International Colloquium on Stability and Ductility of Steel Structures (SDSS'2016)

Timisoara, Romania
30 May – 1 June 2016

Volume 1 of 2

Editors:

Dan Dubina
Viorel Ungureanu

ISBN: 978-1-5108-5770-4

Printed from e-media with permission by:

Curran Associates, Inc.
57 Morehouse Lane
Red Hook, NY 12571

Some format issues inherent in the e-media version may also appear in this print version.

Printed by Curran Associates, Inc. (2018)

For permission requests, please contact European Convention for Constructional Steelwork at the address below.

European Convention for Constructional Steelwork
32, av. des Ombrages, bte 20
1200 Brussels, Belgium

Phone: +32-2/762 04 29
Fax: +32-2/762 09 35

eccs@steelconstruct.com

Additional copies of this publication are available from:

Curran Associates, Inc.
57 Morehouse Lane
Red Hook, NY 12571 USA
Phone: 845-758-0400
Fax: 845-758-2633
Email: curran@proceedings.com
Web: www.proceedings.com

Proceedings of the
International Colloquium on
Stability and Ductility of Steel Structures

SDSS'2016

Edited by
Dan DUBINA and Viorel UNGUREANU

WILEY
Ernst & Sohn
A Wiley Branc

SDSS'2016

The International Colloquium on Stability and Ductility of Steel Structures

30 May – 01 June 2016, Timisoara, Romania

Proceeding of the conference edited by: Dan Dubina and Viorel Ungureanu

TABLE OF CONTENTS

Connections

Members' behavior: tension, compression, beams, beam-columns

Plate, shell and space structures

Robustness

FOREWORD

It was in 1971 when Professor H. Beer from University of Graz, who together with his colleague, Dr. G. Schulz, had proposed the first version of European (ECCS) Buckling Curves equations (Construction Metallique, No. 3, 1970), suggested to organize an International Colloquium, as a forum of discussion and comparison between such equations and those of Eastern Europe, the United States and Japan. The event, entitled "International Colloquium on Column Strength", was organized in November 1972, in Paris, with the support of IABSE, but, sadly, after the sudden death of Professor Beer. This is the event which may be considered the starting point of a long series of International Colloquia dedicated to the Stability of Steel Structures. However, the *real* organization of the series of International Colloquia, thought of as to *globally cover the instability problems of steel structures,* emerged two years later, in London, as a common initiative of Dr. D. Sfiintesco, on behalf of ECCS, and Professor L. Beedle, on behalf of SSRC. The basic idea was to organize several series of *travelling* colloquia, with *sessions* located in different areas of the world. The first series was organized in 1976-1977, with sessions in Tokyo, Liege and Washington, and, further on, the world wide circuit continued.

In October 1982, the Department of Steel Structures and Structural Mechanics of the Politehnica University of Timisoara organized one of the East-European Sessions of the 3rd International Colloquium on Stability of Steel Structures; it was followed by other two sessions, organised in 1983 in Toronto and Paris.

In July 1997, at the Asian Session of the 6th Colloquium, the name of the events was changed into *International Colloquium on Stability and Ductility of Steel Structures - SDSS.* Next, the European Session was organized in September 1999, again in Timisoara. It was followed by those in Budapest (2002), Lisbon (2006) and Rio de Janeiro (2010).

The 2016 edition of SDSS is jointly organized by the *Politehnica University of Timisoara, Department of Steel Structures and Structural Mechanics* in co-operation with the *Romanian Academy - Timisoara Branch*, with the support of the *European Convention of Constructional Steelwork (ECCS),* through the *Structural Stability Technical Committee (TC8),* and the *Structural Stability Research Council (SSRC).*

270 authors from 26 countries on 5 continents are contributing with 115 scientific papers and 4 Keynote Lectures, covering 9 topics, i.e.

1. *Advanced structural design;*
2. *Connections;*
3. *Lightweight steel constructions;*
4. *Members' behaviour: tension, compression, beams, beam-columns;*
5. *Plate, shell and space structures;*

6. *Robustness;*
7. *Seismic-resistant structures;*
8. *Steel-concrete composite members and structures;*
9. *Tubular constructions.*

Since the revision of structural Eurocodes has been already started in Europe, and both stability and ductility of steel structures issues are on the desk of the drafters, this event enables the exchange of knowledge and experience in the field and the debate around the best solutions offered, and is therefore a useful opportunity for all the participants.

On behalf of the SDSS'2016 Organizing Committee, we express our gratitude to Keynote Speakers and authors, to the members of the International Scientific Committee and to all the participants in this international scientific event.

We also appreciate the support of the *Technical Committee 8* within the *European Convention of Constructional Steelwork (ECCS)* and of the *Structural Stability Research Council (SSRC)*.

Thanks are also due to *Joana Albuquerque*, from *CMM – Associação Portuguesa de Construção Metálica e Mista* and also to *Ernst and Sohn and Wiley Company*, for their support and the printing quality of the Proceedings.

Acad. Dan Dubina
Chairman of SDSS'2016

Prof. Viorel Ungureanu
Scientific Secretary of SDSS'2016

Timisoara, May 2016

ORGANIZATION

CHAIRMAN

D. Dubina, *PU Timisoara*

SCIENTIFIC SECRETARY

V. Ungureanu, *PU Timisoara*

SCIENTIFIC COMMITTEE

G.A. Altay	*Turkey*	R. Leon	*USA*
I. Balaz	*Slovakia*	J.R. Liew	*Singapore*
C. Baniatopoulos	*Greece*	J. Loughlan	*UK*
A. Bureau	*France*	J. Machacek	*Czech Republic*
E.M. Batista	*Brazil*	M. Mahendran	*Australia*
R. Beale	*UK*	F. Mazzolani	*Italy*
R. Bjorhovde	*USA*	M. Nakashima	*Japan*
M.A. Bradford	*Australia*	D. Nethercot	*UK*
B. Brune	*Germany*	J. Packer	*Canada*
L. Calado	*Portugal*	J. Paik	*South Korea*
D. Camotim	*Portugal*	A. Plumier	*Belgium*
S.L. Chan	*Hong Kong, China*	N. Rangelov	*Bulgaria*
R. Casciaro	*Italy*	K.J.R. Rasmussen	*Australia*
K.F. Chung	*Hong Kong, China*	J. Ricles	*USA*
C. Chiorean	*Romania*	J.M. Rotter	*UK*
H. Degée	*Belgium*	F. Roure	*Spain*
F. Dinu	*Romania*	C.A. Rogers	*Canada*
L. Dunai	*Hungary*	R. Sause	*USA*
S. Easterling	*USA*	B.W. Schafer	*USA*
A. Elghazouli	*UK*	L.S. Silva	*Portugal*
M. Fontana	*Switzerland*	N. Silvestre	*Portugal*
D. Frangopol	*USA*	H. Snijder	*Netherlands*

L. Gardner *UK*
M. Garlock *USA*
G. Garcea *Italy*
P. Gonçalves *Brazil*
F. Guarracino *Italy*
J. Hajjar *USA*
G.J. Hancock *Australia*
M. Heinisuo *Finland*
M. Hjiaj *France*
B. Izzuddin *UK*
M. Iwata *Japan*
J.P. Jaspart *Belgium*
K. Kasai *Japan*
A. Kawano *Japan*
Z. Kolakowski *Poland*
M. Kotelko *Poland*
U. Kuhlman *Germany*
D. Lam *UK*
R. Landolfo *Italy*
R. Stroetmann *Germany*
J.G. Teng *Hong Kong, China*
R. Tremblay *Canada*
D. Ungermann *Germany*
H. Unterweger *Austria*
B. Uy *Australia*
I. Vayas *Greece*
M. Veljkovic *Sweden*
P. Vellasco *Brazil*
P. Vila Real *Portugal*
A. Wada *Japan*
F. Wald *Czech Republic*
N. Yardimci *Turkey*
B. Young *Hong Kong, China*
R. Zandonini *Italy*
B. Zhao *France*
R. Ziemian *USA*
A. Zingoni *South Africa*
Y.B. Yang *Taiwan, China*

LOCAL ORGANIZING COMMITTEE

I. Both, *Romania*
A. Ciutina, *Romania*
A. Crişan, *Romania*
A. Dogariu, *Romania*
M. Georgescu, *Romania*
D. Grecea, *Romania*
C. Neagu, *Romania*
D. Pintea, *Romania*

Keynote papers

RESEARCH, THEORY AND PRACTICE OF SECOND ORDER DIRECT ANALYSIS FOR DESIGN OF STEEL AND COMPOSITE STRUCTURES

SL Chan, YP Liu and SW Liu

Department of Civil and Environmental Engineering, The Hong Kong Polytechnic University

Abstract: The terms "Second-order Analysis" and "Direct Analysis" appear in many modern design codes but this use is still relatively new to engineers. In the past, second order analysis was mainly to account for sway effects and defined only as the second-order "indirect" analysis, P-Δ-only analysis or simplified second order analysis which needs to be used with the effective length assumption. The modern second-order direct analysis or P-Δ-δ analysis allowing for both sway and member buckling effects with imperfections provides a reliable and proficient tool for practical design. Their underlying principle also carries a very different philosophy to the conventional design approach. This paper outlines the differences between the two design concepts. Several built structures designed completely without effective length assumption are also discussed. The era for re-thinking our design and analysis philosophy seems to have arrived.

1. Background

In the last century, design of steel structures relies on the effective length method (ELM) in conjunction with the first-order linear analysis. This conventional design constitutes the major design exercise since one needs to first determine the safety and stability of a structure under imposed, permanent, wind and seismic loads in ultimate limit state design. This traditional design approach using the first-order linear analysis in conjunction with ELM is not based on a rational theory and it assumes every member buckles simultaneously with its own effective length at its undeformed position. This method is a member-based design and it is, strictly speaking, incompatible with the computer analysis for a frame system. Because of this assumption of drastic buckling in ELM, deflections and stresses cannot be calculated from elastic buckling analysis or the ELM but they are approximated separately from the first-order linear analysis. In practice, this mode of buckling at undeformed position is unrealistic and valid only for non-existent perfect structures since material and geometrical imperfections change the structural geometry once load is applied to a structure.

An alternative, better and more rational design is to simulate the response of a structure directly using a nonlinear analysis. Theories of nonlinear analysis of structures have been studied by numerous researchers. Recent work related to practical applications include works by Liew et al. (2000), Izzuddin and Smith (1996), Chen and Chan (1995), and White and Hajjar (1997), Dubina and Ungureanu (2014), among others. The works have also been summarised and reported in detail Chan and Chui (2000). Applications to special structures like transmission towers and scaffoldings are also demonstrated by Albermani and Kitipornchai (2003), Chandrangsu and Rasmussen (2011) and Chan and Cho (2005) and performance of various nonlinear numerical methods is also studied by Clarke and Hancock (1990). The theoretical works have been extensively studied with encouraging results reported. Actual behaviour is further investigated by tests by Dubina et al. (2009) and Lope et al. (2005). Practicing engineers most commonly use the elastic second-order "indirect" analysis for finding of P-Δ sway moment which in fact has limited use to an improved design in terms of safety, economy and design efficiency because it considers only partly the nonlinear and buckling effects.

In modern design codes, the requirements for seismic design and progressive collapse normally go beyond the elastic limit. In many steel codes, the Second-order Direct Analysis (SODA) has been more widely coded as a preferred method. The new Eurocode-3 (2005), LRFD (2010) and Code of practice for the structural uses of steel Hong Kong (2011) name these methods as direct analysis, second-order analysis, second-order "direct" analysis, second order "indirect" analysis and the first-order linear analysis.

This paper introduces the theory and development of the new second order direct analysis method of design and employs three examples to demonstrate its applicability and outlines the advantages of this new method. The detailed element formulation can be found in a two-part paper by Chan and Zhou (2004) and the theory has been coded in the software NIDA (2015, www.nidacse.com) for research at university and practical use at design institutes.

2. Problems with the traditional effective length method (ELM)

Second-order buckling effects exist whenever a structure is in compression or when it deforms. The effective length method to reduce the buckling resistance is to account for this effect. Fairly speaking, this approach is indispensable in the pre-computer age, but becomes more difficult to handle design of contemporary structures of irregular geometry under various scenarios. The fundamental concept of the method is incorrect since no practical structures are perfect and buckle without deformation at "undeformed" configuration.

In the first-order linear analysis used with the effective length (ELM) method, the critical problem will be on the assessment of the buckling strength and the assumption of effective length. Also the linear analysis is unreliable near the ultimate limit state that its bending moment and force distribution are insufficiently accurate.

The member-based ELM design is only for member design under isolation so its use with a frame analysis software is incompatible. To account for system or frame stiffness, some design codes like the Eurocode-3 (2005) use the classification method of limiting the application of the method to structures with elastic critical factor (α_{cr}) less than 3 or 10 for validity of the first-order linear analysis.

A building frame is classified as non-sway, sway sensitive and sway-ultra-sensitive frames, according to the value of elastic critical load factor, α_{cr}, as,

$$\alpha_{cr} = \frac{H_{Ed}}{V_{Ed}} \frac{h}{\delta_{H,Ed}} \tag{1}$$

in which the meanings of symbols are the same as in the Eurocode-3-1-1(2005) summarised as α_{cr} is the elastic buckling load factor, H_{Ed} is the design horizontal reaction, V_{Ed} is the vertical load, $\delta_{H,Ed}$ is the horizontal displacement at top of storey and h is the storey height.

For different ranges of elastic buckling load factor α_{cr}, the structure is classified as sway and non-sway frames and various charts can be used to determine the effective length factor.

In the Hong Kong Steel Code (2011), an amplification factor as defined in Eqn. (2) is further required to ensure the moment amplified along a member is properly accounted for. Here, the Code uses λ_{cr} in place of α_{cr}.

$$\lambda_{cr} = \frac{\pi^2 EI}{F_c L_e^2} \tag{2}$$

where, L_e is the effective length, EI is the flexural constant and F_c is the member compression force due to design loads.

Selecting a smaller λ_{cr} from Eqns. (1) and (2), the amplified moment, M_{amp}, is calculated as,

$$\mathrm{M}_{amp} = \frac{\lambda_{cr}}{\lambda_{cr} - 1} \mathrm{M}_{\mathrm{L}} \tag{3}$$

Once we know the effective length, we can determine the slenderness ratio (L_e/r) and then we can use the charts, tables or formulae in a code to find the buckling strength p_c and buckling resistance P_c as,

$$P_c = p_c A \tag{4}$$

in which A is the cross sectional area.

The above ELM with and without second-order P-Δ-only analysis is, strictly speaking, suitable for design of isolated members since it needs to assume an effective length for each member and these members are assumed to buckle simultaneously and independently which is far from the actual behaviour. Further, it is not applicable for design of slender frames because "effective length" is based on undeformed geometry and the end nodes of columns in general move with the structure when under loads. For argument sake, if the structure does not deform, it has no displacement, no strain, no stress and no resistance against external loads and therefore the assumption of using the original geometry for checking is not rational.

The stiffness or sway sensitivity of a frame is then used for assessment of the effective length factor (L_e/r) and amplified moment in Eqn. (3), but the actual amplification for bending moment is linked to the deflection of the frame in a nonlinear manner and therefore the method is limited to rather stocky and regular structures which may not be the case in practice.

The underlying principle for the effective length method is to reduce the resistance of a structure and this is conceptually unacceptable since buckling is due to external load and deflection. The mixed use of reducing buckling resistance p_c of a member in Eqn. (4) and amplifying the action or moment in Eqn. (3) is another inconsistency.

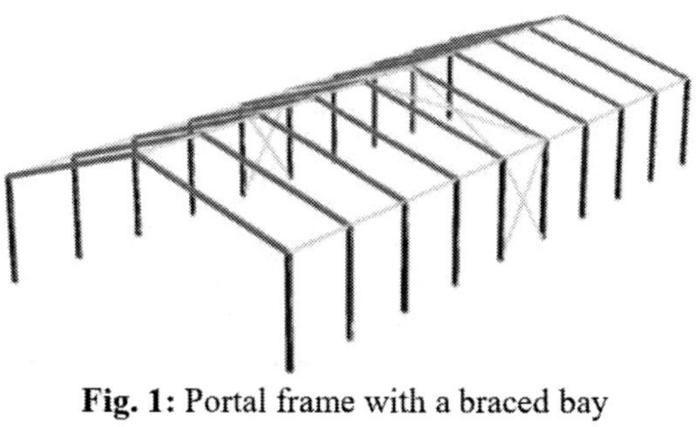

Fig. 1: Portal frame with a braced bay

As can be seen in a simple portal in Fig. 1, the stiffness of one bay increases significantly the buckling strength of the complete frame and the buckling resistance of all vertical columns including the one far from the braced bay which is affected by the "system" stiffness. Thus, the stability design of individual columns should not be assessed in isolation.

3. Types of analysis

It is interesting to note that the Eurocode-3 (2005) and other codes have a special chapter on analysis which the older codes did not have. The simple reason is the old code uses only one method of analysis which is the linear analysis whilst the new codes allow use of several types of analysis. To this, an overview of design and analysis method is needed and shown graphically in Fig. 2 which is extended from a figure in Chan (2001). Various types of buckling loads and analysis represent different levels of accuracy in modelling. Naturally, a more simple method gives a less accurate solution and leaving the part of safety check to the manual design which contains assumption and assessment with lesser theoretical verification.

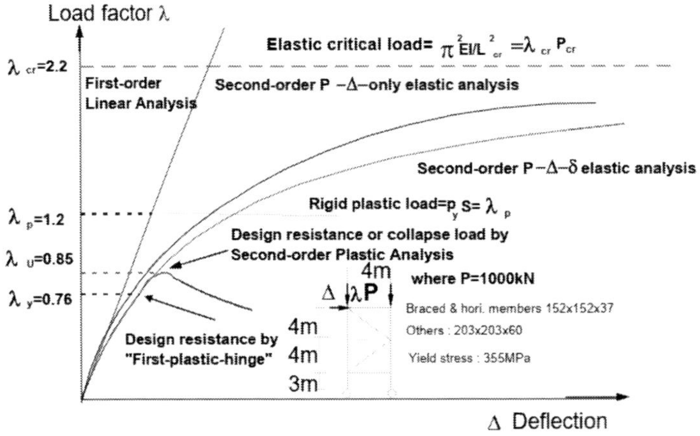

Fig. 2: Common types of analysis for design

4. IMPERFECTIONS

A valid second-order analysis must consider imperfections at both element and frame levels. Ignoring either one of them will provide an under-designed solution to a design exercise. To this, the following two aspects of imperfections are required for consideration in an analysis and design software to meet code requirement.

4.1 Global *P-Δ* imperfection

The linear analysis uses the moment amplification to enlarge the linear moment for sway effect, which can be due to wind load or notional force normally taken as 0.5% for permanent structures and 1% for temporary structures.

In second-order analysis, wind load or notional force can still be used, but an alternative and more reliable and convenient method is to use the elastic buckling mode as the imperfection mode with amplitude set equal to the out-of-plumbness normally taken as one-200th of the building height for permanent structures or other justified values. Other imperfection mode assumption such as use of post-buckling shape can be used. When designing a structure of unconventional shape, the second method of using elastic buckling mode as imperfection mode is more rational since the locations and direction of application of notional force become difficult to determine. There have been arguments on which buckling mode should be used and, in general, the actual imperfection mode is random. Using the elastic buckling mode in the plane of buckling recommended in Eurocode-3 (2005) as imperfection mode is adequate in general and studies by the authors show that the difference is generally small otherwise one could treat each imperfection mode as a separated load case. To this, the program NIDA (2015) contains options of using a selected elastic buckling mode and using the displacement mode as imperfection mode.

4.2 Member *P-δ* imperfection

Linear analysis uses various buckling curves to represent different values of member imperfections and this leads to the production of different design tables in various notional codes, with the exception of the LRFD (2011) which uses only one buckling curve.

In SODA or *P-Δ-δ* elastic analysis, the initial curvature can be pre-set to the code values which can be calibrated from the buckling curves. The member curvature can be varied with an increasing axial force such that the *P-δ* effect can be considered correspondingly (see example 1 of this paper).

4.3 Structural resistance by section capacity check

In the codified linear analysis and design, a member is required for checking against member buckling and sectional strength. In the second-order analysis for design, only the section capacity check is required and checking can be conducted in the following expression.

$$\frac{P}{p_y A}+\frac{(M_y+P\Delta_y+P\delta_y)}{M_{Cy}}+\frac{(M_z+P\Delta_z+P\delta_z)}{M_{Cz}}=\phi\le 1 \qquad (5)$$

in which p_y is the design strength, A is the cross sectional area, M_y M_z M_{Cy} and M_{Cz} are respectively the design moment and moment capacity about the principal axes, P is the axial force, *P-Δ* and *P-δ* are the second-order delta moments and ϕ is the section capacity factor.

In dealing with members with possibility of lateral torsional buckling, the minor axis moment can include the component of the major axis moment (Trahair and Chan, 2003) and the twist angle.

5. First plastic hinge analysis vs. full plastic analysis

In the first plastic hinge design approach where the design load is limited to the load causing the formation of the first plastic hinge, the checking of Eqn. (5) is carried out for all members and the design resistance for the complete structure is then checked and monitored. The design load should be taken as the load causing the formation of the first plastic hinge in a member of the structure. When using the Second-order Direct Plastic Analysis, an incremental –iterative analysis is used with small increments of design load applied to the structural model. The analysis is continued after the first plastic hinge and the design load is indicated by the load causing the complete structure to collapse, indicated as a limit point, a flat curve or a peak in the load vs. deflection plot. Eqn. (5) is used again for checking and insertion of plastic hinge in a member. Nowhere along a member is allowed to have a sectional capacity factor greater than 1 in Eqn. (5) as strain hardening is not considered.

The use of one-element per member with a smooth extension of elastic to plastic analysis represents a consistent design procedure from the first-order linear elastic analysis to second-order plastic analysis since the analysis model using one element per member is unchanged. The consistent design philosophy allows engineers to compare the results by the new and the old method, which allows a simpler evaluation of the advantages of the new approach. As shown in Fig. 3 for the Ziemian's two-story calibration frame, the maximum load is indicated by the load vs deflection curve being horizontal.

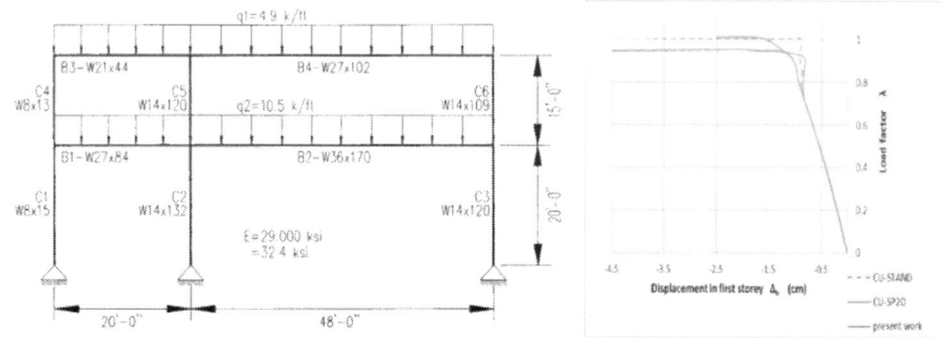

Fig. 3: Comparison of the first story load-deflection response by Ziemian (1992) and NIDA (2015)

6. Recent research

Research on design of pinned or rigid steel frames and composite structures has been basically completed and currently the authors and their research team is working on refined design of semi-rigid frames, tapering members and structures under impact loads. It is envisaged that the design concept for SODA is sustainable that it could be applied to various scenarios like fire engineering, time-history seismic design and advanced structural systems like landslide flexible barrier systems shown in Fig. 4. The new structural form allows large deflection to absorb impact and dynamic energy to reduce the maximum loads on the structure and its supports. ELM is believed to be non-sustainable as it could not be extended to design of similar systems rationally.

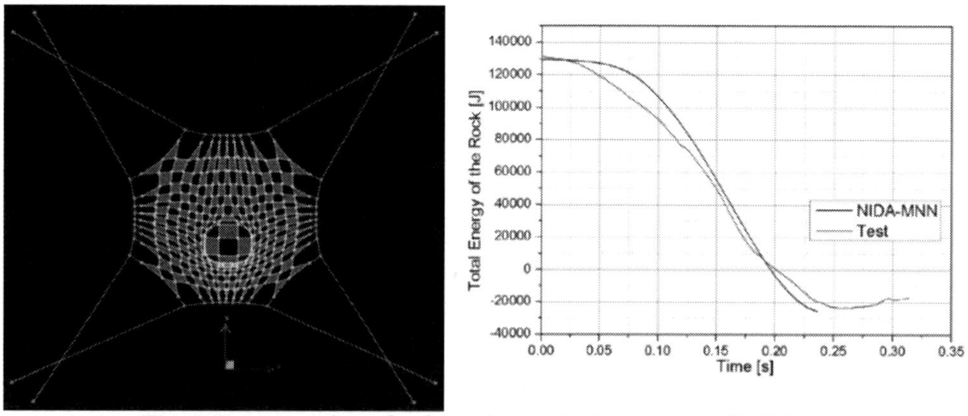

Fig. 4: Energy vs time plot for an impacting body onto a flexible net

7. Verification examples

Before the design method is used in design of practical structures, it is tested against experimental results as well as code recommendation. Example 1 shows the comparison with code design for the simplest single column with free and pinned at one end and fixed at the other end. Example 2 is about the practical application with a comparison of experimental result for buckling of composite columns and design of a composite truss. In example 3, a medium size steel platform of about 1000 tons of steel supporting heavy mobile cranes is designed without assumption of effective length and constructed.

Example 1. Benchmarking with code result

The first example is to demonstrate the performance of using a single element to determine the buckling strength and behaviour of a column under compression. The success of the element in capturing the buckling behaviour of a column implies a consistent extension on the use of a structural model from linear analysis to second-order direct analysis since linear analysis allows use of a single element per member. Modelling of a member by several elements to simulate initial imperfection results not only in unnecessary requirement of longer computer time, but also in a complication in modelling and an inconsistency in linear and nonlinear computer models for structural analysis. To this, the authors proposed the concept of one single element per member two decades ago (Chan and Zhou, 1995).

Methods of analysis	Euler's buckling load (kN)	Design Load Capacity (kN)
NIDA	131	111.5
Analytical $\pi^2 EI/(0.7L)^2$	130.8	-
HK Steel Code (2011)	-	114.6

This example is a column of cross section CHS88.9×3.2, length of 5 m and under an axial force. The analytical Euler's buckling load and design resistance from the Hong Kong Steel Code (2011) are respectively 131 kN and 109 kN, compared well with the solutions for buckling and design resistance by the proposed method and program NIDA (2015) of 131 kN and 108 kN using only one element in the model.

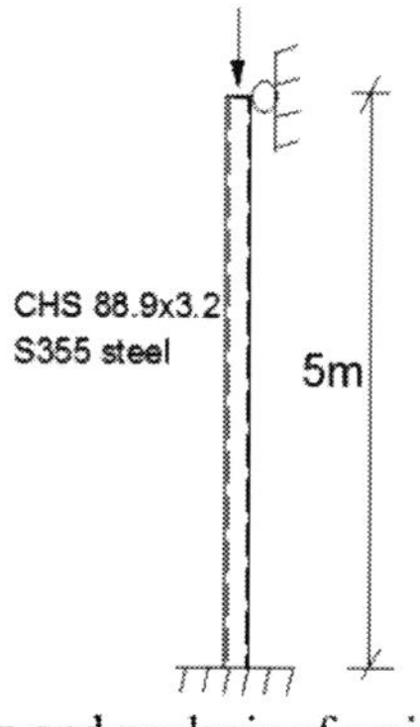

Fig. 5: Design and analysis of a pin-fix column

In the design by the present method, no effective length has been assumed and it gives a similar solution as ELM assuming an accurate effective length factor which is only possible for single columns but not for complex and practical frames. For complex structures, the ELM is unreliable whereas the SODA is still applicable as it is free from effective length assumption. Note that, when using the imperfection given in a code, the analysis result should give a similar value of buckling resistance as given by that particular code.

Example 2. Design of composite columns and composite truss

The same concept can be applied to analysis and design of composite frames. However, the yield and failure surfaces for a composite section are more complicated and allowance must be made for weak tension stress in concrete. The following is typical yield and failure surfaces of a composite column used to control the formation of plastic hinges.

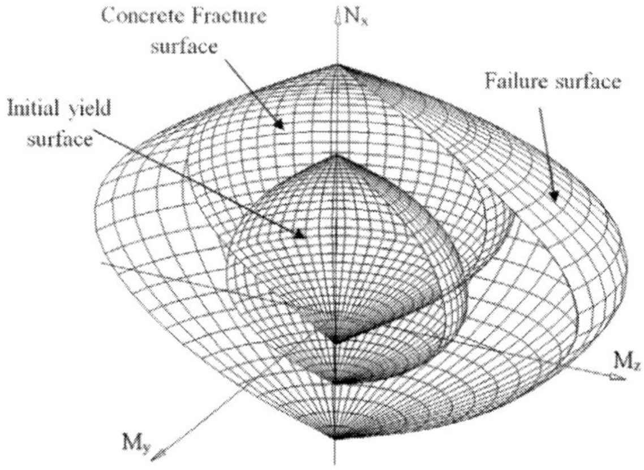

Fig. 6: The concrete crushing, yield and failure surfaces

The results using the proposed method agree well with test, see Fig. 5. Below is a typical load deflection plot of a composite concrete infilled square column and the full set of comparison is reported by Liu, Liu and Chan (2012). The application was further carried out in the design of a 4 level composite truss at Tamar Government Headquarters in Hong Kong with significant saving of material, see Fig. 8.

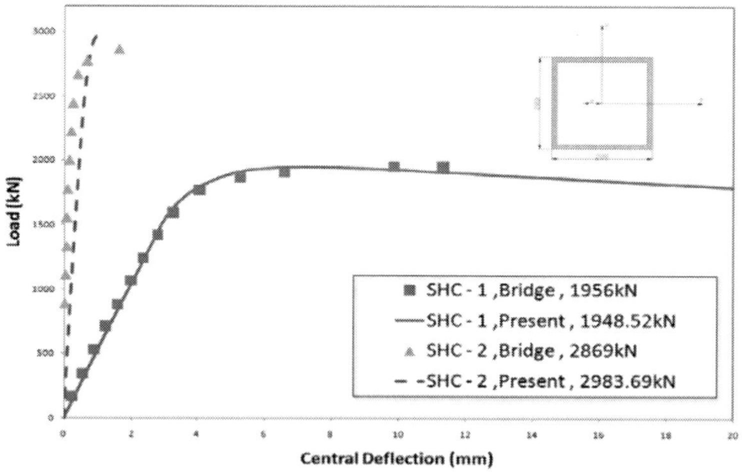

Fig. 7: The load vs deflection plot of a composite column with validation with experiment by Bridge (1976)

Fig. 8: The computer model and completed structure with the top 4 level composite truss designed by SODA without effective length

Example 3. Design of an approx. 1000 ton steel platform

It is common to construct high rise buildings along steep hill slope in Hong Kong so cranes for bore pile foundation are required to be supported by platform. The steel platform shown in Fig. 9 was designed and constructed for support of mobile cranes for bore piling in a high-rise building in Hong Kong Island. The structure was originally designed by the ELM and it was later re-designed by SODA because of the slenderness of the structure and to requirement of the local code since some load cases have α_{cr} less than 10 which requires the consideration of second-order effects. Note that the proposed second order direct analysis has no such a limit and it can be used for all values of α_{cr}. Currently, the authors even do not bother to check this α_{cr} for a structure to be designed by second-order direct analysis.

The outcome of the SODA showed some members were under-designed and most were over-designed by the ELM. The dimensions of the members were altered and the tonnage of the structure is reduced from 1346 tons to 983 tons. The platform was used for about ten months and dismantled after completion of the piling works.

Fig. 9: The computer model and built platform

7. Conclusions

This paper reports the research, background theory and applications of the method of Second-order Direct Analysis (SODA). Apart from a number of research topics such as development of energy absorbing systems against impact loads, fire engineering and refined semi-rigid frame design, the concept of SODA needs to be introduced into curriculum of undergraduate and graduate courses for students in university and our next generation. With this successful implementation of the design concept in codes and in teaching, our engineering profession could then build more reliable, sustainable and safer structures against human-made or natural hazards.

Acknowledgement

The authors are grateful to the financial supports by the Research Grant Council of the Hong Kong SAR Government on the project "Second-order and Advanced Analysis of Steel Frames Second-order and Advanced Analysis of Arches and Curved Structures (PolyU 152012/14E)" and "Second-Order Analysis of Flexible Steel Cable Nets Supporting Debris (PolyU 152008/15E)" and The Hong Kong Polytechnic University for the project "Stability Design of Composite Structures by Direct Analysis" .

References

[1] Albermani FGA, Kitipornchai S. "Numerical simulation of structural behaviour of transmission towers", *Thin-walled Structures*, 41(2-3), 167-177, 2003.

[2] Bridge RQ. *Concrete filled steel tubular columns*, Research Report No. R283, Sydney University, 1076.

[3] Chan SL. "Non-Linear behaviour and design of steel structures", invited review paper, Journal of Construction Steel Research, 57(12), 1217-1232, 2001.

[4] Chan SL, Chui PPT. "Non-linear Static and Cyclic analysis of semi-rigid steel frames", Elsevier Science, 336, 2000.

[5] Chan SL, Zhou ZH. "Elastoplastic and large deflection analysis of steel frames by one element per member. Parts 1 and 2, *Journal of Structural Engineering, ASCE*, 130(4), 538-553, 2004.

[6] Chan SL, Cho SH. "Second-order P-delta analysis and design of angle trusses allowing for imperfections and semi-rigid connections", *International Journal of Advanced Steel Construction*, 1(1), 157-172, 2005.

[7] Chan SL, Zhou ZH. "Second order analysis of frame using a single imperfect element per member", *Journal of Structural Engineering, ASCE*, vol. 121, No. 6, June, pp. 939-945, 1995.

[8] Chandrangsu T, Rasmussen K. "Investigation of geometric imperfections and joint stiffness of support scaffold systems", *Journal of Constructional Steel Research*, 67(4), 576-584, 2011.

[9] Chen WF, Chan S.L. "Second Order Inelastic Analysis of Steel Frames using Element with Mid-span and End Springs", 121(3), *Journal of Structural Engineering, ASCE*, 530-541, 1995.

[10] *Code of practice for structural uses of steel 2011*, Buildings Department, Hong Kong SAR Government.

[11] Clarke MJ, Hancock GJ. "A study of incremental-iterative strategies for nonlinear analysis", *International Journal for Numerical Methods in Engineering*, 29, 1365-1391, 1990.

[12] Eurocode 3-1-1, *Design of steel structures*, European Standard, CEN, 2005.

[13] Dubina D, Ungureanu V. "Instability mode interaction: From Van Der Neut model to ECBL approach", *Thin-Walled Structures*, 81, 39–49, 2014.

[14] Dubina D, Stratan A, Nagy Zs. "Full-scale tests on cold-formed steel pitched-roof portal frames with bolted joints", *Advanced Steel Construction*, 5(2), 175-194, 2009.

[15] Izzuddin BA, Smith DL. "Large displacement analysis of elasto-plastic thin-walled frames Parts 1 and 2", *Journal of Structural Engineering, ASCE*, 122(8), 905-925, 1996.
[16] Liew JYR, Chen WF, Chen H. "Advanced inelastic analysis of frame structures", *Journal of Constructional Steel Research*, 55(1-3), 245-265, 2000.
[17] Liu SW, Liu YP, Chan SL. "Advanced analysis of hybrid steel and concrete frames, Part 2: Refined plastic hinge and advanced analysis", *Journal of Constructional Steel Research*, 70, 337-349, 2012.
[18] *Load and resistance factor design specification for structural steel buildings*, American Institute of Steel Construction Chicago, AISC, 2011.
[19] Lopes F, Santiago A, Silva LS, Iqbal N, Veljkovic M, Silva JGS. "Sub-frames with reverse channel connections to CFT composite columns – experimental evaluation", *Advanced Steel Construction*, 11(1), 111-129, 2015.
[20] *NIDA. Non-linear Integrated Design and Analysis user's manual*, NAF-NIDA series, Version 9, (http://www.nidacse.com), 2015.
[21] Trahair NS, Chan SL. "Out-of-plane advanced analysis of steel structures", *Engineering Structures*, 25, 1627-1637, 2003.
[22] White DW, Hajjar JF. "Design of Steel Frames without consideration of effective length", *Engineering Structures*, 19(10), 797-810, 1997.
[23] Ziemian RD. *Advanced Methods of Inelastic Analysis for in the Limit States Design of Steel Structures*, Ph.D. Thesis, Cornell University, Ithaca, NY, 1992.

FURTHER DEVELOPMENT OF EUROCODE 3: IMPROVEMENT AND HARMONIZATION

Ulrike Kuhlmann, Adrian Just and Antonio Zizza

University of Stuttgart, Institute of Structural Design, Stuttgart, Germany

Abstract: The presented paper shows an extract of recent work regarding the revision of Eurocode 3 Design of steel structures on European and national level including selected scientific and technical issues and a summary of the activities executed within the European Standardization committee CEN/TC 250/SC 3 "Design of Steel Structures" under the chair of Prof. Dr.-Ing. Ulrike Kuhlmann. After a general overview on the further development of Eurocodes in general and Eurocode 3 in specific in the frame of Mandate M/515, more detailed information is given on the development of the stability relevant parts EN 1993-1-1, EN 1993-1-3 and EN 1993-1-5.

1 Introduction

The Eurocodes form the basis of design for all types of building and civil engineering works, all over Europe. When they were published, prior to 2007, the first generation of EN Eurocodes were the most comprehensive and technically advanced set of standards for structural and geotechnical design in the world. It is widely recognised that long-term confidence in the codes requires the Eurocodes to evolve in an appropriate manner. Specifically the planned new work programme of Mandate M/515 aims at ensuring the standards remain fully up to date through comprising new methods, new materials, and new market requirements. Furthermore, it focuses on further harmonization and a major effort to improve the ease of use of the whole of standards for practical users, see [2], [3]. Further harmonization is needed, leading to a reduction in Nationally Determined Parameters and the number of alternative design methods.

2 Mandate M/515

2.1 General

In 2010, the Eurocodes, comprising 10 standards in 58 parts, were introduced according to the directives of the EU in many countries replacing the national standards. Thereby an important first step for a harmonised European standardization in the field of civil engineering was made.

The further evolution of the Eurocodes is planned in the frame of the Mandate M/515 [2]. Since December 2012, CEN/TC 250 has developed a work programme for Mandate M/515 which was agreed between the European Commission and CEN (the European Committee for Standardization) [3]. In December 2014 the European Commission awarded a grant agreement for Phase 1 of the proposed work to CEN. NEN (the Dutch Foundation for Standardization) responsible for the financial and operational management of Mandate M/515 launched a first call for tenders for the members of the 25 Project Teams of the first phase in April 2015.

2.2 Organization of the Mandate M/515

The mandate has started in 2015 and has a term of 5 years. It is planned that the results of the mandate find their way immediately into the first revision of the Eurocodes in 2020.

The CEN/TC 250 work programme has been split into four over-lapping phases, as illustrated in Fig. 1. This has been done to enable the interdependencies between activities to be effectively managed, and ensure that the work is undertaken as efficiently as possible.
Phase 1 includes those parts of the work programme upon which other activities are primarily dependent for reasons of overall coordination, technical scope or because they are essential for achieving the target dates for delivery of the next generation of Eurocodes.

Project Teams (PT) are responsible for the technical work within the mandate, similar as it was organized for the conversion from the ENV versions to the EN versions of the Eurocodes. Each of these Project Teams consists of a maximum of six members under the leadership of a project team leader. Each of the Project Team has to fulfil a specific technical task in the development of the future Eurocodes.

For each of these tasks a number of subtasks are defined according to the specific needs of the referred parts of Eurocode 3, see Fig. 2. Common sub-tasks which are included in all tasks relating to existing Eurocode parts are

- Reduction in the number of Nationally Determined Parameters (NDPs)
- Refinement to improve the 'ease of use' of Eurocodes by users in practice.

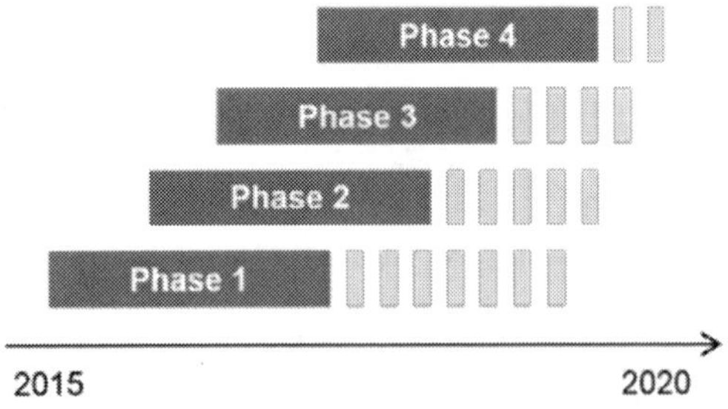

Fig. 1: Phasing of the mandate work after [3]

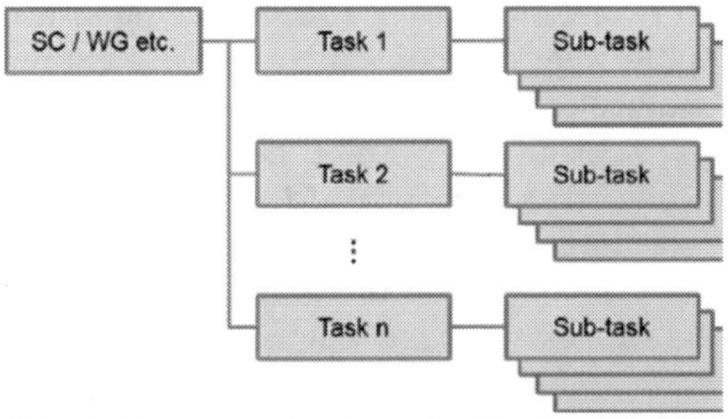

Fig. 2: Structure of tasks and subtasks after [3]

3 Procedure for the Revision of Eurocode 3

3.1 Organization

Of all Eurocodes (EN 1990 – EN 1999), Eurocode 3 (EN 1993) with its 20 parts and approximately 1.450 pages is the largest one. Fig. 3 gives an overview on the structure of Eurocode 3 showing the "Application parts", which refer to the "General parts" within Part 1, see Fig. 3.

EN 1993 "Design of Steel Structures", consisting of all in all 20 parts, presents a considerably extensive and partially complex code system that is highly important for the steel construction in practice. First experiences of application of the series of standards and the elaboration of the National Annexes (NAs) showed that, in terms of a practice-oriented application, a certain revision and harmonization is required and still existing ambiguities should be erased by corrections as soon as possible.

So the requirement is to work towards a coordinated approach in revising Eurocode 3 especially because of the high amount of different types of application covered by the various parts of Eurocode 3. Only thus, the user-friendliness of the Eurocode 3 as required by the principles of the EU-Mandate and CEN/TC 250 can be achieved in the longer run. The following principles have been defined for the further evolution of Eurocode 3, see TC 250/SC 3 Decision 4/2013:

- keep the overall structure of EN 1993 and its parts,
- improve the clarity,
- simplify and harmonise rules (same format, structure, notations, …),
- harmonise different parts of Eurocode 3 and if possible also with other relevant Eurocodes,
- reduce the overall volume (e.g. by avoiding informative annexes) and
- reduce the number of alternatives.

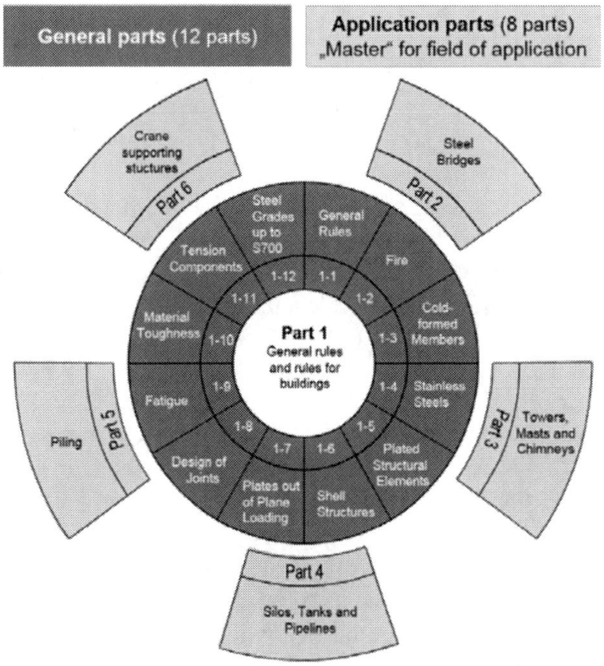

Fig. 3: Structure and overview of Eurocode 3 after [10]

For the future work which is affected by the general revision under the scope of the call for the "Systematic Review" and the works for the EU-Mandate M/515, it is essential to pursue this approach.

In the frame of the SC 3 meeting in Stuttgart in April 2010 the approach shown in Fig. 4 for the revision and harmonization of Eurocode 3 was developed. Significant elements are the so-called Working Groups (technical working groups, until March 2014 called Evolution Groups), which support the work of the SC 3 in scientific and technical terms.

The members of the Working Groups are experts for the particular area of expertise and are nominated by the member countries. The questions relating to the revision and harmonization of Eurocode 3 are solved in cooperation between the CEN/TC 250/SC 3 and the Working Groups and are elaborated in the form of proposals for amendments. These proposals originally were then sent to CEN for approval to be finally introduced into the existing Eurocode 3. Now with the new Mandate M/515 [2] the amendments will be "hold in the basket", until they are integrated into the new versions of Eurocode 3. Furthermore, for the application of Eurocode 3 in practice some open questions exist which need to be answered. In addition recent research offers to some extend new insights on various problems and new results for improved design, which however are not yet available in practice.

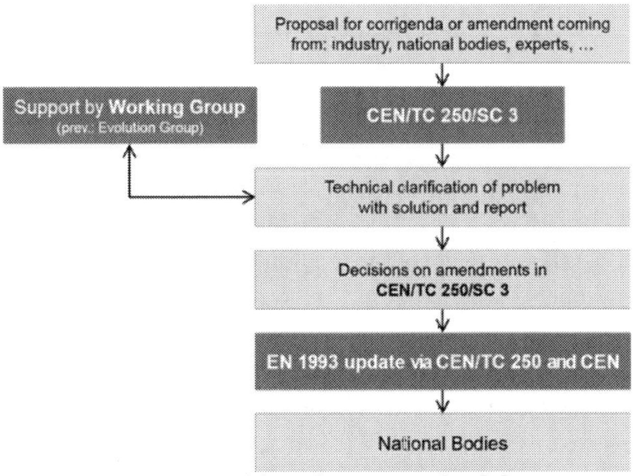

Fig. 4: Revision of EN 1993 after [9]

The SC 3 submitted its last technical review in form of amendments to the already existing parts of Eurocode 3 within the meeting of the CEN/TC 250 in November 14th/15th, 2013 in Delft. For the future, technical modifications are to be carried out in the frame of the mandate work. Exception is made just for safety-relevant amendments, which may still be submitted and decided for the existing versions of Eurocode 3.

For the work within the mandate the 20 parts of Eurocode 3 have been subdivided into 13 tasks. For these tasks the technical contents were developed in form of so-called 'Project Proposals' in collaboration with the respective Working Groups and coordinated within the SC 3, see [3].

As EN 1993-1-1 and EN 1993-1-8 are the basic parts of Eurocode 3, there is the necessity of harmonization in a number of issues with other parts, so that these two parts are dealt right at the beginning, in Phase 1. Furthermore four SC 3-tasks, mainly stability parts, are assigned to the early Phase 2 of the mandate. The material specific parts of Eurocode 3, e.g. EN 1993-1-4 and EN 1993-1-10 are assigned to Phase 3, while in Phase 4 of the mandate primarily the application parts, such as e.g. EN 1993-2 for steel bridges, are covered. The assignment of the tasks to the phases is shown in Table 1.

Table 1: Tasks of Mandate M/515

Task-Ref.	Task-Phase	Corresponding Part of EN 1993	Task-Name
SC 3.1	1	EN 1993-1-1	Design of sections and members according to EN 1993-1-1
SC 3.2	1	EN 1993-1-8	Joints and connections according to EN 1993-1-8
SC 3.3	2	EN 1993-1-3	Cold-formed members and sheeting. Revised EN 1993-1-3
SC 3.4	2	EN 1993-1-5	Stability of plated structural elements. Revised EN 1993-1-5
SC 3.5	2	EN 1993-1-6, -1-7	Harmonization and extension of rules for shells and similar structures. Revised EN 1993-1-6 and EN 1993-1-7
SC 3.6	2	EN 1993-1-2	Fire design of steel structures. Revised EN 1993-1-2
SC 3.7	3	EN 1993-1-4	Stainless steels. Revised EN 1993-1-4
SC 3.8	3	EN 1993-1-9	Steel fatigue. Revised EN 1993-1-9
SC 3.9	3	EN 1993-1-10	Material and fracture. Revised EN 1993-1-10
SC 3.10	4	EN 1993-2,-1-11	Steel bridges and tension components. Revised EN 1993-2 and EN 1993-1-11
SC 3.11	4	EN 1993-3	Consolidation and rationalisation of EN 1993-3
SC 3.12	4	EN 1993-4	Harmonization and extension of rules for storage structures. Revised EN 1993-4-1 and EN 1993-4-2
SC 3.13	4	EN 1993-1-12, -4-3, -5, -6	Evolution of existing parts of EN 1993 that are not included in the other tasks

3.2 Revision and future development of Eurocode 3 in context with Mandate M/515

The first revised version of the Eurocodes altogether is expected to be published in 2020. Fig. 5 shows the planned time schedule for the revision and the further evolution of the Eurocodes, which can be subdivided into the following two activities:

1) General revisions and maintenance of the Eurocodes:
   This is the usual procedure for a code revision according to CEN, which is launched in form of a call for "systematic reviews" to the national standardization bodies. The evaluation and implementation of the suggestions and comments is then carried out by the CEN/TC 250 Subcommittees and their Working Groups.

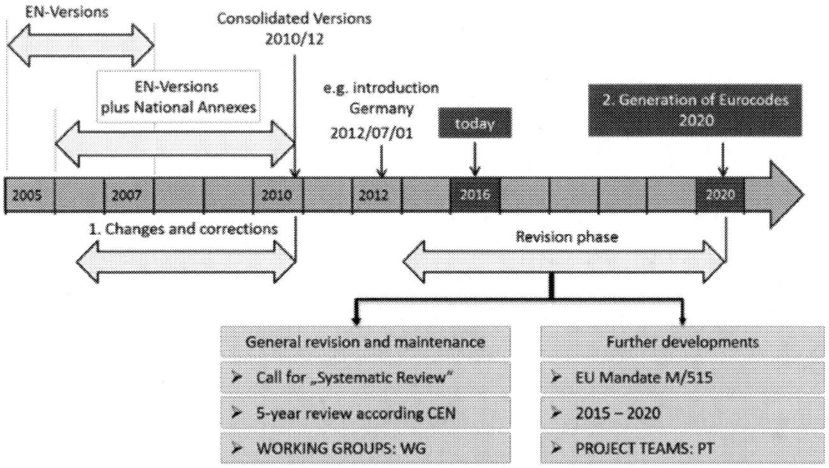

Fig. 5: Revision time-table of the Eurocodes

2) Technical enhancements of the Eurocodes in the frame of the Mandate M/515:
   The further development takes place simultaneously to the general revision within Mandate M/515. Similar to the transfer of the ENV-versions into the EN-version, the realisation is conducted by project teams. Their task is to include all changes and amendments into a complete new text version of the respective part of Eurocode 3.

4 Technical review of selected code rules

4.1 General

The following sections show the recent development separated for some of the stability relevant parts of Eurocode 3 and tries to explain the technical review of some selected rules. Due to the limited space of this paper, only a small extract is shown here.

4.2 Eurocode EN 1993-1-1

4.2.1 Stability rules – member buckling

For the stability rules of member buckling in EN 1993-1-1 there is a clear need of clarification and harmonization. Counting all alternatives for verification given in EN 1993-1-1, one may sum up at least eight alternatives or even more, see Fig. 6.

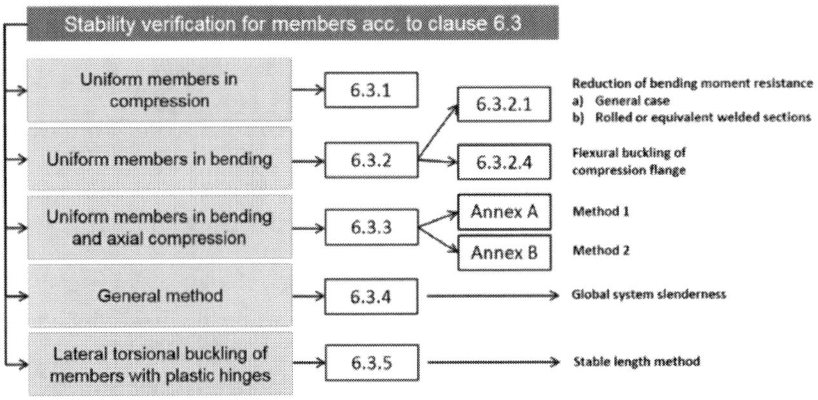

Fig. 6: Stability verification for members and alternatives according to EN 1993-1-1

Recent investigations have shown that there is a chance of improvement and harmonization of the stability checks in EN 1993-1-1, which have led to a set of proposals that will be included in the Eurocode. For example currently two simplified methods exist for lateral-torsional buckling resistance of beams: the "General case" (clause 6.3.2.2) and the specific case of "Rolled and equivalent welded sections" (clause 6.3.2.3). A new common approach has now been developed so that the redundancy between different formulae dealing with the same sections and problems is removed. Also for the case of torsional and torsional-flexural buckling a specific formula has been developed no longer referring conservatively only to the reduction curves of flexural buckling. At the same time the rules of Uniform members in bending and compression of Method 2 as given in Annex B have been extended for mono-symmetric cross-sections, which are often used in practice. The proposed and agreed amendments were partially developed in the PhD thesis of A. Taras [12] and presented in a series of ECCS TC8 and CEN/TC 250/SC 3/WG 1 meetings which led to a decision in TC 250/SC 3 for an amendment, see also [13]. This will be implemented into the code by the meanwhile established Project Team SC 3.T1.

Discussing intensively on member buckling rules in EN 1993-1-1, following a proposal by the Working Group CEN/TC 250/SC 3/WG 1 responsible for the evolution of EN 1993-1-1, CEN/TC 250/SC 3 had taken several important decisions in October 2015:

- Decision 10/2015: CEN/TC 250/SC 3 agreed that a CEN Technical Report giving elastic critical buckling solutions for flexural buckling and lateral torsional buckling in terms of F_{cr}, L_{cr} and M_{cr} respectively shall be prepared by WG 1 and cited in EN 1993-1-1.
- Decision 11/2015: CEN/TC 250/SC 3 agreed that Method 2 in Annex B will remain in EN 1993-1-1 for determining the stability of beam columns. Method 1 in Annex A will be moved to a Technical Specification and will be cited in EN 1993-1-1.
- Decision 16/2015: CEN/TC 250/SC 3 agreed that the contents of Informative Annex BB.3 will be moved to a Technical Specification and cited in EN 1993-1-1.

So in view of the new development in [13] it was decided to keep Method 2, see Fig. 6, in the main text and reduce the number of alternatives and the overall amount of pages by transferring two of the parallel methods into Technical Specifications (TS). Usually Technical Specifications are used for the development of new parts of Eurocodes, before an agreement can be achieved, in order to allow those member states that already want to apply these new rules to do so on a legal basis. So as it can also be seen in Table 2, a TS is a fully normative document with the same procedure for development and agreement as a Eurocode. The only difference to a Eurocode is that national standards may exist in parallel. In the case of the two methods described above (Method 1 in Annex A and Stable length method) which will be transferred to Technical Specifications, this is a way to continue applying the methods on nearly the same legal level, even if they are no longer part of EN 1993-1-1. Maybe that in future the use of these methods become less frequent because they are limited in scope and spread, so that they may be skipped then. But at the moment this procedure allows for a smooth transition period. This was the basis for a positive decision on the necessary reduction of alternatives leading in future to a clear direct stability verification in EN 1993-1-1.

Table 2: Comparison by characteristics of Technical Specification and Technical Report following CEN/CENELEC Internal Regulations Part 2 [1]

TS	TR
normative document	for information and transfer of knowledge
to be announced and made available at national level	availability at national level is voluntary
conflicting national standards may continue to exist	conflicting national standards may continue to exist
no standstill obligation exists except if the Technical Board has specifically decided to do so	no standstill obligation exists
approved by a weighted vote of CEN/CENELEC national members	approved by simple majority of CEN/CENELEC national members
to be reviewed every 3 years at the latest	no time limit is specified for the lifetime of TRs and it is recommended that TRs are regularly reviewed

As a very difficult issue it was also decided not to include extensive information on elastic critical buckling forces, buckling length and elastic critical lateral torsional buckling moments in the main document of EN 1993-1-1 but to collect these rules in a Technical Report which will be fully authorized by CEN/TC 250/SC 3 and cited in EN 1993-1-1. This will suffice the need of some member states of complete information without extending the volume of EN 1993-1-1 by textbook information on "applied mechanics".

4.2.2 Accepted amendments

In a number of cases pre-decisions have been taken by TC 250/SC 3. The proposals of amendments have been drafted following a specific format containing the following elements: topic, clause in EN 1993, reasons for amendment and the background information on which the amendment is based. This gives a perfect basis for the implementation of these changes and additions into EN 1993-1-1 by the Project Team. It also always the follow-up of decisions taken in TC 250/SC 3, which for the decisions in the past at the time of conversion or drafting of ENV 1993 has been rather difficult. A number of reasons have been explained for the accepted amendments in EN 1993-1-1 in more details also by the chairman of the Working Group 1, H.H. Snijder in [11]. The amendments for EN 1993-1-1 so far decided are given in Table 3.

Table 3: Accepted amendments for EN 1993-1-1, Status October 2015 after [9]

SC 3-Document-No. Decision Amendment-No.	Relevant clause	Subject
N1895, Decision 05/2013, AM-1-1-2012-08-V2	6.2.6 (3) 6.2.6 (6)	Shear resistance of cross-sections and influence of buckling
N1898, Decision 08/2013 AM-1-1-2012-01 to 05	1.6, 6.2.5, 6.2.9, 6.3.2, 6.3.3 and Annex B, 5.5, Table 5.2 Annex B – Tables B.1 and B.2 Annex B, Table B.3; 6.3.1 (6.3.1.2 to 6.3.1.4) and 6.3.2 (6.3.2.2 and 6.3.2.3)	Amended rules for member buckling and semi-compact cross-section design (1) Uniform buckling rules for flexural, torsional-flexural and lateral-torsional buckling. (2) Strength of Class 3 sections with linear transition between class 2 and class 4. (3) Corrected c/t limits for "internal" cross-sectional components, esp. rectangular hollow sections (4) Modification of Tables B.1 and B.2 of Annex B (beam-column buckling, method 2) – Clarification/Simplification of k_{ij} factors and inclusion of Circular Hollow Sections (CHS) (5) Clarification of Cm factors in Annex B (Method 2)
N1986 Decision 20/2014 AM-1-12-2013-01 to -08	1.1.1, 1.2.2, 3.1(2) and 3.2.1(1), 3.2.2, 3.2.3, 5.4.1 (3) and (4)B, notes to 5.4.3 (1), 6.2.3 (2) and 6.2.3 (3), 6.3.1.2, Table 6.2	Integration of various clauses of EN 1993-1-12 into EN 1993-1-1, as EN 1993-1-12 Additional rules for the extension of EN 1993 up to steel grades S 700 will be dissolved
N2017 Decision 23/2014 AM-1-2014-01	6.3.1.2 / Table 6.2	Buckling curves for L-sections
N2031 Decision 23/2014 AM-1-2014-01	6.3.1.2 / Table 6.2	Buckling curves for heavy wide flange rolled sections
N2164 Decision 19/2015 AM-1-1-2015-001	Table 6.2, clause 6.3.1	Classification of rolled I-Profiles fabricated in steel grade S460 within Table 6.2 of EN 1993-1-1
N2191 Decision 20/2015 AM-1-1-2011-01	1.1.2(1)	The limit on the material thickness of hollow sections

4.3 Eurocode EN 1993-1-3

For EN 1993-1-3 a high number of proposals of amendments have been already accepted in TC 250/SC 3 in view of the future implementation of these changes into the new version of EN 1993-1-3. Aside of small editorial corrections the various amendments already decided in TC 250/SC 3 concern e.g. the following items:

- Material, addition of steels S390GD, S420GD, S450GD;
- Effect of flange curling;
- Calculation of edge and intermediate stiffeners;
- Purlins with semi-continuity given by overlaps or sleeves;
- Reduction factors for pull-through resistance;
- Mixed effective width/effective thickness method for outstand elements (Annex D);
- Elastic-plastic resistance with yielding at the compressed flange;
- Elastic-plastic resistance of trapezoidal sheeting webs;
- Stresses in compressed flange;
- Resistance of self-tapping screws;
- Effective width to calculate second moment of area of stiffeners and sheeting with large U stiffeners in the flange;
- Concentrated line load and line load.

A crucial issue forms the existing Annex A (normative) – Testing procedures, which should be enhanced and also increased in volume. Part of the discussion was the idea to transfer the content of Annex A into a Technical Specification because these testing procedures are usually applied only by a small number of specialized users of laboratory testing and do not form a matter of the general design. However as shown also in Table 2, a Technical Specification may allow a parallel national rule which would of course undermine the intended harmonization.

4.4 Eurocode EN 1993-1-5

The current activities of the Working Group WG5 for the revision of EN 1993-1-5 focus on the harmonization with other parts of the Eurocode 3 as well as the development of additional rules to extend the scope of use. Extension of the code rules for example concern various interaction formulae, patch loading and corrugated web girders and hybrid girders.

A European research project COMBRI [6] can be considered as a stepping stone for a large number of research projects in the field of plate buckling and for a systematic development of EN 1993-1-5. The main outcomes from [6] are summarized in [7]. More background information and references are also given in [8] with a focus on the extended code rules. During the last TC 250/ SC 3-meetings several proposals for amendments concerning EN 1993-1-5 have been decided. They will be implemented into the revised EN 1993-1-5 by the relevant Project Team of Task SC 3.4 that probably will take up its work in autumn 2016. Table 4 gives an overview on various modifications of EN 1993-1-5 which have been successfully adopted by decisions in TC 250/SC 3.

An example for the extension of the code rules is the new interaction formula for the case of combined patch loading, bending moment and shear (F-M-V interaction), see Eq. (1).

$$\bar{\eta}_1^{3.6} + \left[\bar{\eta}_3 \cdot \left(1 - \frac{F_{Ed}}{2 \cdot V_{Ed}}\right)\right]^{1.6} + \eta_2 \leq 1.0 \qquad (1)$$

where: $\bar{\eta}_1 = \frac{M_{Ed}}{M_{pl,Rd}}$ and $\bar{\eta}_3 = \frac{V_{Ed}}{V_{bw,Rd}}$ and η_2 utilization ratio transverse force F.

Table 4: Accepted amendments for EN 1993-1-5, Status October 2015 after [9]

SC 3-Document-No. Decision Amendment-No.	Relevant clause	Subject
N1881 Decision 06/2013 AM-1-5-2011-02	4.5.4	Resistance of longitudinally stiffened plates subjected to direct stresses; Interaction between plate and column buckling
N1824 Decision 04/2012 AM-1-5-2011-03a	5.3 (1)	Shear resistance of longitudinally stiffened girders; Requirement of hinged boundary conditions for the use of shear buckling curves
N1881 Decision 06/2013 AM-1-5-2011-03b	5.3 (2)	Shear resistance of longitudinally stiffened girders; Closed section stiffeners
N1950 Decision 06/2013 AM-1-5-2011-03c	5.3 (4)	Shear resistance of longitudinally stiffened girders; 1/3-reduction for open-section stiffeners
N1824 Decision 04/2012 AM-1-5-2011-04	4.5.1 (4)	Resistance of longitudinally stiffened plates subjected to direct stresses; Effective area
N1824 Decision 04/2012 AM-1-5-2011-06	6.4, 6.5	Resistance of steel plate girders subjected to patch loading
N1881 Decision 06/2013 AM-1-5-2011-08	1.1 (2), Note 4	Allowable panel curvature
N1881 Decision 06/2013 AM-1-5-2011-09	1.4 (1)	Definition of cross-sectional area of longitudinal stiffener A_{sl}
N1881 Decision 06/2013 AM-1-5-2012-13	Annex D.2.4	Corrugated web girders; Patch loading resistance
N1933 Decision 06/2013 AM-1-5-2013-16	Section 7	Interaction between transverse force, bending moment and shear force, see Eq. (1)
N2040 Decision 05/2015 AM-1-5-2011-07	Annex D.2.1	Corrugated web girders and transverse bending
N2040 Decision 05/2015 AM-1-5-2013-14	9.1 (2) 9.2.1 (5) 9.2.1 (6) 9.2.4 (5) and Figure 9.5	Definition of stiffeners cross-sections in chapter 9

Recently an amendment regarding the reduced stress design of plates under biaxial compression has been submitted, as Braun [4] has shown that the current formulation may lead to unsafe results for this case. Therefore a modification has been proposed by introducing a V-factor to the third term in the interaction formula for the case of biaxial compression, see Eq. (2). Together with a second safety relevant amendment concerning the patch loading resistance formulae these modifications have meanwhile be accepted by TC 250 as safety relevant issues to become introduced as changes already in the existing code EN 1993-1-5.

$$\left(\frac{\sigma_{x,Ed}}{\rho_x \cdot \frac{f_y}{\gamma_{M1}}}\right)^2 + \left(\frac{\sigma_{z,Ed}}{\rho_z \cdot \frac{f_y}{\gamma_{M1}}}\right)^2 - V \cdot \left(\frac{\sigma_{x,Ed}}{\rho_x \cdot \frac{f_y}{\gamma_{M1}}}\right) \cdot \left(\frac{\sigma_{x,Ed}}{\rho_x \cdot \frac{f_y}{\gamma_{M1}}}\right) + 3 \cdot \left(\frac{\tau_{Ed}}{\chi_w \cdot \frac{f_y}{\gamma_{M1}}}\right)^2 \leq 1 \qquad (2)$$

where: $V = \rho_x \cdot \rho_z$ when $\sigma_{x,Ed}$ and $\sigma_{z,Ed}$ are both compression
$V = 1.0$ in all other cases

The investigations leading to this proposal have been conducted on unstiffened plates. Current investigations are dealing with the case of stiffened panels as well as with the positive effect of tensile stresses [14] in the design of multiaxial loaded panels.

Table 5: Accepted amendments for EN 1993-1-5, Status October 2015 after [9]

SC 3-Document-No. Decision Amendment-No.	SC 3-Document-No. Decision Amendment-No.	SC 3-Document-No. Decision Amendment-No.
N2040 Decision 05/2015 AM-1-5-2014-01	Annex A.3	Shear buckling coefficients
N2040 Decision 05/2015 AM-1-5-2014-02	E.1	Calculation of effective width for stress levels below the yield strength
N2040 Decision 05/2015 AM-1-5-2014-03	9.2.2	Welding of the longitudinal stiffeners
N2127 Decision 06/2015 AM 1-5_2015-01	6.5 (3)	Effective loaded length for patch loading type c) safety related amendment
N2128, Decision by correspondence AM-1-5-2015-02	Clause 10 (5b) Note, Equation (10.5)	Reduced stress design of plates under biaxial compression safety related amendment

5. Summary and Conclusions

As shown in the beginning of this paper, EN 1993 „Design of Steel Structures“ consists of all in all 20 parts making it a considerably large and partially complex code system that however is important for the steel construction practice. First experiences of application of the series of standards and the elaboration of the National Annexes (NAs) showed that, in terms of a practice-oriented application, a certain revision and harmonization is required and still existing ambiguities should be erased by corrections as soon as possible.

For the further evolution of Eurocodes the Mandate M/515 has been agreed between European Commission and CEN. After a general overview on the mandate and its organisation the procedure for the revision of Eurocode 3 is presented. Until the second generation of Eurocodes will presumably appear in 2020, the development of Eurocode 3 will be based on one hand on a general revision and maintenance initiated by systematic reviews and realised by the "Working Groups". On the other hand technical enhancements of the various parts will be realised in the frame of the Mandate M/515 by "Project Teams".

Furthermore within a technical review of some selected code rules insight has been given in recent developments and decisions taken in TC250/SC3 concerning the stability relevant parts EN 1993-1-1, EN 1993-1-3 and EN 1993-1-5. Major aims of these measures are the improvement of "ease of use" for the practitioners and a harmonisation by reducing NDPs and number of alternatives.

Aside of an overview on the concept for the future development of Eurocode 3 this paper should also encourage engineers to participate in the revision process by contributions e.g. to the national systematic reviews. Experts as well practitioners will realise that the input given in that evolution process of Eurocode 3 will turn into benefit for them by a much more clearer, user-friendly and harmonised code.

References

[1] CEN/CENELEC Internal Regulations Part 2, *Common Rules For Standardization Work*, June 2015.

[2] Commission European, "Mandate M/515 – Mandate for amending existing Eurocodes and extending the scope of structural Eurocodes", *Document CEN/TC 250 – N953*, Brussels, 2012.

[3] Commission European, "Response to Mandate M/515: Towards a second generation of EN Eurocodes". *Document CEN/TC 250 – N993*, Brussels, 2013.

[4] Braun B. "Stability of steel plates under combined loading". *Dissertation No. 2010-3*, Institute of Structural Design, University of Stuttgart, 2010.

[5] Greiner R, Kettler M, Lechner A, Jaspart JP. "SEMI-COMP: Plastic Member Capacity of Semi-Compact Steel Sections". *Final Report of RFCS-Project RFS-CT-00044*, 2008.

[6] Kuhlmann U. et al. "Competitive steel and composite bridges by improved steel plated structures (COMBRI)". *Final Report of RFCS-Project RFS-CR-03018*, 2007.

[7] Kuhlmann U, Braun B, Degée H, Zizza A. "New chances and developments of Eurocode 3 Part 1.5 – Bridge design aspects". *Steel Construction*, 4(4), 224–231, 2011.

[8] Kuhlmann U, Zizza A, Braun B, Chacon R, Kövesdi B, Sinur F. "Safety and stability of slender plated elements - new chances and developments of Eurocode 3 Part 1.5", *Proceedings of the 8th International Conference on Thin Walled Structures, Recent research advances and trends*, editors Dubina D, Ungureanu V, Timisoara, Romania, September 2011, 13-28.

[9] Kuhlmann U, Günther HP. "Anpassung und Koordination der Überarbeitung der Normenreihe EN 1993 Stahlbau", *Final report, Project P 52-5-16.142-1375/11 DIBt*, 2014.

[10] Kuhlmann U, Zizza A. "Concept for the Future Development of Eurocode 3 - Design of Steel Structures", *IABSE Conference – Structural Engineering: Providing Solutions to Global Challenges*, 5-7 September 23-25 2015, Geneva, 693-700.

[11] Snijder HH. "The next version of part 1-1 of Eurocode 3 on steel structures: needs for an update and amendments". *A 14-A Conferinta Nationala de Constructii Metalice: "Construieste cu STEEL"*, 19-20 November 2015, Cluj-Napoca, Romania, Document CEN/TC 250/SC 3/WG 1 – N0087, 2015.

[12] Taras A. "Contribution to the Development of Consistent Stability Design Rules for Steel Members". *PhD thesis*, Graz University of Technology, 2010.

[13] Taras A, Greiner R, Unterweger H. "Proposal for amendment rules for member buckling and semi-compact cross-section design". *Document CEN/TC 250/SC 3 – N1898*, 2013.

[14] Zizza A. "Buckling behaviour of Unstiffened and Stiffened Steel Plates under Multiaxial Stress State". *Dissertation (in preparation)*, Institute of Structural Design, University of Stuttgart, publication expected 2017.

STRUCTURAL STABILITY AND DUCTILITY OF DRIVE-IN RACKS UNDER SEISMIC LOADING

Tanim Ahmed, Kim J.R. Rasmussen and Hao Zhang

School of Civil Engineering, The University of Sydney, Sydney, Australia

Abstract: This paper presents the findings regarding the ductile response of a drive-in rack structure obtained from shake table tests carried out on two full scale rack sub-assemblies. The racks were subjected to the El Centro (1940) ground motion record with an increasing scale, with peak ground accelerations ranging from 0.017*g* to 0.34*g*. Plastic response of the rack was identified from visible local deformations of the rack uprights and residual overall drift of the structure. Ductility factors are assessed from two nominally identical tests carried out on frames with the same bracing configuration. Conclusions are drawn about the initiation of plastic deformation of uprights and the effectiveness of energy dissipating spine bracing.

1. Introduction

Driven by maximum space utilization and efficient engineering design requirements, industrial storage rack members have evolved into structural components with highly utilized material and section capacity. Steel storage racks are usually fabricated from cold-formed sections that are perforated to facilitate connection of other component members. Unlike other common structures, steel storage racks are designed to carry live loads much higher than their self-weight. In addition, they are slender structures that can reach heights up to 25 m, with clad racks that can go up to 45 m. In addition, their structural behaviour is characterized by local, global and distortional instability, non-linear beam-column and base plate joint behaviour, and limitation in capacity to develop plastic hinges at beam-column joints. The design standards for regular steel structures are hence not sufficient to incorporate the characteristics of steel storage rack systems. Moreover, the increased usage of storage racks that are accessible to public makes it important to design these structures with proper guidelines that address their special characteristics.

The two most commonly used types of racks are selective racks and drive-in racks. Selective racks store goods in an accessible manner, at a cost of space. On the other hand, drive-in racks achieve better space utilization by sacrificing accessibility and following a first in-last out principle. Drive-in racks allow the forklift truck to drive into the rack to store goods, which limits the rack to have any bracing at the entry side. Bracings can only be placed at the back of the rack, known as spine bracings, and at the top of the rack, known as plan bracings.

Available design standards for steel storage racks include RMI 2008, FEMA 2005, FEM 2010 and AS 4048 [1-4]. These standards provide design approaches that often require a significant amount of experimental data, specific to the rack to be designed. Design of racks for high seismic areas become more complicated and the lack of available standards makes it a challenging task for the designer. Existing design codes for seismic design are based on force-controlled design or capacity design [5]. The total base shear developed for the design seismic event is used for calculating the member forces. For ultimate limit state design, the structure is designed for seismic forces smaller than those corresponding to the linear elastic response, taking into consideration the structure's capacity to dissipate energy by plastic deformation [6]. Eurocode-8 [6] allows the design of a structure using the results of a linear analysis, where the analysis is carried out based on the "design response spectrum", which is a reduced form of the elastic response spectrum. The reduction is done using a behavioural factor, which takes into account the global and local ductile capacity of the structural system. While different codes have similar provisions for determining the reduction factor, different terms are used including ductility factor, R-factor, q-factor and behaviour factor. Herein, the factor will interchangeably be referred to as the R-factor and the q-factor. In general, the factor determines the energy dissipation capacity of the structure by means of factors accounting for ductility, damping and over-strength of members. Hence, a complete definition of the R-factor can be given as:

$$R=R_S.R_\mu.\ R_\zeta$$

where, R_S is the over-strength dependent component, R_μ is the ductility dependent component and R_ζ is the damping dependent component [5]. The seismic design forces are directly related to the reduction factor, which makes it an important parameter in seismic design. Depending on the structural type, the code specifies the value to be taken for the reduction factor. Many studies have been carried out to assess the behaviour factor of various types of structures. A review of results from the past 30 years on strength reduction factors can be found in the study by Miranda and Bertero [7]. Ozmen and Inel [8] carried out a study on the evaluation of the R-factor to be used for existing mid-rise reinforced concrete buildings. Kappos [9] studied the ductility and over-strength part of the behaviour factor.

Here, the objective is to find the ductility dependent component of the reduction factor, or the q-factor. In general, the ductility factor can be defined as the ratio between the elastic and inelastic responses of a structure. According to Eurocode 8 [6], the behaviour factor (q) is an approximation of the ratio of the seismic forces that the structure would experience if its response was completely elastic with a fixed amount of viscous damping, to the design seismic forces. In the literature, several methods can be found for determining the q-factor [10]:

- Methods based on the ductility factor theory.
- Methods based on an extrapolation of inelastic dynamic response analysis of single-degree-of-freedom systems.
- Methods based on an energy approach.
- Methods based on the accumulation of damage.

Hereafter, the approach to assess the q-factor is based on its definition as the ratio of the elastic response and inelastic response at collapse. The definition of collapse can depend on the failure criteria defined by the design code. Here, the peak response (maximum base shear) is considered for assessing the q-factor.

2. Full scale tests

2.1 Test specimen

Tests were carried out on a four-storey high, three-bay wide and three pallets deep drive-in racking system. The structural response of a drive-in rack system is affected by the type of spine bracing used in the rack system. Prior to the destructive test reported in this paper, a number of tests were carried out on the drive-in rack system, with a varied range of spine bracing configurations, to study their effects on the three-dimensional response of the rack system [11]. The configuration that had bracings on the top two rows showed optimum performance and was chosen for the destructive test to assess the ductility factor. Figure 1 shows the tested rack configuration and numbering used for identifying the uprights.

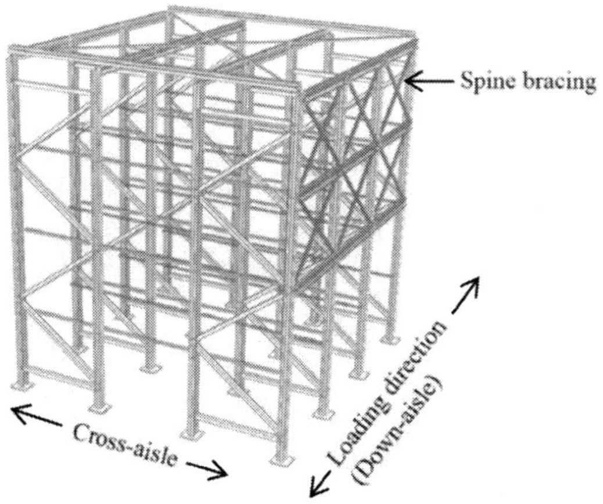

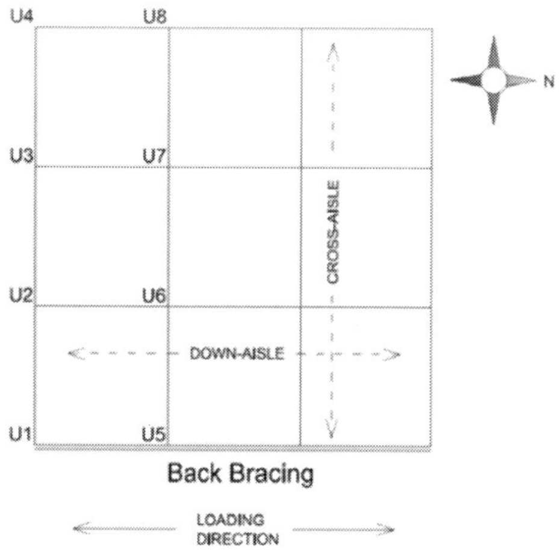

(a) (b)

Fig. 1: (a) Test frame and (b) location wise upright numbering

2.2 Test rig

The testing facility has been designed and commissioned in the Structures and Materials Laboratory at the University of Sydney. Details of the design and components of the testing facility are described in [12]. The facility is a uniaxial shaking table, designed to test full scale, 3×3 bay wide and four storey high drive-in racks. The complete system consists of a frame, able to house the rack both in down-aisle and cross-aisle directions and a 600 kN dynamic rated hydraulic jack attached to the frame. Below the test frame, four parallel I-beams are anchored directly to the concrete slab of the structures laboratory. The test frame sits on the I-beams using a system of rollers that allow the whole frame to move front and back, restraining both uplift and lateral (cross-aisle) movements. Figure 2 shows a schematic diagram of the different components of the test rig and the test frame.

2.3 Instrumentation

Location-wise, three types of instrumentation were used in the testing facility. Type-1 was instrumentation attached directly on the rack, which included (i) strain gauges to calculate axial forces developed in the uprights, spine braces and plan braces, (ii) accelerometers placed on the RHS support frame to measure the base acceleration, and (iii) accelerometers placed on the rack and concrete blocks to detect any occurrence of concrete block sliding. The second type of instrumentation was used for measuring the down-aisle displacements at different locations of the rack. A separate instrument frame, placed on the opposite side of the hydraulic jack attachment, as shown in Fig. 2, was built for housing a total of 10 LVDTs, on three different levels. On the top and mid-level, LVDTs were placed at four different locations with equal intervals, whereas the bottom level had two LVDTs at the outer corners. The third type of instrumentation measured the cross-aisle movements. An instrument frame in the plane parallel to the spine bracing plane was constructed with three horizontal beams. Laser sensors placed on the rack provided cross-aisle movement data, using the horizontal beams of the instrument frame as reference plane.

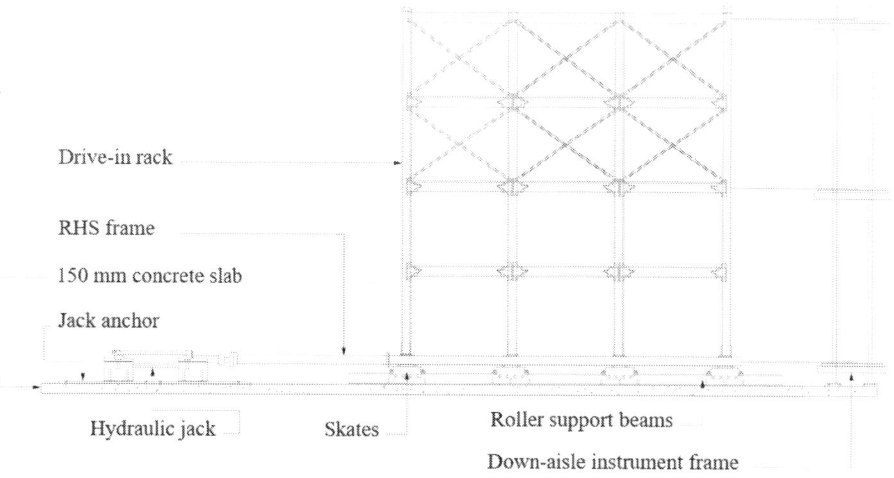

Fig. 2: Schematic diagram of test setup

2.4 Time history

Three different ground motion records were used during the different stages of the seismic testing, which included El Centro (1940), Duzce (1999) and Tabas (1978) earthquake events, sourced from PEER strong motion database. The destructive tests were carried out adopting the ground motion record of El Centro earthquake, having an unscaled peak ground acceleration of 0.31g. The acceleration time history of El Centro earthquake is shown in Fig. 3.

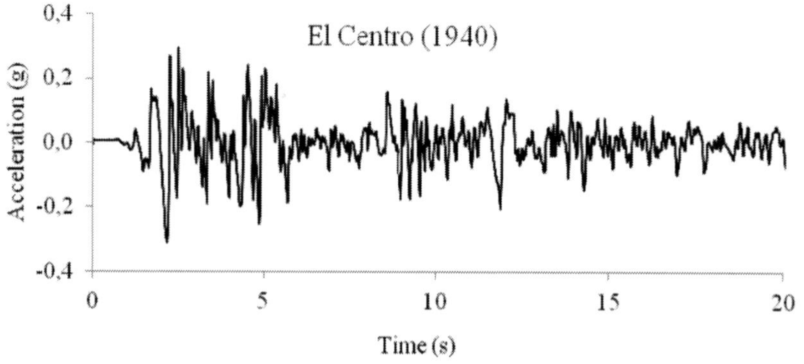

Fig. 3: Acceleration time history of El Centro (1940) earthquake

2.5 Tests

The procedure for assessing the q-factor involved application of the selected earthquake in the down-aisle direction of the rack with an increasing scale, starting with 10% of the design level earthquake. In this case, El Centro-180 earthquake signal has been scaled for the down-aisle direction of the rack, following NZS 1170 [13] and considering the zone coefficient of Wellington. In this test, the applied signals were 10%, 20%, 30%, 60% 90% and 100% of the scaled earthquake. During the test, after each increment, the peak ground acceleration (PGA) was plotted against the top displacement at the front of the rack. At the last increment, i.e. at 100%, a drop in the PGA was observed. At the same time, local deformations of the uprights were also visible on the back uprights featuring the spine braces. The uprights failed at mid-height, where the stiffness changes due to the absence of spine bracings in the lower half of the rack. The outer upright at the back experienced the maximum deformation. The next row of uprights, adjacent to the back uprights, also had visible deformation. In total, eight uprights at the back, out of the total of 16, experienced permanent deformation. Figure 4 shows the location of deformation on the upright and the deformation from different angles.

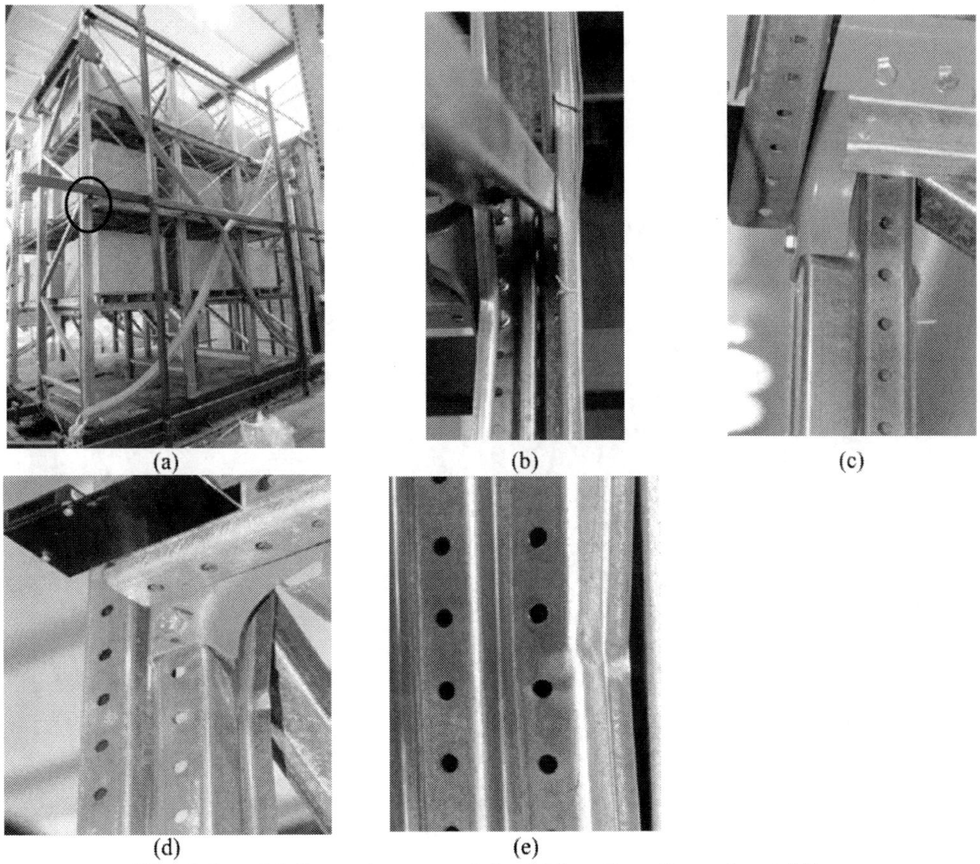

(a) (b) (c) (d) (e)

Fig. 4: Local deformation on upright: (a) location of prominent failure, (b, c, d, e) view of localized deformations from different angles

In order to accurately define the initial linear elastic zone of the rack response, a second destructive test was carried out on a newly setup rack with same spine bracing configuration. In this test, the earthquake signal was applied in smaller steps with 5% increments all through. The test started with 5% of the scaled earthquake signal, with the last signal being 95%. The test series was stopped when the base shear dropped and failure of the uprights was visible. The rack started showing signs of plastic deformation of the upright at the back of the rack, initiated by the pallet runner brackets, at 80% of the scaled earthquake. At the next increment of 85%, the deformation increased with a drop in the base shear curve.

In seismic design, both natural frequency and damping characteristics play significant roles. While all the system components are fixed, the structure's natural period can change depending on the adopted spine bracing configuration. As presented in [11], seven different spine bracing configurations were investigated to study their seismic response. In addition to those, a different type of energy dissipating bracing system [14, 15] was also investigated as reported in this paper. As shown in Fig. 5, the spine bracing system was modified by introducing an energy dissipating component in the bracing system. Instead of the regular pipe braces, flat bars were used and the damper component was placed in the middle of the bracing system. Four different rack configurations were investigated to study the dynamic characteristics and the seismic response.

- *Setup-1*: Rack with regular pipe braces at all levels of the middle bay only (Configuration-5 in [11]).
- *Setup-2*: Setup-1 with energy dissipating bracing system replacing pipe braces.
- *Setup-3*: Spine bracing only on top two levels, covering all three bays (Configuration-3 in [11]) and
- *Setup-4*: Setup-3 with an additional energy dissipating cross brace, spanning the bottom two levels and all three bays.

Figure 6 shows the two setups having the damper elements. The dynamic characteristics of the racks were obtained from free vibration tests, where the rack was given a sudden displacement at the base and stopped. The vibration of the rack, followed by the applied displacement was measured at different locations, in the loading direction. To compare the seismic response, 60% of the scaled El Centro earthquake was applied to all four setups.

Fig. 5: Spine bracing system with damping element

3. Test results

3.1 Ductility factor

Table 1 summarises the rack response at different stages during the incremental loading for Test-2. During the incremental tests of the first frame, the rack response in relation to the

PGA was monitored. In the tests of the second frame, both the base shear and the PGA were plotted as the tests progressed under increasing amplitude. Analysis of the recorded data from the tests showed that in order to follow the displacement path defined by the acceleration control signal, in some cases, the jack produced higher accelerations than the control signal, attributed by the presence of high frequency components in the acceleration time history. These high peaks of accelerations are in the high frequency zone in the earthquake response spectrum; whereas the rack's natural frequency is much lower in the down-aisle direction (0.84 Hz for Test-2). As a result, the rack does not respond directly to those sudden peaks of accelerations.

(a) (b)

Fig. 6: Racks with energy dissipating spine bracing system (a) Setup-2 and (b) Setup-4

Figure 7 shows the fast Fourier transform of the acceleration time history for both the control signal and the measured accelerations. It can be seen that the control signal has a higher content of high frequency components, especially around 2.5 Hz and higher. Figure 8 contains the response spectrum curves for both signals, indicating that the measured acceleration would produce higher responses for structures with periods less than 0.7 s. The tested rack had a natural period of 1.18 s in the down-aisle direction.

Figure 9 shows the peak ground acceleration vs rack displacement curve. As seen in the Figure, the initial part of the curve is non-linear with an increasing slope but the response becomes nearly linear after a PGA of about 0.1g. The time histories of ground acceleration and rack displacement show that the PGA does not occur at the same time as the maximum top displacement (Fig. 10). On the other hand, the base shear and top displacement generate a more meaningful curve (Fig. 11), as the rack response, in terms of top displacement and base shear follow a synchronized phase (Fig. 12).

Both destructive tests produced similar base shear versus top displacement curves. Up to 20% of the scaled earthquake, the rack response stayed linear and afterwards it started to show nonlinear behaviour (Fig. 11). For the second test, the first four data points in the linear range were extended assuming linear response to calculate the ductility factor as shown in Fig. 13.

Table 1: Rack response during incremental tests, Test-2

Test (% of scaled earth-quake)	Rack response	Residual dis-placement (mm)	Observations
5%	First increment.	0	
10%, 15%, 20%	Response is linear.	0	Rack is still in elastic range.
25%	Non-linear response initiates.	0	No visible deformation.
30%	Sudden drop in base shear curve.	0	Possible reason: bolt slippage in rack joints.
35%-75%	Response is nonlinear, following a similar backbone curve.	3	No visible deformation.
80%	Following same backbone curve.	8	First sign of permanent deformation: indentation at mid-height of outer back upright, caused by brackets of pallet runners.
85%	Drop in base shear curve.	20	Increase in upright deformation.
90%	Base shear curve goes up.	55	Safety cables are engaged, limiting the rack displacement.
95%	Base shear curve keeps going up.	133	Engagement of safety cables confirmed both from response curve and from captured video.

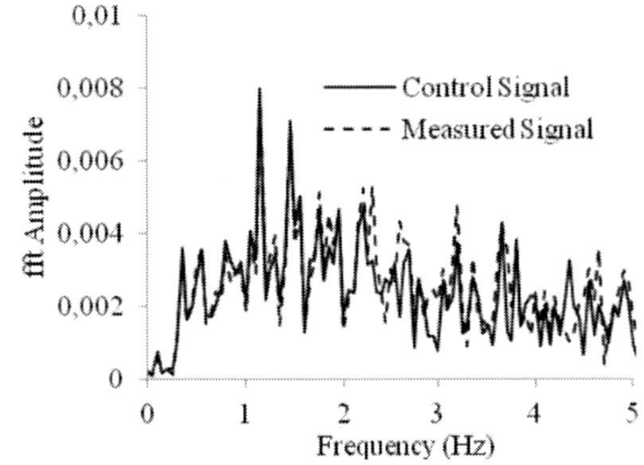

Fig. 7: Fast Fourier transform of acceleration signal

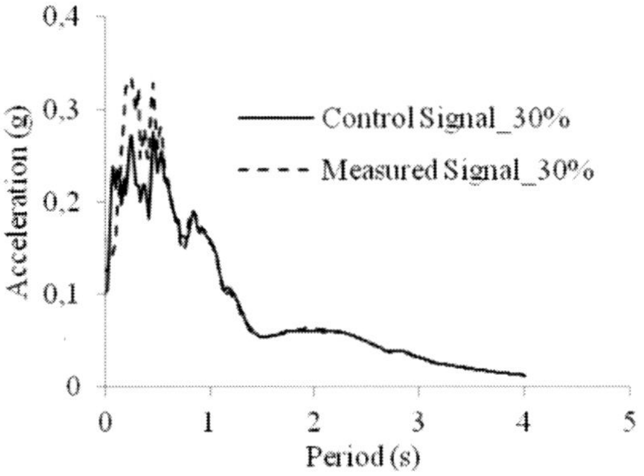

Fig. 8: Response spectra of control signal and measured signal

The q-factors obtained from the tests are presented in Table 2. Although the same earthquake signal was used for both the tests, the main difference was the increment size causing a small variation in the peak response. For the first test, the step after 60% of the signal was 90%. The test with 90% of the scaled signal started with the rack having almost no damage, resulting in a higher base shear. In case of the second test, where increments of 5% were applied all through, the rack started accumulating damage from the test with 75% scaled signal, which resulted in lower peak base shear than in the previous test.

Table 2: q-factors obtained from the destructive tests

Test No.	Maximum base shear, V_{max} (kN)	Displacement at V_{max} (mm)	Displacement from elastic response (mm)	Scaled earth-quake% at V_{max}	q-factor
1	72	270	83	90%	3.25
2	68	258	75	80%	3.45

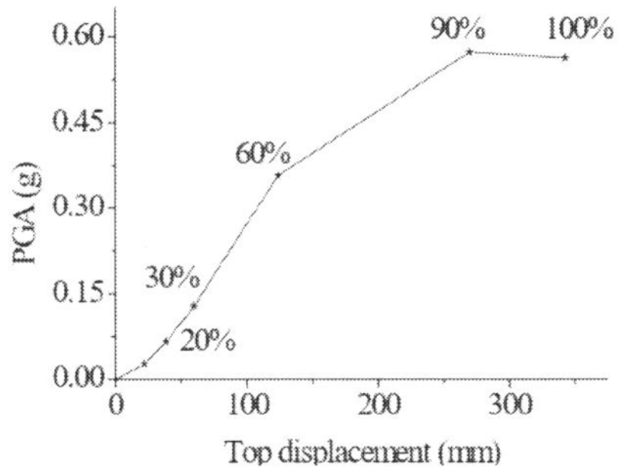

Fig. 9: PGA vs top displacement, Test-1

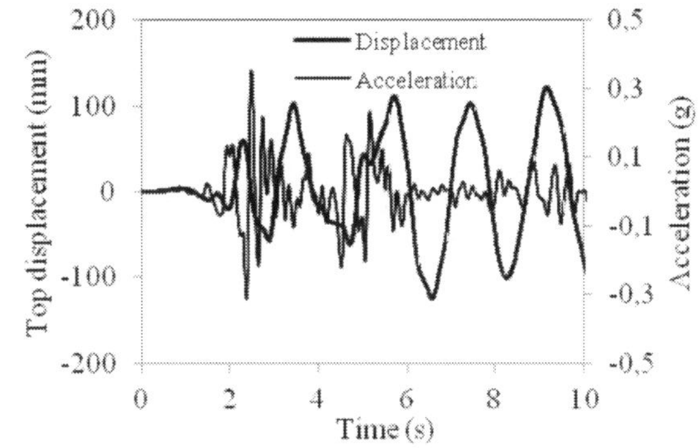

Fig. 10: Top displacement and acceleration time histories for 60% of the scaled earthquake, Test-1

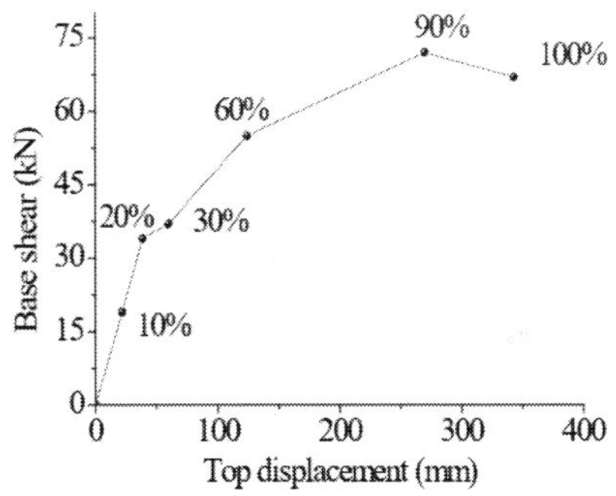

Fig. 11: Base shear vs. top displacement, Test-1

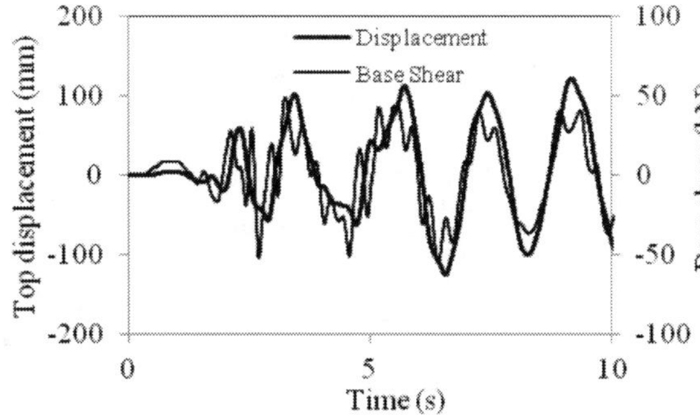

Fig. 12: Top displacement and base shear time histories for 60% of the scaled earthquake, Test-1

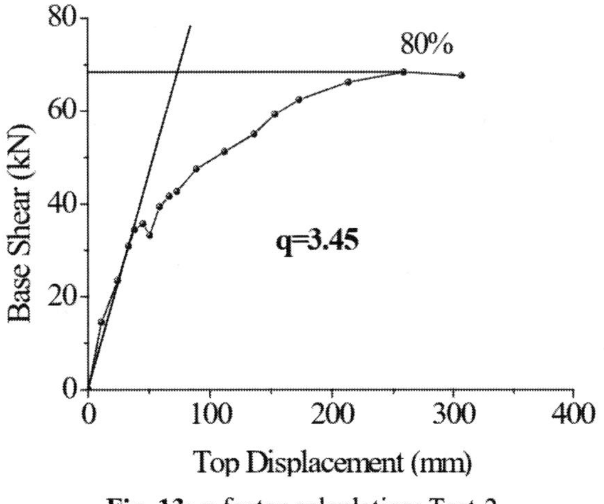

Fig. 13: q-factor calculation: Test-2

3.2 Energy dissipating bracing system

The attenuation of vibration amplitude reflects the damping characteristics of the structure. The critical damping ratio of the structure was calculated from the displacement response at the top of the rack. The decay of motion has been related to amplitudes 10 cycles apart, starting from the initial maximum displacement. Figure 14 shows the top displacement of the rack with Setup-1 under free vibration.

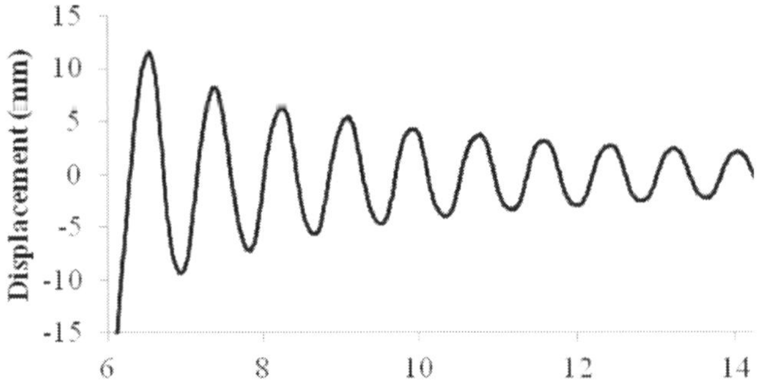

Fig. 14: Top displacement at the front of rack Setup-1

In Setup-2, the use of the damper braces changed the natural period of the structure from 0.81 second to 1.21 second. There was also significant improvement of the critical damping ratio, which changed from 2.9% to 6%. The change in the spine braces made the rack more flexible, which in addition to the damping element, was also attributed to the use of flat bars instead of the stiffer pipe braces. Figure 15 shows the top displacement response of the rack with Setup-1 and Setup-2.

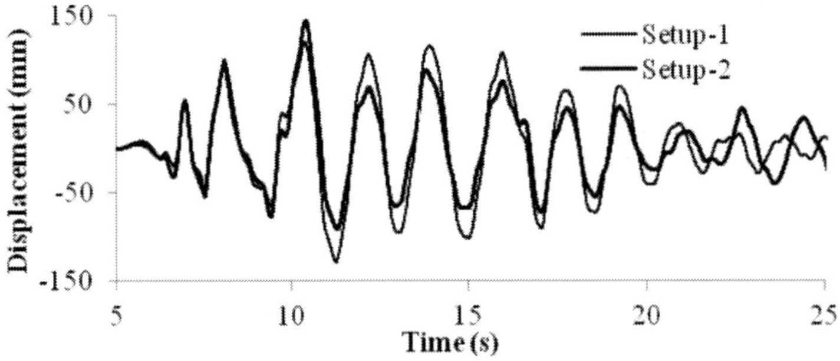

Fig. 15: Top displacement time history of rack Setup-1 and rack Setup-2

As can be seen from the figure, the initial sway amplitude was almost the same at the beginning, for both the racks. During the subsequent large sway motions between 10s and 20s, the rack with damping braces (Setup-2) performed better than the one with regular braces, by reducing the sway amplitudes.

In Setup-4, the additional bracings increased the seismic axial forces in the connected uprights. Among the tested four cases, Setup-4, with the added damper brace, experienced the least top displacement. The added damping element experienced maximum deformation and lost capacity near the end of the seismic signal. Table 3 summarises the test results.

Table 3: Summary of results for racks with damper braces

Test #	Bracing Configuration	Period of Structure (s)	Damping Ratio %	Base Shear (kN)	Max Top Displacement (DI) (mm)	
					Front	Back
1	Setup-1	0.81	2.9	65	112	80
2	Setup-2	1.21	6.0	56	145	126
3	Setup-3	1.18	2.3	53	128	93
4	Setup-4	1.13	3.5	59	97	69

The seismic axial forces developed in the different uprights under El Centro-60% loading are presented in Table 4 and Table 5. The locations of the uprights (U1, U2, ... , U8) are shown in Fig. 1. It follows from Table 4 and Table 5 that for Setup-1 and Setup-2, the use of damper braces has mixed effects on the compression forces in the uprights. Rather, the effectiveness is evident from the tension response of the uprights. There is about 25% reduction in tension forces in uprights U1 and U2. The tension force in the upright featuring the spine braces (U5) is reduced by 37.5% and in upright U6, the reduction is about 56%. In Setup-4, unlike Setup-1 and Setup-2, the energy dissipating element does not replace any existing bracing member of Setup-3; rather it comes as an additional member. Hence, in most of the uprights, both tension and compression forces are somewhat increased. The improvement achieved from Setup-4, is the reduced top displacement of the rack.

Table 4: Maximum seismic compression in uprights

Rack Configuration	U1 (kN)	U2 (kN)	U3 (kN)	U4 (kN)	U5 (kN)	U6 (kN)	U7 (kN)	U8 (kN)
Setup-1	-26	-11	-26	-25	-72	-15	-19	-18
Setup-2	-23	-9	-19	-11	-77	-14	-24	-9
Setup-3	-45	-5	-24	-19	-11	-19	-26	-12
Setup-4	-65	-11	-32	-29	-13	-11	-27	-17

Table 5: Maximum seismic tension in uprights

Rack Configuration	U1 (kN)	U2 (kN)	U3 (kN)	U4 (kN)	U5 (kN)	U6 (kN)	U7 (kN)	U8 (kN)
Setup-1	12	27	26	17	32	32	11	8
Setup-2	9	21	22	16	20	14	7	7
Setup-3	11	16	25	15	7	1	4	9
Setup-4	12	24	28	18	14	3	11	12

4. Conclusions

Seismic tests have been performed on full scale and fully loaded drive-in rack sub-assemblies. The tests were carried out on a testing facility recently designed and commissioned at the University of Sydney as part of this project. Two frames were tested, inducing plastic deformations in both by step-wise increasing the selected ground motion signal. The effect of increment size has been identified in terms of maximum scale of the earthquake signal required to reach ultimate response.

The failure initiated at mid-height of the upright members featuring spine braces, where the absence of spine bracing in the bottom two levels caused abrupt change in upright stiffness in the down-aisle direction. The inelastic deformation process was initiated by indentations caused by the brackets of pallet runners.

Ductility factors were determined from the two tests, producing values of 3.25 and 3.45. The small difference in the values may be attributed to (i) the different increment sizes used for scaling the applied time history signal and (ii) the fact that the first frame underwent a number of tests with a varied range of spine bracing configurations and earthquake signals prior to the destructive test, as reported in [11].

Although not visible from the direct comparison of the maximum top displacements of the racks, the displacement response time history does show improved performance when energy dissipating components are installed in the spine bracing system. Under seismic action, the damper elements experience deformation and attract a major portion of the seismic force, which reduces damage to the other components of the rack. The advantage of such damper elements is that, unlike rack uprights, they are easily replaceable after damage.

Acknowledgments

This research was supported under Australian Research Council's Linkage Projects funding scheme (project grant LP120100458). The authors are grateful to Dematic Pty. Ltd. for supplying all testing materials at no cost.

References

[1] RMI, *Specification for the design, testing and utilization of industrial steel storage racks*, Rack Manufacturers Institute: Charlotte (USA), 2008.

[2] FEMA, *Seismic Considerations for Steel Storage Racks Located in Areas Accessible to the Public*, in *Current Storage Rack Seismic Design Practices*, National Institute of Building Sciences: Washington, DC, 2005

[3] FEM, *10.2.08, Recommendations for the design of static steel pallet racking in seismic conditions (version 1.02)*, Federation Europeenne de la Manutention (FEM) racking and shelving product group (European Racking Federation): West Midlands, UK, 2010.

[4] AS 4084, *AS 4048 Steel Storage Racking*, Standards Australia, Sydney, 2012.

[5] Ferraioli M, Lavino A, Mandara A. "Behaviour factor of code-designed steel moment-resisting frames", *International Journal of Steel Structures*, 14(2), 243-254, 2014.

[6] *Eurocode 8 Design of structures for earthquake resistance - Part 1: General rules, seismic actions and rules for buildings*. 2004.

[7] Miranda E, Bertero VV. "Evaluation of strength reduciton factors for earthquake-resistant design", *Earthquake Spectra*, 10(2), 357-379, 1994.

[8] Ozmen HB, Inel M. "Evaluation of strength reduciton factor for existing mid-rise RC buildings", *The 14th World Conf. on Earthquake Engineering*, Beijing, China, 2008.

[9] Kappos AJ. "Evaluation of behaviour factors on the basis of ductility and overstrength studies", *Engineering Structures*, 21(9), 823-835, 1999.

[10] Castiglioni CA, Zambrano A. "Determination of the behaviour factor of steel moment-resisting (MR) frames by a damage accumulation approach", *Journal of Constructional Steel Research* 66(5), 723-735, 2010.

[11] Ahmed T, Rasmussen KJR, Zhang H. "Down-aisle seismic tests of drive-in rack systems", *Proceedings of the International Colloquium on Stability and Ductility of Steel Structures (SDSS2016)*, Timisoara, Romania, 2016.

[12] Ahmed T, Rasmussen KJR, Zhang H. "Development of a seismic testing facility to test full scale steel storage racks", *Proceedings of the International Colloquium on Stability and Ductility of Steel Structures (SDSS2016)*, Timisoara, Romania, 2016.

[13] *NZS 1170.5:2004 Structural design actions - Part 5: Earthquake actions - New Zealand*, Standards New Zealand, 2004.

[14] Li G, Li HN. "Experimental Study and Application in Steel Structure of 'Dual Functions' Metallic Damper", *Advanced Steel Construction*, 9(3), 247-258, 2013.

[15] Comini R, Schafer BW. "Stability of energy-dissipating steel fuses in an innovative seismic system for cold-formed steel structures", *Annual Stability Conference*, Structural Stability Research Council, Toronto, Canada, 2014.

MODELS FOR STABILITY ANALYSIS AND DESIGN OF STEEL AND COMPOSITE PLATE GIRDERS

Ioannis Vayas

National Technical University of Athens, Athens, Greece

Abstract: Various numerical models for analysis and design of steel and composite plate girders are introduced. They include six and seven degree of freedom beam models, models in which plate girders are represented by an equivalent truss and shell element FE models. Linear and non-linear analysis methods are employed for stability problems of members that include lateral torsional buckling modes. The applicability of 6 and 7 DOF beam models and truss models are studied. 3D models for complete bridge decks in which the steel girders are represented by truss models are employed. The validity of these models is studied for orthogonal, skewed and curved decks in the construction stage and the final service state.

1. Introduction

Steel and composite plate girders are usually represented in structural analysis by means of beam elements. This is because of their simplicity and due to the fact that most Codes are structured on the basis of beam elements analysis output when defining required design verifications. For example the main parts of Eurocodes 3 and 4 expect from analysis internal forces and moments N_{Ed}, M_{Ed} to compare and verify against the relevant plastic, elastic or effective resistances and their interactions [1], [2]. Enhanced computational capacities and the easier access to more sophisticated software tools allow a wider application of refined models that may comply with the design principles. Recognizing this fact, EN 1993-1-1 [1] defines in its Principles that "*Analysis shall be based upon calculation models of the structure that are appropriate for the limit state under consideration*" and that "*The method used for the analysis shall be consistent with the design assumptions*".

The most usual representation for a steel or composite plate girder is the six (6) degree of freedom (DOF) beam element that correspond to three displacements (u_x, u_y, u_z) and three rotations (θ_x, θ_y, θ_z) per node, based on Euler-Bernoulli or Timoshenko beam theories that maintain the assumption that cross sections remain plane after deformation. With the inclusion of the first derivative of either the total angle of twist (θ'_x) [3] or its primary part [4] as an additional unknown, the seven (7) DOF beam element is created that allows the direct determination of the warping bimoment M_w (or B_w in terms of [1]) and the primary and secondary torsion moments, ignoring or taking into account the secondary torsional shear deformation effects, respectively. Beam elements have been further developed by adding more DOFs in the stiffness matrix, based on the so-called "Higher-Order Beam Theories", in order to include shear lag effects due to flexure or torsion. Thus, employing two independent warping parameter in beam transverse directions for the flexural problem and two more for the torsional one, a ten (10) DOF beam element arises that allows for the direct inclusion of shear lag effects in the beam model [5]. Refined beam elements are more accurate in stress analysis and more suitable for fatigue limit state design.

In order to overcome problems that were encountered in bridge design, truss models were successfully introduced to represent steel or composite plate girders [6]. The truss is composed of chords and posts of beam elements representing the flanges and part of the web and truss elements as bracings that represent the web. For composite girders, the top chord may be linked to additional beam elements that represent the concrete slab. This truss representation allows more accurate analysis for complicated topologies, like in skew or curved bridges, as well as coverage of stability problems at construction stages and at service conditions [7], [8], [9], [12].

Plate girders may also be represented by shell or volume finite elements. However, they are normally employed in design practice for investigations of complicated construction details or joints, other local areas, where other models fail to provide accurate solutions as required by the limit state under consideration, or for calibration purposes with the results of other more simple models [5]-[9].

This paper gives an overview for the application of various models and methods for analysis and design of I-shaped girders. In order to provide guidance for the design practice, it illustrates the use of linear and non-linear analyses with and without imperfections that may be employed in the context of the Eurocodes.

2. Beam models

As discussed before, beam models are the simplest way to represent steel or composite plate girders. This section illustrates their use for a simple case study that refers to analysis and design of a simply supported class 1 girder subjected to compression and biaxial bending, Fig. 1. The girder is represented by a 6 or 7 DOF beam element and alternative methods of analysis according to [1] are employed. All calculations were done with the commercial software SOFISTIK [10] widely used in engineering practice.

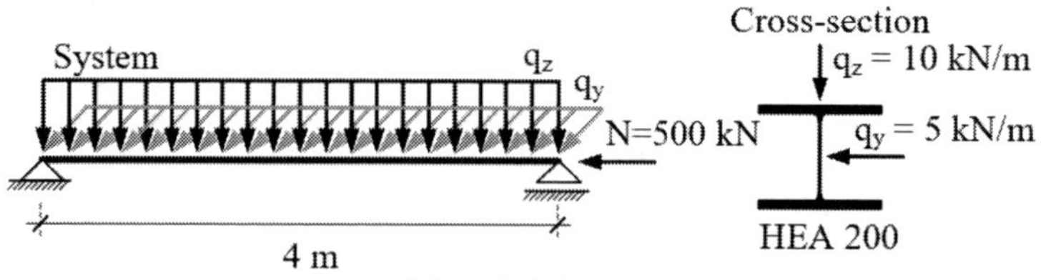

Fig. 1: Examined girder under compression and biaxial bending

Following methods of analysis were considered:

- **LA** Linear analysis + member check according to [1], 6.3.3.

- **GNIA** Geometric non-linear elastic analysis with imperfections + cross section check.
- **GMNIA** Geometric and material non-linear analysis with imperfections.
- **General Method** This method is proposed in [1], 6.3.4. It does not strictly apply to the specific problem due to biaxial bending but is employed here for comparison reasons.

Equivalent geometric imperfections were considered as following:

- **I1** Initial bow weak axis imperfections $k \cdot e_{0,d}$ as [1] 5.3.4, with $k = 0.5$. For weak axis buckling curve c is relevant. $e_{0,d}$ is then according to [1] Table 5.1 equal to $L/200$ for GNIA (elastic analysis) and $L/150$ for GMNIA (plastic analysis).
- **I2** Imperfections following the shape of the elastic critical buckling mode η_{cr} as in [1] 5.3.2 (11).

Cross section check was done according with appropriate interaction formulae. For analysis with the 6 DOF element the interaction relation as proposed in [1], 6.2.9.1 (6) was used, which is written as:

$$\frac{M_{y,Ed}}{M_{N,y,Rd}}^{\alpha} + \frac{M_{z,Ed}}{M_{N,z,Rd}}^{\beta} \leq 1 \qquad (1)$$

Analysis with the 7 DOF element delivers in addition a bimoment M_w. Cross section resistance is then checked by following relation as proposed in [11], which may is expressed by Eqn. (2), where all quantities express the ratio between internal forces and moments divided by the corresponding plastic resistances, α_f, α_w are the ratios between the cross section areas of the flange and correspondingly the web to the total section area and p is the sign of the product $(m_z \cdot m_w)$:

$$\left[\frac{m_y \cdot (1+\alpha_f) - 2 \cdot (\alpha_w + n)}{2 \cdot \alpha_f}\right]^2 + |m_z - p \cdot m_w| \leq 1 \qquad (2)$$

Figures 2, 3 and 4 show exemplary results for the analysis with the 7 DOF element. Fig. 2 shows transverse displacements of the centroid, Fig. 3 moments M_z around the weak axis and Fig. 4 bimoments all at mid-span. It may be seen that each method starts with different equivalent imperfections and that the imperfections for elastic analysis (GNIA method) are smaller than those for plastic analysis (GMNIA method).

Table 1 summarizes the achieved load factors to failure for all analyses employed. Following observations can be made and conclusions drawn:

- Non-linear material behavior results in significant stiffness reductions at higher loads and therefore lower limit loads. Opposite to the provisions of [1] initial bow imperfections should be accordingly lower for plastic analysis than for elastic analysis.
- The critical buckling modes of the 6 DOF beam element do not "recognize" a lateral torsional but a flexural buckling problem. So does the shape of geometrical imperfections if selected to correspond to the 1st buckling mode.
- The limit loads from GMNIA analyses are similar for both 6 and 7 DOF beam element. However they differ when GNIA analysis and cross section plastic capacity is the selected design method. This is due to the fact that 6 DOF elements do not deliver bimoments and so their influence in the capacity check is ignored.
- The "General Method" delivers acceptable accurate limit loads. However, this applies when 7 DOF beam elements are used due to the inclusion of the bimoment in the determination of the ultimate load amplifier. It should be examined if its field of application could be extended to problems where members are subjected to both in- and out-of-plane loading.
- Linear analysis in combination with member capacity checks following the existing provisions of [1] provides results on the safe side.

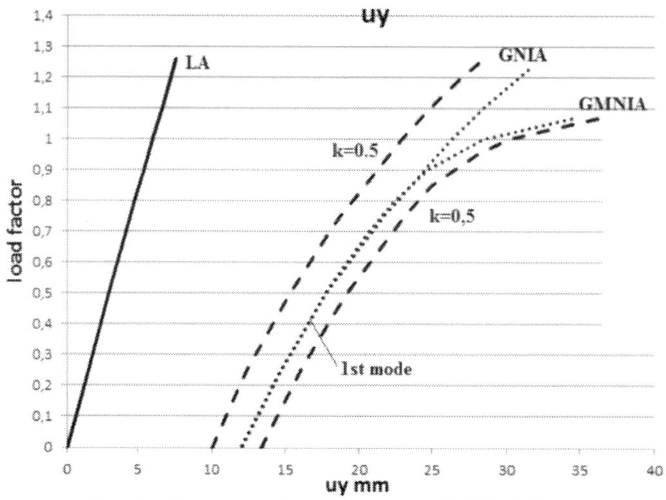

Fig. 2: Transverse displacements - 7 DOF element

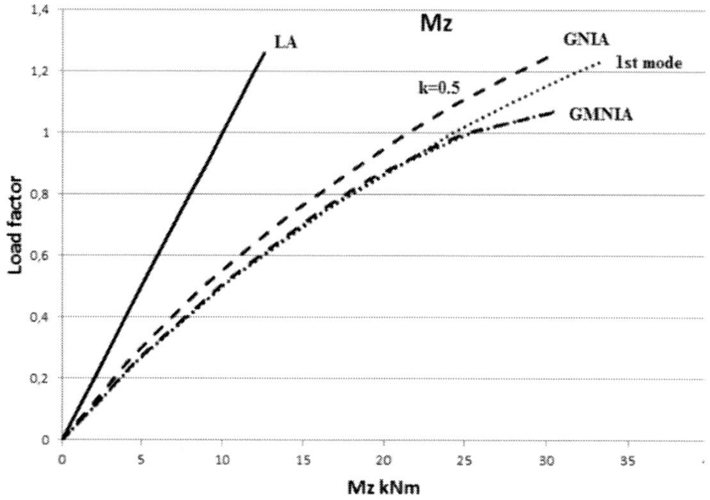

Fig. 3: Bending moments around the weak axis - 7 DOF element

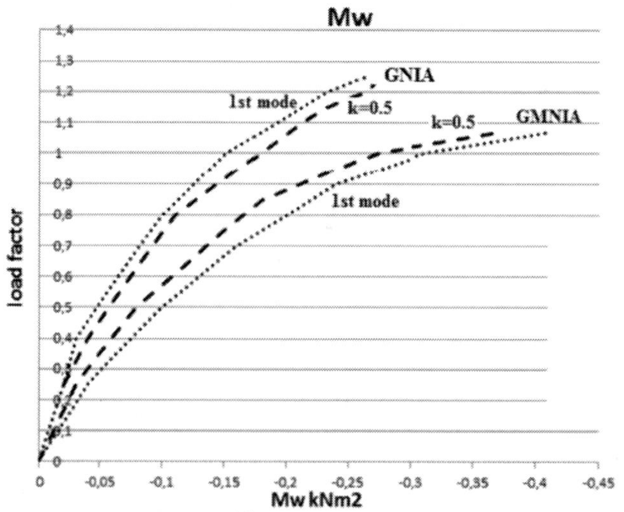

Fig. 4: Bimoment - 7 DOF element

Table 1: Load factors for beam models

Type of analysis	Type if imperfections	6 DOF	7 DOF
General method	-	1.28	1.06
GNIA + cross section check	I1 k=0.5	1.37	1.26
	I2 1st mode	1.34	1.24
GMNIA	I1 k=0.5	1.08	1.07
	I2 1st mode	1.09	1.07
LA+ member check		1.0	1.0

3. Truss models

Truss models were successfully introduced for analysis and design for composite bridge decks at service and at construction phases [6]. The flanges of the truss are beam elements with cross section composed of the flange and part (1/3) of the web of the steel girder. The flanges are connected by a "hybrid combination" of truss and beam elements that represent the web of the steel girder. The concrete flange is represented by another beam element rigidly connected with the top steel flange of the truss with an offset as shown as shown in Fig. 5. Despite its simplicity, the 1/3 rule has been proven to provide accurate deflections and stresses for a wide range of geometric properties.

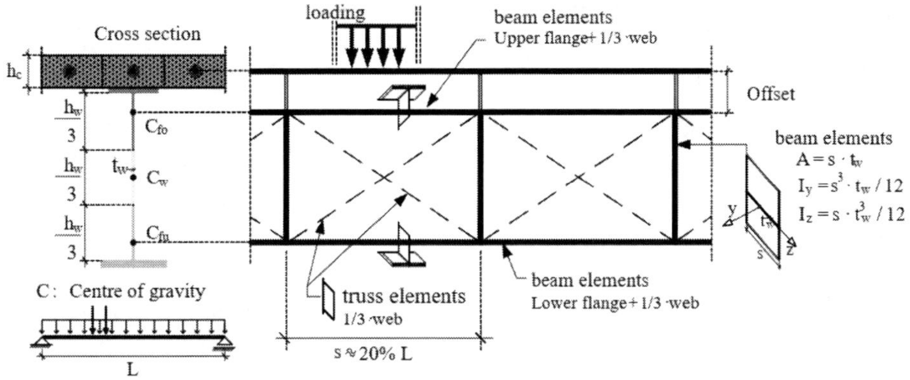

Fig. 5: Truss idealization for a composite plate girder

This representation allows for the introduction of horizontal or transverse bracings and accordingly the study of lateral stability during construction stages. Fig. 6 shows the truss representation of a four plate girder simply supported bridge deck with a light horizontal bracing scheme. The construction sequence adopted includes two construction stages, one where the two end regions are concreted followed by concreting of the mid-span area. The girders are all steel in the first construction stage. In the second construction stage the girders are composite in the end regions and only steel in at mid-span. The concrete slab at end regions is represented in this example by a grillage of interconnected beams.

Buckling modes and corresponding load factors may be determined by linear buckling analysis. Fig. 6 shows the shape of the 1st buckling mode. It may be seen that the critical buckling modes correspond to lateral buckling of the steel girders in the end regions at construction stage 1 and in the mid-span area at construction stage 2.

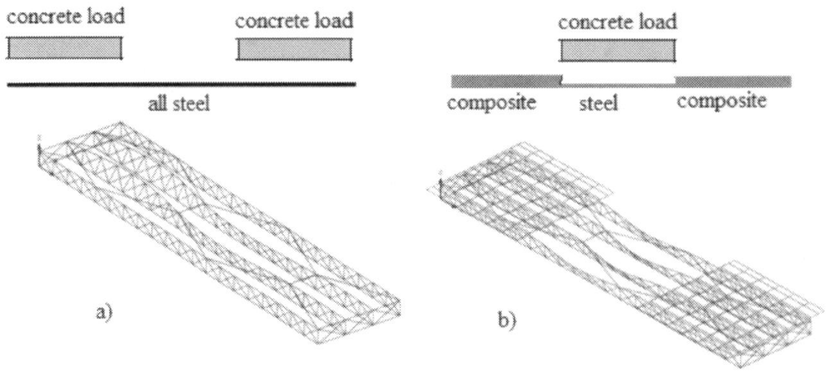

Fig. 6: Shape of 1st buckling mode of a bridge deck at two construction stages

Table 2 summarizes the values of the critical load factors for the first two buckling modes and compares the results of the truss model with the corresponding ones for an FEM model using shell elements with ABAQUS Software [13]. A good agreement between the two models at both construction stages can be observed. It is mentioned that if the entire deck had been concreted in one stage the critical buckling factor would be 1.5, i.e. half of that at concreting at two stages.

Table 2: Critical buckling load factors

Type of model	Buckling mode	Stage 1	Stage 2
Truss model	1st	3.003	5.434
	2nd	3.017	5.671
FEM	1st	3.103	5.298
	2nd	3.115	5.528

4. Truss models for skewed bridge decks

In skewed bridges special phenomena, like twisting and out of plane rotation of the main girders during concreting, concentration of reaction forces and fatigue problems due to out-of-plane web distortion occur that make their analysis and design intricate. Cross-frames are placed transversally, either skewed or normal, in order to prevent lateral torsional buckling of the steel girders during concreting of the deck. Cross-frames restrain deflections causing girders to rotate out of plumb and lateral stresses to be developed into the flanges. The out of plane rotation varies, depending on the location along the girder span.

3D models with steel girder representation by a truss model are very efficient to represent the ensemble of skewed bridges and may be used for both strength and stability analysis [8]. This is illustrated in the example of a simply supported composite bridge as shown in Fig. 7. For the stability analysis of this bridge in the construction phase, the steel girders are modelled by equivalent trusses.

Fig. 8 shows the shape of the 1st buckling mode during concreting with cross bracings only at the end sections and with additional intermediate cross bracings every 8 m. It may be seen that in both cases lateral buckling between cross bracings occurs. Table 3 gives the corresponding critical load factors. It may be seen that these factors are increased by the factor of 10 when intermediate bracings are used. Fig. 6 and Table 3 show also that the truss model delivers almost identical results compared to the more sophisticated shell FEM model.

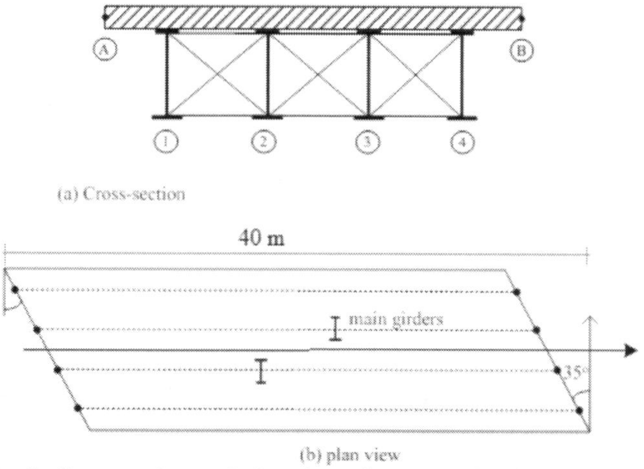

Fig. 7: Cross section and plan view of a composite skewed bridge

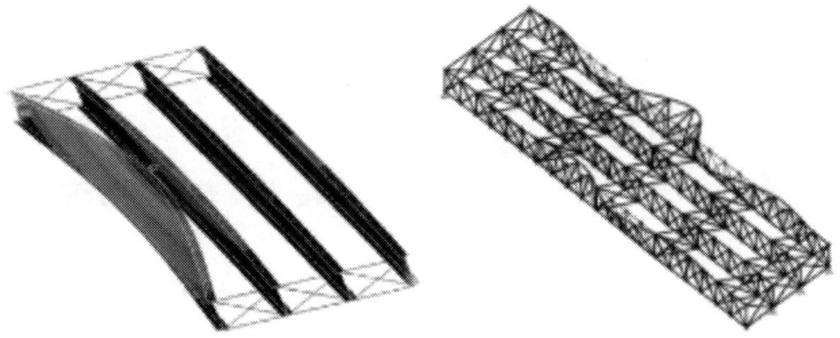

Fig. 8: Shape of the 1st buckling mode for the shell element FEM and the truss model [8]

Table 3: Critical buckling load factors for the skewed bridge [8]

Type of model	Buckling mode	No intermediate bracings	With intermediate bracings
Truss model	1st	0.390	3.370
	2nd	0.390	3.419
FEM	1st	0.367	3.472
	2nd	0.367	3.498

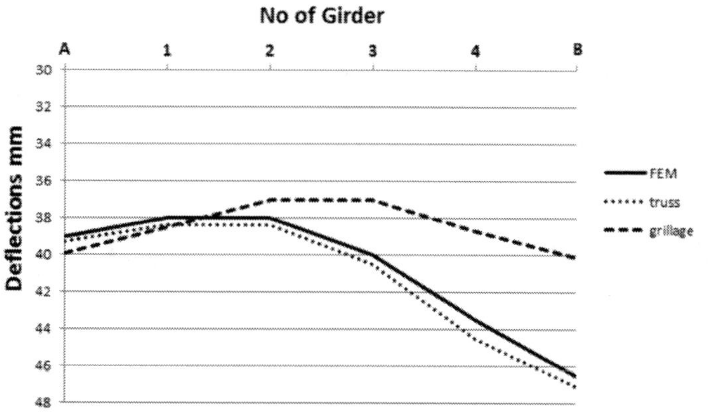

Fig. 9: Mid-span deflections of the deck of Fig. 7 due to UDL loading

Unlike grillage models, truss models are also appropriate for analysis of skewed plate girder decks in the final stage. Fig. 9 shows for the deck of Fig. 7 the deflections at mid-span, transverse to the traffic direction, due to UDL loading as calculated by grillage analysis, the proposed truss model and a FEM model using shell elements. It may be seen that the truss model delivers results very close to those of the reference FEM model, while the grillage model does not adequately reflect the behaviour.

5. Truss models for curved bridge decks

Curved composite bridges display unique behavior characteristics, some of which are not immediately obvious. The presence of curvature affects the geometry and as a consequence, the behavior of the structure. Curved bridges are subjected to coupled torsion and bending because of the curvature and their analysis is more complex than straight bridges. In addition to simple vertical flexure behavior, there can be significant torsional loading and twisting of the girders that cause lateral stresses to the flanges. Such phenomena are difficult to investigate by grillage analysis. For this reason, 3D truss models as introduced before may be employed [7]. This is illustrated in the example of a simply supported curved composite plate girder bridge shown in Fig. 10. The plate girders are connected with cross bracings at intermediate positions and at bridge ends.

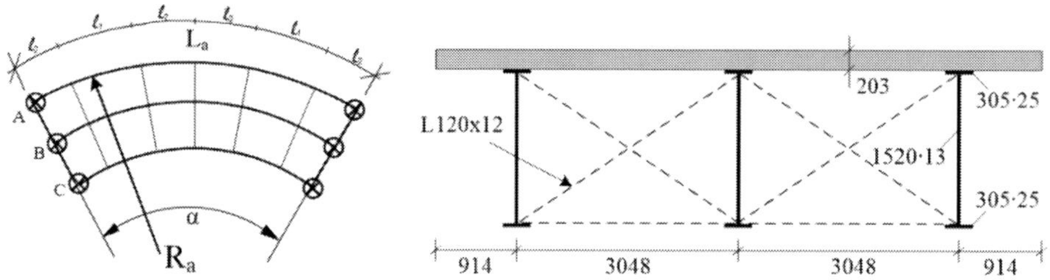

Fig. 10: Cross section and plan view of a composite curved bridge, L_a= 30.5 m

This bridge was analysed at the constructions stage and the final stage by shell element FE model and a truss model. Fig. 11 gives for the concreting stage the shape of the 1st buckling mode and Table 4 the corresponding buckling factors as determined by linear buckling analysis. The similarity between buckling shapes and buckling factors is obvious. Other results for the final stage are referenced in [7]. The main advantage of the truss model is its ability to simulate the whole structure using fewer members, offering at the same time a faster solution and an easier interpretation of results.

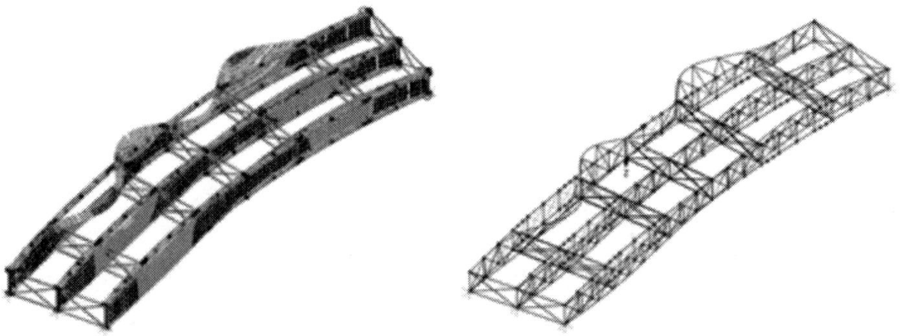

Fig. 11: 1st buckling mode at the concreting stage for the shell element and the truss model

Table 4: Critical buckling load factors for the curved bridge

Buckling modes	Truss model	FEM
1st	3.849	3.779
2nd	4.808	4.581
3rd	5.880	5.903

6. Truss models for inelastic stability analysis

Truss models have been employed not only for strength and linear buckling analysis but also for non-linear inelastic stability analysis including imperfections (GMNIA analysis). Fig. 1 shows the geometric properties and the loading conditions for the plate girder studied. The girder has been analysed in the ABAQUS Software [13] by two models, a shell element FE model and a truss model. Non-linear analysis was employed under consideration of both geometric and material non-linear effects. Equivalent geometric imperfections corresponding to the 1st buckling mode were considered. The imperfections were scaled appropriately so that the initial lateral displacement of the top flange at mid span ranged from 13.3 mm to 40 mm, i.e. from $L/300$ to $L/100$. Failure patterns of the last step for the system with imperfection 13.3 mm are shown in Fig. 12. Fig. 13 shows the development of the mid span deflections up to and after failure [9]. It may be seen that the proposed truss model correlates very well with the detailed shell element model up to failure. The softening behavior after failure is similarly captured in both models. The truss model is able to proceed with non-linear analysis in higher values of displacements indicating a more stable numerical behavior.

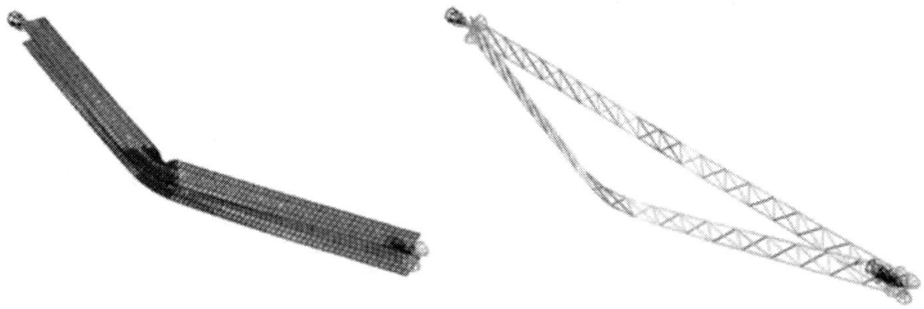

Fig. 12: Failure pattern of the shell FE model and the truss model with initial imperfection 13.3 mm

7. Conclusions

Various numerical models to represent steel and composite plate girders were presented. They include six and seven degree of freedom beam models, equivalent truss models and shell element FE models. It was shown that the classical 6 DOF beam models do not correctly predict lateral torsional buckling modes and should be substituted by 7 DOF elements. It was found that equivalent geometric imperfections for elastic GNIA analysis should be larger than those for plastic GMNIA analysis, oppositely to the proposals of [1]. Models in which the plate girders are represented by equivalent trusses are appropriate to represent isolated girders and complete composite bridge decks in association with the concrete slabs. They were shown to correctly predict the structural response in the construction phase and the final service state for orthogonal, skewed and curved bridges.

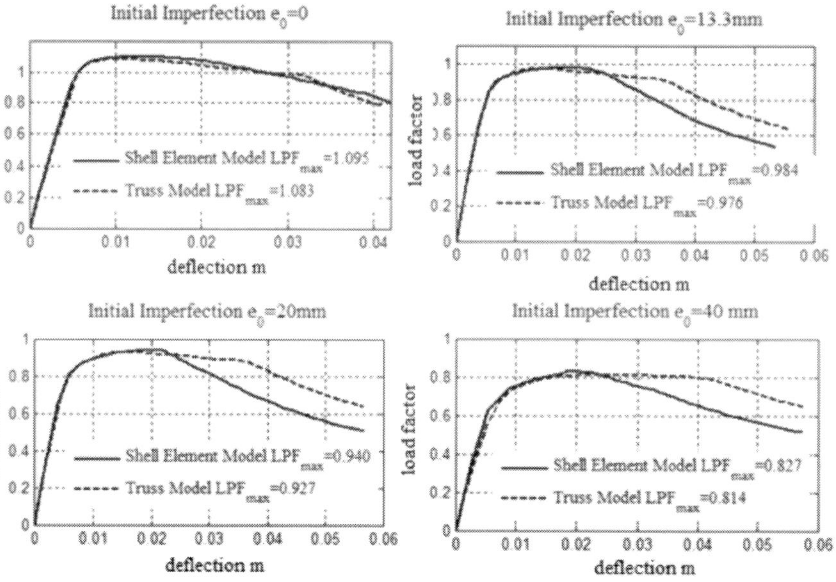

Fig. 13: Load factor vs. mid-span deflection for various lateral top flange imperfections

References

[1] EN 1993-1-1. *Eurocode 3: Design of steel structures – Part 1-1: General rules and rules for buildings*, CEN, 2005.

[2] EN 1994-1-1. *Eurocode 4: Design of composite steel and concrete structures – Part 1-1: General rules and rules for buildings*, CEN, 2005.

[3] Sapountzakis EJ, Mokos VG. "Warping shear stresses in nonuniform torsion by BEM". *Computational Mechanics,* 30, 2003.

[4] Tsipiras VJ, Sapountzakis EJ. "Bars Under Nonuniform Torsion - Application to Steel Bars, Assessment of EC3 Guidelines", *Engineering Structures,* 60, 133-147, 2014.

[5] Sapountzakis EJ, Dikaros IC. "Advanced 3–D Beam Element of Arbitrary Composite Cross Section Including Generalized Warping Effects", *International Journal for Numerical Methods in Engineering*, 2015.

[6] Vayas I, Iliopoulos A, Adamakos Th. "Spatial systems for modelling steel-concrete composite bridges – Comparison of grillage systems and FE models", *Steel Construction Design and Research,* 3, 100-111, 2010.

[7] Adamakos Th, Vayas I, Petridis St, Iliopoulos A. "Modeling of curved composite I-girder bridges using spatial systems of beam elements", *Journal of Constructional Steel Research*, 67(3), 462-470, 2011.

[8] Vayas I, Adamakos Th, Iliopoulos A. "Three dimensional modelling for steel-concrete composite bridges using systems of bar elements – Modeling of Skewed Bridges", *International Journal of Steel Structures*, 11(2), 157-169, 2011.

[9] Smyrnaios SV, Iliopoulos A, Vayas I. "Truss models for inelastic stability analysis and design of steel plate girders", *Engineering Structures*, 105, 165-173, 2015.

[10] SOFISTIK, Finite Element Software, www.sofistik.com.

[11] Vayas I. „Interaktion der plastischen Grenzschnittgrößen doppelsymmetrischer I-Querschnitte", *Stahlbau 69*, H. 9, 693-706, 2000.

[12] Vayas I, Iliopoulos A. *Design of Steel-Concrete Composite Bridges to Eurocodes*, CRC Press, 2014.

[13] Abaqus 6.11: Online Documentation, Dassault Systèmes, Providence, Rhode Island, 2014.

Advanced structural design

BUCKLING SOLUTIONS FOR THIN-WALLED MEMBERS BY USING THE CONSTRAINED FINITE ELEMENT METHOD: DEMONSTRATIVE EXAMPLES

Sándor Ádány, Dávid Visy and Róbert Nagy

Department of Structural Mechanics, Budapest University of Technology and Economics

Abstract: In this paper a novel method, the constrained finite element method is applied to solve linear buckling problems of thin-walled steel members. First the method is briefly presented, then numerical examples are provided demonstrating the modal decomposition within the shell finite element method. Linear buckling solutions are shown for pure global, distortional and local-plate buckling with various loading and boundary condition, also for member with holes, and also for members with cross-section changes.

1. Introduction

A widely used practical approach to understand and analyse the complex behaviour of a structural members is to decompose the complex phenomenon into simpler ones, or, to interpret the complex phenomenon as a superposition of simpler phenomena. This is the reason why the deformations of a thin-walled beam or column member are frequently categorized into deformation classes: global, distortional, local-plate and other modes, based on some characteristic features of the deformations.

Though the knowledge of pure buckling modes, especially the values of the associated critical loads are essential in the design of thin-walled members, still only two methods are known to able to do the modal decomposition in a general case: the generalized beam theory (GBT) and the constrained finite strip method (*c*FSM), see [1,2]. In this paper a novel method is used. The proposed method follows the logic of cFSM, however, polynomial longitudinal shape functions are applied together with a longitudinal discretization. Thus, strips are transformed into multiple shell finite elements. The longitudinal shape functions are selected in such a way that modal decomposition similar to cFSM can be realized, therefore, the new method can readily be described as constrained finite element method (cFEM), possessing all the modal features of cFSM, but with significantly more flexibility in its applications.

The cFEM method is using shell finite elements, therefore, various engineering problems can be solved. If the modal features are to be utilized, a highly regular mesh is necessary. This required regularity of the mesh means an important practical limitation, but otherwise the method is general: first- and second-order static analysis as well as dynamic analyses can be performed, for arbitrary loading and boundary conditions. Holes can easily be handled, too, once they fit into the regular mesh. Due to the modal nature of the method, mode identification is also possible, but here buckling problems are presented only, solved by cFEM to demonstrate the applicability and flexibility of the method.

2. From FSM to FEM

In finite strip method a member is discretized into longitudinal strips, instead of finite element method, which applies discretization in both the longitudinal and transverse directions. In Fig. 1 a single strip is highlighted, along with the local coordinate system and the degrees of freedom (DOF) for the strip.

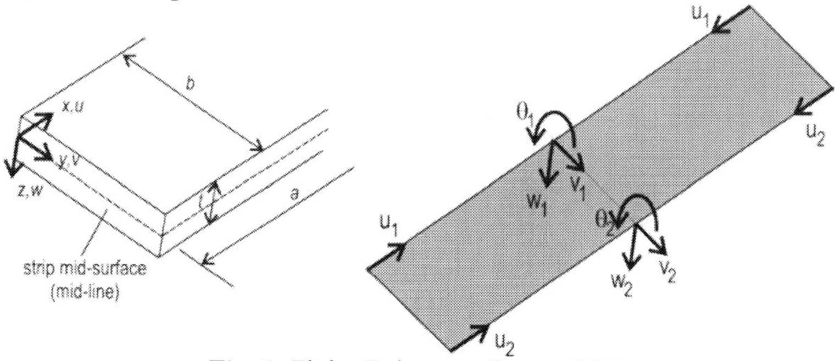

Fig. 1: Finite Strip, coordinates, DOF

In FSM the displacements are approximated as follows.

$$u(x,y)=\begin{bmatrix}N_{y1,1} & N_{y1,2}\end{bmatrix}\begin{bmatrix}u_1\\u_2\end{bmatrix}\cos\frac{m\pi x}{a} \quad v(x,y)=\begin{bmatrix}N_{y1,1} & N_{y1,2}\end{bmatrix}\begin{bmatrix}v_1\\v_2\end{bmatrix}\sin\frac{m\pi x}{a} \tag{1}$$

$$w(x,y)=\begin{bmatrix}N_{y3,1} & N_{y3,2} & N_{y3,3} & N_{y3,4}\end{bmatrix}\begin{bmatrix}w_1\\\vartheta_1\\w_2\\\vartheta_2\end{bmatrix}\sin\frac{m\pi x}{a} \tag{2}$$

with the linear (N_{y1}) and cubic (N_{y3}) transverse shape functions as follows:

$$N_{y1,1}=1-\frac{y}{b} \quad \text{and} \quad N_{y1,2}=\frac{y}{b} \tag{3}$$

$$N_{y3,1}=1-\frac{3y^2}{b^2}+\frac{2y^3}{b^3} \; N_{y3,2}=-y+\frac{2y^2}{b}-\frac{y^3}{b^2} \; N_{y3,3}=\frac{3y^2}{b^2}-\frac{2y^3}{b^3} \; N_{y3,4}=\frac{y^2}{b}-\frac{y^3}{b^2} \tag{4}$$

where a is the member length, and b is the strip width.

The above formulae represent pinned-pinned boundary conditions. An important feature of the longitudinal shape functions is that (i) the same longitudinal functions are used for u and w, and (ii) the longitudinal function for v is the first derivative of that used for u and w. By keeping these above characteristics, a significant generalization is proposed: instead of using trigonometric functions (or function series), polynomial longitudinal shape functions are used, together with a longitudinal discretization, as typical in finite element method. The interpolations are as follows:

$$u(x,y)=\begin{bmatrix}N_{y1,1} & N_{y1,2}\end{bmatrix}\begin{bmatrix}u_1\\ u_2\end{bmatrix}\times\begin{bmatrix}N_{x2,1} & N_{x2,2} & N_{x2,3}\end{bmatrix}\begin{bmatrix}c_{u1}\\ c_{u2}\\ c_{u3}\end{bmatrix} \tag{5}$$

$$v(x,y)=\begin{bmatrix}N_{y1,1} & N_{y1,2}\end{bmatrix}\begin{bmatrix}v_1\\ v_2\end{bmatrix}\times\begin{bmatrix}N_{x3,1} & N_{x3,2} & N_{x3,3} & N_{x3,4}\end{bmatrix}\begin{bmatrix}c_{v1}\\ c_{v2}\\ c_{v3}\\ c_{v4}\end{bmatrix} \tag{6}$$

$$w(x,y)=\begin{bmatrix}N_{y3,1} & N_{y3,2} & N_{y3,3} & N_{y3,4}\end{bmatrix}\begin{bmatrix}w_1\\ \vartheta_1\\ w_2\\ \vartheta_2\end{bmatrix}\times\begin{bmatrix}N_{x3,1} & N_{x3,2} & N_{x3,3} & N_{x3,4}\end{bmatrix}\begin{bmatrix}c_{w1}\\ c_{w2}\\ c_{w3}\\ c_{w4}\end{bmatrix} \tag{7}$$

with the second-order (N_{x2}) and third-order (N_{x3}) longitudinal shape functions as:

$$N_{x2,1}=1-\frac{3x}{a}+\frac{2x^2}{a^2}\quad N_{x2,2}=\frac{4x}{a}-\frac{4x^2}{a^2}\quad N_{x2,3}=-\frac{x}{a}+\frac{2x^2}{a^2} \tag{8}$$

$$N_{x3,1}=1-\frac{3x^2}{a^2}+\frac{2x^3}{a^3}\quad N_{x3,2}=-x+\frac{2x^2}{a}-\frac{x^3}{a^2}\quad N_{x3,3}=\frac{3x^2}{a^2}-\frac{2x^3}{a^3}\quad N_{x3,4}=\frac{x^2}{a}-\frac{x^3}{a^2} \tag{9}$$

With the above longitudinal shape functions the 'strip' is transformed into 'shell finite element'. This shell element is unusual, since longitudinal and transverse directions are distinguished, and different interpolation functions are used for the various displacements. The details of this shell element are not discussed here, but it can be understood that the resulted shell element has 6 nodes, some with 7- DOF and some with 1-DOF, so that the total number of DOF of the element is 30.

3. cFEM analysis

Modal decomposition is made possible by introducing constraint equations. The constraints are determined by mechanical criteria. These criteria are theoretically independent of the longitudinal shape functions, thus, the same criteria can be used as in cFSM, as summarized in [3,4]. It is to note that even though mechanical criteria are the same, as well as the cross-section deformation modes are (can be) the same, the deformation modes of the whole member are necessarily different from those of cFSM.

There is another important practical difference, as a direct consequence of the different longitudinal interpolation functions. If the classical and simplest sine-cosine shape functions are used in FSM, they easily lead to the so-called signature curves of the member, which is the critical load (typically to the first mode) in the function of the wave length (i.e. buckling length). On the other hand, if FEM linear buckling analysis is performed, typically one given structural configuration is solved (i.e., given length and given boundary conditions), however, multiple critical loads and associated buckling modes are determined. The cFEM buckling analysis is essentially a FEM analysis, but can be performed to any pure mode (i.e. pure global, pure distortional, etc.) or any combination of the pure modes.

As discussed e.g. in [5], linear buckling analysis can be carried out in many slightly different ways, depending on (i) which second-order terms of the strain-displacement relationship are considered, and (ii) whether simplifications are introduced e.g. in the integration through the plate thickness. The actual FEM formulation is programmed in such a way that all these options (and some more options) can easily be changed so that the effect of the various options can be studied.

4. Demonstrative examples

In Fig. 2 the pure global and local buckling solutions of a simple column problem is presented (Example #1). The considered column is 500 mm long, doubly symmetrical I-shaped cross section (40×10 mm flanges, 200×4 mm web). The ends are locally and globally pinned. The material is isotropic steel (E = 210 GPa, ν = 0.3). The column is loaded by a concentric axial loading, uniformly distributed over the end cross-sections (represented by nodal forces, as required by FEM technique). The global buckling solutions coincide the well-known analytical (or numerical) solutions, namely: the first mode is a minor-axis flexural buckling with one single wave, the second mode is a pure torsional mode with one single wave, while higher modes consist of multiple waves along the member length. The first three modes of pure local-plate buckling are also shown.

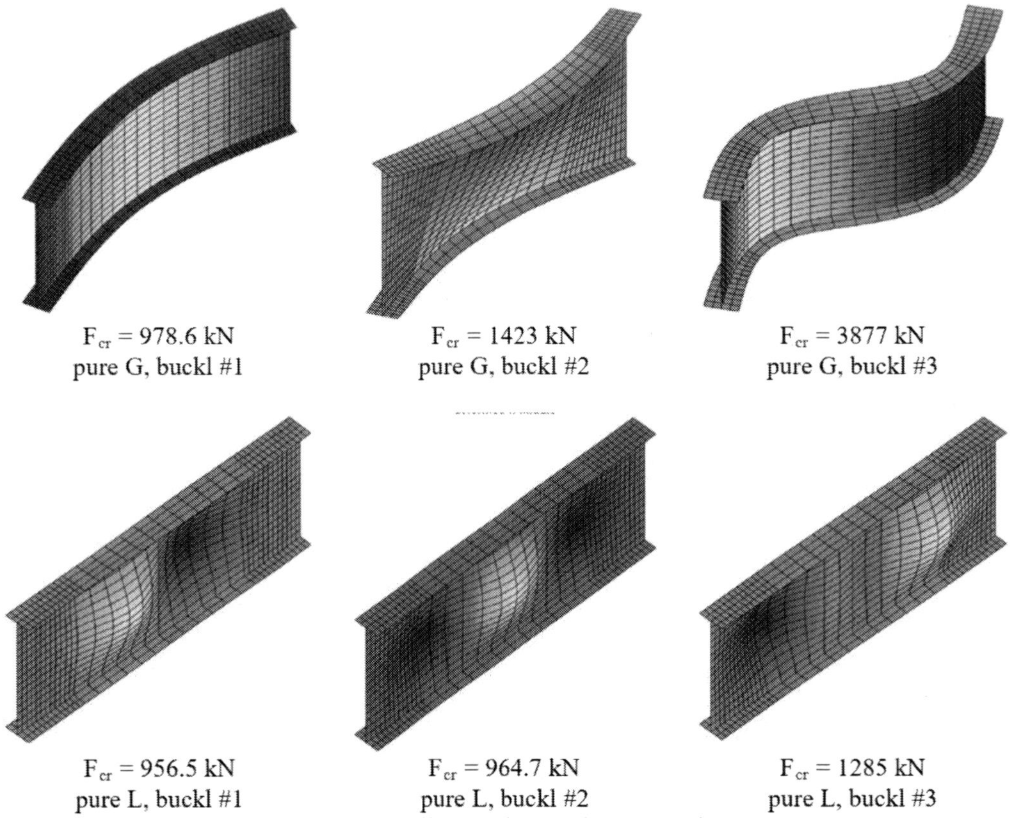

Fig. 2: Buckling mode samples: Example #1

Example #2 is to demonstrate that various loading can be applied. In fact, the loading can be arbitrary as far as it can be handled by shell finite element analysis, which means that there is no real practical limitation for the loading. Example #2 is similar to Example #1, however, the member is loaded by a transverse force (instead of the axial force). The load is a concentrated vertical force at the middle of the beam, acting at either the top flange, or the centroid, or the bottom flange. Note, in all the cases the load is acting in the plane of the symmetry of the cross-section. As well known, the load position has significant effect on the lateral-torsional (global) critical moment. The results show that the load application point has pronounced influence on the local behavior and associated critical values, too, see Fig. 3.

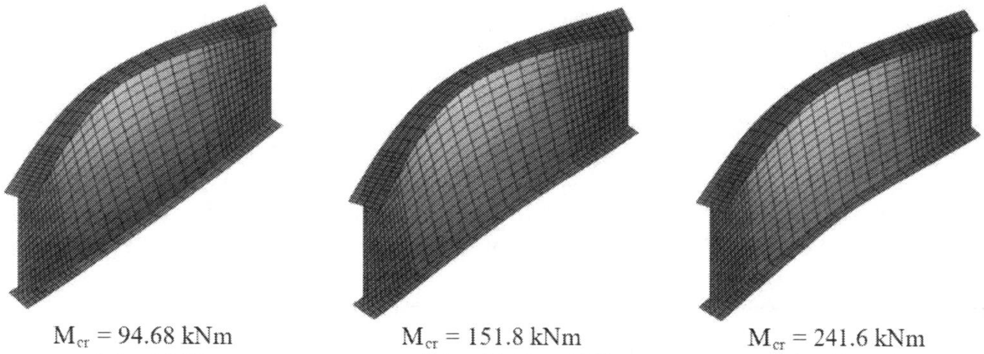

M_{cr} = 94.68 kNm
pure G, buckl #1, top

M_{cr} = 151.8 kNm
pure G, buckl #1, middle

M_{cr} = 241.6 kNm
pure G, buckl #1, bottom

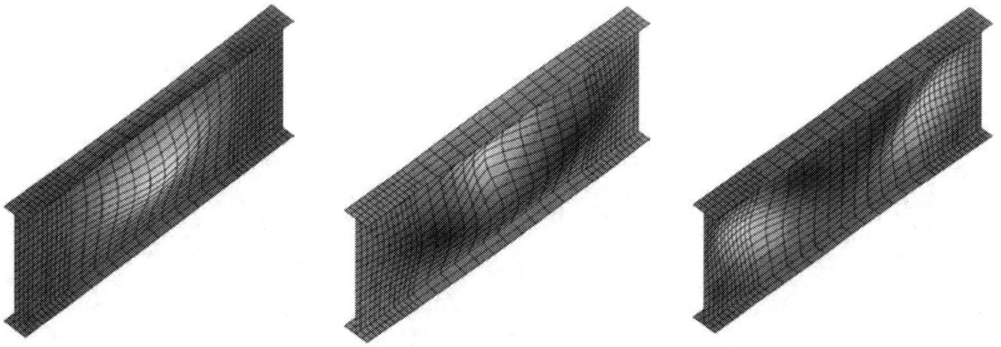

M_{cr} = 39.73 kNm
pure L, buckl #1, top

M_{cr} = 100.3 kNm
pure L, buckl #1, middle

M_{cr} = 151.4 kNm
pure L, buckl #1, bottom

Fig. 3: Buckling mode samples: Example #2

Since cFEM is based on a shell finite element, various support conditions can easily be modelled. Example #3 is similar to Example #2 (i.e., simply supported beam with a concentrated transverse force), but the end supports are somewhat released. Namely: only the web-to-flange junction points are supported against transverse translations. This change in the supports has little effect on the global modes, but has significant effect on the local-plate modes and on the associated critical load values, see Fig 4. Note, the movements of the flanges toward each other are not restrained by the applied supports, that is why the whole web can buckle. The critical values are only slightly influenced by the load position, but the critical values are much smaller compared to the classic "fork"-support case of Example #2.

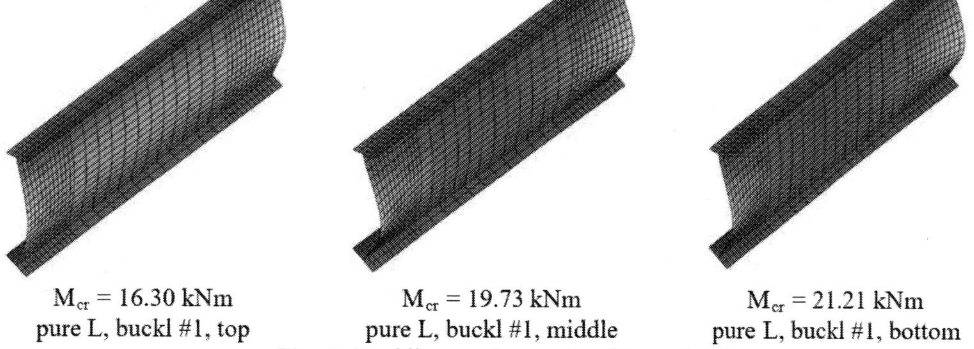

M_{cr} = 16.30 kNm
pure L, buckl #1, top

M_{cr} = 19.73 kNm
pure L, buckl #1, middle

M_{cr} = 21.21 kNm
pure L, buckl #1, bottom

Fig. 4: Buckling mode samples: Example #3

In Example #4 the effect of the second-order strain terms (i.e., the elements of the Green-Lagrange strain matrix) is demonstrated. The problem itself is identical to Example #2 (i.e., simply supported beam with a concentrated transverse force), the load being applied at the junction of the web and the top flange. Pure local-plate buckling solutions are shown in Fig. 5. The buckling calculation is completed by systematically switching on and off the second-order terms of the longitudinal strain, of the transverse strain, and of the shear strain. The buckled shapes as we all as the critical values are greatly affected by the considered terms. (Note, the results can directly be compared to the corresponding case of Example #2, where all the second-order strain terms are considered.) The results show that for the actual problem the transverse effect is dominant, i.e., the actual buckling could be classified as web crippling which, however, is in interaction with plate buckling and shear buckling.

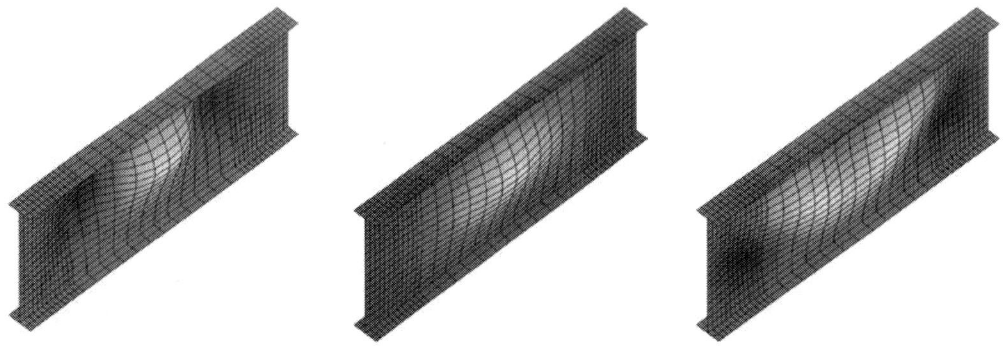

M_{cr} = 330.7 kNm
pure L, buckl #1, longit

M_{cr} = 38.40 kNm
pure L, buckl #1, transv

M_{cr} = 126.7 kNm
pure L, buckl #1, shear

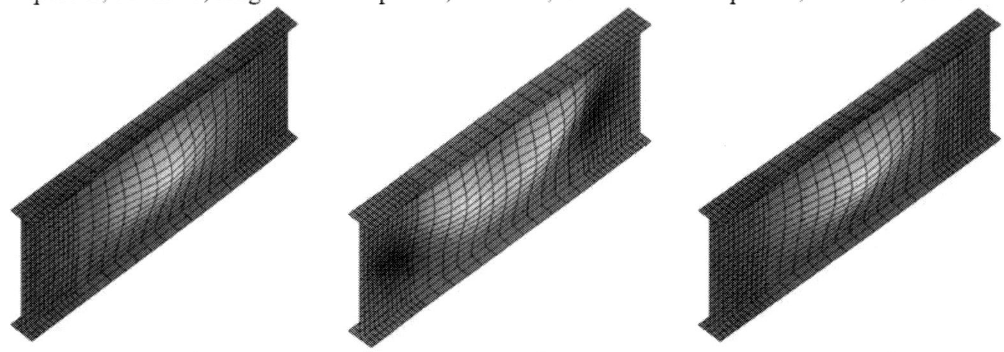

M_{cr} = 36.38 kNm
pure L, buckl #1, long+transv

M_{cr} = 114.1 kNm
pure L, buckl #1, long+shear

M_{cr} = 41.52 kNm
pure L, buckl #1, transv+shear

Fig. 5: Buckling mode samples: Example #4

Though modal decomposition is worked out only for prismatic members, members with different cross-sections can be joined. Thus, if the member is built up from prismatic segments, the proposed cFEM works. Example #5 is similar to the previous examples, i.e., it is a simply supported column with a concentrated axial force, or a simply supported beam with a transverse force acting at the junction of the web and the top flange. Here, however, the middle 300-mm-long segment of the member is replaced by a stronger, mono-symmetric I-section: top flange 80×12 mm, bottom flange 40×12 mm, web 200×6 mm. Some results are given in Fig. 6. In case of the column problem now the first mode is a flexural-torsional buckling, unlike in Example #1, and with an increased critical value. In case of the beam problem the lateral-torsional mode and the first local mode is similar to the prismatic case (see Example #2, Fig. 3), but the critical values are significantly increased.

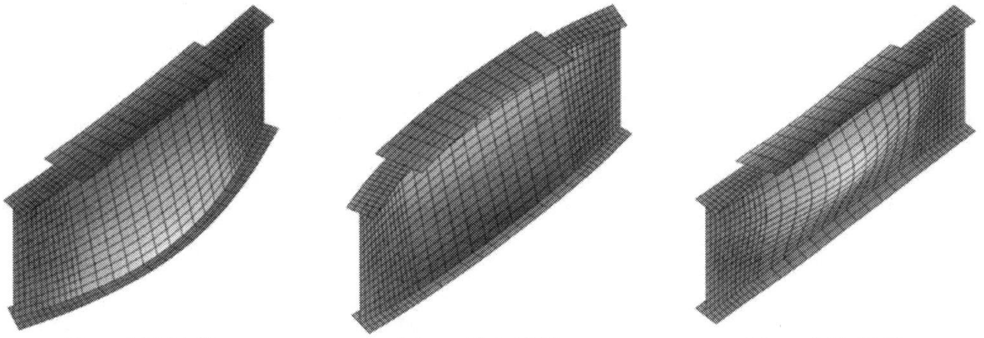

F_{cr} = 1422 kN | M_{cr} = 506.4 kNm | M_{cr} = 109.4 kNm
pure G, buckl #1 | pure G, buckl #1 | pure L, buckl #1

Fig. 6: Buckling mode samples: Example #5

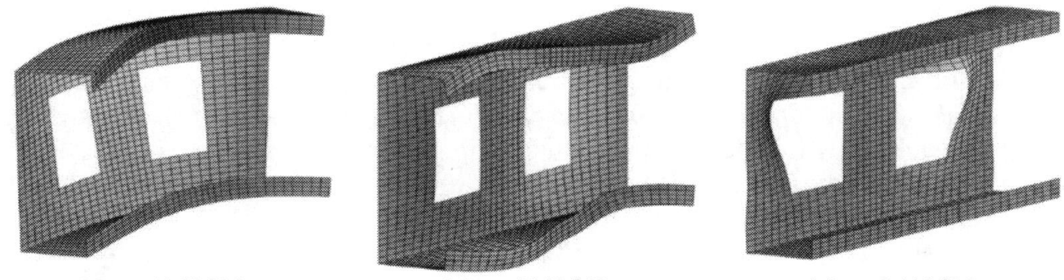

M_{cr} = 1352 kNm | M_{cr} = 1089 kNm | M_{cr} = 5.411 kNm
pure G, buckl #1 | pure D, buckl #1 | pure L, buckl #1

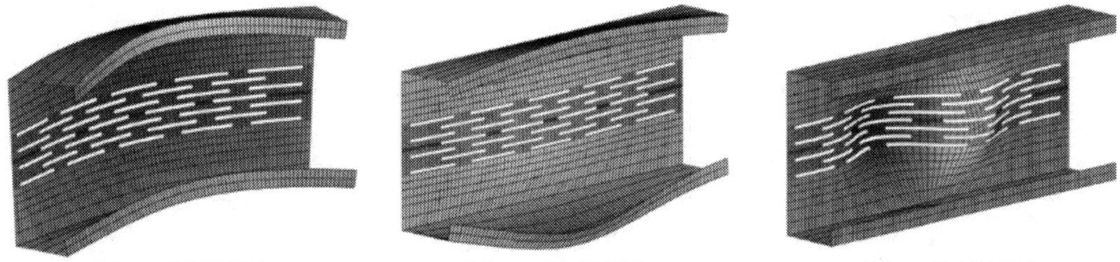

M_{cr} = 1549 kN | M_{cr} = 69.61 kNm | M_{cr} = 2.354 kN
pure G, buckl #1 | pure D, buckl #1 | pure L, buckl #1

M_{cr} = 1160 kN | M_{cr} = 1624 kNm | M_{cr} = 11.34 kN
pure G, buckl #1 | pure D, buckl #1 | pure L, buckl #1

Fig. 7: Buckling mode samples: Example #6

The last example (Example #6) is a simply supported beam with C-shaped cross-section (web, flange and lip widths are 200, 80 and 20 mm, respectively, plate thickness is 2 mm). The beam is loaded by mid-point concentrated transverse force, acting at the junction of the web and bottom flange. Pure global, distortional local-plate modes are calculated with considering holes: either two large rectangular holes, or slotted web, or a large elliptical opening is considered. In Fig. 7 the first (practically meaningful) modes are shown. The presence, the dimensions and arrangements of the holes greatly influences the results, especially local and distortional buckling. Surely, the investigation of the effect of holes on the buckling behavior of thin-walled members (in the light of modal decomposition) is a research topic for the future.

5. Summary and conclusions

In this paper the applicability of the newly proposed constrained finite element method (cFEM) is demonstrated by numerical examples. The method uses rectangular shell finite elements, thus, requires a highly regular mesh, but otherwise it is generally applicable to any problem that can reasonably be modeled by shell finite elements. Thus, various boundary conditions (e.g., intermediate and/or elastic restraints) as well as practically arbitrary loading can be considered. Members with holes can readily be handled as far as the holes are rectangular or can be approximated by multiple rectangles. Also, the cross-section might change along the member length if the member is built up from prismatic segments. The cFEM is developed in such a way that the calculations can be performed in various options, i.e., many details of the calculations (e.g., the considered second-order strain terms) can easily be adjusted by the user. It is believed that the proposed cFEM is the most practically flexible among the other methods with modal decomposition feature.

Acknowledgments

The work was conducted with the financial support of the OTKA K108912 project of the Hungarian Scientific Research Fund.

References

[1] Ádány S, Schafer B.W. „A full modal decomposition of thin-walled, single-branched open cross-section members via the constrained finite strip method", *Journal of Constructional Steel Research*, 64(1), pp. 12-29, 2008.

[2] Silvestre N, Camotim D, Silva NF. „Generalized Beam Theory revisited: from the kinematical assumptions to the deformation mode determination", *Int. Journal of Structural Stability and Dynamics*, 11(5), Oct 2011, pp. 969-997.

[3] Ádány S, Schafer BW. „Generalized constrained finite strip method for thin-walled members with arbitrary cross-section: Primary modes", *Thin-Walled Structures* (2014), Vol 84, pp. 150-169.

[4] Ádány S, Schafer BW. „Generalized constrained finite strip method for thin-walled members with arbitrary cross-section: Secondary modes, orthogonality, examples", *Thin-Walled Structures* (2014),Vol 84, pp. 123-133.

[5] Ádány S. "Global Buckling of Thin-Walled Columns: Analytical Solutions based on Shell Model", *Thin-Walled Structures* (2012), Vol 55, pp 64-75.

STATICS AND KINEMATICS IN STRUCTURAL DYNAMIC RESPONSE TO WIND ACTION

Florin Blaga and Pavel Alexa

Technical University of Cluj-Napoca, Romania

Abstract: Accounting for dynamic action of wind on multi-story structures may reveal aspects of structural behavior that cannot be inferred from static analysis. Also, dynamic analysis to wind leads to relevant differences in the values of the few parameters that may be computed via static analysis. Present contribution deals with time – history type analysis of multistory steel structures dynamically acted upon by wind. Wind action is accounted for by recorded and processed speedo – grams at several altitudes. Computed parameters are lateral level displacements and base shear force. The numerical results are associated to a 21 level steel frame considered in several cases of its lateral stiffness and added viscous damping levels.

1. Introduction

As in the case of earthquake dynamic action, the wind loading on shear type structures becomes more and more a dynamic problem [1], [2], [3]. The traditional static approach to wind action is rather complex in its arithmetic mainly due to the necessity of taking into account the dynamic effects of wind loading [4], [5], [6]. The coefficients change / increase, indeed, the value of the wind pressure that, still, remains a static phenomenon. Generally, dynamic effects do not reduce to values of some parameters such as the traditional lateral displacements. Dynamic analysis of structures goes beyond the values of static and kinematic parameters associated to limit states. The benefit of the dynamic analysis of structures has to be viewed in terms of not so classical components of the mechanical state of the structure. Let the static state (stress – resultants, unit stresses) and kinematic state (displacements, deformations) be considered such classical concepts. The design limit states impose limit values of several parameters that are related to static behaviour. An efficient tool in the hands of structural engineers to adequate the structural conceiving and sectional sizing is the stiffness (overall and sectional) of the structure and of its members. The modern, simple and versatile means to control the structural behaviour / response under dynamic loading is the level of inherent or, rather, added damping [7], [8], [9]. Damping is a "hidden state" that can only express itself during the motion of the structure. The effect of the damping is the dissipation of induced energy that is, in its turn, intimately connected to the motion. Therefore, mitigating or just assessing earthquake and / or wind response via damping can only be performed by dynamic analyses. Here lies the importance of dynamic analysis of earthquake or wind acted upon multi-story structures - much more than in the values of behavioural parameters. In the same time, the very acceptance of motion in the structural analysis paves the way to the involvement of other parameters in evaluating the structural performances under dynamic loadings. Several such parameters are the earthquake / wind input energy and its structural components (kinetic, strain dissipated energies) that are able and strong enough to accurately describe the mechanical state of the structure and emphasize the effectiveness of earthquake / wind protection of structures. Dynamic analysis open the door of the more comprehensive structural assessment even larger by allowing for evaluating the contribution of not only the damping level but, also, of inertial / mass distribution to structural response [10], [11] and of the modal participation to the overall response [12].

Present contribution is part of a larger study that involves dynamic (of time – history modal type) analysis of multi-story planar structures acted upon by wind. Transition from static to dynamic in the case of wind action involves a mathematical phase (stochastic, probabilistic) required by the numerical model (as a time function) of the velocity of wind [13], [14[15]. The numerical results included in the contribution refer to absolute lateral displacements at story levels and to the base shear force induced by wind action on a 21 levels steel planar frame (Fig. 1). A base reference velocity of wind of 27.00 m/s is considered. Several states of added linear viscous damping expressed in the values of ratio of critical damping (of 2%, 5%, 8%, 10% and 15%) are considered. Lateral stiffness is, also, taken into account in the form of the braced structure (X braces located in the two lateral spans – Fig. 2). The total height of the structure (76.00 m) leads to two different intensities of the wind velocity (Fig. 3, Fig. 4). The unbraced structure of 5% ratio of critical damping is considered as reference structure allowing for assessment and comparison of computed results.

2. Structures and wind actions

The 21 level planar structure in its two stiffness states (reference unbraced state – Fig. 1 and braced state – Fig. 2) are design – in what regards the sizing of their cross sections – such that they fulfil the ultimate limit state provisions [16].

Dynamic wind actions are considered in the forms of two different winds – in what regards the time variation of their velocities: wind V1 (Fig. 3) and wind V2 (Fig. 4), each of them associated to two zones of the height of the structure (up to 30.00 m and between 30.00 m and 76.00 m).

3. Computed results

An analytical a multi-degree-of-freedom system (Fig. 5) has been adopted as the analytical model of the analysed structure. The usual lateral level displacements have been selected as the degrees of freedom. A simple transformation of wind action allowed the transition from velocity to nodal applied forces (Fig. 5). The modal component of the dynamic analysis lead to the values of the natural modes of vibrations in the case of reference structure and in the case of braced structure. The values of the first three eigen periods of the reference structure are: $T1$ = 4.13 sec.; $T2$ = 1.42 sec.; $T3$ = 0.80 sec. The first three eigen periods of the braced structure are: $T1$ = 2.90 sec.; $T2$ = 0.81 sec.; $T3$ = 0.39 sec. The large differences between the two stiffness cases in the values of eigen periods ensure – on an acceptable extent - the validity of the conclusions in terms of structural response to dynamically applied wind action.

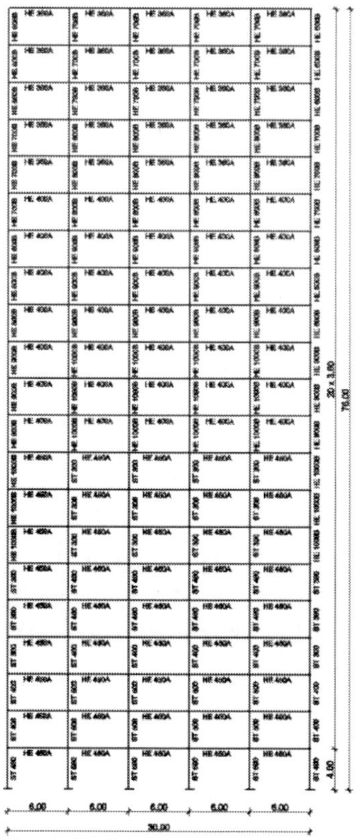

Fig. 1: Reference structure

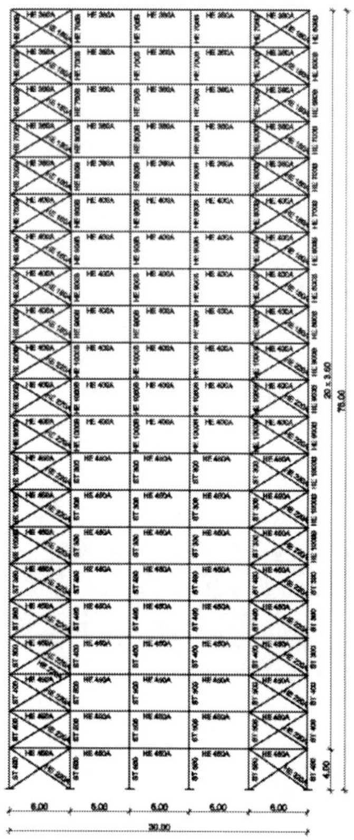

Fig. 2: Braced structure

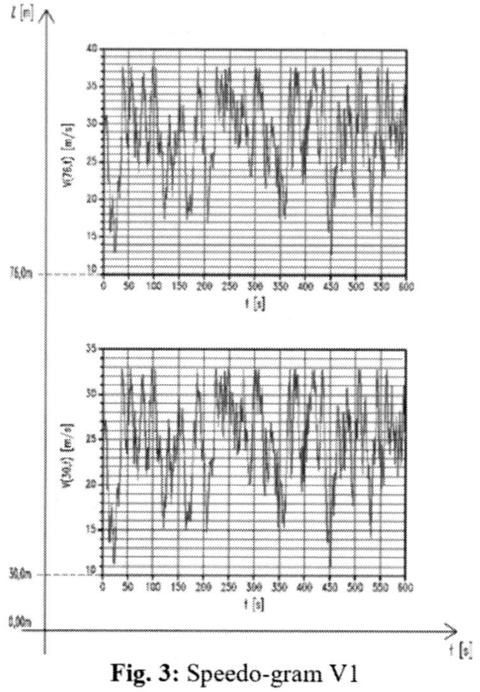

Fig. 3: Speedo-gram V1

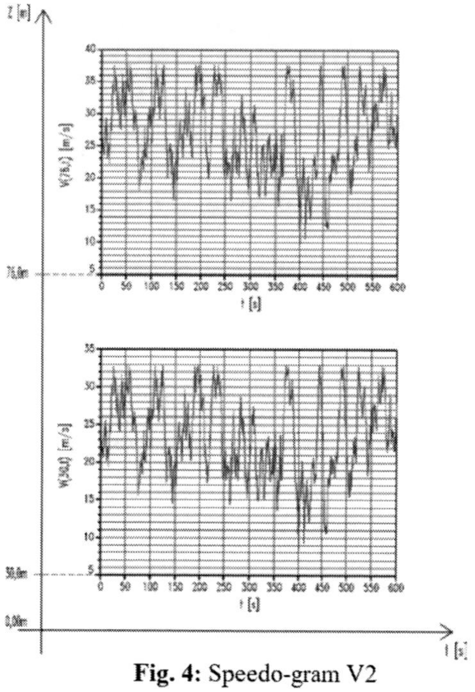

Fig. 4: Speedo-gram V2

Top level lateral displacements in the seven cases of equipping with added damping and lateral stiffness are presented in both, the case of wind action V1 (Fig. 6) and wind action V2 (Fig. 7). The diagrams include also the results of static analyses performed on reference structure (referred as AS-REF) and on braced structure (referred as AS-CV) respectively.

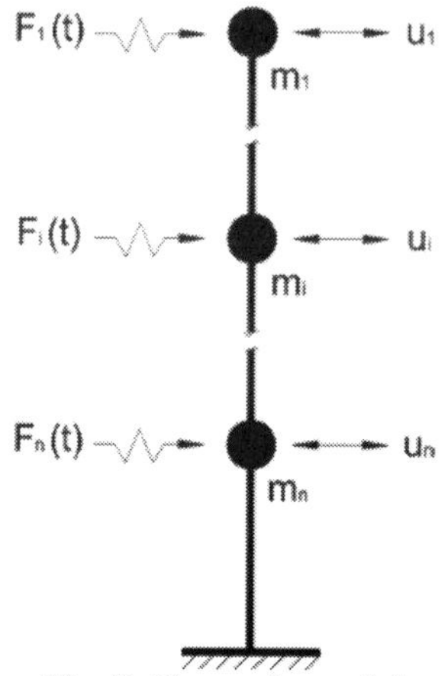

Fig. 5: Dynamic model

Base shear force induced by the wind actions has been, also, computed in the same cases of equipping and wind actions and the two cases of static analyses. Under wind action V1, the base shear diagram is presented in Fig. 8, while for the case of wind action V2, the results are emphasized in Fig. 9.

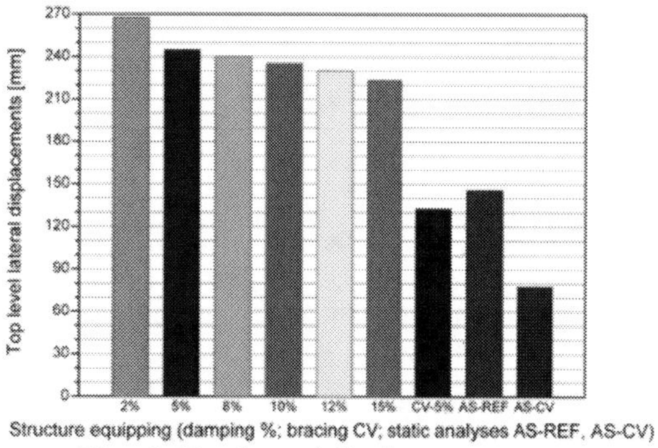

Fig. 6: Top lateral displacements – wind action V1

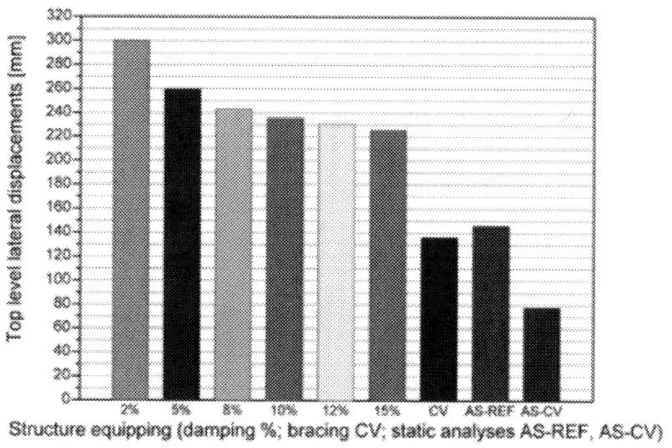

Fig. 7: Top lateral displacements – wind action V2

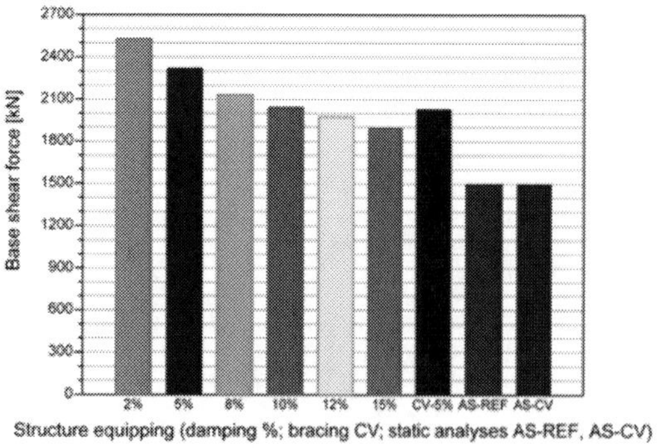

Fig. 8: Base shear force – wind action V1

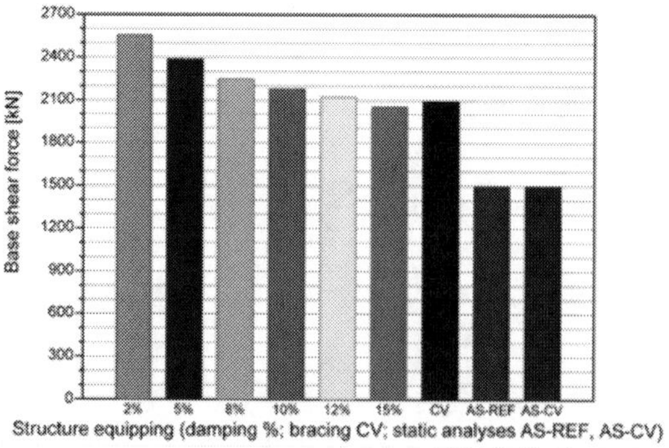

Fig. 9: Base shear force – wind action V2

A simple synthesis of the results referring to both parameters (top lateral displacements and base shear force) is presented in in the case of wind action V1 (Fig. 10) and in the case of wind action V2 (Fig. 11) as well.

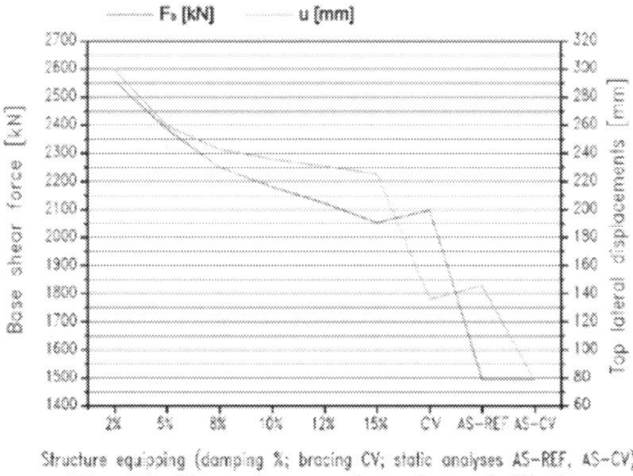

Fig. 10: Base shear and top lateral displ. Wind action V1

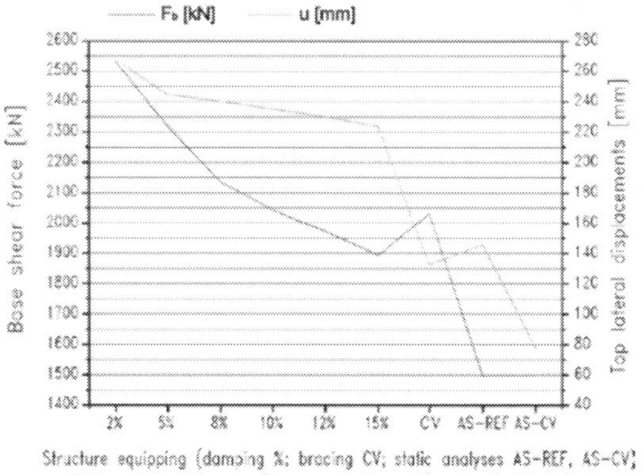

Fig. 11: Base shear and top lateral displ. Wind action V2

Another comparison is given by the induced kinematics (Fig. 12) and statics (Fig. 13) of the two different wind actions V1 and V2, respectively.

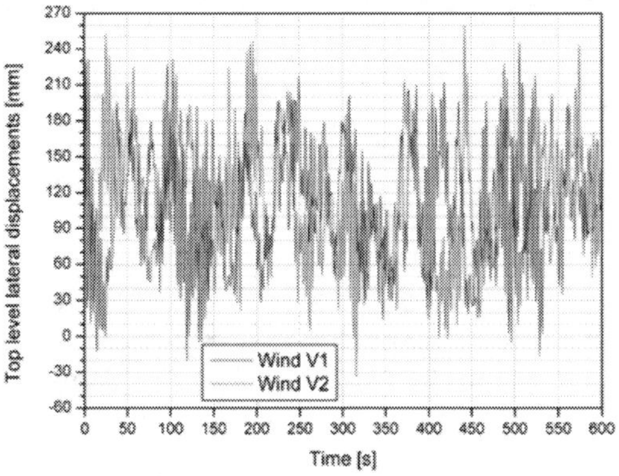

Fig. 12: Top lateral displ. induced by V1, V2

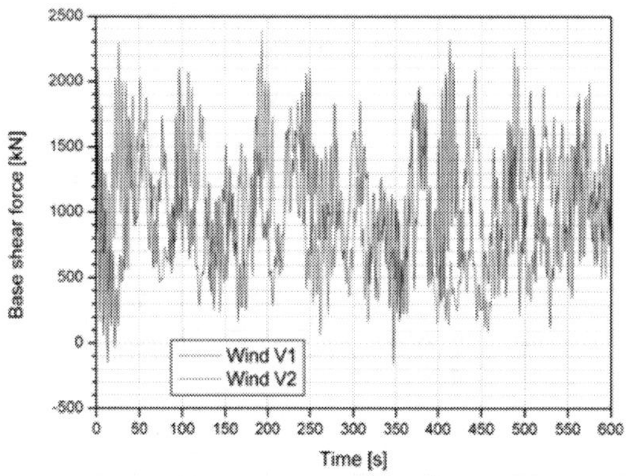

Fig. 13: Base shear forces induced by V1, V2

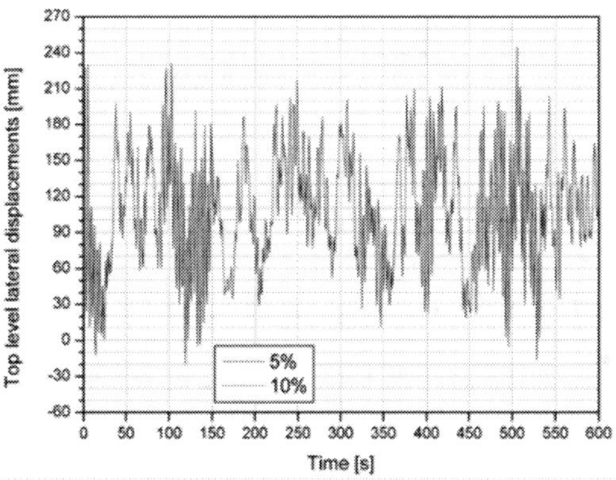

Fig. 14: Top lat. displ. versus added damping-V1

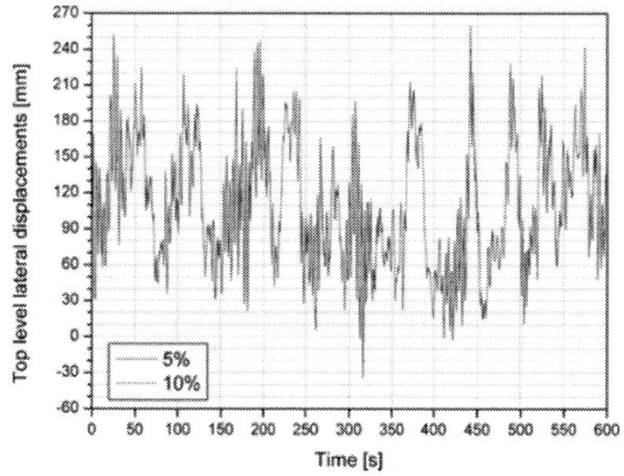

Fig. 15: Top lat. displ. versus added damping-V2

The wind mitigation by added viscous damping is exhibited by a selected set of the computed results that are expressed in top lateral displacements for both wind actions (Fig. 14, Fig. 15) and in base shear forces (Fig. 16, Fig. 17) for two levels of ratio of critical damping (5% and 10%, respectively).

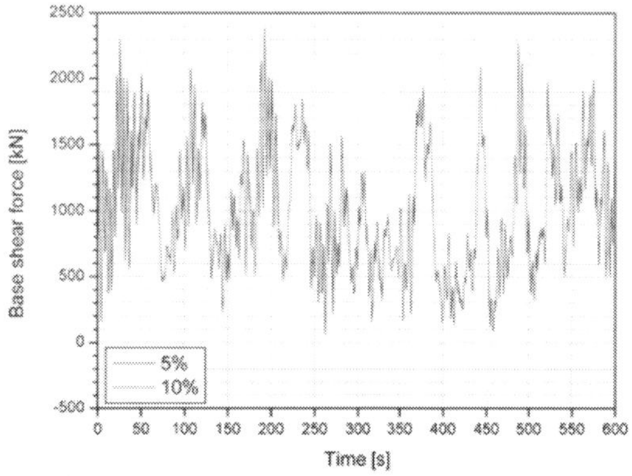

Fig. 16: Base shear force versus added damping-V1

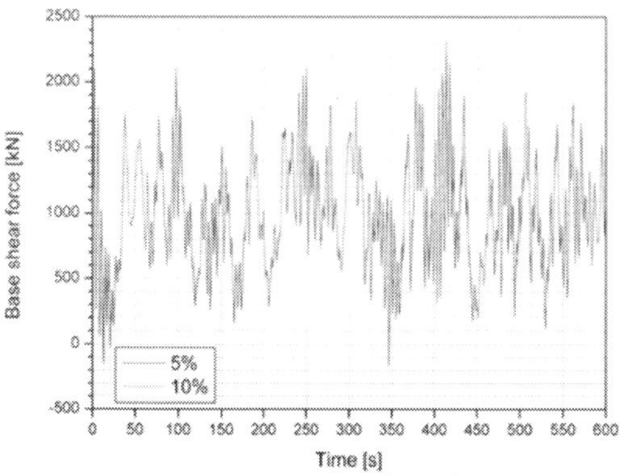

Fig. 17: Base shear force versus added damping-V2

4. Conclusions

A set of time - history modal analyses have been conducted on a 21 level steel frame acted upon by wind actions. Computed results involve a kinematic parameter (top lateral displacement) and a static parameter (base shear force). A comparative basis of the results is built up by several techniques of equipping the reference frame: added lateral stiffness via bracing and added viscous damping. Performed analyses emphasize not only the possibility of carrying out such analysis but, also, the specificity of computed numerical results. Clear differences in numerical values of these results may be noticed in the several cases involved such as between the two levels of added damping, the two situations of lateral stiffness, the two wind actions. The contribution also pleads for the time – history type analyses of structures acted upon by wind actions as the only way of emphasizing the mitigation of wind response via added damping. The contribution intends to constitute just a step in a larger study that involves other mechanical aspects associated to wind response of multi-story structures that can only be exhibited by dynamic analyses such as the influence (on wind response) of added damping, of mass distribution, of added mass. The energy state will be deeply dealt with as an effective tool in assessing the mitigation of wind response of multi-story structures.

References

[1] Davenport AG. "The Relationship of wind structure to wind loading", *Proc. of the Symposium on Wind Effect on Building and Structures.* London, vol. 1, pp. 54-102, 1965.

[2] Wilson Ed. "Dynamic Analysis With Emphasis On Wind and Earthquake Loads". 22, 1999. http://seabc.ca/documents/seminars/dynamic_analysis_basics.pdf

[3] Chen JW, Liang Li. "Investigation on Dynamic Response of Steel Tower Structure under Time-History Wind Load", *Applied Mechanics and Materials,* Vol. 166-169, pp. 699-707, 2012.

[4] EN 1991-1-1 "Eurocode 1: Actions on structures-Part 1-4: General actions, Wind actions", CEN, Brussels, 2005.

[5] ASCE7 "Minimum design loads for buildings and other structures".

[6] CR 1-1-4/2012 "Design code. Assessment of wind action on structures", (in Romanian).

[7] Prodan O, Ladar I, Alexa P. "Seismic risk mitigation via supplemental viscous damping", *Acta Tehnica Napocensis: Civil Engineering & Arhitecture*, Vol 54, No. 1, pp. 13-27, 2011.

[8] Prodan O, Ladar I, Alexa P. "Seismic performances of steel structures with supplemental damping. Numerical studies", *14th European Conference on Earthquake Engineering*, Ohrid, Republic of Macedonia, pp. 476.

[9] Woodhouse J. "Linear damping models for structural vibration", *Journal of Sound and Vibration*, 215(3), pp. 547-569, 1998.

[10] Ladar I. "Seismic response of steel multi-story structures. An energy approach", PhD Thesis, Technical University of Cluj – Napoca, 2013 (in Romanian).

[11] Suciu D. "Steel multi-storey structures with added mass. An energy approach", PhD Thesis, Technical University of Cluj – Napoca, 2013 (in Romanian).

[12] Ladar I. "Inertial distribution of seismic input energy into multi-storey structures", *5th National Conference on Earthquake Engineering and 1st National Conference on Earthquake Engineering and Seismology*, Bucharest, Romania, 2014.

[13] Barszcz T, Bielecka M, Bielecki A, Wójcik M. „Wind speed modeling using Weier strass function fitted by a genetic algorithm", *Journal of Wind Engineering and Industrial Aerodynamics*, Vol. 109, pp. 68-78, 2012.

[14] Bielecki A, Barszcz T, Wójcik M. „Modelling of a chaotic load of wind turbines drivetrain" *Mechanical Systems and Signal Processing*, Vol. 54-55, pp. 491-505, 2015.

[15] Bielecka M, Barszcz T, Bielecki A, Wójcik M. Fractal Modelling of Various Wind Characteristics for Application in a Cybernetic Model of a Wind Turbine. *Artificial Intelligence and Soft Computing. Lecture Notes in Computer Science*, Vol. 7268, pp. 531-538, 2012.

[16] EN 1993-1-1 "Eurocode 3: Design of steel structures-Part 1-1: General rules and rules for buildings", CEN, Brussels, 2005.

ELASTIC CRITICAL BUCKLING MODE AS IMPERFECTION ON FRAME WITH NON-UNIFORM CROSS-SECTION ALONG THE MEMBERS WITH IMPERFECTION ACCORDING EN 1993-1-1 CLAUSE 5.3.2 (11)

Ján Brodniansky

Department of Steel and Timber Structures, Faculty of Civil Engineering, Slovak University of Technology, Bratislava, Slovakia

Abstract: This paper focuses on practical application of EN 1993-1-1, EN 1999-1-1 clause 5.3.2(11) in calculation of frames with non-uniform cross-section along the members. Due to complexity of the solved problem, a FEM-based computer program has been developed. The presented example has been calculated in the mentioned computer program and documented by partial results. According to the mentioned clause, it is possible to obtain the imperfection amplitude in the shape responding to the elastic critical buckling mode. Calculated **E**quivalent **U**nique **G**lobal and **L**ocal **I**nitial imperfection (**EUGLI** imperfection) is then used as imperfection in the verifying calculation of structures. This paper compares two possible imperfections, one according to the clause 5.3.2(11) and the other according to the clause 5.3.2(3) a) b).

1. Introduction

According to the standards [1,2] and clause 5.3.2(11), it is possible to use the elastic critical buckling mode as an imperfection and determine the amplitude of this imperfection. Imperfection is called **E**quivalent **U**nique **G**lobal and **L**ocal **I**nitial imperfection (**EUGLI** imperfection). The whole calculation procedure according to the 2nd order theory with EUGLI imperfection is called EUGLI method. The whole theoretical background of EUGLI imperfection was published in [3,4,5,6]. Because of complexity of the solved problem, a computer program was developed in MATLAB programming language [7]. Some descriptions that include computer program background can be found in [8,9]. This paper presents a relatively new matrix calculation procedure of EUGLI method, presented on a frame structure with non-uniform cross-section along the members. This relatively new matrix procedure is very well suited for computer program calculations. Iteration procedure and matrix formation of EUGLI method are described on the following example of a frame structure.

Method 5.2.2 a) [1,2] is used for calculation and both imperfections according to the clause 5.3.2(11) [1,2] (EUGLI imperfection) and clause 5.3.2(3) a) b) (global initial sway imperfection and local initial bow imperfection) [1,2] are implemented in the frame structure.

2. Frame with non-uniform cross-section along its members

Frame scheme can be found on "Fig. 1", a static scheme with distribution of second moment of area is on "Fig. 2" and loading on "Fig. 3". Welded I sections, steel grade S355, taken as 3rd cross-sectional class, are considered, and imperfection buckling curve b is considered as well. All frame members are bonded against lateral-torsional buckling and out of plane flexural buckling. Each frame member is divided into 20 FEM elements (160 together - included frame edges).

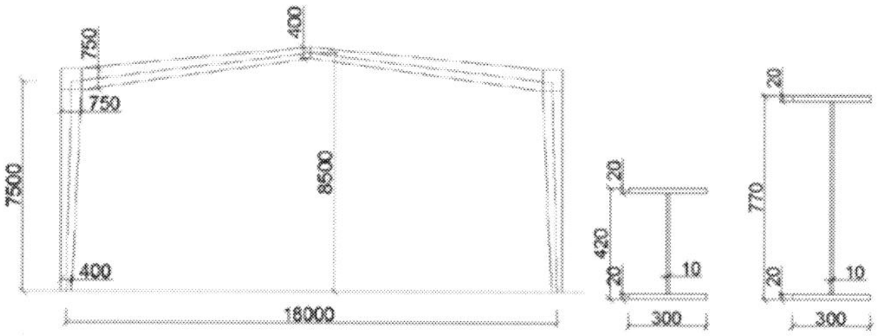

Fig. 1: Frame geometry and edge cross sections

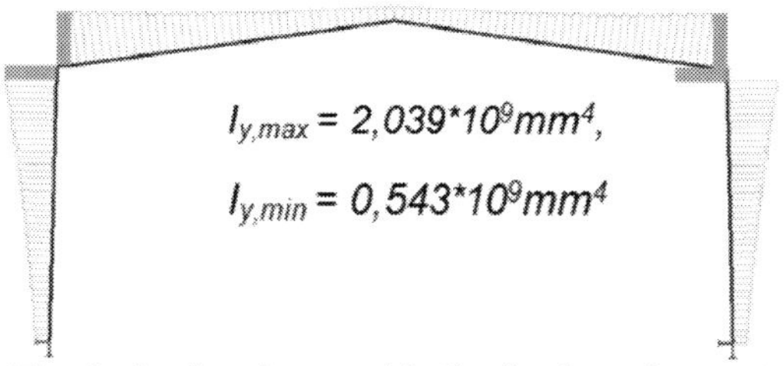

Fig. 2: Static scheme with distribution of second moment of area

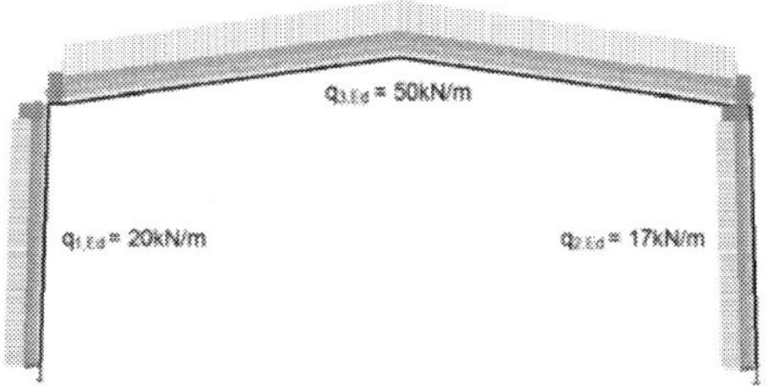

Fig. 3: Frame loading

2.1 Imperfection considered according to the clause 5.3.2(11) in [1,2]

In EUGLI imperfection method, the iterative approach is used to determine the critical cross-section on the structure. In the first iteration step, it is necessary to estimate the initial position of the critical cross-section. In the calculation of EUGLI imperfection amplitude (equations (10a) [3], (3) [4], (5.13) [5]), cross-sectional parameters of this particular cross-section are used. Calculation then continues with 2nd order analysis and evaluation of utilization ratio along the structure. In case the position of cross-section with maximal utilization factor calculated by means of EUGLI method is different from the estimated position, the next iteration step is needed. The iteration procedure is finished when the position of the cross-section with maximal utilization percentage (by means of EUGLI method) calculated by the program is the same in two consecutive iterations. This cross-section is called the critical cross-section. However, there is a possibility in automatic computer calculations that this iterative approach does not reach convergence – the program enters a never-ending cycle. To avoid cycling of the iteration procedure, it is advisable to use the so-called 'brute force method'. In calculation procedure of critical cross-section, which parameters are used for EUGLI imperfection amplitude, in every cross-section on the structure is considered. The best way to align obtained results is to use the matrix form (EUGLI method utilization matrix "Fig. 13"). The values in cells along the rows represent the EUGLI method utilization ratios of the structure in cross-sections *j*, where the row *i* shows the imperfection amplitude considered on the base of parameters from *i* cross-section (*i* and *j* define the position in matrix). Values along the matrix diagonal (*i=j*) represent the utilization factors in cross-sections which define the EUGLI imperfection amplitude. If maximal utilization ratio in row lies on the matrix diagonal, *i=j* defines the position of critical cross-section. When programming, it is necessary to ensure that the amplitude (on base of its parameters) is set up for each cross-section *i*. For the amplitude, the structural utilization of cross-sections *j* is evaluated and is written into row *j*. Part of the computer program designated for finding the critical cross-section stands on finding the matrix cell with maximal utilization factor on matrix diagonal (by means of EUGLI method). All rows must be checked.

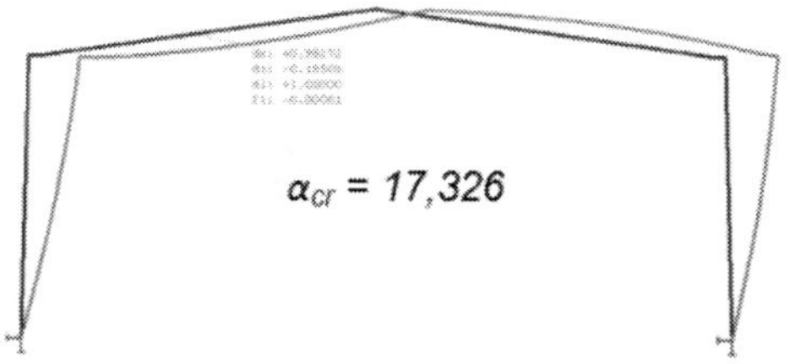

Fig. 4: Buckling mode

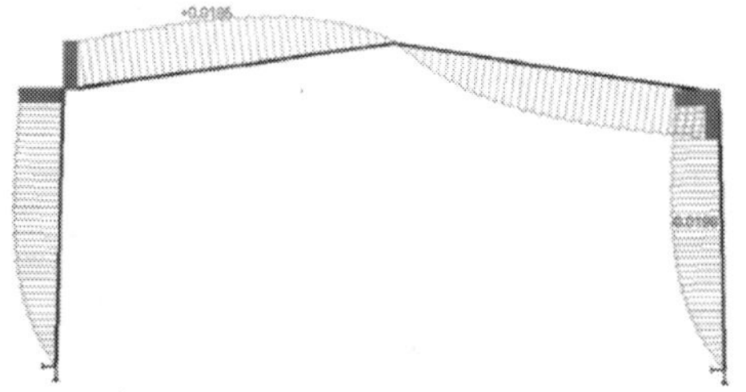

Fig. 5: 2nd derivation of buckling mode (curvature)

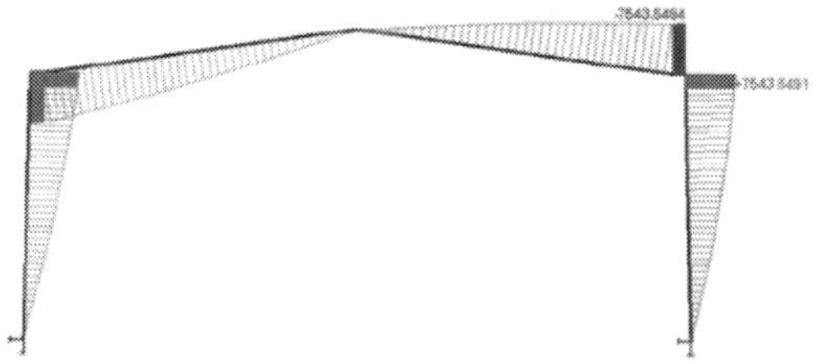

Fig. 6: Bending moment from imperfection with amplitude 1m (*7543,6491 kNm*)

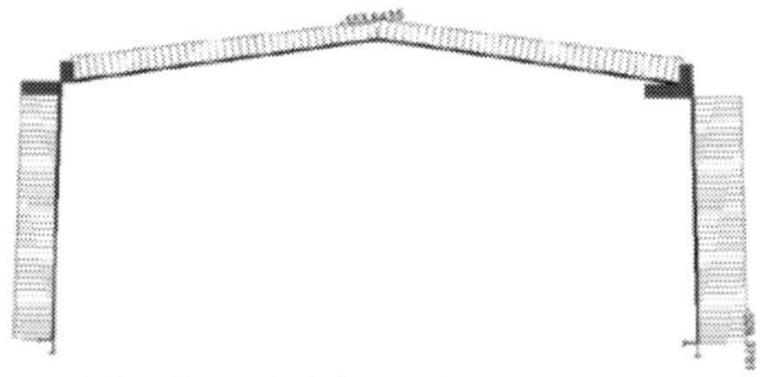

Fig. 7: Axial force (*509,0791 kN*)

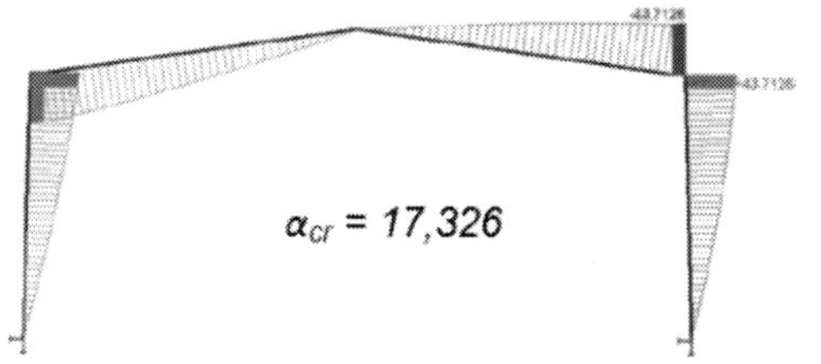

Fig. 8: EUGLI imperfection bending moment (*43,7126 kNm*) for α_{cr}

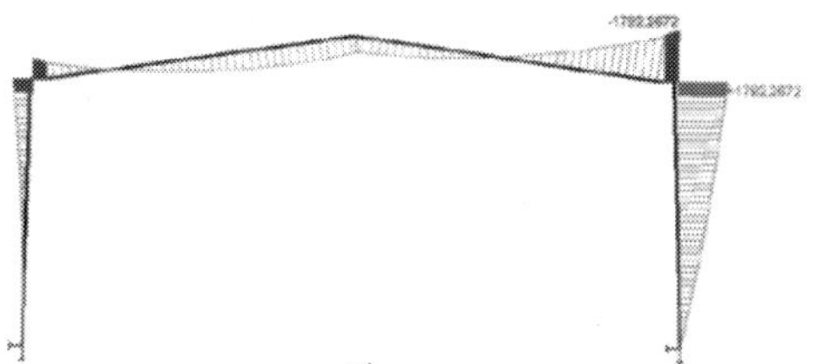

Fig. 9: Resulting 2nd order bending moment (*1792,2672 kNm*)

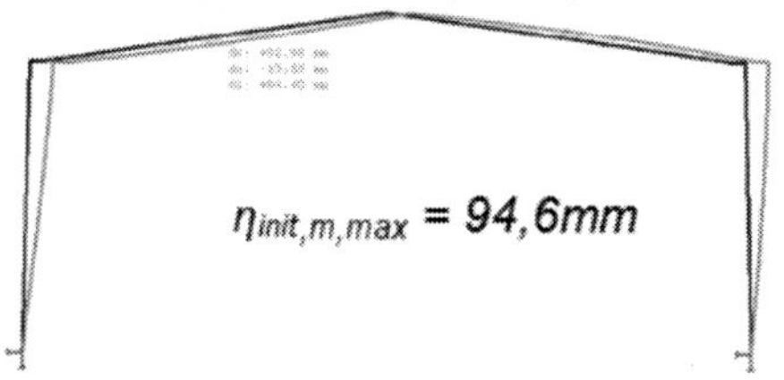

Fig. 10: Imperfection with amplitude value

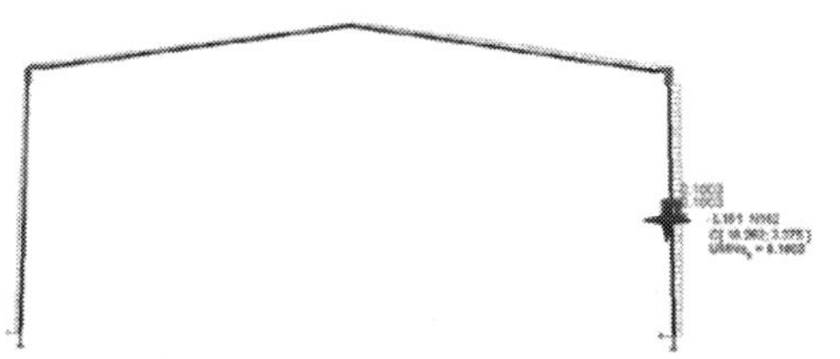

Fig. 11: Critical cross-section and EUGLI method utilization ratio (*0,1003*)

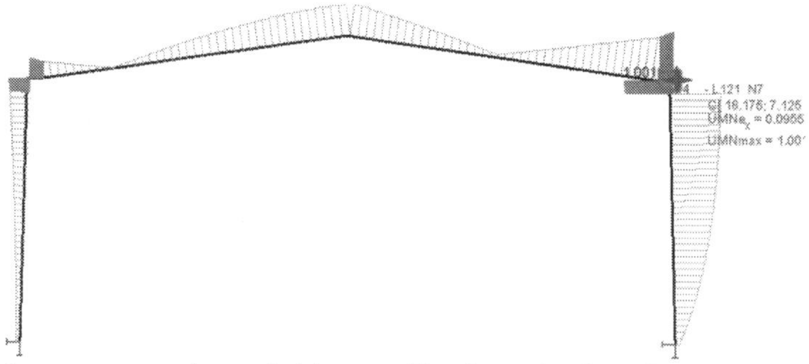

Fig. 12: Ultimate cross-section and ultimate utilization ratio along frame members (1.0018)

	j									
i	1	2	143	144	145	149	150	151	152	153
1	0,08733	~~0,10274~~	~~0,22194~~	0,22195	0,22147	0,21373	0,21001	0,20542	0,19989	0,19333
2	0,07883	0,08655	0,15204	0,15239	0,15249	0,1499	0,14834	0,14633	0,14384	0,14081
143	0,07173	0,07303	0,09367	0,0943	0,09488	0,0966	0,09684	0,09699	0,09703	0,09696
144	0,07175	0,07307	0,09382	0,09445	~~0,09503~~	~~0,09674~~	~~0,09698~~	~~0,09712~~	0,09715	0,09707
145	0,07178	0,07312	0,09404	0,09467	0,09525	0,09694	0,09717	0,0973	0,09733	0,09724
149	0,07199	0,07353	0,09583	0,09645	0,09702	0,09858	0,09875	0,09882	0,09877	0,09858
150	0,07209	0,07372	0,09662	0,09724	0,0978	0,0993	0,09945	0,09949	0,0994	0,09918
151	0,07221	0,07395	0,09763	0,09824	0,09879	0,10022	0,10034	0,10034	0,10021	0,09994
152	0,07237	0,07425	0,09892	0,09953	0,10007	0,1014	0,10148	~~0,10143~~	0,10124	0,10091
153	0,07257	0,07464	0,1006	0,1012	0,10173	0,10293	0,10296	0,10285	0,10259	0,10217

Fig. 13: Part of the EUGLI method utilization matrix (critical cross-section 151)

The whole structure consists of 160 elements (161 cross-sections). In the "Figs. 4-12" the results generated by computer program are presented in graphic form. "Fig. 4" represents the calculated buckling mode with critical factor. In the "Fig. 5", 2nd derivation of buckling mode is displayed. "Fig. 6" displays the bending moment distribution from deformation into imperfection with amplitude 1m, bending moment distribution ("Fig. 6") is obtained by multiplying the "Fig. 5" by bending stiffness (EI_y). "Figs. 7,8,9" represent the internal forces, "Fig. 7" the axial force distribution, "Fig. 8" the distribution of EUGLI imperfection bending moment and "Fig. 9" the distribution of the resulting 2nd order bending moment from external loading on imperfect structure. In the "Fig. 10", there is the EUGLI imperfection with amplitude value and in the "Fig. 11" the position of critical cross-section and EUGLI method utilization ratio (value 0.1003 can be found on "Fig. 13" too) is displayed. Possible iteration cycles are displayed in "Fig. 13" by red arrows. These cycles have not been done by the computer program. They only show how the classic iteration procedure may look. In first iteration step, the cross-section no. 1 is chosen as the critical one. Maximal utilization ratio given by EUGLI method is in the cross-section no. 144. If the cross-section no. 144 (second iteration) is taken as the critical cross-section along with its initial parameters for the next iteration cycle, maximal utilization calculated by EUGLI method is in the cross-section no. 152. If the cross-section no. 151 is taken into consideration as the initial parameter (i.e. the initial critical cross section), the next iteration calculation cycle will calculate the critical cross-section as cross-section no. 151 and this means the end of the iteration cycle.

Because of in-plane uniform transverse load distribution, 2nd order bending moment from this external loading is calculated. Critical cross-section is searched for only by means of 2nd order axial force and 2nd order bending moment caused by deformation of the structure into imperfect shape. 2nd order theory is used in all stages of the calculation and therefore it is possible to add the bending moment caused by external loading and the bending moment caused by the structure being deformed into imperfect shape at the end of the calculation procedure (details are in [3,5]). Critical cross-section obtained by means of stability calculation does not need to be the cross-section with maximal resulting utilization on the structure ("Fig. 11" and "Fig. 12" positions of cross-sections are not the same). Resulting ultimate utilization ratio is in the "Fig. 12" with its maximal value of 1.0018. In the presented frame example, an approach of deforming the structure to the imperfect shape and calculation of internal forces according to 2nd order theory from external loading of the structure is used.

2.2 Imperfection considered according to the clause 5.3.2(3) a) b) in [1,2]

By using the imperfection according to the clause 5.3.2(3) a) b) in [1,2], it is necessary to check large amount of imperfection shapes, respectively to find the imperfection shape that produces the highest utilization ratio. Both sway and bow imperfections together are considered as imperfections. In agreement with the previous chapter, the frame is considered in its imperfect shape and under external loading. In the Table 1, some of the imperfection shapes of frame from "Fig. 1" are displayed (for better imagination, they are magnified). Global sway imperfection is considered only on the right side, in agreement with the effect of external loading. For each imperfection shape, the resulting value of utilization ratio is shown. The real values of used imperfections are $\phi = 0.00297$; $e_0 = 27$ mm for columns; $e_0 = 34.8$ mm for rafter. Red star means the position of cross-section with maximal utilization and the rectangle highlights the imperfection shape which produces maximal utilization on the frame structure.

Table 1: Imperfection shapes

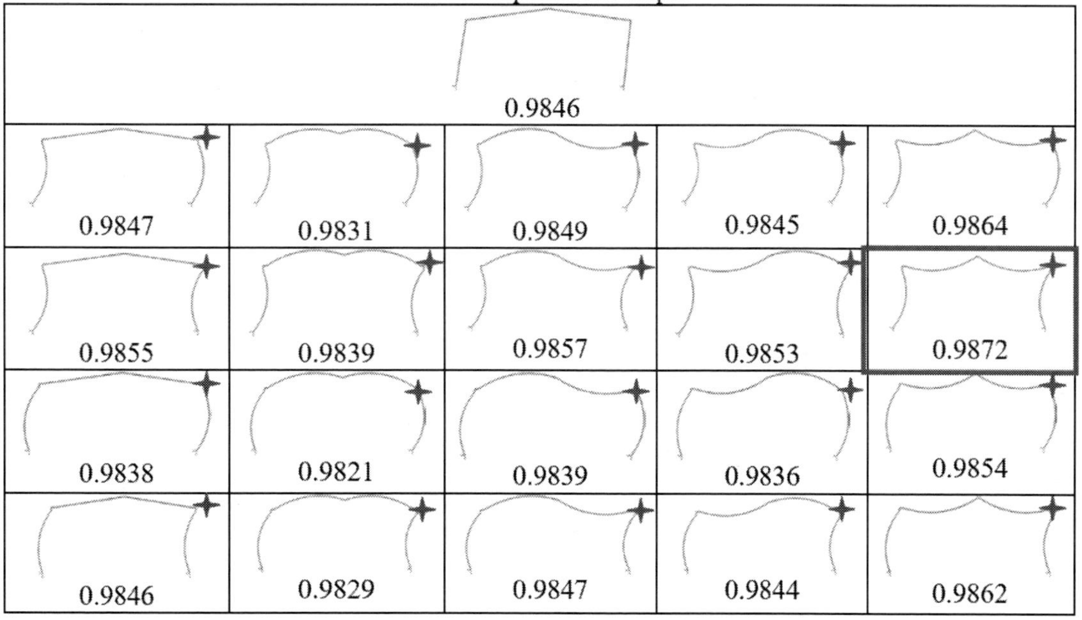

		0.9846		
0.9847	0.9831	0.9849	0.9845	0.9864
0.9855	0.9839	0.9857	0.9853	0.9872
0.9838	0.9821	0.9839	0.9836	0.9854
0.9846	0.9829	0.9847	0.9844	0.9862

All partial frames (frames with established imperfections) were modelled in the same computer program mentioned in the introduction of this paper, but EUGLI iteration procedure was turned off.

3. Conclusions

In the Table 2, there is a short recapitulation of the results of verified frame structure. On the base of the achieved results, it can be concluded that by considered imperfections (5.3.2(11) in [1,2] and 5.3.2(3) a) b) [1,2]), maximal utilization factors are basically the same. One of the reasons why this happens is the ambition to use the real proportion of loading. Critical coefficient α_{cr} of presented frame is relatively high (17.33).

Table 2: Results

Frame		
Imperfection according to:	5.3.2(11) in [1,2] (EUGLI imperfection)	5.3.2(3) a) b) in [1,2]
Maximal utilization value:	1.0018	0.9872
No. imperfection shapes:	2*	21**

* By imperfection acc. to 5.3.2(11) [1,2] are considered two inverse imperfection shapes

** By imperfection acc. to 5.3.2(3) a) b) [1,2] are not evaluated all possible imperfection shapes

The unquestionable advantage of EUGLI imperfection method is the universality of its use. There are only two imperfection shapes in comparison with imperfection 5.3.2(3) a) b) in [1,2]. EUGLI imperfection method is not implemented into commercial civil engineering computer programs, and 2nd order theory calculation is basically the standard procedure in such programs, implementation of EUGLI method represents only a small amount of programming code. Some of these programs allow the application of buckling mode shape as imperfection, but calculation of EUGLI imperfection amplitude is not implemented. Use of the EUGLI method utilization matrix gives one of the easiest ways to implement this procedure into computer programs. On "Fig. 14" is shown computer program interface for obtaining graphical and numerical results.

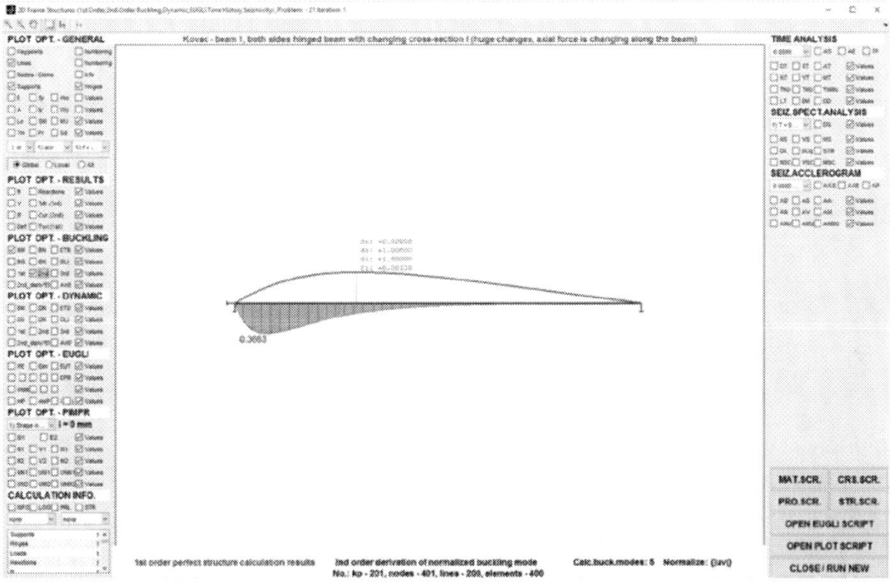

Fig. 14: Computer program graphical results interface

Universality of EUGLI imperfection method is shown in Table 3, where some other structural examples calculated by the author of this paper are displayed. In each of them, it is difficult to set the imperfection in any other way than using the EUGLI imperfection.

Table 3: Some of the calculated structures

1. Upper chord of open truss bridge [3] 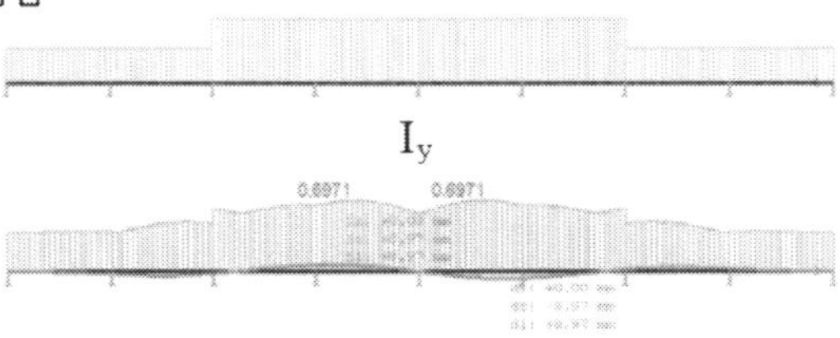 Utilization and imperfection	2. Arc structure – Žďákovský most [5] 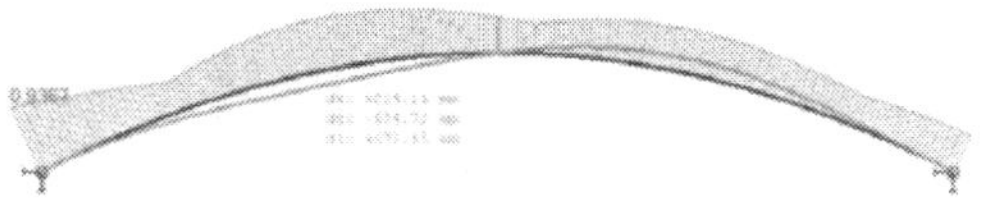 Utilization and imperfection
3. Arc with changeable cross-section 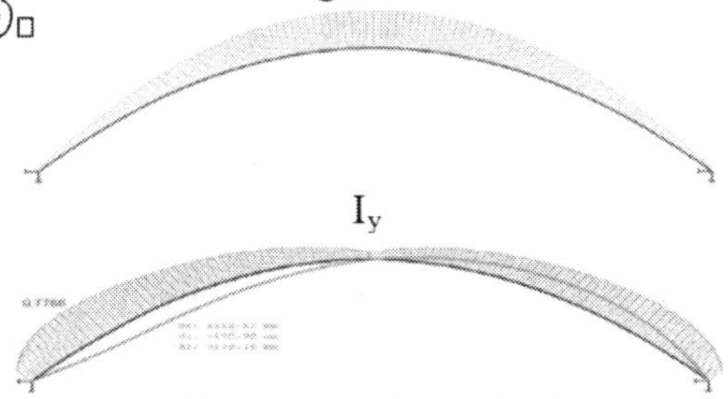 Utilization and imperfection	4. Special shape frame (half raindrop shape) 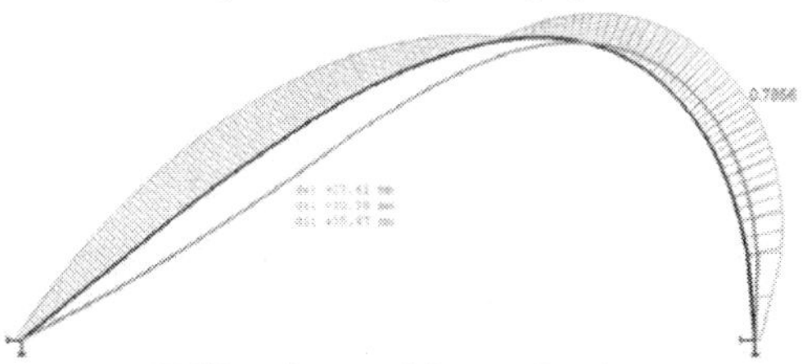 Utilization and imperfection
5. Simple beam on spring base 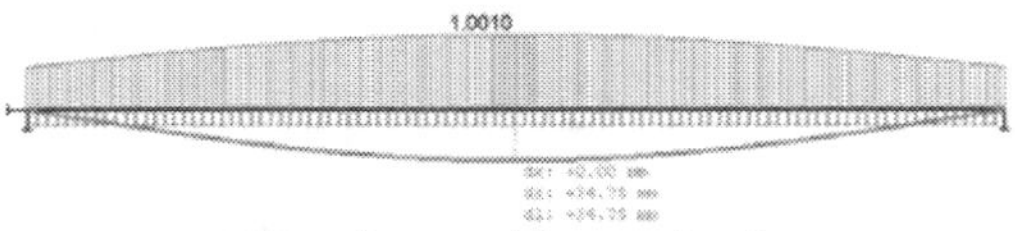 Utilization and imperfection	6. Fixed-fixed beam temperature load 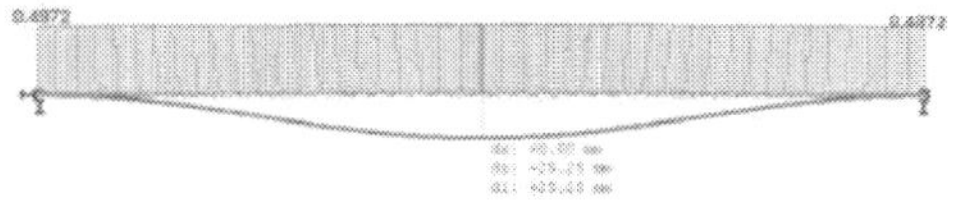 Utilization and imperfection

The first row of Table 3 presents examples which can be found in the following - 1st example in [3] and 2nd example in [5]. The results were additionally verified by the presented computer program, and were in very good agreement. The second and third row shows the structures which were chosen by the author and they represent other options where it is possible to use the EUGLI imperfection and the whole EUGLI method.

Acknowledgement

This paper has been supported by the Slovak Research and Development Agency (SRDA) – grant from the research program APVV No. 0236-12.

References

[1] EN 1993-1-1: 2005, *Eurocode 3: Design of steel structures – Part 1-1: General rules and rules for buildings.*

[2] EN 1999-1-1: 2005, *Eurocode 9: Design of aluminium structures – Part 1-1: General structural rules.*

[3] Chladný E, Štujberová M. "Frames with Unique Global and Local Imperfection in the Shape of Elastic Buckling Mode (Part 1, Part2)", *Stahlbau*, Vol. 82, No. 8, 609-617, No. 9, 684-694, 2013.

[4] Baláž I. "Determination of the Flexural Buckling Resistance of Frames with Members with Non-Uniform Cross-sections and Non-Uniform Axial Compression Forces", *Proceedings of the 34th Meeting of Experts on Steel Structures*, Pezinok, October 16-17, 2008, 17-22, ISBN 978-80-227-2950-5.
[5] Dalemulle M. *"Vzperná odolnosť oblúkových konštrukcií v ich rovine" (Buckling Resistance of Arches in their Plane)*, Dissertation thesis, Slovak University of Technology in Bratislava, 2013 (in Slovak).
[6] Kovač M. *"Vzperná odolnosť kovových prútov a prútových konštrukcií. Aplikácie nových metód z Eurokódov" (Buckling Resistance of Metal members and Frames. Applications of New Methods from Eurocodes)*, Bratislava, Slovak University of Technology in Bratislava, 2012, ISBN 978-80-227-3681-7 (in Slovak).
[7] MATHWORKS: *Programming language MATLAB v. 2009*, www.mathworks.com.
[8] Brodniansky J, Kovač M. "Application of EUGLI Imperfection for Assessment of Members and Frames with 4th Class Cross-Section", *International Masaryk Conference for Ph.D. Students and Young Researchers*, December 2014, Volume V, ISBN 978-80-87952-07-8.
[9] Brodniansky J. "2nd Order Structural Analysis with Imperfection According 5.3.2(11) EN 1999-1-1 (Method for Computer Analysis)", *eConference of Doctoral Students and Young Researchers*, July 20-24, 2015, ISBN 978-80-972051-0-2.

ANALYSIS OF STEEL MEMBERS USING FEM AND BIG DATA

Rolando Chacón, Enrique Mirambell and Esther Real

Department of Civil and Environmental Engineering. Technical University of Catalonia, Barcelona, Spain

Abstract: This paper describes trends concerning the analysis of steel structures using FEM and Big Data from two perspectives: i) the increase of the computing capacity and ii) the design guidelines. For the former, computers allow performing considerably vast studies for frames, members or cross-section verification. For the latter, guidelines follow a particularly careful pace when it comes to standardizing new methodologies related to the use of FE in the analysis of steel structures. Both trends move at a rather different speed, with consequences in steel design in a broad sense. The paper presents emphasis of the discussions in the frame of TC8 encompassing "Numerical Modeling"- "Structural Design"- "Statistics" and "Big Data" .

1. Introduction

The potential of the numerical methods in countless fields of the engineering world is nowadays recognized by the whole scientific and engineering community with practically no exceptions. The numerical simulations of the most complex coupled phenomena are increasingly feasible and provide extremely realistic phenomenological insights as well as accurate predictions of those behaviors. This increasing trend is relentless and as well as other fields associated with computing, will face a considerable boost in the years to come.

The construction industry, however, is lagging behind the advent of numerical methods when it comes to standardization and design codes. Recognizably, even if the construction processes as well as the static/dynamic behavior of real structures may be realistically simulated with the available computing capabilities and Software, few design guidelines provide to the designer a systematic and comprehensive way of presenting standard calculations using FEM [1]. While the types of analysis have been shifted from hand calculations to linear and second order analysis over the last decades, the design based upon structural analysis and member verification has prevailed for longer than a century.

The standardization (and benchmarking) of these guidelines is being currently treated within the core of TC8 and TWG 8.3 of ECCS. In particular, design manuals and research aimed at defining the initial conditions that must be assumed by the designer are being developed in recent times. Even in the simplest cases, the type of elements, the meshing, the material modeling, the boundary conditions, the analyses and the initial structural and geometrical imperfections are parameters that must be assessed thoroughly with engineering judgment, especially, those which are designer-assumed.

This paper describes the state-of-the-art of such guidelines as well as of investigations performed in the core of both groups and their treatment towards these initial conditions when it comes to generating large amounts of simulations that are subsequently used in standardization. The paper presents a particular emphasis in new trends discussed in the frame of TC8 and TWG 8.3 encompassing "Numerical Modeling"- "Structural Design"- "Statistics" and "Big Data".

2. Computers and Steel

Steel and computers grew up together. The computing steelmaking industry experienced simultaneous booms in the last century and as a matter of fact, the steel industry played a pioneering role in developing applications for the new computer technology. Capital equipment devoted to controlling steelmaking processes with state-of-the-art sensors and data acquisition technologies performed according to the computing capacity of the moment [2]. By the mid-1960s, almost a fifth of the world's process control computers were installed in the steel industry [3].

Computers played a primary role in structural control when steel came into research labs in Europe giving birth to the European buckling curves from 1960 onwards. An international programme was carried out by ECCS to study the behaviour of standard columns. More than 1000 buckling tests on various types of members with varying slenderness values were tested and a rational, probabilistic approach developed the basis of design of compressed steel members. Nowadays, this basis is used not only for standard sections but for a variety of geometries and metallic materials. The vast experimental programme provided significantly enough results for erecting a frame of structural design that is still used nowadays.

Decades after, personal computers have allowed designers to analyse steel structures from various perspectives: i) structural analysis (linear and subsequently, nonlinear), ii) automation in member verifications (formerly performed by hand calculations), iii) simulation of physical phenomena by means of advanced FE-based computations and more recently iv) systematic usage and control of construction processes in the building information modeling (BIM) sector. Steel design is particularly prone to automation for several reasons such as the nature of the members (generally standard profiles) and the elastic-plastic behavior of the material. For these reasons, standard structural analysis and design software as well as advanced FE-based simulations have emerged first within the steel design sector rather than in other construction materials with a more complex, uncertain and less understood material nonlinearity. Figure 1 displays four images related to the aforementioned perspectives: a) standard design and structural analysis, b) recently launched ECCS member verification mobile applications [4], c) advanced simulations of 3D joints and d) complex 4D analysis of construction processes in steel buildings with BIM [5].

In recent years, the increase in computing capacity have relentlessly decreased the computing time of similar analysis of members, frames, shells, or any given simulation. On the other hand, software (and generally, know-how) have evolved to a more user-friendly approach and a myriad of tutorials can be found over the Internet. Consequently, the nature and complexity of simulations have increased with computing capacity but these processes are more feasible than years ago. As a result, in present days, with the same computing time and standard equipment used in design offices, the amount of simulations that can be done today is ten- to hundredfold the amount of simulations performed fifteen years ago. In addition, the accuracy of the results, the ease of use of software, the greater trial-and-error interaction of the design and the boom of sharing information online have increased in a way that not only the amount of simulations may be multiplied ten to hundred times but also, their quality. Presently, researchers and designers alike are exploiting this computing capacity quite considerably. Expectedly, in only fifteen years, the computing capacity as well as the complex nature of the simulations will reach unprecedented levels in structural design history.

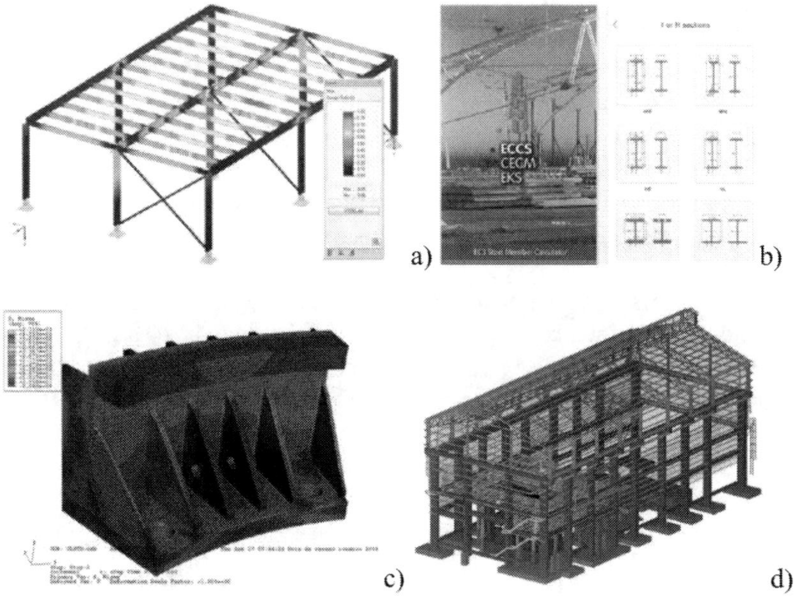

Fig. 1: Computer-aided analysis in steel structures a) Structural analysis b) Member verifications c)FEM analysis d) BIM analysis

3. Computer-aided analysis and Design guidelines

The main core of all structural design guidelines around the world is nowadays based upon member verifications (cross-sectional capacity, stability of members and/or frames and joints) and linear structural analysis (to a certain extent, nonlinear). Admittedly, some codes include guidelines related to FE-based simulations but broadly speaking, it can be stated that chapters devoted to such technologies represent a minor part of the structural design guidelines. In Europe, the general rules for steel design EN 1993-1-1 [6] includes sections related to types of analysis that may and/or must be performed as a function of the structural problem but when it comes to the usage of FE-based simulations, it refers to the informative Annex C found on the rules for plated structures EN 1993-1-5 [7] as well as to the particular rules for shell structures EN 1993-1-6 [8]. Numerically speaking, EN 1993 defines different types of analysis that may/must be performed as a function of the nature of the structure and the corresponding limit state:

- Linear analysis LA
- Linear Bifurcation Analysis LBA
- Geometrically Nonlinear Elastic Analysis GNA
- Materially Nonlinear Analysis MNA

- Geometrically and Materially Nonlinear Analysis GMNA
- Geometrically Nonlinear Elastic Analysis with imperfections GNIA
- Geometrically and Materially Nonlinear Analysis with imperfection GMNIA

3.1 EN 1993-1-5. Annex C

EN 1993-1-5-C includes recommendations for the modeling of plated structures. Firstly, the designer must select a type of analysis to be performed. The type of nonlinearity for the material and the geometry must be chosen accordingly. Depending on the type of analysis, the use of imperfections is defined as compulsory. On the other hand, EN 1993-1-5-C states that the user must account carefully for three types of considerations: i) adequate modeling (meshing) with an adequate Software, ii) adequate definition of loads and boundary conditions and iii) clear documentation of the Limit State criteria and the partial factor. For the particular case of imperfections, EN 1993-1-5-C defines two alternatives. The modeling may be based upon a refined analysis or upon equivalent geometric imperfections. For the former, an initial geometrical eigenmode-based imperfection may be used together with a residual stresses pattern (geometric and structural imperfections). For the latter, a combination of geometrical initial imperfections defined by the recommendations may be used. In principle, this only-geometric combination is equivalent to the use of geometric-structural imperfections and should include both effects. There is a decision to be taken by the user in this definition. The designer-assumed conditions defined by EN 1993-1-5-C are displayed in Table 1. These conditions together with the above mentioned initial considerations erect the frame of the current EN 1993-1-5-C recommendations for the usage of FE-analyses in plated structures.

Table 1

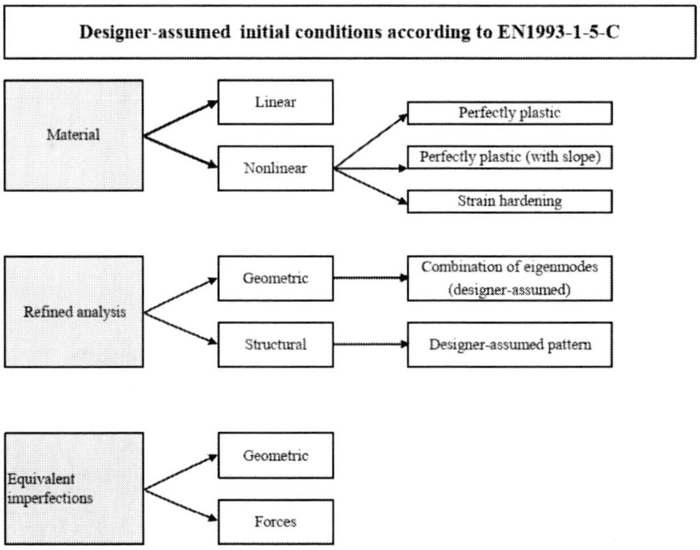

3.2 EN 1993-1-6

The complexity of the structural analysis of shell structures implies that obtaining stress-strain-displacement fields in a given model often requires numerical modeling. Closed-form solutions and mechanically-based verifications are seldom available in shell structures with arbitrary loading, boundaries and geometries. Consequently, guidelines concerning the strength and stability of shells have included numerical alternatives for years.

Particular emphasis is given in EN 1993-1-6 to the stress analysis in circumferential, normal and meridional directions. Boundary conditions of substructures are particularly complex and recommendations concerning this matter are succinctly provided. The basis of design state that the modeled shells shall sustain all actions and satisfy the requirements related to overall equilibrium, limitation of cracks due to cyclic plastification and due to fatigue. A window is open for the design assisted by testing similarly to EN 1993.

4. Design assisted by testing (DAT)

The design assisted by testing is covered in the Annex D of EN 1990 [9]. This annex includes sections encompassing a broad sense in which testing methodology, statistical treatment and general basis of design are presented. The guidelines are not limited to a specific material but are presented in a generic way. The design assisted by experimental tests in steel structures have been used recently for specific geometries and steel performance [10]. Broadly speaking, the statistical treatment of the design assisted by testing depicted in EN 1990 involves:

- A set of tests that allows defining an experimental value R_{exp};
- A mathematically engineered model that may reproduce the analyzed phenomenon that allows defining an calculated value R_{cal};
- The coefficient of variation (CoV) of the geometrical and material magnitudes of the analyzed phenomenon.
- A statistical procedure for determining the partial safety factor associated to R_{exp} / R_{cal}

By using vast experimental databases and verified design models, the procedure has also been used for calibrating partial safety factors for different verifications within the design of steel structures [11],[12],[13],[14],[15]. Recently, in the frame of a vast RFCS European project (SAFEBRICTILE), a team of researchers has been gathering relevant information associated with tests of steel members subjected to compression and flexural loads in which material, geometrical and other relevant data from the industry are systematically used. The aim of the project is to establish standardized rules for the safety assessment stability-prone steel members.

It is worth pointing out that in most of the statistical evaluations performed in recent years, sets of numerical databases have also been used within the developed calibrations. Experimental specimens tested in laboratories include geometrical as well as structural imperfections that are random but limited to a certain value associated with fabrication tolerances guaranteed by steel makers. Numerical tests, however, should be designed in such a way both geometrical and structural imperfections are included in order to reproduce at least, similar conditions when compared to real experimental tests. In addition, some authors have adapted the statistical procedure depicted in Annex D by adding variability of the results if numerical specimens are used within the calibrations (in order to penalize the difference between numerical and experimental results). Furthermore, the randomness of the initial conditions and its effect on both the results and on the potential calibration is still a matter of debate.

5. Towards a design assisted by numerical testing (DANT)

The convergence of several trends shows a potential future in which the structural design and verification of new steel members, new metallic materials and new methodologies will be performed via a design assisted by numerical testing. Trends such as the listed below may open a new array of possibilities:

- The computer-aided control at the industry levels, providing better and continuous data related to the initial conditions of the steel members. The generated amount of data was untreatable in past years but with the advent of computing capacity (and storage), this is no longer a critical issue.
- The possibility of performing hundreds to thousands of simulations with random or deterministic nature in unprecedented short time spans with full account of nonlinearity. The amount of generated data may exceed by far the data used for developing the European buckling curves.
- The capacity and ease of use of Software that allow the users implementing complex coupled structural phenomena at both research and design levels.

Within the ECCS Technical Committee TC8, the stability analysis of members subjected to flexural and compression loads has been a matter of scrutiny and study for decades. Special attention has been paid to the interaction of forces in compressed members since they represent a major verification to be included in designs of all steel structures. The general methods that are currently implemented in EN1993-1-1 for these interactions include sets of sequential calculations that provide accurate results but in some cases, with a costly numerical development. Consequently, design alternatives are continuously discussed and evaluated for the sake of enhancement not only in safety terms but also, in terms of clarity, conciseness and practice. Some recent studies [16] aimed at providing more concise design methods for the cross-sectional resistance of hollow sections are based upon more than 40000 numerical tests with rigorously developed initial conditions concerning geometric and structural imperfections. This type of research works may be systematically found in the years to come. As long as the initial conditions are systematically documented, these research works may continuously feed with data new calibrations and new insights of standard problems [17].

Advanced design manuals aimed at establishing standard recommendations for the use of FE-models in stability analysis are under preparation in the core of TC8 and TWG 8.3. Two key aspects of this potential design involve a standard definition of large parametric studies and, a sound implementation of the potential variability of the random/deterministic nature of the simulations.

5.1 Large parametric studies

Large parametric studies have generally been performed by generating an initial model with a parametric nature (scalable geometries and parametric abilities for other variables). Subsequently, short scripts run the simulations with either random or deterministic approaches following user-designed functions and user-defined arrays. As a result, a large amount of simulations must be post-processed and for this purpose, macros and/or additional codes are needed. With the use of proper statistical procedures, a design assisted by numerical testing (DANT) is on the verge of being used. Table 2 summarizes the general path a designer follows for developing such procedure.

Table 2

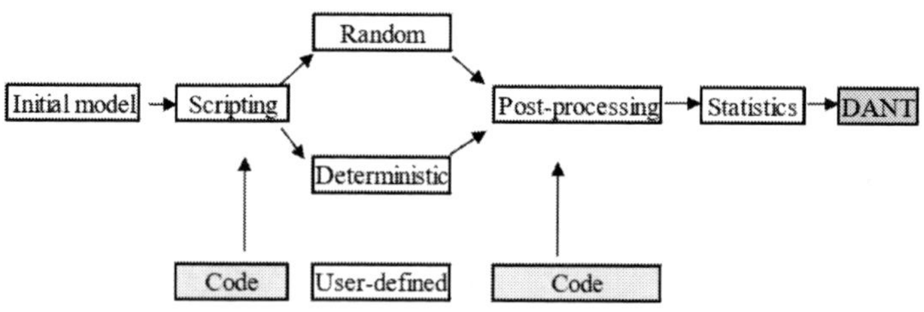

5.2 Statistical Evaluation

The statistical procedure employed in DAT depicted in Annex D of EN 1990 [9] follows rational assumptions concerning statistical significance, statistical distributions and hypothesis tests. The procedure relies on industry data for defining variability of material and geometries and, in the potential use of numerical tests, additional variability associated with the FEM models should be included. Attempts for using additional FEM variations V_{FEM} are found in the literature but further studies concerning this matter are needed. Table 3 displays succinctly the parts involved in this evaluation.

Table 3

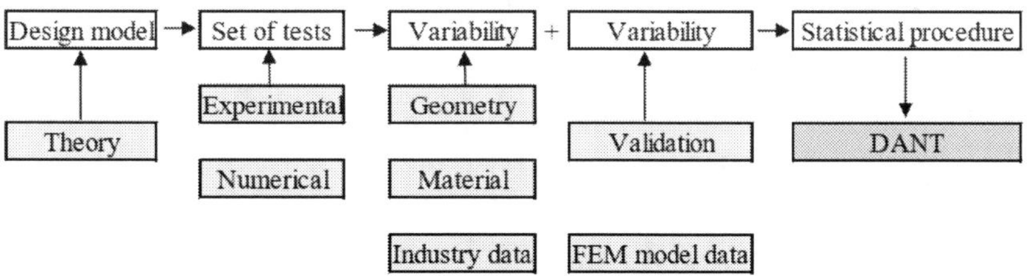

6. Conclusions

Finally, it is worth pointing out that the number of tests that may be generated and used with DANT involves a twofold conclusion: i) the amount of potential simulations may exceed by orders of magnitudes the amount of experimental tests used in developing the present design verifications and ii) the quality of simulations in stability analysis is high and accurate for the potential definition of member verification not only with formulae but with FEM design. As a result, both analysis and verification may be shifted from hand calculations to computer-aided design and automation if proper standardization is foreseen and implemented in the years to come.

Acknowledgments

The authors acknowledge the financial support provided by ASCEM (Asociación para la Construcción de Estructuras Metálicas) to the first author.

References

[1] Shayan S, Rasmussen K, Zhang H. "Direct design of 2D steel frames by GMNIA analysis", *Proceedings of the 7th European Conference on Steel and Composite Structures* (Eds.: R. Landolfo, F. Mazzolani), Naples, Italy, 181-182, 2014.

[2] Stout T, Roberts S. "Some applications of computer control in the iron and steel industry". *Iron Steel Engineering*, 37, 101-110. 1960.

[3] Aylen J. "Megabytes for metals: development of computer applications in the iron and steel industry". *Ironmaking and Steelmaking*. Vol. 31 (6), pp. 465-478. 2004.

[4] http://www.steelconstruct.com/site/

[5] http://www.tekla.com

[6] EN1993-1-1. *Eurocode 3. Design of steel structures – Part 1-1. General rules*, CEN.

[7] EN1993-1-5. *Eurocode 3. Design of steel structures – Part 1-5. Plated Elements*, CEN.
[8] EN1993-1-6. *Eurocode 3. Design of steel structures – Part 1-6: Shells*, CEN.
[9] EN1990. *Basis of Structural Design*, CEN, 2006.
[10] Dubina D. "Structural Analysis and Design Assisted by Testing of Cold-Formed Steel Structures". *Thin-Walled Structures*. 46(7-9), 741-764. 2008.
[11] Chacón R, Braun B, Kuhlmann U, Mirambell E. "Statistical evaluation of the new resistance model for steel plate girders subjected to patch loading", *Steel Construction*, 5(1), 10-15, 2012.
[12] Bock M, Miriada F, Real E. "Statistical evaluation of a new resistance model for cold-formed stainless steel cross-sections subjected to web crippling", *International Journal of Steel Structures*, 15(1), 227-244, 2015.
[13] Simões da Silva L, Rebelo C, Nethercot D, Marques L, Simões R, Vila Real P. " Statistical evaluation of the lateral-torsional buckling resistance of Steel I-beams, part 1: variability of the Eurocode 3 resistance model", *Journal of Constructional Steel Research* 65, 818-831, 2009.
[14] Simões da Silva L, Rebelo C, Nethercot D, Marques L, Simões R, Vila Real P. " Statistical evaluation of the lateral-torsional buckling resistance of Steel I-beams, part 2: variability of the steel properties", *Journal of Constructional Steel Research*, 65, 832-849. 2009.
[15] Tankova T, Simões da Silva L, Marques L, Rebelo C, Taras A. "Towards a standardized procedure for the safety assessment of stability design rules", *Journal of Constructional Steel Research*, 103, 290-302, 2014.
[16] Nseir J. "Development of a new design method for the cross-section capacity of steel hollow sections". *Doctoral Thesis*. Université de Liège. 2015.
[17] Chacón R, Mirambell E. "Numerical modelling for plated structures-Proposal for a standard methodology for benchmarking according to EN1993-1-5-Annex C", *Proceedings of the 7th European Conference on Steel and Composite Structures* (Eds.: R. Landolfo, F. Mazzolani), Naples, Italy, 1009-1010, 2014.

STEEL BRIDGE LAUNCHING ANALYSIS USING ADVANCED NUMERICAL SIMULATIONS

Rolando Chacón, Rubén Zorrilla, Nicolás Uribe and Carles Duñó

School of Civil Engineering. Universitat Politècnica de Catalunya, Barcelona, Spain

Abstract: In this paper, realistic nonlinear 3D simulations of incrementally launched steel bridge girders are presented. Doubly-symmetric I-girders, girders with corrugated webs and curved girders are analyzed by means of similar numerical modeling. The simulation is presented in a two fold fashion since the generated data information may be exploited at both design and construction stages. For the former, the model provides predictive capabilities for inferring potential failure due to the combination of loads (bending, shear and patch loading altogether). For the latter, the results obtained are presented in a way that data may be used for planning a structural health monitoring (SHM) deployment.

1. Introduction

In this paper, realistic nonlinear 3D simulations of incrementally launched steel bridge girders are presented. Several cases such as: Doubly-symmetric I-girders, girders with corrugated webs and curved girders are analyzed by means of similar numerical modeling. The analysis accounts for three sources of nonlinearity: geometry, material and boundary conditions. The principal feature of this type of analysis is the realistic reproduction of the launching. The girder is fully assembled and then pushed longitudinally along the launching path. As the girder reaches intermediate supports, a full-contact formulation is activated and reactions are transmitted between both bodies. Reactions, displacements, stresses and strains are available throughout the launching process in every point of the girders. The simulations are presented in a twofold fashion since the generated data information may be exploited at both design and construction stages. For the former, the model provides predictive capabilities for inferring potential failure due to the combination of loads (bending, shear and patch loading altogether). For the latter, the results obtained are presented in a way that data may be used for planning a structural health monitoring (SHM) deployment. Relevant figures concerning results as well as detailed information concerning the numerical modeling of the process are provided.

2. Steel bridge launching

The incremental launching method (ILM) is a bridge construction procedure in which the superstructure is set on an assembling yard, typically placed in one side of the landform to be crossed, and then pushed longitudinally towards its final position. The launching is typically performed statically in a series of increments from one side to another. It is estimated that thousands of concrete and/or steel bridges have been incrementally launched worldwide [1-3]. The ILM has gained popularity since it may guarantee minimal disturbances to the surroundings which may be of a great concern during construction. Careful analysis, though, is compulsory at design stages, since a thorough time-stepped solution is required. The process involves a continuous change of the structural schemes of the bridge with varying geometries, loads and boundary conditions. Furthermore, at construction stages, a considerably specialized construction equipment and expertise are needed. Figure 1 displays an example of the continuous change of the structural scheme in a steel girder which is incrementally launched.

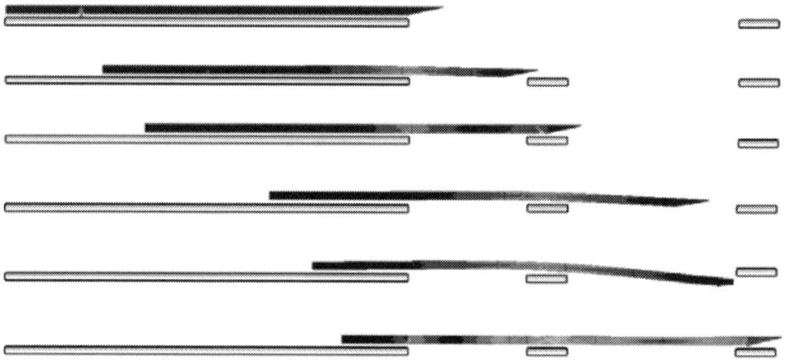

Fig. 1: Bridge launching in a steel girder with varying structural schemes

Studies related to the analysis of the structural behavior of straight doubly-symmetric steel plate girders when launched are abundant [4] [5] [6] [7]. The girders are generally launched in pairs or in groups since these arrangements provide a considerable rotational stiffness. Lateral torsional buckling (LTB), patch loading and the interaction of bending, shear and concentrated forces are generally the most prominent conditions for the steel girders design. Bridge launching in which the steel girders are assembled with corrugated webs are also found and analyzed in the literature [8]. The design is generally based upon the stress analysis (shear and secondary moment about the minor axis of the loaded flange M_z) on a discrete basis. The girders are analyzed at precise unfavorable points at which launching is expected to generate the worst case scenario. Theoretical approaches aimed at determining the non-uniform torsion of curved, incrementally launched girders have been recently launched [9] but numerical analysis of such constructive process are not yet generalized in design offices due to, mainly, their computational cost and complexity associated.

This paper presents an overall description of three numerical examples of bridge launching developed by the authors, namely, doubly-symmetric, web-corrugated and horizontally-curved steel girders. These examples are developed and exploited in such a way that: i) the phenomenological insight concerning the structural analysis is studied from the construction perspective, ii) extraction of results that may be useful during the construction processes for the sake of comparison with data obtained with SHM deployments.

3. Numerical simulations

A FE-based numerical model is used as a simulation tool. The model is implemented in the commercial Software Abaqus-Simulia [10]. The model reproduces realistically the movement of one steel girder over a launching platform towards the intermediate and end supports. This movement is applied through a large series of small increments. The size of the increment is a function of the total length of the girder (ranging approximately from L/1500 to L/3000). As a result, the incremental procedure is static but nonlinear. The bodies are simulated with shell elements. Shell elements allow developing 3D geometries realistically for cases in which one dimension is considerably smaller than the others. The material nonlinearity for metallic materials is based upon an ideal elastic-plastic constitutive equation. For multi-axial stresses, the uniaxial constitutive equation includes the von Mises criterion. Hardening is accounted for by including an isotropic formulation.

The boundary conditions are quite particular. The numerical model was expected to reproduce a multi-body physical problem that involved a mechanical interaction between the steel plate and the support conditions. The steel plate girders were modeled with first-order shell elements. The supports were modeled as analytical, rigid and frictionless surfaces on which the steel plates were able to slide and/or transmit contact stresses but conversely, were not able to penetrate through. These analytical surfaces were geometrically defined as objects rigidly connected to the ground.

The process consisted of a sequential movement (static) of the girder along a given path. In this particular case, the path is defined as a straight horizontal line but may also be defined as circular, for the case of horizontally curved bridges. The process encompasses full nonlinearity for the plate geometry, the material (elastic-plastic with isotropic hardening) and the boundary conditions (a contact-based formulation between bodies). Table 1 summarizes the numerical features of the performed simulations. The models were validated by comparing deflections, strains and vibrations with a scale-reduced test of a steel profile launched incrementally in the lab using both beam elements and shell elements [11] [12] [13].

Table 1

NUMERICAL SIMULATION			
Software	Abaqus 6.10	Interaction properties: · *Normal behaviour: No penetration and separation after contact* · *Tangential behaviour: Frictionless*	
Solver	Abaqus-Standard		
Procedure	Geometrically nonlinear		
Analized structure	Two spans (20 and 30 m respectively) bridge launching process	Bearing modelling	Analytical rigid surfaces
Cross-section	Transversally stiffened plate girder	Bearing shape	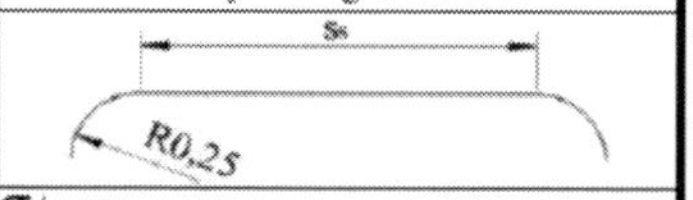
Load	Only self-weight		
Material properties:		Constitutive equation	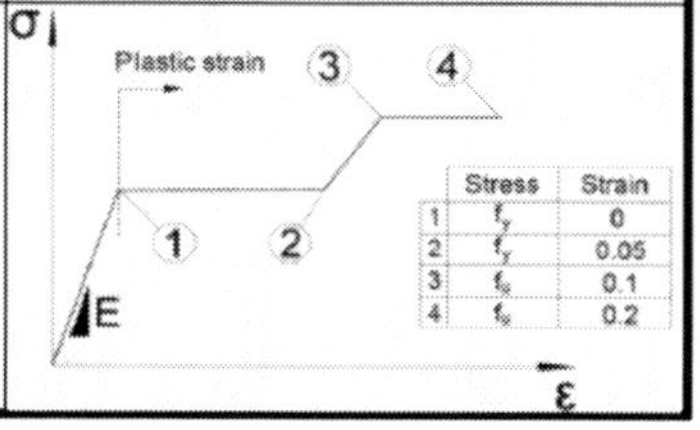
· *Steel class*	*S355*		
· *E (N/mm²)*	*210000*		
· *Poisson ratio*	*0,3*		
· *Density (kg/m³)*	*7850*		
· f_{yf} *(N/mm²)*	*355*		
· f_{yu} *(N/mm²)*	*490*		

4. Incremental launching in a doubly-symmetric, straight steel I-girder

The first example corresponds to the simplest case: doubly-symmetric, straight steel I-girder. The focus of the analysis was on the understanding of the patch loading phenomena not only for the worst-case scenario, that has been studied in last decades (worst panel with a concentrated load in the middle), but also for travelling loads during launching. Detailed analysis of the investigation is given in [13].

Figure 2 displays the evolution of the vertical stresses in web panels for travelling loads when reaction moves from one stiffener to another. The provided information is compared with analytical patch loading models developed in recent years [14][15].

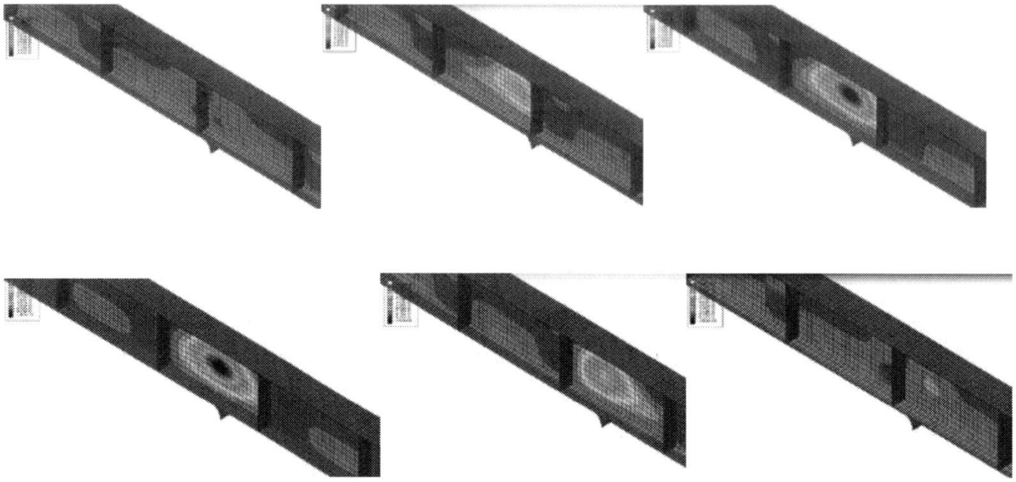

Fig. 2: Vertical stresses in the directly loaded panel

On the other hand, information related to displacements and strains is extracted from key points of the girder and presented in a potential SHM form (figure 3).

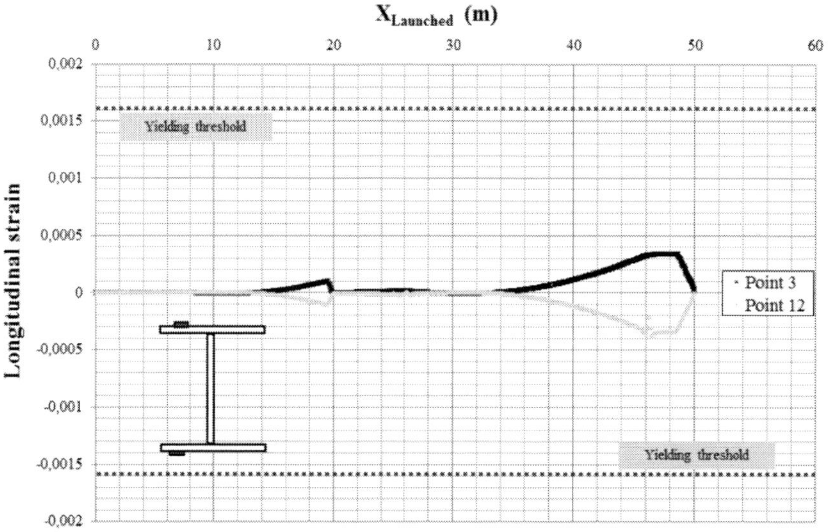

Fig. 3: Longitudinal strains in the top flange for example 1

5. Incremental launching in steel I-girder with corrugated webs

The second example corresponds to a case in which the web of the girder is corrugated. The focus of the analysis was on the understanding of the interaction of stresses and the moment distribution on the web, not only for the worst-case scenario that has been studied in last decades [16] [17] but also, for travelling loads during launching. Detailed analysis of the investigation is given in [18]. A model of a monotonically increasing web-corrugated panel loaded up to failure and found in [16] was developed for calibration purposes.

Figure 3 displays the evolution of the von Mises stresses in web panels for travelling loads when reaction moves from one stiffener to another. It is worth pointing out how the load generates stress distributions in accordance with the web corrugation. Expectedly, the magnitude and sign of the generated moment distribution may be inferred. Furthermore, Fig. 4 shows the vertical displacement of the launching nose during construction (a typical magnitude measured in SHM deployments).

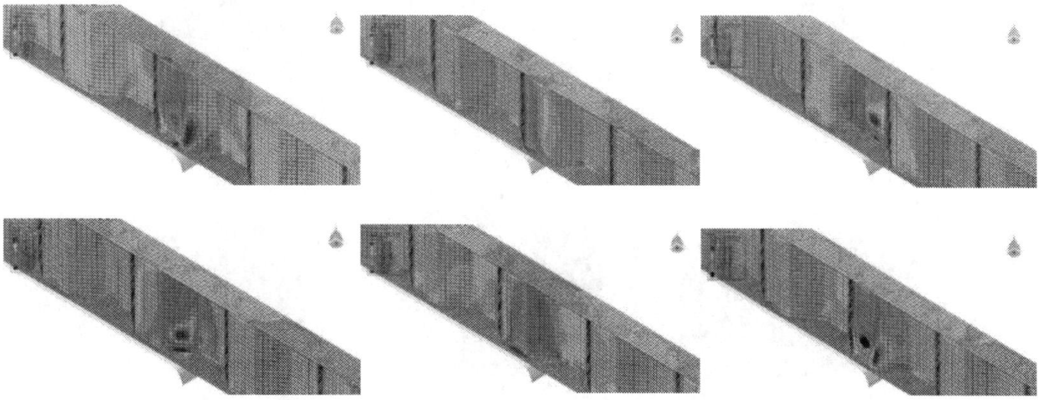

Fig. 3: Von Mises stresses in the directly loaded panel as the reaction is introduced variably

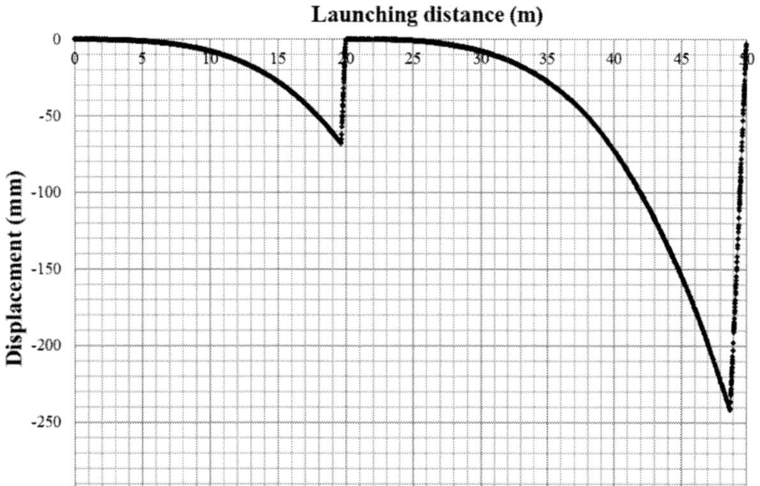

Fig. 4: Vertical displacement in a node located at the launching nose for example 2

6. Incremental launching in an horizontally-curved steel I-girder.

The third example presents an I-shaped cross section, but in this case, considering a curved plan alignment. In this particular case, full elasticity is considered since the study is focused on reactions and deflections. Furthermore, geometric imperfections are not taken into account for the sake of limiting the amount of variables. The focus of this analysis is on assessing the specific phenomena associated to the curved nature of the problem, such as the torsion. In addition, the model can be further improved for the sake of studying the design of lateral braces and the necessary equipment for recovering the launching misalignment of the girder.

Figure 5 depicts a few snapshots of a control cross-section close to the launching nose throughout the procedure. Noticeably, a torsion phenomenon is observed (unsymmetrical stress distribution). Finally, figure 6 shows first attempts of plots that show the evolution of radial forces throughout the launching process. Radial inwards reactions showing torsion loads are observable and expected. It has to be said that this ultimate version is still being validated and under preparation [19].

Fig. 5: Control cross section evolution during the two spans launching (not-to-scale)

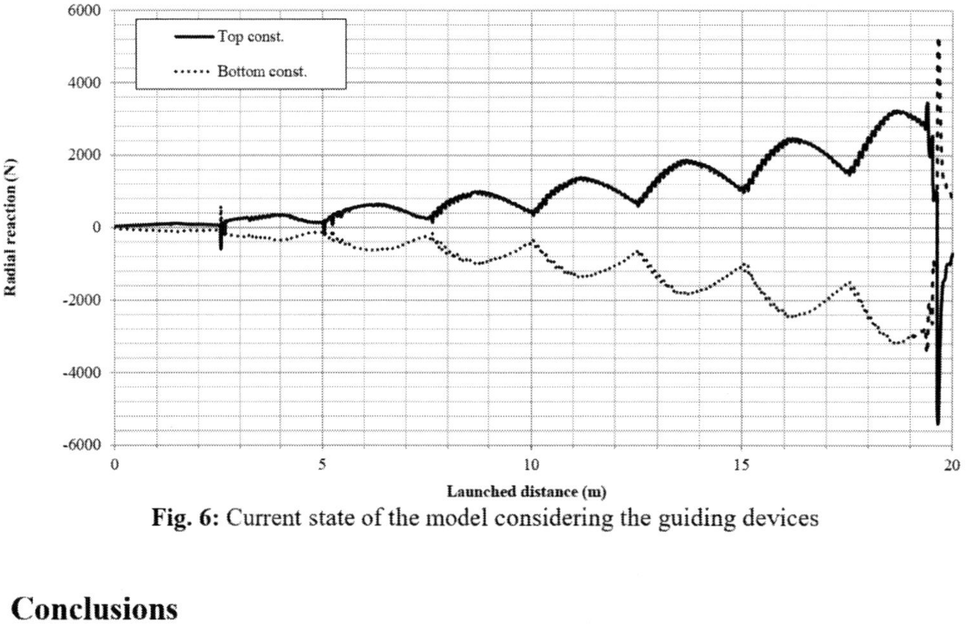

Fig. 6: Current state of the model considering the guiding devices

7. Conclusions

In this paper, realistic 3D numerical models aimed at reproducing the mechanical behavior of a steel plate girder during bridge launching are presented. Doubly symmetric steel I-girders, web corrugated elements as well as curved plan aligned beams are analyzed during a hypothetical process of incremental launching. The numerical simulation provides useful information at design stages (for obtaining phenomenological insights of the processes) as well as at construction stages, since key values related to displacements and strains may be used in SHM deployments. During this process, a continuous monitoring of displacements (vertical and horizontal), strains and stresses as well as contact forces is obtained as a result. These magnitudes can be extracted from the model at any increment (or position of the girders) and at any point of the whole model.

The presented numerical simulations have been studied in a manifold fashion:

- For detecting a potential failure due to concentrated loading, uneven moment distribution or undesired displacement during launching.
- For understanding the travelling patch loading phenomenon in a steel girder at serviceability and/or at failure.
- For predicting numerically the results concerning strains and displacements. The model might be useful at construction stages when comparing the numerical results with values obtained *in situ*.
- For designing SHM deployments. At this point, the numerical model may feed the SHM project, which may also feedback the results for further calibrations.

References

[1] LaViolette M, Wipf T, Lee Y, Bigelow J, Phares B. *Bridge construction practices using incremental launching*, American Association of State Highway and Transportation Officials, AASHTO, 2007.

[2] Rosignoli M. *Bridge Launching*, Thomas Telford, 2002.

[3] Gohler B, Pearson P. *Incrementally Launched Bridges. Design and Construction*, Ernst and Sohn, Berlin, Germany, 2000.

[4] Chacón R, Zorrilla R. "Structural Health Monitoring in Incrementally Launched Steel Bridges: Patch Loading Phenomena Modeling", *Automation in Construction*. 58, 60-73, 2015.

[5] Martins OP, Sampaio AZ. "The application of virtual reality technology in the construction of bridge: The cantilever and incremental launching methods", *Automation in Construction*, 37(1), 58-67. 2011.

[6] Marzouk M, El-Dein H, El-Said M. Application of computer simulation to construction of incrementally launching bridges, Journal of Civil Engineering and Management, Vol. 13 (1), 27-36. 2007.

[7] Granath P, Thorsson A, Edlund B. "I-shaped steel girders subjected to bending moment and travelling patch loading", *Journal of Constructional Steel Research*, 54(3), 409-421. 2000.

[8] Jung K, Kim K, Sim C, Kim J. "Verification of Incremental Launching Construction Safety for the Ilsun Bridge, the World's Longest and Widest Prestressed Concrete Box Girder with Corrugated Steel Web Section", *Journal of Bridge Engineering*, 16(3), 453-460, 2011.

[9] Granata M. "Analysis of non-uniform torsion in curved incrementally launched bridges", *Engineering Structures*, 75, 374-387, 2014.

[10] Abaqus FEA, Simulia© V6.10.3. Dassault Systèmes. 2013.

[11] Chacón R, Uribe N, Oller S. "Numerical and experimental study of the incremental launching method of a steel bridge", *Experimental Techniques*, 2013.

[12] Zorrilla R. "Reproducción experimental de puente por empujes sucesivos empleando FEM (Validation of FEM shell model of incremental launching process using experimental data)", *Eduacero*, 2(5), 2013, http://eduacero.com/revista/pdf/2013-2-5.pdf.

[13] Zorrilla R. "Simulación numérica a escala real de proceso de lanzamiento de puente metálico por empujes sucesivos", *Bachelor Thesis*, Universitat Politècnica de Catalunya, 2014, http://hdl.handle.net/2099.1/24672.

[14] Graciano C, Ayestarán A. "Steel plate girder webs under combined patch loading, bending and shear", *Journal of Constructional Steel Research*, 80(1), 202-212, 2013.

[15] Chacón R, Mirambell E, Real E. "Transversally stiffened plate girders subjected to patch loading. Part 1. Preliminary study", *Journal of Constructional Steel Research*, 80(1), 483-491, 2013.

[16] Kövesdi B, Braun B, Kuhlmann U, Dunai L. "Patch loading resistance of girders with corrugated webs", *Journal of Constructional Steel Research*, 66(12), 1445-1454, 2010.

[17] Jáger B, Dunai L, Kövesdi B. "Girders with trapezoidally corrugated webs subjected by combination of bending, shear and path loading", *Thin-Walled Structures*, 96, 227-239, 2015.

[18] Duñó C. "Estudio del comportamiento estructural de una viga armada de alma corrugada durante su proceso constructivo", *Bachelor Thesis*, Universitat Politècnica de Catalunya, 2014.

[19] Zorrilla R, Chacón R. "Structural analysis of an horizontally-curved steel girder when launched incrementally". Manuscript under preparation.

ASSESSMENT OF THE OVERALL CAPACITY OF TAPERED MEMBERS WITH NON-SYMMETRICAL RESTRAINTS

João D. R. Ferreira, Paulo J. M. M. F. Vila Real and Carlos A. S. Couto

RISCO, Department of Civil Engineering, University of Aveiro, Portugal

Abstract: Part 1-1 of Eurocode 3 presents the General Method to assess the out-of-plane stability of steel members, for cases that fall outside the scope of the interaction formulae. However, there is still a lot of uncertainty regarding the safety and accuracy of this method, particularly when dealing with non-uniform members. In this paper, a comparison between results obtained with the General method and by numerical analyses, for tapered members with non-symmetrical restraints, is provided. It is seen that for cases commonly found in practice, the method returns over-safe results, and that for some of these cases only the assessment of the in-plane resistance would be sufficient to guarantee its safety.

1. Introduction

Part 1.1 of Eurocode 3 presents the so-called General Method (GM) to assess the out-of-plane stability of steel members when the application of the expressions for flexural buckling (FB), lateral-torsional buckling (LTB) or their interaction, given in clauses 6.3.1, 6.3.2 and 6.3.3 respectively, is not possible, such as the case of tapered members and members with complex support conditions. Although this method has already been object of some studies, there is still a lot of uncertainty regarding its validation and proper application. One of the aspects that stills needs further clarification is how the presence of complex restraint conditions affects the design against buckling with this methodology.

Regarding the influence of lateral restraints on the design of members against buckling, several studies exist in the bibliography. Here, it is important to mention [1], where the influence of the vertical position (along the cross-section) and the horizontal position (along the length) of lateral restraints in prismatic beam-columns was investigated, and it was seen that the interaction formulae of Eurocode is well suited for this kind of members, provided that proper values for the critical loads are considered.

As for the study of tapered members in the presence of lateral restraints, few studies can be found and consequently the knowledge regarding their influence on the safety and accuracy of the GM is limited. In [2], a wide comparison of the method to results from experimental tests, which sought for guaranteeing the safety of application of the method to a large scope of cases, does not fully address this subject. In [3], a comparison between numerical results and the GM for a frame with lateral restraints was performed, though only for a single case.

Due to a still scarce validation of the GM, along with the fact that its application to complex support conditions has receive no attention apart from the two mentioned papers (at least to the knowledge of the authors), provided the motivation for the study here presented. Therefore, it is the goal of this paper to provide some additional insight into the application of the GM to these members, as well as to provide indications on how to improve the accuracy of the method.

The latter is done by presenting a numerical study of the application of the GM for prismatic and tapered members that are normally found in practice, with symmetrical and non-symmetrical distributions of lateral restraints. The in-plane and out-of-plane behaviour of single members is analysed using a FEM software for a geometrically and materially nonlinear analysis with imperfections (GMNIA), in order to calculate the global capacity of the members, and the results are compared to the application of the GM considering the European buckling curves from Eurocode 3.

2. The General Method

The GM is presented in clause 6.3.4 of Part 1-2 of Eurocode 3, and allows to assess the overall stability of steel members by ensuring that:

$$\chi_{op} \cdot \alpha_{ult,k} / \gamma_{M1} \geq 1.0 \qquad (1)$$

where $\alpha_{ult,k}$ is the minimum load amplifier of the design loads to reach the characteristic resistance of the most critical cross section of the structural component, considering its in-plane behavior without taking lateral or lateral torsional buckling into account however accounting for all effects due to in plane geometrical deformation and imperfections, global and local, where relevant and χ_{op} is the reduction factor to take account of lateral and lateral-torsional buckling, for the generalized non-dimensional slenderness $\bar{\lambda}_{op}$ given by:

$$\bar{\lambda}_{op} = \sqrt{\alpha_{ult,k} / \alpha_{cr,op}} \qquad (2)$$

where $\alpha_{cr,op}$ is the minimum amplifier for the in plane design loads to reach the elastic critical resistance of the structural component with regards to lateral or lateral torsional buckling without accounting for in plane flexural buckling.

The out-of-plane reduction factor χ_{op} may be taken as the minimum or by an interpolation of the values χ_z for out-of-plane FB and χ_{LT} for LTB, both calculated for the slenderness $\bar{\lambda}_{op}$.

3. Numerical modelling

With the purpose of running GMNIA simulations, the finite element program SAFIR [4] was used, wherein each shell element was defined by 4 nodes, with 6 degrees of freedom in each node. The material law considered was a bi-linear stress-strain relationship with a perfect elastic-plastic behaviour, as indicated in Part 1-1 of Eurocode 3.

A mesh with 10 elements per flange width, 18 elements per web height and 100 elements per each 20m of length were considered, after a mesh sensitivity analysis and a validation of the results obtained with this mesh has been performed. The loads were applied to the model by means of nodal forces, and fork support conditions were defined by imposing null displacements of the nodes at the extremities of the structural members that were analysed (see Fig. 1).

Geometric global imperfections were introduced in the model by changing the node coordinates according to the shape given by the first eigenmode given by a linear buckling analysis (LBA). Since SAFIR does not possess the capability of performing this type of analysis, the software Cast3M [5] was used for that purpose, whereas the coupling between SAFIR and Cast3M models was automated by the use of the software RUBY [6]. The amplitude of the geometric imperfections were set to follow the ones used for the determination of the European buckling curves. Therefore, the node with the maximum displacement was scaled to $L/1000$, with all the other nodes scaled accordingly, where L is the member length.

As for the material imperfections, they were introduced in the model by considering the residual stresses pattern for welded profiles described in ECCS [7], according to the "S235" steel grade, which was adopted in this study.

Fig. 1: Numerical model

4. Parametric study

4.1 Cases covered

In Table 1 are described the cases covered in the numerical study. Tapered members are usually obtained by welding plates, and thus two welded profiles with dimensions equivalent to the European hot-rolled profiles HEB300 and IPE200 were considered.

Table 1: Cases covered in the parametric study

Profiles	Bending moment diagram	Lengths (m)	Loading case	Taper ratio - γ_h	Distribution of lateral restraints
IPE200 welded "IPE200w"	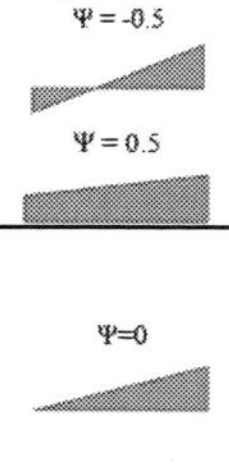	3m 5m 8m	Column Interm. beam-column Beam	$\gamma_h = 1$ $\gamma_h = 2$ $\gamma_h = 3$	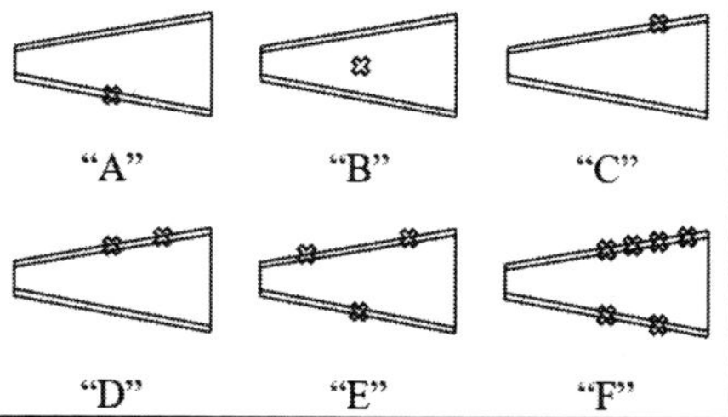
HEB300 welded "HEB300w"					

The bending moment diagram and the lengths were chosen in order to consider cases that are likely to be found in practice. Numerical simulations for all the possible combinations between the items in Table 1 were performed, meaning that a total of 486 cases were analysed.

4.2 Buckling curves for each member

One of the problems that arise when dealing with the application of the GM to tapered members is the choice of the buckling curve, which depends on the cross-section geometry that varies along the member. In order to illustrate this, in Table 2 the appropriate buckling curves for the different geometries involved in this study are shown, in agreement to Tables 6.2 and 6.4 of EC3-1-1, according to each type of buckling phenomenon.

Table 2: Buckling curves for out-of-plane flexural buckling and lateral-torsional buckling for the profiles analysed, according to Tables 6.2 and 6.4 of Part 1-1 of Eurocode 3

Taper ratio	IPE200w 	HEB300w 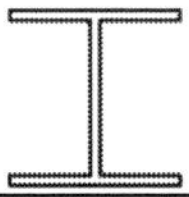
$\gamma_h = 1$	h_{const} = 200mm FB: Curve c LTB: Curve d	h_{const} = 300mm FB: Curve c LTB: Curve c
$\gamma_h = 2$	h_{min} = 200mm, h_{max} = 400mm FB: Curve c LTB: Curve d	h_{min} = 300mm, h_{max} = 600mm FB: Curve c LTB: Curve c
$\gamma_h = 3$	h_{min} = 200mm, h_{max} = 600mm FB: Curve c LTB: Curve d	h = 600mm h_{min} = 300mm, h_{max} = 900mm FB: Curve c LTB: Curve c, Curve d

Another issue found when dealing with these members has to do with how to obtain χ_{op}, as it was seen in previous studies that there is a lack of consistency in the interpolation formulae provided by the Eurocode (see [8]). Due to this aspect, [9] strongly recommends the use of the minimum value between the reduction factors χ_z and χ_{LT}. This means that for the IPE200w cases the curve "d" should be used, whereas for the HEB300w, the curve "c" should be used for the taper ratios $\gamma_h = 1$ and $\gamma_h = 2$, but it is not clear whether for $\gamma_h = 3$ one should use curve "c" or "d". To overcome these uncertainties, both buckling curves "c" and "d" will be considered in the remaining part of this study.

5. Analysis of results

5.1 Effect of the distribution of lateral restraints

In Fig. 2, it can be seen the effect of different configurations for the lateral restraints, for four members analysed with the IPE200w, grouped according to their geometry and the loads applied.

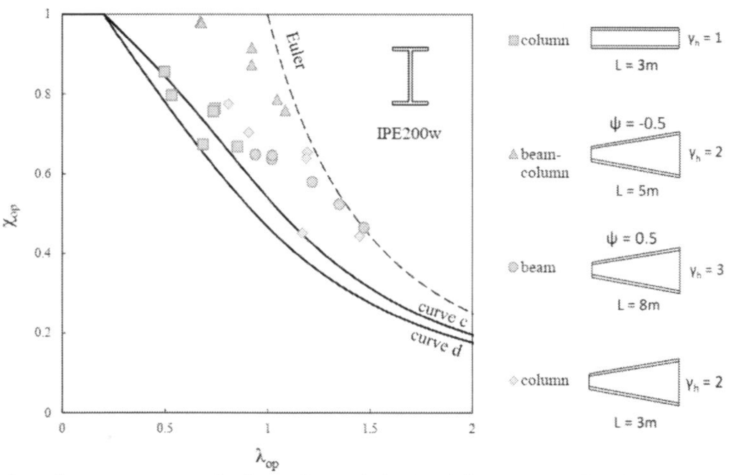

Fig. 2: Comparison between numerical results and the buckling curves for the IPE200w, considering all the distributions of lateral restraints found in Table 1, for four of the cases analysed

It is observed that for the beams and beam-columns it is verified a somewhat consistent behaviour of the numerical results when varying the type of restraint conditions, as is exemplified by the beam and beam-column case in Fig. 2. The same is not true for the column cases, where, for instance, for a $\bar{\lambda}_{op}$ around 1.25, considerably different reduction factors were obtained for the tapered column example shown in Fig. 2.

This may be related to the possibility of purely torsional modes being captured in the LBA analysis, but probably means that, contrary to what was seen in [1], a proper value of the critical load may not be sufficient to guarantee consistent results in the verification of these members with the GM, when considering different restraint conditions.

5.2 Safety according to each lateral restraint distribution

In Fig. 3 and Fig. 4, results for the IPE200w and HEB300w profiles are respectively plotted, for three of the lateral restraint configurations studied.

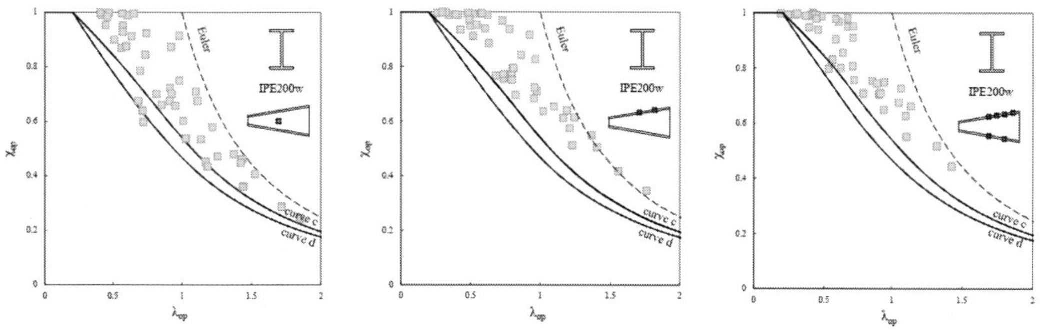

a) "B" b) "D" c) "F"

Fig. 3: Comparison for the IPE200w, for three of the lateral restraint configurations

For the IPE200w, only four of the cases analyzed fall below the buckling curve "c". These cases below curve "c" correspond to columns, which are not likely to be built in the form of tapered members. On the other hand, curve "c" yields safe but over conservative results, especially for the cases where the bending moment is considerably high. For such cases, the beneficial influence of the bending moment diagram is not accounted for by the reduction factor for LTB, which was calibrated for the uniform bending moment (ψ=1) for a prismatic member. Moreover, the types of bending moment diagrams found in tapered members are not that severe, since it would not make sense to use a tapered member for a uniform bending moment case. Thus, the geometry of tapered members along with the pattern of the bending moment diagrams for which they are normally subjected could be exploited in the formulae in order to have more accurate and economic results.

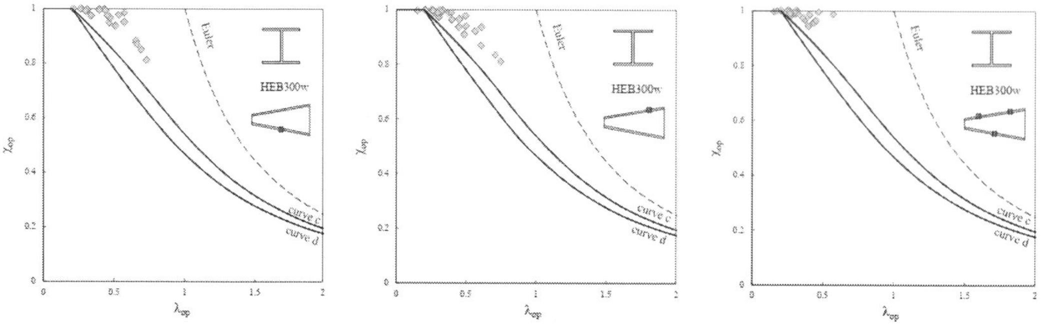

a) "A" b) "C" c) "E"

Fig. 4: Comparison for the HEB300w, for three of the lateral restraint configurations

The nature of the bending moments diagram considered for the HEB300w cases, the fact that the lengths analyzed correspond to critical values that are much larger than the cross-section resistance, and the high proximity between the in-plane and out-of-plane resistances for this kind of profile, resulted in the ultimate resistance obtained being very close to the in-plane resistance for most of the cases, which led to reduction factors very close to the unity, and even equal to 1 for cases where the in-plane displacements completely governed the collapse of the member (see Fig. 4).

5.3 Overall safety

In Fig. 5 and Fig. 6, comparisons between the reduction factors obtained numerically ($\chi_{op,GMNIA}$) and the ones obtained with the GM ($\chi_{op,GM}$) are shown, which include all the cases analysed.

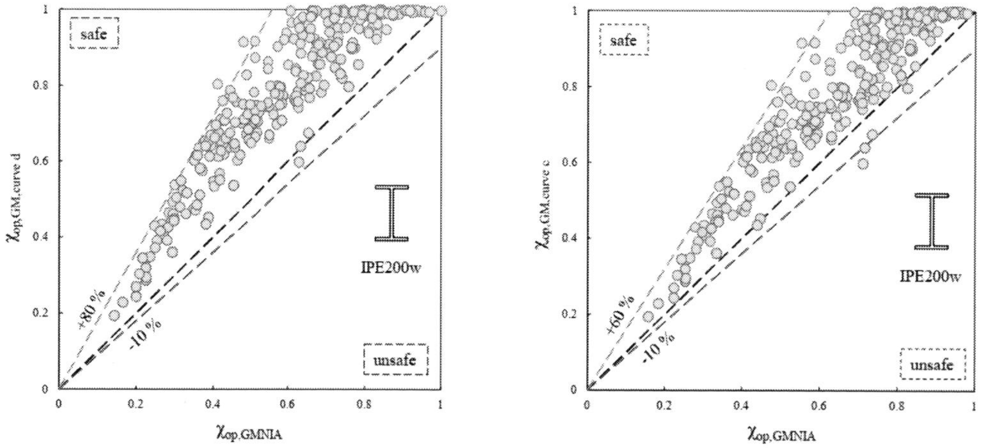

a) Curve "d" b) Curve "c"

Fig. 5: Comparison between the reduction factors obtained by the numerical results and the buckling curves "c" and "d", for the IPE200w

As reported before, it is seen that buckling curve "d" is too conservative for almost all the cases analysed for the profile IPE200w (see Fig. 5), with members frequently achieving close to 80% more resistance than what is considered by affecting the in-plane resistance by the reduction factors according to the buckling curve "d".

If the buckling curve "c" was used instead, a better utilization of the characteristics and benefits of tapered members would be achieved. This is in line with results found in the project "TaperSteel", where other type of non-uniformities were analysed.

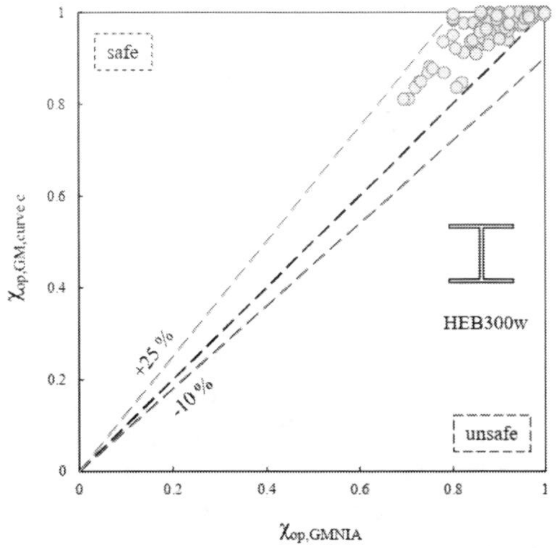

Fig. 6: Comparison between the reduction factors obtained by the numerical results and the buckling curve "c", for the HEB300w

As for the plots for the HEB300w in Fig. 6, the results show that the curve "c" is well-adjusted, as almost all the cases fall between 25% on the safe side and 10 % on the unsafe side.

6. Conclusions

The main conclusions are:

1. The influence of different lateral support conditions is not consistently portrayed by the European buckling curves when applied with the General Method. This lack of consistency is more pronounced for members under pure compression;
2. The results for the tapered beam-columns indicate that, due to the nature of the bending moment diagrams applied to these members, the curves preconized in the Eurocode 3 for the LTB phenomenon in welded profiles result frequently in over-conservative results, and that probably a consideration of the buckling curve "c" for the majority of the cases would be more adjusted;
3. Due to the proximity between the in-plane and out-of-plane resistances for the H-profiles, the collapse analysis for laterally restrained tapered members with this profiles often result in values close or equal to their in-plane resistance, which mean that for some of these cases an in-plane check would be sufficient to guarantee its safety.

Acknowledgments

The work here present was developed in the scope of the project TAPERSTEEL - PTDC/ECM-EST/1970/2012, "Stability design of non-uniform steel members", funded by the Fundação para a Ciência e Tecnologia (FCT).

References

[1] Aswandy, Greiner R. "Design of members under bending and axial compression with intermediate lateral restraints," *Eurosteel 2005*, 2005.

[2] Bijlaard F, Feldmann M, Naumes J, Sedlacek G. "The 'general method' for assessing the out-of-plane stability of structural members and frames and the comparison with alternative rules in EN 1993 - Eurocode 3 - Part 1-1," *Steel Constructions*, 3(1), 19–33, 2010.

[3] Greiner R, Ofner R. "Comparison of General Method with Traditional Methods - Example of a sway frame with lateral restraints," 2007.

[4] Franssen J-M, "SAFIR: A thermal/structural program for modelling structures under fire," *Eng. J.*, 42(3), 143–158, 2005.

[5] CEA, "CAST 3M is a research FEM environment; its development is sponsored by the French Atomic Energy Commission <http://www-cast3m.cea.fr/>." 2012.

[6] Couto C, Real PV, Lopes N. "Ruby - an interface software for running a buckling analysis of SAFIR models using Cast3M," 2013.

[7] ECCS, *Ultimate Limit State Calculation of Sway Frames with Rigid Joints*. European Convention for Constructional Steelwork, 1984.

[8] Marques L, Rebelo C. "Review of the General Method procedure in EC3-1-1 as a global stability verification procedure," in *Eurosteel 2014*, 2014.

[9] CommitteE TC8, "Resolution of ECCS/TC8 with respect to the General Method in EN1993-1-1."

STIFFNESS REDUCTION METHOD FOR THE IN-PLANE DESIGN OF STEEL FRAMES

Merih Kucukler[a,b], Leroy Gardner[a] and Lorenzo Macorini[a]

[a] *Department of Civil and Environmental Engineering, Imperial College London, London, UK*
[b] *Department of Civil Engineering, Bilecik Seyh Edebali University, Bilecik, Turkey*

Abstract: A stiffness reduction method for the in-plane design of steel frames is proposed in this paper. The proposed method is performed by (i) reducing the flexural stiffnesses of the members of a frame on the basis of the first-order forces they withstand and (ii) carrying out Geometrically Nonlinear Analysis. The ultimate capacity of the structure is defined as the point at which the ultimate strength of the most heavily loaded cross-section is reached. Owing to the full consideration of the deleterious influence of the spread of plasticity and imperfections through stiffness reduction, the proposed approach eliminates the need of using member design equations or modelling member out-of-straightnesses; only cross-section strength checks are required. The verification of the proposed approach against results from the nonlinear finite element modelling of a series of benchmark frames is presented.

1. Introduction

The load-bearing capacity of steel structures is limited by the detrimental influence of the spread of plasticity, instabilities and imperfections. In traditional design, these adverse effects are generally considered through semi-empirical member design equations and effective length or notional load concepts. However, this type of a design approach is rather indirect, and requires lengthy calculations for the design of every individual member in the structure. Design by Geometrically Nonlinear Analysis (i.e. second-order elastic analysis) with stiffness reduction, on the other hand, represents a more direct and practical approach. In this approach, instabilities and the influence of boundary conditions and restraint from adjoining members are accounted for by conducting a Geometrically Nonlinear Analysis, while load level and the corresponding influence of geometrical imperfections, residual stresses and the spread of plasticity are allowed for through stiffness reduction. This type of a design approach was previously recommended by Liew et al. [1] and Kim and Chen [2]. However, the stiffness reduction methods put forward in these studies can only be applied through specifically developed structural analysis software that is not readily available to structural engineers. Surovek-Maleck and White [3] recommended a method that can be performed through conventional structural analysis software; this method has now been included in the two most recent versions of AISC 360 specifications, including AISC 360-10 [4]. It should be noted though that the Direct Analysis Method of AISC 360-10 [3,4] still requires the use of column strength equations in design, and can lead to overly-conservative results in some cases due to the neglect of the influence of bending moment shape on the development of plasticity [5].

With the aim of extending these previous studies, this paper presents a stiffness reduction method for the in-plane design of steel frames. In the proposed approach, the flexural stiffnesses of members of a frame are reduced through the developed stiffness reduction functions on the basis of the first-order forces they resist, then Geometrically Nonlinear Analysis (i.e. second-order elastic analysis) of the frame is performed. Owing to the full consideration of the deleterious influence of plasticity and imperfections through the developed stiffness reduction functions, the proposed approach eliminates the need of using member design equations or modelling member out-of-straightnesses, but only requires cross-section strength checks, thus offering a practical means of structural steel design. The presented stiffness reduction approach is intended to be used in a design framework where the in-plane and out-of-plane design of steel frames under in-plane loading are performed separately. The finite element modelling approach adopted in this study is described in Section 2 of this paper, while the development, application and validation of the proposed stiffness reduction method are described in Sections 3 and 4.

2. Finite element modelling

This section describes the finite element modelling (FE) approach adopted in this study. The results of the Geometrically and Materially Nonlinear FE Analyses with Imperfections (GMNIA) are used in Section 4 to validate the proposed stiffness reduction approach. Owing to their computational efficiency, beam finite elements were used to develop the finite element models through the finite element analysis software Abaqus [6]. Specifically, the linear Timoshenko beam element referred to as B31OS in the Abaqus element library, which is able to consider shear deformations, is employed. The tri-linear elastic-plastic stress-strain relationship illustrated in Fig. 1 is used, where E is the Young's modulus, E_{sh} is the strain hardening modulus, f_y and ε_y are the yield stress and strain respectively, ε_{sh} is the strain value at the onset of strain hardening and f_u and ε_u are the ultimate stress and strain respectively. E_{sh} was assumed to be 2% of E, which was taken as 210 GPa, and ε_{sh} was assumed as $10\varepsilon_y$, complying with the ECCS [7] material model recommendations for hot rolled steel. In all numerical models S235 steel (f_y = 235 MPa) was used.

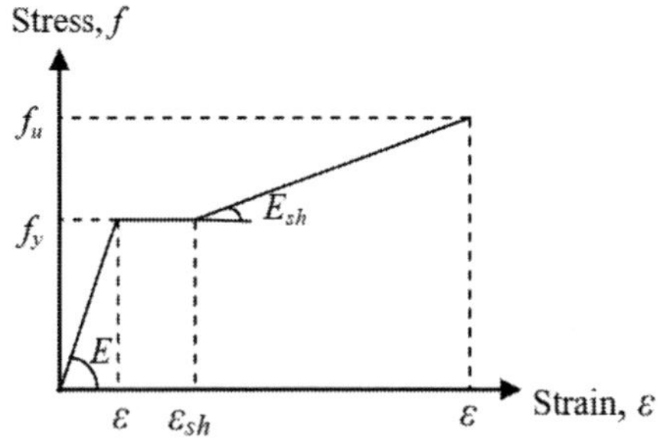

Fig. 1: Material stress-strain curve used in finite element models

The ECCS [7] residual stress patterns, shown in Fig. 2, were applied to the models through the SIGINI subroutine [6]. The initial member out-of-straightness was assumed to be half-sine wave in shape and 1/1000 of the corresponding member length in magnitude, while the initial frame out-of-plumbness (i.e. initial sway imperfection) was assumed to be 0.002 rad [8]. The geometrical imperfections were modelled in the most unfavourable directions considering the loading conditions of the investigated frames.

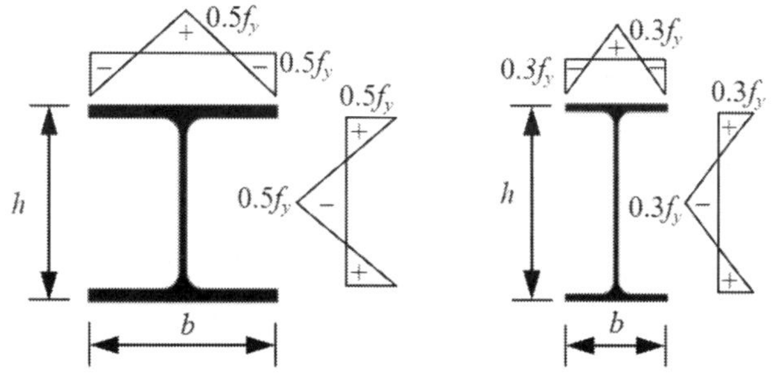

Fig. 2: Residual stress patterns applied to finite element models (+ve = tension, −ve compression)

3. Description of the stiffness reduction method

In this section, the application of the proposed stiffness reduction method is explained. Stiffness reduction functions previously developed by Kucukler et al. [9] are presented, the adopted ultimate cross-section strength equation is provided, and the overall design procedure is then described.

3.1 Stiffness reduction functions

3.1.1 Stiffness reduction function for axial loading τ_N

Using the European column buckling curves [10], Kucukler et al. [9] derived a stiffness reduction function for axial loading τ_N, which is expressed as:

$$\tau_N = \frac{4\psi^2}{\alpha^2 N_{Ed}/N_{pl}\left[1+\sqrt{1-4\psi\dfrac{N_{Ed}/N_{pl}-1}{\alpha^2 N_{Ed}/N_{pl}}}\right]^2} \quad \text{but} \quad \tau_N \le 1.0$$

$$\text{where } \psi = 1+0.2\alpha\frac{N_{Ed}}{N_{pl}}-\frac{N_{Ed}}{N_{pl}}, \tag{1}$$

in which N_{Ed} is the axial load applied to the member, N_{pl} is the axial yield load equal to the cross-sectional area times yield stress f_y (i.e. $N_{pl} = Af_y$) and α is the imperfection factor. The values of imperfection factors for hot-rolled I section members in the case of major axis failure are given in Table 1. It should be noted that use of the stiffness reduction function τ_N in a Linear Buckling Analysis (LBA) provides an exact match to the inelastic flexural buckling strengths obtained through the European column buckling curves. Nevertheless, herein, an imperfection factor α equal to 0.26 (i.e. α = 0.26) is recommended for the case of I-section members whose aspect ratios are smaller than 1.2 (i.e. $h / b \le 1.2$) as Kucukler [11] observed that this value leads to more accurate and less conservative results in comparison to α=0.34, as adopted in EN 1993-1-1 [10].

Table 1: Parameters for the stiffness reduction functions for axial loading τ_N and bending τ_M

	τ_N	τ_M			
Aspect ratio	α	τ_{MI}	ϕ_y	ξ_m	β_m
$h / b \leq 1.2$	0.26	0.04	$0.5W_{el,y}/W_{pl,y}$	0.98	1.5
$h / b > 1.2$	0.21	0.08	$0.7W_{el,y}/W_{pl,y}$	0.95	1.5

3.1.2 Stiffness reduction function for bending τ_M

In Kucukler et al. [9], a stiffness reduction function for bending τ_M was recommended, which is given by the following expression:

$$\tau_M = 1.0 \quad \text{if} \quad C_m M_{y,Ed} / M_{y,pl} \leq \phi_y, \tag{2a}$$

$$\tau_M = (1-\tau_{MI})\left[1-\left(\frac{C_m M_{y,Ed}/M_{y,pl}-\phi_y}{\xi_m-\phi_y}\right)^{\beta_m}\right]^{1/\beta_m}+\tau_{M1} \quad \text{if} \quad \phi_y \leq C_m M_{y,Ed}/M_{y,pl} \leq \xi_m, \tag{2b}$$

$$\tau_M = \tau_{M1}\left[1-\left(\frac{C_m M_{y,Ed}/M_{y,pl}-\xi_m}{1-\xi_m}\right)\right] \quad \text{if} \quad \xi_m \leq C_m M_{y,Ed}/M_{y,pl}, \tag{2c}$$

in which $M_{y,Ed}$ is the maximum absolute value of bending moment along the length of a member, $M_{y,pl}$ is the major axis plastic bending moment resistance equal to the plastic section modulus $W_{pl,y}$ multiplied by the yield stress f_y (i.e. $M_{y,pl} = W_{pl,y} f_y$), and ξ_m, β_m, ϕ_y and τ_{MI} are auxiliary coefficients proposed by Zubydan [12], the values of which are provided in Table 1. In the table, $W_{el,y}$ is the elastic section modulus of the cross-section. To consider the influence of the bending moment shape on the development of plasticity, the maximum absolute bending moment along the length $M_{y,Ed}$ is multiplied with a moment gradient factor C_m, which is determined as given by:

$$C_m = \frac{-1.5M_{y,Ed}+4M_A+6M_B+4M_C}{12.5M_{y,Ed}} \geq 0.12, \tag{3}$$

where M_A, M_B and M_C are the absolute bending moment values at the quarter, centre and three-quarter points along the length of the member.

3.1.3 Stiffness reduction function for combined bending and axial loading τ_{MN}

Kucukler et al. [9] derived a stiffness reduction function for combined bending and axial loading τ_{MN} which can be expressed as follows:

$$\tau_{MN} = \tau_M \tau_N \left[1-\left(\frac{N_{Ed}}{N_{pl}}\right)^{0.8}\left(\frac{C_m M_{y,Ed}}{M_{y,pl}}\right)\right]. \tag{4}$$

As can be seen from the equation, the maximum absolute bending moment along the member length $M_{y,Ed}$ is factored by the moment gradient factor C_m to consider the effect of bending moment shape on the development of plasticity in the expression for τ_{MN}. For the case of pure compression, τ_{MN} degenerates into τ_N, while τ_{MN} is equal to τ_M for pure bending.

3.2 Ultimate cross-section strength equations

In view of their accuracy and simple form, the continuous ultimate strength interaction equations proposed by Duan and Chen [13] were chosen in this study. This expression is given by:

$$1/\alpha_{ult,c} = \left(\frac{N_{Ed}}{N_{pl}}\right)^{1.3}+\frac{M_{y,Ed}}{M_{y,pl}} = 1, \tag{5}$$

where $1/\alpha_{ult,c}$ is the cross-section utilisation factor.

3.3 Steps followed in the application of the stiffness reduction method

The following steps are adopted in the application of the proposed stiffness reduction method:

1. Carry out Linear Elastic Analysis (i.e. first-order elastic analysis) of a frame assuming the out-of-plumbness to be 0.002 rad and in the most unfavourable direction.
2. Calculate the stiffness reduction factors τ_{MN} of each member considering the member forces from the Linear Elastic Analysis.
3. Determine the upper bound stiffness reduction factor of the frame $\tau_{MN,lim}$ by taking the greater of 0.8 or the minimum stiffness reduction factor calculated for all the members $\tau_{MN,min}$, i.e. $\tau_{MN,lim}$=max(0.8; $\tau_{MN,min}$).
4. Reduce the flexural stiffness (i.e. second moment of area I_y) of each member using its corresponding stiffness reduction factor τ_{MN}. If the stiffness reduction factor of a member τ_{MN} is larger than the upper bound stiffness reduction factor of the frame $\tau_{MN,lim}$, use $\tau_{MN,lim}$ to reduce the flexural stiffness of the member.
5. Perform Geometrically Nonlinear Analysis of the frame with reduced stiffness.
6. Make cross-section strength checks using Eq. (5). Provided the utilisation factors for all the members $1/\alpha_{ult,c}$ are less than or equal to 1.0, the design is satisfactory.

It is worth noting that the influence of the out-of-plumbness can be taken into account by direct modelling or through the application of notional loads to the structure with perfect geometry as explained in EN 1993-1-1 [10] and AISC 360-10 [4].

4. Application of the stiffness reduction method to non-redundant Kanchanalai [14] frames

In this section, the proposed stiffness reduction method (SRM) is applied to non-redundant benchmark frames sensitive to the deleterious influence of plasticity and imperfections, which were previously studied by Kanchanalai [14]. Additionally, comparisons of the proposed approach against the direct analysis method of AISC 360-10 [4] and the notional load method (NLM) of EN 1993-1-1 [10] are also presented. Unlike the stiffness reduction method presented in this paper and the direct analysis method of AISC 360-10 [4], the nominal (unreduced) elastic stiffnesses are considered in the notional load method of EN 1993-1-1 [10]. Nevertheless, to consider the influence of the plasticity on the response a structure, the initial out-of-plumbness is assumed to be equal to 0.005 rad in lieu of 0.002 rad, which is the maximum out-of-plumbness value permitted by the relevant European standard with respect to the fabrication and erection of steel structures [15]. Since this amplified (i.e. equivalent) out-of-plumbness is taken into account through notional horizontal loads, this approach is referred to as the notional load method in the literature, which is also adopted herein.

The geometrical properties and loading conditions of the considered Kanchanalai [14] frames are shown in Figs 3 (a)-(d), where the frames are labelled as FP, FR, LFP and LFR considering the use of base beams and leaning columns (i.e. the columns pinned at the both ends) within them. As can be seen from the figures, the frames are subjected to horizontal loading H_{Ed} and vertical loads N_{Ed} applied at the tops of the columns. For the columns, HEB 400 and HEB 200 sections were used. The beam sections were specified such that the parameter G_R, determined by dividing the ratio of the second moment of area of the column I_c to its length L to the ratio of the second moment of area of the beam I_b to its length L_b (i.e. $G_R=(I_c/L)/(I_b/L_b)$), were equal to 0, 2.0 and 3.0. Different column slenderness values λ_y, equal to the square root of the ratio of the column length L to the radius of gyration of the column section about the major axis i_y (i.e. $\lambda_y = \sqrt{L/i_y}$), are considered.

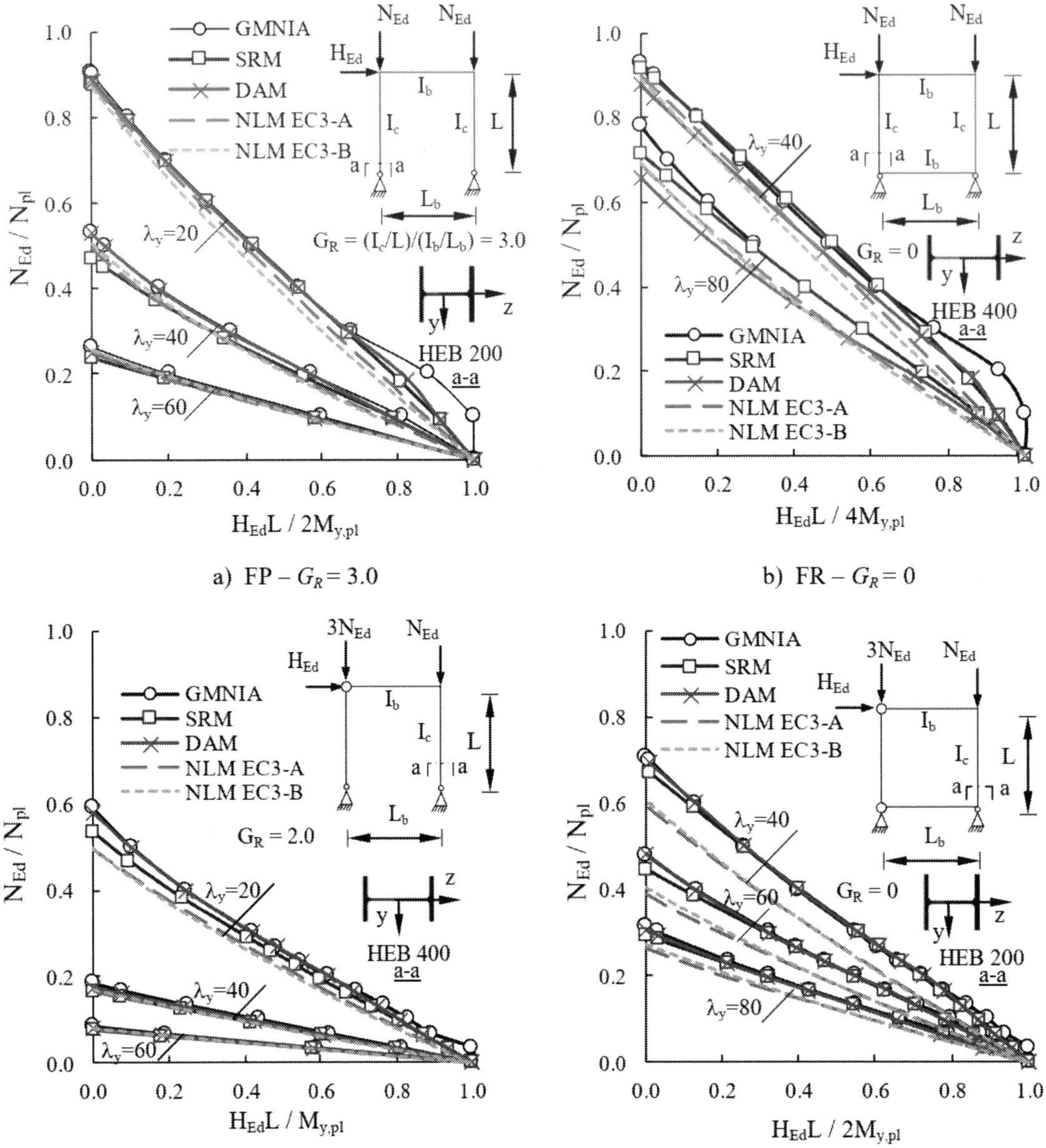

a) FP – G_R = 3.0 b) FR – G_R = 0

c) LFP – G_R = 2.0 d) LFR – G_R = 0

Fig. 3: Comparison of the proposed stiffness reduction method (SRM) against GMNIA, the direct analysis method (DAM) of AISC 360-10 [4] and the notional load method with the beam-column design method provided in Annex A (NLM EC3-A) and Annex B (NLM EC3-B) of EN 1993-1-1 [9] for the non-redundant Kanchanalai [14] frames (L_b>>L).

Figs 3 (a)-(d) show the comparison of the ultimate strength predictions obtained through the proposed SRM against those obtained from GMNIA for different frame configurations, column slenderness values and column-to-beam stiffness ratios. Note that N_{pl} and $M_{y,pl}$ correspond to the axial yield load and major axis plastic bending moment resistance of the columns respectively. It is seen from the figures that in all the considered non-redundant frames sensitive to the detrimental influence of plasticity and imperfections, the proposed SRM provides accurate and safe ultimate strength predictions. The presence of leaning columns within the LFP and LFR frames creates a further destabilising effect on the columns that are providing the lateral stability. The larger the axial compression withstood by the leaning column, the larger the destabilising effect. As can be seen from Fig. 3, the proposed SRM is able to capture accurately the leaning column effects, which are quite significant as the axial compression withstood by the leaning columns are three times that of the columns providing the lateral stability. The ultimate strength values obtained through the DAM are also shown in Figs 3 (a)-(d), where it is seen that the DAM also provides accurate strength predictions. Nevertheless, for the case of the FR frames whose columns are under double curvature bending, the proposed SRM provides more accurate results than the DAM owing to the consideration of the influence of the bending moment shape on the development of plasticity within its stiffness reduction functions through the moment gradient factors C_m. It should be noted that in Kucukler et al. [5], a much wider range of cases are considered to verify the proposed SRM, where it is shown that the proposed SRM leads to considerably more accurate results in comparison to the DAM for sway inhibited steel members under double curvature bending.

The proposed SRM is also compared against the NLM of EN 1993-1-1 [10] in Figs 3 (a)-(d). For the case of the NLM, the ultimate strength values determined using both the beam-column design methods provided in the Annex A (NLM EC3-A) and Annex B (NLM EC3-B) are shown, where exact equivalent moment factors were calculated and used in the calculations, following the recommendations of Boissonnade et al. [16]. As can be seen from Figs 3 (a)-(d), the proposed SRM provides more accurate results in comparison to both the NLM EC3-A and NLM EC3-B for the great majority of the cases. Particularly, for the case of the LFR frame, the proposed approach is significantly more accurate. It is worth noting that the overconservatism of the NLM EC3-A and NLM EC3-B observed in this paper is in line with the observations of Boissonnade et al. [16], where it is stated that the beam-column design methods of EN 1993-1-1 [10] were developed for simply-supported beam-columns and may provide less accurate, but safe, results for other cases.

5. Conclusions

A stiffness reduction method for the in-plane design of steel frames has been proposed in this paper. In the application of the proposed method, the flexural stiffnesses of the members of a steel frame are reduced on the basis of the first-order forces they resist through the developed stiffness reduction functions. Geometrically Nonlinear Analysis (GNA) of the frame is then performed and cross-section strength checks are made. The proposed method was verified against GMNIA results obtained through finite element modelling of a series of non-redundant Kanchanalai [14] frames, which are sensitive to the deleterious effects of plasticity and imperfections; some of the frames also included leaning column effects. The proposed method was also compared against the direct analysis method of AISC 360-10 [4] and the notional load method of EN 1993-1-1 [10]. It was observed that the proposed approach leads to more accurate results than the direct analysis method for the cases where the columns are under double curvature bending and provides more accurate ultimate strength predictions in comparison to the notional load method of EN 1993-1-1 [10] in the great majority of cases.

Owing to the elimination of the need of using member design equations or modelling member out-of-straightnesses, the proposed stiffness reduction method, which only requires cross-section strength checks, provides a practical means of structural steel design. Since it does not require the use of column strength equations, unlike the direct analysis method, the proposed approach can be extended to frames comprising members with irregular geometry such as curved, tapered or stepped members. Future research will focus on the application of the proposed approach to frames with irregular members and geometries and the extension of the proposed approach to the out-of-plane design of steel frames utilising the methods developed by Kucukler et al. [17,18].

References

[1] Liew JYR, White, DW, Chen WF. "Second-order refined plastic hinge analysis for frame design. Part I.", *Journal of Structural Engineering, ASCE*, 119(11), 3196-3216, 1993.

[2] Kim SE, Chen WF. "Practical advanced analysis for unbraced frame design", *Journal of Structural Engineering, ASCE*, 122(11), 1259-1265, 1996.

[3] Surovek-Maleck AE, White DW. "Alternative approaches for elastic analysis and design of steel frames. Part I: Overview", *Journal of Structural Engineering, ASCE*, 130(8), 1186-1196, 2004.

[4] AISC 360-10. *Specifications for structural steel buildings*, American Institute of Steel Construction (AISC), Chicago, 2010.

[5] Kucukler M, Gardner L, Macorini L. "In-plane design of steel frames through a stiffness reduction method", *Journal of Constructional Steel Research*, Submitted for publication.

[6] Abaqus v. 6.10. *Reference Manual*, Simulia, Dassault Systemes, 2010.

[7] ECCS. *Ultimate limit state calculation of sway frames with rigid joints*, Tech. Rep., No: 33, Technical Committee 8 of European Convention for Constructional Steelwork (ECCS), 1984.

[8] AISC. *Code of standard practice for steel building and bridges*, American Institute of Steel Construction (AISC), Chicago, 2010.

[9] Kucukler M, Gardner L, Macorini L. "A stiffness reduction method for the in-plane design of structural steel elements", *Engineering Structures*, 73, 72-84, 2014.

[10] EN 1993-1-1. *Eurocode 3 Design of steel structures. Part 1-1: General rules and rules for buildings*, European Committee for Standardization (CEN), Brussels, 2005.

[11] Kucukler M. *Stiffness reduction approach for structural steel design*, Ph.D. thesis, Imperial College London, 2015.

[12] Zubydan AH. "A simplified model for inelastic second order analysis of planar frames", *Engineering Structures*, 32(10), 3258-3268, 2010.

[13] Duan L, Chen WF. "A yield surface equation for doubly symmetrical sections", *Engineering Structures*, 12(2), 114-119, 1990.

[14] Kanchanalai T. *The design and behavior of beam-columns in unbraced frames*, AISI project, No. 189., Tech. Rep., The University of Texas at Austin. 1977.

[15] EN 1090-2. *Execution of steel structures and aluminum structures-Part 2:Technical requirements for steel structures*, European Committee for Standardization (CEN), Brussels, 2008.

[16] Boissonnade N, Greiner R, Jaspart JP, Lindner J. *Rules for member stability in EN 1993-1-1: Background documentation and design guidelines*, No. 119. European Convention for Constructional Steelwork (ECCS), 2006.

[17] Kucukler M, Gardner L, Macorini L. "Lateral-torsional buckling assessment of steel beams through a stiffness reduction method", *Journal of Constructional Steel Research*, 109, 87-100, 2015.

[18] Kucukler M, Gardner L, Macorini L. "Flexural-torsional buckling assessment of steel beam-columns through a stiffness reduction method", *Engineering Structures*, 101, 662-676, 2015.

ASYMPTOTIC AND PATH-FOLLOWING ANALYSES OF THIN WALLED STRUCTURES USING A MIXED SOLID FINITE ELEMENT

Domenico Magisano, Leonardo Leonetti and Giovanni Garcea

DIMES, Università della Calabria, Italy

Abstract: The reasons of the better performances of mixed, stress--displacements, 3D solid finite elements in the analysis of slender elastic structures are explained. It will be shown that mixed or compatible description, also when derived from the same finite element and then completely equivalent from the discretization point of view, behave very differently when implemented in both asymptotic and path-following solution strategies due to the occurrence of a pathological locking phenomenon in the compatible formulation. The notable advantages of the used of a 3D mixed solid finite element in Koiter asymptotic analysis are also highlighted.

1. Introduction

A growing number of researchers is aimed in developing 3D solid finite elements [1] for linear and nonlinear elastic analysis of thin shells. This is essentially due to a series of important advantages that 3D models show with respect to technical ones. In particular, they allow the use of the simple continuum mechanics strain and stress measures as function of translational degrees of freedom only and then make possible to avoid the use of complicated and expensive rules for updating the rotations and to reduce the order of nonlinearity of the problem.

Among the other advantages of 3D solid elements it is interesting to recall [1-3]: the possibility in describing the strain energy and its variations using simpler expression with benefits in computational efficiency with respect to shell elements; the possibility to exactly obtain the complete 3D stress state and then simplicity in applying nonlinear constitutive laws; the possibility to simply consider structures with solid-like and thin-walled-like regions. Furthermore the model is objective but, using the Green-Lagrange strain measure has a quadratic dependency only from the displacement field avoiding the use of complex geometrically exact shell formulations [4,5].

Nevertheless, formulating robust solid-shell elements is more demanding than formulating 2D-shell elements. To maintain acceptable the number of finite element variables, almost all the element proposed are based on a low order displacement interpolation. Consequently they are plagued not only by shear and membrane locking phenomena, as in the classical shell models, but also from trapezoidal and thickness locking [1,2,6,7]. The locking is usually sanitized by applying the concepts of Assumed Natural Strain, Enhanced Assumed Strain [2,7] and mixed and hybrid formulation [1,3,6].

Successful attempts to eliminate the locking phenomena in 3D solid elements by using mixed and hybrid formulations can be found in [1-3,6,7]. In comparing mixed (hybrid) and compatible finite elements many authors (see for example [1,2]) conclude that the mixed models are more robust and allows larger step in path-following analysis, due to the better description of the stress field. This is, in our opinion, not the true reason and in this paper a deeper investigation is addressed to solve this misunderstood.

The advantages in using mixed finite element have been exhaustively explained in two papers [8,9], that we recommend for further details, regarding path-following and Koiter [10,11] asymptotic analyses of 2D framed structures. Mixed or compatible finite elements even based on equivalent interpolations have completely different performances in geometrically nonlinear analysis using both asymptotic and path-following solution strategies. This is a crucial, even if frequently misunderstood, point in developing numerical algorithms as already widely discussed in [8,9,11]. Both numerical strategies require the strain energy to be smooth enough in the description variables, allowing, by keeping the first few terms in its Taylor expansion, to reduce as much as possible the truncation error. In path-following analysis, this ensures a fast convergence of the Newton iterative process; in asymptotic analysis, it implies that the higher-order energy terms neglected in the Taylor series expansion used by the formulation [10,11] be really irrelevant, allowing the accurate recovery of the equilibrium path. As the smoothness of a nonlinear function is strictly related to the choice of the primary variables, the setup of the best description is the key aspect to be considered in approaching the solution of a nonlinear problem. In this paper, to show the true reasons of the different observed behaviour, we use a mixed and compatible element based on the same interpolation fields. In this way they describe exactly the same equilibrium paths but sometimes with extremely different computational efforts. Nevertheless the two description are characterized by a different smoothness in the energy description and then they behave differently, when used within a numerical solution process. In particular, to represent the equilibrium path by using only the displacements as primary variables has to be considered the least suitable choice. It will be shown, in fact, as the truncation error of a Taylor expansion, for the compatible description, pathologically depends on the stiffness coefficients ratio while this is not the case of a mixed format. In presence of large displacements (rigid rotations) and high membranal/flexural stiffness ratios of the elements, the truncation error of the compatible description becomes large and a very slow convergence rate is observed in path following analysis while an erratic estimate of the bifurcation point and the post-critical coefficient occurs in the asymptotic analysis. It is important to note as the occurrence of this phenomenon, that we call "extrapolation locking", is not directly related to the order of the strain energy representation in its description variables and it is important, in our opinion, to well point out and split these aspects. This is made evident by the simple 3D solid finite element used in this paper in which the energy has the lowest dependency from the control variables, 3th and 4th order for the mixed and the compatible description respectively.

Moreover the polynomial order of the discrete energy representation based on 3D solid shells has in the asymptotic context of analysis a further tremendous impact in the efficiency, robustness and coherence of the formulation. In fact all the strain energy variations higher than the third are zero and, consequently, the formulation is simplified and improved. Its coherency and efficiency make the algorithms computationally more accurate and efficient and well suited for imperfection sensitivity analysis of structures with coincident or almost coincident buckling loads with respect to those based on 2D shell finite elements. For simplicity reasons the discrete model is based on the PT18β element [2],that for all the test presented here the element gives correct results. More sophisticated finite element existing or new formulations can be simply implemented in the proposed framework without any substantial modification.

2. Mixed and compatible description

In this section mixed and compatible variational formulations, equivalent from the point of view of the finite element approximation and referred from now on as mixed or compatible description are presented.

2.1 The discrete nonlinear equations

We consider a slender hyperelastic structure subject to conservative loads p growing linearly with the amplifier factor λ. The equilibrium is obtained by the stationary condition of the following functional

$$\Pi[\mathrm{u}] \equiv \Phi[\mathrm{u}] - \lambda \mathrm{pu} = \mathrm{stat}_{(\mathrm{u})} \quad \mathrm{u} \in \mathcal{U} \tag{1}$$

where $u \in \mathcal{U}$ is the field of configuration variables, $\Phi[u]$ denotes the strain energy. It is worth mentioning that when the mixed format is adopted the configuration u collects both displacement and stress fields. The strain energy can, as usual in a FE context, be expressed as sum of element contributions that, using a mixed interpolation for the displacement and the stress field $\boldsymbol{\sigma}$ and letting with Ω_e the element domain, becomes

$$\Phi[\mathrm{u}] \equiv \sum_e \int \left(\boldsymbol{\sigma}^T \boldsymbol{\varepsilon} - \frac{1}{2}\boldsymbol{\sigma}^T \mathrm{C}^{-1} \boldsymbol{\sigma}\right) d\Omega_e = \boldsymbol{\beta}^T \left(\mathbf{L} + \frac{1}{2}\mathbf{Q}[\mathbf{d}]\right)\mathbf{d} - \frac{1}{2}\boldsymbol{\beta}^T \mathbf{H}^{-1} \boldsymbol{\beta} \tag{2}$$

where $\boldsymbol{\varepsilon}$ is the Green-Lagrange strain measure $\mathbf{L}$ and $\mathbf{Q}[\mathbf{d}]$ are the linear and quadratic discrete compatibility matrices, $\mathbf{d}$ is the vector collecting all the finite element displacement variables while $\boldsymbol{\beta}$ is the vector collecting all the stress parameters, $\boldsymbol{C}$ and $\boldsymbol{H}$ are the continuum and discrete elastic matrices. It is worth to note as, by using the Green strain measure, the mixed strain energy (2) has a third order nonlinearity with respect to the configuration variable vector $\mathbf{u} = \{\boldsymbol{\beta}, \mathbf{d}\}$.

The element can also be described in a compatible format by a priori solving the discrete form of the constitutive laws, obtained from the stationary condition with respect $\boldsymbol{\beta}$

$$\boldsymbol{\beta}_c = \mathbf{H}^{-1}\left(\mathbf{L} + \frac{1}{2}\mathbf{Q}[\mathbf{d}]\right)\mathbf{d} \tag{3}$$

Substituting Eq.(3) in Eq.(2) we obtain the compatible description of the strain energy

$$\Phi[\mathrm{u}] = \mathbf{d}^T \left(\mathbf{L} + \frac{1}{2}\mathbf{Q}[\mathbf{d}]\right)^T \mathbf{H}^{-1} \left(\mathbf{L} + \frac{1}{2}\mathbf{Q}[\mathbf{d}]\right)\mathbf{d} \tag{4}$$

that has a 4th order dependence on the displacement variables only.

Similarly a fourth order dependence of the strain energy from the displacements parameters, is obtained with any compatible finite element interpolation when a 3D solid formulation is employed. In this paper we intentionally use the compatible description derived from the mixed finite element in order to have the same discrete approximation for both the description, i.e. the finite element is the same but the format of the problem changes. This allows us to focus of the way as the problem description affects its solution in large deformation problems (see also [8,9]) and then to show the reasons of the better performance of the use of a mixed description. The conclusions are general and hold for any other compatible or mixed finite element. In the numerical experiments the PT18β element [1] is used.

3. The extrapolation locking phenomena and its reduction using 3D mixed solid shell element

In this section we try to give an exhaustive enlightenment on the increasing of robustness and efficiency by using a mixed description of the nonlinear problem. The amazing improvements with respect to both efficiency and accuracy due to the joined use of a 3D solid element and a mixed description in Koiter asymptotic analysis is also discussed. We refer [10-12] and references therein for a review of path-following and Koiter analysis.

3.1 Mixed versus compatible description in nonlinear analysis of slender structures

In [8,9] has been shown that the slow convergence rate or failure of path-following analyses or the low accuracy of Koiter analysis occurs when the strain energy and then its Hessian are not smooth enough. The phenomenon becomes pathological (extrapolation locking) when compatible descriptions are adopted. The Hessian of strain energy in Eqs. (2) and (4) are respectively

$$\mathbf{K}[\mathbf{u}] = \begin{bmatrix} -\mathbf{H} & \mathbf{B}[\mathbf{d}] \\ \mathbf{B}[\mathbf{d}]^T & \mathbf{\Gamma}[\boldsymbol{\beta}] \end{bmatrix}, \qquad \mathbf{K}[\mathbf{u}] = \mathbf{B}[\mathbf{d}]^T\mathbf{H}^{-1}\mathbf{B}[\mathbf{d}] + \mathbf{\Gamma}[\boldsymbol{\beta}_c] \tag{5}$$

where $\mathbf{B}[\mathbf{d}] = \mathbf{L} + \mathbf{Q}[\mathbf{d}]$ and $\mathbf{\Gamma}[\boldsymbol{\beta}]$ is defined from $\boldsymbol{\beta}^T\mathbf{Q}[\mathbf{d}]\mathbf{d} = \mathbf{d}^T\mathbf{\Gamma}[\boldsymbol{\beta}]\mathbf{d}$. To show the different smoothness of the two description we compare the estimated stiffness $k_j[\mathbf{u}] = \mathbf{u}^T\mathbf{K}[\mathbf{u}_j]\mathbf{u}$ in a given direction $\mathbf{u}$ for the stiffness matrix assembled in two point $\mathbf{u}_j$ and $\mathbf{u}_0$. This value is strictly related to the convergence of path-following analyses or to bifurcation locking in Koiter formulation [8,9]. For the mixed description we have

$$\begin{cases} k_0[\mathbf{u}] \equiv \mathbf{u}^T\mathbf{K}[\mathbf{u}_0]\mathbf{u} = -\boldsymbol{\beta}^T\mathbf{H}\boldsymbol{\beta} + 2\boldsymbol{\beta}^T\mathbf{Q}[\mathbf{d}_0]\mathbf{d} + \mathbf{d}^T\mathbf{\Gamma}[\boldsymbol{\beta}_0]\mathbf{d} \\ k_j[\mathbf{u}] \equiv \mathbf{u}^T\mathbf{K}[\mathbf{u}_j]\mathbf{u} = -\boldsymbol{\beta}^T\mathbf{H}\boldsymbol{\beta} + 2\boldsymbol{\beta}^T\mathbf{Q}[\mathbf{d}_j]\mathbf{d} + \mathbf{d}^T\mathbf{\Gamma}[\boldsymbol{\beta}_j]\mathbf{d} \end{cases} \tag{6a}$$

Letting $\Delta k = k_j - k_0$ we have that the difference is

$$\Delta k_m = 2\boldsymbol{\beta}^T\mathbf{Q}[\mathbf{d}](\mathbf{d}_j - \mathbf{d}_0) + (\boldsymbol{\beta}_j - \boldsymbol{\beta}_0)^T\mathbf{Q}[\mathbf{d}]\mathbf{d} \tag{6b}$$

where we exploit the linear dependence from the configuration variables of matrices $\mathbf{Q}$ and $\mathbf{\Gamma}$ and the equivalence $\mathbf{Q}[\mathbf{d}_1]\mathbf{d}_2 = \mathbf{Q}[\mathbf{d}_2]\mathbf{d}_1$ for each $\mathbf{d}_1$ and $\mathbf{d}_2$. Eq.(6b) show as Δk_m is independent from the "hard" terms in $\mathbf{H}^{-1}$ and estimate the difference according to the quadratic matrix $\mathbf{Q}$. For the compatible description using similar arguments we obtain

$$\Delta k_m = (\mathbf{d}_j + \mathbf{d}_0)^T\mathbf{Q}[\mathbf{d}]\mathbf{H}^{-1}\mathbf{Q}[\mathbf{d}](\mathbf{d}_j - \mathbf{d}_0) + (\boldsymbol{\beta}_{cj} - \boldsymbol{\beta}_{c0})^T\mathbf{Q}[\mathbf{d}]\mathbf{d} \tag{6b}$$

that, on the contrary, is strongly affected from $\mathbf{H}^{-1}$. In particular the two estimated stiffness k_0 and k_j could be very different in presence of directions of $\mathbf{H}^{-1}$ with very different stiffness ratio, as typically occurs in slender structures, coupled with also small displacements. Conversely (see [5,9,13]) the mixed description makes the strain energy and its Hessian smoother than the correspondent compatible one, making the truncation errors negligible with respect to the retained terms. In path-following analysis the relevance of the locking is directly related to the length of the incremental steps and has, as a consequences, a pathological reduction in the step size when compatible description (elements) are used. As the Koiter method is based on Taylor expansion the use of a mixed description affects both efficiency and accuracy. The use of a solid finite element model, furthermore, allows to zeroing all the fourth order terms of the strain energy variations with a great improvement of both the computational performances and coherence of the asymptotic expansion.

4. Numerical results

In this section the effectiveness and reliability of both the method of analysis are tested in a series of benchmark problems. For path-following analyses we will show the efficiency of the mixed description which allows very large steps. For the asymptotic formulation we present the increasing in accuracy. In the following we denote with M and C the results obtained using the mixed and compatible description while with F and D the buckling results obtained respectively by zeroing or maintain in a linearized form, the pre-critical displacements (linear buckling analysis).

4.1 Simple tests

To emphasize the locking phenomenon in a Riks analysis we present the simple case of the Euler beam, with the data reported in Fig. 1.

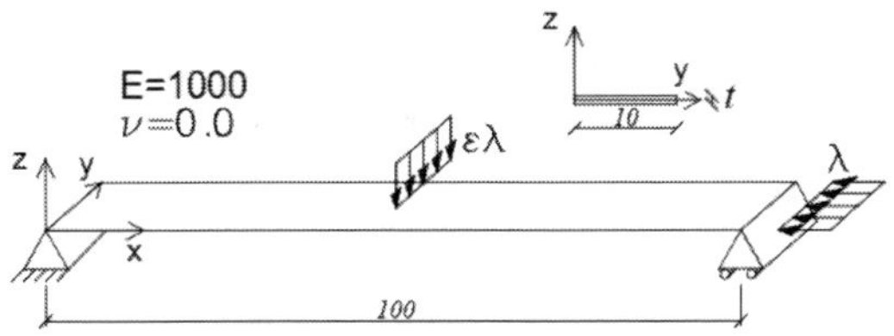

Fig. 1: Euler beam geometry and analysis evolution for increasing k

In Table 1 the number of steps and loops required to reach a maximum displacement in z direction equal to 39, directly related to the CPU time consuming are presented. The results of the mixed formulation, are unaffected by the coefficient $k = (\ell/t)^2$ while the compatible ones are pathological dependent from it. In Table 2 for the same Euler beam we present, for various aspect ratios k and imperfection load ϵ, the buckling loads.

Table 1: Number of steps and loops required to reach a displacement in z direction equal to 39 for the Euler beam

k	10^4	10^5	10^6	10^7	Mixed (unaffected by k)
Steps	38	43	67	failure	27
Loops	133	166	328		75

It is worth to note as the Frozen configuration hypothesis sanitizes the locking but neglecting the pre-critical displacements leads to inaccurate results for structures with an high pre-critical behaviour as for the shallow arc in Fig. 2. In this case the frozen configuration is not capable to produce the correct bifurcation load. Finally it is possible to observe the accuracy of the proposed mixed formulation in the reconstruction of the whole path by comparing the solution with that obtained with the path following analysis.

Table 2: Buckling loads of the Euler beam: normalized values with respect to the analytical one ($\varepsilon=0$)

	$\varepsilon=0.01$				$\varepsilon=0.005$				$\varepsilon=0.001$			
k	C	F	D	M	C	F	D	M	C	F	D	M
10^4	failure	1.001	0.673	1.001	1.112	1.001	0.866	1.001	1.004	1.001	0.994	1.001
10^5	failure	1.001	0.309	1.001	failure	1.001	0.517	1.001	1.040	1.001	0.938	1.001
10^6	failure	1.001	0.111	1.001	failure	1.001	0.209	1.001	failure	1.001	0.673	1.001
10^7	failure	0.999	0.036	0.999	failure	0.999	0.071	1.000	failure	0.999	0.308	0.999

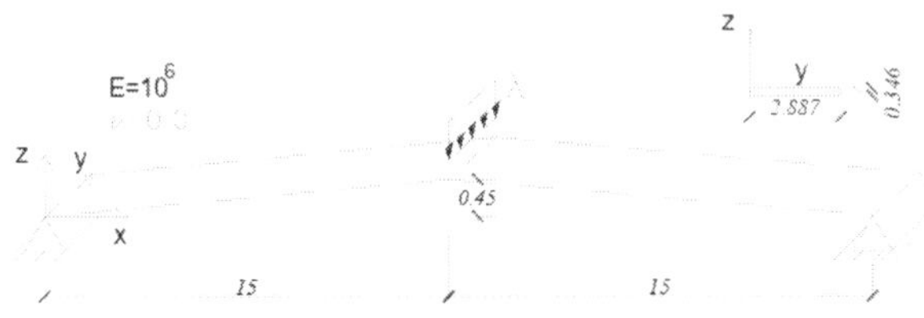

Fig. 2: Geometry and bifurcation values for the shallow arc

Table 3: Buckling loads for the shallow arc

	M	*F*
Mode 1	21.828	30.653
Mode 2	30.667	47.065

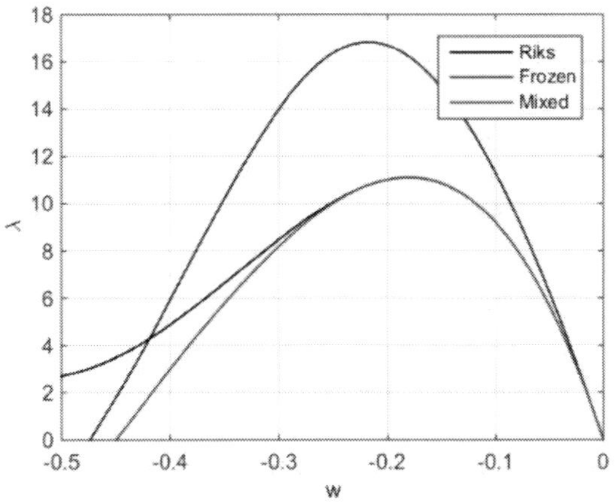

Fig. 3: Shallow arc: equilibrium paths

4.1 Simply supported C-shaped beam

The simply supported beam depicted in Fig. 4 has been analysed under the action of a simple compression axial load plus two small vertical force simulating a torsional load in the middle of the beam span. The buckling modes and values, obtained by using a mesh of (9+16+9)×50 elements are reported in Fig. 5 and Table 4.

The accuracy of the mixed strategy in the evaluation of the limit load and of the initial postcritical behaviour is highlighted in Fig. 6. The equilibrium path, obtained by both M and F Koiter strategies using the first four buckling modes, shows a strong modal interaction between the third local mode and the first two flexural-torsional global modes. The results are in good agreement with the path-following ones using both ABAQUS and our implementation. 24 steps and 73 iterations are required to evaluate the equilibrium path in the case of a mixed description with respect to 74 steps and 175 iterations required by the compatible one, with a clear advantage in terms of computational cost.

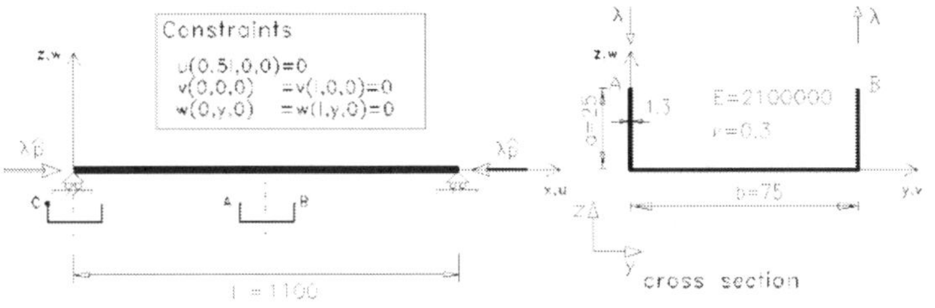

Fig. 4: Channel section: geometry and loads

Fig. 5: Channel section: buckling modes

Table 4: Channel section: Buckling values

	Mode 1	Mode 2	Mode 3	Mode 4
C	1291.49	1718.99	2949.70	2970.55
M	1266.78	1828.07	3092.32	3114.75

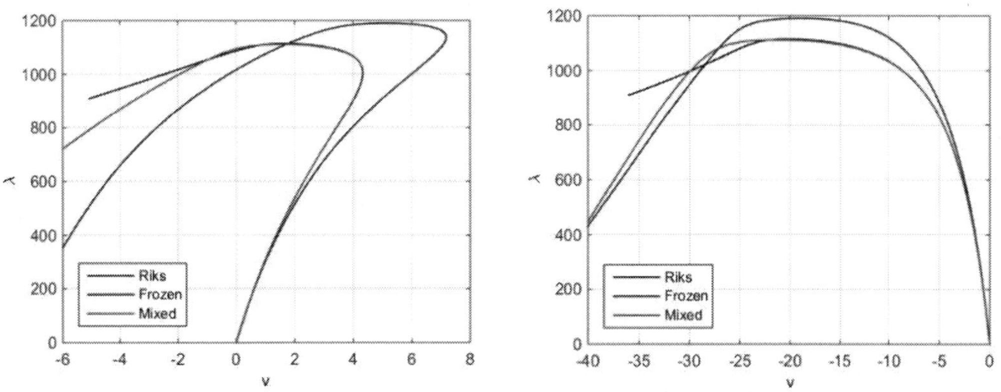

Fig. 6: Channel section: Equilibrium paths $\lambda - w_A, \lambda - w_B$

5. Conclusion

The advantages of using a mixed 3D solid finite element which use both stresses and displacements as primary variables are presented. It sanitizes the extrapolation locking phenomenon allowing very large steps for path-following analyses and increases the accuracy and the efficiency in Koiter formulations. In the latter case it also make possible to naturally zeroing the fourth order variations of the strain energy with notable giving greater efficiency and coherence to the method.

References

[1] Sze K, Chan W, Pian T. "An eight-node hybrid-stress solid-shell element for geometric non-linear analysis of elastic shells", *International Journal for Numerical Methods in Engineering*, 55(7), 853-878, 2002.

[2] Klinkel S, Gruttmann F, Wagner W. "A robust non-linear solid shell element based on a mixed variational formulation", *Computer Methods in Applied Mechanics and Engineering*, 195(1-3), 179-201, 2006.

[3] Vu-Quoc L, Tan X. "Efficient Hybrid-EAS solid element for accurate stress prediction in thick laminated beams, plates, and shells", *Computer Methods in Applied Mechanics and Engineering*, 253, 337-355, 2013.

[4] Garcea G, Madeo A, Casciaro R. "The implicit corotational method and its use in the derivation of nonlinear structura models for beams and plates", *J. Mech. Mater. Struct.*, 7(6), 509-539, 2012.

[5] Garcea G, Madeo A, Casciaro R. "Nonlinear FEM analysis for beams and plate assemblages based on the implicit corotational method", *J. Mech. Mater. Struct.*, 7(6), 539-574, 2012.

[6] Sze K, Ghali A. "Hybrid hexahedral element for solids, plates, shells and beams by selective scaling", *International Journal for Numerical Methods in Engineering*, 36(9), 1519-1540, 1993.

[7] Schwarze M, Reese S. "A reduced integration solid-shell finite element based on EAS and the ANS concept: Large deformation problems", *International Journal for Numerical Methods in Engineering*, (85), 289-329, 2011.

[8] Garcea G, Trunfio G, Casciaro R. "Mixed formulation and locking in path-following nonlinear analysis", *Computer Methods in Applied Mechanics and Engineering*, 165(1-4), 247-272, 1998.

[9] Garcea G, Salerno G, Casciaro R. "Extrapolation locking and its sanitization in Koiter's asymptotic analysis", *Computer Methods in Applied Mechanics and Engineering*, 180(1-2), 137-167, 1999.

[10] Garcea G, Madeo A, Zagari G, Casciaro R. "Asymptotic post-buckling FEM analysis using corotational formulation", *International Journal of Solids and Structures*, 46(2), 377-397, 2009.

[11] Casciaro R. "Computational Asymptotic Post-Buckling Analysis of Slender Elastic Structures", *CISM Courses and Lectures*, No. 470, 2005.

[12] Riks E. "An incremental approach to the solution of snapping and buckling problems", *International Journal of Solids and Structures*, 15(7), 529-551, 1979.

ADAPTATION OF THE AYRTON-PERRY FORMULA FOR THE GLOBAL-DISTORTIONAL-LOCAL BUCKLING OF THIN-WALLED MEMBERS

Mihai Nedelcu

Technical University of Cluj-Napoca, Faculty of Civil Engineering, Romania

Abstract: The paper presents an original adaptation of the Ayrton-Perry formula using the Generalized Beam Theory. This enhanced formula provides the maximum carrying capacity in elastic domain for simply-supported thin-walled members with initial geometric imperfections under various loading conditions. Both buckling deformations and geometric imperfections are considered as pure deformation modes of Global, Distortional or Local type, with sinusoidal variation along the member's length.

1. Introduction

Despite its many simplifying assumptions, the Ayrton-Perry formula (APF) represents a very popular tool for the buckling analysis of thin-walled members considering the initial geometric imperfections, due to its clear mechanical background, simplicity and flexibility. It has been introduced in Eurocode 3 Chapter 1-1 (EC3) [1] for basic cases as flexural and torsional-flexural buckling of columns, and lateral-torsional buckling of beams, all of them belonging to the Global buckling type.

The goal of the research partially presented here is to create a Generalized APF (GAPF) for all deformation types: Global, Distortional and Local. The main instrument is the Generalized Beam Theory (GBT), a specialized theory capable to perform a wide variety of structural analyses, the most important being the study of stability behaviour of thin-walled members [2-[5]. GBT has the ability to decompose the member deformation into a linear combination of pure deformation modes, each one belonging to a separate type (Global, Distortional or Local). A conventional GBT analysis has two steps: (i) *a cross-section analysis* leading to the GBT pure cross-section deformation modes and the corresponding modal mechanical properties and (ii) *a member stability analysis* to obtain the critical loading factors and the associated mode shapes. Only the first step is used in this paper, and next, the pure cross-section deformation modes together with the corresponding mechanical properties are used by GAPF.

This GAPF was first presented in [6], covering Global and Distortional buckling of simply-supported thin-walled members. In this paper, the formula covers the entire range (i.e. also the Local buckling), and illustrative examples are presented for Distortional and Local buckling analysis. Both buckling deformations and geometric imperfections are considered as GBT pure deformation modes with sinusoidal variation and arbitrary number of halfwaves along the member's length. For the time being the GAPF covers one or a combination of two deformation modes [6]. In this paper the GAPF is presented for one deformation mode (but of any type). The development of GAPF for arbitrary combinations of deformation modes is currently under work.

2. GBT - short review

The local coordinate system *x-s-z* and the corresponding displacements *u-v-w* are shown in Fig. 1 for an arbitrary thin-walled member.

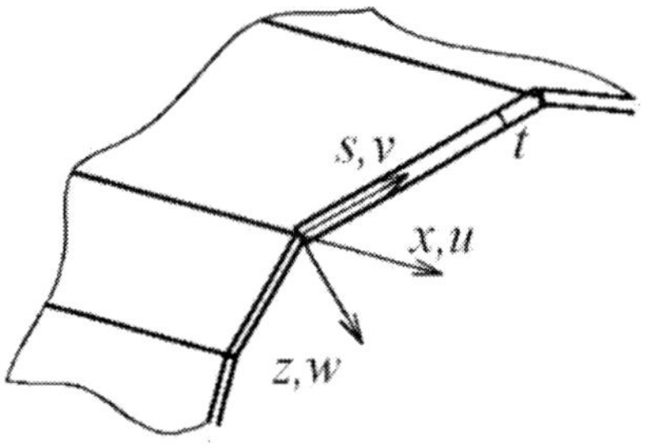

Fig. 1: Thin-walled member - local axes and displacements

In this study, one assumes the classical Kirchhoff-Love hypotheses and also Vlasov's simplifying hypotheses (null membrane transverse extensions and membrane shear strains). According to GBT, any displacement is considered as a linear combination of *m* pure deformation modes of Global-Distortional-Local type, using the product formulation, as follows:

$$u(s,x)=\sum_{k=1}^{m}u_k(s)\phi_k'(x) \quad v(s,x)=\sum_{k=1}^{m}v_k(s)\phi_k(x) \quad w(s,x)=\sum_{k=1}^{m}w_k(s)\phi_k(x) \tag{1}$$

where $u_k(s)$, $v_k(s)$, $w_k(s)$ are the components of the *pure cross-section deformation modes* and $\phi_k(x)$ are the *amplitude functions* describing their longitudinal variation, and the symbol $(\,)'=d(\,)/dx$.

The GBT cross-section analysis provides the pure cross-section deformation modes ($u_k(s)$, $v_k(s)$, $w_k(s)$) and the corresponding modal mechanical properties in the shape of stiffness matrices. Next, the member equilibrium equations are written based on the principle of virtual work applied in its variational form. The GBT system of differential equations in modal formulation has the expression [3]:

$$C_{ik}\phi_k^{IV}-D_{ik}\phi_k''+B_{ik}\phi_k+W_pX_{pik}\phi_k''=0 \tag{2}$$

where *C, D, B* are the cross-section linear stiffness matrices, namely the warping stiffness matrix, the torsion stiffness matrix and the transverse stiffness matrix, respectively. X_p is the geometric stiffness matrix taking into account the degradation due to the membrane longitudinal stresses. The vector W_p (p = 1..4) contains the resultants of the applied pre-buckling stresses, namely: (i) axial force ($W_1 = N$), (ii) major/minor axis bending moments ($W_2 = M_y$, $W_3 = M_z$), and (iii) bimoment ($W_4 = B$). The expressions of the stiffness matrices C, D, B and X_p together with the GBT cross-section analysis are well described in literature.

3. The generalised Ayrton-Perry formula for one deformation mode

The GBT differential equations system given by Eq. (2) is reduced to one equation for the deformation mode k:

$$C_{kk}\phi_k^{IV} - D_{kk}\phi_k'' + B_{kk}\phi_k + W_p X_{pkk}\phi_k'' = 0 \tag{3}$$

One assumes a sinusoidal amplitude function with n halfwaves:

$$\phi_k(x) = a_k \sin(\alpha_n x), \quad \alpha_n = \frac{n\pi}{L} \tag{4}$$

where L is the member length. For linear buckling, introducing Eq. (4) into Eq. (3) it yields as follows:

$$\alpha_n^4 C_{kk} + \alpha_n^2 D_{kk} + B_{kk} + \alpha_n^2 W_{cr,p-k} X_{pkk} = 0 \tag{5}$$

and the critical load from load p by deformation mode k, is found:

$$W_{cr,p-k} = \frac{\alpha_n^2 C_{kk} + D_{kk} + \dfrac{B_{kk}}{\alpha_n^2}}{X_{pkk}} = \frac{P_k}{X_{pkk}} \tag{6}$$

where P_k is the abbreviation of the fraction's numerator. Eq. (6) is a generalized formula for linear buckling, therefore it covers all classical formulas of Global buckling. Minimizing the critical load, one finds out the halfwave length for Distortional and Local modes and also the number of halfwaves at minimum critical load:

$$l_{hw,k} = \pi \sqrt[4]{\frac{C_{kk}}{B_{kk}}} \Rightarrow n_{hw,k} = \frac{L}{l_{hw,k}} \tag{7}$$

Now, if one introduces the initial geometric imperfections, Eq. (5) becomes:

$$C_{kk}\varphi_k^{IV} - D_{kk}\varphi_k'' + B_{kk}\varphi_k + W_p X_{pkk}(\varphi_k'' + \varphi_k^{0''}) = 0 \tag{8}$$

where the geometric imperfection is also a sine function $\phi_k^0(x) = a_k^0 \sin(\alpha_n x)$ and using the abbreviation P_k from Eq. (6), Eq. (8) becomes:

$$P_k a_k - W_p X_{pkk}(a_k + a_k^0) = 0 \tag{9}$$

The maximum value of the displacement amplitude function depending on the loading level is obtained as follows:

$$a_k = \frac{W_p}{W_{cr,p-k} - W_p} a_k^0 \tag{10}$$

The kinematic equations for the longitudinal normal strains, considering their Membrane and Flexural Linear components, are as follows:

$$\varepsilon_x^{M,L} = u' \qquad \varepsilon_x^{F,L} = -zw'' \tag{11}$$

The Membrane longitudinal strains play the main role for Global and Distortional buckling. As for the Local buckling, the Flexural longitudinal strain must be considered. Neglecting the effect of the Flexural transverse extensions and also the Poisson's effect for the Membrane components, the longitudinal normal stresses are obtained as follows:

$$\sigma_x^L = \frac{E}{1-\mu^2}(\varepsilon_x^{M,L} + \varepsilon_x^{F,L}) \tag{12}$$

where E and μ are the Young's modulus and the Poisson's ratio, respectively. Using the GBT product formulation (Eq. (1)) and a new notation, the longitudinal normal stresses given by deformation mode k, have the expression:

$$\sigma_x^L \rightarrow \sigma_k(s,x) = \frac{E}{1-\mu^2}\left(u_k(s) - z w_k(s)\right)\phi_k''(x) \tag{13}$$

Introducing the maximum of the cross-section warping displacement as $\bar{u}_k = \max(|u_k(s)|)$, the corresponding transversal displacement as $\bar{w}_k = w_k(\bar{s})$, and the maximum of z as half the thickness t, the maximum normal stress along the cross-section is found as follows:

$$\sigma_k = -E\left(\bar{u}_k + \frac{t}{2(1-\mu^2)}\bar{w}_k\right)\alpha_n^2 a_k = -E d_k \alpha_n^2 a_k \tag{14}$$

where d_k is the notation of the expression in brackets. Introducing the maximum value of the amplitude function from Eq. (10), the maximum normal stress on the entire member is found:

$$\sigma_k = -E\alpha_n^2 d_k a_k^0 \frac{W_p}{W_{cr,p-k} - W_p} \tag{15}$$

Now, in GBT the Membrane Linear normal stresses can be written as follows:

$$\sigma_x^{M,L}(s,x) = E\sum_{k=1}^{m}\frac{W_k u_k(s)}{C_{kk}} \tag{16}$$

The maximum normal stresses from the initial loading W_p, can be written:

$$\sigma_p = E\frac{W_p \bar{u}_p}{C_{pp}} \tag{17}$$

where $\bar{u}_p = \max(|u_p(s)|)$. The first yield criterion is now written, using also Eq.(15):

$$\sigma_p + \sigma_k = f_y \quad \Rightarrow \quad \frac{\sigma_p}{f_y} + \frac{E\alpha_n^2 d_k a_k^0}{f_y}\frac{W_p}{W_{cr,p-k} - W_p} = 1 \tag{18}$$

One introduces now the critical normal stress $\sigma_{cr,p-k}$ by adapting Eq. (17), the new generalized imperfection factor η^*, the member slenderness λ and the buckling reduction factor χ with the expressions given below:

$$\sigma_{cr,p-k} = E\frac{W_{cr,p-k}\bar{u}_p}{C_{pp}} \qquad \eta^* = \frac{E\alpha_n^2 d_k a_k^0}{f_y}$$
$$\lambda = \sqrt{\frac{f_y}{\sigma_{cr,p-k}}} \qquad \chi = \frac{\sigma_p}{f_y} \tag{19}$$

With these new notations Eq.(18) becomes the GAPF for one deformation mode:

$$\chi^2 + \chi\left(-1 - \frac{1}{\lambda^2} - \eta^*\right) + \frac{1}{\lambda^2} = 0 \tag{20}$$

The smallest solution provides the buckling reduction factor as follows:

$$\chi = \frac{1}{\Phi + \sqrt{\Phi^2 - \lambda^2}} \qquad \text{where} \qquad \Phi = \frac{1}{2}\left(1 + \eta^*\lambda^2 + \lambda^2\right) \tag{21}$$

4. ILLUSTRATIVE EXAMPLES

GAPF is applied on a symmetric steel lipped channel, length L = 1500 mm, thickness t = 2 mm for all the walls, yield stress f_y = 350 MPa. All the other dimensions and material properties are given in Fig. 2. The member is simple supported at both ends, meaning pinned–pinned (both locally and globally) and free to warp end supports, sometimes also referred to as "fork" supports. Moreover, the longitudinal translations are restrained at the closest cross-section to the mid-span where the warping displacements are minimal (this depends on the initial geometric imperfections). In the presented examples, the loading is the axial force (W_1 = N). The geometric imperfections were chosen as sine functions with various amplitudes and number of halfwaves. The cross-section discretization is made with 7, 11 and 1 intermediate nodes between the corners, respectively in the flanges, web and flange lips. A Matlab [7] application was written for the GBT cross-section analysis, which provided a number of 35 pure cross-section deformation modes (the first 14 are presented in Fig. 2) together with the corresponding stiffness matrices. Same application provided the solution of GAPF given by Eq. (21).

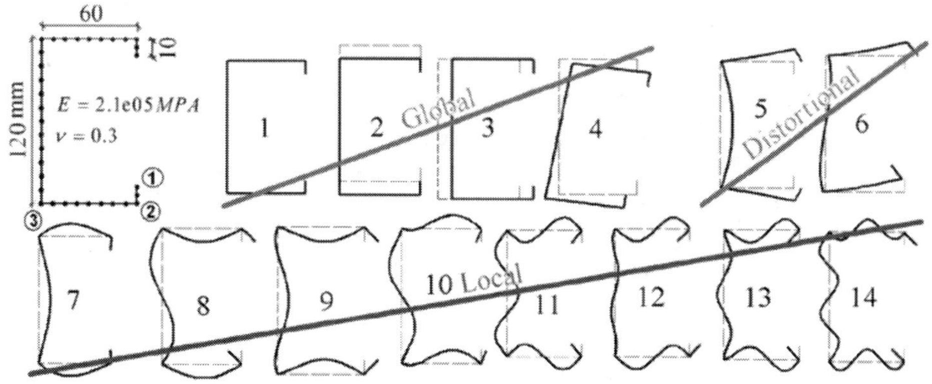

Fig. 2: C section - the first 14 GBT cross-section deformation modes

The semi-analytical results given by the proposed GAPF are compared with the numerical results provided by the Finite Element Analysis (FEA). The numerical simulations were done using Abaqus Standard [8] – Riks analysis. The initial axial loading is introduced as an edge uniform load with the resultant of 1kN. Shell elements of type S4 were considered, in a highly regular rectangular mesh. The cross-section discretisation is constant along the member length and identical with the one used by the GBT cross-section analysis. In longitudinal direction, all the elements have the same size of 10 mm. The coordinates of each node are given by the initial geometric imperfections, and they were introduced in the input file of Abaqus, using another Matlab application. After performing the analysis, the first yield is searched for the normal longitudinal stresses and the corresponding axial loading is used for the calculation of the buckling reduction factor χ.

4.1 Example 1: Distortional buckling

In this case, both the geometric imperfection and the buckling deformation were considered as the pure Distortional mode 5 (see Fig. 2). 15 amplitudes of the initial geometric imperfections were considered, and they were calculated in order to obtain the maximum in-plane deviation (i.e. $d_{\max} = \max(|v_0|_{\max}, |w_0|_{\max})$ equal with $(0.1, 0.2, ..., 1.5)t$. For each amplitude, the number of halfwaves n varied from 1 to 10, giving a total number of 150 buckling analyses. The differences between GAPF and FEA are given in Table 1, the differences greater or equal 5% being highlighted. Fig. 4 presents the buckling reduction factor for an amplitude of 50% of the thickness. It can be seen that the differences are quite big for 1 halfwave, or 2 halfwaves with small amplitudes, and also for 8 to 10 halfwaves with an amplitude of 10% of the thickness. The explanation is that for these cases, the buckling in FEA is not following the deformation shape of mode 5, but it suddenly jumps to different modes, mainly to the Local mode 7 with $n_{hw,7}$ = 15 halfwaves (showed in Fig. 5) – a number which corresponds to the minimum critical load for linear bifurcation analysis and which can be found using Eq. (7). For the other cases, the deformation follows the geometric imperfection and the best results can be seen for a number of halfwaves close to $n_{hw,5}$ = 4 (see Fig. 3), and this value can also be found using Eq. (7). For amplitudes greater than the thickness, the differences exceed 5%, even if the deformation follows the geometric imperfection; it can be concluded that the simplifying assumptions of this formulation are no longer valid.

Table 1: Differences between GAPF and FEA (%) for Distortional buckling

Amplitude	No of halfwaves									
	1	2	3	4	5	6	7	8	9	10
0.1t	24.86	23.08	0.17	1.36	2.09	3.00	3.84	5.00	6.19	7.49
0.2t	24.91	17.59	0.34	0.23	0.70	1.35	1.86	2.74	3.42	4.08
0.3t	24.96	14.28	1.01	0.53	0.28	0.22	0.47	1.05	1.38	1.68
0.4t	24.99	11.83	1.58	1.14	1.06	0.59	0.66	0.24	0.23	0.16
0.5t	25.00	9.01	2.08	1.62	1.74	1.30	1.59	1.30	1.48	1.60
0.6t	25.00	6.48	2.54	2.09	2.33	1.88	2.38	2.18	2.56	2.77
0.7t	24.98	4.10	2.98	2.47	2.90	2.40	3.12	2.98	3.52	3.77
0.8t	24.96	1.86	3.39	2.88	3.43	2.91	3.80	3.68	4.32	4.67
0.9t	24.96	0.33	3.78	3.22	3.95	3.33	4.42	4.31	5.08	5.40
1.0t	24.87	2.48	4.17	3.59	4.47	3.76	5.03	4.92	5.81	6.09
1.1t	24.77	2.77	4.54	3.91	4.95	4.19	5.61	5.48	6.43	6.75
1.2t	24.74	2.96	4.91	4.26	5.47	4.57	6.17	6.01	7.06	7.33
1.3t	24.65	4.63	5.28	4.57	5.94	5.00	6.76	6.56	7.69	7.87
1.4t	24.51	4.89	5.62	4.91	6.44	5.39	7.29	7.07	8.24	8.42
1.5t	24.43	5.17	5.98	5.23	6.95	5.77	7.85	7.55	8.80	8.96

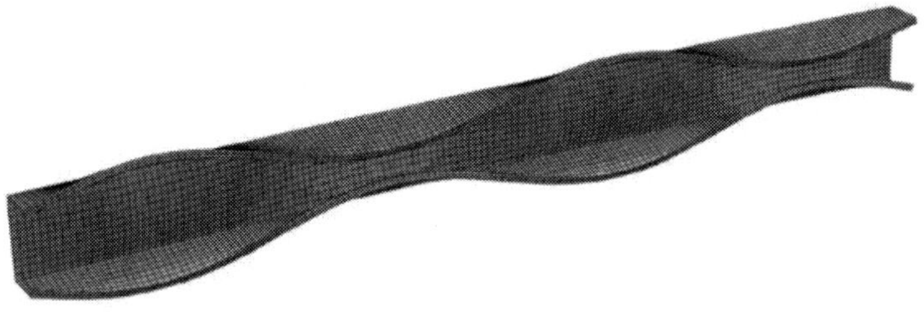

Fig. 3: The Distortional mode 5 with 4 halfwaves – deformations and normal longitudinal stresses

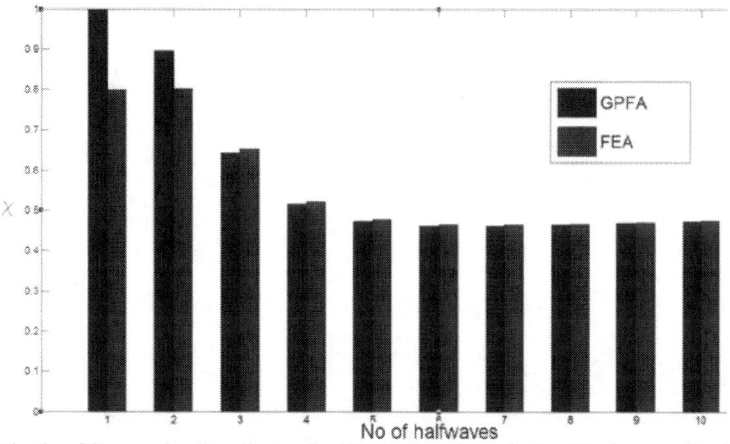

Fig. 4: The buckling reduction factor for Distortional mode 5 with the amplitude = $0.5t$

4.2 Example 2: Local buckling

In this case, both the geometric imperfection and the buckling deformation were considered as the pure Local mode 7 (see Fig. 2). 18 amplitudes of the initial geometric imperfections were considered, with the maximum in-plane deviation previously described $d_{\max}$ equal with $(0.01, 0.025, ..., 0.1, 0.15, ..., 0.5, 0.6, ..., 1.0)t$. For each amplitude, the number of halfwaves n varied from 1 to 35, but only the odd numbers, giving a total number of 324 buckling analyses. The differences between GAPF and FEA are given in Table 2, the differences greater or equal 5% being highlighted. It can be seen that the differences are quite big for 1 to 11 halfwaves with small amplitudes and also for 27 to 35 halfwaves with an amplitude of 1% of the thickness. In these cases, the buckling in FEA is following the deformation shape of mode 7 but at one point, the number of halfwaves jumps to $n_{hw,7}$ = 15 halfwaves, number which corresponds to the minimum critical load for linear bifurcation analysis, (see Eq. (7)). For the other cases, the deformation follows the geometric imperfection and the best results are found for a number of halfwaves close to $n_{hw,7}$, a deformation shape presented in Fig. 5. Fig. 6 presents the buckling reduction factor for an amplitude of 10% of the thickness. For amplitudes greater than 20% of the thickness, the differences usually exceed 5%, even if the deformation follows the geometric imperfection shape; it can be concluded that the simplifying assumptions of this formulation are no longer valid.

Table 2: Differences between GAPF and FEA (%) for Local buckling

Amplitude	No of halfwaves													
	1…	11	13	15	17	19	21	23	25	27	29	31	33	35
0.01*t	28.6	17.2	4.2	3.5	3.0	2.8	1.9	3.8	4.2	7.5	7.9	10.1	10.2	11.7
0.025*t	28.7	12.5	4.4	3.7	3.1	2.9	2.6	2.3	1.9	1.7	0.8	1.7	1.7	2.6
0.05*t	28.8	12.1	4.2	3.5	2.8	2.6	2.2	2.0	1.6	1.4	0.9	0.7	0.3	0.1
0.075*t	28.9	7.1	3.7	3.0	2.1	1.9	1.4	1.3	0.8	0.7	0.1	0.0	0.5	0.6
0.1*t	28.9	6.8	3.1	2.2	1.2	1.0	0.4	0.3	0.2	0.3	0.9	1.0	1.6	1.7
0.15*t	27.9	5.5	1.2	0.2	1.1	1.5	2.2	2.4	3.0	3.1	3.8	3.8	4.5	4.5
0.2*t	27.0	3.5	1.2	2.5	4.0	4.6	5.4	5.7	6.5	6.5	7.3	7.4	8.2	8.1
0.25*t	27.7	1.0	3.9	5.5	7.2	8.0	9.0	9.4	10.3	10.5	11.4	11.4	12.3	12.2
>0.25*t	> 5													

Fig. 5: The Local mode 7 with 15 halfwaves – deformations and normal longitudinal stresses

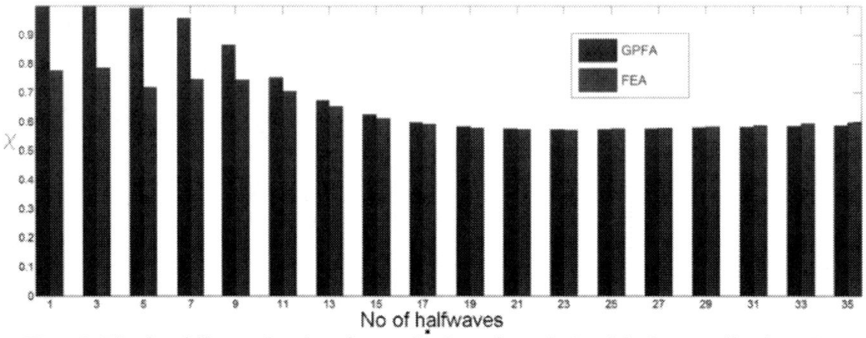

Fig. 6: The buckling reduction factor for Local mode 7 with the amplitude = $0.1t$

5. Conclusions

The paper presents a GBT-based generalization of the Ayrton-Perry formula for simply-supported thin-walled members. The presented theoretical developments cover the continuous elastic buckling with initial geometric imperfections, characterized by one imperfection/deformation mode of Global, Distortional or Local type. The validation of this method was done for axially compressed members with very promising results, and the analyses concerning other types of loading as well as the development of the proposed APF for arbitrary combinations of deformation modes in interactive buckling is currently under work.

References

[1] European Standard, EuroCode 3. *Design of Steel Structures – Part 1.1: General rules and rules for buildings*, EN 1993-1-1, 2005.

[2] Schardt R. *Verallgemeinerte Technische Biegetheorie*, Springer Verlag Berlin, 1989.

[3] Schardt R. "Generalized beam theory – An adequate method for coupled stability problems", *Journal of Thin-Walled Structures*, 19, 161-180, 1994.

[4] Silvestre N, Camotim D. "Second-order generalised beam theory for arbitrary orthotropic materials", *Journal of Thin-Walled Structures*, 40, 791–820, 2002.

[5] Camotim D, Basaglia C, Silvestre N. "GBT buckling analysis of thin-walled steel frames: A state-of-the-art report", *Journal Thin-Walled Structures*, 48, 726–743, 2010.

[6] Nedelcu M. "A GBT-based adaptation of the Ayrton-Perry formula for the Global-Distortional buckling", *Proceedings of the 8th International Conference on Advances in Steel Structures*, Lisbon, Portugal, 22-24 July, 2015.

[7] Matlab Version 7.1.0246 *Documentation*, The Mathwork Inc., 2005.

[8] N Hibbit, Karlsson and Sorensen Inc. *ABAQUS Standard* (Version 6.3), 2002.

FINITE ELEMENT MODEL WITH SEMI-RIGID CONNECTIONS FOR VIBRATION ASSESSMENT OF STEEL MOMENT RESISTING FRAMED STRUCTURES

Halil F. Ozel, Afsin Saritas and Tayseer Tasbahji

Department of Civil Engineering, Middle East Technical University, Ankara, Turkey

Abstract: In this paper, a mixed formulation frame finite element with semi-rigid connections is developed. Consistent mass matrix of the element is obtained such that determination of vibration frequencies of members with varying geometry and material distribution as well as the presence of semi-rigid connections at any section on the element is accurately captured without the need to specify displacement shapes. An accurate shear correction coefficient for wide flange sections is taken into account in order to get closer match with exact solutions. Numerical examples on a portal frame and a multi-story steel moment resisting frame verify the accuracy of proposed element with and without semi-rigid connections.

1. Introduction

Semi-rigidity at connection regions of steel structures greatly influences the vibration characteristics of steel moment resisting framed structures. Finite element models should address modelling of the mass and stiffness matrices for both beam and column members in order to accurately capture transverse shear deformations and rotary inertia along a member's length, as well as the partial fixity introduced by the presence of these connections. Researchers have studied dynamic behavior of steel framed structures and suggested to consider the effect of semi-rigid connections in the last two decades. Chui and Chan [1] and Nader and Astaneh-Asl [2] conducted tests on flexible jointed steel frames accompanied with numerical analyses, and found out the importance of taking into account joint flexibility in structural models. Besides consideration of semi-rigidity at connection region, it is also important to take into account possible inelastic behavior and nonlinear geometric effects on frame members [3-5] in carrying out dynamic analysis. The literature contains significant amount of research work on the investigation of the dynamic behavior of steel framed structures with semi-rigid connections through the use of finite element method [6-10]. Available design codes also try to provide the influence of semi-rigid response for steel structures under dynamic actions, and research studies try to evaluate the code suggestions on this front. In one of such studies, Sophianapoulos [11] compared Eurocode 3 approach with closed form solutions, where the results from both approaches showed that in the fundamental modes of vibration responses are very close to each other; however, in the higher modes, the influence of the flexible joints became apparent and results differed from each other.

The amount of the studies on trying to get better predictions on capturing vibration characteristics of steel framed structures with semi-rigid connections clearly reflects the importance of coming up with more accurate and robust modeling approaches. To this end, a mixed formulation frame finite element with semi-rigid connections is developed in this study, where the element formulation bases on the use of three-field Hu-Washizu-Barr principle. Consistent mass matrix of the element is obtained such that determination of vibration frequencies of members with varying geometry and material distribution as well as the presence of semi-rigid connections at any section on the element is accurately captured without any need for specification of different displacement shape functions for each individual case. The element response does not necessitate further discretization due to the presence of semi-rigid connections. An accurate shear correction coefficient for wide flange sections is taken into account in order to get closer match with exact solutions. A commercial program, SAP2000 [12], is used to validate the results of the proposed model and the previous benchmark solutions in [9].

2. Frame Element Formulation

2.1 Kinematic Relations

Displacements on a material point on the section of a beam that deforms in *xy*-plane can be obtained by calculating Timoshenko beam theory as follows;

$$\begin{Bmatrix} u_x(x,y) \\ u_y(x,y) \end{Bmatrix} = \begin{Bmatrix} u(x) - y\theta(x) \\ v(x) \end{Bmatrix} \tag{1}$$

where $u_x(x,y)$ and $u_y(x,y)$ are the displacements in x and y directions, respectively of any point in the section. $u(x)$ is the displacement of the point $(x,0)$ along x-axis. $v(x)$ is the transverse deflections of the point $(x,0)$ from x-axis in y direction. $\theta(x)$ is the small rotation of the beam cross section around z-axis.

The non-zero strain components $\boldsymbol{\varepsilon}$ include the normal strain in the x direction and shear strain with xy component, where these are calculated from section deformations as follows;

$$\boldsymbol{\varepsilon} = \begin{Bmatrix} \varepsilon_{xx} \\ \gamma_{xy} \end{Bmatrix} = \begin{Bmatrix} u'(x) - y\theta'(x) \\ -\theta(x) + v'(x) \end{Bmatrix} = \begin{Bmatrix} \varepsilon_a(x) - y\kappa(x) \\ \gamma(x) \end{Bmatrix} = \mathbf{a}_s(y,z)\,\mathbf{e}(x) \tag{2}$$

where $\mathbf{e}(x)$ is the section deformation vector given as follows;

$$\mathbf{e}(x) = \begin{bmatrix} \varepsilon_a(x) & \gamma(x) & \kappa(x) \end{bmatrix}^{\mathrm{T}} \tag{3}$$

In Equation (3), $\varepsilon_a(x)$ is the axial strain of the reference axis, $\gamma(x)$ is the shear deformation along y-axis and κ is the curvature about z-axis. Section deformations can be easily calculated from section reference displacements as clearly visible from a one to one comparison of the terms of Equation (2). Furthermore, section compatibility matrix, $\mathbf{a}_s(y,z)$ introduced in Equation (2) is written as follows;

$$\mathbf{a}_s(y,z) = \begin{bmatrix} 1 & 0 & -y \\ 0 & 1 & 0 \end{bmatrix} \tag{4}$$

2.2 Basic System without Rigid Body Modes and Force Interpolation Functions

Element formulation is proposed in *xy*-plane, where the formulation considers two end nodes and relies on a transformation from complete system to basic system. In the whole structure, the element has 3 degrees of freedom (dof) per node, resulting in 6 dofs, where the nodes are placed at element ends. The complete system is proposed such that the axis of the element is aligned with horizontal *x*-axis. The basic system is prescribed for the purpose of removing rigid body modes of motion, and the basic system is chosen as the cantilever beam as shown in Fig. 1, where the fixed and free ends are the left and right ends, respectively. The transformation matrix, **a** for an element with length L is used to relate element end forces in complete system to basic element forces as follows;

$$\mathbf{p}=\mathbf{a}^{\mathrm{T}}\mathbf{q};\quad \text{where}\quad \mathbf{a}=\begin{bmatrix}-1 & 0 & 0 & 1 & 0 & 0\\ 0 & -1 & -L & 0 & 1 & 0\\ 0 & 0 & -1 & 0 & 0 & 1\end{bmatrix} \tag{5}$$

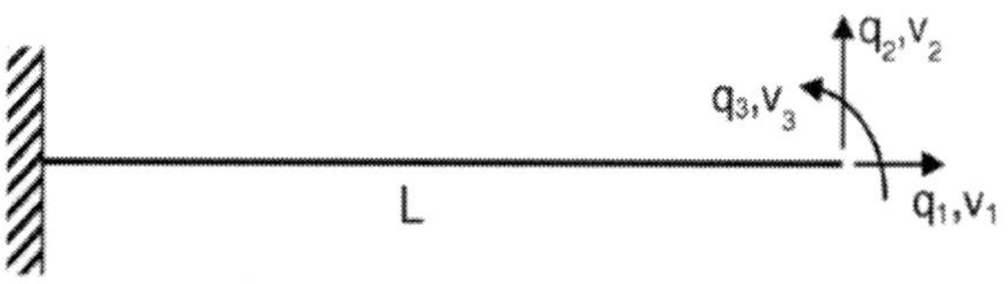

Fig. 1: Cantilever basic system forces and deformations

It is also possible to relate basic element deformation vector v to displacements in complete system by separating 3 rigid body modes and keeping only the basic deformation modes for the element. By this way, it is feasible to derive flexibility matrix that would have been impossible to get in the complete system because of the singularity caused by rigid body modes. Basic element deformations **v** can be calculated from nodal displacements **u** in complete system as follows;

$$\mathbf{v}=\mathbf{a}\,\mathbf{u} \tag{6}$$

Basic element forces at free end, **q** are shown in Figure 1 and given in Eqn. (5). These forces can be related to internal section forces, $\mathbf{s}(x)$ by using the force interpolation matrix $\mathbf{b}(x,L)$ for the cantilever beam configuration as follows;

$$\mathbf{s}(x)=\begin{bmatrix}N(x) & V(x) & M(x)\end{bmatrix}^{\mathrm{T}}=\mathbf{b}(x,L)\mathbf{q}+\mathbf{s}_p(x)$$

$$\mathbf{b}(x,L)=\begin{bmatrix}1 & 0 & 0\\ 0 & 1 & 0\\ 0 & (L-x) & 1\end{bmatrix}\quad\text{and}\quad \mathbf{s}_p(x)=\begin{bmatrix}L-x & 0\\ 0 & L-x\\ 0 & (L-x)^2/2\end{bmatrix}\begin{Bmatrix}w_x\\ w_y\end{Bmatrix} \tag{7}$$

By using Equation (7), it is possible to attain exact equilibrium between the forces at free end of the element and forces at any section that is x units away from the fixed end. Section forces are axial force $N(x)$, shear force in *y* direction $V(x)$, and moment about *z*-axis $M(x)$. In above equation $\mathbf{s}_p(x)$ is the particular solution for uniformly distributed loads in the axial and transverse directions, i.e. w_x and w_y, respectively. By the way, with this approach, it is easy to calculate the particular solution under arbitrary inter element loads that are concentrated or distributed.

2.3 Variational Base and Finite Element Formulation of the Element

Variational form of the element is written by considering independent element nodal displacements $\mathbf{u}$, element basic forces $\mathbf{q}$, and section deformations $\mathbf{e}$ by using three-fields Hu-Washizu functional and implemented as part of beam finite elements by Taylor et al. [13] and Saritas and Filippou [14]. Extension to dynamic case is achieved through introduction of inertial forces $\mathbf{m}\ddot{\mathbf{u}}$ acting at nodes by considering D'Alembert's principle to get the following variational form of the element

$$\delta\Pi_{\mathrm{HW}} = \int_0^L \delta\mathbf{e}^{\mathrm{T}}\left(\hat{\mathbf{s}}(\mathbf{e}(x)) - \mathbf{b}(x,L)\mathbf{q} - \mathbf{s}_p(x)\right)dx - \delta\mathbf{q}^{\mathrm{T}}\int_0^L \mathbf{b}^{\mathrm{T}}(x,L)\mathbf{e}(x)\,dx + \delta\mathbf{q}^{\mathrm{T}}\mathbf{a}_g\mathbf{u} + \delta\mathbf{u}^{\mathrm{T}}\mathbf{a}_g{}^{\mathrm{T}}\mathbf{q} + \delta\mathbf{u}^{\mathrm{T}}\mathbf{m}\ddot{\mathbf{u}} - \delta\mathbf{u}^{\mathrm{T}}\mathbf{p}_{app} = 0 \tag{8}$$

Above equation can also be obtained by considering the general Hu-Washizu variational form with extension to dynamic case by Barr [15]. Eqn. (8) should hold for arbitrary $\delta\mathbf{u}$, $\delta\mathbf{q}$ and $\delta\mathbf{e}$, thus the following three equations should be satisfied in order for the Hu-Washizu-Barr variational to be zero.

$$\mathbf{m}\ddot{\mathbf{u}} + \mathbf{p} \equiv \mathbf{p}_{app}; \quad \text{where} \quad \mathbf{p} = \mathbf{a}_g{}^{\mathrm{T}}\mathbf{q} \tag{9}$$

$$\mathbf{v} \equiv \int_0^L \mathbf{b}^{\mathrm{T}}(x,L)\mathbf{e}(x)dx; \quad \text{where} \quad \mathbf{v} = \mathbf{a}_g\mathbf{u} \tag{10}$$

$$\hat{\mathbf{s}}(\mathbf{e}(x)) \equiv \mathbf{b}(x,L)\mathbf{q} + \mathbf{s}_p(x) \tag{11}$$

Eqn. (9) is the equation of motion that holds for linear or nonlinear material response, and this equation can be collected for each element to get structure's equation of motion. A numerical time integration scheme can be employed to get a solution. Consequence of viscous damping can be simply achieved by adding $\mathbf{c}\dot{\mathbf{u}}$ to the left hand side of the equation, where c is the damping matrix. It is also possible to determine resisting forces $\mathbf{p}$ not only in terms of displacements $\mathbf{u}$ but also as a function of velocities $\dot{\mathbf{u}}$ through the use of a material model that considers time-dependent effects, such as viscoelastic or viscoplastic material models.

For linear elastic material response, section deformations can be calculated as $\mathbf{e}=\mathbf{k}_s^{-1}\hat{\mathbf{s}}$ to obtain the section deformations from section forces through the use of section stiffness matrix $\mathbf{k}_s$. Substitution of section deformations $\mathbf{e}$ to Equation (10) gives:

$$\mathbf{a}_g\mathbf{u} = \mathbf{v} = \mathbf{f}\,\mathbf{q}; \quad \text{where} \quad \mathbf{f} = \int_0^L \mathbf{b}^{\mathrm{T}}(x,L)\mathbf{f}_s(x)\,\mathbf{b}(x,L)dx \tag{12}$$

In above equation $\mathbf{f}$ is the flexibility matrix of the element in the basic system. $\mathbf{f}_s$ is the section flexibility matrix that can be calculated from the inversion of the section stiffness matrix $\mathbf{k}_s$. Further substitution of above equation for linear elastic response in Equation (9) results in

$$\mathbf{m}\ddot{\mathbf{u}} + \mathbf{k}\mathbf{u} = \mathbf{p}_{app}; \quad \text{where} \quad \mathbf{k} = \mathbf{a}^{\mathrm{T}}\mathbf{f}^{-1}\mathbf{a} \tag{13}$$

where $\mathbf{k}$ is the 6×6 element stiffness matrix in the complete system. At this point in the element formulation, presence of semi-rigid connections will be introduced through the following extended version of above equation for the calculation of element end deformations:

$$\mathbf{v} = \mathbf{v}_{\mathrm{Frame}} + \mathbf{v}_{\mathrm{Con}}; \quad \text{where} \quad \mathbf{v}_{\mathrm{Frame}} = \int_L \mathbf{b}^{\mathrm{T}}(x)\mathbf{e}(x)dx; \quad \mathbf{v}_{\mathrm{Con}} = \sum_{i=1}^{nSC}\mathbf{b}^{\mathrm{T}}(x_i)\mathbf{\Delta}_{SC,i} \quad \text{and} \quad \mathbf{\Delta}_{SC} = \left[\delta_{SC}^{axial} \quad \theta_{SC} \quad \delta_{SC}^{shear}\right]^{\mathrm{T}} \tag{14}$$

The first integral along the length of the frame element can be numerically calculated by using a quadrature rule to capture spread of inelastic behavior and n_{SC} is the total number of semi-rigid connections discreetly located along element length; Δ_{SC} is the vector of semi-rigid connection deformations. Introduction of semi-rigid connections along element length in Fig. 1

does not alter the force field under small deformations. Element flexibility matrix is similarly discretized as follows:

$$\mathbf{f} = \mathbf{f}_{\text{Frame}} + \mathbf{f}_{\text{Con}}; \quad \text{where} \quad \mathbf{f}_{\text{Frame}} = \int_L \mathbf{b}^{\text{T}}(x)\mathbf{f}_s(x)\mathbf{b}(x)dx; \quad \text{and} \quad \mathbf{f}_{\text{Con}} = \sum_{i=1}^{nSC} \mathbf{b}^{\text{T}}(x_i)\mathbf{f}_{SC,i}\mathbf{b}(x_i) \tag{15}$$

As a remark, Eqns. (10) and (11) are related to the element state determination, i.e. these equations can be solved independent of Eqn. (9), and then the solution can be condensed out into Eqn. (9) such that the equations of motion can be assembled for all elements. This process was demonstrated above for the linear elastic case. In general, state determination of the element requires an iterative solution in the case of nonlinear behavior, where Eqns. (9) to (11) are needed to be solved. Detailed derivation of this element and its validation towards carrying out nonlinear static analysis of steel framed structures with semi-rigid connections was published in Saritas and Koseoglu [16].

2.4 Section Response

Section response can be obtained by the basic assumption that plane sections before deformation remain plane after deformation along the length of the beam by the use of following section compatibility matrix as $\mathbf{a}_s$ given in Eqn. (2), where the section compatibility matrix now contains the shear correction factor κs as follows

$$\mathbf{a}_s = \mathbf{a}_s(y) = \begin{bmatrix} 1 & 0 & -y \\ 0 & \kappa_s & 0 \end{bmatrix} \tag{16}$$

Shear correction factor κ_S is taken as the inverse of the form factor suggested by Charney et al. [17] for an I-section:

$$\kappa_s = 1/\kappa; \quad \text{where} \quad \kappa = 0.85 + 2.32\frac{b_f t_f}{d\, t_w} \tag{17}$$

The section forces are obtained by integration of the stresses that satisfy the material constitutive relations $\boldsymbol{\sigma} = \boldsymbol{\sigma}(\boldsymbol{\varepsilon})$ according to

$$\mathbf{s} = \int_A \mathbf{a}_s^{\text{T}} \boldsymbol{\sigma}\, dA; \quad \text{where} \quad \boldsymbol{\sigma} = \begin{pmatrix} \sigma_{xx} \\ \sigma_{xy} \end{pmatrix} \tag{18}$$

The derivative of section forces from (18) with respect to the section deformations results in the section tangent stiffness matrix

$$\mathbf{k}_s = \frac{\partial \mathbf{s}}{\partial \mathbf{e}} = \int_A \mathbf{a}_s^{\text{T}} \frac{\partial \boldsymbol{\sigma}(\boldsymbol{\varepsilon})}{\partial \mathbf{e}} dA = \int_A \mathbf{a}_s^{\text{T}} \mathbf{k}_m \mathbf{a}_s\, dA \tag{19}$$

The material tangent modulus km is obtained from the stress-strain relation according to $\mathbf{k}_m = \partial\boldsymbol{\sigma}(\boldsymbol{\varepsilon})/\partial\boldsymbol{\varepsilon}$. Gauss-quadrature, the midpoint or the trapezoidal rule can be used for the numerical evaluation of the integrals in (18) and (19).

2.5 Force Based Consistent Mass Matrix

The derivation of the consistent mass matrix requires the determination of the section mass matrix, where the mass is considered like a distributed load along the length of the beam in cantilever basic system. The section mass matrix is easily computed by following equation through the use of section compatibility matrix as given in Equation(4):

$$\mathbf{m}_s(x) = \int_A \mathbf{a}_s^{\text{T}} \rho(x,y)\, \mathbf{a}_s\, dA; \tag{20}$$

Mass matrix of the force-based element, which will be used in Equation (9), is written in a 6×6 dimension by the method discussed in depth by [18], i.e. in the complete system with 3 degrees of freedom per node, as follows:

$$\mathbf{m} = \begin{bmatrix} \mathbf{m}_{00} & \mathbf{m}_{0L} \\ \mathbf{m}_{L0} & \mathbf{m}_{LL} \end{bmatrix} \tag{21}$$

where the components of element mass matrix are calculated from sub-matrices

$$\begin{aligned}
\mathbf{m}_{LL} &= \mathbf{f}^{-1}\int_0^L \mathbf{b}^{\mathrm{T}}(x,L)\mathbf{k}_s^{-1}(x)\left(\int_x^L \mathbf{b}^{\mathrm{T}}(x,\xi)\mathbf{m}_s(\xi)\mathbf{f}_p(\xi)\mathbf{f}^{-1}d\xi\right)dx \\
\mathbf{m}_{L0} &= \mathbf{f}^{-1}\int_0^L \mathbf{b}^{\mathrm{T}}(x,L)\mathbf{k}_s^{-1}(x)\left(\int_x^L \mathbf{b}^{\mathrm{T}}(x,\xi)\mathbf{m}_s(\xi)\left(\mathbf{b}^{\mathrm{T}}(0,\xi)-\mathbf{f}_p(\xi)\mathbf{f}^{-1}\mathbf{b}^{\mathrm{T}}(0,L)\right)d\xi\right)dx \\
\mathbf{m}_{0L} &= \mathbf{m}_{L0} = -\mathbf{b}(0,L)\mathbf{m}_{LL} + \int_0^L \mathbf{b}(0,x)\mathbf{m}_s(x)\mathbf{f}_p(x)\mathbf{f}^{-1}dx \\
\mathbf{m}_{00} &= -\mathbf{b}(0,L)\mathbf{m}_{L0} + \int_0^L \mathbf{b}(0,x)\mathbf{m}_s(x)\left(\mathbf{b}^{\mathrm{T}}(0,x)-\mathbf{f}_p(x)\mathbf{f}^{-1}\mathbf{b}^{\mathrm{T}}(0,L)\right)dx
\end{aligned} \tag{22}$$

In above equations, element flexibility matrix $\mathbf{f}$ is obtained as given in Eqn. (12). The partial flexibility matrix $\mathbf{f}_p$ is calculated as follows:

$$\mathbf{f}_p(x) = \int_0^x \mathbf{b}^{\mathrm{T}}(\xi,x)\mathbf{k}_s^{-1}(x)\mathbf{b}(\xi,x)d\xi \tag{23}$$

3. Numerical Examples

The first example is a portal frame retrieved from [9], where semi-rigid connections are defined at both ends of the beams in this structure. British UB254×146×37 section is defined to the beam and UC203×203×60 sections are defined for the columns. The length of the beam is 2.9 m and the height of the columns is 3.0 m (see Fig. 2). The results are compared with the SAP2000 analyses results for 1 and 32 elements of each member of the portal frame. This comparison is conducted to verify the accuracy of the SAP2000 model with increased number of the elements. Since SAP2000 uses lumped mass approach, increasing mesh size tends to give a closer match with consistent mass matrix results as evident in Fig. 2.

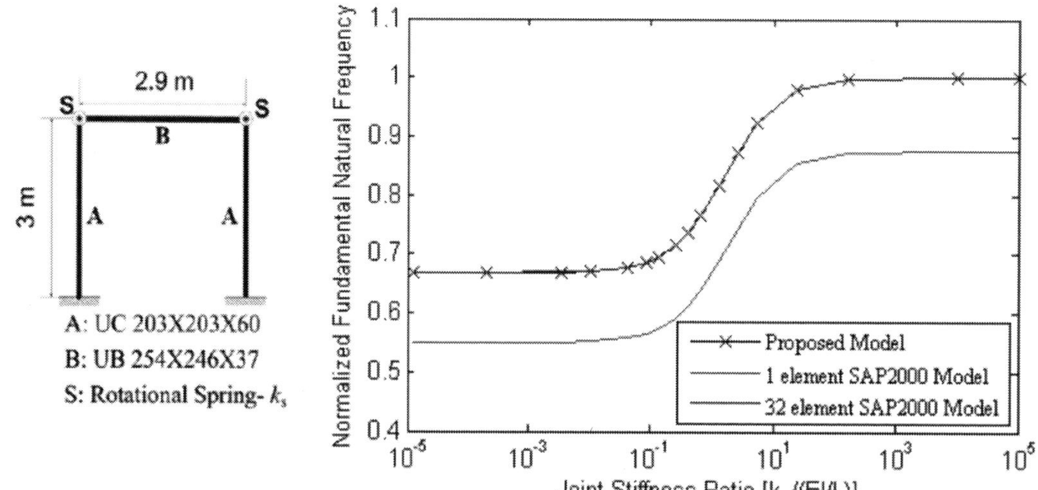

Fig. 2: Portal Frame from the study of [9] and
Fundamental Natural Frequency versus Joint Stiffness Ratio for Portal Frame

The second example is a 3 bay and 6 stories steel frame with European sections HEB260 for columns and IPE300 for beams as given in [9]. The length of the beams is 6.0 m and the height of the columns is 3.75 m. The results of the proposed model are compared with SAP2000 models and results from the study of [9]. On the contrary to the former example, SAP2000 model is prepared with four elements for each member. Since the structural system is a much larger one compared to the previous example, lumped mass approach presents better convergence with 4 elements per member for SAP2000 as compared to the results by ABAQUS model from [9] (see Fig. 3); however, the results of SAP2000 could not perfectly capture the fundamental mode of vibration due to the deficiency of SAP2000 in capturing an accurate shear correction factor for these IPE sections. It is apparent from below figure that proposed model showed very close results with ABAQUS by the use of single element discretization per member.

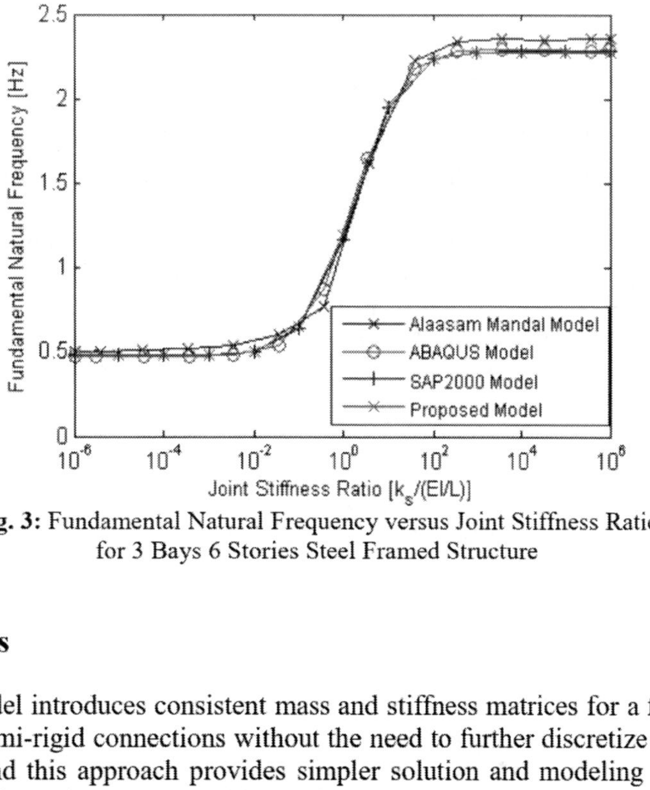

Fig. 3: Fundamental Natural Frequency versus Joint Stiffness Ratio for 3 Bays 6 Stories Steel Framed Structure

4. Conclusions

The proposed model introduces consistent mass and stiffness matrices for a frame member in the presence of semi-rigid connections without the need to further discretize the member into more elements, and this approach provides simpler solution and modeling strategy towards analysis of steel framed structures with semi-rigid connections. Validation studies clearly showed the robustness of the proposed model compared to other finite element programs widely used in research and practice.

Acknowledgments

The authors thank for the support provided by Scientific and Technological Research Council of Turkey (TUBITAK) under Project No: 113M223.

References

[1] Chui PPT, Chan SL. "Vibration and deflection characteristics of semi-rigid jointed frames." *Engineering Structures*, 19(12), 1001-1010, 1997.

[2] Nader MN, Astaneh-Asl A. "Shaking table tests of rigid, semirigid, and flexible steel frames." *Journal of Structural Engineering*, 122, 589-596, 1996.

[3] Nguyen PC, Kim SE. "Nonlinear elastic dynamic analysis of space steel frames with semi-rigid connections." *Journal of Constructional Steel Research*, 84, 72-81, 2013.

[4] Sekulovic M, Salatic R, Nefovska M. "Dynamic analysis of steel frames with flexible connections." *Computers and Structures*, 80, 935-955, 2002.

[5] Sekulovic M, Danilovic MN. "Contribution to transient analysis of inelastic steel frames with semi-rigid connections." *Engineering Structures,* 30, 976-989, 2008.

[6] Galvao AS, Silva ARD, Silveira RAM, Goncalves PB. "Nonlinear dynamic behavior and instability of slender frames with semi-rigid connections." *International Journal of Mechanical Sciences*, 52, 1547-1562, 2010.

[7] Minghini F, Tullini N, Laudiero F. "Vibration analysis of pultruded FRP frames with semi-rigid connections." *Engineering Structures*, 32, 3344-3354, 2010.

[8] da Silva JGS, de Lima LRO, da S. Vellasco PCG, de Andrade SAL, de Castro RA. "Nonlinear dynamic analysis of steel portal frames with semi-rigid connections." *Engineering Structures*, 30, 2566-2579, 2008.

[9] Al-Aasam HS, Mandal P. "Simplified procedure to calculate by hand the natural periods of semirigid steel frames." *Journal of Structural Engineering*, 139, 1082-1087, 2013.

[10] Suarez LE, Singh MP, Matheu EE. "Seismic response of structural frameworks with flexible connections." *Computers and Structures*, 58, 27-41, 1996.

[11] Sophianopoulos DS. "The effect of joint flexibility on the free elastic vibration characteristics of steel plane frames." *Journal of Constructional Steel Research*, 59, 995-1008, 2003.

[12] SAP2000, Structural Analysis Program SAP 2000, CSI Berkeley, CA., 2011.

[13] Taylor RL, Filippou FC, Saritas A, Auricchio F. "Mixed finite element method for beam and frame problems." *Computational Mechanics*, 31(1-2), 192-203, 2003.

[14] Saritas A, Filippou FC. "Inelastic axial-flexure–shear coupling in a mixed formulation beam finite element." *International Journal of Non-Linear Mechanics,* 44(8), 913-922, 2009.

[15] Barr ADS. "An Extension of the Hu-Washizu variational principle in linear elasticity for dynamic problems." *Journal of Applied Mechanics*, June, 465, 1966.

[16] Saritas A, Koseoglu A. "Distributed inelasticity planar frame element with localized semi-rigid connections for nonlinear analysis of steel structures", *International Journal of Mechanical Sciences*, 96-97, 216-231, 2015.

[17] Charney FA, Iyer H, Spears PW. "Computation of major axis shear deformations in wide flange steel girders and columns", *Journal of Constructional Steel Research*, 61, 1525–1558, 2005.

[18] Soydas O, Saritas A. "Free vibration characteristics of a 3d mixed formulation beam element with force-based consistent mass matrix", *Journal of Vibration and Control (Accepted)*, 2016.

USING ADVANCED NUMERICAL MODELS TO ESTIMATE THE FIRE SAFETY OF TRUSS FRAME STRUCTURES MADE OUT OF COLD-FORMED STEEL

Luc Schueremans[a,b], Leen Driesens[b], Jonathan Meylemans[b] and Johan De Cock[a]

[a] *Frisomat NV, Belgium*
[b] *KU Leuven, Department of Civil Engineering, Materials Science and Materials Technology, Belgium*

Abstract: The overall goal of this contribution is to demonstrate to what extent numerical modelling using Safir® is useful to design structures out of cold-formed steel with a minimum fire resistance of 15 minutes: R15 as often requested in several European countries. First a comparison is made in between the simple numerical tools available from standards (EN1993-1-2) and advanced numerical models for fire analysis. Secondly, a parametric study is performed. From the batch of structural analyses made on frame level, a systematic overview of the calculations and a critical appraisal of both the design methodology and obtained results are given.

1. Introduction

A fire resistance of 15 minutes (R15) is often required as minimum fire resistance for load-bearing structural elements in industrial buildings. This for example is the case in Belgium. Although by law specified as "R non-determined", the law indicates that for slender elements it is advised to check the fire resistance to be at least 15 minutes [1].

To achieve this R15, advanced numerical tools look very promising. Therefore, in this paper an extensive numerical program is set up, first to compare the results of these advanced numerical models with the simple numerical tools available from standards (EN 1993-1-2). The advanced numerical tools will enhance much better the specific nature of the failure mechanisms. Secondly, the impact of several parameters on the resulting fire resistance is verified for truss frames, including: the size of the frame (height, span), the loads (roof loads, wind, snow) and the boundary conditions (hinged versus fixed foundation) to be taken into account. These are truss frame types that are typically produced by Frisomat NV.

2. Simplified analysis using EN 1993-1-2 and EN 1993-1-3

In the following, an example is given to illustrate the type of industrial buildings meant in this paper, the way frames are built out of cold-formed steel and the typical fire resistance that is obtained using simplified analyses as available from European standards [2].

In this article a truss frame type of industrial building is treated, named *Astra*. Parallel, 2 types of industrial buildings have been treated elsewhere [3], having a 22° and 10° roof angle, named *Sigma* and *Astrigma* respectively.

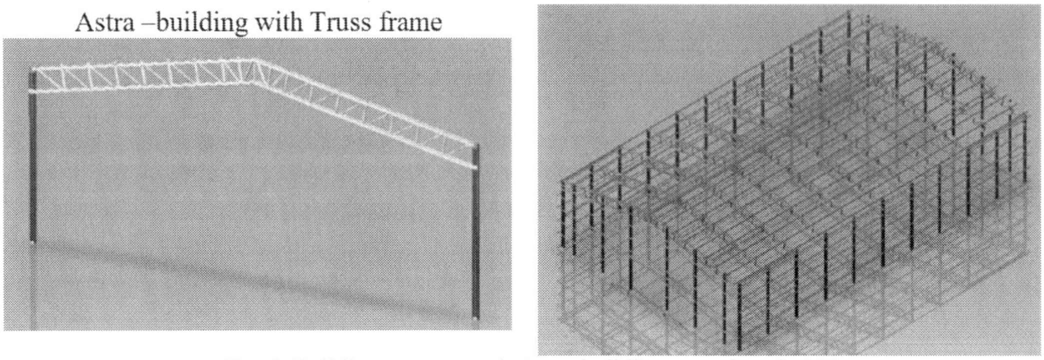

Fig. 1: Building type treated: Astra building – roof angle 10°

The frames are built up from cold-formed sections, namely:

- Double Σ320 for the columns (C) thickness: 2.0 mm till 5.0 mm;
- Double C160 for the lower and upper chord (L and U), thickness: 1.25 mm till 4.0 mm;
- Single C130 for the diagonal and vertical elements (D and V), thickness 1.25 mm till 3.0 mm.

The steel grade equals: S390+GD+Z275, with yield strength of f_y = 390 MPa and an ultimate strength of f_u = 460 MPa (EN 1993-1-3, Table 3.1b). The thickness of the sections varies in steps of 0.25 mm. As a result, these cross-sections are class IV cross-sections. The structural design at ambient temperature is performed in accordance with EN 1993-1-3 [4]. In case of fire, the reduction factor is the proportionality factor $k_{p,0.2,\theta}$, which is more conservative compared to the yield strength reduction $f_{y,\theta}$ used for hot-rolled profiles such as IPE and HEA beams and columns. In the standard, the critical temperature for cold-formed steel structures θ_a is suggested to be taken maximum 350°C, as a conservative assumption, when no calculation is performed. Taking into account the ISO 834 standard fire curve, this temperature is reached after about 3.15 minutes for 2Σ320/2.0 (double sigma profiles with a thickness of 2.0 mm). Does this means that in real cases the fire resistance is limited to 3 minutes? Certainly not. This critical temperature is based on a conservatively taken utilization degree of η_{fi} = 0.70, Eqn. (1). One of the advantages of cold-formed structures is the low utilization degree in case of fire. This is mainly because (1) the dead weight of the structure is low compared to the applied loading, and (2) the part of applied load in case of fire (ψ_{fi}) is limited. In the majority of European countries, as indicated in the national annexes to EN 1990, no snow load needs to be taken into account and the part of wind to be taken into account in case of fire, is limited to ψ_{fi} = 0.2-0.3. In this paper, the methodology is illustrated for the application of the Belgian Application Document to EN 1990 and EN 1991-1-2 [5-6], but the methodology and major findings remain general applicable for other European countries.

$$\eta_{fi} = \frac{G_k + \psi_{fi} Q_{k,w}}{\gamma_g G_k + \gamma_q Q_{k,w}} = \frac{G_k + 0.2 Q_{k,w}}{1.35 G_k + 1.50 Q_{k,w}} = \frac{1 + 0.2 \dfrac{Q_{k,w}}{G_k}}{1.35 + 1.50 \dfrac{Q_{k,w}}{G_k}} \tag{1}$$

3. Parametric survey

The following parameters are altered within the analysis, Table 1. The distance in between the frames is a fixed parameter and always equals: $a = 4.0$ m. This results in 72 analyses for *Astra* frames.

Table 1: Overview of parameters within the outlined fire analyses

Geometry of *Astra* frame	Height (at gutter) (h)	5.0 m; 6.0 m; <u>7.0 m</u>
	Span (L)	17 m; 21 m; <u>25 m</u>; 29 m
Boundary conditions	Foundation	<u>Hinged</u> versus fixed
Dead weight of the frame	Non-insulated building (sheeting TR45 on purlins Z120)	<u>$g_{k,wall}$ = 0.10 kN/m²</u>
		<u>$g_{k,roof}$ = 0.10 kN/m²</u>
Wind load (basis wind velocity v_b and terrain category)	High wind	v_b = 26 m/s – II
	Moderate wind	<u>v_b = 24 m/s – III</u>
	Low wind	v_b = 23 m/s – IV
Snow load (s_k)	Altitude = 0 m;	s_k = 0.50 kN/m²
	Altitude = 200 m;	<u>s_k = 0.62 kN/m²</u>
	Altitude = 400 m;	s_k = 0.85 kN/m²

4. Optimization procedure for design at ambient temperature

The different frames are optimized for the persistent design situation, to correctly respond to the design criteria in both ULS (ultimate limit state) and SLS (serviceability limit state). In that, an optimization procedure is performed. The minimum weight or minimum thickness that complies with the performance requirements is taken as optimal result.

As an example, *Astra* analyses no. 32 is taken. The parameters used are underlined in Table 1.

Within the simplified analysis, the reduced loads acting on the structure are processed with a linear-elastic frame model. The major element that counts is that there is no redistribution of internal forces as function of time or increased temperature. The results for the simplified analysis are outlined within Table 2.

5. Advanced analysis using Safir

These frames are also modeled within Safir [2-3, 8], in a 2.5D model using beam elements although the cross-sections are class 4 sections and not shell elements. In a 2.5D model, a portal frame with restraints perpendicular to the frame is considered. Warping, frame imperfection, bow imperfection and element instability such as lateral torsional buckling are included. This finite element tool contains a thermal and a mechanical part. For the thermal part, there is only a slight difference in thermal outcome for the cross-sections at hand. This is logic: due to the high thermal conductivity, the temperature evolution is nearly constant along the cross-section. Within the numerical analyses of the temperature, a uniform distribution of the temperature along the cross-section is taken as starting point. The difference is negligible. For the thermal part of the analysis, this is performed stepwise as a function of time (and thus implicitly temperature) to enable to cover the changes in the structure due to the increased temperature). Several phases are encountered within the mechanical part of the analysis [8-9], Fig. 2:

- An initial phase 0 where the loads are applied on the frame in ambient conditions:
- A phase 1 where the heating of the profiles results into elongation of the elements and additional internal forces (*N, M, V*) due to hindered deformations, Fig. 2. These internal forces are higher compared to the values at time $t = 0$ s, where the accidental load acts on the structure at ambient temperature (20°C), [3];
- A phase 2 where the capacity of some of the elements is insufficient to withstand the internal forces. As a result, a redistribution of the internal forces occurs in such a way that the frame keeps on acting at its maximum capacity given a certain time and thus remaining yield strength.

This can also be seen from the evolution of the normal forces at ridge and gutter as function of time. On the same diagram the vertical displacements are drawn. After the initial phase one clearly indicates the positive displacements or uplift movement of the building, Fig. 3. Similar, the normal forces at the gutter and ridge change sign. After the maximum capacity is reached (at about 375 s), the capacity starts to decrease and the uplift movement at the ridge retards and eventually at about 800 s starts to drop. Starting from that moment clearly the ridge starts to steadily move downwards, reaching failure at about 1448 s .

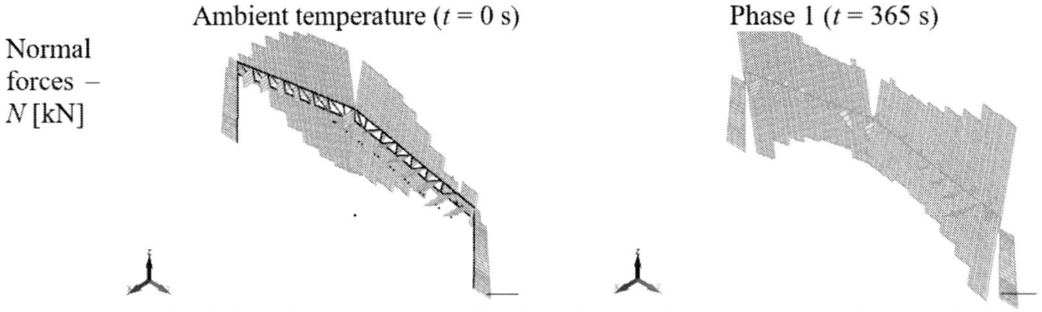

Fig. 2: Evolution of normal forces as function of time – Astra structure analysis no. 32

Table 2: Fire resistance – comparison of simplified analysis according to Eurocode – versus Safir

Frame type	*Astra* – analysis no. 32 Optimized thickness	*Astra* – analysis no. 32 Maximized thickness
Column cross-section (C)	2Σ320/5	2Σ320/5
Truss frame (U_i-L_i)	C160/1.5-2.5	C160/3.0
Vertical elements (V_i)	C130/1.25-3.0	C130/3.0
Diagonal elements (D_i)	C130/1.25-3.0	C130/3.0
Mass of frame [kg]	929 kg	1880 kg
Fire resistance *R* [min] - simplified analysis EN	9 min θ_a = 650°C, η=0.22	17 min 30 sec θ_a = 720°C
Fire resistance *R* [min] - advanced analysis Safir	692 sec or 11 min 32 sec	1448 sec or 24 min 8 sec

The impact of an advanced analysis is clearly visible from Table 2. There is a significant gain in fire resistance that is neglected within a simplified analysis. That is mainly due to the lack of possibility to take into account the redistribution of internal forces. This can only be done within a finite element framework that takes into account the deformed structure in the stiffness matrix, as such a 2nd order analyses is needed therefore.

There thus is a clear positive effect of the internal force redistribution. From a certain point on, indicated as phase 2 in Fig. 3, the roof truss starts to act according to their maximum capacity. The redistribution of the internal forces in the structure is such that they are in line with the maximum capacity at that moment. As such the normal force capacity drops steadily and normal forces are also adapted. In quantities, this gain is significant. For the examples treated the gain in fire resistance yields more than 6 minutes. This is a gain of over 35%.

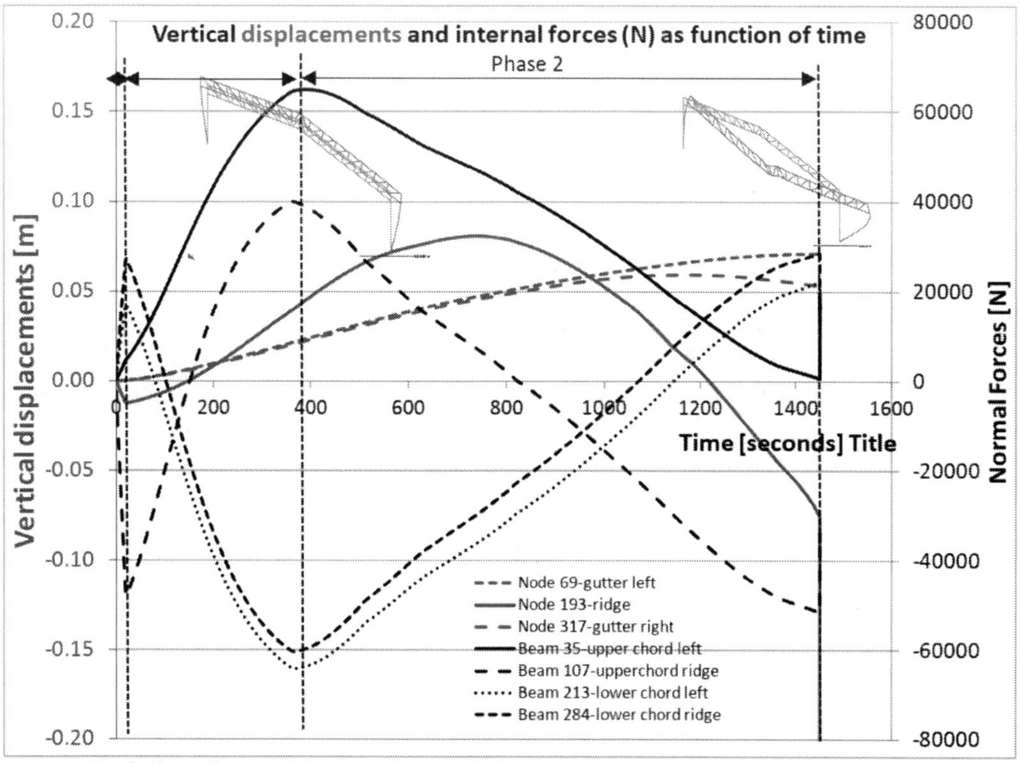

Fig. 3: Evolution of normal force and vertical displacements – analysis no. 32 - maximized thickness

6. Parametric study

The analyses results for different *Astra* frames are outlined in Fig. 4. Following elements are highlighted:

- The required R15 is only reached for part of the smaller spans. The larger the span (and the higher the frame) the more limited the fire resistance;
- There is a clear increase of capacity when fixed boundary conditions can be assumed (analyses 37-72). The impact is largest for the smallest frames and is even adverse for the larger spans. The adverse effect is less when the frame height is larger. This is logic. Within the first phase of the fire, the elements elongate. Because of the fixed foundation, this causes hindered deformations. The larger the span, the more the truss frame needs to deal with additional compressive forces from hindered deformations and thus instability phenomena. For the higher frames this effect is smaller because of smaller bending stiffness of the column, in line with their increased length ($k=EI/L$).

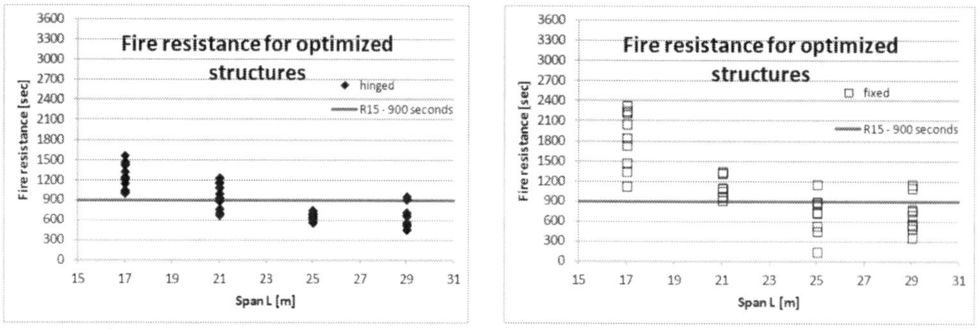

Astra: 16/36 < R15 *Astra*: 11/36 < R15

left: fire resistance for hinged foundations - right: fire resistance for fixed foundation) red line at R15 (900s) threshold line.

Fig. 4: Impact of boundary conditions on fire resistance

7. Increase of fire resistance

There are different options to increase the fire resistance. In this type of cold-formed frames, it is straightforward to increase the thickness till maximum thickness, being 5.0 mm for the columns for example. When this is done within the same height of section, a significant amount of frames tend to fulfil the R15 requirement, Fig. 5. Of course, this will affect the frames for which in ambient temperature the maximum thickness is not yet reached, otherwise, no difference will be obtained. As such, one can conclude from Fig. 5:

- When combining maximization in the element thickness with hinged foundations, nearly the full range of *Astra* frames reaches R15;
- When combining maximization in the element thickness with fixed foundations, the full range of *Astra* frames reaches R15. It is clear that the maximization of thickness is not needed for the smaller spans. As a remark: the analyses were cut-off at 3600s.

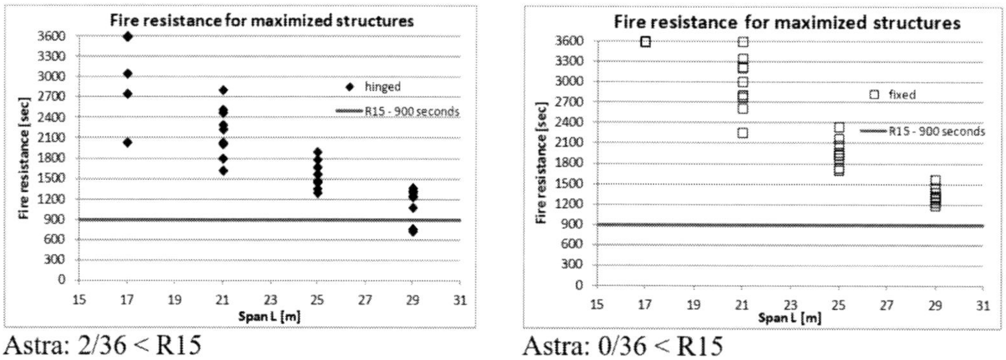

Astra: 2/36 < R15 Astra: 0/36 < R15

left: increased thickness and hinged foundation; right: increased thickness and fixed foundation; red line at R 15 (900s) threshold line

Fig. 5: Impact of over dimensioning to increase the massivity of the section and the capacity reserve

8. General discussion based on the parametric study

The maximum thickness of the sections sets a clear upper limit to both span and gutter height of a frame. Of course, if larger thicknesses would be within range, this would be beneficial both for the section massivity and capacity. For an identical range of dimensions, R15 would be reachable for an increased number of cases and or range of dimensions.

The degree of utilisation plays an important role and is in favour of this type of light-weight structures. Overall, an utilisation degree in between η_{fi} = 0.15-0.20 is found whereas in typical concrete or hot-rolled steel structures a significantly higher degree of utilization generally is present. Because the self-weight of the structure is low in relation to the applied loading and snow does not need to be taken into consideration for the fire analysis, the wind is the main factor affecting the outcome. One can clearly see the impact of higher wind pressure. Because the combination factor for wind equals ψ_{fi} = 0.20, a theoretical utilisation degree of 0.20/1.50 = 0.13 can almost be obtained in practice.

From the analyses made it is clear that within the range of buildings that fulfil the requirements for the persistent design situation, only a relatively small portion – 50% – still fulfil under ISO 834 fire conditions. This is logic. Because the optimization concept at ambient temperature is such that the structure is optimized towards minimum thickness of the sections. These structures are designed to have very small additional conservatism in their design. After 15 minutes of ISO 834 fire, the temperature ranges in between 692°C and 723°C for the C-sections. As such, the remaining capacity of the yield strength, $f_{p.0.2.\theta}$ varies in between 0.13 and $0.11f_y$. At the same time, there also lies the solution.

There is a simple and economical solution to increase the fire resistance. Whenever possible, the thickness of the sections can be increased. As such both, additional massivity is combined with additional margin in cross-sectional capacity and element stability. The steel temperature will lower and there will be gain in capacity mechanically ($f_{p.0.2.\theta}$) and geometrically (W_{eff}, A_{ef}, I_{eff}).

Similar, the use of fixed foundations results systematically in increased fire resistance due to increased capacity and reduction of buckling lengths. Even in case a hinged foundation system is designed at ambient temperature, assuming a fixed foundation at fire conditions seems to be valid, as also confirmed by a real case experiment conducted in 2009 [9-11]:

- In reality, the compartment gas temperature will be horizontally layered. As such the column base will benefit from a lower temperature and thus increased capacity;
- The actual layout of the connections often might have sufficient bending capacity for the accidental fire situation. This bending moment often is limited to 15% - 20% of the bending moment capacity at ambient temperature.

The analysis performed here are not solely valid for ISO 834 standard fire curve. This type of analysis can also be made for natural fire curves, without any alteration in the methodology.

9. Conclusions

This paper demonstrated to what extent numerical modelling using Safir ® is useful to design structures out of cold-formed steel with a minimum fire resistance of 15 minutes:

- The major gain is obtained from the mechanical modelling part. The redistribution of internal forces during the heating of the different sections is positively affecting the time to failure;
- This gain in fire resistance is substantial compared to simplified analyses performed along Eurocode, applying a cross-sectional verification based on a linear-elastic internal force distribution obtained from 1st order analysis;
- The overall analysis using advanced techniques give a better inside into the failure mechanisms that occur during failure in fire conditions.

From the batch of structural analyses on frame level it can be concluded that a significant increase in fire resistance is obtained when the section thickness is maximized and, or fixed boundary conditions are considered.

From the range of geometrical parameters tackled, one can conclude that the largest spans reach a fire resistance R15. For practical purpose one can conclude that reaching R15 - except of the smaller frames (span) - cannot be achieved without additional cost. The additional cost is within the over dimensioning of the frames (thickness of the material) and a change of boundary conditions towards fixed foundations.

References

[1] KB of 01/03/2009. "Bijlage 6: Industriegebouwen", (E: Annex 6: Industrial Buildings), *Belgian State Papers*, 2nd edition, 15/07/2009.

[2] Driessens L, Meylemens J. "Eenvoudige rekenmethode voor constructief brandveilig ontwerpen", (E: simplified design methodology for fire safety engineering), *MSc Thesis*, KULeuven, 2013-2014.

[3] Schueremans L, De Cock J, Furtado AC. "Benefits of advanced numerical models to design fire resistant frames made out of cold-formed steel", *X Congresso de Construçao Metalica e Mista*, Coimbra, 26-27 November 2015, accepted for publication, 2016.

[4] NBN EN1993-1-3, "Eurocode 3 - Design of steel structures - Part 1-3: General rules - Supplementary rules for cold-formed members and sheeting", *European Committee for Standardization*, 2011

[5] NBN EN 1990, "Eurocode 0: Basis of structural design", *European Committee for Standardization*, 2002.

[6] NBN EN1991-1-2 ANB, "Eurocode 1: Actions on structures - Part 1-2: General actions - Actions on structures exposed to fire, National Application Document", *European Committee for Standardization*, 2008.

[7] NBN EN1993-1-2, "Eurocode 3: Design of steel structures - Part 1-2: General rules - Structural fire design (+ AC:2005, + AC:2009)", *European Committee for Standardization*, 2005.

[8] Franssen JM. *User's manual for safire 2007a, a computer programme for analysis of structures subjected to fire*, Ulg, Université de Liège, Belgium, 2008.

[9] Pyl L, Schueremans L. "Brandveilig constructief ontwerp - cursustekst" (E: structural fire design-tutorial notes), *tutorial notes made in the framework of the IWT-TETRA research project*, KULeuven, 2008-2010, IWT090163 "Dunwandig koudgevormde profielen: van theorie naar praktijk" (E: thin-walled cold-formed sections: from theory towards practice, ISBN 978-94-6018-270-9, 2010.

[10] Pyl L, Schueremans L. "Brandveilig constructief ontwerp - voorbeeldoefeningen" (E: structural fire design-example applications), *tutorial notes made in the framework of the IWT-TETRA research project*, KULeuven, 2008-2010, IWT090163 "Dunwandig koudgevormde profielen: van theorie naar praktijk" (E: thin-walled cold-formed sections: from theory towards practice, ISBN 978-94-6018-269-3, 2010.

[11] Pyl L, Schueremans L, Dierckx W, Georgieva I, *Carroccio M, Trippaers R*. "Fire safety analysis of a 3D frame structure consisting of cold-formed sections; Numerical modeling versus experimental behaviour based on a full-scale fire test", *Proceedings of the 6th International Conference on Thin Walled Structures - ICTWS2011*, September, 5 - 7, 2011, Timisoara, Romania, Ed. Dan Dubina and Viorel Ungureanu, ISBN 978-92-9147-102-7, 447-454, 2010.

BENCHMARK PROBLEMS FOR DESIGN BY ADVANCED ANALYSIS - MEMBERS SUBJECT TO MAJOR- AND MINOR-AXIS FLEXURE

Ronald D. Ziemian and Jean C. Batista Abreu

Department of Civil and Environmental Engineering, Bucknell University, Lewisburg, PA, USA

Abstract: Steel design codes continue to be expanded to permit the use of more advanced methods of nonlinear analysis. Designers looking to employ such methods need to validate their analysis software and, just as importantly, verify their ability to properly utilize it. The literature contains many benchmark problems and results to help achieve this, but nearly all are limited to two-dimensional behaviour. This paper is intended to contribute a new set of benchmark problems, towards satisfying the need for a database of examples in which the accurate modelling of three-dimensional or spatial behaviour is essential.

1. Introduction

More and more specifications worldwide are providing designers with opportunities to take advantage of the ever-increasing capabilities of nonlinear frame analysis software. One such example is the American Institute of Steel Construction's *Specification for Structural Steel Buildings*. In 2010, this Specification [1] included completely revised provisions for design by inelastic analysis. With an emphasis on taking advantage of the potential redistribution of member and connection forces and moments that can occur as a result of localized yielding, these provisions represent a logical extension of the AISC's direct analysis method, which first appeared in the AISC Specification in 2005 [2]. Most recently, the AISC Specification [3] has again been expanded to provide opportunities for designers who employ a second-order elastic analysis that is more rigorous than the traditional P-Δ and P-δ methods expected for application of the original direct analysis method. By using an analysis that also accounts for twist, when formulating equilibrium equations on the deformed shape, and includes the explicit modelling of system sway and member out-of-straightness imperfections as well as an inelastic stiffness reduction factor, the compression strengths of members may now be taken as their cross-section strength. Although this new approach is based on elastic analysis and therefore does not permit an inelastic redistribution of internal forces and moments, it can be particularly useful for design situations in which the definitions of the unbraced lengths or effective lengths are not readily apparent. Such cases include, but are not limited to, arches and unbraced Vierendeel trusses, in which the member may be curved and/or the compressive axial force varies along a member's unbraced length.

Key requirements for applying a design method that relies heavily on a rigorous second-order analysis are the confirmation of the software's capabilities and, just as importantly, the designer's competency in using such software. Given that the literature now contains many planar frame examples that can be used for such verifications, the purpose of this paper is to provide results for three benchmark problems in which significant twists develops primarily as a result of second-order effects. These three-dimensional verification studies, which focus on beams and beam-columns, are intended to represent only a few of the array of benchmark problems that an engineer should exercise before using sophisticated levels of analysis to design steel structures.

2. Why a little twist hurts...

Consider the beam-column shown in Fig. 1. The member is subjected to a compressive force P_u and uniformly distributed load ω_{uy} that is assumed to track the direction of gravity, and therefore always remain vertical. For now, it is further assumed that no out-of-plane distributed load and end moments are applied ($\omega_{ux} = 0$, $M_X = 0$ and $M_Y = 0$). The analytical model also includes an initial out-of-plane imperfection with amplitude of δ_{ox} at mid-span. In applying the given loads, the member will not only bend in- and out-of-plane, but it will also twist. Based on the resulting mid-span deflections of δ_x, δ_y, and θ, an equilibrium analysis on the deformed shape will result in major- and minor-axis bending moments at mid span of

$$\begin{aligned} M_{ux'} &= M_{st}\cos\theta - M_{wk}\sin\theta \\ M_{uy'} &= M_{st}\sin\theta + M_{wk}\cos\theta \end{aligned} \tag{1}$$

in which, $M_{st} = \omega_{uy}L^2/8 + P_u\delta_y$ and $M_{wk} = P_u(\delta_{ox} + \delta_x)$.

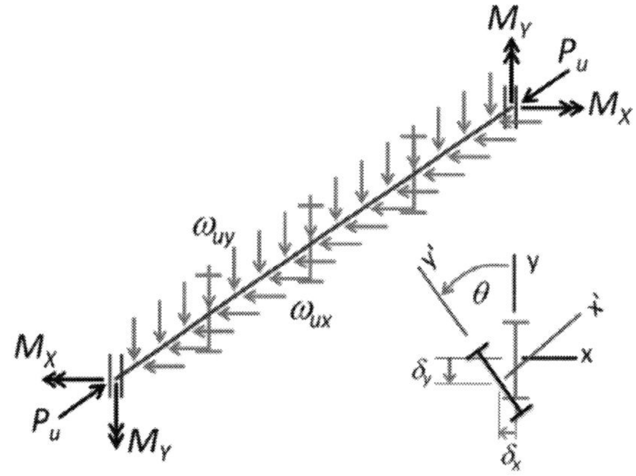

Fig. 1: General description of member investigated

It is important to note that the moments $M_{ux'}$ and $M_{uy'}$, which reference the local cross-section axes at mid-span, are the moments that are intended to be used in any interaction equation design checks.

To gain some insight as to the magnitude of the twisting rotation θ, let's consider the case in which the compressive axial force is inconsequential, $P_u \approx 0$. Substituting the moments from Eq. (1) into AISC's governing interaction equation, given by

$$\frac{M_{ux'}}{\phi M_{nx}} + \frac{M_{uy'}}{\phi M_{ny}} \leq 1.0 \quad (2)$$

and, assuming θ is small ($\cos\theta = 1$ and $\sin\theta = \theta$), yields

$$\frac{M_{st}}{\phi M_{nx}} + \frac{M_{st}\,\theta}{\phi M_{ny}} \leq 1.0 \quad (3)$$

Solving Eq. (3) for an inherent limitation on θ gives

$$\theta \leq \left[\left(\frac{M_{st}}{\phi M_{nx}}\right)^{-1} - 1\right] \Big/ \frac{M_{nx}}{M_{ny}} \quad (4)$$

For the case of no axial force and a range of major-axis to minor-axis available strength ratios M_{nx}/M_{ny}, Fig. 2 provides a plot of the required-to-available strength ratio for major axis bending $M_{st}/\phi M_{nx}$ versus the maximum values of twist θ permitted per Eq. (4). The curves indicate that attempts to fully utilize a member's major-axis bending capacity will require that the amount of twist resulting from an initial out-of-plane member imperfection (or some other perturbation, such as a small amount of out-of-plane eccentricity in the applied load) will need to be limited. It is also observed that using cross sections with large major-axis to minor-axis available strength ratios further exasperates this limitation on θ; a situation that, for efficiency purposes, may often be the case for members being designed to primarily resist major-axis bending. Again, the reason for this behavior is that assessing equilibrium on the deformed shape (reality!) requires that large major-axis demands be resolved into minor-axis demands as the section twists. If the member has a comparatively smaller amount of minor-axis available strength, such minor-axis demands can have a significant impact on the design interaction equation checks.

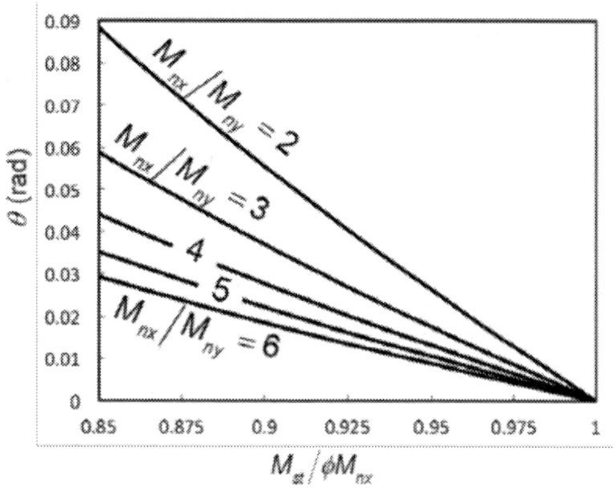

Fig. 2: Upper bounds on permitted twist angle θ

3. Analysis software employed

Several commercially available analysis programs were used to produce the results presented in this paper. In general, all of these programs are known to possess rigorous three-dimensional second-order or geometric nonlinear analysis capabilities. Given that most of today's structural engineering designs are completed using finite element analyses that employ elastic line or framework elements, the focus of this paper is to establish such benchmark results. In regard to torsional resistance, analysis programs often differ in their ability to model uniform and/or nonuniform torsion. Programs employed in this study that only account for the continuous flow of shear stresses on the section, commonly called uniform or St. Venant torsional resistance, include SAP2000 [4] and Strand7 [5]. Analysis software used that also models warping restraint torques included the line elements within ADINA [6], ANSYS [7], and ABAQUS [8]. In all cases, analysis results are provided from models in which the cross-section properties were provided; in other words, the general or integrated properties, such as the second-moment of area, were defined as input. Although not reported herein, nearly identical results are obtained when the dimensions of the sections are provided as input, with the software computing the integrated section properties. To provide comparisons with results obtained using second-order inelastic or geometric material nonlinear analyses (GMNIA), complete shell and brick element models were analyzed using Strand7 [5]. With the exception of the shell and brick element results presented, so-called shear deformations were not included in any of the analyses that employed line elements.

4. Common Properties and Support Conditions of Members Investigated

All of the beams and beam-columns studied in this paper have the same geometric properties, material properties, and support conditions. The member is comprised of a W460×97[1] (W18×65) cross-section profile with a span and unbraced length of 6.1m (240 inches). The material stiffness and strength are E = 200 GPa (29000 ksi), G = 77 GPa (11154 ksi), and F_y = 345 MPa (50 ksi), respectively. When results are provided that are in accordance with AISC's *Design by Advanced Elastic Analysis* method, all section properties (A, I_x, I_y, J, and C_w) are reduced by a factor of 0.8 in computing required strengths. When employing AISC's *Design by Advanced Inelastic Analysis* method, the original section properties and material yield strength are multiplied by a reduction factor of 0.9.

All members are simply supported (pin-support at one end and roller support at the other end) with longitudinal rotation or twist at the member ends restrained. Member ends are assumed warping free with cross-flange bending not resisted at the supports. The axial, flexural, and uniformly distributed loads provided below are assumed to be factored and are applied proportionally. All applied loads track the original global coordinate system. For example, a uniformly distributed gravity load is applied in the vertical direction throughout the loading history; in other words, the applied loads do not track the element axes.

For this compact section, unbraced length, and support conditions, the design capacities of the member are computed to have an axial strength of ϕP_n = 952 kN (214 kip), major-axis flexural strengths of ϕM_{nx} = 380.9 kN-m (3371 kip-in) with C_b =1 for a uniform moment distribution and ϕM_{nx} = 434.2 kN-m (3843 kip-in) with C_b =1.14 for a moment gradient produced by a uniformly distributed load, and a minor-axis flexural strength of ϕM_{ny} = 114.4 kN-m (1013 kip-in). With these values, the major-axis to minor-axis available strength ratios are 3.33 with C_b =1.0 and 3.79 with C_b =1.14. It is also noted that the cross-section axial strength is ϕP_{ns} = ϕP_y = 3825 kN (860 kip).

5. Benchmark Problems and Results

All of the below problems were created and analysed using imperial units (kips and inches). Results presented below were obtained by converting moments and deflections from imperial units to SI units.

[1] d = 467 mm (18.2 in), b_f = 193 mm (7.59 in), t_w = 11.4 mm (0.45 in), t_f = 19.1 mm (0.75 in), A = 12300 mm^4 (19.1 in^2), I_x = 445x10^6 mm^4 (1070 in^4), I_y = 22.8x10^6 mm^4 (54.8 in^4), J = 1140x10^3 mm^4 (2.73 in^4), and C_w = 1140x10^9 mm^6 (4240 in^6)

5.1 Beam subject to bi-axial bending, but no axial force

Most structural engineers and design codes associate second-order effects with additional moments produced by eccentricities from laterally displacing gravity loads in frames (P-Δ sway moments) and axial loads in members (P-δ member moments). The primary purpose of this example is to provide numerical results for a situation in which no axial force exists but there are still significant differences between results obtained ensuring equilibrium on the deformed shape versus undeformed shape; in other words, second-order effects can still be significant even in the absence of axial force.

Using the member properties and support conditions defined above and with reference to Fig. 1, applied loading includes only the equal and opposite end moments of M_X = 323.7 kN-m (2865 kip-in) and M_Y = 17.17 kN-m (152 kip-in). The axial force and uniformly distributed loads shown in Fig. 1 will be taken as $P_u = 0$ and $\omega_{ux} = \omega_{uy} = 0$, and no initial imperfections, such as member out-of-straightness, are included.

Given that there is no axial force, most engineers and design codes would assume that a first-order elastic analysis is adequate. For those referencing the AISC Specification, the reduction factor of 0.8 associated with the direct analysis method for frame design would not be applied. With this, the beam would be shown to deflect vertically in-plane and horizontally out-of-plane, but no twist would be computed along its length. With $\theta = 0$, and per Eq. (1), the resulting mid-span moments would be equal to the applied end moments. It is further noted that the specific values defined for the applied end moments were purposely selected such that the beam would be assessed as adequate in meeting the requirements of AISC's interaction equation Eq. (2). For reference, the major- and minor-axis demands equal 85% and 15% of the member's available major- and minor-axis moment capacities respectively; $M_{ux'}/\phi M_{nx}$ = 0.85 and $M_{uy'}/\phi M_{ny}$ = 0.15. Given that the interaction equation results in a value of unity for an applied load ratio ALR = 1.0, the member proportion could even be defined as optimal. A summary of these results is provided in the second column of Table 1.

Table 1: Results for beam subject to bi-axial bending and no axial force

Effects at mid-span	1st-order elastic analysis	Rigorous 2nd-order elastic analysis $J \neq 0, C_w = 0, E$	$J \neq 0, C_w \neq 0, E$	$J \neq 0, C_w \neq 0, 0.8E$
ALR	1.0	1.0	1.0	1.0
$M_{ux'}$ kN-m (kip-in)	323.7 (2865)	293.7 (2599)	321.3 (2843)	304.2 (2692)
$M_{uy'}$ kN-m (kip-in)	17.2 (152)	137.3 (1215)	43.8 (388)	112 (992)
δ_y mm (in)	16.9 (0.665)	53.2 (2.09)	19.5 (0.768)	50.1 (1.97)
δ_x mm (in)	17.5 (0.689)	107 (4.20)	38.4 (1.51)	111 (4.39)
θ rad (deg)	0.00	0.384 (22.0)	0.0828 (4.74)	0.300 (17.2)
AISC Int. Eq.	1.00	1.97	1.23	1.78
ALR, Int. Eq ≈ 1	1.00	0.80	0.88	0.81
$M_{ux'}$ kN-m (kip-in)	323.7 (2865)	256.4 (2269)	283. 7 (2511)	260.0 (2301)
$M_{uy'}$ kN-m (kip-in)	17.2 (152)	39.0 (345)	29.5 (261)	36.7 (325)
δ_y mm (in)	16.9 (0.665)	16.3 (0.641)	16.0 (0.628)	20.0 (0.787)
δ_x mm (in)	17.5 (0.689)	33.8 (1.33)	26.7 (1.05)	40.1 (1.58)
θ rad (deg)	0.00	0.0978 (5.60)	0.0505 (2.89)	0.0874 (5.01)
AISC Int. Eq. (2)	1.00	1.01	1.00	1.00

A rigorous second-order analysis that accounts for twist and thereby assesses equilibrium on the deformed shape provides significantly different results. The third column of Table 1 provides results for an analysis that models only uniform or St. Venant torsion. Finally, the fourth and the fifth columns of Table 1 provide results for analyses that model both St. Venant and warping torsion, with the fifth column also including the 0.8 reduction on stiffness required by AISC's *Design by Advanced Elastic Analysis* method. Although a summary of the observations that can be made from these results will be provided at the conclusion of this paper, it is noted that second-order inelastic analyses, which employ either 4080 shell elements or 9120 brick elements with both explicitly modelling yielding accentuated by the presence of residual stresses, shows that the member would fail at an applied load ratio of approximately ALR = 0.78 or 78% of the applied end moments. Note that these latter analyses included a reduction factor of 0.9 on both the material stiffness and yield strength ($0.9E$, $0.9F_y$), per the provisions of AISC's *Design by Advanced Inelastic Analysis* method.

5.2 Beam-column with in-plane distributed load

Using the same geometry, member properties, and support conditions defined above, the applied loadings shown in Fig. 1 include a range of combinations of in-plane uniformly distributed load ω_{uy} and axial force P_u. In all cases, the out-of-plane distributed load and applied end moments are taken as $\omega_{ux} = 0$, $M_X = 0$, and $M_Y = 0$. For each combination of loading, three sets of results are provided[2] in Table 2. All results are based on assuming that AISC's direct analysis method is being applied, and thus, include a stiffness reduction factor of 0.8. Included in these results are values obtained for AISC's governing interaction equation for assessing members subject to flexure and axial force, which is

$$\frac{P_u}{\phi P_n} + \frac{8}{9}\left(\frac{M_{ux'}}{\phi M_{nx}} + \frac{M_{uy'}}{\phi M_{ny}}\right) \leq 1.0 \quad \text{when } \frac{P_u}{\phi P_n} \geq 0.2$$
$$\frac{P_u}{2\phi P_n} + \left(\frac{M_{ux'}}{\phi M_{nx}} + \frac{M_{uy'}}{\phi M_{ny}}\right) \leq 1.0 \quad \text{when } \frac{P_u}{\phi P_n} < 0.2 \qquad (5)$$

For this specific member and loading, AISC's available major- and minor-axis moment strengths would be taken as ϕM_{nx} = 434.2 kN-m (3843 kip-in) with C_b =1.14 for a moment gradient produced by a uniformly distributed load, and ϕM_{ny} = 114.4 kN-m (1013 kip-in). The available axial strength would be a function of the design method being employed.

In rows labelled as (a), the member is assumed as nominally straight and a second-order elastic analysis would result in only in-plane behaviour; there would not be any twist, out-of-plane deflection, or minor-axis bending moments computed. In applying AISC's direct analysis method provisions, Eq. (5) would be assessed using an available compressive strength that is based on an effective length equalling the unbraced length of the member, ϕP_n = 952 kN (214 kip).

In accordance with AISC's *Design by Advanced Elastic Analysis* method, additional results are provided in rows (b) and (c). In this case, a more rigorous elastic analysis procedure is employed that ensures equilibrium and compatibility are satisfied on the deformed shape of the member, and thereby includes second-order effects that can be attributed to both P-δ and twist effects. This design method further requires an initial out-of-plane member imperfection, which in this study is modelled by distorting the original geometry of the member in the shape of a sine curve with an amplitude of $L/1000$ at mid-span. The analyses used to produce the results in rows (b) and (c) differ only with regard to the method used for modelling torsional resistance. The analytical results given in row (b) includes only the torsional resistance provided by the St. Venant stiffness GJ, whereas results in row (c) come from an analysis that also includes the warping resistance of the section EC_w produced by cross-flange bending along the length of the member. In applying this design method, Eq. (5) would be assessed using an available compressive strength that is based on the member's cross-section strength, ϕP_n = 952 kN (214 kip).

[2] It should be noted that AISC has adopted significant portions of Table 2 for inclusion in their Commentary on the 2016 *Specification for Structural Steel Buildings* [3].

Table 2: Results for beam-column subject to in-plane distributed load

Load Combination #		1	2	3	4
P_u kN (kip)		0	333.6 (75.0)	556.0 (125.0)	778.4 (175.0)
ω_{uy} kN/m (kip/ft)		58.4 (4.0)	43.8 (3.0)	29.2 (2.0)	14.6 (1.0)
	(a)	271.2 (2400)	207.1 (1833)	139.8 (1237)	70.8 (626)
M_{ux}, kN-m (kip-in)	(b)	269.6 (2386)	206.3 (1826)	139.5 (1235)	70.5 (624)
	(c)	271.1 (2399)	207.0 (1832)	139.8 (1237)	70.8 (626)
	(a)	0	0	0	0
M_{uy}, kN-m (kip-in)	(b)	29.1 (258)	26.4 (234)	21.7 (192)	34.9 (309)
	(c)	6.3 (56)	11.7 (104)	15.8 (140)	32.1 (284)
	(a)	14.7 (0.580)	11.2 (0.443)	7.58 (0.299)	3.84 (0.151)
δ_y mm (in)	(b)	17.6 (0.694)	13.3 (0.524)	8.69 (0.342)	5.11 (0.201)
	(c)	15.0 (0.589)	11.7 (0.460)	8.09 (0.318)	4.71 (0.186)
	(a)	0	0	0	0
δ_x mm (in)	(b)	24.6 (0.967)	24.2 (0.951)	21.2 (0.833)	35.5 (1.40)
	(c)	5.44 (0.214)	11.0 (0.435)	15.6 (0.616)	32.8 (1.29)
	(a)	0	0	0	0
θ rad (deg)	(b)	0.108 (6.19)	0.0790 (4.53)	0.0471 (2.70)	0.0358 (2.05)
	(c)	0.0233 (1.33)	0.0290 (1.66)	0.0266 (1.52)	0.0260 (1.49)
	(a)	0.62	0.77	0.87	0.96
AISC Int. Eq. (5)	(b)	0.87	0.75	0.58	0.62
	(c)	0.68	0.62	0.53	0.60
	(a)	1.61	1.29	1.15	1.04
ALR, Int. Eq ≈ 1	(b)	1.03	1.07	1.13	1.09
	(c)	1.28	1.26	1.25	1.11
ALR_{ult} GMNIA		1.23	1.15	1.16	1.10

Note: (a) traditional GNA, $\delta_{ox} = 0$; (b) rigorous GNIA, $J \neq 0$, $C_w = 0$; (c) rigorous GNIA, $J \neq 0$, $C_w \neq 0$

5.3 Beam-column with in-plane and out-of-plane distributed loads

This benchmark problem is nearly the same as the previous problem, except that (i) an out-of-plane uniformly distributed load ω_{ux} is applied along with the in-plane distributed load ω_{uy} and axial force P_u, and (ii) the initial member out-of-straightness imperfection is never included. Although AISC's *Design by Advanced Elastic Analysis* would require modelling of a member imperfection, this provision has been purposely disregarded in order to provide a benchmark problem with a geometry that is relatively easy to create. In Table 3, results are provided with rows (a), (b), and (c) corresponding to the analysis assumptions presented in the previous section.

6. Conclusions

Using a laterally unbraced I-shaped member subject to various combinations of in-plane and out-of-plane loading effects as well as applied axial force, results intended for benchmarking and verifying nonlinear frame analysis software are provided. In addition to providing numerical results from commercial programs that vary by the degree to which torsional resistance is modelled, several general observations can be made, including

1. For situations in which the major-axis flexure capacity of an I-shaped member is closely approached, only a small amount of twist need occur before such major-axis demands are resolved into minor axis demands;
2. Second-order or geometrical nonlinear analysis is required in order to assess such demands, even for cases in which there is no axial force.

3. The beneficial effects of employing nonlinear analysis programs that account for both uniform and nonuniform (warping) torsional resistance result in reductions in deflections, twist, and minor-axis bending moment demands.

These benchmark problems are meant to be valuable in verifying the proper use of advanced methods of nonlinear analysis, with a variety of software packages, for situations in which the accurate modelling of spatial behaviour is important. The problems also highlight important effects associated with twist that are relevant as the design profession moves towards employing three-dimensional nonlinear analysis.

Table 3: Results for beam-column subject to in-plane and out-of-plane distributed loads

Load Combination #		1	2	3	4
P_u kN (kip)		0	222.4 (50.0)	444.8 (100.0)	667.2 (150.0)
ω_{uy} kN/m (kip/ft)		58.4 (4.0)	43.8 (3.0)	29.2 (2.0)	14.6 (1.0)
ω_{ux} kN/m (kip/ft)		5.84 (0.4)	4.38 (0.3)	2.92 (0.2)	1.46 (0.1)
	(a)	271.2 (2400)	205.9 (1822)	138.9 (1229)	70.3 (622)
$M_{ux'}$ kN-m (kip-in)	(b)	248.3 (2198)	199.6 (1767)	137.4 (1216)	70.1 (620)
	(c)	267.0 (2363)	203.8 (1804)	138.2 (1223)	70.2 (621)
	(a)	27.1 (240)	26.6 (235)	25.4 (225)	22.1 (196)
$M_{uy'}$ kN-m (kip-in)	(b)	111.6 (988)	66.4 (588)	41.8 (370)	26.8 (237)
	(c)	54.2 (480)	44.1 (390)	34.9 (309)	25.5 (226)
	(a)	14.7 (0.580)	11.2 (0.440)	7.54 (0.297)	3.81 (0.150)
δ_y mm (in)	(b)	42.4 (1.67)	19.6 (0.773)	10.0 (0.394)	4.34 (0.171)
	(c)	19.2 (0.755)	13.7 (0.539)	8.76 (0.345)	4.19 (0.165)
	(a)	28.7 (1.13)	27.9 (1.10)	26.7 (1.05)	23.1 (0.909)
δ_x mm (in)	(b)	97.0 (3.82)	62.7 (2.47)	41.7 (1.64)	27.4 (1.08)
	(c)	52.1 (2.05)	43.7 (1.72)	35.6 (1.40)	26.4 (1.04)
	(a)	0	0	0	0
θ rad (deg)	(b)	0.323 (18.5)	0.158 (9.05)	0.0700 (4.01)	0.0232 (1.33)
	(c)	0.101 (5.79)	0.0685 (3.92)	0.0404 (2.31)	0.0165 (0.945)
	(a)	0.86	0.86	0.95	1.02
AISC Int. Eq. (5)	(b)	1.55	1.07	0.74	0.48
	(c)	1.09	0.88	0.68	0.47
	(a)	1.16	1.15	1.05	0.99
ALR, Int. Eq $\approx$ 1	(b)	0.84	0.97	1.12	1.12
	(c)	0.92	1.05	1.18	1.22
ALR_{ult} GMNIA		0.93	1.00	1.09	1.18

Note: (a) traditional GNA, $\delta_{ox} = 0$; (b) rigorous GNIA, $J \neq 0$, $C_w = 0$; (c) rigorous GNIA, $J \neq 0$, $C_w \neq 0$

References

[1] AISC, *Specification for Structural Steel Buildings*, ANSI/AISC 360-10, American Institute of Steel construction, Chicago, Il, 2010.

[2] AISC, *Specification for Structural Steel Buildings*, ANSI/AISC 360-05, American Institute of Steel construction, Chicago, Il, 2005.

[3] AISC, *Specification for Structural Steel Buildings*, ANSI/AISC 360-16, American Institute of Steel construction, Chicago, Il, 2016.

[4] SAP2000 (v. 17.2), Computers and Structures, Inc., USA, www.csiamerica.com

[5] Strand7 (rel. 2.4.6), Strand7 Pty Ltd, Australia, www.strand7.com

[6] ADINA (v. 9.0.1), ADINA R&D, Inc., USA, www.adina.com

[7] ANSYS Mechanical APDL (rel. 16.0), USA, www.anysis.com

[8] ABAQUS/CAE (v. 6.14), Dassault Systèmes Americas Corp., USA, www.abaqus.com

Connections

STRENGTH OF COLD-FORMED STEEL FRAME BASE CONNECTIONS

Mohamed A. El Aghoury[a], Essam A. Amoush[b] and Amr M. El Hady[b]

[a] *Department of Structural Engineering, Ain Shams University, Abassia, Egypt*
[b] *Department of Structural Engineering, Higher Technological Institute, Tenth of Ramadan City, Egypt*

Abstract: The behaviour and strength fixed base connections of cold-formed steel frames have been investigated in the present study. A new form of base assemblage to suite fixed, partially fixed and hinged base is presented. The proposed base configuration is tested using numerical simulation. A finite element model is used to investigate different parameters such as; double back-to-back cold-formed lipped channels, welded built-up I-shape stub with triangular stiffener plates and base plate thicknesses. The column base behaviour is presented along with different modes of failure; while base stiffness is predicted via moment-rotation curves of the column base. Eventually, comparison with Euro Code (EC3) is carried out.

1. Introduction

Cold-formed steel structural members play an important role in modern steel structures due to their high strength and light weight. Steel frames made of cold-formed members assembled with bolted connections are widely used recently. K.F. Chung and L. Lau [1], presented an experimental investigation on the structural performance of cold formed steel members with bolted moment connections. A total of sixteen specimens, two different lipped channel sections back-to-back with interconnections were used. The moment resistance of bolted moment connections with four bolts per member ranged from 42% to 84% from the moment capacities of the connected members. Thus, it was demonstrated that the moment connections among cold formed steel members were structurally feasible and economical through rational design. Lixin Li et al. [2], carried out an experimental study for beam-column galvanized cold-formed steel lipped channel with different lengths and different slenderness ratios. The experimental results of axial load-bending moment interaction diagrams for the specimens were constructed and compared with current North American Design Code (AISI, 2002), and there was a good agreement. Bayan Anwer Ali et al. [3] presented an experimental investigation of cold-formed steel frames. The beam and column members were formed from single lipped channel column and rafter connected back-to-back with others. A total of 10 frames were tested. They found that, the stiffness and performance of the column base connections had a significant effect on the structural frame behaviour. M. Dundu [4], presented an investigation of the behaviour of base connections, that formed from cold-formed channels and hot-rolled angle cleats. A total of 32 column bases were tested. He concluded that, the failure of the base occurred in angle cleats when a cold-formed angle cleat was used. However, the failure occurred in the column channels when hot-rolled angle was used. A simple formula was suggested to design these types of bases.

In the present paper, a new form of base assemblage configuration to suite fixed, partially fixed and hinged base of the steel frame cross section is presented. A verified finite element model is developed using a software program (ANSYS 12.0) [5]. Different results and failure modes are noticed. The finite element results are compared with Euro Code (EC3) [6].

2. Parametric Study Variables

This study is concerned with a parametric study to investigate the behaviour of the fixed base connection, which consists of back-to-back cold-formed lipped channels, welded built-up I-shape stub with triangular plate stiffeners and base plate as shown in "Fig. 1". The stiffener plate width-to-height ratio is equal to 1/5 of I-shape stub height ($H_{conf.}$). The double cold-formed channels are connected with the proposed welded built up I-shape stub via transfer bolts, while the welded built-up I-shape stub is welded with the base plate. Moreover, the base plate is connected to the concrete base via anchor bolts (A) as presented in "Fig. 1". Different parameters such as; cold-formed column sections with different width-to-depth ratios B/D= (0.4, 0.8 and 1.2), width-to-thickness ratio of their flanges, b/t, and depth-thickness ratios of their webs, D/t, are chosen, as listed in "Table 1". Moreover, the I-shape stub is chosen to have different flanges width-to-thickness ratio (b_I/t_I) and webs depth-thickness ratio (D_I/t_I). In addition, the base plate thickness and the arrangement of the transfer web bolts also are listed in "Table 1". Two patterns of the transfer web bolts are considered. The first pattern is type (H) and the second one is type (T) as shown in "Fig. 1". The adopted naming convention for specimens is "C4-80AI6-40" where: the letter C indicates that the shape of cold-formed section is channel, the second digit "4" indicates that the thickness of cold-formed channel is 4 mm, the number "80" indicates that the nominal width of the channel is 80 mm, the letter "A" indicates that the anchorage type is anchor bolts, the letter and number "I6" indicates that the welded built-up I-shape stub thickness is 6 mm and the number "40" indicates that the cold-formed column section width-to-depth ratio is 40 %.

3. Finite element model

The aforementioned parametric study is conducted using the FEA package "ANSYS 12.0" [6]. Thin-shell element with four nodes and six degrees of freedom at each node, SHELL181, is used to simulate the cold-formed column, steel plates and concrete base layers. BEAM4 is used to simulate the bolt shank as spring element. The spring element has two nodes and six degrees of freedom at each node is used to represent the anchor bolts embedded concrete. LINK10 and BEAM44 elements with two nodes have three and six degrees of freedom at each node; respectively. These elements with full bending stiffness are used to simulate the head and nut of the bolts. The model of bolts was calibrated by Jakab [7], to provide the stiffness of the connection. CONTA178, element with two nodes and three degrees of freedom at each node, is used to model the sliding between the shell elements.

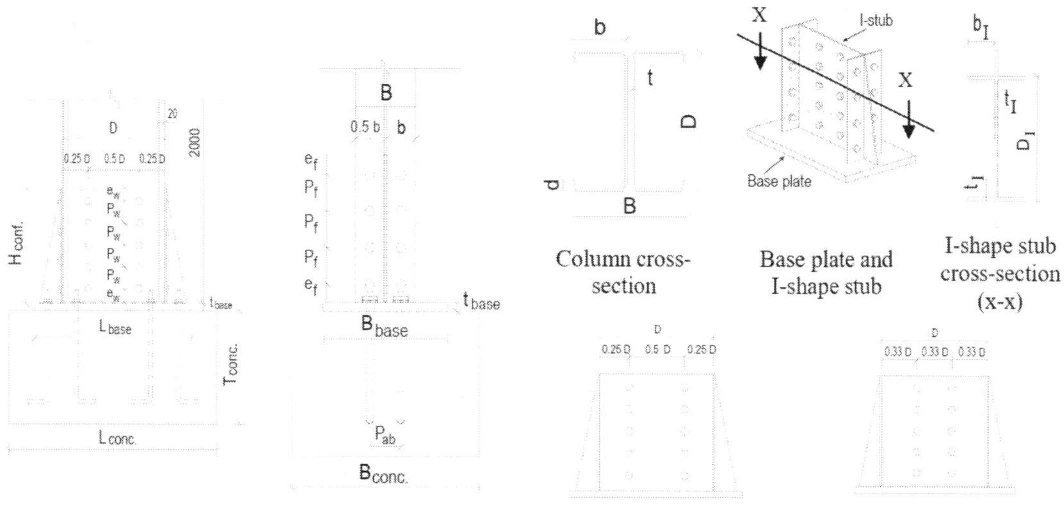

Type (H) Type (T)

Base views Arrangement of web transfer bolts for I-shape stub

Fig. 1: Proposed column base details

SOLID185 element with eight nodes and three degrees of freedom at each node is used for load bearing plate modelling. Both large deflection and elasto-plastic material behaviour have been incorporated in the non-linear finite element model. The mesh density is chosen to keep the element aspect ratio approximately equal to one. Young's modulus, E and the yield stress, F_y, of the steel material are considered as 210000 MPa and 360 MPa; respectively. The yield stress, Fy, of the transfer bolts (M16) is considered as 640 MPa with bolt grade (8.8). Young's modulus of elasticity, E, and cubic strength F_{cu}, of the concrete material are considered as 22000 MPa, 25 MPa; respectively. The shear modulus is taken as 81000 MPa. A bilinear stress-strain curve is adopted for steel material modelling while linear stress-strain curve for the concrete material as shown in "Fig. 2-a". The end condition of the column base is treated as fixed-free condition. The column loaded end is prevented from both rotations about X so as Y axes and translation in Z direction. On the other hand, the unloaded end is prevented from translations and rotations in the three directions X, Y and Z, as shown in "Fig. 2-b". The model is loaded with two proportional loads as shown in "Fig. 2-b". The loads consisted of horizontal load and vertical load act through the C.G. of the column cross-section. At the mid-height of the column, both rotations about X & Y axes and translation in Z direction are prevented. The load is incrementally increased through successive load steps. Newton-Raphson iteration technique is used in solving the nonlinear equations.

Table 1: Cold-formed column, I-shape stub and base plate dimensions

Model	Column Dimensions						t_I (mm)	t_{base} (mm)	W_{tb} type
	D (mm)	b (mm)	B (mm)	B/D	b/t	D/t			
C2-80AI4-40	400	80	165	0.4	40	200	4	24	H & T
C4-80AI6-40					20	100	6		
C2-100AI4-80	250	100	205	0.8	50	125	4	22	
C4-100AI6-80					25	62.5	6		
C2-100AI4-120	170	100	205	1.2	50	85	4	20	
C4-100AI6-120					25	42.5	6		

* t_I = Thickness of welded built-up I-shape stub
* t_{base} = Thickness of base plate
* W_{tb} = Web transfer bolts arrangements

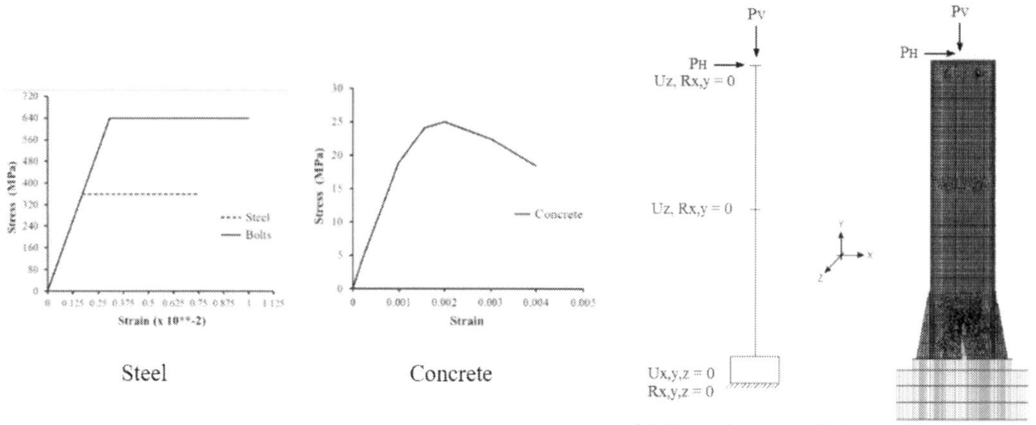

Steel Concrete

a) Material stress-strain relationship

b) Boundary condition and load for the model

Fig. 2: Finite element model

4. Results

4.1 Moment-Rotation relationships

The results of the FE models are listed in "Table 2". The behaviour of the fixed base connection has linear moment-rotation relationship also the models with higher width-depth ratio of the column section, *B/D* = 1.20 the moment-rotation relationship is approximately linear. However, the nonlinear moment-rotation relationship is observed for models with lower width-depth ratio of the column section, *B/D* = 0.40, as shown in "Fig. 3".

Table 2: Finite element results for models with width-depth ratio of *B/D* = 40%, 80% and 120%

Model name	Type of web transfer bolts	Moment resistance M_u (kN.m)	Rotation capacity θ (radian)	Initial stiffness S_{init} (kN.m/rad)	Failure Mode
C2-80AI4-40	Type (H)	95.42	0.000658	172676.0	LYBs
C4-80AI6-40		221.66	0.001527	183802.7	BFs & LYB_I
C2-80AI6-40	Type (T)	93.54	0.000598	189172.7	LYBs
C4-80AI4-40		175.21	0.001301	186525.0	LYB_I
C2-100AI6-80	Type (H)	45.93	0.000936	64214.4	BFs
C4-100AI4-80		64.94	0.001410	64311.3	LYB_I
C2-100AI4-80	Type (T)	54.24	0.001150	66337.1	BFs
C4-100AI6-80		80.70	0.001574	73185.9	LYB_I
C2-100AI4-120	Type (H)	27.96	0.001387	25022.9	BFs & BF_I
C4-100AI6-120		37.20	0.001665	26849.2	LY_I
C2-100AI6-120	Type (T)	25.32	0.001083	27925.9	LYBs
C4-100AI4-120		31.54	0.001342	27296.5	BF_I

where:
- * BFs: Bearing failure in cold-formed channels section around bolt hole
- * LYBs: Local yielding and Bearing failure in cold-formed channels section member
- * LY_I: Local yielding in welded built-up I-shape stub
- * BF_I: Bearing failure in welded built-up I-shape stub around bolt hole
- * LYB_I: Local yielding and bearing failure in welded built-up I-shape stub

For models with B/D = 1.2 and b/t = 50, the initial stiffness, S_{init}, is ranged between 25022.9 and 33225.0 kNm/rad. Also for models with b/t = 25, the initial stiffness is approximately equal as listed in "Table 2". The stiffness of the base connection is inversely proportional to the I-shape stub depth-thickness ratio (D_I/t_I). For models with B/D equals 0.4, the stiffness of the connection is higher than the others. Moreover, the stiffness of the base connections is inversely proportional to width-depth ratio of the cold-formed section B/D as shown in "Fig. 3". Eventually, the arrangements of web transfer bolts types (H and T) approximately have the same effect on the stiffness of the base connection for different models, as shown in "Fig. 3".

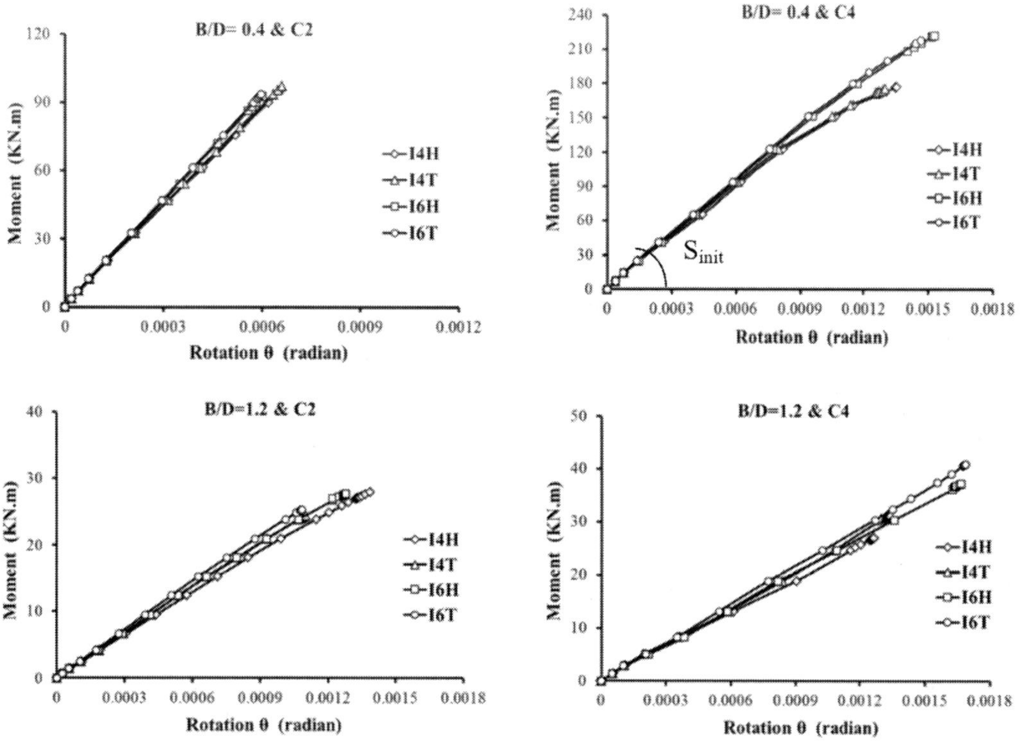

Fig. 3: Moment-rotation relationships for models having B/D = 0.4 and 1.2

4.2 Load-Displacement relationships

The horizontal load-lateral displacement relationships for the studied models are plotted as shown in "Fig. 4". The lateral displacement is measured at the loaded end point. For models with B/D = 1.2, the ductility of the connection is the highest one. Moreover, it is clear that the ductility of the base connection is directly proportional to the width-depth ratio, B/D, and depth-thickness ratio, D/t, of the cold-formed section as shown in "Fig. 4". However, for models with lower width-depth ratio of column section B/D = 0.4, the ductility of the base connection for lower depth-thickness ratio is higher than the other one as shown in "Fig. 4". For models having transfer web bolts types (H and T) with width-depth ratios of column section B/D equal 0.4 and 0.8 approximately have the same ductility. However, for models having B/D = 1.20 and transfer web bolts type (H), the ductility is inversely proportional to the cold-formed thickness as shown in "Fig. 4".

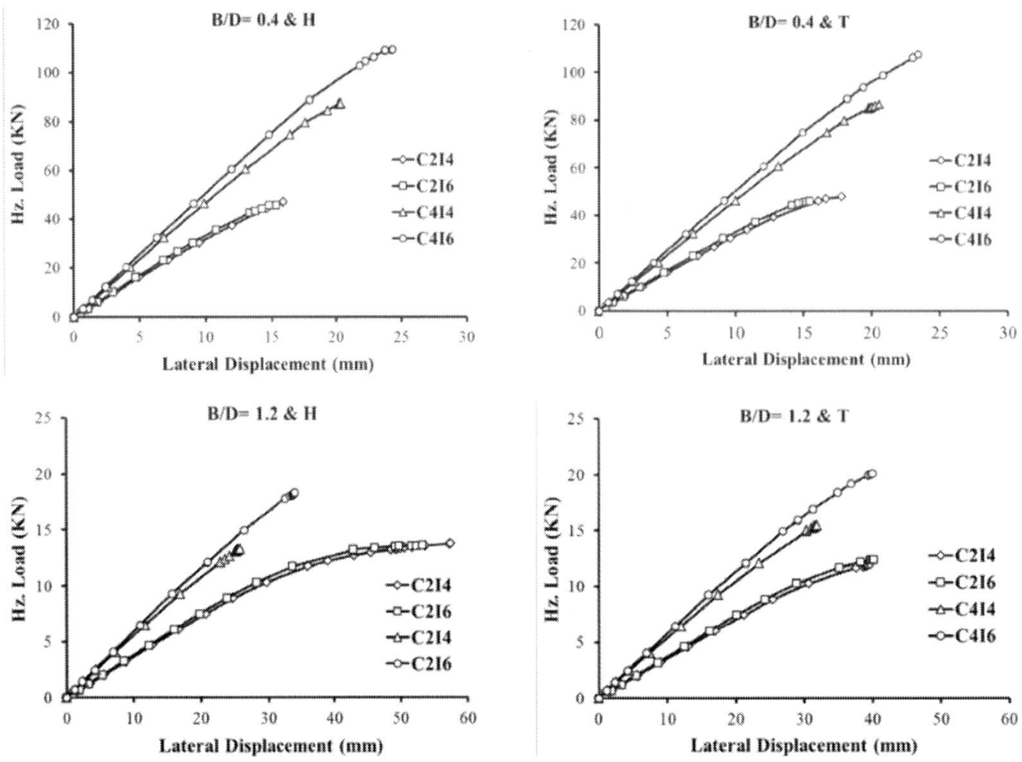

Fig. 4: Load-Displacement relationships for models having B/D = 0.4 and 1.2

4.3 Stress distribution and Failure modes

In cold-formed base connections, there are five basic failure modes, namely; the local yielding (LY), bearing failure (BF) around bolt holes, a couple of local yielding with bearing failure (LYB) in cold-formed sections or/and in the I-shape stub, tension failure (TF) in the tension side due to bending in the base plate and shear failure (SF) in anchor bolts. The stress distributions (Von Mises yield criterion) and failure mode shapes for various models are plotted and analysed, as shown in "Fig. 5". For models with $B/D = 0.8$ and slender cold-formed cross-section, the failure modes are local yielding and bearing failure in cold-formed column section, together with bearing failure in I-shape stub. However, for models with non-compact cold-formed sections the failure modes are local yielding and bearing failure in the I-shape stub.

5. Comparison between Finite Element results and Eurocode

Joint behaviour has a significant effect on the response of structural steel frames and must be included in global analysis and design. The types of joint modelling, with respect to their stiffness and resistance, are classified into rigid, semi-rigid and pinned joints. The stiffness of the column base is defined in terms of stiffness of the connected column. According to the Euro code formula, the limiting stiffness for column bases are 12 EI_c/L_c and 30 EI_c/L_c for braced and un-braced frames, respectively. "Fig. 6" shows comparison, between Euro Code (EC3) [6] and the finite element results, the cold-formed base connections are considered as semi-rigid base connections. The base plate shape and dimensions along with the pattern and number of anchor bolts play an important role in the behaviour of the base.

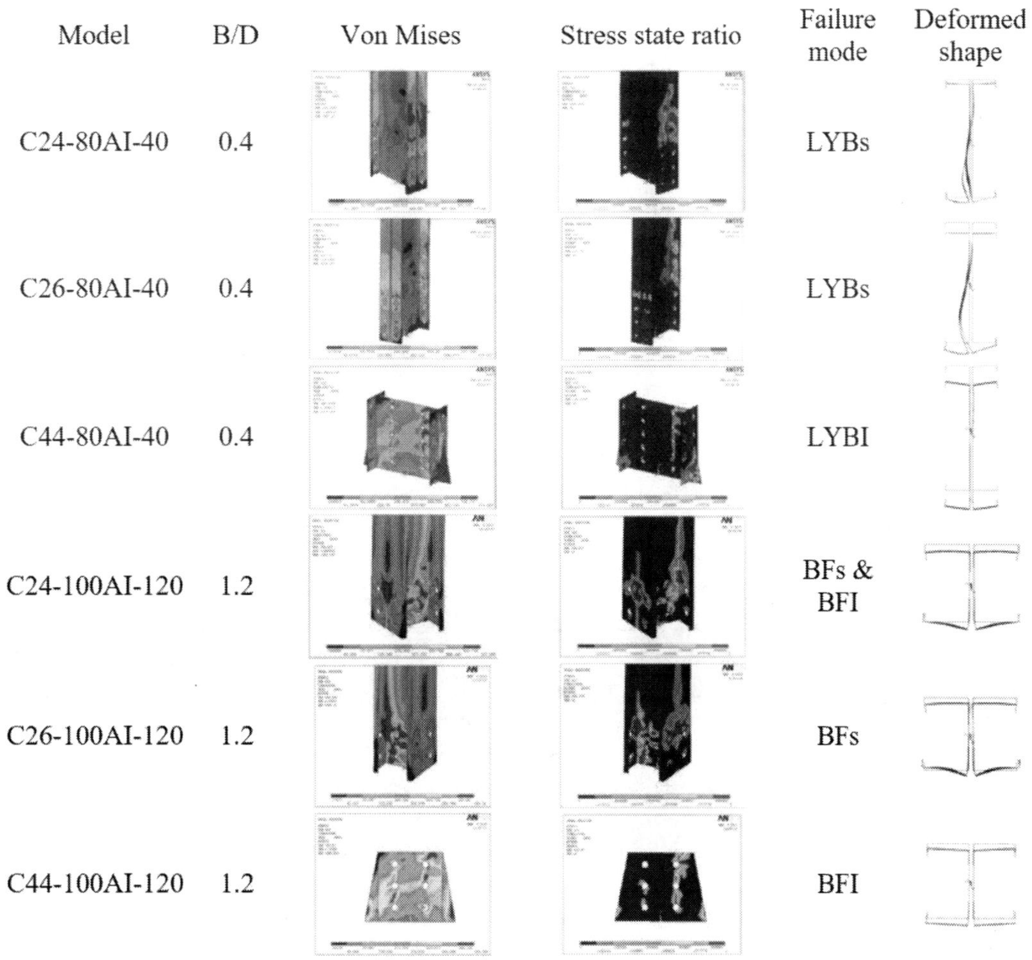

Fig. 5: Stress state ratio and failure mode shapes

Therefore, different base plate thicknesses are used to investigate the behaviour of the cold-formed base connection. "Fig. 7" shows the moment-rotation relationship for different models and different base plate thickness. It is clear that, the models with a base plate of thickness up to 16 mm are classified as semi-rigid base connections as shown in "Fig. 7". The rigidity of base connection is directly proportional to the base plate thickness.

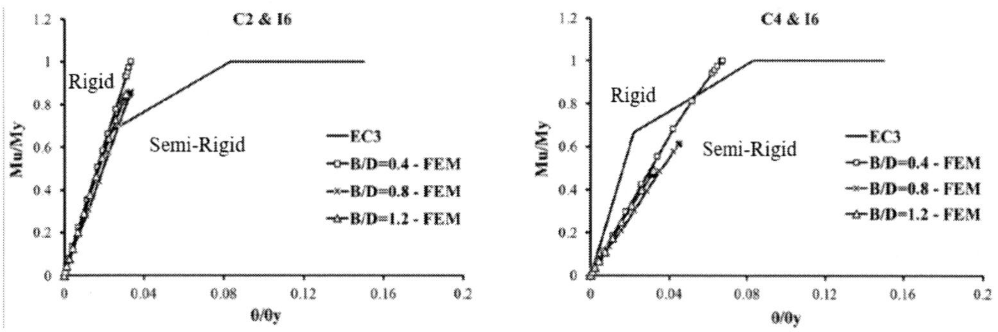

Fig. 6: (M_u/M_y) - (θ/θ_y) relationship for models having B/D= 0.4, 0.8 and 1.2

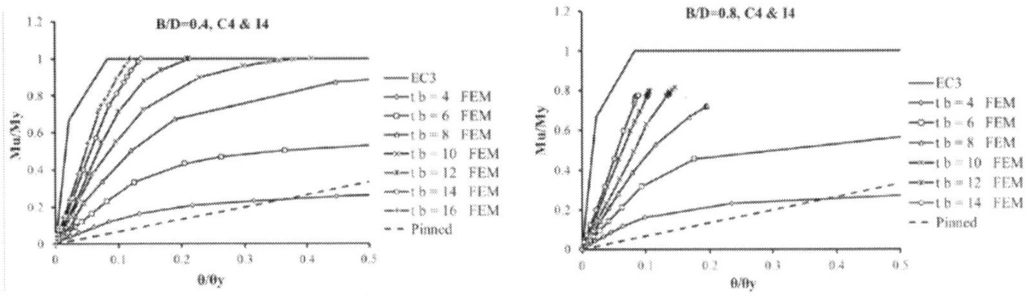

Fig. 7: (M_u/M_y) - (θ/θ_y) relationship for models having B/D= 0.4, 0.8 with different base plate thickness

6. Conclusions

A numerical investigation of the strength and behaviour of cold-formed steel frame base connection is conducted using a finite element software program (ANSYS 12.0). It is noticed that, the stiffness of the base connection is inversely proportional with width-depth ratio of cold-formed column section B/D as well as the depth-thickness ratio of I-shape stub (D_I/t_I). Also, the ductility of the base connection is directly proportional with width-depth ratio of column section B/D and the depth-thickness ratio of column section D/t. However, the ductility of base connection is inversely proportional to depth-thickness ratio of cold-formed column section D/t for models having B/D = 0.4. The effect of transfer web bolts arrangement type (H) or type (T) for models having width-depth ratios equal 0.4, 0.8 and 1.2 is approximately similar regarding the stiffness of the base connection. The base plate shape and dimensions along with the pattern and number of anchor bolts play an important role in the behaviour of the base. In the present parametric study to avoid the base plate failure, the thickness of the base plate in the models having B/D = 0.4, 0.8 and 1.2 should be greater than 16 mm, 14 mm and 12 mm; respectively. A comparison between FEM results and EC3 [6], illustrate that the cold-formed base connections are classified as semi-rigid base connections.

References

[1] Chung KF, Lau L. "Experimental investigation on bolted moment connections among cold formed steel members", *Engineering Structures*, 21, 898–911, 1999.

[2] Li L, Sivakumaran KS. "Beam-Column Behaviour of Cold-Formed Steel Lipped Channel Sections", *4th Structural Specialty Conference of the Canadian Society for Civil Engineering*, June 5-8, 1-10, 2002.

[3] Bayan AA, Sariffuddin S, Mohd HO. "Cold-Formed Steel Frame with Bolted Moment Connections", *International Journal of Civil And Structural Engineering*, 1(3), 534-544, 2010.

[4] Dundu M. "Base connections of single cold-formed steel portal frames", *Journal of Constructional Steel Research*, 78, 38–44, 2012.

[5] ANSYS User Manual Revision 12.0, ANSYS®, Inc., Canonburg, Pennsylvania, USA, 2009.

[6] Eurocode 3 (EC3), ENV 1993-1-1, Design of Steel Structures - *General rules and roles for buildings.*

[7] Jakab G. "Analysis and design of cold-formed C-section members and structures", *PhD Dissertation,* Budapest University of Technology and Economics, 2009.

ANALYTICAL PREDICTION OF THE BEARING RESISTANCE OF BOLTED CONNECTIONS FOR TUBULAR RACKING STRUCTURES

Vittorio Armenante[b], Marina D'Antimo[a], Jean-Françoise Demonceau[a], Jean-Pierre Jaspart[a], Massimo Latour[b] and Gianvittorio Rizzano[b]

[a] *University of Liège, ArGenCo Department, Belgium*
[b] *University of Salerno, Department of Civil Engineering, Italy*

Abstract: The aim of this paper is to provide a formulation able to predict the resistance of simple shear connections realized by fastening a plate to a hollow square section by means of long bolts. The research is motivated by the fact that the rules currently provided by EC3 for defining the bearing resistance of steel connections are limited to the case of stacks of plates, where every plate composing the connection is always restrained out of plane by the adjoining plate or by the bolt head or nut. This is a condition not verified in case of connections realized with long bolts fastening hollow sections. The work is organized in two sections. First, the results of an experimental campaign are presented and successively, the results of a parametric analysis carried out by means of a Finite Element model calibrated on the experimental data are reported. Afterwards, an analytical approach able to include the local buckling effects arising in these kinds of connections is presented, proposing a coefficient for the reduction of the resistance, to be included in the EC3 formulation.

1. Introduction

In shelving structures realized by means of hollow section members, a particular system of connection, that has been proposed only recently, consists in bending and drilling the end of the tube composing the beam in order to fasten it to the squared hollow columns with long bolts. Nevertheless, even though such a kind of connection is already used in constructions, it is not among the cases regulated by the Eurocode 3 Part 1.8 [1]. In fact, the rules currently provided by EC3 for defining the bearing resistance of steel connections are limited to the case of stacks of plates, where every plate composing the connection is always constrained out of plane by the adjoining plate or by the bolt head or nut (Fig. 1). This is a condition not verified in case of connections realized with long bolts fastening hollow sections [2]. In fact, in such connections, the tubes' faces are constrained out of plane by another plate only in the outward direction, but they are free to buckle inward under the localized action applied by the bolt in shear. These joints have already been subject of a preliminary study of the same authors [3], in which it has been pointed out that their behaviour in shear is characterized by a particular failure mode, which is not accounted for in the current formulation provided by EC3. In fact, due to the adoption of tubes with small thickness of the plates and high strength steel, these connections may be characterized by a failure occurring due to the arisement of local buckling phenomena (Fig. 1). Within this framework, the main goal of this paper is to investigate the behaviour of this joint typology, with the aim of providing an analytical formula able to predict the shear resistance with an approach in line with the current approach codified by EC3.

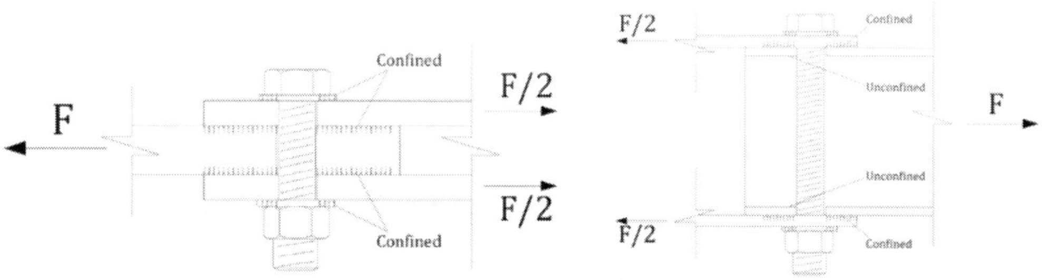

Fig. 1: Layout of the tested specimens

Therefore, in order to provide an approach able to extend the formulation proposed by Eurocode 3 to connections made with tubular elements and long bolts, a model for determining the value of the critical load leading to the local buckling of the plate composing the steel tube is presented. Such a load is evaluated by studying the behaviour of a sub-structure, idealized by extracting from the whole connection only the end portion of the tube's plate subjected to a compression load, which is used to simulate the localized bolt action. Furthermore, in order to introduce in the model the constraining action provided by the remaining part of the tube on the considered sub-structure, particular boundary conditions are applied. Such a sub-model is used to evaluate the critical elastic buckling load by performing buckling analyses in SAP 2000. Afterwards, the value of the critical elastic load obtained from the FE analysis is corrected by adopting Winter's approach [4] in order to account for the fact that the local buckling occurs in plastic field and to account for the effects related to the geometric imperfections. Exploiting the results of the parametric analysis, a new coefficient to be included in the current EC3 formulation to account for local buckling phenomena arising in case of tubular members fastened with long bolts is proposed. The new coefficient is expressed as a function of the ratio between tube's thickness and hole diameter.

2. Evaluation of the buckling load

The procedure applied in this work to assess the value of the critical load leading to the local buckling of the tube is carried out starting from the analysis of a sub-structure extracted from the whole connection representative of the zone of the tube where the local buckling effects are more significant. In particular, the sub-structure is idealized by extracting from the tube's plate only the end portion, whose dimensions are the width of the tube (*H*) and the distance between the center of the hole and the free edge of the tube (e_1) (Fig. 2). This is the zone that, as also demonstrated by the experimental analysis, is more significantly influenced by the local buckling phenomena.

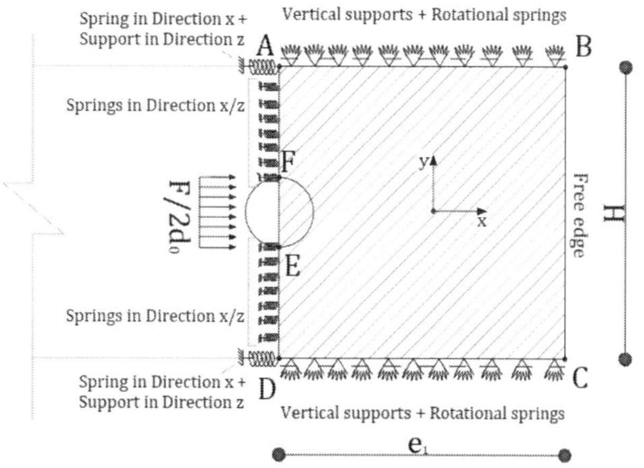

Fig. 2: Boundary conditions applied

Dealing with the modelling of buckling phenomena, it is well known that the definition of the restraining conditions applied at the edges of the plate is of paramount importance, due to the influence that the constraints have on the value of the critical buckling load. This is the reason why, even though in the present work only a small portion of the tube is studied, the continuity of the analysed plate with the remaining parts of the tube is recovered by considering appropriate boundary conditions applied on the edges of the element (Fig. 2).

In particular, in the following, the considerations made to define the stiffness of appropriate springs able to account for the rotational, shear and axial behaviour of the remaining parts of the tube's plates are described. In particular, in order to account for the remaining portions of the tube the following constraints are applied at the edges of the analysed plate (Fig. 3):

- at the intersection of the edges of the plate with the lateral walls of the tube rotational springs and vertical supports are applied (edges A-B and D-C);
- on the edge loaded under compression by the bolt, in-plane and out-of-plane springs are introduced both on the portions of the edge next to the hole (edges A-F and D-E) and in the corners of the plate (points A and D). Furthermore, in the zone of the hole, a uniform load modelling the shear action transferred by the bolt is introduced;
- at the free edge of the tube no boundary conditions are introduced (edge B-C).

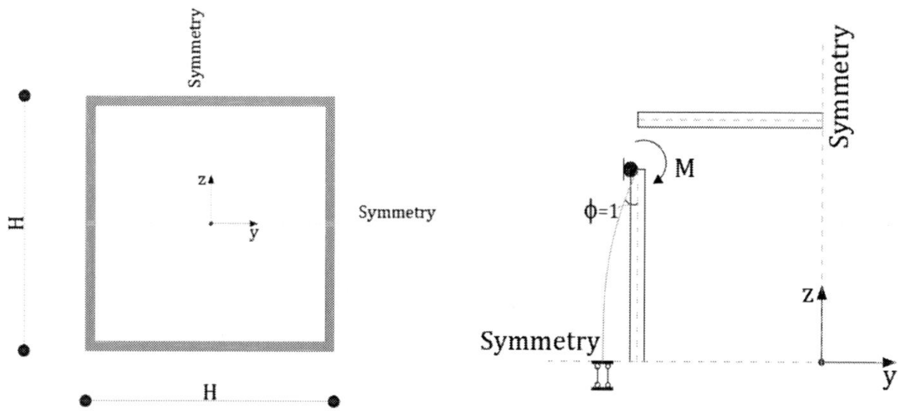

Fig. 3: Scheme used to define the rotational stiffness due to the lateral plates of the tube

Therefore, on the sides of the plate parallel to the load (from A to B and from D to C, Fig. 3), out-of-plane supports and rotational springs are applied in order to model the axial and rotational stiffness of the lateral plates of the tube. In particular, in order to model the axial stiffness of the lateral plates of the tube, on these edges, the out-of-plane supports are defined as rigid elements. Conversely, the stiffness of the rotational springs is determined by defining the value of the bending moment needed to generate a unitary rotation on the lateral plates of the tube (Fig. 3). By adopting this approach, the following expression is obtained:

$$k_{\phi,lateral} = \frac{2EI}{H} \quad (1)$$

where E is the modulus of elasticity of steel, I is the moment of inertia of the lateral tube's plate per unit of length and H is the height of the tube's plate.

The side that goes from B to C is considered free; a similar assumption is made for the portion of the edge on which the load is applied that goes from E to F. In addition, in order to account for the stiffness of the remaining part of the tube behind the analysed plate, various considerations are made. In particular, at points A and D of the model (the corners of the plate) extensional springs in the x direction accounting for the stiffness of the portion of the lateral plate behind the bolt are applied. The stiffness of such springs is easily calculated as:

$$k_{x,corners} = \frac{E(A_{lp}/2)}{L} \quad (2)$$

where L is the length of the tube and A_{lp} is the area of the lateral wall of tube calculated as the product of the tube thickness t and $H/2$. In the same way, in the zone going from A to F and from E to D extensional springs are spread along the edges both in the x and in the y direction. As already made for the stiffness $k_{x,corners}$, the value of the stiffness of the springs spread on the loaded edge is determined in order to account for the axial deformability of the remaining part of the top plate located behind the bolt:

$$k_{x,edge} = \frac{E(A_{tp}/2)}{L} \quad (3)$$

where A_{tp} is the area of the top plate of the tube calculated as the product of the tube thickness t and H. Conversely, the stiffness of the springs in the y direction is determined in order to account for the out-of-plane deformability of the portion of the tube behind the bolt hole. To this scope, the simple scheme of plate fully restrained on three edges loaded with a unitary force applied on the free edge represented in Fig.4 has been studied. Therefore, in order to determine the value of the stiffness of the springs to be applied on the holed side of the plate several schemes with the force applied in different points of the free edge have been solved.

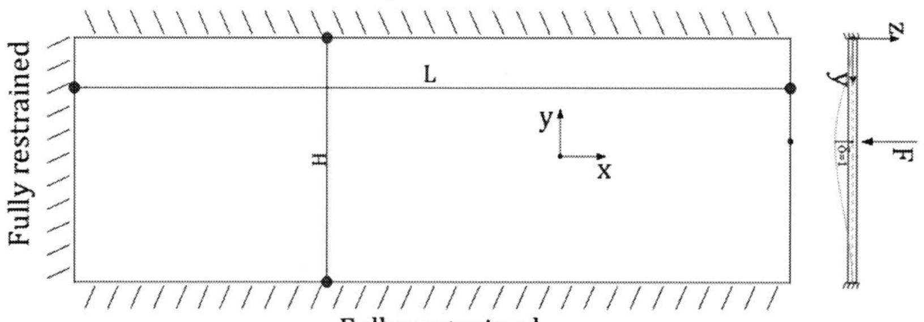

Fig. 4: Scheme used to evaluate the out-of-plane stiffness

In order to maintain a simple structure of the model and to avoid to introduce a variable value of the springs' stiffness, the out-of-plane stiffness of the springs has been assumed constant and equal to the value of the force needed to obtain a unitary displacement at the mid-point of the plate's edge. Conversely, at the corners of the plate, as already said, a fixed support has been applied. This choice is justified by the fact that, from some preliminary analyses, no appreciable differences were found in the results considering a variable value of the springs' stiffness or a constant value. Under the reported hypotheses, the value of the elastic buckling load of the plate with the above described boundary conditions has been determined by developing an appropriate FE model in SAP2000 v.16 software package (Fig.5).

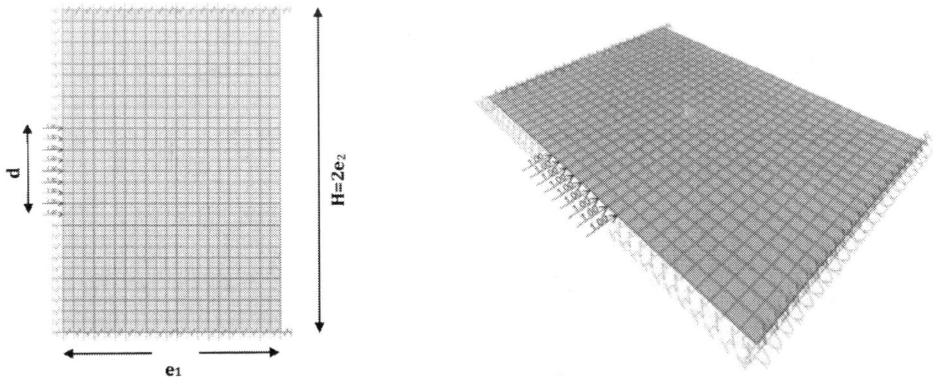

Fig. 5: SAP2000 model of the plate

An example of the results of one of the analyses presented in the next section is delivered in Fig.6, where the shapes of the first two buckling modes of one on the analysed cases are represented. As it is possible to verify, the 1st buckling shape, i.e. the buckling shape characterized by the lowest value of the load multiplier, is the one typically observed also in the experimental analysis.

e_1=48 mm, H=60 mm, t = 0.50 mm, d=16 mm

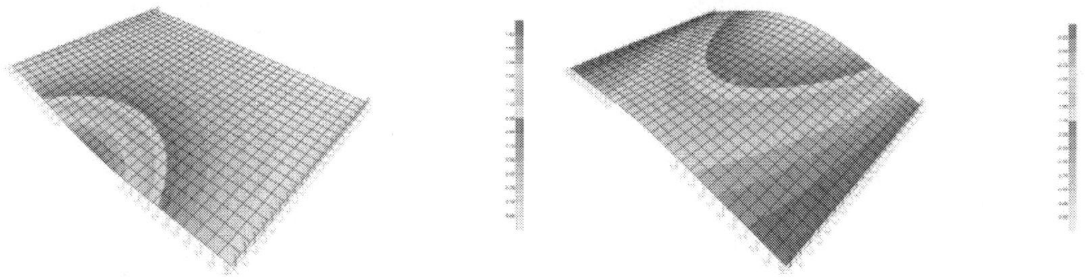

1st mode – α_{cr} =105.3 2nd mode – α_{cr} =306.8

Fig. 6: Buckling modes for a plate with thickness equal to 0.50 mm

In order to obtain a value of the critical buckling load representative of the actual conditions of the plate it is necessary to account for two additional parameters that normally provide a significant influence on such a value: the imperfections and the non-linear behaviour of steel. In fact, both these factors, as widely recognized in technical literature, may provide a significant reduction to the value of the buckling resistance calculated by means of an elastic analysis and, therefore, they need to be accounted for in order to obtain a formulation able to predict the ultimate buckling load of the analysed connection. In this work, in order to account for these factors, the empirical approach of Winter [4], already proposed by EC3 [1] and by the AISI specifications [5], is followed in order to account for the influence of imperfections and of the non-linear behaviour of the material. Such an approach is based on the application of the effective width concept and provides to correct the value of the plastic load that the plate can carry by means of the following expression:

$$N_{cr} = N_{pl}\beta = N_{pl}\left[\left(1-\frac{0,22}{\overline{\lambda_p}}\right)\frac{1}{\overline{\lambda_p}}\right] \leq N_{pl} \tag{4}$$

where $\overline{\lambda_p}$ is a non-dimensional slenderness parameter expressed as:

$$\overline{\lambda_p} = \left(\frac{N_{pl}}{F_{cr}}\right)^{1/2} \tag{5}$$

where F_{cr} is the value of the elastic buckling load, evaluated in this work by means of the buckling analyses described previously, and N_{pl} is the value of the plastic load. It is worth noting that, according to the Winter's methodology, the two coefficients $\frac{1}{\overline{\lambda_p}}$ and $\left(1-\frac{0,22}{\overline{\lambda_p}}\right)$ reported in Eq. 4 account for the influence of the mechanical non-linearity of the material and for the imperfections respectively [4,6] The value of the plastic load N_{pl} is determined according to the formulation currently proposed by EC3 part 1.8 under the hypothesis of full plasticization of the plate:

$$F_{b,Rd} = 2.5 f_u t d \tag{6}$$

The accuracy of the model herein describer has been checked against some of the experimental and FE results currently available and described in another work of the same authors (D'Antimo et al., 2016). In Fig. 7 the values of the maximum loads calculated in the experimental and numerical activity developed in D'Antimo et al. (2016) are compared with those calculated by means of the proposed model and with those calculated by means of the EC3 approach. In particular, in the picture, with a black dot are reported the values of the maximum load calculated in D'Antimo et al. (2016), with a continuous line are represented the values of the bearing resistance of the plate calculated according to the EC3 model, with a dashed and dotted line are represented the values of the bolt resistance calculated according to EC3 model and with a simple dashed line are represented the values of the buckling load calculated with the model proposed in this paper. From such a figure it is easy to observe that in the buckling range the proposed model is able to predict the load values more accurately than EC3. Conversely, for higher values of the *t/d* ratio other failure modes become more significant and, therefore, EC3 provides a higher accuracy in the prediction.

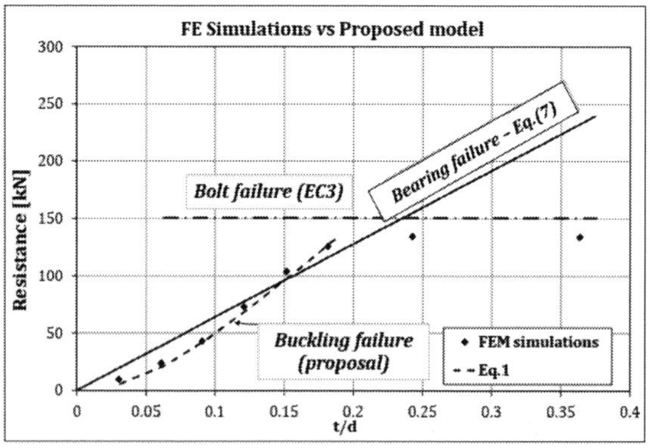

Fig. 7: Mathematical model vs. FE simulation

In addition, it is also possible to observe that the resistance of the connection is accurately predicted in the whole range of analysed values if the resistance is calculated as minimum of the resistance of the three possible failure mechanisms. Therefore, starting from this observation, in the next section a proposal of improvement of the EC3 approach including the local buckling failure mode of the plate by means of the approach proposed in this paper, is described.

3. Proposal of an improvement to EC3 approach

According to Eurocode 3, the bearing resistance of a shear connection is provided by the following relationships:

$$F_{b,Rd} = k\alpha_b f_u td \tag{7}$$

$$\alpha_b = min\left(\frac{e_1}{3d_0};\frac{f_{ub}}{f_u};1\right) \tag{8}$$

where, f_u and f_{ub} are respectively the plate and bolt ultimate strength, d is the bolt diameter, e_1 is the distance of the hole from the plate free edge and k is a parameter accounting for the influence of the lateral distance on the tear-out failure mode. Typically, such a factor, for a flat plate, depends on the distance between the hole and the lateral edge e_2 and is equal to 2.5 provided that e_2 is greater than $1{,}5d_0$. In case of tubes, as already demonstrated in D'Antimo et al. (2016), in order to apply the EC3 formulation, the k factor has to be directly assumed equal to 2.5. In fact, in case of tubular sections, the presence of the side walls allows to neglect the influence of the reduction of resistance arising for small values of the distance of the hole from the free edge ($e_2<1.5d_0$). Eqs. 7-8, account for all the basic failure modes normally arising in a simple bolted connection and, in particular, include the tear-out, bearing and shear failure mechanisms. Basing on the procedure for determining the critical buckling load of plates of tubular members reported in previous sections, the aim of this section is to provide an additional coefficient to be included in Eqs.7-8 to account for the possibility of local buckling failures mode (Fig.1). In particular, with the aim to keep the same structure of the code equation, the idea is to propose to include in coefficient α_b of Eq. 8 an additional factor to be properly calibrated by means of a parametric analysis. To this scope, equating the ratio between the ultimate bearing resistance provided by Eq. 5 and that calculated by means of Eq. 1 to one a new factor $\alpha_{b,st}$ to be added in the classical bearing resistance formula can be defined:

$$\frac{F_{b,Rd}}{N_{cr,Rd}} = \frac{2.5\alpha_{b,st}dtf_u}{2.5\beta dtf_u} = 1 \tag{9}$$

$$\alpha_{b,st} = \beta \tag{10}$$

where $\beta = \left(1-\frac{0,22}{\lambda_p}\right)\frac{1}{\lambda_p}$ is the buckling factor evaluated according to Winter's formula.

Therefore, following this approach, in order to find a simple formula to be used in design, the main issue is to define a correlation between β and the main geometrical parameters of the analysed connection. According to what is reported before, the factors that may influence significantly the load in case of buckling failure mode are t/d, e_1/d and e_2/d ratios. Therefore, the parametric study proposed in this section has been mainly devoted to investigate the influence of these three parameters on the value of β.

Table 3: Results of the parametric analysis

t/d	e_1/d	e_2/d	d [mm]	t [mm]	e_1 [mm]	e_2 [mm]	$N_{crit,FEM}$ [kN]	f_y/f_u [MPa]	N_{pl} [kN]	λ_p [Eq.2]	β	$N_{cr,rd}$ [kN]
1st group of analyses – influence of t/d												
0.05	3	2.5	**16**	0.8	48	40	3.51	443/499	15.97	2.134	0.420	6.711
0.075	3	2.5	**16**	1.2	48	40	11.84	443/499	23.95	1.422	0.594	14.235
0.1	3	2.5	**16**	1.6	48	40	28.08	443/499	31.94	1.067	0.744	23.767
0.125	3	2.5	**16**	2	48	40	54.84	443/499	39.92	0.853	0.870	34.724
0.15	3	2.5	**16**	2.4	48	40	94.77	443/499	47.90	0.711	0.971	46.529
0.05	3	2.5	**12**	0.6	36	30	1.86	443/499	8.98	2.195	0.410	3.682
0.075	3	2.5	**12**	0.9	36	30	6.30	443/499	13.47	1.462	0.581	7.829
0.1	3	2.5	**12**	1.2	36	30	14.95	443/499	17.96	1.096	0.729	13.099
0.125	3	2.5	**12**	1.5	36	30	29.20	443/499	22.46	0.877	0.854	19.182
0.15	3	2.5	**12**	1.8	36	30	50.46	443/499	26.95	0.731	0.956	25.773
2nd group of analyses – influence of e_1/d												
0.1	**3**	2.5	16	1.6	48	40	26.976	443/499	31.94	1.088	0.733	23.417
0.1	**3.5**	2.5	16	1.6	56	40	27.365	443/499	31.94	1.080	0.737	23.542
0.1	**4**	2.5	16	1.6	64	40	27.558	443/499	31.94	1.077	0.739	23.604
0.1	**4.5**	2.5	16	1.6	72	40	27.684	443/499	31.94	1.074	0.740	23.644
0.2	**3**	2.5	16	3.2	48	40	212.934	443/499	63.87	0.673	1.000	63.882
0.2	**3.5**	2.5	16	3.2	56	40	218.957	443/499	63.87	0.673	1.000	63.882
0.2	**4**	2.5	16	3.2	64	40	220.512	443/499	63.87	0.673	1.000	63.882
0.2	**4.5**	2.5	16	3.2	72	40	221.516	443/499	63.87	0.673	1.000	63.882
0.3	**3**	2.5	16	4.8	48	40	718.672	443/499	95.81	0.673	1.000	95.823
0.3	**3.5**	2.5	16	4.8	56	40	739.028	443/499	95.81	0.673	1.000	95.823
0.3	**4**	2.5	16	4.8	64	40	744.221	443/499	95.81	0.673	1.000	95.823
0.3	**4.5**	2.5	16	4.8	72	40	747.638	443/499	95.81	0.673	1.000	95.823
3rd group of analyses – influence of e_2/d												
0.031	3	**1.9**	16	0.5	48	30	0.948	443/499	9.98	3.245	0.287	2.867
0.063	3	**1.9**	16	1	48	30	7.59	443/499	19.96	1.622	0.533	10.638
0.094	3	**1.9**	16	1.5	48	30	25.59	443/499	29.94	1.082	0.736	22.051
0.125	3	**1.9**	16	2	48	30	60.69	443/499	39.92	0.811	0.899	35.870
0.16	3	**1.9**	16	2.5	48	30	118.53	443/499	49.90	0.673	1.000	49.908
0.0313	3	**2.5**	16	0.5	48	40	0.698	443/499	10.00	3.784	0.249	2.488
0.05	3	**2.5**	16	0.8	48	40	3.51	443/499	15.97	2.134	0.420	6.711
0.075	3	**2.5**	16	1.2	48	40	11.84	443/499	23.95	1.422	0.594	14.235
0.1	3	**2.5**	16	1.6	48	40	28.08	443/499	31.94	1.067	0.744	23.767
0.125	3	**2.5**	16	2	48	40	54.84	443/499	39.92	0.853	0.870	34.724
0.15	3	**2.5**	16	2.4	48	40	94.77	443/499	47.90	0.711	0.971	46.529
0.0313	3	**3.5**	16	0.5	48	56	0.51	443/499	10.00	4.420	0.215	2.149
0.05	3	**3.5**	16	0.8	48	56	2.402	443/499	15.97	2.578	0.355	5.665
0.075	3	**3.5**	16	1.2	48	56	7.115	443/499	23.95	1.835	0.480	11.489
0.1	3	**3.5**	16	1.6	48	56	16.774	443/499	31.94	1.380	0.609	19.455
0.125	3	**3.5**	16	2	48	56	32.77	443/499	39.92	1.104	0.725	28.961
0.15	3	**3.5**	16	2.4	48	56	56.62	443/499	47.90	0.920	0.827	39.624
0.05	3	**4.5**	16	0.8	48	72	1.651	443/499	15.97	3.110	0.299	4.771
0.075	3	**4.5**	16	1.2	48	72	5.908	443/499	23.95	2.013	0.442	10.596
0.1	3	**4.5**	16	1.6	48	72	13.209	443/499	31.94	1.555	0.552	17.633
0.125	3	**4.5**	16	2	48	72	25.820	443/499	39.92	1.243	0.662	26.425
0.15	3	**4.5**	16	2.4	48	72	44.617	443/499	47.90	1.036	0.760	36.416
0.2	3	**4.5**	16	3.2	48	72	105.766	443/499	63.87	0.777	0.923	58.923
0.225	3	**4.5**	16	3.6	48	72	150.584	443/499	71.86	0.691	0.987	70.892
0.25	3	**4.5**	16	4	50	72	206.493	443/499	79.84	0.673	1.000	79.853

The analyses have been performed by carrying out three groups of simulations. A ***first group***, devoted to investigate the influence of the t/d ratio. In this case the analyses have been carried out by fixing the values of e_1/d equal to 3 and of e_2/d equal to 2.5 and by varying the bolt diameter d and plate thickness t singularly. In particular, in this first group of analyses the values of d have been assumed equal to 12 and 16 mm and the t/d ratio has been varied in the range 0.05/0.15. The ***second group*** of analyses has been carried out in order to investigate the influence of the e_1/d ratio. In this case the analyses have been carried out by fixing the value of e_2/d equal to 2.5 and by varying t/d in between 0.1 and 0.3 and e_1/d in between 3 and 5. The ***third group*** of analyses has been carried out with the aim to investigate the dependence of the buckling resistance on the parameter e_2/d. In this case e_2/d has been varied in between 1.88 and 3.5, t/d has been varied in the range 0.03 and 0.15 and e_1/d has been fixed as equal to 3. For each of the analysed cases, the procedure previously reported to determine the value of the inelastic buckling load has been applied and, subsequently, the value of the critical elastic buckling load and the value of the reduction factor calculated by means of the Winter's approach have been calculated. The results of these analyses are summarized in Table 3.

4. Conclusions

In this work the behaviour of connections made with SHS and long bolts has been investigated. The analysis performed in this paper has been devoted to develop a design formula able to account for the particular buckling failure mode arising in these connections. First of all, in order to define this formula, a procedure able to predict the critical load of the tube's plate under the action applied by the bolt has been set up. This procedure has been based on the study of the portion of the plate, which is mainly influenced by the buckling phenomena by defining, in order to recover the restraining action due to the remaining parts of the tube, appropriate boundary conditions. The buckling load of the studied plate has been evaluated by first performing a buckling analysis in SAP2000 and by after correcting the value of the critical load by means of the Winter's approach in order to account for imperfections and for the plastic behaviour of steel. This procedure has been validated against the available data retrieved from the tests and analysis performed in [3]. Afterwards a parametric analysis has been carried out in order to find a correlation between the reduction of resistance due to the buckling failure mode and the geometrical properties of the tube's plate. The analysis has shown that the local buckling effects, in the analysed connection, may be significant depending on the values of t/d and of e_2/d demonstrating that the buckling check is required if t/d is lower than $0.018e_2/d+0.124$. Using the results of the parametric analysis a coefficient able to account for the bucking failure mode to be included in EC3 has been defined.

References

[1] CEN (2005). *Eurocode 3: Design of steel structures - Part 1-8: Design of joints.* European Committee for Standardization, CEN, Brussels.

[2] Hoang VL, Jaspart JP, Demonceau JF. "Extended end-plate to concrete- filled rectangular column joint using long bolts", *Journal of Constructional Steel Research,* 113, 2015.

[3] D'Antimo M. "Experimental and theoretical analysis of shear bolted connections for tubular racking structures", under Review on *Engineering Structures.*

[4] Winter G. "Strength of thin steel compression flanges", Reprint No. 32, Ithaca, NY, USA: Cornell University Engineering Experimental Station.

[5] AISI (1996). *Specification for the design of cold formed steel structural members*. New York: American Iron and Steel Institute.

[6] Von Karman T. "Festigheitsprobleme im Maschinenbau". *Encyclopaedie der Mathematischen Wissenschaften*.

THE EFFECT OF M/V RATIO ON LEDGER-FRAMING PERFORMANCE OF COLD-FORMED STEEL BUILDINGS

Deniz Ayhan and Benjamin W. Schafer

Department of Civil Engineering, Johns Hopkins University, Baltimore, MD, USA

Abstract: The objective of this paper is to present test results of joist-to-ledger connection to examine the effect of applied moment/shear ratio on their behavior. In recent full-scale shake table tests, the cold-formed steel building where ledger framing was a key feature, exceeded the predictions based on sub-system level design. The additional lateral strength and stiffness may have been provided by the stiffness of joist-to-ledger connections. Therefore six connection tests are conducted here. It is observed that their stiffness, strength and limit states are highly dependent on not only the connection details but also the applied loading type. These tests are a part of a broader experimental study on ledger framing construction.

1. Introduction

The seismic behavior of light-framed cold-formed steel (CFS) buildings was recently investigated with experimental characterization in different scales: fastener, member and full-scale in CFS-NEES project [1]. The ledger-framed CFS-NEES building has used ledger framing for wall-to-floor connections as shown in Fig. 1. Ledger framing consists of floor and roof joists which are connected to the interior flange of the load-bearing studs via a combination of a rim-track and clip angles. The ledger transfers floor joist loads to the vertical stud framing and therefore, the joists are not required to be aligned with the load bearing studs.

The designer assumes all lateral loads flow to and through the shear walls in a light-framed CFS building. However, full scale tests of the CFS-NEES building indicated that non-traditional load paths were utilized to transmit significant lateral loads. Also the necessity of new approaches were shown since the observed performance exceeded design expectations [2]. The contribution of other structural and even non-structural members; such as wall-to-floor connections, to the lateral load resisting system should be considered along with the shear walls that are particularly designed for lateral seismic demands.

a) Joist on stud b) Joist between studs

Fig. 1: Examples of joist-to-ledger framing connections used in CFS-NEES building [2]

This paper presents a set of sub-assemblage experiments on joist-to-ledger connection to quantify their stiffness and strength. Notably, the code guidance for connection design is mostly based on the shear capacity of fasteners and is not enough to understand the connection behavior. The testing program is designed to evaluate the loading position (applied moment-to-shear (M/V) ratio) with different joist locations relative to studs (mid studs, close and on stud) under monotonic loading. These tests are a part of the broader comprehensive experimental study on ledger framing construction to support future 3D seismic building analysis and design.

2. Testing Program for CFS wall-to-floor connections

The tests conducted on CFS joist-to-ledger connections aim to examine their stiffness and strength, with several configurations. The whole test matrix comprises of 30 specimens to examine different parameters: 1. presence of floor sheathing as an oriented strand board (OSB), 2. presence of top and bottom screws connecting joist and ledger flanges, 3. load location, and 4. clip angle location, depending on joist locations relative to studs (Fig. 2), depending on joist location relative to studs under monotonic and cyclic loadings. Four test series are defined for each parameters. Table 1 shows only the part of the test matrix explored in scope of this paper.

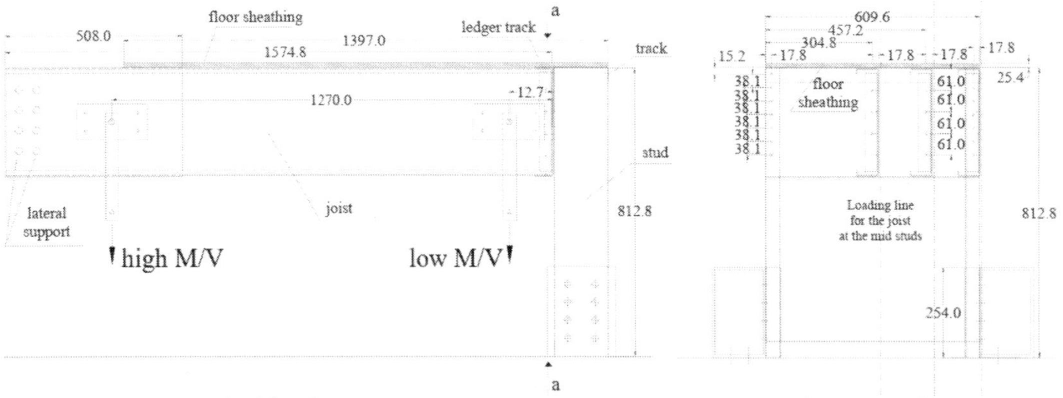

a) side view b) a-a section

Fig. 2: Specimen details including three joist and two loading locations (dimensions in mm)

Table 1: Test Matrix

	test no	joist location	loading location
first series	T4- mid studs (low M/V)	mid studs	connected end
	T5- near stud (low M/V)	near stud	
	T6- on stud (low M/V)	on stud	
third series	T19- mid studs (high M/V)	mid studs	free end
	T20- near stud (high M/V)	near stud	
	T21- on stud (high M/V)	on stud	

Monotonic tests of first and third series are presented here to evaluate effects of applied M/V ratio depending on the joist location. In the first series, the load location is placed at a distance of 12.7 cm (5 in.) from the connected end; and it is moved to a distance of 127 cm (50 in.) towards the free end in the third series to increase applied M/V ratio, see Fig.2. The load location distances are determined based on the potential inflection point location in a joist of the CFS-NEES building, i.e. the farthest location from the connection is where the plastic hinge would be for fully fixed connection and the closest to the connection is almost pin ends. So, essentially we bracket the potential connection response by testing at these two extrema.

Current specimens have floor sheathing (OSB), top and bottom screws to connect the joist and ledger flanges together and a clip angle on the inside face of the joist section to connect the joist web to the ledger web. These tests are conducted under monotonic loading but the same specimens will be tested under cyclic loading in the next phase of this experimental program. The joist-to-ledger connection sub-assemblage are designed according to CFS-NESS building details. Correspondingly, members are selected following the building structural system [2].

The instrumentation consists of well-arranged 12 position transducers (PTs) to measure the necessary deformations [3] to calculate the absolute rotations of connection ($\theta_{con} = \theta_J - \theta_S$), joist ($\theta_J$), stud frame ($\theta_S$) and ledger track ($\theta_L$), and the relative rotations of the joist to the ledger track (θ_{JL}) and clip angle leg (θ_{JA}) as seen in Fig. 3. The stud frame rotation (θ_S) is calculated with linear interpolation of stud rotations (θ_{S1} for stud away from joist and θ_{S2} stud close to joist) depending on joist location.

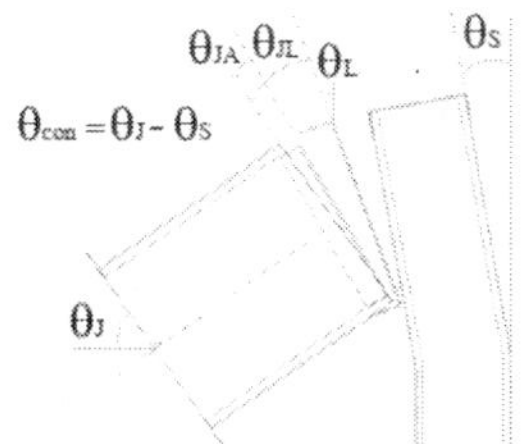

Fig. 3: Rotation components of the connection

3. Experimental Results

The failure modes of tests are depicted in the Fig. 4. The limit state of T4 having the joist at the middle of the ledger is the ledger bottom flange buckling (Fig. 4a). However, the stud web crippling is the failure mode for T5 and T6 when the joist is close to or aligned with the stud, respectively (Fig. 4b,c). The first series is useful to understand that limit states are highly sensitive to the joist location, even if there is no change in the load location (12.7 cm away from the connection). There is a lack of design guidance for these limit states in the current codes.

a) T4 - mid studs (low *M/V*)

d) T19 - mid studs (high *M/V*)

b) T5 - near stud (low *M/V*)

e) T20 - near stud (high *M/V*)

c) T6 - on stud (low *M/V*)

f) T21 - on stud (high *M/V*)

Fig. 4: Primary limit states observed in tests

In the third series, the geometrically identical specimens to the ones of the first series are tested by moving the load location at a distance of 127 cm to create high *M/V* ratio. The change in the load location did not affect the limit states of specimens where joist is placed at mid span of ledger (T19) and aligned with the vertical stud (T20), as seen in Fig. 4d,e. However, the failure of the specimen with the joist near the vertical stud (T21) became due to ledger bottom flange buckling (Fig. 4f).

The partial pull-out of the screws connecting the ledger web to the stud flange, is observed before stud web crippling for the specimens having the joist aligned with the stud (T6 and T21). For all of specimens, the screws connecting sheathing to the stud frame top track and the joist were tilted prior to the failure of specimen. The primary and secondary limit states observed in the tests presented Table 2.

Table 2: All limit states observed in the tests

test no	Ledger flange buckling	Stud web crippling	Fastener pull-out	Ledger section deformation	Fastener tilting	OSB delamination or pull out
T4- mid studs (low *M/V*)	P				S	
T5- near stud (low *M/V*)		P		S	S	
T6- on stud (low *M/V*)		P	S		S	S
T19- mid studs (high *M/V*)	P				S	
T20- near stud (high *M/V*)	P				S	
T21- on stud (high *M/V*)		P	S		S	

P: primary limit state, S: secondary limit state.

The moment-joist rotation behavior of the six tested specimens including stud frame rotation is depicted in Fig. 5. The initial stiffness of T19 is slightly higher than T4, but more remarkably, the high applied *M/V* ratio caused the bottom ledger to buckle early in T19 with a lower capacity. For T5 and T20, the limit state changed with the increase in the *M/V* ratio. This resulted in the increase of stiffness. However their moment capacities are still almost the same. The high applied M/V ratio provided a considerable increase in the strength and stiffness of T21 joist behavior compared to T6. It is clear that the change in the applied *M/V* ratio has different influences on the moment-joist rotation behavior of the specimens depending on the joist location.

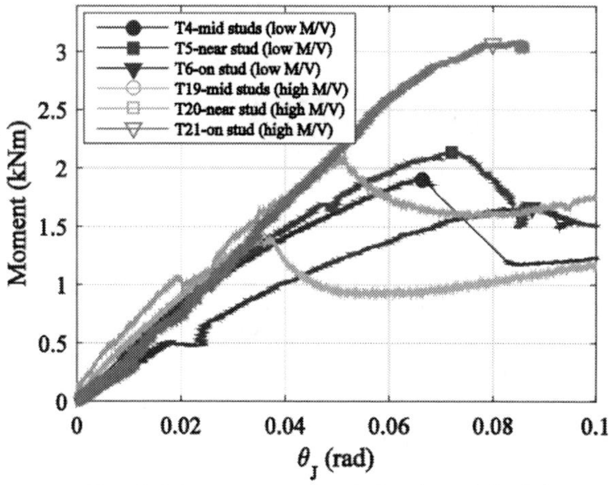

Fig. 5: Moment-rotation behavior of the joist

The connection rotation is the difference between the joist rotation and the rigid body rotation of the stud frame from joist rotation. So, the actual moment-rotation curves of the joist-to-ledger connections are obtained as shown in Fig. 6. The stud frame stiffness would be greater with upward continuity of the studs, however; the connection stiffness is independent than this. The initial stiffness of the connection is observed to be almost the same for all of the specimens except T21. It should be also noticed that for specimens with the high applied *M/V* ratio, the rigid body stiffness of the stud frame is higher than the others. The specimens with the joist at a quarter of the space between studs provided always the most rigid connection, whatever their failure mode is. Also, the moment-connection rotation curves of T5 and T20 have similar pre-failure shape. T21 having the joist on stud with the high applied *M/V* ratio provided the highest connection capacity but the lowest stiffness due to the plastic hinge mechanism of the failure.

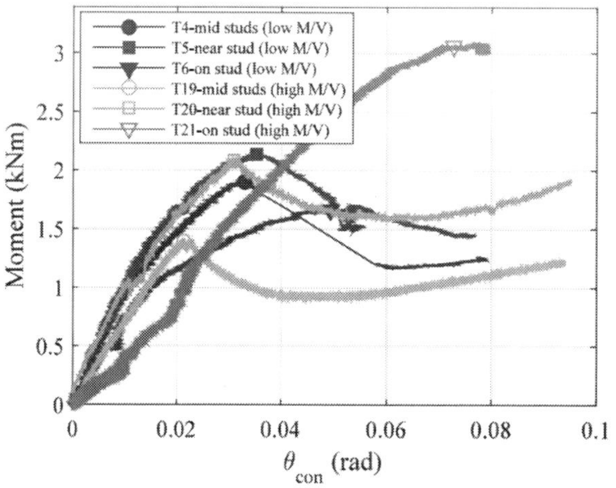

Fig. 6: Moment-rotation behavior of the connection

Table 3 summarizes the maximum applied moment (M_{act}), connection rotation (θ_{con}) corresponding to M_{act} and the connection stiffness (k_1 and k_2) observed up to M_{act}. k_1 and k_2 represent that the moment-rotation behavior of the joist-to-ledger connection up to the failure point has been characterized with a bilinear curve having an inflection at the 40% of M_{act}. Moreover, the ledger rotation (θ_L), rotations between joist and ledger (θ_{JL}), between joist and angle (θ_{JA}), and stud rotations (θ_{S1} and θ_{S2}) corresponding to M_{act} are provided in Table 3. The maximum connection moment and rotation capacity is observed in T21, having the joist aligned with the stud. However, the maximum stiffness of the connection is measured in the specimens T5 and T20 having the joist at a quarter of the space between studs. Even this stiffness is about 2% of a fixed connection stiffness (=5400kNm). It is worth noting that joist-to-connection rotations of the specimens are in a wide range varying between 0.021 and 0.070 radian prior to failure.

Table 3: Measured quantities for joist-to-ledger specimens

test no	M_{act} kNm	θ_{con} rad	k_1 kNm	k_2 kNm	θ_L rad	θ_{JL} rad	θ_{JA} rad	θ_{S1} rad	θ_{S2} rad
T4- mid studs (low *M/V*)	1.902	0.033	98.612	45.112	0.028	0.001	0.001	0.031	0.036
T5- near stud (low *M/V*)	2.134	0.035	103.422	47.163	0.032	0.002	0.000	0.034	0.038
T6- on stud (low *M/V*)	1.650	0.048	73.589	25.945	0.033	0.007	0.008	0.015	0.038
T19- mid studs (high *M/V*)	1.393	0.021	70.675	61.575	0.017	0.001	0.000	0.014	0.013
T20- near stud (high *M/V*)	2.079	0.031	104.580	54.139	0.024	0.002	0.001	0.019	0.021
T21- on stud (high M/V)	3.038	0.070	51.406	40.305	0.048	0.005	0.007	0.007	0.012

The moment-ledger rotation of the specimens is obtained by removing the rigid body rotation contribution of the stud frame, and shown in the Fig. 7. Similar to the connection behavior, all specimens provided the same initial stiffness. This phenomenon suggests that the local effects of the joist location and applied M/V ratio has no significant effect on the initial stiffness of the ledger since the floor sheathing provide the consistency between the stud frame and the attached ledger. The difference in shear capacities can be clearly seen in Fig. 8.

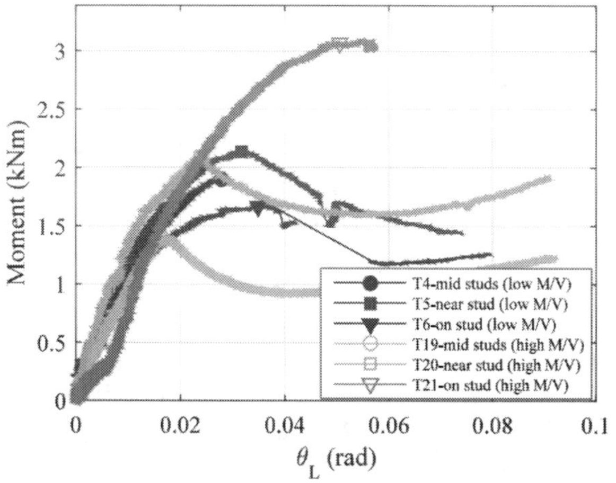

Fig. 7: Moment-rotation behavior of the ledger

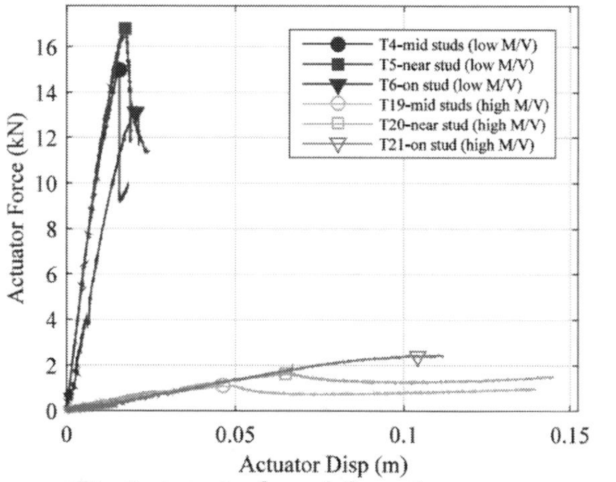

Fig. 8: Actuator force-deformation response

4. Discussion

The effects of applied M/V ratio on the joist-to-ledger connection has been studied herein depending on the joist location. The results of this study show that any of these two connection properties may drive the connection to fail with a different limit state and/or moment-rotation behavior. Consequently, other connection parameters; such as, presence of floor sheathing, presence of top and bottom screws connecting the joist flange to the ledger flange and clip angle location need to be investigated under both monotonic and cyclic loadings. The current specimens are designed to keep these parameters fixed. The effect of floor sheathing has already been investigated [3]. In the future tests, all these parameters will be altered to fully characterize the static and seismic behavior of the ledger framing to support future 3D seismic building analysis and design.

5. Conclusions

Six sub-assemblages of joist-to-ledger connection used in the CFS-NEES archetype building have been tested under monotonic shear load applied either close to the connection or to the joist free end to control *M/V* ratio. Three different joist locations (mid studs, close stud and on stud) are considered. These tests are a part of a broader experimental study on ledger framing construction, which consists of 30 tests. Although the ledger flange buckling and stud web crippling are the primary limit states which control joist-to-ledger connection behavior, there is no guidance for their design check in the codes. The increase in the *M/V* ratio caused always a change either in the moment-rotation response, limit state or both, depending on loading location. The connection ultimate strength decreased when the limit state of ledger bottom flange buckling remained the same, and increased when web crippling and screw pull-out kept driving the failure. The pre-failure stiffness degradation of the connection becomes less dramatic when the *M/V* ratio is high in both cases. Besides, when the limit state changed from stud web crippling to ledger bottom flange buckling with the increase of *M/V* ratio, there is no remarkable change in the connection capacity, except the rigid body rotation of stud frame. The stud frame rigid body stiffness increased due to plastic mechanism when specimens are tested under a load with high *M/V* ratio. Also, the change in the rigid body rotations of the stud frame highly correlate with the joist location in all of the tests, since load path is sensitive to the joist location.

Acknowledgments

This project was partially funded by the U.S. National Science Foundation (NSF-CMMI #1041578). Cold-formed steel products were donated by ClarkDietrich, and fasteners from Simpson Strong-Tie. Laboratory assistance from John Boutillier and Nick Logvinovsky was much appreciated. The views expressed in this work are those of the authors and not those of NSF, or any of the participating companies.

References

[1] Schafer BW, Ayhan D, Leng J, Liu P, Padilla-Llano D, Peterman KD, Stehman M., Buonopane SG, Eatherton M, Madsen R, Manley B, Moen CD, Nakata N, Rogers C. Yu C. "The CFS-NEES effort: Advancing cold-formed steel earthquake engineering", *Proceedings of the 10th National Conference in Earthquake Engineering*, Earthquake Engineering Research Institute, Anchorage, AK., 2014.

[2] Peterman KD. "Behavior of full-scale cold-formed steel buildings under seismic excitations", *Dissertation*, Johns Hopkins University, May, 2014.

[3] Ayhan D, Qin Y, Torabian S, Schafer BW. "Characterizing joist-ledger performance for cold-formed steel light frame construction", *Proceedings of 8th International Conference on Advances in Steel Structures*, Lisbon, Portugal, 2015.

SIMPLIFIED METHODOLOGY OF STRESS CALCULATION IN SLIP-JOINT CONNECTIONS

Michal Botló and Ján Brodniansky

Department of Steel and Timber Structures, Faculty of Civil Engineering, Slovak University of Technology, Bratislava, Slovakia

Abstract: This paper presents an introduction to the topic of slip-joint connections. It describes the principle of slip-joint connections, simplified method of internal forces calculation (single-axis bending moment, axial force) and a 3D model of such connection. By comparing the results of simplified calculation with the data from FEM-based software (ANSYS), it shows that the simplified method generates reliable and safe results. In the conclusion part, the paper identifies the key aspects of the solution which require a more complex examination.

1. Slip-joint connections

1.1 Outline of the research area

Slip-joint connections offer an alternative to the traditionally used flanged ones which are the most common solution in structures of this kind. The principle of a slip-joint is often described as being similar to a pair of inverted plastic cups [3,4].

Despite many advantages, slip-joint connections are rarely used in engineering practice. The reasons for this are various, but it is certain that the lack of design methods is an important one. There is no mention of these connections in the Eurocodes describing the design of steel structures and their connections, the only notion can be found in standards dealing with overhead electrical lights. Even in this case, rather than providing a simplified design methodology, it only offers a recommendation regarding the overlapping length, stating the minimum equals to 1.5 multiple of the tower's diameter.

This solution has been more thoroughly analysed in the Netherlands, where they were commonly used for wind turbine towers. For example, the company WindMaster held many patents concerning production manuals for whole structures including the segment connections [4].

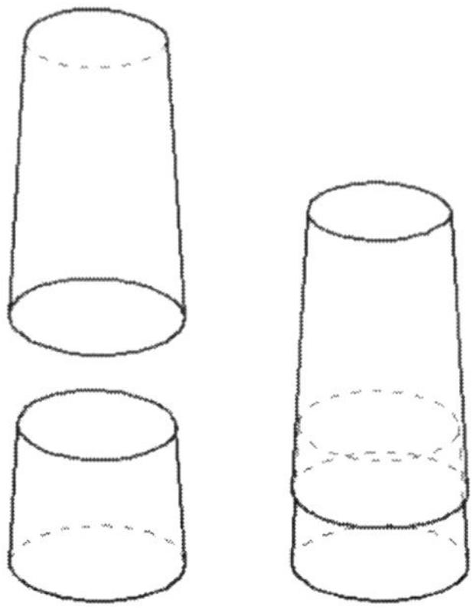

Fig. 1: The basic principle of a slip-joint

1.2 Simplified design of a slip-joint connection

In general, the load in this kind of connection acts in the form of an axial force, two bending moments and torsion moment. For the purpose of verifying the simplified model, only the interaction of axial force with one bending moment was considered (see Fig. 2).

The following section describes the simplified design methodology of a slip-joint connection resistance.

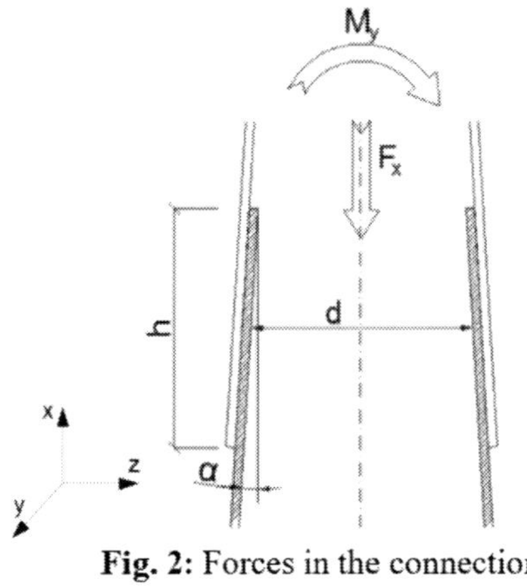

Fig. 2: Forces in the connection [4]

The axial force F_x distributes into two components - one parallel with the angle α of the tower web and one orthogonal to it, referred to as F_k. The latter one incurs inner pressure directed from the diameter centre towards the web of the tower, resulting in a tangent stress in the web itself. Furthermore, the principle of the connection is based on the segments being in direct surface-to-surface contact along the whole length of the connection. It is therefore possible to achieve significant increase in the connection's resistance due to friction, which can be implemented into the calculation in the form of axial force reduction.

The inner pressure on the wall of the structure can be calculated as:

$$P = \frac{F_k}{hd\pi} = \frac{F_x}{hd\pi \tan\alpha} \tag{1}$$

Stress resulting from the action of axial force:

$$\sigma_t = P.\frac{d}{2t} = \frac{F_x}{2ht\pi \tan\alpha} \tag{2}$$

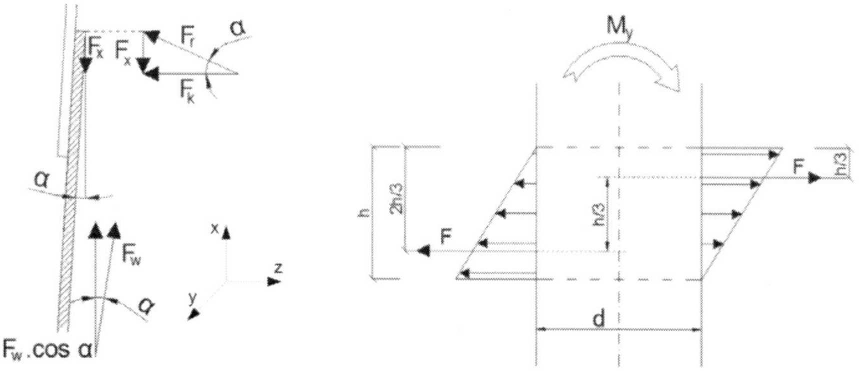

a) Action of the axial force b) Action of the bending moment

Fig. 3: Distribution of forces in the connection [4]

The friction force can be expressed by means of the friction coefficient μ as:

$$F_w = F_r \mu \tag{3}$$

The friction can be included into the axial force in the following way:

$$F_x = F_w \cos\alpha + F_r \sin\alpha \tag{4}$$

By substituting Eqn. (3) into the Eqn. (4), the formula is:

$$F_x = \nu F_r \cos\alpha + F_r \sin\alpha \approx F_r(\tan\alpha + \mu) \tag{5}$$

Stress resulting from the action of axial force (with the contribution of friction):

$$\sigma_t = \frac{F_x}{2ht\pi(\tan\alpha + \mu)} \tag{6}$$

The action of bending moment M_y is, in its simplified form, a pair of forces – one pressing against the upper edge of the connection (i.e. the upper edge of the lower segment), the other pressing against the lower edge of the connection (i.e. tearing the lower edge of the upper segment). The distribution of the bending moment is illustrated in the Fig. 3b.

First of all, it is necessary to determine the tangential stress the bending moment generates. Its value can be calculated as:

$$P_{\max} = \frac{6M_y}{h^2 d} \tag{7}$$

The Eqn. (7) is based on two assumptions:

$$M_y = F\frac{h}{3} \tag{8}$$

$$F = \frac{1}{2} P_{\max} hd \tag{9}$$

Stress resulting from the action of bending moment:

$$\sigma_t = P_{\max} \cdot \frac{d}{2t} = \frac{6M_y d}{2th^2 d} = \frac{3M_y}{h^2 t} \tag{10}$$

The overall stress resulting from the ionteraction of bending moment and axial force can be calculated by a simple addition of Eqn. (6) and Eqn. (10). The final formula for calculation is:

$$\sigma_t = \frac{F_x}{2ht\pi(\tan\alpha + \mu)} + \frac{3M_y}{h^2 t} \tag{10}$$

2. Analysis of a slip-joint connection of a selected structure

A tower of 440 kV electrical transmission line was selected for the purpose of verifying the methodology described above. The tower‘s material is steel S355, it has hexadecagonal cross-sectional shape and is 41.40 m high. The geometry, as well as the loads and other parameters, was extracted from [2] (this research continues on the basis established by [2]). In order to ensure the analysis is accurate, it is necessary to simulate real structure under real conditions, i.e. to sustain the ratio between its parameters and the parameters of loads it withstands.

The concerned connection involves two segments without apertures and side brackets. The average diameter of the connection is 1.555 m. The length of the upper segment is 10.6 m, the length of the lower one is 9.6 m. The input data is shown in the Table 1.

Table 1: Input data summary

Input data	Diameter	$D =$	1.555	[m]
	Conicity angle	$\alpha =$	1.000	[°]
	Web thickness	$t =$	0.010	[m]
	Friction coefficient	$\mu_1 =$	0.10	(steel-steel)
		$\mu_2 =$	0.50	(zinc-coated steel-steel)
Loads	Bending moment	$M_y =$	3088.86	[kNm]
	Axial force	$F_x =$	470.606	[kN]

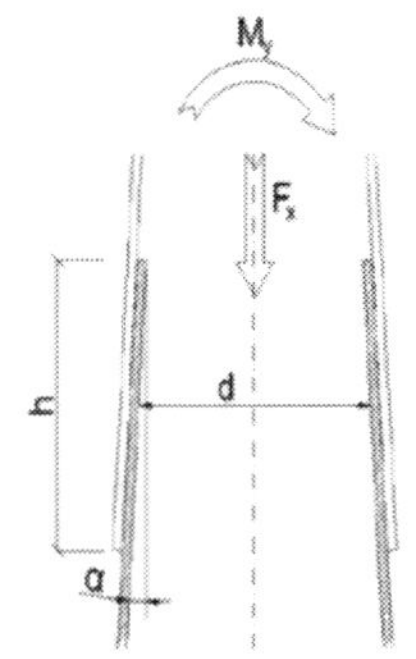

The stresses were calculated for various overlapping lengths (ranging from 1.0×D to 1.5×D). In order to demonstrate the contribution of friction, two sets of results are shown – one for μ = 0.1 (steel-steel, i.e. no surface treatment) and μ = 0.5 (zinc-coated surface). The results can be seen in tables and graphs below.

Table 2: Comparison of calculated stress (friction coefficient μ = 0.1)

	Overlapping length		μ = 0.1 (steel-steel)		
			Calculation result	Software result	Deviation
	[m]		[MPa]	[MPa]	[%]
1	$h_1 = 1.0 \times$ D	1.555	424.259	379.427	11
2	$h_2 = 1.1 \times$ D	1.711	354.018	326.738	8
3	$h_3 = 1.2 \times$ D	1.866	300.323	282.412	6
4	$h_4 = 1.3 \times$ D	2.022	258.324	251.046	3
5	$h_5 = 1.4 \times$ D	2.177	224.832	224.018	0
6	$h_6 = 1.5 \times$ D	2.333	197.677	209.873	-6

Table 3: Comparison of calculated stress (friction coefficient $\mu = 0.5$)

	Overlapping length		$\mu = 0.5$ (zinc-coated surface)		
			Calculation result	Software result	Deviation
	[m]		[MPa]	[MPa]	[%]
1	$h_1 = 1.0 \times D$	1.555	392.542	248.709	37
2	$h_2 = 1.1 \times D$	1.711	325.185	215.381	34
3	$h_3 = 1.2 \times D$	1.866	273.892	197.651	28
4	$h_4 = 1.3 \times D$	2.022	233.927	174.863	25
5	$h_5 = 1.4 \times D$	2.177	202.177	165.084	18
6	$h_5 = 1.5 \times D$	2.333	176.533	144.589	18

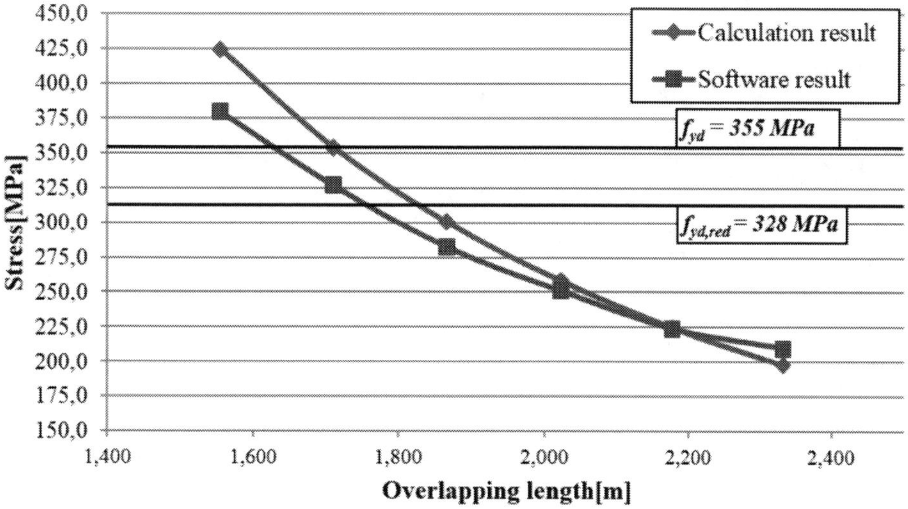

Fig. 4: Relationship between the overlapping length and stresses (friction coefficient $\mu = 0.1$)

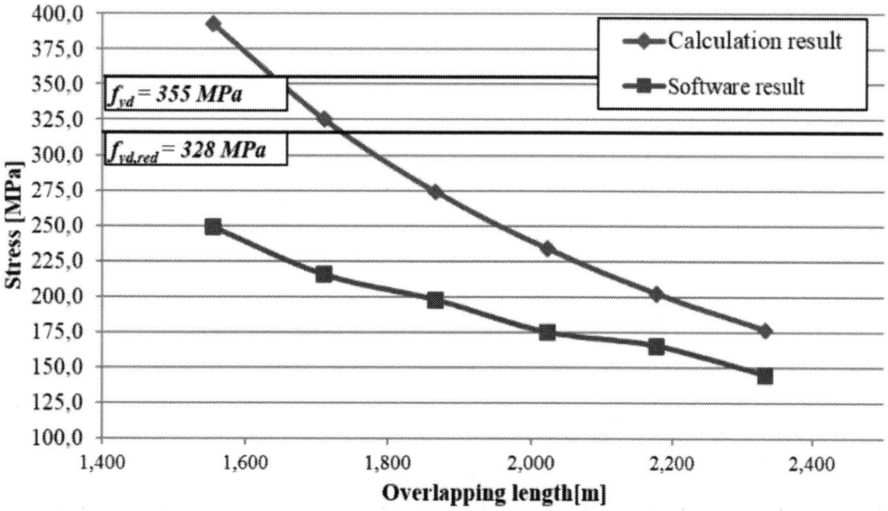

Fig. 5: Relationship between the overlapping length and stresses (friction coefficient $\mu = 0.5$)

3. Conclusion

First of all, the article presented a methodology for simplified examination of a slip-joint connection. The results the methodology provided were then compared to the results from a FEM software. Based on the comparisons presented in Tables 2 and 3 (and the related graphs), the following conclusions were drawn:

1. In both cases, the numerical analysis provided safe results. The only value of stress that exceeded result of FEM software analysis was of overlapping length equal to 1.5 multiple of the section diameter. This, however, is not relevant, as the value of stress is significantly lower than material resistance.
2. As the value of friction coefficient grows, the calculation becomes increasingly non-linear. The effect can be seen in deviations which are more significant for the friction coefficient μ=0.5. Nevertheless, the results of the numerical analysis are safe.
3. The results of both analysis methods assume that the most suitable overlapping length ranges between 1.1 and 1.2 multiple of section diameter. The calculation did not consider the effects of bending moment around the *z*-axis, but its value is considerably lower than that of M_y.
4. The presented methodology provides safe results of the stresses in the structure's webs. A more detailed analysis of friction can bring even more precise results.

Notation

The formulas Eqn. (1) to Eqn. (11) contained the following parameters:

F_x	Axial force
M_y	Bending moment
h	Overlapping length
d	Average diameter of the inner fragment
t	Thickness of the wall
α	Conicity angle
μ	Friction coefficient

References

[1] Kozák J. *Ocelové stožáry a věže,* Prague, Czech Republic, 1990.

[2] Recký J. *Teoretické a konštrukčné problémy oceľových stožiarov*, Bratislava, Slovakia, 2013.

[3] Segeren M, Lourens E. "Investigation of a slip joint connection between the monopile and tower of an offshore wind turbine", *Renewable Power Generation,* 422-432, 2013.

[4] Van Der Tempet J, Schipholt B. "The Slip-Joint Connection - Alternative connection between pile and tower", *Journal of Physics Conference Series*, 2003.

[5] STN EN 1993-3-1: Eurocode 3: Design of Steel Structures. Part 3-1: Towers, masts and chimneys. Towers and masts.

[6] STN EN 50341-1-2006: Overhead electrical lines exceeding AC 45 kV. Part 1: General requirements. Common specifications.

INFLUENCE OF LOCAL BUCKLING ON THE SHEAR RESISTANCE OF BOLTED CONNECTIONS FOR TUBULAR RACKING STRUCTURES

Marina D'Antimo[a], Jean-Françoise Demonceau[a], Jean-Pierre Jaspart[a], Massimo Latour[b] and Gianvittorio Rizzano[b]

[a] *University of Liège, ArGenCo Department, Belgium*
[b] *University of Salerno, Department of Civil Engineering, Italy*

Abstract: The competition in the field of storage racking structures is so that companies are always looking for new structural solutions aiming at containing construction time and expenses. This paper focuses the bearing failure of shear joint using long bolts through thin tubular members with the purpose to check the reliability of Eurocode 3 formulation. The used hollow sections are free to buckle inward leading to a bearing resistance lower than the one predicted using EC3. In fact EC3 does not account for the possible occurrence of out-of-plane deformations in shear connection assuming that the connected plates are restrained by the bolt head and the nut. In order to investigate this phenomenon, an experimental campaign has been developed, a FE model Abaqus/CAE has been validated and a parametric study has been performed. The present paper introduces the conducted investigations.

1. Introduction

Storage racking structures are often considered less important than traditional steel structures mainly due to their use in industrial warehouse and due to the mass production with standardized components [1]. Nevertheless the competition in the field of storage racking structures leads the companies to search for new solutions to streamline the structural elements through the use of thin wall members, with the aim of optimising, in term of time and expenses, the fabrication and assembly procedure. For these structures, connections play a key role on the overall behaviour, even more significant compared to normal steel frames. The actual formulation in Eurocode 3 covers shear connections between two or more plates, assuming that the plates are always laterally restrained by the bolt head and the nut (Fig. 1a). Using long bolts passing through hollow section members, only one side of the section walls is in contact with the bolt head or nut; thus, only this side can be assumed as laterally restraint. In the herein studied connections, the tubular profile is free to buckle inward (Fig. 1b); this could lead to a decrease of resistance, in particular, the bearing one. Hence it is important to understand if the formulations present in the current formulation of EC 3, for the bearing resistance of steel plates, can also be applied to tubular members in racking structures. The bearing resistance of bolted connections has been often investigated with the aim to extend the application of the EC3 rules to cases not covered by the code, such as the case of high strength steel [2, 3, 4] or stainless steel [5, 6] or the case of connections of cold formed strips and hot rolled steel plates [7]. The approach followed in this paper is similar to that provided in [3, 5], in which the accuracy of the application of the codified approach to new cases is firstly investigated by means of a parametric analysis based on FE simulations calibrated on the results of experimental tests.

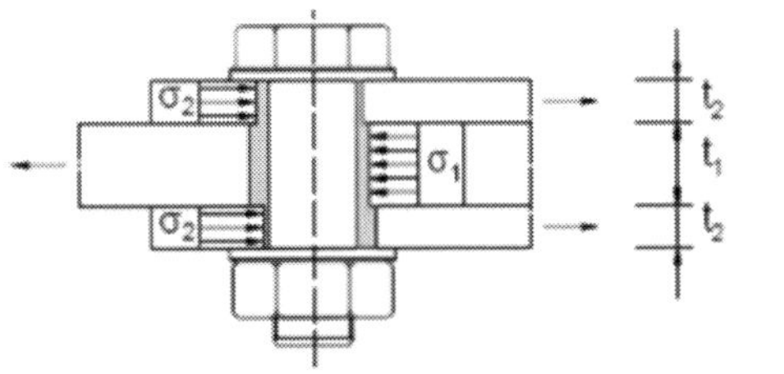

a) Plate connections

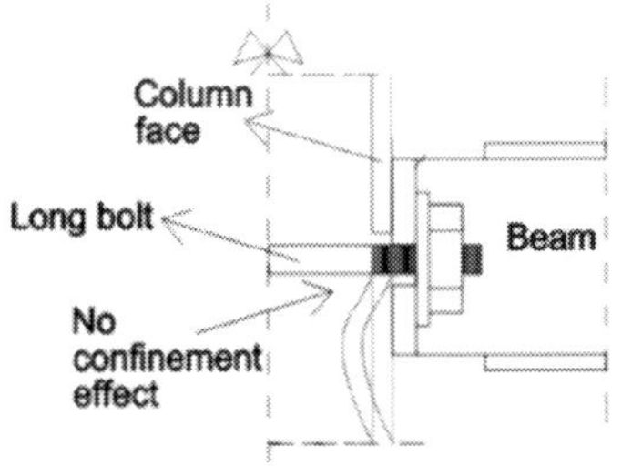

b) Tube-to-plate-connections

Fig. 1: Shear connections

2. Experimental Campaign

Monotonic tensile tests on 24 assembly, divided into six groups, according to material and geometry properties, were carried out at the Laboratory of the University of Liège. Two different materials for the tube, namely HX420LAD (f_y=min.420/max.520 MPa) and S235 (f_y=235 MPa), three different tube thicknesses (2 mm, 2.5 mm and 4 mm) and two bolt diameters (M12 and M16) were tested. All the bolts were 8.8 class ones and were not preloaded before executing the tests. The specimens were connected to the loading machine by means of steel plates, on one side welded to the tube and, on the other side, bolted to the specimen (Fig. 2). The tube's walls were marked with letters, the hole sides were called A and C and the two other sides were called B and D. In Table 1, the material grade and the actual dimensions of all the specimens are summarized. In particular, in this table, the values of the parameters d_A and d_C, which are the minimum dimensions of the hole diameters (on side A and C respectively) and of the parameter d_N, which is the nominal dimension of the hole, are given. Also some information about the holing procedure (drilling or punching) are given.

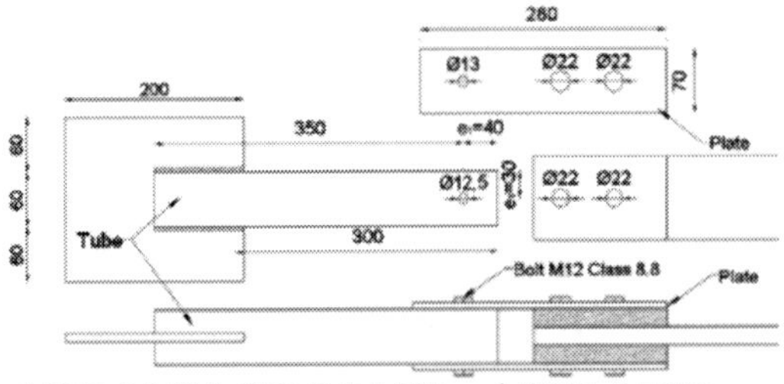

a) HX-2-M12, HX-2.5-M12 and S235-4-M12

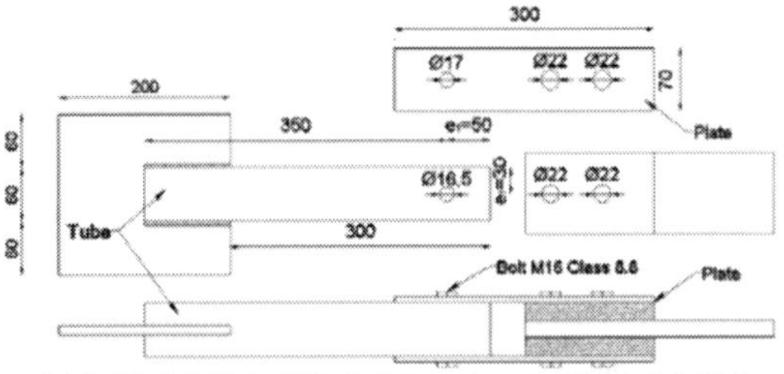

b) HX-2-M16, HX-2.5-M16 and S235-4-M16

Fig. 2: Layout of the tested specimens

Table 1: Specimens' actual geometrical and nominal mechanical properties

	TUBE		L[mm]	B[mm]	d_N[mm]	*Holing procedure*	d_A[mm]	d_C[mm]	t[mm]	e_1[mm]	e_2[mm]
HX-2-M16	HX420LAD	1	399	60.15	16.5	Punched	16.59	16.66	2.04	50.26	29.67
	Thickness	2	398.8	60.16	16.5	Punched	16.50	16.51	2.11	48.66	29.68
	2mm, bolts	3	399.5	60	16.5	Punched	16.50	16.53	2.06	49.02	29.72
	M16	4	399.3	59.8	16.5	Punched	16.53	16.50	2.07	49.11	29.76
HX-2-M12	HX420LAD	5	389.9	60	12.5	Punched	13.28	12.9	2.03	39.27	29.92
	Thickness	6	390.2	59.8	12.5	Punched	12.89	13.05	2.06	40.00	30.21
	2mm,bolts	7	390.1	59.9	12.5	Punched	12.84	12.93	2.09	39.93	30.15
	M12	8	389.6	59.9	12.5	Punched	12.97	12.93	2.04	39.86	30.15
HX-2.5-M12	HX420LAD	9	391.2	60	12.5	Punched	12.55	12.64	2.57	39.60	29.84
	Thickness	10	388.7	60.1	12.5	Punched	12.46	12.56	2.55	40.00	29.99
	2.5mm,bolts	11	390.4	60	12.5	Punched	12.46	12.85	2.51	41.94	29.70
	M12	12	390.3	60.1	12.5	Punched	12.66	12.30	2.55	40.82	29.69
HX-2.5-M16	HX420LAD	13	399	59.9	16.5	Punched	16.59	16.49	2.57	49.58	29.93
	Thickness	14	399	60.1	16.5	Punched	16.98	16.45	2.54	51.67	30.02
	2.5mm,bolts	15	398.9	60.1	16.5	Punched	16.99	16.50	2.62	50.87	30.00
	M16	16	399.1	60	16.5	Punched	16.76	16.45	2.55	49.45	30.00
S235-4-M12	S235	1	391.7	60.5	12.5	Drilled	12.39	12.53	3.85	33.17	30.30
	Thickness	2	391.3	60.4	12.5	Drilled	12.41	12.51	3.87	33.36	30.39
	4mm,bolts	3	391.6	60.4	12.5	Drilled	12.57	12.56	3.85	33.05	30.49
	M12	4	391.5	60.4	12.5	Drilled	12.46	12.53	3.79	34.07	30.35
S235-4-M16	S235	5	391.4	60.3	16.5	Drilled	17.24	17.29	3.8	41.48	30.39
	Thickness	6	391.5	60.6	16.5	Drilled	17.27	17.40	3.8	37.61	30.40
	4mm,bolts	7	391	60.5	16.5	Drilled	17.31	17.37	3.8	40.61	30.49
	M16	8	390.1	60.6	16.5	Drilled	17.26	17.28	3.8	41.27	30.25

2.1 Results

The tests on the connections have been performed using a Schenck RME 600 under monotonic loading conditions. The load has been applied under displacement control mode following a quasi-static protocol aimed at defining the initial stiffness of the connection and the force-displacement response up to failure. In Fig. 3 the experimental results of all the tests have been represented in terms of force and displacement calculated at the external transducer directly applied in correspondence of the tube holes.

3. Application of EC3

In order to evaluate the possibility to simply extend EC3 formulation to the connections analyzed in this work, a comparison between the bearing resistance provided by the code and the one obtained from the experimental tests is given herein. According to Eurocode 3, the resistance of the connection is provided by the minimum value between bolt and bearing resistances.

The bolt resistance can be calculated according to Eq. (1), while the bearing resistance can be evaluated in Eq. (2),

$$F_{v,Rd} = \frac{\alpha_v f_{ub} A_b}{\gamma_{M2}} \tag{1}$$

$$F_{b,Rd} = \frac{k \alpha_b f_u t d}{\gamma_{M2}} \qquad \alpha_b = Min\left(\frac{e_1}{3d_0}; \frac{f_{ub}}{f_u}; 1\right) \tag{2}$$

where, f_u and f_{ub} are respectively the plate and bolt ultimate strength, d and d_0 are respectively the bolt and hole diameter, e_1 is the distance of the hole from the plate free edge and k is equal to 2,5 provided that the distance of the hole from the lateral edge of the plate e_2 is greater than 1,5d. These last equations, account for all the basic failure modes normally arising in a simple bolted connection. The comparison between the resistance provided by the experimental tests and the one predicted by EC3 formulation is reported in Fig. 4, where the ratio between the predicted and the experimental values of the bearing resistance (respectively $F_{b,EC3}$ and $F_{b,exp}$) are represented for each test.

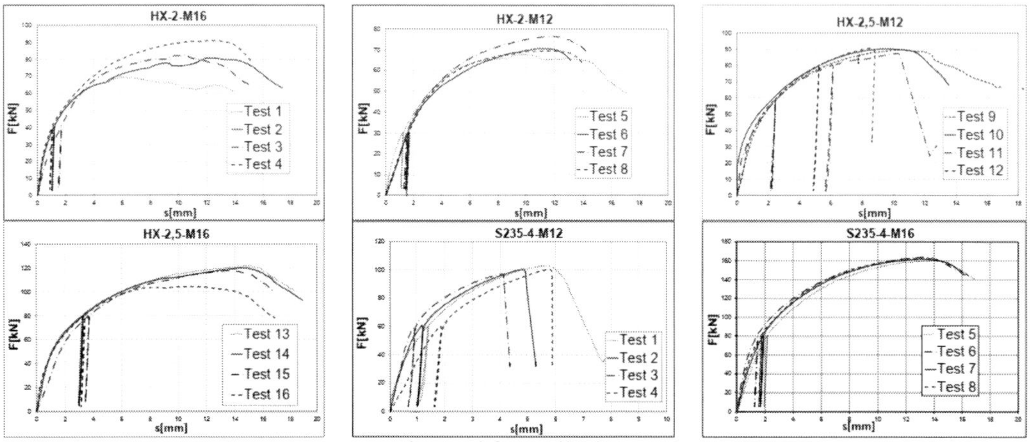

Fig. 3: Experimental force-displacement curves

On average a good prediction is observed (average value of the ratio is equal to 0.9 and 0.74 and the standard deviation is equal to 0.08 and 0.13 for HX420LAD and S235 respectively), nevertheless the comparison highlights a lower safety of EC3 formulation for the specimens characterized by a low value of thickness/hole diameter ratio (Fig. 4). This result can be due to the fact that the EC3 formulation does not account for the local buckling failure mode especially when the local slenderness of the tube's plate is high. Therefore this first comparison points out the need of a further investigation when low values of tube's thicknesses are used.

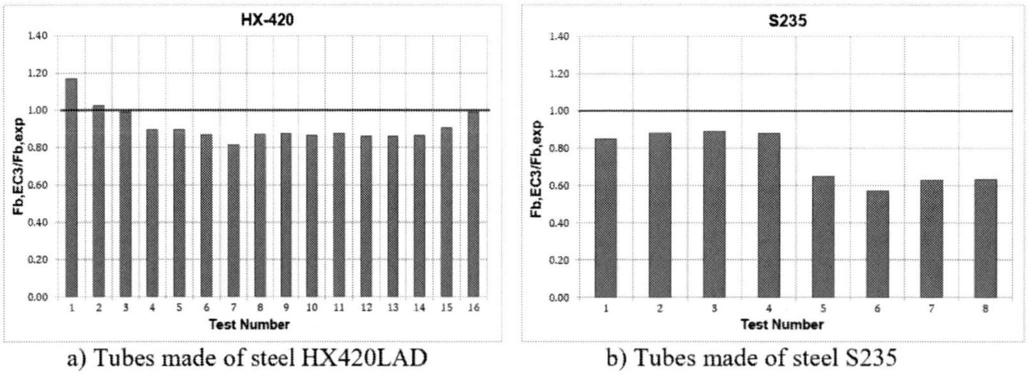

a) Tubes made of steel HX420LAD   b) Tubes made of steel S235

Fig. 4: Accuracy of EC3 model

4. Application of EC3

In order to further investigate the accuracy of EC3 formulation and to extend the results obtained through the experimental program, a Finite Element model has been developed in

Abaqus/CAE software. The required steps to develop the model are the followings: geometrical characterization of the parts, definition of the material properties, definition of the elements' interaction, definition of the boundary conditions and choice of the elements and size of the mesh. To decrease computational time, a quarter of the assembly has been modelled keeping into account the symmetry conditions (Fig. 5) and following a three-dimensional approach. In particular, the geometric non linearity has been properly considered, in order to include the typical buckling phenomena of the plate composing tube arising at high displacements. Regarding the ***element type***, eight-node brick elements with full integration and first order approximation have been adopted (C3D8). Such elements, as reported in [4], are particularly accurate for analysis where buckling effects are significant. The ***material properties*** of tubes and plates have been described by means of an elastic-plastic isotropic model by adopting simplified equivalent bi-linear or quadri-linear true stress-true strain laws defined using the results from the coupon tensile tests. Conversely, considering the absence of the experimental data for the bolts, the behaviour of the material composing the bolts has been modelled using a simplified bi-linear model based only on literature data for strength and deformation. All the ***interactions*** between the different parts have been defined using the surface-to-surface contact. The ***mesh size*** of the elements composing the model has been defined after performing several preliminary analysis and by following some of the guidelines available on the topic [10]. With the aim to reduce the computational effort, the number of elements has been increased only in areas where great stress gradients were expected. Therefore, in order to obtain accurate and stable results the following mesh average geometry has been adopted: the tube mesh starts from 0.8 mm in the zone of the tube subjected to high stress concentrations up to 8 mm and by defining at least six elements within thickness of the plates (Fig. 6); the bolt has been meshed with elements from 0.8 mm up to 2mm; the loading plate has been modelled with a larger mesh adopting only two elements within the thickness of the plate.

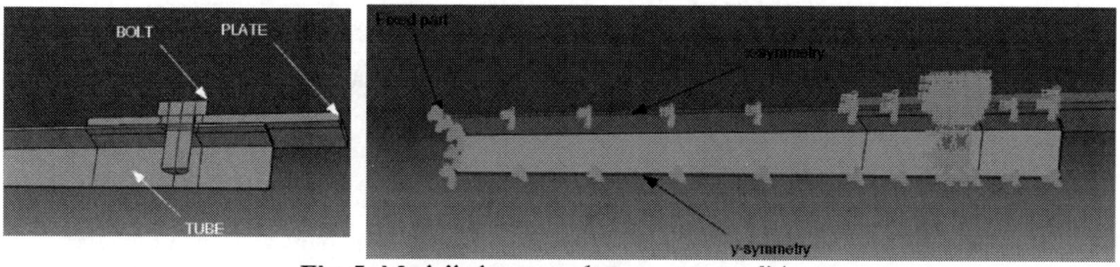

Fig. 5: Modelled parts and symmetry conditions

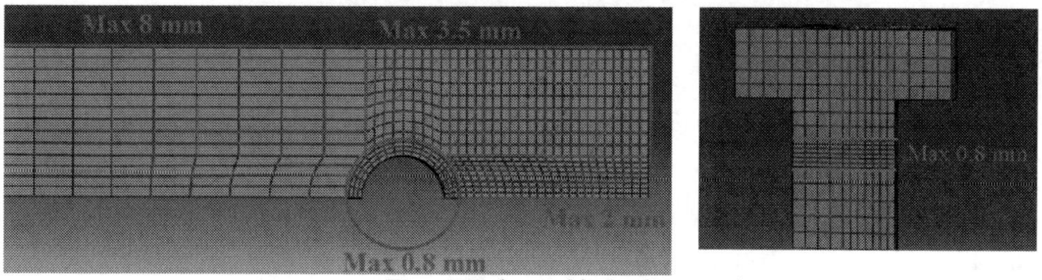

Fig. 6: Tube and bolt element size

4.1 Finite element results vs experimental one

Two different FE models have been developed, one called "ideal", defined according to the nominal geometry of the specimens and considering the nominal diameter of the bolt; and the second one, called "initial", defined accounting for the geometrical imperfections of the holes related to the manufacturing process and considering the net diameter of the bolt (EN1090-2).

In Fig. 7 the comparison between the two FE models is given. The "ideal" model provides an overestimation of the initial stiffness and a good accuracy in predicting the resistance. The "initial" model instead reproduces with high accuracy the stiffness accounting for the hole imperfections. The comparison between the FE model and the experimental tests, in terms of deformation at the end of the test, is shown in Fig. 8.

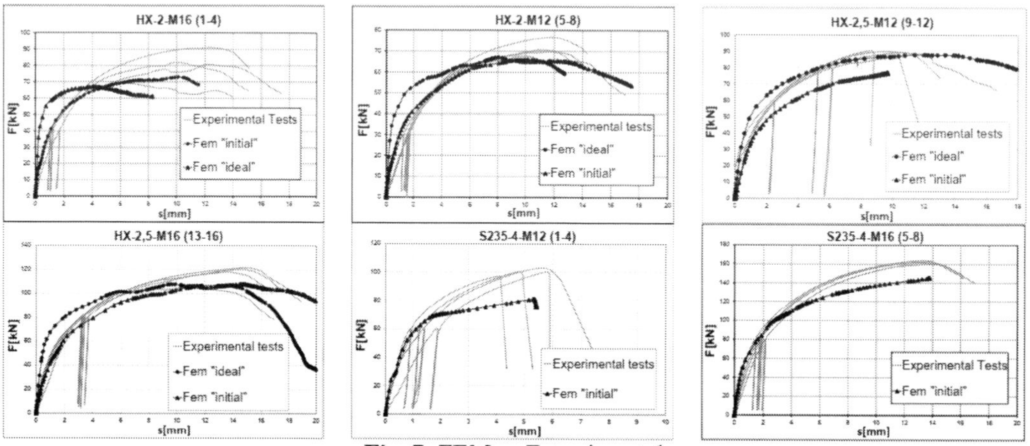

Fig. 7: FEM vs Experimental

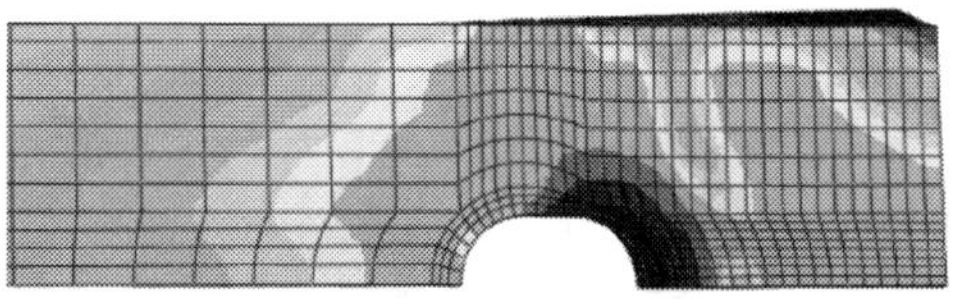

a) Hole elongation in FE model

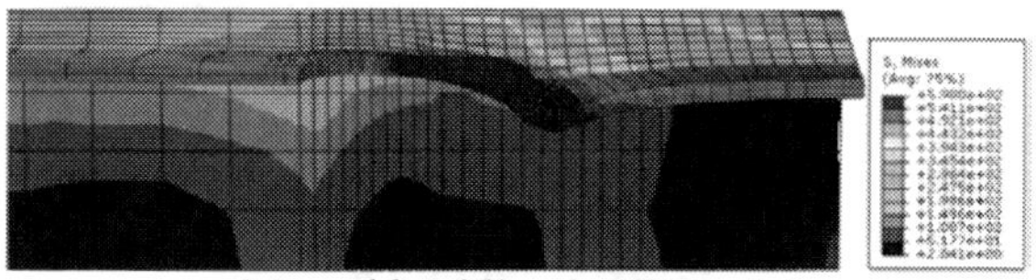

c) Local buckling in FE model

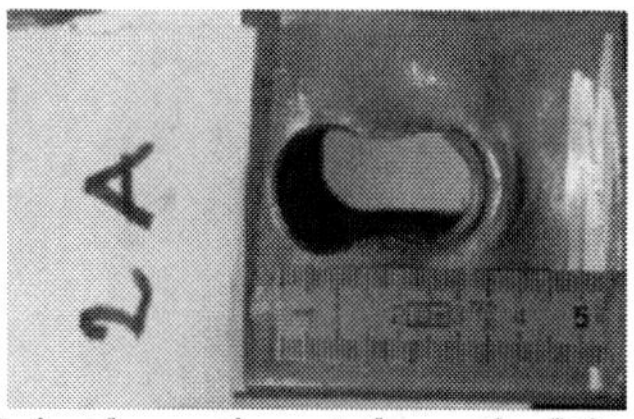

b) Hole elongation at the end of the test

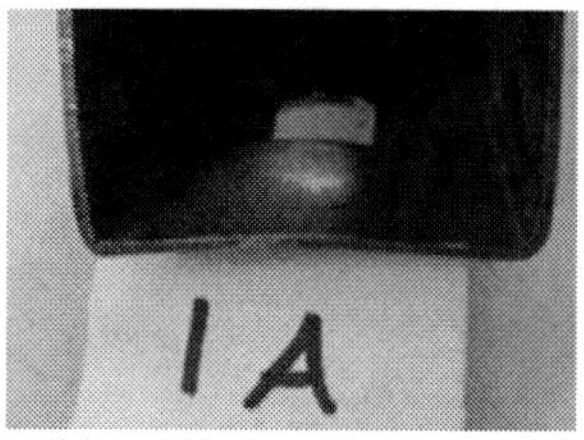

d) Local buckling at the end of the test

Fig. 8: Results series 1-4 HX420LAD

5. Parametric Study

The bearing resistance of the steel connections is mainly influenced by the geometrical parameters e_1, e_2, d, d_0 and t. A parametrical study has been performed starting from the FE model HX420-2.5-M16, by varying the three considered geometrical parameters one by one.

5.1 Influence of e_1 and e_2

Due to the presence of the tube lateral walls, the influence of the edge distance e_2 is less important in this study. On the contrary, e_1 influences the bearing resistance, and, in this work, eight values have been investigated in a range from 16.5 up to 90 mm (Fig. 10).

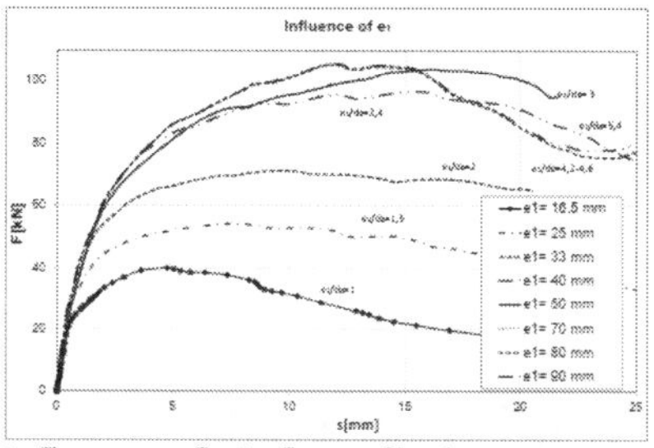

a) Influence of e1: force displacement curve

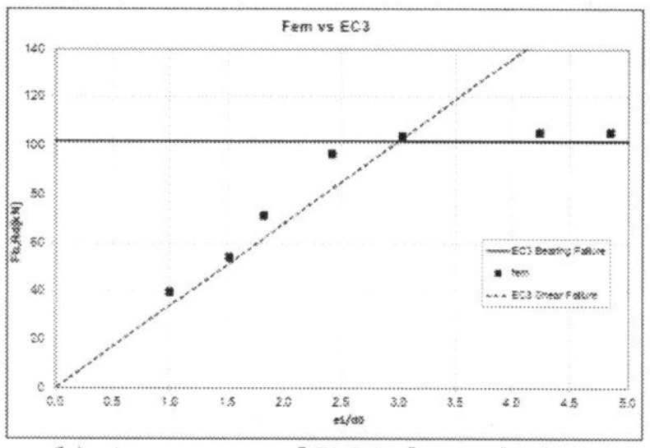

b) Accuracy of EC3 formulation

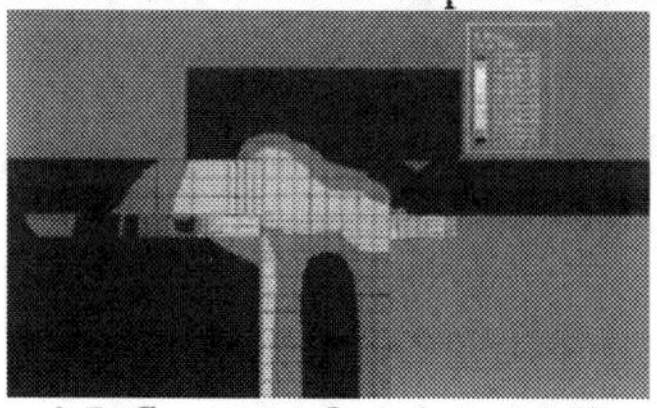

c) Influence of e1 (e1=16.5);

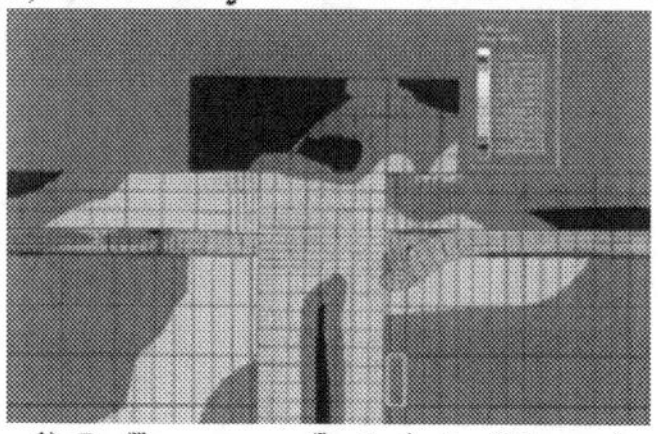

d) Influence of e1 (e1=90mm)

Fig. 9: Influence of e_1 and Accuracy of EC3 formulation

The results of simulations, in terms of load-displacement curves, for each value of e_1/d_0, are depicted in Fig. 9(a) while in Fig. 9(b) the application of the design formulation, provided by EC3, is represented. It is possible to observe that the code equation is slightly conservative in the shear failure mode range, while it is very accurate in the bearing range. In addition, it can be also recognised that the boundary value separating the two failure mode ranges appears well predicted by EC3 formulation.

5.2 Influence of *t*

Clearly, the thickness of the plate also plays a significant role on the bearing resistance. In particular, the decrease of the tube's thickness leads to a decrease of the connection resistance with the progressively appearance of buckling phenomena (Fig. 10(a)). In the performed analysis, eight values of the thickness varying from 0.5 mm up to 6 mm have been considered. In Fig. 10(b) the resistance provided by the model is compared with the resistance predicted by EC3. Above all, in such a graph the FE and EC3 resistance values have been charted versus the t/d_0 ratio. For values of the t/d_0 ratio higher than 0.15 the EC3 formulation provides an adequate prediction of the resistance, on the contrary, for values of t/d_0<0.15, EC3 formulation provides an overestimation of the resistance since it does not account for the local buckling (Fig. 10(c)).

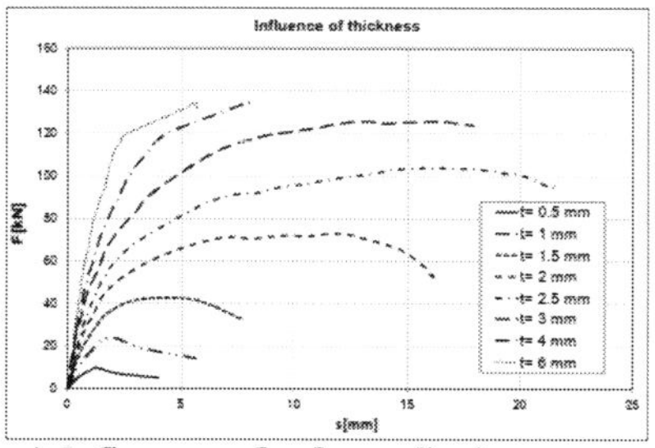

a) Influence of t, force displacement curve;

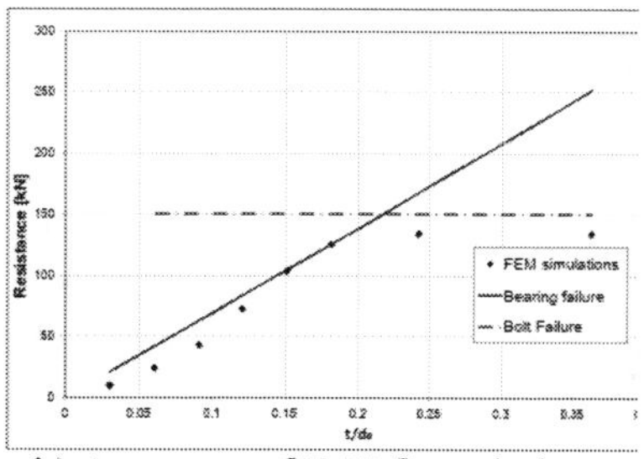

b) Accuracy of EC3 formulation

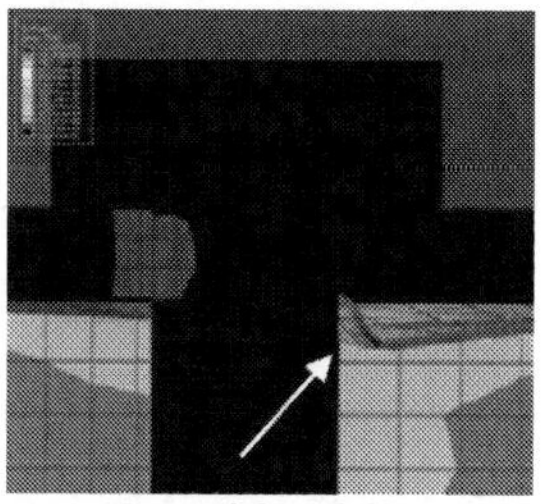

c)Local buckling (e_1=16.5mm)

Fig. 10: Influence of geometrical parameters and accuracy of EC3

6. Conclusions

The main conclusions are:

1. The numerical analyses, conducted in ABAQUS, can accurately predict the ultimate resistance if an appropriate mesh is used, while the prediction of the initial stiffness is strongly influenced by the hole's imperfection and the actual bolt's diameter.
2. The parametric study allowed to recognize that the distance e_1 and the thickness t of the tube influence the prediction of the connection resistance while the distance e_2 is not significant in the examined case. In particular, the developed parametric analyses evidenced that in case of low values of tube thickness and hole diameter ratio a more accurate formulation for predicting bearing resistance is advisable, by properly accounting for the influence of the local buckling of the tube plate in contact with the bolt shank.

Notation

L	Length of the specimen	d_C	Hole diameter (on the side C)
B	Width of the specimen	t	Thickness of the specimen
d_N	Nominal hole diameter	e_1	Distance from the centre of the hole to the edge
d_A	Hole diameter (on the side A)	e_2	Distance from the centre of the hole to the lateral edge

References

[1] Tilburgs K. "Those peculiar structures in cold-formed steel:"racking & shelving", *Journal of steel construction*, 6-2, 95-106, 2013.

[2] Hoang VL, Jaspart JP,Demonceau JF."Extended end-plate to concrete- filled rectangular column joint using long bolts", *Journal of Constructional Steel Research*, 113, 2015.

[3] Puthli R, Fleischer O. "Investigations on bolted connections for high strength steel members", *Journal of Constructional Steel Research,* 57, 313-326, 2001.

[4] Može P, Beg D. "Investigation of high strength steel connections with several bolts in double shear", *Journal of Constructional Steel Research*, 67(3), 333-347.

[5] Salih EL, Gardner L, Nethercot DA. "Bearing failure in stainless steel bolted connections", *Engineering Structures*, Vol. 33, 549-562.

[6] Kiymaz G. "Bearing strength of stainless steel bolted plates in tension", *Nordic Steel Construction Conference*, Malmo, 2-4 September.

[7] Chung KF, Ip KH. "Finite element modelling of bolted connections between cold-formed steel strip and hot rolled steel plates under static shear loading", *Engineering Structures*, Vol. 22, 1271-1284, 2000.

[9] Bursi OS, Jaspart JP. "Calibration of a finite Element model for isolated bolted end-plate steel connections", *Journal of Constructional Steel Research*, 44, 224-262, 1997.

[10] Selamenr S, Garlock M. "Guidelines for modeling three dimensional structural connection models using finite element methods", *International Symposium "Steel Structures: Culture & Sustainability*, Paper No: 14, 21-23 September 2010, Istanbul, Turkey.

[11] D'Antimo M, Jaspart JP, Demonceau JF, Latour M, Rizzano G. "Shear bolted connections for tubular racking structures: Numerical analysis", XXV Congresso CTA, Salerno, 2015.

[12] D'Antimo M, Jaspart JP, Demonceau JF, Latour M, Rizzano G. "Experimental investigation of shear bolted connections for tubular racking structures", *XXV Congresso CTA*, Salerno 2015.

BEHAVIOUR OF OVER-ROOFING STRUCTURES WITH SEMI-RIGID COLUMN-BASE CONNECTIONS

Andra Floricel[a], Viorel Ungureanu[a,b] and Adrian Ciutina[a]

[a] *Department of Steel Structures and Structural Mechanics, Politehnica University of Timisoara, Romania*
[b] *Laboratory of Steel Structures, Romanian Academy - Timisoara Branch, Romania*

Abstract: Renovation of buildings represents the challenge of existing building stock. For this purpose, in case of flat-roof structures, the engineers choose to use over-roofing and over-cladding solutions for updating and rejuvenating outdated buildings. The use of the light-weight steel construction with cold-formed steel structural components leads to important advantages such as strength, lightness, durability, easy adaptation, recyclability and prefabrication. However, the connection between the old and the over-cladded structure often represents a problem, which can limit, in many cases the application of desired over-roofing systems. Present study is focused on the analysis and behaviour of column-base connections in case of over-roofing solutions with cold-formed thin-walled structural elements.

1. Introduction

According to literature [1], the total building stock in Europe is about 196 million dwellings that correspond to 160 million of buildings [2]. In mean, yearly, the new buildings represent about 1.5% of the total building stock [3]. However, the retrofitting rate of existing building stock is much lower than the one needed to future 2050 ambitions: today European buildings are renovated every 30 to 40 years on average, and every 60 to 80 years in the Mediterranean region.

Although flat to pitch conversions have proven a complete transformation solution for the appearance of a building and, in some cases, they are the only effective way of waterproofing the existing roof structure, on the downside, they require detailed structural capacity verification. This is to ensure that the existing roof structure or supporting walls are capable of withstanding the additional loads transmitted by the new roof structure. The conversion creates additional habitable space within the new roof structure, which in turn improves the cost-benefit balance on long-term.

Building conversion by vertical addition can be achieved by using both traditional materials such as timber, reinforced concrete, hot-rolled steel, rectangular hollow steel section and can include innovative ones e.g. glue laminated timber, composite materials with polymeric matrix, cold-formed and/or high strength steel. Of the first category, glue laminated timber is the preferable option nowadays compared to simple timber due to its capacity to obtain sections of general shape with minimal defects and high structural performance. The benefits of using reinforced concrete in some particular cases include the possibility of reducing the dimensions of the vertical addition structure members, though this option is not preferred in high seismic areas and in most cases gives unnecessary extra weight that can endanger the existing buildings structural resistance. This problem was solved by the introduction of modern high strength materials like steel and composite polymeric materials, which lead to a decrease of effects on the existing structure up to ten times. However, the latter can be dismissed due to its high costs. Thus, cold-formed steel becomes an ideal solution for refurbishment as replacement of the existing roofing and for over-cladding solutions. Moreover, due to steel's high adaptability features, each refurbishment solution can be tailored to the building owner's requirements and to the particularities of the project [4].

Depending on the prefabrication level, three types of cold-formed constructive systems can be identified, namely stick-by-stick constructions, panel constructions and modular constructions. Among these, stick-built and panel constructions are certainly the most suitable for vertical addition, mainly for technological reasons related to the easier installation and transportation, which are the most important requirements for super elevation within densely populated areas, as the city centres [4,5] (see Fig. 1).

Fig. 1: Over-roofing of a three storey building: structure and new building lay-out [5]

Cold-formed steel members are efficient in terms of both stiffness and strength. Additionally, the use of thinner sections and high strength steel leads to design problems for structural engineers which may not normally be encountered in routine structural steel design. The design is dominated by specific problems related to (1) stability loss: this problem is specific for cold-formed steel members due to class four cross-sections, (2) connecting technology, which is specific and strongly influences the structural behaviour and detailing. However, lately, the seismic performance (3) of cold-formed steel structures started to be examined [6].

2. Case-study: over-roofing of T744.R apartment building

2.1 Statistical data

According to the Romanian national statistics, there are 493,000,000 m^2 of built floor area, of which residential buildings account for 86%. In terms of construction period, most of these were built during the second half of the 20th century, more specifically 1961-1990 period. More than 90% of the total residential floor area was built before 1989.

According to the 2011 Census of Population and Housing, Romania had a total 5,341,908 of residential buildings representing 8,722,398 households, out of which: (a) 5,107,818 were individual buildings (dwelling type); (b) 57,335 coupled type buildings; (c) 153,886 as apartment building type; (d) 22,858 other type of buildings (see Table 1).

Table 1: Existing building stock and habitants in Romania [7]

Type of buildings	Number of units	Number of residents
Dwelling (individual)	5,107,818	12,409,338
Coupled	24,998	267,386
Duplex	32,347	155,473
Block	153,886	7,061,465

As shown by data presented in Table 1, the last building typology – block buildings, as although they only represent 2.9% from the total building stock, they are housing some 7,061,465 inhabitants, which is more than one third of the total Romanian population. The major part of these buildings is located in urban areas. According to the same census, the apartment building structures have as primary resistance structures concrete (71,802 units) and respectively various types of masonry and other materials (22,313 units). The vast majority of these buildings were constructed at a time when no specific thermal requirements were set, or when such requirements were not demanding. Besides the structural retrofitting of the existing building stock, also the energy retrofitting is needed.

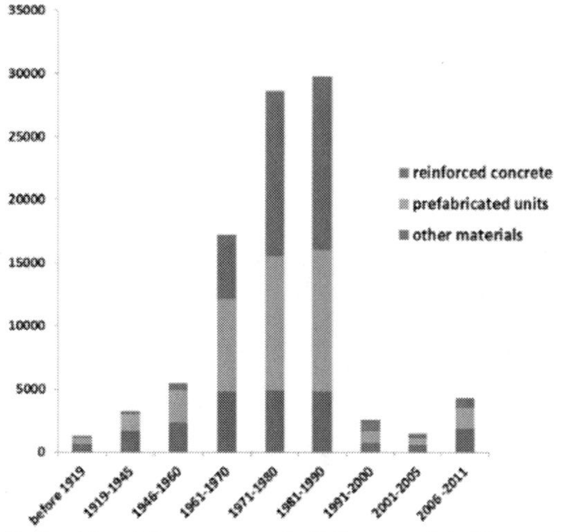

Fig. 2: Residential apartment building stock in Romania [7]

During 1960-1990 period, 48,719 precast multifamily apartment buildings were built which represent more than 85% of the total buildings of this type as shown in Fig. 2. In order to achieve large cost advantages in the construction of these apartment blocks, the National Institute of Design, Research & Technology (IPCT) delivered a series of standardized projects applicable all-over the Romanian territory. The most popular IPCT projects were the low rise 5 storey, as follows: (a) 5-storey IPCT, project types: 770; 744; 994; 1013-1168; 1340; 1586; 2926; 1399; (b) 9-storey IPCT, project type: 772.

2.2 Structural design

The example chosen as case study in present study represents an over-roofing structure for the multifamily apartment building type T744.R. The original structure was built in 1962-1975 period and is considered as representative for large reinforced concrete precast buildings [8,9] - see Fig. 3. The addition of the over-roofing structure consists of two stages: (i) check of the current state of the original structure considering the actual codes and (ii) verification of the structure after including the new-added steel structure.

The state of the original structure was done by 3D analyses using ETABS computer code, by using shell finite elements. The thicknesses of the vertical and horizontal modelled diaphragms were equal to the thickness of the structural layers. The existing concrete building was checked according to the actual standards, proving to possess important reserves of resistance even in the cases of considering different refurbishment strategies for enlarging the interior living space. Details on these aspects are presented elsewhere [10]. In particular, the seismic load was accounted through a spectral analysis using the design ground acceleration value $a_g = 0.20g$ and the control period $T_C = 0.7$s, according to EN 1998-1-1-2004 [11].

The checks performed on the over-roofed structure showed that, by removing the original thermal and hydro insulation layers from the terrace and the addition of the supplementary over-roofing cold-formed structure, the stress distribution is not changing significantly.

The over-roofing systems were analysed by including in the already built models in ETABS, the new steel structures as shown in Fig. 3. For the seismic analysis of the over-roofing systems, a behaviour factor $q = 1.0$, specific to non-dissipative structures, was chosen. Because the existing concrete structure is very rigid, i.e. the first period of vibration is around 0.1 seconds, no amplification effects are introduced to the new structure and the new structure can be designed independently on ground. Fig. 3 presents the FEM model of the existing concrete structure, the model of the over-roofing structure as well as the cross-sections of cold-formed structural elements.

Although in design the connection of over-roofing to the existing structure was considered as pinned, the practical realisation is rather complex, as it should be accommodated with the existing concrete building. Several solutions are analysed in the next section by FEM.

Over-roofing elements:
- Columns: 2C300×2 (built-up members)
- Beams: 2C300×2
- Braces: ϕ16 - roof, ϕ24 - walls

a) IPCT T744R Original Structure b) Cold-formed section (CF)

Fig. 3: Cold-formed over-roofing solutions for the main block type T744.R

2.3 Column base solutions

Two column-base typologies of the over-roofing systems to the existing concrete structures are considered. For both solutions, the columns are composed of two back-to-back cold-formed lipped channels of 300 mm height and a 2 mm thickness:

- base-plate connection, consisting of two rows of M16 bolts, used to connect the webs of column profiles via a steel box section of 6 mm thickness and 150 mm height, placed between the two sections of the column; the base plate is 480×480×20 mm. The column-base connections are M20 bolts, and are fixed on back-side angle profiles, jacketing the concrete horizontal and vertical diaphragms (see Fig. 4a).
- splayed connections, at which three rows of M16 bolts were used to connect the webs of the profiles through a similar steel box section of 6 mm thickness and 220 mm height, placed between the two sections of the column. The vertical stiffeners (220×10 mm) are anchored to

the column flanges by three-rows of M16 bolts (see Fig. 4a). The base plate is 460×500×20 mm. Similar connection of the base-plate to the concrete slab is used.

Both typologies were studied by an alternative anchoring system, by using chemical anchors, drilled and fixed in the concrete slab. The chemical anchors used, were of HILTI type, 140 mm length and a diameter of 20 mm, chosen according to reactions and company catalogue. The solutions are detailed in Fig. 4c and Fig. 4d, respectively.

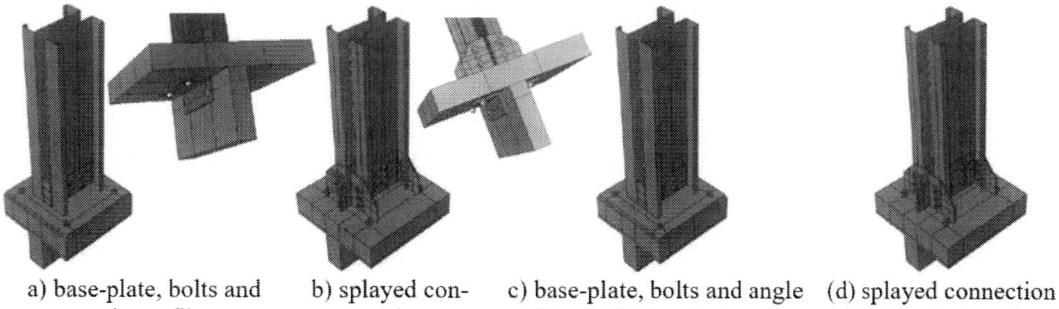

a) base-plate, bolts and angle profiles   b) splayed connection   c) base-plate, bolts and angle profiles with chemical anchors   (d) splayed connection with chemical anchors

Fig. 4: Cold-formed steel column-base connections

3. Column-base FEM numerical investigation

3.1 Description of models

In order to prove the efficiency of the column-base connections chosen for this study, numerical simulations were performed using FE software ABAQUS 6.7. SHELL elements of S4R type, with 4 nodes, reduced integration, 6 DOF per node were used for modelling the cold-formed steel sections, while for all other sections, BRICK elements of C3D8R type, with 8 nodes, reduced integration, 6 DOF per node, were used. In order to obtain accurate results the dynamic explicit solver was used.

The load was applied in displacement control through a maximum horizontal load of 100 mm displacement at the top free-end of the column. The steel elements' material was modelled by using an elastic - perfectly plastic model using the nominal value of yield strength S355. Bolts of class 6.6 have been considered in analyses, modelled by elastic-plastic behaviour with strain hardening as presented by former studies [8,9]. The concrete was considered in the analysis with elastic-plastic behaviour [12]. The option used to simulate concrete plasticity was Concrete Damaged Plasticity. A particular attention was given to the resin in which the anchors are embedded for second fixing solution: an elastic modulus of 3240 N/mm^2 characteristic to Mapefix EP 385 material [13] was considered according to producer.

3.2 Numerical results

In Fig. 5 the global moment-rotation responses of analysed column-base connections are presented. The moment is derived from the lateral force applied to column tip, while the rotation is computed as the ratio between the top lateral displacement and column height.

As expected, the web column-base connections present a limited capacity in bending, generally half as compared to the splayed typologies. The same trend is observed for initial stiffness. However, a relatively small bending capacity is recorded for column-base connections with anchors (see Fig. 5b). Besides that, both typologies develop high-level of rotation of 30 to 50 mrad, without regard to the concrete back-connection system: angle profiles or concrete anchor bolts.

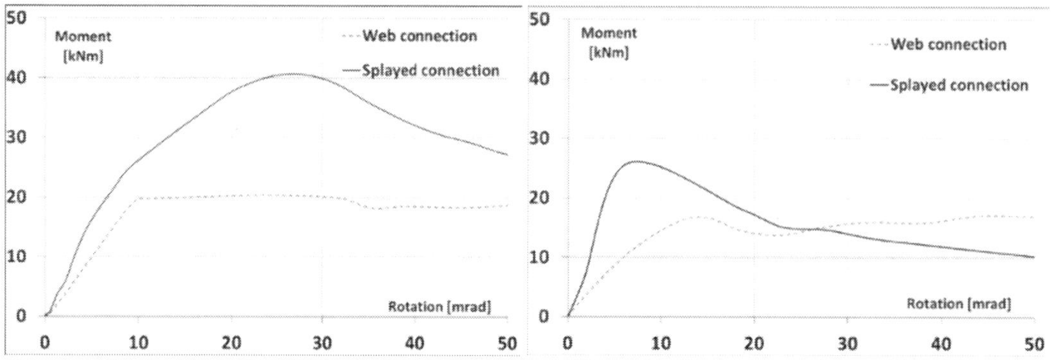

a) column-base connections with angle connectors b) column-base connections with anchors

Fig. 5: Moment-rotation curves of cold-formed column-based connections

Fig. 6 presents the failure mechanisms of column-base connections, both as plastic deformations and plastic stresses. In both models with bolted web connection, the failure is developed in two steps: in a first phase, high values of rotation are developed important rotation stresses around web bolt-holes. In a second step, the compressed flanges get in contact with the base-plate, developing additional resistance, but eventually leading to local buckling of these. The splayed connection typology reaches the target of driving the plastic hinge in column element at higher resistances in case of angle connection. The anchored splayed connection fails by pulling the anchors from concrete, forming traction concrete cones. In all other cases, the concrete slabs proved only elastic stresses with limited plasticity at full loading conditions and recorded locally in bolt holes.

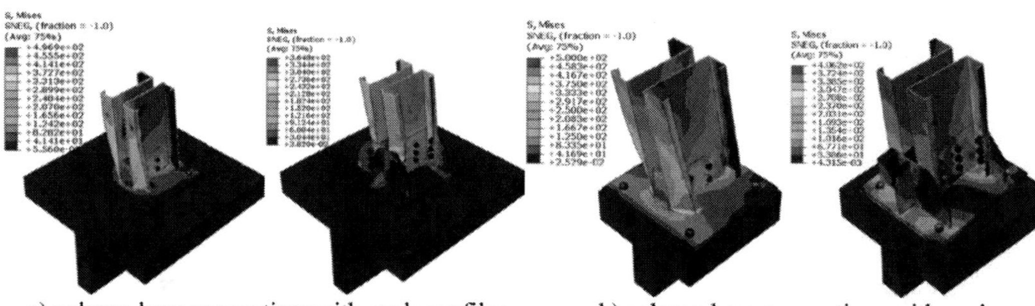

a) column-base connections with angle profiles b) column-base connections with anchors

Fig. 6: Failure mechanisms of column-base connections: deformed shapes / stress distributions

3.3 Joint classification according to EN 1993-1-8

In order to classify the results of the FEM simulations with regard to the EN 1993-1-8 [11] continuous / semi-continuous conditions, the values of initial stiffness and elastic moment capacity were evaluated on the moment-rotation characteristic response curves presented in Fig. 5. The joint stiffness characteristics were computed in accordance to the moment-resisting frame conditions while the resistance was compared to the column plastic value ($M_{pl,c}$). The results are shown in Table 2 in comparison with the resistance and stiffness limits.

All column-base connections are semi-rigid with low values of stiffness, smaller than 10% of the limit for rigid connections. However, the web connections are close to pinned conditions; these typologies exhibiting large rotations and could be considered as ductile.

Splayed column-base connections develop significant resistances and important values of stiffness. Their classification is partial-resistant and semi-rigid. The connection anchoring in concrete slab does not change significantly the joint characteristics as long as the concrete slab remains in elastic range and the base-plate is sufficiently rigid.

Table 2: Initial stiffness and moment resistance of base connections and their classification

	Typology	Angle profile connections		Anchor connections		Limits for resistance and stiffness:
		web	splayed	web	splayed	
Stiffness	$S_{j,ini}$ [kNm/rad]	1864.7	3787.6	1774.3	4302.9	$S_{j,rig}$ =46730 kNm/rad;
	Classification	Semi-rigid	Semi-rigid	Semi-rigid	Semi-rigid	
Resistance	M_j [kNm]	20	35	16.1	33.3	$S_{j,pin}$ = 934 kNm/rad;
	Classification	Partial resistant	Partial resistant	Partial resistant	Partial resistant	$M_{pl,col}$ = 81.6 kNm

4. Structural behaviour integrating behaviour of joints

In order to find the influence of semi-rigid column-base connections on the overall response of the designed frame, the stiffness values given in Table 2 were introduced as column base rotational stiffness (spring stiffness) in ETABS computer program. Table 3 summarises the maximum element forces for beams, columns and braces as comparison with the ideal pinned joints, used in real design of elements.

The results of static analyses show that the values of internal forces are dependent on column-base characteristics especially for beams and columns. In case of beams, the beam bending moments in the span increase by more than 10% when real web bolting column-base characteristics are integrated. However, this variation do not change the beams cross-sections, neither the connection details. In case of columns, the axial load is not affected by the integration of real column-base stiffness, while the bending moment decreases in case of web bolted connection with more than 20%. In these conditions the column cross-sections are slightly over-designed.

Table 3. Stresses in members after integrating behaviour of joints

Members	Stresses	Ideal pin connections	Angle profile connections		Anchor connections	
			web	splayed	web	splayed
Beam	M [kNm]	56.71	62.25	54.53	62.52	55.3
Column	M [kNm]	55.81	41.68	48.63	41.06	49.56
	N [kN]	54.15	54.18	54.16	54.18	54.15
Braces	N [kN]	32.60	32.90	33.28	34.00	33.20

5. Conclusions

For many existing multi-residential apartment block buildings with flat roof, the over-roofing represents a good alternative for retrofitting. The creation of additional spaces by over-roofing is an added value and can economically sustain the retrofitting costs. Probably one of the best solutions for over-roofing is represented by the use of cold-formed steel elements, due to their lightness and high levels of industrialisation that can be used. However, special attention should be payed to connection of the new steel structure with the existing concrete diaphragm structure. Present paper offers two column-base typologies by using column web bolts and respectively a splayed connection anchored to concrete diaphragms in two solutions, by back angle profiles and by anchor bolts.

The FEM simulation of these solutions proved a semi-rigid/partial-strength behaviour, with stiffness values close to nominally pinned connections in case of web fixing. The usual failure modes, in case of web connection, are by column web plasticization followed by contact local buckling of steel flanges. The latter represents also the failure mode of splayed connection with back angle anchoring, while in case of anchoring bolts, an early failure is recorded due to tearing of anchoring bolts.

The integration of real base-connection stiffness values into the structural response lead to variation of internal forces in case of beams and columns of up to 20-25% as compared to pinned-base structure. In consequence the design could be optimised.

References

[1] Norris M, Shiels P. "Regular National Report on Housing Development in European Countries – Synthesis Report", *The housing unit*, Dublin, Ireland, 2004.

[2] *Mettering international issue 1,* 2005.

[3] "Energy-efficient Buildings PPP beyond 2013. Research & Innovation Roadmap", *http://www.ectp.org/cws/params/ectp/download_files/36D2263v2_E2B_Roadmap_Info days_V.pdf,* 2012.

[4] Lorenzo G, Formisano A, Landolfo R, Mazzolani FM, Terracciano G. "On the use of cold-formed thin walled members for vertical addition of existing masonry buildings", - *SDSS'Rio 2010 Colloquium*, Rio de Janeiro, Brazil, 945-952, 2010.

[5] Dubina D, Ungureanu V, Georgescu M, Fülöp L. "Innovative Cold-Formed Steel Structure for Restructuring of Existing RC or Masonry Buildings by Vertical Addition of Supplementary Storey", *Proc. of the 3rd International Conference on Thin-Walled Structures*, Cracow, Poland, 187-194, 2001.

[6] Dubina D, Ungureanu V, Fülöp L, Nagy Zs, Larsson H. "LINDAB Cold-Formed Steel Structures for Small and Medium Size Non-Residential Buildings in Seismic Zones", *NSCC2001 Conference,* Helsinki, Finland, 463-470, 2001.

[7] INNSE. *The official census regarding population and dwellings made on October 20-31, 2011,* The National Institute of Statistics, Romania, http://www.insse.ro/.

[8] Floricel A, Zagari G, Ungureanu V. "Solutions for Column-Base Connections for Steel Over-roofing systems. Numerical investigations." *Proc. of the Eighth International Conference on Advances in Steel Structures*, Lisbon, Portugal, 2015.

[9] Floricel A, Benzar S, Zagari G, Ungureanu V. "Steel column-base solutions for over-roofing of existing buildings. A numerical study", *Proc. of the XIII International Conference on Metal Structures,* Zielona Gora, Poland, 15-17 June 2016 (in print).

[10] Botici AA, Ungureanu V, Ciutina A, Botici A, Dubina D. "Sustainable retrofitting solutions for precast concrete residential buildings". *Proc. of the IALCCE 2012 Symposium*, CRC Press, 357 + CD, 1722-1729, Vienna, Austria, 2012.

[11] EN1993-1-8. *Eurocode 3: Design of steel structures. Part 1-8: Design of joints.* ECS, Brussels, Belgium, 2005.

[12] Korotkov V, Poprygin D, Ilin K & Ryzhov S. "Determination of dynamic reaction in concrete floors of civil structures of nuclear power plant in accidental drops of heavy objects", *ABAQUS Users' Conference*, Boston, 25-27 May, 399–408, 2004.

[13] MAPEI. *Mapefix EP 385 - Pure epoxy, resin-based chemical anchor for structural loads*, http://www.mapei.com/.

TO PREDICTION OF THE CONNECTION DEFORMATION CAPACITY BY COMPONENT BASED FINITE ELEMENT METHOD

Lukáš Gödrich[a], František Wald[a], Luboš Šabatka[b], Marta Kurejková[a] and Jaromír Kabeláč[b]

[a] *Czech Technical University in Prague, Faculty of Civil Engineering, Prague, Czech Republic*
[b] *IDEA RS, Brno, Czech Republic*

Abstract: This paper introduces the prediction of deformation capacity of structural steel connections by the component based finite element method (CBFEM). The distribution of internal forces in connection is analysed by FEA. The connectors' behaviour is modelled by analytical models as components. The prediction of deformation capacity is verified on behaviour of the T-stub, the column web panel in shear, and the beam splices. The influence of the limits of materials strain and the estimation of the strain hardening and upper values material yield stress is described on the beam to column welded connection. The sensitivity study of connection deformation capacity shows applicability for practical design solutions.

1. Introduction

The prediction of deformation capacity of connections is currently offered by component method (CM), which builds up on standard procedures of evaluation of internal forces in connections and their checking. Zoetemeijer [1] was the first who equipped this model with prediction of stiffness and deformation capacity. The elastic stiffness was improved in the work of Steenhius, see [2]. Basic description of components behaviour in major structural steel connections was used by Jaspart for beam to column connections [3] and by Wald et al for column bases [4]. Method, which is implemented in the current European structural standard for steel and composite connections see [5] and [6] can be applied in majority of software for structural steel used in Europe. The model was generalised by da Silva [7]. Procedure starts with decomposition of a joint to components followed by their description in terms of normal/shear force deformation behaviour. After that, components are grouped to examine joint moment-rotational behaviour and classification/representation in a spring/shear model and application in global analyses. Advantage of the component model is integration of current experimental and analytical knowledge of connections components behaviour, bolts, welds and plates. This provides accurate prediction of behaviour in elastic and ultimate level of loading for simple configurations of joints. Disadvantage of component method is that experimental evaluation of internal forces distribution is available only for limited number of joint configurations. In temporary scientific papers, description of atypical components is either not present or has low validity in description of background materials. The CM´s is complex and not developed for hand calculation but for preparation of the design tables or tools.

Development of modern general-purpose software and decreasing cost of computational resources facilitate the trend of prediction of behaviour of structures by FEA. As the computational tools become more readily available and easier to use, even to relatively inexperienced engineers. Steel and steel and concrete connections are one of the last parts of structural design where the FEA is not commonly use. Finite element analyse (FEA) for connections is used from 70s of last century for research purposes. Their ability to express real behaviour of connections is making it a valid alternative to testing and current source of knowledge of connection's behaviour. Material model of structural steel for the research FEA uses true strain stress-strain diagram. Component based finite element method (CBFEM) is a multilevel FEA model developed to analyse and design connections of steel structures with features advanced analyses [8]. The distribution of internal forces in connection is analysed by FEA method. The resistance of steel plates design is evaluated by limiting the strain to 5% as it is recommended in cl. C.8(1) EN 1993-1-5 [9]. The proper behaviour of connectors, of bolts, welds etc., is treated by introducing them as components representing well its behaviour in term of initial stiffness, ultimate resistance and deformation capacity, see [10]. Standard procedure with partial safety factors for material/connections is applied.

The study is prepared in CBFEM IDEA RS, which uses for steel plates the most common MITC4 quadrangular element given by four nodes. Each node has all 6 degrees of freedom in translation and rotation. Deformation along the element are divided into membrane and flexural components. For membrane behaviour are also contemplated rotation perpendicular to the plane of the element. This provides full 3D formulation of element. Bolts are modelled as three sub-components. The first sub-component is the bolt shank, which is represented as a bilinear spring transmitting the compression force only. Its force-deformation diagrams in tension is prepared based on the experiments and standardised values for resistance. Initial stiffness, design tensile resistance, initialisation of yielding and deformation capacity are the main characteristics required of its tensile behaviour. Second subcomponent is connecting the bolt shank to the plates by restrains taking into account the bolt shear resistance. Third subcomponent modelled by restrains is limiting the bearing resistance.

Ductility with strength and stiffness belongs to the three basic parameters describing the behaviour of connections. In moment resistant connections, the ductility is achieved by a sufficient rotation capacity. Compare to well accept methods for determination of the initial stiffness and resistance of many types' structural joints, there are no generally accepted standardised procedures for the determination of the rotation capacity. The deemed to satisfy criteria are selected to help the engineers in EN 1993-1-8 [5]. The estimation of the rotation capacity is important in many applications namely in connections exposed to seismic, see [11] and [12], and extreme loading, see [13] to [15]. The deformation capacity of components has been studied from end of last century [16]. Faella et al [17] carried out tests on T-stubs and derived for the deformation capacity the analytical expressions. Kuhlmann and Kuhnemund [18] performed tests on the column web subjected to transverse compression at different levels of compression axial force in the column. Da Silva et al [19] predicted deformation capacity at different levels of axial force in the connected beam. Based on the test results combined with FE analysis deformation capacities are established for the basic components by analytical models by Beg at al [20]. In the work are represented components by non-linear springs, and appropriately combined in order to determine the rotation capacity of the joint for the end-plate connections, with an extended or flush end-plate, and welded connections. For these connections, the most important components that may significantly contribute to the rotation capacity column were recognised as the web in compression, column web in tension, column web in shear, column flange in bending, and end-plate in bending. Components related to the column web are relevant only when there are no stiffeners in the column that resist compression, tension or shear forces. The presence of a stiffener eliminates the corresponding component, and its contribution to the rotation capacity of the joint can be therefore neglected. End-plates and column flanges are important only for end-plate connections, where the components act as a T-stub, where also the deformation capacity of the bolts in tension is included. The questions and limits of deformation capacity of connections of high strength steel was studied in [21].

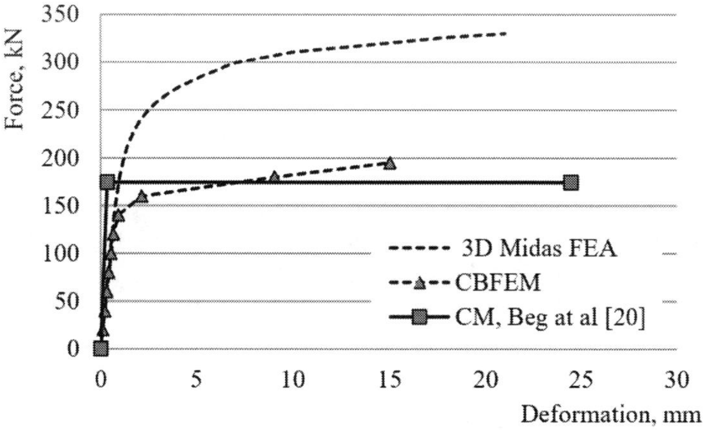

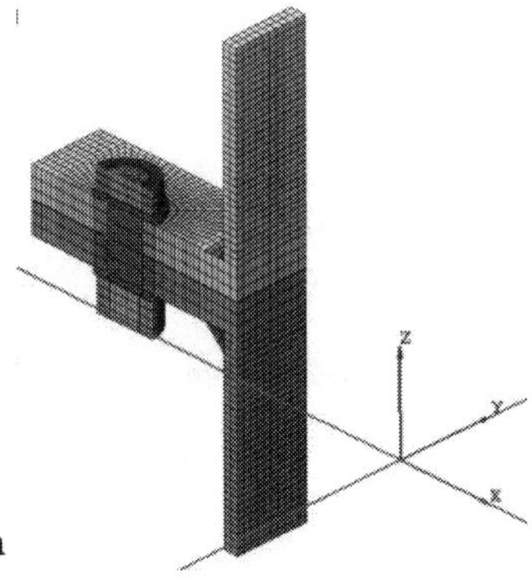

Fig. 1: a) Prediction of deformation capacity for a T stub t_f = 20 mm,
b) mesh and bolt representation for research FEA model by FEM MIDAS code

2. Verification

The detailed procedure for verification of proposed method was prepared [22]. The procedure contains Benchmark studies for connectors and major connections. In cases, where gives the CBFEM method higher resistance, initial stiffness or deformation capacity, the advanced FEM model validated on experiments is used, to approve physically good results of modelling. The quality of prediction of deformation capacity was validated on the T stub behaviour and on the column web panel in shear, which was calculated by CM according to [20].

In Fig. 1 is compared the prediction of the deformation capacity by CM and CBFEM to FEM research model [8]. The mesh of research model is includes. A T stub with flanges thickness t_f = 20 mm and width b_f = 300 mm, web thickness t_w = 10 mm, radius r = 14.1 mm for steel S235 is modelled. The limits for principal stain are expected 20% for plates and 5% for bolts. By CM is calculated the initial stiffness as 485 kN/mm and deformation capacity as 24.47 mm, see Table 1. By the research model using FEM MIDAS validated on experiments and using measured material characteristic, see [10], is the deformation capacity 21.07 mm. Using CBFEM is the deformation capacity predicted as only 15.02 mm. The Table 1 shows the influence of the flanges thickness t_f for the prediction of deformation capacity.

For flange thickness of 10 and 12 mm, the plastification of flange only, is the ed deformation capacity by CBFEM higher then by CM. In CBFEM is taken into account the membrane behaviour of the plate. For flange thickness 15 to 25 mm limits the bolt deformation capacity the behaviour of the T stub already and the CBFEM in these cases is more conservative. The prediction by CM according to [20] seems to be not realistic. For flange thicker than 30 mm has the limiting strain no influence to the deformation capacity of the T stub due to failure of the bolts.

Table 1: Deformation capacity of the T stub

Flange thickness t_f (mm)	Deformation capacity by CBFEM δ_{Cd} $\delta_{Cd, CBFEM}$ (mm)	Deformation capacity by CM $\delta_{Cd, CM}$ (kN)
10	37.50	24.47
12	35.10	24.47
15	23.77	24.47
20	15.02	24.47
25	11.16	13.86
30	3.99	15.71
35	3.66	17.55
40	3.84	19.40
45	4.05	11.50
50	2.59	12.50

Fig. 2 shows the plastic strain of the unstiffened welded beam-to-column connection, which was selected for comparison of prediction of deformation capacity by CBFEM to CM for beam IPE330 and column HEB260 in Fig. 3. The values for model by CM according to [20] were reached by FE analyses and validated on experiments for strain 10%. In cl. 6.4.3(2) of EN 1993-1-8 [5] is for an unstiffened welded beam-to-column joint assumed a rotation capacity φ_{Cd} at least 15.0 mrad. Conservative prediction by estimation in standard is hence confirmed.

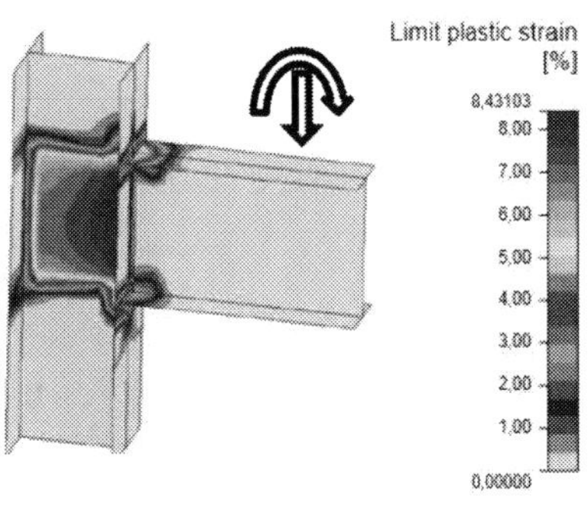

Fig. 2: Distribution of strains in welded connection

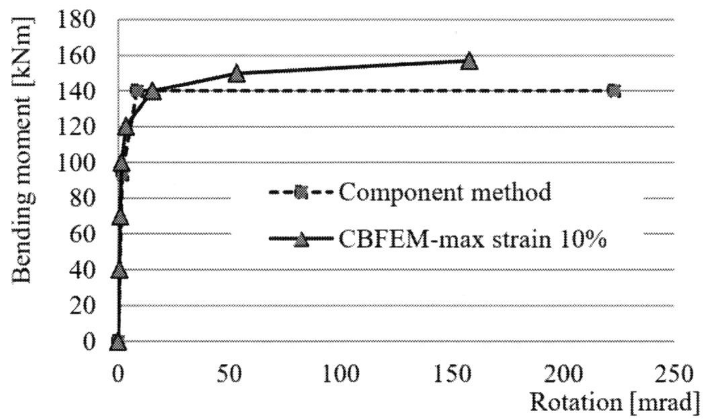

Fig. 3: Moment rotation diagram for beam to column welded unstiffened connection by CBFEM and CM

The bolted beam splice connection of two IPE300 in Fig. 4 shows the influence of the T stub to behaviour of the connection, which exhibits a limited deformation capacity. The end plates with steel S355 were here designed 12 mm thick with bolts M16 4.8. The sensitivity of the end plate thickness to the resistance and changes in the failure modes are demonstrated in Fig. 5. In the moment rotational diagram in Fig. 6 is visible the difference of the prediction of the rotational capacity by CBFEM and CM according to [20].

EN 1993-1-8 [17] in cl. 6.4.2(2) limits the plastic distribution between the bolt rows, for joints with a bolted end-plate connection provided that the design moment resistance of the joint is governed by the design resistance of the column flange or the beam end-plate in bending or the thickness t of either the column flange or the beam end-plate or tension flange cleat satisfies

$$t \leq 0.36d\sqrt{f_{u.b}/f_y} \qquad (1)$$

where d and $f_{u.b}$ are the diameter and strength of the bolt and f_y is the yield strength of the relevant plate. This criterion is in CBFEM taken into account naturally by checking/limiting the deformation capacity of each connector. The development of plastic zones round the bolts and limited plastification till failure of the second bolt row in tension is shown in Fig. 6b.

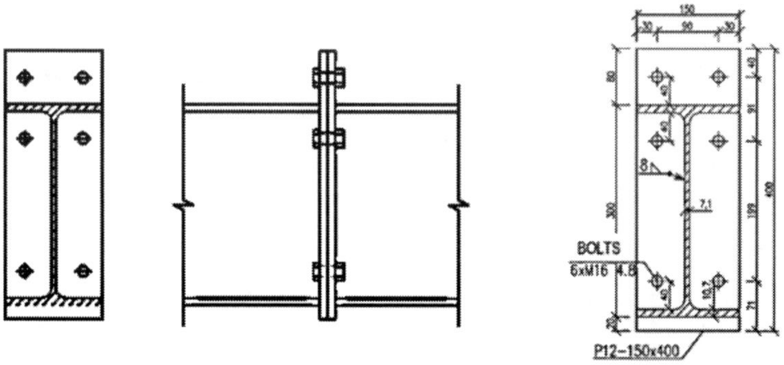

Fig. 4: Geometry of the studied bolted end-plate connection, IPE300, S355

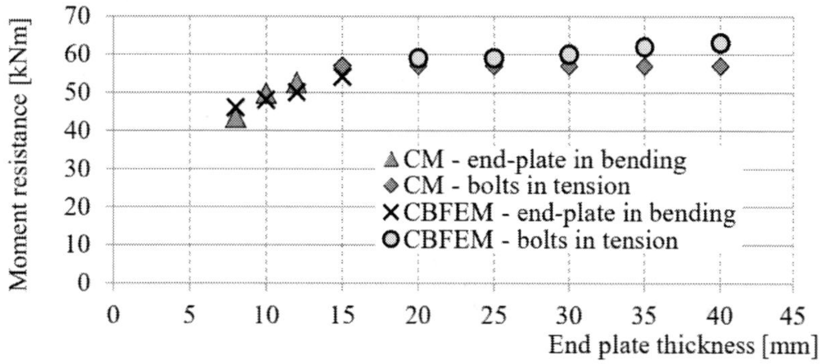

Fig. 5: Comparison of moment resistance predicted by CBFEM and CM for different plate thicknesses

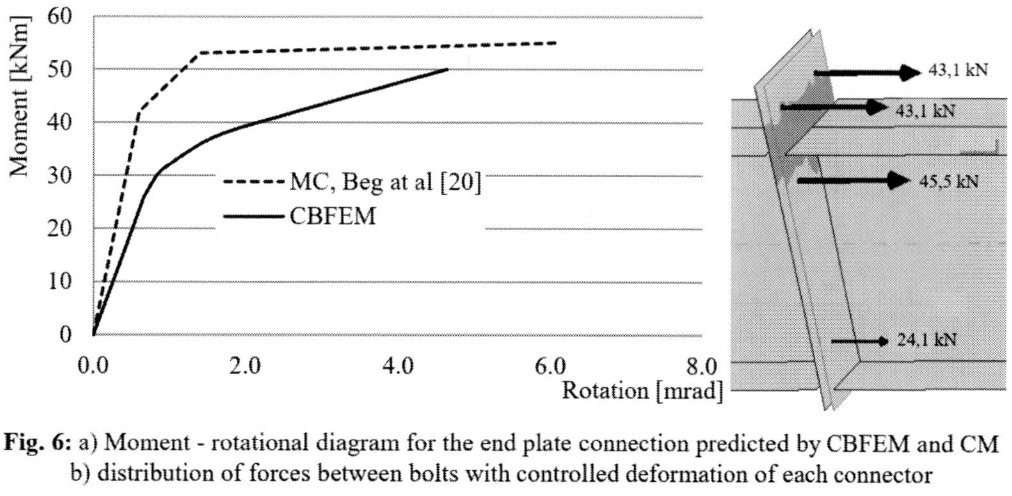

Fig. 6: a) Moment - rotational diagram for the end plate connection predicted by CBFEM and CM b) distribution of forces between bolts with controlled deformation of each connector

3. Sensitivity study

It is commonly known, that the relative strength of individual components is of great importance for ductility of connections. The already available knowledge and good engineering practice in seismic design [23] introducing the parameters for overstrength of parts of joints may be utilised in the methodology for prediction of deformation capacity of connections.

The influence of not guaranteed values of yield strength of the structural steel to the ductility of connections is demonstrated in Fig. 7 for beam to column welded unstiffened connection in Fig. 2 described in previous paragraph. The rotation capacity reduces for example from 170 mrad for column with f_y = 235 MPa to 7 mrad for column with f_y = 1.5 · 235 MPa. For bolted connection it was shown in previous paragraph the good and safe estimation of 5% limit of strain. For column panel zone in shear is the limiting strain changed from 3% till 20 %. Currently is guaranteed for structural steel 15% strain. Such or specific value of steel strain for particular steel may be recommended for prediction of ductility of connection.

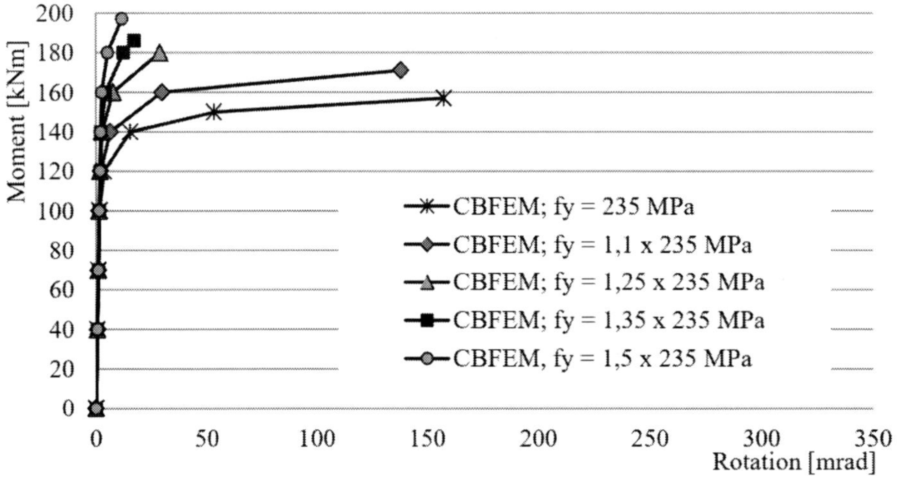

Fig. 7: Moment - rotational diagram for beam to column welded unstiffened connection calculated by CBFEM for changed material properties of column only

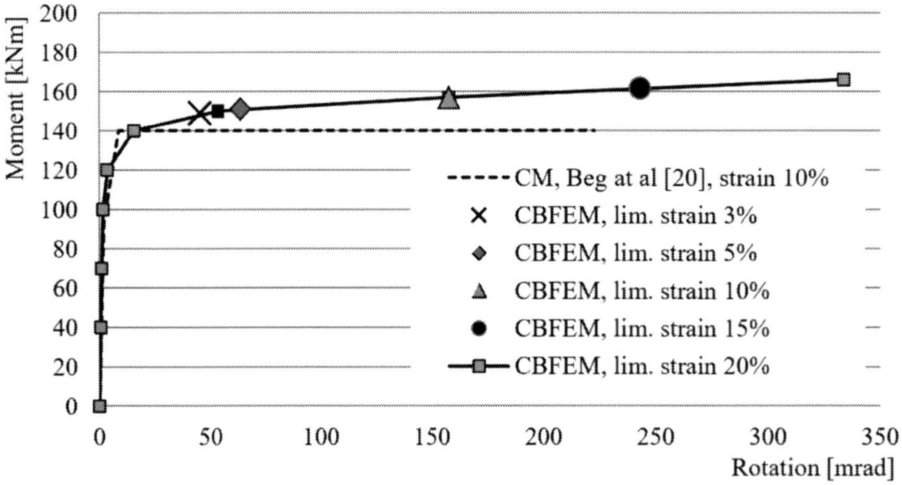

Fig. 8: Moment - rotational diagram for beam to column welded unstiffened connection calculated by CBFEM with limits in strain

4. Conclusions

In current models for prediction of the rotational capacity of structural steel connections by component method is the behaviour of steel plates implemented from finite element analyses FEA. The quality of prediction limits also dissimilarity of assembly procedures for stiffness and resistance, the estimation of the lever arm of internal forces, and neglecting the interaction of internal forces.

The multilevel FEA, component based finite element method CBFEM, gives naturally a realistic prediction of ductility of connections, represented by its deformation and rotational capacity, based on model of steel plates including its membrane action and analytical model of connectors behaviour. The changes of position of internal forces during loading of connection is taken into account. A comparison with the test results shows good agreement.

This contribution is not focusses to welds ductility, which may be of course simply examined by CBFEM, where this component has its stiffness, strength and deformation capacity. The study of this question is under preparation.

Next necessary step for the further studies is validation of CBFEM for prediction of ductility of different types of connections to experiments. For correct application of advanced models in everyday practice is necessary to prepare benchman cases and to standardise the advanced models of connections similar to already published cases in another parts of structural engineering, like in plated structures, fire engineering or structural dynamics.

Acknowledgments

The work was prepared under the project MERLION supported by Technology Agency of the Czech Republic, project No. TA02010159.

References

[1] Zoetemeijer P. "Summary of the Researches on Bolted Beam-to-Column Connections", *Report 6-85-7*, University of Technology, Delft, 1985.

[2] Steenhuis M, Gresnigt N, Weynand K. "Pre-Design of Semi-Rigid Joints in Steel Frames", *Proceedings of the Second State of the Art Workshop on Semi-Rigid Behaviour of Civil Engineering Structural Connections*, COST C1, Prague, 1994, 131-140.

[3] Jaspart JP. "Design of structural joints in building frames", *Prog. Struct. Engng Mater.*, 4 (2002), 18–34.

[4] Wald F, Sokol Z, Steenhuis M, Jaspart JP. "Component Method for Steel Column Bases", *Heron,* 53(2008), 3-20.

[5] EN 1993-1-8, *Eurocode 3: Design of steel structures, Part 1-8: Design of joints*, CEN, Brussels, 2006.

[6] EN 1994-1-1, *Eurocode 4: Design of composite steel and concrete structures, Part 1-1: General rules and rules for buildings*, CEN, Brussels, 2010.

[7] da Silva SL. "Towards a consistent design approach for steel joints under generalized loading", *Journal of Constructional Steel Research,* 64(2008), 1059-1075.

[8] Wald F, Gödrich L, Šabatka L, Kabeláč J, Navrátil J. "Component Based Finite Element Model of Structural Connections", *Proceedings Steel, space and composite structures,* 337-344, 2014.

[9] EN 1993-1-5, *Eurocode 3: Design of steel structures, Part 1-5: Plated Structural Elements*, CEN, Brussels, 2005.

[10] Šabatka L, Wald F, Kabeláč J, Kolaja D, Pospíšil M. "Structural analysis and Design Using Component-Based Finite Element Model", *J. Civ. Eng. and Arch,* 9(2015) 895-901.

[11] Gioncu V, Mazzolani F. *Ductility of seismic resistant steel structures*, Spon Press, 2002.

[12] Grecea D, Stratan A, Ciutina A, Dubina D. "Rotation capacity of the beam-to-column joints under cyclic loading", *Connections in Steel Structures V*, Amsterdam, 2004.

[13] Sherbourne AN, Bahaari MR. "3D simulation of end-plate bolted connections", *Journal of Structural Engineering,* 120(1994), 3122–3136.

[14] Sherbourne AN, Bahaari MR. "Simulation of bolted connections to unstiffened columns T-stub connections", *Journal of Constructional Steel Research,* 40(1996), 169–87.

[15] Sherbourne AN, Bahaari MR. "3D simulation of bolted connections to unstiffened columns—II. Extended endplate connections", *Journal of Constructional Steel Research,* 40(1996), 189–223.

[16] Foley CM, Vinnakota S. "Toward design office moment–rotation curves for end-plate beam-to-column connections", *Journal of Constructional Steel Research,* 35(1995), 217–753.

[17] Faella C, Piluso V, Rizzano G. "Plastic deformation capacity of bolted T-stubs: theoretical analysis and testing", *Moment resistant connections of steel frames in seismic area, design and reliability*, E&FN Spoon. 2000.

[18] Kuhlmann U, Kuhnemund F. "Rotation capacity of steel joints", *Verification Procedure and Component Tests*, 2000.

[19] da Silva LS, Lima L, Vellasco P, Andrade S. "Experimental behaviour of end-plate beam-to-column joints under bending and axial force, Database reporting and discussion of results", *Report on ECCS-TC10 Meeting in Ljubljana*, 2002.

[20] Beg D, Zupancic E, Vayas I. "On the rotation capacity of moment connections", *Journal of Constructional Steel Research,* 60(2004), 601-620.

[21] Girao AMC, Bijlaard FSK, da Silva LS. "Experimental assessment of the ductility of extended end plate connections", *Journal of Engineering Structures,* 26(2004), 1185–1206.

[22] Wald F, Kwasniewski L, Gödrich L, Kurejková M. "Validation and verification procedures for connection design in steel structures", *Proceedings Steel, space and composite structures,* 111-120, 2014.

[23] EN 1998-1, *Eurocode 8: Design of structures for earthquake resistance – Part 1: General rules, seismic actions and rules for buildings*, CEN, Brussels, 2004.

EXPERIMENTAL STUDY ON MECHANICAL BEHAVIOR OF BEAM TO SQUARE CFT COLUMN CONNECTIONS WITH BUILT-IN REINFORCING BARS

Shintaro Matsuo

Faculty of Human-Environment Studies, Kyushu University, Fukuoka, Japan

Abstract: This study presents the development of a new connection of beam to square CFT column. The connection is characterized by the use of built-in reinforcing bars in the CFT members, which joint the column to column without a welding work. Therefore, it can be easily applied on the structure using super high strength steel that has been developed in recent years. In this study, the applicability of this connection is investigated experimentally with several specimens. As a result, it reveals that the hysteretic behaviors are very stable in the case of the appropriate anchored length of the bars. And the proposed strength evaluation method is reasonable for the practical use.

1. Introduction

Recently, a lot of researches have been done on super high strength steel (SHSS), and structural performance of concrete filled steel tubular (CFT) columns or beam to column connections with exterior diaphragms have been revealed by loading tests in literature [1,2,3]. In general, the SHSS requires much higher technique in welding work to assure the weld quality in comparison with conventional steel. Since welding is inevitable in beam to CFT column connections, developments of new construction methods are expected in order to spread the use of structural systems made of SHSS. Therefore, in this paper, it is proposed a new connection, which is composed of beam to column joint made of conventional steel, and CFT column made of SHSS. Then, fundamental mechanical behavior of the connection is investigated by the loading test.

2. Outline of the proposed connection

The details of the connection proposed in this paper are shown in Fig. 1. Exterior diaphragm type is adopted in the beam to column joint. The joint is connected to the upper and lower CFT columns through steel plates for fixing and built-in reinforcing bars. The fixing plates are used in order to fix the built-in reinforcing bars. The projections on the steel plates play a role to prevent columns move in horizontal direction under construction. Shear connectors are welded by fillet weld near the end of inner surfaces of steel tubes, which enables stress transfer from steel tubes to built-in reinforcing bars. This stress transfer mechanism has been already investigated in some previous researches (for example, literature [4]). Anchor plates are installed at the end of built-in reinforcing bars so that anchored length of the bars is able to be made shorter because of the bearing resistance of the anchor plates. In this paper, the author mainly discusses the behavior of the joint between the high strength CFT column and the CFT column made of conventional steel.

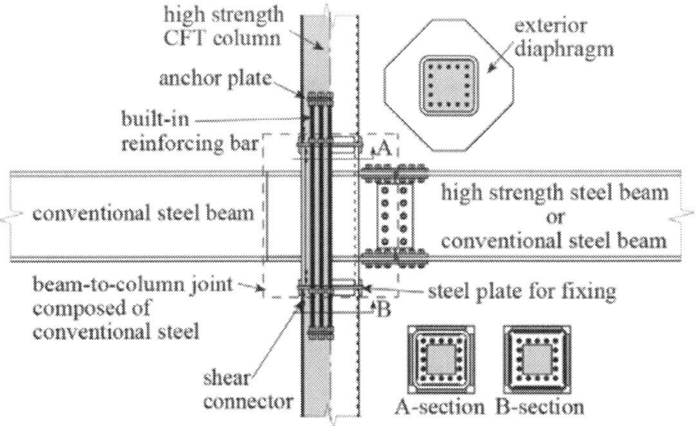

Fig. 1: Details of the connection method proposed in this paper

3. Outline of experiment

3.1 Specimens

An example of specimens is shown in Fig. 2, and the list of specimens is shown in Table 1. The width to thickness ratio of column (*B/t*), distance between the surface of the diaphragm plate and the end of the column (projecting dimension), anchored length of the built-in reinforcing bars and compression strength of filled concrete are adopted as experimental parameters. The dimensions of cross-section of beam, column and diaphragm are shown in Table 1. The steel grade of conventional steel column is BCR295, which is approved by the Minister of land, infrastructure and transportation in Japan, and those of beam and diaphragm are SN490B specified by Japanese Industrial Standard (JIS). Since the objective of the test is to investigate the fundamental mechanical behavior of the connection, the conventional steel (BCR295) is used to steel tubes for column shafts. Pull-out failure mode in the joint is expected in No. 1 specimen. As for the others, full plastic strength of the joint is expected to be reached. The compression strength of the filled concrete is shown in Table 1, and steel material properties are shown in Table 2.

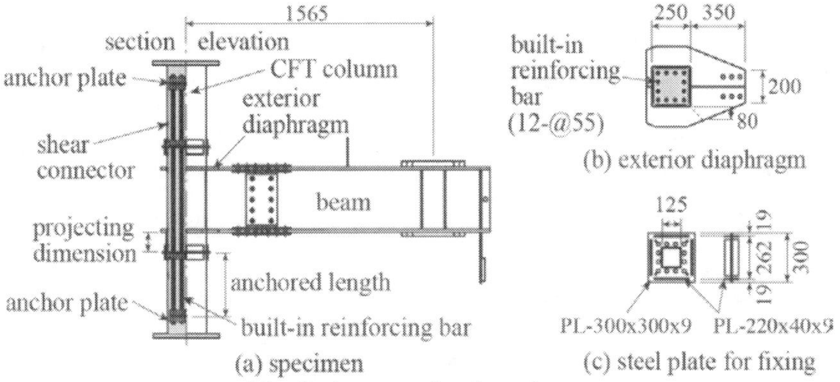

Fig. 2: An example of specimens

Table 1: List of specimens

No.	Steel tube (BCR295) Cross-section (mm)	Steel tube (BCR295) Projecting dimension (mm)	Wide flange beam (SN490B)	Exterior diaphragm (SN490B)	Built-in reinforcing bar (SD490) Bar arrangement	Built-in reinforcing bar (SD490) Anchored length (mm)	Concrete (N/mm^2)
1	250×6	125	H-400×200×9×16	PL-16 See Fig. 2	12-D19 (*D* is the nominal diameter of bar)	10*D*	38.3
2	250×6	125				20*D*	74.8
3	250×9	125					78.0
4	250×6	63					77.9

Table 2: Steel material properties

Grade	Size / Thickness Diameter (mm)	No.	Young's modulus (N/mm^2)	Yield stress (N/mm^2)	Tensile stress (N/mm^2)	Yield strain (%)
SN490B	16	No. 1, Beam	202,528	372	516	0.18
		No. 2~4	208,622	362	539	0.17
BCR295	250×6	No. 1	209,728	404	511	0.19
		No. 2, 4	204,230	378	460	0.19
	250×9	No. 3	201,939	415	452	0.21
SD490	*D*19	No. 1, 3, 4	182,440	526	710	0.29
		No. 2	185,951	538	722	0.29

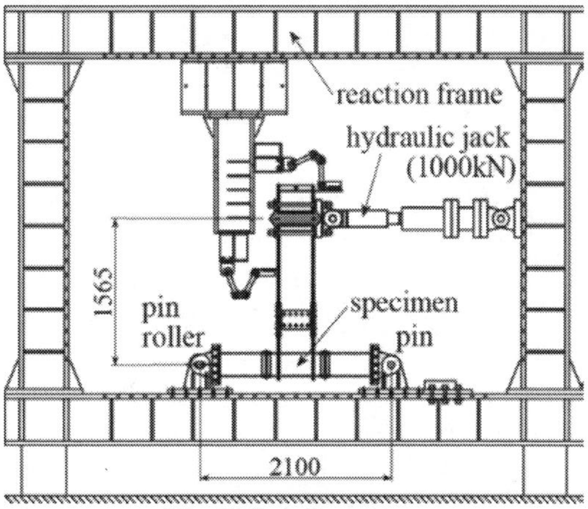

Fig. 3: Loading set up

3.2 Loading plan

The loading set up is shown in Fig. 3. The left side end plate of the column is bolted with a pin roller, and the right one with a pin. The pin roller of the left side permits the horizontal displacement. Specimens are horizontally loaded by an hydraulic jack at the beam end. The horizontal loading is controlled by the story drift angle (R). The amplitude of R is increased by ±0.01 rad from ±0.01 rad to ±0.04 rad after ±0.005 rad. Thereafter, the loading tests are monotonically continued up to the limit of jack stroke.

4. Test results and considerations

4.1 Relations between joint moment and joint rotational angle

Fig. 4 shows relations between joint moment (M_j) and joint rotational angle of the pin roller side (θ_j), where M_j and θ_j are defined as the bending moment and the rotational angle at the joint part, respectively. All specimens represent the hysteresis loops which are influenced by slip characteristics. No. 1 reaches the maximum strength at R of 0.03 rad. The other specimens keep increasing to their strength without strength deteriorations until the end of the loading. The strength of No. 1 is much smaller than those of the others, which failure mode is the pull-out failure of built-in reinforcing bars at the joint. On the other hand, the other specimens are considered to reach to plastic state at the joint, so that no pull-out failure occurs. The strength and absorbed energy of No. 3 with B/t of 28 are larger than those of No. 2 and No. 4 with B/t of 42. This indicates that the strength of the joint is affected by B/t, and that the absorbed energy is almost proportional to the quantity of plastic deformation of the built-in reinforcing bars.

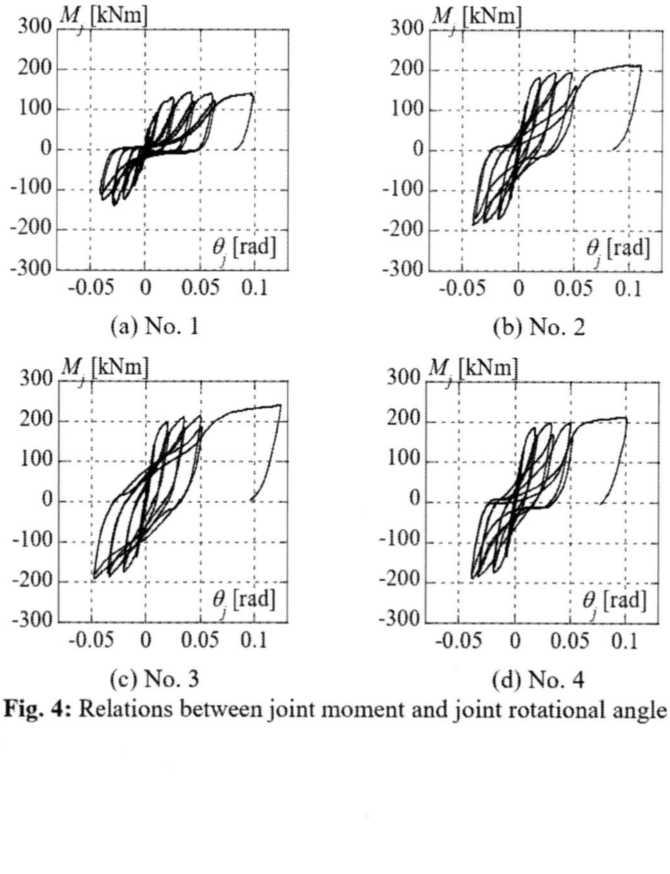

Fig. 4: Relations between joint moment and joint rotational angle

4.2 Relations between strains in steel tube and story drift angle

Fig. 5 shows relations between the strains in a steel tube (ε) and the story drift angle (R). Strain gauges are installed as shown in Fig. 6. 'a' and 'c' in the name of the strain gauges mean longitudinal and circumferential direction of the steel tubes. Positive value of the strain is tension. From Fig. 5, it can be found that circumferential strains of both specimens show tensile strain throughout the loadings. This is because plate of the steel tubes bends in out-of-plane direction, which is caused by the formation of splitting and compression struts of the filled concrete. From Fig. 5(a) and (c), longitudinal strains of No. 1 are larger than those of No. 2, which means that the steel tube of No. 1 resist large tensile force around the joint because of the pull-out failure of built-in reinforcing bars.

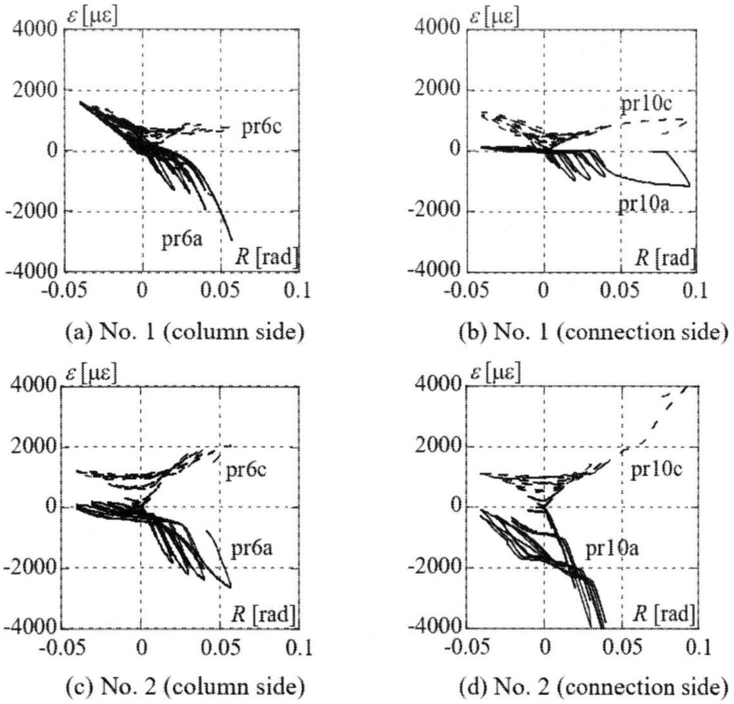

(a) No. 1 (column side) (b) No. 1 (connection side)

(c) No. 2 (column side) (d) No. 2 (connection side)

Fig. 5: Relations between strains in steel tube and story drift angle

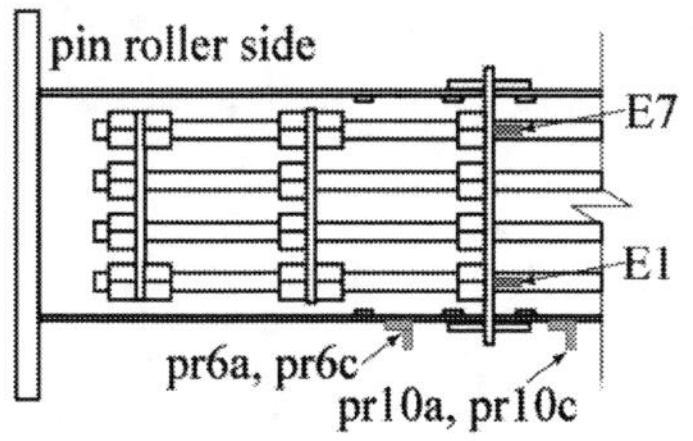

Fig. 6: Arrangement of strain gauges

4.3 Relations between strains of built-in reinforcing bars and story drift angle

Fig. 7 shows relations between the strains of built-in reinforcing bars (${}_r\varepsilon$) and the story drift angle (R). Dashed lines mean the yield strains of the built-in reinforcing bars. Arrangement of the strain gauges are shown in Fig. 6. As for No. 1, the built-in reinforcing bars never reach to the yield strain through the loading process. However, with regard to No. 2, the bars of the outermost edge of the tension side (E1 or E7) exceed the yield strain at R of 0.02 rad., which means that the bar yielding approximately corresponds to the yield strength of the joint (${}_eM_y$). ${}_eM_y$ is the strength when the tangential stiffness in M_j-θ_j relations becomes 1/3 of the elastic stiffness.

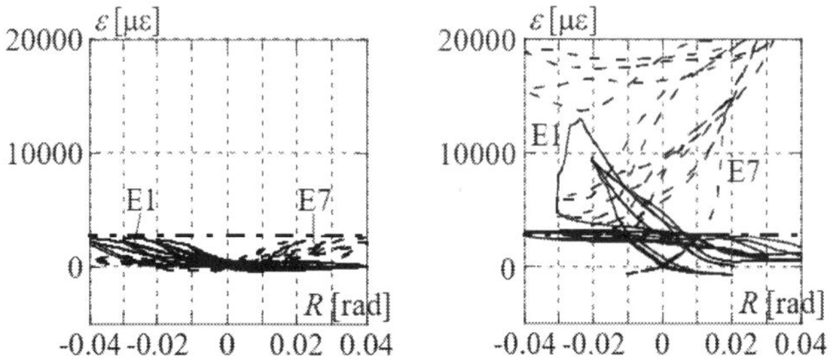

(a) No. 1 (b) No. 2

Fig. 7: Relations between strains of built-in reinforcing bars and story drift angle

5. Estimation of joint strength

5.1 Ultimate strength by the generalized superposed strength method

The ultimate strength of the joint is estimated by assuming full plastic state at the joint as shown in Fig. 8. In this model, compressive force is defined as positive value, and it is assumed that the steel tube and the filled concrete do not resist tensile stress. From this model, the ultimate axial strength (N_u) and the ultimate bending strength (M_u) are obtained by the equations as follows;

$$N_u = \{{}_sD \cdot {}_st + 2(x_n - {}_st){}_st\}{}_s\sigma_y + \sum {}_r\sigma_{yj} \cdot a_j + {}_c\sigma_B \cdot x_n \cdot {}_cD \tag{1}$$

$$\begin{aligned} M_u = &\{\frac{1}{2}{}_sD({}_sD - {}_st) + (x_n - {}_st)({}_sD - x_n + {}_st)\}{}_st \cdot {}_s\sigma_y \\ &+ \sum {}_r\sigma_{yj} \cdot a_j(\frac{{}_cD}{2} - d_j) + \frac{1}{2}({}_cD - x_n + {}_st)x_n \cdot {}_cD \cdot {}_c\sigma_B \end{aligned} \tag{2}$$

where, ${}_sD$: column width; ${}_st$: column thickness; ${}_s\sigma_y$: yield stress of steel tube; ${}_c\sigma_B$: compressive strength of concrete; ${}_cD$: width of concrete part; ${}_r\sigma_{yj}$: stress of the reinforcing bar in the jth row; a_j : cross-sectional area of the bar in the jth row; d_j : distance between compressive edge and the center of the bar in the jth row. Eqn. (1) or Eqn. (2) is superposed by strength of three portions which are composed of steel tube (the first term of each equation), the built-in reinforcing bar (the second term of each equation) and concrete (the third term of each equation).

5.2 Comparison between strength by calculation and that by experiment

The calculated ultimate strengths by Eqn. (2) (${}_cM_u$) are shown in Fig. 4 by horizontal dashed lines. As for the positive loading, it can be found that the calculated results are nearly equal to the experimental results which are shown by circles (${}_eM_u$) except for No. 1, whose failure mode was the pull-out failure. The ${}_eM_u$ is defined as the strength when the tangential stiffness in M_j - θ_j relation becomes 1/6 of the elastic stiffness. On the other hand, the experimental results of the negative loading are somewhat overestimated. These differences are regarded as the influence of repeated loading. The deformation of the joint at the strength of ${}_cM_u$ approximately corresponds to θ_j of 0.02 rad. It can be considered that the generalized superposed strength method is reasonable in order to predict the strength of the new connection proposed by Author.

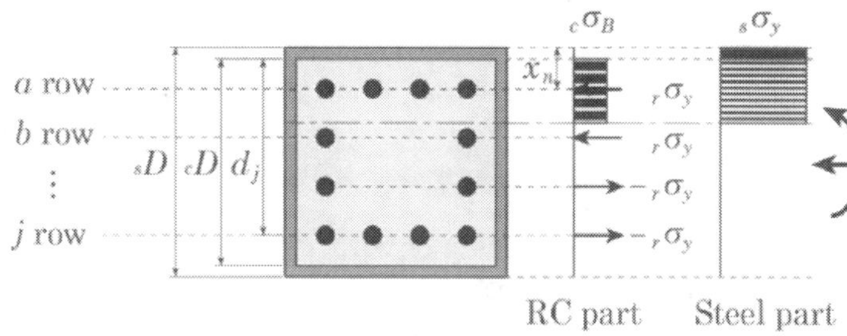

Fig. 8: Stress state of the joint at the full plastic strength

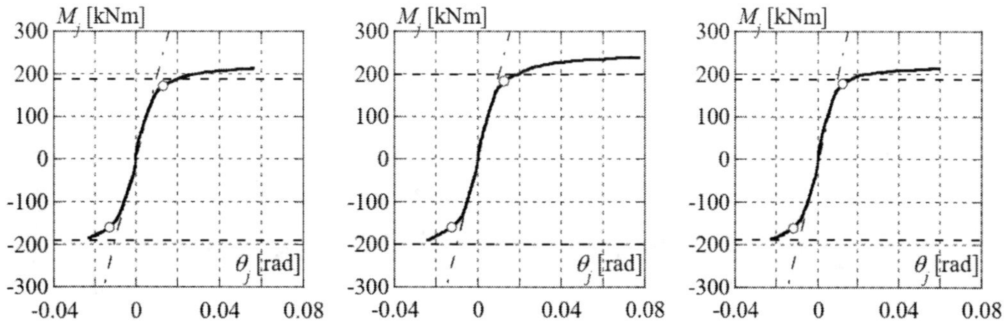

(a) No. 2 (b) No. 3 (c) No. 4

Fig. 9: Skeleton curves and the ultimate strengths
(dashed line: calculated strength ($_cM_u$), circle : experimental strength ($_eM_u$))

6. Conclusions

On the basis of experimental investigation, the conclusions are summarized as follows:

1. A new connection of beam to square CFT columns using built-in reinforcing bars was proposed. The connection was proved to be effective for practical CFT structures.
2. Full plastic state of the joint could be reproduced by adopting the appropriate anchored length of the built-in reinforcing bars. As a result, the hysteretic behavior of the joint is very stable up to the maximum deformation in spite of including the slip characteristics.
3. In this connection method, a part of stress between joint and column is intended to be directly transmitted to the steel tube via the fixing steel plate. In the loading test, it was confirmed that remarkable differences on the strength and hysteresis loop occurred due to the width-thickness ratio of square steel tube. This means that the fixing steel plates have worked as intended.
4. The strength estimation equation has been derived being based on the generalized superposed strength method. The good agreements between the calculated strength values and the experimental strength values were observed in the loading test. It can be considered that the strength estimation method is reasonable for prediction of the strength of the new connection proposed by Author.

Acknowledgments

This research was supported by The Japan Iron and Steel Federation (JISF). I would like to show my appreciation to Prof. Kawano of Kyushu University who gave me insightful comments and suggestions. And I would particularly like to thank Mr. Ueoka, K., Mr. Togawa, D. and Mr. Kashiwagi, K. who are students in Kyushu University.

References

[1] Terasawa T. et al. "Experimental Study on Structural Performance of CFT Column Using Super High Strength Steel (Part 1 ~ Part 4)", *Summaries of technical papers of annual meeting architectural institute of Japan*, C-1 Structures III, 1153-1160, 2007 (in Japanese).

[2] Hirade T. et al. "Experimental Study on Structural Performance of CFT Column Using Super High Strength Steel (Part 5 ~ Part 7)", *Summaries of technical papers of annual meeting architectural institute of Japan*, C-1 Structures III, 1111-1116, 2008 (in Japanese).

[3] Takenaka H. et al. "Elasto-Plastic Behavior of Circular-Column-to-H-shaped-Beam Connections Employing Super High Strength Steel (Part 1)", *Summaries of technical papers of annual meeting architectural institute of Japan*, C-1 Structures III, 755-756, 2008 (in Japanese).

[4] Matsuo S. et al. "Experimental Study on Mechanical Behavior of Exposed-type Square CFT Column Base with Built-in Reinforcing Bars", *Behavior of steel structures in seismic areas (STESSA2012)*, 645-651, 2012.

ASSESSMENT OF LOCKING DEVICES FOR BOLTED CONNECTIONS – EXPERIMENTAL STUDY AND EVALUATION FOR PRACTICAL APPLICATION

Thomas Misiek

Breinlinger Ingenieure, Tuttlingen, Germany

Abstract: EN 1933-1-8 and EN 1998-1 give recommendations for applications where bolts with locking devices or preloaded bolts should be used to avoid self-loosening. This includes shear-loaded joints under impact load, seismic load or vibrations. The paper gives comments on these recommendations. With increased requirements, locking devices are often used with preloaded bolts, leading to modified tightening procedures ("modified torque method" or "modified combined method"). Based on the example of wedge lock washers,–assessment procedures for locking devices are presented. These wedge lock washers secure preloaded bolted joints with tension instead of friction: wedge-shaped cams lead to an increase of preload during untightening. The graphs obtained in the tests and given in the paper allow for a detailed study of the tightening process, both for conventional assemblies and assemblies with wedge lock washers. Based on these tests, the equivalent friction coefficient for the contact interface between washer and bearing surface can be determined. Taking into account the scatter of the equivalent friction coefficient well as the minimum strength of the bolts, nominal preload and nominal tightening torque can be defined. This procedure is explained based on the aforementioned tests.

1. Self-loosening of bolted connections

Vibrations, impact or dynamic loading may cause unlocking and self-loosening of bolted connects: Self-loosening is caused by small slip in shear-loaded lap joints (bearing type connections of Category A according to EN 1993-1-8). At first, the displacement corresponds only to bending of the bolt, no relative displacement between nut or bolt and connected members occur. With increased bending of the bolt, friction is overcome by the bending forces and slip in the connection will allow a small rotation of the nut. This causes a reduction in preload. After several cycles and a sufficient loss of preload, friction forces are low enough to be overcome by (cyclic) shear forces and the nut will finally loosen rapidly. Locking devices prevent self-loosening and reduction of preload by blocking rotation of the nut.

Small clamping lengths increase the problem significantly, as the formula for the theoretical threshold slip $s_{G,th}$ in a connection shows:

$$s_{G,th} = \frac{F_{p,C} \cdot \mu_b \cdot \left(\sum t\right)^3}{12 \cdot E \cdot I} \tag{1}$$

A small clamping length leads to higher bending stresses in the bolt with the same amount of slip. Self-loosening occurs if the bending stiffness $E{\cdot}I$ of the bolt is larger than the friction force in the bearing surface which is provided by the preload (or $F_v \geq F_{p,C}{\cdot}\mu_b$).

2. Self-loosening of bolted connections and treatment in standards

Eurocodes and corresponding execution standards give advice in which cases locking devices shall be used: both EN 1993-1-8 and EN 1090-2 state that if bolted connections are subject to significant vibrations or impact, locking devices should be used. EN 1993-1-8 limits the problem to applications with small clamping lengths Σt. which is certainly an over-simplification

According to EN 1993-1-8, if preload is not explicitly used in the design calculations for slip resistances, but is required for execution purposes or as a quality measure (e.g. for durability) then the level of preload can be specified in the National Annex. For example, the German National Annex defines

$$F_{p,C}{*} = 0.7 \cdot f_{yb} \cdot A_S \tag{2}$$

as a level of preload for such purposes.

There are also requirements for structures for earthquake resistance: EN 1998-1 requires that Categories B and C of bolted joints in shear should be used. This cannot be generalized as a general requirement for slip-resistant connections: As a seismic design situation is an ultimate limit state, but bolted Category B connections (slip-resistant at serviceability) work as bearing type connections and not as a slip-resistant connection in this design situation. In fact, preload is used as a locking method. And other locking methods may also be used. This corresponds to the requirements given in national design codes: The German DIN 4149 – which is based on EN 1998-1 – generally states that all bolts shall be secured against unlocking.

Generally a preload F_p is regarded as a locking method in EN 1090-2. The German National Annex to EN 1993-1-8 also notes, that 50% of $F_{p,C}{}^*$ is sufficient to prevent unlocking. But preload can be significantly reduced by the aforementioned vibrations or impact, causing small shear displacements in the connections or by embedding. So the assumption is in contrast to the results of [1] and [2].

The use of locking devices is a possible way out of this dilemma. Locking devices in steel structures need to be specified. EN 1090-2 gives a list of products which may be used, for example prevailing torque nuts (all-metal or with non-metallic inserts). Nevertheless there are still several products on the market which do not deliver what they promise, such as spring washers, serrated lock washers and self-locking counter nuts. Most of them are based on old DIN standards which were withdrawn in because of "proven ineffectiveness" [3].

Junker's test developed in the 60's [4] allows a qualitative comparison of characteristics of locking methods for bolts under transverse loading conditions. The setup consists of a preloaded assembly connecting two members and a load-cell. The member adjacent to the nut is subjected to repeated transverse displacement of constant amplitude (i.e. displacement controlled test). Between the members, no friction occurs. The reduction in preload is measured continuously. Fig. 1 shows a comparison of two different locking methods: One assembly was locked only by preload, using a standard washer. The other one was locked with an additional locking device, in this case a wedge washer. Wedge washers are not defined in a standard but have shown their suitability. The geometry corresponds to washers according to EN 14399-6, allowing the use with assemblies according to EN 14399-4. Compared with the assembly with the standard washer, the wedge washer shows just a slight reduction in preload, proving its effectiveness and – in contrast – the ineffectiveness of preload only.

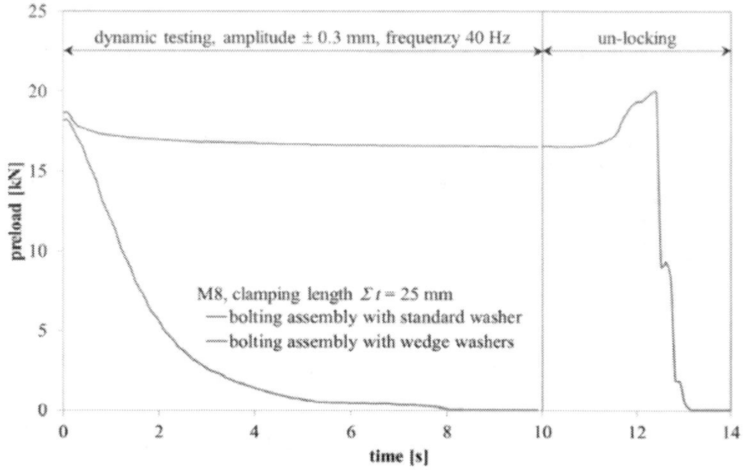

Fig. 1: Assemblies going through Junker's test (Source of data: Nord-Lock GmbH)

3. Wedge-washers and their mode of action

A wedge washer is a two-piece washer with cams between the pairs, as shown in Fig. 2 for the Nord-Lock SC wedge washers.

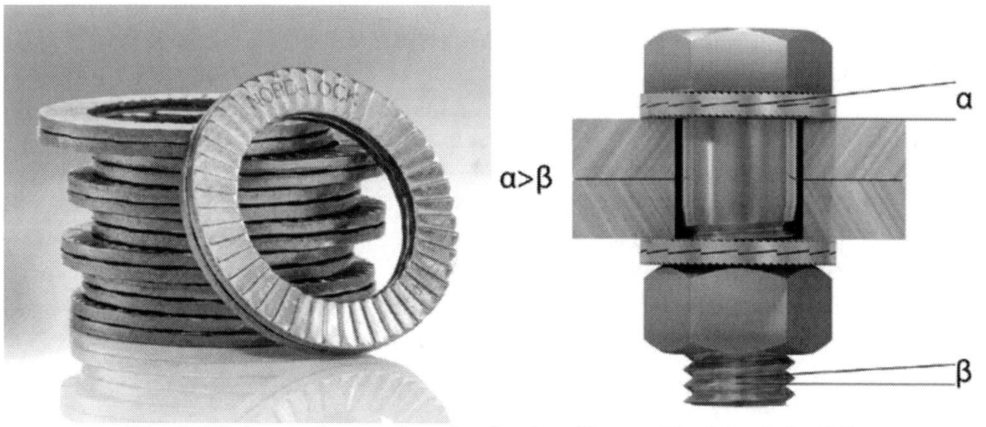

Fig. 2: Wedge washers and mode of action (Source: Nord-Lock GmbH)

The mode of action of wedge washers is based on the principle of the difference between the angle α of the cams and the angle β of the thread. Teeth on the surface of the washer are gripping into the bearing surfaces (Fig. 3), requiring the hardness of the surfaces being lower than the hardness of the wedge washers. The teeth lock the washer in place, allowing movement only in the layer of the cams. But unlocking of the bolt requires an increase of force in the bolt: As the angle α of the cams and the angle β of the thread, the two-piece washer expands more than the corresponding pitch of the thread provides increase of length: Unlocking automatically leads to an increase in preload, in turn preventing unlocking. This effect can also be found in controlled unlocking the nut. Right part of Fig. 1 shows unlocking of the nut after the test. The increase of preload is clearly visible.

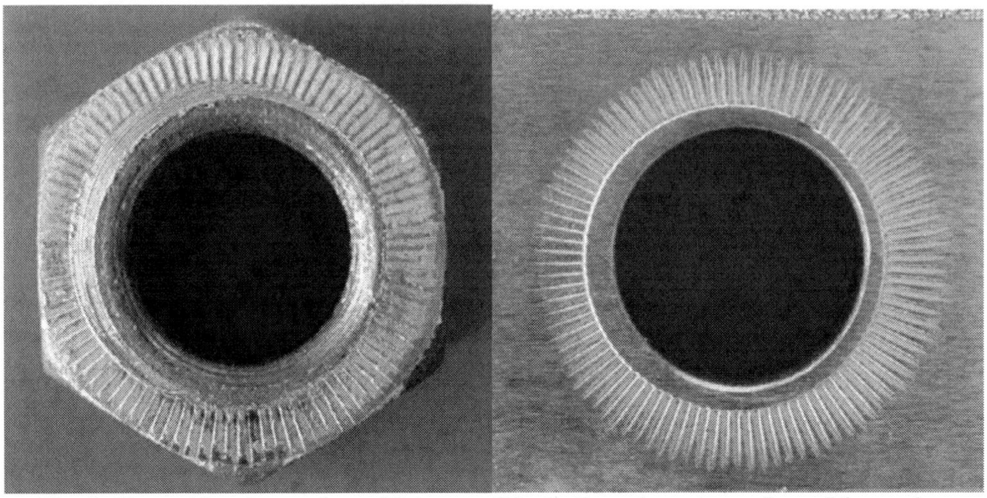

Fig. 3: Nut and bearing surface with traces of the teeth on the outer surface of the washer

4. Tightening of and preload in standard assemblies

EN 14399-1 and EN 1090-2 are working with k-classes and values k_i or k_m which define the ratio between torque and preload $F_{p,C}$, defined in EN 1090-2:

$$k_m = \frac{M_r}{d \cdot F_{p,C}} \tag{3}$$

Assemblies are grouped into k-classes K0, K1 and K2, in which K0 is not relevant here. The classification is based on the quality of lubrication, especially the scatter in friction coefficients. Assemblies of k-class K1 require all test results within the range of $0.10 \leq k_i \leq 0.16$. For calculation of required torque, $k_m = 0.13$ can be assumed. Assemblies of k-class K2 have batch specific k-values k_m with $0.10 \leq k_m \leq 0.23$ and variance $V_k \leq 0.1$. For calculation of required torque, the specific value k_m of the batch has to be used.

Fig. 4 shows the tightening of an assembly EN 14399-4 - M20×80/37 - 10.9/10 - HV - tZn with washers EN 14399-6 - 20 - tZn. The relation between preload and rotation or torque is shown based on an experimental test: In the figure, preload is given on the vertical axis, both moment and rotation are given on the horizontal axis, so increase of reload with increasing moment or rotation can be shown by two curves.The procedures for tightening according to EN 1090-2 for Category B or C connections where preload is intended to transfer shear forces in the connection via friction (with surface treatment in the friction surfaces) are shown by the dotted lines, both for torque method and for combined method.

Torque method requires k-class K2, here with $k_m = 0.15$. $M_{r,2} = 516$ Nm is calculated using Eq. (3). The first tightening step is done with $0.75 \cdot M_{r,2} = 387$ Nm, the second one with $1.1 \cdot M_{r,2} = 568$ Nm, whereas the factor 1.1 considers the usual scatter in k-values. Both steps are shown in Fig. 4. The calculated torque values would be rounded in practice, but are given unrounded here for easier retracing. Application of the combined method can be done with assemblies of k-class K1 or K2. For this example we assume k-class K1: According to EN 1090-2, $k_m = 0.13$ can be assumed simplifying, leading to $M_{r,1} = 447$ Nm. The first tightening step is done with $0.75 \cdot M_{r,1} = 335$ Nm. The second step means turning the nut with a specified part turn ϑ_{CM}, which is depending on the ratio of Σt over diameter d, as given in EN 1090-2 (here $\vartheta_{CM} = 90°$). With this example, both methods lead to quite similar preload.

Tightening procedures and specific parameters given in EN 1090-2 are clearly related to standard assemblies according to EN 14399-1 and the friction conditions defined there. Locking devices such as wedge washers affect friction conditions in the bearing surface, so specific procedures have to be developed.

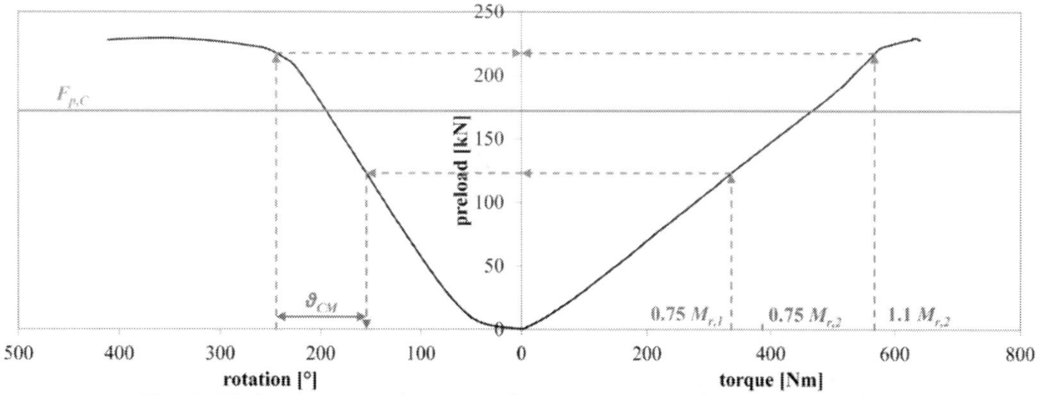

Fig. 4: Tightening procedure according to EN 1090-2 for an assembly in a bolted Category B or C connection

Fig. 5 shows the same experimental test curve with the procedure given in the German National Annex DIN EN 1993-1-8/NA if tightening is intended as a quality measure in bolted Category A connections, for example to prevent unlocking of the connection. As the level of preload is lower (see Eq. (1)), scatter in friction coefficients has a lower impact: Assemblies of k-class K2 can also be used for tightening by torque method, because danger of twisting of the bolt is significantly reduced as the bolt remains still in the linear elastic stress range. Only tightening by modified combined method might lead to small plastic stresses in the bolt. Both torque values for both methods (M_A = 450 Nm for torque method and M_A = 340 Nm for first step of modified combined method) and additional part turn for modified combined method (here $\vartheta_{MCM} = 60°$) are laid down in the National Annex and are not batch specific.

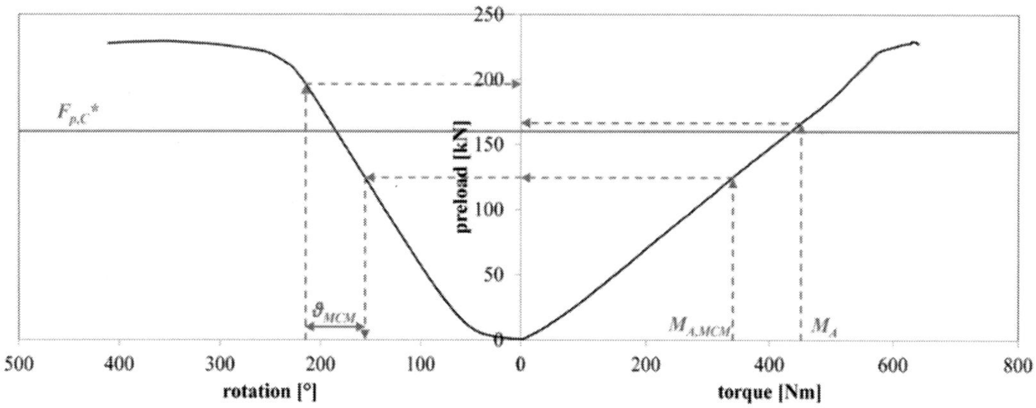

Fig. 5: Tightening procedure according to DIN EN 1993-1-8/NA for an assembly with preload intended as a quality measure

Because of its insensitivity, this tightening procedure served as a basis for the derivation of torque values for connections with wedge washers: Torque given in aforementioned systems of codes always refers to friction coefficients of standard assemblies EN 14399-4 with washers according to EN 14399-6. The geometry of the wedge washer affects the friction coefficient in the bearing surface of the nuts. Both the torque for the torque method and the combined method tightening has to be defined. The additional rotation for the second step of tightening with the combined method is not affected.

Typical friction coefficients for standard assemblies can be calculated by means of Eq. (1) on the basis of the k-values $0.10 \leq k_i \leq 0.16$ defined for k-class K1. The upper and lower bound of k_i lead to minimum preload $F_{p,min}$ and maximum preload $F_{p,max}$. Applying

$$\mu_{tot} = \frac{\frac{M_A}{F_p} - \frac{P}{2\cdot\pi}}{0.577\cdot d_2 + 0.5\cdot D_d} = \frac{d\cdot k_i - \frac{P}{2\cdot\pi}}{0.577\cdot d_2 + 0.5\cdot D_d} \quad (4)$$

for both boundary values leads to the range $\mu_{tot,min} = 0.065$ to $\mu_{tot,max} = 0.115$ for coefficient of total friction for assemblies M12 to M36. There are only two values with significant discrepancies form this range: diameter M12 gives $\mu_{tot,min} = 0.060$ and diameter M36 gives $\mu_{tot,max} = 0.117$. The necessary parameters are pitch P, pitch diameter d_2 and diameter D_b of the friction torque in the bearing surface of the nut, defined as

$$D_b = \frac{D_o + d_h}{2} = \frac{d_{w,\min} + d_h}{2} = \frac{d_{w,\min} + d_1 + 2\cdot c}{2} \quad (5)$$

with d_{wmin}, d_1 and c according to EN 14399-4 and EN 14399-6.

Tests allow a more detailed view in friction conditions: On the basis of test results, were applied torque and torque M_b in the bearing surface of the nuts were measured separately, both friction coefficient in the bearing surface of the nut and in thread can be calculated, using $F_{p,C}*$ as reference preload. 42 tests were performed with assemblies according to EN 14399-4 and diameters M12 to M36, 22 with standard washers according to EN 14399-6, 20 with Nord-Lock SC wedge washers. Based on the test results, the friction coefficient in the bearing surface under the nut was be calculated with

$$\mu_b = \frac{M_b}{0.5\cdot D_b \cdot F_{p,C}*} \quad (6)$$

and the friction coefficient between threads with

$$\mu_{th} = \frac{\frac{M_{th}}{F_{p,C}*} - \frac{P}{2\cdot\pi}}{0.577\cdot d_2} \quad (7)$$

The results are given in Table 1.

Table 1: Friction coefficients in the thread and in the bearing surface of the nut

	mean value	*standard deviation*	*minimum value*	*maximum value*	*5%-fractile*	*95%-fractile*
μ_{th}	0.110	0.032	0.045	0.155	0.055	0.165
$\mu_{b,EN}$	0.101	0.033	0.066	0.182	0.044	0.159
$\mu_{b,NL}$	0.163	0.031	0.109	0.212	0.110	0.217

5. Derivation of tightening parameters for assemblies with wedge washers

Torque for tightening of assemblies with wedge washers must be defined considering a possible scatter in friction conditions. Torque should be high enough to assure defined preload F_P in case of friction conditions in the upper range of the scatter band, but should not be too high to avoid bolt failure in case of friction conditions in the lower range. For latter requirement, the upper values of the ratios F_p/F_y and F_p/F_u associated with Eq. (2) and the tightening procedure given in DIN EN 1993-1-/8NA were used as acceptance criteria. Values of 1.08 and 0.97, respectively, were considered as acceptable. A first attempt to adopt the preload $F_{p,C}*$ as defined in DIN EN 1993-1-8/NA for wedge washers by simply adjusting the torque M_A using

$$M_{A,NL} = F_{p,C} * \cdot \left[\frac{1}{2} \cdot \frac{P + 1{,}154 \cdot \pi \cdot \mu_{th,NL} \cdot d_2}{\pi - 1{,}154 \cdot \mu_{th,NL} \cdot \frac{P}{2}} + \mu_b \cdot \frac{D_0 + d_h}{4} \right] \quad (8)$$

for taking into account the deviating friction coefficient $\mu_{b,NL}$ did not succeed for this reason: Bolt failure could not be excluded in every case. This led finally to the definition of

$$F_{p,C,NL}* = 0.9 \cdot F_{p,C}* = 0.63 \cdot f_{yb} \cdot A_S \quad (9)$$

for nominal preload. This value is still high enough to secure deserved properties (locking). Torque calculated in the aforementioned first attempt was adjusted accordingly by multiplication with 0.90. Statistical evaluation proved that $F_{p,C,NL}*$ will be safely reached when applying $M_{A,NL}$. Compared to the torque M_A given in the German National Annex DIN EN 1993-1-8/NA for an assembly with preload intended as a quality measure (leading to the preload defined by Eq. (2)), $M_{A,NL}$ is approx. 30 % higher than M_A.

Exemplarily for the Nord-Lock SC, Tables 2 and 3 give all necessary information for the torque method and the modified combined method. The torque values for the first tightening step were obtained by multiplying $M_{A,NL}$ with the factor 0.75. The further rotation to be applied, during the second step of tightening were taken from DIN EN 1993-1-8/NA. The values depend on total nominal thickness Σt of parts to be connected (including all packs and washers). Larger rotation would have been acceptable, too.

Table 2: Preload and torque for Nord-Lock SC wedge washers

	M12	*M16*	*M20*	*M22*	*M24*	*M27*	*M30*	*M36*
$F_{p,C,NL}*$ [kN]	45	90	145	170	200	260	315	460
$M_{A,NL}$ [Nm]	130	330	660	850	110	1600	2150	3750
$M_{A,MCM,NL}$ [Nm]	100	250	490	640	825	1220	1650	2800

Table 3: Further rotation for assemblies with Nord-Lock SC wedge washers

Total nominal thickness	$\vartheta_{MCM,NL}$	
	Degrees	*Part turn*
Σt < 2 d	45°	1/8
2 d ≤ Σt ≤ 6 d	60°	1/6
6 d ≤ Σt ≤ 10 d	90°	1/4
10 d ≤ Σt	no recommendation	

Fig. 6 shows the tightening procedure for an assembly EN 14399-4 - M20×80/37 - 10.9/10 - HV - tZn with wedge washers, again both by torque method and by combined method.

6. Conclusions

In case of vibrations, impact or dynamic loading, preload alone is not sufficient to prevent self-loosening of assemblies in a Category A connection. In such cases, locking devices such as wedge washers are of great help. The required torque and specified part turn for proper tightening are given exemplarily, including background information on the derivation of the parameters.

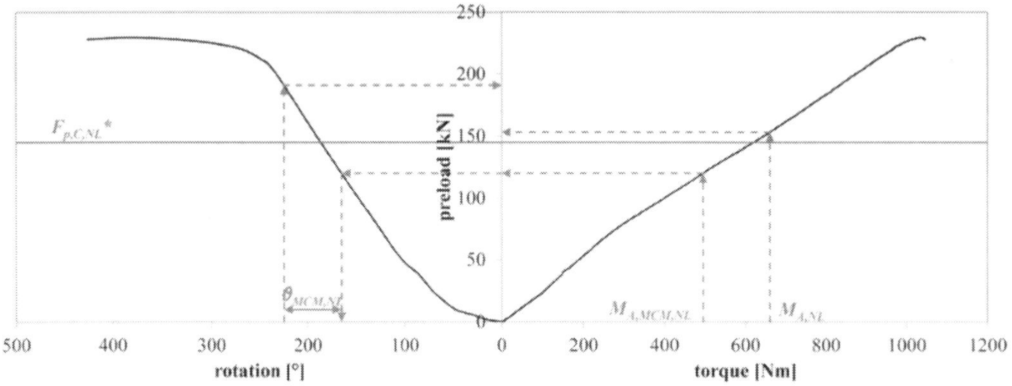

Fig. 6: Tightening procedure for an assembly with wedge washers

Acknowledgments

The investigations presented in chapter 5 were realised while the author was working at the Versuchssanstalt für Stahl, Holz und Steine at Karlsruhe Institute of Technology, Germany. Thank you to Frank Hüssing, Nord-Lock GmbH, Westhausen, Germany for permission of partial publication of the test results. The testes were performed by Dabid Urionabarrenetxea, AZTERLAN - Centro de Investigación Metalúrgica, Durango, Spain.

References

[1] Friede R, Lange J. "Self loosening of prestressed bolts", *Proceedings of the Nordic Steel Construction Conference 2009*, Malmö, Sweden, 272-279, 2009. (http://www.nordicsteel2009.se/pdf/106.pdf).

[2] Friede R, Lange J. "Loss of preload in bolted connections due to embedding and self loosening", *Proceedings of SDSS'Rio 2010 International Colloquium Stability and Ductility of Steel Structures* (Eds.: E. de M. Batista, P. C. G. da S. Vellasco, L. R. O. de Lima), Rio de Janeiro, Brazil, 287-294, 2010 (http://www.labciv.eng.uerj.br/sdss2010/files/sdss_rio_2010_4_15.pdf).

[3] Esser J, Hellwig G. "Wirkungslose Schraubensicherung - Konsequenzen für die Normung", *DIN-Mitteilungen*, 83(4), 14-16, 2004.

[4] Junker GH. "New Criteria for Self-Loosening of Fasteners Under Vibration" *Transactions of the Society of Automotive Engineers*, 78, 314-335 1969.

MONOTONIC AND CYCLIC BACKBONE RESPONSE OF SINGLE SHEAR COLD-FORMED STEEL SCREW-FASTENED CONNECTIONS

Cristopher D. Moen, Fannie Tao and Robert Cole

Virginia Polytechnic and State University, Blacksburg, VA, USA

Abstract: Monotonic and cyclic backbone load-deformation response models for single shear cold-formed steel screw-fastened connections are developed with support from an experimental program. Connection strength and stiffness are correlated to fastener bearing strength of the two connected plies. Cyclic strength and stiffness degradation was negligible. Cyclic excursions resulted in increased connection stiffness from the screw threads locking in the plies.

1. Introduction

Monotonic and cyclic backbone load-deformation response models are developed with experiments in this paper to provide the capability to simulate studs to stud, stud to ledger frame, and stud to track connections for performance-based design (e.g., FEMA P695 [1]) of cold-formed steel framed buildings. Screw fasteners serve as the primary connectors in light steel framing (Fig. 1a) and their discrete behavior drives lateral and gravity system response (Fig. 1b). Whole building seismic analysis, i.e., modeling every cold-formed steel stud, track, shear wall, floor diaphragm, and fastener (Fig. 1c), becomes feasible when connection response, including strength and stiffness degradation and energy dissipation, is accurately predicted.

Fig 1: Light steel framing: (a) 5 story Virginia Tech residence hall (AECOM) constructed with (b) panelized load bearing cold-formed steel framing connected by (c) self-drilling screw fasteners

Extensive test data on cold-formed steel fasteners is available (see [2] for a summary and reference list), and most of these tests focus on monotonic ultimate strength and failure limit states. For example, it is well documented that a fastener will tilt at the connection if the moment created by the single shear eccentricity cannot be resisted by the plies, which occurs when both steel plies are relatively thin (e.g., ply thicknesses t_1 and t_2 are 0.88 mm where Ply 1 is in contact with the fastener head). If one of both plies are relatively thick (e.g., t_1 and/or t_2 are 2.57 mm), then fastener bearing is the controlling strength limit state. These tilting and bearing limit states are represented in codes and standards, for example, AISI-S100-12 Section E4 [3].

More recent work has focused on documenting complete fastener load-deformation response, including cyclic strength and stiffness degradation of typical cold-formed steel framing connections [4, 5]. Initial stiffness and post-peak response is known for different fastener head types [6] and steps have been taken to parameterize the monotonic backbone response [7, 8]. Cyclic strength and stiffness degradation at the fastener level can also be considered in whole building models with newly coded pinching material models [9] that accommodate accurate nonlinear dynamic time history analyses [10-16].

This paper builds on existing research to derive monotonic and cyclic load-deformation backbone parameter models for single shear cold-formed steel-to-steel screw-fastened connections. These models will be used to inform pushover or cyclic seismic simulations of light steel framed subsystems (shear wall, diaphragm) and whole building analysis [e.g., 17] where the connections are input as nonlinear springs or hysteretic elements. Building serviceability calculation and retrofit procedures can also be developed from the models, where connection stiffness at a specific fastener load is required. The parameterization is implemented with a single shear steel-to-steel fastener experimental program introduced in the next section.

2. Screw-fastened connection experimental program

Single shear steel-to-steel screw fastened connection response is experimentally documented in this section. Both monotonic and cyclic fastener tests were performed which are used in Section 3 to define load-deformation backbone response models.

2.1 Test matrix

The t_1 and t_2 steel Ply1 and Ply 2 thicknesses were varied in the following combinations: 3333, 4333, 4343, 4354, 4368, 4397, 5433, 6833, and 9733 where 33, 43, 54, 68, and 97 correspond to nominal ply thickness in thousands of an inch [18] which converts to design thicknesses of 0.88 mm, 1.15 mm, 1.44 mm, 1.81 mm, and 2.57 mm respectively. Each thickness combination was studied for #8, #10, and #12 hex-head self-drilling fasteners and for both monotonic and cyclic loadings totaling 9 thickness combinations x 3 tests per combination x 3 fastener sizes x 2 loading types = 162 tests. The test naming convention denotes the Ply 1 and Ply 2 thicknesses (t_1 and t_2), fastener size, test type (M=monotonic, C=cyclic), and trial number within a test series. For example, '4333-8-M1' represents the test combination with t_1=1.15 mm (43 thousandths of an inch), t_2=0.88 mm (33 thousandths of an inch), a #8 hex-washer head fastener, and the first of three monotonic trials.

2.2 Test setup

A single screw is driven from Ply 1 into Ply 2 as shown in Fig. 2. Aluminum fixtures provide a 102 mm by 102 mm window that supports the plies on 3 sides. This test setup evolved over a few iterations [19] and was inspired by [20]. The window was designed to prevent the two connected plies from separating as moment develops in the single shear connection. This separation occurs in a standard connection test, i.e., [21], which is inconsistent with typical framing connection behavior where the framing members (studs and tracks) have enough flexural stiffness to prevent ply separation [6].

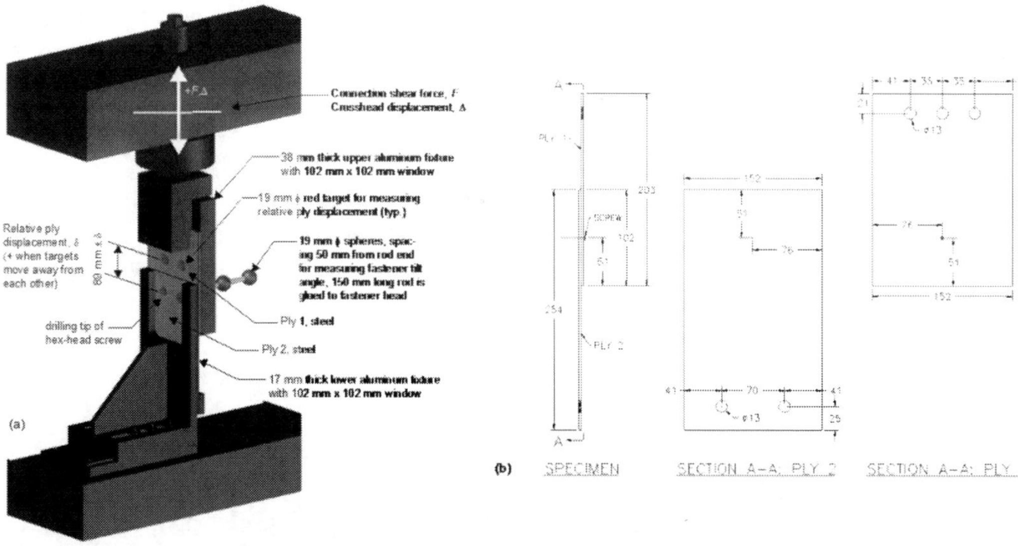

Fig 2: Single fastener connection (a) test setup; (b) specimen dimensions and construction

2.3 Specimen construction, installation, and fastener details

Each specimen consists of two steel plies as shown in Fig. 2b. The typical installation procedure was to drive the fastener (#8, #10, or #12) from Ply 1 into Ply 2, slide the fastened Ply 1 and Ply 2 specimen onto the bolts in the upper and lower aluminum fixtures, and then tighten the 13 mm diameter bolts at the top and bottom fixtures. The load cell reading was zeroed before tightening the bolts to measure any pretension or precompression force. The drilled bolt hole and bolt size are closely matched to prevent ply slip at the bolted connections.

The #10 and #12 hex-head self-drilling fasteners were provided by Simpson Strong-Tie and had order numbers X1S1016 and X1S1216, respectively. The #10 fastener is 25 mm long with a zinc coating, head diameter of 10.2 mm, major thread diameter (out-to-out), D, of 4.74 mm, a #3 drilling tip, and 0.62 teeth per mm (TPI), i.e., pitch P=1/TPI=1.6 mm, and the fastener shear strength F_{ss}=8.2 kN. The #12 fastener is the same except it has a major thread diameter D=5.40 mm and F_{ss}=11.1 kN. The #8 hex-head self-drilling fasteners were provided by ELCO Construction Products with order number EDA285. They are 25 mm long with a zinc coating, head diameter of 8.3 mm, major thread diameter D=4.20 mm, a #2 drilling tip, 0.71 teeth per mm (pitch P=1.4 mm), and F_{ss}=4.9 kN.

2.4 Test measurements and instrumentation

Relative movement between the plies was measured with a custom optical non-contact measurement system designed and validated at Virginia Tech. Relative ply displacement is recorded by following four red sticker targets (two on each ply) with a Microsoft Lifecam cinema HD 720p video camera recording at 30 frames per second. The area and location of each target is identified in every video frame using custom code provided in [22] utilizing the Matlab [23] Computer Vision System Toolbox. The relative displacement between Ply 1 and Ply 2 (δ in Fig. 2a) is calculated by taking the average target area in pixels for the first 50 data points measured in Matlab and using this value and the measured target area (285 mm^2) to convert the distance between targets into mm from pixels. Measurement accuracy is ±0.10 mm [22].

The connection force, measured with a 150 kN load cell (accuracy of ± 0.10 kN), was synchronized to the relative ply displacement δ with common time stamps in the video and data acquisition files. The crosshead displacement, Δ (Fig. 2a), was also recorded using the test machine internal LVDT (accuracy of ± 0.10 mm), and Δ is used to implement the monotonic and cyclic loading protocols discussed later in this paper.

2.5 Specimen material properties

Steel sheet yield stress (f_y) and ultimate stress (f_u) were measured for each ply thickness from tensile coupons tested in accordance with ASTM E8M [24]. The mean (μ) and coefficient of variation (c_v) statistics in Table 1 are calculated from three tests per ply sheet thickness (specimens of the same thickness came from the same sheet). All specimens exhibited a yield plateau followed by strain hardening except for the 43 plies which gradually yielded and fractured with approximately 60% less elongation that the other plies.

Table 1: Steel ply thickness, yield and ultimate strength, and elongation at fracture

Ply	*t* (mm)		*f*$_y$ (MPa)		*f*$_u$ (MPa)		% elongation at fracture	
	μ	c_v	μ	c_v	μ	c_v	μ	c_v
33	0.90	0.01	325	0.03	376	0.02	34.1	0.03
43	1.11	0.01	590	0.01	615	0.01	7.4	0.07
54	1.43	0.00	393	0.01	493	0.00	24.4	0.10
68	1.80	0.00	390	0.04	510	0.03	20.3	0.10
97	2.56	0.00	379	0.00	505	0.02	21.9	0.07

2.6 Specimen loading

Monotonic tests were conducted under crosshead displacement control at Δ=0.025 mm/sec which is consistent with the recommended loading rate in [21]. The cyclic loading protocol is adapted from FEMA 461 quasi-static cyclic deformation-controlled testing protocol [25] as shown in Fig. 3a which was developed to obtain fragility data and hysteretic response characteristics of building components for which damage is best predicted by imposed deformations. The protocol comprises steps of increasing amplitude with two cycles per step and symmetric displacement amplitudes. Each step displacement amplitude is 40% larger than the previous, i.e., $\Delta_i=1.4\Delta_{i-1}$ (see Fig. 3).

The cyclic loading protocol is anchored at the beginning of the second step (i.e., the 3th cycle) with the crosshead displacement Δ_a assumed to define the end of the linear portion of a fastener's load-deformation response and beginning of the fastener damage state [8]. Assuming crosshead and relative ply displacements are equivalent, the anchor displacement Δ_a is calculated as the predicted relative ply displacement at $0.40F_c$, i.e., $\Delta_a=\delta_a=0.4F_c/K_e$, where F_c is the peak (cap) load and K_e is the elastic stiffness of the monotonic response shown in Fig. 3b. FEMA 461 suggests to perform 6 deformation cycles before the first damage limit state, however this was not practical for these fastener tests where plasticity initiates at low load levels through bearing deformation in the thin plies.

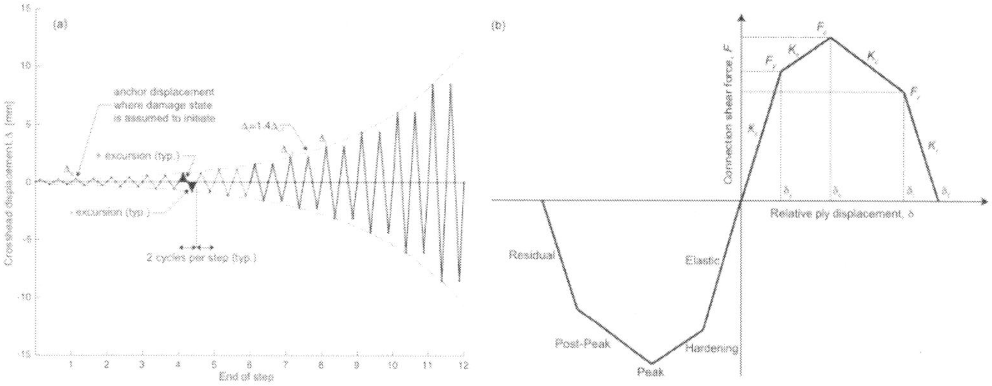

Fig 3: (a) Cyclic fastener loading protocol for test 3333-HEX08; (b) backbone curve notation

3. Monotonic and cyclic backbone characterization and trends

3.1 Backbone nomenclature

The monotonic and cyclic fastener load-deformation responses are characterized as backbone curves to facilitate hysteretic modeling developing including strength and stiffness degradation. The backbone configuration and notation are motivated by [26], and each curve is composed of five regions – elastic, hardening, peak, post-peak, and residual as shown in Fig. 3b. These regions in Fig. 3b are defined by load-displacement points (F_y, δ_y), (F_c, δ_c), (F_r, δ_r), and $(0, \delta_f)$ with related stiffnesses (slopes) K_e, K_s, K_c, and K_r.

3.2 Backbone construction

Monotonic backbone construction was performed by visually selecting (F_c, δ_c) as the first peak load after hardening, then visually selecting (F_y, δ_y) to match the elastic and hardening leg slopes K_e and K_s. The point (F_r, δ_r) is obtained so that the post-peak backbone segment from (F_c, δ_c) to (F_r, δ_r) is a linear fit (i.e., the average) of the tested response, and then δ_f is calculated to equate areas under the load-displacement curve for the tested and backbone response.

The backbone of the positive cyclic response ($+\Delta$ in Fig. 2) was obtained by first identifying the response outline with the Matlab *Boundary* command [13] with the 'shrink factor'=1, and then following the same procedures discussed previously for the monotonic backbone.

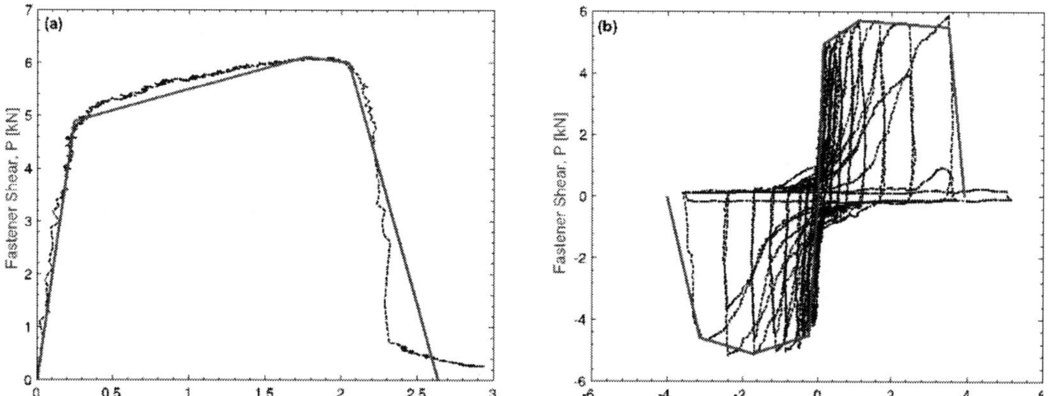

Fig. 4: Backbones for fastener configuration 4354-8-1: (a) monotonic and (b) cyclic loading

The negative excursion cyclic response was consistent with the positive response except when Ply 1 was thin (e.g., 33, 43) resulting in buckling deformation that created excessive friction at the edges of the test window. An example set of backbones is shown in Fig. 4.

3.3 Backbone trends and observations

Backbone segment parameters are extracted for each of the ply combinations and loadings (monotonic, positive cyclic excursion). Table 2 summarizes the results, where M represents the monotonic loading, and C represents the positive cyclic loading. The three trials for each specimen were averaged to give the statistics in Table 2.

Models for the backbone loads (F_y, F_c, F_r) and stiffnesses (K_e, K_s, K_c, and K_r) are proposed based on the tested trends and observations. The fastener load model is $F_{[y,c,r]}=\alpha\psi^{\beta}F_{ss} \leq F_{ss}$ and the fastener stiffness model is $K_{[e,s,c,r]}=\alpha\psi^{\beta}K_a$ where $\psi=[F_{ss}/(t_1Df_{u1})][(F_{ss}/t_2Df_{u2})]$, F_{ss} is the fastener shear strength (typically determined experimentally by manufacturers), $K_a=[1/(E_1t_1)+1/(E_2t_2)]^{-1}$ which is the axial stiffness of two square plies calculated as springs in series assuming a rigid fastener connection, E_1 and E_2 are the ply elastic moduli, and α and β in Table 3 are calculated with the Matlab function *fminsearch* minimizing the difference in error between test and model. Test-to-predicted statistics are provided in Table 3. The backbone model framework is designed to accommodate high strength steel plies [27] and other sheathing materials (OSB, plywood, gypsum) in the near future.

Ply bearing strengths t_1Df_{u1} and t_2Df_{u2} are key contributors to connection strength and stiffness. Decreasing *either* of the ply thicknesses results in decreased strength and stiffness which is reflected in the backbone models. For both monotonic and cyclic tests, the typical response has a linear region where the connection shear stiffness (K_e) is defined by the amount of fastener tilting controlled by bearing deformation in the plies. Changing fastener sizes from #8 to #12 increases K_e because the bearing stress and associated deformation in each ply decreases. The yielding load F_y and peak (cap) load F_c experience minimal cyclic degradation as the fasteners repeatedly bear on the plies (Fig. 5a), and it is hypothesized that the positive cyclic (C+) elastic connection stiffness (K_e) is approximately double the monotonic stiffness (Fig. 5b) because the fastener threads lock into the plies during the first few excursions. The monotonic and cyclic backbone load and stiffness models can be used to calculate δ_y , δ_c , δ_r and δ_f in Fig. 3b, e.g., $\delta_y=F_y/K_e$.

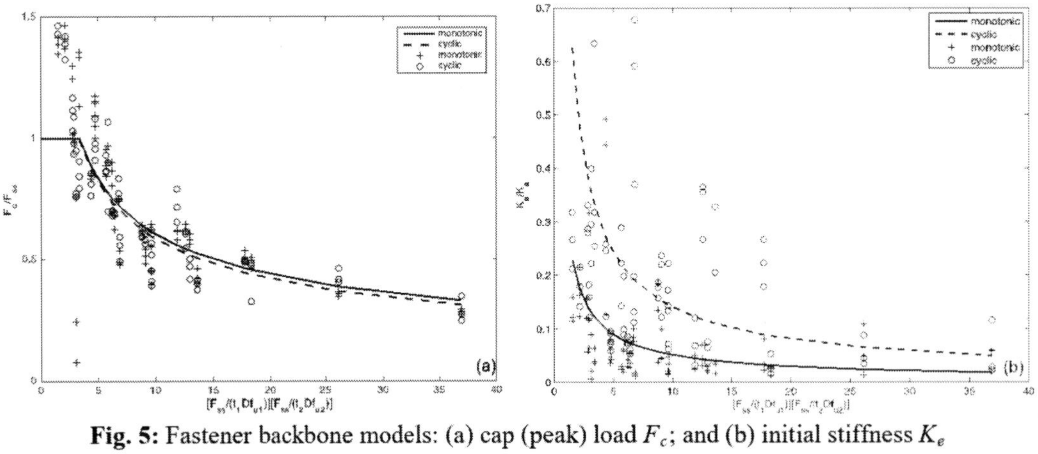

Fig. 5: Fastener backbone models: (a) cap (peak) load F_c; and (b) initial stiffness K_e

Table 2: Monotonic and cyclic backbone parameter summary and statistics

Test	P_y (kN)		P_c (kN)		P_r (kN)		δ_y (mm)		δ_c (mm)		δ_r (mm)		δ_f (mm)		K_e (kN)		K_s (kN/mm)		K_c (kN/mm)		K_r (kN/mm)	
	μ	c_v	μ	c_v	μ	c_v	μ	c_v	μ	c_v	μ	c_v	μ	c_v	μ	c_v	μ	c_v	μ	c_v	μ	c_v
3333-8-M	2.2	0.06	3.0	0.03	1.9	0.08	0.7	0.35	4.01	0.15	12.2	0.11	23.3	0.13	3.6	0.28	0.25	0.34	-0.1	0.22	-0.2	0.21
3333-8-C+	1.9	0.19	3.5	0.08	1.7	0.07	0.4	0.64	4.75	0.11	10.4	0.05	23.7	0.06	6.7	0.50	0.39	0.36	-0.3	0.14	-0.1	0.15
3333-10-M	2.1	0.13	2.9	0.02	1.8	0.09	0.8	0.77	6.33	0.15	15.3	0.04	30.5	0.04	5.2	0.69	0.16	0.44	-0.1	0.13	-0.1	0.12
3333-10-C+	2.3	0.11	3.5	0.06	1.8	0.18	0.6	0.42	5.61	0.20	12.1	0.21	19.8	0.22	5.1	0.41	0.25	0.58	-0.3	0.43	-0.3	0.39
3333-12-M	2.7	0.02	3.2	0.03	2.2	0.09	0.8	0.54	8.40	0.04	14.4	0.05	30.1	0.06	4.2	0.40	0.06	0.39	-0.2	0.21	-0.1	0.18
3333-12-C+	2.2	0.04	3.2	0.15	2.2	0.10	0.7	0.54	5.18	0.35	10.4	0.01	16.6	0.22	5.1	0.77	0.24	0.28	-0.2	0.74	-0.6	0.69
4333-8-M	2.6	0.20	4.5	0.03	3.6	0.13	0.8	0.74	4.39	0.06	5.0	0.11	5.3	0.14	4.7	0.40	0.53	0.21	-2.1	0.68	-15.2	0.36
4333-8-C+	2.6	0.12	4.3	0.17	4.2	0.20	0.2	0.22	4.16	0.43	5.8	0.09	6.9	0.16	13.0	0.38	0.55	0.43	-0.3	0.70	-7.2	0.63
4333-10-M	3.0	0.24	4.8	0.03	2.2	0.24	1.3	0.37	6.85	0.08	9.9	0.03	21.3	0.27	2.4	0.15	0.31	0.34	-0.9	0.25	-0.3	0.60
4333-10-C+	3.0	0.07	3.8	0.08	3.1	0.16	0.6	0.33	6.23	0.16	11.1	0.11	12.8	0.10	6.0	0.27	0.14	0.15	-0.1	0.54	-1.9	0.19
4333-12-M	2.9	0.04	5.4	0.04	3.0	0.04	1.4	0.37	9.25	0.14	14.1	0.13	34.6	0.22	2.3	0.33	0.33	0.11	-0.5	0.01	-0.2	0.57
4333-12-C+	3.3	0.14	4.7	0.17	3.2	0.21	1.2	0.35	5.91	0.49	13.9	0.17	24.9	0.03	3.2	0.48	0.31	0.08	-0.2	0.63	-0.3	0.14
4343-8-M	3.2	0.12	4.9	0.02	3.8	0.09	0.3	0.51	2.96	0.08	3.9	0.14	4.5	0.17	18.7	0.66	0.65	0.23	-1.5	0.50	-7.5	0.51
4343-8-C+	2.6	0.14	4.9	0.06	3.4	0.23	0.2	0.29	3.41	0.41	5.7	0.22	6.6	0.19	17.4	0.16	0.79	0.34	-0.7	0.82	-8.1	1.01
4343-10-M	4.7	0.05	5.5	0.04	2.9	0.16	1.2	0.26	4.97	0.18	14.1	0.21	26.3	0.07	4.4	0.29	0.21	0.18	-0.3	0.13	-0.3	0.28
4343-10-C+	3.4	0.07	5.8	0.03	3.3	0.09	0.5	0.18	4.53	0.29	9.4	0.16	13.9	0.13	7.2	0.17	0.66	0.32	-0.6	0.48	-0.8	0.36
4343-12-M	4.8	0.11	5.7	0.04	3.2	0.10	0.8	0.39	3.70	0.19	13.2	0.08	25.8	0.05	7.1	0.41	0.29	0.45	-0.3	0.18	-0.3	0.13
4343-12-C+	4.1	0.18	6.5	0.03	2.8	0.36	0.2	0.49	2.22	0.23	11.1	0.26	17.9	0.18	21.7	0.26	1.26	0.21	-0.4	0.12	-0.4	0.31
4354-8-M	4.6	0.13	5.8	0.10	4.4	0.20	0.4	0.31	3.77	0.54	4.9	0.59	11.0	1.01	14.1	0.37	0.48	0.44	-1.5	0.40	-18.3	0.89
4354-8-C+	4.5	0.09	5.4	0.05	4.9	0.08	0.1	0.17	1.52	0.34	3.1	0.38	3.5	0.40	38.8	0.10	0.73	0.25	-0.9	1.19	-17.1	0.59
4354-10-M	4.6	0.07	7.0	0.05	4.1	0.05	0.9	0.40	6.93	0.09	13.5	0.04	24.2	0.05	6.2	0.39	0.41	0.24	-0.5	0.23	-0.4	0.16
4354-10-C+	4.7	0.05	5.7	0.01	5.3	0.02	0.5	0.08	2.3	0.07	7.5	0.03	13.3	0.07	9.7	0.08	0.5	0.18	-0.1	0.31	-0.9	0.16
4354-12-M	6.1	0.02	7.0	0.02	4.3	0.21	0.6	0.74	4.4	0.39	12.9	0.13	24.4	0.08	15.1	0.51	0.3	0.27	-0.4	0.56	-0.4	0.27
4354-12-C+	4.4	0.07	6.7	0.01	5.9	0.05	0.2	0.00	2.3	0.06	6.7	0.11	10.1	0.13	22.1	0.07	1.1	0.19	-0.2	0.21	-2.4	0.49
4368-8-M	5.7	0.07	6.9	0.03	5.4	0.10	0.3	0.25	1.3	0.20	1.8	0.23	1.9	0.22	23.3	0.23	1.2	0.23	-4.0	0.45	-41.5	0.16
4368-8-C+	5.8	0.04	6.8	0.03	5.8	0.09	0.2	0.16	1.4	0.15	1.7	0.10	2.0	0.10	24.9	0.17	0.9	0.10	-3.7	0.54	-21.5	0.14
4368-10-M	5.7	0.14	8.8	0.07	4.6	0.08	0.7	0.25	11.4	0.01	14.5	0.10	25.8	0.04	8.2	0.30	0.3	0.41	-1.7	0.40	-0.4	0.16
4368-10-C+	3.9	0.14	8.0	0.01	4.4	0.54	0.4	0.06	7.4	0.12	10.4	0.09	15.4	0.20	10.4	0.18	0.6	0.06	-1.0	0.26	-1.2	0.75
4368-12-M	7.0	0.00	8.3	0.01	6.9	0.07	0.6	0.12	2.9	0.07	14.5	0.01	28.2	0.10	12.1	0.12	0.6	0.11	-0.1	0.34	-0.5	0.13
4368-12-C+	5.4	0.02	8.7	0.04	4.6	0.09	0.3	0.21	8.0	0.13	12.9	0.01	13.9	0.03	20.5	0.25	0.4	0.18	-0.9	0.20	-5.0	0.36
4397-8-M	5.3	0.10	6.8	0.02	5.4	0.04	0.3	0.13	1.2	0.41	1.6	0.30	1.9	0.28	20.7	0.15	1.8	0.30	-3.3	0.19	-25.5	0.48
4397-8-C+	5.8	0.04	7.1	0.01	6.3	0.05	0.1	0.20	0.7	0.13	1.0	0.14	1.4	0.13	41.8	0.16	2.1	0.07	-3.3	0.48	-21.5	0.30
4397-10-M	6.1	0.08	10.4	0.08	8.8	0.25	0.9	0.27	8.5	0.12	10.4	0.27	10.9	0.27	7.3	0.26	0.6	0.17	-2.0	0.98	-27.8	0.60
4397-10-C+	4.7	0.20	6.9	0.05	6.3	0.05	0.1	0.20	1.4	0.18	8.3	0.17	10.1	0.25	63.2	0.41	1.7	0.29	-0.1	0.13	-7.2	0.86
4397-12-M	7.7	0.03	12.6	0.02	9.8	0.13	1.1	0.07	9.5	0.06	27.0	0.09	54.6	0.15	6.9	0.05	0.6	0.07	-0.2	0.61	-0.4	0.55
4397-12-C+	9.0	0.00	11.3	0.08	7.5	0.03	0.7	0.12	12.4	0.08	26.3	0.23	45.0	0.20	12.8	0.13	0.2	0.36	-0.4	0.61	-0.4	0.22
5433-8-M	2.5	0.05	4.4	0.05	3.3	0.10	0.8	0.07	5.5	0.05	5.9	0.06	6.4	0.07	3.1	0.10	0.4	0.20	-3.3	0.20	-7.4	0.35
5433-8-C+	2.9	0.11	4.3	0.04	4.0	0.06	0.1	0.32	4.9	0.21	6.6	0.15	7.0	0.16	24.5	0.27	0.3	0.12	-0.6	0.67	-17.9	0.75
5433-10-M	3.0	0.01	5.2	0.02	3.0	0.16	0.6	0.30	5.2	0.10	8.3	0.17	25.2	0.15	5.5	0.33	0.5	0.02	-0.7	0.12	-0.2	0.28
5433-10-C+	2.9	0.04	4.8	0.05	2.4	0.09	0.1	0.17	5.7	0.11	13.3	0.24	19.6	0.17	37.0	0.13	0.3	0.23	-0.4	0.55	-0.6	0.69
5433-12-M	3.0	0.07	5.7	0.03	3.4	0.20	0.8	0.50	8.3	0.17	13.1	0.23	35.5	0.07	5.2	0.56	0.4	0.19	-0.5	0.17	-0.2	0.05
5433-12-C+	2.8	0.08	5.5	0.01	2.9	0.25	0.1	0.20	11.4	0.05	18.8	0.16	37.8	0.22	25.0	0.16	0.2	0.12	-0.4	0.33	-0.2	0.43
6833-8-M	2.9	0.11	4.1	0.01	3.1	0.18	0.1	0.56	4.2	0.11	5.9	0.31	9.4	0.62	43.2	0.46	0.3	0.16	-0.8	0.61	-3.9	0.66
6833-8-C+	3.0	0.06	4.0	0.05	2.8	0.33	0.1	0.31	5.6	0.17	8.5	0.53	9.9	0.46	25.5	0.29	0.2	0.37	-1.8	0.67	-16.4	1.30
6833-10-M	4.0	0.00	5.0	0.07	2.9	0.21	1.9	0.13	5.0	0.13	8.8	0.14	21.9	0.18	2.1	0.14	0.3	0.36	-0.5	0.23	-0.2	0.45
6833-10-C+	3.7	0.11	4.7	0.07	2.9	0.07	0.2	0.00	4.3	0.24	9.9	0.13	15.2	0.05	18.3	0.11	0.3	0.30	-0.3	0.16	-0.5	0.07
6833-12-M	3.4	0.09	4.7	0.06	3.2	0.08	1.1	0.49	5.8	0.20	16.2	0.08	35.3	0.14	3.7	0.34	0.3	0.47	-0.1	0.13	-0.2	0.20
6833-12-C+	3.2	0.19	4.4	0.04	3.1	0.05	0.1	0.19	6.3	0.26	18.3	0.03	24.5	0.14	35.0	0.20	0.2	0.37	-0.1	0.12	-0.8	0.68
9733-8-M	1.4	0.81	1.8	0.80	1.2	0.70	0.3	0.14	1.9	0.95	2.5	0.78	3.4	0.52	5.2	0.93	0.4	0.45	-0.8	0.94	-1.7	0.82
9733-8-C+	2.7	0.12	4.0	0.10	3.6	0.25	0.1	0.32	3.1	0.27	3.5	0.15	3.6	0.16	41.3	0.24	0.5	0.10	-1.1	0.60	-26.0	0.32
9733-10-M	3.3	0.04	4.1	0.05	2.5	0.01	1.7	0.17	5.1	0.32	10.4	0.14	21.5	0.11	2.1	0.19	0.3	0.42	-0.3	0.32	-0.2	0.14
9733-10-C+	2.7	0.04	4.5	0.08	2.1	0.34	0.0	0.23	6.2	0.03	11.5	0.13	14.4	0.09	73.9	0.24	0.3	0.22	-0.4	0.27	-1.2	0.45
9733-12-M	3.4	0.06	4.8	0.06	3.6	0.02	0.8	0.29	2.9	0.04	13.7	0.06	35.0	0.11	4.6	0.32	0.7	0.10	-0.1	0.20	-0.2	0.15
9733-12-C+	2.9	0.19	4.6	0.06	3.2	0.18	0.2	0.53	3.0	0.10	18.9	0.13	24.2	0.02	16.0	0.62	0.6	0.07	-0.1	0.20	-0.7	0.38

4. Conclusions

This paper makes an important step towards light-steel framed whole building seismic analysis with single shear steel fastener connection backbone models. The models were derived with data from an extensive experimental program considering monotonic and cyclic loadings, where bearing strengths of the connected plies were the primary parameters influencing pre-peak connection stiffness and strength. This work also provides updates and improvements to the typical screw fastener connection test setup, defines a means for generalizing cyclic loading protocols for fasteners, and validates a non-contact computer-vision measurement system that conveniently and accurately measures relative slip and fastener tilt throughout the load-deformation response.

Table 3: Backbone model parameters and test-to-predicted statistics

Backbone parameter	*Loading*	*Trend parameters*		*Test/Predicted*	
		α	β	μ	c_v
F_y	monotonic	1.37	-0.52	1.00	0.22
	cyclic	1.55	-0.61	1.00	0.22
F_c	monotonic	1.76	-0.46	1.00	0.22
	cyclic	1.79	-0.48	1.00	0.17
F_r	monotonic	1.52	-0.59	1.00	0.29
	cyclic	1.78	-0.67	1.00	0.29
K_e	monotonic	0.32	-0.80	1.00	0.99
	cyclic	0.89	-0.80	1.00	0.84
K_s	monotonic	0.013	-0.68	1.00	0.51
	cyclic	0.022	-0.80	1.00	0.73
K_c	monotonic	-0.035	-0.87	1.00	0.97
	cyclic	-0.055	-1.23	1.00	1.19
K_r	monotonic	-0.52	-1.80	1.00	1.64
	cyclic	-0.40	-1.40	1.00	1.50

Acknowledgments

The authors are thankful to the American Iron and Steel Institute for their support, especially the project monitoring task group, Simpson Strong-Tie and ELCO for donating the fasteners, and ClarkDietrich Building Systems for donating the steel plies. The authors are also grateful for the support of Virginia Tech graduate students Hong Pham, David Padilla-Llano, and Chu Ding and Princeton undergraduate student Felicia Jiang.

References

[1] FEMA P695 - *Quantification of Building Seismic Performance Factors*, Federal Emergency Management Agency (FEMA), Document No. FEMA 965, Washington, D.C., 2009.

[2] Peköz T. "Design of cold-formed steel screwed connections." *Proceedings of the Tenth International Specialty Conference on Cold-formed Steel Structures*, Missouri S&T, St. Louis, Missouri, U.S.A., 1990.

[3] AISI-S100-12. *North American Specification for the Design of Cold-Formed Steel Structural Members*. American Iron and Steel Institute, Washington, D.C. (USA), 2012.

[4] Peterman K, Nakata N, Schafer BW. "Hysteretic Characterization of Cold-Formed Steel Stud-to-Sheathing Connections." *Journal of Constructional Steel Research,* 101: 254-264, 2014.

[5] Ayhan D, Schafer BW. "The Effect of M/V Ratio on Ledger-Framing Performance of Cold-Formed Steel Buildings", *Proceedings of the International Colloquium on Stability and Ductility of Steel Structures – SDSS2016*, Timisoara, Romania, 2016.

[6] Haus A, Moen CD. "Energy Dissipation of Cold-Formed Steel Connections", *Virginia Tech Research Report No. CE/VPI-ST-14/02,* Blacksburg, VA.

[7] Moen CD, Padilla-Llano DA, Corner S, Ding C. "Towards Load-Deformation Models for Screw-Fastened Cold-Formed Steel-to-Steel Shear Connections", *Proceedings of the*

22nd International Specialty Conference on Cold-Formed Steel Structures, St. Louis, Missouri (USA), 2014.

[8] Pham H, Moen CD. "Stiffness and Strength of Single Shear Cold-Formed Steel Screw-Fastened Connections", *Virginia Tech Research Report No. CE/VPI-ST-15/07*, Blacksburg, VA, 2015.

[9] Ding C. "Monotonic and Cyclic Simulation of Screw-Fastened Connections for Cold-Formed Steel Framing", *M.Sc. Thesis*, Blacksburg, Virginia (USA): Department of Civil and Environmental Engineering, Virginia Tech, 2015.

[10] Leng J, Schafer BW, Buonopane S. "Modeling the seismic response of cold-formed steel framed buildings: model development for the CFS-NEES building", *Proceedings of the Annual Stability Conference-Structural Stability Research Council*, St. Louis, Missouri (USA), 2013.

[11] Niari SE, Rafezy B, Abedi K. "Numerical Modeling and Finite Element Analysis of Steel Sheathed Cold-Formed Steel Shear Walls", *Proceedings of the 15th World Conference on Earthquake Engineering*, Lisbon (Portugal), 2012.

[12] Bian G, Buonopane SG, Ngo HH, Schafer BW. "Fastener-based computational models with application to cold-formed steel shear walls", *Proceedings of the 22nd International Specialty Conference on Cold-Formed Steel Structures,* St. Louis, Missouri (USA), 2014.

[13] Buonopane SG, Tun TH, Schafer BW. "Fastener-based computational models for prediction of seismic behavior of CFS shear walls", *Proceedings of the 10th National Conference in Earthquake Engineering*, Anchorage, Alaska (USA), 2014.

[14] Ngo HH. "Numerical and experimental studies of wood sheathed cold-formed steel framed shear walls", *M.Sc. Thesis*. Baltimore, Maryland (USA): Department of Civil Engineering, Johns Hopkins University, 2014.

[15] Tun TH. "Fastener-Based Computational Models of Cold-Formed Steel Shear Walls", *Honors Thesis*. Lewisburg, Pennsylvania (USA): Department of Civil and Environmental Engineering, Bucknell University, 2014.

[16] Fülöp L, Dubina D. "Design Criteria for Seam and Sheeting-to-Framing Connections of Cold-Formed Steel Shear Panels", *J. Struct. Eng.*, 132:4, 582-590, 2006.

[17] Padilla-Llano DA. "A Framework for Cyclic Simulation of Thin-Walled Cold-Formed Steel Members in Structural Systems", *PhD dissertation*. Blacksburg, Virginia (USA): Department of Civil and Environmental Engineering, Virginia Tech, 2015.

[18] SFIA, Technical Guide for Cold-Formed Steel Framing Products, Steel Framing Industry Association, <https://sfia.memberclicks.net/sfia-technical-publications>, February 2, 2016.

[19] Corner S. "Screw-Fastened Cold-Formed Steel-to-Steel Shear Connection Behavior and Models", *M.Sc. Thesis*. Blacksburg, Virginia (USA): Department of Civil and Environmental Engineering, Virginia Tech, 2014.

[20] Okasha AF. "Performance of steel frame/wood sheathing screw connections subjected to monotonic and cyclic loading", *M.Sc. Thesis*. Montreal (Canada): Department of Civil Engineering and Applied Mechanics, McGill University, 2004.

[21] AISI S905-13, *Test Standard for Cold-Formed Steel Connections*, American Iron and Steel Institute, Washington, D.C., 2013.

[22] Tao F, Cole R, Moen CD. "Monotonic and cyclic response of single shear cold-formed steel-to-steel and steel-to sheathed connections." *Virginia Tech Research Report No. CE/VPI-ST-16/01*, Blacksburg, VA, 2016

[23] MATLAB, Version 8.6.0.267246. The Mathworks, Inc., www.mathworks.com, 2015.

[24] ASTM. E8M-13, *Standard Test Methods for Tension Testing of Metallic Materials (Metric)*, ASTM International, West Conshohocken, PA. 2004.

[25] FEMA 461 - *Interim protocols for determining seismic performance characteristics of structural and nonstructural components through laboratory testing*, Federal Emergency Management Agency (FEMA), Document No. FEMA 461. 2007.
[26] Ibarra LF, Medina RA, Krawinkler H. "Hysteretic models that incorporate strength and stiffness deterioration", *Earthquake Engineering & Structural Dynamics*, 34(12), 1489-1511, 2005.
[27] Rogers CA, Hancock GJ. "Screwed connection tests of thin G550 and G300 sheet steels", *J. Struct. Eng.*, 125(2), 128-136, 1999.

STIFFNESS OF WELDED BRACE TO LIPPED CHANNEL COLUMN JOINTS: AN EXPERIMENTAL AND NUMERICAL APPROACH

Francesc Roure[a], Teoman Peköz[b], Jordi Bonada[a], M. Rosa Somalo[a], M. Magdalena Pastor[a] and Miquel Casafont[a]

[a] *Universitat Politècnica de Catalunya, Barcelona, Spain*
[b] *Cornell University, Ithaca, USA*

Abstract: Upright frames of the pallet racks fabricated in the U.S. typically have columns welded to the braces. Analysis of the upright frames assuming the brace to column joints as rigid indicates significantly larger stiffness than experimentally tested actual frames. The main cause of this is the flexibility of the brace – column joints. This paper presents an analysis of the flexibility of these joints experimentally and numerically.

1. Introduction

The most common upright frames for the pallet rack structures in the USA have braces welded to the columns. The columns are usually lipped channel sections. The braces are introduced between the lips of the lipped channel sections and welded. A typical example of an upright frame and its details are shown in Figs. 1 to 4.

In general, cross-aisle stability of a rack structure relies on the in-plane stiffness of the upright frames. Analysis of the upright frame, using beam elements and assuming rigid joints between the columns and the braces overestimates the stiffness of upright frame very significantly. This can be observed with an experimental verification of the behaviour of the frame.

The main cause of this difference are the local deformations in the joints between bracing and columns. A similar conclusion was drawn by S. Sambasiva & alter in 2004 [1], and for the down-aisle stiffness of shelving structures by F. Roure & alter in 2015. For analysis and design of upright frame joints Paragraph 6.3.4 of the ANSI code MH 16.1: 2012, for the design of rack structures, requires the consideration of the deformation of the member legs, lips and stiffening elements that make up the section. Thus, the aim of this paper is to analyse these local deformations in the joints, both experimentally and numerically by simulation with

FE, to be able later to introduce them in the global model of the frame, with the objective to obtain a more accurate behaviour of the FE model, closer to the behaviour of the actual frame.

The geometry of the frames is repetitive, and in this study all the joints of the braces to the columns can be classified, with respect to the geometry, into two groups: joints with an angle between brace and column of 90°, and joints with an angle of 45°. The behaviour of the joint is different when it is subjected to tension or to compression. So combining both variables, there are four different situations to analyse for each combination of columns and braces: 90° in tension, 90° in compression, 45° in tension and 45° in compression.

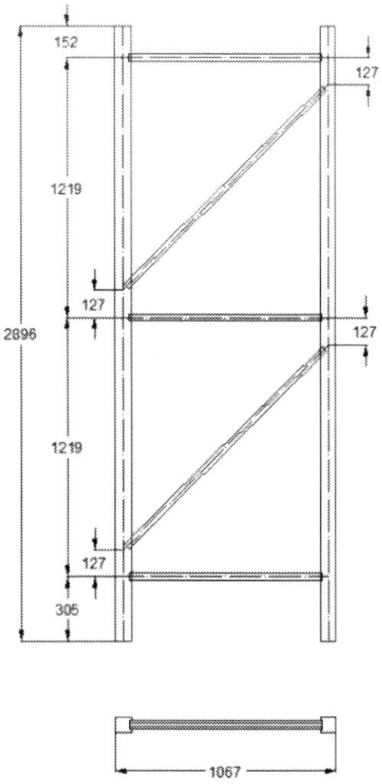

Fig. 1: Frame geometry

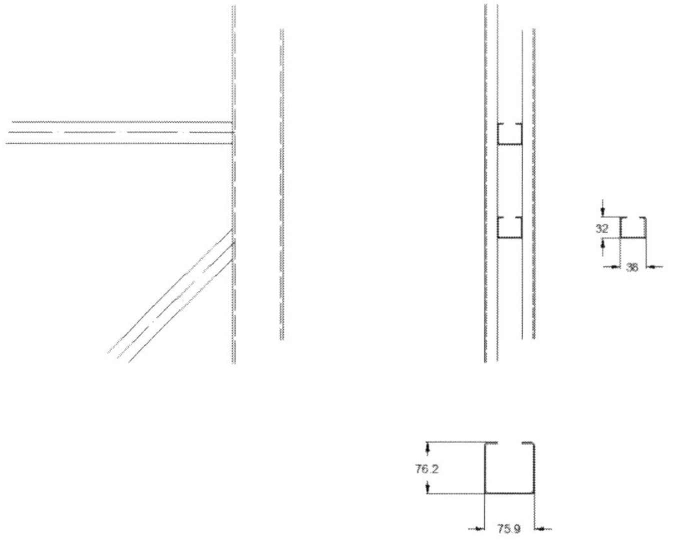

Fig. 2: Brace – column joints

Fig. 3: 90° brace-column joint

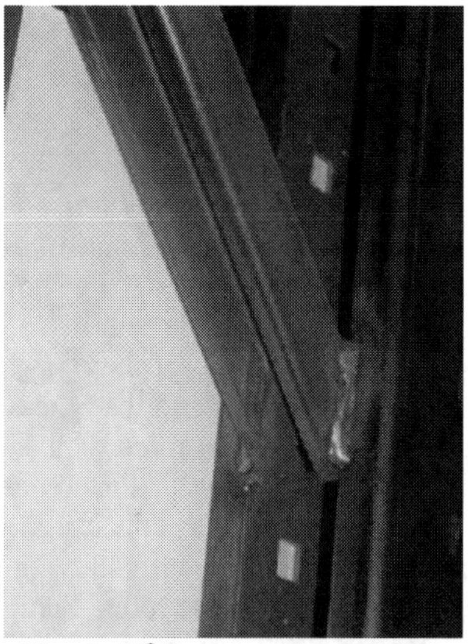

Fig. 4: 45° Brace – column joint

In this paper these four situations are analysed, experimentally and numerically with FE, for two different frames. The frames have the same geometry, but the columns are different: frame A is made with columns having a lipped C section 75.95 × 76.20 × 1.78 mm, and frame B a lipped C section 75.95 × 76.20 × 2.67 mm. For both frames the bracing is the same: a lipped C section 38.10 × 31.75 × 1.52 mm.

2. Description of the tests

2.1 The samples

The following sections provided by Mecalux S.A. were used in this study:

- Columns A: material thickness 1,78 mm, yield stress 515 MPa, ultimate stress 591 MPa;
- Columns B: material thickness 2,67 mm, yield stress 427 MPa, ultimate stress 479 MPa;
- Braces: material thickness 1,52 mm, yield stress 445 MPa, ultimate stress 515 MPa.

The braces are welded at the midsection of the column at an angle of 90° and 45° (see Figs. 3 and 4).

2.2 The experimental setup

The experimental setup can be seen in Figs. 5 to 7. In all the cases two bases are mounted at both ends of the column. The bases have a pin, passing through the centre of gravity of the section. One end is pinned and the other end is sliding pinned. With this setup it is guaranteed that the column is subjected always to statically determinate bending.

The brace is always vertical. When the angle is 90° the column is mounted horizontally. When the angle is 45° the column is mounted diagonally at 45°. The load is applied to the brace by means of a hydraulic cylinder, through a spherical bearing. When a tension load is applied, the cylinder is pinned at both ends. When a compression load is applied, the far end of the cylinder is fixed and the brace is laterally restrained, so that only vertical movement is allowed.

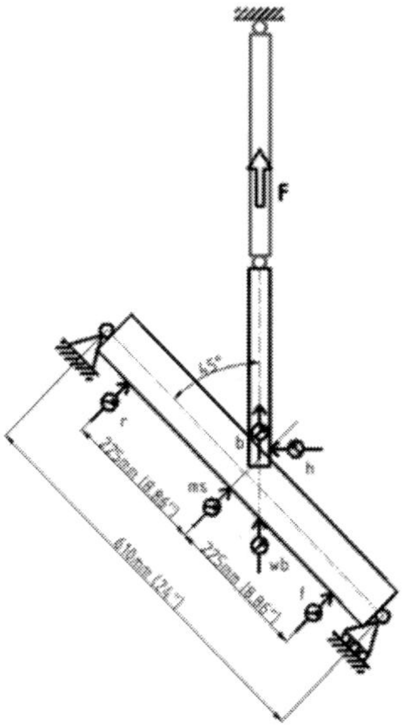

Fig. 5: Test setup for 90°

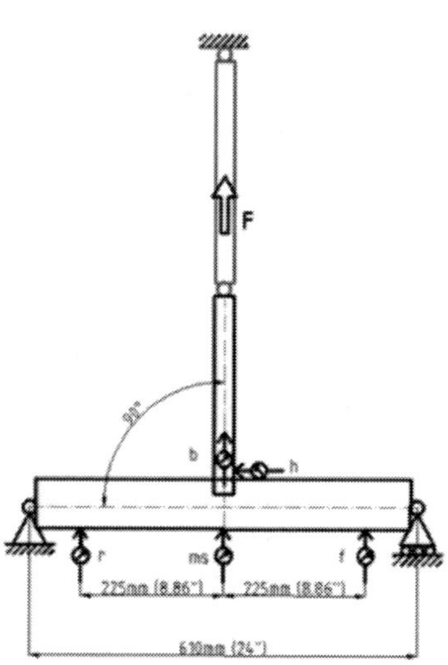

Fig. 6: Test setup for 45°

2.3 Instrumentation

The force applied to the brace is measured by means of a force transducer. The displacement is measured perpendicular to the web of the column (see Fig. 7) at the midspan (ms) and at both sides at 225 mm from the midspan (f and r). The vertical (b) and the horizontal (h) displacement of the brace are also measured. When the angle is 45°, also the vertical displacement of the column (wb) on the action load line is measured. The displacements are measured with potentiometric displacement transducers.

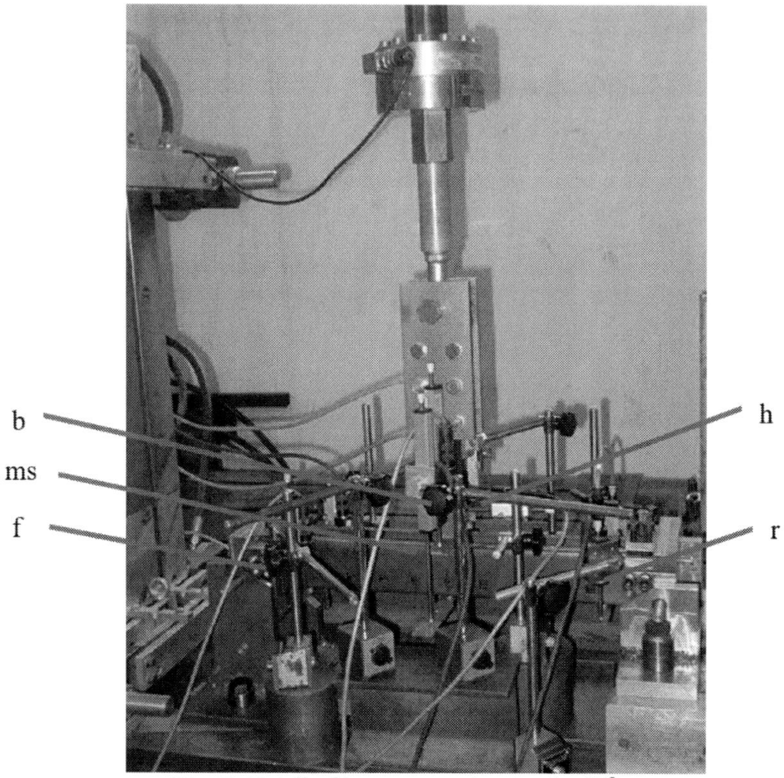

Fig. 7: Detail of instrumentation for the 90° test

2.4 Modus operandi

Initially, the displacement transducers are set to zero. Then the load is gradually increased until the joint reaches the final failure load. The values of the force and the displacements are continuously registered.

3. Experimental results

Three samples have been tested for each situation, and the mean has been calculated.

3.1 Plots

From the experimental data the column bending (midspan) and the column – brace joint deformation (spring) are calculated:

$$\text{midspan} = \text{ms} - (\text{f} + \text{r})/2$$
$$\text{spring} = \text{b} - \text{ms} \quad (\text{for } 90^\circ)$$
$$\text{spring} = \text{b} - \text{wb} \quad (\text{for } 45^\circ)$$

These values are plotted and the initial stiffness of the spring curve is determined (see Fig. 8).

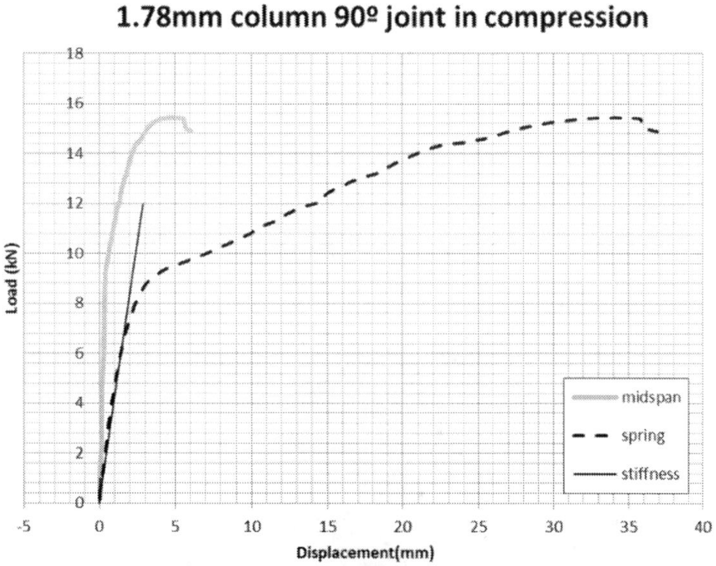

Fig. 8: 90° joint, compression test

3.2 Tables

The experimental results are summarized in the Table 1.

Table 1

Column	Angle	Test	F_u (kN)	midspan (mm)	stiffness (kN/mm)
A (1.78 mm)	90°	tension	25.97	0.73	5.75
		compression	15.23	3.60	3.44
	45°	tension	35.00	1.03	7.65
		compression	21.62	2.63	5.67
B (2.67 mm)	90°	tension	35.07	1.82	14.77
		compression	21.45	3.37	8.97
	45°	tension	49.90	2.90	20.31
		compression	27.64	1.55	14.79

3.3 Behaviour of the joints

The failure modes of the joints are summarized in Table 2.

Table 2

Column	Angle	Test	Failure mechanism
A (1.78 mm)	90°	tension	column lips – brace weld breaking
		compression	excessive deformation of column lips
	45°	tension	column lip tearing
		compression	column lip tearing and brace buckling
B (2.67 mm)	90°	tension	column lip tearing or brace breaking
		compression	excessive deformation of column lips
	45°	tension	brace breaking
		compression	column lip tearing and brace buckling

4. Finite element model

A finite element model has been created, using the ANSYS V14 software. The joint has been modeled with a fine mesh of Shell 181 elements (see Fig. 9). It has been realized that the welds between column lips and braces have to be reproduced in a realistic manner. It is not enough to connect the corresponding nodes of both components, it is necessary to reproduce in the connection the actual thickness of the weld, because it affects the stiffness of the column lip, and finally affects the stiffness of the joint. The weld has been reproduced by connecting nodes of the brace to nodes of the lip, at a distance of 3 mm of the corner, simulating so a 3 mm welding fillet. The effects of the residual stresses due to the welding process are not considered in this model. The model includes material non linearity (see f_y and f_u of the materials in paragraph 2.1) and large displacements.

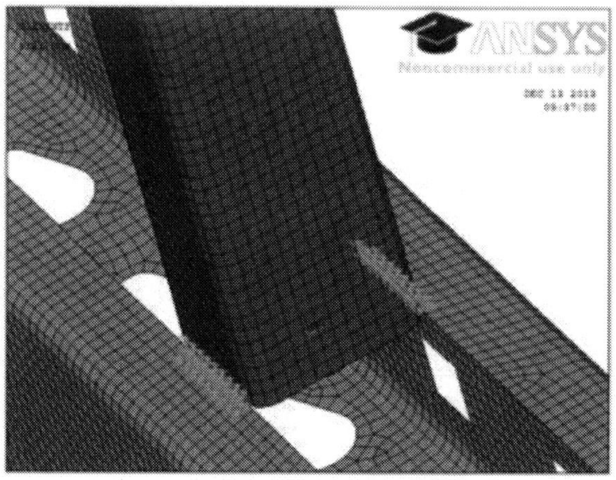

Fig. 9: FE model: detail of the welds

In Fig. 10 the *F-d* curve obtained with the FE model is compared to the experimental *F-d* curves. It can be seen that in the initial part of the curves the agreement between both methods is very good. In fact this is the part interesting for the determination of the cross-aisle stiffness of the frame. There is also good agreement in the ultimate load.

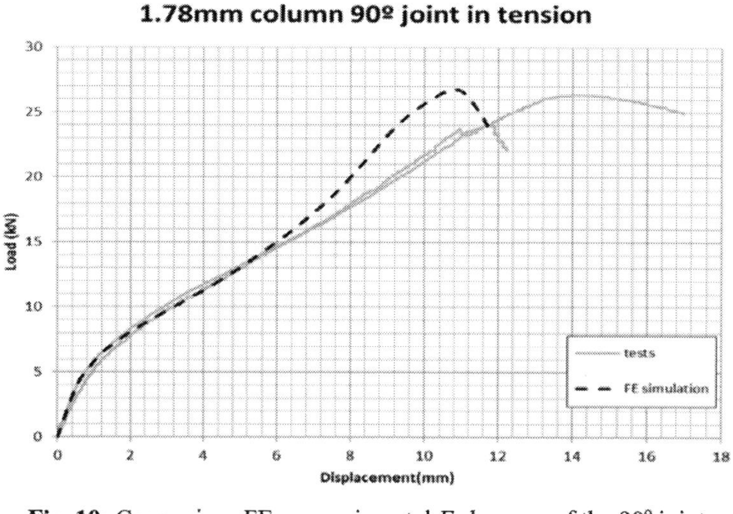

Fig. 10: Comparison FE – experimental *F-d* curves of the 90° joint

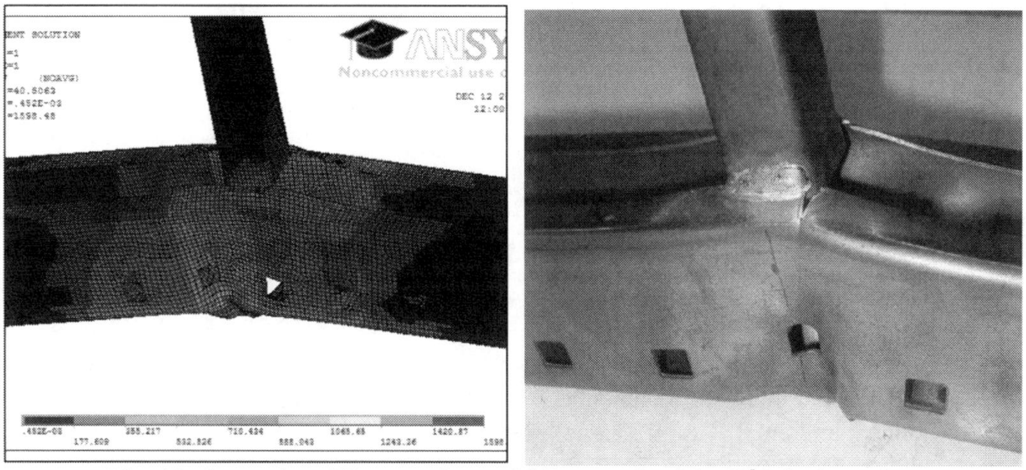

Fig. 11: Comparison FE – experimental failure mode of the 90° joint in tension

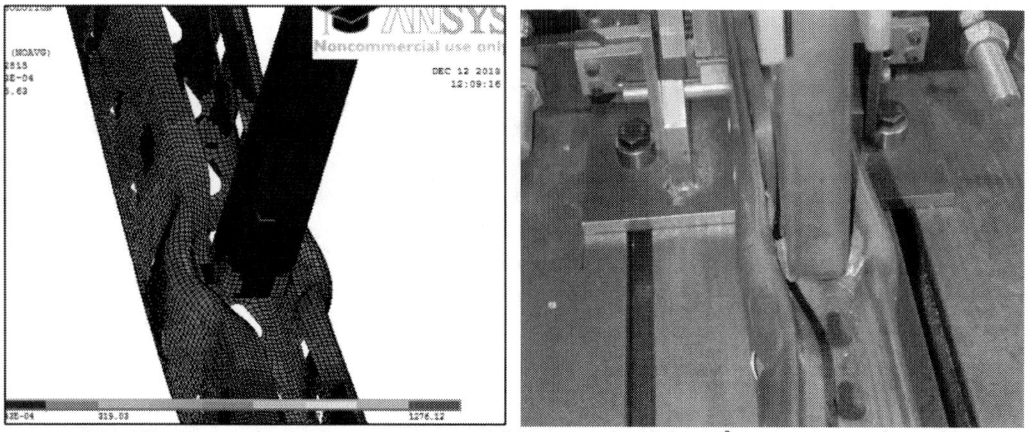

Fig. 12: Comparison FE – experimental failure mode of the 90° joint in compression

The FE model also reproduces very well the failure mode of the joints, as can be seen in Figs 11 and 12.

5. Conclusions

The main conclusions are:

1. The stiffness of the brace-to-column joint can be measured experimentally with precision.
2. The experimental setup has to be designed very carefully, to measure the effects desired, and to cancel those not desired.
3. The behaviour and the stiffness in tension and compression is different.
4. The behaviour and the stiffness for the joints at 90° and at 45° is different.
5. In all the cases the stiffness of the joint is of an order of magnitude (3.44 to 20.31 kN/mm) in the same range of the stiffness of the brace (24.0 to 34.1 kN/mm). In consequence it can be predicted that taking or not in account this stiffness in the model of the whole frame will have an important influence.

6. The stiffness and the behaviour of the joints can also be analysed with a FE model. The model has to include all the geometric details (including the actual thickness and geometry of the weld), and reproduce material non linearity and large displacements.
7. The agreement between the numerical and experimental *F-d* curves of the joints is very good in the first part of the curves. This part is also where the stiffness analysis of the frame is applied.
8. The FE model reproduces also very well the failure mode of the joints.
9. Taking into account all these conditions, further geometries and combinations of brace-column joints can be analysed with FE models, without need of experimental testing.

Acknowledgments

The authors wish to thank the technicians Juan Espada and Francesc J. García for their skilled work in the experimental testing. The authors wish also to thank Mecalux S. A. for providing the samples for the tests.

References

[1] Sambasiva S, Beale RG, Godley MHR. "Shear stiffness of pallet rack frames", *Proceedings of the 7th International Speciality Conference on Cold-Formed Structures*, Orlando, Florida, USA, 295-311, 2004.

[2] Roure F, Somalo MR, Bonada J, Pastor MM, Casafont M. "Down-aisle stiffness of steel shelving structures: Experimental and FE approach", *Proceedings of the 8th International Conference on Advances in Steel Structures*, Lisbon, Portugal, 2015.

[3] RMI, "ANSI MH 16.1: 2012, Specification for the Design, Testing and Utilization of Industrial Steel Storage Racks" *Materials Handling Industry of America (MHIA)*, Charlotte, USA, 2012.

FINITE ELEMENT MODEL VS. EXPERIMENTAL TEST RESULTS FOR COMPOUND COLUMNS AND THEIR EXPOSED BASE PLATE CONNECTIONS

Gianluca Sarracco, Gaetano Della Corte and Raffaele Landolfo

Department of Structures for Engineering and Architecture, University of Naples Federico II, Italy

Abstract: This paper describes finite element modelling of two specimens of compound columns and their exposed base plate connections. The two specimens represented samples of real compound columns extracted from an existing industrial building. They were formerly subjected to physical tests in the laboratory of the Department of Structures at the University of Naples Federico II. Firstly, experimental test results are shortly summarized. Secondly, finite element models are described, illustrating the numerical predictions and comparing them to the experimental results.

1. Introduction

Exposed base plate connections are important structural components in steel constructions, especially in inverted pendulum or moment resisting frame structures. Both schemes are typical of industrial buildings, where, in case of seismic actions, column base connections significantly contribute to the energy dissipation capacity. Past seismic events caused damages in base plate connections, such as plate fracture, excessive bolt elongation, unexpected early anchor bolt failure or grout and concrete crushing [1]. Complex interactions govern the response of this type of connections, involving the column, the base plate, the foundation anchors and the grout joint between the base plate and the foundation. Moreover, industrial buildings are often characterised by compound columns, consisting of two I-shaped profiles connected together by means of battens, to form a single column member. Due to flexibility of connections, shear deformations significantly affect the behaviour of a compound column, considered as a single element.

Two specimens of compound battened columns and relevant exposed base plate connections were tested formerly, in the laboratory of the Department of Structures at the University of Naples Federico II [2]. Subsequently, detailed finite element (FE) models were developed and analysed. The numerical predictions are compared to the experimental results in this paper, trying to highlight critical modelling issues that should be considered when approaching similar problems.

2. Description of experimental tests

Experimental tests were carried out on two full-scale specimens reproducing the geometry of battened columns with exposed base plate connections, whose geometry reproduced real columns of an existing industrial building. Both specimens were characterized by two I-shaped profiles with battens and a stiffened base plate connection anchored by means of round steel rods to a concrete foundation. Fig. 1 reports detailed drawings; Della Corte and Landolfo [2] provides further information, about the experimental tests.

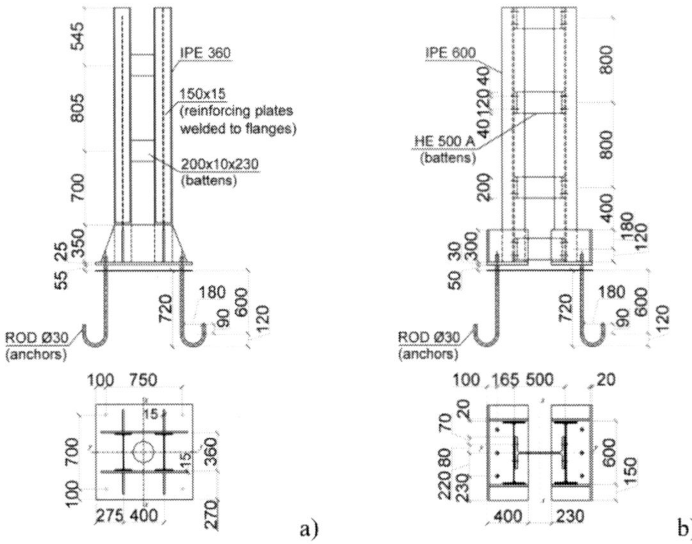

Fig. 1: Geometry of specimen 1 (a) and specimen 2 (b)

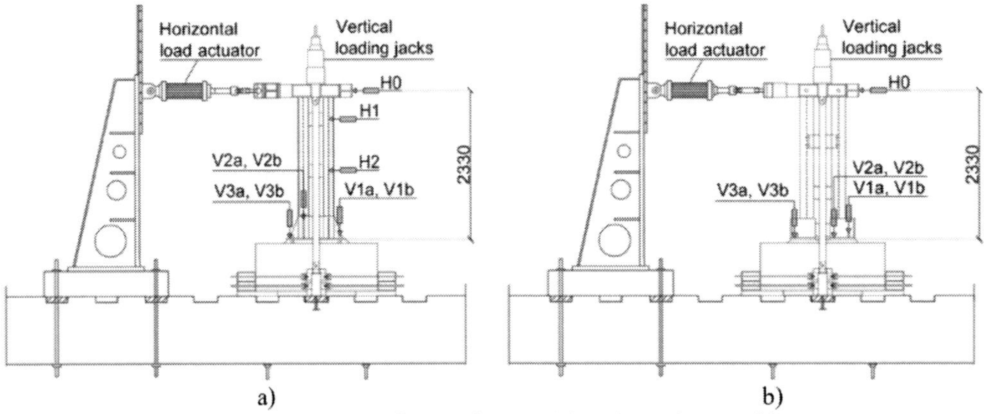

Fig. 2: Test setup for specimen 1 (a) and specimen 2 (b)

Firstly, both specimens were submitted to monotonic loading tests in the elastic range of response, in order for measuring initial specimen stiffness; secondly, they were subjected to cyclic loading tests, in order to observe plastic mechanisms and hysteretic behaviour. The cyclic loading protocols were in accordance with ECCS provisions [3].

Fig. 2 shows the test setup, consisting of a horizontal load actuator, used to apply lateral forces at the column tip, and a couple of vertical jacks, which exerted the compressive axial load by means of vertical ties connected to the laboratory strong floor. Several displacement sensors completed the test setup. Fig. 2 reports those displacement sensors adopted to calibrate the numerical models described in the current paper. Section 4 and section 5 will present experimental results for specimen 1 and specimen 2, respectively, in comparison with numerical responses.

3. Finite element model

3.1 Geometry and mesh definition

Numerical analyses were performed using the software ABAQUS [4]. Both specimens are doubly symmetric allowing half-models to be built up. Fig. 3 shows the meshed half-model of specimen 1 and specimen 2 respectively; the mesh density was increased where stress concentrations were expected to occur.

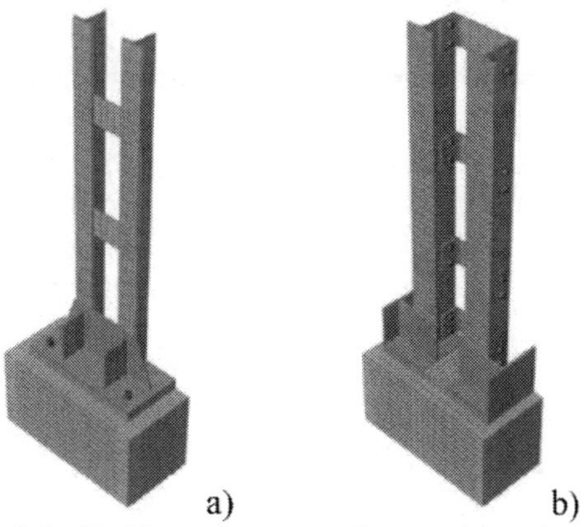

Fig. 3: FE model of half-specimen: specimen 1 (a) and specimen 2 (b)

All the parts constituting the models were meshed with first order linear reduced integration hexahedral elements (C3D8R). This finite element is affected by hour glassing; nevertheless, an adequately fine mesh (*i.e.* at least two elements through the thickness of a plate) and the artificial hourglass stiffness have provided acceptable results. Element distortion was reduced increasing local mesh density along the edges of circular parts and holes. The adopted mesh was checked by means of sensitivity analyses.

It is worth mentioning that the model did not include hooks of anchor rods, while it assumed a fixed restraint at the end of their straight portion (i.e., no slip occurring below this portion in the hooked region).

3.2 Material constitutive models

Material properties were determined by means of laboratory tests and through information provided by the manufacturer. An overview of the main properties in terms of yielding (f_y) and failure (f_u) stresses, as well as ultimate strains (ε_u) is reported in Table 1. Experimental data are available in [2]. Concrete of the foundation was also tested to check its strength. Using results of three cubic specimens, the mean compression strength was equal to 46 MPa.

All steel components were modelled using the Von Mises yield criterion and kinematic strain-hardening rule.

Concrete and mortar were modelled as frictional and pressure dependent Drucker-Prager materials. The unidirectional compressive behaviour of the grout, including post failure behaviour, was calibrated by means of six experimental tests [2]. The tensile strength was fixed at 1/10 of the maximum compressive strength. Parameters of the 3-axial stress response were fixed as values extracted from the literature for concrete [5].

Table 1: Steel average yield strength (f_y), ultimate strength (f_u) and ultimate strain (ε_u)

ELEMENT	f_y (MPa)	f_u (MPa)	ε_u (%)
PLATE 25mm	301	474	30.2
PLATE 30mm	326	504	33.2
IPE 600	320	461	38.6
HE 500 A	389	487	26.7
IPE 360	289	429	37.5

3.3 Boundary conditions and analysis procedure

The bottom section of the foundation block was fixed in the space, since the experimental evidence highlighted the absence of significant displacements at the base of the foundation block.

Interactions between different parts were modelled by means of surface-to-surface contact with normal behaviour defined as "hard" contact and allowing separation. The tangential behaviour was frictionless except for the base plate and grout, where an isotropic friction formulation was adopted, with a constant friction coefficient of 0.45 [6, 7].

Axial compressive loads and lateral displacements were assigned to a master joint at the column top cross section. The specimen self-weight was automatically computed by assigning a gravity acceleration field and the appropriate unit mass. Horizontal loads were applied through displacement control, following the displacement measured by sensor H0 (Fig. 2). Numerical analyses accounted for geometrical non-linearity.

4. Specimen 1

4.1 Monotonic loading test

Fig. 4*a* compares column experimental displacements, measured at the locations shown in Fig. 2*a* to corresponding numerical predictions (vertical force N=150 kN). The figure shows a generally good agreement of numerical and experimental results. Pre-stress in the foundation anchor rods was a model parameter calibrated through back analysis of experimental results. In fact, the anchor rods were hand-tightened at the beginning of the physical tests without control of the applied force. Using the displacement response measured by sensor V1 (Fig. 4*b*), an average pre-tension force equal to 20% of the anchor yielding resistance was estimated. As shown in Figs. 4*a* and 4*b*, the model was able to catch the interaction between the base plate and the grout, showing stiffness reduction due to the transition from a fully compressed column base plate to separation on the tension side of the plate. Therefore, the FE model is able to simulate the change of the neutral axis position at the column base cross section. The plots in Figs. 4*c* and 4*d* also testify this aspect, illustrating that the numerical model acceptably reproduces the displacements monitored by sensors V2 and V3.

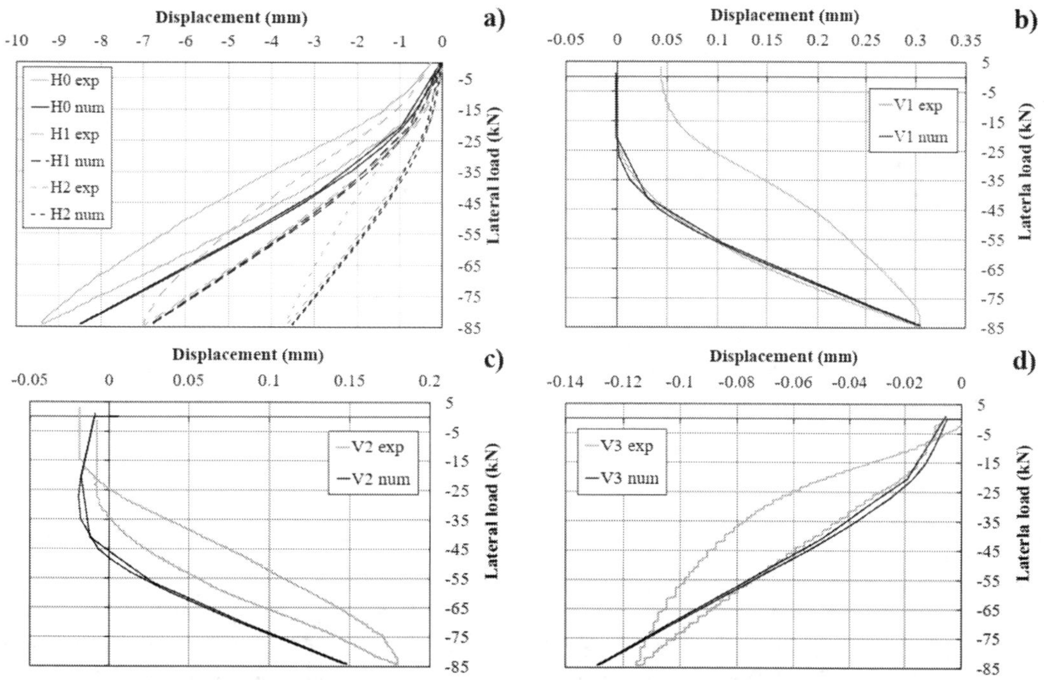

Fig. 4: Experimental vs. numerical results: monotonic loading test of specimen 1 (N=150kN)

4.2 Cyclic loading test

Fig. 5 shows experimental vs. numerical results in case of cyclic loading (N=150 kN).

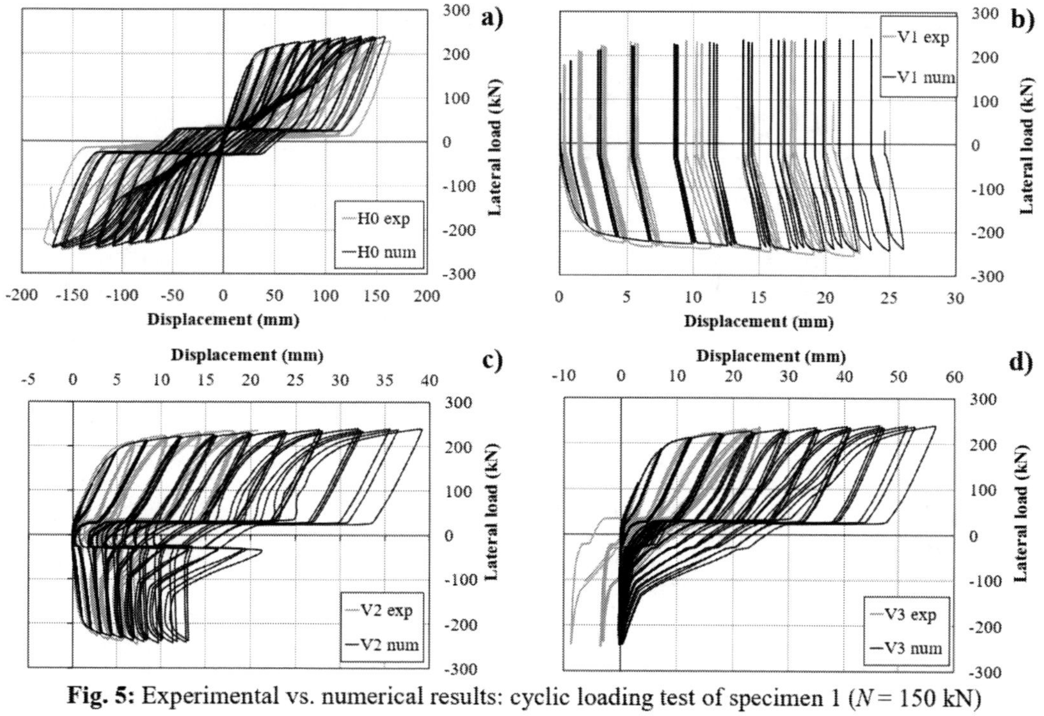

Fig. 5: Experimental vs. numerical results: cyclic loading test of specimen 1 (N = 150 kN)

The tested specimen exhibited fracture of one anchor bolt along with large plastic deformations of the base plate and widespread cracks in the grout. The numerical model captures all these phenomena.

Fig. 5*a* shows that displacements H0 obtained from the numerical model are in good agreement with experimental results at small-to-medium displacement amplitudes. However, a mismatch between numerical and experimental curves is observed when the lateral load reverses at relatively large displacements. During this phase, the FE model shows a premature rise of strength caused by an early contact between the base plate and the grout (Fig 5*d*). Improvements in the cyclic response model of the mortar are deemed useful for eliminating the mismatch. Figs. 5*b* and 5*c* highlight a good agreement between numerical and experimental results for displacements measured by sensors V1 and V2. Displacement sensors V1, V2 and V3 had to be removed before ending the test, because they reached their maximum displacement capacity. Therefore, the corresponding experimental response curves are interrupted at displacement amplitudes smaller than the maximum values of the numerical response curves.

5. Specimen 2

5.1 Monotonic loading test

The monotonic loading test conducted on specimen 2 (N=150 kN) is reported in Fig. 6, where it is compared with numerical results. Fig. 6*a* shows an initial mismatch between numerical and experimental results. However, at displacements larger than approximately 12 mm, the numerical and experimental response curves are almost parallel. Therefore, the initial mismatch could be attributed to effects, such as initial sliding of the physical specimen and sensors, which are rather difficult to be incorporated in the numerical model.

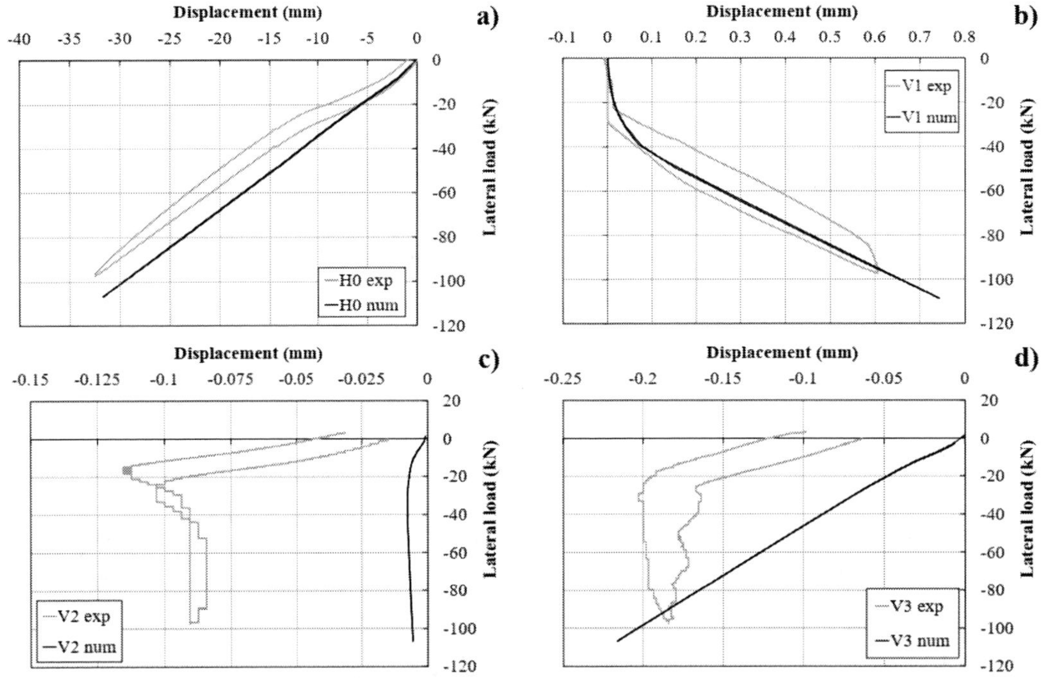

Fig. 6: Experimental vs. numerical results: monotonic loading test of specimen 2 (N=150 kN)

Fig. 6*b* illustrates comparison of numerical predictions and experimental results for displacements measured by sensors V1 (Fig. 2*b*), showing good matching. In this case, back-analysis of experimental results, led to estimating an anchor pretension force equal to 30% of its yielding force resistance, a value calibrated essentially looking at results in Fig. 6*b*. Finally, Figs. 6*c* and 6*d* report displacements measured by sensors V2 and V3 (Fig. 2*b*), showing mismatching with numerical results. However, the measured displacement amplitude was very small (max value 0.20 mm approximately) and the difference in terms of this local behaviour was considered acceptable in view of the good prediction of the global response (see also results for cyclic tests, section 5.2). One should consider also difficulties in predicting the initial response of mortar based on mono-dimensional stress-strain relationships obtained from standard material tests. Besides, the initial mortar specimen response usually shows some nonlinear behaviour, while the model assumed a linear initial response (*i.e.* a constant Young's modulus).

5.2 Cyclic loading test

Fig. 7 compares results obtained from experimental and numerical cyclic loading tests of specimen 2 (N=150 kN). As for specimen 1, measurements of sensors V1, V2 and V3 had to be stopped before ending the test, otherwise exceeding their displacement capacity. The whole test had to be stopped when the displacement capacity of the horizontal load actuator was attained, prior to the occurrence of any abrupt failure of the specimen. However, the peak displacement demand was sufficiently large to make the test significant. The specimen exhibited large plastic deformations of foundation anchor rods and crushing of the grout, as well as significant plastic deformations at connections between the column web panel and the battens.

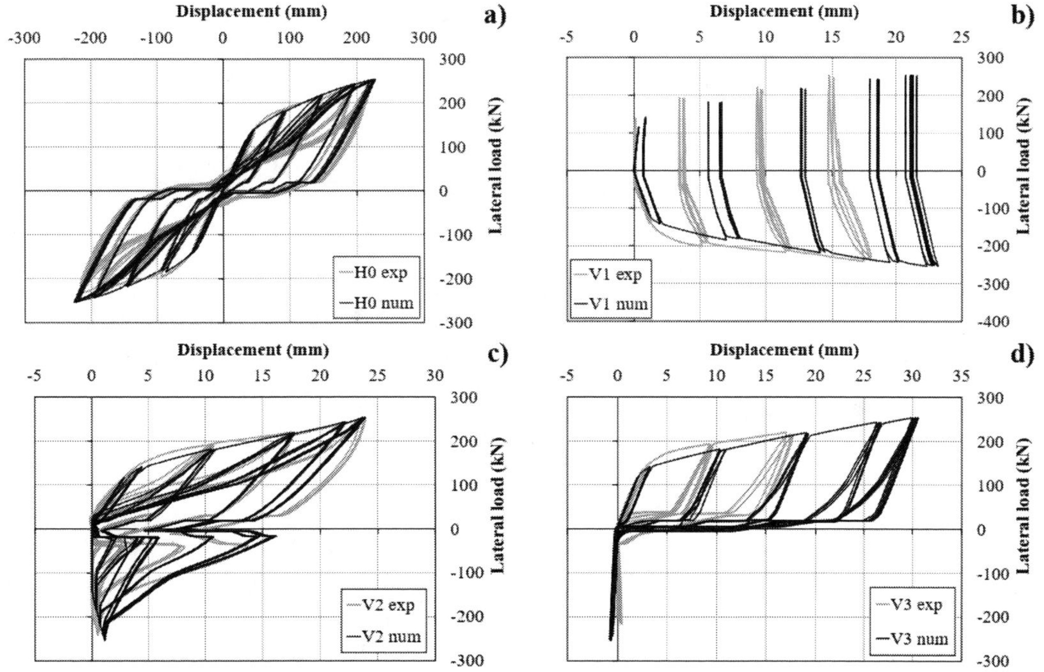

Fig. 7: Experimental vs. numerical results: cyclic loading test of specimen 2 (N=150 kN)

Fig. 7*a* shows that numerical displacements H0 are in good agreement with experimental measurements, at small and medium displacement amplitudes. Nevertheless, at large displacements the numerical model presents hysteretic loops slightly larger than experimental ones. Absence in the model of a clearance between bolts and holes might explain such type of behaviour. However, at this stage of the study, the introduction of clearance caused solver problems and unstable results, and this is the subject of further on-going investigations. Figs. *7b*, *7c* and *7d* highlight generally good predictions also for displacements measured by sensors V1, V2 and V3, respectively.

6. Conclusions

Finite element modelling of two specimens of compound columns and their exposed base plate connections, formerly tested in laboratory, was carried out. The model predictions were compared to experimental results for both monotonic and cyclic tests. Monotonic tests were useful to validate specimen boundary conditions and to estimate the initial pre-stress of anchor rods. Cyclic tests were utilized for checking the implementation of material plastic properties (*i.e.* kinematic strain hardening for steel and crack damage for concrete). Generally, both monotonic and cyclic response models have shown good agreement with experimental results, especially for small-to-medium displacements. Further developments are required in order to achieve a more suitable assessment of local response phenomena, such as grout degradation and effects of clearances between bolts and holes. However, such phenomena had some, but small, effects only at large displacements under cyclic loading.

Acknowledgments

This research was financially supported by the ReLUIS consortium (Italian network of University laboratories for research on earthquake engineering), whose contribution is gratefully acknowledged.

References

[1] Lee D, Goel SC, Stojadinovic B. "Exposed Column-Base Plate Connections Bending About Week Axis: I. Numerical Parametric Study", *International Journal of Steel Structures*, 8, 11-27, 2008.

[2] Della Corte G, Landolfo R. "Experimental tests of compound columns and their base-plate connection subjected to axial and horizontal forces", *The 8th International Conference on Advances in Steel Structures (ICASS 2015)*, Lisbon, Portugal, July 22-24, 2015.

[3] ECCS (European Convention for Constructional Steelwork), *Recommended testing procedure for assessing the behaviour of structural steel elements under cyclic loads,* Technical Committee 1, TWG 1.3-Seismic Design, Publ. No. 45, 1986.

[4] Hibbitt, Karlsson, Sorensen "ABAQUS Analysis User's Manual, Version 6.12", Dassault Systèmes Simulia Corp., Providence, RI, USA, 2012.

[5] Lubliner J, Oliver J, Oller S, Oñate E. "A Plastic-Damage Model for Concrete", *International Journal of Solids and Structures*, vol. 25, no.3, pp. 229–326, 1989.

[6] Jaspart JP, Vandegansb D. "Application of the component method to column bases", *Journal of Constructional Steel Research,* 48, 89–106, 1998.

[7] Kanvinde AM, Jordan SJ, Cooke RJ. "Exposed column base plate connections in moment frames - Simulations and behavioral insights", *Journal of Constructional Steel Research*, 84, 82–93, 2013.

BOLTED BEAM TO COLUMN JOINTS WITH HAUNCHES: EXPERIMENTAL AND NUMERICAL INVESTIGATIONS

Cristian Vulcu[a], Keunwoo Lee[a], Aurel Stratan[a], Cosmin Maris[a] and Dan Dubina[a,b]

[a] *Politehnica University of Timisoara, Dept. of Steel Structures and Structural Mechanics, Romania*
[b] *Laboratory of Steel Structures, Romanian Academy, Timisoara Branch, Romania*

Abstract: The current paper outlines the experimental program on bolted extended end-plate connections with haunches investigated at the Politehnica University of Timisoara in the framework of the European EQUALJOINTS project. The design of the joints was performed aiming to obtain full-strength and rigid joints. Further, the paper presents a set of results from the experimental investigations, and the outcomes of the numerical investigation program. Consequently, the calibration of the numerical model is presented, as well as the parametric study, which was performed in order to assess the influence of: member size, haunch geometry, imperfection, and response of internal / external joint configurations.

1. Introduction

The use of pre-qualified joints is a common practice in U.S. and Japan. Nevertheless, the standard joints pre-qualified according to codified procedures in U.S. and Japan cannot be extended to Europe, due to differences in materials and section shapes. Moreover, the beam-to-column joint types usually adopted in U.S. and Japan are not commonly used in Europe. As a result, the existing scientific and technical background on pre-qualification may not be directly extended to European context. For this reason, a European research project entitled European pre-QUALified steel JOINTS (EQUALJOINTS) [1], is currently underway and aims at seismic pre-qualification of several beam-to-column connection typologies common in the European practice, and which are illustrated in Fig. 1. In particular, the research activities will focus on the standardization of design and manufacturing procedures based on different geometric and mechanical parameters of selected joint typologies. Hence, a large experimental program supported by theoretical and numerical analyses is proposed to provide reliable standard joints that can be easily used by designers.

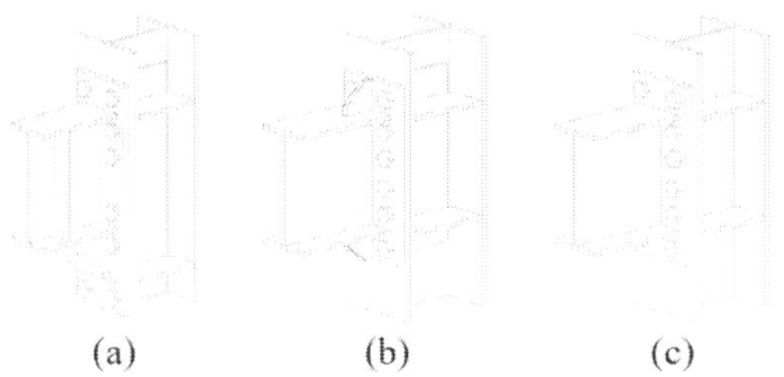

(a) (b) (c)

Fig. 1: Connection typologies with: a) haunches, b) stiffened end-plate, c) un-stiffened end-plate – [1]

The current paper outlines the experimental program on bolted extended end-plate connections with haunches that will be investigated at the Politehnica University of Timisoara in the framework of EQUALJOINTS project [1]. It is to be mentioned that the design objective was to get full-strength and rigid joints. Furthermore, the outcomes of the ongoing experimental and numerical investigations are presented. In particular, the following sections are related to the description of test results, calibration of the numerical model, and parametric study – aimed to assess the influence of: member size, haunch geometry, imperfection, response of internal/external joint configuration.

2. Experimental program

The aim of the experimental program, at the Politehnica University of Timisoara, is to investigate the behavior of bolted connections with haunches under monotonic and cyclic loading. A set of 24 beam-to-column joint assemblies are being tested in order to evaluate different parameters that are specific for this type of connection. Single sided and double sided joints were selected from moment resisting frames (MRF) as illustrated in Fig. 2.

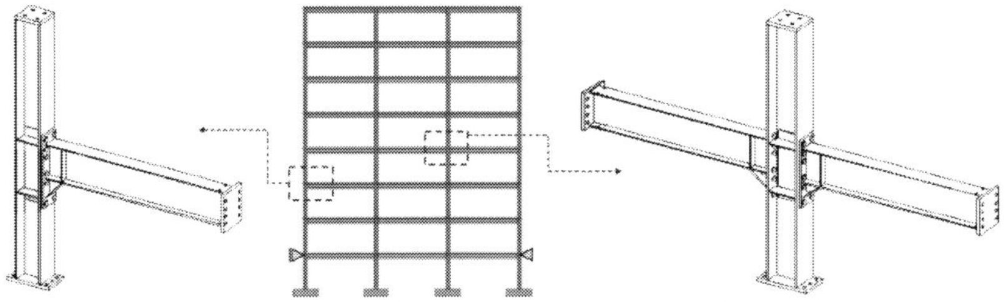

Fig. 2: Single-sided and double-sided joints selected from MRF structure

The experimental program, summarized in Table 1, contains three groups of specimens:

- Group 1: full strength & rigid joint, shallow haunch (35° angle), strong web panel;
- Group 2: full strength & rigid joint, steep haunch (45° angle), strong web panel;
- Group 3: full strength & semi–rigid joint, shallow haunch (35° angle), balanced web panel.

Group 1 and Group 2 serve for qualifying two alternative haunch geometries (lower and upper limit of reasonable haunch angle) for considered range of beam size. Group 3 investigates joints with balanced panel zone strength, which also leads to a semi–rigid classification of the joint (connection and panel zone). Two supplementary web plates are used for the joints of Group 1 and Group 2, while for Group 3 only one supplementary web plate is used. Additionally, larger column depth increases the range of prequalified column sizes. The complete parameters considered within the experimental program are: loading protocol (monotonic and cyclic), member size, single-sided and double sided connections, strong panel zone / balanced panel zone, strong beam and haunch geometry. Table 2 makes an overview of the parameters considered within the experimental program and describes the labelling of the specimens. As can be observed, the experimental program covers 24 tests on joint specimens, out of which three tests are to be performed under monotonic loading: EH2-TS-35-M, EH2-TS-45-M, and EH2-XB-35-M, while the rest are subjected to cyclic loading.

Table 1: Experimental program on bolted beam-to-column joints with haunches

Group 1: single-sided joints	Group 2: single-sided joints	Group 3: double-sided joints
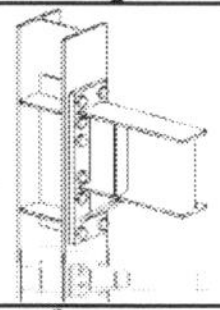 EH1-TS-35 IPE 360 HEB 280	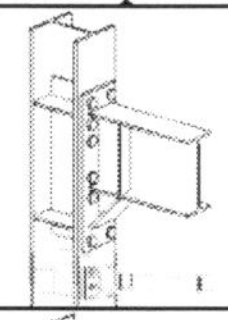 EH1-TS-45 IPE 360 HEB 280	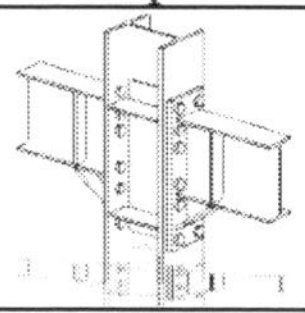 EH1-XB-35 IPE 360 HEB 340
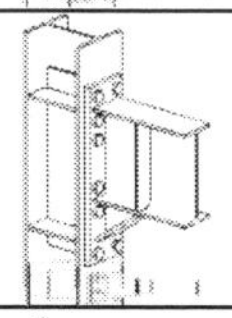 EH2-TS-35 IPE 450 HEB 340	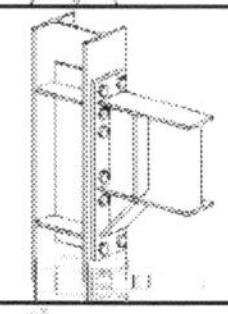 EH2-TS-45 IPE 450 HEB 340	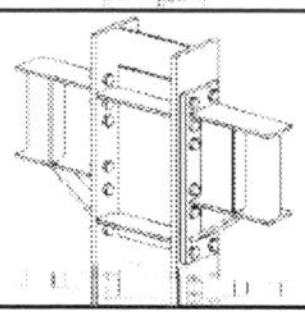 EH2-XB-35 IPE 3450 HEB 500
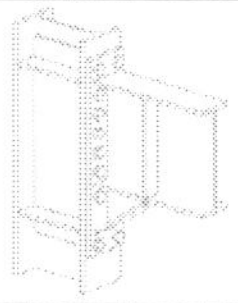 EH3-TS-35 IPE 600 HEB 500	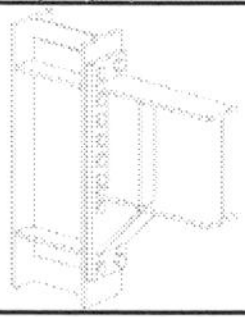 EH3-TS-45 IPE 600 HEB 500	

Table 2: Description of joint configuration and labelling

Gr.	Conn. type	Joint config.	Haunch geom.	Loading protocol	Beam/column depth 1	2	3
1	EH	TS	35°	M	-	EH2-TS-35-M	-
	EH	TS	35°	C1	EH1-TS-35-C1	EH2-TS-35-C1	EH3-TS-35-C1
	EH	TS	35°	C2	EH1-TS-35-C2	EH2-TS-35-C2	EH3-TS-35-C2
	EH	TS	35°	CA	EH1-TS-35-CA	EH2-TS-35-CA	EH3-TS-35-CA
	EH	TSO	35°	C	EH1-TSO-35-C	-	EH3-TSO-35-C
2	EH	TS	45°	M	-	EH2-TS-45-M	
	EH	TS	45°	C1	EH1-TS-45-C1	EH2-TS-45-C1	EH3-TS-45-C1
	EH	TS	45°	C2	EH1-TS-45-C2	EH2-TS-45-C2	EH3-TS-45-C2
3	EH	XB	35°	M	-	EH2-XB-35-M	-
	EH	XB	35°	C1	EH1-XB-35-C1	EH2-XB-35-C1	-
	EH	XB	35°	C2	EH1-XB-35-C2	EH2-XB-35-C2	-

Note:
- Connection type: Haunched beam to column connection (EH)
- Joint configuration and panel zone: exterior joint with strong column web panel (TS), exterior joint with strong column web panel/strong beam (TSO), interior joint with balanced column web panel (XB)
- Haunch geometry: angle of haunch 35° (35), angle of haunch 45° (45)
- Loading protocol: monotonic (M), cyclic (C1, C2), alternative cyclic protocol (CA)

The analytical design procedure of the bolted connections with haunches was developed based on the component method implemented in EN 1993-1-8 [2]. The design of the joints was performed considering the formation of the plastic hinge in the beam, at the end of the haunch. Further, the components of the joint were designed for an equal or higher capacity in comparison to the fully yielded and strain hardened plastic hinge. Furthermore, based on pre-test numerical investigations [3], the assumptions considered in design were as follows: (i) the active bolt rows were those situated near the flange in tension, (ii) the compression centre was located at a distance equal to 60% of the haunch depth measured from the bottom flange of the beam.

3. Experimental investigations

The experimental investigation program on bolted beam-to-column haunched joints is currently underway. The first test was performed for the EH2-TS-35-M joint assembly, and was followed by the experimental investigation of EH2-TS-45-M joint assembly. Both specimens were subjected to monotonic loading. Further experimental investigations on joint assemblies were conducted under cyclic loading. However, the current paper presents the results obtained from the monotonic loading conditions.

3.1 Test set-up, instrumentation and material properties

The experimental test set-up (conceptual scheme and illustration) for the bolted beam-to-column joints is presented in Fig. 3. A hydraulic actuator connected at the tip of the column served as loading device. The support at the bottom end of the column was pinned (horizontal and vertical displacements were restrained), while the tip of the beam was connected to a vertical link pinned at both ends – and therefore the vertical displacement at the tip of the beam was restrained, but the horizontal displacement was free. The beam out of plane deformations were blocked using a lateral support system.

The global instrumentation consisted in measuring: the force in the actuator, displacement at the tip of the column, horizontal and vertical displacement at column base and tip of the beam. The local instrumentation was aimed at measuring the deformations within the dissipative zone, connection zone, and column web panel. Additionally, with the aim to identify the yielding sequence, the connection zone was prepared by whitewashing. The material properties were obtained from tensile tests of steel samples representative for the main joint components: beam flange and web, end-plate, column flange and web.

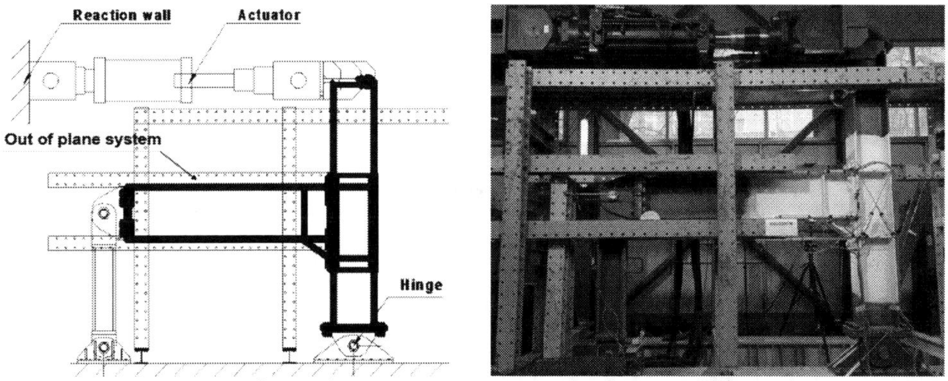

Fig. 3: Experimental test set-up – conceptual scheme and illustration

3.2 Test results

The primary test results are shown in terms of moment-rotation curve (rotation computed at column centreline, bending moment computed at column face) and illustration of the failure mode. Fig. 4 shows the response of EH2-TS-35-M and EH2-TS-45-M joint assemblies under monotonic loading. As can be observed in Fig. 5, the yielding was initiated in the beam flanges near the ending of the haunch, and was followed by large plastic deformations – local buckling of flanges and web. No damage was observed within end-plate and column web panel. Consequently, the plastic mechanism considered in the design was confirmed, i.e. formation of the plastic hinge in the beam.

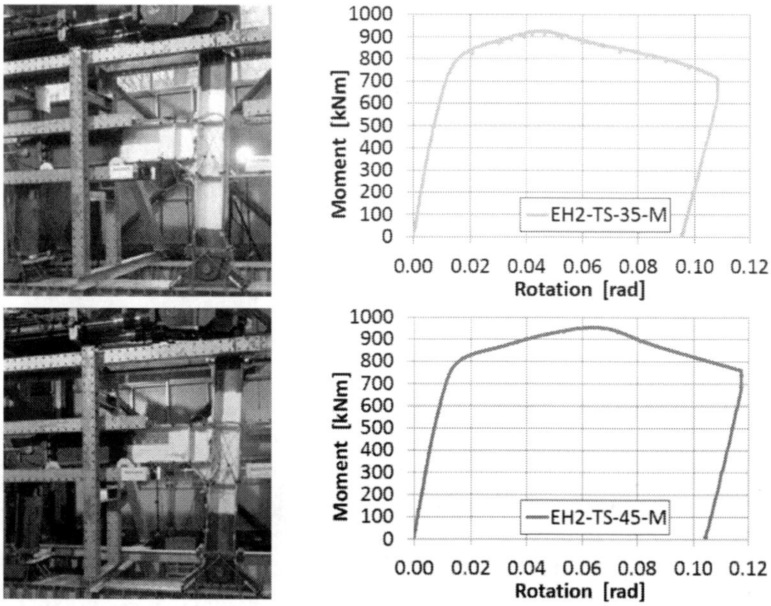

Fig. 4: Response of EH2-TS-35-M and EH2-TS-45-M joint assemblies under monotonic loading: illustration of failure mode, moment-rotation curve

Fig. 5: Yielding sequence corresponding to EH2-TS-35-M joint assembly

4. Numerical investigations

In addition to the experimental program, numerical investigations were performed. The pre–test numerical investigations were realized with the aim to obtain an accurate prediction for response of the beam-to-column joint assemblies. The numerical models of the joints developed at this stage were further used for the calibration based on experimental data. The numerical simulations were performed with the finite element modelling software Abaqus [4].

4.1 Calibration of the material model

Based on the tensile tests conducted on material samples, a set of numerical simulations were performed for the calibration of the material model. In particular, the engineering stress-strain curve obtained from the test was computed into the true stress – true strain curve using equation C.1 from EN 1993-1-5 [5]. Beyond the ultimate strength, trial values needed to be used for stress and strain in order to obtain a good comparison with the experimental force-displacement curve, see Fig. 6. The calibration procedure was used for the following parts:

beam flange and web (IPE450 – S355), end-plate (S355), column flange and web (HEB340 – S355), plates for reconstructed beam (IPE360 / IPE600 – S460). The detailed information regarding the values of the material model for each part can be found in [6]. Additionally, the material properties for bolts, see [3], were calibrated based on data from a past project [7], in which experimental investigations were performed on T-stubs.

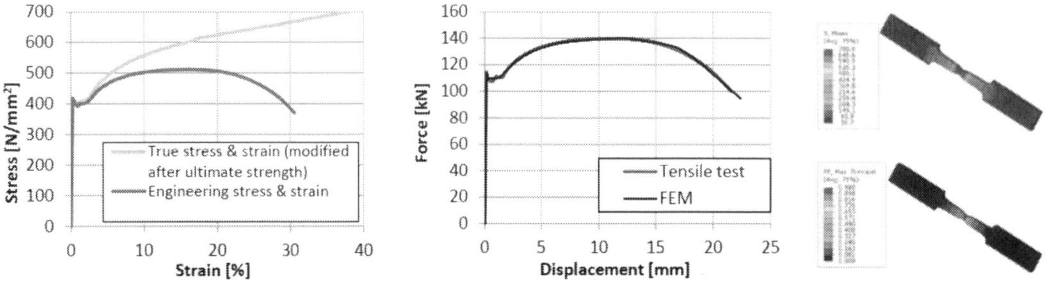

Fig. 6: Calibration of the material model (e.g. IPE450 beam flange): engineering and true stress/strain, test vs. simulation, FE model of steel samples illustrating the failure mode

4.2 Calibration of the joint model

The modelling procedure of the beam-to-column joint models is further described. All the components of the steel haunched joint were modelled using solid linear hexahedral elements of type C3D8R. The nominal geometry was used, except for the thickness of some components (e.g. beam, end-plate). A dynamic explicit type of analysis was used due to large contact surfaces within joint models. The tie constraint was used for welded connections, while the contact interaction between different parts of the joint assembly was defined using a normal (hard contact) and tangential (penalty with friction) interaction law. It is to be mentioned that the out-of-plane support system was modelled through a set of elements placed at the level of the beam flanges according to the test set-up, and in addition a small gap was considered based on the observations from the test. Frictionless contact between beam flanges and the lateral support system was applied. The boundary conditions were adopted according to the test set-up (see Fig. 7a), i.e. column base was pinned, a double-hinged link was connected to the tip of the beam, and the load was applied through a displacement control at the tip of the column. The pre-stressing of bolts was performed by defining and linking a connector to the bolt nut, respectively bolt rod, and specifying a connector force. Fig. 7b illustrates the mesh discretization of the joint assembly.

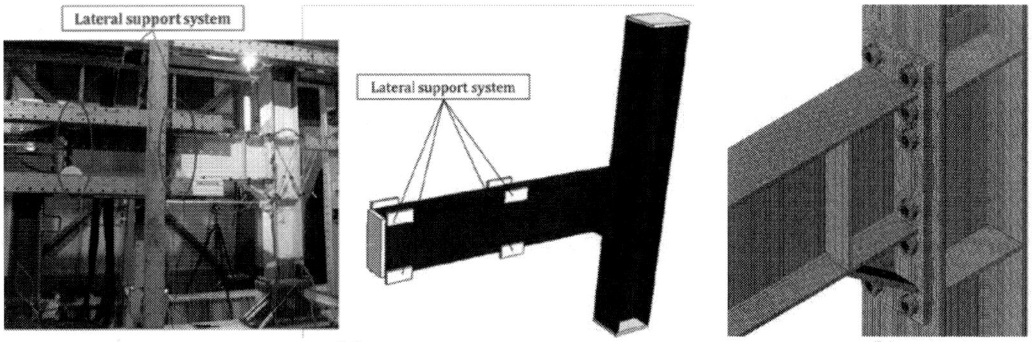

(a) (b)

Fig. 7: Beam-to-column joint model (EH2-TS35-M): (a) lateral support system, (b) mesh discretization of joint assembly

From the calibration process, a set of numerical models were obtained which were capable of reproducing with a good accuracy the response of the joints considering the moment-rotation curve and the failure mechanism, see Fig. 8. Consequently, the comparison between test and simulation is further shown in terms of moment rotation curve (with the rotation computed at column centreline, and the bending moment computed at column face). Furthermore, the response for the two beam-to-column joint configurations (EH2-TS-35-M, EH2-TS-45-M) is presented in terms of stress distribution (von Misses) and plastic strain. As can be observed in Fig. 8, the plastic hinge developed in the beam near the haunch, while the connection zone experienced only elastic deformations. Additionally, the numerical investigations confirmed the yielding sequence observed during the test, namely yielding of the top and bottom beam flanges close to the haunch, followed by large plastic deformations of flanges and web (elongation / local buckling).

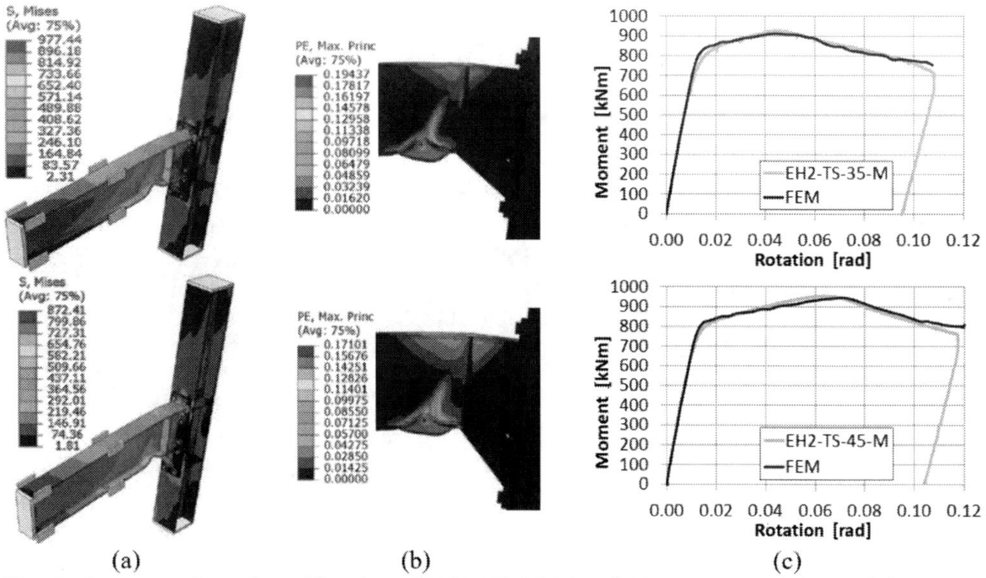

Fig. 8: Outcomes from the calibration of EH2-TS-35-M and EH2-TS-45-M joint models: (a) von Misses stress distribution, (b) plastic strain / failure mechanism, (c) comparison between test and simulation in terms of moment-rotation curve

4.3 Parametric study

The calibration of the joint models allowed the extension of the experimental program with a parametric study, aimed to investigate the following cases: (i) response of joint assemblies with different member size, (ii) response of interior joint configuration (with two opposite beams), (iii) response of joint assemblies with higher steel grade within the beam (S460), (iv) influence of geometrical imperfections. The outcomes of the parametric study are illustrated in Fig. 9 to Fig. 12.

Regarding the influence of member size, Fig. 9 shows the response of the joint specimens with smaller and respectively larger cross sections (e.g. IPE360 / IPE600 beams) in terms of moment rotation curve and plastic strain. As can be observed, the beam size showed two main effects on joint response. Firstly, larger plastic strains were attained for the larger beam size. Secondly, strength degradation was more pronounced for the larger beam size. The designed failure mechanism was confirmed for all cases, i.e. formation of the plastic hinge in the beam close to the haunch. All joint components of three models have minor plastic deformation (characterized mainly by an elastic response).

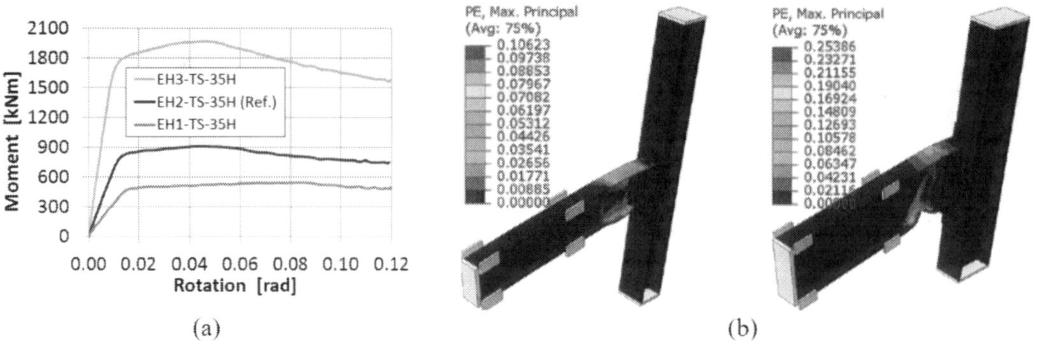

(a) (b)

Fig. 9: Influence of member size: (a) moment-rotation curve, (b) plastic strain for EH1-TS-35H and EH3-TS-35H

The response of the interior joint assembly is represented in Fig. 10 in terms of moment-rotation curve and plastic strain. Compared to the reference model, a small difference can be observed, i.e. higher capacity corresponding to rotations above 0.05 rad. This is due to the different response under hogging and sagging bending moment. Consequently, the response in terms of initial stiffness, capacity, and failure mechanism was similar compared to the reference model. Furthermore, the column web panel was not affected by the shear force developed due to the presence of the two opposite beams.

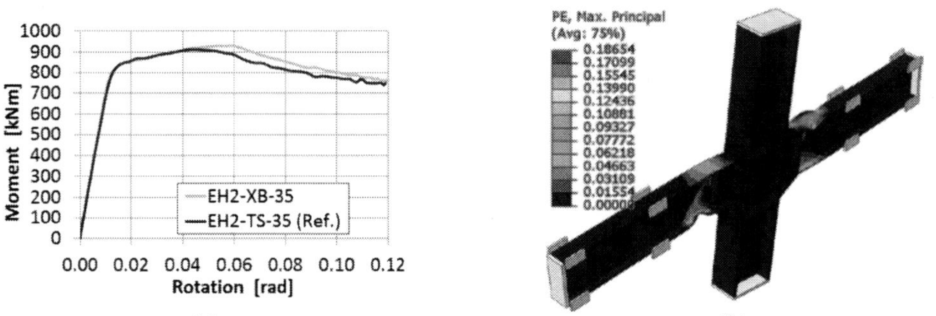

(a) (b)

Fig. 10: Response of interior joint (with two opposite beams): (a) moment-rotation curve, (b) plastic strain for EH2-XB-35

The influence of higher steel grade within the beam is illustrated in Fig. 11. The effect of material overstrength in the beam was investigated by using a beam from a higher steel grade (S460, with measured yield strength of 452 N/mm^2). This results in a material overstrength of 1.27 with respect to the nominal strength of 355 N/mm^2. This is only slightly larger than the overstrength considered in design (1.25). Both hogging and sagging bending moment loading were analysed for the joint assemblies with IPE360, and respectively IPE600 beam sections. The failure mechanism was characterized by the formation of the plastic hinge in the beam. Using a higher steel grade within the beam lead to the increase of the capacity (higher bending moment within the plastic hinge), and therefore the connection zone / joint components were subjected to higher strength demands. Consequently, the joint components suffered plastic deformations. As can be observed, hogging vs. sagging bending moment showed differences only for rotations above 80 mrad. Additional information is available in [6].

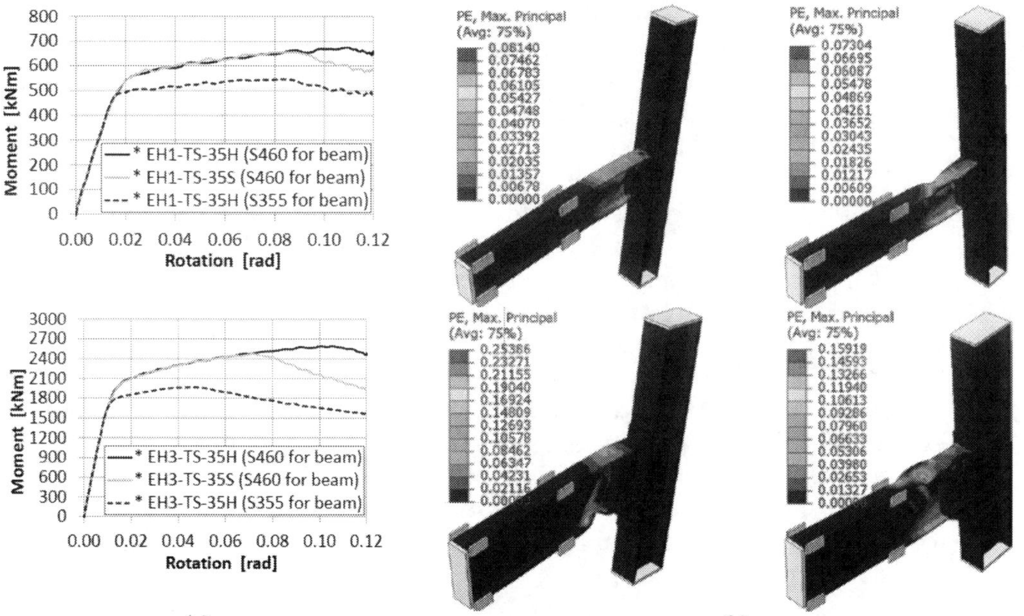

(a) (b)

Fig. 11: Influence of higher steel grade within the beam: (a) moment-rotation curve, (b) plastic strain for *EH1-TS-35 and *EH3-TS-35 joint models subjected to hogging and sagging bending moment

The influence of the geometrical imperfection is illustrated in Fig. 12. Within the reference model (EH2-TS-35), geometrical imperfections were not considered. Within three other models, out of square imperfections (see Fig. 12b), were applied with the following amplitudes: 0.65%, 1%, 2% of the beam width. Large geometrical imperfections of the beam within the plastic hinge location lead to the decrease of the capacity (degradation of the maximum bending moment). It is to be mentioned that EN 10034 [8] specifies a maximum out of square tolerance of 1% of the beam width. As can be observed, only larger imperfection (2%) leads to a noteworthy change of the response.

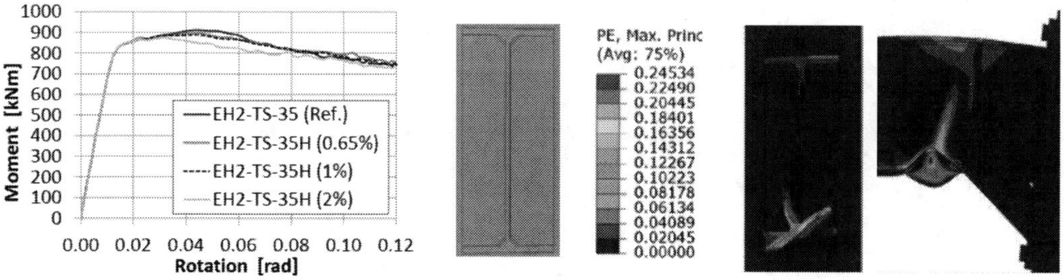

(a) (b) (c)

Fig. 12: Influence of geometrical imperfection: (a) moment-rotation curve, (b) shape of the geometrical imperfection considered at plastic hinge location, (c) plastic deformations within the plastic hinge (EH2-TS-35 – 2%)

5. Conclusions

Experimental investigations evidenced a good response of the bolted beam-to-column joints with haunches. The plastic mechanism considered in design was confirmed, i.e. formation of the plastic hinge in the beam. The experimental data from material and joint tests were further used for the calibration of the material and joint models. The comparison between test and simulation evidenced a very good correlation in both moment-rotation curve and failure mechanism (development of plastic hinge in the beam). This allowed for a better understanding of the joint response. Further, a parametric study was performed for the investigation of different parameters on the behaviour of the bolted beam-to-column joint assemblies. The main outcomes of the parametric study are as follows:

1. Plastic mechanism was characterised by beam yielding, with only minor plastic strains in the joint components. The effect of larger material overstrength in the beam did not change the plastic mechanism, but lead to more important plastic deformations in joint components (bolts, end plate, haunch flange).
2. For the adopted solver (Abaqus/Explicit) modelling of geometrical imperfections of the beam may be omitted.
3. Interior joint has similar initial stiffness, capacity, and failure mechanism compared to exterior joint. Only incipient yielding in the column web panel was evidenced.
4. The designed steel haunched bolted beam-to-column joint assemblies confirmed the designed assumptions (response / failure mechanism) under monotonic loading, Also, the experimental and numerical investigations lead to a (preliminary) validation of the analytical design procedure.

Acknowledgments

The present work was supported by the funds of European Project EQUALJOINTS: "European pre-QUALified steel JOINTS", Grant No. RFSR-CT-2013-00021.

References

[1] Landolfo R, et al. *European pre-QUALified steel JOINTS – EQUALJOINTS*, RFSR-CT-2013-00021, project website: http://dist.dip.unina.it/2013/12/09/equaljoints/

[2] EN 1993-1-8. *Eurocode 3: Design of steel structures - Part 1-8: Design of Joints*. CEN – European Committee for Standardization, 2005.

[3] Maris C, Vulcu C, Stratan A, Dubina D. "Experimental program and numerical simulations of bolted beam-to-column joints with haunches". *Proceedings of the 8th International Conference on Behaviour of Steel Structures in Seismic Areas*, Shanghai, China, 2015.

[4] *Abaqus Analysis User's Manual*, Version 6.14. Abaqus, Inc., Dassault Systèmes Simulia Corp., Providence, RI, USA, 2014.

[5] EN 1993-1-5. *Eurocode 3: Design of steel structures - Part 1-5: Plated structural elements*. CEN – European Committee for Standardization, 2005.

[6] Lee K. *Numerical investigation on seismic behaviours of steel haunch connections*, (Dissertation/Erasmus Mundus Programme) Politehnica University of Timisoara, 2016. (http://ct.upt.ro/suscos/files/2014-2016/Keunwoo_Lee_Dissertation.pdf)

[7] Dubina D, et al. *High Strength Steel in Seismic Resistant Building Frames (HSS-SERF)*. Final Report. Directorate-General for Research and Innovation, Unit D.4 – Coal and Steel, RFCS Publications, European Commission, ISBN 978-92-79-44081-6, doi:10.2777/725123, 2015.

[8] EN 10034:1993. *Structural Steel I and H Sections – Tolerances on Shape and Dimensions*. CEN – European Committee for Standardization, 1993.

COLUMN WEB BUCKLING COMPONENT ANALYSIS

Chen Zhu and Kim J.R. Rasmussen

School of Civil Engineering, The University of Sydney, Sydney, Australia

Abstract: In beam-to-column joints of steel frames, the column web is a major component resisting compression when the joint is subject to bending. If unstiffened, it may buckle and the buckling behaviour may cause additional rotation of the joint. The web can therefore be treated as an additional component when applying the Component Method. However, the behaviour of this component is not well understood. For instance, in Eurocode 3, it is considered as an infinitely stiff link. This paper presents an analysis of the buckling of column webs subject to concentrated transverse compression forces. It derives a force-displacement relationship covering the elastic, inelastic and post-ultimate ranges. The presented model is shown to be in reasonable agreement with experiments.

1. Introduction

The column web is a major component of steel joints. It bears most of the compression force transmitted from the beam flange (Fig.1a) when the steel joint is subject to bending. Therefore, the behaviour of the steel joint will be greatly affected by the behaviour of the column web component, which has not been well understood and is often not considered in design. For example, in Eurocode 3 Part 1-8, this component is essentially considered as an infinitely stiff link. The Code suggests stiffening this component if prone to buckling.

While thus Eurocode 3 favours a strong or stiffened column web, in practice, a weaker unstiffened column web may be suitable when it can sustain sufficient large compression forces in the post-buckling range (Fig. 2). In this range, the resistance may increase significantly depending on the plate slenderness. Moreover, it can provide larger ductility in the post-ultimate range which is usually desirable in structural design. Therefore, it is necessary to devise a new model which can predict the full range behaviour of the column web component.

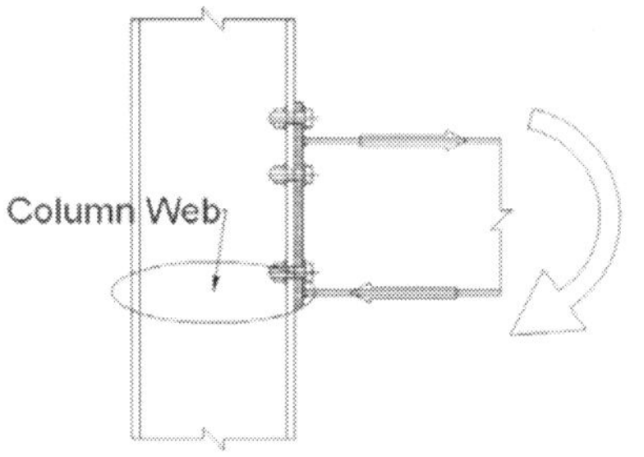

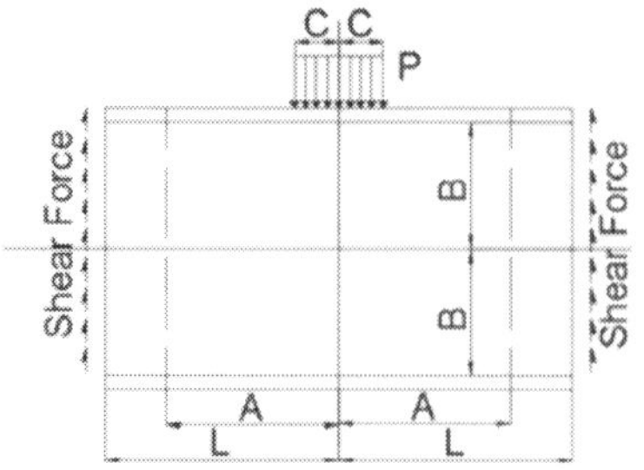

a) Column Web Component b) Loading Condition of the Component

Fig. 1: The Column Web Component and Its Idealised Loading Condition

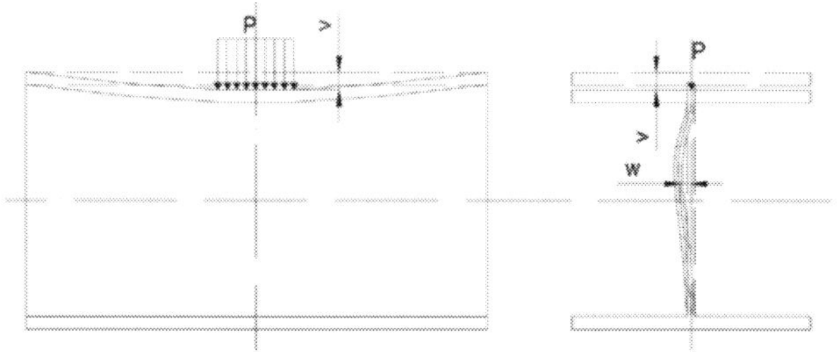

Fig. 2: Buckling Behaviour of the Column Web Component Subject to Patch Loading

2. Related literature

Since existing research has focused on the bearing load of strong or stiffened columns, few techniques for modelling the detailed behaviour of unstiffened column webs can be found in the literature. However, a closely related research field is steel girders subject to patch loading (Fig. 1b), for which several applicable models, such as the elastic buckling analysis and methods for predicting ultimate strength, can be found.

For the elastic buckling analysis, Khan and Walker [1] devised a viable analytical method for the plate subject to patch loading. Their elastic buckling analysis can be extended to the elastic and inelastic post-buckling ranges by adding the second order stress function and inelastic Young's modulus.

A slender column web will reach its ultimate load in the post-buckling range. The most robust design method for predicting the ultimate load is given in Eurocode3 [2]. It adopts a plate buckling approach and is readily understood by practicing engineers. Compared to other methods, it is the most accurate method for predicting the Univ. of Sydney (USyd) tests [3].

For the post-ultimate range, Roberts' mechanism model can be employed [4]. The model assumes the girder collapses due to web folding at the ultimate load (Fig. 2) and that web buckling is the major contributor to the post-ultimate deformations. With modifications to suit the column web component, as described in this paper, it can accurately predict the post-ultimate behaviour compared with experiments.

The column web component is usually subject to axial compressive load, which produces a bi-axial compressive stress state in the web near the compression flange of the beam. However, according to Beg et. al. [5], the deformation capacity of columns with medium to high slenderness, as are typical for the kind of frameworks this paper is intended for, is not significantly affected by a longitudinal compressive axial load. Consequently, the effect of axial load in the column is ignored in the presented model.

3. The Proposed Model

The proposed model includes three parts: a) Elastic and inelastic post-buckling analysis, b) Determination of ultimate resistance and corresponding displacement, and c) Determination of post-ultimate force-displacement relationship.

3.1 Elastic and Inelastic Post-Buckling Analysis

3.1.1 Elastic Post-Buckling Analysis

The total energy of the column web component can be written as:

$$\Pi = U_b + U_m + U_f - W_P \tag{1}$$

The web plate bending energy U_b can be obtained by classical plate theory (Eq. (2)).

$$U_b = \int_{-B}^{B}\int_{-A}^{A} \frac{D}{2}\left\{\left(\frac{\partial^2 w_1}{\partial x^2}+\frac{\partial^2 w_1}{\partial y^2}\right)^2 + 2(1-\nu)\left[\left(\frac{\partial^2 w_1}{\partial x \partial y}\right)^2 - \frac{\partial^2 w_1}{\partial x^2}\frac{\partial^2 w_1}{\partial y^2}\right]\right\} dxdy \tag{2}$$

The web plate membrane energy U_m consists of three parts (Eqs. (3-6)) in which $U_m^{'}$ is an additional term suggested by Khan and Walker [1] when a statically admissible stress field (Eqs. (10-16)) is applied.

$$U_m = U_{m1} + U_{m2} + U_m^{'} \tag{3}$$

$$U_{m1} = \int_{-B}^{B}\int_{-A}^{A} \frac{t_w}{2}\left[\frac{\partial^2 F}{\partial y^2}\left(\frac{\partial w}{\partial x}\right)^2 + \frac{\partial^2 F}{\partial x^2}\left(\frac{\partial w}{\partial y}\right)^2 + 2\frac{\partial^2 F}{\partial x \partial y}\frac{\partial w}{\partial x}\frac{\partial w}{\partial y}\right] dxdy \tag{4}$$

$$U_{m2} = \int_{-B}^{B}\int_{-A}^{A} t_w\left[\frac{\partial^2 F}{\partial y^2}\frac{\partial u}{\partial x} + \frac{\partial^2 F}{\partial x^2}\frac{\partial v}{\partial y} + \frac{\partial^2 F}{\partial x \partial y}\left(\frac{\partial v}{\partial x}+\frac{\partial u}{\partial y}\right)\right] dxdy \tag{5}$$

$$U_m^{'} = -\int_{-B}^{B}\int_{-A}^{A} \frac{t_w}{E}\left[\left(\frac{\partial^2 F_0}{\partial y^2}+\frac{\partial^2 F_0}{\partial x^2}\right)\left(\frac{\partial^2 F_2}{\partial x^2}+\frac{\partial^2 F_2}{\partial y^2}\right)\right] dxdy \tag{6}$$

In Eqs. (4-5), F is the stress function which can be expressed as $F = F_0 + F_2$, where F_0 is the fundamental term of the stress function and F_2 is the second order term of the stress function.

The strain energy of the flange U_f can be obtained from:

$$U_f = \int_{-A}^{A} \frac{EI_f}{2}\left\{\frac{\partial^2 v_f(x)}{\partial x^2}\right\}^2 dx \tag{7}$$

where v_f is the deflection of the flange plate. The work done by the applied patch load can be written as:

$$W_P = Pv \tag{8}$$

where v is the average displacement at loading area in y-axis direction.

In Eqs. (1-8), *F*, *v* and *w* are the unknown terms which will be determined in Sections 3.1.1.1-3.1.1.3. Then, in Section 3.1.1.4, the total energy Π (Eq. (1)) will be minimised to obtain the buckling load and the force-displacement relationship in the elastic post-buckling range.

3.1.1.1 Out-of-Plane Displacement *w*

The out-of-plane displacement, *w* (Fig. 2), is adopted from Khan and Walker's method [1] and given in Eq. (9). Its shape is close to the experimental observations made during the Usyd tests.

$$w = w_1 = q_1 t_w \cos\left[\pi x/(2A)\right]\left\{\cos\left[\pi y/(2B)\right] + 0.25\sin(\pi y/B)\right\} \tag{9}$$

3.1.1.2 Stress Function *F*

In order to achieve sufficient accuracy in the post-buckling analysis, the stress function will not only include the fundamental term, F_0, but also the second order term, F_2. For F_0, Khan and Walker [1] suggested the following statically admissible stress field (Eqs. (10-16)).

$$\frac{\partial^2 F_0}{\partial y^2} = \frac{P}{t_w}\left(-\frac{3L}{8B^3}\right)y\left(2 - \frac{C}{L} - \frac{x^2}{CL}\right), \quad |x| \le C \tag{10}$$

$$\frac{\partial^2 F_0}{\partial y^2} = \frac{P}{t_w}\left(-\frac{3L}{4B^3}\right)y\left(1 - \left|\frac{x}{L}\right|\right), \quad |x| > C \tag{11}$$

$$\frac{\partial^2 F_0}{\partial x^2} = \frac{P}{t_w}\left(-\frac{1}{8C}\right)\left(2 + \frac{3y}{B} - \frac{y^3}{B^3}\right), \quad |x| \le C \tag{12}$$

$$\frac{\partial^2 F_0}{\partial x^2} = 0, \quad |x| > C \tag{13}$$

$$\frac{\partial^2 F_0}{\partial x \partial y} = \frac{P}{t_w}\left(\frac{3}{8BC}\right)x\left(1 - \frac{y^2}{B^2}\right), \quad -C \le x \le C \tag{14}$$

$$\frac{\partial^2 F_0}{\partial x \partial y} = \frac{P}{t_w}\left(\frac{3}{8B}\right)x\left(\frac{y^2}{B^2} - 1\right), \quad x < -C \tag{15}$$

$$\frac{\partial^2 F_0}{\partial x \partial y} = \frac{P}{t_w}\left(\frac{3}{8B}\right)x\left(1 - \frac{y^2}{B^2}\right), \quad x > C \tag{16}$$

The second order field stress function F_2, can be obtained from a perturbation analysis as set out by Budiansky [6]. The governing equation can be written as:

$$\nabla^4 F_2 = E\left[\frac{\partial^2 w_1}{\partial x^2}\cdot\frac{\partial^2 w_1}{\partial y^2} - \left(\frac{\partial^2 w_1}{\partial x \partial y}\right)^2\right] \tag{17}$$

By using Eq. (9) for the buckling displacement (w_1), F_2 can be obtained as,

$$F_2 = \frac{5Eq_1^2 t_w^2 A^2}{128B^2}\cos\frac{\pi x}{A} + \frac{Eq_1^2 t_w^2 B^2}{32A^2}\cos\frac{\pi y}{B} + \frac{-Eq_1^2 t_w^2 B^2}{2048A^2}\cos\frac{2\pi y}{B} + \frac{Eq_1^2 t_w^2 B^2}{162A^2}\sin\frac{3\pi y}{2B} + \frac{Eq_1^2 t_w^2}{32B^2 A^2}\left(\frac{1}{A^2} + \frac{1}{4B^2}\right)^{-2}\cos\frac{\pi x}{A}\sin\frac{\pi y}{2B} \tag{18}$$

3.1.1.3 The Average Displacement at Loading Area

By integrating the y-axis strain, the displacement of the web plate at the top flange junction can be found as,

$$v_f(x) = \int_{-B}^{B}\left(\frac{1}{E}\left(\frac{\partial^2 F_0}{\partial x^2} + \frac{\partial^2 F_2}{\partial x^2} - \upsilon\frac{\partial^2 F_0}{\partial y^2} - \upsilon\frac{\partial^2 F_2}{\partial y^2}\right) - \frac{1}{2}\left(\frac{\partial w_1}{\partial y}\right)^2\right)dy \tag{19}$$

So, the average y-axis displacement at loading area, v, can be obtained as,

$$v = \frac{1}{2C}\int_{-C}^{C} v_f(x)\,dx = \frac{B}{2CEt_w}P + \left(\frac{10\pi t_w^2 A\sin(\pi C/A)}{64BC} + \frac{5\pi^2 t_w^2}{64B}\right)q_1^2 \tag{20}$$

By re-arranging Eq. (20), q_1^2 can be expressed in terms of v and P,

$$q_1^2 = 32B\left[-vB/(Et_w) + P(2C)\right]/\left\{5\pi t_w^2\left[2A\sin(\pi C/A) + \pi C\right]\right\} \tag{21}$$

3.1.1.4 Total Energy and Force-Displacement Relationship

By substituting Eq. (9) into Eq. (2), the web plate bending energy can be obtained:

$$U_b = q_1^2 \frac{E\pi^4 t_w^5 \left(17B^4 + 40B^2A^2 + 32A^4\right)}{6144B^3A^3\left(1-\upsilon^2\right)} = C_{Ub}q_1^2 \tag{22}$$

By substituting Eqs. (9-16, 18) into Eqs. (3-6) and integrating Eq. (3) by parts, the web plate membrane strain energy can be simplified into:

$$U_m = C_{Um1}P^2 + C_{Um2}Pq_1^2 + C_{Um3}q_1^4 \tag{23}$$

$$C_{Um1} = a_{Cum}\begin{pmatrix} \left(1+\upsilon-C^2-C^2\upsilon/3-\upsilon CL\right)\left(10702A^3B^6\pi^3+5351A^5B^4\pi^3+668.9A^7B^2\pi^3\right) \\ -13377A^5B^4CL\pi^3-6689A^7B^2CL\pi^3+6625A^3B^8\pi^3+3312A^5B^6\pi^3 \\ +414.1A^7B^4\pi^3+111.5A^7C^4\pi^3+278.7A^{10}C\pi^3+10702\left(1+\upsilon\right)A^4B^6C\pi^3 \\ +5351\left(1.833+\upsilon\right)A^6B^4C\pi^3+668.9\left(4.332+\upsilon\right)A^8B^2C\pi^3-278.7A^7C^3L\pi^3 \\ +836.1A^8CL^2\pi^3+1784\left(C+2.5L\right)A^3B^4C^3\pi^3+891.8A^5B^2C^4\pi^3 \\ -836.1A^9CL\pi^3+13377A^4B^4CL^2\pi^3-2230A^5B^2C^3L\pi^3+6689A^6B^2CL^2\pi^3 \end{pmatrix} \tag{24}$$

$$C_{Um2} = a_{Cum}\begin{pmatrix} -26754\left(\left(1+\upsilon\right)\sin\left(\pi C/A\right)+\upsilon\right)A^6B^4Et_w^3+293.6\left(1+\upsilon\right)AB^8CEt_w^3\pi \\ +146.8\left(1+\upsilon\right)A^3B^6CEt_w^3\pi+13396\left(1+\upsilon+2\left(1+\upsilon\right)\cos\left(\pi C/A\right)\right)A^5B^4CEt_w^3\pi \end{pmatrix} \tag{25}$$

$$C_{Um3} = a_{Cum}\begin{pmatrix} 3.402A^8CE^2t_w^6\pi^7+41.85B^8CE^2t_w^6\pi^7 \\ +20.93A^2B^6CE^2t_w^6+74.47A^4B^4CE^2t_w^6\pi^7+27.22A^6B^2CE^2t_w^6\pi^7 \end{pmatrix} \tag{26}$$

where, $a_{Cum} = 1/2230A^3B^3CEt_w\pi^3\left(4B^2+A^2\right)^2$. By substituting Eq. (19) into Eq. (7), the flange bending energy can be obtained:

$$U_f = q_1^4\left(25b_fE\pi^8t_w^7\right)/\left(24576B^2L^3\right) = C_{Uf}q_1^4 \tag{27}$$

By substituting Eq. (20) into Eq. (8), the work done by the applied patch load can be obtained:

$$W_P = \frac{B}{2CEt_w}P^2 + \left(\frac{5\pi t_w^2 A\sin\left(\pi C/A\right)}{32BC} + \frac{5\pi^2t_w^2}{64B}\right)q_1^2P = C_{WP1}P^2 + C_{WP2}q_1^2P \tag{28}$$

By substituting Eqs. (22-28) into Eq. (1), the total energy of the system can be re-written as:

$$\Pi = C_{Ub}q_1^2 + C_{Um1}P^2 + C_{Um2}Pq_1^2 + C_{Um3}q_1^4 + C_{Uf}q_1^4 - C_{WP1}P^2 - C_{WP2}q_1^2P \tag{29}$$

By applying the principle of stationary energy for Π with respect to q_1 and then substituting Eq. (21) for q_1^2, the buckling load and the elastic post-buckling force-displacement relationship can be obtained:

$$P = K_{p-b}v + P_{cr} \tag{30}$$

$$K_{p-b} = -\frac{2\left(C_{Um3}+C_{Uf}\right)C_{q12}}{C_{Um2}-C_{WP2}+2\left(C_{Um3}+C_{Uf}\right)C_{q11}} \tag{31}$$

$$P_{cr} = -\frac{C_{Ub}}{C_{Um2}-C_{WP2}+2\left(C_{Um3}+C_{Uf}\right)C_{q11}} \tag{32}$$

3.1.2 Inelastic Post-Buckling Stiffness

In the inelastic post-buckling range, the post-buckling stiffness gradually reduces due to material yielding. This gradual yielding bending behaviour can be modelled using a simplified model in which a reduced Young's modulus (Eq. (33)) is substituted for the elastic Young's modulus in the elastic post-buckling analysis.

$$E_y = 0.278E \tag{33}$$

3.2 Ultimate Strength and Corresponding Displacement

The provisions of Eurocode 3 allow the ultimate strength P_u to be easily predicted with adequate accuracy. Its corresponding displacement can then be calculated using the elastic and inelastic post-buckling stiffnesses. In the proposed method, it is assumed that elastic post-buckling dominates the first two thirds of the post-buckling range. The point at the end of the elastic post-buckling range is defined as the yield point. By this assumption, the displacement at the yield point v_y can be obtained as,

$$v_y = 2\left(P_u - P_{cr}\right)/\left(3K_{p-b}\right) \tag{34}$$

After the yield point, the inelastic post-buckling stiffness applies to the remaining one third of the post-buckling range. The displacement corresponding to the ultimate load point v_u can be obtained as,

$$v_u = v_y + \left(P_u - P_{cr}\right)/\left(3K_{p-b,y}\right) \tag{35}$$

where $K_{p-b,y}$ is the inelastic post-buckling stiffness.

3.3 Post-Ultimate Behaviour

After the applied patch load reaches the ultimate load of column web component, the web will collapse under decreasing load. This collapse behaviour usually progress gradually and is associated with folding of the column web and bending of the top flange (Fig. 3). In order to model this behaviour, a new model, which is based on Roberts' mechanism model [4], is proposed (Fig. 3).

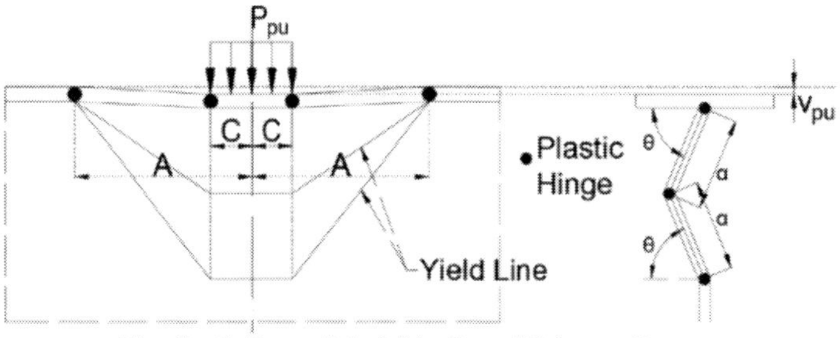

Fig. 3: Collapse Model in Post-Ultimate Range

Based on the virtual work method introduced by Roberts et. al. [4], the incremental post-ultimate resistance and displacement can be expressed in terms of the angle of web folding, θ (Eqs. (36-37) and Fig. 3).

$$P_{pu} = P_u + \left(4A + 4t_f\right)M_w\left(\frac{1-\eta(\theta+\Delta\theta)}{\alpha\cos(\theta+\Delta\theta)} - \frac{1-\eta(\theta)}{\alpha\cos\theta}\right) \tag{36}$$

where, $\eta(\theta) = t_w/\left(t_w + 2\alpha\cos\theta\right)$.

$$v_{pu} = v_u + 2\alpha\left[\sin\theta + \sin(\theta+\Delta\theta)\right] \tag{37}$$

Based on the experimental observations made during the USyd tests, the half depth of the folding web, α, can be defined as the distance from the top edge of the web to the maximum deflection point. Therefore it can be obtained by equating the first derivative of the deflection function (Eq. (37)) to zero,

$$\alpha = 0.76143667B \approx 0.75B \tag{38}$$

The initial angle of the folding web θ_0 can be defined as:

$$\theta_0 = \operatorname{asin}\left(1 - v_u / 2\alpha\right) \tag{39}$$

Starting from $\theta = \theta_0$, the post-ultimate curve can be obtained by solving Eqs. (36-37).

4. Comparison with Test Results

Three sets of experimental data are compared to predictions obtained using the presented model. The tests include those conducted at the University of Sydney [3] and those reported by Granath and Lagerqvist [7] and Chacón, et al. [8]. The presented model is seen to be in reasonable agreement with the experiments (Fig. 4).

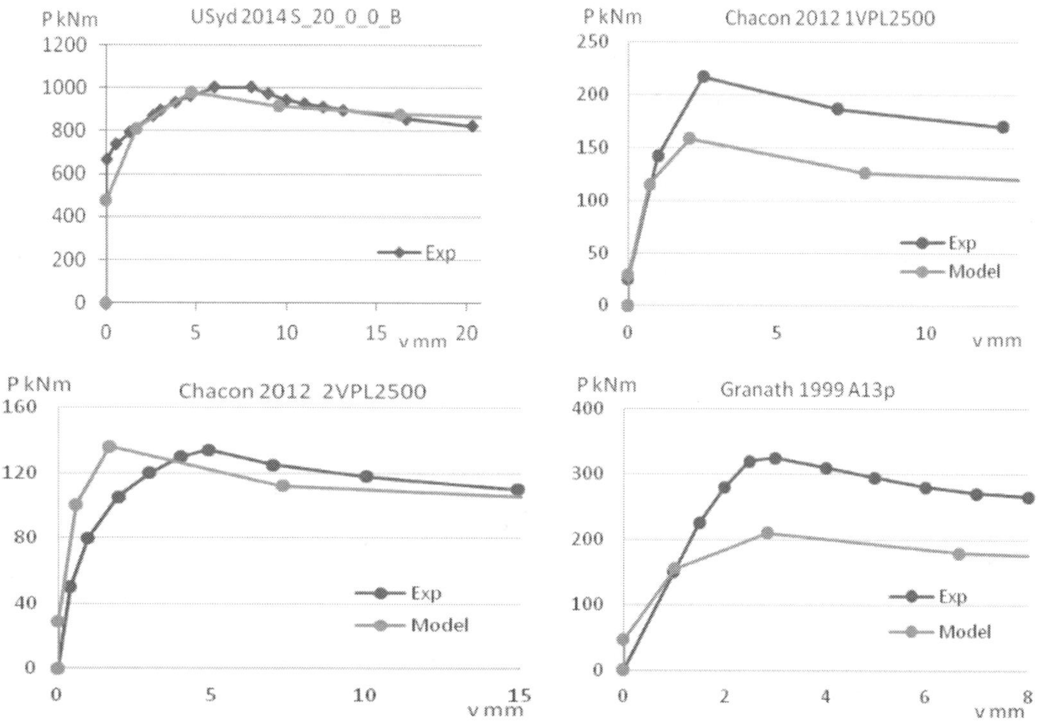

Fig.4: Comparison between the presented model and experiments

5. Conclusion

The paper presents an analysis of the buckling and post-buckling of column webs subject to transverse compression forces. It derives a force-displacement relationship, which can be incorporated in the Component Method, covering the elastic, inelastic and post-ultimate stages of a buckled web in compression. The presented model is shown to be in reasonable agreement with experiments.

Notation

A	Half length of the buckling area (Fig. 2), as per [1].
B	Half depth of web plate
C	Half length of applied patch load
L	Half length of the web plate
P	Total applied patch load
P_{cr}	Buckling load
P_u	Ultimate load
P_{pu}	Resistance in the post buckling range
v	Average displacement at loading area in y-axis direction
w	Out-of-plane displacement (Fig. 2)
ν	Poisson's ratio
E	Young's modulus
D	Rigidity of web plate
t_w	Thickness of web plate
t_f	Thickness of flange
q_1	Amplitude factor for deflection w
K_{p-b}	Elastic post-buckling stiffness
$K_{p-b,y}$	Inelastic post-buckling stiffness
E_y	Reduced Young's modulus for inelastic buckling range
v_y	y-axis average displacement at loading area at yield point
v_u	y-axis average displacement at loading area at ultimate load
v_{pu}	y-axis average displacement at loading area in the post-ultimate range
θ	Angle of the web folding (Fig. 3)

References

[1] Khan MZ, Walker AC. "Buckling of plates subjected to localised edge loading," *The Structural Engineer,* vol. 50, 225-232, 1972.

[2] Eurocode 3, "EN1993, Part 1.5: Plate Structure Elements," *Eurocode 3: Design of steel structures,* 2006.

[3] Zhu C. "Full Moment Rotation Behaviour of Steel Structural Joints," *Ph.D. Thesis, School of Civil Engineering, University of Sydney, NSW, Australia,* 2016 [In Preparation].

[4] Shahabian F, Roberts TM, "Buckling of slender web plates subjected to combinations of in-plane loading," *Journal of Constructional Steel Research,* 99-122, 1999.

[5] Beg D, Zupancic E, Vayas I, "On the rotation capacity of moment connections," *Journal of Constructional Steel Research,* vol. 60, 601-620, 2004.

[6] Budiansky B, "Theory of Buckling and Post-Buckling Behavior of Elastic Structures," *Advances in Applied Mechanics,* vol. 14, 1-65, 1974.

[7] Granath P, Lagerqvist O, "Behaviour of girder webs subjected to patch loading," *Journal of Constructional Steel Research,* vol. 50, 49-69, 1999.

[8] Chacón R, Mirambell E, Real E, "Transversally stiffened plate girders subjected to patch loading. Part 1. Preliminary study," *Journal of Constructional Steel Research,* vol. 80, 483-491, 2013.

Lightweight steel constructions

STABILITY ANALYSIS OF THIN-WALLED MEMBERS WITH CURVED CROSS-SECTION PARTS: ELASTIC BEHAVIOUR

Sándor Ádány[a], Dávid Jobbágy[a], Attila L. Joó[b] and Dávid Visy[a]

[a] *Department of Structural Mechanics, Budapest University of Technology and Economics*
[b] *Department of Structural Engineering, Budapest University of Technology and Economics*

Abstract: In this paper the buckling behaviour of thin-walled members with cross-sections with curved parts is investigated. Due to the curved parts, shell-like buckling is a potential mode of failure. The objective of the research is to understand whether shell-like buckling behavior might be governing in practical cold-formed steel members. For this aim, numerical studies have been carried out, involving linear buckling analysis as well as nonlinear analysis with imperfections, by considering various cross-sections. Based on the results it is concluded that shell-like behaviour might be critical in certain cases. In this paper only the elastic calculations are presented, inelastic behavior will be discussed in subsequent papers.

1. Introduction

As linear cold-formed steel profiles have become everyday solutions in many applications (e.g., purlins, rafters), several research activities started with aiming to develop more efficient cross-sections. These research and/or innovation activities led to more refined cross-section shapes, e.g. with multiple longitudinal stiffeners. Lately, attempts for a more formal mathematical optimization have been reported by various research groups [1-4]. Though the strategies followed by these attempts are sometimes different (e.g., different target functions and constraints are used, also, various optimization techniques are used, etc.), it is a general observation that the found optimal cross-section shapes tend to consist of curved parts rather than flat parts, at least if no special constraints are used to avoid the formation of curved parts.

One problem is that the highly curved cross-section shapes are impractical. However, it is reasonable to assume that some combination of flat and curved parts might be advantageous, e.g., by assuming some classical cross-section, but with unusually large corner radii. Another problem is that the behaviour of thin-walled members with such curved cross-sections is not yet investigated in a comprehensive manner, therefore it is questionable whether the reported optimal cross-sections are properly analyzed by considering all possible failure modes. Namely: since curved cross-section parts mean cylindrical surfaces, shell-like behaviour is possible, but shell-like behavior is not considered in current cold-formed steel member design.

In the paper the results of numerical parametric studies are presented. The calculations are completed by shell finite element analysis. Both column and beam members are investigated, considering two cross-section topologies, but a large number of curved and non-curved cross-sections, by systematically changing the corner radii in a wide range. In this paper elastic calculations are presented only, including linear buckling analysis, and geometrically nonlinear analysis with imperfections. The preliminary results suggest that in certain cases shell-like behaviour should be considered in predicting the capacity.

2. Overview, solution strategy

The objective of the research is to check whether shell-like buckling can or cannot be critical in case of thin-walled (cold-formed steel) members. The aim is not to investigate specific product(s), but to analyze buckling phenomena. Therefore, only simple cross-section geometries are selected. One single cross-section topology is chosen for pure compression, and another one for pure bending. The topology for compression is a doubly-symmetrical hollow section shape, (with a maximum dimension of 100 mm,) while the topology for bending is a C-like singly-symmetrical open cross-section shape (with 100 mm width and 130 mm height). Since the emphasis is on the curved parts, within the given topology the corners are rounded with variable corner radius, the radius being varying in between zero and the physically possible maximum (i.e., 50 mm). In case of the hollow section, therefore, the increasing radius transforms the shape from a square hollow section (SHS) to a circular hollow section (CHS), as shown in Fig. 1a. Similarly, Fig. 1b shows the considered C-like shapes with the changing corner radius. The thickness is also a variable, changing in between 0.4 mm and 3.0 mm.

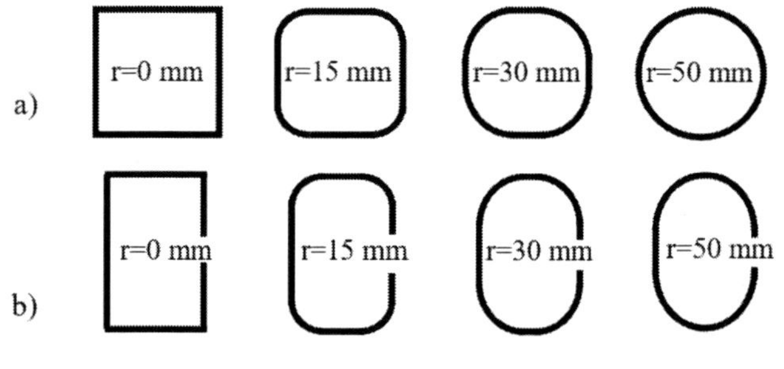

a) Hollow sections b) C-like sections
Fig. 1: Cross-sections

Since the main question here is the possible importance of shell-like buckling, and since shell-like buckling is associated with small buckling waves, only short members are considered, with a length equal to 200 or 300 mm, which is roughly twice as much as the maximum cross-section dimension for the SHS type and the C-like section respectively. The selection of short member length automatically eliminates the global buckling phenomena. It is also to mention that distortional buckling is practically also eliminated by the selection of member length and cross-section shapes. In case of hollow sections distortional buckling mode theoretically exists, however, the associated critical load is much larger than those belong to local-plate buckling, hence, it is reasonable to assume that the effect of distortional buckling for the considered column problems is negligible. In case of C-like cross-sections distortional buckling is typically important, however, in our cases the flange lips are relatively large, and if such a cross-section is subject to bending, the lips are lightly compressed, hence distortional buckling has minor role. We can conclude, therefore, that our study investigates local-plate buckling and shell-like buckling of thin-walled beam and column members.

The final goal of the numerical studies is to estimate the load-bearing capacity of the members with (and without) significant curved parts. In the lack of real experiments, the load bearing estimation is carried out by finite element analysis, using shell finite elements, and considering material and geometric nonlinearity with imperfections (i.e. GMNI analysis). For the imperfection geometric imperfections are intended to use, taken as properly scaled buckling shapes. The major steps of the research work therefore are as follows:

- parametric model definition,
- linear buckling analysis for a large number of cases, by systematically changing the model parameters in wide range of the parameters,
- development of a method to numerically and automatically characterize the buckling modes,
- imperfection sensitivity analyses by using elastic material and geometric nonlinear analysis with geometric imperfections,
- load bearing capacity estimation with GMNI analysis. .

Based on the results of the GMNI analyses the load bearing capacity of the members can be assessed and conclusion can be drawn.

In this paper only the elastic analyses are presented, while the inelastic analyses will be discussed in subsequent papers.

3. Linear buckling analysis

In order to be able to make large number of analyses, a parametric finite element model has been built in Ansys [5]. The model uses shell finite elements. Various shell element types have been tested, and it was concluded that there is no significant difference in between the considered element types if sufficiently dense discretization is applied. Finally, SHELL281 is selected for further analyses, which is an 8-node element. The advantage of this element is that the curved parts can smoothly be modelled since the element edges can be curved.

The member ends are supported as follows: a globally and locally hinged-hinged support was created, but with restrained warping. One may think of this arrangement as if a thick plate were welded to both ends, and the plate is supported in one point by a hinge (i.e., by restraining translations and twisting rotation around the longitudinal axis of the member, while allowing the rotations around the other axes). Practically: a master node is defined at each end to which each end cross-section node is linked by rigid constraint equations (CERIG command in ANSYS). It is to note that some slightly different end support arrangements have also been tested, but it was concluded that the local behaviour (local-plate buckling or shell-like buckling) is hardly influenced by the small variation of the supports.

Linear buckling analyses are performed for both cross-section topologies, with 0-50 mm corner radii, within the thickness range of 0.4-3.0 mm. In each case the first 200-300 critical loads and corresponding buckled shapes are calculated. Some of the modes are shown in Fig. 2. In general, if the deformations are concentrated to the flat parts of the member (while the curved parts are subject to much smaller deformations), the buckled shape is most likely plate-like buckling. On the other hand, if significant deformations appear at the curved parts, the buckled shape is most likely shell-like buckling. By the visual inspection of the buckling modes it can be concluded that:

(i) in case of small corner radius (r<25 mm) the first few hundred buckling modes can be classified as (classic) plate-like modes,

(ii) in case of larger corner radius the first buckling modes are plate-like buckling, but shell-like buckling appears among the higher modes,

(iii) the larger the corner radius, the sooner the shell-like buckling appears,

(iv) both "shell-type" (see Fig. 2b #110) and "axisymmetric-like" (see Fig. 2b #158) modes appear, however, axisymmetric modes are found only as very high modes in case of very large corner radius,

(v) the increasing tendency of the critical loads are dependent on the cross-section shape: the larger the corner radius, the slower the increasing of the critical loads (e.g., in case of a hollow section with r=5 mm and t=1 mm, the ratio of the 200th and the 1st critical load is 16.8, while the same ratio is 4.3 if the radius is 40 mm).

#1 #100 #200

a) Hollow section, r=5 mm, t=1 mm

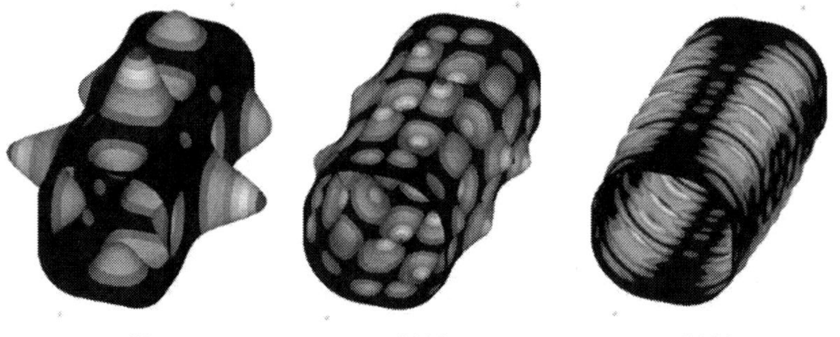

#1 #110 #158

b) Hollow section, r=40 mm, t=1 mm

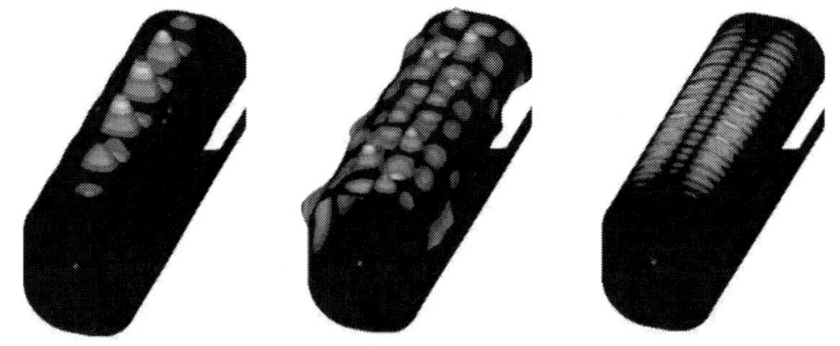

#1 #67 #89

c) C section, r=40 mm, t=1 mm

Fig. 2: Buckling mode samples

4. Spectral analysis of buckling modes

Buckled shapes are intended to use as geometric imperfections in nonlinear analyses for a large number of cases. It is known that the various buckling modes have very different post-critical behaviour. It is expected, therefore, that the member will show significantly different imperfection sensitivity depending on the nature of the imperfection, i.e., depending on the nature of the buckling mode which is used as geometric imperfection. Since we have many different cross-section shapes, and hundreds of buckling modes for each case, it is highly beneficial to be able to numerically characterize the buckled shape, which numerical characterization might later be connected to the imperfection sensitivity (or: post-critical behaviour).

Here a simple and automatic characterization is proposed and used, which can be summarized as follows: (i) longitudinal straight lines are defined at some characteristic points of the member, namely: two in the flat part of the cross-section (f1,f2), and two in the curved part (c1,c2), (ii) the displacements along the lines are collected, (iii) the displacement function along each line is approximated by Fourier series, (iv) the Fourier coefficients are normalized.

Since in most of the cases only a few coefficients have non-zero values, the few non-zero coefficients show the characteristic buckling length(s), as well as show whether the deformations are dominant in which part of the member. Also, it can be observed that shell-like buckling is characterized by smaller buckling lengths compared to plate-like buckling.

As an example, Table 1 shows the normalized Fourier coefficients for 4 lines of a hollow section (r=40mm and t=1mm), for the 3 buckling modes shown in Fig. 2b. It can be observed that (i) mode #1 is clearly plate-like buckling since the numbers (and thus the deformations) in the corner region (c1,c2) are much smaller than those in the flat parts (f1,f2), (ii) mode #158 is clearly shell-like buckling, (iii) while mode #110 shows both plate-like and shell-like characteristics.

Table 1: Spectral decomposition of selected buckling modes

Hollow section r=40 mm t=1 mm												
Buckl. shape #1				Buckl. shape #110				Buckl. shape #158				Half-wave length [mm]
Node sets				Node sets				Node sets				
f1	f2	c1	c2	f1	f2	c1	c2	f1	f2	c1	c2	
0	0	0	0	0	4	1	1	0	0	0	0	200
1	1	0	0	0	0	0	0	1	1	0	1	100
0	0	0	0	0	42	18	19	0	0	0	0	67
3	3	0	0	0	0	0	0	1	0	3	0	50
0	0	0	0	0	100	32	38	0	0	0	0	40
100	85	2	2	0	0	0	0	0	0	1	0	33
0	0	0	0	0	16	1	6	0	0	0	0	29
71	61	1	1	0	0	0	0	0	0	1	0	25
0	0	0	0	0	9	1	2	0	0	0	0	22
0	0	0	0	0	0	0	0	0	0	1	1	20
0	0	0	0	0	7	1	1	0	0	0	0	18
0	0	0	0	0	0	0	0	0	0	1	3	17
0	0	0	0	0	7	0	0	0	0	0	0	15
0	0	0	0	0	0	0	0	0	0	4	7	14
0	0	0	0	0	20	1	1	0	0	0	0	13
0	0	0	0	0	0	0	0	2	1	17	61	13
0	0	0	0	0	6	0	0	0	0	0	0	12
0	0	0	0	0	0	0	0	4	2	37	100	11
0	0	0	0	0	1	0	0	0	0	0	0	11
0	0	0	0	0	0	0	0	1	0	2	14	10

5. Imperfection sensitivity analysis

In order to learn the sensitivity of the members to various geometric imperfections, through which the post-critical behaviour can also be explored, various buckling modes are used as geometric imperfections in nonlinear analyses. Both elastic and materially nonlinear analyses have been performed, here, however, only the elastic analyses are discussed.

In the frame of the elastic imperfection sensitivity both hollow and C sections are considered with various corner radius. In order to reduce the number of cases, one single thickness is assumed (t=1 mm), and for each radius only three imperfection shapes are applied, namely a plate-like, a mixed, and a shell-like. Finally, the buckled shapes are scaled so that the maximum (out-of-plate) displacement would be equal to 0.05, 0.1, 0.15, 0.2, 0.4, 0.5, 0.75, 1.0, 1.5, 2.0, 3.0, and 4.0 mm, respectively.

By using the above parameters, geometric non-linear analyses are carried out (with elastic material), during which load-displacement curves are established. The maximum point of the load-displacement curve then is considered as the elastic load-bearing capacity of the member. If the geometric nonlinear analysis is carried out for a given member and a given imperfection shape, but with increasing the initial value of the imperfection, the change in the elastic capacity characterizes the sensitivity of the member to the given imperfection type.

It is known from classic textbooks that the elastic capacity can be quite different from the elastic critical load, the difference being dependent on the nature of buckling: in case of plate buckling significant post-buckling reserve might exist, (i.e. the capacity can be larger than the critical load), while in case of shell buckling the capacity is typically much smaller than the critical load. However, the actual capacity is also dependent on the initial magnitude of the imperfection.

By having the load-displacement curves for the various cases, the relationship in between the imperfection shape (i.e. buckling mode), imperfection magnitude and elastic capacity can be established.

Though the analyses can technically be fully automated, the detailed evaluation of the results highlights the difficulties and limits of this type of finite element calculation. In many cases a sudden change of the deformed configuration occurs as the load is increasing. If this snap occurs, the calculated capacity (though technically correct) does not belong to the original imperfection, therefore, the calculated capacity cannot be used to characterize the member sensitivity to the initially considered imperfection shape. It is also observed that this snap phenomenon is typical if the initial imperfection magnitude is small, and rarer and rarer as the imperfection magnitude increases. In Fig. 3 this phenomenon is illustrated, where load-displacement curve is plotted for a hollow section with r=0 mm and t=1 mm, by using buckling mode #10 as imperfection but with a small initial magnitude (0.005 mm). Some typical deformed shapes are also presented, belonging to various points of the curve, as marked.

As a consequence of this snapping phenomenon, automatic evaluation of the results is questionable, since it is practically impossible (or: at least it would require an unreasonably huge effort) to properly assign the elastic capacities and initial imperfections. In the lack of a reliable automatic evaluation process the imperfection sensitivity is established only for a few characteristic cases, by manually checking the non-linear behaviour during the whole load incremental process. In Fig. 4 the imperfection sensitivity is illustrated for a hollow section with r=40 mm and t=1 mm. It can be observed that (i) plate-like imperfection leads to elastic capacities larger than the critical load, (ii) if considerable shell-like deformation exists in the imperfection, the elastic capacity is much smaller than the critical load, (iii) the larger the amplitude of the initial imperfection is, the smaller the elastic capacity is, and (iv) the elastic capacity is not too sensitive to the magnitude of the geometric imperfection.

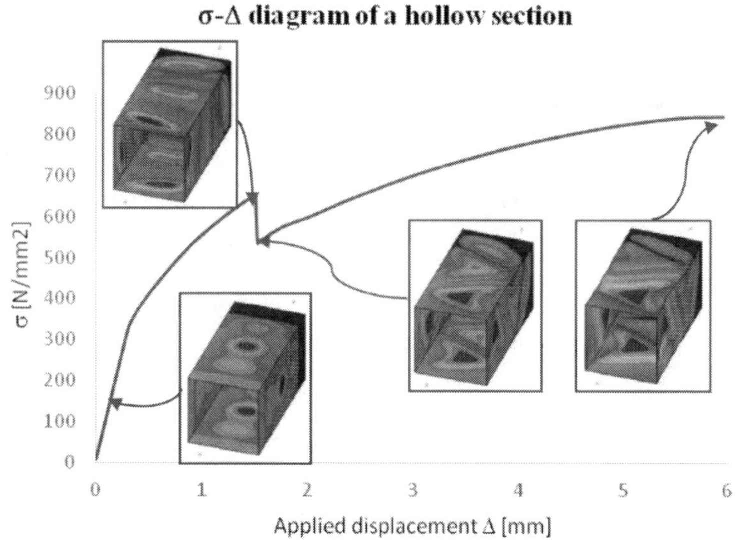

Fig. 3: Geometric nonlinear behaviour of a hollow section (r=0 mm, t=1.0 mm) with buckling mode #10 as initial imperfection (imperfection magnitude 0.005 mm)

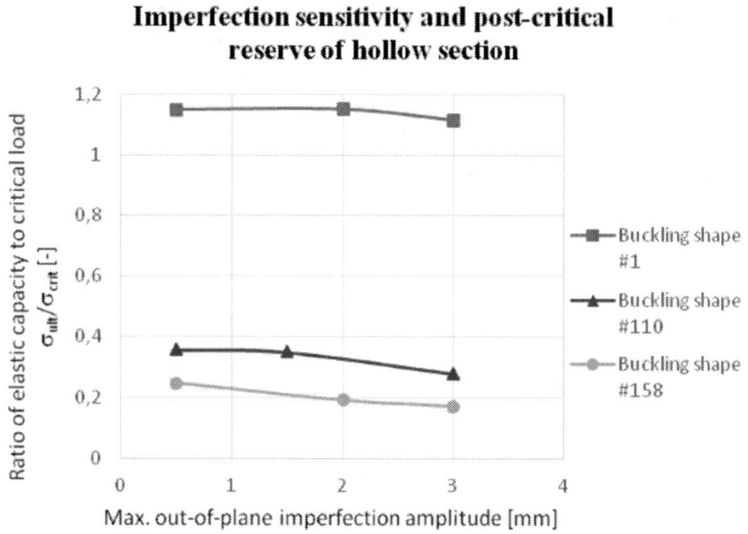

Fig. 4: Imperfection sensitivity of a hollow section (r=40 mm, t=1.0 mm) with various buckling modes and amplitudes as initial geometric imperfection

6. Summary and conclusions

In this paper the first phase of a research on the buckling behaviour of thin-walled members with cylindrically curved parts has been presented. The focus is on the local buckling behaviour, including plate-like and shell-like buckling. Parametric numerical studies have been completed on two selected cross-section types, namely hollow section and C-like section, by systematically varying the radius of the curved parts. Elastic buckling modes have been determined first, which characterized numerically, then used as geometric imperfections in non-linear elastic analyses. Based on the results the following conclusions can be drawn.

- If the curved parts are significant in the cross-section, shell-like buckling is possible. In linear buckling analysis the shell-type modes are among the higher modes. The corresponding critical load values are typically multiples of the lowest critical load value with the tendency as follows: the smaller the cylindrical part of the cross-section, the larger the ratio of the shell-type critical load to the lowest critical load.
- By using the buckled shapes as initial geometric imperfections, elastic (or inelastic) capacities can be calculated. The calculation of the elastic capacity can be numerically challenging, especially if the initial magnitude of the imperfection is small. The results justify the well-known fact that the ratio of the elastic capacity to the corresponding critical load is highly dependent on the imperfection shape. When the initial imperfection shape is taken as a plate-like buckling mode in case of a hollow section with r=40 mm, t=1 mm, the ratio of the elastic capacity to the critical load is 1.2, while if the initial imperfection shape is taken as a shell-like buckling mode this ratio could be as small as 0.2. However, when a section, having small radius (r=5 mm, t=1 mm), is analyzed the ratio of the elastic capacity to the critical load might reach 8 as well.
- It is also observed that the elastic capacity is influenced by the initial imperfection amplitude. The general tendency is that the larger the initial magnitude, the smaller the predicted elastic capacity.
- It is also observed that if various buckling modes are considered for the same member, the predicted minimal capacity might belong to plate-like imperfections, but might also belong to shell-like imperfections. The observation is that shell-like imperfections might be governing only in case of members with significant curved parts: in the considered cases shell-type imperfections are critical if the corner radius is more than 30-35 mm, which means that most parts of the member is cylindrical, though plane parts also exist.

Since the real behaviour is greatly influenced by the material nonlinearity, the above conclusions are not the final ones. The inelastic behaviour, thus, must also be studied, the results of which will be reported in subsequent papers.

Acknowledgments

The work work was conducted with the financial support of the OTKA K108912 project of the Hungarian Scientific Research Fund.

References

[1] Leng J, Guest JK, Schafer BW. "Shape optimization of cold-formed steel columns", *Thin-Walled Structures,* vol. 49, pp. 1492-1503, 2011.

[2] Leng J, Zhanjie L, Guest JK, Schafer BW. "Shape optimization of cold-formed steel columns with fabrication and geometric end-use constraints", *Thin-Walled Structures*, vol. 85, pp. 271-290, 2014.

[3] Moharrami M, Louhghalam A, Tootkaboni M. "Optimal folding of cold-formed steel cross-sections under compression", *Thin-Walled Structures*, vol. 76, pp. 145–156, 2014.

[4] Gilbert BP, Savoyat TJM, Teh LH. "Self-shape optimization application: optimization of cold-formed steel columns", *Thin-Walled Structures*, vol. 60, pp. 173–184, 2012.

[5] ANSYS 16.2-Mechanical APDL.

COMPARISON OF DIFFERENT EUROCODE VERIFICATIONS OF A THIN-WALLED MEMBER PORTAL FRAME

Rudolf Ároch and Ján Brodniansky

Department of Steel and Timber Structures, Slovak University of Technology in Bratislava, Slovakia

Abstract: The paper is focused on the different approaches for verification of a portal frame composed of thin-walled members according to Eurocodes. The main emphasis is devoted to the calculation approach using a unique global and local imperfection described in clause 5.3.2 (11) EN 1993-1-1 or clause 5.3.2(11) EN 1999-1-1. The paper is devoted to single-storey portal frames composed of class 4 2C cross-sections. The above mentioned standards give us more possibilities of calculation procedures for frame verification. They differ in complexity and level of difficulty. Common analysis programs can be used or specialised iterative computer programs based on the FEM analysis, as the one which was developed at the Bratislava University, are needed. The paper compares the complexity of the different procedures with the obtained utilization level.

1. Introduction

The standards [1, 2, 3, 4] allow the usage of more possibilities of calculation procedures for frame verification. A frame structure can be calculated according to clause 5.2.2(3) (Structural stability of frames) by three procedures a), b) and c). In the calculation according to 5.2.2(3) a) imperfections defined in clause 5.3.2(3) a) and b) or imperfections according to clause 5.3.2(11) can be used. Imperfection from clause 5.3.2(11) is presented in both standards [1, 2] and in the Slovak National Annex to [1] as well. The Slovak National Annex to [1] and [2] give a more detailed description compared to [1]. In this paper the notation given in [2] or in the Slovak National Annex to [1] is used. According to this verification procedure it is possible to use the elastic critical buckling mode as the imperfection shape and to determine also the amplitude of this imperfection. This imperfection is called the **E**quivalent **U**nique **G**lobal and **L**ocal **I**nitial imperfection (**EUGLI** imperfection). The whole calculation procedure according to the 2nd order theory with the EUGLI imperfection will be denoted as the EUGLI method in the following text. Because of the use of structural members containing class 4 2C cross-sections we needed a specialized iterative computer program based on the FEM analysis. This program uses a calculation procedure that includes standards [3, 4] which describe calculation of effective cross-sectional characteristics. The whole theoretical background of the EUGLI imperfection is published in [5, 6, 7 and 8]. The computer program developed by the authors is written in MATLAB programming language [9]. All of the partial graphical and numerical results were obtained by the use of this computer program. Some of the descriptions which include details on the computer program can be found in [10, 11]. This paper is focused on the different approaches for verification of a portal frame composed of thin-walled members according to Eurocodes and for lack of space there is only a basic explanation of the EUGLI method iteration procedure.

2. Thin-walled frame

The scheme of the investigated frame is on "Fig.1", the static scheme with the distribution of cross-sectional area and second moment of area (of the gross cross-section) is on "Fig.2 and 3", and three loading types (A, B, C) on "Fig.4, 5 and 6". 2C cross-sections of class 4, a steel grade S390 and the imperfection buckling curve ***a*** ($\alpha = 0,21$) is considered.

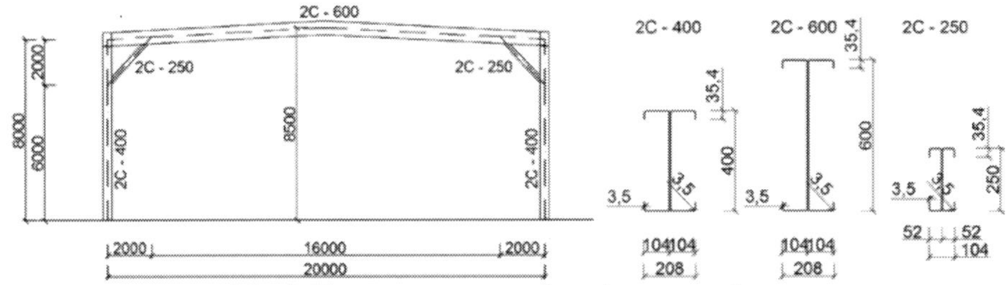

Fig. 1: Frame geometry and used cross-sections

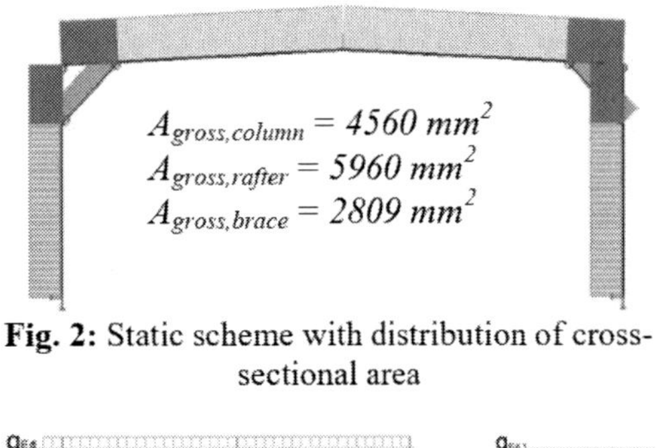

Fig. 2: Static scheme with distribution of cross-sectional area

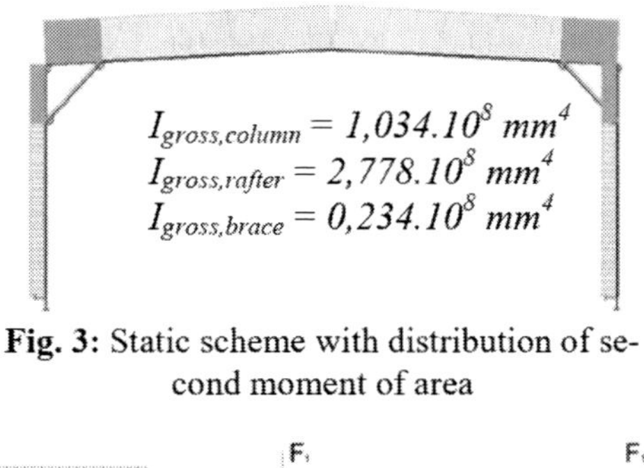

Fig. 3: Static scheme with distribution of second moment of area

Fig. 4: Loading type **A** Uniform load distribution on rafter

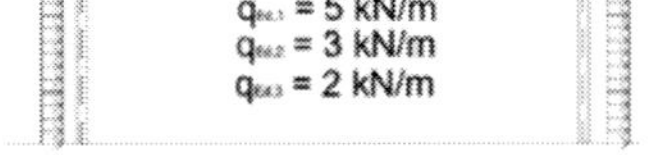

Fig. 5: Loading type **B** Uniform load distribution on rafter and column

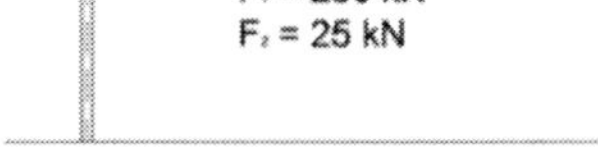

Fig. 6: Loading type **C** Point loads on top of the columns

All frame members are restrained against lateral-torsional buckling and out-of-plane flexural buckling. Each frame member is divided into 50 FEM elements. Other characteristics of the structure are shown on the following figures.

Loading type **A** ("Fig.4") represents a real structural loading on the frame, this load combination can be regarded as dead and snow load (symmetric load combination). Loading type **B** ("Fig.5") represents a real structural loading on the frame, this load combination can be regarded as dead, snow and wind load (non-symmetric load combination). Loading type **C** ("Fig.6") is not considered as a real loading combination, this combination of loads is for demonstrating the behaviour of a frame structure with a low critical factor α_{cr}. Loading types **A** and **B** are considered to achieve a maximal utilization factor approximately equal to 1.

3. Imperfection according 5.3.2(11) [1, 2] (EUGLI imperfection)

An iterative approach is used in the EUGLI imperfection method for the determination of the critical cross-section of the structure. It is necessary to estimate the position of this cross-section in the first iteration step. In the calculation of EUGLI imperfection amplitude (equations (10a) [5], (3) [6], (5.13) [5]) the cross-sectional parameters of this cross-section are used. For Class 4 cross-sections we have to use the effective cross-sectional characteristics ($A_{eff}, I_{eff}, W_{eff}$) that change along the member lengths due to the stressing (axial load and bending moment acting simultaneously). The elastic critical buckling mode of the effective structure can be denoted as $\eta_{cr,eff}$. This procedure is described in [12]. The design value of the maximum amplitude of the EUGLI imperfection is then

$$\eta_{init,m,max} = e_{0d,eff,m} \frac{N_{cr,eff,m} \left|\eta_{cr,eff}\right|_{max}}{EI_{eff,m} \left|\eta''_{cr,eff}\right|_{m}} \tag{1}$$

where

$e_{0d,eff,m} = \alpha\left(\bar{\lambda}_{eff,m} - 0.2\right)\dfrac{M_{Rk,m}}{N_{Rk,m}} \dfrac{1 - \dfrac{\bar{\lambda}_{eff,m}^{2}\chi_{eff}}{\gamma_{M1}}}{1 - \bar{\lambda}_{eff,m}^{2}\chi_{eff}}$ is the design value of the bow imperfection for the equivalent member;

$\bar{\lambda}_{eff,m} = \sqrt{\dfrac{A_{eff,m} f_y}{N_{cr,eff,m}}} = \sqrt{\dfrac{N_{Rk,m}}{N_{cr,eff,m}}}$ is the relative slenderness;

$N_{Rk,m} = A_{eff,m} f_y$ is the characteristic resistance to normal force of the critical cross-section;

$M_{Rk,m} = W_{eff,m} f_y$ is the characteristic moment resistance of the critical cross-section;

χ_{eff} is the reduction factor for the relevant buckling curve and the relative slenderness;

$N_{cr,eff,m} = \alpha_{cr,eff} N_{Ed,m}$ is the elastic critical force, i.e. the axial force in the critical cross-section m under the critical loading of the structure based on the effective cross-sectional properties.

To avoid the iterative procedure, it is possible to use the so called "brute force method". First every cross-section of the structure is considered as the critical cross-section. The EUGLI imperfection amplitude is calculated in turn and the utilizations of the cross-sections of the structure are calculated. The best way how to show the obtained results is to use a matrix form (EUGLI method utilization matrix "Fig.7"). The matrix values in each row i represent the

EUGLI method utilization ratios of the cross-sections *j* obtained with the imperfection amplitude for the critical cross-section in *i* (*i* and *j* are defining the position in the matrix). The final critical cross-section is that one which has the maximal value in a row on the diagonal.

		j					
		1	2	49	50	51	52
i	1	0,060196	0,073141	0,010[illegible]	0,641494	0,49528	0,484327
	2	0,053721	0,060196	0,334483	0,3[illegible]7359	0,247716	0,242237
	49	0,047552	0,04786	0,065382	0,0[illegible]7073	0,011806	0,011545
	50	0,047542	0,047841	0,064959	0,066633	0,011436	0,011183
	51	0,047243	0,047243	0,051916	0,053048	1,65E-06	1,62E-06
	52	0,047243	0,047243	0,051916	0,053048	1,69E-06	1,65E-06

Fig. 7: Part of the EUGLI method utilization matrix (critical cross-section $i = j = 50$)

The critical cross-section is found with the axial forces from the external loading and from the bending moments caused by the deformation of the structure into the imperfect shape only (no bending moments from the external loading). 2nd order theory is used in all stages of calculation. The bending moments caused by the external loading are then added at the end of the calculation procedure (details are in [5, 7]). The critical cross-section does not need to be the cross-section with the maximal final utilization of the structure. For the verification of the presented frame example we have deformed the structure to the imperfect shape and calculated the internal forces and moments according to the 2nd order theory under all external loading.

For obtaining correct values of cross-sectional parameters, more iteration cycles have to be done. Effective cross-sectional parameters are calculated according to the actual stressing in each structural cross-section. Iteration cycling is needed until the utilization of the structure in two following iterations does not change.

Results obtained in the 1st and the last (10th) iteration are presented on "Fig.8 to 12" of Table1. Only one set (Loading type **A**) of detailed results is shown, results of the other combinations are summarized in Table 2. The thick lines are showing the differences between the 1st and the 10th iteration (1st iteration – gross cross-sections, 10th iteration – effective cross-sections).

The EUGLI method utilization matrix on "Fig.7" displays the result for the last (10th) iteration cycle of Loading type **A** combination. The final ultimate utilization ratio is on "Fig.12b" with its maximal value of *1,0264*. All calculations are made in specialised iterative computer program developed by the authors.

4. Imperfection considered according clause 5.3.2(3) a) b) in [1, 2]

By using the imperfections according clause 5.3.2(3) a) b) in [1, 2] it is necessary to check a large number of imperfection shapes and it is necessary to find the imperfection shape that produces the highest utilization ratio. Both sway and bow imperfections are considered as imperfections. In agreement with the previous chapter the frame is considered in the imperfect shape with external loading applied. In Table 3 some of the imperfection shapes of the frame on "Fig.1" are displayed (for better imagination they are magnified).

Table 1: Results - Loading type A

1st iteration results	10th iteration results
 **Fig. 8a:** Second moment of area distribution (gross cross-section)	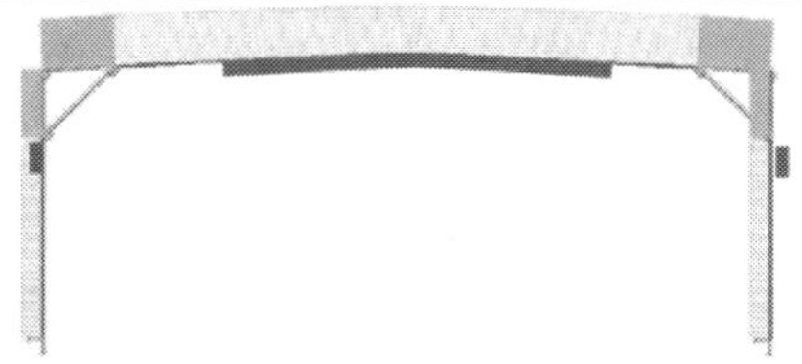 **Fig. 8b:** Second moment of area distribution (effective cross-section)
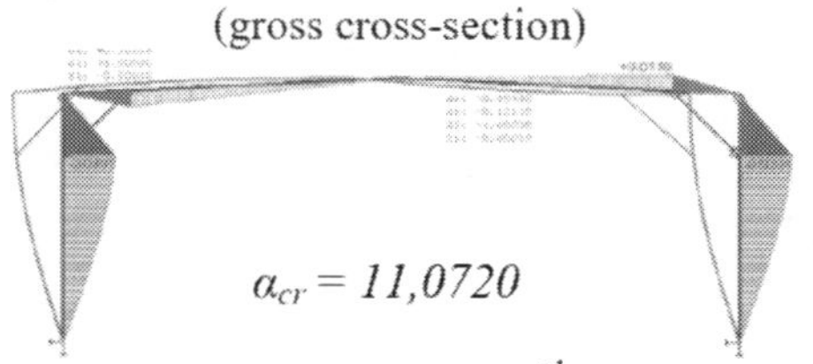 **Fig. 9a:** Buckling mode and 2nd derivation of buckling mode (curvature)	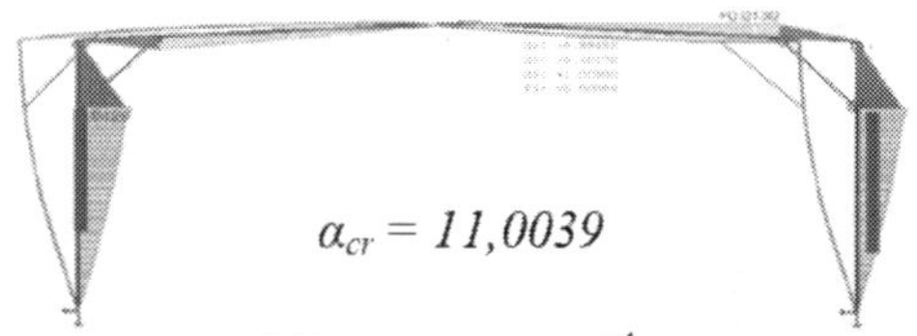 **Fig. 9b:** Buckling mode and 2nd derivation of buckling mode (curvature)
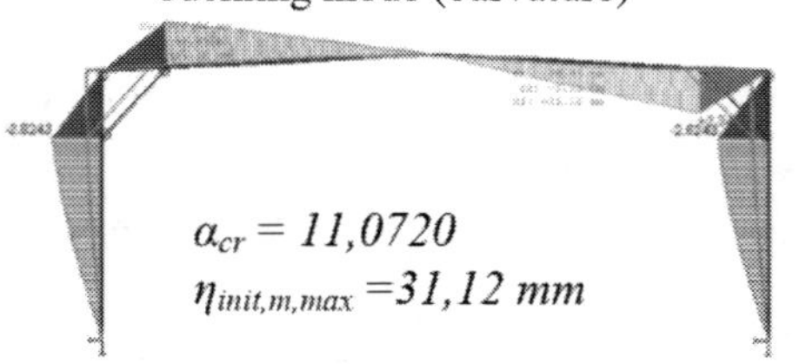 **Fig. 10a:** EUGLI imperfection bending moment for α_{cr} and imperfection with amplitude value	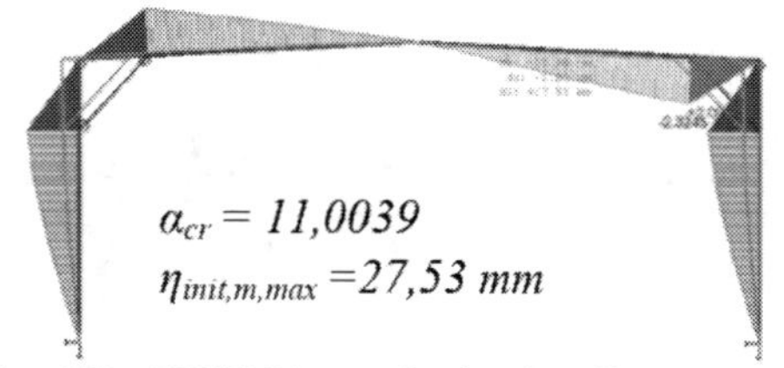 **Fig. 10b:** EUGLI imperfection bending moment for α_{cr} and imperfection with amplitude value
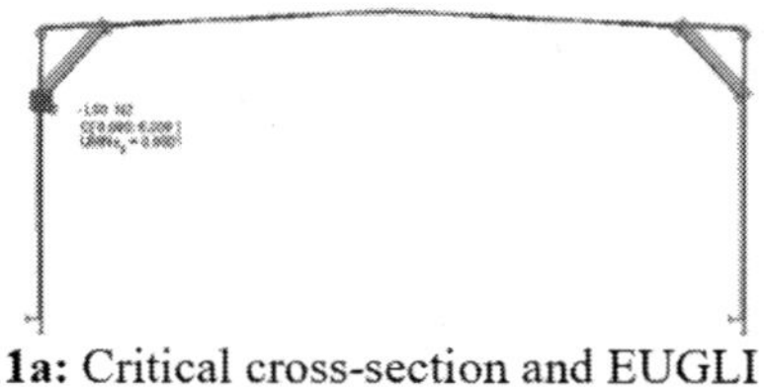 **Fig. 11a:** Critical cross-section and EUGLI method utilization ratio (*0,0601*)	 **Fig. 11b:** Critical cross-section and EUGLI method utilization ratio (*0,0666*)
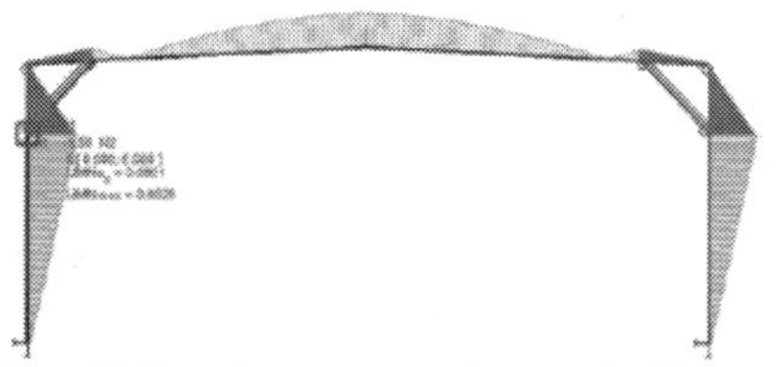 **Fig. 12a:** Ultimate cross-section and ultimate utilization ratio (*0.8531*)	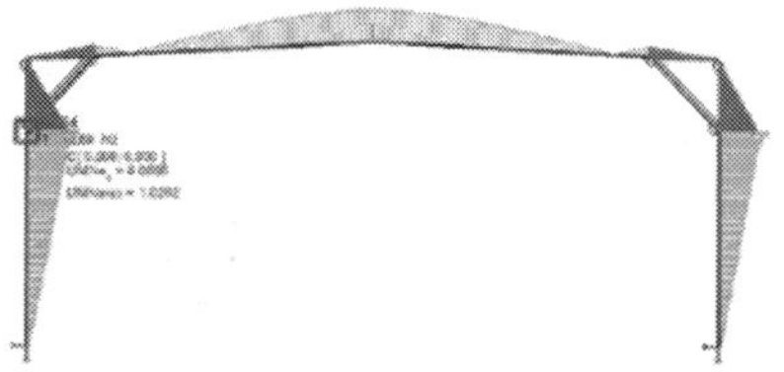 **Fig. 12b:** Ultimate cross-section and ultimate utilization ratio (*1,0264*)

Table 2: Results for all loading types

Loading type	Iter.	α_{cr}	$\eta_{init,m,max}$	EUGLI util. ratio	Ultimate util. ratio
A	1st	11,0720	31,12 mm	0,0601	0.8531
	10th	11,0039	27,53 mm	0,0666	1,0264
B	1st	18,5455	32,37 mm	0,0408	0,8914
	10th	18,4224	28,88 mm	0,0450	1,0527
C	1st	3,7460	33,74 mm	0,2233	0,7309
	10th	3,7048	32,71 mm	0,2468	0,8168

Table 3: Imperfection shapes with results

A	1,0211	A	1,0211	A	1,0194	A	1,0222	A	1,0192
B	1,0505	B	1,0439	B	1,0506	B	1,0416	B	1,0504
C	0,7919	C	0,7096	C	0,7857	C	0,7038	C	0,7849

A	1.0222	A	1,0222	A	1,0192	A	1.0183	A	1,0228
B	1,0505	B	1,0414	B	1.0415	B	1,0493	B	1,0510
C	0,7992	C	0,7028	C	0,7165	C	0,8003	C	0,7835

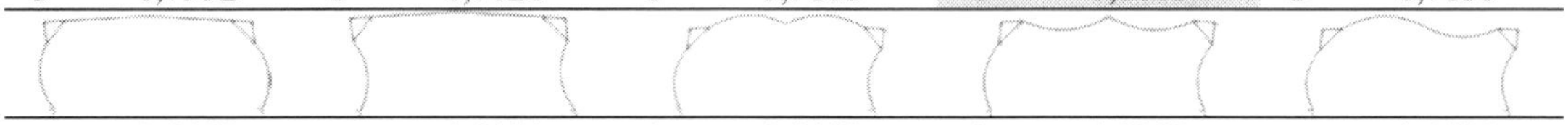

A	1,0183	A	1,0059	A	1,0183	A	1,0203	A	1,0032
B	1,0403	B	1,0420	B	1,0500	B	1,0509	B	1,0506
C	0,7178	C	0,7015	C	0,7854	C	0,7839	C	0,7840

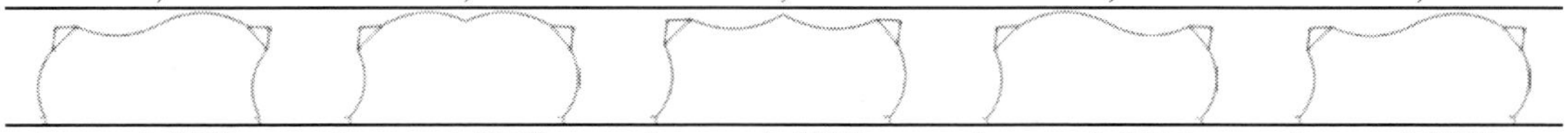

A	1,0193	A	1,0279	A	1,0298	A	1,0286	A	1,0288
B	1,0502	B	1,0526	B	1,0534	B	1,0531	B	1,0527
C	0,7852	C	0,7965	C	0,7950	C	0,7951	C	0,7960

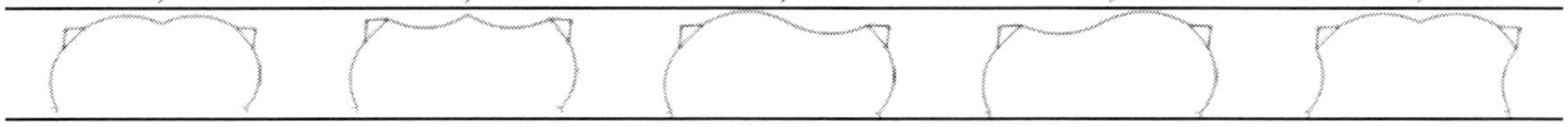

A	1,0246	A	1,0265	A	1,0252	A	1,0254	A	1,0285
B	1,0512	B	1,0524	B	1,0521	B	1,0515	B	1,0530
C	0,7975	C	0,7964	C	0,7965	C	0,7974	C	0,7809

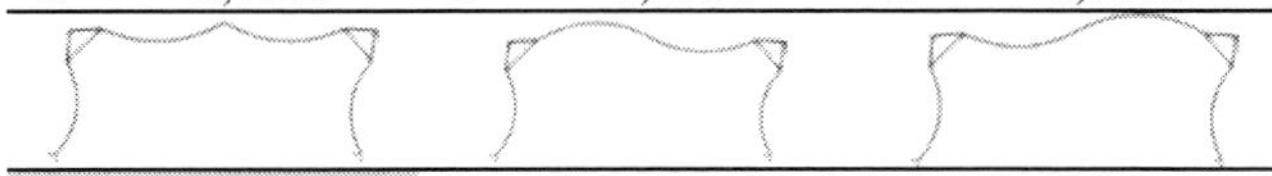

NO IMPERFECTION

A	1,0305	A	1,0297	A	1,0299	A			0,9926
B	1,0541	B	1,0535	B	1,0532	B			1,0460
C	0,7792	C	0,7794	C	0,7803	C			0,7505

Real values of the used imperfections are ϕ=0,00297 (in the frame peak a maximal sway value 25,2 mm); e_0=20 mm for the bottom part of the columns; e_0=6,6 mm for the upper part of the columns; e_0=26,7 mm for the rafter.

The displayed imperfection shapes are recalculated using the developed program (EUGLI method calculation is turned off). The same iterative calculation approach for obtaining effective cross-sectional parameters as in chapter 3 of this paper (10 iterations) is used. Under each imperfection shape in table 3 the maximal utilization factors calculated for all three loading types (combinations) are shown. There are not all possible imperfection shapes in table 3, only some of them are shown (the variety of these shapes is significant). The last table cell is reserved for the results of calculation without imperfections (for comparison purposes only).

5. 1st order analysis procedures in [1, 2, 3]

Parameters for the reduction factor calculation (χ) for this verification procedure are displayed in Table 4. In this calculation gross cross-section parameters of the column (A_{gross}= 4560 mm^2, I_{gross} = 1,034.10^8 mm^4, W_{gross}= 5,217.10^5 mm^3) are used and the bottom part (h = 6 m) of the column is investigated. All other necessary values are taken from the 1st order analysis made by the presented computer program.

Table 4: Parameters for reduction factor for buckling calculation

Loading type	α_{cr}	N_{Ed}	L_{cr}*	β*	χ
A	11,0720	84 kN	15,18 m	2,53	0,426
B	18,5455	58 kN	14,12 m	2,35	0,478
C	3,7460	260 kN	14,84 m	2,47	0,443

* L_{cr} – buckling length of column, β – buckling length factor ($L_{cr} = \beta.h$)

Calculation parameters and results according clause 6.3.3(4) [1] are displayed in Table 5. In this calculation the effective cross-sectional parameters ($A_{eff(axial)}$ = 2644 mm^2, $I_{eff(bending)}$ = 0,879.10^8 mm^4, $W_{eff(bending)}$ = 4,435.10^5 mm^3) are used.

Table 5: Parameters and results for 6.3.3(4) [1]

Loading type	C_{my}	k_{yy}	M_{Ed}	χ_{LT}	Util. ratio
A	0,9	1.003	158,7 kNm	1	1,1113
B	0,9	0,963	167,6 kNm	1	1,0511
C	0,9	1,208	75,9kNm	1	1,0995

Calculation results according clause 6.2.5(2) [3] are displayed in Table 6.

Table 6: Results for 6.2.5(2) [3]

Loading type	Util. ratio
A	1,1993
B	1,1554
C	1,1548

6. Conclusions

There is a short recapitulation of results of the verified frame structure in Table 7. On the base of the achieved results it can be concluded, that by considering imperfections (5.3.2(11) in [1, 2] and 5.3.2(3) a) and b) [1, 2]) the maximal utilization factors are basically the same. According to the results obtained in table 3 differences between the imperfect structures and the perfect structure are very low (Loading type A – 2%, Loading type B – 1% and Loading type C – 5%). This is mainly caused due to the selected loading which should represent real loading cases. Classical 1st order calculations according clauses 6.3.3(4) [1] and 6.2.5(2) [3] are relatively conservative.

The unquestionable advantage of the EUGLI imperfection method is the universality of its use. There are only two imperfection shapes in comparison with the imperfections 5.3.2(3) a) and b) in [1, 2] where there are plenty.

Table 7: Results

Method	Maximal utilization ratios				
	No imperfection	5.2.2(3) a) [1, 2]		5.2.2(3) c) [1, 2]	
Load. type		5.3.2(11) [1, 2]	5.3.2(3) a) b) [1, 2]	6.3.3(4) [1]	6.2.5(2) [3]
A	0,9926	1,0264	1,0305	1,1113	1,1993
B	1,0460	1,0527	1,0541	1,0511	1,1554
C	0,7505	0,8168	0,8003	1,0995	1,1548

The EUGLI imperfection method is not implemented into commercial civil engineering computer programs until now. The 2nd order theory calculation is basically standard procedure in such programs, implementation of the EUGLI method represents only a small amount of additional programming code. Use of both imperfections 5.3.2(11) [1, 2] (EUGLI imperfection) and 5.3.2(3) a) b) [1, 2] together with the combination with the effective cross-sectional parameters calculation is a new procedure and therefore further work is planned to calculate more cases of similar examples for comparison purposes.

Acknowledgement

This paper has been supported by the Slovak Research and Development Agency (SRDA) – grant from the research program APVV No. 0236-12.

References

[1] EN 1993-1-1: *Eurocode 3: Design of steel structures – Part 1-1: General rules and rules for buildings*, 2005.

[2] EN 1999-1-1: *Eurocode 9: Design of aluminium structures – Part 1-1: General structural rules*, 2005.

[3] EN 1993-1-3: *Eurocode 3: Design of steel structures – Part 1-3: General rules – Supplementary rules for cold formed members and sheeting*, 2006.

[4] EN 1993-1-5: *Eurocode 3: Design of steel structures – Part 1-5: Plated structural elements*, 2008.

[5] Chladný E, Štujberová M. "Frames with Unique Global and Local Imperfection in the Shape of Elastic Buckling Mode (Part 1, Part2)". *Stahlbau*, Vol. 82, No. 8, pp. 609-617, No. 9, pp. 684-694. 2013.

[6] Baláž I. "Determination of the Flexural Buckling Resistance of Frames with Members with Non-Uniform Cross-sections and Non-Uniform Axial Compression Forces". Proceedings of the 34th Meeting of Experts on Steel Structures, Pezinok, Oct. 16-17, 2008, pp. 17-22.

[7] Dalemulle M. "Vzperná odolnosť oblúkových konštrukcií v ich rovine" (Buckling Resistance of Arches in their Plane), *Dissertation thesis*, Slovak University of Technology in Bratislava, 2013 (in Slovak)

[8] Kovač M. *Vzperná odolnosť kovových prútov a prútových konštrukcií. Aplikácie nových metód z Eurokódov (Buckling Resistance of Metal members and Frames. Applications of New Methods from Eurocodes)*, Bratislava, Slovak University of Technology in Bratislava, 2012, ISBN 978-80-227-3681-7 (in Slovak)

[9] MATHWORKS: *Programming language MATLAB v.2009.* www.mathworks.com

[10] Brodniansky J, Kovač M. "Application of EUGLI Imperfection for Assessment of Members and Frames with 4th Class Cross-Section". *International Masaryk Conference for Ph.D. Students and Young Researchers*, December 2014, Volume V.

[11] Brodniansky J. "2nd Order Structural Analysis with Imperfection According 5.3.2 (11) EN 1999-1-1 (Method for Computer Analysis) ". *eConference of Doctoral Students and Young Researchers*, July 20-24, 2015.

[12] Brodniansky J, Aroch R, Belica A. "Unique Global and Local Initial Imperfection in the Shape of the Elastic Buckling Mode. Application for Frames with Class 4 Cross-Sections" *Eurosteel 2014: 7th European Conference on Steel and Composite Structures.* Abstracts book + Proceedings (USB), Napoli, Italy, Sept. 10-12, 2014, Brussels, ECCS, 2014, p. 6.

STABILITY AND LOAD-BEARING CAPACITY OF THE LIGHT GAUGE COLD-FORMED PURLINS AT THE SLOPE ROOFS

Eduard L. Ayrumyan, Nikolay I. Kamenshchikov and Maxim A. Liplenko

Melnikov Central Research and Design Institute of Steel Structures, Moscow, Russia

Abstract: This report presents the test results of behaviour of the steel thin-wall purlins made of single C-shaped and Z-shaped cold-formed sections at the building roof structures with a pitch up to 22 degrees.

1. Introduction

One of well-known types of light gauge steel structures (LGSS) are the roof purlins made of single or coupled cold-formed profiles with a C-shaped, Z-shaped or sigmoid cross-section.

The purlins are used as bearing elements supporting light weight cladding (e.g. corrugating sheets, sandwich panels or metal tiles) at the roof steel framework with pitch from 6° to 30°. Nowadays the results of some experimental research of LGSS purlin actual work are known [see 1, 2, 3, 4]. As a rule the test technique and design of these purlins were based on vertical load (mainly, snow or wind) excluding the pitched load component. But building experience shows: stability and capacity of these purlins depend on the roof pitch - the greater the pitch the larger influence it on the purlin behaviour. Therefore investigations of the behaviour of the thin-wall cold-formed purlins at two roof systems with different slopes were carried out by Melnikov Steel Institute (Russia) in 2013.

2. Test technique

The test samples were the single-span and double-span continuous purlin systems with 6 m span made of C-shaped and Z-shaped single sections (Fig.1).

The purlins were supported on oblique beams, which simulated the upper truss chord or the pitch roof frame. The roof sandwich panels were located and fixed with screws at the purlins (see Table 1). Two types of sandwich panels were used for experiment purpose: those with foamed polyisocyanurate insulant thickness of 50 or 100 mm (further PIR) and the panels with the mineral wool insulant thickness of 120 mm (further MW).

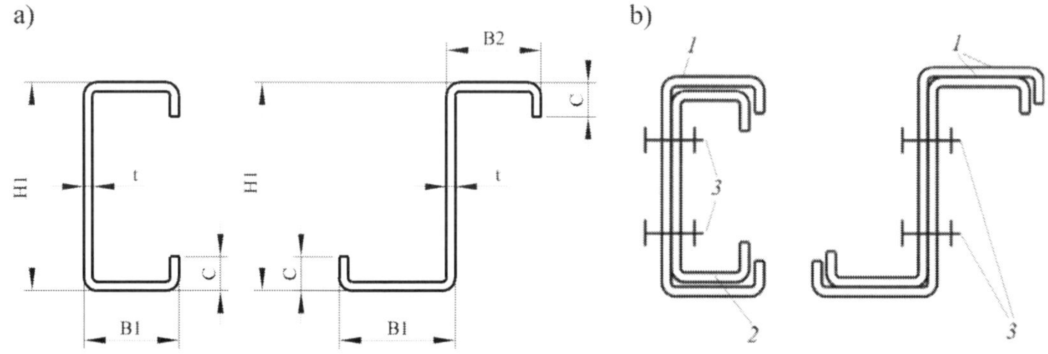

a – cross sections of the experimental samples; b – types of connection purlins at the middle support of the two-span system: 1 – purlin; 2 – connecting profile; 3 - bolt

Fig. 1: Experimental samples and connection types of two-span purlins

Table 1: Test configuration (section sizes are showed in mm)

Section Type C - H1 × B1 × C × t Z - H1 × B1 × B2 × C x t	Span System of Purlins	Purlins Step, [m]	Panel Sandwich	Roof Pitch
C-160×80×26×3	Continuous Two-span	1.5	PIR-50	22° 6°
C-200×95×21×3			PIR-100	22° 6°
Z-200×100×92×16×3				22°
Z-280×105×97×17×3	Single span	2.5	PIR-100	22°
C-280×100×21×3			MW-120	
Z-280×105×97×17×3			PIR-100	6°
C-280×100×21×3			PIR-100	

The test stand was consisted of three triangular frames located parallels at 6.0 m distance (Fig. 2). Oblique girts were joined with lower runner by ball-and-socket joint and were freely supported by upright post made of double rolled channels. It allowed to installation of girts at specified angle to contour line. The purlins were fastened to the girts with cold-formed angles reinforced with the stiffening rib. In case of double-span system, the purlins abutted at the middle support. C-shaped sections were joined with smaller C-shaped section. The Z-shaped sections were joined by means of overlapping. Joints were connected with M16 bolts. Sandwich panels were fastened to the purlins with self-drilling and thread-cutting screws.

Load upon the sample consisted of gaged sand sacks 30 kg each and 120 cm long concrete tiles (cross-section 50×50 cm, average weight 350 kg). The sacks and tiles were laid on the sandwich panels. To avoid tiles' sliding down along the slope, usual ice rakes were fastened to panels. The loading was gradual, at least 15 minute interval was between loading steps. The concrete tiles were laid during first steps. It was made with overhead crane. The sand sacks were laid during final loading steps. Applied loading and displacements were registered during the research. Displacements were measured with the mechanical deflection indicators. For convenience, the latters were located in each purlin's mid-span.

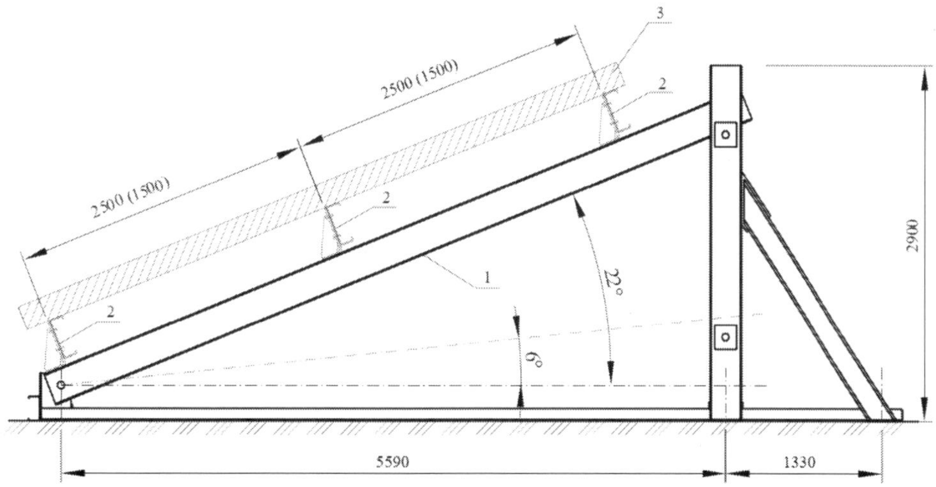

1 – test stand; 2 – tested purlin; 3 – roof sandwich panel;
Fig. 2: Experimental set

Fig. 3: Test of two-span purlins

3. Test results and data analysis

Basing on research results, the vertical mid-span displacement/load relationship curves were plotted.

Double-span system research results are shown at Fig.4. The curves also show purlin's theoretical deflection due to applied experimental load. Cross-section's geometrical parameters were calculated basing on Corporate Standard [5] developed by the Melnikov Steel Institute for calculating and designing of cold-formed structures.
The purlins were made of 350 grade steel. Tensile testing result showed: average yield strength 385 MPa, instantaneous strength 450 MPa, tension set 32 %.

Due to gradual loading, deflection of the double-span purlins enlarged linearly up to loading's certain value (Fig. 4).

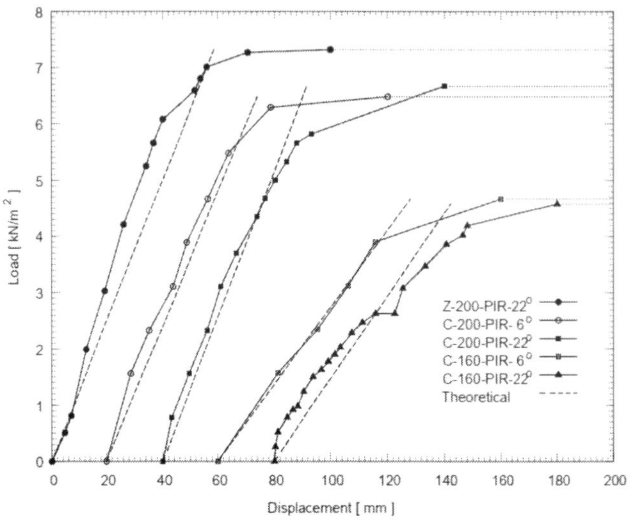

Fig. 4: Load vs vertical middle span displacements of the average purlin for two-span system

It should be noted that 3-day time exposure took place during testing of purlin C-160-PIR-22°, load being 2.63 kN/m^2. As result, deflection increased for 6.77 mm, load being unchanged. Rotation of purlin's cross-section took place during loading. Lower flange of C-shaped purlin had arc displacement towards the slope while lower flange of Z-shaped purlin rotated vice versa (i.e. from the slope). Before loss of purlin's sustaining capacity, self-drilling and thread-cutting screws faced destruction which started at the middle of the purlin span and continued to its supports. Bearing capacity of the samples was lost due to torsional-flexural buckling. During destruction of purlin C-200-PIR-22 it twisted towards the slope despite of the fact that direction of rotation was opposite (Fig. 5a). Other C-shaped purlins destroyed according to usual pattern with twisting from the slope (Fig. 5b).

a)

a) the turn of purlin towards the slope at the inclination of 22°

b)

b) the turn of purlin against the ramp at the inclination of 6°

Fig. 5: The support assembly of purlin and the type of collapse

Unique feature of double-span purlins with PIR sandwich panels is their sustaining capacity's weak dependence on roof pitch. The purlin bearing capacity for extreme load values difference proved to be 3 %, roof pitch being 6° and 22°. It was so because PIR sandwich panels are of great relative stiffness and they release the purlins; which are fastening to the girts by angles reinforced with stiffening rib to prevent rotation of the purlin cross-section on the support. In case of high roof pitch, load's pitch component increases therewith impacting section's upper web. It influences purlins' work and quite changes fracture pattern. Nevertheless, testing of Z-200 purlin at pitch 22° showed that its sustaining capacity is 11 % higher than that of C-200 purlin.

Results of single-span systems' tests are shown at Fig.6. Sustaining capacity of single-span C-shaped and Z-shaped purlins (pitch being 22°) was lost due to purlin's rotation towards the slope as well as during testing of C-200-PIR-22° double-span purlin.

At pitch equal to 6°, purlins kept losing bearing capacity as result of local buckling of the wall and top flange in mid-span (Fig. 7). When pitch changed from 6° to 22°, then critical load for Z-shaped purlin decreased by 27 %. But if pitch is 6°, then sustaining capacity of C- and Z-shaped purlins and that of PIR sandwich panels is practically the same (difference does not exceed 5 %).

The flexural rigidity of MW panels and the rigidity of their screw connections with purlins are much lower than for PIR panels (Fig. 8 a). Also the mineral insulant of the continues double-span MW panels are crushed at the average support and their lower metal sheeting are scaled off under much lower test load, than limit design load (Fig. 8 b). As the result of this load-bearing of the purlins with MW panels is less, than with PIR panels.

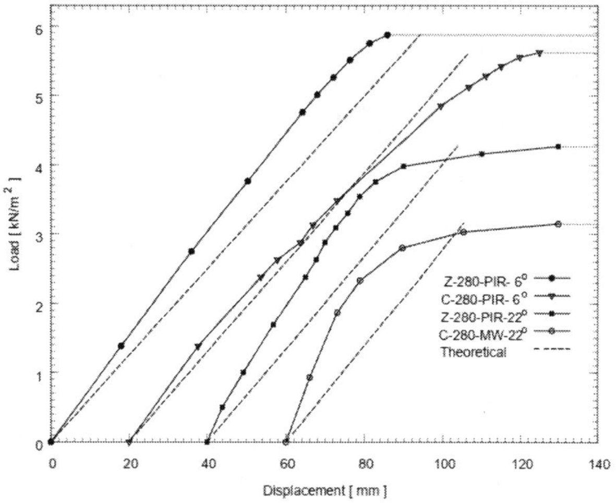

Fig. 6: Load vs vertical middle span displacements of the average purlin for single-span system

a) b)

a) general view of Z-purlins

b) local buckling of wall and top flange of the Z-purlin

Fig. 7: Z-purlins at the slope 6° after testing

a) b)

a) panel failure b) panel destroy at the average support

Fig. 8: Test of purlins with MW sandwich panel

4. Conclusions

The tests were allowed to research the work of the single-span and multi-span purlins in conditions most closely resembling natural ones.

Obtained data was resulted in the following conclusion:

- Load-bearing capacity of the C-shaped double-span purlins at pitch 6° to 22° is practically unchanged.
- Double-span and single-span Z-shaped purlins showed higher load-bearing capacity than the C-shaped purlins.
- PIR sandwich panels provide much better cooperation with the purlin than MW panels and can be used as bracing diaphragms.
- If roof pitch is more up 6°, these purlins should be fastened to beams by means of more rigid device than a cold-formed support angle.

Acknowledgments

The authors would like to express appreciation to the specialists of JSC "Samara plant Electroshield-Stroyindustria" for the assistance in carrying out this test.

References

[1] Ayrumyan EL, Galstyan VG. "Investigation of the actual work of thin-walled cold-formed purlins of galvanized steel", *Journal of Industrial and Civil Engineering*, ISSN 0869-7019, 2002, Vol. 6, pp. 31-34.

[2] Dubina D, Ungureanu V. "Behaviour of multi-span cold-formed Z-purlins with bolted lapped connections", *Thin-Walled Structures*, 2010, 48, 866-871.

[3] Vrany T, Braham M, Belica A. "Restraint of purlins for various roof systems", *11th Nordic Steel Construction Conference NSCC 2009*, pp. 422-429.

[4] Pekoz TB, Soroushian P. *Behavior of C- and Z-purlins under uplift*, Report No. 81-2, Department of Structural Engineering, Cornell University, Ithaca, NY, 1981.

[5] STO 97217441-001-2013, "Construction of the steel frame of cold-formed profiles for low-rise buildings of various purpose", *Standard of Organization*, Moscow, Russia, 2013, C. 99.

ANALYTICAL EVALUATION OF LATERAL RESPONSE OF CFS SHEAR WALLS

Nadia Baldassino[a], Matteo Accorti[a], Riccardo Zandonini[a],
Federica Scavazza[b] and Colin A. Rogers[c]

[a] *Department of Civil, Environmental and Mechanical Engineering, University of Trento, Italy*
[b] *Cogi s.r.l., Calliano, Trento, Italy*
[c] *Department of Civil Engineering and Applied Mechanics, McGill University, Montreal, Canada*

Abstract: This paper focuses on a study of the analytical evaluation of the shear performance of CFS walls. At this aim, the available analytical methods developed for wooden shear walls were considered. The first part of the paper is devoted to an overview of such methods, which are revisited to account for the non-regular pattern of fasteners typical of CFS shear walls. In the second part of the paper, the comparison between experimental and analytical results allows an appraisal of the reliability of such hand methods as design tools for a preliminary evaluation of the CFS shear walls performances.

1. Introduction

The use of cold-formed steel profiles and systems in residential buildings of small and medium dimensions has recently become a viable alternative to traditional framing systems. Lightweight, high structural efficiency, durability, easy transportation, rapidity and simplicity of installation of the building equipment make it a competitive constructional system.

The University of Trento has recently been involved in a research project focusing on the development of an innovative industrialised housing system composed of cold-formed steel profiles. The system makes use of trussed beams and wall's panels as principal structural components. The innovative content of the project is associated with the use of a single section for building up all the structural components. The study required hence a preliminary phase devoted to the identification of the most suitable sectional geometry, which was selected on the basis of both industrial and structural requirements. The response of 2D and 3D subassemblies was then investigated. In this framework, particular attention was devoted to the response of shear walls, which carry out the task of transferring to the foundations the vertical loads of the flooring systems and the horizontal forces due to wind and earthquake actions. In the study developed in Trento [1], several walls' configurations were tested. The possible influence of the sheathing to the wall's response was also considered. Aiming to provide a better understanding of the walls' response, ancillary tests on the main wall components, i.e., the sheathings, the sheathing connections and on hold-down devices, were also conducted. The response of steel walls under lateral loads was extensively investigated in many countries. Extensive experimental studies allow an in-depth understanding of the complex mechanisms, which govern the wall's response, and identification of the key parameters affecting their behaviour. Various tests were first performed in the field of wooden shear walls allowing researchers to establish simplified relationships between the sheathing connections and the overall shear wall response to lateral loads. In the literature, various formulations are available which mainly focus on the wall's response in the elastic range. The availability of such methods and their effectiveness in estimating the stiffness and strength of wooden shear walls, suggested to verify their reliability also for the steel framed shear walls. An attempt to apply these existing formulations to the cases investigated in Trento was performed.

This paper focuses on the available analytical methods (shortly named hand methods). The first part of the paper is devoted to an overview of such methods, which are revisited by considering the non-regular pattern of fasteners, which typically characterise CFS shear walls. In the second part of the paper, the comparison between experimental and hand method results allows an appraisal of the reliability of such hand methods as design tools for a preliminary evaluation of the CFS shear walls performances.

2. Analytical simple-formula methods

Extensive research activities in the field of the wooden shear walls and diaphragms subject to lateral loads have been carried out since late 1920's. The experimental activities performed in this field allowed the researchers to observe the deformation pattern and to identify the main parameters affecting the walls' response. Analytical models for hand calculations were also developed providing simplified computational tools for evaluating stiffness and strength of walls. In these formulations, parameters such as dimensions and properties of the sheathing, number and position of the sheathing-wall frame connections, stiffness and resistance of the bare connection are considered while the contribution of a framed support is disregarded. Experimental results of the walls tested in Trento [1] clearly showed the substantial influence of the sheathing on the overall wall response. The contribution of the bare steel skeleton can be neglected and the wall performance attributed to the sheathing and its connections.

Between the available formulations proposed in the literature, in this study the methods of Tuomi & McCutcheon [2,3], Easley & al. [4], Kallsner & Girhammar [5] are considered. They are based on the two following hypothesis:

1. sheathings assumed homogeneous and isotropic and with rigid behaviour;
2. sheathing to steel skeleton fasteners characterised by a linear load-displacement relationship.

The application of the methods of Tuomi & McCutcheon [2,3], Easley & al. [4], as proposed by the authors, requires that fasteners are positioned with a regular and symmetric pattern. A generalization of these methods to the case of irregular pattern of the fasteners, which characterized the cases investigated on Trento, was hence performed. Moreover, additional assumptions were made:

1. the overall resistance of the CFS wall is evaluated as sum of the resistance associated to the single sheathing panel;
2. the contribution offered to the wall performance by continuous tracks is neglected;

3. the wall performance is not affected by the distance between the fasteners and the edge of the sheathing (i.e., the bearing resistance is not critical).

In the following the methods are briefly recalled.

2.1 Tuomi & McCutcheon method

This method [2,3] assumes that:

1. the framing systems distorts as a parallelogram while the sheathing retains its rectangular shape (Fig. 1);
2. sheathing is applied vertically to the frame and is continuous between the upper and lower chords;
3. loading is static;
4. distortions and deformations are small.

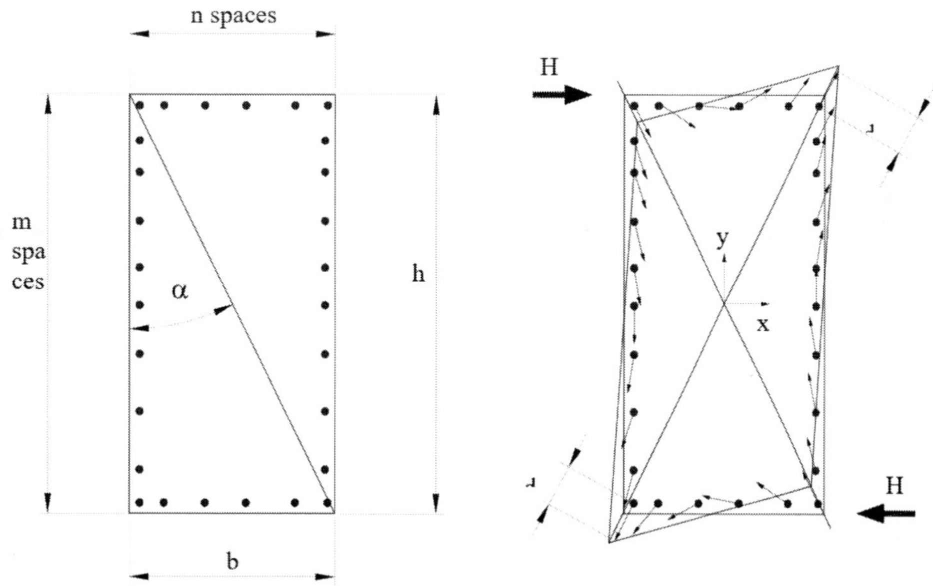

Fig. 1: Framing and sheathing deformation

Wall's strength and stiffness are computed through a balance between the internal energy associated with the fasteners' deformation and the external energy associated with the lateral force H. In case of non-regular pattern of fasteners, the lateral resistance H is evaluated as:

$$H = kd\left\{\sum_{i=1}^{n}\left[\sin^2\alpha + \left(-1+2\frac{is}{b}\right)^2\cos^2\alpha\right] + \sum_{j=1}^{m}\left[\cos^2\alpha + \left(-1+2\frac{js}{b}\right)^2\sin^2\alpha\right]\right\} \quad (1)$$

where:

- k fastener's stiffeness;
- d corner distortion shown in Fig. 1;
- s distance between the first and i-th fastener;
- m,n number of spaces between fasteners in the vertical and horizontal direction, respectively;
- b,α panel dimension and angle defined in Fig. 1.

In case of internal fasteners Eqn. (1) has to be implemented more than once: once for the 'external framing', and the others for the 'internal framings'.

The lateral displacement at the top of the wall Δ is computed as sum of the displacement associated to the fasteners' deformation (Δ_c) and the shear deformation of the sheathing (Δ_t). By assuming a linear force (*f*) - displacement (*x*) relationship for the fasteners

$$f = kx \quad (2)$$

the lateral displacement Δ is evaluated as:

$$\Delta = \Delta_c + \Delta_t = \frac{4H}{k \sin^2 \alpha \sum x^2} + \frac{Hh}{Gtb} \quad (3)$$

where:

x fastener's elongation;
h,b,t height, width and thickness of the sheathing;
G sheathing shear modulus.

2.2 Easley & al. method

This method [4] assumes that:

1. the forces acting of the fasteners at the top and at the bottom of the panel are characterised by a component F_{ex} in the x-direction and a component F_{ey} in the y-direction. F_{ex} is assumed constant while F_{ey} is proportional to the distance of the fastener from the centre of the connection (Fig. 2);
2. the force on the internal fasteners acts only in vertical direction and it is proportional to the distance from the centre of the connection (Fig. 2);
3. the maximum force F_s in vertical direction acts on the fasteners of the sides of the sheathing (Fig. 2).

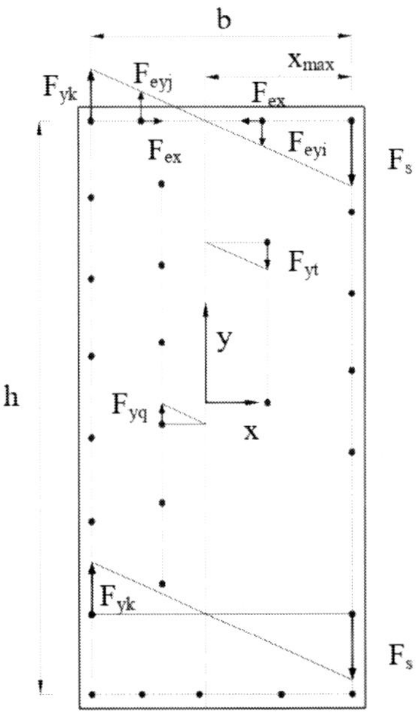

Fig. 2: Sheathing fasteners forces

Under the above mentioned hypothesis, from the rotational equilibrium the maximum force F_s is evaluated as:

$$F_s = \frac{Nbh}{\beta x_{max}} \quad (4)$$

where:

$$\beta = \left[n_s + \frac{1}{x^2_{\max}} \left(\sum_{i=1}^{m} x_i^2 n_j + 2\sum_{i=1}^{ne} x_{ei}^2 \right) \right] \tag{5}$$

n_s number of fasteners at a distance x_{max} from the centre of the fasteners;
x_i x- coordinate of the i-th fastener;
x_{ei} x-coordinate of the i-th fastener at the top and at the bottom of the panel;
x_{max} maximum x-coordinate of the fastener;
m number of columns of fasteners with the exclusion of the one at the distance x_{max};
n_j number of fasteners on the single column;
n_e number of fasteners at the top and at the bottom of the panel;
N wall's lateral resistance for unit length;
b, h dimensions as in Fig. 2.

If the most stressed fasteners are considered, i.e., the ones at the corners, and their resistance is obtained from tests, Eqn. (4) allows evaluating the wall resistance N per unit length. Consequently, the wall lateral resistance H is evaluated as

$$H = Nb \tag{6}$$

The shear deformation (γ) is calculated by considering the shear deformation due to the fasteners (γ_1) and the sheathing shear deformation (γ_2). By assuming a linear force-displacement relationship for the fasteners, the lateral top displacement Δ is computed as:

$$\Delta = \gamma\, h = (\gamma_1 + \gamma_2)h = Nh\left(\frac{2h}{k\beta\, x_{\max}} + \frac{1}{Gt} \right) \tag{7}$$

where k identifies the fastener's stiffness and G and t identify the sheathing' shear modulus and the panel thickness, respectively.

2.3 Källsner & Girhammar method

The fundamental hypotheses of this method [5] are:

1. sheathing are free to rotate;
2. framing joints act as hinges;
3. small displacements if compared to the dimensions of the sheathings.

In the assessment of the wall resistance the framing is assumed to deform along the diagonals with an inclination γ with respect to the vertical direction, while the sheathing rigidly rotates of an angle φ (Fig. 3). The angles γ and φ, determined by considering the minimum of the potential energy, enable evaluation of the force acting on the i-th fastener (f_i) as a function of its coordinates:

$$f_i = \sqrt{F_{x,i}^2 + F_{y,i}^2} = Hh\sqrt{\left(\frac{x_i}{\sum_{i=1}^{n} x_i^2} \right)^2 + \left(\frac{y_i}{\sum_{i=1}^{n} y_i^2} \right)^2} \tag{8}$$

where
H lateral force;
h panel height;
k stiffness of the sheathing to framing fastener;
x_i; y_i coordinates of the i-th fastener with respect to the centroid of the connections.

The failure of the most stressed fasteners, the resistance of which obtained from tests is f_y, leads to Eqn. (9) for the wall lateral capacity H:

$$H = \frac{f_y}{h\left[\sqrt{\left(\frac{x_i}{\sum_{i=1}^{n} x_i^2}\right)^2 + \left(\frac{y_i}{\sum_{i=1}^{n} y_i^2}\right)^2}\right]_{max}} \tag{9}$$

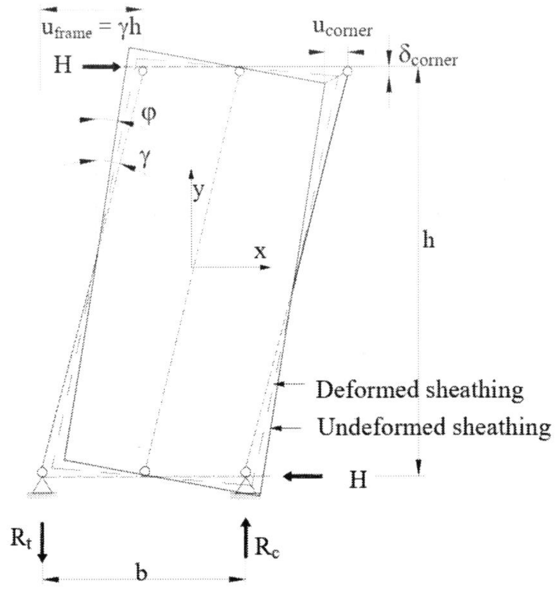

Fig. 3: Wall's deformation

The fasteners' deformation (Δ_c) and the shear sheathing deformation (Δ_s) allow then to evaluate the wall lateral displacement (Δ):

$$\Delta = \Delta_c + \Delta_s = \frac{Hh^2}{k}\left[\frac{1}{\sum_{i=1}^{n} x_1^2} + \frac{1}{\sum_{i=1}^{n} y_1^2}\right] + \frac{Hh}{Gtb} \tag{10}$$

where k and G identify the fastener's stiffness and the sheathing' shear modulus, respectively and t and b the thickness and the width of the panel.

3. Experimental vs. hand methods

The study of the response of CFS shear walls performed in Trento [1] comprises tests, under vertical and lateral loads, of single storey shear walls with dimensions of 2400 mm × 3018 mm. The steel framing elements (e.g. chords and studs) were built up using the same C-like cold-formed section. The steel framing systems consisted of configurations characterized by different bracing systems. The case of wall with vertical studs without bracing system was also considered. For most of the walls two different configurations were tested: the first with sheathing and the second without sheathing. Four different types of cement board and one gypsum board were considered in the study. Tests were performed both in monotonic and cyclic regime.

Complementary tests aimed at characterizing the sheathing material and the connection between the sheathing and the steel skeleton in shear were also carried out.

In order to verify the reliability of the hand methods, the cases of sheathed walls tested in monotonic regime (Fig. 4) were recalculated according to the methods presented in the previous sections. Table 1 summarises the results of the tests considered in terms of maximum lateral load (H_{max}), force ($H_{y,conv}$) and displacement ($\Delta_{y,conv}$) associated to the conventional yielding point.

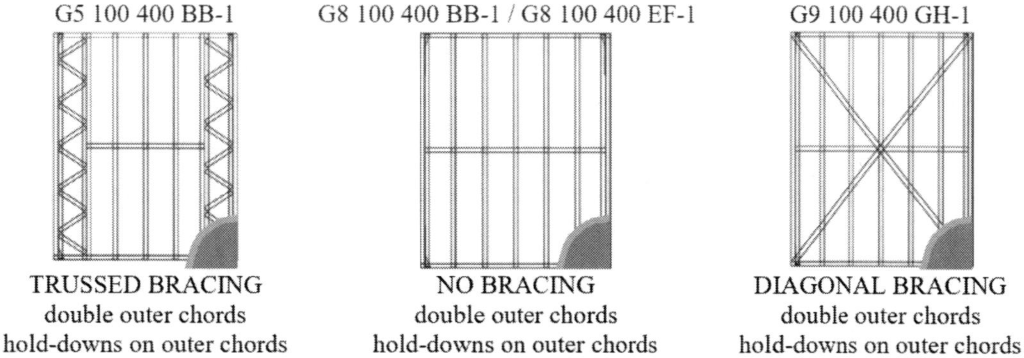

Fig. 4: Walls considered for the comparison

Table 1: Experimental results

Specimen	Material Sheathings	H_{max} (kN)	$H_{y,conv}$ (kN)	$\Delta_{y,conv}$ (mm)
G5 100 400 BB-1	Fibreboard/ Fibreboard	64.20	54.57	10.44
G8 100 400 BB-1	Fibreboard/ Fibreboard	66.48	56.51	10.96
G8 100 400 EF-1	Cement board / Wood-fibre cement sheet	70.04	59.53	15.80
G9 100 400 GH-1	Cement-bonded panels / Gypsum fibreboard	76.92	65.38	16.56

Tables 2-5 compare the lateral resistance (H_{hand}) and the lateral displacement (Δ_{hand}) evaluated by the hand methods with the experimental results of Table 1. For the comparison, force and displacement associated with the experimental conventional yielding point were considered.

Table 2: Experimental vs. hand methods results for wall G5 100 400 BB-1

Hand method	H_{hand} (kN)	$H_{hand}/H_{y,conv}$	Δ_{hand} (mm)	$\Delta_{hand}/\Delta_{y,conv}$
Tuomi & McCutcheon	54.67	1.00	13.16	1.26
Easley & al.	33.22	0.61	7.97	0.76
Kallsner & Girhammar	44.37	0.81	11.14	1.07

Table 3: Experimental vs. hand methods results for wall G8 100 400 BB-1

Hand method	H_{hand} (kN)	$H_{hand}/H_{y,conv}$	Δ_{hand} (mm)	$\Delta_{hand}/\Delta_{y,conv}$
Tuomi & McCutcheon	54.67	0.97	13.16	1.20
Easley & al.	33.22	0.59	7.97	0.73
Kallsner & Girhammar	44.37	0.79	11.14	1.02

The results of Tables 2-5 show that only the Tuomi & McCutcheon approach leads to a reliable estimation of the elastic lateral force for the cases G5 100 400 BB-1 and G8 100 400

BB-1. On the contrary, the associated lateral displacement is overestimated. In all the remaining cases, hand methods generally underestimate both strength and lateral wall's displacement. The simplified assumptions at the base of the different methods appear not able to describe the complex mechanisms of force transmission between structural components. This is particularly apparent for wall G9 100 400 GH-1, i.e. the wall with diagonal bracing system where the contribution of the steel skeleton is significant.

Table 4: Experimental vs. hand methods results for wall G8 100 400 EF-1

Hand method	H_{hand} (kN)	$H_{hand}/H_{y,conv}$	Δ_{hand} (mm)	$\Delta_{hand}/\Delta_{y,conv}$
Tuomi & McCutcheon	47.09	0.79	16.82	1.06
Easley & al.	28.61	0.48	10.19	0.64
Kallsner & Girhammar	38.22	0.64	14.27	0.90

Table 5: Experimental vs. hand methods results for wall G9 100 400 GH-1

Hand method	H_{hand} (kN)	$H_{hand}/H_{y,conv}$	Δ_{hand} (mm)	$\Delta_{hand}/\Delta_{y,conv}$
Tuomi & McCutcheon	28.26	0.43	8.48	0.51
Easley & al.	17.17	0.26	5.13	0.31
Kallsner & Girhammar	22.93	0.35	7.17	0.43

4. Concluding remarks

This paper summarises the results of a study focusing on the response of CFS shear wall under lateral loads. The reliability of existing hand methods developed for wooden shear wall and diaphragms as design tools for a preliminary evaluation of CFS shear walls performances is investigated. The methods are first presented and then used for evaluating the response of four of the walls tested by the Authors. The comparison between experimental and hand method results show that the simplified hypotheses at the base of these methods prevent a reliable evaluation of both strength and stiffness.

Acknowledgments

This project was funded by Autonomous Province of Trento and Cogi s.r.l.. The Authors gratefully acknowledge the work by the laboratory technicians Stefano Girardi, Marco Graziadei and Alessandro Banterla.

References

[1] "SteelMax: Experimental and numerical analysis of the performance of CFS profiles and subassemblies under elementary and complex states of stress", *Research report*, Dept. of Civil, Environmental and Mechanical Engineering, University of Trento, 2014.

[2] Tuomi RL, McCutcheon WJ. "Racking Strength of light-frame nailed Walls", *Journal of Structural Division*, ASCE 104 (ST7), 1131-1140, 1978.

[3] McCutcheon WJ. "Racking deformation in Wood Shear Walls", *Journal of Structural Engineering*, 111, 257-269, 1985.

[4] Easley JT, Foomani M, Dodds RH. "Formulas for wood shear walls", *Journal of Structural Division*, ASCE 108 (ST11), 2460-2478, 1982.

[5] Kallsner B, Girhammar UA. "Analysis of fully anchored light-frame timber shear walls-elastic model", *Materials and Structures*, 42, 301-320, 2009.

GBT CALCULATION OF ELASTIC BUCKLING LOADS OF COLD-FORMED STEEL RACK COLUMNS

Miquel Casafont, Jordi Bonada, M. Magdalena Pastor and Francesc Roure

Department of Strength of Materials and Structural Engineering,
Universitat Politècnica de Catalunya, Barcelona, Spain

Abstract: The article shows the performance of a beam finite element procedure recently developed to carry out linear buckling analysis of perforated rack columns. The procedure implements the Generalized Beam Theory in such a way that it can handle members with perforations. The theoretical bases of the method are described elsewhere in a previous paper by the authors. The investigation included herein is mainly focused on practical issues, such as the sensitivity of the results to the finite element size, or the accuracy of the procedure when applied to columns with complex cross-sections and holes. The final conclusion of the study is that the new method is efficient and rather accurate.

1. Introduction

The article presents a numerical procedure for carrying out linear buckling analysis of cold-formed steel columns with multiple perforations distributed along their length, such as those used in typical steel storage rack structures. The Generalised Beam Theory (GBT) is used in the analysis by means of a beam finite element that has been adapted to perforated members.

The proposed approach consists in meshing the perforated and non-perforated parts of the beam model with two different finite elements (Figs. 1 and 2), the perforated and non-perforated finite element, respectively. The GBT mechanical properties of the perforated element have been reduced to take into account the holes. Afterwards, the perforated and non-perforated parts are linked by means constraint equations on the degrees of freedom of the finite elements. The buckling loads are finally determined by solving a constrained eigenvalue problem.

The bases of the new method are briefly summarized in Section 2 of the present article. For those readers interested in the theoretical aspects of the procedure, it is recommended to go to reference [1], which includes a rather comprehensive explanation on how the standard GBT beam finite element can be adapted to consider perforations (an alternative method is also presented in [2]). The actual aim of the article is to show the practical performance of the proposed procedure when applied to real rack columns. This is done in Sections 3 and 4. The concluding remarks of Section 5 close the paper.

2. Theoretical basis for the analysis of perforated members via GBT

Generally speaking, the method presented in this article follows the standard GBT procedure when deriving the cross-section mechanical properties and creating the finite element stiffness matrices. The presence of holes only involves small changes into these two operations. The proposed approach differs from the usual GBT procedure in the process of assembly of the element stiffness matrices to obtain the global stiffness matrix. This particular matrix assembly is the main topic of this Section.

The first step in any GBT calculation is the cross-section analysis, from which the deformation modes and the modal GBT cross-section properties are determined [3]. The deformation modes are usually grouped into four different classes [4]: (i) conventional modes, including the axial extension mode, global modes (G), distortional modes (D) and local modes (L); (ii) natural shear modes (NS); (iii) transverse-extension modes (TE); and (iv) local shear modes (LS). Each mode is defined by means of a set of three functions $u(s)$, $v(s)$ and $w(s)$, corresponding to the mid-plane wall displacements of the cross-section in the local x, s and z coordinate system (see Fig. 1.a).

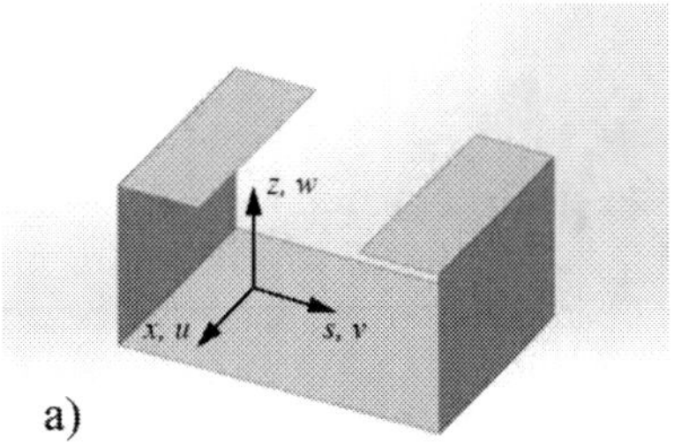

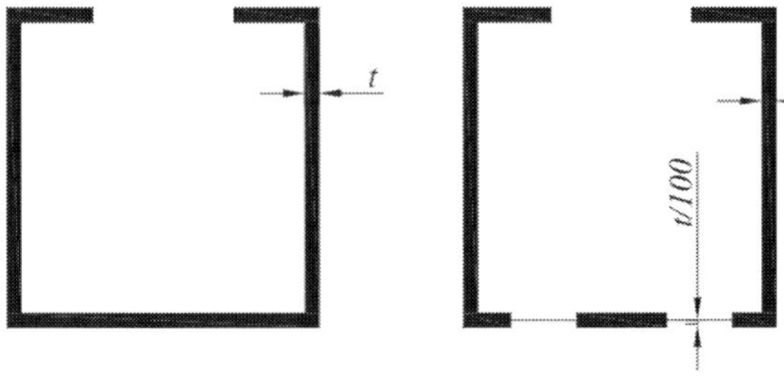

Fig. 1: a) GBT Local coordinate system; b) Unperforated and perforated channel cross-sections

In the cross-section analyses of the present investigation, the TE modes are not calculated. The other three groups of modes are determined, but only some of them are used when carrying out the member finite element analysis, as discussed in Sections 3 and 4.

In the GBT calculation procedure proposed herein for perforated members, the cross-section analysis is carried out twice, on the non-perforated and perforated cross-sections. In Fig. 1.b, it can be observed that both cross-sections are the same, but for the thickness of the perforated segment. All the analyses of this investigation are performed considering that the thickness of this segment is $t/100$, where t is the gross sheet thickness.

Sub-indexes np and p will be used for the parameters corresponding to the non-perforated and perforated cross-sections, respectively. For instance, the vectors of the deformation modes are designated as: $u_{k,np}(s)$, $v_{k,np}(s)$ and $w_{k,np}(s)$; and $u_{k,p}(s)$, $v_{k,p}(s)$ and $w_{k,p}(s)$.

Finally, once the deformation modes have been determined, the displacement field of a member is expressed using a modal vector space:

$$u(x,s) = u_k(s)\phi_{k,x}(x), \quad (1.a)$$

$$v(x,s) = v_k(s)\phi_k(x), \quad (1.b)$$

$$w(x,s) = w_k(s)\phi_k(x), \quad (1.c)$$

where the summation convention applies to subscript k, and $\phi_k(x)$ is the amplitude function of mode k in the longitudinal member direction (x direction in Fig. 1.a).

In the second step of the GBT procedure, the amplitude functions are determined by means of a modal beam finite element analysis. Two types of shape functions are commonly used for the modal amplitudes $\phi_k(x)$ when developing the GBT finite element, depending on the type of mode considered. Hermite cubic polynomials are used for the amplitudes of the L, G, D, NS and TE modes, and Lagrange cubic polynomials are used for the amplitudes of the LS modes (see [1]). The degrees of freedom of the finite element problem are the parameters $\phi^H{}_{k,x}(x)$, $\phi^L{}_{k,x}(x)$ and $\phi^H{}_k(x)$ involved in the shape functions used to define the amplitude functions:

Hermite mode: $\phi_k{}^H(x) = \phi_{k,x}{}^H(0)\cdot\psi_1{}^H + \phi_k{}^H(0)\psi_2{}^H + \phi_{k,x}{}^H(L_e)\psi_3{}^H + \phi_k{}^H(L_e)\cdot\psi_4{}^H$ (2.a)

Lagrange mode: $\phi_k{}^L(x) = \phi_{k,x}{}^L(0)\cdot\psi_1{}^L + \phi_{k,x}{}^L\left(\frac{L_e}{3}\right)\cdot\psi_2{}^L + \phi_{k,x}{}^L\left(\frac{2L_e}{3}\right)\psi_3{}^L + \phi_k{}^L(L_e)\cdot\psi_4{}^L$ (2.b)

where functions $\psi^H{}_i$ and $\psi^L{}_i$ can be found in the literature (for instance in [1]). The element displacement vector in a GBT analysis has the following form:

$$\{d\}^{(e)} = \left\{\{d^1\}^{(e)} \quad \cdots \quad \{d^N\}^{(e)}\right\}^T; \tag{3.a}$$

with:

$$\text{Hermite mode: } \{d^j\}^{(e),H} = \left\{\phi_{j,x}{}^H(0) \quad \phi_j{}^H(0) \quad \phi_{j,x}{}^H(L_e) \quad \phi_j{}^H(L_e)\right\}^T; \tag{3.b}$$

$$\text{Lagrange mode: } \{d^j\}^{(e),L} = \left\{\phi_{j,x}{}^L(0) \quad \phi_{j,x}{}^L\left(\frac{L_e}{3}\right) \quad \phi_{j,x}{}^L\left(\frac{2\cdot L_e}{3}\right) \quad \phi_{j,x}{}^L(L_e)\right\}^T \tag{3.b}$$

In the equations above, j is the mode number, and N the maximum number of modes.

In the proposed procedure, the beam finite element mode is composed of a sequence of, minimum, two parts: perforated and unperforated parts (see Fig. 2.a). Each part can be meshed with a different number of elements, as it will be seen in Section 3. The assembly of the element stiffness matrices within each part is performed according to the standard GBT finite element procedure, i.e., elements are connected one to each other by simply locating the element stiffness matrices in the proper place (somewhat similar to the standard finite element procedure).

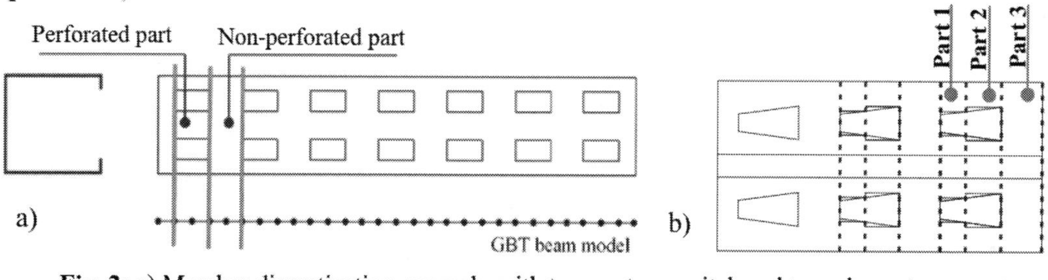

Fig. 2: a) Member discretization example with two parts per pitch and two elements per part; b) Example with three parts per pitch

The key point of the new procedure is the assembly of the perforated parts with the non-perforated parts. There are two coincident nodes, and sets of degrees of freedom, on the edge of each part. The connection between the parts is established by means of constraint equations on the degrees of freedom of each pair of coincident nodes, i.e., on the amplitude values on the edges of the parts.

Two constraint equations per connection should be set for Hermite modes, since degrees of freedom $\phi^I{}_k$ and $\phi^I{}_{k,x}$ are present in their displacement vector; while just one equation is needed for Lagrange modes, because they only have $\phi'^{l}_{k,x}$ terms. Some examples of constraint equations can be:

-Hermite modes:

$$u(x_l,s)-u(x_{l+1},s)=u_{k,np}(s)\phi^l_{k,x}-u_{k,p}(s)\phi^{l+1}_{k,x}=0,\tag{4.a}$$

$$w(x_l,s)-w(x_{l+1},s)=w_{k,np}(s)\phi^l_{k}-w_{k,p}(s)\phi^{l+1}_{k}=0,\tag{4.b}$$

-Lagrange modes:

$$u(x_l,s)-u(x_{l+1},s)=u_{k,np}(s)\phi^l_{k,x}-u_{k,p}(s)\phi^{l+1}_{k,x}=0.\tag{4.c}$$

where summation convention applies to k; $\phi'^{l}_{k}=\phi_k(x_l)$ and $\phi^{l+1}_{k}=\phi_k(x_{l+1})$; and l is a node number of the member discretization. For instance, node l can belong to the cross-section of the non-perforated part, and node $l+1$ to the cross-section of the perforated part (or vice versa).

Constraint equations (4) are introduced into the finite element procedure via the Lagrange multiplier's method. Finally, the buckling loads and buckling shapes are obtained by solving the eigenvalue problem of the constrained (linked) model.

3. Validation and illustrative examples for distortional and global buckling

In the above mentioned paper [1], the procedure proposed was validated by carrying out linear buckling analyses on 4 different channel columns, S1 to S4 in Fig. 3 (the sheet thickness of S1 and S3 is 1.5 mm, while the sheet thickness of S2 and S4 is 3 mm). These cross-sections are very similar to those produced by the North American rack industry.

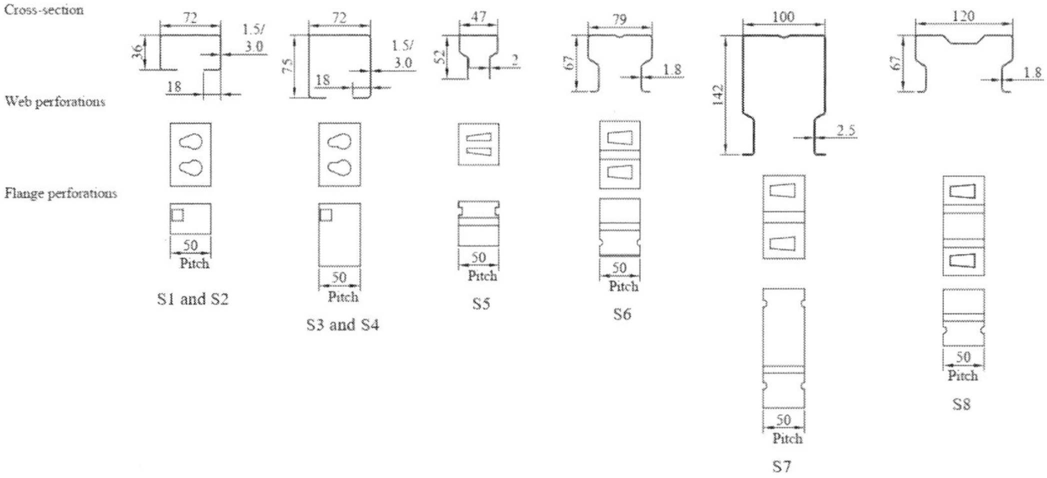

Fig. 3: Columns investigated

The aim of the analyses carried out in [1] was (i) to find a set of constraint equations producing a robust procedure, and (ii) to select which deformation modes have to be included in the analyses to get accurate buckling loads for distortional and global buckling modes. Finally, it was concluded that the best option is to combine conventional D and G deformation modes, together with SL modes; and that the best constraint equations are:

$$u_{k,np}(s)\phi^l_{k,x}-u_{k,p}(s)\phi^{l+1}_{k,x}=0,\tag{5.a}$$

$$u_{k,np}(s)\phi^l_{k}-u_{k,p}(s)\phi^{l+1}_{k}=0,\tag{5.b}$$

where D and G, and SL modes are considered when defining (5.a), and only D and G are considered in (5.b), because there are not ϕ^d_k degrees of freedom for the SL Lagrangian modes. Fig. 4.a shows the accuracy achieved in [1] for columns S1 to S4. In this figure, the values on the vertical axis are rations between the GBT buckling loads and shell finite element buckling loads. It can be observed that the distortional buckling loads are slightly underestimated (especially for simply supported columns); while the torsional-flexural buckling loads may be overestimated for some cross-sections (especially for short columns).

In the present article, the accuracy and applicability of the procedure is further investigated. First, a short study on the finite element mesh discretization is carried out. Afterwards, the performance of the method when applied to more complex cross-sections is also assessed. Finally, the results of the first analyses carried out to determine local buckling loads are presented.

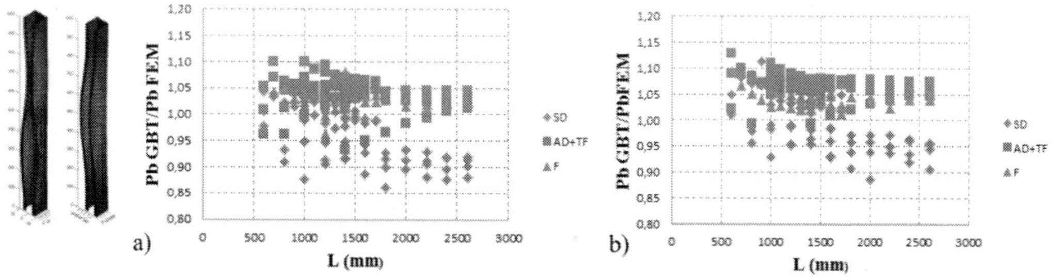

Fig. 4: Accuracy for American rack columns without (a) and with (b) flange holes

3.1 Sensitivity with respect to cross-section and member discretization

A study on the sensitivity of the results of the proposed procedure with respect to the cross-section and member discretization was carried out. Firstly, the study was focused on the effect of the number of cross-section nodes on the resulting buckling loads. S4 columns were analysed using five different cross-section discretizations, as shown in Figs. 5.a to 5.e. 10 is the lowest number of nodes considered, because it is the minimum number of nodes needed to generate a channel cross-section with two perforations (Fig. 5.a). Afterwards, nodes are added to both the non-perforated and perforated segments, until a maximum number of 23.

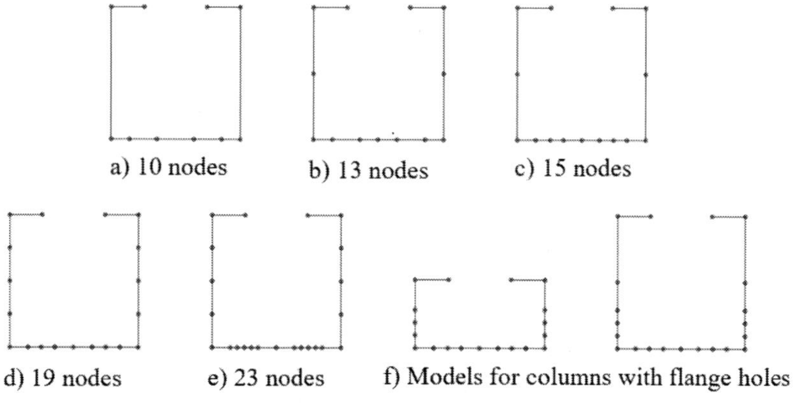

Fig. 5: Cross-section models investigated

The results of the analyses performed on three different lengths of a S4 column, 1000 mm, 1600 mm and 2200 mm, are plotted in Fig. 6. It can be observed that rather good estimates of the FEM buckling loads are already obtained with the poorest discretization, 10 nodes. If the number of nodes is increased, the GBT/FEM ratios only decrease a little bit. They become almost constant after the first intermediate node is placed in the middle of the perforations, i.e., when 15 nodes are used. Therefore, it seems that from a practical point of view, it is not necessary to use a large number of cross-section nodes. 10 or 13 nodes would be enough for a S4 column. In the present investigation, however, all analyses are performed using always a node in the middle of each perforation. All channel sections are modelled with 15 nodes distributed as shown in Fig. 5.c.

In the second analysis, the number of elements to be used in each part of the member, perforated and non-perforated, is investigated. The study is also carried out working on column S4. The same three column lengths used in the cross-section discretization analysis are investigated, considering three different number of GBT finite elements per part: 1, 2 and 3 elements. The results of the study are included in Table 1, where it can be clearly observed that 1 element per part is the optimum option. The use of more elements does not change the accuracy of the procedure. Nevertheless, it is important to note that rack columns usually show a short pitch, around 50 mm. Such a small pitch also produces small perforated and non-perforated parts, which finally results in rather dense meshes, even if only one finite element is used in each part.

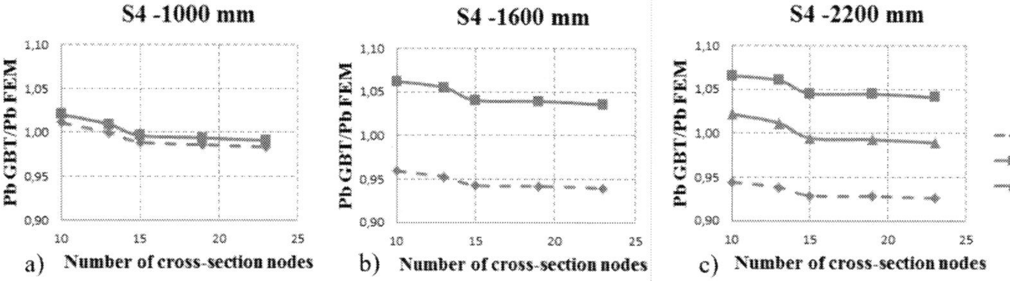

Fig. 6: Sensitivity to the cross-section discretization

Table 1. Sensitivity to the number of finite elements per part (1 element means 1 element per part). SD: symmetric distortional; AD: antisymmetric distortional; TF: torsional-flexural; F: flexural.

Mode	L (mm)	P_{bFEM} (kN)	P_{bGBT} (kN) *1element*	P_{bGBT} (kN) *2elements*	P_{bGBT} (kN) *3elements*	*GBT/FEM 1element*	*GBT/FEM 2elements*	*GBT/FEM 3elements*
SD	1000	485.19	479.79	479.75	479.75	0.99	0.99	0.99
SD	1600	405.16	382.05	382.02	382.01	0.94	0.94	0.94
SD	2200	379.87	352.80	352.77	352.77	0.93	0.93	0.93
AD+TF	1000	563.11	561.04	560.99	560.98	1.00	1.00	1.00
TF	1600	302.00	314.16	314.13	314.12	1.04	1.04	1.04
TF	2200	177.79	185.84	185.82	185.82	1.05	1.05	1.05
F	2200	815.10	810.71	810.62	810.61	0.99	0.99	0.99

3.2 Columns with large flange perforations

There are some rack columns that contain small flange perforations, usually circular, that can be ignored when performing the calculation of the distortional and global elastic buckling loads (see columns S5 to S8 in Fig.3). For instance, in [5] it was possible to accurately predict the buckling loads of some columns without taking into account such flange holes.

However, the flange perforations that can be found in typical American channel rack columns are usually large, see columns S1 to S4 in Fig. 3. Their effect on the buckling loads cannot be neglected [5]. The next step of the investigation was to verify whether the proposed procedure can be applied to these columns. Cross-sections S1 to S4 were analysed using the cross-section models presented in Fig. 5.f. The resulting accuracy for all analysed columns is shown in Fig. 4.b. The results are not as good as those corresponding to columns without flange perforations, but they are still acceptable.

3.3 Validation for typical rack sections with multiple stiffeners

The performance of the proposed analysis procedure when applied to rack sections with multiple lips was assessed by carrying out analysis on columns S5 to S8 of Fig. 3 (more columns will be investigated in the near future). They are very similar to columns produced by European manufacturers.

The results of the analyses are presented in Table 2. It shows the buckling loads for two different lengths of columns, for which the symmetric distortional buckling mode (SD) or the torsional-flexural buckling mode (TF) is critical. According to the ratios presented in the last column of the table, it seems that the procedure works rather well. The trend is similar to what can be seen in Fig. 4.

In the analyses discussed in Section 3.1 and in [1], perforations were always modelled in GBT as rectangular holes circumscribing the actual perforations. It is worth to point out that the Matlab code developed to implement the procedure has been recently improved. Nowadays it allows the user to introduce three different parts per pitch. Perforations can be better modelled if the three parts are used, as can be seen in Fig. 2.b for column S5.

Table 2.Buckling loads of sections with multiple lips (See Table 1 for the meaning of SD and TF)

Column	Mode	L *(mm)*	P_{bFEM} (kN)	P_{bGBT} (kN)	*GBT/FEM*
S5	SD	1000	137.77	131.06	0.95
S5	TF	1500	76.18	78.80	1.03
S6	SD	800	217.99	210.89	0.97
S6	TF	2000	112.41	110.81	0.99
S7	SD	1300	227.35	205.74	0.90
S7	TF	3000	174.02	159.02	0.91
S8	SD	800	275.21	266.13	0.97
S8	TF	3000	128.36	130.27	1.01

4. First validation and illustrative examples for local buckling

The first results of local buckling calculations are shown to close the article. The proposed procedure is applied to determine the local buckling loads of columns S1 to S4 without flange holes. The column lengths considered in the analyses range from 250 mm to 1200 mm.

The analyses are carried out including conventional GBT L, D and G deformation modes. Since the longitudinal displacements u_k of the local modes are null, their constraint equations cannot be directly added to Eq. (5). A new set of constraint equations is used for these modes:

$$w_{k,np}(s)\phi_{k,x}^{l} - w_{k,p}(s)\phi_{k,x}^{l+1} = 0 , \tag{6.a}$$

$$w_{k,np}(s)\phi_{k}^{l} - w_{k,p}(s)\phi_{k}^{l+1} = 0 . \tag{6.b}$$

These constraints are applied together with Eqn. (5), which are the expressions used to link the D and G modes of the perforated and non-perforated parts. When calculating local buckling loads, SL modes are not considered.

Fig. 7 shows the accuracy of the computations performed using the cross-section discretization shown in Fig. 5.c (15 nodes). The analyses are carried out twice, for two different member meshes: 1 element per part, and 2 elements per part. It can be observed that slightly better results are obtained when using 2 elements per part. This is because this mesh can reproduce in a more accurate way the short longitudinal buckling waves of the local modes. For the columns analysed, the accuracy does not improve if more than 2 elements are used.

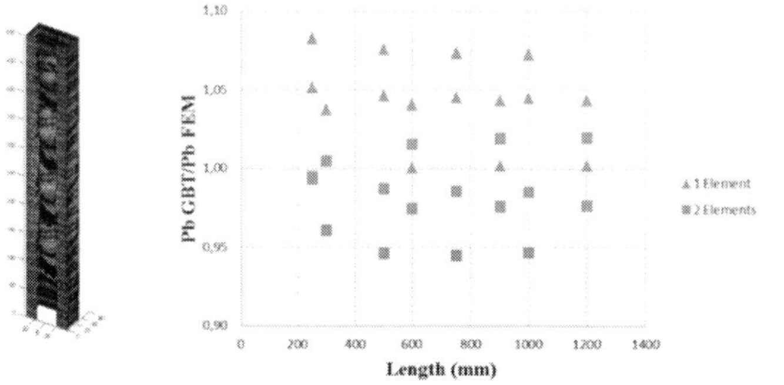

Fig. 7: Accuracy for local buckling of American rack columns

5. Conclusions

The paper confirms the good results obtained in a previous investigation on the calculation of distortional and global buckling loads of perforated rack columns by means of the GBT. It has been shown that the proposed procedure works rather well when applied to complex cross-sections with different perforation patterns (even with perforated flanges). Guidance has also been given on how the GBT finite element models should be meshed to get an efficient procedure. Finally, encouraging results have been obtained for local buckling loads.

Future research will be focused on (i) further verifying the method with new cross-sections, and (ii) improving its accuracy. For instance, it is necessary to improve the predictions of the global buckling loads of short members, and the distortional buckling loads of pinned columns.

References

[1] Casafont M, Bonada J, Pastor M.M., Roure F. "GBT calculation of distortional and global buckling loads of cold-formed steel channel columns with multiple perforations", *8th International Conference on Advances in Steel Structures*, Lisbon, Portugal, 2015.

[2] Cai J, Moen C. "Generalized beam theory buckling analysis for members with holes", *8th International Conference on Advances in Steel Structures*, Lisbon, Portugal, 2015.

[3] Bebiano R, Silvestre N, Camotim D. "GBT formulation to analyse the buckling behavior of thin-walled members subjected to non-uniform bending", *International Journal of Structural Stability and Dynamics*, 7(1), 23-54, 2007.

[4] Gonçalves R, Ritto Corrêa M, Camotim D. "A new approach to the calculation of cross-section deformation modes in the framework of generalised beam theory", *Computational Mechanics*, 46, 759-781, 2010.

[5] Casafont M, Pastor MM, Bonada J, Roure F, Peköz T. "Linear buckling analysis of perforated steel storage rack columns with the Finite Strip Method", *Thin-Walled Structures*, 61, 71-85, 2012.

OPTIMIZATION OF TRAPEZOIDAL SHEET GEOMETRY BY IMPROVING STIFFENERS DISTORTIONAL BEHAVIOUR

Mircea Georgescu[a,b] and Viorel Ungureanu[a,b]

[a] *Politehnica University of Timisoara, Department of Steel Structures and Structural Mechanics, Romania*
[b] *Romanian Academy - Timisoara Branch, Laboratory of Steel Structures, Romania*

Abstract: Trapezoidal steel sheets having the depth of corrugation larger than 50 mm represent a frequently used product in industrial buildings as well as in other types of steel and concrete construction. The usual depth ranges of such products extend up to 200 mm, over a large variety of cross-sectional geometries and wall thicknesses. Being involved into a process of evaluation of load tables for a set of deep corrugation trapezoidal sheets, the authors have observed a number of drawbacks of the existing products, mostly related to the inappropriate geometry of their cross-section, leading to an inadequate distortional behaviour of these elements. Thus, correction measures were identified, leading to an optimized geometry of the section and a considerable increase of resistance in bending.

1. Introduction: role and advantages of deeper corrugation trapezoidal sheets in industrial building

Deeper corrugation trapezoidal steel sheeting represent nowadays an extremely frequent thin-walled structural element used to build modern roofing systems. The decision taken very often by consultants, investors or designers to adopt such a solution in industrial building (or similar type) roofing is actually due to some well-known advantages of such products. These advantages would be the lower value of permanent load per high value of bending resistance, allowing for a total elimination of the classic purlin systems and for a considerable simplification of the erection operations. In case of large surface roofing, this solution is leading to superior productivity levels when installing the sheeting, which motivates the high interest of constructors. Furthermore, light roofing steel systems induce much lower seismic load even in case of prefabricated concrete structures, which strongly recommends them to replace traditional caisson elements used in such cases.

Considering all these features, a large number of suppliers in the field of thin-walled cold-formed steel elements are offering such products in a large variety of cross-section geometry, sheet thicknesses and steel grades. This also leads to a very strong competition on the market between similar products, calling for permanent innovation and improvement. Being involved into the process of issuing load tables (according to the new European codes in the field) for a set of deep corrugation trapezoidal sheets fabricated by a supplier acting on the market, the authors have observed a number of drawbacks of the existing products. The observed drawbacks were mostly related to the inappropriate geometry of the cross-section, leading to a lesser behaviour as far as local instability modes specific to thin-walled profiles are concerned. This has also led to some modification proposals for the existing geometry, some of them with opposite effects, which are discussed in the paper and finally applied onto the cross-section with beneficial results.

2. Set of trapezoidal sheeting under investigation

The studied set of trapezoidal sheeting describing the geometry of each specimen wall thickness and steel grade of mother coil is presented in Table 1. The authors consider the depth range described below as very relevant for the products in current use on the market.

Table 1: Geometry of the studied specimens

Symbol of trapezoidal sheet	Depth of corrugation [mm]	Pitch of corrugation (centres) [mm]	Width of the sheet (centres) [mm]	Nominal wall thickness [mm]	Steel grades as per [1]
58.210.840	58	210	840	0.60-0.75-0.88-1.00	S250GD S280GD
85.280.1120	85	280	1120	0.75-0.88-1.00-1.25	S280GD S350GD
135.310.930	135	310	930	0.75-0.88-1.00-1.25	S280GD S350GD
153.280.840	153	280	840	0.75-0.88-1.00-1.25	S280GD S350GD
158.250.750	158	250	750	0.75-0.88-1.00-1.25	S280GD S350GD
200.420.840	200	420	840	0.75-0.88-1.00-1.25	S280GD S350GD

As a principle, all trapezoidal sheets are installed with their wider flange in upper position while their narrower flange comes into direct contact with the main frame supporting element (see Fig. 1). This is the normal position of the product, designed to support (at wider flange level) all other layers of the roofing system, plus distributed loads acting on the roof (i.e. permanent load, imposed load, snow load and wind load). A consequence of such position adopted when installing the product, is that the wider flange normally remains within the compression zone of the cross-section area under gravity loading combinations and thus it is prone to local instability modes. This affects both the plane elements of the flange and the intermediate stiffeners possibly existing within upper flange width (usually one or even two intermediate stiffeners symmetrically located on the flange).

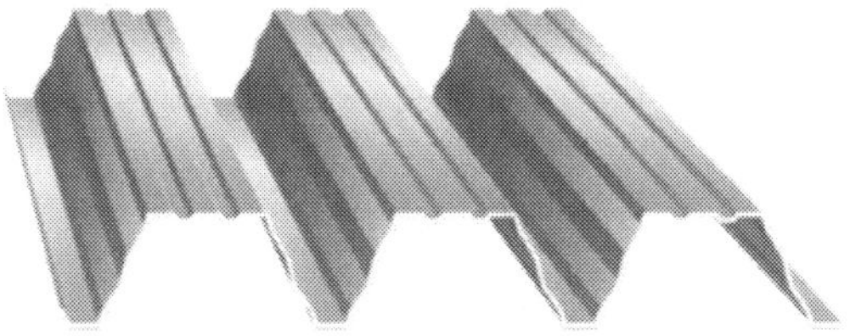

Fig. 1: Example of trapezoidal sheet with the wider flange in upper position

3. Collapse modes and minimum resistance of the sheet

The procedure used to create load tables for any type of trapezoidal sheet implies an identification of possible collapse modes of the sheet associated with each static scheme (single span, two span or three span scheme) followed by the assessment of the final result given by the minimum load value.

A brief review of the procedure is given below for single span and two span static scheme in order to identify the geometrical characteristics of interest for this paper.

3.1 Single span trapezoidal sheet

The relevant collapse modes are:

a) Collapse by bending at mid span, as per 6.1.4.1(1) in [1] (in the following simple intermediate steps of the demonstrations are skipped):

$$\frac{M_{Ed}}{M_{c,Rd}} = 1 \quad \Rightarrow \quad q_1 = \frac{8M_{c,Rd}}{b_u L^2} \tag{1}$$

b) Collapse by web crippling on the end support, as per 6.1.7.1 (1) in [1] (also simple intermediate steps of the demonstration skipped):

$$F_{Ed} = R_{w,Rd} \quad \Rightarrow \quad q_2 = \frac{2R_{w,Rd}}{b_u L} \tag{2}$$

In upper relations (and in the next ones), the following notations have been used, i.e. $M_{c,Rd} = W_{eff} f_{yb}/\gamma_{M0}$ = design resistance of the effective cross-section (class 4); W_{eff} = effective section modulus; M_{Ed} = maximum bending moment from loads; $R_{w,Rd}$ = local transverse resistance of the web as per 6.1.7.3 in [1]; F_{Ed} = maximum local transverse force (support reaction); L= span of the sheet; b_u = actual panel width of the sheet (multiple of corrugation pitch); q_i = load value corresponding to subsequent collapse mode (i = 1, 2 etc.).

The design load (in the table) for single span static scheme is established as the minimum value between the upper two, i.e.:

$$q_{all}^{ULS} = min\{q_1; q_2\} \tag{3}$$

3.2 Two span trapezoidal sheet

The use of a two span static scheme leads to the presence of more collapse modes than in the previous case. These relevant collapse modes are:

a) Collapse by bending in the first span at distance x = 0.4215L from the left support as per 6.1.4.1(1) in [1]:

$$\frac{M_{Ed}}{M_{c,Rd}} = \frac{0.07 q_4 b_u L^2}{M_{c,Rd}} = 1 \Rightarrow q_4 = \frac{M_{c,Rd}}{0.07 b_u L^2} \tag{4}$$

b) Collapse by bending on the intermediate support as per 6.1.4.1(1) in [1]:

$$\frac{M^B_{Ed}}{M^B_{c,Rd}} = \frac{0.125 q_5 b_u L^2}{M^B_{c,Rd}} = 1 \Rightarrow q_5 = \frac{M^B_{c,Rd}}{0.125 b_u L^2} \tag{5}$$

c) Collapse by web crippling on the intermediate support as per 6.1.7.1 (1) in [1]:

$$\frac{F_{Ed}}{R^B_{w,Rd}} = \frac{1.25 q_6 b_u L}{R^B_{w,Rd}} = 1 \Rightarrow q_6 = \frac{R^B_{w,Rd}}{1.25 b_u L} \tag{6}$$

d) Collapse by web crippling on the end support as per 6.1.7.1 (1) in [1]:

$$\frac{F_{Ed}}{R^A_{w,Rd}} = \frac{0.375 q_7 b_u L}{R^A_{w,Rd}} = 1 \Rightarrow q_7 = \frac{R^A_{w,Rd}}{0.375 b_u L} \tag{7}$$

e) Collapse by combined negative bending moment and local transverse force (support reaction) on intermediate support, as per 6.1.11(1) in [1]:

$$\frac{M^B_{Ed}}{M^2_{c,Rd}} + \frac{F^B_{Ed}}{R^B_{w,Rd}} = \frac{0.125 q_8 b_u L^2}{M^2_{c,Rd}} + \frac{1.25 q_8 b_u L}{R^B_{w,Rd}} = 1.25 \tag{8}$$

In the upper relation the resistance moment on the intermediate support was determined using the section modulus calculated for the case of the effective aria with upper flange in tension, i.e.:

$$M^2_{c,Rd} = \frac{W^2_{eff} f_{yb}}{\gamma_{M0}} \tag{9}$$

The subsequent load value writes:

$$q_8 = \frac{1.25}{b_u \left(\frac{0.125 L^2}{M^2_{c,Rd}} + \frac{1.25 L}{R^B_{w,Rd}} \right)} \tag{10}$$

The design load (in the table) for two span static scheme is established as the minimum value between the upper five values, i.e.:

$$q^{ULS}_{all} = min\{q_4; q_5; q_6; q_7; q_8\} \tag{11}$$

As visible from upper described procedures, the moment resistance and the transverse force resistance on supports (each one alone or in interaction) actually govern the obtained load values. This is the reason why, present paper focusses on these values, trying to improve the resulting design load by improving cross-section performance via its geometrical characteristics determined either for bending or for transverse force on support (web crippling).

Actually, the authors have observed that relatively minor changes at the level of cross-section existing geometry generate quite consistent improvement as far as upper resistances are concerned. This proposed optimizing procedure is extremely important for series products (as trapezoidal sheets) where even small improvements of product characteristics might help to beat competition of similar items.

4. Effective cross-section of trapezoidal sheet according to [1]

Trapezoidal sheet is usually built of a number of wider upper flanges providing the required support for gravity or uplift loading, a certain number of sloping webs and some narrower bottom flanges also used for a proper contact with external or intermediate supports. The geometry of this cross-section and especially in terms of angles between the plain elements and width-to-thickness ratios is prescribed by Table 5.1 in [1]. In order to improve the behaviour of this cross-section (framed into Class 4 sections and thus prone to local buckling) intermediate stiffeners are provided on the upper flange and / or on the web. The shape of these stiffeners is also recommended, see 1.6.2 in [1]. Depending on the depth of the trapezoidal cross-section (web depth) one, two or even three intermediate stiffeners may be provided on the web. A typical example of deeper corrugation trapezoidal sheet is presented in Fig. 2 for one of the specimens in the studied set, i.e. sheet 153.280.840. The drawing is representing half of a corrugation with all intermediate stiffeners provided in this particular case. This pattern is repeated for a suitable number of times in order to form the subsequent full sheet section.

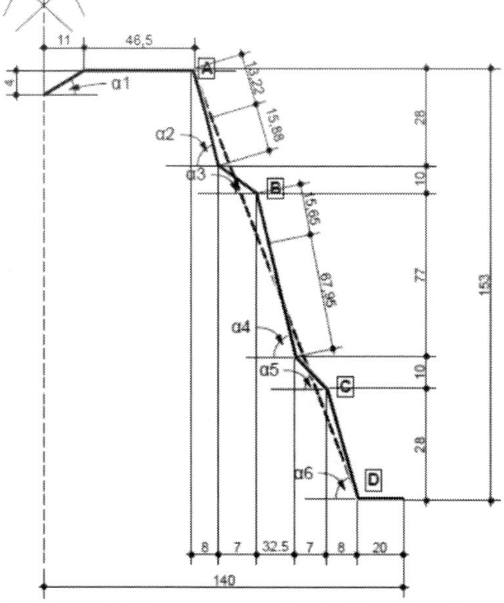

Fig. 2: Existing geometry for half-corrugation

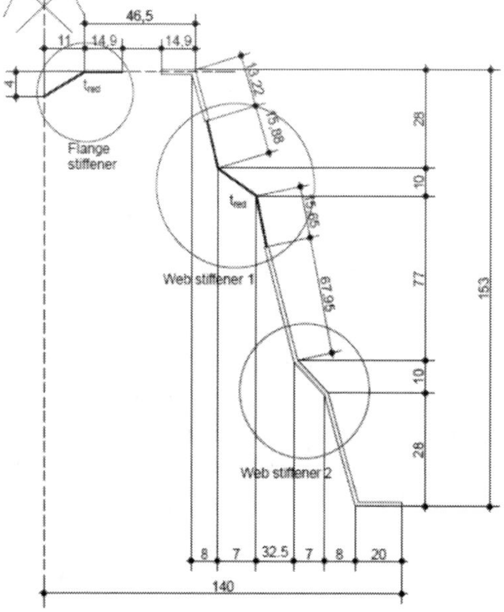

Fig. 3: Effective cross-section for half-corrugation

By applying the prescribed procedure of [1] on upper gross cross-section, the effective cross-section for sheet 153.280.840 results as presented in Fig. 3. The method of obtaining the final geometry of the effective cross-section is leading (in the zone of the intermediate stiffeners) to a reduced wall thickness in comparison with the nominal thickness of the sheet. This reduced thickness value depends on stiffener characteristics and performance as far as its distortional buckling resistance is concerned. As observed by the authors, variations of the reduced thickness lead to relevant variations of geometrical characteristics, allow for an optimization process and thus justify the discussion below.

5. Optimization of the cross-section geometry in view of increasing the bending resistance

The prescribed procedure to obtain a suitably reduced wall thickness for the intermediate stiffeners on the flange and on the web includes a number of steps, i.e.:

- Calculation of the reduced wall thickness for the flange stiffener in compression using the elastic critical stress for flexural or distortional buckling of the stiffener ($\sigma_{cr,s}$);
- Calculation of the reduced wall thickness for the web stiffener in compression using the elastic critical stress for flexural or distortional buckling of the stiffener ($\sigma_{cr,sa}$);
- Considering the interaction between flange and web stiffener and calculation of a unique reduced wall thickness by using both upper values of critical stress to obtain a modified

value for the stress ($\sigma_{cr,mod}$). This unique thickness is further on used to define the effective cross-section of the sheet and calculate allowable loads.

The elastic critical stress values are determined in this paper for the particular case of 153.280.840 sheet, as an example of application for the proposed optimization. Similar applications may be performed for any other specimen of the studied set or for similar other products. The elastic critical stress for the flange intermediate stiffener as per 5.5.3.4.2(2) in [1] writes:

$$\sigma_{cr,s} = \frac{4.2k_w E}{A_s}\sqrt{\frac{I_s t^3}{4b_p^2\left(2b_p + 3b_s\right)}} \tag{12}$$

where: A_s is the area of flange stiffener; I_s is the second moment of area of the flange stiffener section about its centroid axis parallel to the flange; k_w is the coefficient of rotational stiffness, safely taken as unity; t is the design thickness of stiffener section elements; b_p and b_s are the geometrical lengths defined by [1].

The elastic critical stress for the web stiffener in compression (closest to the flange) as per 5.5.3.4.3(7) in [1] writes:

$$\sigma_{cr,sa} = \frac{1.05k_f E\sqrt{I_{sa}t^3 S_1}}{A_{sa}S_2\left(S_1 - S_2\right)} \tag{13}$$

where: A_{sa} is the area of web stiffener; I_{sa} is the second moment of area of the web stiffener section about its centroid axis parallel to the web; k_f is the coefficient of rotational stiffness, safely taken as unity; t is the design thickness of stiffener section elements; S_1 and S_2 are the geometrical lengths defined by [1].

In the final stage, the interaction between flange and web stiffener local buckling is considered by calculating a modified value of the critical stress based on the previous two, i.e.:

$$\sigma_{cr,mod} = \frac{\sigma_{cr,s}}{\sqrt[4]{1+\left[\beta_s\dfrac{\sigma_{cr,s}}{\sigma_{cr,sa}}\right]^4}} \tag{14}$$

where β_s is the coefficient depending on geometrical parameters explained in [1].

As a final result, the reduction factor, χ_d, and the reduced wall thickness in the stiffener zone are obtained using previous critical stress, i.e.:

$$t_{red} = \chi_d \cdot t \tag{15}$$

Thus, the effective cross-section area of the trapezoidal sheeting, as defined by [1] is derived. By applying the upper procedure, the authors have observed the relatively low values of the elastic critical stress (i.e. less than 100 N/mm^2) which further on led to low values of the reduced thickness in the stiffener zone (i.e. t_{red} = 0.35t...0.40t) and which finally penalized the bending resistance of the effective cross-section. This called for an optimization of the sectional geometry in order to avoid such effects. To this aim, equations (13) and (14) were re-examined, thus observing the important role of the second moment of area I_s or I_{sa} in a possible increase of the σ_{cr} values. If somehow these values could be increased, an increase of the critical stress and consequently of the reduced thickness would be possible.

6. Optimization of the cross-section geometry to increase trapezoidal sheet bending and transverse force resistance

A first conclusion derived when examining the geometry of the studied specimens (in view of a possible optimization) is the shallow depth of the flange and web intermediate stiffeners.

This feature of the existing products, as designed by the fabricator, is directly related to the angle of the stiffeners with the horizontal. These angles (denoted α_1, α_3 and α_5 in Fig. 2) are presented in Table 2 with their values, resulting from the product catalogue. Trying to increase the reduced thickness used in the zone of the stiffeners via an increase of the elastic critical stress ($\sigma_{cr,mod}$) obtained in its turn by increasing I_s and I_{sa} values, the authors have modified the values of angles α_1, α_3 and α_5 (see "Proposed" values in Table 2 and Fig. 4):

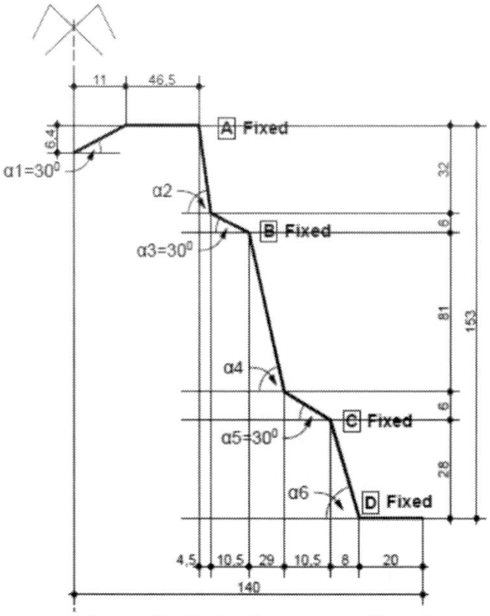

Fig. 4: Optimized geometry Step 1

Table 2: Angles of the intermediate stiffeners with horizontal (existing versus proposed)

Specimen symbol	Angle α_1 [°]		Angle α_3[°]		Angle α_5[°]	
	Existing	Proposed	Existing	Proposed	Existing	Proposed
58.210.840	17.7	30	-	-	-	-
85.280.1120	45	30	45	30	-	-
135.310.930	13	30	38.7	30	41.6	30
153.280.840	20	30	55	30	55	30
158.250.750	20	30	63.4	30	63.4	30
200.420.840	16.4	30	48.4	30	48.4	30

After determining the effective geometrical characteristics for the new proposed geometry of the 153.280.840 trapezoidal sheet (see Fig. 4), a considerable increase of all characteristics including the bending resistance ($M_{c,Rd}$) are visible, as pointed out in Table 3.

Table 3: Improvement of bending characteristics via optimization (Step 1)

Characteristic symbol	Initial geometry value	Proposed geometry value	Relative increase +Δ[%]
χ_d	0.382	0.534	+40%
A_{eff} [mm^2]	127.42	138.07	+8.4%
$I_{y,eff}$ [mm^4]	2091120	2308740	+10.4%
$W_{eff(-)}$ [mm^3]	25914	29658	+14.5%
$M_{c,Rd}$ [Nmm]	7255920	8304240	+14.5%

However, this encouraging result obtained by the Step 1 of the optimization needs to be checked by determining its influence on the transversal force resistance of the web which also influence the load value (see eqns. (2) to (8)).

Table 4: Improvement of transverse force resistance via optimization (Step 2)

Characteristic symbol	Initial geometry	Optimized geometry: Step 1	Relative variation [%]	Optimized geometry: Step 2	Relative variation [%]
$R_{w,Rd1}$ [N] (1st support)	4096	3179	-22%	4590	+12.1%
$R_{w,Rd2}$ [N] (2nd support)	17376	13484	-22%	19472	+12.1%

When calculating the values of the specimen resistance to transverse force (on the end support and on the intermediate support respectively) for the trapezoidal sheet 153.280.840 using the geometry resulting from Step 1 of the optimization, a drastic decrease of these values result as visible in Table 4.

This decrease imposes some new measures of optimization included in Step 2, i.e.:

- changing the values of the stiffeners angle with horizontal to: $\alpha_1=\alpha_3=\alpha_5=45°$;
- aligning the centres of the web stiffeners on the straight line uniting points A and D on Fig. 2, in order to minimize eccentricities of web stiffener corners which strongly influence the web capacity in shear and transverse force.

A new geometry of the cross-section thus results, which fulfils both bending and transverse force demands (see Table 4 final column). Though increasing angle α_3 from 30° to 45° apparently diminishes the value of I_{sa}, this is only a minor compromise as the governing value in determining the elastic critical stress is in fact I_s (see eqns. (12) and (14)) which actually increases via the new measures adopted in Step 2 of the optimization. Similar procedures may be applied for any similar specimen. Actually, the proposed geometrical modifications are minor changes in respect with the initial shape, with considerable influence over the resulting section resistances, very important for a series product.

7. Conclusions

By studying a set of deep corrugation trapezoidal sheets, the authors have observed some drawbacks of their geometry (presented in the product catalogues) which lead to a lower performance of the flange and web intermediate stiffeners in bending and distortion. These drawbacks mostly refer to the inappropriate angle of the intermediate stiffener with horizontal. Some optimization measures are proposed in the paper, which lead to a considerable increase in bending and transverse force resistance of the samples. These measures are proposed for adoption into the fabrication of the series products.

References

[1] EN 1993-1-3, *Eurocode 3: Design of steel structures. Part 1-3: General Rules. Supplementary rules for cold-formed thin gauge members and sheeting*, ECS, 2006.

CRITICAL STRESSES OF STEEL PLATE ELEMENTS ON ELASTIC FOUNDATION SUBJECTED TO COMPRESSION

Péter Hegyi, László P Kollár and László Dunai

Department of Structural Engineering, Budapest University of Technology and Economics, Budapest, Hungary

Abstract: A new direction of development in the field of light-gauge building systems is when the load bearing elements made of cold-formed steel sections are encased in an ultra-lightweight material. These structural systems can be used beneficially in the case of residential buildings. To be able to derive design formulae for these structures closed-form solutions are needed for the different buckling modes of cross-sections. In this paper the calculation of critical buckling stresses of plate elements having three or four simply supported edges resting on elastic foundation, representing the bracing material is presented.

1. Introduction

A new direction of development in the field of light-gauge building systems is when the load bearing elements made of cold-formed steel (CFS) sections are encased in an ultra-lightweight material (density of 100-200 kg/m^3). This material can act as heat and moisture insulation as well as fire protection. Besides these features the material can also provide bracing effect to the CFS elements therefore increases the resistance against stability failures. Such lightweight material is the polystyrene aggregate concrete (PAC), which contains no gravel or sand only polystyrene granules and admixtures to improve material properties [1].

An experimental research focused on structural elements was conducted on flexural [2] and compression [3] elements. The aim of these tests was to gain information on the behaviour of PAC encased CFS columns and beams, as well as to investigate the load increment due to the bracing effect provided by the PAC. It was found that the PAC has significant effect on both behaviour and resistance, which means that this novel structural system can be used beneficially in residential buildings. For this purpose, design procedures are required which can take into account the bracing effect of PAC. Our final goal is to derive these design formulae based on the Direct Strength Method (DSM) already standardised in the US, Australia and New Zealand [4].

The characteristic parameters required for the DSM are the critical stresses for local, distortional and global buckling, respectively. For this end we need to have a procedure for determining these stresses in the case of cross-sections with PAC bracing. As a first step in this procedure the buckling stress of compressed plate elements resting on elastic foundation is investigated in this paper.

2. Elastic foundation

To be able to take into account the bracing effect of PAC on the plate buckling, the material is treated as an elastic half-space. This idea comes from the theory of sandwich beams, where a similar stability phenomenon, i.e. the face wrinkling may occur [5]. During loading, as the face material of the sandwich panel gets in-plane compression, the core material restrains the buckling, resulting in a problem of buckling of plate resting on solid foundation. This solid foundation is treated as a Winkler-type model in the sandwich beam theory [5], assuming elastic material behaviour. Although the real material behaviour of the core is not elastic in the full range, the results obtained by this assumption were found to be in good correlation with the experimental data. For steel plates encased in PAC the same assumption is applicable, since the structural problem and the material behaviour are similar. The original equation of the foundation modulus shows a dependence on the half wavelength l [5]:

$$k_w^* = \frac{2 \cdot \pi \cdot E_c}{(3-\upsilon_c)\cdot(1+\upsilon_c)} \cdot \frac{1}{l} \tag{1}$$

where

E_c and ν_c are the elastic modulus and Poisson's ratio of the foundation material, respectively;
k_w* is the foundation stiffness assuming unidirectional waves.

Eqn. (1) is derived with the assumption that the buckling shape of the face is a sine wave in the direction of the loading while in the perpendicular direction it remains straight. For the case of plate buckling, where the buckling shape is composed of two sine waves perpendicular to one another, a modified foundation modulus is derived:

$$k_w = \frac{2 \cdot \pi \cdot E_c}{(3-\upsilon_c)\cdot(1+\upsilon_c)} \cdot \sqrt{\frac{1}{l_x^2} + \frac{1}{l_y^2}} \tag{2}$$

where l_x and l_y are the buckling lengths in x and y direction, respectively;
k_w is the foundation stiffness assuming two directional waves.

To prove that Eqn. (2) is valid the following trigonometric identity can be used:

$$w(x,y) = \sin\frac{\pi x}{l_x} \cdot \cos\frac{\pi y}{l_y} = \frac{1}{2}\left(\sin\left(\frac{\pi x}{l_x} - \frac{\pi y}{l_y}\right) + \sin\left(\frac{\pi x}{l_x} + \frac{\pi y}{l_y}\right)\right) \tag{3}$$

The left side of Eqn. (3) is the assumed load pattern. The right side of the equation should be modified according to the following expressions:

$$l_x = \frac{l}{\cos\alpha} \tag{4a}$$

$$l_y = \frac{l}{\sin\alpha} \tag{4b}$$

Substituting Eqn. (4) into the Eqn. (3) the terms can be simplified to the sum of two unidirectional sine waves in a rotated coordinate system $\xi - \eta$:

$$w(x,y) = \frac{1}{2}\left(\sin\left(\pi\frac{\cos\alpha \cdot x - \sin\alpha \cdot y}{l}\right) + \sin\left(\pi\frac{\cos\alpha \cdot x + \sin\alpha \cdot y}{l}\right)\right) = \frac{1}{2}\left(\sin\left(\frac{\pi\eta}{l}\right) + \sin\left(\frac{\pi\xi}{l}\right)\right) \tag{5}$$

Thus the two-directional load pattern can be converted to two unidirectional waves. The foundation modulus therefore can be obtained by superimposing two unidirectional foundation moduli calculated by Eqn. (1). Based on Eqn. (4) the corresponding wavelength l can be expressed as:

$$\frac{1}{l^2} = \frac{1}{l_x^2} + \frac{1}{l_x^2} \tag{6}$$

which is what had to be proven.

Eqn. (2) can predict the foundation modulus for different buckling lengths: the smaller the buckling length the stiffer the foundation. This interaction between the buckling phenomenon and the restraining effect characterises the plate buckling behaviour. In the next chapters two types of rectangular plate will be investigated taking into account the effect of elastic foundation.

3. Buckling stress of rectangular plate elements having four simply supported edges

3.1 Analytical solution

The differential equation of buckling for plates resting on elastic foundation is as follows:

$$\frac{D}{t}\left(\frac{\partial^4 w}{\partial x^4} + 2\frac{\partial^4 w}{\partial x^2 \cdot \partial y^2} + \frac{\partial^4 w}{\partial y^4}\right) - \sigma_x \frac{\partial^2 w}{\partial x^2} - \sigma_y \frac{\partial^2 w}{\partial y^2} - 2\tau_{xy}\frac{\partial^2 w}{\partial x \partial y} = \frac{k_w}{t} w \tag{7}$$

where D is the stiffness of the steel plate, $\frac{E_s t^3}{12(1-\nu_s)}$;

E_s and ν_s are the elastic modulus and Poisson's ratio of steel, respectively;

t is the plate thickness;

σ_x, σ_y and τ_{xy} are the membrane stresses in the steel plate.

In most practical cases σ_y and $\tau_{xy} = 0$, and only σ_x acts on the plate. The boundary conditions for simply supported edges in the y direction are (the other is the same but with index replacement):

$$w = 0 \tag{8a}$$

$$\frac{\partial^2 w}{\partial x^2} = 0 \tag{8b}$$

Eqn. (7) can be solved analytically. As the critical buckling stress of long plates is of interest, we will assume the buckling shape in the form of ($l_y = b$):

$$w = A_0 \cdot \sin\frac{\pi x}{l_x} \cdot \sin\frac{\pi y}{b} \tag{9}$$

Substituting Eqn. (9) into Eqn. (7) after straightforward algebraic manipulation we arrive at:

$$\sigma_x = \frac{D}{t}\left(\frac{\pi}{l_x}\right)^2 + 2\nu_s \frac{D}{t}\left(\frac{\pi}{L_y}\right)^2 + \frac{D}{t}\left(\frac{\pi}{L_y}\right)^4 \left(\frac{l_x}{\pi}\right)^2 + 2(1-\nu_s)\frac{D}{t}\left(\frac{\pi}{L_y}\right)^2 + \frac{k_w}{t}\left(\frac{l_x}{\pi}\right)^2 \tag{10}$$

To find the actual σ_{cr} Eqn. (10) needs to be minimised with respect to l_x. As k_w itself depends on l_x, no closed form solution can be derived directly.

Three main parameters characterise Eqn. (10): the b/t ratio of the plate and the material properties of the foundation material (E_c and ν_c). As no data is available for the Poisson's ratio of the PAC material yet, it is assumed to be 0.2 below. For the two remaining parameters a wide range parametric study was performed, the plate b/t ratio was varied between 30 and 300 (using fixed t = 1 mm and varying b), while the elastic modulus of the PAC was set between 0 and 200 MPa based on previous measurements (see [2] and [3]). Eq. (10) was minimised using numerical tools.

Fig. 1 shows the critical stresses as a function of b/t ratio for the case of several different elastic moduli. It is obvious that significant increment can be reached in the buckling stress in the case of more slender plate elements. As the b/t increases the critical stress reaches a limit value. This behaviour represents that at very large plate slenderness the resistance against buckling is provided solely by the PAC bracing. However, as the b/t ratio decreases the increment in σ_{cr} decreases, too.

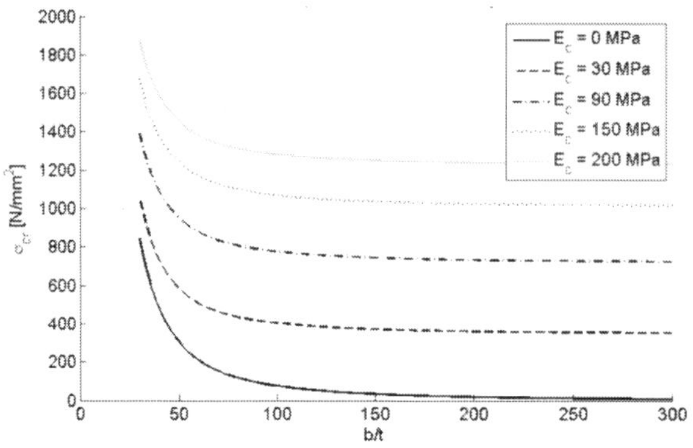

Fig. 1: Critical plate buckling stress as a function of b/t and E_c

Fig. 2a) shows the relationship of buckling length and critical stress for different foundation stiffnesses. It is visible that the critical stress increases significantly while the buckling length decreases. The typical buckling length is several centimetres in the studied range. This finding is in good correlation with previous test results [2], where C-section columns were investigated, and the buckling length for local failure was found to be in the same range (see Fig. 2b)).

(a)

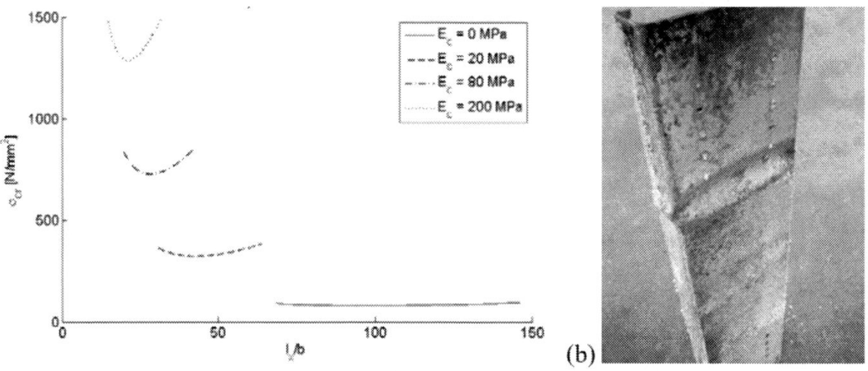

(b)

Fig. 2: a) $l_x - \sigma_{cr}$ relationship for different E_c with b/t = 98; b) Local failure

Based on the results obtained for σ_{cr}, a curve-fitting process was carried out using neural network based software Eurequa [6]. For the investigated parameter range the proposed equation which describe the phenomenon is:

$$\sigma_{cr} = 4 \cdot \frac{\pi^2 E_S}{12(1-\upsilon_S^2)(b/t)^2} - \frac{4680\,MPa}{b/t} + 2.35E_c + \sqrt{E_c \cdot 3025\,MPa} \tag{11}$$

3.2 Finite strip solution

To be able to verify the results obtained by analytical calculations a finite strip analysis was also carried out using CUFSM software [7]. The parametric range covered by CUFSM was the same as detailed before. As an initial guess (l_{x0}) for buckling length the numerical solution of Eqn. (10) was used. Plate elements were investigated in the range [0.7 l_{x0}; 1.5 l_{x0}], the minimum buckling stress was searched in this interval. The plate element was divided into 20 equal parts for the analysis while the buckling length range consisted of 50 parts.

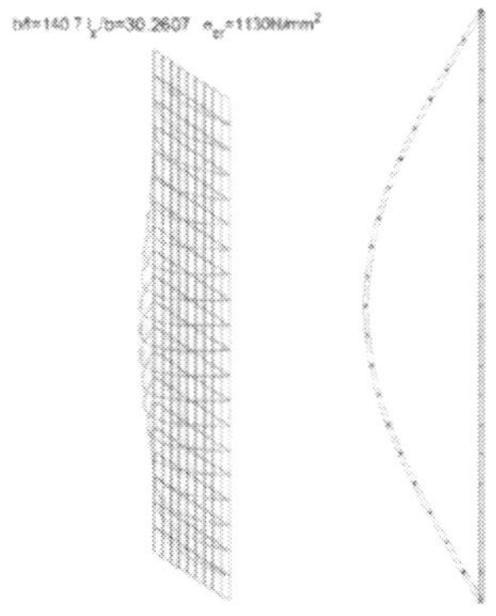

Fig. 3: Buckling shape of finite strip analysis

The results showed that the buckling shape perpendicular to the loading direction is close to the assumed sine wave (Fig. 3). As a conclusion the relative difference between the finite strip results and Eqn. (11) are quite small. Eqn. (11) can be used in the range of b/t [30;300] and E_c [30;200] MPa with error less than 4%. Eqn. (11) gives greater error as E_c reaches 0 MPa, as can be seen in Fig. 4.

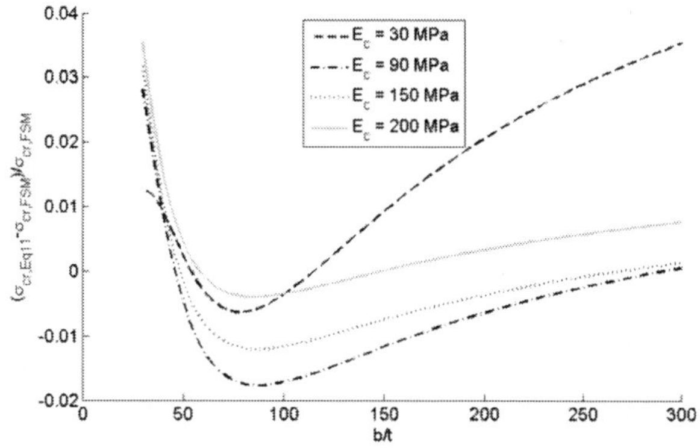

Fig. 4: Relative difference between finite strip results and Eqn. (11)

4. Buckling stress of rectangular plate elements having three simply supported edges and one free edge

4.1 Analytical solution

The main principles of the analytical analysis of this section are similar to those in Section 3. Here we apply the Rayleigh-Ritz method and assume the buckling shape based on [8] as:

$$w = A_0 \cdot y \cdot \sin\frac{\pi x}{l_x} \tag{12}$$

Going through the steps of Rayleigh-Ritz method the result is:

$$\sigma_x = \frac{D}{t}\left(\frac{\pi}{l_x}\right)^2 + 6(1-\nu_s)\frac{D}{t}\frac{1}{b^2} + k_w\left(\frac{l_x}{\pi}\right)^2 \tag{13}$$

The last term of the equation is the same as the last term in Eqn. (10), which predicts that the effect of the foundation is similar to that found in Section 3. In the current case for calculating the foundation modulus, the buckling length l_y was approximated by $2b$. A parametric study was conducted again on the same parameter range as it was described before.

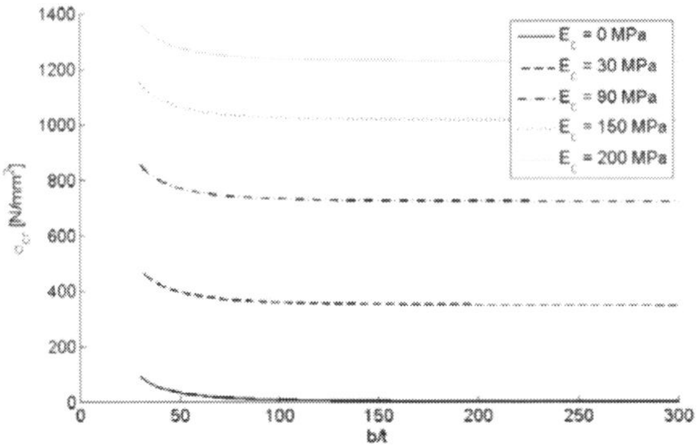

Fig. 5: Critical plate buckling stress as a function of b/t and E_c

The results for slender plates in Fig. 5 are similar to Fig. 1 which is based on the fact that the critical stress is provided mainly by the foundation. For the case of less slender plates the difference between the three and four simply supported edge solution gets larger which is due to the effect of the free edge (see Table 1).

Table 1: Critical stress of plates with different boundary conditions (b/t = 150)

E_c [MPa]	3 ss + 1 free edge	4 ss edge
0	3.59 MPa	33.74 MPa
100	777.58 MPa	797.21 MPa

In the unbraced case the last term in Eqn. (13) is zero, therefore the critical buckling length is infinite. The actual critical stress asymptotically reaches this value as the length of the plate increases. On the other hand, in the case of the braced plates, an exact minimum exists for buckling. This change in the behaviour is visible in Fig. 6.

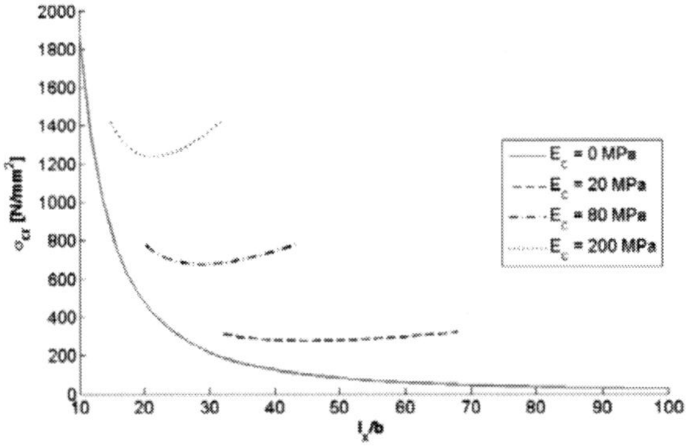

Fig. 6: $l_x/b - \sigma_{cr}$ relationship for different E_c, with b/t = 98

Based on these results a curve-fitting was performed which resulted in the following equation:

$$\sigma_{cr} = 0.43 \cdot \frac{\pi^2 E_s}{12(1-\upsilon_s^2)(b/t)^2} + \frac{1200MPa}{b/t} + 2.22E_c + \sqrt{E_c \cdot 3390MPa} - 40MPa \quad (14)$$

4.2. Finite strip solution

A finite strip analysis was conducted to check for the approximations in the previous section. The same procedure was carried out as before by using CUFSM program. For small slenderness the assumed and calculated buckling shapes were closely matched (Fig. 7a)).

As the slenderness increases the agreement between the two shapes gets less adequate. Moreover, the increasing E_c has the same effect (Fig. 7b) and c)). This discrepancy does not appear in the relative difference between the critical stresses calculated by CUFSM and Eqn. 14 (Fig. 8). This is explained by the fact that where the buckling shape is different from the assumed shape, the main contribution in the critical stress is made by the foundation and not by the plate itself. The simplification $l_y = 2b$ should be further investigated by FE analysis.

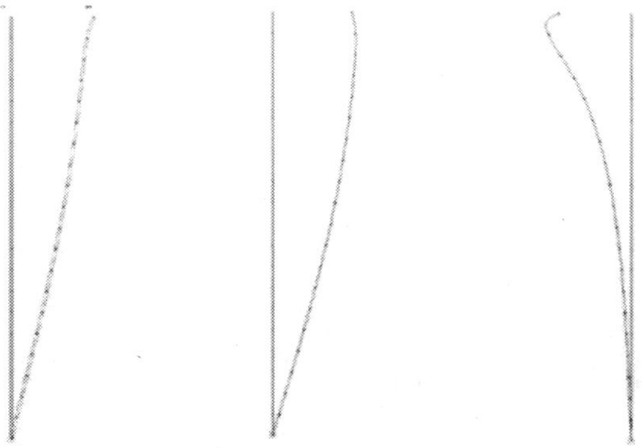

Fig. 7: Buckling shapes: a) b/t = 159.6, E_c = 80 MPa; b) b/t = 294.6, E_c = 56 MPa; c) b/t = 297.3, E_c = 194 MPa

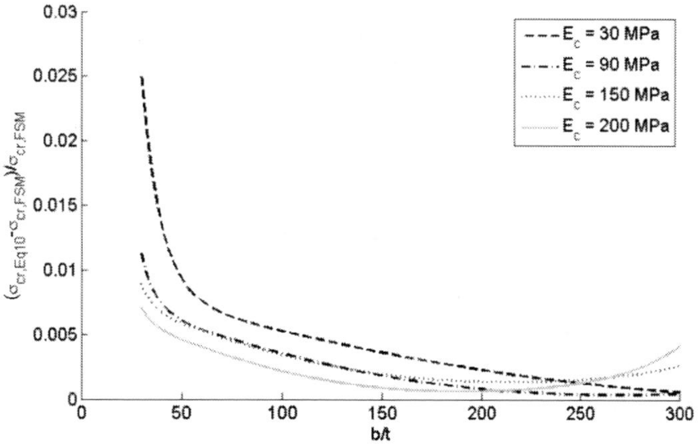

Fig. 8: Relative difference between finite strip results and Eqn. (14)

5. Conclusion

A new direction of development in the light gauge construction induced experimental and theoretical investigations on ultra-lightweight-concrete encased CFS elements. The final aim of the research is to derive design formulae for such kind of structures based on the Direct Strength Method. In this paper the local buckling behaviour of plate elements subjected to compression was introduced. The main conclusions of the paper are as follows:

1. a new formula was derived and verified which can predict the foundation modulus of an elastic solid half-space loaded in the shape of two-directional sine-waves;
2. the local buckling behaviour of plate elements with four simply supported edges were investigated both analytically (by means of Rayleigh-Ritz method) and numerically (CUFSM);
3. the investigation was conducted to plate elements with three simply supported edges and one free edge;
4. it was found that significant increment can be reached by the elastic foundation;
5. closed form equations were proposed which can be reliably used within a wide range of parameters.

The results obtained should be further investigated by FE models. Based on the extension of these results the local buckling stress of a cross section can be determined which is one of the key parameters of DSM design.

Acknowledgments

The work reported in this paper has been developed partly in the framework of the project „Talent care and cultivation in the scientific workshops of BME" project, supported by the grant TÁMOP-4.2.2.B-10/1--2010-0009. The reported R&D project is part of the program of the co-operating partners in the frame of "Funding market-oriented R+D activities in Central Hungary" (KMR_12-1-2012-0236); the help and support of WM21 Hungária Ltd., TDT ALTENERGIA Co. and Horizont Globál Ltd. are highly appreciated.

References

[1] Lublóy É, Balázs LGy, Kopecskó K, Tóth E, Dunai L, Hegyi P, Drávucz O. "Thermal insulation capacity of concretes by expanded polystyrene aggregate", *Proceedings of the Fourth International fib Congress*, 750-751, Paper No. 214, Mumbai, India, 2014.

[2] Hegyi P, Dunai L. "Experimental study on ultra-lightweight-concrete encased cold-formed steel structures. Part I: Stability behaviour of elements subjected to bending", *Thin-walled Structures,* Vol. 101, 75-84, 2016.

[3] Hegyi P, Dunai L. "Experimental study on ultra-lightweight-concrete encased cold-formed steel structures. Part II: Stability behaviour of elements subjected to compression", *Thin-walled Structures,* Vol. 101, 100-108, 2016.

[4] Schafer BW. "Review: The Direct Strength Method of cold-formed steel member design", *Journal of Constructional Steel Research* Vol. 64, 766-778, 2008.

[5] Hegedűs I, Kollár LP. "Application of sandwich beam theory in structural stability analysis" in *Special problems in stability* editor Kollár L, 2006 (in Hungarian).

[6] www.nutonian.com.

[7] Schafer BW, Ádány S. "Buckling analysis of cold-formed steel members using CUFSM: conventional and constrained finite strip methods", *18th International Specialty Conference on Cold Formed Steel Structures*, Florida, 2006.

[8] Kollár LP, Springer G. *Mechanics of composite structures*, Cambridge, UK, 2003.

SHEAR TESTS OF RIVET FASTENED RECTANGULAR HOLLOW FLANGE CHANNEL BEAMS WITH WEB OPENINGS

Poologanathan Keerthan, Mahen Mahendran and Somadasa Wanniarachchi

Science and Engineering Faculty, Queensland University of Technology, Brisbane, Australia

Abstract: The rivet-fastened rectangular hollow flange channel beam (RHFCB) is a new cold-formed hollow section proposed as an alternative to welded hollow flange steel beams. No research has been undertaken on the shear behaviour and strength of rivet fastened RHFCBs with web openings. Hence a detailed experimental study involving 30 shear tests was undertaken to investigate the shear behaviour and strength of rivet fastened RHFCBs with web openings. Experimental results showed that the current design rules are inadequate for the shear design of rivet fastened RHFCBs with web openings. Improved design equations have been proposed for the shear strength of rivet fastened RHFCBs with web openings.

1. Introduction

In the past, hollow flange sections (HFS) including the Hollow Flange Beam section and the LiteSteel Beam (LSB) [1] have been widely used in residential, industrial and commercial buildings, mainly as flexural members, due to their improved structural performance and light weight. However, these HFSs are no longer manufactured today due to the expensive dual electric welding process used in their manufacturing process, as well as other factors. The rivet-fastened rectangular hollow flange channel beam (RHFCB) shown in Fig. 1 is a new type of cold-formed HFS, proposed as an alternative to the welded HFS. The RHFCB is fabricated by intermittently rivet-fastening two cold-formed rectangular hollow flanges to a web plate. Unlike other conventional cold-formed sections, the HFS family including the rivet-fastened RHFCB, has no unsupported edges. Previous HFS beams are made from single strip of high strength steel through the use of combined cold-forming and dual electric resistance welding process. The rivet-fastened RHFCB uses the much more affordable rivet-fastening system and gives the flexibility of using different combinations of flange and web steel thicknesses and grades due to the way that it is being assembled. It also has lips, possibly contributing to additional strength.

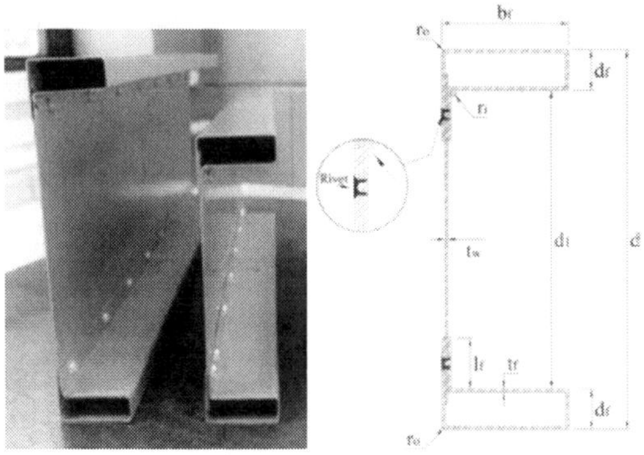

Fig. 1: Rivet Fastened RHFCB

Current practice in flooring systems is to include openings in the web of floor joists or bearers so that building services can be located within rivet fastened RHFCBs. Without web openings, building services have to be located under the joists or bearers leading to increased floor height. Although various shapes can be used for these web openings in floor joists, the most common shape is circular.

The use of web openings in a beam section significantly reduces its shear capacity due to the reduced web area. The reduction in the primary shear resisting area will lead to a significant reduction in the shear capacity of RHFCBs. However, the effect of web openings on their flexural capacity is insignificant as the web openings are normally located at the centre of web. There are many variables that affect the shear capacity of members containing web openings. They include the shape, size and location of the web openings and also the slenderness of the web element. The main aim of this research is to investigate the effect of circular web openings of varying diameters on the shear capacities of rivet fastened RHFCB sections using a detailed experimental study. This paper presents the details of a series of shear tests of RHFCBs with circular web openings, and the results. Experimental shear capacities are compared with the predicted shear capacities using the current design rules in the Australian/New Zealand cold-formed steel structures standard AS/NZS 4600 [2] and the North American Specification [3].

2. Shear Capacity of Rivet Fastened RHFCBs without Web Openings

Keerthan and Mahendran [4] proposed new shear capacity equations for the rivet fastened RHFCBs without web openings based on experimental and finite element analysis results. They found that the current shear design rules are very conservative. This is because they do not include the observed post-buckling strength and the effects of increased fixity at the web-flange juncture of RHFCBs. However, Test and FEA results show that there is considerable amount of post-buckling strength for rivet fastened RHFCBs subjected to shear. Hence post-buckling shear strength can be taken into account in the design of rivet fastened RHFCBs with large clear web height to thickness ratios (d_1/t_w). Rivet-fastened RHFCB also has lips, possibly contributing to additional shear capacity.

Keerthan and Mahendran [4] proposed new design equations for the shear capacity of rivet fastened RHFCBs in a similar manner to those of the section moment capacity of beams subject to local buckling (Equations 1 and 2). In these equations the Direct Strength Method based nominal shear capacity (V_v) is proposed using V_{cr} (elastic buckling capacity in shear) and V_y (shear yield capacity) with a power coefficient of 0.30.

$$V_v = V_y = 0.6 \times d_1 \times t_w f_{yw} \text{ for } \frac{d_1}{t_w} \leq 0.86\sqrt{\frac{Ek_{RHFCB}}{f_{yw}}} \text{ (Shear yielding)} \tag{1}$$

$$V_v = \left[1 - 0.15\left(\frac{V_{cr}}{V_y}\right)^{0.30}\right]\left(\frac{V_{cr}}{V_y}\right)^{0.30} V_y \text{ for } \frac{d_1}{t_w} > 0.86\sqrt{\frac{Ek_{RHFCB}}{f_{yw}}} \tag{2}$$

(Elastic and inelastic shear buckling)

$$V_{cr} = \frac{k_{RHFCB}\pi^2 E d_1 t_w}{12(1-\upsilon^2)\left(\frac{d_1}{t_w}\right)^2} \tag{3}$$

$$k_{RHFCB} = k_{ss} + 0.80(k_{sf} - k_{ss}) \tag{4}$$

$$k_{ss} = 5.34 + \frac{4}{(a/d_1)^2} \text{ for } \frac{a}{d_1} \geq 1 \tag{5}$$

$$k_{ss} = 4 + \frac{5.34}{(a/d_1)^2} \text{ for } \frac{a}{d_1} < 1 \tag{6}$$

$$k_{sf} = \frac{5.34}{(a/d_1)^2} + \frac{2.31}{(a/d_1)} - 3.44 + 8.39(a/d_1) \text{ for } \frac{a}{d_1} < 1 \tag{7}$$

$$k_{sf} = 8.98 + \frac{5.61}{(a/d_1)^2} - \frac{1.99}{(a/d_1)^3} \text{ for } \frac{a}{d_1} \geq 1 \tag{8}$$

$$\lambda = \sqrt{\left(\frac{V_y}{V_{cr}}\right)} = 0.815\left(\frac{d_1}{t_w}\right)\sqrt{\left(\frac{f_{yw}}{Ek_{RHFCB}}\right)} \tag{9}$$

where k_{ss}, k_{sf} = shear buckling coefficients of plates with simple-simple and simple-fixed boundary conditions, a = shear span of web, d_1 = clear height of web, f_{yw} = web yield stress.

3. Experimental Study into the Shear Behaviour of RHFCBs with Web Openings

It is vital that key parameters are chosen carefully in the design of a test program. In order to fully understand the shear behaviour of RHFCB sections with web openings several important issues were considered when deciding these parameters such as the ratios of the depth of web openings to clear height of web (d_{wh}/d_1) and the depth of web openings to thickness of web (d_{wh}/t_w). In this experimental study, test specimens of RHFCBs with circular web openings were designed to fail in shear prior to reaching other section capacities.

All RHFCB specimens were tested using an Instron testing machine. The experimental set-up is shown in Fig. 2. The RHFCB test sections were fabricated by connecting two rectangular hollow flanges to a web plate with rivets spaced at 100 mm. Two RHFCB sections were bolted back to back using three T-shaped stiffeners and web side plates located at the end supports and the loading point in order to eliminate any torsional loading of test beams and possible web crippling of flanges and flange bearing failures. In order to simulate a primarily shear condition, relatively short test beams of span based on two aspect ratios (shear span a/ clear web height d_1) of 1.0 and 1.5 were selected. A 30 mm gap was included between the two RHFCBs sections (see Fig. 2) to allow the test beams to behave independently while remaining together to resist torsional effects. High strength steel bolts (M16 8.8/S) were used to avoid bolt failure during testing. Flanges were restrained by straps to eliminate flange distortion. The support system was designed to ensure that the test beam acted as a simply supported beam with pinned supports at each end. The measuring system was set-up to record the applied load and associated test beam deflections. Two laser displacement transducers were located on the test beam under the loading point and web panel to measure the vertical and lateral deflections, respectively (see Fig. 2). The cross-head of the testing machine was moved at a constant rate of 0.7 mm/minute until the test beam failed. Table 1 shows the RHFCB specimens tested in this experimental study. The shear force induced in each RHFCB section is equal to the applied load (P) divided by 4 in the back to back RHFCB test arrangement. In Table 1, RHFCB specimens are specified by their overall depth (d) x flange width (b_f) x flange depth (d_f) x flange thickness (t_f) x web thickness (t_w) (refer to Fig. 1). They present the measured flange and web thicknesses (t_f and t_w), clear web heights (d_1) and yield stresses (f_{yw}) of the web elements of tested rivet fastened RHFCBs. The lip length was maintained at 20 mm for all the specimens.

Fig. 2: Experimental Set-up

Fig. 3 shows the load-deflection curves for the shear test of 250×51×16×1.25×1.25 RHFCB section with 100 mm web openings (aspect ratio =1.5) while Figs. 4 (a) and (b) show the failure modes of 200×51×16×1.25×1.25 RHFCB with 60 mm and 100 mm web openings, respectively (aspect ratio = 1.5). Table 1 shows the shear capacities of RHFCBs with web openings obtained from this experimental study.

The shear tests conducted in this study did not show any unacceptable separation between the two flange lips and the web of RHFCB test specimens (see Fig. 4). There was noticeable separation of flange lips and web element only after the ultimate load was reached with test specimens retaining the torsional rigidity provided by the hollow flanges. The lack of continuity along the web to flange juncture did not appear to cause any major changes to the expected shear buckling and failure modes as seen Fig. 4. Their shear failure modes were similar to those observed for welded RHFCBs with web openings (LSBs) and included tension field action. However, the effect of rivet spacing should be further investigated using numerical studies.

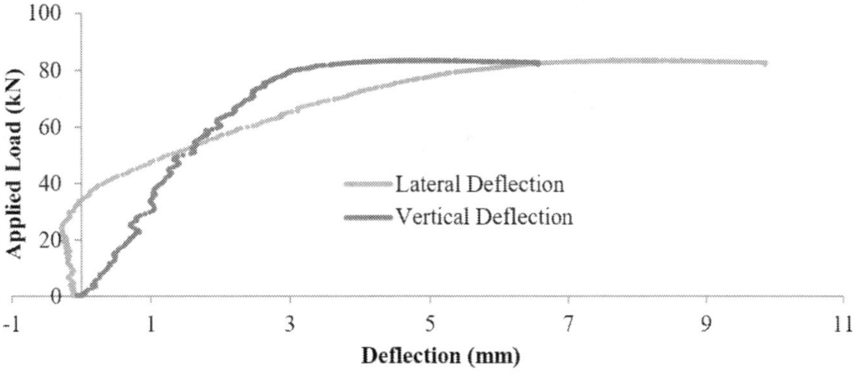

Fig. 3: Applied Mid-span Load versus Deflection for 250×51×16×1.25×1.25 RHFCB with 100 mm Web Openings (Aspect Ratio = 1.5)

Table 1: Details of Shear Tests and their Results

No.	RHFCB Section $(d \times b_f \times d_f \times t_f \times t_w)$	Aspect Ratio	f_{yw} (MPa)	t_w (mm)	d_1 (mm)	d_{wh} (mm)	d_{wh}/d_1	V_v (kN)	q_s Test	q_s Eqs. 10 to 12	q_s AS/NZS 4600(2005)
1	150×52×16×1.25×1.25	1.0	352	1.25	115.0	0	0.00	31.8	1.00	1.00	1.00
2	150×52×16×1.25×1.25	1.0	352	1.25	115.3	50	0.43	25.0	0.79	0.71	0.59
3	150×52×16×1.25×1.25	1.0	352	1.25	117.5	60	0.51	22.8	0.72	0.65	0.56
4	150×52×16×1.25×1.25	1.0	352	1.25	117.4	75	0.64	20.1	0.63	0.55	0.48
5	200×52×16×1.25×1.25	1.0	352	1.25	170.0	0	0.00	35.4	1.00	1.00	1.00
6	200×52×16×1.25×1.25	1.0	352	1.25	167.0	60	0.36	27.9	0.79	0.77	0.92
7	200×52×16×1.25×1.25	1.0	352	1.25	168.5	80	0.47	23.9	0.68	0.68	0.83
8	200×52×16×1.25×1.25	1.0	352	1.25	170.0	100	0.59	20.2	0.57	0.59	0.74
9	200×52×16×1.25×1.25	1.0	352	1.25	168.7	110	0.65	19.0	0.54	0.54	0.67
10	200×52×16×1.25×1.25	1.0	352	1.25	171.2	120	0.70	17.1	0.48	0.50	0.64
11	250×52×16×1.25×1.25	1.0	352	1.25	218.8	0	0.00	38.1	1.00	1.00	1.00
12	250×52×16×1.25×1.25	1.0	352	1.25	218.0	60	0.28	31.6	0.83	0.83	1.00
13	250×52×16×1.25×1.25	1.0	352	1.25	218.4	100	0.46	24.8	0.65	0.69	1.00
14	250×52×16×1.25×1.25	1.0	352	1.25	219.0	125	0.57	21.0	0.55	0.60	0.97
15	250×52×16×1.25×1.25	1.0	352	1.25	219.0	150	0.68	16.8	0.44	0.51	0.84
16	150×52×16×1.25×1.25	1.5	352	1.25	119.6	0	0.00	27.0	1.00	1.00	1.00
17	150×52×16×1.25×1.25	1.5	352	1.25	119.4	50	0.42	21.7	0.80	0.72	0.62
18	150×52×16×1.25×1.25	1.5	352	1.25	119.6	60	0.50	20.2	0.75	0.66	0.57
19	150×52×16×1.25×1.25	1.5	352	1.25	118.1	75	0.64	17.7	0.66	0.55	0.48
20	200×52×16×1.25×1.25	1.5	352	1.25	167.0	0	0.00	29.5	1.00	1.00	1.00
21	200×52×16×1.25×1.25	1.5	352	1.25	169.0	60	0.36	23.7	0.80	0.78	0.94
22	200×52×16×1.25×1.25	1.5	352	1.25	170.0	80	0.47	20.2	0.69	0.68	0.84
23	200×52×16×1.25×1.25	1.5	352	1.25	167.0	100	0.60	17.7	0.60	0.58	0.71
24	200×52×16×1.25×1.25	1.5	352	1.25	167.7	110	0.66	15.9	0.54	0.54	0.67
25	200×52×16×1.25×1.25	1.5	352	1.25	167.0	120	0.72	14.6	0.50	0.49	0.61
26	200×52×16×1.25×1.25	1.5	352	1.25	217.4	0	0.00	32.0	1.00	1.00	1.00
27	250×52×16×1.25×1.25	1.5	352	1.25	218.2	60	0.27	26.9	0.84	0.84	1.00
28	250×52×16×1.25×1.25	1.5	352	1.25	219.0	100	0.46	20.8	0.65	0.69	1.00
29	250×52×16×1.25×1.25	1.5	352	1.25	219.6	125	0.57	17.5	0.55	0.60	0.97
30	250×52×16×1.25×1.25	1.5	352	1.25	220.4	150	0.68	14.7	0.46	0.52	0.85

(a) Depth of Web Opening = 60 mm (b) Depth of Web Opening = 100 mm

Fig.4: Failure Modes of 200×51×16×1.25×1.25 RHFCBs with Web Openings (Aspect ratio = 1.5)

4. Proposed Equations for the Shear Capacity of RHFCBs with Web Openings

Since the preliminary studies showed that the currently available shear capacity reduction factor equations are either conservative or unsafe, new equations are proposed to predict the shear capacity of RHFCBs with web openings based on experimental results (see Table 1). It is proposed that the shear capacity of RHFCBs with web openings (V_{nl}) can be calculated using a reduction factor q_s applied to the shear capacity of RHFCBs without web openings (V_V). This approach is similar to that used in the current cold-formed steel design codes AS/NZS 4600 [2] and AISI S100 [3] and by other researchers [5,6]. The presence of web openings lead to reduced shear capacities for RHFCBs. Therefore the use of a shear capacity reduction factor (q_s) to the shear capacities of RHFCBs without web openings is considered adequate as a simple design method. Section 2 presents the shear capacity equations for RHFCBs without web openings (V_V), as Eqs. (1) to (9). Eqs. (10) to (12) show the proposed design equations for the shear capacity of RHFCBs with web openings (V_{nl}). It is proposed that Eqs. (10) to (12) can be used in combination with Eqs.(1) to (9) to predict the shear capacity of RHFCBs with web openings.

$$V_{nl} = q_s V_v \qquad d_{wh}/d_1 \le 0.70 \tag{10}$$

$$q_s = 1.0 - 0.6\left[\frac{d_{wh}}{d_1}\right] \qquad 0 < \frac{d_{wh}}{d_1} \le 0.30 \tag{11}$$

$$q_s = 1.06 - 0.8\left[\frac{d_{wh}}{d_1}\right] \qquad 0.30 < \frac{d_{wh}}{d_1} \le 0.70 \tag{12}$$

Table 1 shows the shear capacity reduction factor ($q_s = V_{nl}/V_V$) and the depth of web openings to clear height of web ratio (d_{wh}/d_1) from the shear tests conducted in this study while Figs. 5 and 6 show the non-dimensional curve of q_s versus d_{wh}/d_1. In order to assess the accuracy of the proposed design equations for the shear capacity of RHFCBs with web openings (Eqs. 10 to 12), their predictions are compared with the experimental shear capacity reduction factors in Table 1 and Fig. 5. It shows that the shear capacity reduction factors predicted by Eqs. 10 to 12 agree well with the experimental shear capacity reduction factors. The mean value of test to predicted shear capacity reduction factor ratio is 1.01 while the corresponding coefficient of variation (COV) is 0.075.

Fig. 6(a) shows the shear capacity reduction factors (q_s) of 150×51×16×1.25×1.25 RHFCBs as a function of depth of web opening to clear height of web ratio in comparison to those proposed by AS/NZS 4600 [2]. AISI S100 [3] also present the same equations for q_s as a function of d_{wh}/d_1. It shows that the AS/NZS 4600 [2] design equations are conservative for

150×51×16×1.25×1.25 RHFCB sections. Fig. 6 (b) shows the shear capacity reduction factors of 200×51×16×1.25×1.25 RHFCBs as a function of depth of web opening to clear height of web ratio in comparison to those proposed by AS/NZS 4600. It shows that the AS/NZS 4600 [2] design equations are unconservative for 200x51×16×1.25×1.25 RHFCB sections (slender sections). However, the proposed design equations are able to predict the experimental shear capacities of RHFCBs with web openings.

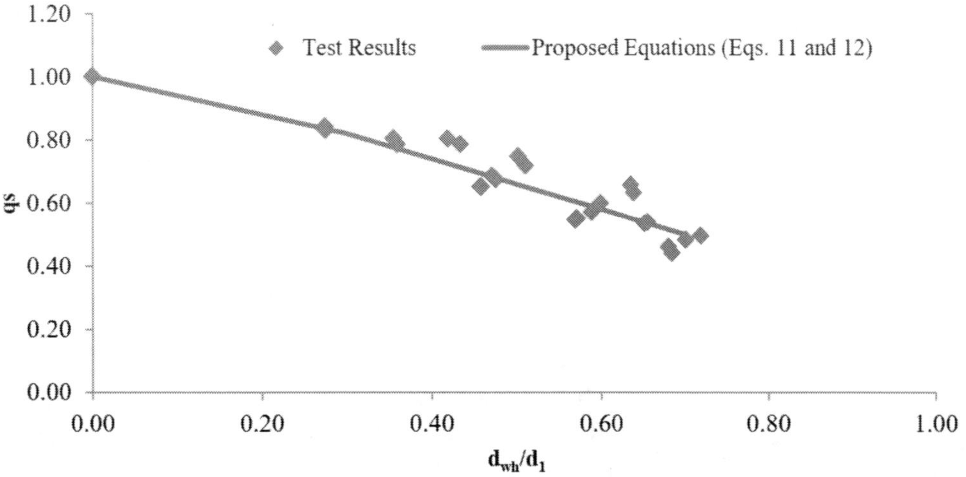

Fig. 5: Comparison of Shear Capacity Reduction Factors for RHFCBs with Proposed Equations

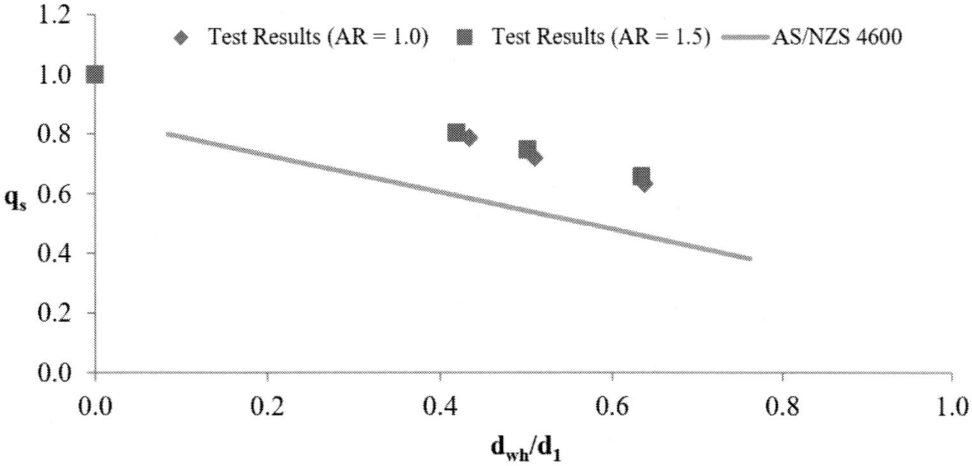

(a) 150×51×16×1.25×1.25 RHFCBs

Fig. 6: Comparison of Shear Capacity Reduction Factors of RHFCBs with AS/NZS 4600 Equations

5. Conclusions

This paper has presented the details of an experimental study of 30 shear tests into the shear behaviour of rivet-fastened rectangular hollow flange channel beams (RHFCBs). Comparison of ultimate shear capacities from tests showed that AS/NZS 4600 [2] and AISI [3] design equations are inadequate for the shear design of Rivet fastened RHFCBs with web openings. Appropriate improvements have been proposed in the form of modified shear capacity reduction factors to determine the shear capacity of RHFCBs with web openings based on experimental results. This experimental study was limited to 100 mm rivet spacing and hence further research is currently under way to investigate the effect of different rivet spacing on the shear capacities of RHFCBs with web openings.

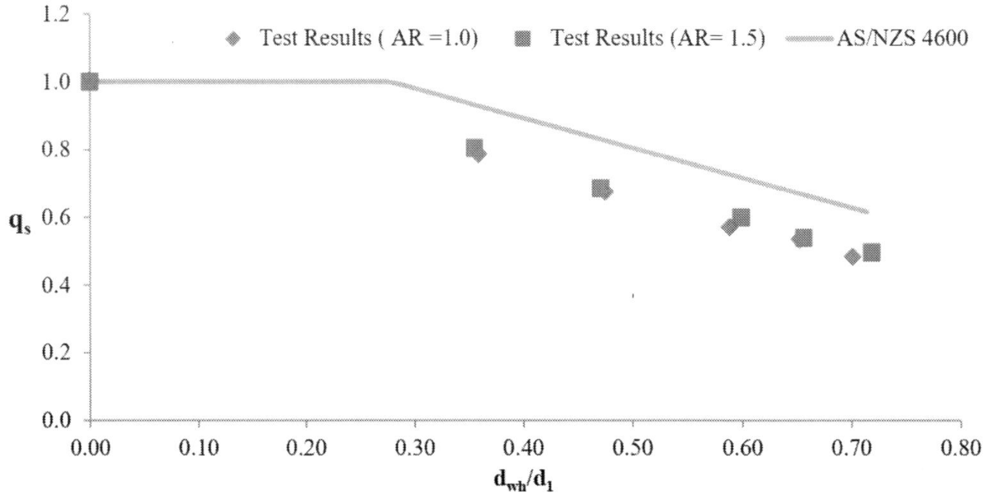

(b) 200×51×16×1.25×1.25 RHFCBs

Fig. 6: Comparison of Shear Capacity Reduction Factors of RHFCBs with AS/NZS 4600 Equations

References

[1] OneSteel Australian Tube Mills, (OATM) "Design of LiteSteel Beams", Brisbane, Australia, 2008.

[2] Standards Australia/Standards New Zealand (SA) "Australia/New Zealand Standard AS/NZS4600 Cold-Formed Steel Structures", Sydney, Australia, 2005.

[3] American Iron and Steel Institute (AISI) "North American Specification for the Design of Cold-formed Steel Structural Members", AISI, Washington, DC, USA, 2007.

[4] Keerthan P, Mahendran M. "Numerical Modelling of Rivet Fastened Rectangular Hollow Flange Channel Beam Subject to Shear", *8th International Conference on Advances in Steel Structures (ICASS'2015)*, Lisbon, 2015.

[5] Eiler MR. "Behaviour of Web Elements with Openings Subjected to Linearly Varying Shear", *Masters Thesis*, University of Missouri-Rolla, Missouri-Rolla, USA, 1997.

[6] Shan MY, LaBoube RA, Langan JE, Yu WW. "Cold-formed Steel Webs with Openings: Summary report", *Thin-Walled Structures*, Vol. 27, 79 – 84, 1997.

INFLUENCE OF THE DISTORTIONAL-LATERAL BUCKLING MODE ON THE INTERACTIVE BUCKLING OF SHORT CHANNELS

Zbigniew Kolakowski and Tomasz Kubiak

Department of Strength of Materials, Lodz University of Technology, Lodz, Poland

Abstract: The paper deals with an influence of the distortional-lateral buckling mode on the interactive buckling of thin-walled short channels section beam subjected to the bending moment. Initial imperfection and the shear lag phenomenon and distortional deformations are taken into account. A plate model (2D) is adopted for the channel section. The structure is assumed to be simply supported at the ends. A method of the modal solution to the coupled buckling problem within the Koiter's theory, using the semi-analytical method (SAM) and the transition matrix method, was applied. The calculations and experimental preliminary tests were carried out for short channels made of steel and laminate.

1. Introduction

Beams are the fundamental element in steel structures that carry loads mainly via bending. In the majority of cases, a possibility to manage comparatively high loads with thin-walled beams is limited not by their strength only but mainly by stability.

The major development of research on stability of thin-walled isotropic structures took place in the 1970s and the 1980s. Since the late 1980s, the Generalized Beam Theory (GBT) was developed extensively. At that time numerous studies employing such theories as: GBT, FSM (Finite Strip Method) and DSM (Direct Strength Method). In [1] a method for calculation of critical forces for pure bending of thin-walled beams, implemented on the basis of the Finite Strip Method (FSM), is presented, whereas decomposition of the buckling mode was conducted on the basis of assumptions taken from the GBT. This new method was referred to as the constrained Finite Strip Method (cFSM). In [2], models employed in the finite strip method are studied. The strength and local buckling of channel section and Z-section beams under bending were investigated numerically and experimentally in [3]. In [4,5], special attention was paid to the distortional-global interaction buckling.

Within the research project entitled "Experimental and numerical investigations of nonlinear stability of thin-walled composite structures" (founded by the National Centre for Sciences, Poland), experimental investigations were carried out for short composite channels subject to bending in the web plane as there is a lack of investigations devoted to the buckling analysis of such channel sections. The results pointed out especially to an effect of the global distortional-lateral buckling mode on postbuckling equilibrium paths. It was decided to explain that phenomenon in composite channels first for a simpler case referring to steel channel sections and then for composite beams under bending. An influence of the distortional-lateral buckling mode on the interactive buckling of short steel channels is presented. The results of calculations are compared to the results of experimental tests.

The concept of interactive buckling (i.e., coupled buckling) that involves the general asymptotic Koiter's theory of stability is fundamental for theoretical considerations. The nonlinear stability of thin-walled channels in the first order approximation of Koiter's theory is solved with the modified analytical-numerical method (ANM) [6] and second order postbuckling coefficients were estimated with the semi-analytical method (SAM) [7].

In the present study, a plate model (2D) of the short channel is adopted to describe all buckling modes. Instead of the finite strip method, the exact transition matrix method and the numerical method of the transition matrix using Godunov's orthogonalization is used. The interaction between all the walls of structures have been taken into account, the shear lag phenomenon and also the effect of cross-sectional distortions were included.

2. Formulation of the problem

A prismatic thin-walled channels built of plates connected along longitudinal edges were considered. The girders were simply supported at its ends. It was assumed that the material structures were made of obeyed Hooke's law.

The nonlinear problem of stability was solved with the asymptotic perturbation method. Let λ be a load factor. The displacement fields U and the sectional force fields N (Koiter's type expansion for the buckling problem) were expanded into power series with respect to the dimensionless amplitude of the r-th mode deflection ζ_r (see [6,7]):

$$\begin{aligned} U &\equiv (u,v,w) = \lambda U_0 + \zeta_r U_r + \zeta_r^2 U_{rr} + \ldots \\ N &\equiv (N_x, N_y, N_{xy}) = \lambda N_0 + \zeta_r N_r + \zeta_r^2 N_{rr} + \ldots \end{aligned} \tag{1}$$

where the prebuckling (i.e., unbending) fields are U_0, N_0, the first nonlinear order fields are U_r, N_r (eigenvalues problems) and the second nonlinear order fields – U_{rr}, N_{rr}, respectively. The range of indices is $[1, J]$, where J is the number of interacting modes.

If the structure contains the geometric imperfections in following form:

$$\bar{U} = \zeta_r^* U_r \tag{2}$$

then the equilibrium equations can be written in the following way:

$$\left(1 - \frac{M}{M_r}\right)\xi_r + a_{pqr}\xi_p\xi_q + b_{rrrr}\xi_r^3 = \frac{M}{M_r}\xi_r^* \qquad r = 1, \ldots, \mathrm{J} \tag{3}$$

where: M is a magnitude of the applied bending moment, M_r, ζ_r, ζ_r^* – the buckling moment instead of the load parameter λ_r of the r-th buckling mode, the dimensionless amplitude of the r-th buckling mode and the dimensionless amplitude of the initial deflection corresponding to the r-th buckling mode, respectively. The coefficients a_{pqr} and b_{rrrr} are described for example in [6].

On the basis of Castigliano's theorem, a relative angle of rotation of the girder in bending on the support as a function of the M/M_1 load was determined through differentiation of the expression for the potential energy corresponding to (3) with respect to M/M_1:

$$\frac{\alpha}{\alpha_1}=\frac{M}{M_1}\left[1+\frac{M_1}{Ma_0}\sum_{r=1}^{J}\frac{M_1}{M_r}a_r\zeta_r\left(0.5\zeta_r+\zeta_r^*\right)\right] \quad (4)$$

where α_1 – minimal critical angle of rotation of the beam under pure bending, corresponding to the minimal value of the critical moment M_1.

3. Analysis of the calculation results

A detailed analysis of the calculations is conducted for thin-walled steel channels of two thicknesses (see Fig. 1): $b_1 = b_3 = 40$ mm, $b_2 = 80$ mm, $l = 275$ mm, $t = 0.5$ mm and 1.2 mm, $E = 200$ GPa, $\nu = 0.3$; that are subject to bending in the web plane (i.e., the upper flange is compressed, whereas the lower one is subject to tension).

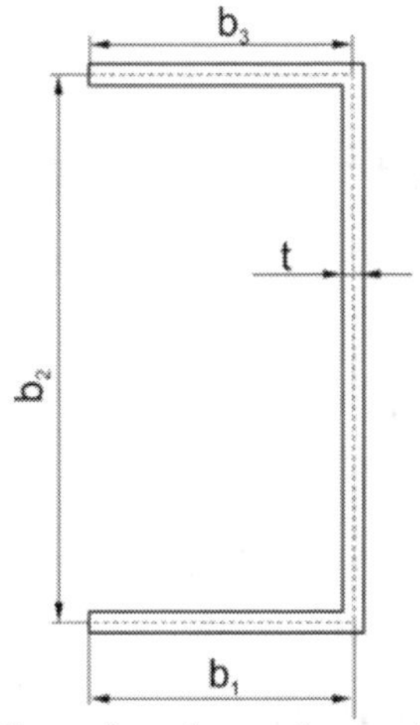

Fig. 1: Channel section under consideration

In Table 1, values of critical loads with the corresponding number of halfwaves m of buckling along the longitudinal direction of the channel for the two cases of thickness (i.e. t = 0.5 and 1.2 mm) are presented. The following r index notations are introduced: 1 – the lowest value of the critical load corresponding to the local buckling mode for $m \neq 1$, 2 – the value of the critical load corresponding to the local buckling mode for m = 1, 3 – the value of the critical load corresponding to the global buckling mode for $m = 1$. Additionally, values of critical loads of the global lateral buckling mode, corresponding to the beam model (i.e., 1D model), are given in brackets for the global load (i.e., r = 3).

Table 1: Buckling stresses σ_r for channels

t [mm]	index r	m	σ_r [MPa]
0.5	1	3	25.7
	2	1	64.0
	3	1	1782 (5463)
1.2	1	3	148.1
	2	1	360.9
	3	1	4394 (5480)

As can be easily seen, the values of global loads σ_3 for the plate model (2D) are significantly lower than for the beam model (1D). For the beams the buckling modes for r =1; 2; are distortional-local modes, whereas for r = 3 is a global distortional-lateral mode (see Fig 2).

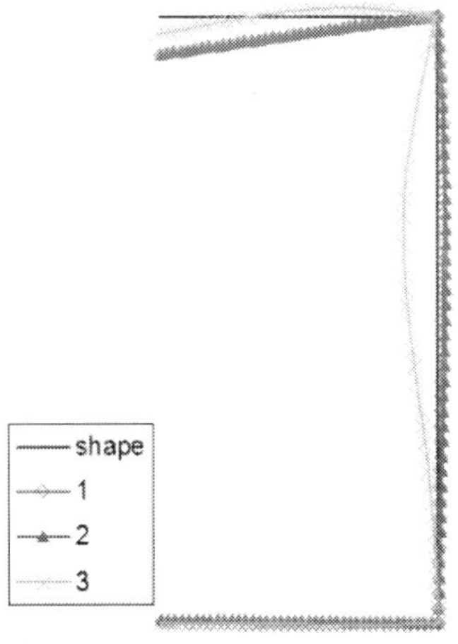

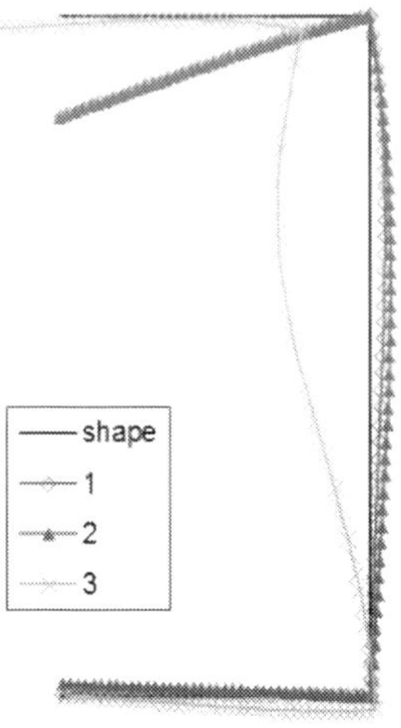

a) t = 0.5 mm b) t= 1.2 mm

Fig. 2: Buckling modes of channel section beams with different wall thickness t

Detailed calculations for the nonlinear problem were carried out as well. The postbuckling equilibrium paths for the channels under consideration were determined. The imperfections (3) assumed as:$\zeta_1^* = |0.2|$, $\zeta_2^* = |0.2|$, $\zeta_3^* = |1.0|$.

In Figs. 3 and 4, results in form of the postbuckling equilibrium path in the coordinate system representing the ratio of the bending moment to the minimal value of the critical moment $M/M_1 = \sigma/\sigma_1$ as a function of the angle of rotation of the channel on the support to the critical angle of rotation corresponding to the minimal value of the moment α/α_1, determined from relationship (4), are depicted.

Each figure shows the curves for the following approach:

- three-modal (J=3 in (3) and (4)), i.e., when an interaction of all three possible modes is considered;
- two-modal (J=2) for two cases under consideration, accounting for an interaction of two modes:
  - a) for r=1 and 2;
  - b) for r=1 and 3;
- one-modal (J=1) for r=1.

Each curve is described with the respective indices for the given buckling modes. In Fig. 3 the curves obtained for the one-, two- and three-modal approaches are plotted for beam with wall thickness t = 0.5 mm. Fig. 4 shows more complex plots of the postbuckling equilibrium paths. The curve corresponding to the one-modal approach has the highest position, whereas for the two cases of the two-modal approach, the curves are very close to each other. For $M/M_1 \leq 5$, the curve for the case (a) lies below, whereas for $M/M_1 > 5$ – above the curve representing the case (b). Attention should be paid to the fact that for the case (a) the relation of buckling stress for corresponding modes is as follow: $\sigma_2/\sigma_1 = 2.44$, whereas for the case (b) - $\sigma_3/\sigma_1 = 29.7$, respectively.

In the case of an interaction of three modes (J=3), unsteady postbuckling equilibrium paths were obtained. To presents them, a nonlinear system of equations describing the postbuckling equilibrium paths (3) was solved with two methods. The first method consisted in solving the system via controlling an increment in external load (the curve denoted as 1,2,3-L in Fig. 4), in the second one, an increment in the local displacement amplitude ζ_1 was controlled (the curve marked as 1,2,3-D in Fig. 4). One can easily notice that an interaction of two modes corresponding to the number of halfwaves $m = 1$ (i.e., $r = 2$ and $r = 3$) and the local mode ($r = 1$) makes unsteady equilibrium paths occur, which is an unexpected effect in the conducted analysis. It follows from much higher internal sectional forces, which affects directly values of the coefficient a_{pqr} in Eq. (3) than in the beam with wall thickness $t = 0.5$ mm. In the case ($t = 0.5$ mm), an effect of nonsymmetrical equilibrium paths was been observed (Fig. 3).

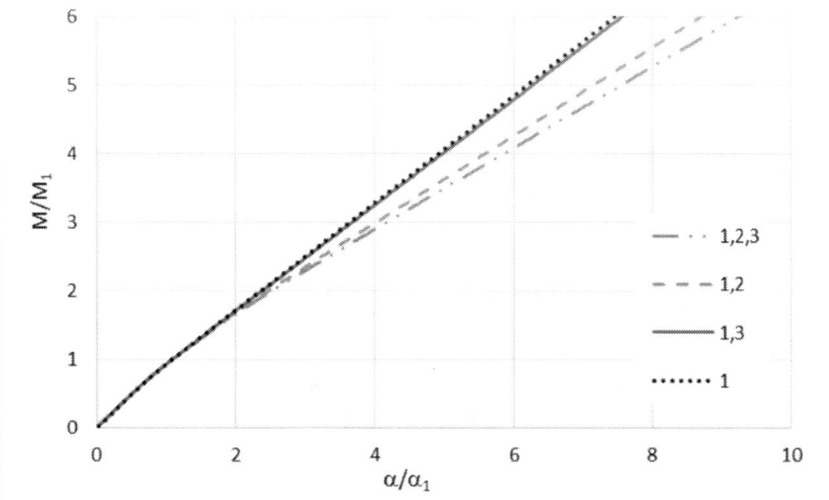

Fig. 3: Postbuckling paths for beam with wall thickness t =0.5 mm

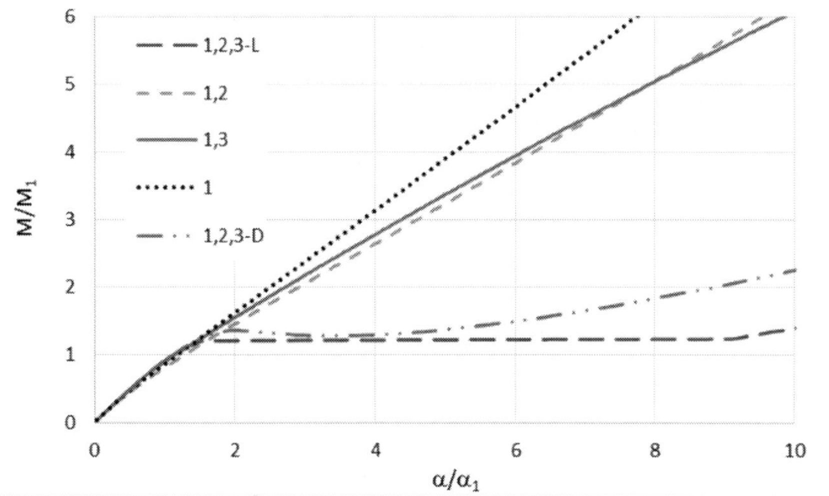

Fig. 4: Postbuckling paths for beam with wall thickness t =1.2 mm

The method employed to solve the nonlinear stability problem is based on the modal approach within the elastic range, which makes an interpretation of the attained results much easier. These issues caused that it was decided to conduct preliminary experimental investigations for the channels to confirm the assumptions taken in the theoretical considerations and the numerical models.

4. Experimental investigations

The channels under considerations were investigated experimentally. The tests were conducted on an universal testing machine, equipped with specially designed grips. The investigated steel and composite beams were subjected to pure bending. This type of load was applied in a four-point bending test. A scheme of the performed bending test with dimensions describing the span of support and the span of load is presented in Fig. 5. A special grip was designed and manufactured to avoid stress concentration in the place where the beams were supported and loaded. The grips were used to ensure the load corresponding to pure bending. The tests were performed at a constant velocity of the cross-bar equal to 1.5 mm/min.

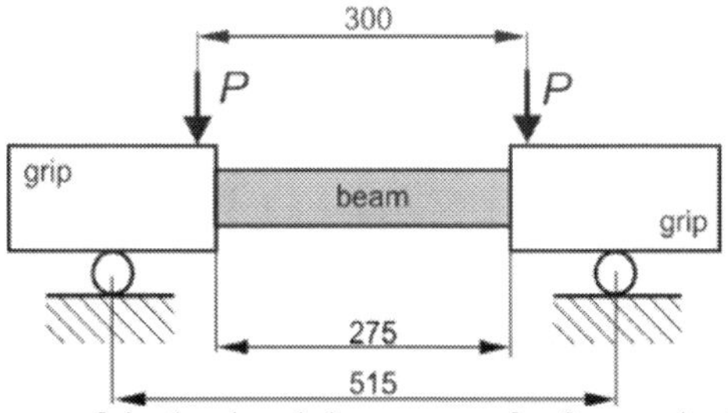

Fig. 5: Scheme of the load and the support for four-point bending test

The values of the loading force applied to the system and the displacement in the points where the load was applied were obtained directly from machine sensors.

The models were subjected to testing until failure. The values of loads and vertical displacements of the machine traverse were recorded. These quantities were then recalculated into values of the bending moment M [Nm] and the angle of rotation α [deg] on the support of the channel investigated.

4.1 Steel channels

Two models for each thickness of steel plates were made (i.e., $b_1 = b_3 = 40$ mm, $b_2 = 80$ mm, $l = 275$ mm, $t = 0.5$ mm and 1.2 mm). For each sheet, the average values of material constants were determined, namely: E = 197 GPa, σ_y = 475 MPa. The sheet plate was cut with the waterjet method, and the models of beams were cold formed.

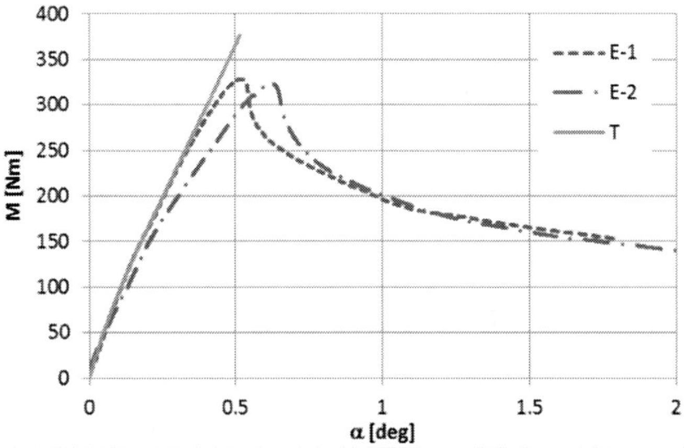

Fig. 6: Bending moment vs. angle of rotations for steel beams with $t = 0.5$ mm

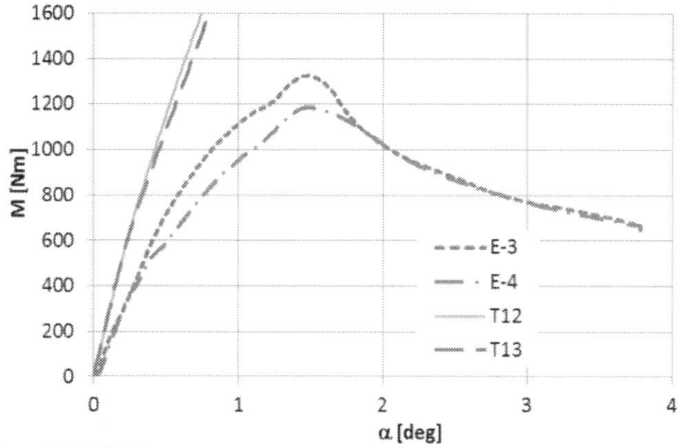

Fig. 7: Bending moment vs. angle of rotations for steel beams with t = 1.2 mm

In Figs. 6 and 7 the collective results of the experimental investigations (denoted as E-1, E-2, E-3 and E-4) of the channel section beam under consideration are shown. The results of the analytical-numerical calculations (denoted as T, T12 and T13) for the assumed values of imperfections are also presented to facilitate an interpretation of the results obtained with these two methods. The results of the theoretical calculations of the bending moment M and the angle of rotation α were transformed to the distance equal to 515 mm (Fig. 5) between the supports of the model of beam. In Fig. 6 (for the thickness of the beam walls case t = 0.5 mm), the theoretical curve for the three-modal approach agrees well with the experimentally obtained curves. In Fig. 7, only the theoretical curve obtained for the two two-modal approaches (a) and (b) is depicted. In this case (beams wall thickness t = 1.2 mm), one can see a lack of conformity of the results obtained theoretically and experimentally. The experimentally obtained curves are situated below the theoretical ones, which may indicate a strong effect of the distortional-lateral buckling mode on the interactive buckling, and, thus, on the post-buckling equilibrium paths.

4.2 Composite channels

Experimental investigations have been also performed for channel section beams made of eight layers GFRP laminate. Dimensions of composite beams under considerations were as follow: b_1=b_3= 40 mm, b_2= 80 mm, l = 275 mm, $t \approx 2$ mm (8 layers × thickness of layer c.a. 0.25 mm). The material properties of one ply have been determined according to standard and they are: E_1 = 38.5 GPa; E_2 = 8.1 GPa; ν_{12} = 0.27 and G_{12} = 1.92 GPa.

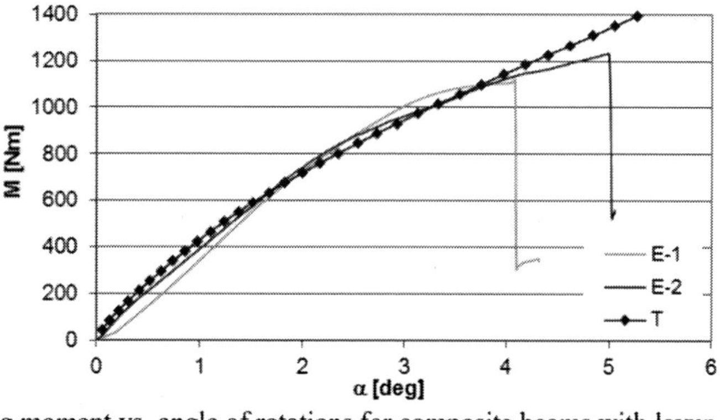

Fig. 8: Bending moment vs. angle of rotations for composite beams with layup $[0/-45/45/90]_S$

Six different symmetrical lay-ups have been consider. They were as follow: $[0/-45/45/90]_S$; $[90/-45/45/0]_S$; $[90/0/90/0]_S$; $[0/90/0/90]_S$; $[45/-45/45/-45]_S$; $[45/-45/90/0]_S$. Exemplary results presenting a comparison between theoretical calculations and experimental tests for composite beam with layer arrangement $[0/-45/45/90]_S$ are shown in Fig. 8. The rest results for composite beams will be presented during SDSS16 Conference.

5. Conclusion

An influence of the global distortional-lateral buckling mode on the interactive buckling of short channels subjected to bending in the web plane is presented. A nonlinear stability problem was solved with the SAM using Koiter's asymptotic theory in the modal description. Special attention was paid to theoretical unsteady postbuckling equilibrium paths. Experimental investigations were performed for steel and composite channels. An effect of the buckling mode for $m=1$ on the results of investigations was observed.

The theoretical and experimental investigations should be treated as an introduction to a more thorough analysis both in terms of theoretical considerations, employing the analytical-numerical methods and the FEM, as well as experimental investigations.

References

[1] Adany S, Schafer BW. "Buckling mode decomposition of single-branched open cross-section members via finite strip method", *Thin-Walled Structures*, 44, 563-584 and 585-600, 2006.

[2] Camotim D, Silvestre N. "Distortional buckling before and after Greg Hancock", *Proceedings of the 5th International Conference on Coupled Instabilities in Metal Structures,* Vol. 2, (Eds.: K. Rasmussen, T. Wilkinson), Sydney, Australia, 155-167, 2008.

[3] Yu C, Schafer BW. "Simulation of cold-formed steel beams in local and distorsional buckling with applications to the direct strength method", *Journal of Constructional Research*, 63, 581-590, 2007.

[4] Niu S, Rasmussen KJR, Fan F. "Distortional-global interaction buckling of stainless steel C-beams: Part I – Experimental investigation", *Journal of Constructional Steel Research*, 96, 127-139, 2014.

[5] Niu S, Rasmussen KJR, Fan F. "Distortional-global interaction buckling of stainless steel C-beams: Part II – Numerical study and design", *Journal of Constructional Steel Research*, 96, 40-53, 2014.

[6] Kolakowski Z, Krolak M. "Modal coupled instabilities of thin-walled composite plate and shell structures", *Composite Structures*, 76, 303-313, 2006.

[7] Kolakowski Z, Mania JR. "Semi-analytical method versus the FEM for analysis of the local post-buckling", *Composite Structures*, 97, 99-106, 2013.

TEST BASED FINITE ELEMENT DEVELOPMENT FOR DIAPHRAGM ACTION

Anita Lendvai and Attila L. Joó

Budapest University of Technology and Economics, Department of Structural Engineering

Abstract: Lightweight roof and wall structural systems are widely used for industrial type steel buildings and it has been recognised that assembled systems of profiled sheeting and purlins show strong resistance to loads acting in their own plane. In our experimental research the purpose was to determine the stiffness and strength in widely used structural configurations of trapezoidal sheeting and purlins without additional stiffeners. To improve the Eurocode formulas a series of panel experiments (18 tests altogether) were carried out where 3 m × 3 m diaphragms were examined in different configurations for static load. Based on the test series a shell finite element based modelling method has been developed in ANSYS environment to be able to follow the real behaviour.

1. Introduction

In the past few years numerous research work [7] [8] has been carried out internationally and by a research group at the Budapest University of Technology and Economics: Young Scientific Conference work [1] and diploma work [2] as well, in order to amend the inaccuracy of the existing Eurocode formulas calculating the stressed skin effect, the stiffening effect of diaphragms built up of purlins and trapezoidal sheets [3] [4] [5]. The inaccuracy might cause underestimation of the whole building's stiffness in earthquake related areas [6].

In order to be able to examine the diaphragm stiffness first a test series has been carried out from which we could get exact panel stiffness data for each specimen, as well as the load bearing capacity and failure types. Our aim was to develop and verify a shell finite element based numerical model in ANSYS upon these results, so we will be able to follow the real behaviour of the diaphragm, and extend the results to any different, non-examined section sizes which might be relevant in engineering practice.

2. The test series

2.1 The test set-up

The 3-dimensional view of the test set up can be seen on Fig. 1, the loading frame is built up from horizontal HEA180 beams fixed to the vertical SHS columns. Between the SHS columns 2U160 bracings have been applied in order to provide a stable loading frame to which we have built up the diaphragms. The purlins were fixed to the HEA beams, and the trapezoidal sheets were fixed straight to the purlins. There were no additional bracing applied in the diaphragms. A static horizontal load has been applied by a hydraulic jack to the right side of the bottom HEA180 beam until the total failure mechanism of the diaphragm.

The overall size of the test series is also included in Fig. 1. The distance of the horizontal HEA180 beams is 3 m, to which the vertical purlins has been fixed in 1.5 metres distance. The overall size of the trapezoidal sheet is 3 m × 3 m. Measured parameters were the horizontal deflection and the horizontal force loaded by the hydraulic jacks, besides that the failure types of the panels has been examined during the tests.

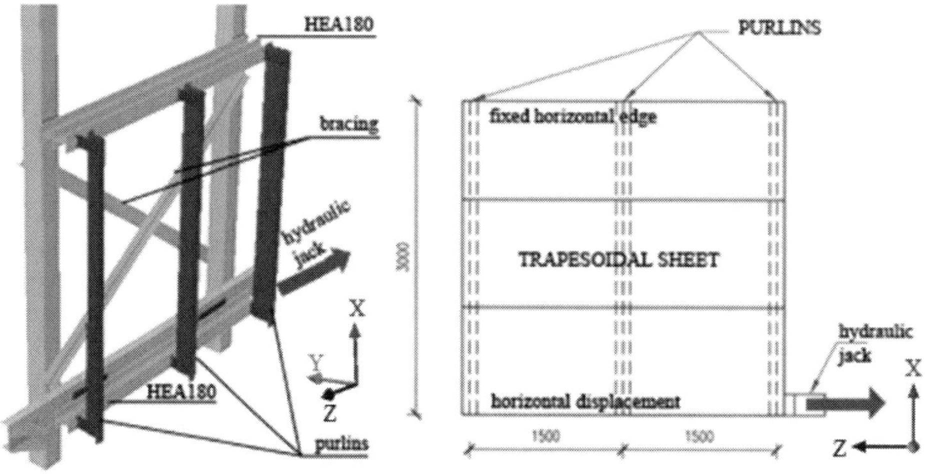

Fig. 1: Test set-up 3D and front view

2.2 Execution of test series

During the design of the test series the most important parameters were the examination of the effect of the purlin's size and trapezoidal sheet's profile height, the number of fixings and the effect of sheeting used only the outer or both side of the purlins (inner and outer side).

For this reason we have planned 18 different configurations, the below structural elements have been used in our test series:

- trapezoidal sheets: LTP20/0.5; LTP45/0.5 (see Fig. 2 for sheet profiles);
- purlins: Z200/1.5; Z250/2.0; Z300/2.0 (see Fig. 2 for purlin profiles);
- number of fixings: in every through or in every alternate through;
- usage of trapezoidal sheeting: only one side (outside) or both sides of the purlins (sheets fixed only to purlins, no fixings applied between two sheets).

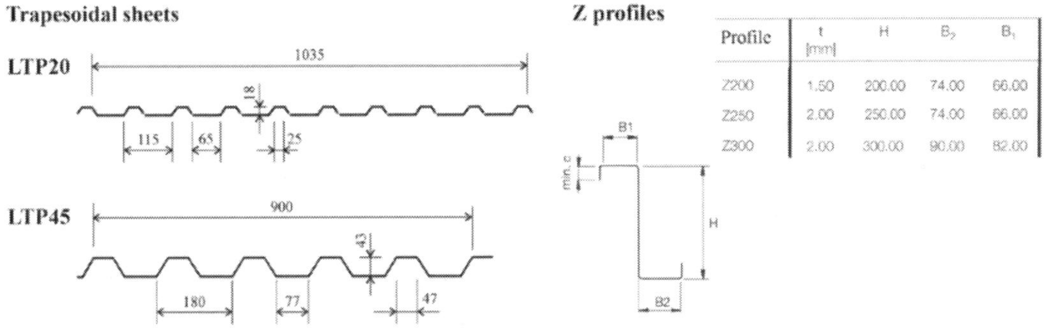

Profile	t [mm]	H	B_2	B_1
Z200	1.50	200.00	74.00	66.00
Z250	2.00	250.00	74.00	66.00
Z300	2.00	300.00	90.00	82.00

Fig. 2: Lindab trapezoidal sheet and purlin profiles applied in the test series

3. Evaluation of test results

3.1 Method of evaluation

During the test series the horizontal displacements of the bottom HEA beam has been measured, and the corresponding force values. These data ranges have been illustrated on a typical force-displacement diagram.

In Fig. 3 the test results of the panel experiment No. 7 (see Table 1) can be seen: in this configuration LTP20 sheets outside has been fixed to the Z250 purlins in every alternate through. The panel has been loaded 8-times, in which the load has been increased in every load-step until the failure of panels (exceeding the plastic range). In Fig. 4 the appearance of local failure types are marked, and according to the thick trend line the stiffness of the panels has been determined. Types of failures are shown in Fig. 4.

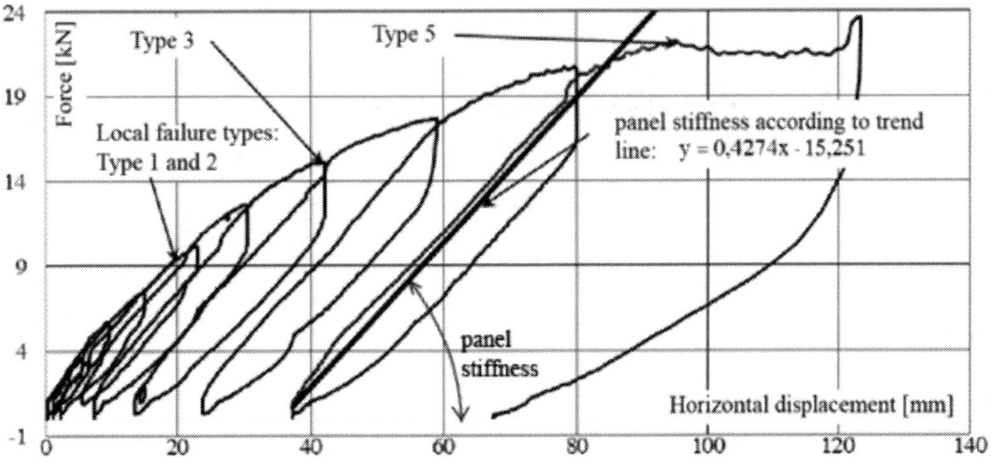

Fig. 3: Diagram of the panel test No. 7

During the panel experiments we have defined the typical local failure types with description and photo documentation, which will be presented in detail in the next chapter.

3.2 Typical failure types

The failure types have been numbered in order to later references. In most cases the combination of failure types has been determined, see Fig. 4 for pictures of the failure modes.

The bottom line of the panels has been subjected to an increasing horizontal load, and in most cases the failure of the panel has been started with the increase of the opening between the attached sheets, after – depending on panel strength – other local failure types has been appeared. In stronger panel constructions – for example in panels with fixings in every through – another typical local failure type appeared: distortional buckling of the purlin's upper end. During the tests we have experienced horizontal and vertical curvature of the sheet edge as well. Typically the end failure of the panel was caused by hole elongation around the self-drilling screws. As a conclusion, depending on the stiffness of the specified construction, the local failure type started with either the purlin's distortional buckling or the increasing opening between the sheets, and the horizontal or vertical distortional of sheet edges. The end failure type was mostly the hole elongation around fixings. The failure of the self-drilling has not occurred during testing.

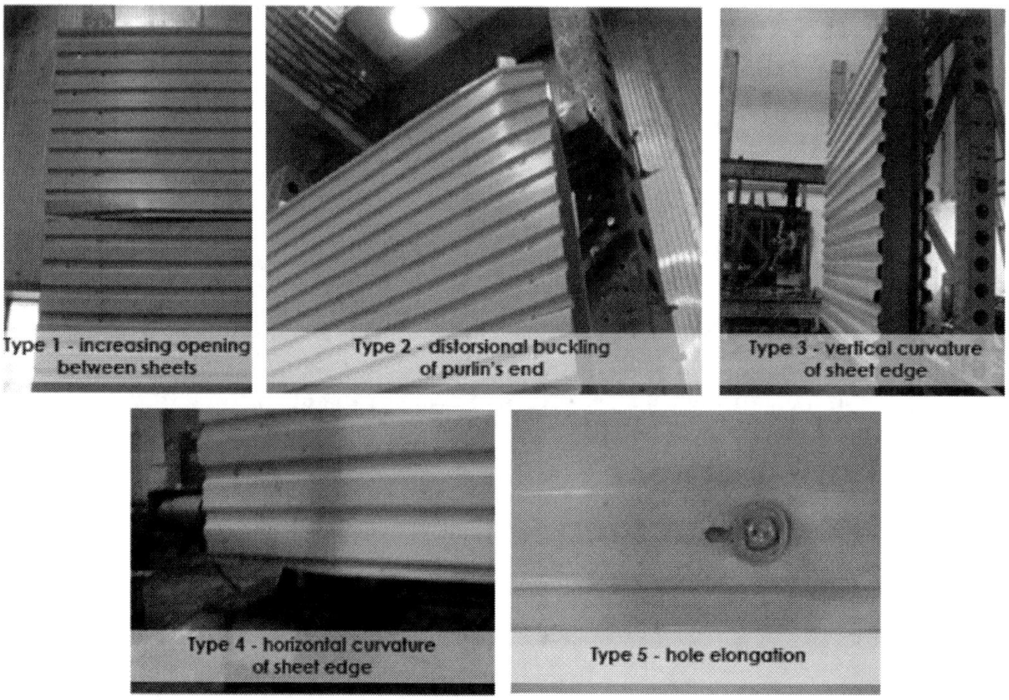

Fig. 4: Typical failure types from left to right: increasing opening between sheets, distortional buckling of purlins, horizontal- and vertical curvature of sheet edges, hole elongation

3.3 Panel stiffnesses

In Table 1 we have summarized the load carrying capacities and calculated panel stiffnesses, which are based on the trend line, shown in Fig. 3.

In general the conclusion is that for the most stiff constructions, including double sided sheets (inner and outer side fixed sheets), and the number of fixing has a great effect on panel stiffness. For example in the tests made of the smallest purlin section and trapezoidal sheets (see Table 1., test No. 1-3), the simple and in every through fixed construction (No. 1) is still stiffer than the construction made of double sheets, but which are fixed only in every alternate through.

Table 1: Load carrying capacities and stiffnesses from the test series

Test No.	Purlins	Trapezoidal sheets	Number of fixings	Application of sheeting	Panel stiffness (kN/mm)
1	Z200/1.5	LTP20/0.5	In every through	Single	0.4327
2			In every alternate through	Single	0.1875
3				Double	0.3812
4		LTP45/0.5	In every through	Single	0.2281
5			In every alternate through	Single	0.1459
6				Double	0.2826
7	Z250/2.0	LTP20/0.5	In every through	Single	0.4274
8			In every alternate through	Single	0.3605
9				Double	0.5688
10		LTP45/0.5	In every through	Single	0.3187
11			In every alternate through	Single	0.2056
12				Double	0.2945
13	Z300/2.0	LTP20/0.5	In every through	Single	0.4086
14			In every alternate through	Single	0.3842
15				Double	0.5947
16		LTP45/0.5	In every through	Single	0.3800
17			In every alternate through	Single	0.1967
18				Double	0.3260

4. Finite element analysis

4.1 Finite element model

In order to examine the stiffness of the diaphragms, a non-linear finite element analysis has been performed using ANSYS. The frame itself with the HEA180 beams and the different purlins have been modelled by the use of SOLID187 element type, the trapezoidal sheets were modelled by SHELL181 element type. On the top HEA180 beam fixed supports have been applied, while on the bottom HEA180 beam fixed support have been applied in axis 'X' and 'Y', while free displacement has been allowed in axis 'Z'.

The complexity of the semi-rigid connection due to self-drilling screws between the purlins and sheeting has been established by laboratory tests, where the stiffness of a connection were determined by tension tests on the purlin and sheeting materials and self-drilling screws. From these laboratory tests we could achieve the connection's stiffness, which has been used for BUSHING JOINT element type, which has six degrees of freedom, three translations and three rotations, all of which can potentially be characterized by their rotational and translational degrees of freedom as being free or constrained by stiffness. Depending on the type of the diaphragm each node in the middle of the trapezoidal sheet's through has been connected to the purlin's flange node.

The purlin and hot rolled steel connections are modelled with CONTA178 elements, which represent contact and sliding between any two nodes of any types of elements). The horizontal load has been applied on the edge of the hot rolled HEA180 section with SURF156 element (3-D Structural Surface Line Load Effect).

Table 2: Properties of the finite elements used, see Fig. 1 for x, y, z axes

	SHELL181	SOLID187	CONTA178
Description	Four-node element with six degrees of freedom at each node	3-D 10-Node Tetrahedral Structural Solid	Two node element with three degrees of freedom at each node with translations in the X, Y, and Z directions
Number of nodes	4	10	2
Degrees of freedom	x, y, and z translational and rotational displacements	x, y, and z translational displacements	x, y, and z translational displacements

For the material properties we have applied 235 MPa yield stress for the hot rolled steel sections, 350 MPa yield stress for the trapezoidal sheets, and 360 yield stress for the purlins. The applied Young modulus is 210000 MPa.

For the examination of panel stiffnesses' linear elastic material model has been used. The entire assembly and the details along with meshing is shown in Fig. 5. The load is horizontally applied by steps on the side of the bottom HEA180 beam, and the deflection of the bottom HEA180 beam has been monitored besides the local failure types in comparison with the laboratory results. As elastic range has been examined, imperfections are not included which have small effect on panel stiffness.

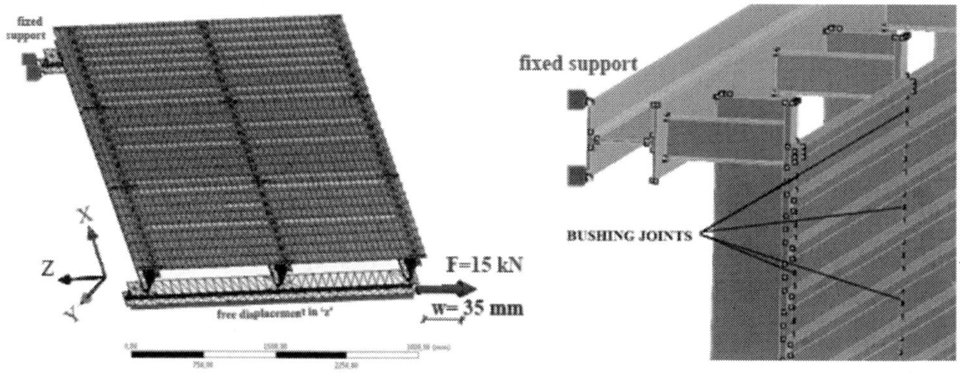

Fig. 5: Meshed diaphragm (No. 1 test set up)

4.2 Model behavior

During numerical modelling the typical model behaviour in the elastic range has been examined. Compared to our laboratory experiments we have experienced all the failure types described in 3.2 except the hole elongation (see Fig. 6), which is a typical plastic failure type and needs more detailed modelling of the surroundings of the self-drilling screws. Therefore further examination will be executed to define the load bearing capacity of the sheeting. So we concentrated on the stiffness comparison with the test results.

In stiffer configurations we have experienced Type 1 and 2 behaviour modes in Fig. 4 (increasing opening between sheets, and distortional buckling of purlin's end), though in less stiff diaphragms Type 3 and 4 has been experienced (vertical and horizontal curvature of sheets edges), similarly to laboratory experiments.

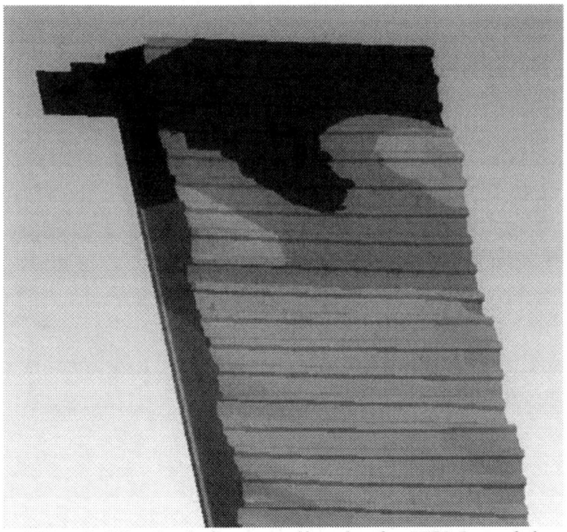

Fig. 6: Vertical sheet curling in the FEM model

4.3 Comparison of the stiffnesses of the experimental tests and FEA

In Fig. 7 the results of test No. 2 can be seen, where the panel stiffnesses in elastic range have been compared. The laboratory test result has given the result of 0.1875 kN/mm stiffness, compared to the ANSYS numerical model value, which is 0.1807 kN/mm. The difference between the two results is ~4%, which is a reasonable difference in respect of the complex behavior. All other models showed similar behavior.

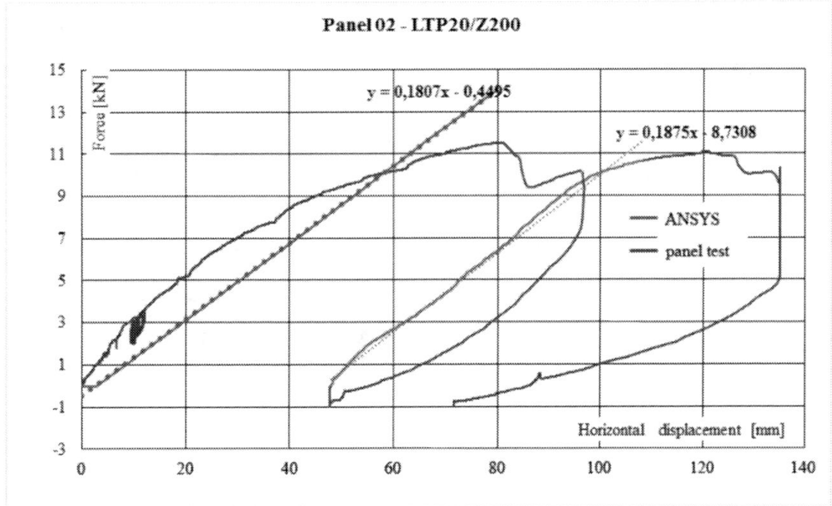

Fig. 7: Comparison of diaphragm stiffnesses: ANSYS numerical model and laboratory test

5. Summary

In our panel experimental test series and FE-modelling of them we have got experiences as follows: the typical failure types in FE-model showed similar types as in tests: firstly the deformation of the sheets, and secondly the distortional mode of the purlin ends. The failure of the self-drilling screws and its surroundings has not been modelled in any cases. In our numerical analyses the calculated stiffness has been verified with the laboratory tests.

Summarizing the panel stiffness results the stiffest constructions were the configurations with double side applied sheets in higher profiled purlins configurations (Z250 and Z300). During application of small profiled sheet and purlin this effect is not valid (combining Z200 and LTP20), in this case the single sides sheets in the every through fixed version was the stiffest construction.

In general it can be determined that the application of smaller profiled sheets (LTP20) increases the panel stiffness, compared to the higher profiled ones (LTP45). This is valid for all the tests.

After verification of ANSYS model in the elastic range our next aim is to improve the numerical model in plastic range around the self-drilling screws to be able to follow the diaphragm's real behaviour, during which the model will be verified upon the diaphragm's load bearing capacity. After the verification we are planning to generalize the test series, in which by the help of virtual tests we can determine the panel stiffness and load bearing capacity for other configurations as well.

References

[1] Kenéz Á, Rácz A. "Stiffening effect of trapezoidal sheetings on steel halls", *Young Scientific Conference* (in Hungarian), supervisors: L. Dunai, A. L. Joó, BME Department of Structural Engineering, 2014.

[2] Schaul P. "Stiffening effect of trapezoidal sheetings on the behaviour of steel halls", *MSc. diplomawork* (in Hungarian), supervisor: A. L. Joó, I. Kotormán, BME Department of Structural Engineering, 2013.

[3] Yiu PKA. "BRANZ study report. Profiled sheet steel claddings as diaphragms – General review", *Building Research Association*, New Zealand, 1987.

[4] Luttrell LD. *Strength and behaviour of light-gage steel shear diaphragms*, Cornell University, Ithaca, New York, 1967.

[5] Davies JM. *Developments in stressed skin design*, The University of Manchester, Manchaster, UK, 2007.

[6] Höglund T. *Stabilisation by stressed skin diaphragm action*, Swedish Institute of Steel Construction, Stockholm, 2002.

[7] Wrzeslen A., Lim JBP, Xu Y, MacLeod IA, Lawson RM. "Effects of stressed skin action on the behaviour of cold-formed steel portal frames", *Engineering Structures*, 105(2015), 123-136.

[8] Ilveskoski O. *Stabilization of steel structures by sandwich panels*, HAMK University of Applied Sciences, Hämeenlinna, Finland, 2015.

MECHANICS OF THE LOCAL-DISTORTIONAL INTERACTION IN FIXED-ENDED LIPPED CHANNEL COLUMNS

André Dias Martins, Dinar Camotim and Pedro Borges Dinis

CEris, ICIST, DECivil, Instituto Superior Técnico, University of Lisbon, Portugal

Abstract: This paper addresses an investigation on the post-buckling response of fixed-ended thin-walled lipped channel columns experiencing local-distortional (L-D) interaction. The results presented and discussed are obtained through GBT analyses and provide the evolution, along given equilibrium paths, of the column deformed configuration (expressed in modal form) and relevant displacement profiles. Particular attention is devoted to interpreting the quantitative and qualitative differences exhibited by the column post-buckling behaviours associated with the occurrence of L-D interaction due to either the closeness between the L and D critical buckling stresses ("true interaction") or a secondary bifurcation phenomenon ("secondary-bifurcation interaction"). Moreover, a comparison with similar results recently obtained for simply supported lipped channel columns is also presented.

1. Introduction

Traditionally, the non-linear behaviour of prismatic thin-walled members (*e.g.*, cold-formed steel ones) could only be rigorously assessed by resorting either (i) to time-consuming and computation-intensive shell finite element simulations or, to a smaller extent, (ii) to spline finite strip analyses, both providing outputs not easy to apprehend/interpret. However, in the last few years Generalised Beam Theory (GBT) has emerged as a very promising alternative to obtain similarly accurate results in a much more efficient and (mostly) clarifying manner (only a few structurally meaningful d.o.f. are required), mainly due to the genuine geometrically non-linear GBT formulation, valid in the large deformation range, developed by Silvestre and Camotim [1]. By performing GBT geometrically non-linear imperfect analyses (GNIA) of prismatic thin-walled members, it becomes possible to unveil and quantify the contributions of the various deformation modes to the member structural response under consideration. This feature enables the acquisition of much deeper insight on that response, thus providing a much clearer picture of the mechanical aspects involved – *i.e.*, makes GBT GNIA an ideally suited numerical tool to investigate complex coupled problems, like the one addressed in this work – the local-distortional interactive behaviour of thin-walled lipped channel columns.

As far as isotropic thin-walled members affected by local-distortional (L-D) interaction are concerned, the only GBT post-buckling results available were obtained for simply supported lipped channel (LC) columns by Silvestre and Camotim [2] and, recently, Martins *et al.* [3]. However, fixed-ended columns exhibit a much higher distortional post-critical strength reserve than their simply supported counterparts, thus leading to more pronounced and relevant L-D interaction effects. Therefore, the aim of this work is to present and discuss GBT-based numerical results concerning the geometrically non-linear response of fixed-ended thin-walled LC columns experiencing L-D interaction, in order to shed fresh light on the mechanics underlying this coupling phenomenon – in particular, it will be possible to assess the marked differences between the interactive behaviours of simply supported and fixed-ended LC columns. The results presented and discussed were obtained through GBT elastic post-buckling analyses and provide the evolution, along given equilibrium paths, of the column deformed configuration (expressed in modal form) and relevant displacement profiles. Particular attention is devoted to interpreting/explaining the quantitative and qualitative differences exhibited by the column post-buckling behaviours associated with the occurrence L-D interaction due to either the closeness between the L and D critical buckling stresses ("true interaction") or to a secondary bifurcation phenomenon ("secondary-bifurcation interaction") – *e.g.*, [4].

2. GBT buckling behaviour

In order to investigate the mechanics of the geometrically non-linear response of thin-walled columns experiencing L-D interaction, it is indispensable to begin by selecting column geometries (cross-section dimensions and length) with "appropriate" (L and D) critical buckling loads. In order to shed new light on this coupling behaviour, three fixed-ended columns exhibiting it are considered, namely (i) one experiencing "true L-D interaction", with $R_{DL}=P_{crD}/P_{crL}\approx1.00$ (P_{crD} and P_{crL} are the critical D and L buckling loads), and (ii) two affected by "secondary-bifurcation L-D interaction", (ii$_1$) one exhibiting a "secondary L bifurcation" ($R_{DL}\approx0.60$) and (ii$_2$) the other a "secondary D bifurcation" ($R_{DL}\approx1.60$). Table 1 shows the cross-section dimensions (b_w, b_f, b_l, t – web-flange-lip widths and wall thickness) of the three selected steel columns (E=210GPa, ν=0.3). The curves in Figs. 1(a)-(c) provide the variation of P_{cr} (critical buckling load) with the column length L (logarithmic scale) for the fixed-ended LP columns. Marked on these curves are the selected lengths adopted to ensure the desired

Table 1: Fixed-ended LP column selected geometries, local, distortional and global buckling loads and buckling mode critical half-wave numbers (dimensions in mm and loads in kN)

Column	b_w	b_f	b_l	t	L	P_{crD}	P_{crL}	P_{crG}	R_{DL}
LP1	120	110	10	2.50	800	200.6(1)	333.3(6)	2932.4(1)	0.60
LP2	120	110	10	1.40	900	57.1(1)	59.0(7)	1636.9(1)	1.00
LP3	120	80	10	1.10	800	44.7(1)	27.9(8)	1269.2(1)	1.60

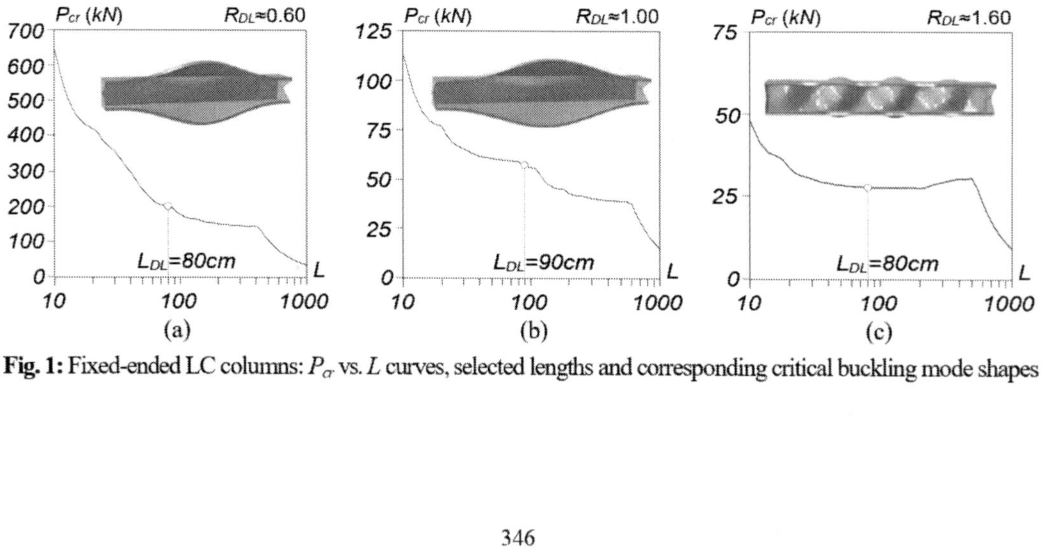

Fig. 1: Fixed-ended LC columns: P_{cr} vs. L curves, selected lengths and corresponding critical buckling mode shapes

interaction "levels". The critical L and D buckling loads and the corresponding buckling mode half-wave numbers (n_L and n_D) are also given in Table 1 (the buckling mode shapes are depicted in Figs. 1(a)-(c)). A possible interaction with global buckling is avoided in these columns by ensuring that the corresponding critical buckling loads are much higher than their L and D counterparts.

3. GBT post-buckling behaviour under L-D interaction

The previously selected columns are used to investigate the influence of L-D interaction on their structural response, by means of GBT GNIA analyses, with the aim of shedding fresh light on the mechanical aspects underlying the post-buckling behaviour of fixed-ended columns (with warping-prevented end sections). The results concerning the column experiencing "true L-D interaction" are addressed first (Section 3.1), followed by those dealing with the columns affected by "secondary-bifurcation L-D interaction" (Section 3.2). All the columns analysed contain (i) critical-mode initial geometrical imperfections with small amplitudes (10% of the wall thickness t), obtained through a preliminary buckling analysis, and (ii) are longitudinally discretised into 20 GBT-beam finite elements[1]. Fig. 2 shows the GBT discretisation adopted, which comprises 17 (6 natural and 11 intermediate) nodes and leads to (i) 17 conventional deformation modes (4 global, 2 distortional and 11 local), (ii) 14 shear modes and (iii) 14 transverse extension modes. The main features of the most relevant ones (5 conventional, 2 shear and 1 transverse extension – numbered sequentially in this order), *i.e.*, with significant contributions to the member structural response, are also displayed. The GBT results are compared with values yielded by shell finite element analyses (SFEA) carried out in ABAQUS [5], employing models already adopted in previous L-D interaction studies [4].

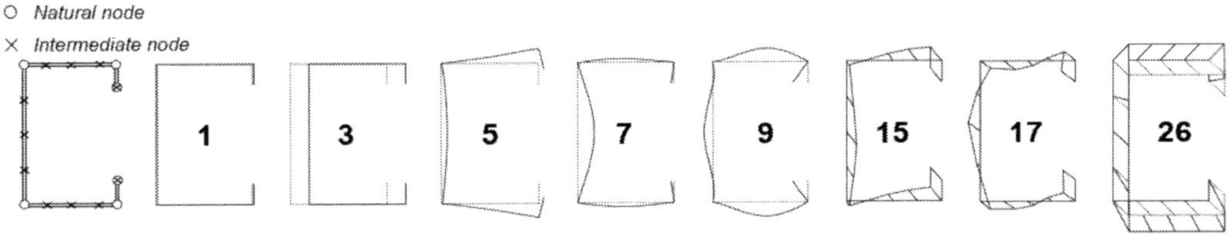

Fig. 2: Lipped channel (i) in-plane deformed configurations of the most relevant conventional and transverse extension modes, and (ii) warping displacements of the most relevant shear modes

3.1 True L-D interaction

Fig. 3(a)-(b) display the L-D post-buckling equilibrium paths P/P_{cr} vs. v/t (v is the mid-span top flange-lip corner displacement parallel to the web caused by the applied load, *i.e.*, excluding the value stemming from the initial geometrical imperfection) of two fixed-ended LP columns differing in the shape of the initial geometrical imperfections considered. The columns contain "pure" D (Fig. 3(a)) and L (Fig. 3(b)) initial geometrical imperfections, with the former involving outward flange-lip motions (as depicted in Fig. 1(b)) – they are hereafter termed R_{DL}=1.00+D and R_{DL}=1.00+L column, respectively. The preliminary GBT buckling analysis yields the modal participation factors of the two above buckling modes: (i) **5**-88%, **7**-10%, **9**-2% (D critical mode with one inner half-wave and two outer quarter-waves) and (ii) **5**-9%, **7**-82%, **9**-7% (L critical mode with seven half-waves). The relevance of the different deformation modes for the column structural response is assessed through the performance of several "approximate analyses", namely those including the deformation mode sets (i) **1+5**, (ii) **1+3+5**, (iii) **1+3+5+7+9**, and (iv) **1+3+5+7+9+15+17+26**. The curves in Figs. 4(a)-(c) provide the evolution, as P/P_{cr} increases from P/P_{cr}=0.75, of the mid-web transverse displacement ($w(x)$) profiles due to the contributions of modes **5**, **7+9** and **3+5+7+9+26** (Figs. 4(a)-(c) - $w_5(x)$, $w_{7+9}(x)$ and $w_{3+5+7+9+26}(x)$≡$w(x)$), for the two imperfect columns - Figs. 4(a_1)-(c_1)

[1] Similar results were obtained employing 30 GBT-beam finite elements – differences below 1%.

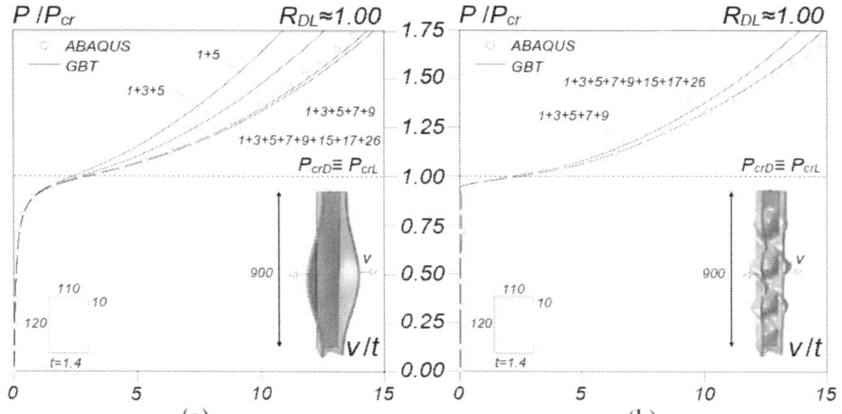

(a) (b)

Fig. 3: GBT post-buckling equilibrium paths of the lipped channel column undergoing true L-D interaction and obtained with different deformation mode sets and for (a) D and (b) L initial geometric imperfections

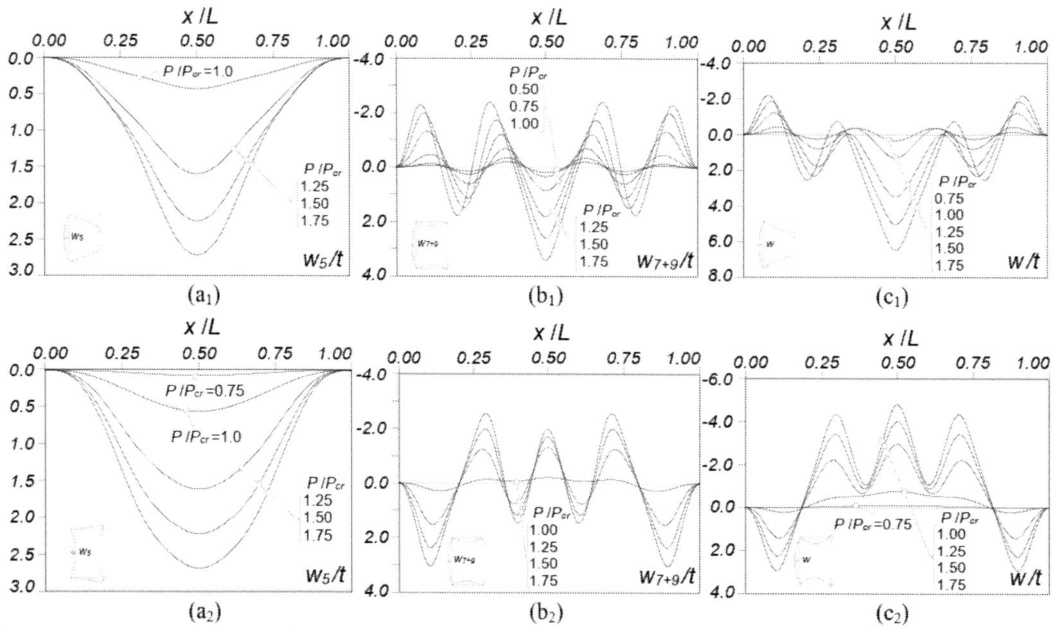

Fig. 4: Evolution of the web transverse bending displacement profiles due to deformation modes (a) **5** ($w_5(x)$), (b) **7+9** ($w_{7+9}(x)$) and (c) **3+5+7+9+26** ($w_{3+5+7+9+26}(x)\equiv w(x)$), for the ($_1$) R_{DL}=1.00+L and ($_2$) R_{DL}=1.00+D columns

and Figs. 4(a$_2$)-(c$_2$) concern the R_{DL}=1.00+D and R_{DL}=1.00+L columns, respectively. On the other hand, Fig. 5 shows column deformed configurations at several advanced loading/post-buckling stages, namely P/P_{cr}=1.00; 1.25; 1.50; 1.75, for the R_{DL}=1.00+L column. The analysis of all these post-buckling results prompts the following remarks:

(i) The GBT equilibrium paths obtained with the **1+3+5+7+9+15+17+26** deformation modes virtually coincides with the SFEA ("exact") ones up to P/P_{cr}=1.75, for both imperfect columns (see Figs. 3(a)-(b))[2] – note that the shear and transverse extension modes are significantly less relevant than in the simply supported columns recently analysed by the authors [3].

[2] Rigorously speaking, the virtual coincidence between the GBT and SFEA equilibrium paths only occurs up to around P/P_{cr}=1.50. For higher applied loads, the two pairs of equilibrium paths start to diverge slightly, even if the deformed configurations remain practically identical. Measures to eliminate this slight divergence are currently under investigation.

(ii) Recalling that the L critical mode exhibits seven half-waves, it is just logical to expect that the mid-web transverse displacement profile $w_{7+9}(x)$ of the R_{DL}=1.00+L column retains this shape as loading progresses (see Fig. 4(b_1)) – indeed, it is just an amplification of the initial imperfecion component $\bar{\phi}_{7+9}(x)$. Conversely, the fact that, at P/P_{cr}=1.0, the mid-web transverse displacement profile $w_5(x)$ has a shape akin to the column critical D buckling mode (one inner half-wave and two outer quarter-waves), as shown in Fig. 4(a_1), constitutes clear evidence of L-D interaction, since the initial imperfecion component $\bar{\phi}_5(x)$ has seven half-waves – $w_5(x)$ gradually moves away from this shape as loading increases. The "exact" mid-web transverse displacement profile $w_{3+5+7+9+26}(x)$≡$w(x)$ (see Fig. 4(c_1)), just reflects the combination of these two tendencies. Similar observations can be made for the R_{DL}=1.00+D column, by switching the roles played by $w_5(x)$ and $w_{7+9}(x)$. Indeed, the shape of $w_5(x)$ always remains akin to the critical D buckling mode (see Fig. 4(a_2)), while that of $w_{7+9}(x)$ moves gradually towards seven half-waves (see Fig. 4(b_2)), again providing clear evidence of L-D interaction. This gradual emergence of deformation patterns akin to the L and D buckling modes at early loading stages, regardless of the initial geometrical imperfection shape, characterises the "true L-D interaction". Naturally, this coupling phenomenon also occurs in elastic-plastic columns, provided, that the squash load P_y is not well below P_{cr}. Fig. 5, concerning the R_{DL}=1.00+L column, clearly shows the occurrence of this coupling phenomenon, which is already visible at P/P_{cr}=1.00.

(iii) Figs. 4(a_1)-(c_1) shows that the R_{DL}=1.00+L column displays *inward* mid-span flange-lip motions (see also Fig. 5). This feature, which is a bit surprising (the column is more flexible w.r.t. *outward* motions), stems from the fact that the local deformations at mid-span are associated with outward web bending motions, thus "forcing" the distortional deformations to follow that same trend. It is wort mentioning that a similar behavioural feature was observed in simply supported columns [4], a context in which it is explained by the presence of minor-axis bending caused by the effective centroid shift stemming from the stress redistribution due to local buckling – an explantion not holding for fixed-ended columns (there are no effectice centroid shift effects).

(iv) Figs. 4(c_1)-(c_2) show the web transverse displacement longitudinal profiles $w(x)$≡$w_{3+5+7+9+26}(x)$ and provide clearly evidence that the L-D interaction effects are much more relevant in these fixed-ended columns than in the similar simply supported columns analysed in [3], which is due to the much higher distortional post-critical strength reserve obtained by fixing the end sections. For instance, at P/P_{cr}=1.25 both the L and D deformations are clearly visible (see Fig. 5). However, in a simply supported column with the same level of L-D interaction (R_{DL}≈1.00), such clear visibility is only achieved at P/P_{cr}=1.75 [3].

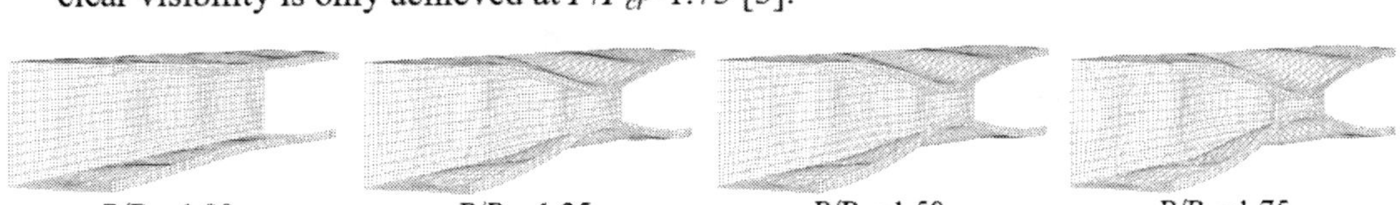

P/P_{cr}=1.00 P/P_{cr}=1.25 P/P_{cr}=1.50 P/P_{cr}=1.75

Fig. 5: ABAQUS deformed configurations at P/P_{cr}=1.00; 1.25; 1.50; 1.75 (R_{DL}=1.00+L column)

3.2 Secondary-bifurcation L-D interaction

The results presented in this section concern R_{DL}=0.60 and R_{DL}=1.60 columns and are qualitatively similar to those reported in the previous section. Therefore, the corresponding descriptions and discussions are abbreviated whenever possible.

3.2.1 Secondary local bifurcation

Fig. 6(a) shows the equilibrium paths P/P_{crD} vs. v/t and the initial geometrical imperfection shape included in the GBT analysis, akin to the critical D critical buckling mode shape with outward flange-lip motions (see Fig. 1(a)). A preliminary GBT buckling analysis showed that this mode exhibits visible contributions from deformation modes **5**-92%, **7**-6%, **9**-2%. As in Section 3.1, (i) Fig. 6(a) also shows equilibrium paths yielded by different "approximate analyses" (including distinct deformation mode sets) and (ii) Figs. 7(a)-(c) and 8 concern the R_{DL}=0.60+D column (only the critical-mode initial imperfection is considered). On the basis of these results, the following conclusions can be drawn:

(i) Until P/P_{crD}=1.00, $w(x)$ is clearly akin to the critical D buckling mode (see Fig. 7(c)), with a dominant contribution from mode **5** – at this loading stage, the impact of the remaining modes is very small. On the other hand, $w_{7+9}(x)$ exhibits one inner half-wave and two outer quarter-waves (see Fig. 7(b) for P/P_{crD}=1.00 and 1.25), *i.e.*, consists of the tiny contribution of these deformation modes to the column critical (D) buckling mode (amplification of $\bar{\phi}_{7+9}(x)$). Therefore, the column exhibits a typical distortional post-buckling behaviour [6] up to P/P_{crD}≈1.00.

(ii) However, as the applied load increases beyond P/P_{crD}=1.25 the shape of $w_{7+9}(x)$ gradually switches into a seven half-wave pattern (see Fig. 7(b) for P/P_{crD}=1.50 and 1.75 – note that the secondary-

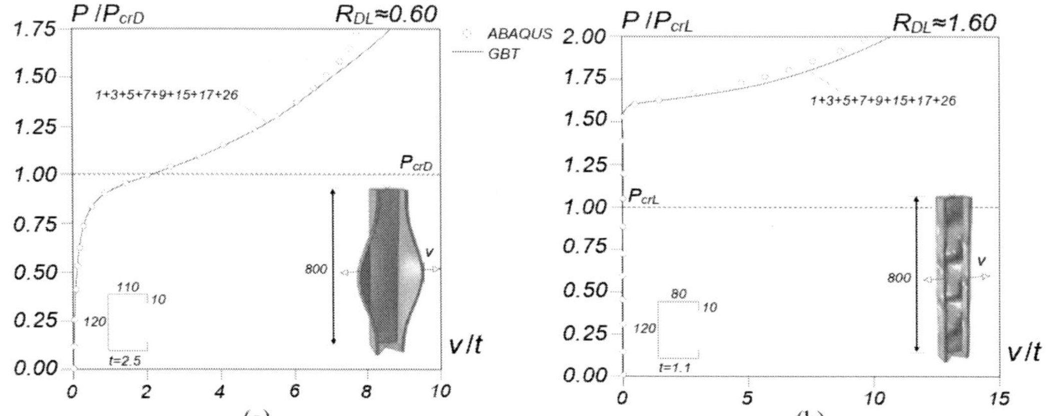

(a) (b)

Fig. 6: GBT post-buckling equilibrium paths of the column undergoing secondary (a) local and (b) distortional bifurcation L-D interaction

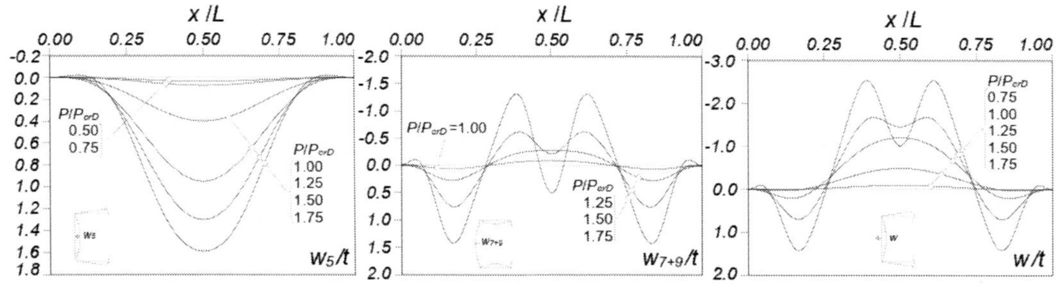

Fig. 7: Evolution of the web transverse bending displacement profiles due to deformation modes (a) **5** ($w_5(x)$), (b) **7+9** ($w_{7+9}(x)$) and (c) **3+5+7+9+26** ($w_{3+5+7+9+26}(x)$≡$w(x)$), for the R_{DL}=0.60+D column

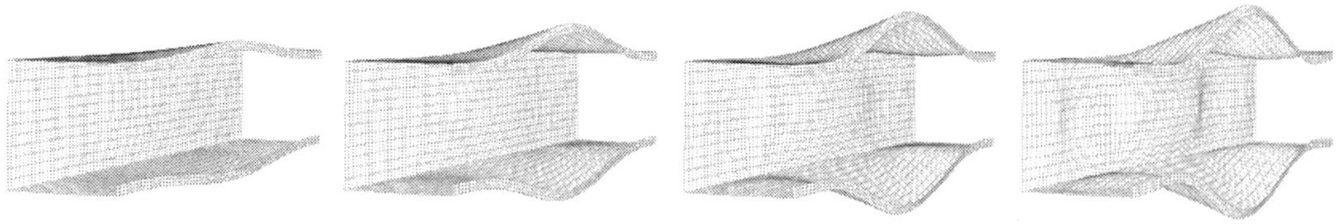

P/P_{crD}=1.00 P/P_{crD}=1.25 P/P_{crD}=1.50 P/P_{crD}=1.75

Fig. 8: ABAQUS deformed configurations at P/P_{cr}=1.00; 1.25; 1.50; 1.75 (R_{DL}=0.60+D column)

bifurcation occurs at P/P_{crD}=1/0.6=1.66), akin to the critical L buckling mode[3]. The emergence of this local deformation pattern at advanced loading stages characterises the "secondary local bifurcation L-D interaction".

3.2.2 Secondary distortional bifurcation

Fig. 6(b) shows the equilibrium path P/P_{crL} vs. v/t and the initial geometrical imperfection shape included in the GBT GNIA (in this case, exhibiting eight half-waves). The preliminary GBT buckling analysis showed that the columns critical (L) buckling modes combines contributions from modes **7**-88%, **9**-9%, **5**-2%. The close observation of the results presented in Figs. 9 and 10, again similar to those shown in the previous sections, leads to the following comments:

(i) Fig. 9(c) shows the mid-web displacement longitudinal profile $w(x) \equiv w_{3+5+7+9+26}(x)$ consists almost exclusively of the eight half-wave contribution from local deformation modes **7+9** (akin to the critical buckling mode) for P/P_{crL}<1.25. However, for higher applied loads this picture changes, as the participation from mode **5** gradually grows in amplitude and its shape switches from eight half-waves to a single one (see Fig. 9(a))[4] – consequently, the amplitudes of the eight half-waves exhibited by $w(x)$ become progressively more unequal, particularly in the column central region. The emergence of this distortional deformation pattern akin to the critical D buckling mode deformations at fairly advanced loading stages (P/P_{crL}≈1.25) characterises the "secondary distortional bifurcation L-D interaction". Concerning the elastic-plastic behaviour of this column, it may be either (i_1) practically local, if the squash load P_y does not exceed $1.25P_{crL}$, or (i_2) affected by L-D coupling, if the squash load is higher than that value (*e.g.*, [6]) – naturally, the relevance of the L-D interaction effects increases with the squash load (yield stress).

(ii) Once more, the column deformed configuration displays inward flange-lip motions – see item (iii) in Section 3.1, concerning the R_{DL}=1.00+L column, and note that the local deformations at mid-span are again associated with outward web bending motions.

(iii) Comparing the results obtained in this work and those concerning simply supported columns, reported in [3], it is readily concluded that the effects of the "secondary distortional bifurcation L-D interaction" are much more relevant in the fixed-ended columns – this same comment also applies to the columns experiencing "secondary local bifurcation L-D interaction", addressed in Section 3.2.1. In fact, it was found that the simply supported columns fail elastically very soon after the "secondary-bifurcation" load level, which does not happen in the fixed-ended columns – they exhibit a considerable strength reserve beyond that load level. Therefore, squash loads (yield stresses) considerably higher than the "secondary-bifurcation" load level will often lead to the development of significant L-D interaction prior to the onset of yielding in elastic-plastic

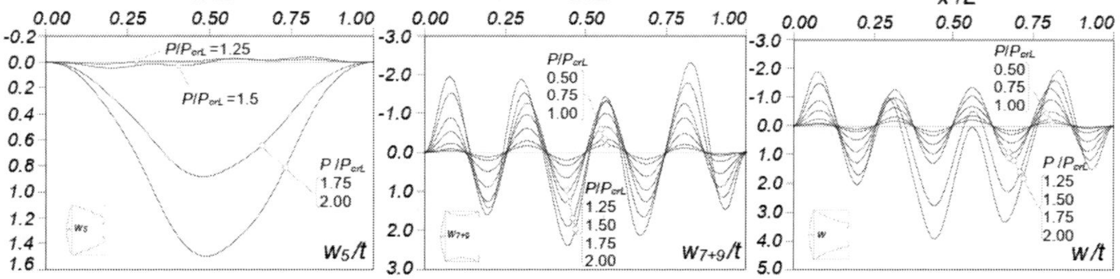

Fig. 9: Evolution of the web transverse bending displacement profiles due to deformation modes (a) **5** ($w_5(x)$), (b) **7+9** ($w_{7+9}(x)$) and (c) **3+5+7+9+26** ($w_{3+5+7+9+26}(x) \equiv w(x)$), for the R_{DL}=1.60+L column

[3] Note that the n_L=6 was given in Table 1. However, in fixed-ended columns the identification of the "pure" critical modes is difficult to achieve – in this case, the critical L buckling mode contains a non-negligible participation of mode **5**. By performing an "approximate buckling analysis" with only local deformation modes, it was found that the number of local-half-waves is indeed seven.

[4] Naturally, the load level at which the distortional deformations emerge depends on the column R_{DL} value and support conditions.

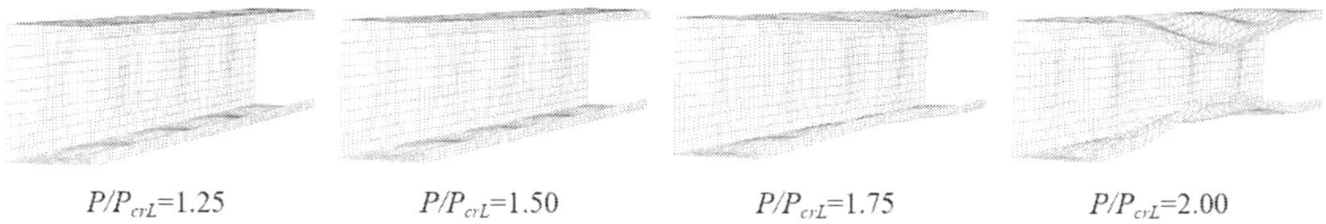

P/P_{crL}=1.25 P/P_{crL}=1.50 P/P_{crL}=1.75 P/P_{crL}=2.00

Fig. 10: ABAQUS deformed configurations at P/P_{cr}=1.25; 1.50; 1.75; 2.00 (R_{DL}=1.60+L column)

fixed-ended columns, thus causing a non-negligible failure load erosion [4] – such consequences only very rarely arise in elastic-plastic simply supported columns under the same circumstances.

4. Conclusion

A GBT-based numerical investigation concerning the geometrically non-linear response of fixed-ended lipped channel experiencing various levels of L-D interaction was reported. The main objective was to shed fresh light on the mechanical aspects underlying this coupling phenomenon. Among the various conclusions drawn from this investigation, the following ones deserve to be specially noted:

(i) Three types of L-D interaction were identified and mechanically characterised: (i) true interaction and (ii) secondary local/distortional bifurcation interaction. In the first case, deformations akin to both the L and D critical buckling modes appear at early loading stages (combining contributions from GBT L and D deformation modes). Conversely, in the last two cases deformations akin to the non-critical buckling mode only emerge at advanced loading stages.

(ii) There are very strong similarities between the L-D interactive behaviour of the fixed-ended and simply supported columns (the latter studied in [3]), in the sense that deformation patterns akin to both the local and distortional critical buckling modes emerge along the equilibrium path (the "core business" of L-D interaction). However, the extent and relevance of the coupling effects is considerably more pronounced in the fixed-ended columns, due to the much higher distortional post-critical strength reserve. Thus, the impact on the failure loads of elastic-plastic columns is also more relevant, which has non-negligible design implications.

Acknowledgments

The first author gratefully acknowledges the financial support of FCT (*Fundação para a Ciência e a Tecnologia – Portugal*), through the doctoral scholarship SFRH/BD/87746/2012.

References

[1] Silvestre N, Camotim D. "Nonlinear generalized beam theory for cold-formed steel members", *International Journal of Structural Stability and Dynamics*, 3(4), 461-490, 2003.

[2] Silvestre N, Camotim D. "GBT-based analysis of the local-plate/distortional buckling mode interaction in lipped channel columns", *Proceedings of 4th International Conference on Coupled Instabilities in Metal Structures* (CIMS'04 – Rome, 27-29/09), M. Pignataro, J. Rondal, V. Gioncu (eds.), Editura Orizonturi Universitare (Timisoara), 449-462, 2004.

[3] Martins AD, Dinis PB, Camotim D, Providência P. "On the mechanics of local-distortional interaction in lipped channel thin-walled columns", *Website Proceedings of SSRC Annual Stability Conference*, (Nashville, 24-27/3), 2015.

[4] Martins AD, Dinis PB, Camotim D, Providência P. "On the relevance of local-distortional interaction effects in the behaviour and design of cold-formed steel columns", *Computers & Structures*, 160(November), 57-89, 2015.

[5] Simulia Inc. (2009), *ABAQUS Standard* (version 6.9-3).

[6] Silvestre N, Camotim D. "Local-plate and distortional postbuckling behavior of cold-formed steel lipped channel columns with intermediate stiffeners", *Journal of Structural Engineering*, 132(4), 529-540, 2006.

EXPERIMENTAL INVESTIGATION OF THE CROSS-SECTIONAL STABILITY OF COLD-FORMED STEEL BUILT-UP COLUMNS

Francisco Meza, Shanshan Cheng, Jurgen Becque and Iman Hajirasouliha

Department of Civil and Structural Engineering, the University of Sheffield

Abstract: This paper describes an experimental investigation into the structural performance of 12 built-up cold-formed steel stub columns, aiming to study the buckling interaction between individual components. The built-up sections were assembled by bolting channel and plate sections together with thicknesses ranging from 1.2 mm to 2.4 mm. Various connector spacings (ranging from 100 mm to 333 mm) were investigated. The experiments revealed a pronounced extent of restraint and interaction between the individual components of the section while buckling, with the connector spacing having a significant effect on the observed buckling modes, but a much less pronounced effect on the ultimate capacity.

1. Introduction

Cold-formed steel (CFS) structural members are widely used in modern structures as they are typically light-weight, are relatively easily manufactured, possess a high strength-to-weight ratio and allow a rapid speed of construction. The CFS industry is currently experiencing a drive towards larger spans and higher loads. Examples are multi-storey buildings constructed entirely out of cold-formed steel [1], and cold-formed steel portal frames [2]. Built-up members, obtained by joining two or more sections together using bolts, rivets or screws, are increasingly used in practice. However, the fundamental understanding of the behaviour of these sections is lagging behind practice. The influence of various parameters, such as the type of connector used, the spacing between the connectors, the degree to which the individual components work as one cross-section, and the occurrence of individual as well as coupled instabilities needs more study.

A number of experiments have previously been carried out on built-up sections [3-8], involving various section types and connecting methods. However, it should be noted that all these tests have only accounted for built-up sections made of two identical components, in which both components buckle at the same time.

The purpose of the experimental program described in this paper is to investigate the various instabilities which occur in built-up CFS stub columns, the interaction between the cross-sectional components during buckling and the effect of the connector spacing on the buckling modes and the ultimate capacity of the columns.

2. Specimen geometry

The specimens were constructed by assembling four individual sections into two different geometries using M6 bolts (Fig. 1). The lengths of the type 1 and type 2 columns were 1100 mm and 800 mm, respectively, in accordance with the specifications given in [9]. Built-up column 1 was designed with 2, 3 or 5 equally spaced connectors, while built-up column 2 was designed with 2, 4 or 6 connectors.

The symbol 'SC' is here used to label the built-up cross-sections, followed by a '1' or a '2' to indicate the cross-sectional configuration (Fig. 1), and again followed by the number of rows of intermediate connectors (i.e. not counting the connectors in the end sections). Since each test was repeated, the letters 'a' and 'b' indicate the first and the second test of twin columns. Regarding the cross-sectional components, the letters 'T' and 'P' refer to channels and plate sections, respectively, followed by the nominal width of the web of the channel (or the width of the plate section) in mm, and the nominal thickness of the section in mm multiplied by 10.

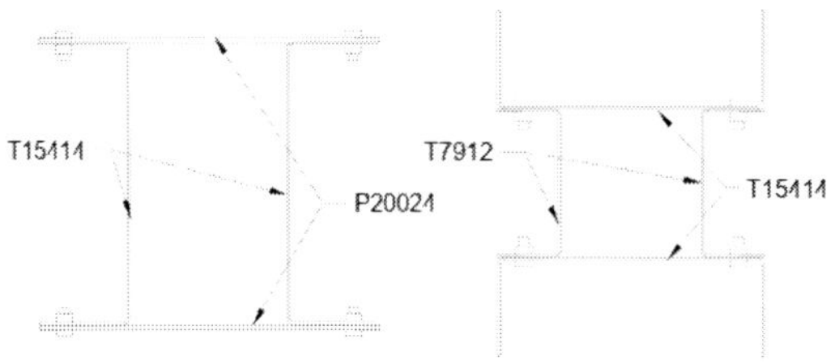

Fig. 1: Built-up cross-sections

All the individual sections were measured prior to their assemblage. The measured (outer) dimensions and the spacing between the connectors are listed in Table 1 and Table 2, for built-up columns of type 1 and 2, respectively.

Table 1: Measured dimensions of built-up columns 1

Column	Channel			Plate		Connector spacing (mm)
	Component	Web (mm)	Flange (mm)	Component	Width (mm)	
SC1-2a	T15414-1	154.00	53.92	P200-1	199.83	333
	T15414-2	154.23	53.62	P200-2	199.73	
SC1-2b	T15414-3	154.13	53.85	P200-3	199.67	333
	T15414-4	154.13	53.45	P200-4	199.83	
SC1-3a	T15414-5	154.23	53.75	P200-5	199.53	250
	T15414-6	154.10	53.67	P200-6	199.27	
SC1-3b	T15414-7	154.17	53.43	P200-7	199.43	250
	T15414-8	154.10	53.75	P200-8	200.40	
SC1-5a	T15414-9	154.17	53.60	P200-9	200.10	167
	T15414-10	154.17	53.63	P200-10	199.47	
SC1-5b	T15414-11	154.13	53.37	P200-11	198.93	167
	T15414-12	154.23	53.62	P200-12	198.73	

Table 2: Measured dimensions of built-up columns 2

Column	Channel Component	Channel Web (mm)	Channel Flange (mm)	Channel Component	Channel Web (mm)	Channel Flange (mm)	Connector spacing (mm)
SC2-2a	T15414-1	154.07	53.45	T7912-1	78.97	36.28	233
	T15414-2	154.10	53.30	T7912-2	79.03	36.27	
SC2-2b	T15414-3	154.00	53.48	T7912-3	78.87	36.42	233
	T15414-4	154.00	53.42	T7912-4	79.17	36.32	
SC2-4a	T15414-5	154.07	53.60	T7912-5	79.20	36.42	140
	T15414-6	154.00	53.28	T7912-6	79.13	36.48	
SC2-4b	T15414-7	154.07	53.72	T7912-7	79.10	36.52	140
	T15414-8	154.17	53.53	T7912-8	79.17	36.48	
SC2-6a	T15414-9	154.07	53.53	T7912-9	79.07	36.53	100
	T15414-10	154.33	53.37	T7912-10	79.10	36.43	
SC2-6b	T15414-11	154.20	53.75	T7912-11	79.07	36.18	100
	T15414-12	154.27	53.70	T7912-12	79.03	36.32	

To correct the slight relative difference in length between the individual components, both ends of each built-up column were manually filed. Much care was put into this process to ensure that a completely flush bearing surface was obtained between the specimen ends and the end plates. Endplates with dimensions of 250x300 mm^2 and a thickness of 20 mm were attached to both ends of the columns using a Sikadur 31 FC Normal 2-part thixotropic epoxy resin. In addition to hand filing the ends, this provided a second means to ensure a uniform introduction of the load into all components.

3. Material properties

Tensile coupon tests were carried out to obtain the material properties of the test specimens. Two flat coupons were taken along the centre line of the web of each type of channel section and along the centre line of the plate section. In addition, two corner coupons were also cut from the web-flange junction of each type of channel section. All flat coupons had a nominal width of 12.5 mm, while the corner coupons had a nominal width of 6 mm. Corner coupons were tested in pairs in order to avoid the introduction of unwanted bending moments due to their asymmetrical shape. All coupons were tested following the specifications given in [10]. Table 3 lists the values of the material properties obtained for each coupon, as well as the average values, in which E is the Young's modulus, $\sigma_{0.2\%}$ is the 0.2% proof stress, σ_u is the tensile strength and ε_f is the elongation after fracture, measured over a gauge length of 50 mm for the flat coupons and 25 mm for the corner coupons.

Table 3: Material properties of tensile coupons

Type	Section	E (GPa) Ind.	E (GPa) Avg.	$\sigma_{0.2\%}$ (MPa) Ind.	$\sigma_{0.2\%}$ (MPa) Avg.	σ_u (MPa) Ind.	σ_u (MPa) Avg.	ε_f (%) Ind.	ε_f (%) Avg.
Flat	Plate2.4-1	195	195	437	437	-	519	-	16
Flat	Plate2.4-2	-		436		519		16	
Flat	T15414-1	207	207	620	609	709	705	10	11
Flat	T15414-2	206		597		700		12	
Flat	T7912-1	197	195	437	419	530	537	13	14
Flat	T7912-2	193		401		544		15	
Corner	T15414-1	226	231	617	617	-	733	-	11
Corner	T15414-2	236		616		-		-	
Corner	T7912-1	211	217	487	487	-	605	-	17
Corner	T7912-2	222		486		-		-	

4. Imperfection measurements

Imperfections may have a significant influence on the stability of thin-walled structural members, particularly when coupled instabilities are involved. The imperfections of all test specimens were therefore recorded using a laser displacement measuring device before testing. The measurements were performed after the built-up columns were assembled into their final configurations, as joining the single sections together might somewhat modify their geometric imperfections.

The imperfections of built-up column 1 were measured along three longitudinal lines on each face of the column, as illustrated in Fig. 2a), while in built-up column 2 the imperfections were measured along twenty different longitudinal lines, as shown in Fig. 2b). In built-up column 1, a maximum out-of-plane imperfection of 0.87 mm was observed in the flat plate sections, while an average of 0.27 mm was obtained. The webs of the channel sections showed a maximum out-of-plane imperfection of 0.64 mm and an average value of 0.12 mm. In built-up column 2, the largest out-of-plane imperfection (0.69 mm) of the web occurred in the outer channels, while the average imperfection was 0.10 mm. In the inner channels the maximum out-of-plane imperfection of the web was 0.58 mm and the average was 0.17 mm. In the flanges of the outer channels, the maximum out-of-plane deviation was 0.67 mm, with an average of 0.16 mm.

5. Test set-up

All the specimens were tested between fixed end supports. To capture the axial deformation of the specimens and the out-of-plane deformations of their components, 14 LVDTs were used. Twelve of them were used to measure the out-of-plane deformations of the component sections and were placed at mid-distance between connectors at different heights to capture the localization of the deformations in the later stages of the test (Figs. 3, 4). The other two LVDTs were placed on each side of the column to measure the axial deformation of the column. It was found that the readings given by both of these LVDTs were in close agreement for all specimens.

The specimens were tested in an EHS universal testing machine with a capacity of 1000 kN. The tests were carried out by applying a strain rate of $1.7\text{x}10^{-6}$ /s to all specimens, resulting in a displacement rate of 0.11 mm/min for built-up column 1 (with a length of 1100 mm) and 0.08 mm/min for built-up column 2 (with a length of 800 mm).

6. Test results

6.1 Deformed shape and critical stresses

As shown in Fig. 3, all test specimens of type 1 failed due to the interaction of a 'global-type' flexural buckling of the individual plate sections between connectors and local buckling of the channels. Distinctively different buckling patterns were observed for different connector spacings. In the case of columns SC1-5, the channel sections and the plates buckled with approximately equal buckle half-wave lengths, equal to the distance between the connectors (Fig. 3 c). The flanges of the channel buckled outward in almost perfect sympathy with the plate sections, while a small gap opened up between the flanges and the plates in places where the flanges buckled inwards. This, of course, was due to the fact that the webs of the channels prevented the plates from buckling inwards. It is thereby noted that the 'natural' local buckle half-wavelength of the individual unrestrained channel was determined to be 170 mm, which is close to the connector spacing of 166 mm in SC1-5. In columns SC1-2 and SC1-3, with connector spacings of 333 mm and 250 mm, the buckle half-wave length of the channels was observed to be approximately half the distance between connectors (Fig. 3 a) and 3 b)). The channels thereby received a significant amount of restraint from the plates against outward buckling. The plates, in turn, were prevented by the channel webs from buckling in-wards and therefore developed a pattern whereby they mostly remained straight in one field, followed by outwards buckling in the next field, in a process which necessarily required a finite amount of bolt slip.

a) Built-up column 1 b) Built-up column 2

Fig. 2: Imperfections measurements in built-up columns

a) SC1-2a b) SC1-3a c) SC1-5a

Fig. 3: Deformed shape in columns

Each specimen of geometry 2 failed due to local buckling of its components. In columns SC2-2, the buckling half-wave length of the outer channels coincided with half the distance between connectors (Fig. 4a)). This could be expected since the 'natural' half-wave length of the individual unrestrained channel (170 mm) is relatively close to half the connector spacing (117 mm). In columns SC2-4, on the other hand, the connector spacing was 140 mm and the channels buckled with a half-wave length equal to the distance between connectors (Fig. 4b)). In columns SC2-6 (with a connector spacing of 100 mm), interestingly, a somewhat irregular buckling wave pattern was observed in the outer channels. While, along most of the column length, the buckle half-wave length coincided with the connector spacing, occasionally half a wave would span over an intermediate connector (Fig. 4c)). In almost all columns of geometry 2, however, the cross-sections carrying the connectors coincided with the nodal lines of the outer channel buckling pattern. The flanges of the outer channel did not receive any direct restraint from other components. The flanges of the inner channel sections, however, were prevented from buckling outwards by the presence of the outer channel webs. Therefore, they buckled with a buckle half-wave length equal to half the distance between the connectors in all columns.

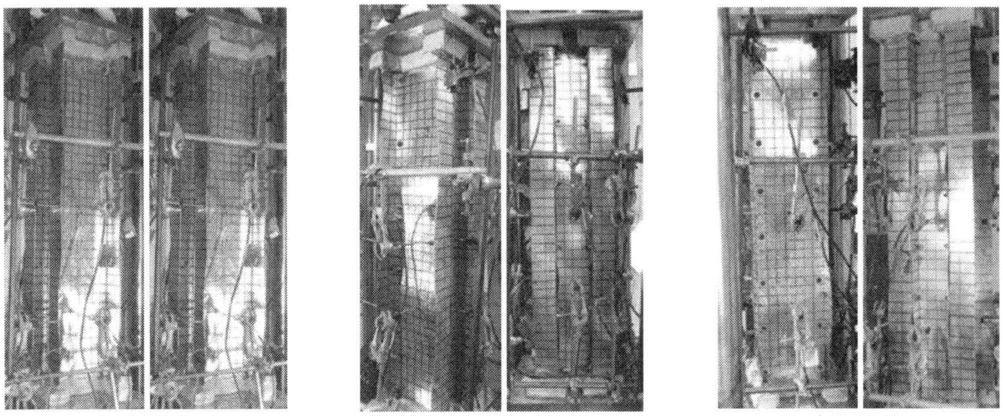

a) SC2-2a b) SC2-4a c) SC2-6a

Fig. 4: Deformed shape in columns

An estimate of the critical buckling stresses of the various components was obtained from the LVDT readings in the test. Initially, up to the point of first buckling, the load was transmitted uniformly to all the components (as indicated by the strain gauges). In built-up column SC1-2, the plate sections buckled in a flexural mode before buckling of the channel sections occurred, while simultaneous buckling of the channel and plate sections was observed in columns SC1-3 and SC1-5. It is a reasonable assumption, due to the known lack of post-buckling capacity in the global flexural mode, that after buckling of the plates, any increase of load would be resisted entirely by the channels. This allowed an estimate of the load carried by the channels at buckling to be established. Buckling loads were obtained from the LVDT readings by plotting the square of the lateral displacement readings against the load and extending the linear trend line to the load axis.

In built-up column 2, the outer channels buckled before the inner channels in all the test specimens. As these channels buckled in a local mode, they possessed significant post-buckling capacity and the distribution of the load over the various components of the cross-section could no longer be easily determined past the point of first buckling. Therefore, no estimate for the buckling stress of the inner channels is listed in Table 4.

The buckling stresses obtained from the tests were compared with theoretically determined buckling stresses. These were obtained while considering the individual channels in isolation, without any interaction with the rest of the cross-section, and while assuming a buckle half-wave length equal to the distance between the connectors, or half the distance between the connectors, as observed in the test. The theoretical critical buckling stress of each channel section was obtained under those assumptions using the CUFSM 4.05 software [11]. For the plate sections, a lower and an upper bound of the critical buckling stress was determined (using Euler's equation), corresponding to assuming one or two half-waves in between the connectors, respectively.

Table 4 shows that for columns SC1-2, where the plate sections buckled before the channels, the stresses at which the plate sections buckled were close to the theoretical upper bound, indicating that the channel in its pre-buckled state provided a significant amount of restraint to the plates, forcing them to buckle outwards. The channels, on the other hand, buckled at a stress somewhat lower than the theoretically predicted value. This is thought to be a result of the additional forces exerted by the plates on the channels while they were trying to extend their buckle half-wave length beyond half the connector spacing (Fig. 3a), a process which necessarily required a certain amount of rotation and translation at the connector points. In columns SC1-3, simultaneous buckling of the channel and the plate sections was experimentally observed. Again, the experimentally derived buckling stress of the channels is slightly lower than the theoretically predicted value, while the measured buckling stress in the plates is about 17% lower than the theoretical upper bound. In columns SC1-5, the channels and the plate sections were seen to buckle simultaneously with roughly the same half-wave length. In this case, the channels actually buckled at a stress well above the theoretical value. This could possibly be due to the buckling pattern mobilizing the shear stiffness of the connectors.

Regarding built-up column 2, Table 4 shows that the buckling stresses of the outer channels obtained from the test are about 14% to 38% larger than the theoretical values. This indicates that the buckling stress of these channels was substantially increased by the restraint provided by the pre-buckled inner channels through 1. necessary compatibility of the deformations at the connector points, and 2. the surface contact between the inner channel flanges and the outer channel webs in locations where the outer channel webs wanted to buckle inward.

Table 4: Buckling stresses of the different component sections of built-up column 1

Column 1	Theoretical buckling stress (MPa)			Buckling stress from test (MPa)		Column 2	Theoretical buckling stress (MPa)		Buckling stress from test (MPa)	
	Channel	Plate		Channel	Plate		Inner Channel	Outer Channel	Inner Channel	Outer Channel
		Lower	Upper							
SC1-2a	63	10	38	52	41	SC2-2a	146	71	-	-
SC1-2b	63	10	38	52	41	SC2-2b	146	71	-	81
SC1-3a	67	17	68	58	57	SC2-4a	156	65	-	89
SC1-3b	67	17	68	56	56	SC2-4b	156	65	-	89
SC1-5a	63	38	152	81	81	SC2-6a	201	78	-	108
SC1-5b	63	38	152	83	83	SC2-6b	201	78	-	108

6.2 Ultimate loads

Fig. 5 shows the load vs. axial displacement curves of all specimens. For geometry 1, twin columns showed a maximum difference in ultimate load of 8.97%. The tests show a modest increase of the ultimate load of the built-up columns as the spacing between the connectors is reduced. More specifically, halving the spacing between connectors from 333 mm to 167 mm produced an increase of the ultimate load of 10.81%. For geometry 2 the largest difference in ultimate load between twin columns was 6.28%. The tests showed, however, that in this case reducing the spacing between connectors does not necessarily result in a noticeable increase of the ultimate load. For example, columns with a connector spacing of 140 mm (columns SC2-4) showed a comparable (and actually slightly higher) ultimate load than columns with connectors spaced every 100 mm (columns SC2-6).

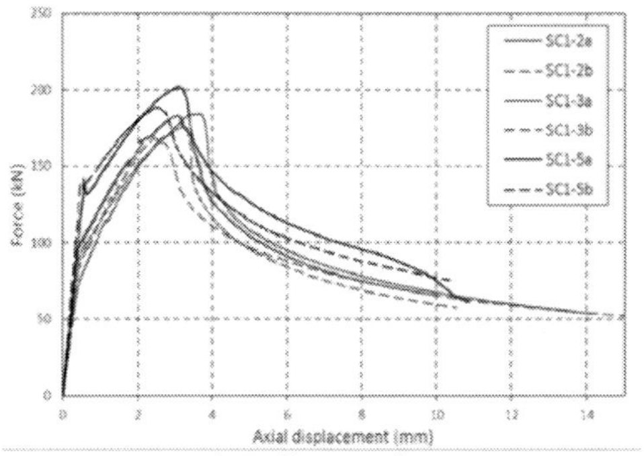

a) Built-up column 1

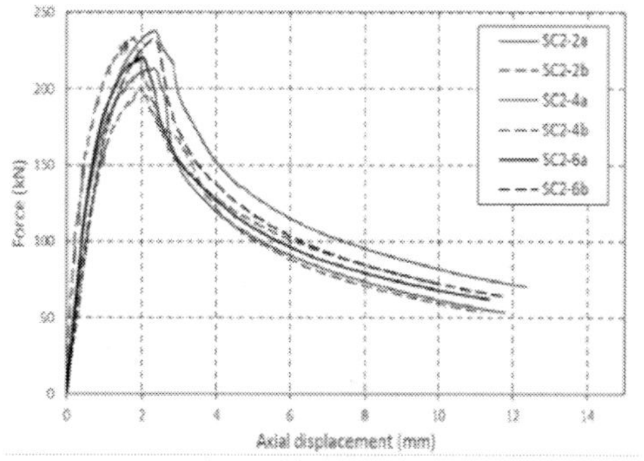

b) Built-up column 2

Fig. 5: Axial load vs axial deformation curve

7. Conclusions

A detailed experimental program of 12 built-up stub columns with two different geometries has been described. The columns were compressed between fixed supports. The investigation included determining the material properties of the specimens through coupon tests, measuring the specimen out-of-plane imperfections, recording the deformed shapes and ultimate loads of the test specimens, and determining the buckling stress of each section component. By comparing the theoretical predictions of the buckling stresses to the experimentally measured values, it is concluded that a significant amount of interaction takes place between components, which in a particular component either results in restraint relative to the preferred buckling shape, or additional action promoting instability.

While the buckling pattern in the specimens under consideration appeared highly dependent on the connector spacing, the ultimate capacity showed less sensitivity. For instance, an increase in the ultimate strength of 10.81% was observed in built-up column 1 when halving the spacing between connectors. Moreover, reducing the spacing between connectors did not always results in a noticeable increase of the cross-sectional capacity, as observed in columns SC2-4 and SC2-6.

References

[1] Li Y, Shen Z, Yao X, Ma R, Liu F. "Experimental Investigation and Design Method Research on Low-Rise Cold-Formed Thin-Walled Steel Framing Buildings", *Journal of Structural Engineering*, 139, 818–836, 2013.

[2] Dundu M. "Design Approach of Cold-formed Steel Portal Frames", *International Journal of Steel Structures*, 11(3), 259–273, 2011.

[3] Young B, Chen J. "Design of Cold-Formed Steel Built-Up Closed Sections with Intermediate Stiffeners", *Journal of Structural Engineering*, 134(5), 727–737, 2008.

[4] Zhang JH, Young B. "Compression tests of cold-formed steel I-shaped open sections with edge and web stiffeners", *Thin-Walled Structures*, 52, 1–11, 2012.

[5] Lau HH, Ting TCH. "An Investigation of the Compressive Strength of Cold-formed Steel Built-up I Sections", *Proceedings of the Sixth International Conference on Advances in Steel Structures. ICASS'09*, 441–449, 2009.

[6] Whittle J, Ramseyer C. "Buckling capacities of axially loaded, cold-formed, built-up C-channels", *Thin-Walled Structures*, 47(2), 190–201, 2009.

[7] Piyawat K, Ramseyer C, Kang TH. "Development of an Axial Load Capacity Equation for Doubly Symmetric Built-Up Cold-Formed Sections", *Journal of Structural Engineering - ASCE*, 139(12), 1–13, 2013.

[8] Georgieva I, Schueremans L, Pyl L, Vandewalle L. "Experimental investigation of built-up double-Z members in bending and compression", *Thin-Walled Structures*, 53, 48–57, 2012.

[9] Ziemian RD. "Appendix B - Technical Memoranda of Structural Stability Research Council", 2010.

[10] BS EN ISO 6892-1, "Metallic materials — Tensile testing Part 1 : Method of test at ambient temperature", British Standard Institution, 2009.

[11] Schafer BW. "Buckling analysis of cold-formed steel members using CUFSM : conventional and constrained finite strip methods". Orlando, Florida, 2006.

BUCKLING ANALYSIS OF STEEL RACK-SECTION MEMBERS BY THE GBT-BASED ECBL APPROACH

Stefano de Miranda[a], Domenico Melchionda[a], Viorel Ungureanu[b,c] and Dan Dubina[b,c]

[a] *Department of Civil, Chemical, Environmental and Materials Engineering – DICAM, University of Bologna, Italy*
[b] *Department of Steel Structures and Structural Mechanics, Politehnica University of Timisoara, Romania*
[c] *Laboratory of Steel Structures, Romanian Academy – Timisoara Branch, Romania*

Abstract: The Erosion of Critical Bifurcation Load (ECBL) approach is a practical tool to characterize the instability behaviour of thin-walled cold-formed steel members. In this framework, the Generalized Beam Theory (GBT) is used to find (1) the interaction point of simultaneous buckling modes and (2) the erosion of the critical bifurcation load referred to this interactive buckling of two or more buckling modes associated with the same critical load. In this sense a modified version of the ECBL approach is proposed. Some numerical results involving pallet rack uprights in compression are presented and compared with experimental ones, classical ECBL approach and EN 1993-1-3.

1. Introduction

Coupled instabilities represent a characteristic of thin-walled steel members in compression or bending. The actual buckling curves included in the design codes, i.e. EN 1993-1-3 [1], are based on experimental tests carried out on hot-rolled sections. For the purpose of practical use these curves have been adapted in order to cover the stability design problems of thin-walled cold-formed steel members. In case of compact sections the erosion is due to the imperfections, while for thin-walled members, supplementary erosion is induced by the phenomenon of coupled instabilities.

Given a member in compression, it is assumed that two simultaneous buckling modes may occur. If N_u is the ultimate load, and N_{cr} the ideal critical one, the following relation may be written:

$$N_u = (1-\psi)N_{cr} \quad (1)$$

The *erosion factor* ψ was introduced as a *measure* of erosion of critical load. Gioncu [2] classified the interaction types by means of this erosion factor, as follows:

class I: weak interaction (W), $\psi \leq 0.1$;
class II: moderate interaction (M), $0.1 < \psi \leq 0.3$;
class III: strong interaction (S), $0.3 < \psi \leq 0.5$;
class IV: very strong interaction (VS), $\psi > 0.5$.

Obviously, an appropriate framing of each type of coupled instability into the relevant class is very important because the methods of analysis used for design have to be different from one class to another. In case of weak or moderate interaction, structural reliability will be provided by simply using of design code safety coefficients, while in case of strong or very strong interaction, special methods are needed. In [3] Dubina summarised the main coupled instability cases, which may appear within the bar members.

Based on the concept of Erosion of Critical Bifurcation Load (ECBL), Dubina [3] proposed an approach to evaluate the ultimate strength in local/distortional-global interactive buckling. This approach enables to use the Ayrton-Perry format of European buckling curves [1,4] to calibrate appropriate buckling curves for any kind of interactive local/distortional-global buckling.

In the spirit of [5], the Generalized Beam Theory (GBT) [6,7,8] is used to find the interaction points of buckling modes and, based on that, to evaluate the erosion coefficient, to propose a modified version of the ECBL approach and to apply it to the analysis of a pallet rack uprights in compression. The analysis consists in two steps. In the first step, the instability mode interaction points and their related imperfections are identified using GBT bifurcation analyses. In the second step, imperfection factors associated to the modified version of the ECBL approach are calculated in terms of the erosion coefficients computed in the first step. The buckling curves related to the modified and classic ECBL approaches are thus obtained and compared with experimental results. Also, for comparison, the buckling curves proposed by EN 1993-1-3 are presented.

2. Erosion of Critical Bifurcation Load

To illustrate the ECBL approach, let consider Fig. 1 showing the typical theoretical buckling curves of a thin-walled member in compression (black curves) and the actual one (green curve).

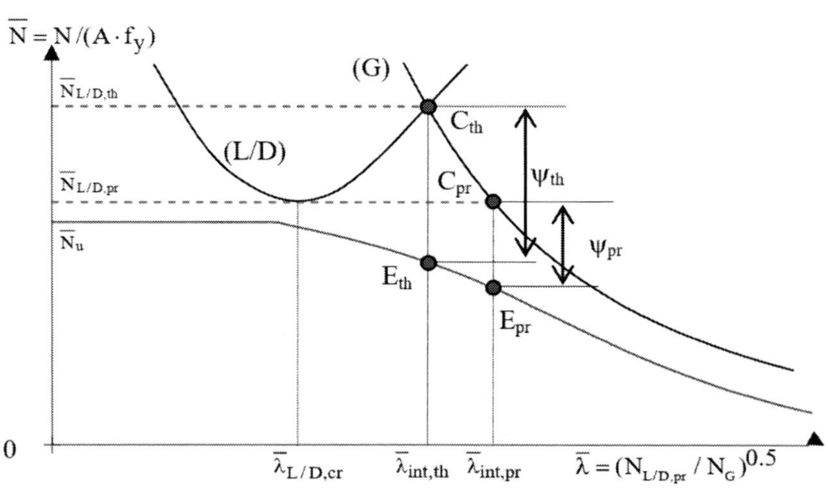

Fig. 1: Interactive buckling model based on the ECBL approach

In Fig. 1, curve G denotes the quantities associated to global buckling modes, while curve L/D those associated to local/distortional ones. Moreover, $\overline{N}$ is the dimensionless axial force (where N is the axial force and Af_y the plastic strength of full cross-section) and $\overline{\lambda}$ is the dimensionless slenderness (where $N_{L/D,pr}$ is the smallest value between the local or distortional critical buckling load and N_G is the critical global buckling load).

As it can be noted, global mode (G) interacts with local/distortional (L/D) one. The ECBL approach proposed by Dubina in [3], to evaluate the ultimate strength in global-local/distortional interactive buckling, hereinafter called ECBL$_{pr}$, is based on the main hypothesis that the reduction of the buckling load is most significant when the global buckling load is almost equal to the minimum value of local/distortional one [9,10]. In particular, in the ECBL$_{pr}$ the "practical interaction point", marked as C_{pr} in Fig. 1, is assumed as the point where the erosion of the critical load, ψ_{pr}, is maximum. On the other hand, in the same graph it is also possible to identify the "theoretical interaction point", marked as C_{th}, corresponding to the intersection of the two buckling modes, and its related erosion ψ_{th}. The erosions can be expressed as follows:

$$\psi_* = 1 - \overline{N}_u / \overline{N}_{L/D,*} \tag{2}$$

where * can be *pr* or *th* and $\overline{N}_u$ is the dimensionless ultimate load.

In the framework of the ECBL$_{pr}$ approach, the Ayrton-Perry formula can be written as

$$\left(\overline{N}_{L/D,pr} - \overline{N}\right)\left(1 - \overline{N}\overline{\lambda}^2\right) = \alpha_{pr}\overline{N}\left(\overline{\lambda} - 0.2\right) \tag{3}$$

where α_{pr} is the imperfection factor. Substituting the coupling point E_{pr} in the above equation leads to

$$\alpha_{pr} = \frac{\psi_{pr}^2}{1-\psi_{pr}} \frac{\sqrt{\overline{N}_{L/D,pr}}}{1-0.2\sqrt{\overline{N}_{L/D,pr}}} \tag{4}$$

Following the same path outlined above, a new version of the ECBL approach, hereinafter called ECBL$_{th}$, is now proposed. It mainly consists in using the theoretical erosion, ψ_{th}, instead of the practical one, ψ_{pr}, to evaluate the buckling strength curves. The starting point is still Eq. (3), but now it is solved considering the coupling point E_{th} instead of E_{pr}. The following new expression for the imperfection factor, renamed α_{th}, is obtained:

$$\alpha_{th} = \frac{\psi_{th}^2 + \left(\dfrac{\overline{N}_{L/D,pr}}{\overline{N}_{L/D,th}} - 1\right)\psi_{th}}{1-\psi_{th}} \frac{\sqrt{\overline{N}_{L/D,th}}}{1-0.2\sqrt{\overline{N}_{L/D,th}}} \tag{5}$$

The results related to the application of the ECBL$_{th}$ approach and the comparison with those coming from the ECBL$_{pr}$ one are shown in the following.

3. The Generalized Beam Theory

In this section, the Generalized Beam Theory with shear deformation recently proposed in [7,8] is briefly summarized. The interested reader can refer to [7,8] for further details.

The fundamental idea of the GBT is to consider a thin-walled beam as an "assembly" of thin plates and to assume the displacement field of the beam as a linear combination of predefined cross-section deformation modes (which are known beforehand) multiplied by unknown functions depending on the beam axial coordinate, that can be called generalized displacements. Deformation modes are subdivided into fundamental flexural-shear modes, local flexural modes and nonlinear warping modes. Fundamental flexural-shear modes are the same as those of the classical GBT proposed by Schardt [6]: one fundamental flexural-shear mode can be defined for each vertex of the cross-section midline. Local flexural modes and nonlinear warping modes, based on nodes internal to the walls of the cross-section, can be added in order to enrich, respectively, the local bending behavior and the warping description. In the following, for simplicity, nonlinear warping modes are not considered.

As extensively discussed in [8], an important step of the deformation modes definition is the modal decomposition. It allows, by a modal transformation, to give a mechanical meaning to the generalized deformation parameters and also to distinguish the classical parameters of standard beam theories. For example, the in-plane displacements associated to the flexural-shear modes of a rack section RSB125×3.2 [11,12] expressed in the modal base are shown in Fig. 2. As it can be noted, the first four modes correspond to the modes of a classical shear deformable Vlasov beam with rigid cross-section. The other modes engender cross-section distortion.

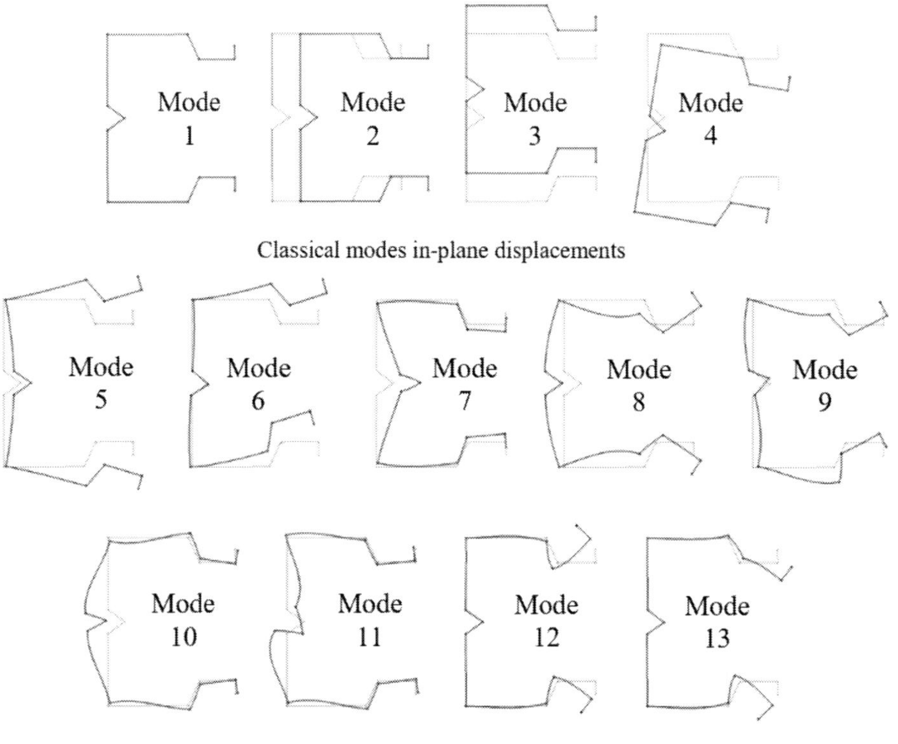

Distortional modes in-plane displacements

Fig. 2: Deformation modes of RSB125×3.2 section after modal decomposition

4. Numerical results

In this section, the numerical results obtained for the rack-section-member RSB125×3.2 in compression both with the $ECBL_{pr}$ approach and the $ECBL_{th}$ one are shown and compared with experimental results. Experimental results come from an extensive campaign carried out at the CEMSIG Research Center (http://cemsig.ct.upt.ro/) within Civil Engineering Faculty of Politehnica University of Timisoara [11,12]. Also, for comparison, the curve obtained with EN 1993-1-3 [1] is included.

The theoretical buckling curves and the coupling points have been obtained using the GBT. In particular, GBT results have been obtained using the CAPS finite element code, developed at the Laboratory of Computational Mechanics, DICAM-University of Bologna (www.unibo.it/LAMC). The numerical ultimate loads have been obtained with GMNIA analysis using shell finite elements implemented in the software ABAQUS/CAE [11]. Material properties are assumed as in [9,10].

4.1 Interaction points

Figure 3 shows the theoretical buckling curves obtained using GBT, of the rack-section-member RSB125×3.2 in compression with simple support-simple support boundary conditions. A distinction between Global-Local (GL) and Global-Distortional (GD) interaction points is made. The results show that the GD interaction points are characterized by lower critical loads with respect to GL interaction points. Accordingly, only the GD interaction points are considered in the following.

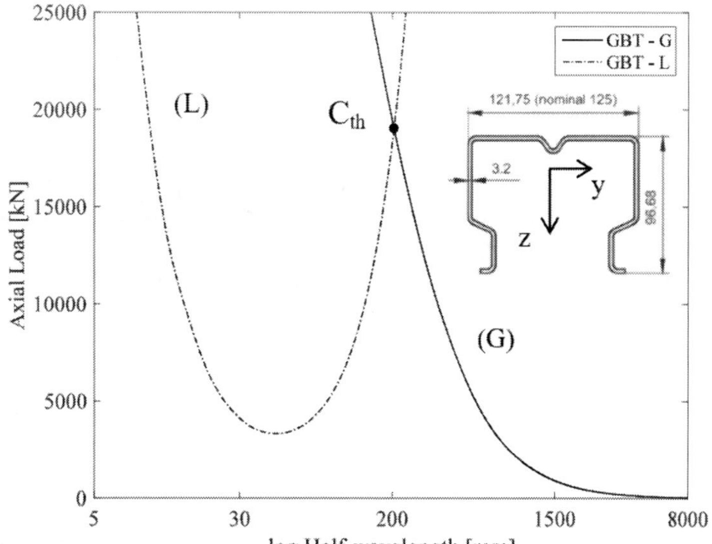

a)

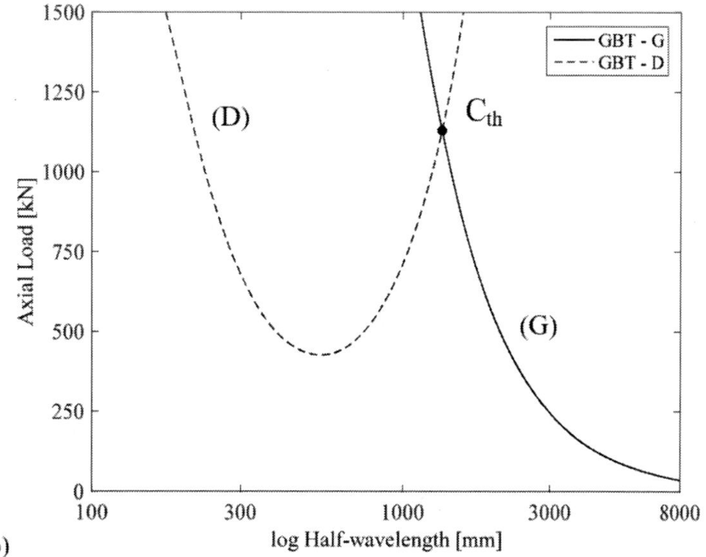

b)

Fig. 3: Theoretical buckling curves for RSB125×3.2: a) GL and b) GD

4.2 Erosion of buckling strength and ECBL buckling curves

In order to evaluate the ultimate load, GMNIA shell finite element analyses of the rack section member RSB125×3.2 in compression have been carried out using ABAQUS software [13]. Rectangular 4-noded shell elements with reduced integration (S4R) have been used, with mesh size of about 5 mm×5 mm. Two types of geometrical imperfections have been considered: overall geometric imperfection, with maximum size at the mid-length equal to 1/750 of column length (as proposed by ECCS Recommendations [14]), and local/distortional imperfection according to the Schafer and Peköz codification [15]. Restraints have been set to reproduce the simple supported boundary condition with warping and section distortion restrained at both beam ends. Only rotations about y-axis and z-axis were allowed on beam ends, torsion being prevented. The lengths of the specimen, were the lengths corresponding to (1) C_{th} interactive point to determine the theoretical erosion, ψ_{th} and (2) C_{pr} interactive point to determine the practical erosion, ψ_{pr}, as presented in Fig. 1.

The analyses have been conducted into two steps. The first step consists in a GBT linear buckling analysis useful to find a buckling mode or combination of buckling modes affine with the relevant imperfections. Two buckling analyses for each case were performed: the first one using only classical global GBT modes to find a global imperfection shape and a second buckling analysis using only distortional GBT modes to find a distortional imperfection shape. After imposing the initial geometric imperfection, obtained as a scaled linear combination of buckling modes from the previous step, a GMNIA analysis with arc-length solver was performed to determine the ultimate load. The residual stresses, induced by the cold-forming process, are not considered because they have no significant influence on the ultimate strength of the member [12].

Once the ultimate loads, and hence ψ_{pr} and ψ_{th}, have been evaluated, the ECBL buckling curves have been obtained as described in Section 2. Figure 4(a) shows the theoretical buckling curves obtained by the GBT, while the buckling curves calculated with both the $ECBL_{pr}$ and the $ECBL_{th}$ approaches are shown in Fig. 4(b) and compared with the one obtained using the design according to EN15512:2009 [16] code. The very good agreement between $ECBL_{th}$ and experimental results can be observed. On the other hand, $ECBL_{pr}$ and EN15512 results are not always on the safe side.

It should be noted that the curve related to the EN15512:2009 [16] code has been obtained by using the stub-column strength. It is well known that the wavelength for distortional buckling is significantly longer than that for local buckling. This means that distortional buckling is not usually identified by a conventional stub-column test. Furthermore, if a stub column test exhibits a distortional failure mode, it is unlikely that the length is sufficient to determine the minimum distortional buckling load.

5. Concluding remarks

The buckling analysis of steel rack-section members by the GBT-based ECBL approach has been addressed in this work. In particular, the GBT has been used to find the mode interaction points and the related imperfections and a modified version of the ECBL approach has been proposed and applied to the analysis of the rack systems member RSB125×3.2. Numerical results coming from both the classical and the modified ECBL approach are compared with experimental ones, showing that the modified ECBL approach is in very good agreement with experimental results, while results from the classical ECBL approach are not always on the safe side.

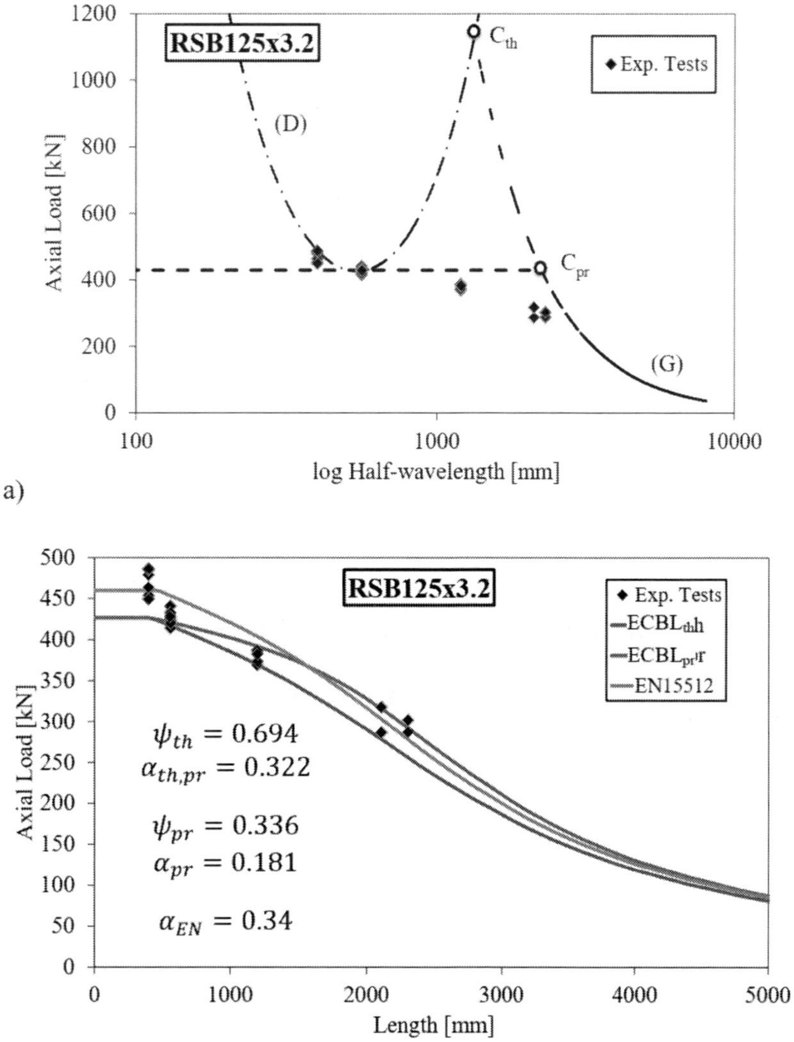

a) b)

Fig. 4: a) Theoretical and practical interaction point for RSB125×3.2 and b) Buckling strength curves for RSB125×3.2

References

[1] EN1993-1-3, "Eurocode 3: Design of steel structures. Part 1-3: General Rules. Supplementary rules for cold-formed thin gauge members and sheeting", *CEN*, Brussels, 2006.

[2] Gioncu V. "General Theory of Coupled Instabilities", *Thin-Walled Structures*, 19(2-4), 81-127, 1994.

[3] Dubina D. "The ECBL approach for interactive buckling of thin-walled steel members", *Steel & Composite Structures*, 1(1), 75-96, 2001.

[4] Rondal J, Maquoi R. "Formulation d'Ayrton-Perry pour le flambement des barres métalliques", *Construction Métallique*, 4, 41-53, 1979.

[5] de Miranda S, Melchionda D, Ungureanu V, Dubina D. "GBT-based buckling analysis of rack systems in the framework of ECBL approach", *Stability of Structure 14th Symposium SSK*, Zakopane, Poland, 8-12 June 2015, 81-82.

[6] Schardt RP. "Verallgemeinerte Technische Biegetheorie", *Springer*, Berlin, 1989.

[7] de Miranda S, Gutierrez A, Miletta R, Ubertini F. "A generalized beam theory with shear deformation", *Thin-Walled Structures*, 67, 88-100, 2013.

[8] de Miranda S, Madeo A, Miletta R, Ubertini F. "On the relationship of the shear deformable Generalized Beam Theory with classical and non-classical theories", *International Journal of Solids and Structures*, 51, 3698–3709, 2014.

[9] van der Neut A., "The interaction of local buckling and column failure of thin-walled compression members", *Proc. of the 12th Int. Congress on Applied Mechanics*, Heteny M.K., Vincenti W.G. (eds.), Springer-Verlag, Berlin, 389-399, 1969.

[10] Dubina D, Ungureanu V. "Instability mode interaction: From Van Der Neut model to ECBL approach", *Thin-Walled Structures*, 81, 39-49, 2014.

[11] Crisan A, Ungureanu V, Dubina D. "Behaviour of cold-formed steel perforated sections in compression. Part-1: Experimental investigations", *Thin-walled Structures*, 61, 86-96, 2012.

[12] Crisan A, Ungureanu V, Dubina D. "Behaviour of cold-formed steel perforated sections in compression. Part-2: Numerical investigations and design considerations", *Thin-walled Structures*, 61, 97-105, 2012.

[13] ABAQUS, "Theory manual", Hibbit, Karlson and Sorenson Inc.; 2007.

[14] ECCS, "European recommendations for the design of light gauge steel members". ECCS, Brussels, 1987.

[15] Schafer BW, Peköz T. "Computational modelling of cold-formed steel characterising geometric imperfections and residual stresses", *J. Construct. Steel Res.,* 47(3), 193-210, 1998.

[16] EN15512, "Steel static storage systems - adjustable pallet racking systems - principles for structural design", *CEN*, Brussels, 2009.

THE "COMBINED APPROACH" FOR THE DESIGN OF SHEAR DIAPHRAGMS MADE OF TRAPEZOIDAL SHEETING

Thomas Misiek[a], Gerhard Huck[b] and Saskia Käpplein[a]

[a] *Breinlinger Ingenieure, Tuttlingen, Germany*
[b] *Ingenieurbüro für Leichtbau, Karlsruhe, Germany*

Abstract: Diaphragms made of trapezoidal sheeting are often used for the stabilization of members or work as a bracing to transfer horizontal loads (wind loads or seismic loads) to the ground. Currently in Europe there are two approaches for the design of these shear diaphragms, both often denominated by the names of the corresponding developers Schardt & Strehl and Bryan & Davies respectively. While the mechanical background for the calculation of the diaphragm stiffness of both approaches is more or less identical, there are differences in the level of detailing, means in the number of parameters regarded as significant for practical design. And generally, Schardt & Strehl's approach is much more reworked and therefore easier to use in practical design. On the other hand, Bryan & Davies' approach is much more realistic regarding the failure modes (shear buckling, failure of fasteners) and load-bearing capacity. Most recently a new approach was elaborated based on work previously done by Baehre & Wolfram and taking into account new developments and findings. This "combined approach" presented in the paper combines the advantages of both approaches from Schardt & Strehl and Bryan & Davies as given above. Detailed information is given how to elaborate the sheeting-related parameters in tables, allowing an easy use in practical design.

1. Introduction

EN 1993-1-3 gives general advice for the design of diaphragms, but does not specify a design approach. In Europe, both the design approaches according to Schardt & Strehl [1], [2] and according to Bryan & Davies [3], [4] can be used. Bryan & Davis' approach was predominantly used in the UK and Scandinavia, while Schardt & Strehl's approach was used in Central Europe. While Bryan & Davis' approach claims to be more realistic in terms of diaphragm stiffness, Schardt & Strehl's approach is much easier to use. Most recently, an approach combining the advantages of both approaches was published [5] which will be presented and discussed here.

It should be highlighted that application of non-European approaches would also be possible, as long as they follow the general provisions in EN 1993-1-3. After some small adaptions, this applies for example for the approaches published by SDI [6] (basically developed by Luttrell), by the US Army Corps of Engineers [7] (the so-called Tri-Services Manual) or in the AISI standard [8].

2. Basic definitions

Beside the general comparison of the different European approaches, this paper focuses on the diaphragm stiffness S, namely on the part of sheeting distortion (warping) and on global buckling strength. To allow for easier understanding, it should be noted that different definitions or different units for stiffness are applied with the verification of diaphragms. The following definitions may also be found:

- S [kN/m]: shear stiffness of the diaphragm;
- S_V [kN/m]: "shear stiffness of the diaphragm, per unit length of the span" (see EN 1993-1-3, section 10.3.5);
- S [kN]: "portion of the shear stiffness provided by the sheeting for the examined member" (see EN 1993-1-3, section 10.1.1);
- S [kN/m]: "shear stiffness (per unit of beam length)" (in EN 1993-1-1, Annex BB).

We recommend the following definitions:

- S [kN]: shear stiffness of the diaphragm;
- S_i [kN]: shear stiffness related to the individual member;
- S_V [kN/m]: equivalent shear modulus.

In some cases, the reference length is the span L_{Si} (or s) instead of the diaphragm length L_S, see Fig. 1 for explanation of dimensions.

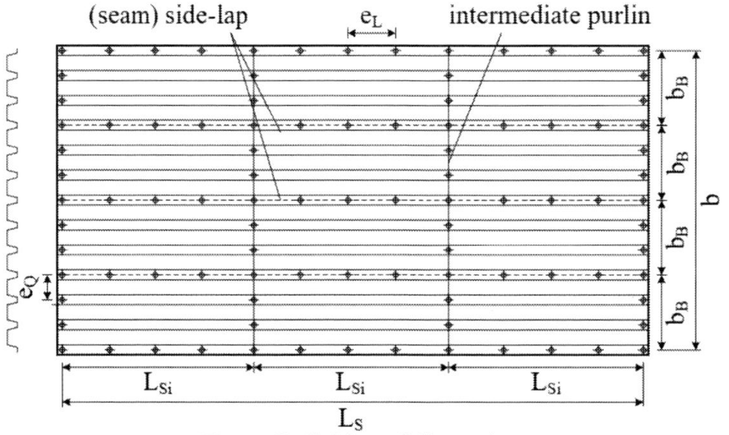

Fig. 1: Definition of dimensions

3. Introduction of current European approaches

3.1 General

All approaches are based on the following verifications to be made:

- Verification of the resistance of the diaphragm at ultimate limit state, usually done on the basis of limiting design value $T_{V,Rd}$ in [kN/m] of shear flow.
- Limitation of the angle of shear deformation γ_S at serviceability limit state, usually done on the basis of limiting value $T_{V,Cd}$ in [kN/m] of shear flow.
- Verification of the load-bearing capacity of the connections.
- Verification of the load-bearing capacity at sheeting panel's supports.

The last two verifications are not discussed in this paper.

3.1 Schardt & Strehl's approach

Schardt & Strehl's approach is based on a model assuming fully linear elastic behaviour of the diaphragm. Ultimate limit state is therefore defined by yielding in the of corner radius at transition of flange to web. Resistance therefore is rather low, relationship to reality questionable. Serviceability limit state is defined by the angle of shear deformation $\gamma_S = 1/750$. Verification is done on the basis of shear flow:

$$T_{V,Cd} = S_V \cdot \gamma_{S,\max} = \frac{S}{L_S} \cdot \gamma_{S,\max} \tag{1}$$

With Schardt & Strehl's approach, an additional verification at serviceability limit state is stipulated for adhesively bonded bituminous membranes.

Calculation of diaphragm stiffness is rather simple since flexibility of connections is neglected. This may even be reasonable with the background of a linear elastic model. The stiffness equation is composed by two terms related to shear deformation and sheeting distortion.

$$S = S_V \cdot L_S = \frac{L_S}{K_1 + \frac{K_2}{L_S}} \tag{2}$$

Necessary parameters K_1 and K_2 are usual given in tables provided by the manufacturer which makes application easy.

3.2 Bryan & Davies' approach

With Bryan & Davis' approach, ultimate limit state is governed by buckling of the diaphragm. Global buckling resistance of the diaphragm is calculated using

$$T_{Rk,g} = 1.1 \cdot T_{cr,g} = \frac{36}{L_S^2} \cdot \sqrt[4]{D_{x,g} \cdot D_{y,g}^3} \tag{3}$$

with orthogonal bending stiffness $D_{x,g}$ and $D_{y,g}$, see also EN 1993-1-5, Annex D. The equation is virtually identical to the equation for shear flow corresponding to the linear elastic buckling stress, only the constant factor is slightly increased by 10% (36 instead of 32.4) to take into account of post-critical buckling resistance. For multi-span systems, [3] and [4] give adapted equations, leading to increased resistance.

Local buckling resistance is calculated as linear elastic buckling stress using

$$T_{Rk,l} \equiv T_{cr,l} = k_\tau \cdot \frac{\pi^2 \cdot E}{12 \cdot (1 - \nu^2) \cdot L_S} \cdot \left(\frac{t}{a_{\max}}\right)^2 = 4{,}83 \cdot \frac{E}{L_S} \cdot \left(\frac{t}{a_{\max}}\right)^2 \tag{4}$$

with a_{max} being the width of the plane cross-section elements of the sheeting.

Interaction of local and global buckling has to be taken into account for

$$a > 2.9 \cdot t \cdot \sqrt{E/f_y} \tag{5}$$

with

$$T_{Rk} = \frac{T_{Rk,g} \cdot T_{Rk,l}}{T_{Rk,g} + T_{Rk,l}} \tag{6}$$

There are no defined verifications for serviceability limit state in [3] or [4], but limitation of angle of shear deformation is a suggested one. If this is desired, calculation may be based on the shear stiffness

$$S = b \Big/ \sum c_{i,j} \tag{7}$$

with

$$c = c_{1,1} + c_{1,2} + c_{2,1} + c_{2,2} + c_{2,3} + c_3 \tag{8}$$

The different component flexibilities $c_{i,j}$ refer to the sheeting deformation, the flexibility in the connections and strain in the edge members. Even though the approach is quite realistic and allows accurate capture of boundary conditions, calculation is cumbersome and preparation of easy-to-use tables is nearly impossible. A first attempt to simplify this approach was therefore presented in [9], proposing the formula

$$S = S_V \cdot L_S = \frac{L_S}{K_1 \cdot \alpha_2 + K_1^* \cdot e_L + \dfrac{K_2 \cdot \alpha_1 \cdot \alpha_4 + K_2^* \cdot \alpha_3}{L_S}} \tag{9}$$

for the shear stiffness. This can be regarded as the first step towards the combined approach as presented in the next chapter. The parameters K_1 and K_2 are calculated according to Schardt & Strehl and further parameters are

$$K_1^* = s_s \Big/ b_B \tag{10}$$

$$K_2^* = 2 \cdot s_p \cdot e_Q \tag{11}$$

with spacing of fasteners e_L at side laps and e_Q at transverse edge (usually equivalent to the pitch) plus flexibility s_s of connections at side laps (seams) and s_p of connections at transverse edges. The spacing of side-laps b_B is also the width of the sheeting panels, see Fig. 1. The coefficients α_i take into account of the number of spans. They were actually set to unity in [9], but often maintained when using this approach. Beside these coefficients there is no difference between Schardt & Strehl's and Bryan & Davies' approaches any more. As at serviceability limit state influence of flexibility of connections should be negligible

$$T_{C,k} = \gamma_{S,\max} \cdot S_V = \frac{1}{750} \cdot \frac{1}{K_1 \cdot \alpha_2 + \dfrac{K_2 \cdot \alpha_1 \cdot \alpha_4}{L_S}} \tag{12}$$

may be used instead of Eqn. (9). Nevertheless Eqn. (9) should be used if shear stiffness is used in ultimate limit state design, for example in stability design of purlins.

4. The "combined approach"

4.1 Basic design procedure

The "combined approach", which basically goes back to Baehre & Wolfram [9], was first presented in [5]. Some more information about preparation for practical use can be found in [10]. As with Bryan & Davies' approach, the ultimate limit state is defined by local buckling of the plane cross-section elements of the sheeting panel or global buckling of the diaphragm. Different from their approach, the buckling resistance is calculated using buckling curves according to EN 1993-1-3 and EN 1993-1-5: For global buckling, this results in taking into account of post-critical buckling resistance. As usual with buckling curves, slenderness

$$\bar{\lambda}_{c,g} = \sqrt{\frac{f_{yb}}{\tau_{cr,g} \cdot \sqrt{3}}} = \sqrt{\frac{f_{yb} \cdot t}{T_{cr,g} \cdot \sqrt{3}}} \tag{13}$$

is controlling input parameter to calculate the reduction factor

$$\chi_{c,g} = \frac{1.5}{0.5 + \bar{\lambda}_{c,g}^2} \leq 1.0 \tag{14}$$

For local buckling, procedure is basically the same, using critical buckling stress equivalent to Eqn. (4), but with buckling curve according to EN 1993-1-3, section 6.1.5 to calculate $\chi_{c,l}$. Buckling resistance is finally calculated using the minimum reduction factor χ_c:

$$T_{R,d} = \chi_c \cdot \frac{f_{y,b}}{\gamma_{M1} \cdot \sqrt{3}} \cdot t \tag{15}$$

Interaction of local and global buckling needs not to be taken into account: In [4], Bryan & Davies also neglected the interaction which is comprehensible when comparing the absolute values for global and local buckling resistance: Resistance to local buckling is often by the power of ten higher than resistance to global buckling. Baehre & Wolfram [9] did also not consider interaction.

Angle of shear deformation should be limited to $\gamma_{s,max}$ = 1/750 in serviceability limit state. This can be easily done on the basis of the limiting shear flow. Following Schardt & Strehl's approach, the limiting shear flow is calculated with

$$T_{C,k} = \gamma_{S,\max} \cdot S_V = \frac{1}{750} \cdot \frac{1}{K_1 + \frac{K_2}{L_S}} \tag{16}$$

not taking into account the flexibility of the connections. The limiting shear flow $T_{b,Ck}$ should be additionally considered for applications with adhesively bonded bituminous membranes.

In case diaphragm stiffness is used for ultimate limit state design, for example in the stability design of purlins, flexibility of connections has to be considered and Eqn. (2) has to be expanded accordingly:

$$S = S_V \cdot L_S = \frac{L_S}{K_1 + K_1^* \cdot e_L + \frac{K_2 + K_2^*}{L_S}} \tag{17}$$

4.2 Preparation of design tables

For ease of use, basic parameters necessary for the design of a diaphragm should be given in tables, provided by the manufacturer of the sheeting panel. Tables 2 and 3 show how this might be done, here for a sheeting type 100/275, direction of lay with seams in trough (decking). More information on use of design tables for sheeting is given in [10].

For serviceability limit state design, basic parameters are the limiting shear flow $T_{b,Ck}$ for applications with adhesively bonded bituminous membranes and the parameters K_1, K_2, K^*_1 and K^*_2 to be used with Eqn. (16) or Eqn. (17). The later ones were calculated with $s_s = s_p$ = 0.35 mm/kN, see [9] and [11].

The shear flow $T_{Rk,g}$ is the global buckling resistance of the diaphragm, calculated for a reference length L_R. In case of the largest span of a diaphragm $L_{Si} > L_R$, the global buckling resistance has to be reduced according to

$$T'_{Rk,g} = T_{Rk,g} \cdot \left(\frac{L_R}{L_{Si}}\right)^2 \tag{18}$$

The shear flow $T_{Rk,l}$ representing the local buckling resistance usually becomes not decisive. Calculation of forces at end supports is done via coefficient K_3. For ultimate limit state design, the partial factor γ_M according to EN 1993-1-3 should be used.

Table 1: Parameters for serviceability limit state design (example for sheeting panels 100/275)

t_N mm	$T_{b,Ck}$ kN/m	K_1 10^{-4}·m/kN	K_2 10^{-4}·m²/kN	K^*_1 10^{-4}·1/kN	K^*_2 10^{-4}·m²/kN
0.75	1.86	0.254	35.192	4.242	1.925
0.88	3.96	0.215	23.115	4.242	1.925
1.00	2.84	0.188	16.555	4.242	1.925
1.25	7.06	0.149	9.282	4.242	1.925
1.50	11.29	0.124	5.804	4.242	1.925

Table 2: Parameters for ultimate limit state design (example for sheeting panels 100/275)

t_N mm	$T_{Rk,g}$ kN/m	L_R m	$T_{Rk,l}$ kN/m	K_3 -
0.75	8.09	7.0	21.46	0.406
0.88	10.41	7.0	35.54	0.442
1.00	12.71	7.0	53.05	0.472
1.25	17.98	7.0	96.80	0.530
1.50	23.82	7.0	116.80	0.582

4.3 Comparison of current approaches

The three European approaches have many similarities, but some distinct differences. Table 3 gives a comprehensive overview.

Table 3: Comparison of the European approaches

	Schardt & Strehl	Bryan & Davies	combined approach
ultimate limit state	yielding due to transverse bending in the radius at transition from flange to web	global and local buckling (elastic critical buckling stresses) with interaction	global and local buckling (with post-critical buckling resistance) without interaction
		limitation of stresses due to shear flow to $0.25 f_y$	limitation of stresses due to shear flow to $0.25 f_y$
serviceability limit state	limitation of global deformation via the angle of shear deformation to $\gamma_{S,max}$ = 1/750, neglecting the flexibility of the connections		
	-	coefficients α_1, α_2, α_4	-
	in case of adhesively bonded roofs: limitation of relative displacement between top flange to bottom flange (crest and trough)		
shear stiffness	neglecting the flexibility of the connections	considering the flexibility of the connections	considering the flexibility of the connections
	-	coefficients α_1, α_2, α_3, α_4	-

Fig. 2 shows the shear flow vs. displacement measured in a test. The diaphragm stiffness calculated according to the different approaches is shown, too. It can be seen that Schardt & Strehl's approach gives a good estimate for the initial stiffness, legitimating neglect of flexibility of connections at serviceability limit state. For ultimate limit state, flexibility has to be taken into account. The calculated diaphragm strength is also shown, in this case governed by strength of the connections at seams. The example is presented in full details in [5].

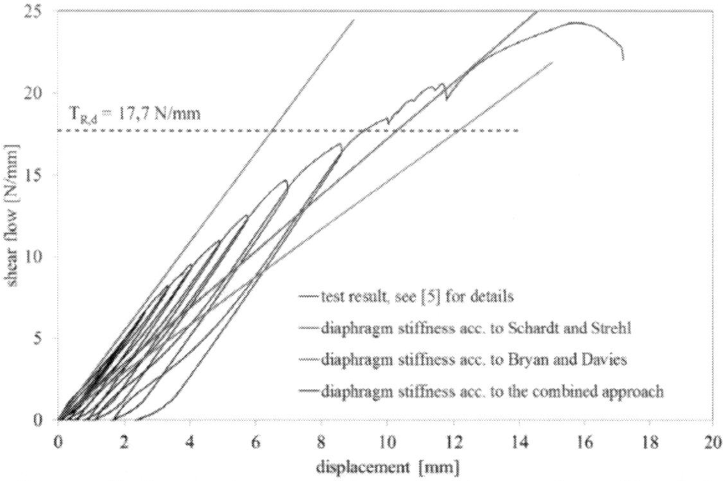

Fig. 2: Shear flow vs. displacement measured in a test and comparison with calculated stiffness

5. Remarks for practical application

5.1 Sheeting distortion and warping

As it was often pointed out that the deformations calculated following Bryan & Davies' approach are higher than measured in tests, this topic is worth some separate considerations. Both in [12] and [13] calculated flexibility due to warping distortion is identified as the reason. This flexibility is due to the sheeting not being continuously fastened, and because the shear centre does not coincide with where the shear flow is in the sheeting. Therefore a short comparative study is presented here, also including Luttrell's approach

$$S = S_V \cdot L_S = \frac{E \cdot t \cdot L_S}{2 \cdot (1+\nu) \cdot \frac{u}{b_R} + \rho \cdot D_n + C} = \frac{E \cdot t \cdot L_S}{2 \cdot (1+\nu) \cdot \frac{u}{b_R} + \rho \cdot \frac{D}{L_S} + C} \tag{19}$$

used in [6] and [8] for calculating the stiffness. As a basis for comparison, the warping parameters (unit m²/kN) were chosen: K_2 in Schardt & Strehl's approach and $D/(E \cdot t)$ in Luttrell's approach. A warping parameter for Bryan & Davies' approach suitable for direct comparison does not exist, but can be calculated from the flexibility $c_{1,1}$ due to sheeting distortion: With

$$c_{1,1} = \frac{b \cdot b_R^{2.5} \cdot K}{E \cdot t^{2.5} L_S^2} = \frac{b}{L_S^2} \cdot K_W \tag{20}$$

we obtain

$$K_W = \frac{b_R^{2.5} \cdot K}{E \cdot t^{2.5}} \tag{21}$$

as a warping parameter. Coefficients α_1 and α_4 are set to unity here. It should be pointed out that different values K are given in [3] and [4] resulting from further improvement of calculation procedures. For webs perpendicular to the plane of the diaphragm ($\phi = 0°$ or $\theta = 90°$), values given in [4] need to be replaced by the ones given in [14].

A parametric study was performed, covering the range of common cross-sections with heights of 35 mm to 165 mm. Both directions of lay were covered (decking with wide flanges in crest, sheeting with wide flanges in trough). The following conclusions could be drawn:

1. Values of warping distortion calculated with Bryan & Davies' approach and Schardt & Strehl's approach are generally similar: Bryan & Davies' approach gives about 90% of the values of Schardt & Strehl's approach. Sole exception are the values for sheeting with wide flanges in trough were calculation according to [4] gives just about 75% of the values of Schardt & Strehl's approach.
2. Luttrell's approach gives much smaller values of warping distortion as he neglects out-of-plane deformation from torsion of the ribs. For decking, warping distortion is just about one third of the values predicted by Schardt & Strehl's approach. For sheeting, difference is even much larger, leading to just about 8% of the values predicted by Schardt & Strehl's approach. Especially the latter values are questionable, because with wide flanges fixed, out-of-plane deformation plays a major role in warping distortion.
3. Theoretical approaches for derivation of warping distortion operate with idealized boundary conditions while in practice, tests and often also in numerical investigations there is an overhang of the sheeting panel over the theoretical line of support defined by the fastener and also a finite length of support. This leads to boundary conditions deviating from the idealized ones by an elastic spring restraint of warping distortion. Numerical investigations working with the idealized boundary conditions assumed in theoretical approaches confirm the values given in [4] with a deviation of 5% in the maximum, see [14] and further investigations of these authors.

5.2 Global buckling

The global buckling curve introduced with the "combined approach" makes demand of post-critical buckling resistance. This raises the question if this post-critical buckling resistance can be activated with shear diaphragms by such an amount: The test data forming the basis of EN 1993-1-5, Annex D, consists of results of tests with plate girders with corrugated webs [15]. Up to now, [3] takes just a small amount of post-critical resistance into account, realized by the constant factor 36 instead of $32.4 = 3.28\ \pi^2$ determined in [16].

To verify this, a survey on tests was done, whose result is shown in Fig. 3. The number of tests in which buckling failure was observed is generally limited. Unfortunately, with this limited number of tests, basic information for re-calculation is often missing. This refers mainly on material properties; for example the yield strength assumed in the re-calculation of the tests [17] and [18] had to be estimated being equal to the values given for specimens from other tests. On the other hand, documented failure loads are often not maximum loads, thus not taking into account post-critical behavior. In [17] differentiation is made between "observed buckling loads" and "theoretical failure loads", leading to the assumption that the first mentioned loads are the ones measured with first buckling when reaching the level of the linear elastic buckling stress. This can be, for example, clearly seen when looking at the load-deflection-curves given in [18] and [19]. Buckling resistance is defined at the point of first bent (end of linear elastic range) which should correspond to the linear elastic buckling stress. The authors assume that the slight increase in linear elastic buckling stress observed in the tests is caused by the boundary conditions, providing some small amount of clamping. The only tests were consideration of post-critical resistance is confirmed are the ones documented in [20]. These tests were performed on aluminium sheeting panels. Interestingly, despite the nonlinear stress-strain-curve associated with aluminium, the results are situated exactly on the buckling curve.

It should also be noted that design against global buckling resistance requires fixing in every trough at end support. Otherwise the constraints adhering to the derivation of the elastic critical buckling stress and the buckling curve are violated. Failure might be observed by local effects. This was obviously the case with some of the tests published in [17] and [21], were in each case three tests were performed with fixing in alternate troughs. In [3] and [4] a reduction of critical buckling stress by half is recommended for fixing in alternate trough which was not applied in this evaluation.

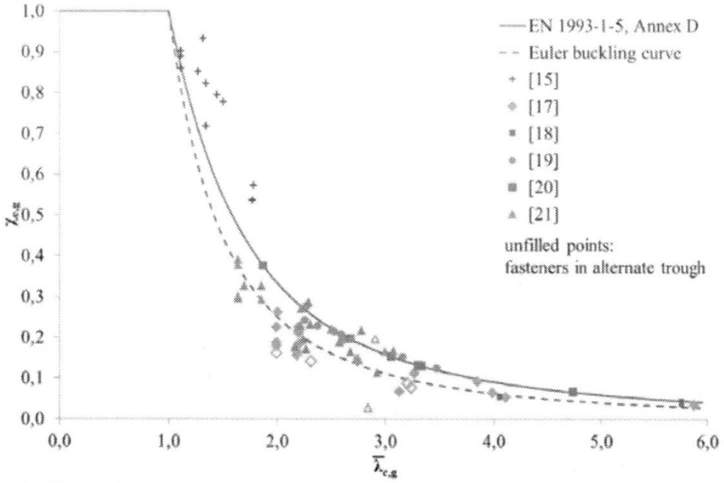

Fig. 3: Global buckling failure of diaphragms – comparison of buckling curve with tests

6. Conclusions

Compared to current approaches for the design of shear diaphragms made of trapezoidal sheeting, the new "combined approach" has significant advantages in practical design. Independent from the approaches, open questions of fundamental nature were identified and discussed, showing the need for further research.

References

[1] Schardt R, Strehl C. "Theoretische Grundlagen für die Bestimmung der Schubsteifigkeit von Trapezblechscheiben – Vergleich mit anderen Berechnungsansätzen und Versuchsergebnissen". *Stahlbau* 45(4), 97-108, 1976.

[2] Schardt R, Strehl C. "Stand der Theorie zur Bemessung von Trapezblechscheiben". *Stahlbau* 49(11), 325-334, 1980.

[3] Davies JM, Bryan ER. *Manual of stressed skin diaphragm design*, London: Granada Publishing, 1982.

[4] ECCS. *European Recommendations for the Application of Metal Sheeting acting as a Diaphragm - Stressed Skin Design (ECCS Publication No. 88)*, Brussels: European Convention for Constructional Steelwork, 1995.

[5] Kathage K, Lindner J, Misiek Th, Schilling S. "Proposal to adjust the design approach for the diaphragm action of shear panels according to Schardt and Strehl in line with European regulations", *Steel Construction – design and research*, 6(2), 107-116, 2013 & 7(4), 281, 2014.

[6] SDI. *Diaphragm Design Manual*, Fox River Grove: Steel Deck Institute, 2004.

[7] Departments of the Army, the Navy, and the Air Force. *Seismic Design for Buildings, TM 5-809-10; NAVFAC P-355; AFM 88-3, Ch 13*, Washington: Military Publications Division, 1982.

[8] AISI. *North American Standard for the design of profiled steel diaphragms panels*, Washington, Detroit, Pittsburgh: American Iron and Steel Institute, 2013.

[9] Baehre R, Wolfram R. "Zur Schubfeldberechnung von Trapezblechen". *Stahlbau*, 55(6), 175-179, 1986.

[10] Huck G, Misiek Th. "Überarbeitete Anlagen der Typenprüfungen für Trapezprofile – Erläuterungen und Hintergründe". *Stahlbau*, 83(12), 873-879, 2014.

[11] Misiek Th, Käpplein S. "Strength and stiffness of shear-loaded fastenings of metal members and sheeting using fastening screws and rivets". *Steel Construction – Design and Research*, 6(1), 14-23, 2014.

[12] Baehre R. "Zur Schubfeldwirkung von Aluminiumtrapezprofilen". *Stahlbau*, 62(3), 81-87, 1993.

[13] Bakhti F, Tremblay R, Rogers CA. "Revisiting the SDI and ECCS method for in-plane shear flexibility of metal roof deck diaphragm using 3D nonlinear finite element analysis", *Proceedings of the 15th World Conference on Earthquake Engineering on Connections*, Lisbon, Portugal, 2012.
(http://www.iitk.ac.in/nicee/wcee/article/WCEE2012_2482.pdf).

[14] Dürr M, Saal H. "Influence of profile distortion on the shear flexibility of profiled steel sheeting diaphragms", *Recent research and developments in cold-formed steel design and construction: Seventeenth International Specialty Conference on Cold-Formed Steel Structures* (Ed.: RA LaBoube), Orlando, Florida, US, 617-624, 2004
(http://scholarsmine.mst.edu/isccss/17iccfss/17iccfss-session8/5)

[15] Johansson B, Maquoi R, Sedlacek G, Müller C, Beg D. *Commentary and worked examples to EN 1993-1-5 "Plated Structural Elements"*, Brussels: European Commission Joint Research Center, European Convention for Constructional Steelwork (Joint Report), 2007 (http://eurocodes.jrc.ec.europa.eu/show_Entity.php?file_id=EC_00000028).

[16] Bergman S, Reissner H. "Neuere Probleme aus der Flugzeugstatik – Über die Knickung von Wellblechstreifen bei Schubbeanspruchung". *Zeitschrift für Flugzeugtechnik und Motorluftschiffahrt*, 20(18), 475-481 1403-1417, 1975

[17] Lawson RM. *The flexibility and strength of corrugated and folded plates*, Department of Civil Engineering, University of Salford, 1976 (http://usir.salford.ac.uk/14884/).

[18] McFarland DE. *An Investigation of the static stability of corrugated rectangular plates loaded in pure shear*, Department of Mechanics and Aerospace Engineering, University of Kansas, 1967 (http://hdl.handle.net/1808/18628).

[19] Easley JT. "Buckling formulas for corrugated metal shear diaphragms". *Journal of the structural Division*, 101(7), 1403-1417, 1975.

[20] Baehre R. *Gutachten über die Schubfeldwirkung von Aluminiumtrapezprofilen*. Karlsruhe, 1985.

[21] Nunna RV. *Buckling of profiled steel diaphragms*, AISI Buckling Task Group, Report No. 11-03, 2011 (http://scholarsmine.mst.edu/ccfss-library/163).

LOCAL AND GLOBAL STABILITY ANALYSIS OF CFS STRUCTURAL MEMBERS WITH PARTICULAR SHAPES USING SPOT WELDING TECHNOLOGY

Zsolt Nagy[a] and Robert Ballok[b]

[a] *Technical University of Cluj-Napoca, Faculty of Civil Engineering, Cluj-Napoca, Romania*
[b] *Gordias LTD, Cluj-Napoca, Romania*

Abstract: The paper analyze the local and global stability of built-up cold formed steel members using spot welding technology. Local and global stability of cold formed steel members is a classical problem which has analytical solutions for basic shapes. In case of non-conventional built-up members this problem exhibits a difficult task. The built-up member presented is composed by four U-profiles, forming a boxed section with back-to-back distanced U members in the web. The U profiles are connected using spot welding technology at the level of the obtained I shaped section. The built-up members are used as secondary column to sustain wall cladding system, for steel and concrete structures with large frame interspace. A non-linear finite element model is used to simulate the behavior of the built-up member subjected to bending and compression. The analysis is performed using Abaqus software package. Initial imperfections of the member are included in the model, considering the buckling mode shapes obtained from preliminary linear perturbation analysis. The paper provide preliminary conclusions on the structural performance of the considered built-up member.

1. Introduction

The use of built-up cold-formed steel sections instead of single profiles may offer greater economy in building construction [1]. Cold formed single profiles are not suitable for large spans and heavier loads, as most of cold profiles are mono symmetric sections and have relatively small torsional rigidity. The basis of research of cold-formed built-up members is based on results obtained on hot-rolled built-up members [2].

Cold-formed built-up members are usually composed of simple sections (C, Z, Sigma, Hat section). The individual sections are connected into the new section by using bolt, screw or weld. Prediction on flexural buckling and torsional buckling resistance between single and a built-up member can give significant differences (20 to 80 %) [3].

Studies concerning built-up members conclude that preventing shear slip in connections of built-up members is a critical factor for strength [4]. The slip in connections leads to additional flexibility, making more complicated the development of a suitable design method for built-up members [5]. Spot-weld connections avoid the slip in connections but these can introduce residual stresses. Moreover, the use of spot-weld connection increases the degree of automation of fabrication process. To use the advantages of cold-formed steel profiles and carry heavier loads or cover larger spans, connection influence in built-up arrangements and flange/web stiffeners placement were also studied by the research community [6], [7].

The built-up member presented in this paper is composed of cold-formed U profiles which is realized using spot-welding technology. The new built-up member will serve as secondary column to support the cladding system for steel or precast concrete structures, in case of increased bays and for relatively high buildings. Usually for such intermediate columns are used hot rolled profiles. The application of the proposed built-up member is presented in Fig. 1.

Fig. 1: Application as secondary column of the proposed built-up member

2. The proposed built-up member

The built-up members presented in this study are obtained by combining of four U-profiles, forming an I-section with back-to-back distanced U members in the web. The U profiles are interconnected at discrete points using spot welding technology at the level of the obtained I shaped section. The obtained boxed shape provides lateral and torsional stiffness. The built-up member is proposed for support of the cladding system, withstanding bending and axial load.

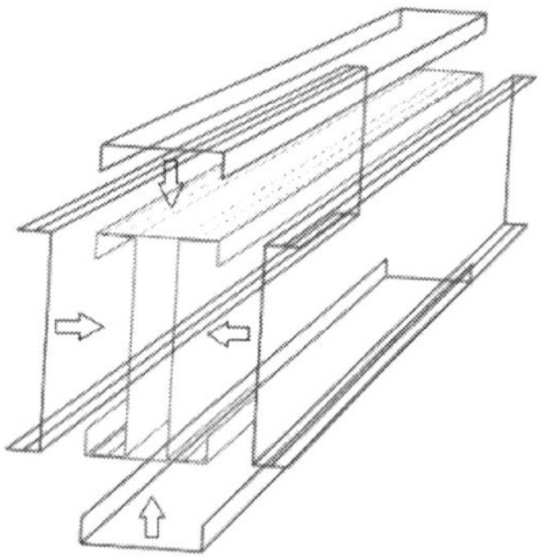

Fig. 2: Four U profiles forming a boxed I-section

Fig. 2 presents the concept of the new built-up section while Fig. 3 give details about the spot-welding machine, the section of the built-up member and the applied prototype column.

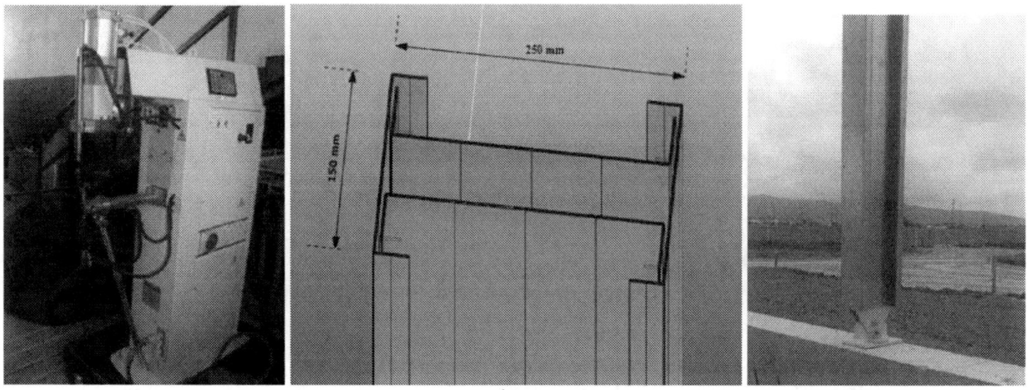

a) b) c)

Fig. 3: a. Spot welding machine, b. Composed section, c. The prototype column

The goal was to obtain built-up members with higher strength-to weight ratio and reduced sensitivity to bucking effects and initial imperfections. The columns in this particular case behave more like a beam, as axial load is almost negligible. The dominant load combination for such a member in a structure will results from the wind load. For wind pressure, the compressed outer flange of the built-up member is stiffened by the cladding system. But in case of wind suction this has unfavorable effect on the column, as the compressed inner flange is laterally not supported.

The used cold-formed steel profiles were selected conducting preliminary FEM simulations. Initially C-profiles were proposed to build the column, but numerical results shows no benefits of the channel section lip. Hence the channel sections were replaced with U profiles.

For the spot welds a nominal spacing of 150 mm was determined. The number of spot-welds along the column was evaluated according to the required shear strength, as well as to undertake the shear flow between two spot-welds. On the other hand, the distance between spot welds is important from local-buckling point of view: previous numerical analyses with larger spot-weld spacing shows local buckling of individual elements between discrete fastenings.

The nominal dimensions of built-up member specimens are presented in Table 1. The length of the column for the required application is 6200 mm. The cross-section depth is 250 mm and the flange width is 150 mm.

Table 1

Column length	6.2 m
Weight	104 kg
Flanges	U150/2
Web	2 × U250/2
Spot welding density	150 mm

The columns are pin ended. As Fig. 4 shows, welded brackets were attached with 8 mm diameter self-tapping screws to the built-up sections, in order to simulate the hinged-end conditions at both supports.

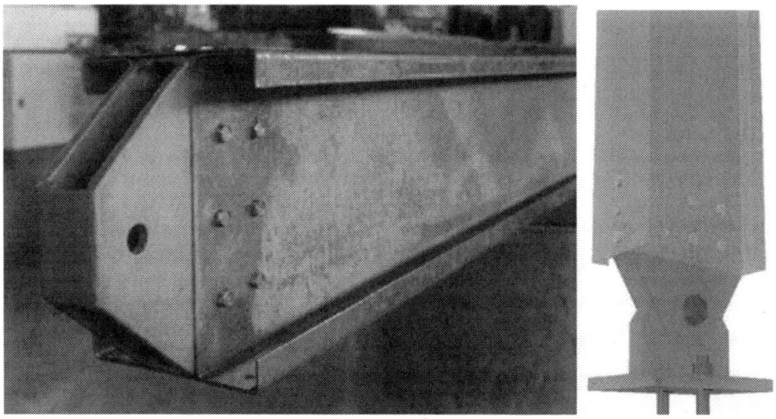

Fig. 4: Column end connection

3. FEM model details of built-up member

To evaluate the performance of the proposed built-up column arrangement, predicting resistance, stiffness and stress distributions as well as interaction of elements of the built-up member, 3D FE models were developed using Abaqus 6.13 [8]. The analysis type conducted is a nonlinear displacement-based simulation using explicit dynamics. The model consists of S4R shell elements. To keep mesh regularity, the rounded corners of the sections were not included in the model [9]. Material plasticity is accounted in the analysis, assigning elastic-perfect plastic material properties with yield strength of 350 N/mm^2. Young modulus and Poisson's ratio were taken 210,000 N/mm^2 and 0.3 respectively.

For the modelling of spot-welds, point-based mesh independent fasteners were disposed using CONN3D2 connector elements, set up to provide a rigid behavior for each directions. To capture local instability phenomena, geometrical imperfection was specified, as it is described below. A good mesh regularity was obtained using structured meshing technique and quad element shapes with 30 mm mesh size.

The displacement is applied to a reference point which in turn is coupled at the end of the column. Fig. 5 shows the meshed FE model of the built-up column.

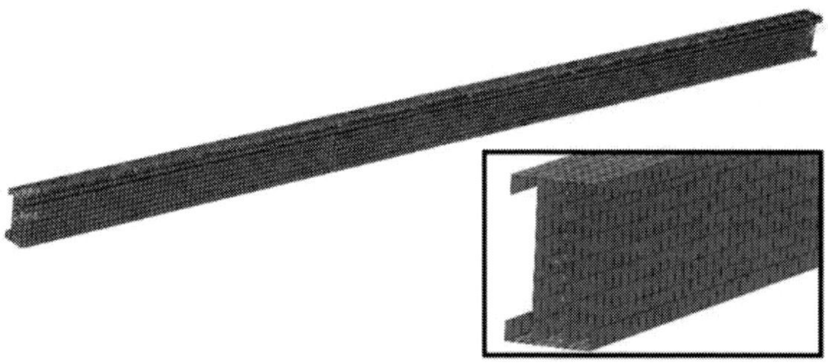

Fig. 5: FE model

The column is loaded initially with an axial load of 12600 N, representing the self-weight of the wall system supported by the column. The self-weight of the column is considered applying gravity load in corresponding direction using gravity load module available in Abaqus 6.13. Maintaining the axial load, the displacement is applied perpendicular on the column with smooth growing. The analysis ran in displacement control. Displacement is distributed using constraints through coupling load surfaces to the reference point as shown in Fig. 6a. Thus, the equivalent load with the applied displacement is uniformly distributed. The adequate step time period for the analysis was found 6 second.

The edges of the column on the load surface level was prevented from lateral displacement, considering the benefits provided by cladding system. Fig. 6 presents the uniformly load application method and the column end conditions set up for the analysis.

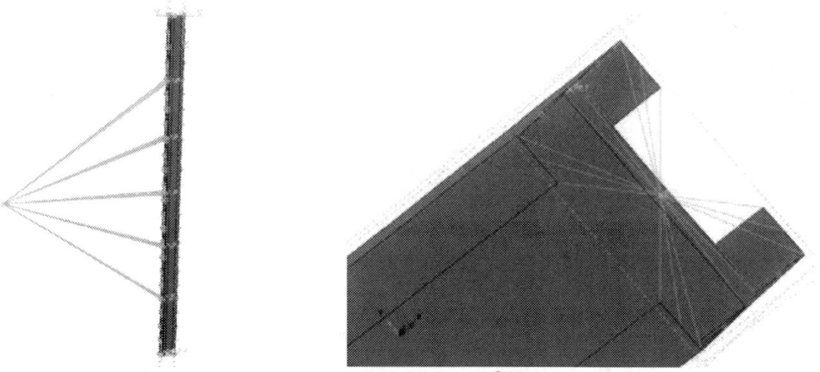

a) b)

Fig. 6: a) Uniform load distribution, b) column end conditions equivalent with pinned end

Because the high slenderness of the cold-formed profiles, geometrical imperfection may have influence on the response of built-up member. To include this imperfection in the model, prior to the non-linear analysis, an eigen-frequency analysis of the perfect model was performed. Thus the buckling modes were obtained. Several numerical approaches using this type of analysis were proposed in the literature to generate geometrical imperfect shapes [10], [11]. The perfect shape of the column was disturbed with the buckling mode shape introducing the imperfection command in the "edit keyword" window (Fig. 7). The command loads an imperfect shape from a selected analysis output database file, specified as first step. The second line of the command specifies the number of the buckled shape selected for the analysis and the amplitude of maximum deflection.

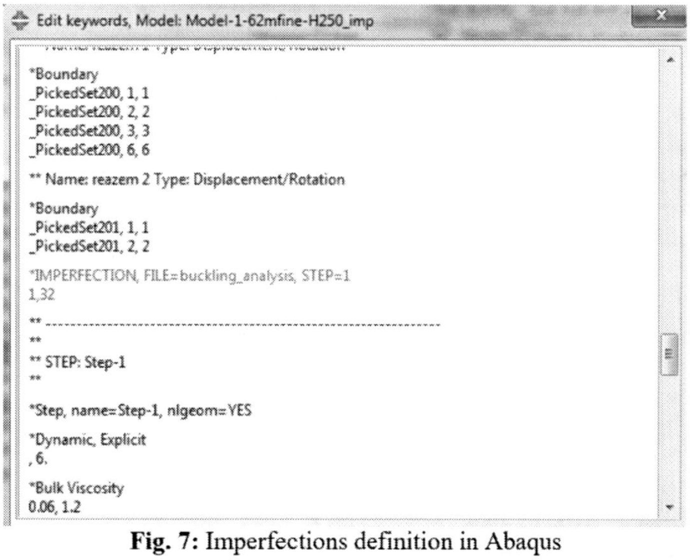

```
Edit keywords, Model: Model-1-62mfine-H250_imp
*Boundary
_PickedSet200, 1, 1
_PickedSet200, 2, 2
_PickedSet200, 3, 3
_PickedSet200, 6, 6
** Name: reazem 2 Type: Displacement/Rotation
*Boundary
_PickedSet201, 1, 1
_PickedSet201, 2, 2
*IMPERFECTION, FILE=buckling_analysis, STEP=1
1,32
** ----------------------------------------------------------------
**
** STEP: Step-1
**
*Step, name=Step-1, nlgeom=YES
*Dynamic, Explicit
, 6.
*Bulk Viscosity
0.06, 1.2
```

Fig. 7: Imperfections definition in Abaqus

The first buckling shape was considered to generate the model with imperfections. The global imperfection amplitude was taken as $L/200$, as initial bow imperfection, corresponding to curve b for plastic analysis, according to Table 5.1 of EN1993-1.1, L being the full length of the column. The model on which the frequency analysis is conducted, must have the same meshing as the non-linear model proposed for analysis. Fig. 8 shows the imperfect shape of the model with the considered scale of imperfection.

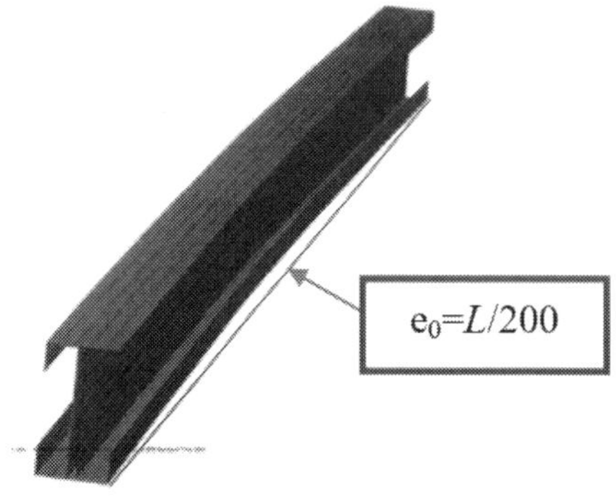

Fig. 8: Imperfect shape using the first buckling mode

4. Analysis results and discussions

The results of the nonlinear analysis confirmed the initial ULS and SLS design results. The column was designed to resist for 3.4 kN/m uniformly distributed wind load and 12.6 kN axial load. The non-linear analysis was performed considering both with and without geometrical imperfections. Figs. 9 and 10 capture the failure mechanism of the analyzed column under pressure and suction loads. Fig. 11 shows the obtained structural response of the specimen in terms of load-deflection behavior.

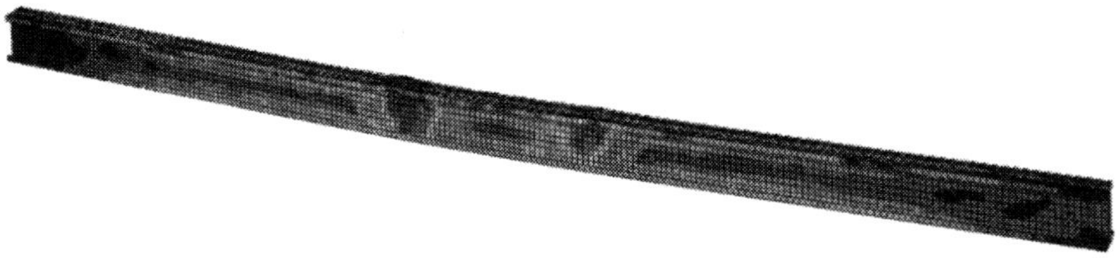

a)

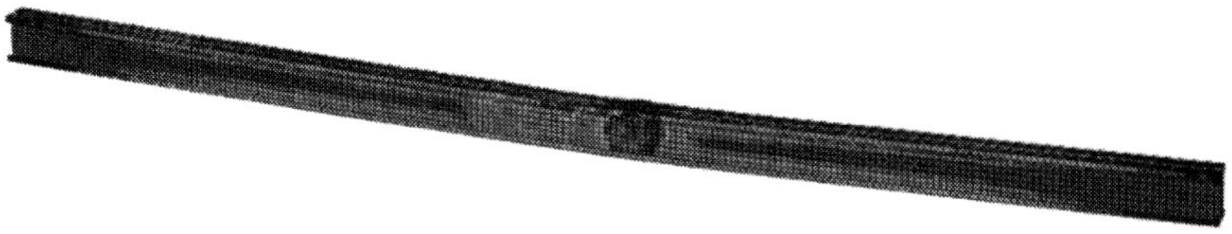

b)

Fig. 9: Failure of column subjected to pressure load (compressed flange laterally restrained): a) model with imperfections, b) model without imperfections

The numerical model without geometrical imperfections show almost no differences to pressure and suction load. The maximum load obtained for perfect model is 10.16 kN/m.

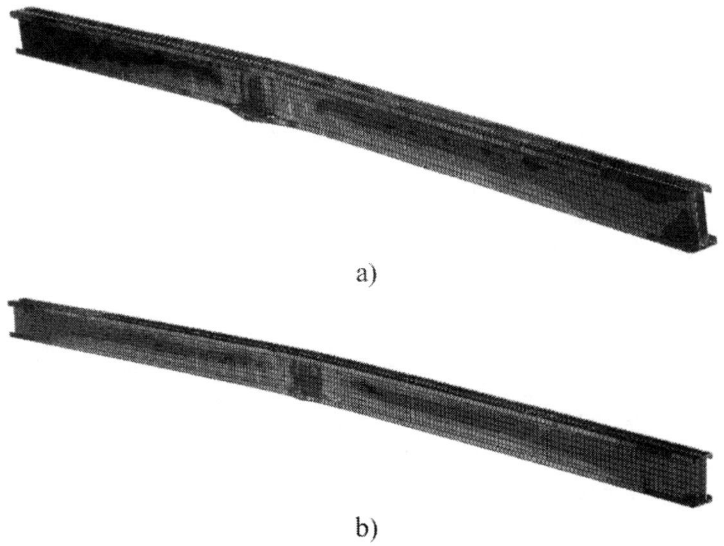

Fig. 10: Local failure of column subjected to suction load (compressed flange laterally not restrained): a) model with imperfections, b) model without imperfections

The load carrying capacity of the model including a bow imperfection from eigen-frequency analysis resulted 8.88 kN/m in case of pressure loads on wall and 8.16 kN/m for suction load, which is more than double of the required design load level. In case of imperfect model there are differences between pressure and suction load behavior, because the laterally unstiffened flange stability is decreased due to imperfections.

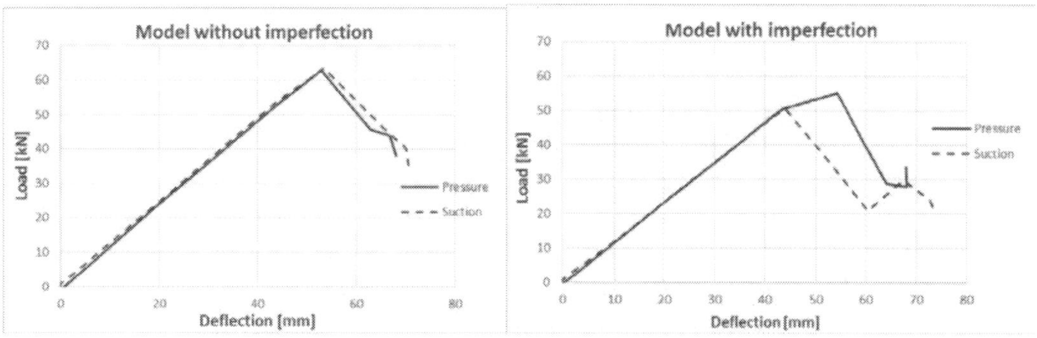

Fig. 11: Load-deflection curves without and with imperfections

5. Conclusions

The structural performance of an innovative spot welded cold-formed built-up steel column have been presented. Fast and easy fabrication of the built-up section leads to realize an efficient structure. The structural performance for the built-up columns made by cold-formed built-up steel profiles are evaluated numerically using non-linear FE simulation. The FE model includes the material and geometrical non-linearity. The analysis accounts for initial geometric imperfections using the eigen-shapes obtained from a non-linear buckling analysis. In case of pressure loads, where compressed flange is stabilized by the cladding, there is a 10 % decrease of load carrying capacity of the analyzed member considering imperfections. In case of suction loads, where compressed flange is not stabilized by the cladding, the load carrying capacity of the analyzed member considering imperfections is much more affected (decrease is about 20 %). The predicted ultimate loads obtained through numerical assessments provide promising results because the resulted specimen capacity is more than double of the required design load level, which confirm such built-up members to be used in construction practice. Research will continue with experimental investigation on the considered column, and several numerical assessments for different column heights and load conditions, as well as spot-weld arrangement along the column.

References

[1] LaBoube RA, Yu WW, Jones ML. "Spacing of Connections in Compression Flanges of Built-up Cold-Formed Steel Beams", *14th International Specialty Conference on Cold-Formed Steel Structures*, St. Louis, Missouri U.S.A., October 15-16, 1998.

[2] Anbarasu M, Kumar PB, Sukumar S. "Study on the capacity of cold-formed steel built-up battened columns under axial compression", *Latin American Journal of Solids and Structures,* 11(12), Rio de Janeiro, Brazil, 2014.

[3] Fadhluhartini M, Mohd Syharul HMS, Shahrin M, Mahmood MdT. "Ultimate load of built-up cold-formed steel column", *ARPN Journal of Engineering and Applied Sciences*, 9(11), 2095-2101, November 2014.

[4] Sherman DR, Yura JA. (1998). "Bolted double angle compression members", *Journal of Constructional Steel Research,* 46, 1998.

[5] Georgieva IB, Schueremans L, Vandewalle L, Pyl L. "Design of built-up cold-formed steel columns according to the direct strength method", *Procedia Engineering - Steel Structures and Bridges,* Vol. 40, 119–124, 2012.

[6] Fadhluhartini M, Mohd Syharul HMS, Shahrin M, Mahmood MdT. "Assessment of Connection Arrangement of Built-up Cold-formed Steel Section under Axial Compression" *Advanced Material Research,* Vol. 1043, 252-257, 2014.

[7] Arunal G, Sukumar S, Karthika V. "Study On Cold-Formed Steel Built-Up Square Sections With Intermediate Flange And Web Stiffeners" *Asian Journal Of Civil Engineering (BHRC)*, 16(7), 919-931, 2015.

[8] ABAQUS. "Theory Manual", Hibbit, Karlson and Sorenson, Inc., 2007.

[9] Georgieva IB, Schueremans L, Pyl L, De Roeck G. "Non-linear finite element analysis of built-up members of cold-formed steel profiles", *Proceedings of the 6th International Conference on Thin-Walled Structures – Recent Research Advances and Trends – Editors: Dubina. D. and Ungureanu V.,* Timisoara, Romania, September 5-7, 2011.

[10] Portioli F, Amico BD, Di Lorenzo G, Landolfo R. "Modeling of geometric imperfections in numerical simulations of built-up cold-formed steel beams", *Proceedings of the 6th International Conference on Thin-Walled Structures – Recent Research Advances and Trends – Editors: Dubina. D. and Ungureanu V.,* Timisoara, Romania, September 5-7, 2011.

[11] Schafer BW, Peköz T. "Computational modelling of cold-formed steel characterizing geometric imperfections and residual stresses", *Journal of Constructional Steel Research,* 47(3), 193-210, 1998.

EXPERIMENTAL INVESTIGATIONS OF COLD-FORMED STEEL TRAPEZOIDAL BEAMS OF SCREWED CORRUGATED WEBS

Zsolt Nagy[a], Viorel Ungureanu[b,c], Dan Dubina[b,c] and Robert Ballok[d]

[a] *Faculty of Civil Engineering, Technical University of Cluj-Napoca, Cluj-Napoca, Romania*
[b] *CMMC Department, Politehnica University of Timisoara, Timisoara, Romania*
[c] *Laboratory of Steel Structures, Romanian Academy, Timisoara Branch, Timisoara, Romania*
[d] *Gordias LTD, Cluj-Napoca, Romania*

Abstract: The paper presents the results of a new experimental investigation on steel trapezoidal beams made by cold-formed steel profiles for flanges (i.e. back-to-back lipped-channel sections) and corrugated sheets for web. The tests were carried out at the Technical University of Cluj-Napoca in co-operation with Politehnica University of Timisoara. Self-drilling screws have been used both for connecting the flanges to the web and as seam fasteners to ensure the continuity of the web. Monotonic tests were performed on two beams of 12 meter span, having different connecting arrangements between the flanges and web.

1. Introduction

The use of corrugated webs instead of plain ones brings many advantages but also several disadvantages. One major disadvantage derived from the production cost of the corrugated sheets and their connection with the flanges. Keeping the web of the beam, the reduction of self-weight is possible by the reduction of the thickness, resulting high slenderness and higher risk of web buckling. The use of corrugated web for the beam increases the shear buckling stability. As a result of its profiling, the web does not participate in the longitudinal transfer of bending stresses. Therefore, in static terms, the corrugated web beam behaves similar to a lattice girder, in which the bending moments and applied forces are transferred only via the flanges, while the transverse forces are only transferred through the diagonals and verticals of the lattice girder, in this case the corrugated web. So, the girder's flanges provide the flexural strength of the girder with no contribution from the corrugated web which provides the girder's shear capacity. The failure of the corrugated web occurs by steel yielding, web buckling (local or distortional or their interaction). Lateral-torsional buckling of the girder and local flange buckling of corrugated web, separately or in interaction, represent other possible failure modes. The Austrian company Zeman supply single story steel structures with large spans, using welded thin corrugated web to thick flanges [1].

Research related to bending moment capacity of corrugated web beams was done by Lindner and Aschinger [2,3]. They demonstrated that web shear flux produces weak axis bending moments, which reduce the overall bending moment capacity of the structural element. Sherman and Fisher [4] were studied the corrugated web and flange connection of such beams. They concluded that is enough to weld only the longitudinal edge of trapezoidal sheeting, to carry the longitudinal slippage between web and flange.

Lindner [5] studied by experimental tests the lateral-torsional behaviour of steel girders with corrugated webs and found that the torsional section constant I_T for a beam with corrugated web doesn't differ from that of a beam with flat web, but the warping section constant I_w is different.

The dimensioning of corrugated web beams is ruled by Annex D of the EN 1993-1-5:2006 [6], together with specific aspects of EN 1993-1-1:2006 [7] and EN 1993-1-3:2006 [8].

A new technological solution of such a system, composed by webs made of trapezoidal cold-formed steel sheets and flanges of built-up cold-formed steel members (back-to-back lipped channels) has been developed at the CEMSIG Research Centre (http://www.ct.upt.ro/en/centre/cemsig) of the Politehnica University of Timisoara [9-10]. The connections between flanges and web were made with self-drilling screws. Five beams with different arrangements for self-drilling screws and shear panels have been experimentally tested. It should be emphasize the new solution, as a whole, is 100% composed by cold-formed steel elements, avoiding the combination of two types of products, i.e. cold-formed for webs and hot-rolled for flanges.

A similar solution has been proposed and analysed in the frame of PRECASTEEL project [11], but using blind rivets as seam fasteners for the corrugated web and bolts for web-to-flange connections. For flanges, back-to-back lipped channel or two types of hat-sections have been used. Deep corrugation web sheeting of longitudinal intermediate stiffeners have been applied in this solution. However, looking to the test results, one observes the sensitivity to distortion of corrugation still remain high.

A structural steel solution used for portal frame configuration has been investigated at the Technical University of Cluj-Napoca [12]. The frame is made by cold-formed steel beams with screwed corrugated web and SHS columns, showing very good structural performance.

By extending the application of the technical solution described in [9-10] for parallel flanges girders, promising experimental results have been obtained on trapezoidal beams made of cold-formed steel profiles and corrugated web [13]. Monotonic test were performed on two beams of 12 meter span, having different connections between the flanges and web. The paper presents the experimental program and the obtained results.

2. Design of beam specimens for testing

In order to obtain realistic specimens for testing the selected geometry of the corrugated web beam was considered similar with the cold-formed steel truss solution tested by Jakab et al. [14]. Fig. 1(a) presents the structural solution [12] based on a pilot project, while Fig. 1(b) shows the geometry for the beam considered for testing and the main dimensions. The loadings considered for the predesign phase of the beam with corrugated web have been selected according to specific loading conditions for Cluj-Napoca / Romania, i.e.: (1) dead load from cladding: 0.20 kN/m^2 (γ_G=1.35); (2) technological loading: 0.15 kN/m^2 (γ_G=1.35); (3) snow loading: 1.5 kN/m^2 (γ_S=1.5).

(a)

(b)
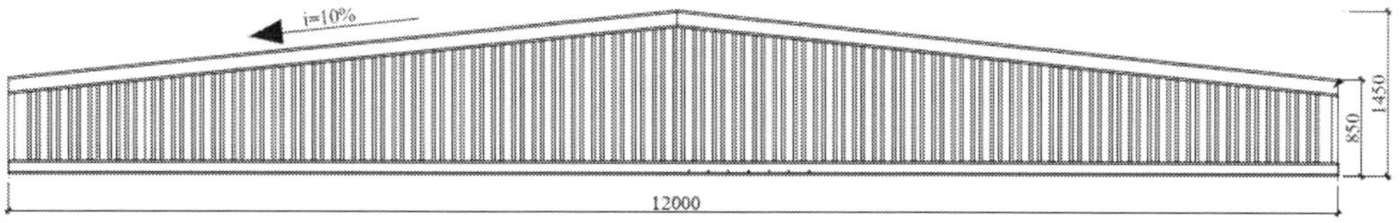

Fig. 1: (a) Structural solution; (b) Geometry of the proposed specimen for testing

In case of 4 m bay, a total load of 10.89 kN/m for ULS stage has been obtained. Purlins at a distance of 1.5 m, equally distributed, have been used, considering the slope of the roof of 10%. For the beam test setup, a similar load distribution was used applying concentrated loads into 8 points, i.e. the geometrical location corresponding to the purlin location.

Two beams with corrugated webs with a span of 12 m have been selected for testing, considering different arrangements for self-drilling screws, denoted as Beam-1 and Beam-2. Fig. 2 presents the components of the beams with corrugated web, i.e.

- back-to-back lipped channel sections for upper flanges - 2×C150/2.0 (grade S350GD+Z);
- back-to-back lipped channel sections for bottom flanges - 2×C120/2.0 (grade S350GD+Z);
- corrugated web with the corrugation depth of 43 mm and the thickness of 0.5 mm - LTP45/0.5 (grade S250GD+Z);
- reinforcing shear panels - supplementary plates of 0.5 mm thickness and 830 mm length, at the beam ends where the shear force is maximum (doubling the corrugated web on both faces) (grade S250GD+Z);
- self-drilling screws of 4.8 mm diameter with EPDM washers for both flange-to-web connection and seam fasteners for corrugated webs;
- bolts M12 class 8.8 for upper flanges to the end support connection.

Due to constructive reasons, the beam assembly includes 3 welded pieces: two identical one over the supports, to be able to distribute the support reaction at the full height of the beam and one central piece for the upper flange joining (see Fig. 2).

Fig. 2: Support and ridge detail of the tested beam

The flat sheet disposed in the support area is fixed to the support element and to the flanges. Supplementary, uniformly distributed local fixing to the web trapezoidal sheeting was applied. The continuity of the web was realized by overlaps and 2 rows of seam fasteners, following the principle of stressed skin action. The number of the seam fasteners was determined by the level of shear forces, increasing in the area of the supports and reducing the number in the span. In Fig. 3 are presented the details of the seam fasteners distribution of the corrugated webs, i.e. 16 pcs. uniformly distributed in the support area and 11 pcs. in the span. The difference between Beam-1 and Beam-2 specimens was the number and the distribution of the flange-to-web fasteners (see Fig. 4) and the thickness of the web in the support area, i.e. Beam-2 specimen had a double web (2×0.5 mm thickness of corrugated sheet) on ¼ of the span. The specimens Beam-1 and Beam-2 had totally 300 kg and 320 kg respectively.

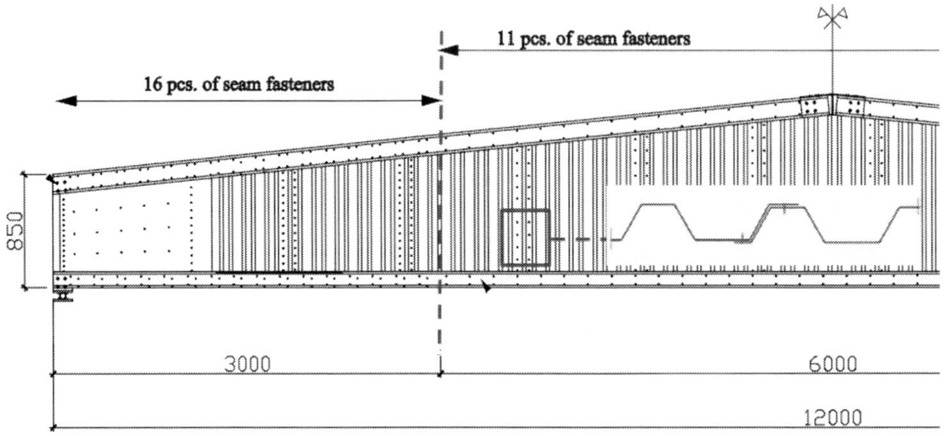

Fig. 3: Seam fasteners used for web sheeting: 16 seam fastener in support area and 11 seam fasteners in span (overlap details)

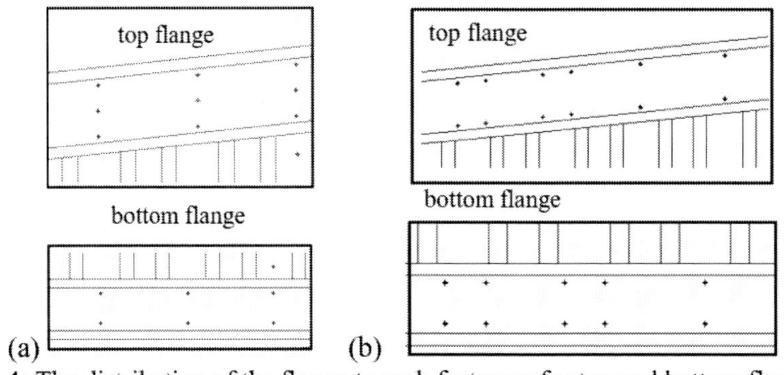

Fig. 4: The distribution of the flange-to-web fasteners for top and bottom flanges. (a) Beam-1; (b) Beam-2

3. Experimental setup

The testing of Beam-1 and Beam-2 specimens was done in the Structural Testing Laboratory of the Technical University of Cluj-Napoca. The main objective of the testing program was to define the load carrying capacity of the specimens and to compare the results with preliminary FEM results, used for the design of the specimens.

The specimens were monotonically loaded, with the load distribution according to Fig. 5 (ten points bending tests), introduced by two parallel devices in 8 points. The load distribution devices were stabilized laterally by the vertical guiding elements, braced in 4 points.

The full-scale testing program was completed with tensile tests to determine both the material properties for beam components and the behaviour of connections.

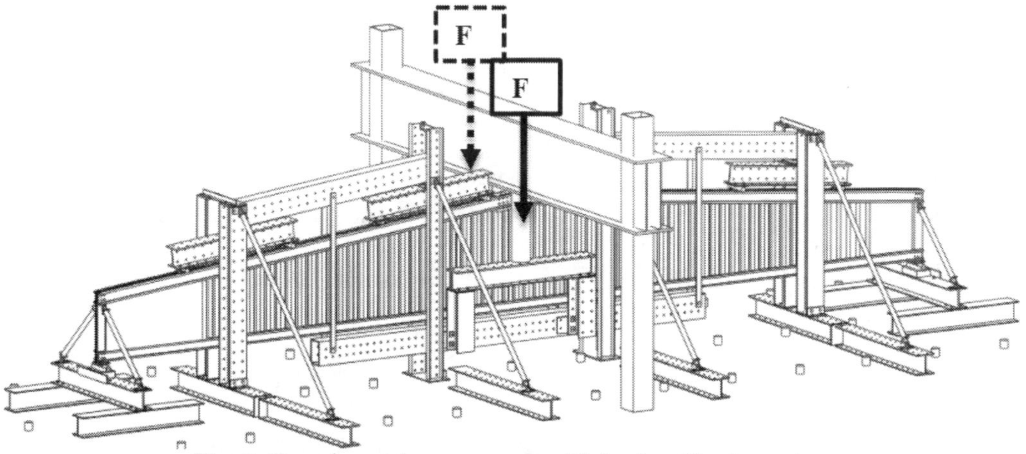

Fig. 5: Experimental arrangement with load application points

To keep the applied load position constant in the mid span, double hinged supports were selected. Because the upper support, point was free to move along the beam line (only the out-of-plane movement was locked by the support brace). Preliminary FEM analyses confirmed that no important differences were recorded between double hinge and simply supported configuration results.

4. Test results

The first tested specimen was Beam-1. In this case the first deformation, which corresponds to the buckling of shear panel, appears for a displacement of 13 mm at 28 kN (see Fig. 6a). At 21 mm small distortions of the corrugated web have been recorded for a load of 55 kN, as shown in Fig. 6b. For a displacement of 40 mm, corresponding to a load of 90 kN, two screws located at the bottom flange near support failed in shear, followed by failure in bearing of web-to-flange connection (see Fig. 6c).

The initial stiffness of $K_{0\text{-}Exp}$ = 1998.3 N/mm has been obtained and the maximum load is reached at F_{max} = 101.1 kN. The collapse mechanism produced at displacement of 64 mm.

(a) (b) (c)

Fig. 6: (a) Deformed shape of the beam end shear panel; (b) distortions of the corrugated web; (c) failure in bearing of web-to-flange connection

In the case of second specimen, i.e. Beam-2, the number and distribution of screws for web-to-flange connections have been change (see Fig. 4b), together with the addition of extra corrugated sheet panels (i.e. 3 widths of sheet, resulting in 2×0.5 mm thickness of corrugated sheet, including the shear panels) near each support.

In this case the first deformation, which corresponds to the buckling of shear panel, appears for a displacement of 14 mm at 51 kN (see Fig. 7a). At 35 mm small distortions of the corrugated web have been recorded for a load of 120 kN. For a displacement of 41 mm, corresponding to a load of 128 kN, local buckling of the top flange has been observed (see Fig. 7b). The initial stiffness of $K_{0\text{-}Exp}$ = 3830.0 N/mm has been obtained and the maximum load is reached at F_{max} = 138.6 kN.

(a) (b)

Fig. 7: (a) Deformed shape of the beam end shear panel; (b) local buckling of top flange

Due to the non-uniform loading (loading devices were controlled manually), out-of-plane stability loss of the upper flange between stabilization points was recorded, resulting in a lower load carrying capacity, than the estimated (without considering load introduction imperfections). The collapse appears suddenly after reaching the maximum force, by developing a local failure mechanism on the top flange located in the third part of the span, together with out-of-plane stability loss of the beam. Fig. 8 presents the deformed shape of the beam at collapse. The improvements in the second test showed an increased load-carrying capacity and also a much higher stiffness (beam stiffness was almost double), as shown in Fig. 9.

Fig. 8: Failure mechanism on the top flange of Beam-2

Finally, Fig. 9 shows comparatively, for the two tested specimens, the load-displacement curves and the ultimate (ULS) and serviceability limit state (SLS) levels.

5. Comparison: corrugated web beam vs. truss girder

The cold-formed steel beams of corrugated web and built-up section for flanges in a direct comparison with truss girders using also cold-formed built-up steel section, tested by Jakab [14], covers some weaknesses of the truss system, i.e. low shear capacity near the supports and high stress concentration in joints. In case of the classical truss arrangement, some members are designed according to connection related requirements – the member has to be large enough to accommodate a desired number of screws [12]. The use of trapezoidal sheeting connected in every corrugation to upper and lower chords increase the stability performance of the compressed chord members and ensure stress distribution. Also, cold-formed steel beams of corrugated webs presents the advantage of a reduced self-weight.

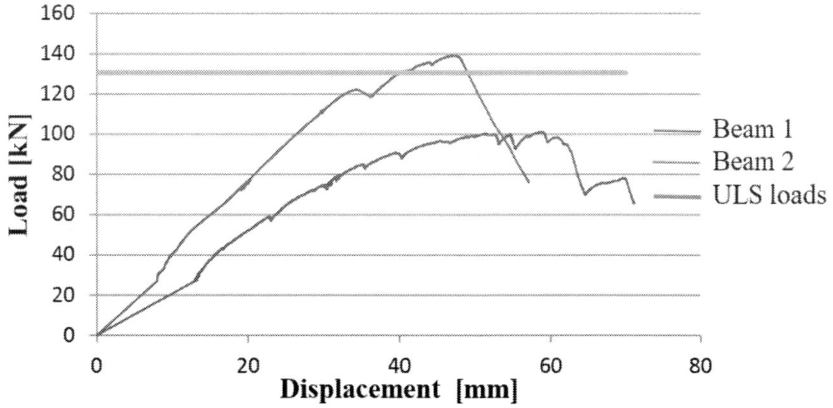

Fig. 9: Load-displacement diagrams for the tested beams

The 12 m span trusses tested by Jakab [14] were characterized by the following failure modes, i.e. interaction of axial compression (including flexural buckling) and bending of the upper chord, interaction of axial compression and bending in brace members and failure of the joints. The components of the truss beams tested in [14] were: 2×C150/2.5 for the upper chord, 2×C150/1.5 for the bottom one and C100/2 for diagonals and vertical members. The weight of the tested trusses was 290 kg, while the corrugated web beam specimens Beam-1 and Beam-2 had 300 kg and 320 kg respectively. It can be observed both systems have the same geometry and steel consumption.

The failure in case of corrugated web girder is more ductile, involving distortions of the corrugated web and shear buckling of web sheeting. The numerical analysis on a 12 m span corrugated web girder shows the structural performance differences compared to the truss test results (see Fig. 10).

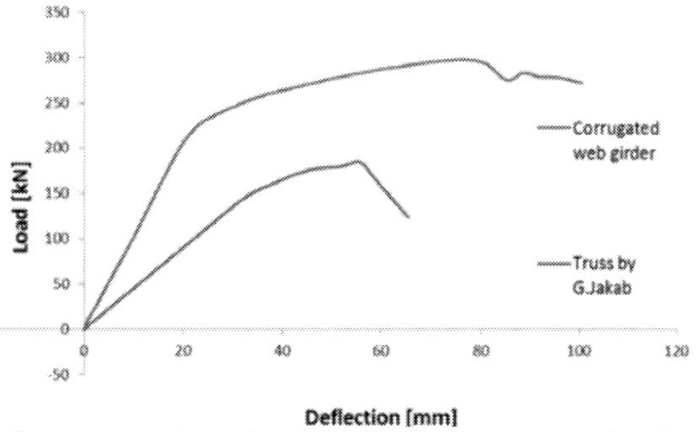

Fig 10: Load-deflection curves: truss (test results) and corrugated web girder (FEA results)

6. Conclusions

The paper presented the results of the experimental program on full scale corrugated web beams with trapezoidal shape made by cold-formed steel components of 12 m span. Two specimens, i.e. Beam-1 and Beam-2, were monotonically tested in a simple support configuration. The test results showed the structural performances of the analysed beams, emphasizing the differences in behaviour and collapse mechanism, when the beam configuration is changed. In case of Beam-1 specimen, where reduced web thickness was applied, the failure was produced under shear, due to the slippage between the web and the flanges, where the connection elements failed. In case of Beam-2 specimen, the web thickness and web-to-flange connections were improved obtaining a much higher load carrying capacity, the failure of the specimen was produced due to stability loss, in which load introduction imperfections took important roles. It is important to observe the recorded stiffness differences between Beam-1 and Beam-2, the obtained stiffness in the second beam test was almost double, emphasizing the importance or the shear stiffness of the beam web near the support.

In a direct comparison of a 12 m span corrugated web beam, with similar geometry and steel consumption with the truss one [14], the girder with corrugated web had almost double capacity than the classical Johnston truss with cold-formed lipped channel chords. Significantly better performances were achieved with increased resistance, stability, ductility and stiffness.

The research will continue by numerical models for calibration and validation of experimental models, optimisation of the number of self-drilling screws used for connections and parametric studies.

References

[1] Zeman & Co GmbH: *SINBEAM - Corrugated Web Beam*, Technical Document, 2012 (www.zeman-stahl.com/).

[2] Lindner J, Aschinger R. *Biegetragfähigkeit von I-Trägern mit trapezförmig profilierten Stegen*, Bericht VR2087, TU Berlin, Institut für Baukonstruktionen und Festigkeit, 1988.

[3] Lindner J, Aschinger R. "Zur Torsionssteifigkeit von Trapezstegträgern", *Stahlbau*, 59(4), 113-120, 1990.

[4] Sherman D, Fisher J. "Beams with corrugated webs", *Proc. of the 1st International Specialty Conference on Cold-formed Steel Structures*, Missouri-Rolla, 198-204, 1971.

[5] Lindner J. "Lateral-Torsional buckling of beams with trapezoidally corrugated webs", *Proc. of the 4th Int. Colloquium on Stability of Steel Structures*, Budapest, Hungary, 305-308, 1990.

[6] EN1993-1-5:2006. *Eurocode 3: Design of steel structures - Part 1-5: Plated structural elements*. CEN, Brussels.

[7] EN1993-1-1:2005. *Eurocode 3: Design of steel structures - Part 1-1: General rules and rules for buildings*. CEN, Brussels.

[8] EN1993-1-3:2006. *Eurocode 3: Design of steel structures. Part 1-3: General Rules. Supplementary rules for cold-formed thin gauge members and sheeting*. CEN, Brussels.

[9] Dubina D, Ungureanu V, Gîlia L. "Cold-formed steel beams with corrugated web and discrete web-to-flange fasteners", *Steel Construction*, 6(2), 74-81, 2013.

[10] Dubina D, Ungureanu V, Gîlia L. "Experimental investigations of cold-formed steel beams of corrugated web and built-up section for flanges", *Thin-Walled Structures*, 90(2015), 159–170.

[11] *Prefabricated steel structures for low-rise buildings in seismic areas (Precasteel)*, Final report, Research Fund for Coal and Steel, RFSR-CT-2007-00038, 2013.

[12] Nagy Zs, Gîlia L, Ballok R. "Romanian application of cold-formed steel beams with screwed corrugated webs", *Steel Construction*, 6(2), 139-143, 2013.

[13] Ballok RI. *Structural systems with trapezoidal beams made of cold-formed steel profiles and corrugated web*. PhD Thesis, Technical University of Cluj-Napoca, Romania, 2014.

[14] Jakab G, Dunai L, Kotorman I. "Development of a new cold-formed truss system". *Proc. of the 5th Int. Conf. on Thin-Walled Structures*, Brisbane, Australia, 2008.

BUCKLING MODE AND ELASTIC BUCKLING STRENGTH ESTIMATION OF CORRUGATED LIGHT-GAUGE PANEL UNDER SHEAR LOADING

Atsushi Sato, Shintaro Ito and Tetsuro Ono

Nagoya Institute of Technology, Japan

Abstract: Buckling strength of Corrugated Light-Gauge Panel (CLGP) is significantly affected by the geometrical configuration. The purpose of this study is to clarify the elastic buckling strength of the CLGP under shear loading. Local buckling that will determine the strength was focused in this study. Finite Element Analysis Method is used to calculate the elastic buckling strength of CLGP. Formula which includes geometric configuration parameters is proposed from the regression analysis of the numerical results.

1. Introduction

Shear wall is used for lateral force resisting component in Steel Framed House (SFH). Corrugated Light-Gauge Panel (CLGP) is one of the option that can be used for the wall. Usually, steel sheet is used for corrugated panel; local buckling phenomenon must be prevented to ensure sufficient strength and ductility. To delay the occurrence of local buckling, plate elements in corrugated panel are designed to have small width-to-thickness ratio. Design recommendation used in Japan only specifies the buckling coefficient where the boundary conditions of the plate elements are in two extremes, i.e., pinned or fixed support. In real practice buckling coefficient will exist between these two, and it shall be determined by the neighbour elements. In current design, conservative buckling coefficient is used [1]. Ikarashi et al. investigated the buckling strength of corrugated panel by FEA and theoretical energy equilibrium method [2, 3]; however, the geometrical configuration was limited. Geometrical configuration of the corrugated light-gauge panel have a lot of parameters, and it will directly decide the amount of steel used in the wall. Moreover, the configuration of the panel will determine the buckling modes. To understand the elasto-plastic behaviour of the CLGP, the characteristic computed from elastic theory will be helpful.

In this paper, eigenvalue analysis were conducted to survey the dominant buckling mode of the corrugated panels for the preliminary study purpose. Inclined angle of the corrugated shape was also included to determine the configuration. Firstly, classification of the buckling modes were investigated. Secondly, evaluation of local buckling mode was conducted, and the index which is useful to evaluate buckling coefficient was proposed. Finally, evaluation formula that can predict the elastic buckling coefficient from the geometrical information was proposed.

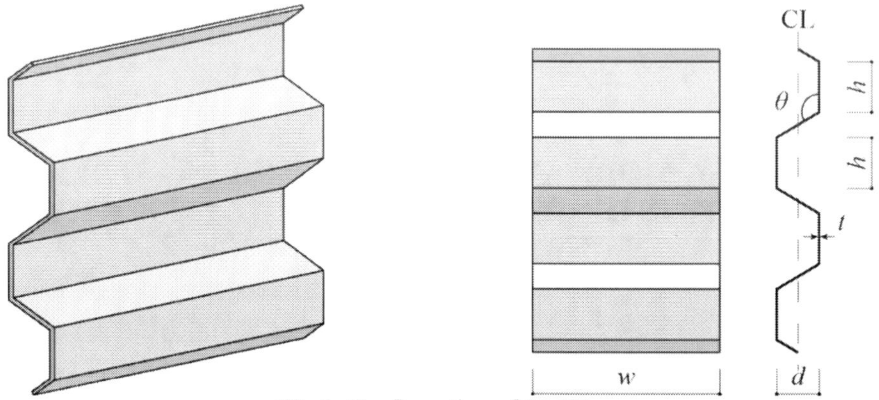

Fig 1: Configuration of CLGP

2. Numerical Simulation

Elastic Finite Element Analysis (E-FEA) was used to conduct the numerical simulation. Commercial software called Marc (ver. 2011) [4], which is produced by MSC Software Corp., was used for this purpose. Fig. 2 shows the Finite Element Model (FEM), and the boundary conditions. Table 1 summarizes the parameters that will determine the corrugated shear wall configuration. Wall width w, angle θ, plate thickness t, and depth-to-height ratio r ($=d/h$) are the configuration parameters (see Fig 1). Shear wall was rotated 90 degree and both sides of the shear wall elements were restrained not to rotate in x-axis. Both sides of the end are simulating that shear wall is connected to studs that have fairly large rigidity. Shell elements are used to model shear wall; the resolution of the element is less than 10mm in each direction. One end of the shear wall is fix-end (bottom side in Fig. 2), all nodes at the loading side end (upper side in Fig. 2) are linked to a primary node that is used to control in-plane behaviour. The primary node have a freedom to move in x-direction; the movement of other directions were restrain. To eliminate the span-to-depth ratio, the model was simulated to have infinite height by setting the boundary condition at the vertical ends (x-direction in the Fig.2). Nodes at both ends in the vertical direction (i.e., lines B, and C in Fig. 2) were linked to nodes at line A in the same location (i.e., in z-direction) to synchronize the movement. Initial imperfection is not introduced in the model. Young's modulus E = 205000 N/mm^2 and Poisson's ratio ν = 0.3 is used in the model. To calculate the elastic buckling strength from eigenvalue analysis, unit force was applied at the primary node that is used to control the movement at panel end (see Fig. 2).

Table 1: Configuration Parameters for Analysis Specimens

	w [mm]:	θ [°]:	t [mm]:	r ($=d/h$):	h [mm]:
Range	320 - 910	90 -150	0.1 – 2.0	0.1 -1.0	100
resolution	65	5	0.1	0.1	-

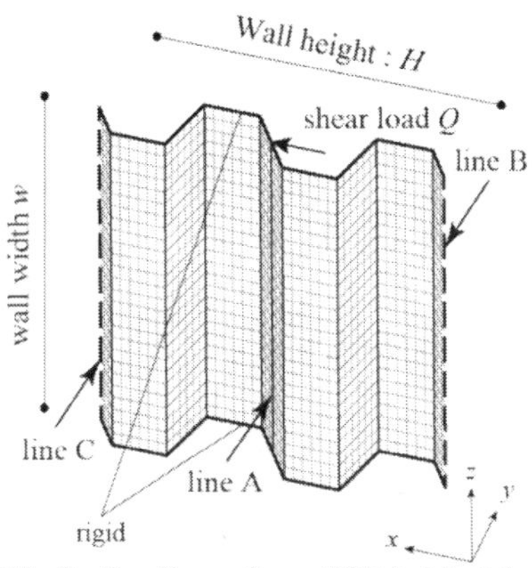

Fig 2: Configuration of FEA-Model

3. Eigenvalue Analysis

3.1 Elastic Buckling Stress

From force equilibrium, shear stress at each component will be equal. Shear stress at the plate components can be computed by following equation where Q_e is elastic buckling strength obtained from eigenvalue analysis. The value four in the denominator is corresponding to the analysis model that have four flat and inclined components in the height direction (see Fig. 2).

$$\tau_e = \frac{Q_e}{4ht\{1 + r\tan(\theta - \pi/2)\}} \tag{1}$$

3.2 Equivalent Height of the CLGP Plate Component

To evaluate local buckling, it is important to evaluate the depth-to-thickness ratio of the plate components. In this study, it is assumed that dominant local buckling will occur in the slender component of the panel (i.e., flat component or inclined component).

The configuration of the corrugated panel will be determined from the height of the flat component (h) and depth of the panel (d) that is including the angle of the inclined component (θ) (see Fig. 3). Parameter r is the depth-to-height ratio ($=d/h$). From the geometric configuration, following equation is obtained to distinguish the dominant plate component. Based on the parameters shown in Table 1, equivalent height of CLGP plate component can be shown in Table 2.

$$\sin\theta \geq r \cdots \text{Flat component is the dominant element} \tag{2-a}$$

$$\sin\theta < r \cdots \text{Inclined component is the dominant element} \tag{2-b}$$

Table 2: Equivalent Height of the CLGP Plate Component

$r\backslash\theta$		θ												
		90	95	100	105	110	115	120	125	130	135	140	145	150
	0.1-0.5	100.0	100.0	100.0	100.0	100.0	100.0	100.0	100.0	100.0	100.0	100.0	100.0	100.0
	0.6	100.0	100.0	100.0	100.0	100.0	100.0	100.0	100.0	100.0	100.0	100.0	104.6	120.0
	0.7	100.0	100.0	100.0	100.0	100.0	100.0	100.0	100.0	100.0	100.0	108.9	122.0	140.0
r	0.8	100.0	100.0	100.0	100.0	100.0	100.0	100.0	100.0	104.4	113.1	124.5	139.5	160.0
	0.9	100.0	100.0	100.0	100.0	100.0	100.0	103.9	109.9	117.5	127.5	140.0	156.9	180.0
	1.0	100.0	100.4	101.5	103.5	106.4	110.3	115.5	122.1	130.5	141.4	155.6	174.3	200.0

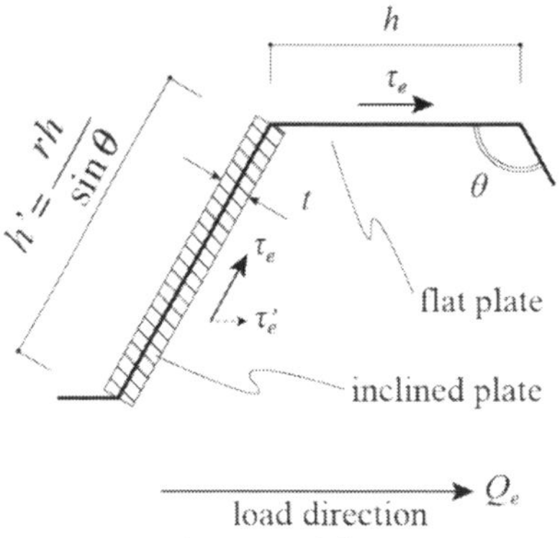

Fig 3: Shear Buckling Stress

3.3 Elastic Buckling Coefficient

Elastic local buckling stress of the plate component can be evaluated by following equation. Coefficient k_e included in the equation will be determined from the boundary condition, and Eq. 4 are used for the evaluation. Coefficient k_e is so-called elastic buckling coefficient.

$$\tau_e = k_e \frac{\pi^2 E}{12(1-\nu^2)(h_{eq}/t)^2} \tag{3}$$

$$k_e = 8.98 + \frac{5.60}{(w/h)^2} \cdots \text{All sides are Fixed support} \tag{4-a}$$

$$k_e = 5.34 + \frac{4.00}{(w/h)^2} \cdots \text{All sides are Pinned support} \tag{4-b}$$

Fig. 4 shows the buckling modes that is computed from the eigenvalue analysis. Fig. 4(a) shows the global buckling for the corrugated panel, i.e., buckling deformation was similar to the mode that can be observed in a plane steel sheet subject to shear. Fig. 4(b) shows the local buckling that appeared in plate components, respectively. Fig. 4(c) shows the coupled buckling behaviour that is the combination of global and local buckling. Corrugated panel which showed similar buckling mode to Fig. 4(b) can be evaluated by Eq. (3).

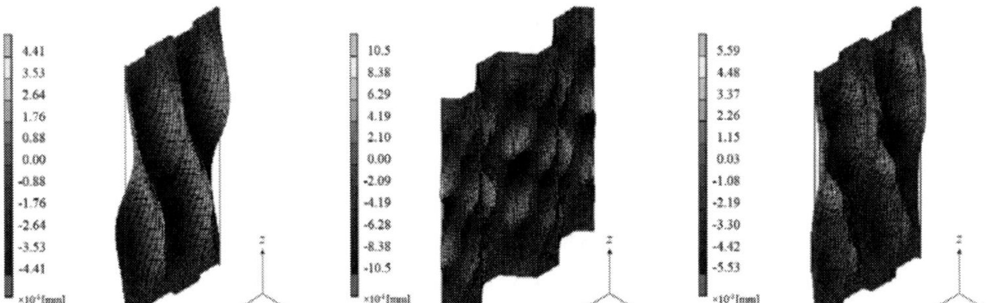

a) Global buckling b) Local buckling c) Coupled buckling

Fig. 4: Classification of Buckling Mode

Fig. 5 shows the relationships between shear stress and corrugated panel parameters. Vertical axis is shear stress computed from Eq. (1), two horizontal axes are depth-to-height ratio r and inclined angle θ. In this figure, buckling modes are express by the difference of legend. As can be recognized from Fig. 5, thickness of the plate provides a difference in buckling modes. Although the shape is the same, global buckling will be the dominant buckling mode in a thicker plate. Once the buckling mode changes to global mode, shear stress will show a different trend. Moreover, it can be recognized that the influence of parameter r is more sensitive than angle θ to the buckling mode.

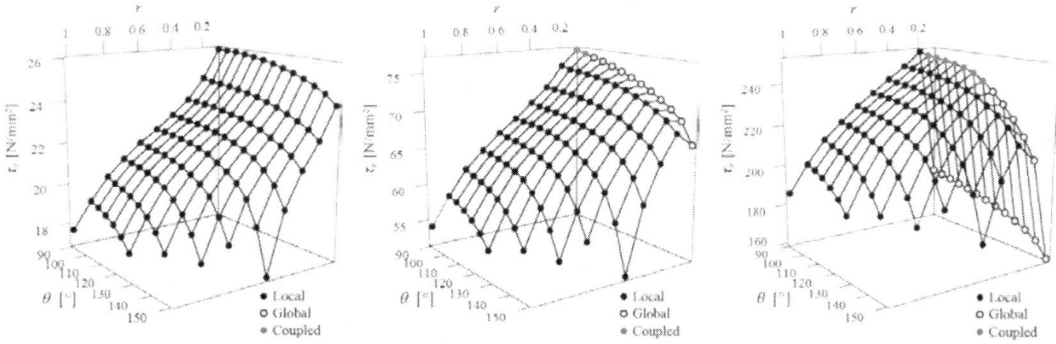

a) t=0.4 [mm] b) t=0.7 [mm] c) t=1.3 [mm]

Fig 5: Relationship between Shear Stress and Buckling Mode

4. Evaluation of Buckling Coefficient

4.1 Normalized Rotational Stiffness

Although two extremes of the boundary condition will be evaluated by Eq. 4, all elements in the real practice will exist between these two. Buckling coefficient will be influenced by the neighbour elements. Local buckling mode that was observed from the eigenvalue analysis appeared in a diagonal direction. (see Fig. 6(a)). Local buckling deformation of the corrugated panel was the repetition of the deformed shape shown in Fig. 6(b). The elastic rotational stiffness of Fig. 6(b) can be calculated as Eq. 5. Normalizing by relative flexural stiffness of the dominant plate component, normalized rotational stiffness k_θ can be obtained. Normalized rotational stiffness is only the function of geometrical configuration. Fig. 7 shows the relationships between normalized rotational stiffness k_θ and depth-to-height ratio r with inclined angle θ.

$$K_\theta = \frac{r+\sin\theta}{r}\cdot\frac{6EI}{h_{eq}} = k_\theta\frac{6EI}{h} \tag{5}$$

$$k_\theta = \begin{cases} \dfrac{r+\sin\theta}{r}\cdots(\sin\theta \geq r) & \text{(6-a)} \\[2ex] \dfrac{r+\sin\theta}{\sin\theta}\cdots(\sin\theta < r) & \text{(6-b)} \end{cases}$$

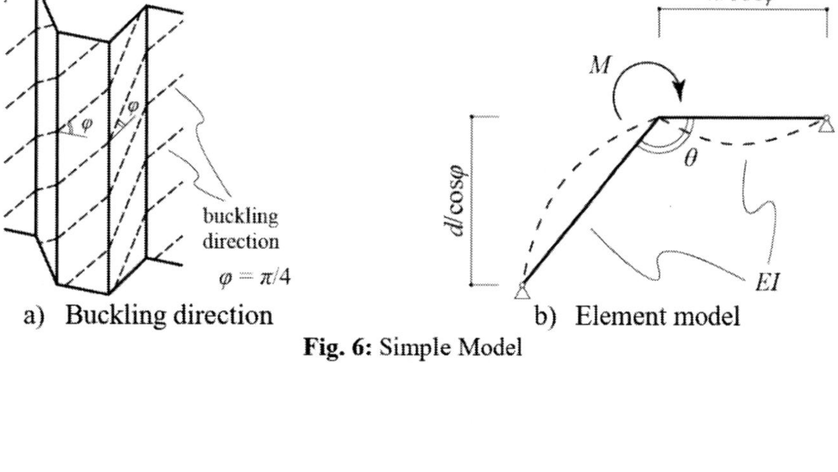

a) Buckling direction b) Element model

Fig. 6: Simple Model

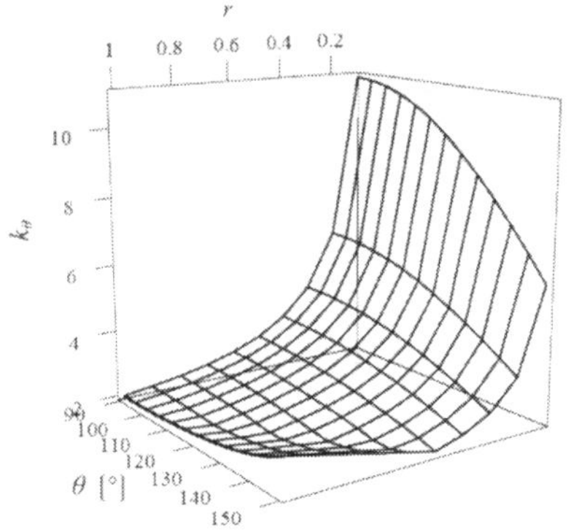

Fig. 7: Normalized Rotational Stiffness

Fig. 8(a) shows the relationship between elastic buckling coefficient k_e and normalized rotational stiffness k_θ where the width-to-height ratio of the plate component is w/h =9.10. The legend corresponds to buckling modes (see Fig. 4). It can be recognized that local buckling mode shows a clear trend, and the buckling coefficient of local buckling will be located between the two extreme coefficients (i.e., Eq. 5). Buckling coefficient of corrugated panel which showed global buckling had a smaller value than local buckling mode. Eliminating global and coupled buckling modes, the results of local buckling mode can be illustrated as Fig. 8(b). Similar clear trends were observed in different width-to-height ratio w/h; all results are illustrated in Fig. 9(a).

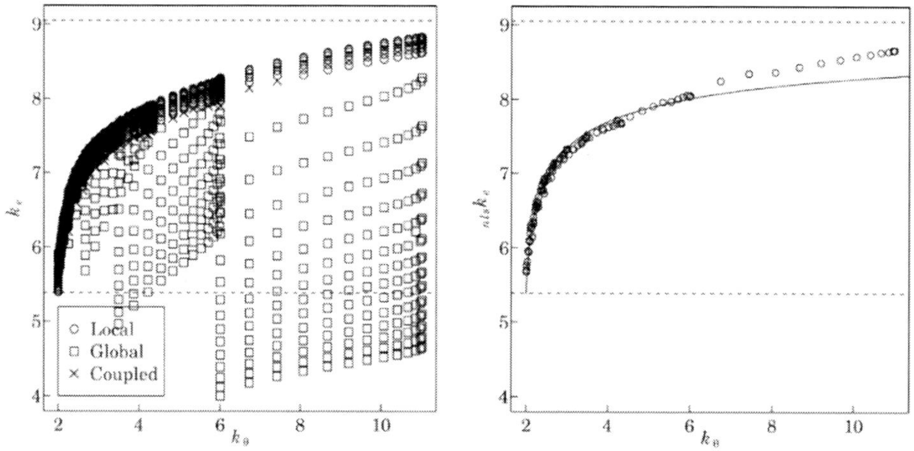

a) k_e-k_θ as all buckling mode b) k_e-k_θ as local buckling mode

Fig. 8: Relations between k_e and k_θ

4.2 Proposal of Evaluation Formula

As illustrated in Fig. 8(b), local buckling coefficient of corrugated panel shows a good correlation between normalized rotational stiffness. To evaluate the buckling coefficient corresponding to local buckling, regression analysis was conducted to obtain an evaluation formula. Following formula was proposed to conduct the regression analysis. Basic idea of this equation was following "*Menegotto-Pinto Model*" [5].

$$k_e = a + \frac{d}{\left\{1+\left(\frac{k_\theta - 2}{b}\right)^c\right\}^{\frac{1}{c}}}(k_\theta - 2) \tag{7}$$

To satisfy two extreme boundary conditions (i.e., Eq. 5), Eq. (8) can be obtained. Eq. (8.a) is obtained from fixed support, and Eq. (8.b) is obtained from pinned support.

$$a + bd = 8.98 + \frac{5.60}{(w/h_{eq})^2} \tag{8-a}$$

$$a = 5.34 + \frac{4.00}{(w/h_{eq})^2} \tag{8-b}$$

Based on these equations, following formula can be obtained from regression analysis.

$$_{cal}k_e = a(w/h_{eq}) + 26.4 \cdot f(k_\theta) \tag{9}$$

Two components is the right side will be computed as follow.

$$\begin{cases} f(k_\theta) = \dfrac{k_\theta - 2}{\left\{1+\left(\dfrac{k_\theta - 2}{0.14}\right)^{0.50}\right\}^{\frac{1}{0.50}}} & \text{(10-a)} \\ a(w/h_{eq}) = 5.34 + \dfrac{4.00}{(w/h_{eq})^2} & \text{(10-b)} \end{cases}$$

Normalized rotational stiffness k_θ, and width-to-height ratio of the dominant plate component w/h_{eq} can evaluate the local buckling coefficient of the corrugated panel. Evaluated buckling coefficients are illustrated in Fig. 8(b) and 9(b). Good agreement can be recognized from both figures.

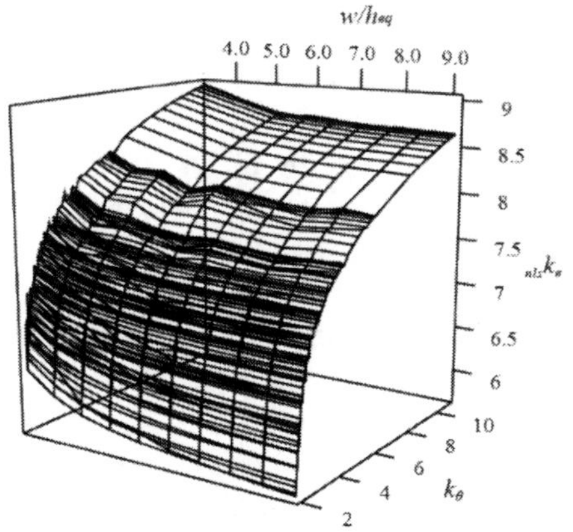

a) Eigenvalue results

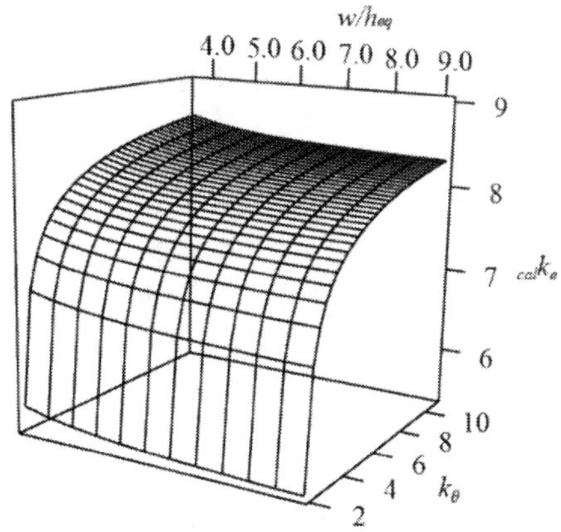

b) Computed Results

Fig. 9: Relation between k_e, k_θ and w/h_{eq}

4.3 Validity of the Proposed Formula

Fig. 10 shows the comparison between eigenvalue results and computed results. Vertical axis is eigenvalue results and horizontal axis is computed results. All corrugated panels that showed local buckling mode are shown in the figure. Dotted lines shown in the figure are ∓ 5% errors. The correlation coefficient was 0.98 and good agreement was confirmed.

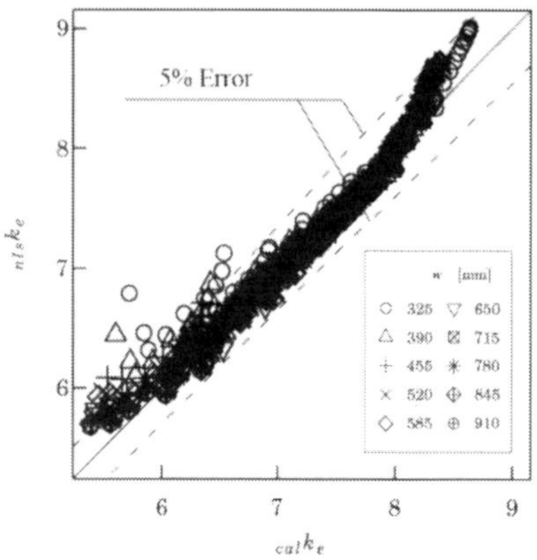

Fig. 10: Validity of the proposed Formula

5. Conclusion

In this paper, E-FEA was conducted for several CLGP configurations to capture the trend of buckling strength and buckling modes for preliminary study purpose.
The main conclusions are:

1. Buckling modes were classified to Local buckling, Global buckling and Coupled buckling. The elastic buckling strength of the CLGP which showed local buckling can be evaluated by the Normalized Rotational Stiffness k_θ. The index k_θ is considering the effect of neighbour elements that is supporting the plate component.
2. Evaluation formula which is composed by configuration parameters r, θ, w, h_{eq}, was proposed. Good agreement of the proposed formula was confirmed with the eigenvalue results.

References

[1] Architectural Institute of Japan. *Recommendations for Stability Design of Steel Structure, Japan*, 2009.
[2] Ikarashi K, Okamoto J, Sato K. "Buckling mode and elastic buckling strength estimation of rectangular corrugated plate under in-plane shear force", *Journal of Structural and Construction Engineering, Japan*, 427-435, 2015.
[3] Ikarashi K, Okamoto J, Sato K. "Large deformation behaviour and collapse mode of rectangular corrugated steel shear wall", *Journal of Structural and Construction Engineering, Japan*, 693-701, 2015.
[4] Marc 2011 User's Guide, *MSC. Software Corporation, USA*, 2011.
[5] Menegotto, Pinto M., P.E., "Method of Analysis for Cyclically Loaded R.C. Plane Frames Including Changes in Geometry and Non-Elastic Behaviour of Elements under Combined Normal Force and Bending", *Proceeding IABSE Symposium, Lisbon*, 15-22, 1973.

BENEFITS OF LOAD REDISTRIBUTION TO THE RELIABILITY OF A COLD-FORMED STEEL JOIST-SUPPORTED FLOOR SYSTEM

Brooks H. Smith[a], Aritra Chatterjee[b], Sanjay R. Arwade[a], Cristopher D. Moen[b] and Benjamin W. Schafer[c]

[a] *Department of Civil & Environmental Engineering, University of Massachusetts, Amherst, MA, USA*
[b] *Department of Civil & Environmental Engineering, Virginia Tech, Blacksburg, VA, USA*
[c] *Department of Civil Engineering, Johns Hopkins University, Baltimore, MD, USA*

Abstract: Using Monte Carlo simulations, this paper investigates the benefits that elastic load sharing and post-elastic residual capacity may have upon the system reliability of a simple cold-formed steel (CFS) floor under gravity load. An algorithm was developed and implemented in MATLAB that is capable of rapid iterative load redistribution with a sheathed CFS floor system and has been validated to an accuracy of approximately 5% against high fidelity finite element simulations. A simple rectangular floor system was simulated using randomized dead loads, live loads, and joist capacities. Benefits to the capacity of the floor system are found to be 3 to 5 percent for elastic load sharing and 20 to 30 percent for post-elastic residual capacity.

1. Introduction

Typical cold-formed steel (CFS) floor systems involve many repetitive joist members laid in parallel, yet the design specifications are based upon the reliability of individual members alone and ignore the benefits imparted by such a system. This paper investigates the benefits that load redistribution and residual capacity have upon the system reliability of cold-formed steel floors by using Monte Carlo simulation to evaluate floor system reliability with and without load redistribution.

The United States National Design Specifications (NDS) for wood construction establishes a repetitive member factor that allows a 15 percent increase in the capacity of wood joists in a floor system that involves several identical members working in parallel [1]. Cold-formed steel construction is similar, but no such factor exists. In the NDS, the repetitive member factor is intended to include the benefits of three effects: (1) elastic load sharing, (2) post-elastic

residual capacity, and (3) composite action between sheathing and joists [2]. The first two of these effects are investigated and quantified for a simple CFS floor system in this paper.
An analysis algorithm was developed and implemented in MATLAB that is capable of rapid iterative load redistribution for an arbitrary sheathing-covered CFS floor system. Joist capacities are calculated based upon the 2012 American Iron and Steel Institute S100 design specification for the United States [3], while demands are determined through finite element analysis using beam elements for sheathing and joists. Within this paper, a simple 7.32 m × 14.63 m [24 ft × 48 ft] floor system is investigated, consisting of 1200S250-97 CFS joists spaced at 0.61 m [24 in] on-center, sheathed with 24/16 oriented strand board (OSB), and having blocking at third points. The simulation program was validated to an accuracy of within 5% against detailed ABAQUS finite element analyses of this floor system. The floor system is roughly based upon the layout of the building investigated and tested extensively in full-scale for the CFS-NEES project [4,5], and future work by the authors will explore the effects of various features of this real floor upon load redistribution.

Monte Carlo simulations were then performed using randomized live loads, dead loads, and joist capacities with the goal of quantifying the failure rate of floor systems both with and without load redistribution. Elastic load sharing is found to provide a benefit of approximately 3 to 5 percent and plastic residual capacity is found to provide a benefit of approximately 20 to 30 percent to the capacity of the floor system.

2. Simulation methodology

The simulation analysis methodology relies upon the assumption that the joists and OSB panels act as Euler-Bernoulli beams with the joists treated as single span beams and the OSB panels treated as multi-span continuous beams over the joists. . This method is essentially a static finite element analysis of the isolated floor using beam elements. The only difference is that the OSB reactions are treated as partially distributed loads (equal to the width of the sheathing board) upon the joists, rather than as point loads. The method gives the same results as a designer should find if they were to treat the sheathing boards as beams. By implementing the method in MATLAB, efficient and repeated simulations such as those required in Monte Carlo simulation are possible. Based upon an assumption of elastic-perfectly plastic joist response, load redistribution effects can be simulated.

2.1 Load redistribution theory

Load redistribution is performed on the level of an individual sheathing board. The method identifies any joists below the board that have an unfactored demand to capacity ratio greater than one, indicating overload, and attempts to move their associated excess load to other joists within the span of the sheathing board. This implicitly assumes an elastic-perfectly plastic overall response for the joists. The process used for this is most easily explained through an instructive example, as given in Section 2.1.1 below.

2.1.1 Example: Multiple over-capacity joists

This example is based upon a simply supported, 1.22 m × 2.44 m [4 ft × 8 ft] sheathing board spanning over five 0.61 m [2 ft] on-center joists – or in other words, a 4-span continuous beam, as shown with labels for each reaction support in Fig. 1.

It is important to note that, for simplicity, this example imagines a floor system in which there is only a single sheathing board. In the actual program, this same process is repeated for every sheathing board in the structure. Effects of this will be discussed after the example in section 2.1.2.

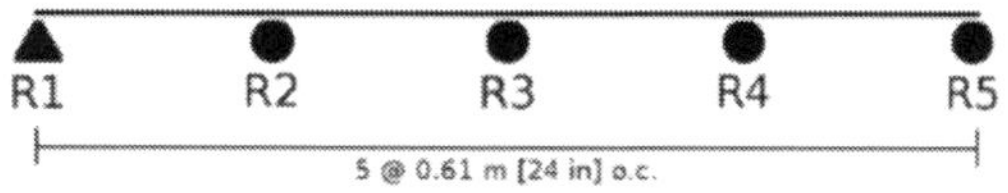

Fig. 1: Labelled beam figure, as referenced in Section 2.1.1. The sheathing board is treated as a beam, and each of the reactions (R#) occur on joists

A hypothetical set of unfactored *D/C* ratios is shown below in Eq. (1), with R2 being 17% over-capacity and R3 being 9% over-capacity. Variability in D/C ratios is due to the different reactions generated for a continuous beam and the random capacities of the joists.

(1)

In this example, 17% of the load must be redistributed off R2, and 9% of the load must be redistributed off of R3. To do this, the solution to the final, redistributed configuration, is a weighted average of three different support conditions: one where all joists are present, one where R2 is completely removed, and one where both R2 and R3 are completely removed, as follows in Eq. (2):

$$\frac{1.0}{1.17}\cdot \qquad + \frac{1.17-1.09}{1.17}\cdot \qquad + \frac{1.09-1.0}{1.17}\cdot$$

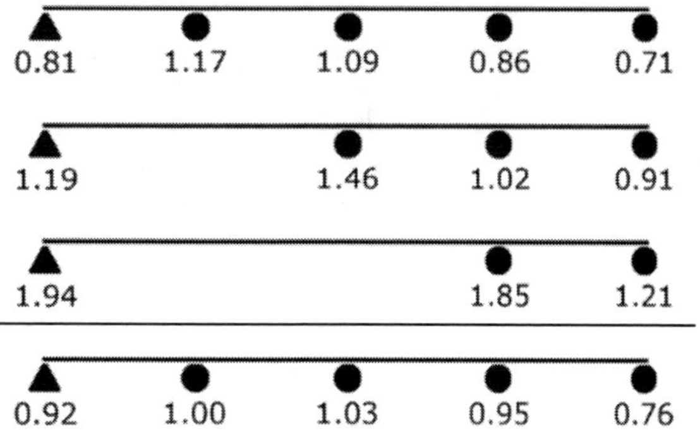

(2)

Note that R3 is still over-capacity. This is due to the load first redistributed from R2, when R3 was still present. A second iteration of load redistribution would then be performed – but with only one joist over-capacity, a weighted average of only two support conditions would be calculated as shown below in Eq. (3):

$$\frac{1.0}{1.01}\cdot \qquad + \frac{1.01-1.0}{1.01}\cdot$$

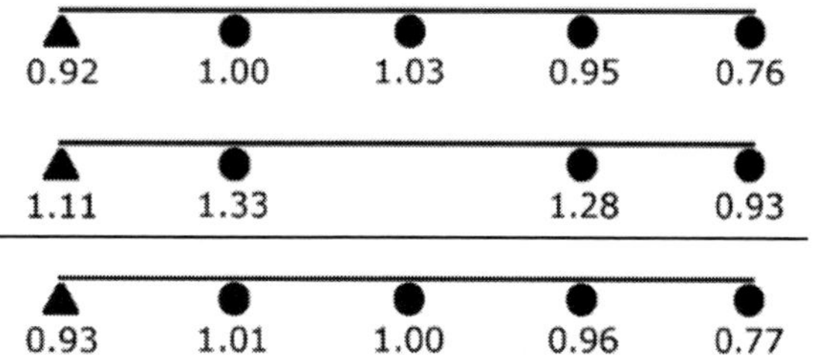

(3)

Again, one joist is over-capacity: R2 this time, and another similar iteration would next be performed. However, load could be passed back and forth between R2 and R3 over infinite iterations. To prevent this, there is a tolerance of 0.005 used, such that the program will stop if a load of less than 0.5% of the capacity is to be redistributed. There would therefore likely only be one more iteration after this.

2.1.2 Implications of multiple sheathing board layers

The above example illustrates the methodology for a single sheathing board. However, a more typical structure with many sheathing boards does affect the procedure somewhat. In particular, joist *D/C* ratios are not recalculated until all sheathing boards have gone through the load redistribution procedure. That is, one sheathing board redistributing its load does not affect how much subsequent sheathing boards that span over the same joist redistribute their loads – until the next iteration begins with new *D/C* ratios and new over-capacities.

Importantly, this means that it may take several iterations to eliminate more than one over-capacity condition, as one sheathing board may move load off a joist, but another may push load from an even more over-capacity joist onto that same joist. Excess load may also be passed back and forth over a distance of several joists, in the same manner as in the example, but staggered sheathing board layouts may spread the distance even further.

2.2 Calculation of demand to capacity ratios

The redistribution algorithm used here relies on the ratio of demands to capacity to determine when load must be redistributed among the joists. To calculate demands, gravity load combinations are based upon Allowable Strength Design (ASD). Joist capacity has been calculated in keeping with current practice in the United States cold-formed steel construction industry (AISI S100-12), by checking for bending, distortional buckling, shear, web crippling, combined bending and web crippling, and combined bending and shear failures.

All code-based design calculations include safety factors, and thus the final design calculations provide D_f and C_f, the factored demand and capacity. Additional calculation is required to obtain D_u and C_u, the unfactored demand and capacity of components. Those calculations fall into one of two categories, based upon the critical limit state.

1. **Gravity system (ASD), single load effect limit state:** Where the failure mode involves a stress resulting from a single, isolated load effect (e.g. bending or web crippling), $C_u = \Omega * C_f$, where Ω is the ASD factor of safety for the given failure mode.
2. **Gravity system (ASD), combined load effect limit state:** Where the failure mode involves combined stresses from multiple load effects (e.g. combined bending and web crippling or combined bending and shear), unfactored demand and capacity cannot be calculated directly. Design standards provide interaction equations in the form of inequalities. The upper limit to the inequality (typically 1.0) plays the role of the factored capacity C_f, whereas the calculated interaction equation is the factored demand. The unfactored demand D_u is obtained by setting all safety factors $\Omega_x = 1.0$ in the interaction equation, and further assuming that $C_u = C_f$. For example, consider joist strength calculations for a combined bending and shear limit state (AISI S100-12, eq. C3.3.1-1), with interaction equation inequality (Eq. (4)).

$$\sqrt{\left(\frac{\Omega_b M}{M_{nxo}}\right)^2 + \left(\frac{\Omega_v V}{V_n}\right)^2} \leq 1.0 \tag{4}$$

where Ω_b is the the bending safety factor, M and M_{nxo} are the moment demand and capacity, Ω_v is the shear safety factor, and V and V_n are the shear demand and capacity. The unfactored demand and capacity are then taken as in Eq. (5).

$$D_u = \sqrt{\left(\frac{1.0\ \ M}{M_{nxo}}\right)^2 + \left(\frac{1.0\ \ V}{V_n}\right)^2}\ , \qquad C_u = 1.0 \tag{5}$$

3. Model validation

The methodology described in Section 2 is based upon temporarily removing elements to simulate load redistribution. Therefore, the goals of validation are to ensure that the load distributions for the virgin structure and for a structure with elements removed are correct and accurate. To this end, a three-dimensional finite element analysis was developed using ABAQUS, with the same floor geometry as described above and 4-node shell elements (S4R) of approximately 2.5 × 2.5 cm [1 × 1 in] size, using linear elastic material properties. The 1200S250-97 joists were attached to 1200T200-97 tracks via spring elements representing screw fasteners, and blocking was added at 1/3 points. Uniformly distributed loads were applied to shell elements above the joists that represent the sheathing boards, and these were pinned to the top flange of the joists.

As the most common failure mode for these joists was bending, the primary output values examined were joist moments, calculated by integrating the axial force of the shell elements with their distance from the neutral axis. The maximum joist moments were compared between ABAQUS and the equivalent MATLAB simulations.

MATLAB moment distribution results were found to agree with ABAQUS moment distribution results to within an accuracy of approximately 5 percent for each of the eight models tested. Example comparisons are plotted in Fig. 2.

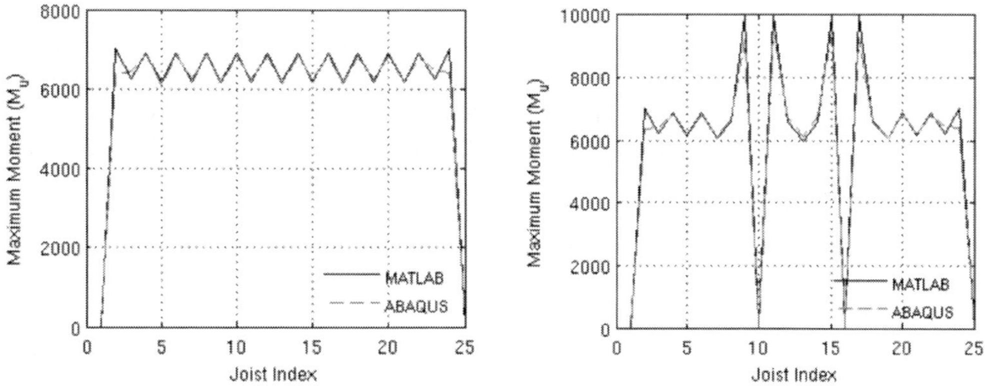

Fig. 2: Validation graphics depicting the maximum moment distribution in each of the 25 joists, represented from left edge to right edge by 'Joist Index', when all joists are present (left) versus when joists with indices 10 and 16 are removed (right)

4. Monte Carlo simulation procedure

Having validated the simulations as accurate, large sets of stochastic simulations may be used to reach conclusions about the benefits of load redistribution. To quantify the extra benefit that redistribution provides, a series of Monte Carlo simulations were performed at progressively decreasing deterministic reductions in the capacity of the joists. That is, a Monte Carlo simulation was performed at full joist capacity, then at 5 percent less capacity, then 10 percent less, and so on. A system which performs equally as well at 5 percent reduction, for example, would imply that there is 5 percent excess reliability in the system. Joist capacity was reduced in increments of 5 percent, from 0% (i.e. full capacity) to 70%. 1000 random runs were performed in each of these 17 increments.

Random input variables for the Monte Carlo simulations were all Gaussian and consisted of:

1. Joist capacity, with a coefficient of variation (COV) of 0.15:

$$COV = 0.15 = \sqrt{V_m^2 + V_F^2 + V_P^2} \tag{6}$$

where V_m, V_F, and V_P are the material, fabrication, and professional factors, respectively, for CFS joist members, as defined in AISI S100-12, table F1 [3]. This random capacity was applied independently to each joist in the system.

2. Live load, with a COV of 0.25, as defined by Meimand and Schafer [6] and based upon AISI calculations. This random live load was applied independently to the tributary area over each joist, which is similar to what the AISI S100-12 code expects for the ideal distribution of random loads.
3. Dead load, with a COV of 0.10, as similarly defined by Meimand and Schafer [6] and based upon AISI calculations. This random dead load was also applied independently to the tributary area over each joist.

5. Results

Two separate benefits are found in the results for the Monte Carlo simulations: (1) elastic load sharing, which is a result of beam effects partially averaging the load distributed to each joist, and (2) post-elastic residual capacity, which is a result of differential deflection redistributing load from a yielding joist. The first is identified by comparing the number of failures that would occur if the entirety of the load within the tributary area above each joist were applied to each joist, with the failures that actually occur on the initial iteration. That is, comparing the calculation that a typical designer would perform with the calculation assuming that sheathing boards are beams. The fact that the sheathing boards span multiple joists smooths fluctuations in the live and dead loads in the joist tributary areas and thereby reduces the variance of the demand on the joists. Reducing the demand variance leads to an increase in the predicted reliability.

Post-elastic residual capacity is identified by comparing the number of failures that occur on the initial iteration with the number of failures that occur on the final iteration of load redistribution. Often, this redistribution is able to redistribute load from joists that would have failed to joists that had excess capacity, thereby preventing system failure (defined by any one joist exceeding its capacity). However, as joist capacity is progressively reduced in Monte Carlo simulations, eventually there is simply not enough capacity available in the system, and redistribution instead pushes more joists over their capacity; that is, causing progressive collapse. This distinction is made in the results, and the difference in benefits is quantified.

5.1 Reliability benefits

The overall benefits of elastic load sharing may be defined as the difference between a demand calculation based purely upon tributary area and one based upon treating the sheathing boards as beams. Similarly, the benefits of post-elastic residual capacity may be defined as the difference between the likelihood of failure at the initial iteration and the final iteration of load redistribution. When system failure is defined as at least one joist having an unfactored *D/C* ratio greater than 1.0, load sharing provides a typical benefit of 3 to 5 percent, while residual capacity provides a benefit of 20 to 30 percent. That is, a system without elastic load sharing is as likely to fail as a system with 3 to 5 percent lower joist capacity and elastic load sharing (see Fig. 3a). The benefits of both effects decrease at higher probabilities of failure.

An alternative method of measuring the benefit of these effects is to examine the reliability index of individual members. The goal in AISI code is to maintain a member reliability index of at least 2.5 (equal to a probability of failure of 0.6%). As shown in Fig. 3b, the member reliability index with code-based tributary area calculations is about 2.75. However, including elastic residual capacity, the reliability index does not drop to 2.75 until about an 8% capacity reduction. Including post-elastic load redistribution, that capacity reduction jumps to almost 43%. These much higher benefits are in line with a measurement of the system reliability benefits at a low probability of failure, when load sharing and residual capacity are most effective.

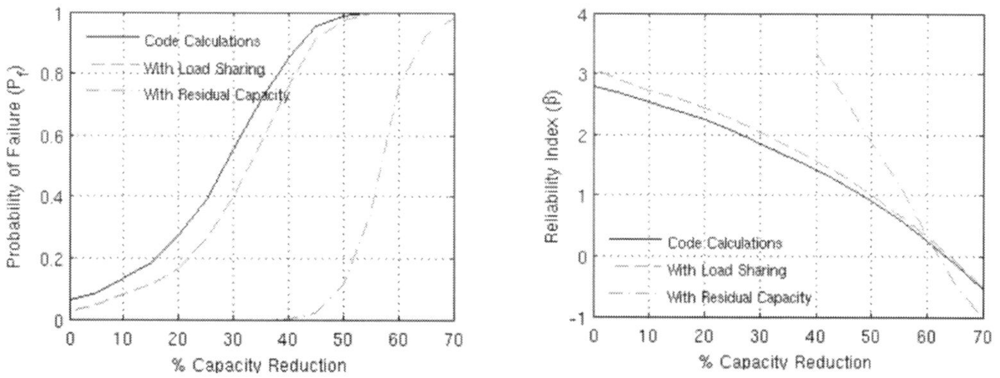

a) Probability of failure of the overall system   b) Reliability index of the individual members

Fig. 3: Failure prevention benefits of elastic load sharing and post-elastic residual capacity, as demonstrated by the probability of system failure (left), and the reliability index of the individual members (right) before and after load sharing and load redistribution

5.2 Load redistribution characteristics

These load redistribution effects may be examined more closely by measuring whether they help reduce failures (productive effect) or if they increase failures (counterproductive) within a given simulation. As shown in Fig. 4a, elastic load sharing is usually beneficial, though there is always a non-zero probability of it being counterproductive. For post-elastic residual capacity (Fig. 4b), the difference caused by redistributing the load from failing joists, this measure is particularly important because counterproductive redistribution is physically equivalent to progressive collapse.

Given that progressive collapse is a catastrophic system failure, the point at which it gains non-negligible probability (in this case, at 35 to 40 percent, as shown in Fig. 4b) should be considered when evaluating the benefits of repetitive systems. There is also a later point, at about 60 percent capacity reduction, at which residual capacity is more likely to hurt than help.

6. Conclusion

These Monte Carlo simulations have examined the benefits to the overall system strength capacity of elastic load sharing and post-elastic residual capacity for a simple, cold-formed steel joist-supported floor system. While it is important to note that this only affects the strength capacity and floor systems may be governed by serviceability considerations, the main conclusions to this computational examination are:

1. Elastic load sharing results in a benefit to the overall system capacity of 3 to 5 percent beyond traditional code-based design equations.
2. Post-elastic residual capacity results in an additional benefit to the system of 20 to 30 percent to the capacity of the floor system.
3. Therefore, the simple, repetitive CFS floor system examined could be designed with approximately 25% percent less capacity and still achieve the same strength reliability intended by the AISI S100-12 design.

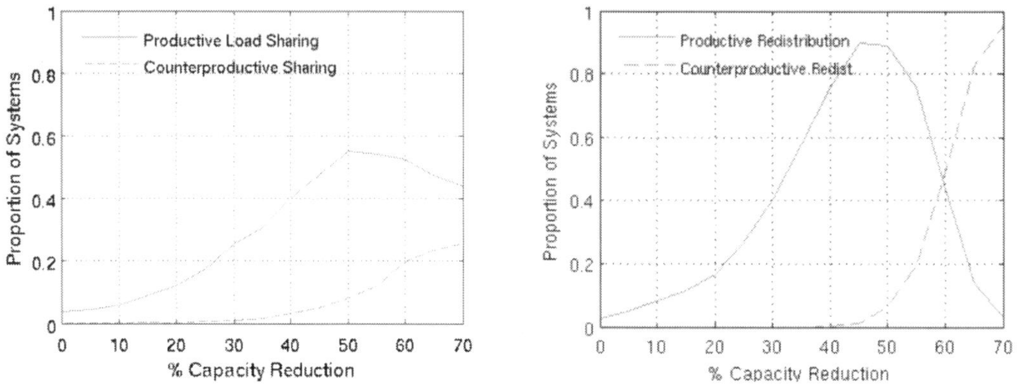

a) Elastic load sharing productivity b) Post-elastic residual capacity productivity

Fig. 4: Plot of whether elastic load sharing (left) and post-elastic residual capacity (right) is productive, reducing the number of member failures within a system, or counterproductive, increasing the number of member failures within a system. Counterproductive residual capacity load redistribution is physically equivalent to progressive collapse

Acknowledgments

The financial and in-kind support of the American Iron and Steel Institute is gratefully acknowledged, as is the financial support provided by the United States National Science Foundation through grants CMMI-1301033, 1301001, and 1300484. Any opinions, findings, and conclusions or recommendations expressed in this material are those of the authors and do not necessarily reflect the views of the American Iron and Steel Institute or the National Science Foundation.

References

[1] American Wood Council. "National Design Specification for Wood Construction, 2015 Edition", *ANSI/AWC NDS-2015*. September 2014.

[2] Verrill SP, Kretschmann DE. "Sources of Confusion in the Determination of ASTM Repetitive Member Factors for the Allowable Properties of Wood Products", *Journal of Structural Engineering*, 130-133. January 2012.

[3] American Iron and Steel Institute. "North American Specification for the Design of Cold-Formed Steel Structural Members, 2012 Edition", *ANSI/AISI S100-12*. June 2013.

[4] Schafer BW. Website for "NEESR-CR: Enabling Performance-Based Seismic Design of Multi-Story Cold-Formed Steel Structures", www.ce.jhu.edu/cfsnees, accessed on 21 January 2016.

[5] Peterman K, Schafer BW, Madsen RL, Buonopane S, Nakata N. "Experimental Performance of Full-Scale Cold-Formed Steel Buildings Under Seismic Excitations", *Network for Earthquake Engineering Simulation (NEES)*, Dataset, DOI: 10.4231/D3DB7VR05, 2014.

[6] Meimand VZ, Schafer BW. "Impact of Load Combinations on Structural Reliability Determined from Testing Cold-Formed Steel Components", *Structural Safety,* 48, 25-32, 2014.

WEB CRIPPLING TESTS OF RIVET FASTENED RECTANGULAR HOLLOW FLANGE CHANNEL BEAMS UNDER TWO-FLANGE LOAD CASES

Edward Steau, Mahen Mahendran and Poologanathan Keerthan

Science and Engineering Faculty, Queensland University of Technology, Brisbane, Australia

Abstract: The rivet-fastened rectangular hollow flange channel beam (RHFCB) is a new cold-formed hollow section proposed as an alternative to welded hollow flange steel beams. To date, no investigation has been conducted on their web crippling behaviour and strengths. Hence an experimental study was conducted to investigate the web crippling behaviour and capacities of rivet fastened RHFCBs under End Two Flange (ETF) and Interior Two Flange (ITF) load cases. Experimental results showed that the current design rules are unconservative for rivet fastened RHFCB sections under ETF and ITF load cases. Hence new equations were proposed to determine the web crippling capacities of rivet fastened RHFCBs.

1. Introduction

In steel construction, cold-formed steel (CFS) hollow flange channel beams are innovative products that represent a modern light weight building technique (Fig. 1). They are thin-walled structural members designed with two rectangular hollow flanges. This is unique from other commonly used open CFS sections such as C-, Z-, or hat-shaped sections. HFCBs have no unstiffened elements (no free edges), hence reducing the local buckling effect in the flanges. They are also designed to have the hollow flanges away from the neutral axis, making them structurally efficient as beams. In doing so, this combines the stability found in hot-rolled sections with the high capacity-to-weight ratio found in cold-formed sections. Hence welded hollow flange channel beams have been manufactured and used in many applications until recently. A rectangular hollow flange channel beam (RHFCB) was proposed using an intermittently rivet fastening process as an alternative to welded beams (see Fig. 1).

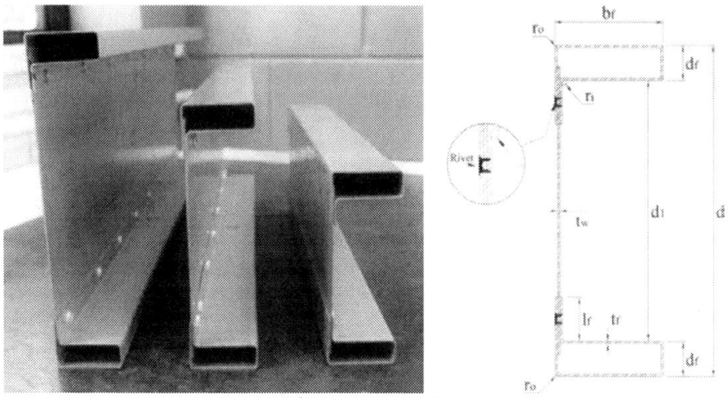

Fig. 1: Rivet Fastened RHFCB

The rivet fastened RHFCB can be easily manufactured using a cold-forming and rivet fastening process. The independent method of joining the web and flange plate elements of the section allows greater optimisation by choosing appropriate combinations of web and flange plate thicknesses. For example, the use of thicker web plate elements will increase the lateral distortional buckling capacity of RHFCB flexural members. The rivet fastened RHFCB also has additional lips, possibly contributing to increased rigidity and strength. The structural behaviour of welded hollow flange channel beams has been investigated by many researchers in the past [1,2]. However, only limited research has been conducted on the structural performance of rivet fastened RHFCB sections.

Since the rivet fastened RHFCB is a thin-walled flexural member, web bearing/crippling instability may occur. Web crippling is a web buckling failure mode that is caused by concentrated loads or support reactions. Rivet fastened RHFCB joists and bearers that are unstiffened against this type of loading are also vulnerable to similar web crippling failures, but no research has been undertaken yet.

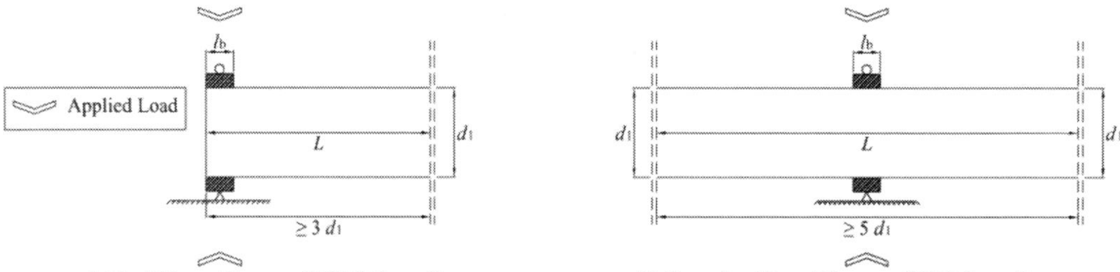

a) End Two Flange (ETF) Loading b) Interior Two Flange (ITF) Loading

Fig. 2: Loading Conditions for Web Crippling Tests [3]

Two-Flange web crippling load cases are defined in AISI S909 [3], which are shown in Fig. 2: End-Two-Flange (ETF) and Interior-Two-Flange (ITF) load cases. Suitable design equations are needed to predict the web crippling capacity of cold-formed steel sections subject to these load cases. However, the web crippling design equations in most cold-formed steel structural design codes are empirical, derived from experimental studies consisting of over 1200 tests of conventional cold-formed steel sections. The current AS/NZS 4600 [4] and AISI S100 [5] web crippling design equations provide a unified web crippling capacity equation. However, it is not applicable to the rivet fastened RHFCB due to the presence of two rectangular hollow flanges. Unlike other open cold-formed steel sections, rivet fastened RHFCB can also be subjected to flange crushing, combined web crippling and flange crushing and lip failures.

Rivet fastened RHFCBs can be used as flexural members in steel building systems. For them to be used as flexural members, their flexural, shear and web crippling capacities must be known. However, no investigation has been conducted into the web crippling behaviour and strength of rivet fastened RHFCBs. In this research web crippling behaviour and strength of rivet fastened RHFCBs under ETF and ITF load cases was investigated using an experimental study. This paper presents the details of this experimental study, and the results. Experimental web crippling capacities are compared with the predicted capacities using the current design rules, based on which appropriate improvements are recommended.

2. Web Crippling Design of Cold-Formed Steel Beams

Recent years have seen the development and introduction of different test methods and design equations for cold-formed thin-walled structures. However, past research is primarily based on experimental studies of open cold-formed steel sections (C-, Z-, hat-shaped and built-up sections). Limited research only has been conducted on the web crippling behaviour of welded hollow flange channel beams (LSB) [2].

The latest AISI S909 [3] presents the details of the suitable test method for conducting web crippling tests of cold-formed steel beams. The AISI test method is applicable to single-web, multiple-web and built-up web sections, and hence it is also applicable to rivet fastened RHFCBs. The AISI S909 recommends the following test specimen lengths for the two different load cases shown in Fig. 2. In this research the rivet fastened RHFCB test specimen lengths are based on this recommendation. Two-Flange Load Case: ETF: $L_{min} = 3.0d_1$ (see Fig. 2 (a)) and ITF: $L_{min} = 5.0d_1$ (see Fig. 2 (b)) where d_1 = depth of the flat portion of the web measured along the plane of the web; and l_b = bearing length.

The unified web crippling design equation (Eq. 1) presented in AS/NZS 4600 [4] and AISI S100 [5] defines the web crippling capacity (R_b) of open cold-formed steel sections. The web crippling design equation takes into consideration the clear height of web to thickness ratio (d_1/t_w), inside bent radius to thickness ratio (r_i/t_w), bearing length to thickness ratio (l_b/t_w), web thickness (t_w) and yield stress (f_{yw}). However, it does not take into consideration the effects of hollow flanges and rivet spacing in rivet fastened RHFCBs.

$$R_b = Ct_w{}^2 f_{yw} sin\theta\left(1 - C_r\sqrt{\frac{r_i}{t_w}}\right)\left(1 + C_l\sqrt{\frac{l_b}{t_w}}\right)\left(1 - C_w\sqrt{\frac{d_1}{t_w}}\right) \qquad (1)$$

where C = coefficient; θ = angle between the plane of the web and the plane of the bearing surface ($45° \leq \theta \leq 90°$); C_r = coefficient of inside bent radius ratio (r_i/t_w); C_l = coefficient of bearing length ratio (l_b/t_w); and C_w = coefficient of web slenderness ratio (d_1/t_w).

3. Experimental Investigation

The web crippling behaviour of rivet fastened RHFCB sections was investigated under two load cases, ETF and ITF load cases. Suitable inside bent radius to thickness ratios (r_i/t_w), bearing length to thickness ratios (l_b/t_w) and clear height of web to thickness ratios (d_1/t_w) were considered to ensure that web crippling failures occurred prior to other failures.

3.1 Test Details

Fifty two RHFCB test specimens with overall heights of 150, 200 and 250 mm were fabricated by rivet fastening two rectangular hollow flanges of 53 mm (width) × 18 mm (height) to a web plate at 100 mm spacing. Different flange and web thicknesses in the range of 0.7 to 1.5

mm and made of two steel grades, G300 and G500, were used. Most test specimens were made of G300 steel except for six. Tables 1 and 2 show the details of the web crippling test specimens used in this investigation on ETF and ITF load cases, respectively. In these tables, RHFCB specimens are specified by their overall depth (d) x flange width (b_f) x flange depth (d_f) x flange thickness (t_f) x web thickness (t_w) (refer to Fig. 1). They present the measured flange and web thicknesses (t_f and t_w), clear web heights (d_1) and yield stresses (f_{yw}) of the web elements of tested rivet fastened RHFCBs. The yield stresses of flange elements (f_{yf}) can also be obtained from Tables 1 and 2 as they were also made from the same steel sheets. They are 290 MPa, 310 MPa and 336 MPa for flange thicknesses of 0.95 mm, 1.15 mm and 1.2 mm, respectively. Since at the web-flange junction, the outside of the inner bent corners (r_i) is combined with the additional lips, the inner bent radius (r_i) of rivet fastened RHFCBs was considered as zero. The lip length was maintained at 20 mm for all the specimens. The specimen lengths of $3d_1$ and $5d_1$ were selected for ETF and ITF load cases, respectively, based on AISI S909 [3] recommendations. The measured overall depth and width were found to be equal to their nominal values.

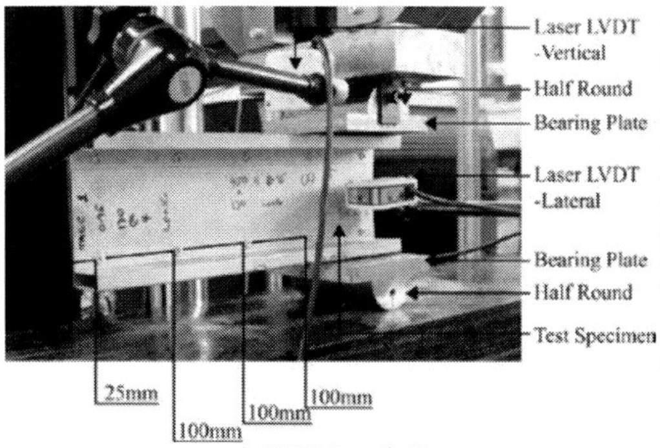

a) ETF Load Case

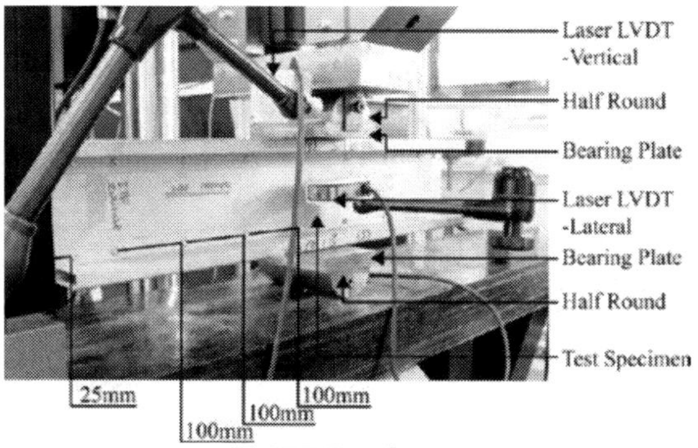

b) ITF Load Case

Fig. 3: Web Crippling Test Set-up

Support conditions were designed to ensure that all the test specimens had pinned supports. Single rivet fastened RHFCB section was used in the tests as used by Keerthan et al. [2] for their LSB tests under ETF and ITF load cases. It should be noted that no signs of twisting were observed during the experiments using single sections. Figs. 3 (a) and (b) show the test set-up used in the web crippling tests for ETF and ITF load cases, respectively. They were built based on the recommended AISI S909 [3] standard test method shown in Figs. 2 (a) and (b). Four different bearing lengths (25 mm, 50 mm, 100 mm and 150 mm) were used for both ETF and ITF load cases.

The web crippling tests were conducted using an Instron testing machine. A measuring system was set-up to record the applied load and associated test beam displacements. The specimen was placed in the Instron testing machine and a small load was applied first to allow the loading and support systems to settle evenly on the bearings. The measuring system was then initialised with zero values and the loading was commenced. The cross-head of the testing machine was moved at a constant rate of 0.7 mm/minute until the test beam failed.

3.2 Test Results and Discussions

A comprehensive summary of all the test results is given in Tables 1 and 2 where the ultimate web crippling capacities of tested rivet fastened RHFCB sections under ETF and ITF load cases are shown. Siahaan et al. [6] recommended 100 mm rivet spacing for RHFCBs based on their section moment capacity studies. Hence this study used the same rivet spacing in all the web crippling tests. Web crippling behaviour observed during these tests (see Figs. 4 (a) and (b)) exhibited no detrimental behaviour due to such intermittent fasting along the web to flange juncture. There was no excessive separation between web and flange plates during the tests. The web crippling test observations also confirm that 100 mm rivet spacing is suitable for RHFCBs.

Fig. 4 (a) shows the web crippling failure modes of rivet fastened RHFCBs under ETF load case while Fig. 4 (b) shows them under ITF load case. As seen in these figures, web crippling failures occurred within the clear height of web (between the two hollow flanges). Slender test specimens began to deflect out of plane and reached their ultimate web crippling capacities. All the specimens displayed significant ductility at failure. Fig. 5 (a) shows the applied load versus vertical deflection curves for the web crippling tests of 150×53×18×1.15×1.15 RHFCB section with 100 mm bearing length under ETF load case. Fig. 5 (b) shows the applied load versus vertical deflection curves for the web crippling tests of 150×53×18×1.15×1.15 RHFCB sections with 100 mm bearing lengths under ITF load case.

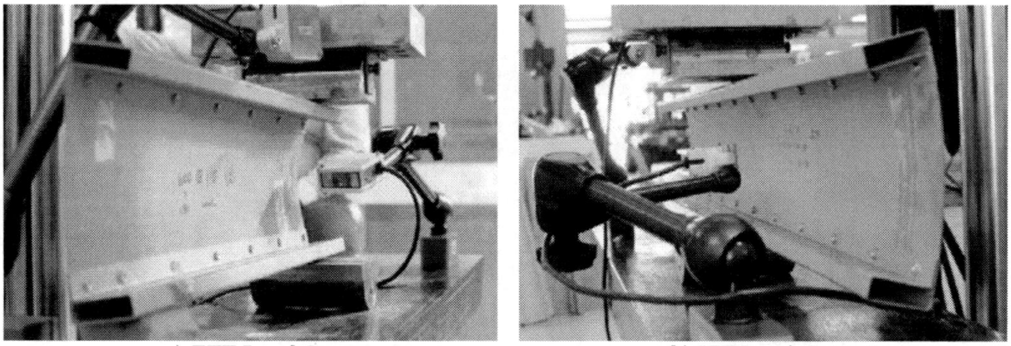

a) ETF Load Case b) ITF Load Case

Fig. 4: Web Crippling Failure Modes of 150×53×18×1.15×1.15 Rivet Fastened RHFCBs (Bearing Length = 100 mm)

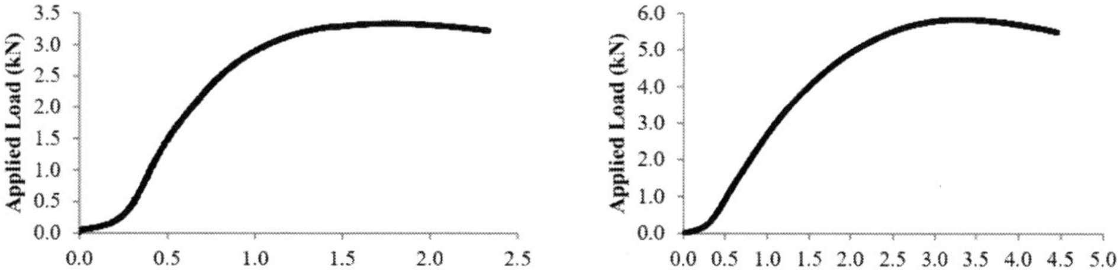

a) ETF Load Case b) ITF Load Case

Fig. 5: Plots of Applied Load versus Vertical Deflection of 150×53×18×1.15×1.15 Rivet Fastened RHFCBs (Bearing Length = 100 mm)

Table 1: Web Crippling Capacities of Rivet Fastened RHFCBs and Comparisons for ETF Load Case

Test No.	Rivet Fastened RHFCB Section ($d \times b_f \times d_f \times t_f \times t_w$)	t_f (mm)	t_w (mm)	d_1 (mm)	d_1/t_w	f_{yw} (MPa)	l_b (mm)	Web Crippling (kN) Test	AS/NZS 4600	Proposed Eq. 2	Test/ (AS/NZS 4600)	Test/ Proposed Eq. 2
***1	150×53×18×1.20×1.20	1.24	1.23	109.1	89.0	336	25	3.38	5.00	N.A.	0.68	N.A.
2	200×53×18×1.20×1.20	1.24	1.22	160.0	131.1	336	25	3.38	4.32	3.28	0.78	1.03
3	250×53×18×1.20×1.20	1.24	1.24	208.3	167.9	336	25	2.67	3.96	2.62	0.67	1.02
4	150×53×18×0.95×0.95	0.91	0.93	119.0	127.9	290	50	1.86	2.44	1.80	0.76	1.03
5	150×53×18×1.15×1.15	1.10	1.11	118.6	106.8	310	50	3.18	3.89	3.08	0.82	1.03
*6	150×53×18×1.15×1.50	1.11	1.58	118.8	75.1	545*	50	4.28	14.80	N.A.	0.29	N.A.
7	200×53×18×1.15×1.15	1.12	1.11	167.6	150.9	310	50	2.38	3.37	2.32	0.71	1.03
8	150×53×18×1.20×1.20	1.25	1.24	108.0	87.4	336	50	4.65	5.50	4.62	0.85	1.01
9	200×53×18×1.20×1.20	1.24	1.24	160.1	129.6	336	50	3.32	4.78	3.55	0.69	0.93
10	250×53×18×1.20×1.20	1.25	1.24	210.0	169.3	336	50	2.62	4.24	2.72	0.62	0.96
11	150×53×18×0.95×0.95	0.92	0.92	118.5	128.8	290	100	1.98	2.65	1.89	0.75	1.05
12	150×53×18×1.15×1.15	1.11	1.12	118.6	105.8	310	100	3.39	4.38	3.37	0.77	1.00
*13	150×53×18×1.15×1.50	1.11	1.56	118.6	76.0	545*	100	7.29	15.72	N.A.	0.46	N.A.
14	200×53×18×1.15×1.15	1.11	1.10	167.4	152.1	310	100	2.53	3.65	2.42	0.69	1.05
15	150×53×18×1.15×0.70	1.11	0.71	118.3	166.6	303	100	1.03	1.53	0.94	0.67	1.10
16	150×53×18×1.15×0.95	1.11	0.90	117.9	131.0	290	100	1.91	2.53	1.78	0.76	1.07
17	150×53×18×1.20×1.20	1.24	1.23	109.7	89.1	336	100	4.74	5.97	4.84	0.79	0.98
18	200×53×18×1.20×1.20	1.24	1.24	159.5	128.8	336	100	3.54	5.29	3.82	0.67	0.93
19	250×53×18×1.20×1.20	1.25	1.24	210.0	170.0	336	100	2.74	4.62	2.87	0.59	0.96
20	150×53×18×0.95×0.95	0.91	0.90	120.2	133.5	290	150	2.01	2.70	1.85	0.74	1.09
21	150×53×18×1.15×1.15	1.18	1.13	118.1	104.5	310	150	3.42	4.79	3.63	0.71	0.94
**22	150×53×18×1.15×1.50	1.11	1.53	118.1	77.1	545*	150	8.75	16.08	N.A.	0.54	N.A.
23	200×53×18×1.15×1.15	1.12	1.11	167.5	150.9	310	150	2.60	3.99	2.60	0.65	1.00
24	150×53×18×1.20×1.20	1.23	1.23	109.8	89.2	336	150	4.90	6.38	5.06	0.77	0.97
25	200×53×18×1.20×1.20	1.23	1.23	158.4	128.7	336	150	3.74	5.60	3.96	0.67	0.94
26	250×53×18×1.20×1.20	1.25	1.24	209.3	169.4	336	150	2.78	4.95	3.02	0.56	0.92
										Mean	**0.68**	**1.00**
										COV	**0.175**	**0.053**

4. Comparisons of Test and Predicted Web Crippling Capacities

Experimental ultimate web crippling capacities from this study were compared with the predictions from the design equation (Eq. 1) based on AS/NZS 4600 [4] and AISI S100 [5] in Tables 1 and 2. Test results were not considered in these comparisons if the test specimens had failed by flange crushing or combined web crippling and flange crushing. For the prediction of web crippling capacities, support and flange conditions were taken as unfastened, stiffened or partially stiffened flanges and Two-flange loading (ETF and ITF) and the corresponding web crippling coefficients are as follows: ETF load case - $C = 13$; $C_r = 0.32$; $C_l = 0.05$; and $C_w = 0.04$. ITF load case - $C = 24$; $C_r = 0.52$; $C_l = 0.15$; and $C_w = 0.001$. The inner bent radii of RHFCBs were taken as zero in these calculations.

For the ETF load case, the mean value of test to predicted web crippling capacity of rivet fastened RHFCB by AS/NZS 4600 [4] and AISI S100 [5] is 0.68 while the corresponding coefficient of variation (COV) is 0.175. For the ITF load case, the mean value of test to predicted web crippling capacity of rivet fastened RHFCB by AS/NZS 4600 [4] and AISI S100 [5] is 0.24 while the corresponding COV is 0.315. Hence it can be concluded that AS/NZS 4600 [4] and AISI S100 [5] design equations are considerably unconservative for rivet fastened RHFCB sections, in particular for the ITF load case.

Since AS/NZS 4600 [4] and AISI S100 [5] design equations were developed for open cold-formed steel sections (C-, Z-, hat-shaped and built-up sections), new web crippling capacity equations should be developed for rivet fastened RHFCBs with rectangular hollow flanges. Details of the proposed web crippling capacity equations for rivet fastened RHFCBs are given in the next section.

Table 2: Web Crippling Capacities of Rivet Fastened RHFCBs and Comparisons for ITF Load Case

Test No.	Rivet Fastened RHFCB Section (d×b_f×d_f×t_f×t_w)	t_f (mm)	t_w (mm)	d_1 (mm)	d_1/t_w	f_{yw} (MPa)	l_b (mm)	Web Crippling Capacity (kN)			Test/ (AS/NZS 4600)	Test/ Proposed Eq. 3
								Test	AS/NZS 4600	Proposed Eq. 3		
**1	150×53×18×1.20×1.20	1.24	1.24	110.2	89.2	336	25	8.13	20.41	N.A.	0.40	N.A.
**2	200×53×18×1.20×1.20	1.23	1.25	161.0	129.3	336	25	7.10	20.66	N.A.	0.34	N.A.
**3	250×53×18×1.20×1.20	1.24	1.24	213.1	171.8	336	25	5.53	20.48	N.A.	0.27	N.A.
**4	150×53×18×0.95×0.95	0.91	0.91	119.8	131.6	290	50	3.50	12.03	N.A.	0.29	N.A.
**5	150×53×18×1.15×1.15	1.10	1.12	117.8	105.1	310	50	7.71	18.49	N.A.	0.42	N.A.
**6	150×53×18×1.15×1.50	1.10	1.54	117.4	76.2	545*	50	10.81	57.03	N.A.	0.19	N.A.
**7	200×53×18×1.15×1.15	1.11	1.11	166.9	150.3	310	50	4.82	18.17	N.A.	0.27	N.A.
**8	150×53×18×1.20×1.20	1.24	1.24	109.4	88.5	336	50	8.75	23.81	N.A.	0.37	N.A.
**9	200×53×18×1.20×1.20	1.23	1.23	159.2	129.3	336	50	6.09	23.60	N.A.	0.26	N.A.
**10	250×53×18×1.20×1.20	1.25	1.24	212.5	171.3	336	50	4.81	23.89	N.A.	0.20	N.A.
11	150×53×18×0.95×0.95	0.91	0.91	119.9	131.7	290	100	3.43	14.66	3.02	0.23	1.14
12	150×53×18×1.15×1.15	1.11	1.15	116.8	101.5	310	100	5.89	23.36	5.81	0.25	1.01
*13	150×53×18×1.15×1.50	1.15	1.57	116.7	74.3	545*	100	9.56	70.23	N.A.	0.14	N.A.
14	200×53×18×1.15×1.15	1.12	1.12	166.5	148.6	310	100	4.93	22.29	4.35	0.22	1.13
15	150×53×18×1.15×0.70	1.12	0.71	117.2	165.0	303	100	1.80	10.06	1.67	0.18	1.07
16	150×53×18×1.15×0.95	1.12	0.90	116.5	129.4	290	100	3.47	14.39	2.99	0.24	1.16
17	150×53×18×1.20×1.20	1.25	1.24	110.0	89.0	336	100	7.17	28.63	7.65	0.25	0.94
18	200×53×18×1.20×1.20	1.23	1.25	161.8	129.9	336	100	5.78	28.97	6.32	0.20	0.91
19	250×53×18×1.20×1.20	1.25	1.25	212.2	170.4	336	100	4.58	28.92	5.11	0.16	0.90
20	150×53×18×0.95×0.95	0.90	0.94	118.2	125.7	290	150	3.56	17.60	3.52	0.20	1.01
21	150×53×18×1.15×1.15	1.12	1.11	118.4	106.6	310	150	5.93	24.89	5.63	0.24	1.05
*22	150×53×18×1.15×1.50	1.15	1.58	116.9	73.9	545*	150	11.59	79.68	N.A.	0.15	N.A.
23	200×53×18×1.15×1.15	1.11	1.11	167.0	150.4	310	150	4.93	24.84	4.50	0.20	1.10
24	150×53×18×1.20×1.20	1.24	1.24	109.3	88.1	336	150	7.28	32.55	8.21	0.22	0.89
25	200×53×18×1.20×1.20	1.24	1.24	160.7	129.6	336	150	5.73	32.48	6.66	0.18	0.86
26	250×53×18×1.20×1.20	1.24	1.24	213.5	172.1	336	150	4.62	32.42	5.32	0.14	0.87
										Mean	**0.24**	**1.00**
										COV	**0.315**	**0.108**

Note 1: Measured values of t_f, t_w and d_1 are given in this table. Other values are nominal. l_f = 20 mm. *G500 steel web element.

Note 2: AS/NZS 4600 and AISI S100 web crippling design equations are identical.

* Section failed due to flange crushing. ** Section failed due to combined web crippling and flange crushing.

*** Section failed due to lip failure. N.A. Not Applicable.

5. Proposed Web Crippling Design Equation

Since the current AS/NZS 4600 [4] and AISI S100 [5] web crippling design equations were found to be unsafe for rivet fastened RHFCBs, new design equations are proposed in this section based on experimental results. Eq. 1 from AS/NZS 4600 [4] and AISI S100 [5] is proposed with modified web crippling coefficients C, C_r, C_l and C_w. Since the outside of the inner bent corners (r_i) is combined with the additional lips at the web-flange junction, the inner bent radius (r_i) of rivet fastened RHFCBs was considered as zero. Therefore, C_r was taken as zero. Tables 1 and 2 show the comparison of the web crippling capacities from tests with the predictions from the proposed Eqs. 2 and 3. Table 3 shows the associated, modified web crippling coefficients. For ETF load case, the mean value of test to predicted web crippling capacity ratio is 1.00 with a COV of 0.053, while these values are 1.00 and 0.108 for ITF load case. It has been shown that the web crippling capacities predicted by Eqs. 2 and 3 agree well with the experimental web crippling capacities of rivet fastened RHFCBs under ETF and ITF load cases.

Test capacity results for flange crushing and combined web crippling and flange crushing failures were not considered in the above proposed equations. However, since the web crippling capacity increased when RHFCBs were subjected to a combined web crippling and flange crushing failure, it is recommended to use the same equations for RHFCBs subject to web crippling and combined web crippling and flange crushing as a conservative option.

Using the capacity reduction factor (ϕ_w) given in the North American Cold-formed Steel Specification AISI S100 [5] with the mean and COV values in Tables 1 and 2 gave capacity reduction factors (ϕ_w) of 0.84 and 0.83 for ETF and ITF load cases, respectively. Therefore it is recommended to use a ϕ_w factor of 0.80 for both ETF and ITF load cases.

$$R_b = 15.58 t_w^2 f_{yw} sin\theta \left(1 + 0.03\sqrt{\frac{l_b}{t_w}}\right)\left(1 - 0.055\sqrt{\frac{d_1}{t_w}}\right) \quad \text{ETF load case (2)}$$

$$R_b = 20.79 t_w^2 f_{yw} sin\theta \left(1 + 0.04\sqrt{\frac{l_b}{t_w}}\right)\left(1 - 0.05\sqrt{\frac{d_1}{t_w}}\right) \quad \text{ITF load case (3)}$$

Table 3: Proposed Web Crippling Coefficients for Rivet Fastened RHFCBs

Load Case	Equations	C	C_r	C_l	C_w	Mean	COV
ETF	AS/NZS 4600	13.00	0.32	0.05	0.04	0.68	0.175
	Proposed	15.58	0.00	0.03	0.055	1.00	0.053
ITF	AS/NZS 4600	24.00	0.52	0.15	0.001	0.24	0.315
	Proposed	20.79	0.00	0.04	0.05	1.00	0.108

6. Conclusions

This paper has presented the details 52 web crippling tests of an innovative thin-walled and intermittently rivet fastened rectangular hollow flange channel beam (RHFCB) under ETF and ITF load cases and the results. The ETF and ITF load tests showed that the new RHFCBs failed by web crippling in most cases. However, some of them failed by flange crushing when thinner flange elements and smaller bearing lengths were used while many others failed by combined web crippling and flange crushing under ITF load case when smaller bearing lengths were used. The web crippling capacity design equations in AS/NZS 4600 [4] and AI-SI S100 [5] were unable to predict the capacities of rivet fastened RHFCB sections under ETF and ITF load cases. This paper has therefore proposed new equations to predict the web crippling capacities of rivet fastened RHFCBs under two-flange load cases. The use of 100 mm rivet spacing was found to be adequate.

References

[1] Keerthan P, Mahendran M. "Experimental studies on the shear behaviour and strength of LiteSteel beams", *Engineering Structures*, 32(10), 3235-3247, 2010.

[2] Keerthan P, Mahendran M, Steau E. "Experimental studies of web crippling behaviour of hollow flange channel beams under two flange load cases", *Thin-Walled Structures*, 85, 207-219, 2014.

[3] American Iron and Steel Institute (AISI S909). "TS-9-05 Standard test method for determining the web crippling strength of cold-formed steel beams", *AISI*, Washington, DC, USA, 2013.

[4] Standards Australia/Standards New Zealand (SA). "AS/NZS4600 Australia/New Zealand standard cold-formed steel structures", *SA*, Sydney, Australia, 2005.

[5] American Iron and Steel Institute (AISI S100). "North American specification for the design of cold-formed steel structural members", *AISI*, Washington, DC, USA, 2012.

[6] Siahaan R, Keerthan P, Mahendran M. "Section moment capacity tests of rivet fastened rectangular hollow flange channel beams", *22nd International Specialty Conference on Cold-Formed Steel Structures*, St Louis, Missouri, USA, 277-294, 2014.

SLENDER STEEL ARCH STABILIZED BY A TEXTILE MEMBRANE

Ondrej Svoboda and Josef Machacek

Czech Technical University in Prague, Faculty of Civil Engineering, Prague, Czech Republic

Abstract: Stabilization effects of textile membranes on in-plane and out-of-plane behaviour of a steel tube arch supporting the membranes is investigated. Model of a large membrane structure using Ferrari® Précontraint 702S fabric was tested in laboratory under various loadings and the resulting stress/deflection values are presented. The stability behaviour of the inner supporting steel tube was of primary interest. The SOFiSTiK software was employed to analyse the behaviour in 3D GNIA with respect to the mesh sensitivity, proper modelling of the membrane prestressing and boundary conditions and validated using test results. Parametric studies cover membranes supported by one central steel arch with various geometries.

1. Introduction

Within last few decades the Tensioned Fabric Structures (TSF) have become popular not only for their visual attractiveness but also for their ability to be a fully load-bearing component in a sophisticated structure. Various membrane materials are available for both single layer double curved shapes (hypars, cones, barrels) and double/more layers pneumatic constructions (inflatable cushions, air supported pneumatic structures). The PVC coated polyester fabric (e.g. Précontraint FERRARI®) seems to be a rational choice for common single layer structures with the lifetime up to 20 years, acceptable cost and good joining possibilities (welding or sewing). More expensive but longer lifetime provides glass fabric coated by PTFE (Teflon), possibly by silicon rubber or titanium dioxide, joined by bonding. Rather expensive but excellent material is expanded PTFE coated 2 sides by fluoropolymer film (TENARA®), joint by welding. The membranes may also be thermally insulated using Nanogel Aerogel™ and 2 sides coated with PTFE (in result translucent, with total thickness 9 mm, CABOT Cor.). High density polyethylene fabrics (HDPE, coated with LDPE) joined by sewing may also be used however, with short lifespan up to 10 years. Plastic foils of low thickness 50÷500 μm are used mainly as inflatable cushions, nowadays predominantly from ETFE (TEXLON®) or THV materials, with versatile use in both facades and roofing.

Proper design of structures with TSF requires geometrically and materially non-linear analysis with imperfections (GMNIA). The essential in such analysis is an appropriate input of the membrane material behaviour. The material is due to its structure non-homogeneous, orthotropic (warp and fill directions) and non-linear. The most sophisticated model with 15 input parameters obtained from laboratory tests was proposed by Kato *et al.* [1]. This model demonstrates excellent agreement with test data but because of its complexity and time requirements is rather unsuitable for practical utilization. Gosling, apart from a simplified approach, suggested in [2] strain-strain-stress approach using response surfaces linking strains to stresses through three dimensional representations for FE analysis. A non-linear material model with 5 parameters based on experimental results and depending on load ratios in warp and fill direction was proposed by Galliot and Luchsinger [3], [4]. Model parameters are: warp and fill Young's modules for the respective load ratio 1:1, the change in warp and fill Young's modules and the Poisson's ration. Recently a simplified stress-strain model for coated plain-weave fabrics was developed by Pargana and Leitao [5]. Model consists of three non-linear elements to model the yarns and an isotropic plate to model the coating.

In accord with recommendation by Tensinet Analysis & Materials working Group [6], [2] a simplified elastic approach may be employed using the simple plane stress theory. The supplied test data provides elastic modules for warp and fill directions and corresponding Poisson's ratio, valid for anticlastic type of structures. Recently Uhlemann *et al.* [7] analysed two approaches concerning simplified elastic constants for design of the membranes (Japanese MSAJ and European acc. to [6]) and found the European approach more general and reasonable for PES/PVC material (but still far from satisfactory).

Outstanding and unique structures using either common or sophisticated textile fabrics were designed in last few decades which indicate a new ways and directions of novel structures. Allianz Arena in Munich (design by Herzog & de Meuron, 2005), SkySong ASU in Scotsdale (FTL Design Engineering Studio, 2009) or Union Station in Denver (Skidmore, Owings & Merrill LLP (SOM), 2014) are just some examples of excellent projects realized in last decade.

For these structures is necessary to use supporting components as slender as possible to follow a concept of the delicate, light and attractive structure with the membrane surface and to avoid any visual intervention of supporting steelworks into the membrane area. While membrane surface is exclusively tensioned, supporting construction is most often exposed to a compression and/or bending. This type of loading, in combination with slender elements, results into stability problems and design must be done with respect to these effects.

Different situations occur, when membrane surface is joined with the supporting steel structure continually. In this case the membrane represents a spring support for the supporting structure, the critical length of individual steel elements is changing, and both parts of the entire system cannot be investigated separately but as one complex structure using proper software package which allows integrated modelling and computing (e.g. EASY [8], FORTEN [9], SOFiSTiK [10], Rhino Membrane [11], NDN [12], etc.). Detailing of membrane structures and erection methods are well described by Seidel [13] and the basic design in European Design Guide [6].

This paper demonstrates significant stabilizing effects of membranes to the respective supporting steelwork, based on numerical parametrical studies validated by tests. The tests relate to structural model representing a concert stage structure with two supporting arches. Influence of various membrane prestressing on the structure stabilization is analysed and compared with the test results. Following parametrical study concerns nonlinear behaviour of tubular arches which support textile membranes of various geometries.

2. Tests

2.1 Description of the test model, places of monitoring, way of prestressing

Model of an outdoor covered stage (with an approx. reduction 1:10) was proposed using formfinder software [14]. The software enables intuitive manipulation with membrane shapes under required stress level in interactive way and export/import to other programs through DXF/DWG files. Supporting steelwork involves two supporting arches (outer and inner CHS tubes), bottom edge steel wire ropes and membrane plates, see Figure 1.

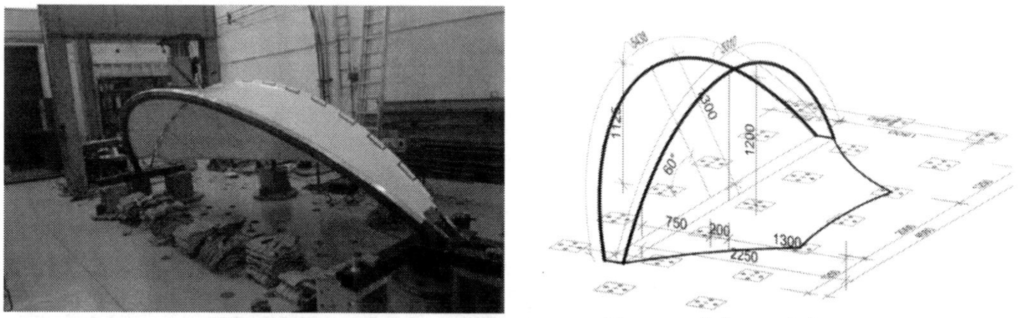

Fig. 1: General view at laboratory and layout of the model

The membrane is PVC coated polyester fabric Ferrari® Précontraint 702S with opaque surface (weight 830 g/m^2). Principal dimensions of the vertical inner tubular arch are LxH = 4500x1200 [mm], while outer tubular arch has inclination of 60° in respect to horizontal.

The fabric of Serge Ferrari group is made of polyester scrim coated both sides with liquid PVC and PVDF topcoat, weldable for joining. The material is due to its structure non-homogeneous, orthotropic (warp and fill directions) and non-linear, with elastic constants presented further at chapter of numerical modelling. Nevertheless, the patented fabrication process ensures similar elongation behaviour in both warp and fill directions and minimum creep. Concerning material characteristics the biaxial test performed by Lab BLUM Stuttgart [15] is available. With respect to these tests both warp and fill braking loadings were considered as $S_{ult} \approx 56$ kN/m, while working loading $S_{max} = S_{ult}/5 \approx 11.2$ kN/m to exclude tearing and prestressing up to $P = S_{max}/5 \approx 2.24$ kN/m, however in the model reality up to $P \leq 0.5$ kN/m.

The steel tubes from grade S355J0 were hot-formed in workshop. The investigated inner tube of ø 26.9x3.2 [mm] and outer one of ø 88.9x3.2 [mm] were welded to steel blocks to form a fixed-in frame acc. to Fig. 1. Coupon tests of the inner steel tube resulted into average yield strength $f_y = 475$ MPa, ultimate strength $f_u = 595$ MPa and elongation 27.1 %. Modulus of elasticity was considered in accord with Eurocode 3 as $E = 210$ GPa.

Membrane was joined with the outer arch using riveted aluminium keder profile while to inner tube via alternating pockets. Common 7x7 wire peripheral rope from CarlStahl Company with diameter of 6 mm in curved cuff fastened the membrane to corner plates and anchors. The investigated inner arch was fitted with transducers (electrical potentiometers) in vertical (V), transverse (H) and longitudinal (L) directions to measure deflections. Their positions and loading points (P) are shown in Fig. 2. In supports and middle of the arch strain gauges denoted (T) were placed, always in four mutually perpendicular positions, for later comparison of stresses with numerical results.

The prestressing of the membrane resulted from the membrane cut as prepared by EASY software [8] considering uniform prestressing of roughly $P = 0.5$ kN/m. During assembly the peripheral ropes were tightened and the membrane checked against wrinkling. 8 strain gauges were placed at various locations within the membrane surface for rough information on prestressing values. The measuring before and after prestressing resulted into average prestrain value of ε_y = 140 μm/m (i.e. $P \approx 0.09$ kN/m) in perpendicular direction with respect to the supporting arches and ε_x = 450 μm/m (i.e. $P \approx 0.30$ kN/m) in parallel direction to the arches.

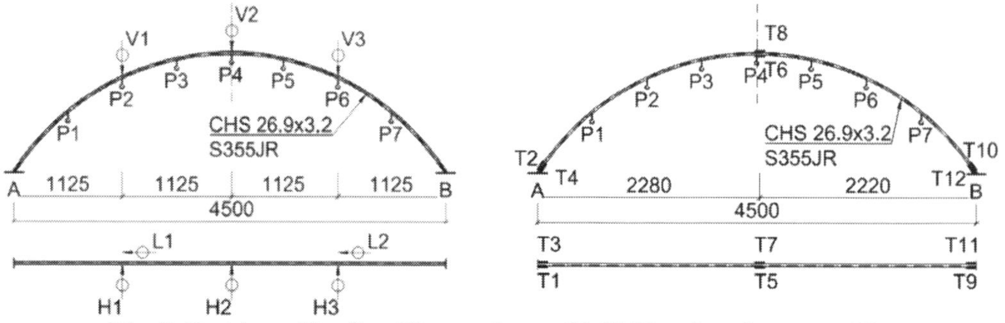

Fig. 2: Positions of loading (P), transducers (V, H, L) and strain gauges (T)

2.2 Test results

The investigation concerned exclusively the inner steel arch to find stabilizing effect of the membrane to its nonlinear behaviour. First the inner arch alone (without fastening the membrane) was loaded and second, after the membrane assembly, the complete membrane structure. For loading calibrated pouches with steel pellets were used and suspended from seven given points *P* at the arch. The loadings were carefully arranged to simulate uniform symmetrical loading and asymmetrical loading corresponding to the first in-plane buckling mode. Maximum loadings for arch alone are in Fig. 3, for complete membrane structure in Fig. 4.

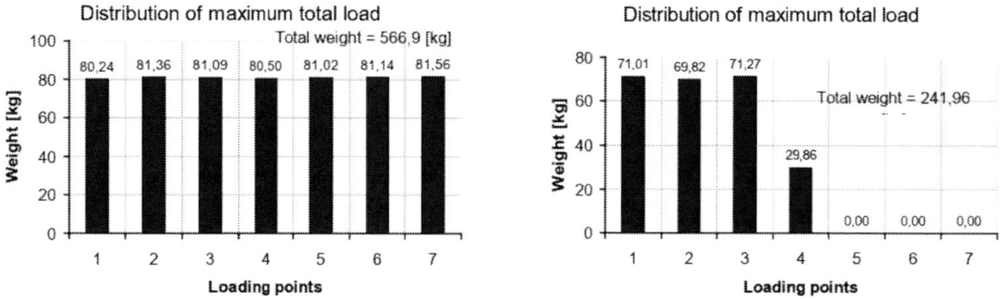

Fig. 3: Loading of the inner steel arch alone: Left-symmetrical, right - asymmetrical

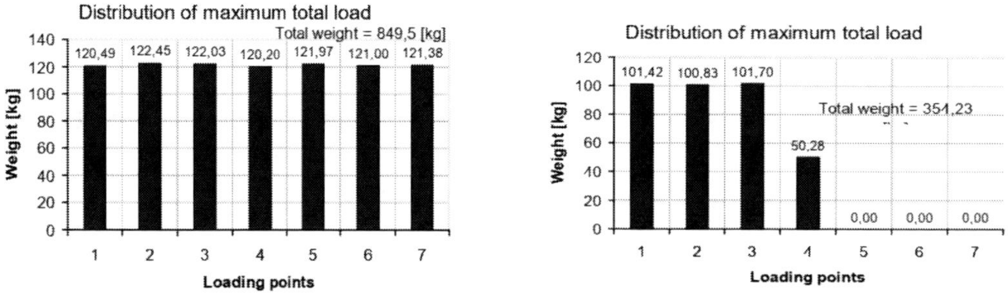

Fig. 4: Loading of the arch joined with the membranes: Left-symmetrical, right - asymmetrical

Individual loading steps amounted for roughly 1/10 of maximum loading, each followed by unloading. The tests were terminated when abnormal deflections out-of-arch plane or in-arch-plane were reached.

2.2.1 Symmetrical Loadings

Deflections of the inner arch under increasing loading in vertical and horizontal directions are shown in Fig. 5 (unloading is not included and generally was fully elastic). The arch without membrane buckled out-of-plane at total loading approaching $F_0 = 5.5$ kN, with associated vertical deflection along all span down. On the other side the test with arch stabilized by the membrane was terminated under total load of $F_M = 8.3$ kN, showing very small and nearly linear increase of the mid span deflection. Stabilizing effect of the membrane is enormous.

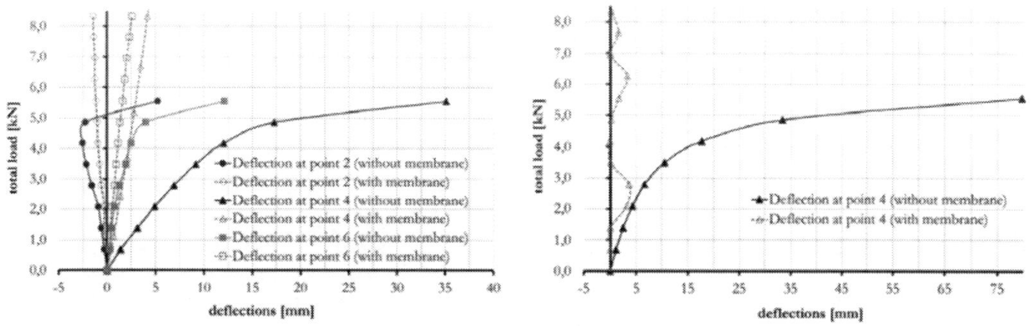

Fig. 5: Symmetrical loadings:
Vertical deflections (left), transverse deflections (right)

2.2.2 Asymmetrical Loadings

Vertical and transverse horizontal deflections under increasing loading are shown in Fig. 6. Testing of the arch without membrane ("0") terminated under total loading $F_0 = 2.37$ kN, giving maximal vertical deflection $\delta_0 = 41.3$ mm and horizontal one $\eta_0 = 3.5$ mm. The arch stabilized by membrane ("M") deflected much less, giving for the same loading $F_M = F_0 = 2.37$ kN values $\delta_M = 18.5$ mm and $\eta_M = 0.5$ mm. The stabilizing effect of the membrane concerning vertical deflection resulted into reduction of 45 %.Difference comparing vertical deflections in positions V1 and V3 (see Fig. 2) gives $\Delta\delta_0 = 68.3$ mm and $\Delta\delta_M = 32.9$ mm.

Measured displacements and stresses (not presented) prove expected enormous effect of the membrane on buckling load and strength of inner supporting arch, particularly in asymmetrical loadings. Test values are used for validation of numerical analyses.

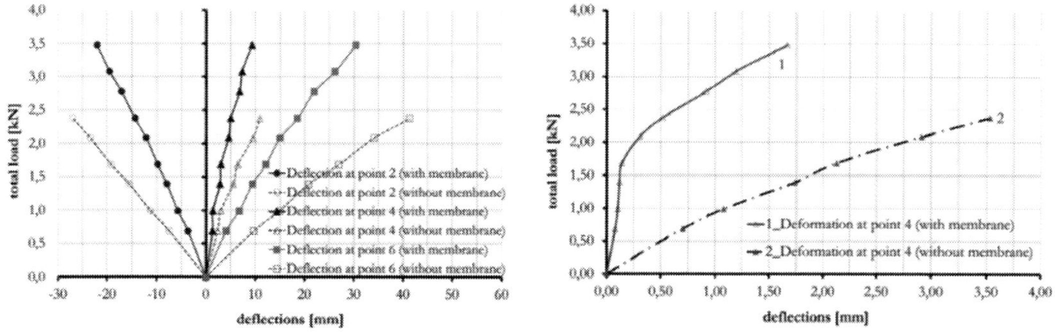

Fig. 6: Asymmetrical loadings:
Vertical deflections (left, positive down), transverse deflections (right)

3. Numerical analysis

3.1 Numerical model and validation of tests

SOFiSTiK software [10] was used to perform GNIA (using N-R iteration) both for the inner arch alone and the complete membrane structure. Various meshing of the membrane was analysed (square sizes of 25, 50, 100 and 200 [mm]) giving nearly identical stress results (differences $\leq 0.2\%$), but with increase of computer time between largest and least mesh up to 107 times. Optimum mesh with 50 mm size was used through the analysis.

Arch alone analysis: The tube arch was introduced as ø 26.9x3.2 [mm] with built-in supports, material S355J0 and Young's modulus E = 210 GPa. Initial geometry as measured was introduced into analysis (theoretical radius R = 2709 mm with out-of-plane (transverse) deflection at the crown w_0 = 17 mm). The elastic buckling analysis of the perfect arch under symmetrical loading in accordance with Fig. 3 (left) resulted into total critical loads $TL_{cr,1}$ = 6.7 kN (out-of-plane single wave buckling), $TL_{cr,2}$ = 15.4 kN (out-of-plane two waves buckling), $TL_{cr,3}$ = 18.7 kN (in-plane two waves buckling), etc.

GNIA transverse deflection curves for symmetrical loading are shown in Fig. 7 left, indicating the first out-of-plane bifurcation earlier but in accordance with test, roughly at 4.9 kN. Vertical deflections for asymmetrical loading correspond also well to test results, Fig. 7 right.

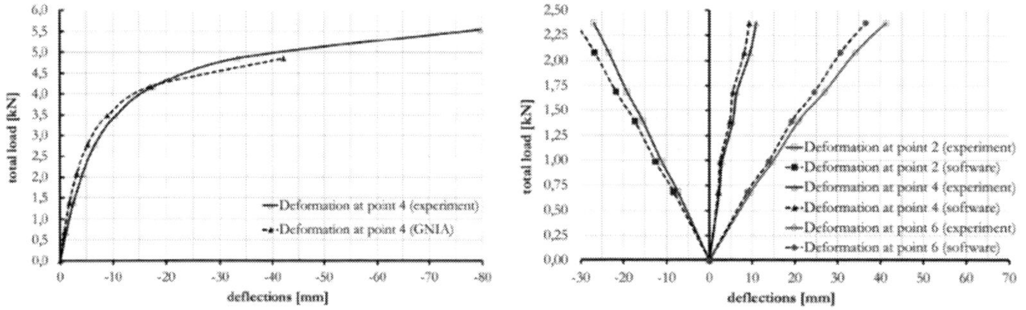

Fig. 7: Arch alone, test and GNIA results. Left: transverse deflections for symmetrical loading at midspan. Right: vertical deflections for asymmetrical loading

Full membrane structure analysis: The membrane FERRARI® 702 was considered with linear elastic orthotropic behaviour taking account for influence of load ratios according to [3]. The necessary five parameters were taken as: warp and fill Young's moduli for 1:1 load ratio $E_w^{1:1}$ = 635.3 kN/m and $E_{wf}^{1:1}$ = 661.9 kN/m, the change in warp and fill Young's moduli ΔE_w = 295 kN/m and ΔE_f = 168 kN/m, the Poisson's ratio ν = 0.196. The warp direction was considered as perpendicular to the arch plane and conversion to stresses due to approximate thickness of the membrane as 0.7 mm.

Initial geometrical imperfections of the arch were measured just before the first loading and were found to be 5.0 mm in horizontal direction at the top of the arch and +/-10.0 mm in vertical direction at the quarters of the arch. Initial shape of the membrane surface was generated automatically by the AutoCAD as a surface with minimal area. The shape was exported into SOFiSTiK software. For each level of the membrane prestress the corresponding value of the perimeter cable pretension was calculated based on formula:

$$S_S = r \cdot S_M \tag{1}$$

where S_S is a tension force in a perimeter cable [kN],
r radius of the cable curvature [m],
S_M prestress in the membrane perpendicular to the cable [kN/m].

The equilibrium state and final unloaded shape of the structure was found under initial SOFiSTiK software calculations. The results of GNIA under various pretension of the membrane P [kN/m] is shown in Fig. 8. The level of pretension is the crucial parameter for the arch stability and resistance capacity. Higher pretension in membrane logically means higher stability and smaller deflection of the investigated arch. It should be recalled, that the real pretension during the testing was influenced by the alternating "pockets" connection of the membrane to the inner arch and due to this fact rather non-uniform and slacked. As mentioned the randomly measured values during the test at several localities (which however can't be considered as a reliable ones) were between 0.09 and 0.3 [kN/m]. This is reflected in numerical analysis with pretension close to zero (0.2 kN/m) which provides results very near to the test ones. Therefore, the received GNIA results with the corresponding prestress 0.2 kN/m justify use of the model for following parametric studies.

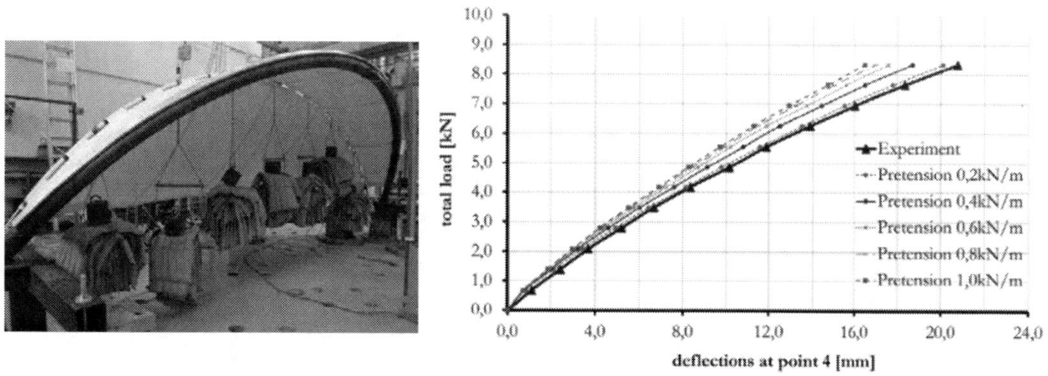

Fig. 8: Symmetrical loading of the membrane model:
Comparison of the test and GNIA vertical deflections under various pretension

3.2 Parametrical study

Inner arches with practical dimensions arranged in a barrel type structure in according with Fig. 9 and their buckling/strength behaviour were studied in a parametrical study. The mid arch in the 5 arch assembly, where the edge arches are continuously supported in all three directions, have been studied with span $L = 6 \div 20$ [m], rise $H = L/10 \div L/2$ and spacing $B = 3 \div 10$ [m]. The full study results will be published soon. E.g. tube arch ø 133x5 [mm], steel grade S355, with $L = 12$ m, $H = 3.6$ m and $B = 3$ m and uniformly loaded vertically under 3 kN/m (snow loading) with mesh size 250 mm, gives following buckling loads: arch alone in-plane $N_{cr.y} = 189$ kN, out-of-plane $N_{cr.z} = 70$ kN; arch with membrane prestressed by $P \approx 5.0$ kN/m gives $N_{cr.y} = 346$ kN and $N_{cr.z} = 542$ kN, i.e. the membrane support to the arch is substantial.

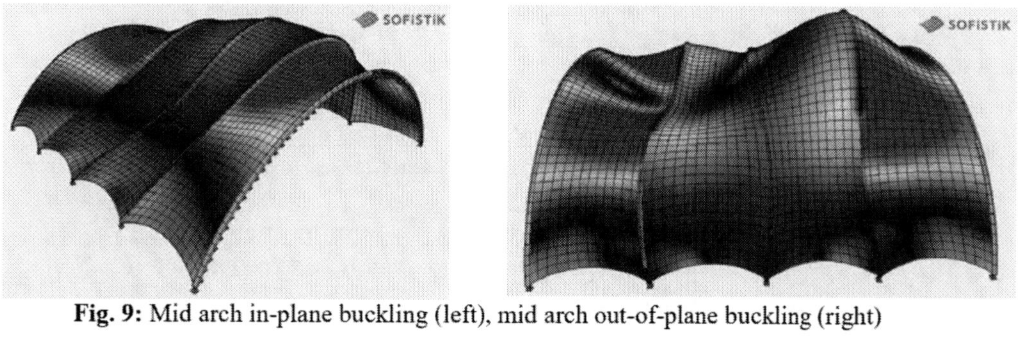

Fig. 9: Mid arch in-plane buckling (left), mid arch out-of-plane buckling (right)

4. Conclusions

The main conclusions are:

1. Tests confirmed decisive effect of textile membranes on both in-plane and out-of-plane supporting arch stability and strength.
2. GNIA with linear elastic orthotropic behaviour of the textile membrane and using SOFiSTiK software proved to be adequate for numerical modelling.
3. The real value of the membrane prestressing substantially influence the deflections and strength of the supporting arch structure (see the sensitivity study in Fig. 8).
4. Parametric studies of large barrel membrane structures supported by a row of steel arches result into huge savings in design of supporting steel arches due to enormous increase of both in-plane and particularly out-of-plane buckling loads in comparison to the ones of an arch alone (e.g. in the shown case $N_{cr,z}$ increased 7.7 times).

Acknowledgments

This work was supported by the Czech Grant Agency, grant GACR No. 105/13/25781S. Thanks are due to company Archtex s.r.o. Prague for help with assembly of the membranes.

References

[1] Kato S, Yoshino T, Minami H. “Formulation of constitutive equations for fabric membranes based on concept of fabric lattice model”, *Engineering Structures*, 21(8), 691-708, 1999.

[2] Gosling P. “Basic Philosophy and calling notice”, *Tensinet analysis & Material working group, Tensinews* (www.tensinet.com), 13, 12-15, 2007.

[3] Galliot C, Luchsinger RH. “A simple model describing the non-linear biaxial tensile behaviour of PVC/coated polyester fabrics for use in finite element analysis”, *Composite Structures*, 90(4), 438-447, 2009.

[4] Galliot C, Luchsinger RH. “Non-linear properties of PVC-coated fabrics used in tensairity structures”, *Proceedings ICCM17*, Edinburgh, UK, 10 p., 2009.

[5] Pargana JB, Leitao WMA. “A simplified stress-strain model for coated plain-weave fabrics used in tensioned fabric structures”, *Engineering Structures*, 84, 439-450, 2015.

[6] Foster B, Mollaert M. *European Design Guide for Tensile Surface Structures*, TensiNet, 354 p., 2004.

[7] Uhlemann J, Stranghöner N, Saxe K. “Stiffness parameters for architectural fabrics: An analysis of two determination procedures”, *Structural Engineering International*, 25(1), 9-19, 2015.

[8] technet gmbh Berlin-Stuttgart. http://www.technet-gmbh.com.

[9] ixForten 4000. http://www.ixforten.com/

[10] SOFiSTiK 2014. http://www.sofistik.de/

[11] Rhino Membrane. http://www.membranes24.com/rhino-membrane.html

[12] membrane NDN software. http://www.ndnsoftware.com/

[13] Seidel M. *Tensile Surface Structures - A Practical Guide to Cable and Membrane Construction*, John Wiley & Sons, 2009.

[14] Formfinder Software GmbH, Wien. http://www.formfinder.at/main/software/.

[15] Laboratorium BLUM Stuttgart, *Report on biaxial tests Précontraint 702*. Web pages of http://arc-ae.net/ARC1010-Pisco/, 2005.

PLASTIC BEHAVIOUR OF THIN-WALLED COLD-FORMED STEEL MEMBERS UNDER ECCENTRIC COMPRESSION

Viorel Ungureanu[a,b], Maria Kotełko[c], Jan Grudziecki[c], Andra Floricel[a] and Dan Dubina[a,b]

[a] *Department of Steel Structures and Structural Mechanics, Politehnica University of Timisoara, Romania*
[b] *Laboratory of Steel Structures, Romanian Academy – Timisoara Branch, Romania*
[c] *Department of Strength of Materials, Łódź University of Technology, Poland*

Abstract: The paper investigates the possibility to use the local plastic mechanisms to characterize the ultimate strength of short thin-walled cold-formed steel members of class 4 subjected to compression and bending. The research is based on previous studies and some new investigations of the authors. The plastic mechanisms for lipped channel members in eccentric compression about minor axis and the evolution of those mechanisms are analysed. Three different types of mechanisms for members in compression with different eccentricities is examined.

1. Introduction

The EN 1993-1-1 [1] code distinguishes the class 4 cross-section, for which the load-capacity corresponds to the maximum compression stresses less than the yield stress. There are members prone to local buckling in the elastic range. Thin-walled cold-formed steel (TWCFS) structures are usually made of thin-walled members of class 4 sections. Since these sections are prematurely prone to local or distortional buckling and they do not have a real post-elastic capacity, a failure of such members is initialized by the local-global interactive buckling of plastic-elastic type, not an elastic-elastic one. Thus, the failure at ultimate stage of those members, either in compression or bending, always occurs by forming a local plastic mechanism. This fact suggests the possibility to use the local plastic mechanism analysis to characterize the ultimate strength of such members. The load-carrying capacity of such members subjected to simple states of loading (pure bending or pure axial compression) is relatively well determined both on the basis of the theory of thin-walled structures and in the code specifications. However, determination of the load-carrying capacity of TWCFS members subjected to combined load, particularly eccentric compression, is still an open question and the code specifications for that case should be improved. The first attempt into the identification of plastic mechanisms in short TWCFS members subjected to eccentric compression has been published in [2].

In this paper buckling loads and buckling modes for the wide range of eccentricities are determined. The attention is focused on plastic mechanisms for extreme cases: small positive eccentricities and large negative eccentricities. Geometry of plastic mechanisms is derived on the basis of buckling modes and FE simulation patterns. Failure curves obtained from YLA (yield line analysis) are compared with FE calculation results.

Subject of the present analysis are TWCFS members subjected to eccentric compression about minor axis, namely lipped channel section columns. Positive and negative eccentricities along the symmetry axis are investigated, as shown in Fig. 1. The dimensions of the cross-section are $h \times b \times c \times t$ = 150×60×20×2 mm and an internal radius r = 3 mm. The columns of length L= 450 mm were investigated, made of structural steel of yield stress f_y = 355 N/mm^2.

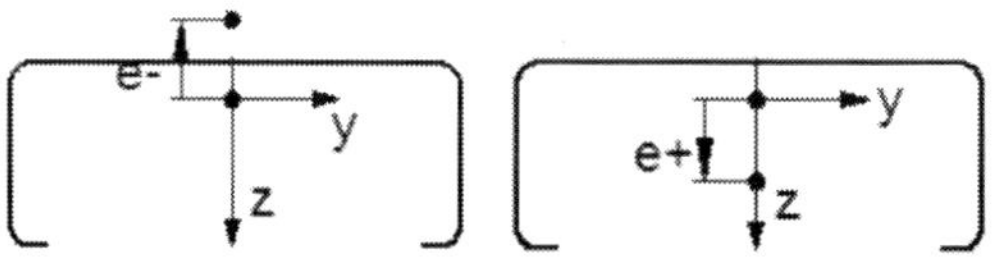

Fig. 1: Lipped channel sections with negative and positive eccentricities

2. Yield line analysis – alternative approaches

The yield line mechanism analysis (YLA) has been widely used to study steel members and connections that involve local collapse mechanisms. This method can be used to study post-elastic behaviour, load-carrying capacity, ductility, rotation capacity and energy absorption. A detailed history of yield line mechanism theory has been presented by Zhao [3]. A state of the art review of the application of yield line analysis to cold-formed members has been presented by Hiriyur and Schafer [4]. Ungureanu et al. [5] make an inventory and classify the geometrical and analytical models for the local plastic mechanisms aiming to characterize the ultimate capacity of some of the most used cold-formed steel sections in structural applications.

The plastic mechanism approach is based on two basic methods [6], named the *energy method (work method)* and the *equilibrium strip method.* A comparative study of these two methods has been carried out by Flockhart et al. [7], who analysed a thin-walled spot welded box-section beam. Further investigation into a comparison of those two approaches has been published by Kotelko and Mania [8].

Using the energy method the Principle of Virtual Velocities of the general following form is applied:

$$P \cdot \dot{\delta} = \int_V \sigma_{ij} \dot{\varepsilon}_{ij}^p \; (\beta, \chi) dV \tag{1}$$

where δ is the global generalized displacement, $\dot{\delta}$ is the rate of change of the global generalized displacement, β is the vector of kinematical parameters of the plastic mechanisms (kinematically admissible displacement), χ is the vector of geometrical parameters of the plastic mechanisms, $\dot{\varepsilon}_{ij}^p$ is the strain rate tensor. As a result, a load - deformation formula is obtained, the graphical representation of which is a *failure curve.*

The equilibrium strip method treats the plastic mechanism as a compatible collection of strips of infinite small or unit width parallel to the direction of applied force. On the basis of free-body-diagram of a separated strip an equilibrium equation is formulated and then, those equations are integrated across the walls of the plastic mechanism, in order to obtain simultaneous equilibrium equation for the mechanism as a whole [2]. An application of this method is restricted to the analysis of local plastic mechanisms build of stationary yield lines only. It is widely used in investigations of plated columns under compression and delivers a direct relation between an applied compressive force and the deflection of the column.

3. Identified plastic mechanisms

Since the buckling mode influences to large extent a failure mode of compressed bar, the buckling behaviour of examined columns was investigated. Buckling loads and modes were determined for the large range of eccentricities, from: e = -60 mm to e = +100 mm. The analysis was carried out using CUFSM code [9] based on Finite Strip Method. For two cases (positive and negative eccentricities) the FSM results were verified using analytical-numerical method based on the asymptotic approach [10]. As shown in Table 1, for positive eccentricities (e = 10, 30, 60, 100 mm) and e = -5 mm a typical distortional buckling takes place. For negative eccentricities e = -10 mm a local-distortional buckling mode was observed. For eccentricities e = -30 mm and e = -60 mm the local buckling takes place.

Table 1: Stress distribution and buckling modes of selected columns under eccentric compression

e [mm]	Stress distribution	Buckling mode	e [mm]	Stress distribution	Buckling mode
10		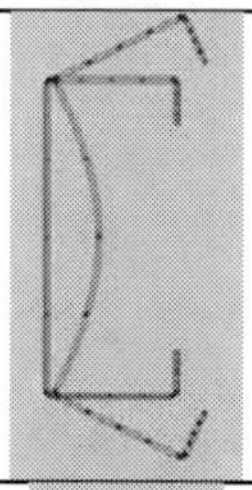	-5	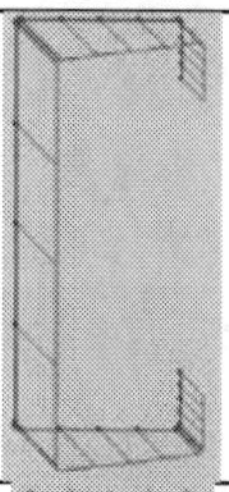	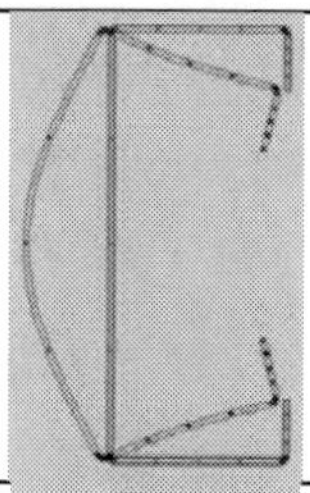
30		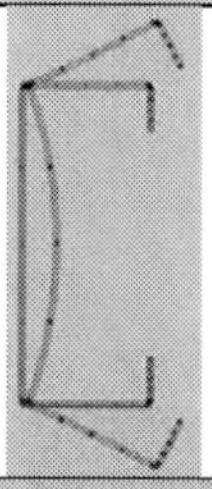	-10	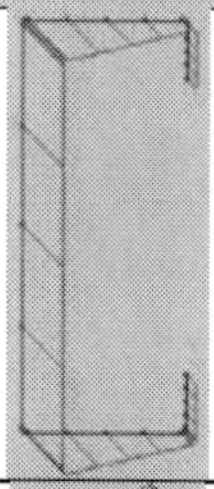	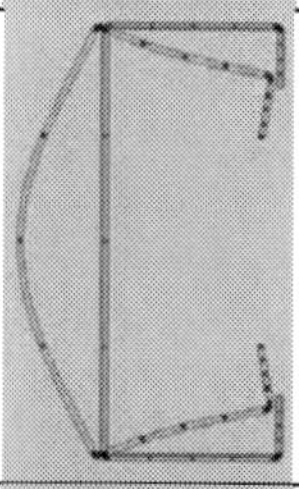
60		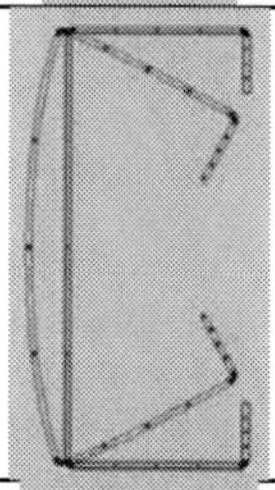	-30	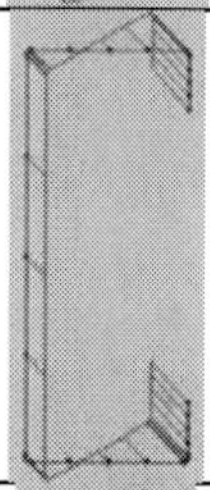	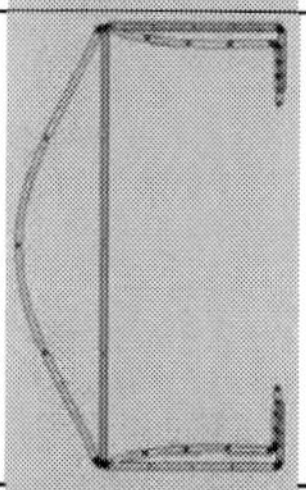
100	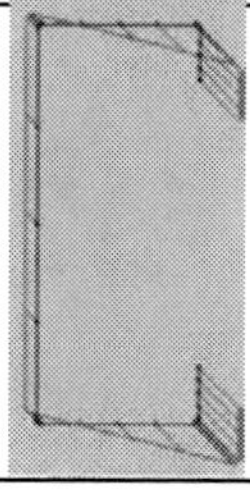	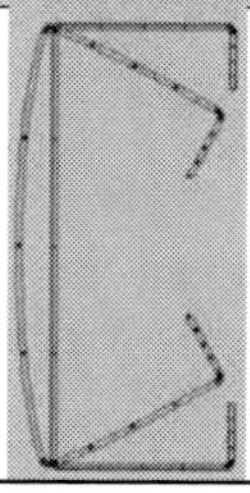	-60	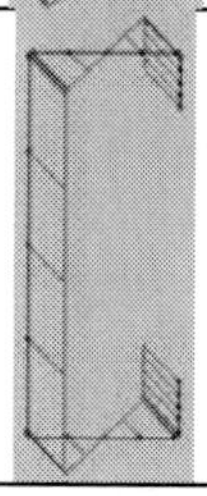	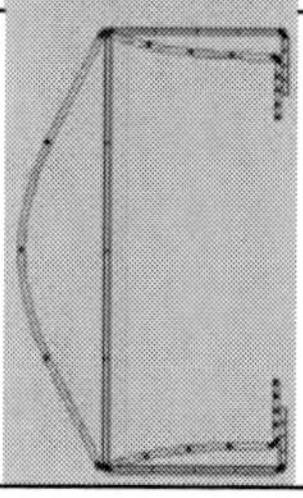

Four simplified types of plastic mechanisms have been identified for lipped channel-section columns subjected to eccentric compression by Ungureanu et al. [2]. In the present paper, as mentioned, the attention is focused on two cases, i.e. the lowest positive and the largest negative eccentricities. For the smallest positive eccentricity, a mechanism similar to CF1 (see Fig. 2) has been derived, originally developed by Murray and Khoo [11]. For this eccentricity a typical distortional buckling mode was determined (see Table 1).

a) FE deformation pattern b) The theoretical plastic mechanism model (CF1)

Fig. 2: Plastic deformation of the column subjected to small positive eccentricity

For large negative eccentricities the mechanism shown in Fig. 3 has been observed. It corresponds to the local buckling mode (see Table 1), where local buckling of the web is a predominant factor. Thus, the mechanism consists of the pitched-roof or roof mechanism in the web and additional yield lines and tension fields in the flanges. However, since the local buckling of the web plays a predominant part, those additional yield lines and tension fields were neglected in the further analysis.

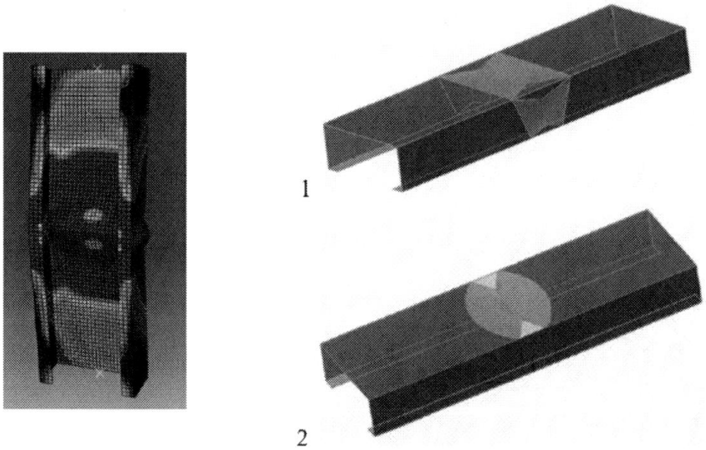

a) FE deformation pattern b) theoretical plastic mechanism model (1-pitched-roof, 2-roof)

Fig. 3: Plastic deformation of the column subjected to large negative eccentricity

4. Yield line analysis

4.1 Positive eccentricities – CF1 mechanism

A detailed geometry of the CF1 mechanism is shown in Fig. 4. It is termed as three-hinge flange mechanism [8], since one can distinguish three different yield lines, namely:

- line 1 of length $l_1 = b$;

- line 2 of length $l_2 = b_1$;
- line 3 of length $l_3 = b_1/\cos\beta$. In this case the total energy of plastic deformation takes form of Eqn. (2).

$$W = m_p \sum_{i=1}^{3} l_i^{'} \beta_i \tag{2}$$

where: $l_2^{'} = 2l_2$, $l_3^{'} = 4l_3$ and β_i is the angle of rotation at i^{th} yield line. After numerical derivation of the energy in Eqn. (2), with respect to the parameter θ, one obtains the current bending moment at the global plastic hinge in terms of θ and – after simple recalculation – the compressive load versus the shortening of the column.

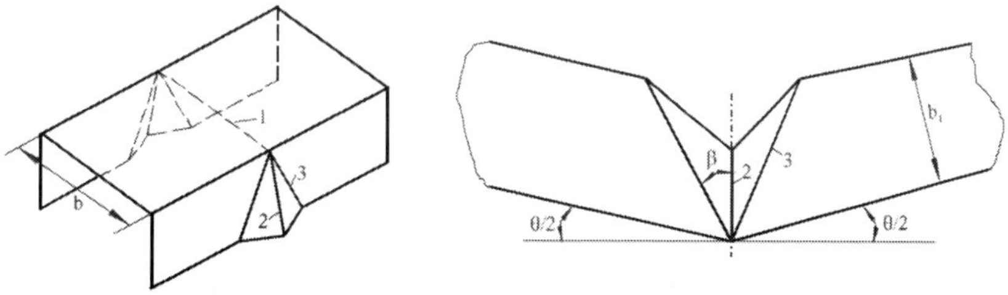

Fig. 4: Mechanism CF1 (three-hinge mechanism)

4.2 Negative eccentricities – roof mechanism and pitched-roof mechanism

As mentioned above, in the case of large negative eccentricities the predominant part plays local buckling of the web and subsequently, the development of plastic mechanism in the web is similar to the mechanism in the plate subjected to uniform compression. In that case, from the principle of virtual velocities we obtain the following variational formulation:

$$\delta W_{ext} = \delta W_b + \delta W_m \tag{3}$$

where δW_{ext} is the variation of work of external forces, δW_b is the variation of the energy of bending plastic deformation, while δW_m is the variation of the energy of membrane plastic deformation.

The variation of the energy of bending plastic deformation dissipated at yield lines is:

$$\delta W_b = \sum_k l_k m_p \delta\beta_k \tag{4}$$

where l_k is a length of the yield-line and β_k is an angle of relative rotation of two walls of the global plastic hinge along that line, m_p is a fully plastic moment in the wall cross-section.

In the pitched roof mechanism the energy of bending deformation takes the form (see Fig. 5a):

$$W_b = W_{AB+CD+\ldots} + W_{AE+\ldots} = \sum_{i=1}^{3} l_i \beta_i \cdot m_p + \sum_{j=1}^{4} l_j \beta_j \cdot m_p \tag{5}$$

where : $l_2 = l_3 = b$, $l_1 = b\text{-}2s$, $\beta_2 = \beta_3 = \theta$, $\beta_1 = 2\theta$, $l_j = a/\sin\gamma$ and $\beta_j = \arctan[\tan\gamma \cdot \cos\gamma]$.

The variation of plastic strain energy dissipated at plastic zones of membrane stresses ACE and BDF is:

$$\delta W_m = (N_x \delta\varepsilon_x^p + N_y \delta\varepsilon_y^p) A_p \tag{6}$$

where: N_x, N_y are membrane forces per unit length, A_p is the area of membrane stresses - plastic zones. The membrane forces N_x, N_y can be determined by using the associated flow rule for Huber-Mises yield criterion [8].

The variation of virtual work of external forces amounts:

$$\delta W_{ext} = 2Nab\sin\theta\cdot\delta\theta = 2Nb\Delta\cdot\delta\theta \tag{7}$$

where N is a force per unit length ($N = P/t$).

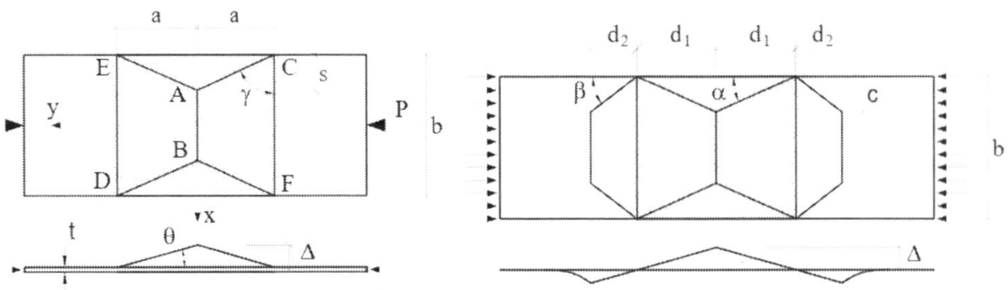

a) pitched-roof b) roof

Fig. 5: Plastic mechanisms models: a) pitched-roof; b) roof

Taking into account Eqns. (4) to (7) in Eqn. (3), a relation of compressive external force N or P in terms of deflection Δ or rotation angle θ (represented graphically as a *failure curve*) is evaluated, i.e.

$$N = 2m_p\frac{b-s}{b\cdot\Delta} + \frac{2a}{b\cdot\Delta\sin\gamma}m_p\cdot\frac{\cos\gamma}{1-\sin^2\theta\cdot\sin^2\gamma} + \frac{1}{2}(a/b)\frac{s}{\Delta}\left(N_x\alpha_x + N_y\alpha_y\right) \tag{8}$$

where: N_x, N_y are given in [8], $N_0 = \sigma_{pl}\cdot t$, $m_p = \dfrac{\sigma_{pl}t^2}{4}$, t is the plate thickness and σ_{pl} is the yield stress of the material.

Mahendran [12] derived a formula, analogous to Eqn. (8) for the roof mechanism shown in Fig. 5b, similar to the mechanism shown in Fig. 3b(2). The formula is based on the equilibrium strip method and takes into account initial imperfections, having the form presented in Eqn. (9).

$$\frac{\sigma}{f_y} = \frac{b-2c}{b}\left[\sqrt{(1+r)^2\left(\frac{\Delta}{t}\right)^2+1} - (1+r)\frac{\Delta}{t}\right] + \frac{c}{b}\left[\sqrt{\Delta_{kt}^2+1} - \Delta_{kt} + \frac{1}{\Delta_{kt}}\ln\left(\Delta_{kt}+\sqrt{\Delta_{kt}^2+1}\right)\right] \tag{9}$$

where $\Delta_{kt} = \dfrac{2(1+r)}{(k_1+k_2)}\dfrac{\Delta}{t}$, $k_1 = \mathrm{cosec}^2\alpha$, $k_2 = \mathrm{cosec}^2\beta$, f_y the yield stress and $r = 0.6$ [12].

5. Numerical results

In order to investigate the applicability of the mechanisms presented in paragraph 4 for members subjected to both small positive and large negative eccentricities comparative numerical calculations have been carried out.

Fig. 6 shows the comparison of load-shortening diagrams for the bar subjected to small positive eccentricities (e = 5, 10 mm) obtained from FE calculations, together with pre- and post-buckling paths derived from analytical-numerical (ANM) algorithm based on the asymptotic method [10] (for the eccentricity e = 10 mm only) and the failure curve, corresponding to CF1 mechanism, derived from energy method in paragraph 4.1.

An analogous comparative diagram for large negative eccentricities (e = -60 mm) is shown in Fig. 7. FE results are compared with the results of ANM calculations for e = -60 mm and two failure curves, corresponding to "roof" mechanism - Eqn. (8) and "pitched-roof" mechanism - Eqn. (9), respectively. In both diagrams an upper-bound load capacity estimation is indicated as intersection point of post-buckling path and failure curve.

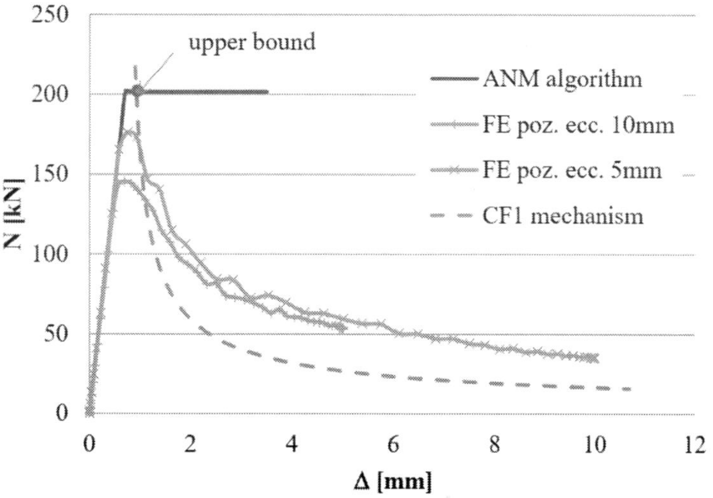

Fig. 6: Load-shortening diagrams for small positive eccentricities

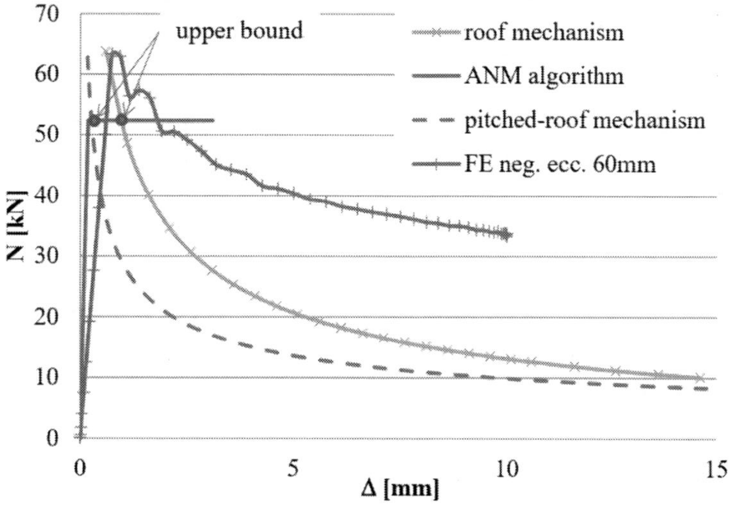

Fig. 7: Load-shortening diagrams for large negative eccentricity, e = -60 mm

6. Conclusions

For small positive eccentricities the mechanism CF1 gives a relatively good prediction of post-ultimate column behaviour. Also an upper-bound estimation of load-carrying capacity obtained from the inter-section of post-buckling path and failure curve is in good agreement with FE prediction for the smallest eccentricity, i.e. e = 5 mm.

For large negative eccentricity the post-ultimate capacity obtained for both mechanisms: “pitched-roof” and “roof” is underestimated. It indicates that additional yield lines and tension fields in the flanges cannot be entirely neglected. Nevertheless, an upper-bound estimation of load-carrying capacity (inter-section of post-buckling path and failure curve) is in relatively good agreement with FE prediction. Thus, obtained results confirm the possibility to use the local plastic mechanism analysis to characterize the ultimate strength of short columns under eccentric compression, although the plastic mechanisms models should be improved.

References

[1] EN1993-1-1 (2005). *Eurocode 3: Design of steel structures - Part 1-1: General rules and rules for buildings*. European Committee for Standardization, Brussels, Belgium (including EN1993-1-1:2005/AC, 2009).

[2] Ungureanu V, Kotełko M, Grudziecki J. "Plastic mechanisms for thin-walled cold-formed steel members in eccentric compression", *Acta Mechanica et Automatica*, 10(1), 33-37, 2016.

[3] Zhao X-L. "Yield line mechanism analysis of steel members and connections", *Prog. Struct. Engng. Mater.*, 5, 252-262, 2003.

[4] Hiriyur BKJ, Schafer BW. "Yield-line analysis of cold-formed steel members", *International Journal of Steel Structures,* 5(1), 43-54, 2004.

[5] Ungureanu V, Kotełko M, Mania RJ, Dubina D. "Plastic mechanisms database for thin-walled cold-formed steel members in compression and bending", *Thin-Walled Structures*, 48(10-11), 818-826, 2010.

[6] Kotełko M. "Load-carrying capacity and mechanisms of failure of thin-walled structures" (in Polish), *WNT*, Warszawa, 2010.

[7] Flockhart CJ, Murray NW, Grzebieta RH. "Comparison of upper-bound rigid-plastic yield line mechanism analysis by the energy and equilibrium strip method", *Proc. of the 2nd Australasia Congress of Applied Mechanics*, Canberra, Australia, 1999, Paper R053.

[8] Kotełko M, Mania JR. "Alternative solution of the problem of load-capacity of thin-walled plated structures", *Mechanics and Mechanical Eng.*, 12(4), 323-336, 2008.

[9] Li Z, Schafer BW. "Buckling analysis of cold-formed steel members with general boundary conditions using CUFSM: conventional and constrained finite strip methods", *Proc. of the 20th Int. Spec. Conf. on Cold-Formed Steel Structures*, St. Louis, MO., USA, November, 2010.

[10] Kolakowski Z. "A semi-analytical method for the analysis of the interactive buckling of thin-walled elastic structures in the second order approximation", *Int. J. Solids Structures*, 33(25), 3779-90, 1996.

[11] Murray NW, Khoo PS. "Some basic plastic mechanisms in thin-walled steel structures", *Int. J. Mech. Sci.*, 23(12), 703-713, 1981.

[12] Mahendran M. "Local plastic mechanisms in thin steel plates under in-plane compression", *Thin-Walled Structures,* 27(3), 245-261, 1997.

OPTIMISATION OF COLD-FORMED STEEL SECTIONS FOR MAXIMUM ENERGY DISSIPATION CAPACITY AND STRENGTH

Jun Ye[a], Iman Hajirasouliha[a], Jurgen Becque[a] and James Lim[b]

[a] *Department of Civil and Structural Engineering, the University of Sheffield, UK*
[b] *Civil & Environmental Engineering Department, the University of Auckland, New Zealand*

Abstract: In this paper, a novel non - linear shape optimization method is presented. In a first step, ten different prototype CFS sections are optimized for maximum flexural strength based on EC 3. All EC 3 requirements, as well as specific manufacturing and construction constraints, are taken into account in the optimisation process. Global optimal solutions are obtained through the Particle Swarm Optimisation (PSO) algorithm. Subsequently, the PSO algorithm is linked to an inelastic post-buckling analysis using ABAQUS and an optimisation is carried out towards maximum energy dissipation. It is shown that, for the same amount of material, both bending capacity and dissipated energy can be significantly increased.

1. Introduction

Structural systems composed of cold-formed steel (CFS) members provide a wide range of advantages. They typically offer a high strength-to-weight ratio, leading to more efficient designs with less redundant material. They result from a straightforward and flexible manufacturing process and are made by bending relatively thin metal sheets into a wide range of cross-sectional shapes using either a cold-rolling or a press braking process. However, the typically large width-to-thickness ratios of cross-sectional plate assemblies (resulting from practical limitations on the sheet thicknesses which can be processed at room temperature) leaves CFS members susceptible to several instabilities (in particular: local, distortional and global buckling modes). The large width-to-thickness ratios of CFS members are also typically outside the limits prescribed by seismic design codes.

Whilst some research has previously been carried out on the seismic behaviour of CFS stud walls [1], it is fair to say that research into the plastic behaviour and the energy dissipation of CFS primary load-bearing elements (i.e. beams and columns) is still limited. Monotonic and cyclic tests [2, 3], however, have been conducted to study the seismic behaviour of CFS channel beams. The results showed a substantial ductility and energy dissipation capacity despite the full plastic moment capacity being reduced by local buckling. Another research project on the development of elements and bolted connections for CFS moment-resisting frames has shown that the ductility and energy dissipation can be significantly improved by gradually adding intermediate folds into the flat plate sections to eventually form curved profiles [4], as show in Fig. 1. While typical CFS cross-sections are unable to satisfy the flat width-to-thickness ratios in practical seismic design, the research indicated that local buckling can be postponed by optimizing the shapes of CFS profiles, while also allowing for the development of large plastic deformations with associated energy dissipation.

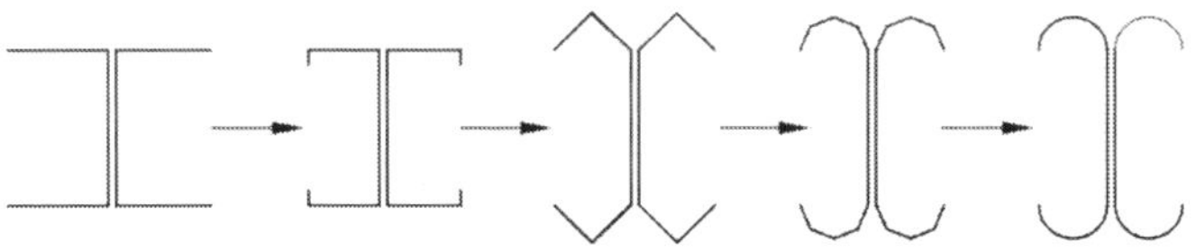

Fig. 1: Development of CFS sections towards curved flanges to increase cross-sectional ductility

The flexibility of CFS members with respect to the manufacturing and construction processes offers great scope and potential to develop cross-sections with high ductility and energy dissipation. However, no research has previously been carried out on the optimization of CFS members based on their post-buckling behaviour, ductility and energy dissipation.

In this paper, a framework is first proposed to optimize cold formed steel sections with respect to their maximum bending capacity. The relative dimensions of the cross-sections, the location and number of intermediate stiffeners and the angle of the inclined lip were considered as the main design variables. The best performing prototypes were subsequently optimized with respect to maximum dissipated energy. For this purpose the commercial FE package ABAQUS[5] was combined with a Particle Swarm Optimization (PSO) algorithm to conduct the optimization and the ensuing results and recommendations are presented.

2. Optimisation for maximum bending capacity

In this paper the CFS sections to be optimized were evaluated according to the stability and strength provisions in EC3 [6], accounting for both local and distortional buckling. In a first step the optimisation procedure was aimed at maximizing the bending capacity while keeping the coil width (or the total developed length of the cross-section) constant:

$$\max\,(M_{c,Rd} = W_{eff}(x)\cdot f_y) \qquad d_{\min} \le x_i \le d_{\max} \;\text{ for }\; i = 1, \dots, n \tag{1}$$

where $M_{c,Rd}$ is the moment resistance of the cross-section about the major axis and $W_{eff}(x)$ is the effective section modulus. For each design variable x_i lower and the upper bounds, $d_{\min}$ and $d_{\max}$, were specified, as summarised in Table 1. All EC3 design constraints were considered in the optimisation process. Manufacturing and practical limitations were also taken into account based on data provided by the UK CFS industry. For example, a minimum lip length of 15 mm (or 5mm for the edge stiffener of a return lip) was maintained in order to make the manufacturing process possible, the dimension *R* was constrained to offer a small flat width between the flange and the intermediate stiffener, and a minimum flange width of 50*mm* was specified to support and connect to floor systems. In addition, the total depth of the section was constrained between a minimum of 100mm (to allow space for two bolts) and a maximum of 350mm to limit the floor depth and the storey-to-storey height. The value of 350mm was, however, turned out not to be critical in the exercise. The optimization was carried out with respect to the cross-sectional capacity (excluding global instability). This situation is representative, for instance, of purlins connected to a steel deck with concrete topping, or even of roof purlins where the lateral and rotational stiffness of the roof diaphragm are sufficient to avoid lateral-torsional buckling. The radius of the rounded corner and the elastic modulus and the Poisson's ratio of the material were taken as 3 mm, 210 GPa and 0.3, respectively. The yield stress of the CFS material was considered to be f_y = 450 MPa. The length of one leg of the intermediate stiffeners was 10 mm and the two legs had an intersecting angle of 60°. The performance of the selected prototypes was compared to that of a standard commercially available channel (with lip=17 mm, flange=75 mm, depth=231 mm and thickness=1.5 mm).

Table 1: Selected prototypes, design variables and constraints (coil width L=415mm)

Prototype	Sections	Design variables	Constraints based on EC3	Manufacturing & practical limitations
①	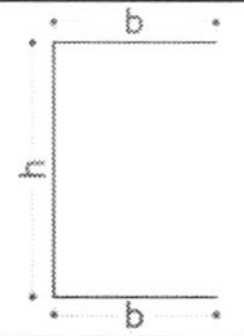	$x=b/L$	$b/t \leq 50$ $h/t \leq 500$	$b \geq 50$ $100 \leq h \leq 350$
②	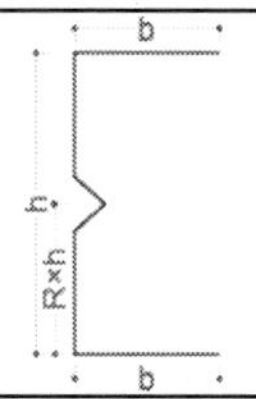	$x_1=b/L$ $x_2=R$	$b/t \leq 50$ $h/t \leq 500$	$b \geq 50$ $100 \leq h \leq 350$ $0.1 \leq R \leq 0.9$
③	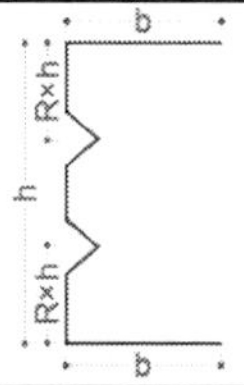	$x_1=b/L$ $x_2=R$	$b/t \leq 50$ $h/t \leq 500$	$b \geq 50$ $100 \leq h \leq 350$ $0.1 \leq R \leq 0.9$
④	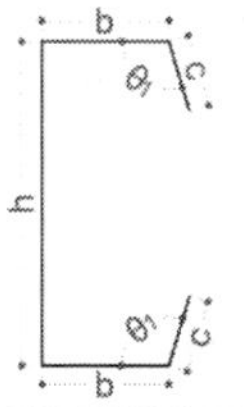	$x_1=c/b$ $x_2=b/L$ $x_3=\theta_1$	$0.2 \leq c/b \leq 0.6$ $b/t \leq 60$ $c/t \leq 50$ $h/t \leq 500$ $\pi/4 \leq \theta_1 \leq 3/4\pi$	$b \geq 50$ $100 \leq h \leq 350$ $c \geq 15$
⑤	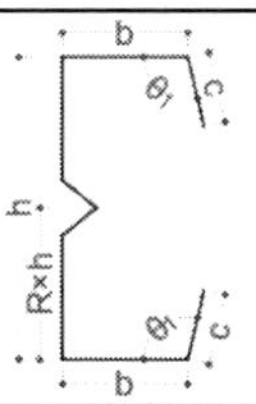	$x_1=c/b$ $x_2=b/L$ $x_3=R$ $x_4=\theta_1$	$0.2 \leq c/b \leq 0.6$, $b/t \leq 60$ $c/t \leq 50$ $h/t \leq 500$ $\pi/4 \leq \theta_1 \leq 3/4\pi$	$b \geq 50$ $100 \leq h \leq 350$ $c \geq 15$ $0.1 \leq R \leq 0.9$
⑥	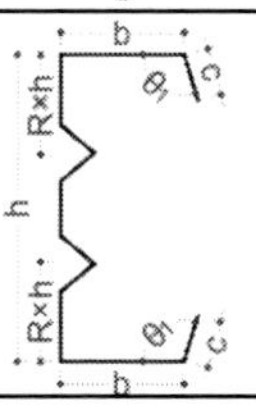	$x_1=c/b$ $x_2=b/L$ $x_3=R$ $x_4=\theta_1$	$0.2 \leq c/b \leq 0.6$ $b/t \leq 60$ $c/t \leq 50$ $h/t \leq 500$ $\pi/4 \leq \theta_1 \leq 3/4\pi$	$b \geq 50$ $100 \leq h \leq 350$ $c \geq 15$ $0.1 \leq R \leq 0.4$

Prototype	Sections	Design variables	Constraints based on EC3	Manufacturing & practical limitations
⑦	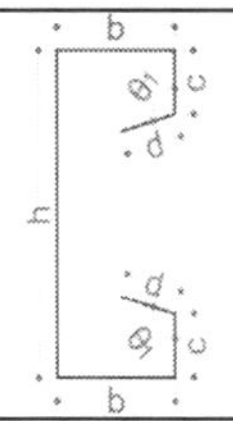	$x_1=c/b$ $x_2=d/b$ $x_3=b/L$ $x_4=\theta_1$	$0.2\leq c/b\leq 0.6$ $0.1\leq d/b\leq 0.3$ $b/t\leq 90$, $c/t\leq 60$ $d/t\leq 50$, $h/t\leq 500$ $\pi/4\leq\theta_1\leq 3/4\pi$	$b\geq 50$ $100\leq h\leq 350$ $c\geq 15$ $d\geq 5$
⑧	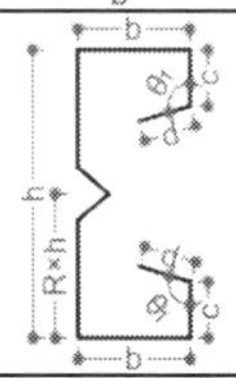	x_1=c/b x_2=d/b x_3=b/L x_4=R $x_5=\theta_1$	$0.2\leq c/b\leq 0.6$ $0.1\leq d/b\leq 0.3$ $b/t\leq 90$, $c/t\leq 60$ $d/t\leq 50$, $h/t\leq 500$ $\pi/4\leq\theta_1\leq 3/4\pi$	$b\geq 50$ $100\leq h\leq 350$ $c\geq 15$ $d\geq 5$ $0.1\leq R\leq 0.9$
⑨	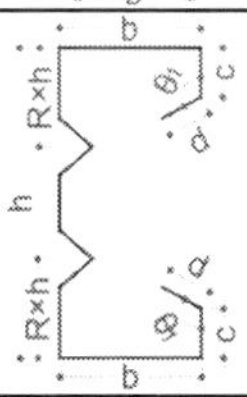	$x_1=c/b$ $x_2=d/b$ $x_3=b/L$ $x_4=R$ $x_5=\theta_1$	$0.2\leq c/b\leq 0.6$ $0.1\leq d/b\leq 0.3$ $b/t\leq 90$, $c/t\leq 60$ $d/t\leq 50$, $h/t\leq 500$ $\pi/4\leq\theta_1\leq 3/4\pi$	$b\geq 50$ $100\leq h\leq 350$ $c\geq 15$ $d\geq 5$ $0.1\leq R\leq 0.4$
⑩	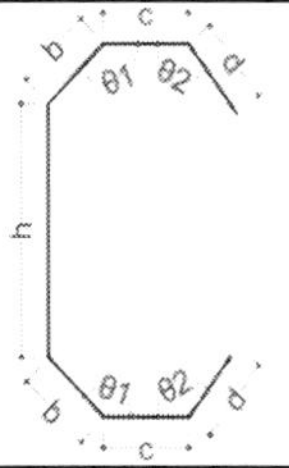	$x_1=\theta_1$ $x_2=\theta_2$ $x_3=b$ $x_4=c$ $x_5=d$	$h/t\leq 500$ $7/12\pi\leq\theta_1\leq 5/6\pi$ $\pi/4\leq\theta_2\leq 3/4\pi$ $30\leq b\leq 48$; $50\leq c\leq 60$; $15\leq d\leq 60$	$100\leq h\leq 350$ $d\geq 5$

The Particle Swarm Optimization (PSO) algorithm was used to solve the problem. This optimization algorithm is population-based and does not need any gradient information, so that it can avoid local optima. To facilitate the optimization process, both the design and the optimisation procedures were formulated in Matlab [7]. The population of the particle swarm was taken as 100 individuals, and 100 iterations were used to obtain the optimum result. Each of the prototypes was optimized 3 times using different initial random particles and the result with the maximum bending capacity was chosen as the optimum section. Table 2 summarises the bending capacities of the optimized cross-sections corresponding to each prototype.

It is clear from Table 2 that, in general, the optimized cross-sections tend to adopt the minimum specified flange width and, hence, have a large depth-to-width ratio. However, prototypes ① to ③, which have the largest depth, still show the lowest bending capacity (one which is even lower than the standard non-optimized section). The optimized lipped channel section ④ shows that a flange edge stiffener can improve the bending capacity by more than 25%. Adding an intermediate stiffener in prototype ⑤ increases the bending capacity of the section by only 2% compared to prototype ④. Even worse, using two intermediate stiffeners in the web in a symmetric arrangement (prototype ⑥) reduces the flexural capacity of the optimum CFS section by 5.2%, compared to prototype ④. This is due to the stiffener in the tension zone being ineffective. The results also indicate that, using the same amount of material, the newly developed folded-flange section (prototype ⑩) provides the maximum flexural strength compared to the other prototypes.

Table 2: Geometrical details and bending capacities of the optimized sections

Section	*h* (mm)	*b* (mm)	*C* (mm)	*d* (mm)	θ_1 (deg)	θ_2 (deg)	*R*	Bending capacity (kN·m)
①	315	50	-	-	-	-	-	9.84
②	305	50	-	-	-	-	0.856	11.08
③	295	50				-	0.186	9.92
Standard	231	75	17	-	90	-	-	10.30
④	270	50	23	-	91	-	-	13.38
⑤	263	50	21	-	92	-	0.79	13.66
⑥	234	50	20	-	90	-	0.223	12.69
⑦	242	50	29	7.5	90	-	-	15.11
⑧	240	50	25	7	135	-	0.9	14.62
⑨	232	50	25	6.5	135	-	0.1	13.41
⑩	185	48	50	17	105	95	-	16.12

3. Optimisation for maximum dissipated energy

In Section 2, the sections were optimized with respect to their maximum bending capacity. The aim of this Section is to achieve optimum shapes with respect to maximum energy dissipation. The optimization process was conducted for prototype ④ and prototype ⑩ (Table 1) since 1. prototype ④ is widely used in practice and 2. prototype ⑩ was previously found to possess the highest bending capacity. Prototype ④ was optimized twice: only while keeping the lip at a straight angle with the flange, and once with the angle as an independent degree of freedom.

3.1 FE models

FE models of CFS cantilever beams were developed using an 8-node quadrilateral shell element with reduced integration (S8R). The stress-strain behaviour of the CFS plate material was simulated using the constitutive model presented by Haidarali and Nethercot [8].

The effect of geometrical imperfections was included in the FE analyses by first conducting an elastic eigenvalue buckling analysis. The mode shape was then scaled and superimposed on the initial model. The amplitude of the imperfection was determined according to the values recommended in [9].

Previous experiments on CFS beams [2, 3] under both monotonic and cyclic loading have shown failure with initial local bottom flange/web buckling, followed by fracture. The damage was typically concentrated in a zone with a length of around one local buckling half-wave [3]. Since the fracture criteria of steel are typically expressed in terms of the equivalent plastic strain, this variable will be used in the optimization process in order to avoid fracture.

3.2 Problem formulation

The optimization procedure aimed to optimize each CFS prototype beam (Fig. 2) with regard to its total energy dissipation throughout the loading history, indicated by *E(X)*. An upper bound $\bar{\varepsilon}_p$ was set for the maximum equivalent plastic strain ε_p in the beam fixed end zone to prevent the fracture in the vicinity of the connection:

$$\max \quad E(X) \tag{2}$$

$$\text{subject to} \quad \varepsilon^p(X) \le \bar{\varepsilon}^p \tag{3}$$

$$d_{min} \le x_i \le d_{max}, \quad (i = 1, \cdots, n) \tag{4}$$

Four levels of plastic strain $\bar{\varepsilon}^p = 0.2, 0.15, 0.1, 0.07$ were used in equation (3) in order to study its effect on the results.

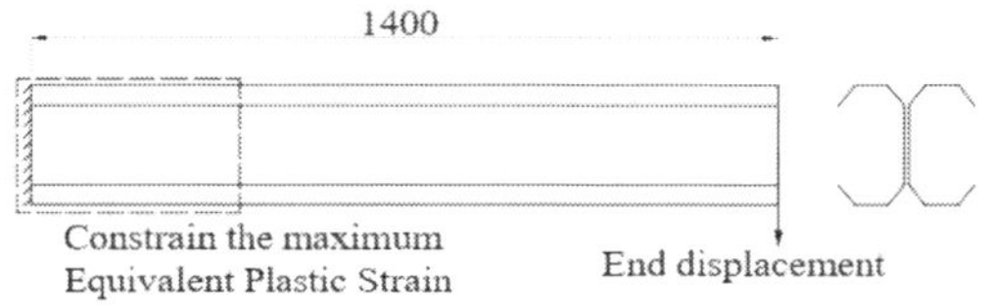

Fig. 2: A cantilever beam model for optimization (length=1400 mm)

3.3 Optimization technique

The Particle Swarm Optimization (PSO) algorithm described in Section 2 was used to solve the problem. The procedure to link ABAQUS to the PSO algorithm is presented in Fig.3. The number of iterations was taken as 100 for all prototypes. To obtain good convergence, the population of the particle swarm was set as 10 for prototype ④ and 15 for prototype ⑩ (in order to accommodate the greater number of design variables).

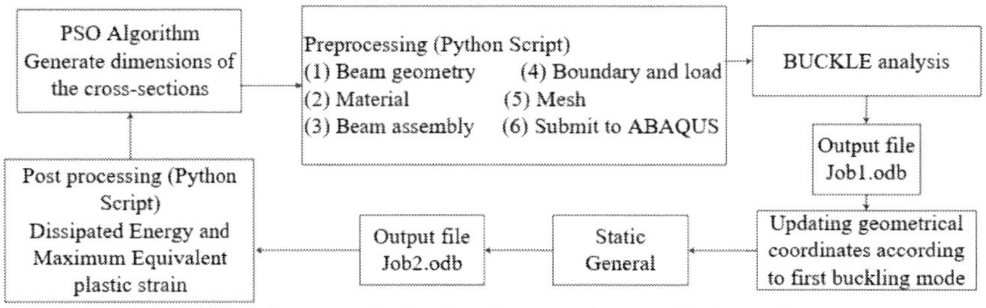

Fig. 3: Flowchart: optimization for maximum dissipated energy

3.4 Optimization results

The prototypes were optimized under monotonic loading. The dimensions of the optimal shapes corresponding to their plastic strain limit $\bar{\varepsilon}^p$ are shown in Table 3. It is seen that the flange width of prototype ④ increases, while the depth of the cross-sections decreases, with a reduction of the equivalent plastic strain limit This affects the energy dissipation of the cross-sections in two ways: (a) It shift the maximum strain demand from the fixed end towards the buckled plates; and (b) It increases the extent of the plastic zone for a given cantilever tip displacement. Logically, the amount of dissipated energy decreases with a decrease of $\bar{\varepsilon}^p$.

Fig. 4 graphically compares the amounts of dissipated energy, subject to various maximum plastic strain limits. It shows that, starting from a lipped channel with the simplest geometry, significant gains in the energy dissipation can be made. The dissipated energy increases by up to 22.9% by adjusting the relative dimensions of the web, the flanges and the lip in prototype ④ with a fixed lip. By relaxing the angle of the lips relative to the flange, a gain of only 0.4% was achieved compared with the optimum channel ④ with a fixed lip. Also, the newly developed folded-flange section (prototype ⑩) dissipated 28%, 27%, 16% and 10% more energy than the optimal lipped channel section (④ with a fixed lip) under plastic strain limits of 0.2, 0.15, 0.1 and 0.07, respectively.

Table 3: Optimized dimensions of the cross-sections

Prototypes	$\bar{\varepsilon}^p$	h (mm)	b (mm)	c (mm)	d (mm)	θ_1	θ_2	E (J)
Standard	0.21	231	75	17	--	--	--	329
④ with $\theta_1 = \pi/2$	0.20	256	50	30	--	--	--	404
	0.15	226	73	22	--	--	--	378
	0.10	187	78	36	--	--	--	323
	0.07	150	83	49	--	--	--	255
④	0.20	228	59	35	--	79	--	405
	0.15	218	62	36	--	130	--	395
	0.10	212	66	36	--	59	--	376
	0.07	149	83	50	--	87	--	255
⑩	0.20	176	48	50	21	135	57	516
	0.15	164	46	51	28	132	104	480
	0.10	155	39	53	39	116	101	473
	0.07	146	33	56	46	102	94	280

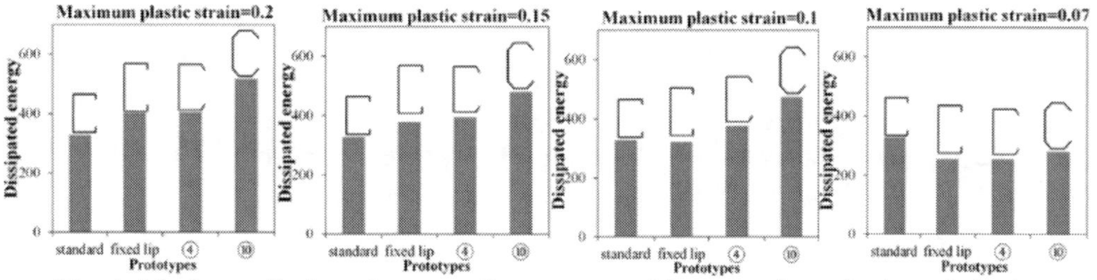

Fig. 4: Maximum dissipated energy of prototypes, subject to various plastic strain limits

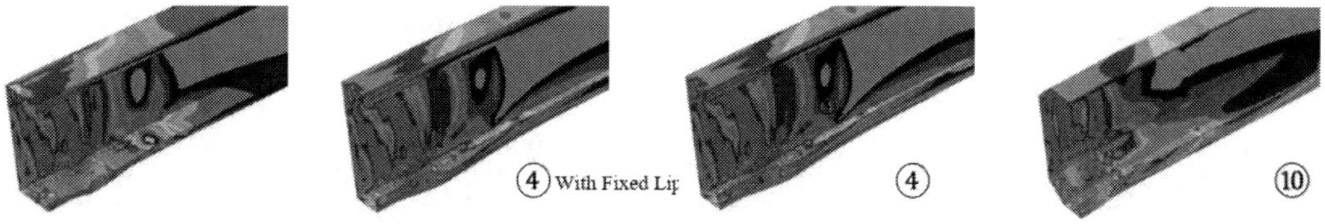

Fig. 5: Distribution of von Mises stress in standard and optimal beams, subject to $\bar{\varepsilon}_p \leq 0.2$ (with stress $\geq 450MPa$ in dark grey area)

The optimal cross-sectional shapes and the distribution of their von Mises stress (for $\bar{\varepsilon}_p \leq 0.2$) is shown in Fig.5, where the dark grey colour represents von Mises stress larger than 450 *MPa* at the fracture stage. By optimising the dimensions of the lipped channel section, it is shown that the extent of the plastic zone is significantly increased in the flange, which leads to an improvement in the dissipated energy. However, when the lip-flange angle was included as a design variable, the plastic region (and the dissipated energy) underwent no apparent change It is also clear from Fig.5 that the plastic region in prototype ⑩ developed very uniformly in the flanges and web, resulting in a higher amount of dissipated energy.

4. Conclusions

A procedure is presented in this paper to obtain optimized channel sections for maximum bending capacity and plastic energy dissipation. A PSO algorithm was linked to the ABAQUS finite element software to carry out an inelastic post-buckling analysis. The following conclusions can be drawn from the results:

(1) The bending capacity of the standard CFS beam was increased by 30% by optimizing the relative dimensions of the flat plates and the inclined angles. The results indicate that flanges with double-fold lips can considerably increase the flexural capacity of CFS beams (by up to 55%), while using intermediate stiffeners in the web does not necessarily increase the capacity.

(2) The folded-flange section was shown to possess the maximum flexural capacity compared to other cross-sectional prototypes, with a 57% and a 20% higher capacity than the standard and the optimum lipped channel sections, respectively.

(3) The dimensions of the optimal shapes with the highest amount of dissipated energy strongly depend on the maximum value of the plastic strain $\bar{\varepsilon}^p$. In general, the flange width of the optimal section increases, while the depth decreases, with any reduction in equivalent plastic strain limit. The amount of dissipated energy also reduces with a decrease of the equivalent plastic strain limit.

(4) The folded-flange section, for the same amount of material, dissipates 28%, 27%, 16% and 10% more energy than the optimal lipped channel section, under plastic strain limitations of 0.2, 0.15, 0.1 and 0.07, respectively.

Acknowledgments

This work was supported by the EPSRC grant EP/L019116/1. The authors would like to thank the EPSRC for their financial support.

References

[1] Nithyadharan M, Kalyanaraman V. "Behaviour of cold-formed steel shear wall panels under monotonic and reversed cyclic loading", *Thin Walled Structures*, 60(2012), 12-23.

[2] Padilla-Llano DA, Moen CD, Eatherton MR. "Cyclic axial response and energy dissipation of cold-formed steel framing members", *Thin Walled Structures*, 78(2014), 95-107.

[3] Calderoni B, De Martino A, Formisano A, Fiorino L. "Cold formed steel beams under monotonic and cyclic loading: Experimental investigation", *J Constr Steel Res*, 65(2009), 219-227.

[4] Sabbagh AB, Petkovski M, Pilakoutas K, Mirghaderi R. "Development of cold-formed steel elements for earthquake resistant moment frame buildings", *Thin Walled Structures*, 53(2012), 99-108.

[5] ABAQUS, in, Hibbitt, Karlsson & Sorensen, Inc, Pawtucket, USA, 2007.

[6] CEN, *Eurocode 3: Design of steel structures, Part 1.3: General rules-Supplementary rules for cold formed members and sheeting*, Brussels: European Comittee for Standardization, 2005.

[7] MATLAB V. 7.10. 0 (R2010a). MathWorks, Natick, MA. 2010.

[8] Haidarali MR, Nethercot DA. "Finite element modelling of cold-formed steel beams under local buckling or combined local/distortional buckling", *Thin Walled Structures*, 49(2011) 1554-1562.

[9] Schafer BW, Pekoz T. "Computational modeling of cold-formed steel: characterizing geometric imperfections and residual stresses", *J Constr Steel Res*, 47(1998), 193-210.

DISTORTIONAL–GLOBAL BUCKLING INTERACTION OF PERFORATED RACK UPRIGHTS: NUMERICAL STUDY AND DESIGN

Xianzhong Zhao[a] and Chong Ren[a,b]

[a] *Department of Structural Engineering, Tongji University, Shanghai, China*
[b] *Department of Civil Engineering, Shanghai University, Shanghai, China*

Abstract: This paper describes a numerical investigation into distortional–global buckling interaction of cold-formed steel perforated rack uprights when subject to axial compression. The numerical modelling strategy is carefully validated against the existing experimental data. The strengths and failure modes of perforated uprights obtained from finite element analysis are compared with design strengths predicted using the direct strength method (DSM). The comparisons indicate that the ultimate strengths predicted by the current DSM are conservative in this study. Hence, modified DSM formulae are presented to consider the interaction of distortional–global buckling for perforated rack uprights.

1. Introduction

In the industry, the rising demand of storage rack requires simple and reliable design approaches and attracts researchers to study its structural behaviour. The role of cold-formed steel rack upright is to take compression load, and the stability is concerned preferentially. The characteristics of storage upright are about complicated cross-sections and continuous perforations along the length, which have significant influences on the structural behaviour.

Experiments are suitable for the determination of design strength of cold-formed steel rack structures. There have been a large number of experimental studies on rack uprights and have published a number of journal articles that reviewed and summarised the structural behaviour of cold-formed steel rack uprights[1]-[3]. However, the practicability and computational efficiency of numerical method offer obvious superiorities on the complex and varied cross sections of rack uprights employed in the industry. Nowadays, numerical method has been increasingly adopted by reason of the practical difficulties of experimental investigations, particularly for extensive parametric studies, see [4][5], for example. Local, distortional and global bucklings are three typical types of buckling which a cold-formed steel member may experience [6]. In the past, the buckling interactions have attracted the attention of many researchers and have been the subject of a number of journal articles. In the period of post-buckling, ultimate strength and failure modes are likely to be strongly affected by the buckling interactions between these three buckling modes, in which the ultimate strength may be eroded significantly and may be unacceptable for safe design when the member experiences buckling interaction phenomena. Some publications reveal this point: the interaction between local buckling and material yield was investigated by Kwon et al. [7]. The interaction between local and lateral-torsional buckling modes was investigated by Becque and Rasmussen [8]. In recently, study on interactions involving distortional buckling has been carried out. For example, Camotim and his colleagues provided a number of journal articles which reviewed and summarised the distortional buckling interactions, including local and distortional interaction [9] and local-distortional-global interaction [10]. In the design propose, the direct strength method (DSM) provides predictions of the strength of columns and beams failing in local, distortional and global buckling modes. Likewise, for interactive mode only local–global buckling interaction is considered in the DSM. Recently, the buckling interaction investigations for improving the DSM approach were extensively conducted, and the focuses of the vast majority outcomes were on the DSM modifications of the local-distortional [11] and local-global [12] buckling interactions. The distortional–global buckling interaction is therefore considered to be sufficiently important to warrant further investigation.

In this paper, a numerical investigation into distortional–global buckling interaction of cold-formed steel perforated rack uprights when subject to axial compression is presented. The focus of this study is to provide practical and effective design method based on the current direct strength method (DSM) which is conservative for predicting the strengths of distortional–global buckling interaction. Furthermore, the finite element modelling strategy is carefully validated against the existing experimental data.

2. Finite element analysis model and validation

The finite element software ABAQUS is employed to investigate the buckling interaction behaviour of members. The members are assigned to specify section type, for example, 100-80-1.8-h-1200 represents the web length (100 mm), the flange height (80 mm), the thickness (1.8 mm) and the overall length (1200 mm) of upright, and h indicates that the upright is perforated. Fig. 1 shows typical perforation locations and dimensions at web and flange, respectively. Geometrical and material nonlinearities need be considered for cold-formed steel member, and therefore the material properties are assumed as Young's modulus E=206 GPa and Poisson's ratio ν = 0.3. The stress-strain curve is assumed to follow that of an elastic-perfectly plastic material and yield stress is 370 MPa. The boundary condition of loading side of a member is restrained at a centroid of cross-section and all the displacements of cross-section at end are tied to the centroid by coupling constraint, and the lateral and vertical components of the displacement at the centroid are assumed to be zero (U1 = U2 = 0). The boundary condition at the other end is assumed to be symmetrical boundary condition (U3 = UR1 = UR2 = 0). The member is subjected to an axial compression, which is acting at the centroid of cross-section of loading side (see Fig. 2 for the boundary conditions and the loading position). Two steps of analyses are performed here. One is the elastic, linear buckling analysis, which is to determine critical buckling load and to provide corresponding buckling mode. The other is the plasticity and geometric non-linear analysis, which is to determine load-displacement response curve and to provide failure mode of the member. In order for the results obtained from the finite element analysis to be accurate enough, the maximum element size used in the finite element mesh is kept to be 5 mm. The four node shell element of reduced integration scheme built in ABAQUS is employed to carry out the static linearity and nonlinearity analyses. The element used is a thin, shear flexible, isometric quadrilateral shell element. In tests of numerical models with different meshes, it can be shown that this kind of mesh is accurate enough as any further reduction (i.e. 2 mm) or enlargement in element size (i.e. 10 mm) has negligible influence on the obtained results. Cold-formed steel are known to be imperfection sensitive, and selection of distribution and magnitude of imperfections is significant. In this study, a set of simplified imperfections are selected. The distributions of imperfection shapes adopted in the model are determined by first elastic eigen-mode in the linear buckling analysis. For the magnitudes of imperfection of finite element analysis, $0.75t$ is selected where t is the thickness of the member.

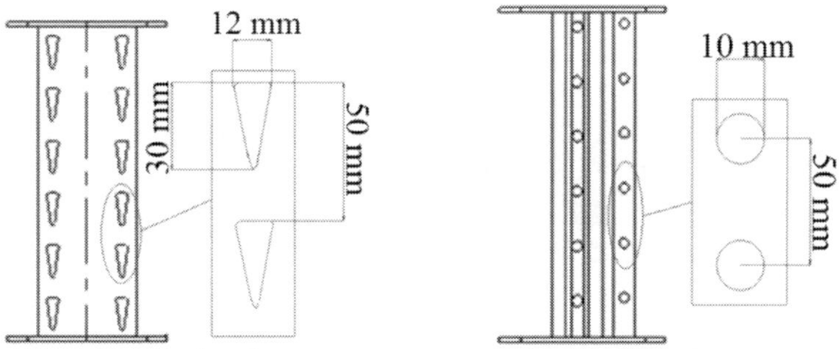

Fig. 1: Web and flange perforation positions and dimensions

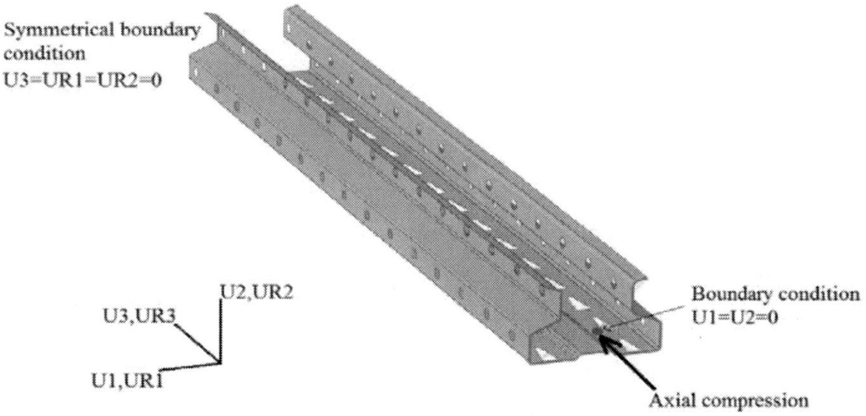

Fig. 2: Boundary conditions and loading position of finite element model

The present finite element models are validated by using available experimental data [3]. The finite element models are then analysed over the full loading range to determine the ultimate strengths and post-ultimate response. Fig. 3 shows the comparisons of the numerical results together with the test results, including load-displacement curve with failure load and failure mode of the member. The good agreements of the present finite element models with experimental data reveal that the present finite element models can provide good predictions.

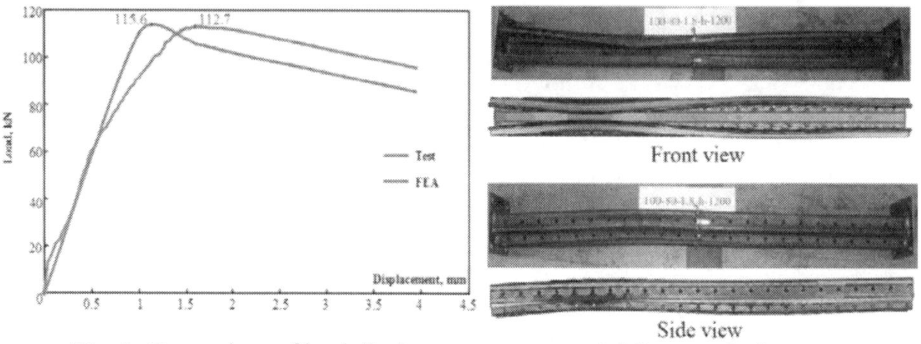

Fig. 3: Comparison of load-displacement curves and failure modes between test and FEA (100-80-1.8-h-1200)

3. Finite element analyses and parametric studies

Following the finite element model validations, an extensive parametric study is carried out to further investigate the buckling behaviour of cold-formed steel perforated rack uprights. A total of 84 finite element models are established and analysed, in which the parameters of finite element model consider varied section dimensions, varied member lengths and the varied thickness, as summarised in Fig. 4 and Table 1.

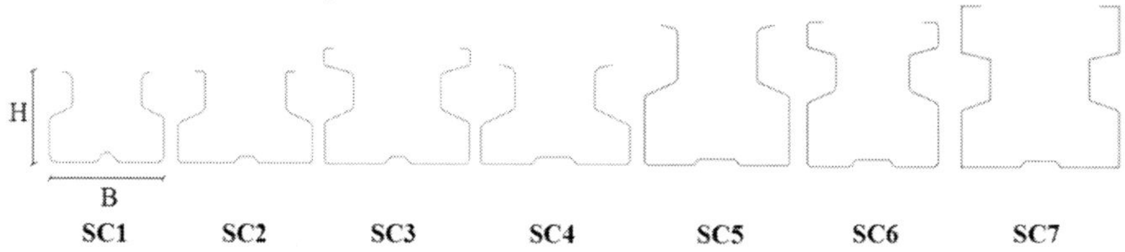

Fig. 4: Cross section shapes

Table 1: Cross section dimensions.

	Cross section	H (mm)	B (mm)	t (mm)	A_{net} (mm^2)	L (mm)
SC-1	75-55-1.8	75	55	1.8	369	L=300, 600, 900, 1500, 2000, 2400, 3000
SC-2	90-56-1.8	90	56	1.8	412	L=300, 600, 900, 1500, 2000, 2400, 3000
SC-3	90-66-1.8 90-66-2.0	90	66	1.8 2.0	495 548	L=300, 600, 900, 1500, 2000, 2400, 3000
SC-4	100-60-1.8 100-60-2.3	100	60	1.8 2.3	477 609	L=300, 600, 900, 1500, 2000, 2400, 3000
SC-5	100-90-2.0 100-90-2.3 100-90-2.8	100	90	2.0 2.3 2.8	654 752 915	L=300, 600, 900, 1500, 2000, 2400, 3000
SC-6	100-110-2.8 100-110-3.0	100	110	2.8 3.0	1080 1158	L=300, 600, 900, 1500, 2000, 2400, 3000
SC-7	120-120-3.0	120	120	3.0	1350	L=300, 600, 900, 1500, 2000, 2400, 3000

For a given upright section dimension, thickness and length, a non-linear finite element analysis is performed using the modified Riks method from which a load-shortening displacement response curve is obtained. Fig. 5 illustrates the load-shortening displacement curve of a 70-55-1.8-h-900 perforated upright and the deformed shape of contour plot of the upright when the load reaches ultimate load point. Additionally, the comparison of deformed and non-deformed cross section at mid-span is also shown in the figure. It can be seen in this figure that the load increases with the displacement until it reaches the ultimate load point. After this point, the load decreases with further increased displacement, indicating that the failure of the member is a typical buckling failure. It also can be found in the figure that the cross section of upright has intermediate twisting deformation at mid-span, and combines with one flange moving inwards and the other moving outwards (O–I mode), indicating that the main failure mode is the distortional and flexural-torsional bucklings interaction mode. In this study, the buckling mode is mainly controlled by distortional buckling where the upright is short, and the distortional–global buckling interaction is occurring mainly where the upright is in intermediate or long length, and the global buckling is the dominant buckling mode where the upright is long. It should be noted that local buckling is occurring rarely in these cross sections of upright, this is due to the strong stiffener at web and flanges and also because the systematic perforations suspend stress distribution to avoid stress concentration.

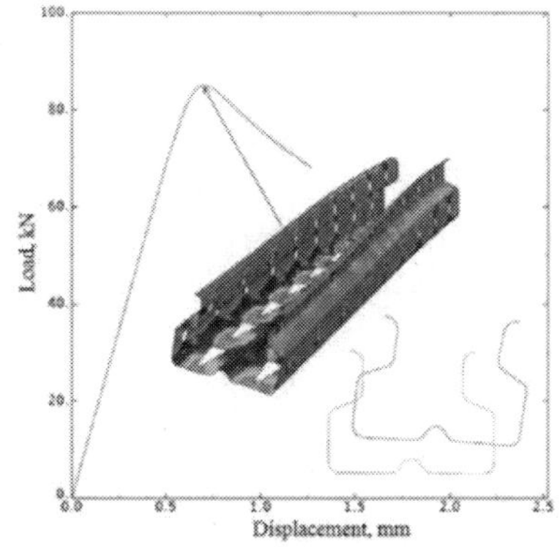

Fig. 5: Load-shortening displacement curve and buckling failure mode of perforated upright (70-55-1.8-h-900)

The distortional–global buckling interaction of perforated upright examined in this investigation is not prescribed in the North American Specification (AISI S100) and the Australian/New Zealand Standard (AS/NZS 4600:2005) where the direct strength method (DSM) for columns is employed. Therefore, the applicability of the distortional–global buckling interaction of perforated upright in the DSM is evaluated. Fig. 6 provides comparisons of the predicted strengths obtained from the current DSM and finite element models for perforated upright of various dimensions and lengths. The slenderness of the upright is represented by $\lambda=\sqrt{P_y/P_{cr,1}}$, where $P_{cr,1}$ is the critical load of buckling which is calculated using the first eigenvalue of buckling from FEA. Three types of buckling failure, distortional buckling, global buckling and distortional-global buckling interaction are contained in the figure. The distortional–global buckling interaction modes are located at the range between distortional buckling and global buckling. In other words, the distortional buckling failure locates at range of the slenderness $\lambda < 1.3$. For slenderness $1 < \lambda < 2$, one can see that the distortional buckling combines with global buckling to dominate the buckling failure mode. When $\lambda > 2$ the buckling mode is only controlled by the flexural-torsional buckling. It can be seen the conservatism of the current DSM to predict ultimate loads for both distortional buckling and global buckling. In the current DSM of column design, the equations aim to predict the nominal axial strength when members subject to distortional buckling and global buckling failures. While the design codes does not contain equations or design curves for predicting strength when members experience the distortional–global buckling interaction. Thus, Yang and Hancock

[13] based on the current DSM and proposed equations for predicting distortional buckling strength considering the distortional–global buckling interaction, and are given as follows,

$$\text{For } \lambda_D \leq 0.561 \qquad P_{nD} = P_{nE} \tag{1}$$

$$\text{For } \lambda_D > 0.561 \qquad P_{nD} = P_{nE}\left[1 - 0.25\left(\frac{P_{crD}}{P_{nE}}\right)^{0.6}\right]\left(\frac{P_{crD}}{P_{nE}}\right)^{0.6} \tag{2}$$

where $\lambda_D = \sqrt{P_{nE}/P_{crD}}$; and P_{crD}= elastic distortional buckling load. The variations of the ratios $P_{nD,FEA}/P_{nD,DSM}$, $P_{nE,FEA}/P_{nE,DSM}$ and $P_{nD+E,FEA}/P_{nD+E,DSM}$ with λ for distortional buckling, global buckling and distortional–global buckling interaction obtained from current DSM and Yang and Hancock' method are shown in Fig. 7 (a)-(c), which also show the associated averages and standard deviations. It can be found in this figure that: (i) The distortional buckling and global buckling of current DSM consistently underestimate the column ultimate strength values. (ii) In global buckling, the extreme high values of the averages and standard deviations and the discrete points prove that the current DSM is over-conservative in predicting the global buckling strength of perforated uprights. (iii) The distortional–global buckling interaction equations proposed by Yang and Hancock provide reasonable ultimate strength estimates, the average and standard deviation equal to 0.90/0.11, which reflects the greater scatter exhibited by the distortional–global buckling interaction ultimate strength predictions.

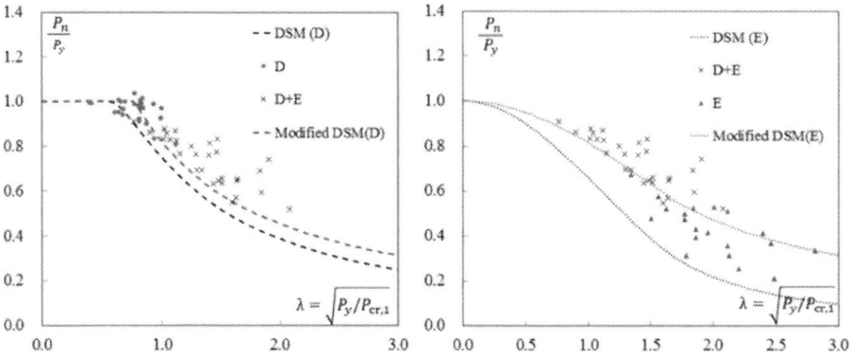

(a) Distortional buckling (b) Global buckling

Fig. 6: The comparisons between DSM and FEA

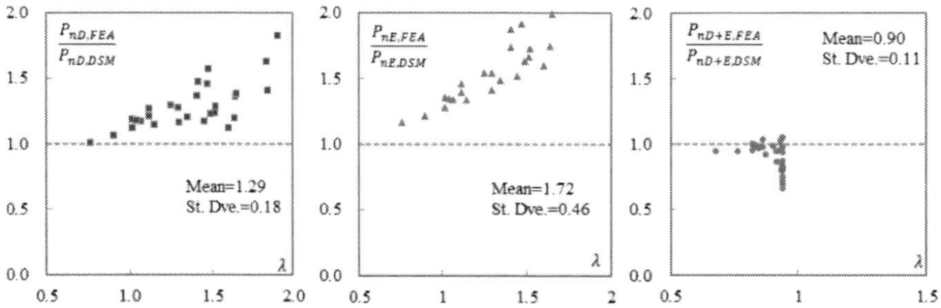

(a) (b) (c)

Fig. 7: Variation with λ of the ratios (a) Distortional buckling, (b) Global buckling, (c) Distortional–global buckling interaction. (Current DSM)

It has been shown in previous figures that the current DSM predictions are over-conservative. Therefore, on the basis of the ultimate loads of the perforated uprights obtained from finite element analysis, the modified distortional and global buckling curves are established and compared with the nominal axial compression strength determined using the current DSM, as shown in Fig. 6. Furthermore, the modified formulae for calculating the nominal axial compression strength subjected to distortional buckling and global buckling are as follows,

1. Distortional buckling $\lambda \leq 0.764$ $\frac{P_{nD}}{P_y} = 1$ (3)

$\lambda > 0.764$ $\frac{P_{nD}}{P_y} = \sqrt{\frac{P_{cr,1}}{P_y}} - 0.18\frac{P_{cr,1}}{P_y}$ (4)

2. Global buckling $\lambda \leq 1.70$ $\frac{P_{nE}}{P_y} = 0.815^{\frac{P_y}{P_{cr,1}}}$ (5)

$\lambda > 1.70$ $\frac{P_{nE}}{P_y} = \frac{0.945}{\left(\sqrt{\frac{P_y}{P_{cr,1}}}\right)}$ (6)

where $\lambda = \sqrt{P_y/P_{cr,1}}$ According to Yang and Hancock method, P_y is replaced by P_{nE} in the distortional buckling expressions. The equations for P_{nD} considering the distortional–global buckling interaction can be expressed as follows,

3. Distortional–global buckling interaction

$\lambda \leq 0.925$ $\frac{P_{nD}}{P_{nE}} = 1$ (7)

$\lambda > 0.925$ $\frac{P_{nD}}{P_{nE}} = \sqrt{\frac{P_{cr,1}}{P_{nE}}} - 0.18\frac{P_{cr,1}}{P_{nE}}$ (8)

where $\lambda = \sqrt{P_{nE}/P_{cr,1}}$. Fig. 8 (a) - (c) schematically illustrates the variations of the ratios $P_{nD,FEA}/P_{nD,DSM}$, $P_{nE,FEA}/P_{nE,DSM}$ and $P_{nD+E,FEA}/P_{nD+E,DSM}$. with λ for distortional buckling, global buckling and distortional–global buckling interaction obtained from modified DSM, respectively. It can be found that the modified DSM for distortional buckling, global buckling and distortional–global buckling interaction bring the averages and standard deviations down considerably, Compare with the current DSM in global buckling, the average and standard deviation of the modified DSM reduce to 1.10 and 0.13, which reveals the lower scatter exhibited by perforated upright ultimate strength predictions. It also can be seen in this figure that the modified DSM formulae provide accurate and reliable ultimate predictions of the distortional–global buckling interaction, the $P_{nD+E,FEA}/P_{nD+E,DSM}$ average and standard deviation are 0.97 and 0.09. It should be noted that the modified formulae concerning the distortional–global buckling interaction is more capable than Yang and Hancock method for complex cross sectional perforated uprights.

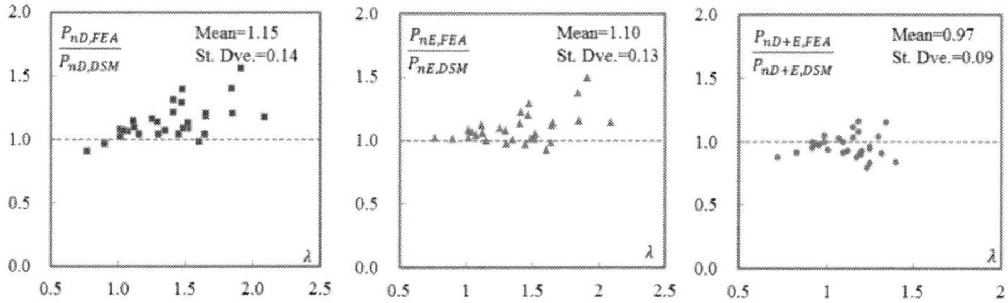

(a) (b) (c)

Fig. 8: Variation with λ of the ratios (a) Distortional buckling, (b) Global buckling, (c) Distortional–global buckling interaction. (Modified DSM)

4. Conclusions

A numerical investigation and parametric studies on the distortional–global buckling interaction of cold-formed steel perforated rack uprights subjected to axial compression have been performed using finite element methods. Some important conclusions drawn from the present study are summarised as follows:

1. Finite element model has been established and the modelling approach has been validated by comparing with test results. The good agreement of the finite element model

with the experimental data has demonstrated that the finite element model can provide a good prediction of steel rack uprights subjected to axial compression.
2. The distortional–global buckling interaction has been discussed, and the conservatism of the current DSM to predict ultimate loads for both distortional buckling and global buckling has been evidenced.
3. The modified DSM formulae of the distortional–global buckling interaction have been proposed. Compare with Yang and Hancock's method, the modified DSM formulae have been proved to be more accurate and reliable in strength predictions.

Acknowledgments

The authors gratefully acknowledge the supports provided by China Postdoctoral Science Foundation (2014M561517) and National Natural Science Foundation of China (51508402).

References

[1] Casafont M, Pastor MM, Roure F, Peköz T. "An experimental investigation of distortional buckling of steel storage rack columns". *Thin-Walled Structures*, 49(8), 933-946. 2011.
[2] Trouncer AN, Rasmussen KJR. "Flexural–torsional buckling of ultra light-gauge steel Storage rack uprights" *Thin-Walled Structures*, 81, 159-174, 2014.
[3] Ren C, Zhao X, Qin R. "Experimental and numerical investigations of the steel storage rack uprights". *Nordic Steel Construction Conference*, Tampere, Finland, 2015.
[4] Casafont M, Pastor M, Bonada J, Roure F, Peköz T. "Linear buckling analysis of perforated steel storage rack columns with the Finite Strip Method". *Thin-Walled Structures*, 61, 71-85, 2012.
[5] Dinis PB, Young B, Camotim D. "Local–distortional interaction in cold-formed steel rack-section columns". *Thin-Walled Structures*, 81, 185-194, 2014.
[6] Hancock GJ. "Local, distortional, and lateral buckling of I-beams", *Journal of Structural Division (ASCE)*, 104(11), 1787-1798, 1978.
[7] Kwon YB, Kim BS, Hancock GJ. "Compression tests of high strength cold-formed steel channels with buckling interaction". *J. of Constr. Steel Research,* 65(2), 278-289, 2009.
[8] Becque J. Rasmussen KJR. "Experimental investigation of the interaction of local and overall buckling of cold-formed stainless steel columns". *Research report R887, School of Civil Engineering, University of Sydney,* 2006.
[9] Martins AD, Dinis PB, Camotim D, Providencia P. "On the influence of local-distortional interaction in the behavior and design of cold-formed steel web-stiffened lipped channel columns". *Twenty-second international specialty conference on cold-formed steel structures*, St. Louis, Missouri, USA, November, 2014.
[10] Santos ES, Dinis PB, Batista EM, Camotim D. "Cold-formed steel lipped channel columns undergoing local-distortional-global interaction: Experimental and numerical investigation". *Twenty-second international specialty conference on cold-formed steel structures*, St. Louis, Missouri, USA, November, 2014.
[11] Dinis PB, Camotim D. "Cold-formed steel columns undergoing local–distortional coupling: Behaviour and direct strength prediction against interactive failure", *Computers & Structures*, 147, 181-208, 2015.
[12] Nandini P, Kalyanaraman V. "Interaction of local and distortional buckling with lateral torsional buckling of cold-formed steel beams", *Fifth International Conference on Coupled Instabilities in Metal Structures*, Sydney, Australia, 2008.
[13] Yang D, Hancock GJ. "Compression Tests of High Strength Steel Channel Columns with Interaction between Local and Distortional Buckling". *Journal of Structural Engineering*, 130(12), 2004.

Members' behavior: tension, compression, beams, beam-columns

ELASTIC STABILITY OF U-SHAPED MEMBERS UNDER BI-AXIAL BENDING

André Beyer[a], Nicolas Boissonnade[b], Abdelouabab Khelil[c] and Alain Bureau[a]

[a] *CTICM, France*
[b] *University of Applied Sciences of Western Switzerland, Switzerland*
[c] *Lorraine University, France*

Abstract: This paper deals with analytical approaches to the elastic stability of U-shaped members under bi-axial bending. The influence of pre-buckling displacements is highlighted and it is shown how they influence elastic instability of the member. Also, it is studied in which cases instability and second order effects should be accounted for when U-shaped members are used.

1. Introduction

Even if the elastic critical loads are of purely theoretical nature, their knowledge is of great practical interest. In fact, the critical load amplifier α_{cr}, i.e. the factor with which the applied loads have to be multiplied to attain the elastic critical load, is an important parameter to judge whether a member (or an entire structure) is affected by second order effects and therefore to determine if member failure occurs by instability (second order effects are important) or by exceeding the (plastic or elastic) resistance of the most loaded section (second order effects are negligible).

The current European standard assessing the stability of steel members, EN 1993-1-1 [1] does neither contain a method to determine the ultimate resistance of U-shaped members under bi-axial bending nor any indication of how their critical loads may be determined. Also, whereas literature relative to critical loads of mono- and double-symmetric I sections in bending is widely spread (see for example [2], [3], [4]), the elastic buckling of U-shaped members is treated much less frequently. Recent studies in this field have been published in [5], [6], [7]. In [8], the case of U sections subjected to bi-axial bending is studied explicitly. However, all of the previously mentioned publications concerning U-shaped members have neglected the influence of pre-buckling displacements. For I-shaped members, it has already been shown in many publications that their influence on the critical loads may be considerable (see for example [9], [10], [11], and [12]). Hence, it seems of interest to study the influence of pre-buckling deflections on the critical load explicitly members of U section. Hereafter, two analytical expressions for the critical load amplification factor are presented, the first one neglecting the influence of pre-buckling displacements and the second one taking them into account. This last expression is validated through Finite Element Analyses and the importance of second order effects arising from minor axis bending is evaluated for U-shaped members.

The theoretical developments presented hereafter are based on the reference system and the displacements defined on Fig. 1.

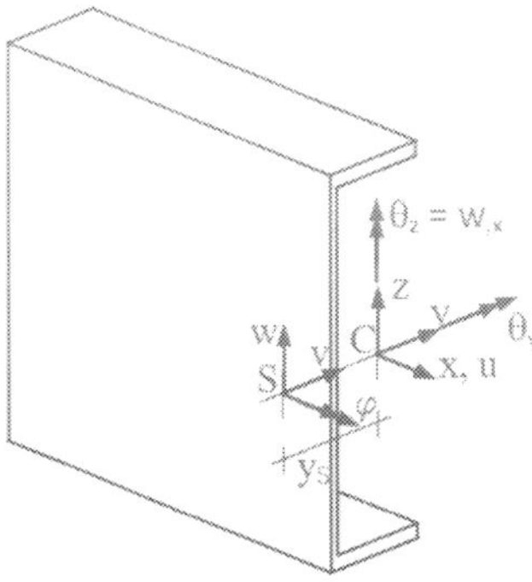

Fig. 1: Definition of the reference system and the associated displacements

2. Analytical approach for the determination of the critical loads

In the following critical loads are determined on the basis of the energy method stipulating that a system is in a state of indifferent equilibrium (i.e. point of elastic instability) if the second variation of its total potential energy Π vanishes as expressed in Eq. (1):

$$\partial^2\Pi = 0 \tag{1}$$

In order to determine the potential energy of the member, an appropriate displacement field has to be adopted. In this study the one given in Eq. (2) is used. A detailed discussion of the quality of different displacement fields to predict the member behaviour is given in reference [11].

$$\begin{aligned} U(x) &= u - y(v_{,x}\cos(\varphi) + w_{,x}\sin(\varphi)) - z(w_{,x}\cos(\varphi) - v_{,x}\sin(\varphi)) - \omega\varphi_{,x} \\ V(x) &= v - (z - z_S)\sin(\varphi) - (y - y_S)(1 - \cos(\varphi)) \\ W(x) &= w + (y - y_S)\sin(\varphi) - (z - z_S)(1 - \cos(\varphi)) \end{aligned} \tag{2}$$

The index ",x" in Eq. (2) means "derivate of the function f with respect to x". The step-by-step derivation of the potential energy is omitted here due to space limitations.

The second variation is directly given in Eq. (3). It should be noted that Eq. (3) supposes that pre-buckling torsional twist is negligible. Also, it is supposed that the studied sections are symmetric about their y-axis (see Fig. 1) and that they are only acted by moments applied at the member ends (the work of external loads is not involved).

The notations used in Eq. (3) are defined in the corresponding section of the present paper. Here, it is only recalled that the displacements associated with the operator "∂" refer to post-buckling displacements whereas v and w refer to pre-buckling deflection. Eq. (3) shows that the influence of pre-buckling displacements is introduced through the terms I_y/I_z and I_z/I_y. If the pre-buckling displacements are neglected, these terms vanish. In this case the terms M_z^2/EI_z and M_y^2/EI_y also vanish: $M_y/EI_y = w_{,xx} \approx 0$ and $M_z/EI_z = v_{,xx} \approx 0$.

$$\begin{aligned}\partial^2\Pi = 0 = &\int_L EI_y \partial w_{,xx}^2 + EI_z \partial v_{,xx}^2 + EI_w \partial \varphi_{,xx}^2 + GI_t\, \partial \varphi_{,x}^2 dx \\ &-2\int_L \left(1-\frac{I_z}{I_y}\right) M_y \partial v_{,xx} \partial \varphi + \frac{1}{2}\left(1-\frac{I_z}{I_y}\right)\frac{M_y^2}{EI_y}\partial\varphi^2 dx \\ &+2\int_L \left(1-\frac{I_y}{I_z}\right) M_z \partial w_{,xx}\partial\varphi - \beta_y M_z \partial \varphi_{,x}^2 - \frac{1}{2}\left(1-\frac{I_y}{I_z}\right)\frac{M_z^2}{EI_z}\partial\varphi^2 dx\end{aligned} \tag{3}$$

3. Elastic member buckling of U sections

3.1 Bi-axial bending

By neglecting the influence of pre-buckling displacements, the authors of reference [8] obtained a formulation for the critical load amplification factor α_{cr} for U-shaped members under bi-axial bending. In case of constant bending moments, their solution reads:

$$\alpha_{cr,1} = \frac{\sqrt{\left(\frac{M_y^2}{N_{cr,z}} + \frac{M_z^2}{N_{cr,y}}\right)\left(GI_T + \frac{\pi^2}{L^2}EI_w\right) + M_z^2\beta_y^2} - M_z\beta_y}{\frac{M_y^2}{N_{cr,z}} + \frac{M_z^2}{N_{cr,y}}} \tag{4}$$

With Eq. (3) for the second variation of the total potential energy of the member, a more complex expression for the critical load amplifier can be obtained. This expression considers explicitly the influence of pre-buckling displacements.

$$\alpha_{cr,2} = \frac{\sqrt{\left[\frac{M_y^2}{N_{cr,z}}\left(1-\frac{I_z}{I_y}\right) + \frac{M_z^2}{N_{cr,y}}\left(1-\frac{I_y}{I_z}\right)\right]\left(GI_T + \frac{\pi^2}{L^2}EI_w\right) + M_z^2\beta_y^2} - M_z\beta_y}{\frac{M_y^2}{N_{cr,z}}\left(1-\frac{I_z}{I_y}\right) + \frac{M_z^2}{N_{cr,y}}\left(1-\frac{I_y}{I_z}\right)} \tag{5}$$

Obviously, Eq. (4) is similar to Eq. (5). The influence of pre-buckling displacements is introduced by the factors "$1-I_z/I_y$" and "$1-I_y/I_z$" associated to the bending moments M_y and M_z, respectively. In order to illustrate the difference between both formulations furthermore, Fig. 2 compares the interaction curve obtained by using Eq. (4) and Eq. (5) for the example of a profile UPE 200 subjected to constant bi-axial bending, for a fixed value of 2.0 for α_{cr}. The member length is 4.0 m. The member is supposed to have fork support conditions at both ends and is assumed to be fabricated of steel S275 (used for the calculation of the plastic bending moments). It should be noted that **positive minor axis bending produces tension in the web of the U section.**

Fig. 2 clearly shows the difference between the two formulations for the critical load amplifier α_{cr}. In particular for high values of negative minor axis bending (or if negative minor axis bending is the leading component of the applied loads), the two formulations diverge. Observing Eq. (4) one may notice that it represents a circle equation (the choice of the graduation in Fig. 2 deforms the shape of the interaction curve). Consequently, the value of the major axis bending moment becomes 0 for two values of the minor axis bending moment: for the chosen example $M_{z,pos}$ = 19,86 kNm and $M_{z,neg}$ = -273,0 kNm. Conversely, when Eq. (5) is used, the value of the major axis bending moment does neither vanish for positive nor for negative minor axis bending. This observation implies that for the studied example, the member is not sensible to elastic instability if it is subjected to minor axis bending only. This conclusion will be justified in the next section, treating the special case of minor axis bending.

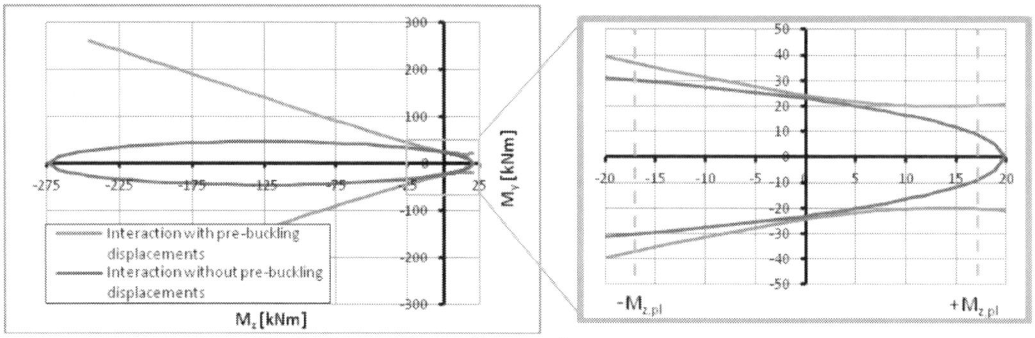

Fig. 2: Comparison of interaction curves – $\alpha_{cr} = 2$

2.2 Particular case of minor axis bending

Based on equations Eq. (4) and Eq. (5), it is possible to derive an expression for the critical minor axis bending moments:

$$M_{cr,z,1} = \pm N_{cr,y}\left(\sqrt{\frac{I_w}{I_y} + \frac{GI_T}{EI_y}\frac{L^2}{\pi^2} + \beta_y^2} \mp \beta_y\right) \tag{6}$$

$$M_{cr,z,2} = \pm\frac{N_{cr,y}}{1-\frac{I_y}{I_z}}\left[\sqrt{\left(1-\frac{I_y}{I_z}\right)\left(\frac{I_w}{I_y} + \frac{GI_T L^2}{EI_y\pi^2}\right) + \beta_y^2} \mp \beta_y\right] \tag{7}$$

Since U sections are not symmetric about their z-axis, the signs "+" and "-" should be kept in Eq. (6) and (7) (the positive and negative critical minor axis bending moments do not have the same value – see Fig 2).

As before, the influence of pre-buckling displacements is introduced by the factor "$1-I_y/I_z$" in Eq. (7). It is interesting to note that the radicand of Eq. (7) may be negative. Consequently, the following condition has to be fulfilled in order to obtain a numerical value of the critical minor axis bending moment:

$$\beta_y^2 > -\left(\frac{I_w}{I_y} + \frac{GI_T}{EI_y}\frac{L^2}{\pi^2}\right)\left(1-\frac{I_y}{I_z}\right) \tag{8}$$

By reorganising Eq. (8), it is possible to determine a maximum length above which elastic instability will not occur for U-shaped members subjected to (constant) minor axis bending:

$$L_{max,ltb.Mz} = \pi\sqrt{\left(-\frac{\beta_y^2}{1-\frac{I_y}{I_z}} - \frac{I_w}{I_y}\right)\frac{EI_y}{GI_T}} \tag{9}$$

If the section UPE 200 is considered again, one obtains a maximum length of 1680 mm. This finding is validated by elastic second order Finite Element calculations including geometrical imperfection about the z- axis (half sine wave; amplitude = length/1000). The calculations are performed with the commercial software ANSYS v.13 using the element Shell181 to model the member. The studied members are subjected to positive minor axis bending (web is in tension). The obtained results are plotted in Fig. 3 and Fig. 4.

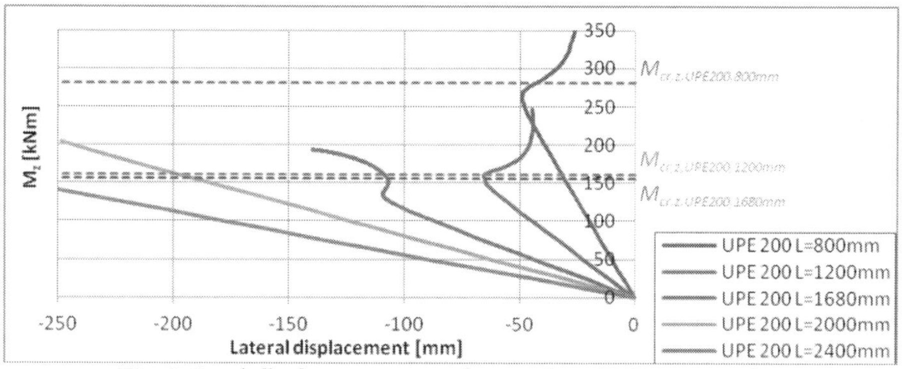

Fig. 3: Load-displacement curve for member of UPE 200 section

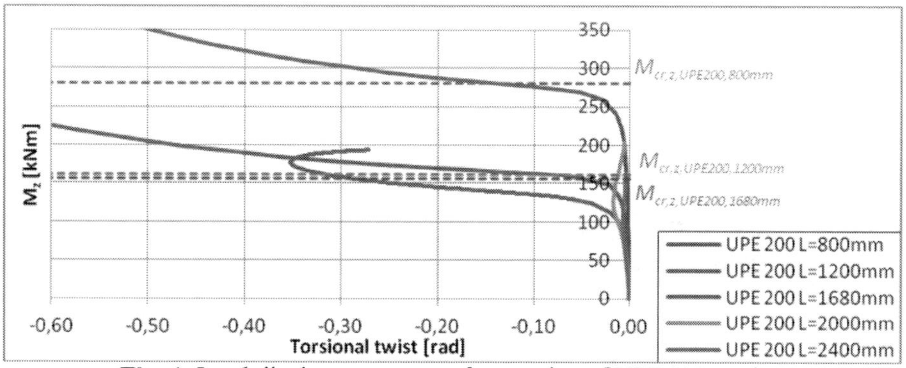

Fig. 4: Load-displacement curve for member of UPE 200 section

The previous figures show the resulting load – lateral displacement and load – torsional twist curves for different member lengths. It should be noted that the lateral displacement and the torsional twist is measured in the reference system. As the studied section is not symmetric about the z-axis, the signs associated to the displacements should therefore be kept (negative lateral displacements indicate a displacement of the gravity centre towards the web).

Obviously, up to a member length of 1680 mm, the member behaviour becomes non-linear at a certain value of the minor axis bending moment. Conversely, for member lengths of 2000 mm and 2400 mm, the behaviour is nearly completely linear (see especially Fig. 3). Hence, second order effects and instability may be considered negligible for these member lengths. Fig. 3 and Fig. 4 also show that the expression given in Eq. (7) for the critical moment $M_{cr,z,2}$ yields satisfactory results and represents well the tendencies observed in the Finite Element calculations (obviously elastic second order calculations do not yield a unique value of the critical moment).

Here before, it has been validated that a member of UPE 200 section subjected to constant minor axis bending is not sensible to second order effects and elastic instability if its length exceeds $L_{max,ltb,Mz}$. This length is given in Table 1 for other commonly used UPE sections.

Table 1: Maximum length $L_{max,ltb,Mz}$ for different U sections

Section	$L_{max,ltb,Mz}$ [mm]	$L_{max,ltb,Mz}/h$
UPE 80	942.9	11.8
UPE 140	1246	8.90
UPE 200	1680	8.40
UPE 270	2146	7.95
UPE 400	2817	7.04

As can be seen, the lengths $L_{max,ltb,Mz}$ are rather short, but they are still at the lower bound of the practical range. To answer the question whether standard U-shaped members under minor axis bending may be sensible to failure by instability, it seems interesting to determine the relative slenderness (see Eq. (10)) resulting from the calculated critical minor axis bending moment. For this example the member of UPE 270 section is supposed to be subjected to positive minor axis bending and made of steel S235.

$$\bar{\lambda}_{LT,Mz} = \sqrt{\frac{M_{z,Rk}}{M_{cr,z}}} \tag{10}$$

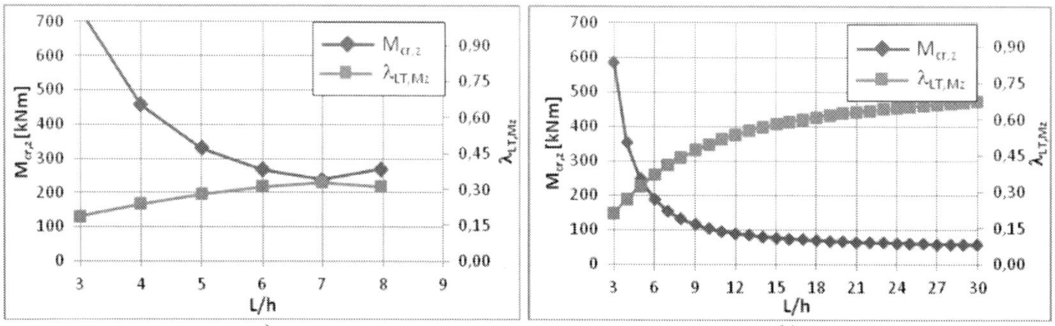

a) b)

Fig. 5: Evolution of the critical moment and the associated relative slenderness if a) pre-buckling displacements are considered and b) if pre-buckling displacements are not considered

One should note that the diagrams represented in Fig. 5 possess a double vertical axis in order to visualize both, the evolution of the critical minor axis bending moment and the associated relative slenderness.

Fig. 5 b) shows that, if the influence of pre-buckling displacement is neglected, the relative slenderness attains values implying that lateral torsional buckling under minor axis bending is to be considered. Conversely, by taking into account the influence of pre-buckling displacements, the relative slenderness is rather small. Also, Fig. 4 a) shows that elastic lateral torsional buckling does not occur for the chosen example if the member length exceeds the maximum value of $L_{max,ltb,Mz} = 2147$ mm ($L/h \approx 7.95$).

Up to now, the present section always considered the case of positive minor axis bending. In order to obtain a critical load for the case of negative minor axis bending, the result of Eq. (7) should be negative. However, for standard U sections, Eq. (7) always yields a positive value (it is recalled that β_y is always positive and "$1-I_y/I_z$" is always negative if the axis definition in section 1 is used).

Hence, it can be concluded that U-shaped members subjected to negative minor axis bending are never sensible to elastic instability. This also explains the form of the interaction behaviour shown in Fig. 2, i.e. with increasing negative minor axis bending the major axis bending moment necessary to cause elastic instability increases, too.

4. Conclusions

The present paper dealt with the elastic instability of U-shaped members subjected to bi-axial bending. Based on the study presented here, the following conclusions may be drawn:

1. For U-shaped members under bi-axial bending, the assumption whether pre-buckling displacements may be neglected or not has a great influence on the form of the inter-

action equation for the critical load amplifier α_{cr} and the expression of the critical bending moments (especially for $M_{cr,z}$);

2. **If pre-buckling displacements are neglected**, the analytical solution implies that the member may be sensible to elastic instability for both positive (web in tension) and negative (web in compression) constant mono-axial minor axis bending;
3. **If pre-buckling displacements are considered**, the analytical solution shows that U-shaped members are never sensitive to elastic instability if they are subjected to negative constant mono-axial minor axis bending. In case of positive minor axis bending, they are only sensible to elastic instability if their length is less than a limit value, noted as $L_{max,ltb,Mz}$ in the present paper;
4. The numerical simulations confirm the analytic expressions for the critical minor axis bending moment considering pre-buckling displacements;
5. Based on the conclusions given in points 1 to 4, it is recommended to take into account the influence of pre-buckling displacements when critical loads and critical load amplifiers are to be calculated for U-shaped members subjected to minor axis bending and a combination of major and minor axis bending. If this is not done, one might draw an erroneous conclusion on the sensitivity to second order effects and instability of U-shaped members in bending.

Notations

I_y	Second moment of area about the major axis of the section
I_z	Second moment of area about the minor axis of the section
I_w	Warping constant of the section
I_t	Torsional constant of the section
y_s	Lateral distance between shear centre S and centroid C
E	Young's modulus of the material
G	Shear modulus of the material
L	Length of the member
$N_{cr,y}$	Critical axial force for major axis buckling
$N_{cr,z}$	Critical axial force for minor axis buckling
$M_{cr,z,1}$	Critical minor axis bending moment without influence of pre-buckling displacements
$M_{cr,z,2}$	Critical minor axis bending moment with influence of pre-buckling displacements
$M_{z,Rk}$	Characteristic value of the minor axis bending resistance
$\alpha_{cr,1}$	Critical load amplification factor without influence of pre-buckling displacements
$\alpha_{cr,2}$	Critical load amplification factor with influence of pre-buckling displacements
β_y	Wagner constant for asymmetry about the minor axis: $\beta_y = \frac{1}{2I_z}\int_A y\left(y^2 + z^2\right)dA - y_s$
$\lambda_{LT,Mz}$	Relative slenderness for lateral torsional buckling under minor axis bending

References

[1] *Eurocode 3: Design of Steel Structures — Part 1-1: General Rules and Rules for Buildings, EN 1993-1-1*, European Committee for Standardization, Brussels, 2005.

[2] Klöppel K, Möll R, Wagner G. "Ein Beitrag zur Bestimmung der Kippstabilität von gedrückten Durchlaufträgern", *Der Stahlbau*, 31, 1962.

[3] Timoshenko SP, Gerer JM. *Theory of Elastic Stability*, International Student Edition, McGraw-Hill, 1961.

[4] Galéa Y. "Moment critique de déversement d'une poutre soumise à des moments d'extrémité et une charge uniforme: proposition d'une formulation analytique de C1 et C2 ", *Revue Construction Métallique* (4), 61-72, 2006.

[5] Seah LK, Khong PW. "Lateral-Torsional Buckling of Channel Beams", *Journal of Constructional Steel Research,* 17(4), 265-282, 1990.

[6] Sudhir Sastry YB, Krishna Y, Budarapu PR. "Parametric studies on buckling of thin walled channel beams", *Computational Materials Science,* 96(B), 416-424, 2015.

[7] Li LY, Chu XT, Kettle R. "Lateral-torsional buckling analysis of partially-laterally restrained thin walled channel-section beams", *Journal of Constructional Steel Research,* 60(8), 1159-1175, 2004.

[8] Cheng SS, Kim, B; Li LY. "Lateral-torsional buckling of cold-formed channel sections subject to combined compression and bending", *Journal of Constructional Steel Research*, 80, 174-180, 2013.

[9] Roberts TM, Azizian ZG. "Influence of pre-buckling displacements on the elastic critical loads of thin walled bars of open cross section", *International Journal of Mechanical Sciences*, 25(2), 93-104, 1983.

[10] Roberts TM, Benchiha M. "Lateral Instability of Monosymmetric Beams with Initial Curvature", *Thin-Walled Structures*, 5(2), 111-123, 1987.

[11] De Ville de Goyet V. "L'analyse statique non linéaire par la méthode des éléments finis des structures spatiales formées de poutres à section non symétrique", Ph.D thesis, Liège University, 1989.

[12] Erkmen RE, Attard MM. "Lateral-torsional buckling analysis of thin-walled beams including shear and pre-buckling deformation effects", *International Journal of Mechanical Sciences*, 53(10), 918-925, 2011.

NUMERICAL INVESTIGATION INTO STRONG AXIS BENDING-SHEAR INTERACTION IN ROLLED I-SHAPED STEEL SECTIONS

R.W.A. Dekker[a], H.H. Snijder[a] and J. Maljaars[a,b]

[a] *Eindhoven University of Technology, The Netherlands*
[b] *TNO, Delft, The Netherlands*

Abstract: Clause 6.2.8 of EN 1993-1-1 covers the design rules on bending-shear resistance, taking presence of shear into account by a reduced yield stress for the shear area. Numerical research on bending-shear interaction by means of the Abaqus Finite Element modelling software is presented. The numerical model is validated against the experimental results. A material model based on various tensile test coupons was used incorporating the actual material properties within the tested cross-section. Strong axis three-point bending tests were simulated by means of continuum solid elements. Good agreement was achieved between numerical and experimental result, both are compared with the EN 1993-1-1 design rule.

1. Introduction

Clause 6.2 of EN 1993-1-1 [1] – known as Eurocode 3 – covers the cross-sectional resistance of steel sections. The bending-shear interaction design rules assign the shear stresses to the shear area, which consists of the web and a part of the flanges. The presence of shear is taken into account by a reduced yield stress for the shear area.

This paper presents the numerical model used to investigate the bending-shear interaction in three-point-bending. This model is validated by tests. The experimental test series executed at Eindhoven University of Technology consisted of rolled HEA280, HEB240, HEM180 and IPE360 beams in steel grades S235J0+M, S355J2+M and S460J0+M bent about their strong axis and HEA280 beams bent about their weak axis. In the test series different beam lengths were used in order to invoke different shear utilization ratios. The simulation of an IPE360 beam in steel grade S355 and HEA280 beam in steel grade S460 bent about their strong axis is presented.

Bending-shear interaction is taken into account by reducing the plastic moment resistance, based on the reduction of the yield stress for the shear area. The presentation of the numerical model is preceded by code requirements and by the results from the experimental test programme.

2. Design rules according to various standards

The design rules on moment-shear interaction result in a check in which the design bending moment M is compared with the design value of the reduced resistance to bending allowing for the presence of a shear force (M_V). For the sake of comparison with the different design rules all partial factors are ignored.

The governing design rules on moment-shear interaction are listed in clause 6.2.8 of EN 1993-1-1 [1]. In case the acting shear force V is less than 50% of the plastic shear resistance V_{pl}, the influence of shear on the bending moment resistance can be ignored. The general method is to reduce the yield stress of the shear area of the section (A_v) by multiplication with the reduction factor (1-ρ), as in Eq. (1).

$$f_{y,r} = (1-\rho)f_y \tag{1}$$

$f_{y,r}$ Reduced yield stress of the shear area

ρ Reduction factor to determine the reduced design values of the resistance to bending moments allowing for the presence of shear forces, see Eq. (2)

f_y Yield stress

$$\rho = \left(\frac{2V}{V_{pl}} - 1\right)^2 \tag{2}$$

V Shear force due to loading

V_{pl} Plastic shear resistance

$$V_{pl} = \frac{A_v f_y}{\sqrt{3}} \tag{3}$$

A_v Shear area, see Fig. 1b for the case of rolled I-shaped sections.

In the case of rolled I-shaped sections Eurocode 3 prescribes an alternative design rule - describing the reduced plastic bending moment allowing for shear force M_v - that can be used, according to Eq. (4), for the case $V > ½V_{pl}$.

$$M_V = \left[W_{pl} - \frac{\rho A_w^2}{4t_w}\right] f_y \tag{4}$$

M_V Reduced bending moment allowing for shear force

W_{pl} Plastic section modulus

A_w Web area, see Fig. 1a

t_w Web thickness

In addition to differences in the design rules, variations in the definition of the shear area are present in different standards as displayed in Fig. 1. This results in a relatively small plastic shear resistance according to DIN 188000 [2] and large shear resistance according to NEN 6770 [3], see Fig. 2.

a) A_w EN 1993-1-1 b) A_v EN 1993-1-1 c) A_v DIN 18800 d) A_v NEN 6770

Fig. 1: Definition of web and shear area in rolled I-shaped sections

The different dimensions of an IPE360 and HEA280 beam result in differences in the moment resistances, graphically presented in Fig. 2. The differences in influence of the shear force on the plastic moment resistance between these two beams are substantial. Fig. 2 a) focuses on the influence of the size of the shear area compared to the total section area, where Fig. 2 b) focuses on the differences in Eurocode 3, DIN 188000 and NEN 6770 design rules.

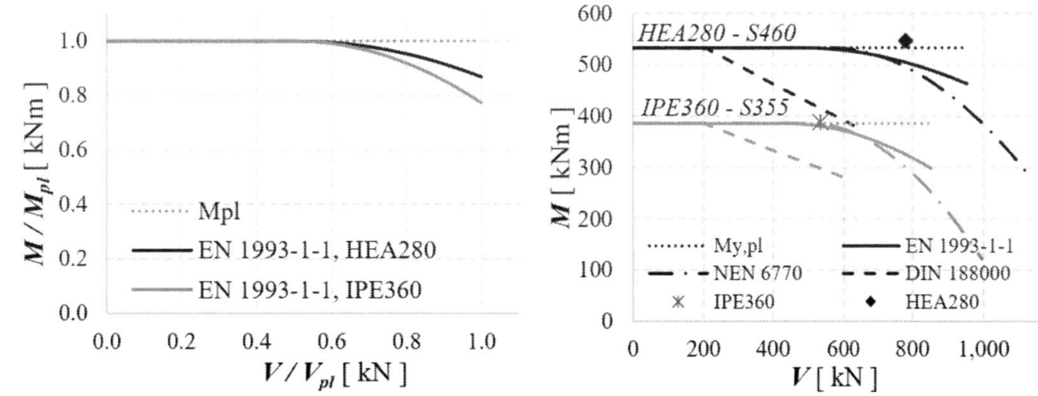

a) Non-dimensional b) Dimensional

Fig. 2: Bending-shear interaction graphs, comparing EN, DIN and NEN

2.1 Yield criterion

The reduced yield strength following clause 6.2.8 of EN 1993-1-1 [1] is in contradiction with the Von Mises yield criterion for a two dimensional stress state prescribed in clause 6.2.1. The Von Mises criterion results in Eq. (5) for combined normal and shear stress after rewriting as reduced yield stress.

$$f_{y,r} = \sqrt{1 - \left(\frac{V}{V_{pl}}\right)^2} f_y \tag{5}$$

Comparison between Eq. (1), (2) and (5) shows that the Von Mises yield criterion always takes shear forces into account, while the EN 1993-1-1 reduced yield stress disregards the presence of shear for $V < 0.5V_{pl}$. For $V > 0.83V_{pl}$ the reduced yield stress following Von Mises is larger than that according to Eq. (1) and Eq (2). For $V < 0.83V_{pl}$ it is the other way around.

3. Experimental test results

3.1 Test program

Tests were carried out in order to determine the cross-sectional resistance under a combination of bending and shear. The test program consisted of a total of 40 three point bending tests in both the strong and weak direction in the test set-up of Fig. 3 a). In total 27 different combinations of sections and shear utilization ratios $n = V/V_{pl}$ were tested, see Table 1. The test results were used for validation of a numerical model. This paper focuses on the test results and numerical model of two beams with a utilization ratio $n > 0.5$, ensuring a substantial influence of shear force present in the section. The HEA280 and IPE360 section selected for this purpose are indicated in bold in Table 1 and had a shear force utilization ratio of respectively $n = 0.83$ and $n = 0.67$.

The specimens were loaded by a 2 MN hydraulic pressure jack about their strong axis. The beams were provided with reusable stiffeners at the supports and a welded stiffener at mid-span. Additional plates were bolted to the mid-span stiffener to provide these in extra stiffness, see Fig 3.b). Strains were measured with 3 rosette strain gauges on the web and in total 6 linear strain gauges. The displacements were monitored on both sides of the web at mid-span and supports by means of Linear Variable Displacement Transducers (LVDTs). The measurement system is described in detail in [4].

Table 1: Test programme, 3-point strong axis bending in strong and weak axis

Section	Steel grade	*Axis*	*# (n)*	$n=V/V_{pl}$	L (mm)
HEA 280	S235J0+M	strong	7 (5)	0.33 - 0.50 - 0.67 - 0.83 - 1.00	3,630 to 1,083
		weak	6 (4)	0.25 - 0.38 - 0.50 - 0.67	983 to 560
	S355J2+M	strong	7 (5)	0.33 - 0.50 - 0.67 - 0.83 - 1.00	3,630 to 1,083
	S460M	**strong**	3 (2)	0.67 - **0.83**	1,790 to 1,400
IPE360	**S355J2+M**	**strong**	7 (4)	0.33 - 0.50 - **0.67** - 0.83	2,996 to 1,095
HEB240	S355J2+M	strong	5 (3)	0.50 - 0.67 - 0.83	2,195 to 1,264
HEM180	S355J2+M	strong	5 (4)	0. 38 - 0.50 - 0.67 - 0.83	2,740 to 1,020

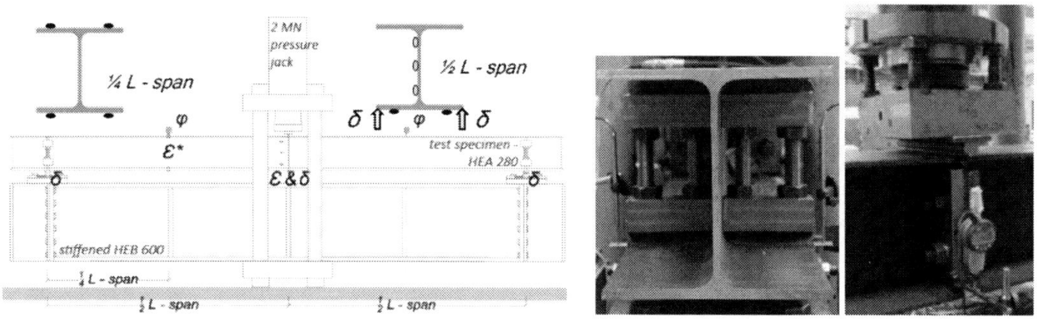

a) Location strain gauges: linear (flange) and rosette (web)

b) Reusable stiffeners at support (left), reinforced stiffener at mid-span (right)

Fig. 3: Strong axis bending test set-up

3.2 Tensile tests

Material properties were measured using tensile coupons in agreement with EN 10002-1 [5], making use of an initial strain rate of 0.00007 s^{-1} which was increased in the plastic branch towards 0.00025 s^{-1}. The tensile tests were paused a number of times in the yield plateau and during hardening in order to obtain the velocity independent yield strength and stress-strain curve, respectively, based on the procedure described by the Technical Memorandum [6]. The Young's modulus E, yield strength f_y and ultimate strength f_u resulting from the tests are provided in Tables 2 and 3, where the final row presents the weighted average values of the entire cross-section. The true stress-plastic strain curves presented in Fig. 4 and Fig. 5 provide the influence of the position of the tensile coupons on the material properties. In both sections the web has a higher yield strength than the flanges.

3.3 Specimen dimension measurements

In order to be able to reproduce the test results in a numerical simulation, the specimen dimensions were measured at mid-span (B) and the beam ends (A, C) before testing, see Fig. 6.

Table 4 presents the measured height h, width b, flange thickness t_f, web thickness t_w and root radius r of the beams discussed in this paper.

Table 2: Material properties IPE360 in S355 – all dimensions in N/mm²

	E	f_y	f_u
I_F0	207,225	366.0	471.4
I_F1	212,090	363.8	581.9
I_F2	203,230	376.7	627.2
I_F3	211,330	366.2	480.4
I_F4	199,690*	357.3*	488.3*
I_W0	205,900	409.6	623.3
I_W1	207,200	425.5	458.1
I_W2	205,360	430.2	615.6
I_W3	206,970	427.2	746.4
I_W4	207,440	424.0	694.9
I_W5	203,190	403.2	615.2
w.av.	**207,346**	**375.5**	**478.6**

** Based on compression coupon and tensile test results*

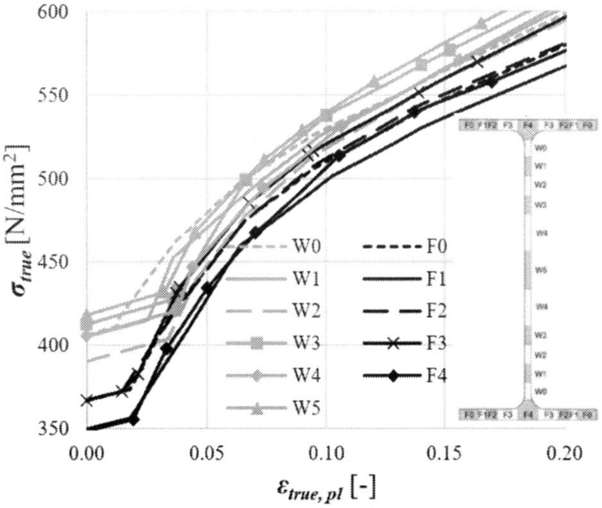

Fig. 4: Stress-strain curves coupons IPE360 in S355

Table 3: Material properties HEA280 in S460 – all dimensions in N/mm²

	E	f_y	f_u
C_F0	217,360	487.9	595.7
C_F1	211,470	483.2	600.0
C_F2	218,698	468.6	605.1
C_F3	212,355	466.1	606.5
C_F4	205,710	456.6	604.5
C_F5	205,090	433.2	613.6
C_W0	206,670	630.4*	678.8
C_W1	225,385	490.9	617.8
C_W2	208,760	505.7	620.2
w.av.	**212,533**	**486.3**	**613.1**

** In absence of a yield plateau the 0.2% proof stress was given*

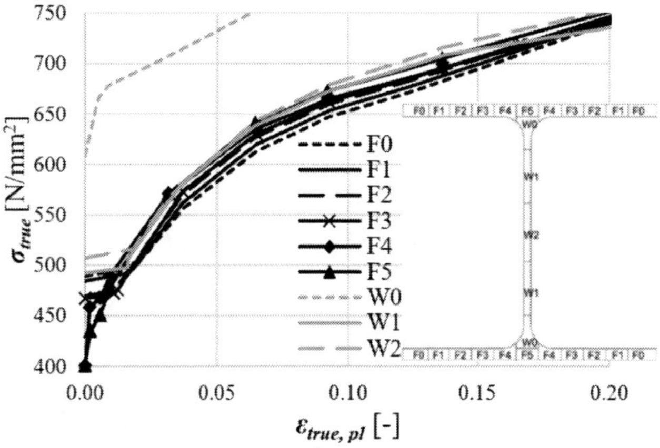

Fig. 5: Stress-strain curves coupons HEA280 in S460

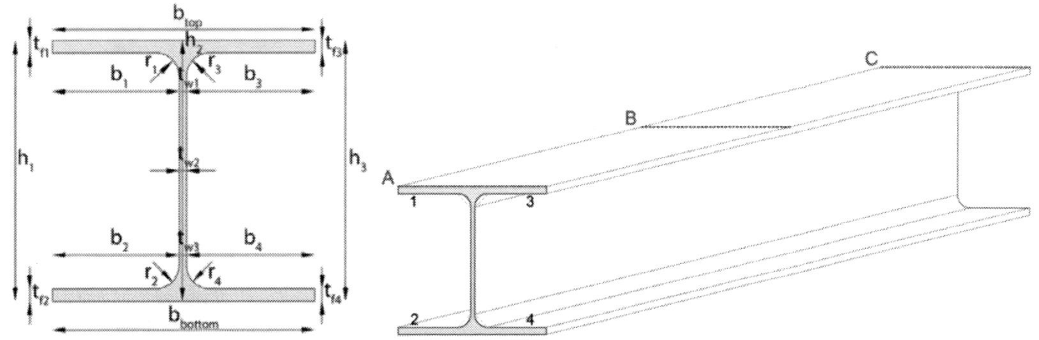

Fig. 6: Specimen measurements: variables at each section (left) and measurement locations (right)

Table 4: Beam measurements of specimen I3 and C2b – dimensions in mm

Beam	L_{beam}		h_1	h_2	h_3	b_{top}	b_{bottom}	t_{f1}	t_{f2}	t_{f3}	t_{f4}	t_{w1}	t_{w2}	t_{w3}	r
		A	360.2	360.2	361.0	171.2	170.6	12.1	12.2	12.2	12.0	7.2	7.3	7.5	18
IPE360	2,178	B	360.8	-	361.0	171.1	170.5	12.1	12.3	12.2	12.1	-	-	-	18
		C	359.8	360.2	361.2	171.0	170.6	12.1	12.3	12.2	12.1	7.2	7.4	7.5	18
		A	275.0	273.6	273.5	281.0	281.8	12.5	12.4	12.3	12.2	8.2	8.2	8.1	23
HEA280	1,959	B	275.4	-	273.8	281.1	281.8	12.4	12.4	12.3	12.2	-	-	-	23
		C	275.7	273.8	273.5	281.1	281.9	12.4	12.4	12.4	12.2	8.2	8.1	8.1	23

3.4 Test results

Beam HEA280, S460 was loaded with an initial cross-head velocity of 1.11 mm/min which was increased to 3.33 mm/min in the plastic branch of the force-displacement curve. The experiment was executed successfully and the beam failed at a static load of 1481 kN at an in plane deflection of 5.6% of the span, see Fig. 8. In the final stages of the test large deformations of the upper flange had presented themselves. Likewise some buckling of one of the mid-span stiffeners was observed, even though additional stiffness was arranged for this stiffener.

Beam IPE360, S355 was loaded with a cross-head velocity of 0.83 up to 2.22 mm/min. This test was executed without lateral torsional buckling constraints, while in hindsight those were required. However, the test result was probably not significantly influenced by lateral torsional buckling, which is substantiated by a reference test of the same nominal dimensions but with constraints, which appeared to have an almost equal ultimate load. Beam IPE360, S355 failed at a static load of 1040 kN at an in plane deflection of 3.6% of the span, see Fig. 8.

4. Numerical model

The Finite Element software Abaqus [7] was used for the numerical simulations. The material properties in the numerical model were based on the true stress-strain properties resulting from the tensile coupons tests as displayed in Fig. 4 and Fig. 5 and the measured dimensions of Table 4 were used for each beam. Fig. 7 shows the boundary conditions applied, with rolling hinges over one line of nodes at the supports and a longitudinal constraint at the mid-section. At the supports the nodes of the bottom flange were tied to each other (over 10 mm length conform the stiffener thickness), creating a rigid body surface which could only rotate around the x-axis. Following the previously described method, the additional stiffness provided by plates bolted to the mid-span stiffener was modeled.

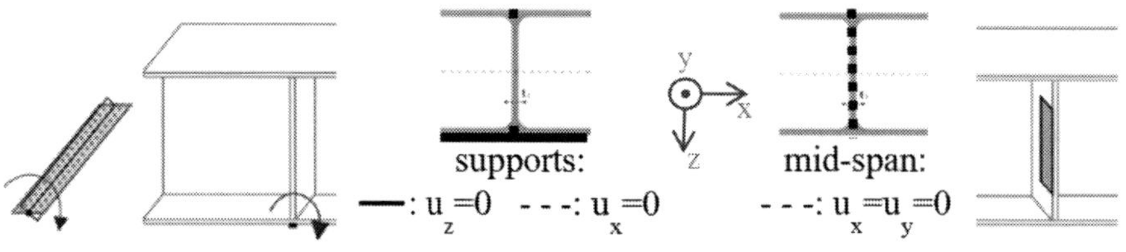

Fig. 7: Boundary conditions of the numerical model

Continuum (solid) elements were used in simulations, because the presence of roots is of large importance for further studies into the stress distribution in the root area. The calculation time of the quadratic elements appeared to be uneconomically large and a model with linear elements with incompatible mode formulation for preventing overstiffness due to parasitic shear (C3D8i) appeared to provide similar resistance as quadratic elements. These elements were thus used in the study.

A mesh refinement study was executed for both sections. Because bending-shear interaction was studied, 2 and 3 elements over the thickness did not suffice and a mesh with 4 elements over flange and web thickness appeared to be required. The element width was approximately 5 times larger than the thickness and the length of the elements varied over the length of the beam. At mid-span and at the supports the length was 5 mm and away from supports and mid-span the length was 40 mm. The remainder elements varied in length, with a default configuration consisting of a gradual transition from 5 mm towards 40 mm and back to 5 mm.

The numerical simulations of the IPE360 lead to a failure load of 1065 kN - which is 102.4% of the experimental failure load - at an in plane displacement of 2.8% of the span, see Fig. 8. The corresponding bending resistance is $0.97M_{pl}$ at a shear force of $0.68V_{pl}$. In the case of the HEA280, the failure load was 1562 kN - 105.5% of the experimental failure load – at an in plane displacement of 5.5% of the span. The corresponding bending moment is $0.92M_{pl}$ at a shear force of $0.89V_{pl}$.

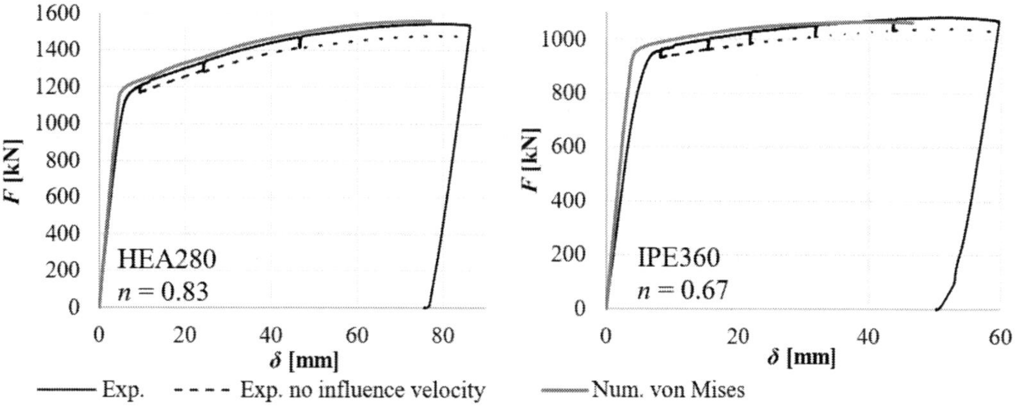

Fig. 8: Tests and numerical results for the HEA280 (left) and IPE360 (right) beam

Fig. 9: Von Mises stresses in IPE360 (top) and HEA280 (bottom) at the ultimate load

At failure, strains of 7.8% in the longitudinal direction of IPE360 were reached on the standard position F1 (bottom flange) according to the numerical model. In HEA280 strains of 8.2% in the longitudinal direction of HEB280 were reached on the standard position F1. At the failure load shear stresses were predominantly present in the web for both sections, whereas normal stresses in longitudinal direction dominated in the root section. The progression of the Von Mises stresses corresponded with the emergence of yield lines in the experimental tests. Fig. 9 displays the Von Mises stresses at the ultimate failure load.

5. Conclusions

This paper presents a numerical model that is capable of simulating various configurations of I-shaped sections in 3-point bending with short spans, where shear and bending moment both influence the resistance. The main conclusions are:

1. the FEM model makes use of C3D8i elements: 4 elements over the web and flange thickness were used; the element size over the flange width and web height corresponds to 5 times the size in corresponding thickness; and in the longitudinal direction elements of 5mm depth were sufficient for the midsection see Fig. 9,
2. the numerically determined failure loads of the two sections analysed were 102.4% and 105.5% of the experimentally determined failure loads,
3. in these two FE models the shear stresses were concentrated in the web,
4. strain hardening should be considered in determination of the bending-shear resistance, since the strains at failure are significantly larger than the strains at first yield,
5. the IPE360 section reached a bending resistance of $0.97M_{pl}$ with a shear force of $0.68V_{pl}$, the HEA section a bending resistance of $0.92M_{pl}$ with a shear force of $0.89V_{pl}$ was reached, in these cases the Eurocode 3, DIN 188000 and NEN 6770 design rules are conservative, see Fig. 2.

In future research the accuracy of the EN 1993 design rule regarding bending-shear resistance will be investigated for a larger range of shear forces and sections. In addition the development of shear stresses and their location will be investigated.

Acknowledgments

This research has received funding from the European community's Research Fund for Coal and Steel (RFCS) under grant agreements no. RFSR-CT-2013-00023. The specimens were provided by ArcelorMittal and tested at the Pieter van Musschenbroek Laboratory, Department of the Built Environment, Eindhoven University of Technology (TU/e).

References

[1] EN 1993-1-1, *Eurocode 3: Design of steel structures- part 1-1: General rules and rules for building,* Brussels, 2011.

[2] DIN 18800-1:2009-11, *Steel structures - Part 1: Design and construction,* Berlin, Deutches Institut für Normung, 2008.

[3] NEN 6770, *TGB 1990 – Steel structures - Basic requirements and basic rules for calculation of predominantly statically loaded structures,* Delft, Nederlands Normalisatie Instituut, 1997.

[4] Dekker, RWA, Snijder HH, Maljaars J. "Experimental Study into Bending-Shear Interaction of Rolled I-shaped Sections", *Proceedings of the 13th Nordic Steel Construction Conference* (Eds.: M. Heinisuo, J. Mäkinen), Tampere, Finland, (pp. 115-116, full 10 pages paper on USB-stick), 2015.

[5] EN 10002-1 - *Metallic materials - tensile testing - part 1: Method of testing at ambient temperature,* Brussels, CEN, 2001.

[6] Zieman, RD. *Guide to Stability Design Criteria for Metal Structures*, John Wiley and Sons, 2010.

[7] *Abaqus 6.14 Online Documentation*, Dassault Systèmes, 2014.

STRUCTURAL EFFICIENCY ASSESSMENT OF HOT-ROLLED STEEL PROFILES

Gianmaria Di Lorenzo, Antonio Formisano and Raffaele Landolfo

Department of Structures for Engineering and Architecture, University of Naples "Federico II", Italy

Abstract: In the current paper the origin of steel *I* beams has been analysed from historical-technological and structural points of view. First, the way leading to discover the most famous structural member of modern steel engineering has been reconstructed. Later on, parameters of structural efficiency have been defined as a function of the difference in terms of performance between the solid billet and the member final shape on the basis of the more or less raw material centrifugation. Finally, the proposed method, framed within a reverse engineering approach, is used to decode the different steel section property tables aiming at understanding the design strategy for their definition. This approach allows to select, among members having the same area and without local buckling phenomena, the best profile for each performance category of steel members used at the international level.

1. Introduction

Widely used as member with either wide or narrow flanges subjected to bending moment and/or axial forces, the undisputed star of the steelwork is certainly the double-T profile, defined according to the code EN 10079:2007 [1] as I (narrow flanges – see §3.4.7.1) or H (wide flanges – see §3.4.7.4.2) beam. This standard defines I-beams each hot-rolled profile with a geometrical configuration similar to the letter *I*, where the flanges have a width (b) not exceeding $0.66{\cdot}h$ (h=depth) and, in any case, less than 300 mm. On the other hand, the H-beams are profiles with flanges larger than $0.66{\cdot}h$ (or 300 mm). Sections with flanges wider than $0.80{\cdot}h$ are sometimes called either columns or profiles with effective wide flanges.

How was the concept born and who invented this range of sections? Unquestionably it can be regarded as one of the innovative products of the industrial revolution, but it is not easy to give an immediate answer to these questions, because its origins are to be found in different areas of emerging modern engineering. In any case, rather than from a theoretical study on the material optimization resulting from the application of methods and models developed by eminent scholars of that time, such as Claude-Louis Navier with his "beam theory" (1826), the idea of the double-T profile and its evolution draws inspiration from both the experience and the technological progress.

In the present paper a contribution on both the genesis of double-T section profiles and the structural efficiency of such members is provided, being justified by the wide use of such sections into the most common seismic-resistant steel typologies [2, 3, 4, 5]. Therefore, some expeditious efficiency parameters, able to both evaluate the section structural performance with respect to the starting semi-finished product and assess the effects produced by the rolling process, have been defined. These parameters will also permit to compare, from the performance point of view, different profiles belonging to the same steel section property table (e.g. comparison among European profiles) or framed within different tables (e.g. comparison between European profiles and American ones). Finally, the same parameters will allow to directly identify the optimal profile related to both a specific performance feature and an use category. Such a procedure can be also applied in the design phase for optimization of either complex tailor-made hot-rolled profiles or cold-formed ones [6, 7, 8].

2. Designation and standardization

The diffusion of hot-rolled sections with double-T profiles into different Engineering fields was such that their production was standardized through predefined types and sizes.

Although already in 1860s there was an extensive range of profiles produced on industrial scale, the unification process leading to the first steel double-T section property table ended only in 1920s with the production of "Normal Profiles", marked with "PN" or "NP".

A direct testimony is contained in the manuals of that time (Colombo, 1877 [9], Boubée, 1880 [10] and Breymann, 1925 [11]), where the first double-T profiles with tapered flanges, called IPN, were mentioned. A rigorous study on materials and products used in steelworks form origins (early 1800s) up to 1950 was conducted by both the American Institute of Steel Constructions [12] and the British Constructional Steelwork Association [13].

Subsequently, from the establishment of the European Community on Coal and Steel (ECSC) (1951), the unification process of hot-rolled profiles was done in Europe, alongside the normal series with the "European Profiles", marked with "PE" or "E".

Since the 1960s, through European standards EU 19-57 and 53-62, respectively, European double T profiles with narrow flanges (IPE) and wide flanges (HE) were born, being discernible with a number identifying their height (h) expressed in millimetres. Since the '70s, using the same rolling mills, but making small movements on the position of the rollers, heavy (marked with M or O) and light (marked with A) series of double-T cross-sections were included in the European list of profiles together with the standard series profiles.

3. Structural prerequisites and efficiency ratios

The structural efficiency of double-T profiles, observed experimentally by Fairbairn, demonstrated by Navier with the "beam theory" and generalized later by Barrè de Saint Venant, is linked to the high "centrifugation" of the material generated by the rolling process. In fact, for equal material used, this process allows to significantly increase the second moment of area of the section with respect to that of the main direction, identified with the strong axis (y-y axis). Starting from a semi-finished product with a square section (billet or bloom), the cross-section area is in fact moved from the centre to the periphery, going to delineate the two typical horizontal lintels (wings or flanges, with an area from 50% to 75 % of the total profile one) and the vertical connection element (web).

The morphological transformations (geometric effects) of semi-finished products deriving from the rolling process contribute to increase, together with the micro-structural modifications (mechanic effects), the structural performance of steel profiles [14, 15].

As it is known, in the double-T beams, the flanges are parts able to withstand the bending moment (over 80%), while the web has the task of absorbing mainly the shear stresses associated with the flexural regime. The passage from a compact section to a thin-walled open one produces, by contrast, a significant reduction in terms of stiffness and resistance to torsion stresses.

A qualitative and quantitative comparative analysis about the performance of profiles can be quickly performed through an appropriate "efficiency ratio" ρ_S, defined as follows:

$$\rho_S = S/S_{SQ} \tag{1}$$

where S is the section geometric property connected to a given structural performance and S_{SQ} is the corresponding geometric property of the square section (starting billet or equivalent square section) with the same area A.

The efficiency ratio expresses the geometrical effect of the working process on both the section geometric properties (S) and the related structural performances, assessed against the starting semi-finished product having the same area, that is the same weight and unit cost.

With respect to the geometric properties to be analysed, from the beam theory it is recalled that the section second moment of area along the strong axis (I_y) is directly correlated to the flexural performance at the Serviceability Limit State (SLS) in terms of vertical (elastic deflection) and horizontal (drift) displacements. The elastic section modulus ($W_y = I_y/(h/2)$) is related to the bending strength, whereas the shear area (A_{vz}), very close to the web area (A_w), is linked to the shear strength. The total strength capacity of flexural steel members against the Ultimate Limit State (ULS) is dependent from these parameters.

For steel profiles, the second moment of area along the vertical axis (*z-z*), the so-called weak axis, is instead indicative of their buckling performance (Ultimate Limit State for instability) in the cases when they are compressed (flexural buckling) or subjected to bending moment (lateral-torsional buckling or lateral-torsional buckling in case of beams). Moreover, the torsion behaviour at the Ultimate and Serviceability Limit States is correlated to the primary torsion second moment of area (I_T) and to the warping constant (I_w), the latter in the case of non-uniform torsion.

Finally, it should be remembered that the section centrifugation in the different directions is graphically represented by the central ellipse of inertia, which can be considered as a sort of polar diagram of the inertia ($I = A \cdot i^2$). The relationship between the two main radii of gyration ($\bar{\imath} = i_y / i_z$) expresses the "flexural anisotropy" of the section, namely the probable inequality in terms of the material disposal between the strong axis and the weak axis. Moreover, the transverse shape ratio ($r_{HB}=h/b$, h and b being the depth and the width of the profile, respectively) is used by the EN 10079 code to differentiate the profiles with narrow flanges ($r_{HB} > 1.5$ and $b < 300mm$) from those with wide flanges ($r_{HB} \leq 1.5$ or $b \geq 300mm$). This latter parameter, together with the flexural anisotropy one, other than being correlated to the section morphology, provides useful indications on the optimal use of compressed profiles subjected to instability (flexural or flexural-torsional) phenomena.

For a given profile having the same buckling length (L_0) in the two main planes ($L_{0,y}=L_{0,z}$), the flexural anisotropy ($\bar{\imath}$) coincides with the ratio between the member geometrical slenderness ($\lambda = L_0/i$) related to the strong axis (λ_y) and that of the weak axis (λ_z).

On the basis of this ratio the profiles can be divided into three conventional categories: sections with low ($\bar{\imath} \leq 2$), medium ($2 < \bar{\imath} \leq 4$) and high ($4 < \bar{\imath}$) anisotropy. In the absence of intermediate supports, e. g. provided by diagonal bars or bracing systems, the use of high anisotropy sections should be avoided for compressed members that, due to onset of premature global instability phenomena, do not allow an adequate exploitation of the base material.

Generally, it is possible to observe that, if the profile grows in homothetic way, maintaining unaltered its proportions, its structural efficiency (ρ_S) does not change, depending on the geometrical characteristics of the original profile. On the contrary, if the profile retains its outer shape, the flexural efficiency (ρ_I; ρ_W) decreases with increasing thickness of both the web (t_w) and the flanges (t_f), tending to degenerate into the starting semi-finished product.

The profiles of the European list of sections are designed, as the height (and/or area) is increasing, with flexural efficiency augmenting according to the strong axis in terms of both strength and deformability. For narrow flanges double-T profiles (Fig. 1) belonging to the IPE series, the efficiency in terms of flexural deformability ($\rho_{Iy}=I_y/I_{y,SQ}$, I_{SQ} being the second moment of area of the equivalent square section) varies from 16 to 45 with increasing height from 80mm (IPE80) to 600mm (IPE600), respectively. This parameter, with increasing height, instead varies from 8 to 48 for wide flanges HEB series profiles.

The increase of the flexural efficiency is even greater for wide flanges HEA profiles that, moving from HE100A to HE1000A, have ρ_{Iy}variable from 9 to 55. Less marked are the efficiency increments in terms of resistance ($\rho_{Wy}=W_y/W_{y,SQ}$), that is the increments in terms of the elastic section modulus which, on average, increased from 4 to 10 times the performance of the equivalent square profile with increasing height. By analyzing the behaviour with respect to the weak axis (Fig. 2), the flexural efficiency ($\rho_{Iz}=I_z/I_{z,SQ}$), related to problems of global flexural and flexural-torsional instabilities, is not strictly increasing, but has a maximum for profiles with height equal to about 300mm. This is particularly true for wide flanges double-T profiles of the HE series. The loss of efficiency around weak axis for medium-large height profiles (h> 300 mm), although to a lesser extent, is also found for IPE series profiles. In all double-T profiles within the section property table, the thinning of the section in the vicinity of the web reduces from 20% to 40% the profile shear strength in comparison to the equivalent square section, with efficiency ($\rho_{Avz}=A_{vz}/A_{v,SQ}$) increasing from 0.6 to 0.8 with the member height (Fig. 3a).

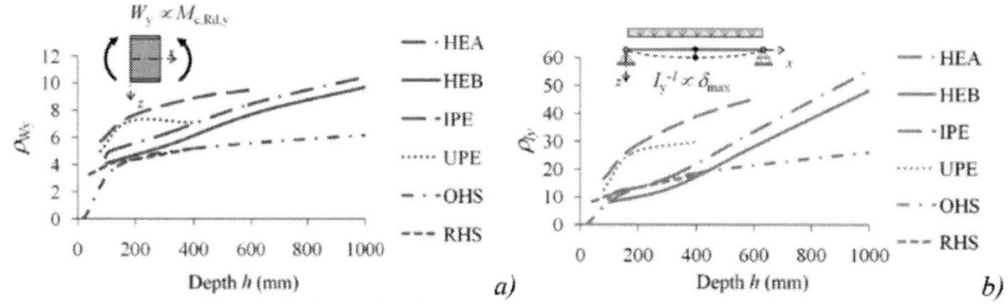

Fig. 1: Bending and flexural efficiencies at the ULS (a) and SLS (b) (*y*-*y* axis)

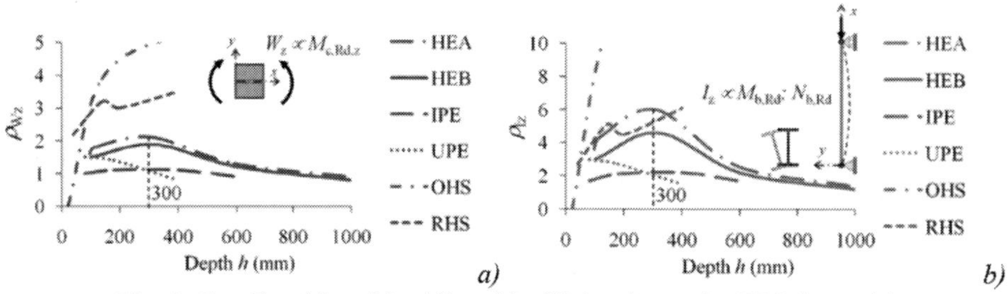

Fig. 2: Bending (a) and buckling (b) efficiencies at the ULS (*z*-*z* axis)

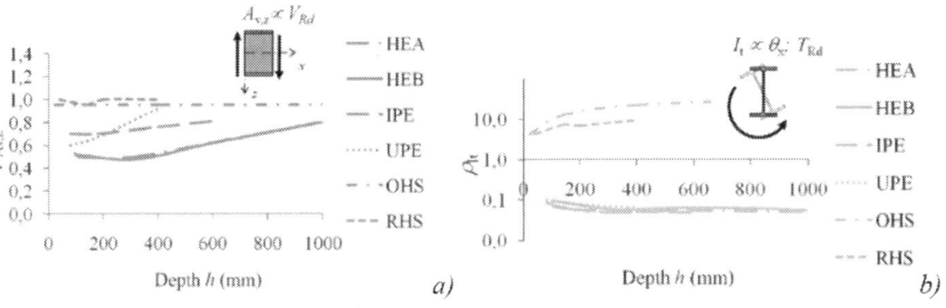

Fig. 3: Shear efficiency at the ULS (z axis) (a) and torsion efficiency (x-x axis) (b)

As further reversal of the medal, the lamination process strongly reduces the torsion efficiency of open shape profiles (Fig. 3b), characterized by primary torsion inertia (I_T) equal to just 5÷10% of that of starting semi-finished products.

Finally, the increase of the transverse shape ratio proportionally to the profile height is reflected directly on the flexural anisotropy parameter. As it is seen in Fig. 4, HE profiles may be considered "actually" as wide flanges cross-sections if flexural anisotropy is medium-low. In fact, members with heights greater than 600 ÷ 650 mm have ratios between radii of gyration larger than 4, which is very different slenderness values along two main axes. Regardless from their size, a low anisotropy is instead shown by both hollow profiles and new HD (wide flanges columns) and HP (wide flange bearing piles) profiles, whose use is particularly recommended for highly compressed members, e.g. located in tall buildings and foundation piles. On the other hand, HE profiles with heights greater than 600mm, similarly to IPE profiles, are particularly suitable for the production of elements subjected to bending moment and located into buildings having floors with medium-large bays. In these cases, also the recent HL (extra wide flanges) profiles may be adopted. In comparison to HE members, these profiles, although having a high flexural anisotropy, ensure a better lateral stability due to the greater width of the flanges, whose size for the entire series is equal to about 400 mm.

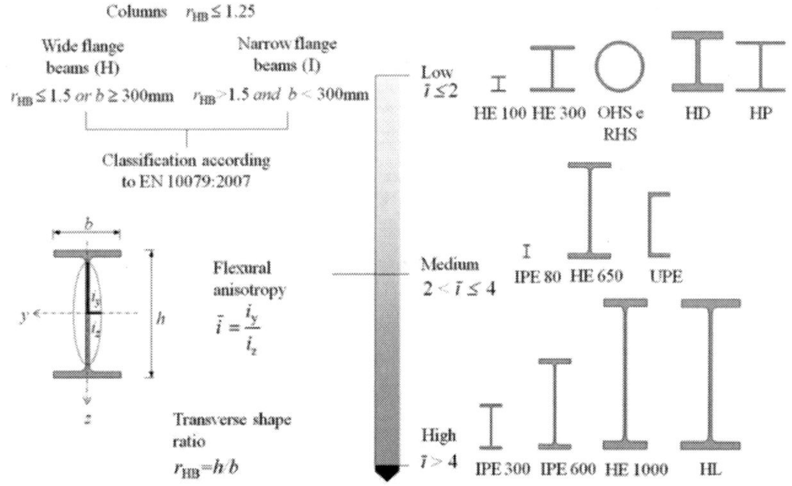

Fig. 4: Flexural anisotropy of profiles

Neglecting the onset of possible local or global buckling phenomena, the increase of flexural efficiency (ρ_{Iy}; ρ_{Wy}) as the cross-section height augments gives rise, within the design phase, to an apparent "paradox", that is the realization of always slenderer members with longitudinal shape ratios ($r_{HL}=h/L$, h and L being the depth and the length of the profile) gradually decreasing with the member length.

As an example, for floor-decking of civil residence buildings (permanent loads g_k=4 kN/m^2 and service loads q_k = 2 kN/m^2) made of steel corrugated trapezoidal sheeting and composite slabs, the secondary beams, having pitch p = 1.5÷2 m, can be pre-designed ($h(L) = r_{HL} \cdot L$) (Fig. 5) with depth/span ratio r_{HL} variable from 1/20 to 1/25 for members of small (L = 3m) and medium-large (L = 7÷10m) lengths, respectively. The shape ratios for sizing main beams posed with pitch p_{TP} variable from 5 m to 7 m, decreased from 1/15 to 1/20 when beams of small (L = 5 m) or large (L = 15 m÷20 m) lengths are of concern, respectively.

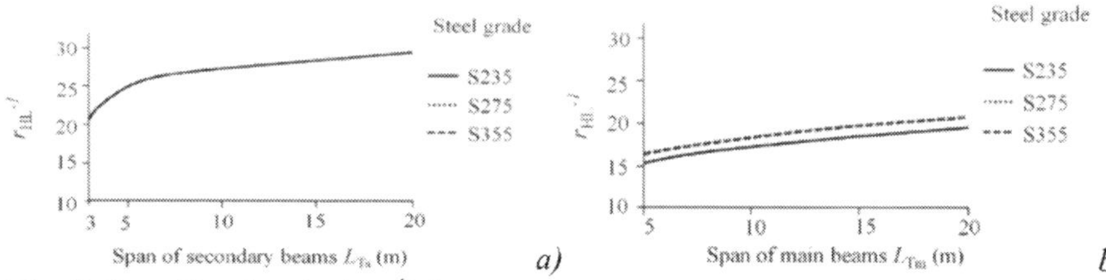

Fig. 5: Span/Depth ratio (r_{HL}^{-1}) for secondary beams T_S (pitch p=1.5 m÷2 m) (a) and for main beams T_m (pitch p=L_{Ts}=5 m÷7 m) (b)

This is explained because the vertical displacements (δ) and bending moments (M_{Ed}) of a steel member in bending subjected to a uniformly distributed load grow with the fourth dimension of the length and with the length square, respectively. As a consequence, the related inertial characteristics (I_y and W_y) grow, for the above-mentioned increase of efficiency (ρ_s), in an even more marked way with the required height necessary to meet the design requirements.

4. Selection of the optimal cross-section for different application fields

The structural efficiency parameters, other than providing useful indication about the effects produced by the rolling process on the cross-section geometrical properties, if applied to members with the same area (A), or the same weight (G_k) and cost per meter linear, enable to directly identify the optimal profile for performance feature and use category (beams under bending moment, beams subjected to torsion, compressed and compressed-flexural columns).

Taking as a benchmark example a billet or a semi-finished shape with square section of 50mmx50mm (cross-section A = 2500 mm^2), the structural performances of hot-rolled profiles HE100B, HE120A, IPE180 and UPE180 can be compared (Fig. 6). When possible buckling phenomena are prevented, for example by adopting specific measures as various kinds of torsion restraints, the IPE series profiles are arguably the best sections in terms of flexural behaviour both in terms of resistance (ULS) and deformability (SLS). Their structural efficiency under bending (ρ_{Wy}) at the ULS is 7.5, that is structural performances larger than 1.5 and 2 times, respectively, those of profiles HEA and HEB.

Even more marked differences are at the SLS, with a performance referred to the strong axis (ρ_{Iy}) greater than about 2 - 3 times that of homologous wide flanges double T profiles. Obviously, the activation of possible flexural-torsional buckling phenomena can significantly affect the structural efficiency of IPE profiles. In fact, in this case, it is convenient to adopt a wide flanges double-T profile. Among these, for equal material used, the HEA series profiles, resulting more centrifuged with respect to the two main directions, are certainly the most efficient cross-sections, offering a structural efficiency ρ_{Iz}=4.3, that is a double value compared to the corresponding IPE profile one.

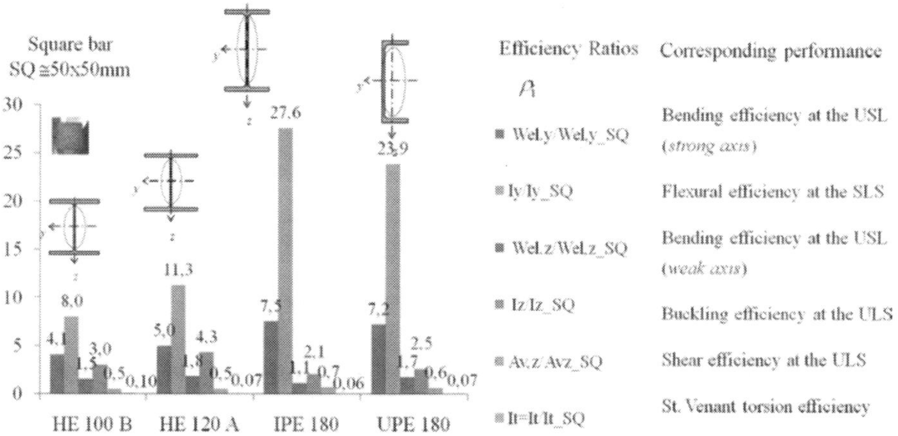

Fig. 6: Comparison of performances among main hot-rolled profiles having the same area

Very well performing as elements under bending are also the UPE profiles, recently introduced in the European list of sections (EN 10279) [16]. These members, if compared to the homologous IPE profiles, although they have not a doubly symmetrical shape (shear-torsion coupling), show a higher centrifugation with respect to the *z-z* axis. This makes advisable their use in the presence of biaxial bending moment, as in the case of the purlins in inclined pitched roofs. Moreover, the presence of a straight lateral surface, without tapers essentially for technological reasons, favours their use as steel beams within staircases.

Finally, as compressed elements, the more efficient profiles are the HEB and HEM series profiles, which, together with a high centrifugation of the section with respect to the two main directions, add a significant compactness useful to prevent local instability phenomena.

5. Concluding remarks

In this paper, after the analysis of the genesis of the hot-rolled steel double-T profiles, some parameters related to the profile structural efficiency have been provided. These parameters have given information about the effects that the material centrifugation generated by the lamination process induces on the structural performance of such profiles. In particular, it has been investigated the efficiency trend of the main geometric properties related to the structural performances at the ULS (in terms of both strength and instability) and at the SLS (in terms of deformability).

To this purpose, by analysing the IPE and HE European double-T profiles, it has been observed that the bending moment efficiencies related to the strong axis, valued in comparison to the starting semi-finished section (benchmark profile), are not constant. In particular, these flexural efficiency parameters grow, as the height increases, up to 10 times in terms of strength and 50 times in terms of deformability. Contrary, the efficiency increases over the weak axis are less pronounced, reaching a maximum value for double-T profiles with height of 300mm.

The proposed methodology, applied to profiles having the same cross-sectional area (same weight and cost), has allowed to rapidly identify the optimal profile for performance features and categories of use. Again with reference to the European section property table, it has been observed under quantitative terms that the IPE series profiles are, in the case of prevented flexural-torsional buckling, the most high-performance sections under bending moment. However, greater efficiency has been found for HE profiles, with particular regard to the lightened series, when both buckling and flexural-torsional buckling phenomena are taken into account.

Finally, the proposed analysis approach can be considered as a reverse engineering technique and, as a further development of the study, it can also be adopted to compare the performance of different steel section property tables used at the international level.

References

[1] EN 10079:2007 "Definition of steel products", CEN.

[2] Formisano A, Mazzolani FM. "On the catenary effect of steel buildings", *Proc. of the COST Action C26 Final Conference "Urban Habitat Constructions under Catastrophic Events"*, Naples, Italy, 16-18 September 2010, 619 – 624, 2010.

[3] Formisano A, Mazzolani FM. "Progressive Collapse and Robustness of Steel Framed Structures", in B.H.V. Topping, (Editor), *Proc. of the Eleventh International Conference on Computational Structures Technology*, Civil-Comp Press, Stirlingshire, UK, Paper 13, doi:10.4203/ccp.99.13, 2012.

[4] Formisano A, Landolfo R, Mazzolani FM. "Robustness assessment approaches for steel framed structures under catastrophic events", *Computers and Structures*, 147, 216-228, 2015.

[5] Faggiano B, Fiorino L, Formisano A, Macillo V, Castaldo C, Mazzolani FM. "Assessment of the design provisions for steel concentric X bracing frames with reference to Italian and European codes", *Open Construction and Building Technology Journal*, 8, 208-215, 2014.

[6] Calderoni B, Formisano A, De Martino A. "Flexural cyclic behaviour and low-cycle fatigue of cold-formed steel members", *Proc. of the Final Conference of COST ACTION C12*, Innsbruck, Austria, 20-22 January 2005, 301-309, 2005.

[7] Calderoni B, De Martino A, Formisano A, Fiorino L. "Cold formed steel beams under monotonic and cyclic loading: experimental investigation", *Journal of Constructional Steel Research*, 65, 219-227, 2009.

[8] Landolfo R, Mammana O, Portioli F, Di Lorenzo G, Guerrieri MR. "Laser welded built-up cold-formed steel beams: Experimental investigations", *Thin-Walled Structures*, 46, 781-791, 2008.

[9] Colombo G. *Manual of the Engineer* (in Italian), 1877.

[10] Boubée FP. *Theoretical-practical elementary treaty of metallic steelwork* (in Italian), 1880.

[11] Breymann GA. *General treaty of Civil Constructions with special hints to the grandiose constructions* (in Italian), 1925.

[12] Herbert WF. *Iron and Steel Beams 1873 to 1952*, American Institute of Steel Construction, New York, 1953.

[13] Bates W. *Historical structural steel work handbook*, British Constructional Steelwork Association, London, 1984.

[14] Formisano A, Faggiano B, Landolfo R, Mazzolani FM. "Ductile behavioural classes of steel members for seismic design", *Proc. of the 5thInt. Conf. on the Behaviour of Steel Structures in Seismic Areas (STESSA '06)*, Yokohama, 14-17 August 2006, 225-232, 2006.

[15] D'Aniello M, Landolfo R, Piluso V, Rizzano G. "Ultimate Behaviour of Steel Beams under Non-Uniform Bending", *Journal of Constructional Steel Research*, 78, 144-158, 2012.

[16] EN 10279. *Hot rolled steel channels. Tolerances on shape, dimension and mass*, 2000.

BEHAVIOUR AND DESIGN OF SHORT-TO-INTERMEDIATE HOT-ROLLED STEEL ANGLE COLUMNS

Pedro Borges Dinis, Dinar Camotim and Virgílio Preto

CEris, ICIST, DECivil, Instituto Superior Técnico, Universidade de Lisboa, Portugal

Abstract: The paper addresses the buckling, post-buckling (elastic and elastic-plastic) and ultimate strength behaviour of fixed-ended short-to-intermediate hot-rolled steel equal-leg angle columns. The results presented and discussed are obtained by means of (i) Generalised Beam Theory (buckling) and (ii) ABAQUS shell finite element simulations (buckling and post-buckling) – particular attention is paid to the determination of column elastic-plastic failure loads, accounting for both initial geometrical imperfections and residual stresses. Then, the numerical failure loads gathered in a limited parametric study are used to assess the quality of their predictions provided by a novel Direct Strength Method approach, recently proposed and validated in the context of cold-formed equal-leg angle columns.

1. Introduction

The buckling, post-buckling, strength and design of equal-leg angle columns have attracted the attention of several researchers in the past (*e.g.*, [1-8]). As far as the design is concerned, some of these publications devoted a fair amount of work to develop rules and procedures aimed at predicting the ultimate strength of short-to-intermediate columns, adopting mostly local buckling concepts. However, the numerical simulations recently carried out by Dinis *et al.* [9-11], concerning fixed and pin-ended short-to-intermediate angle columns with slender legs ($b/t > 20$, where b is the leg width and t the wall thickness), (i) shed new light on key mechanical aspects related to the structural response of such members, namely the fact that it is strongly influenced by the length-dependent interaction between two global buckling modes (major-axis flexural-torsional and minor-axis flexural), and (ii) showed that it is possible to have a single rational approach, based on the Direct Strength Method (DSM – *e.g.*, [12]), that can handle adequately both fixed-ended and pin-ended equal-leg angle columns (angle columns are currently explicitly excluded from the domain of application of that design approach). However, it must be said that no similar investigation was ever carried for hot-rolled steel equal-leg angles, characterised by much stockier legs ($b/t < 20$) and the presence of relevant residual stress effects - the present work aims at providing a first contribution towards filling this gap.

The paper begins by presenting and discussing a brief comparative study between the buckling behaviours of equal-leg steel (E = 210 GPa, ν = 0.3) angle columns with various leg slenderness values, carried out by means of Generalized Beam Theory (GBT) analyses and enabling the characterisation of the columns that are numerically investigated in this work. Then, the paper addresses the post-buckling (elastic and elastic-plastic) and failure behaviour of fixed-ended short-to-intermediate hot-rolled equal-leg angle columns (herein termed "angle columns") - the results are obtained by means of ABAQUS shell finite element analyses (SFEA) including both initial geometrical imperfections and residual stresses – special attention is paid to the modelling of the latter. Finally, a limited parametric study is carried out to gather numerical failure load data that are subsequently used to assess the quality of the estimates provided the DSM approach recently proposed in [11] for the design of cold-formed steel angle columns (with slender legs: b/t > 20) - the proposed approach leads to accurate failure load estimates for short-to-intermediate fixed-ended hot-rolled equal-leg angle columns.

2. Buckling, Post-Buckling and Ultimate Strength Behaviour

The curves in Fig. 1(a) concern angle columns with three cross-section dimensions, namely b = 50; 70; 90 mm and b/t = 7.5; 10; 15; 25. Each curve (i) provides the variation, with the length L (logarithmic scale), of the critical buckling stress f_{cr} and (ii) was obtained by means of GBT buckling analyses performed with the code GBTUL and including 7 deformation modes: 4 global (**1-4**) and 3 local (**5-7**). Figs. 1(b) display the GBT modal participation diagrams associated with each P_{cr} vs. L curve, providing the contributions of each GBT deformation mode to the column buckling modes (*e.g.*, a b/t = 70/7 column with L = 100 cm buckle in a mode involving the participations from modes **2**, **4** with 4.5% + 95.5%, respectively). Fig. 1(c) shows the evolution of the percentage participation of mode **2** in the column buckling modes (p_2). Fig. 1(d) depicts the buckled mid-height cross-sections of the b=70 mm + b/t=10 columns with L=25; 100; 500 cm, as well as the in-plane shapes of deformation modes **2-6** (axial extension not shown). These buckling results prompt the following remarks:

(i) For all columns, (i_1) P_{cr} corresponds to *single half-wave* buckling and decreases monotonically with L, (i_2) the torsion mode **4** has a highly dominant participation in the critical buckling modes (except in the very long columns), and (i_3) the buckling mode switches abruptly from flexural-torsional (FT: **2+4**) to flexural (F: **3**) at L=L_T. The FT buckling stresses may be calculated with the analytical formulae derived in [11] for both fixed-ended and pin-ended angle columns, which only differ in the minor-axis bending support conditions, (G =E/[2 (1+ν)] is the shear modulus)

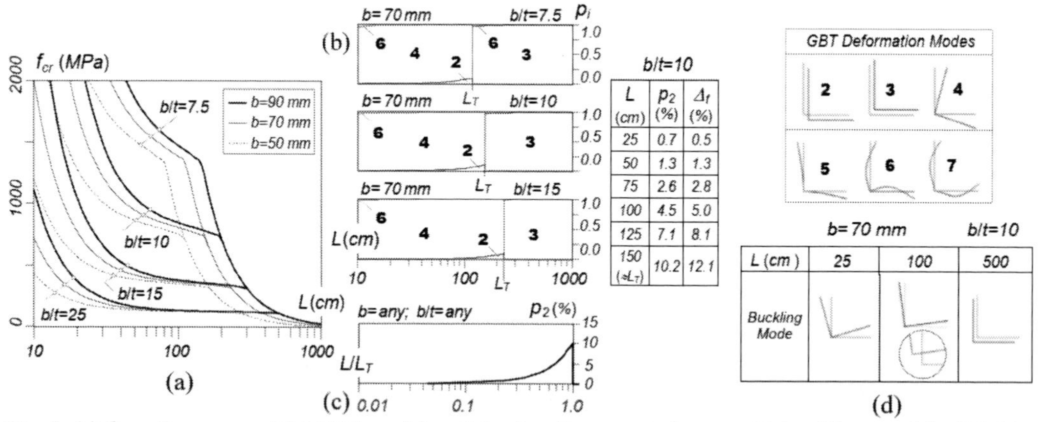

Fig. 1: (a) f_{cr} vs. L curves and (b) GBT modal participation diagrams (columns with b = 70 mm + b/t= 7.5; 10; 15), including a table with a few p_2 values, (c) evolution of p_2 with L (normalised w.r.t L_T), and (d) in-plane shapes of 6 GBT deformations modes and 3 critical buckling modes of columns with b = 70 mm + b/t= 10

$$f_{bt} = G\frac{t^2}{b^2} + \pi^2 \frac{E\,t^2}{12\,(L/2)^2} \qquad f_{crft} = \frac{4}{5}\left(f_{bt} + f_{bf} - \sqrt{(f_{bt} + f_{bf})^2 - 2.5\, f_{bt}\, f_{bf}} \right) \tag{1}$$

where f_{bt} and $f_{bf}=\pi^2 E\, b^2/[6 \times (L/2)^2]$ are the angle column pure torsional and major-axis flexural buckling stresses.

(ii) Since the column flexural buckling load is independent of t, the trios of flexural-torsional curves concerning columns with the same leg width b end in the same flexural curve – see Fig. 1(a).

(iii) The critical stress $f_{cr.LT}$, corresponding to the transition between FT and F buckling (at $L=L_T$) depends only on the steel properties and the b/t ratio – naturally, the value of L_T varies with b. The values of $f_{cr.LT}$ and L_T can be calculated by means of the expressions

$$f_{cr.L_T} = \frac{E}{K} \qquad L_T = b\sqrt{\frac{K\,\pi^2}{6}}\;, \qquad \text{where } K = (1+\nu)\left[2.25\left(\frac{b}{t}\right)^2 - 4.0\right] \;. \tag{2}$$

(iv) The "plateau" of the P_{cr} vs. L curve corresponds to FT buckling – see the b/t =25 column curve in Fig. 1(a). However, Fig. 1(b) shows that the amount of major-axis flexure (**2**) is initially minute and only becomes visible (growingly) as L increases – the table in Fig. 1(b) makes it possible to quantify the variation of p_2 with L, for the b=70 mm + b/t=10 column. It was shown [11] that there is a good correlation between the variations with L of p_2 and the percentage drop of f_{crft} with respect to f_{bt}, given by the parameter $\Delta_f=[(f_{bt} - f_{crft})\, /f_{crft}]\times100$, whose values are also given in the table of Fig. 1(b) – note that, for the b=70 mm + b/t=10 columns, the values of p_2 and Δ_f at L=150 cm ($\approx L_T$) are equal to 10.2% and 12.1%, respectively.

(v) The length of the "flexural-torsional plateau" decreases visibly with b/t – for b/t <10, such plateau does not even exist. However, note that the variation of p_2 (or Δ_f) with $L=L_T$ is the same regardless of the angle b/t and b values – this feature can be clearly observed in Fig. 1(c).

(vi) Finally, note the tiny contribution of the anti-symmetric local mode **6** to the "flexural" buckling mode of columns with b/t <10 – p_6 (vi$_1$) has a maximum value for $L=L_T$, (vi$_2$) decreases as L increases and (vi$_3$) is larger for the stockier legs (for instance, p_6 is equal to 3% and 5%, respectively for b/t =7.5 and b/t =6 – the latter column are not included in Fig. 1).

2.1 Column geometry selection

The hot-rolled steel angle column geometries considered in this work were taken from [13], enforcing that (i) buckling occurs in predominantly torsional modes and (ii) the critical buckling stress for $L=L_T$ ($f_{cr.LT}$) does not exceed 700MPa – since the steel grades covered in [13] exhibit yield stresses 450MPa, the collapse of columns exhibiting critical stresses above this value is clearly governed by plasticity – such columns are stocky. Fig. 2 plots $f_{cr.LT}$ against b/t for the whole set of cross-section dimensions

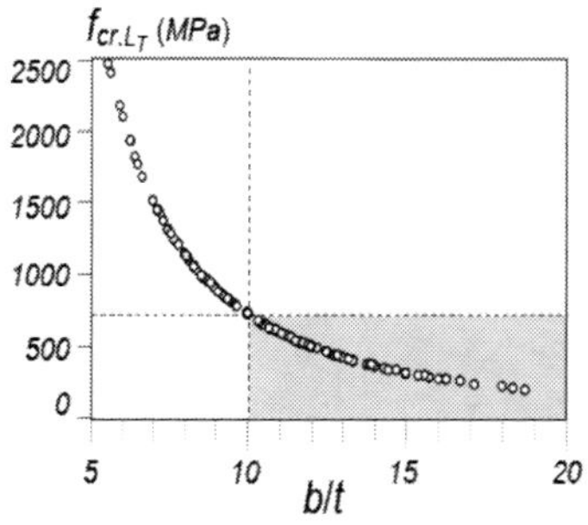

b	t	b/t	L≈0.25 L_T (Δ_f≈0.8) L	L≈0.25 L_T (Δ_f≈0.8) f_{cr}	L≈0.5 L_T (Δ_f≈2.8) L	L≈0.5 L_T (Δ_f≈2.8) f_{cr}	L≈0.75 L_T (Δ_f≈6.5) L	L≈0.75 L_T (Δ_f≈6.5) f_{cr}	L≈L_T (Δ_f≈12.3%) L	L≈L_T (Δ_f≈12.3%) f_{cr}
50	5	10	270	1061.5	540	851.3	810	788.3	1080	736.7
70	7		377.5	1062.2	755	851.5	1132.5	788.5	1510	737.0
90	9		487.5	1059.8	975	850.8	1462.5	787.8	1950	736.1
50	3.33	15	407.5	407.1	815	362.1	1222.5	342.9	1630	323.0
70	4.67		570	407.2	1140	362.2	1710	343.0	2280	323.0
90	6		735	406.9	1470	362.0	2205	342.8	2940	322.8
50	2.5	20	545	216.5	1090	200.5	1635	191.4	2180	180.8
70	3.5		765	216.4	1530	200.5	2295	191.4	3060	180.6
90	4.5		982.5	216.4	1965	200.5	2947.5	191.4	3930	180.7

Fig. 2: Hot-rolled angle column selection: $f_{cr.LT}$ vs. b/t plot, column b, t, L, L_T, f_{cr} and Δ_f values (mm and MPa)

covered in [13] and makes it possible to conclude that only 79 (out of 177) column geometries satisfy the enforced conditions – they correspond to the shaded area in Fig. 2, *i.e.*, are such that $20<b/t\leq 10$. The three column families analysed in this work were selected from these 79 columns and their geometries (b, t, $L\approx L_T$) are given, together with the associated f_{cr} values, in the table included in Fig. 2. In order to make it possible to assess the relevance of the interaction between the FT and F buckling modes on the behaviour of the hot-rolled angle columns, the columns analysed exhibit four normalised length values, namely L=1.0;0.75;0.5;0.25 L_T – the corresponding Δ_f values are given in the table.

2.2 Elastic and elastic-plastic post-buckling behaviour

The elastic post-buckling behaviour of the selected 36 columns, containing critical-mode initial imperfections with small amplitudes (10% of the wall thickness t), was determined through ABAQUS SFEA – all modelling issues can be found in [6]. Figs. 3(a) show the upper parts ($P/P_{cr}>0.4$) of the column equilibrium paths P/P_{cr} *vs.* β (β is the mid-span web chord rigid-body rotation) for columns with (i) b/t= 10; 15; 20 and (ii) L equal to $\approx 0.25\ L_T$ (Fig. 3(a_1)), $0.5\ L_T$ (Fig. 3(a_2)), $0.75\ L_T$ (Fig. 3(a_3)) and L_T (Fig. 3(a_4)). Fig. 3(b) displays the equilibrium paths P/P_{cr} *vs.* d_m/t, where d_m is the mid-span corner displacements due to minor-axis flexure, for b=70 mm + b/t=10 columns with L=0.5 L_T and L=0.75 L_T. The observation of these elastic post-buckling results prompts the following comments:

(i) Angle columns with identical b/t and L/L_T values exhibit practically coincident equilibrium paths, *i.e.*, post-buckling behaviours. Moreover, it is worth noting that these post-buckling behaviours are qualitatively similar to those unveiled for the slender cold-formed angle columns – Figs. 3(a) include equilibrium path concerning columns with b/t=58 (b=70, t=1.2), for comparative purposes.

(ii) The column post-buckling behaviour involves the simultaneous occurrence of cross-section torsional rotations and translations (corner displacements). The relative importance of the latter has strong impact on the column post-buckling response, namely on its post-critical strength reserve – this feature can be clearly observed in Figs. 3(a)-(b).

(iii) Indeed, two equilibrium path sets can be identified: those concerning (iii_1) the shorter column ($L/L_T\approx$0.25; 0.5), involving small corner displacements and clearly stable (fairly high post-critical strength), and (iii_2) the longer column ($L/L_T\approx$0.75; 1.0), associated with significant corner displacements and exhibiting well defined limit points, followed ($L/L_T\approx$0.75) or not ($L/L_T\approx$1.0) by a torsional rotation reversal.

(iv) However, post-buckling behavioural differences are observed when b/t decreases, namely (iv_1) the equilibrium paths become more flexible and (iv_2) the limit points occur for higher P/P_{cr} values – *e.g.*, in the columns with $L/L_T\approx$0.75, one has P_{max}=1.09 P_{cr} (b/t =20) and P_{max}=1.15 P_{cr} (b/t =10).

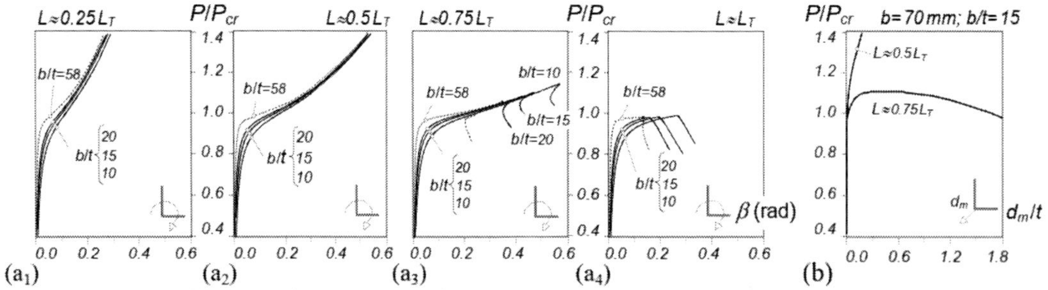

Fig. 3: Elastic equilibrium paths (a) P/P_{cr} *vs.* β ($L/L_T\approx$0.25;0.5;0.75;1.0) and (b) P/P_{cr} *vs.* d_m/t ($L/L_T\approx$0.5;0.75)

Finally, Fig. 4(a) depicts the upper portions (P/P_{cr}>0.4) of typical intermediate ($L\approx 0.5\ L_T$) column elastic-plastic equilibrium paths for columns with b=70 mm and b/t=10; 15; 20: P/P_{cr} *vs.* β concerning yield-to-critical stress ratios $f_y/f_{cr}\approx$1.0, 1.8, 2.5 (elastic equilibrium paths of Fig. 3(a_2) are also shown).

Fig. 4(b) displays three plastic strain diagrams of the b/t=15 column, corresponding to equilibrium states located along the f_y/f_{cr}≈1.8 equilibrium path (indicated in Fig. 4(a)) and including the column collapse mechanism. The observation of all these results leads to the following remarks:

(i) The columns with f_y/f_{cr}>1.0 exhibit some elastic-plastic strength reserve and ductility prior to failure – doubling the yield stress leads to a strength increase over 20%. Moreover, the elastic-plastic strength reserve increases with b/t – see the P_u/P_{cr} values given in the table of Fig. 4(a).

(ii) Diagram *I* in Fig. 4(b) shows that yielding starts at the corner longitudinal edge around the ¼ and ¾-span column regions, where the longitudinal normal and shear stresses are higher [10].

(iii) For the longer columns ($L≈L_T$ – results not shown), there is no benefit in having a yield stress much larger than f_{cr}, since the collapse occurs, abruptly, in the elastic range – see Fig. 3(a_4).

(iv) The markedly different post-buckling behaviours exhibited by the short-to-intermediate and long columns were also observed for the slender cold-formed columns (b/t>20) – this behavioural feature was at the root of the DSM design approach developed specifically for such columns [11].

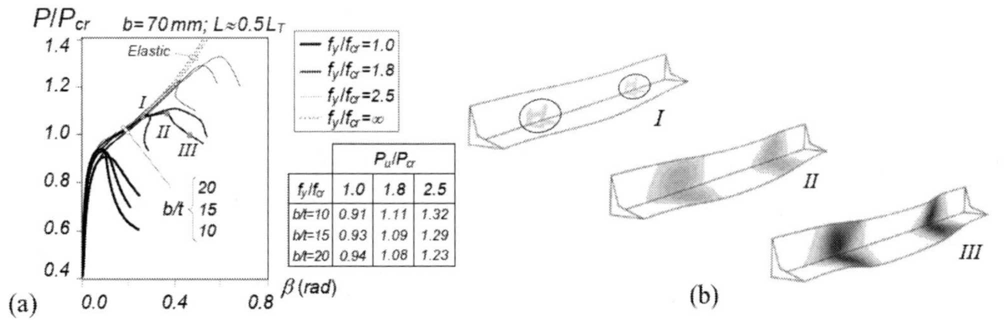

Fig. 4: (a) Elastic-plastic equilibrium paths P/P_{cr} *vs.* β concerning b=70 mm, b/t=10; 15; 20, $L≈0.5L_T$ and f_y/f_{cr}≈1.0; 1.8; 2.5, and (b) plastic strain diagram evolution for b/t=15 and f_y/f_{cr}≈1.8

2.2.1 Residual stress effects

It is well known that initial geometrical imperfections and residual stresses influence the load-carrying capacity of structural members built from hot-rolled steel profiles – in the case of columns, such influence is already "embedded" in the design curves appearing in virtually all existing codes. As far as the residual stresses in angles are concerned, experimental studies carried out in the 70's led to the practically universal adoption of the 3-point linear diagram depicted in Fig. 5(a_1) and included in the current version of Eurocode 3. However, the fast increasing use high-strength steel to fabricate angles with fairly wide legs (*e.g.*, b/t=300/35) prompted further experimental investigations aimed at determining the residual stress distribution in these new profiles [14, 15] – Fig. 5(a_2) shows the 4-point linear distribution recently proposed by Može *et al.* [15]. The numerical study presented next is intended to make a comparative assessment of the influence of the two residual stress distributions displayed in Fig. 5(a) on the failure load of angle columns. Such an assessment is made resorting to the parameter $\chi=[(P_u-P_{u.rs})/P_u]\times 100$, which measures the failure load drop caused by the residual stresses – $P_{u.rs}$ and P_u are the failure loads of angle columns with and without residual stresses. The columns analysed (i) have geometries defined by b=70 mm, b/t=7.5; 10; 15; 20, and $L\approx 0.25L_T$; $0.5L_T$; $0.75L_T$; L_T, (ii) contain critical-mode initial imperfections with 0.1 t amplitude and (iii) exhibit various yield stresses (f_y=300, 500, 700 MPa) – stipulating an unbounded dependency between f_y and the maximum residual stress penalises the high-strength steel columns (it is well known that the residual stresses in hot-rolled steel profiles are bounded – for instance, a maximum value of 70 MPa is suggested in [15]). Fig. 5(b) plots the variation of χ with L/L_T, b/t and $\lambda_{ft}=(f_y/f_{cr.ft})^{0.5}$. The observation of these failure load results prompts the following remarks:

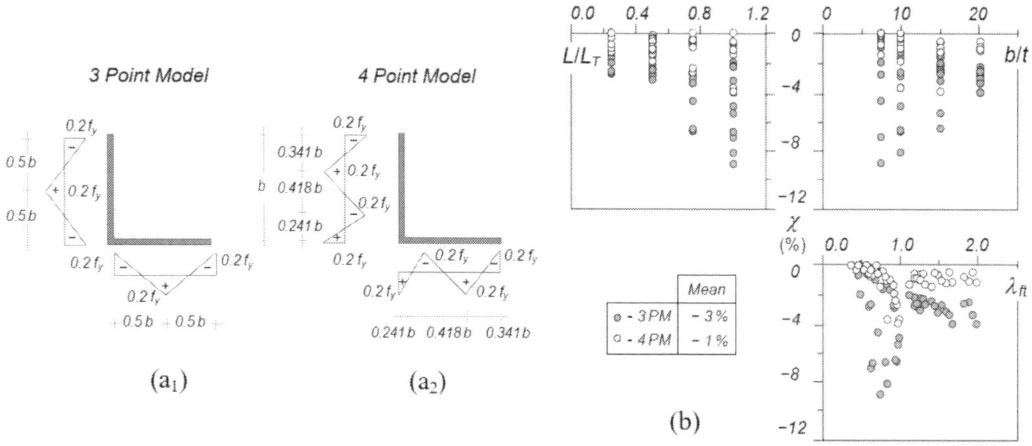

Fig. 5: Hot-rolled angle columns: (a) 3- and 4-point residual stress distributions and (b) plots χ vs. L/L_T, b/t, λ_{ft}

(i) The influence of the residual stresses on the column failure load is only mildly relevant, as the value of χ rarely exceeds 4%. The 3-point distribution causes larger failure load drops than its 4-point counterpart – average failure load drops of 3% and 1%, respectively.

(ii) The nine columns experiencing failure load drops higher than 4% (with the 3-point distribution) exhibit b/t=7.5 or 10 and slenderness below 1.0, which is accordance with current knowledge. In addition, the higher failure load drops occur for columns exhibiting L/L_T=0.75 or 1 values.

In view of the fairly similar failure loads obtained with the two residual stress distributions, and since the 3-point one has been used for a long time (recall that is included in Eurocode 3), it was decided to adopt it in the parametric study presented in the next section. As shown above, such a decision is on the safe side, since it leads to lower failure loads.

3. Parametric Study: Failure Load Data

In order to assess whether the DSM design approach developed in the context of slender cold-formed angle columns is also applicable to their hot-rolled steel (E=210 GPa, ν=0.3) counterparts, a parametric study was carried out with the primary aim of acquiring a relatively large column failure load data. This study involves fixed-ended columns with the cross-section dimensions and lengths given in the table of Fig. 2 – the rounded corners are disregarded. In each case, the columns analysed have lengths and yield stresses (elastic-perfectly plastic material model) selected to ensure (i) buckling in flexural-torsional modes (P_{cr} vs. L curve "horizontal plateaus" – see Fig. 1(a)) and (ii) covering a wide critical slenderness range – the yield stress values adopted were 300, 500, 700 MPa. A total of 108 columns were analysed, all containing (i) 3-point residual stress distributions (see Fig. 5(a)) and (ii) initial imperfections combining (ii$_1$) a critical flexural-torsional component, with amplitude equal to 10% of the wall thickness t, and (ii$_2$) a non-critical minor-axis flexural component, with amplitude equal to $L/1000$ – this value is generally adopted for hot-rolled columns (*e.g.*, [15]).

4. DSM Design Considerations

The design of fixed-ended angle columns has attracted the attention of a few researchers in the past. In the context of cold-formed angle columns, Young [4] and Silvestre *et al.* [6] put forward two DSM-based design approaches that are based on the use of global and local strength curves to estimate the failure load (f_{nle} estimates). Even if these design approaches predicted the available failure load data quite efficiently, the fact they are mostly empirical (the length dependence of the column post-buckling and failure behaviour and failure is not explicitly taken into account) led Dinis and Camotim [11] to look for a more rational design approach for fixed-ended angle columns – pin-ended columns were also dealt with, but they fall outside the scope of this work. They showed that the length-dependence of the column post-buckling behaviour can be linked to the characteristics of the critical buckling mode, namely the amount of major-axis flexure, which can be accurately quantified through parameter Δ_f. Accordingly, it was necessary to develop a set of genuine flexural-torsional strength curves, intended to play the role of the local strength curve in the traditional DSM column design against local-global interactive failures. Therefore, the proposed DSM-based approach for the design of fixed-ended angle columns, termed "DSM-F design approach", can be cast in the form

$$f_{nfte}=\begin{cases}\beta\cdot f_{ne} & \text{if} \quad \lambda_{fte}\leq\left(0.5+\sqrt{0.25-b}\right)^{\frac{1}{2a}}\\ \beta\cdot f_{ne}\left(\dfrac{f_{crft}}{f_{ne}}\right)^{a}\left[1-b\left(\dfrac{f_{crft}}{f_{ne}}\right)^{a}\right] & \text{if} \quad \lambda_{fte}>\left(0.5+\sqrt{0.25-b}\right)^{\frac{1}{2a}}\end{cases} \quad \text{with} \quad \lambda_{fte}=\sqrt{\frac{f_{ne}}{f_{crft}}} \quad , \tag{3}$$

$$a=\begin{cases}0.001\,\Delta_f^3-0.032\,\Delta_f^2+0.250\,\Delta_f+0.400 & \text{if} \quad \Delta_f\leq 5.0\\ 0.001\,\Delta_f+0.970 & \text{if} \quad \Delta_f>5.0\end{cases} \quad , \tag{4}$$

$$b=\begin{cases}0.014\,\Delta_f+0.150 & \text{if} \quad \Delta_f\leq 7.0\\ 0.248 & \text{if} \quad \Delta_f>7.0\end{cases} \quad , \tag{5}$$

where f_{ne} is the DSM nominal global strength, Δ_f=[(f_{bt}–f_{crft})/f_{crft}]×100 (f_{bt},f_{crft} obtained from Eqns. (1)).

4.1 Assessment of the DSM-F design approach

Attention is now turned to assessing the performance of the DSM-F design approach (f_{nfte} estimates) in predicting the numerical failure loads gathered in this work and involving hot-rolled steel angle columns with short-to-intermediate lengths. Figs. 6(a)-(b) plot the f_u/f_y and f_u/f_{nfte} (failure-to-predicted) ratios against λ_{fte} for the whole set of numerical column failure loads (average, standard deviation and maximum/minimum values of f_u/f_{nfte} given in the table of Fig. 6(b)) – the curve in Fig. 6(a) is the DSM local-global interactive design curve. The observation of these results prompts the following remarks:

(i) Most f_u/f_y values fall well below the DSM curve, thus confirming that the current DSM local-global strength curve cannot provide efficient (safe accurate) column failure load predictions.

(ii) The DSM-F procedure leads to quite accurate predictions of the numerical ultimate strengths – the f_u/f_{nfte} average and standard deviation read 1.07/0.06. This result is just logical, since the hot-rolled steel angles columns, characterised by stocky legs (b/t<20), share the behavioural features that were at the root of the DSM-F design approach: interaction between flexural-torsional and minor-axis flexural buckling. Nevertheless, further studies are required to confirm these findings.

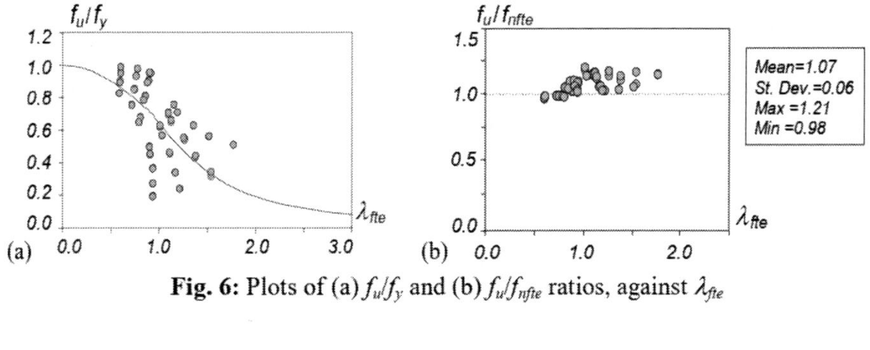

Fig. 6: Plots of (a) f_u/f_y and (b) f_u/f_{nfte} ratios, against λ_{fte}

5. Conclusion

This paper reported the available results of an ongoing investigation on the buckling, post-buckling (elastic and elastic-plastic), strength and design of fixed-ended hot-rolled steel equal-leg angle columns with short-to-intermediate lengths, *i.e.*, buckling in flexural-torsional modes. Initially, numerical results concerning the angle column buckling and post-buckling behaviour were briefly presented and discussed, highlighting (i) the fact that they share key behavioural features with cold-formed steel angle columns (exhibiting much slender legs – b/t>20), namely the strong effect of the length-dependent interaction between major-axis flexural-torsional and minor-axis buckling, and (ii) the reduced impact of the residual stresses on the column failure load. Then, a parametric study involving 108 fixed-ended columns (several cross-section dimensions, lengths and yield stresses) was performed to gather failure load data through ABAQUS SFEA. Finally, a novel DSM design approach recently proposed for cold-formed steel angle columns was briefly reviewed and its validity for hot-rolled steel angle columns assessed. It was shown that this DSM design approach yields accurate and mostly safe failure load predictions for the above 108 columns. Even so, further studies are needed to confirm these findings.

References

[1] Kitipornchai S, Chan SL. "Nonlinear finite-element analysis of angle and tee beam-columns", *Journal of Structural Engineering* (ASCE), 113(4), 721-739, 1987.

[2] Adluri SMR, Madugula MK. "Flexural buckling of steel angles: experimental investigation", *Journal of Structural Engineering* (ASCE), 122(3), 309-317, 1996.

[3] Popovic D, Hancock GJ, Rasmussen KJR. "Axial compression tests of cold-formed angles", *Journal of Structural Engineering* (ASCE), 125(5), 515-523, 1999.

[4] Young B. "Tests and design of fixed-ended cold-formed steel plain angle columns", *Journal of Structural Engineering* (ASCE), 130(12), 1931-1940, 2004.

[5] Rasmussen KJR. "Design of slender angle section beam-columns by the direct strength method", *Journal of Structural Engineering* (ASCE), 132(2), 204-211, 2006.

[6] Silvestre N, Dinis PB, Camotim D. "Developments on the design of cold-formed steel angles", *Journal of Structural Engineering* (ASCE), 139(5), 680-694, 2013.

[7] Shifferaw Y, Schafer BW. "Cold-formed steel lipped and plain angle columns with fixed ends", *Thin-Walled Structures*, 80(July), 142-152, 2014.

[8] Mesacasa Jr. E, Dinis PB, Camotim D, Malite M. "Mode interaction and imperfection-sensitivity in thin-walled equal-leg angle columns", *Thin-Walled Structures*, 81, 138-149, 2014.

[9] Dinis PB, Camotim D, Silvestre N. "On the local and global buckling behaviour of angle, T-section and cruciform thin-walled members", *Thin-Walled Structures*, 48(10-11), 786-797, 2010.

[10] Dinis PB, Camotim D, Silvestre N. "On the mechanics of angle column instability", *Thin-Walled Structures*, 52(March), 80-89, 2012.

[11] Dinis PB, Camotim D. A novel DSM-based approach for the rational design of fixed-ended and pin-ended short-to-intermediate thin-walled angle columns, *Thin-Walled Structures*, 87(February), 158-182, 2015.

[12] Schafer BW. "Review: the direct strength method of cold-formed steel member design", *Journal of Constructional Steel Research*, 64(7-8), 766-778, 2008.

[13] ArcelorMittal. 2015. "ArcelorMittal Sections" http://www.arcelormittal.com/sections/en.

[14] Ban H, Shi G, Shi Y, Wang Y. "Residual stress tests of high-strength steel equal angles", *Journal of Structural Engineering* (ASCE), 138(12), 1446-1454, 2012.

[15] Može P, Sinur F, Rejec K, Cajot L-C, Beg D. "Buckling curves for large angle profiles: residual stress measurements and numerical analysis", *Proceedings of 7th European Conference on Steel and Composite Structures* (EUROSTEEL 2014 – Naples, 10-12/9), R. Landolfo, F. Mazzolani (eds.), ECCS (Brussels), 175-176, 2014. (full paper in USB Key Drive Proceedings).

MEMBER BUCKLING BEHAVIOUR OF HIGH-STRENGTH STEEL IN FIRE

Rebekka Ebel[a], Claudio Scandella[b], Mario Fontana[b] and Markus Knobloch[a]

[a] *Ruhr-Universität Bochum, Institute of Steel, Lightweight and Composite Structures, Germany*
[b] *ETH Zürich, Institute of Structural Engineering, Switzerland*

Abstract: The nonlinear stress-strain response and the stiffness decrease of high-strength steel at elevated temperature affect the member buckling behaviour of structures in fire. Furthermore the steel grade, the slenderness, the profile dimensions, the load case, the load combination and the stability case markedly influence the member resistance. Based on experimental data for the material behaviour of high strength steel in fire this paper focus on the different influence parameters on the interaction curves for beams and beam-columns considering lateral torsional and flexural buckling.

1. Introduction

High-strength steel is increasingly be used for mechanical engineering constructions, bridge structures and traction elements for example. Structural steels with yield strengths above 500 MPa will also be used more often for building structures in the future. The fire design on the material, member and entire structural level is important for the design of buildings. In fires, steel members heat up quickly and the strength and stiffness of steel decreases rapidly. The decreasing stiffness of steel during heating and the non-linear material behaviour strongly influence the stability behaviour of steel members subjected to fire.

A lot of different parameters, like the member length, the cross section type, the structural system, have already a marked influence on the stability behaviour of steel member at ambient temperature and should be considered in simplified analytical methods, (e.g. [1], [2]). Under fire conditions additional factors should be considered, like the thermal creep (strain rate and heating rate effects) and especially the temperature, (e.g. [3], [4]). Up to now commonly used rolled profiles have yield strength up to 550 MPa, welded cross sections may already be made with yield strengths above 550 MPa. In the future also the rolled profiles may be made out of high strength steel. High strength steel indicates a stress-strain response that is often distinct nonlinear already at ambient temperature. A number of recent research topics are related to structures build in high strength steel, (e.g. [5], [6]). However a lack of studies in the fire behaviour of high strength steel members (columns and beam-columns) exists.

The aim of this study is to analyse the fire behaviour of high strength steel members based on material test results. Thus the first part of this paper presents the test results considering the material behaviour of high strength steel in fire. Secondly, the influence of the material behaviour on the member behaviour is analysed.

2. Material behaviour of high strength steel in fire

An experimental study on the material properties of high strength steel at elevated temperatures was performed at the ETH Zurich. Steady-state tests on two steel grades (S690QL and S960QL) were carried out. The test specimens were made out of steel plates with a raw material thickness of 12 mm. Tests were carried out at elevated temperatures of 400°C, 550°C, 700°C and 900°C. In addition, ambient temperature material tests were performed for comparison. The tests were performed with two different strain rates (0.02%/min and 0.10%/min) to analyse the thermal creep effects. Fig. 1 shows the mean stress-strain relationships of three single material tests.

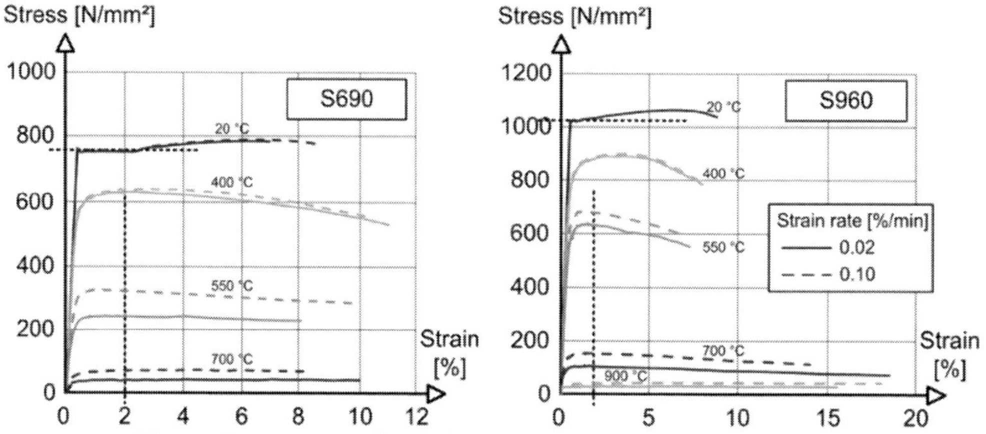

Fig. 1: Mean stress-strain relationships of the material coupon tests on high-strength steel (left: S690; right: S960) at the ETH Zurich

The stress-strain relationships vary for the different temperatures and strain rates. With increasing temperature the influence of the strain rate becomes extensive. For steel grade S960 the material behaviour even for 20°C do not show a yield plateau. The yield strengths of the different steel grades in the case of ambient temperature are bigger than the nominal values (S690: $f_{y,20°C}$ = 760 MPa, S960: $f_{y,20°C}$ = 1026 MPa).

Additionally, the reduction factors for the effective yield strength reached at 2% total strain $k_{y,\theta}$ and the elastic modulus $k_{E,\theta}$ show the influence of the strain rate and the differences for the temperatures and steel grades (Fig. 2). Fig. 2 additionally shows the reduction factors according to EN 1993-1-2 [7].

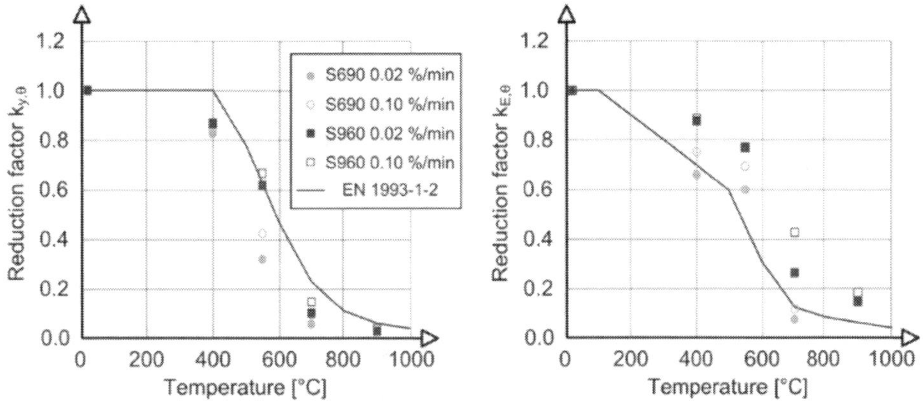

Fig. 2: Reduction factors $k_{y,\theta}$ (left) and $k_{E,\theta}$ (right) for the tested steel grades under different temperature conditions

3. Flexural and lateral torsional buckling of high strength steel members in fire

3.1 Numerical Model for analysing member buckling behaviour

The experimentally determined material behaviour was used as input for the analyses of the stability behaviour of high strength steel members under fire conditions. For this purpose the actual material behaviour for elevated temperatures was used in a finite element model with beam elements. A comprehensive parametric study based on a geometric and material nonlinear analysis considering elevated temperatures with ABAQUS [8] showed the influence of the high-strength steel material behaviour.

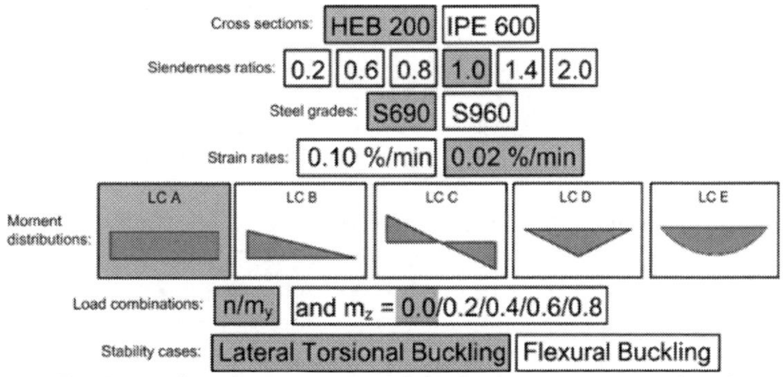

Fig. 3: Analysed parameters and the basic parameter set (grey)

Unless otherwise stated, the grey values were used in the following to analyse the influence of the strain-rate, the steel grade, the temperature, the slenderness and the cross-section type for different load combination and for the stability cases flexural buckling as well as lateral torsional buckling were analysed (Fig. 3).

The finite element model considered the measured non-linear material behaviour as well as geometric imperfections and residual stresses. For the geometric imperfections an eigenmode-affin approach was chosen. Thus, a combination of an initial bow imperfection and a prerotation was used for the case of lateral torsional buckling. The maximum deformation of the initial imperfections was scaled by $L/1000$, [1]. A common residual stress pattern [9] with values related to a yield strength f_y = 235 MPa [10] was considered in the numerical study.

3.2 Influence of the material behaviour for different temperatures and strain rates

Fig. 4 shows the axial compression-uniaxial bending moment interaction curves for the basic configuration and steel grade S690 (left) and S960 (right) considering lateral-torsional buckling. The axial compression and bending moment respectively are normalized by their corresponding plastic resistances at elevated temperatures.

Independent of the steel grade the normalized strengths for pure bending and pure normal force are smaller than 1. That stability effects as well as geometric imperfections and residual stresses avoid reaching the temperature dependent cross-sectional resistance based on the effective yield strength reached at 2% strain.

The graphs for both steel grades show, that for 20 and 400°C nearly the same interaction curves with the same values result. With increasing temperature the difference between the curves for S690 and S960 results bigger. Also the distributions of the stress strain relations differ more for higher than for lower temperatures (Fig. 1). The member of the two different steel grades shows almost similar M-N-interaction behaviour for 20°C and 400°C (Fig. 4).

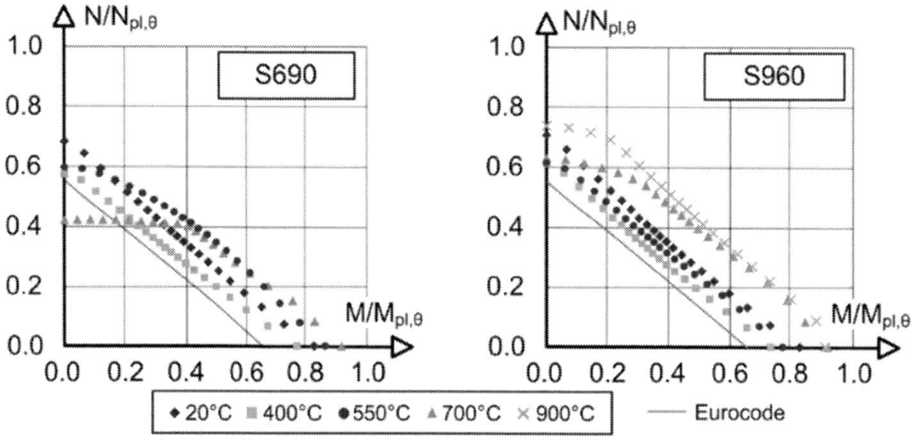

Fig. 4: N-M-Interaction curves for the basic parameter combination considering different temperatures and the two different steel grades

3.3 Influence of the slenderness (beam length)

The stability behaviour is significantly influenced by the beam length and the slenderness respectively. Fig. 5 shows the interaction curves for the basic parameter set and 20°C (to) and 700°C (bottom) for lateral torsional buckling. In addition to the basic profile (HEB 200, left) the results for the profile IPE 600 (right) are shown. For all graphs the flexural buckling slenderness for ambient temperature about the weak axis was fixed and varied between 0.2 and 2.0. The corresponding range of the lateral torsional buckling slenderness is between 0.18 and 1.32 for the profile HEB 200 and for the profile IPE 600 between 0.17 and 1.60. The slender a beam the smaller are the resulting capacities and the stability behaviour and not the cross section behaviour effects the results. On the other side for beams with small slenderness ratios capacities result which are comparable with the cross section capacities. To show this effect three methods to determine the cross section capacity are added – the linear interaction, the capacities according to the plastic theory and the interaction from EN 1993-1-1 [11], all adopted for fire design.

Additionally, the temperature respectively the material behaviour for the different temperature influence the interaction curves. As mentioned before, for elevated temperatures the interaction curves become nonlinear. For the temperature 700°C the influence of the slenderness is bigger than for ambient temperature.

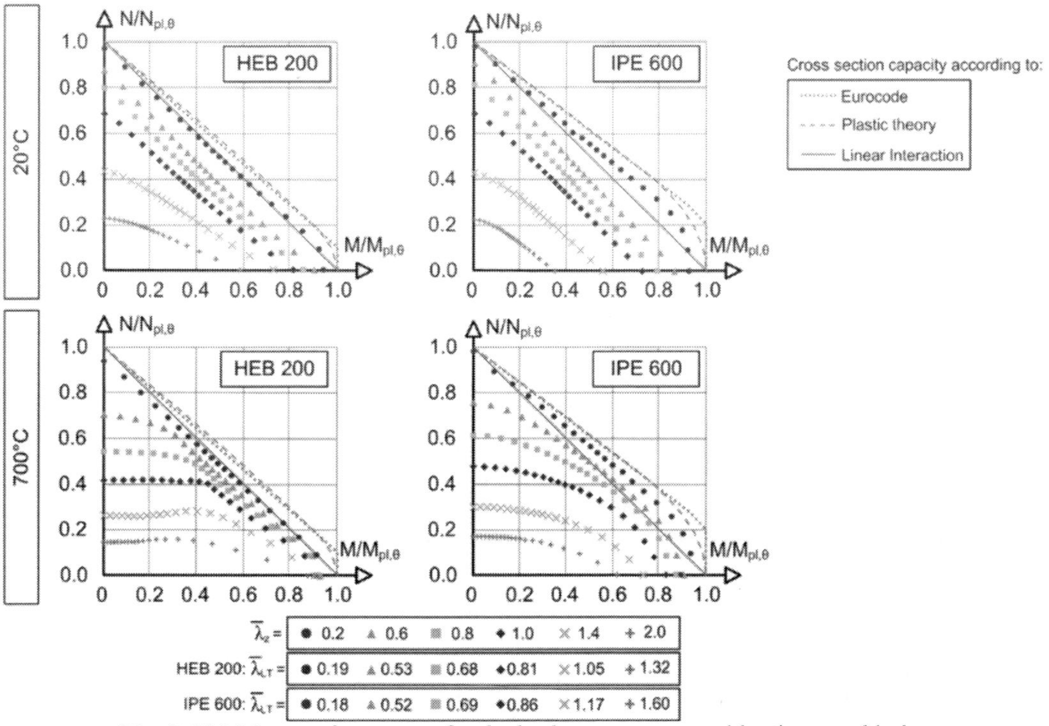

Fig. 5: N-M-Interaction curves for the basic parameter combination considering different slenderness ratios λ_z for flexural and lateral torsional buckling

For load case A, Fig. 6 displays the resulting reduction factors for flexural buckling χ_{fi} (pure compression) and lateral torsional buckling $\chi_{LT,fi}$ (pure bending). The reduction factors are related to the slenderness that considers the temperature. The results show a large scatter in particular for flexural buckling (Fig. 6 left).

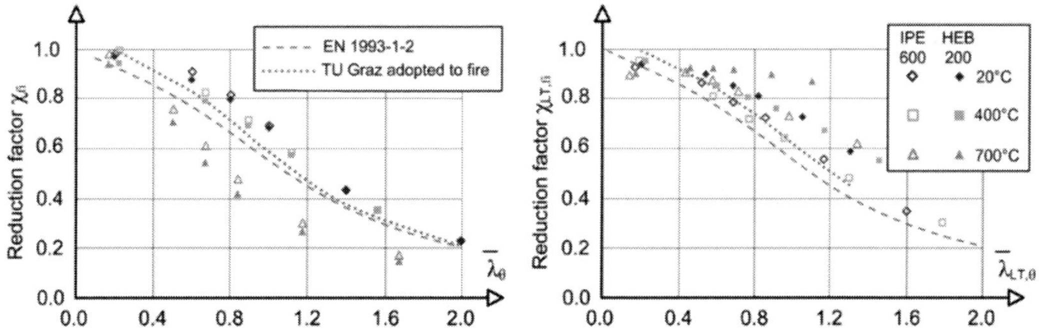

Fig. 6: Reduction factors $\chi_{z,fi,cal}$ (left) and $\chi_{LT,fi,cal}$ (right)

The comparison with the standard approach in EN 1993-1-2 shows that only one buckling curve may have difficulties in taking care of the scatter. Although the related values in Fig. 6

– the reduction factors χ_{fi} and the slenderness – consider the temperature, the resulting reduction factors χ_{cal} show a scatter and can't be determined with one curve.

In addition the buckling curves from the research team from TU Graz are adopted for fire design considering α according to EN 1993-1-2 and added in the diagrams in Fig. 5 [2]. The curve results in a higher reduction factor range than the values for EN 1993-1-2.

3.4 Influence of the load case and the load combination

The capacities are also strongly affected by an existing moment about the weak axis. Fig. 7 shows the interaction curves of the basic parameter set and considering different ratios of m_z. With increasing moment about the weak axis the capacities decrease. For all temperatures it is nearly the same influence.

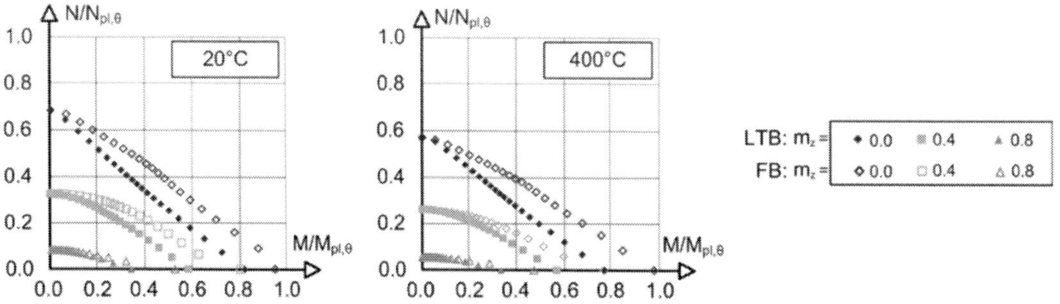

Fig. 7: N-M-Interaction curves for the basic parameter combination considering different ratios for M_z and different temperatures

The stability case (flexural buckling or lateral torsional buckling) affects the distribution of the interaction curves. Increasing bending moments less influence the remaining axial force capacity for flexural buckling in comparison with lateral torsional buckling.

Fig. 3 shows the analysed cases for the moment distributions M_y. All analysed beams are single span beams with end moments (load case A (LC A) to LC C), concentrated load (LC D) or a distributed load (LC E). Distinctive bending moment distributions lead to large plastic zones in the member. Fig. 8 shows the resulting length of this ideal zone for LC D and E considering a linear elastic-perfectly plastic stress strain relation (Fig. 8 right) and no imperfections. To get this plastic zones the strain values at the strain plane has to result infinite.

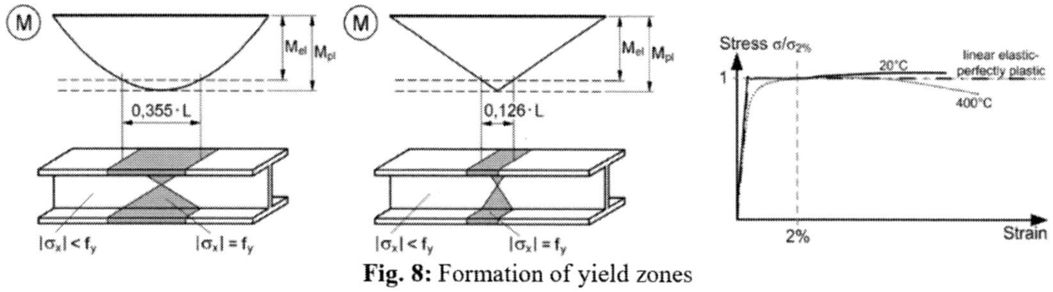

Fig. 8: Formation of yield zones

The formation of the plastic zones influences the stability behaviour and this leads to differences in the capacities, [12]. The different the assumptions for the material behaviour and the imperfections from the ideal approach, the less are the plastic zones as displayed in Fig. 8. The related presentation of the material behaviour shows that for higher temperatures the stress strain response results distinctive nonlinear.

The resulting *N-M* interaction curves for the different load cases are displayed in Fig. 9. For each temperature the capacities are sorted in the same order – top down: C-B-D-E-A. This order coincide with the amplitude of the moment distribution area. For pure bending the curves result in the same point, the calculated reduction factor for flexural buckling. For the load case with small plastic zone length (LC B and C) the resulting interaction curves are more curved than for the other cases. The analysis for the stability case flexural buckling result in comparable relations.

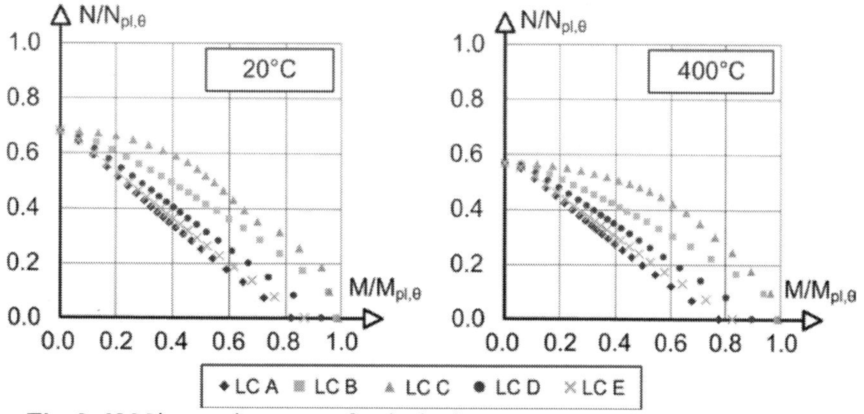

Fig. 9: *N-M* interaction curves for the basic parameter combination considering different load cases and different temperatures

4. Conclusion

The temperature as well as the appropriate nonlinear material behaviour strongly affect the load carrying behaviour of high strength steel members at elevated temperatures and the distributions of the $N\text{-}M_y$-interaction curves. The results of this study can be used to review design methods and to develop suitable analytical models.

References

[1] Ebel R. *Systemabhängiges Tragverhalten und Tragfähigkeiten stabilitätsgefährdeter Stahlträger unter einachsiger Biegebeanspruchung.* (System-dependent bearing characteristics and load bearing capacities of steel beams under uniaxial bending) Dissertation, Bochum 2014, http://www-brs.ub.ruhr-uni-bochum.de/netahtml/HSS/Diss/EbelRebekka/.

[2] Greiner R, Taras A. "New design curves for LT and TF buckling with consistent derivation and code-conform formulation", *Steel Construction*, 3, 176-186, 2010.

[3] Knobloch M, Somaini D, Pauli J, Fontana M. "Stability of steel columns subjected to fire", *Proceedings of Stability and Ductility of Steel Structures*, Rio de Janeiro, 465-472, 2010.

[4] Knobloch M, Fontana M. "Stability of steel structures in fire", *Stahlbau*, 83, 257-264, 2014.

[5] Rasmussen KJR, Hancock GJ. "Tests of high strength steel columns", *Journal of Constructional Steel Research*, 34, 27-52, 1995.

[6] Beg D, Hladnik L. "Slenderness limit of Class 3 I cross-sections made of high strength steel", *Journal of Constructional Steel Research*, 38, 201-217, 1996.

[7] EN 1993-1-2 (2006), Eurocode 3 – Design of steel structures, Part 1-1, General rules – Structural fire design, CEN Brussels, 2006.

[8] FEM Program *Abaqus* Version 6.14, Simulia, 2014.

[9] ECCS-CECM-EKS. Publication No. 33: *Ultimate Limit State Calculation of Sway Frames with Rigid Joints*. Brüssel 1984.

[10] Wolf C. *Tragfähigkeit von Stäben im Baustahl*. Dissertation Shaker Verlag 2006, Bochum 2006.

[11] EN 1993-1-1 (2005), Eurocode 3 – Design of steel structures, Part 1-1, General rules and rules for buildings, CEN Brussels, 2005,

[12] Ebel R, Knobloch M. "Lateral torsional buckling behavior of different static systems", *Proceedings of the Eighth International Conference on Advances in Steel Structures*, Lisbon, Portugal, 2015.

DUCTILITY REQUIREMENTS FOR STRUCTURAL DETAILS WITH STRESS CONCENTRATION AND DIFFUSE NECKING

Petr Hradil[a], Ludovic Fülöp[a], Asko Talja[a] and Petri Ongelin[b]

[a] *VTT Technical Research Centre of Finland Ltd., Finland*
[b] *Ruukki Construction Oy, Finland*

Abstract: Structural steels need to demonstrate certain ductility in order to be used in buildings design. Fulfilling these criteria of design codes may be difficult especially for new high-strength steels where ductile fracture of tensile coupons appears at relatively low deformation. The goal of our study is to develop a calculation method of alternative ductility limits for structural details in a particular design situation. Such limits might be less demanding. Application of the method to a large variety of material models resulted in a simple criterion for the standard coupon test, the minimum distance between the uniform elongation and the coupon failure. This distance is called "necking capacity" of a tensile coupon in our paper.

1. Introduction

Material ductility is one of the most important requirements in structural steel design. It is implicitly used to alleviate stress concentrations in structural details and explicitly required in the plastic design and seismic design situations. Eurocode 3 [1] prescribes the ductility requirements in the form of minimum ultimate-to-yield strength ratio f_u/f_y, uniform elongation ε_u, and elongation at fracture *A5* (see Table 1). When the material yield strength is increased in production by alloying or heat treatment, its ductility is generally decreasing. Therefore more relaxed ductility requirements are applied to high strength steels (S500 to S700) in exchange for certain restrictions in plastic design and semi-rigid connections [2]. This trend continues with development of higher grades such as S960, which is often beyond the current limits.

The purpose of our study is to develop a more rational method to determine the ductility requirements for the structural detail in particular design situations. Such requirements may be easier to satisfy than the general ones. Additionally design situations where high strength steel can be used without limitations are identified. The method relies on standardized material tests according to EN 10002-1 [3] and EN ISO 6892-1 [4] and properly calibrated finite element models (FEM). Numerical calculation is often able to generate acceptable solution for a given design case, but it is also powerful instrument to cover wide range of structural details and material parameters for the further generalization to design recommendations. An example of this approach is presented in the second part of this paper, where the developed method is applied in a large parametric study to formulate generalized ductility requirements for a structural detail.

2. Calculation of ductility requirements

The method presented in this section predicts the minimum required elongation at fracture *(A5)* of tensile coupon, for a given combination of f_u/f_y and ε_u to fulfil the selected design limit in a particular detail. It relies on a standard material coupon test and two numerical models, one of the selected structural detail and one of the coupon itself. The calculation is based on the assumption that strains larger than the uniform elongation of tensile coupon ε_u can be accepted in localized areas of statically loaded structures. However, the load should not cause ductile failure in materials, and therefore the plastic strains and stress triaxiality should remain within acceptable limits. Unfortunately, it is impossible to describe the relation of stress triaxiality and plastic strain at fracture with data from commonly used coupon test, and therefore the present method generates conservative results. A more accurate solution requires a rather complicated testing programme.

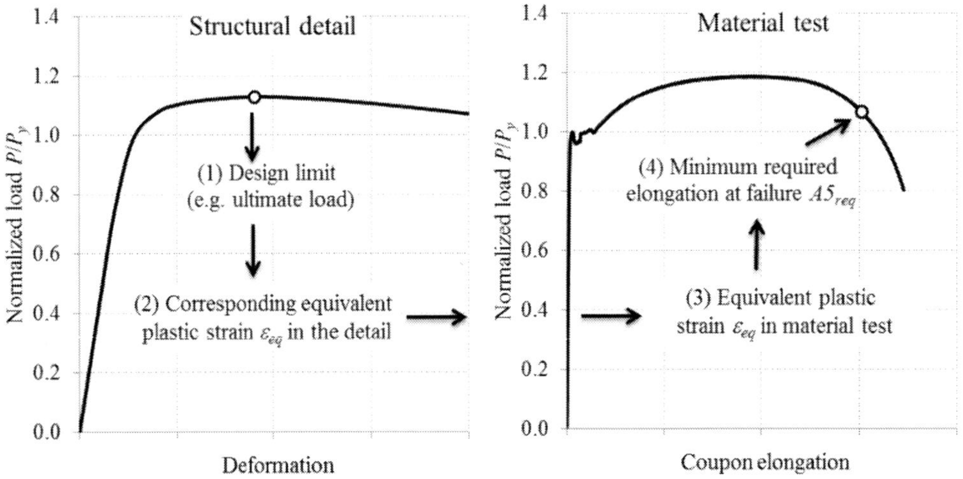

Fig. 1: Prediction of minimum required elongation at failure

The proposed method is based on a single parameter, the maximum equivalent plastic strain ε_{eq} in the critical cross-section of the structural detail. If a few conditions are satisfied, it is safe to assume as required coupon elongation $A5_{req}$ the point when the same equivalent plastic strain ε_{eq} is reached in the coupon during the coupon test (Fig. 1). The required conditions are:

(a) Stress triaxiality at the failure of the structural detail is smaller than the stress triaxiality of tested coupon. This condition is true on details with notches and holes in tension and plain rectangular coupons (see Fig. 2) except for the initial part of loading [5]. Similar findings were reported by Dunaud [6].

(b) The relation between failure strain and stress triaxiality (the damage curve) of a material is monotonic non-increasing function as in Fig. 2. It has been observed that the fracture strains of non-notched specimens are always higher than those of notched samples with lower triaxiality [7], [8].

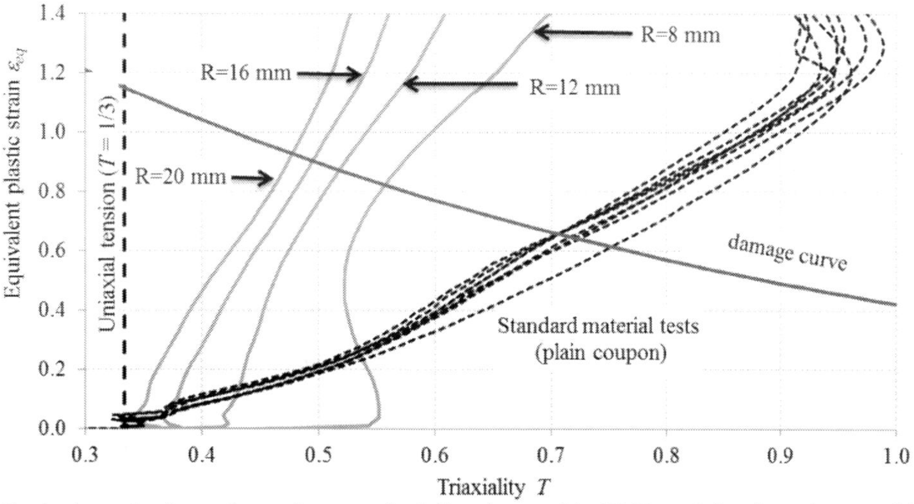

Fig. 2: Equivalent plastic strains and stress triaxiality observed in S960 models of specimens with hole in tension (R=8 to 20 mm) and plain coupons, and example damage curve of the material

The equivalent plastic strains were often found largest in the middle of the coupon cross-section and therefore hard to measure directly. For that reason, finite element models could be efficiently used. However, the numerical approximation brings additional uncertainties in the form of finite element mesh density and initial imperfections. We have observed that smaller elements at the stress concentration areas result in larger strains at the same load levels, and therefore two more conditions have to be satisfied when FEM is employed:

(c) The finite element mesh of the structural detail has at least the same density as the mesh of the simulated coupon test. It is recommended that the same software is used for both numerical models with the same element types and calculation settings.

(d) The surface imperfections of the real coupon are smaller than the imperfections implemented in FEM model of the coupon. This ensures higher stress concentration in the model, and therefore more conservative results.

3. Parametric study

3.1 Material models

Our goal was to use in the parametric study a stress-strain relationship which creates the largest possible plastic strains at the given elongation. Therefore we used, conservatively, ideally plastic material in true stress-strain terms beyond the ultimate load. This assumption excludes metal softening for instance due to micro-voids or dynamic recrystallization [9], but is still valid for the commonly used steel grades and their structural applications. The simplified true stress- true plastic strain models are tri-linear with strain hardening up until the true uniform strain.

The studied range was 1.0 to 1.5 for the f_u/f_y ratio, and 2% to 50% for uniform strain ε_u. Four series of such parametrical studies were carried out with the yield strength 250, 500, 750 and 1000 MPa. We observed that the highest strains at fracture elongation *A5* were always achieved in the series with f_y = 1000 MPa. As high strains correspond to the most conservative case, results presented in this paper relate to this group of materials only.

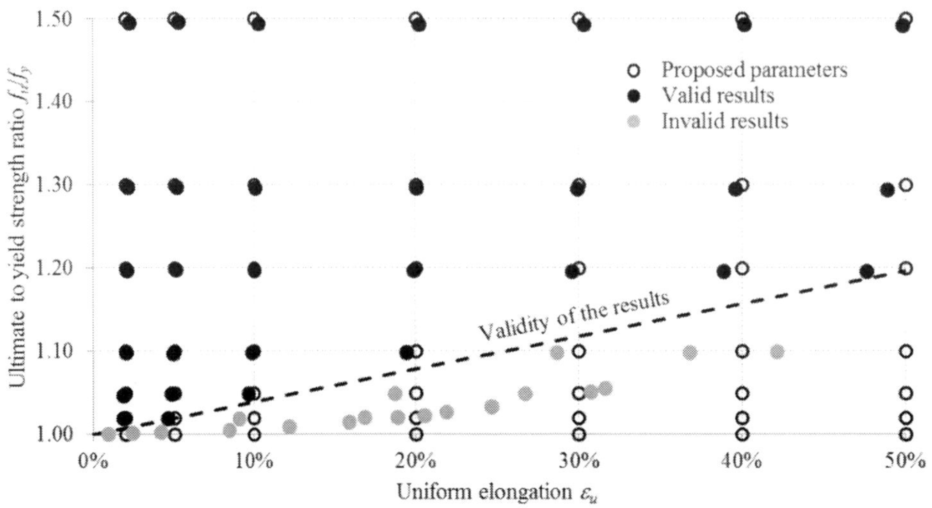

Fig. 3: Selection of material models based on proposed and calculated material parameters

The validity of the material models was verified by numerical simulation of the tensile tests with true stress-strain relation based on the selected range of material parameters f_u/f_y and ε_u. It was observed that in materials with small f_u/f_y ratios and high ε_u, the diffuse necking was initiated earlier than expected. Such materials were excluded from our calculations and the material models used in the parametric study are those, where the uniform elongation ε_u is smaller than $2.55(f_u/f_y - 1)$ as demonstrated on Fig. 3.

The numerical models were calibrated against real material tests of coupons and structural details as described in [5] with several additional steel grades. All of the experiments used in the present paper are summarized in Table 1.

Table 1: Measured material properties and ductility requirements of Eurocode 3 (EC3)

Material	f_y (MPa)	f_u (MPa)	f_u/f_y	ε_u	*A5*	coupons tested	structural details tested	min f_u/f_y	EC3 min ε_u	min *A5*
S275	300	418	1.39	18%	35%	1	-	1.15	2.1%	15%
S355(1)	393	550	1.40	13%	28%	1	-	1.15	2.8%	15%
S355(2)	396	473	1.19	18%	31%	2	5	1.15	2.8%	15%
S700	708	785	1.12	10%	20%	1	5	1.10	5.0%	10%
S960[1)]	1062	1167	1.10	3%	9%	16	5	(1.10)	(7.6%)	(10%)

[1)] S960 is not in scope of current Eurocode 3

3.2 Structural details

The basic model selected for our study was tensile plate with 8x80 mm cross-section and central hole (CHT). For this detail we were able to validate finite element models against the real experiments with the hole diameter *d* from 8 mm to 40 mm [5]. Since the most critical strains were always related to the smallest diameter of the hole, we selected 8 mm for the entire parametric study (see Fig. 4).

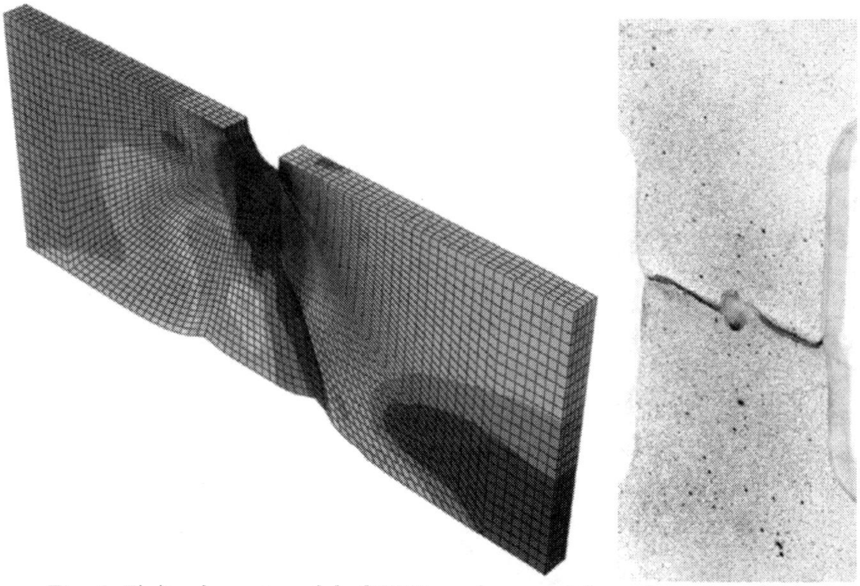

Fig. 4: Finite element model of CHT specimen and the real specimen [11]
(von Mises stress distribution, S960 with 8 mm diameter of the hole)

Simple modification of our FEM models produced another common detail with stress concentration. Central hole was changed into circular side notch (SNT) of the same radius by applying symmetry boundary conditions to the opposite face than in CHT simulations. The results were added to the previously calculated CHT simulations.

3.3 Design criteria

The ultimate limit state resistance of cross-section reduced by holes or notches is related to the ultimate strength of material f_u and the area of net-section A_{net} in Eurocode 3. Therefore our selected design criterion was that the cross-section has to resist the load equal to f_uA_{net}. On the other hand, if the elongation at the ultimate load is not considered, it would be possible that some of the materials would not fulfil minimum deformation requirements. For that reason we have also studied elongation of details in tension. Since the Eurocode 3 does not explicitly state minimum allowable deformation, we have selected several levels of details' elongations on 100 mm gauge: 2 mm, 3 mm, 4 mm and 5 mm. The results from the extreme cases (2 mm and 5 mm) are presented herewith.

4. Results and discussion

The proposed requirement for minimum elongation at failure *A5* of the coupon was obtained by the method described in Fig. 1 for 196 materials and 2 details. The results presented in this section are the envelope of the highest values interpolated between the calculation points. They are compared to real measured *A5* values in four material tests with different steel grades. In most of the cases, the central hole in tension (CHT) was more critical than the side notch in tension (SNT). The results are further limited by the maximum yield strength 1000 MPa, the minimum diameter of the hole or notch $d \geq 8$ mm, and $\varepsilon_u \leq 2.55(f_u/f_y - 1)$ as described in the previous sections.

4.1 Ultimate load in the net-section

Some of the studied steel grades did not fulfil the material requirements given in Eurocode 3 (see Table 1). Nevertheless, all of the selected materials were able to reach the ultimate load $f_u A_{net}$ in both geometrical configurations. This is also confirmed by experiments with CHT [5].Therefore we were always able to find such *A5* that guarantees the corresponding level of equivalent plastic strain before ductile failure of the material. We would like to note that the design load according to Eurocode 3 is always more conservative because it is further reduced by several factors.

It can be observed that the results in Fig. 5 that all five tested materials have adequate elongation *A5* to reach the desired ultimate load. E.g. S275 has available *A5* elongation of 35% far exceeding the demand of ~25%. The *A5* limits form nearly vertical lines, and therefore are almost independent on the f_u/f_y ratio. Moreover, the values of minimum elongation at failure *A5* differ from the uniform strain ε_u by 0.3 to 5.7 percent. This indicates that the difference between *A5* and ε_u could be a suitable parameter to describe the new proposed ductility requirements. We call it the necking capacity of the material.

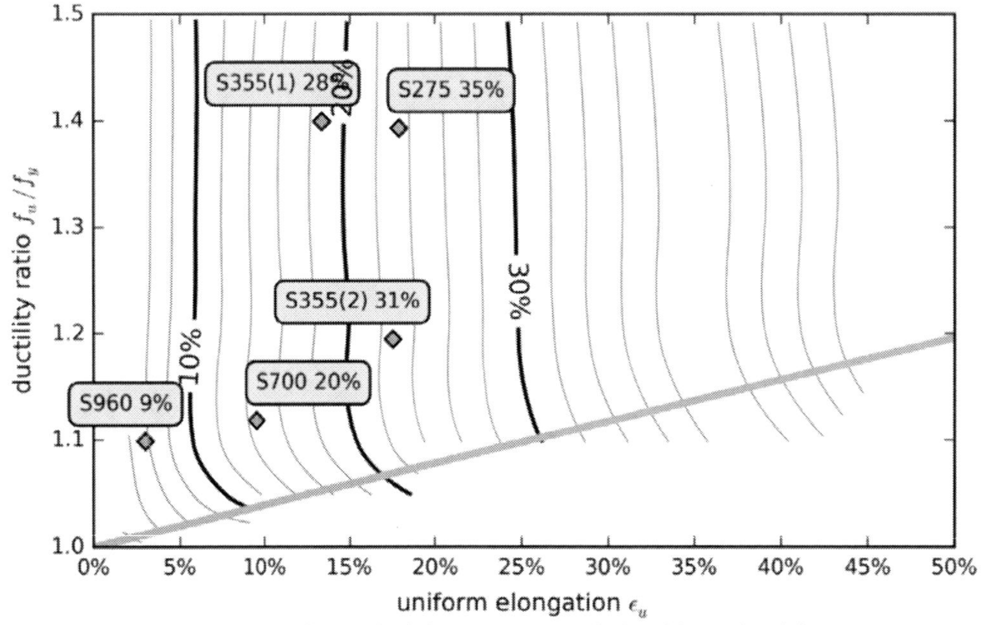

Fig. 5: Contour lines of minimum *A5* to reach the ultimate load $f_u A_{net}$ with measured values of *A5* of five real materials in text boxes

4.2 Elongation 2 to 5 mm measured on 100 mm gauge length

The same evaluation was performed using various deformation levels as the design criterion. Fig. 6 shows the results for the extreme case with 5 mm limit elongation measured on 100 mm gauge length. It can be observed that the measured coupon of S960 failed the 5 mm criterion on Fig. 6. All other materials fulfil the criteria from 2 mm to 5 mm elongation.

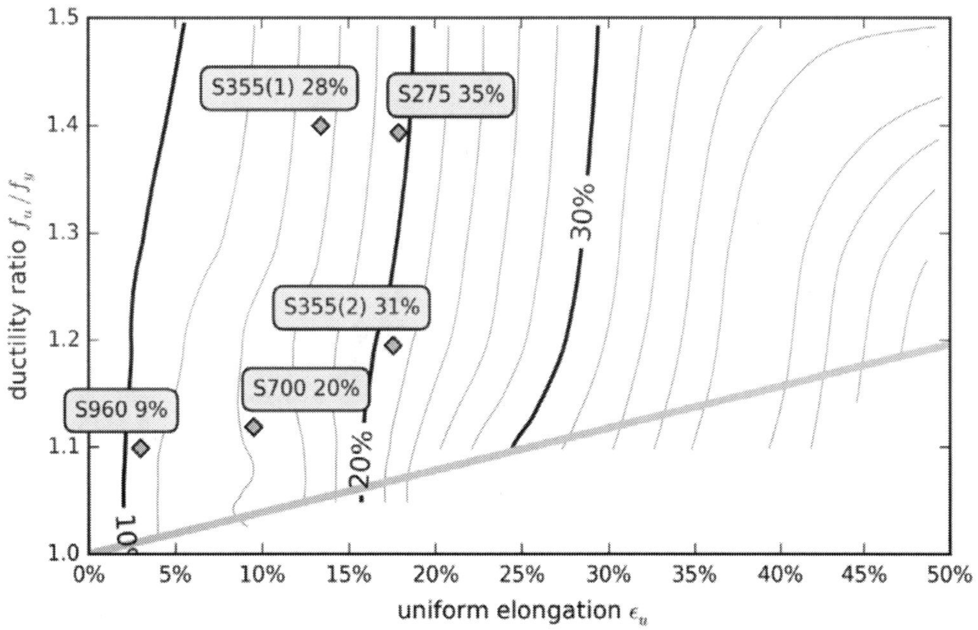

Fig. 6: Contour lines of minimum *A5* to reach the elongation 5 mm

Table 2 shows the maximum difference between the required elongation at failure *A5* and uniform elongation ε_u that was observed for each of the design limits. Should those results be used in real design situation, the labour of finding the real deformation limit of the structural detail lays on the designer. Our intention is to investigate this problem further and propose a single requirement that would take implicitly the allowed deformation into account.

Table 2: Ductility requirements on CHT and SNT details

Allowed elongation on 100 mm gauge	Minimum necking capacity $A5$-ε_u required in coupon tests
2 mm	6%
3 mm	7%
4 mm	8%
5 mm	8%

5. Conclusions

The main conclusions are:

1. The ductility requirements prescribed in current Eurocode 3 in the form of minimum ultimate-to-yield strength ratio f_u/f_y, uniform elongation ε_u and elongation at fracture seem to be questionable.
2. Based on our study, the ductility requirement to reach specific design load or deformation can also be expressed as a "necking capacity", the difference between the elongation at coupon failure *A5* and the uniform strain ε_u.
3. The required necking capacity to reach ultimate load f_uA_{net} is 6% in the present study and the recommended levels of the necking capacity related to certain elongation are summarized in Table 2. Moreover, the required necking capacity seems to be almost independent on the f_u/f_y ratio. It should be noted that these values are only valid for se-

lected CHT and SNT details (8x80 mm plate in tension with central hole or side notches diameter 8 mm).
4. The presented method may be used by designers to validate case-by-case the ability of particular material to reach specific design limits in a given detail.
5. The results can also be further generalized by calculating more structural details in different design situations. Then the necking capacity requirement may work as alternative limit to the existing ductility criteria.

Acknowledgments

The research leading to these results has received funding from Finnish Metals and Engineering Competence Cluster's (FIMECC) program BSA - Breakthrough steels and applications (2014-2018) and its project Design beyond present codes – enabling efficient utilisation of new materials. We would like to thank Ruukki Construction Oy and SSAB Oy for the experimental test results of coupons and CHT specimens.

References

[1] EN 1993-1-1 *Eurocode 3: Design of steel structures - Part 1-1: General rules and rules for buildings*, European Committee for Standardisation, 2005.

[2] EN 1993-1-12 *Eurocode 3: Design of steel structures - Part 1-12: Supplementary rules for high-strength steels*, European Committee for Standardisation, 2007.

[3] EN 10002–1 *Metallic materials - Tensile testing - Part 1: Methods of test at ambient temperature*, European Committee for Standardisation, 2001.

[4] EN ISO 6892–1 *Metallic materials - Tensile testing - Part 1: Method of test at room temperature*, European Committee for Standardisation, 2009.

[5] Hradil P, Talja A. "True stress-strain relationship for finite element simulations of structural details", *Proceedings of the 13th Nordic Steel Construction Conference (NSCC-2015)*, Tampere, Finland, 229-230, 2015.

[6] Dunaud M. *Ductile fracture at intermediate stress triaxialities: Experimental investigations and micro-mechanical modelling*, Massachusetts Institute of Technology, 2013.

[7] Hancock JW, Mackenzie AC. "On the mechanisms of ductile failure in high-strength steels subjected to multi-axial stress states", *Journal of the Mechanics and Physics of Solids* 31(1), 1-24, 1976.

[8] Holland D, Halim A, Dahl W. "Influence of stress triaxiality upon ductile crack propagation", *Steel Research,* 61(10), 504-506, 1990.

[9] Guo-Zheng Q. "Characterization for Dynamic Recrystallization Kinetics Based on Stress-Strain Curves" *Recent Developments in the Study of Recrystallization* (Ed.: P. Wilson), InTech, 2013.

[10] EN 1993-1-3 *Eurocode 3: Design of steel structures - Part 1-3: General rules - Supplementary rules for cold-formed members and sheeting*, European Committee for Standardisation 2006.

[11] Valkonen I. "20th European Conference on Fracture (ECF20): Ultimate limit loads in welded joints and in net sections of high strength steels with yield stress 960 MPa" *Procedia Materials Science*, 3, 720-725, 2014.

AUTHOR INDEX

International Colloquium on Stability and Ductility of Steel Structures (SDSS'2016)

Timisoara, Romania
30 May – 1 June 2016

Volume 2 of 2

Editors:

Dan Dubina
Viorel Ungureanu

ISBN: 978-1-5108-5770-4

Printed from e-media with permission by:

Curran Associates, Inc.
57 Morehouse Lane
Red Hook, NY 12571

Printed by Curran Associates, Inc. (2018)

For permission requests, please contact European Convention for Constructional Steelwork at the address below.

European Convention for Constructional Steelwork
32, av. des Ombrages, bte 20
1200 Brussels, Belgium

Phone: +32-2/762 04 29
Fax: +32-2/762 09 35

eccs@steelconstruct.com

Additional copies of this publication are available from:

Curran Associates, Inc.
57 Morehouse Lane
Red Hook, NY 12571 USA
Phone: 845-758-0400
Fax: 845-758-2633
Email: curran@proceedings.com
Web: www.proceedings.com

Proceedings of the
International Colloquium on
Stability and Ductility of Steel Structures

SDSS'2016

Edited by
Dan DUBINA and Viorel UNGUREANU

WILEY

Ernst & Sohn
A Wiley Branc

SDSS'2016

The International Colloquium on Stability and Ductility of Steel Structures

30 May – 01 June 2016, Timisoara, Romania

Proceeding of the conference edited by: Dan Dubina and Viorel Ungureanu

TABLE OF CONTENTS

Members' behavior: tension, compression, beams, beam-columns

Plate, shell and space structures

Robustness

Steel-concrete composite members and structures

Tubular constructions

FOREWORD

It was in 1971 when Professor H. Beer from University of Graz, who together with his colleague, Dr. G. Schulz, had proposed the first version of European (ECCS) Buckling Curves equations (Construction Metallique, No. 3, 1970), suggested to organize an International Colloquium, as a forum of discussion and comparison between such equations and those of Eastern Europe, the United States and Japan. The event, entitled "International Colloquium on Column Strength", was organized in November 1972, in Paris, with the support of IABSE, but, sadly, after the sudden death of Professor Beer. This is the event which may be considered the starting point of a long series of International Colloquia dedicated to the Stability of Steel Structures. However, the *real* organization of the series of International Colloquia, thought of as to *globally cover the instability problems of steel structures,* emerged two years later, in London, as a common initiative of Dr. D. Sfiintesco, on behalf of ECCS, and Professor L. Beedle, on behalf of SSRC. The basic idea was to organize several series of *travelling* colloquia, with *sessions* located in different areas of the world. The first series was organized in 1976-1977, with sessions in Tokyo, Liege and Washington, and, further on, the world wide circuit continued.

In October 1982, the Department of Steel Structures and Structural Mechanics of the Politehnica University of Timisoara organized one of the East-European Sessions of the 3rd International Colloquium on Stability of Steel Structures; it was followed by other two sessions, organised in 1983 in Toronto and Paris.

In July 1997, at the Asian Session of the 6th Colloquium, the name of the events was changed into *International Colloquium on Stability and Ductility of Steel Structures - SDSS.* Next, the European Session was organized in September 1999, again in Timisoara. It was followed by those in Budapest (2002), Lisbon (2006) and Rio de Janeiro (2010).

The 2016 edition of SDSS is jointly organized by the *Politehnica University of Timisoara, Department of Steel Structures and Structural Mechanics* in co-operation with the *Romanian Academy - Timisoara Branch,* with the support of the *European Convention of Constructional Steelwork (ECCS),* through the *Structural Stability Technical Committee (TC8),* and the *Structural Stability Research Council (SSRC).*

270 authors from 26 countries on 5 continents are contributing with 115 scientific papers and 4 Keynote Lectures, covering 9 topics, i.e.

1. *Advanced structural design;*
2. *Connections;*
3. *Lightweight steel constructions;*
4. *Members' behaviour: tension, compression, beams, beam-columns;*
5. *Plate, shell and space structures;*

6. *Robustness;*
7. *Seismic-resistant structures;*
8. *Steel-concrete composite members and structures;*
9. *Tubular constructions.*

Since the revision of structural Eurocodes has been already started in Europe, and both stability and ductility of steel structures issues are on the desk of the drafters, this event enables the exchange of knowledge and experience in the field and the debate around the best solutions offered, and is therefore a useful opportunity for all the participants.

On behalf of the SDSS'2016 Organizing Committee, we express our gratitude to Keynote Speakers and authors, to the members of the International Scientific Committee and to all the participants in this international scientific event.

We also appreciate the support of the *Technical Committee 8* within the *European Convention of Constructional Steelwork (ECCS)* and of the *Structural Stability Research Council (SSRC)*.

Thanks are also due to *Joana Albuquerque*, from *CMM – Associação Portuguesa de Construção Metálica e Mista* and also to *Ernst and Sohn and Wiley Company*, for their support and the printing quality of the Proceedings.

Acad. Dan Dubina
Chairman of SDSS'2016

Prof. Viorel Ungureanu
Scientific Secretary of SDSS'2016

Timisoara, May 2016

ORGANIZATION

CHAIRMAN

D. Dubina, *PU Timisoara*

SCIENTIFIC SECRETARY

V. Ungureanu, *PU Timisoara*

SCIENTIFIC COMMITTEE

G.A. Altay	*Turkey*	R. Leon	*USA*
I. Balaz	*Slovakia*	J.R. Liew	*Singapore*
C. Baniatopoulos	*Greece*	J. Loughlan	*UK*
A. Bureau	*France*	J. Machacek	*Czech Republic*
E.M. Batista	*Brazil*	M. Mahendran	*Australia*
R. Beale	*UK*	F. Mazzolani	*Italy*
R. Bjorhovde	*USA*	M. Nakashima	*Japan*
M.A. Bradford	*Australia*	D. Nethercot	*UK*
B. Brune	*Germany*	J. Packer	*Canada*
L. Calado	*Portugal*	J. Paik	*South Korea*
D. Camotim	*Portugal*	A. Plumier	*Belgium*
S.L. Chan	*Hong Kong, China*	N. Rangelov	*Bulgaria*
R. Casciaro	*Italy*	K.J.R. Rasmussen	*Australia*
K.F. Chung	*Hong Kong, China*	J. Ricles	*USA*
C. Chiorean	*Romania*	J.M. Rotter	*UK*
H. Degée	*Belgium*	F. Roure	*Spain*
F. Dinu	*Romania*	C.A. Rogers	*Canada*
L. Dunai	*Hungary*	R. Sause	*USA*
S. Easterling	*USA*	B.W. Schafer	*USA*
A. Elghazouli	*UK*	L.S. Silva	*Portugal*
M. Fontana	*Switzerland*	N. Silvestre	*Portugal*
D. Frangopol	*USA*	H. Snijder	*Netherlands*

LOCAL ORGANIZING COMMITTEE

FLEXURAL BUCKLING BEHAVIOUR CONSIDERING DIFFERENT RESIDUAL STRESS APPROACHES

Manfred Käsmaier[a], Rebekka Ebel[a], Matthias Kraus[b] and Markus Knobloch[a]

[a] *Ruhr-Universität Bochum, Institute of Steel, Lightweight and Composite Structures, Germany*
[b] *Bauhaus-Universität Weimar, Faculty of Civil Engineering, Chair of Steel and Hybrid Structures, Germany*

Abstract: Stability failures, such as flexural buckling, are often decisive in the design of steel beam and column members. To provide the background and knowledge for developing and improving corresponding design methods, the influence of different residual stress approaches on the structural behaviour of hinged beams and columns in terms of flexural buckling is being analysed in this paper. Numerical calculations (finite element method) based on the plastic zone theory show, that residual stress distributions have a strong influence on the plastification process and the bearing behaviour. In addition, depending on the effective lengths and residual stresses, different failure criteria emerge.

1. Introduction

Current design models may partially not lead to accurate results for beams, columns and beam-columns compared to results of tests and numerical studies [1], [2]. This paper analyses the fundamental load-carrying behaviour with physical and geometrical nonlinear calculations using FE-methods considering initial deformations and residual stresses. The residual stresses due to rolling and welding processes markedly affect the bearing behaviour. This paper shows the influence of residual stress distributions on the column strength and flexural buckling behaviour about the strong and weak axis due to axial compression as well as combined axial compression and bending.

2. Numerical study

Fig. 1 shows the main assumptions for the numerical analyses carried out with the finite element program ANSYS [3]. The cross section of the investigated system had a height-to-with-ratio of $h/b = 1$.

Three different approaches for residual stress distributions were considered:

Case 1) no residual stresses;

Case 2) residual stresses corresponding to rolled I-profiles (rolling residual stresses);

Case 3) residual stresses corresponding to welded I-profiles (welding residual stresses).

The residual stress distributions due to rolling and welding processes are set as idealized patterns and ordinates according to [4]. Further details of the numerical study are given in [1].

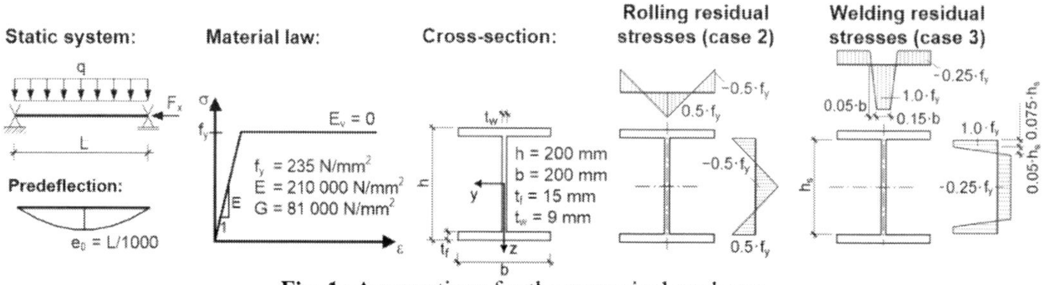

Fig. 1: Assumptions for the numerical analyses

3. Influence of the residual stress distribution on flexural buckling due to a compressive normal force N

3.1 Flexural buckling behaviour about the strong axis

This section deals with flexural buckling about the strong axis. The influence of different approaches for residual stress distributions on the load-carrying behaviour is analysed. Particular attention is drawn to the plastification development in a structural system as well as the characterization of the failure criteria.

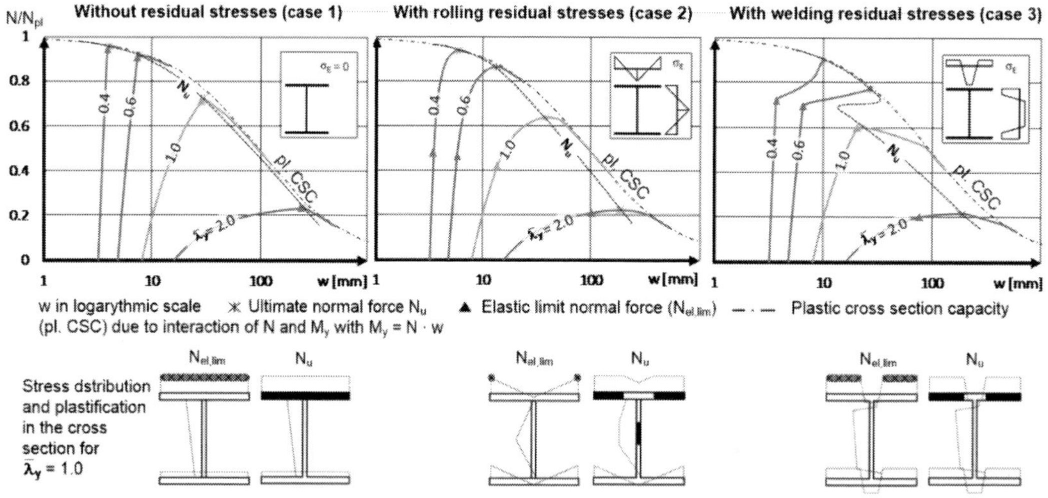

Fig. 2: Load-midspan deflection curves (top) and plastification in the cross-section (bottom) for flexural buckling about strong axis and different residual stress approaches

Principally, load-midspan deflection curves increase uniformly for low axial loads. With the beginning and growing of plastification, the graphs show a nonlinear inclination and after reaching a certain load level a clear drop, in case of a stability problem. Fig. 2 exemplarily shows the load-midspan deflection curves regarding flexural buckling about the strong axis of structural systems with a nondimensional slenderness of $\bar{\lambda}_y$ = 0.4, 0.6, 1.0 and 2.0. Different symbols designate significant stress situations, such as the elastic and the ultimate limits regarding compressive normal forces. For these states, the stress distribution and cross section plastification are shown as well. The plastification due to compressive stresses begins at the inner radius of the curvature regarding the member deflection, for the given systems this is the upper flange of the cross section.

The development of the plastification and hence of the load-midspan deflection curves were strongly affected by the considered residual stress distributions. For systems without residual stresses (case 1), the elastic limit normal force was almost equal to the ultimate normal force. With the entire plastification of the upper flange, the load displacement curves showed a sharp bend going along with a decrease of the bending stiffness EI_y of approximately 88 %. Due to plastification, the centroid of the remaining elastic cross section repositioned, influencing the reference line of the remaining elastic structural system. The centreline offset in conjunction with the compressive load led to an increase of bending moments and corresponding deflections w. As a result, the ultimate normal forces were just slightly higher than the elastic ones.

Considering rolling residual stress patterns (case 2), the elastic limit was reached for a state, where plastification starts at the upper flange tips (maximum of the compressive residual stresses). The corresponding normal force of the elastic limit was smaller than the one calculated neglecting residual stresses (case 1). With increasing load, the plastification continuously spread over the cross section area. The bending stiffness and hence the load displacement curves decreased steadily. The ultimate normal forces were smaller than in case 1.

For structural systems considering welding residual stresses (case 3), the load displacement curves featured several sharp bends as a result of the mainly rectangular distributed residual stresses, for which major cross section areas plastified simultaneously leading to a sudden decrease of bending stiffness. Hereby

- the plastification basically began either in the upper flange due to compressive stress or in the lower flange due to tensile stress, depending on the slenderness,
- depending on the slenderness, the bearing after initiation of the plastification differed.

For further explanation Fig. 3 shows different load-midspan deflection curves regarding nondimensional slenderness of $\bar{\lambda}_y = 0.2$ to 2.0. The stress states leading to an initiation of plastification in cross section areas in the upper or in the lower flange are indicated in these load deflection curves using four boundary lines. Following effects can be distinguished:

- While for structural systems with $\bar{\lambda}_y \leq 1.4$ plastification began in the upper flange due to compressive stresses (of loading) in combination with compressive residual stresses (boundary line 1-1), for systems with $\bar{\lambda}_y \geq 1.6$ it was initiated at the lower flange due to tensile stresses (of loading) and positive residual stresses (boundary line 4-4).
- The structures with $\bar{\lambda}_y \leq 0.4$ showed a strong increase of the load displacement curves with the beginning of lower flange plastification in areas of compressive residual stresses (boundary line 2). The plastification process caused this effect due to which the arrangement of elastic cross section parts became symmetrical leading to a shift of the centroid to its initial position. The centreline offset dissolved and the increase of bending moments as well as the deflections w reduced.
- For $\bar{\lambda}_y \leq 0.7$, the plastification of the upper flange in regions of tensile residual stresses (boundary line 3-3) limited the ultimate normal force.

- For $\bar{\lambda}_y \geq 0.8$, the ultimate normal force was already reached at an earlier stage, after plastification of the upper flange in areas of compressive residual stresses occurred (boundary line 1-1).
- To display the course of the ultimate normal forces N_u in the diagram, a certain shift of the displacement w can be recognized between $\bar{\lambda}_y$ = 0.7 and 0.8.

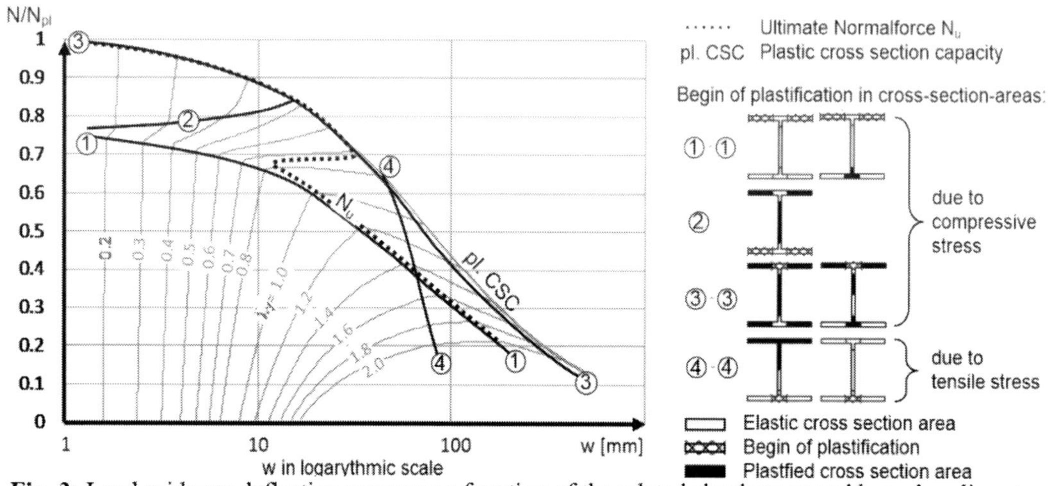

Fig. 3: Load-midspan deflection curves as a function of the related slenderness and boundary lines to indicate the start of plastification for columns in compression with welding residual stresses (case 3)

The ultimate normal forces were in general lower for case 3 (residual stresses due to welding process) than for case 2 (residual stresses due to rolling process). The main reasons were:

- For small $\bar{\lambda}_y$ the spread of the plastified areas in the flanges due to compressive stresses was greater for case 3 than for case 2.
- For higher slenderness, in case 3 the elastic limit normal force was smaller due to the beginning of the plastification in the lower flange in regions with tensile residual stresses (boundary line 4-4).

Depending on $\bar{\lambda}_y$ and the residual stress distributions, different failure criteria could be recognized. The comparison of the connecting line of the ultimate normal forces with the line of the plastic cross section capacity (pl. CSC) in Fig. 2 illustrates, that for small $\bar{\lambda}_y$ the structural systems failed due to a plastic cross section mechanism. For greater slenderness, an eigenvalue drop (drop of critical loading factor) of the partially plastified structural system below 1 was the decisive failure criteria (eigenvalue failure of the partially plastified system). Fig. 4 shows and compares the failure criteria for cases 1, 2 and 3 with regard to $\bar{\lambda}_y$. In cases 2 and 3 the plastic cross section mechanism is decisive for higher slenderness than in case 1. This is a result of higher plastification rates in cases 2 and 3 going along with a decrease of bending stiffness and an increase of bending moments, which enhances the formation of the plastic cross section mechanism.

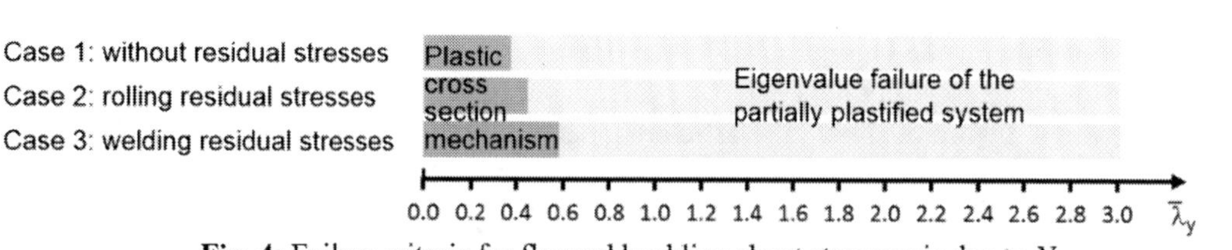

Fig. 4: Failure criteria for flexural buckling about strong axis due to N

3.2 Flexural buckling behavior about the weak axis

Fig. 5 shows the load-midspan deflection curves for flexural buckling about the weak axis as well as stress distributions and the state of plastification regarding $\bar{\lambda}_z = 1.0$. The plastifications of the analysed structures initiated at the inner radii of the curvature regarding the beam deflection, which were the left edges of the cross sections for the given structure. General issues of the load bearing behaviour and the development of plastification were similar to flexural buckling about the strong axis. However, in comparison to the load-deflection curves for buckling about the strong axis (see Fig. 2) the following differences could be identified:

- The limit normal forces $N_{el,lim}$ (elastic limit) and N_u (ultimate load) were lower,
- the load displacement curves decreased more continuous, and the ratio of the ultimate load N_u to the axial load at the elastic limit $N_{el,lim}$ was larger,
- the ultimate load was reached for lower cross-section utilizations (larger distances between the values N_u and pl. CSC).

The lower values of $N_{el,lim}$ and N_u stem from the fact, that the bending stiffness EI_i is lower in case of bending about the weak axis. The ratio of $N_u/N_{el,lim}$ is larger because for buckling about the weak axis no plastification of entire flanges but a successive plastification occurred. This led to a continuous decrease of the bending stiffness and an increase in midspan-deflections v and bending moments M_z. In comparison to buckling about the strong axis, the limit normal force N_u was reached for lower utilization of the plastic cross section capacity due to the fact that the plastic capacity ratio $M_{pl,i}/M_{el,i}$ is greater in case of bending about the weak axis than in case of bending about the strong axis.

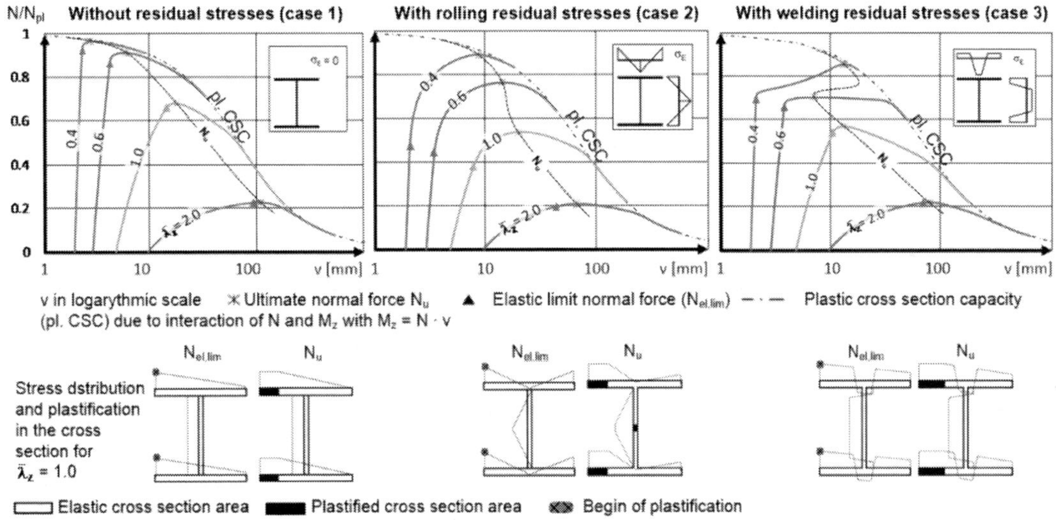

Fig. 5: Load-midspan deflection curves (top) and plastification in the cross-section (bottom) for flexural buckling about the weak axis and different residual stress approaches

The spread of plastification markedly affected the load-carrying behaviour and ultimate axial force. For low slenderness ($\bar{\lambda}_z < 0.8$) the spread of the plastification due to compressive stresses was greater in case 3, which led to smaller maximum normal forces in case 3 than in case 2. For high slenderness ($\bar{\lambda}_z > 0.8$) due to lower elastic limit normal forces $N_{el,lim}$ the maximum normal forces were smaller for case 2 than for case 3.

Further results on the investigation of the load-carrying behaviour regarding the influence of welding and rolling residual stresses can be found in [1].

4. Influence of a bending moment on the flexural buckling behaviour

The behaviour and capacity of a structure in compression (normal force N) is influenced by additional bending moments M. Fig. 6 shows, for flexural buckling about the strong axis, the load-midspan deflection curves and the decisive plastifications of the cross sections at the ultimate limit state for beam-columns with nondimensional slenderness of $\bar{\lambda}_y$ = 1.0 in case 2 (rolled section) and 3 (welded section) as a function of the ratio of the bending moment M_y to the corresponding plastic capacity $M_{pl,y}$. The load-midspan deflection curves are only displayed until the ultimate load is reached, for the purpose of a clear arrangement.

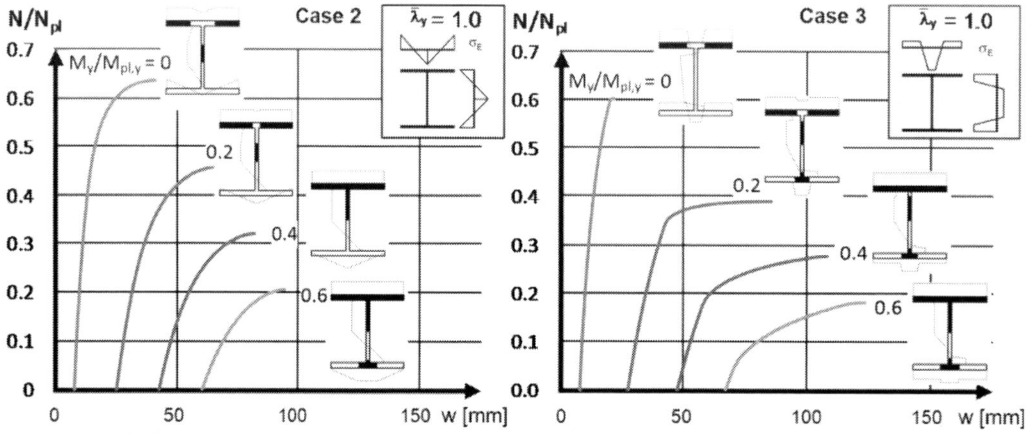

Fig. 6: Load-midspan deflection curves and plastification at ultimate load state for flexural buckling about strong axis due to N and M_y for rolling (left) and welding (right) residual stresses; $\bar{\lambda}_y$= 1.0

Fig. 6 illustrates the influence of the bending moment M_y on the load-carrying behavior of the beam-columns. With an increase of the ratio $M_y/M_{pl,y}$:

- the maximum normal force decreased,
- the deflection at midspan w increased,
- the compressive stresses in the upper flange and tensile stresses in the lower flange and hence the plastification increased,
- the utilization of the plastic cross section capacity increased.

The comparison of the load-midspan deflection curves of case 2 and case 3 shows, that with regard to residual stresses of rolled profiles (case 2), the curves reveal a steadier decrease and smaller deflections w. The plastified regions in the lower flange due to tensile stresses developed for smaller bending moments in case 3 than in case 2.

Fig. 7 shows the interaction of the internal forces N and M_y in terms of bearing capacity. Fig. 7 (left) shows the internal forces related to the plastic cross section resistances, Fig. 7 (right) related to $N_{b,y}$, where $N_{b,y}$ is the ultimate normal force considering flexural buckling with a compression load only, as determined in section 3.1. Additionally, the plastic cross section capacity is given.

Comparing the interaction curves for $\bar{\lambda}_y$ = 1.0 and 2.0 for either case 2 or 3, the graphs on the left side show big differences while on the right side, relating to $N_{b,y}$, the interaction curves are quite similar. In addition, Fig. 7 shows, that for the beam-columns with rolling residual stresses (case 2), the ultimate loads were in general higher than for systems with welding residual stresses (case 3). The differences regarding the ultimate loads were affected by the following influences:

- For equal compression forces, the plastification was distinctly greater in case 3 than in case 2 as shown in section 3.1.
- For case 3, the cross section area with tensile residual stresses in the lower flange plastified due to tension stresses, resulting from a bending moment $M_y > 0$. Fig. 6 shows the plastified areas for $\bar{\lambda}_y = 1.0$. This effect arose at a lower bending moment level than for case 2 and in particular for structures with greater nondimensional slenderness.

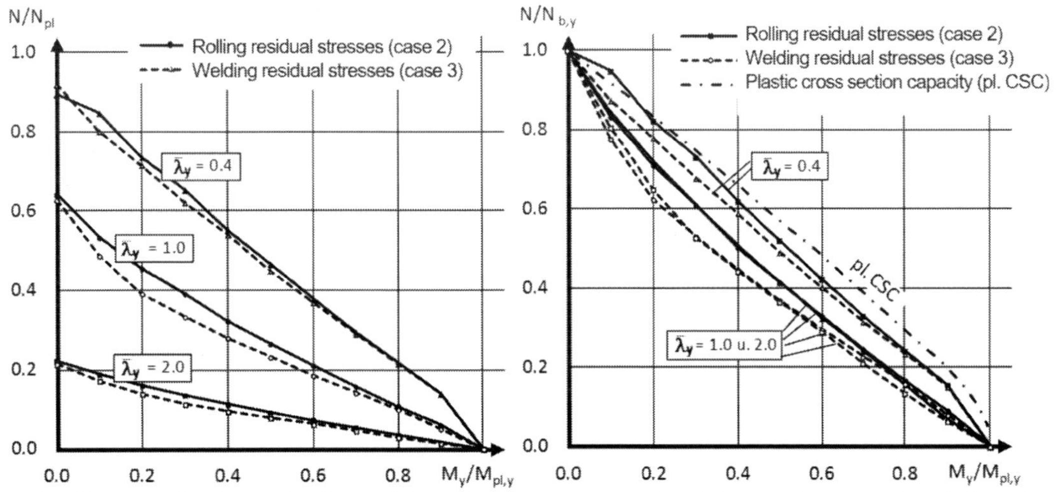

Fig. 7: Interaction N-M_y for flexural buckling about strong axis related on N_{pl} resp. $M_{pl,y}$ (left) and related on ultimate normal force considering flexural buckling $N_{b,y}$ resp. $M_{pl,y}$ (right)

5. Conclusions

Residual stress patterns strongly affect the plastification process and the load-carrying capacity of structural members in compression and bending. The plastification process causes a decrease of member bending stiffness and a repositioning of the centroid of the remaining elastic cross section. The residual stress distribution and the member slenderness are decisive for the location of the onset and sequence of plastification.

Notation

E	Modulus of elasticity
E_v	Modulus of elasticity of strain hardening
f_y	Yield strength
σ	Stress
ε	Strain
$\bar{\lambda}_i$	Nondimensional slenderness for flexural buckling about *i-i* axis
I_i	Moment of inertia about *i-i* axis
v, w	Displacements in y or z-direction
N	Normal force
M_i	Bending moment about *i-i* axis
N_{pl}	Resistance to normal forces of the cross section
$N_{b,i}$	Buckling resistance of a compression member due to buckling about *i-i* axis

$M_{pl,i}$ Plastic resistance to bending moments about *i-i* axis of the cross section
$M_{el,i}$ Elastic resistance to bending moments about *i-i* axis of the cross section

References

[1] Käsmaier M. *Tragverhalten und Tragfähigkeiten von stabilitätsgefährdeten Trägern und Stützen bei kombinierter Beanspruchung*. Dissertation, Ruhr-Universität Bochum 2015.
[2] Ebel R. *Systemabhängiges Tragverhalten und Tragfähigkeiten stabilitätsgefährdeter Stahlträger unter einachsiger Biegebeanspruchung*. Dissertation, Ruhr-Universität Bochum 2014.
[3] ANSYS Software ANSYS Inc., Canonsburg, PA, USA.
[4] ECCS-CECM-EKS, Publication No. 33, *Ultimate Limit State Calculation of Sway Frames with Rigid Joints*. Rotterdam 1984.

COMMENTS ABOUT THE BUCKLING CHECK IN EN 1993-1-1:2005 FOR BUILT-UP COMPRESSED MEMBERS

Helmuth Köber

Technical University of Civil Engineering Bucharest, Romania

Abstract: The present work is intended to point out some aspects in EN 1993-1-1:2005 [1] regarding the buckling check of built-up compression members that could be improved or that could be at least more clearly or rigorously expressed. Certain difficulties of using the general stability checking relations in the European standard EN 1993-1-1:2005 [1] about the z-z axis, (which does not cross the material), were highlighted. Several design procedures were compared used to check the general stability of compression elements with uniform built-up cross-sections formed of two widely distanced chords connected by battens.

1. Introduction

Five different cross-sections were analyzed: two UPN16 hot rolled profiles (buckling curve C) (see Fig. 1a), two hot rolled profiles UPN24 (buckling curve C) (see Fig. 1b), two 160x80x6 rectangular cold formed pipes, (buckling curve C) (see Fig. 1c), two rectangular hot rolled pipes 160x80x6, (buckling curve A) (see Fig. 1c), and two IPE200 profiles (buckling curve B) (see Fig. 1d). The buckling curves in brackets are the ones recommended by EN 1993-1-1:2005 [1].

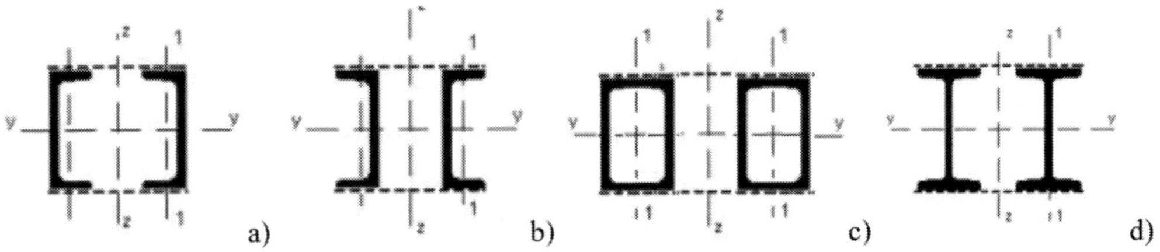

Fig. 1: Analyzed cross-sections: a) cross-section 1 (two UPN16); b) cross-section 2 (two UPN24); c) cross-section 3 and cross-section 4 (two pipes 160x80x6 cold formed, respectively hot rolled); d) cross-section 5 (two IPE 200)

Two steel grades S235 and S355 were considered. In order to avoid the influence of the buckling length values, all the analyses were done on double pinned members in compression. The range of the considered slenderness factor values λ_z was between 60 and 180.

2. Short Presentation of the Considered Procedures

The general stability of the analyzed compressed members, about the axis which does not cross the material, was performed taking into consideration five different procedures in order to compare the obtained results:

-*Procedure 1*: the one recommended by EN 1993-1-1:2005 [1], that uses a single value for the initial bow imperfection $e_0 = L/500$ for all situations.

-*Procedure 2*: the one given in ENV 1993-1-1:1992 [2]; several values of the initial bow imperfection ($e_0 = L/600...L/180$) are used depending on the buckling curve recommended for the chord cross-sections (according to figure 5.5.1 [2] and table 5.5.3 [2]).

-*Procedure 3*: the one proposed in EN 1993-1-1:2005 [1], considering different values of the initial bow imperfection ($e_0 = L/350...L/150$), depending on the buckling curve recommended for the chord cross-sections (according to table 5.1 [1] and table 6.2 [1]).

-*Procedure 4*: the one given in the previous Romanian code for design of steel structures STAS 10108/0-78 [3] with the amendment that the buckling curves were chosen according to EN 1993-1-1:2005 [1]. In this code the stability check about the z-z axis is done on the compressed member as a whole.

-*Procedure 5:* the one indicated in EN 1993-1-1:2005 [1], considering a continuous (smeared) connection between the two chords. The results of procedure 5 were used as reference values for the analyzed procedures.

Table 1: Considered initial bow imperfection values e_0 for procedures 1÷3

Analyzed procedure	Imperfection value for buckling curve		
	a	b	c
Procedure 1	$L/500$	$L/500$	$L/500$
Procedure 2	$L/600$	$L/380$	$L/270$
Procedure 3	$L/300$	$L/250$	$L/200$

2.1 Procedure 1

The buckling check for built-up compressed members formed of two widely distanced chords according to EN 1993-1-1:2005 [1] is performed with the relation:

$$\frac{N_{ch,Ed}}{N_{bch,Rd}} \leq 1 \qquad \text{(EN 1993-1-1:2005 [1], relation (6.71))} \qquad (1)$$

where:

$N_{ch,Ed}$ is the design compression force in a chord at mid-length of the built up member;

$N_{bch,Rd}$ is the design value of the buckling resistance of the chord, taking the buckling length of the chord L_{ch} equal to the distance between two consecutive connection elements,

$N_{bch,Rd} = \chi_{ch} A_{ch} f_y / \gamma_{M1}$,

A_{ch} is the cross-sectional area of one chord,

χ_{ch} is the reduction factor for the buckling of one chord between to consecutive connection elements,

f_y is the yield strength of steel,

γ_{M1} is the partial safety factor for the buckling resistance of members, $\gamma_{M1} = 1.0$.

$$N_{ch,Ed} = 0{,}5N_{Ed} + \frac{M_{Ed}h_0A_{ch}}{2I_{eff}}; \quad M_{Ed} = \frac{N_{Ed}e_0 + M^I_{Ed}}{1 - \frac{N_{Ed}}{N_{cr}} - \frac{N_{Ed}}{S_v}} \quad \text{([1], relation (6.69))} \qquad (2)$$

where:

N_{Ed} is the design compression force in the whole built-up member,

M_{Ed} is the design value of the maximum bending moment in the middle of the built-up member considering second order effects,

h_0 is the distance between the centroids of chords,

A_{ch} is the cross-sectional area of one chord,

I_{eff} is the effective second moment of area of the built-up member about the z-z axis,

$I_{eff} = 0{,}5h_0^2A_{ch} + 2\mu I_{ch}$; where μ is the efficiency factor (see the table in Fig. 2),

Criterion	Efficiency factor μ
$\lambda \geq 150$	0
$75 < \lambda < 150$	$\mu = 2 - \frac{\lambda}{75}$
$\lambda \leq 75$	1,0
where $\lambda = \frac{L}{i_0}$; $i_0 = \sqrt{\frac{I_1}{2A_{ch}}}$; $I_1 = 0{,}5h_0^2A_{ch} + 2I_{ch}$	

Fig. 2: Efficiency factor μ (table 6.8 of EN 1993-1-1:2005 [1])

I_{ch} is the in plane second moment of area of one chord,

e_0 is the initial bow imperfection (according to 6.4.1(1), $e_0 = L/500$ for all situations),

M^I_{Ed} is the design value of the maximum bending moment in the middle of the built-up member without considering second order effects; for pure compression $M^I_{Ed} = 0$,

N_{cr} is the effective critical force of the built-up member, $N_{cr} = \pi^2 E \cdot I_{eff}/L^2$,

n is the distance number of battening planes,

S_v is the shear stiffness of the battened or laced panel.

$$S_v = \frac{24EI_{ch}}{a^2\left[1 + \frac{2I_{ch}}{nI_b}\frac{h_0}{a}\right]} \leq \frac{2\pi^2 EI_{ch}}{a^2} \quad \text{(EN 1993-1-1:2005 [1], relation (6.73))} \qquad (3)$$

where:

I_b is the in plane second moment of area of one batten,

a is the distance between two consecutive battened connection elements.

2.2 Procedure 2

In case of procedure 2 the same buckling check relation was used as for procedure 1, as shown in Eq. (1). Different values of the initial bow imperfection are used (see the values indicated in Table 1), depending on the buckling curve recommended for the chord cross-sections according to figure 5.5.1 of ENV 1993-1-1:1992 [2].

2.3 Procedure 3

The general stability check of the built-up members about the z-z axis was done with the same relation, as in case of procedure 1 (see Eq. 1). Different values of the initial bow imperfection were considered (depending on the buckling curve recommended for the chord cross-sections according to table 5.1 of EN 1993-1-1:2005 [1]).

2.4 Procedure 4

For procedure 4 the general stability check about the z-z axis (which does not cross the material) was performed using the relation:

$$\frac{N_{Ed}}{N_{bz,Rd}} \leq 1 \qquad \text{(EN 1993-1-1:2005 [1], relation (6.46))} \qquad (4)$$

where:
N_{Ed} is the design compression force on the whole built-up member,
$N_{bz,Rd}$ is the design value of the buckling resistance of the whole member about z-z axis,

$$N_{bz,Rd} = \chi_z A\, f_y / \gamma_{M1} \qquad \text{(EN 1993-1-1:2005 [1], relation (6.47))} \qquad (5)$$

where:
A is the area of the entire built-up cross-section, $A = 2 \cdot A_{ch}$,
χ_z is the reduction factor for the buckling of the whole built-up member about z-z axis,

$$\chi_z = \frac{1}{\phi_z + \sqrt{\phi_z^2 - \bar{\lambda}_z^2}}, \text{ but } \chi_z \leq 1{,}0 \qquad \text{(EN 1993-1-1 [1], relation (6.49))} \qquad (6)$$

In case of procedure 4 the relative slenderness factor $\bar{\lambda}_z$ was determined using $\lambda_{z,tr}$ (the transformed slenderness factor from the old Romanian design code for steel structures STAS 101 08-0/78 [3]):

$$\bar{\lambda}_z = \frac{\lambda_{z,tr}}{\lambda_1};\ \lambda_{z,tr} = \sqrt{\lambda_z^2 + \lambda^2};\ \lambda_z = \frac{L_{cr,z}}{i_z};\ \lambda = \frac{a}{i_{z1}} \cdot \sqrt{\frac{\pi^2}{12} \cdot \left(1 + \frac{I_{ch}}{a} \cdot \frac{h_0}{I_b}\right)} \qquad (7)$$

where:
λ_1 is the Euler slenderness factor, used to determine the relative slenderness value,
$\lambda_1 = \pi(E / f_y)^{0.5}$, E is the Young modulus of steel, $E = 2{,}1 \cdot 10^5 N/mm^2$,
$L_{cr,z}$ is the buckling length of the entire built-up member about the z-z axis, $L_{cr,z} = L$,
i_{z1} is the in plane radius of gyration of a single chord, $i_{z1} = \sqrt{\frac{I_{ch}}{A_{ch}}}$,
i_z is the radius of gyration about the z-z axis of the entire cross-section of the built-up member; $i_z = \sqrt{\frac{I_z}{A}}$, where $I_z = 0{,}5 h_0^2 A_{ch} + 2 I_{ch}$ and $A = 2 \cdot A_{ch}$.

2.5 Procedure 5

In case of procedure 5, the buckling check about the z-z axis was performed using Eq. (4) considering the built-up member as a uniform member with a continuous cross-section.
The results of procedure 5 were used as reference values for the considered procedures.
For procedure 5 the reduction factor χ_z is calculated considering the relative slenderness factor $\bar{\lambda}_z$ as follows:

$$\chi_z = \frac{1}{\phi_z + \sqrt{\phi_z^2 - \bar{\lambda}_z^2}};\ \bar{\lambda}_z = \frac{\lambda_z}{\lambda_1};\ \lambda_z = \frac{L_{cr,z}}{i_z} \qquad (8)$$

where:
i_z is the radius of gyration about the z-z axis of the entire cross-section of the built-up member; $i_z = \sqrt{\frac{I_z}{A}}$, where $I_z = 0{,}5 h_0^2 A_{ch} + 2 I_{ch}$ and $A = 2 \cdot A_{ch}$.

3. Results and Comments

3.1 Buckling resistance values

Considering the fact, that the deformability of connection elements aggravates the buckling phenomena of the compressed built-up members about the z-z axis, it would have been expected for the values of the buckling capacity obtained for procedure 5 to be the greatest from all considered procedures.

Procedure 1 (the one recommended in EN 1993-1-1:2005 [1], that considers a single value for the initial bow imperfection $e_0 = L/500$ for all situations) conducts in most cases to greater buckling capacities than procedure 5. The white stripes are greater than the black stripes in the graphics shown in Figs. 3, 4 and 6.

In the graphics presented in Fig. 3, 4 and 6, case 1 represents the situation for $\lambda_z = 60$, case 2 the situation for $\lambda_z = 75$, case 3 the situation for $\lambda_z = 90$, case 4 the situation for $\lambda_z = 105$ and so on till case 9 for $\lambda_z = 180$.

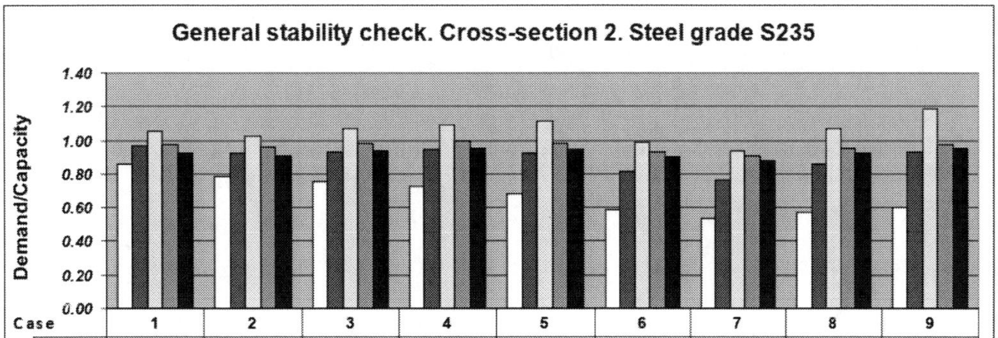

Case	1	2	3	4	5	6	7	8	9
□ Proc.1	0.858	0.786	0.755	0.726	0.679	0.583	0.532	0.569	0.600
■ Proc.2	0.970	0.924	0.934	0.946	0.925	0.815	0.763	0.854	0.934
□ Proc.3	1.054	1.029	1.069	1.095	1.112	0.991	0.939	1.071	1.188
■ Proc.4	0.977	0.960	0.983	0.996	0.984	0.929	0.907	0.950	0.974
■ Proc.5	0.924	0.910	0.936	0.954	0.947	0.899	0.882	0.927	0.954

Fig. 3: Cross-section 2; S235; Demand/capacity ratio for different values of the slenderness factor λ_z

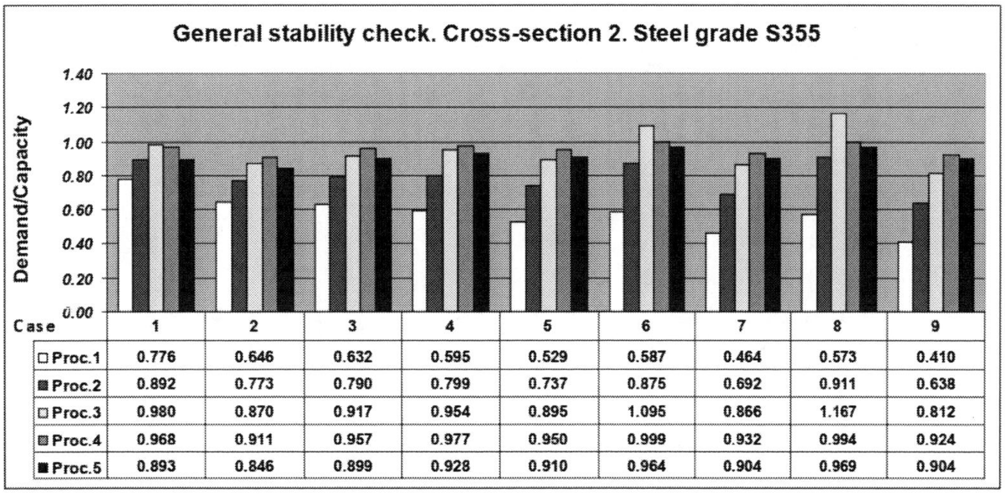

Case	1	2	3	4	5	6	7	8	9
□ Proc.1	0.776	0.646	0.632	0.595	0.529	0.587	0.464	0.573	0.410
■ Proc.2	0.892	0.773	0.790	0.799	0.737	0.875	0.692	0.911	0.638
□ Proc.3	0.980	0.870	0.917	0.954	0.895	1.095	0.866	1.167	0.812
■ Proc.4	0.968	0.911	0.957	0.977	0.950	0.999	0.932	0.994	0.924
■ Proc.5	0.893	0.846	0.899	0.928	0.910	0.964	0.904	0.969	0.904

Fig. 4: Cross-section 2; S355; Demand/capacity ratio for different values of the slenderness factor λ_z

The following remarks can be made about the values of the buckling resistances noticed in the five considered procedures by analyzing the graphics in Fig. 3 and 4:

1. For cross-section 2 the differences between procedures 1 and 5 are almost always on the unsafe side. In case of steel grade S235 the differences are from about 6% for $\lambda_z = 60$, to 41% for $\lambda_z = 165$. For steel grade S355 the differences are varying between 13% for $\lambda_z = 60$ and 55% for $\lambda_z = 180$.

2. Procedure 2, that considers an initial bow imperfection $e_0 = L/270$ for cross-section 2, is generally unsafe compared to procedure 5 for both considered steel grades (in most cases the stripes for procedure 2 are smaller than the ones for procedures 5, see Fig. 3 and 4).

3. In case of procedure 3 an initial bow imperfection value $e_0 = L/200$ was considered for cross-section 2. Procedure 3 appeared to be the safest when using steel grade S235 (compared to procedure 5 up to 25% smaller buckling resistances were obtained, see Fig. 3). In case of using steel grade S355 procedure 3 is unsafe compared to procedure 5 with up to 12%, for λ_z = 120, 150 and 180. For the other considered values of λ_z procedure 3 is on the safe side when using S355.

4. Procedure 4, that takes into consideration the prescriptions of the former Romanian code [3] conducts always to smaller buckling capacities compared to procedure 5. In case of steel grade S235 the buckling resistances are between 2% and 6% smaller than the one obtained in case of procedure 5. For the steel grade S355 the differences between the buckling capacities values are varying among 2% and 8%.

5. For the steel grade S235 procedure 3 appears to be the safest. In case of steel grade S355 it is difficult to decide which one of the procedures 3 or 4 is the safest. For $\lambda_z = 60$, 135 and 165 procedure 3 conducts to the smallest buckling resistances, whilst for $\lambda_z = 75$, 90, 105, 120, 150, 180 procedure 4 is safer.

3.2 Comments about the use the efficiency factor μ in EN 1993-1-1:2005 [1]

The limits for using the efficiency factor μ should be more clearly defined. The following example illustrates this. A battened column is considered (see Fig. 5a). If $\lambda_z \leq 75$, $\mu = 1{,}0$ and the effective second moment of the area of the cross-section about the z-z axis would be:

$$I_{eff} = 0{,}5h_0^2 A_{ch} + 2\mu I_{ch} \quad \text{(EN 1993-1-1:2005 [1], relation (6.74))} \tag{9}$$

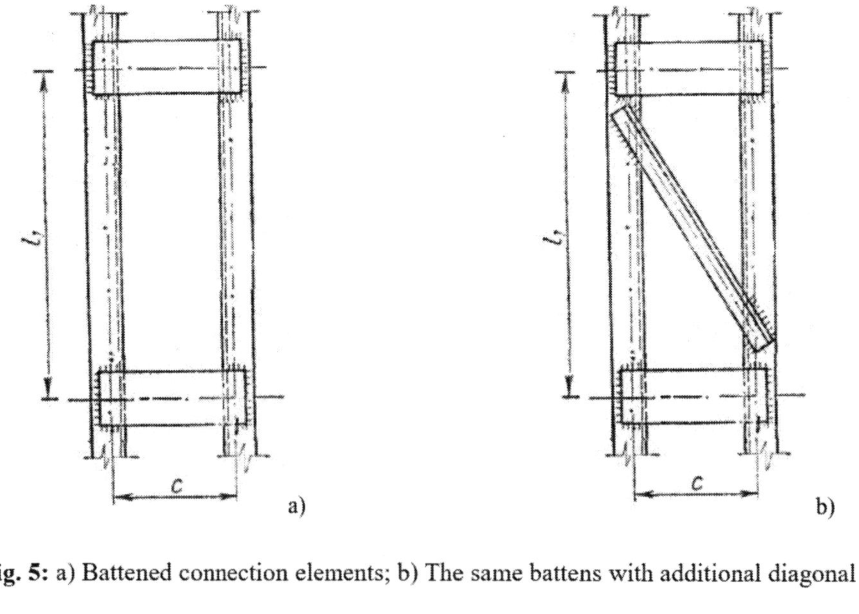

Fig. 5: a) Battened connection elements; b) The same battens with additional diagonal laces

Let us presume now that laces are added (it is not a usual case but the mathematical relations should cover it properly, as it could be a retrofitting solution) and a laced system results (see Fig. 5b). In this new situation, irrespective of the values of λ_z, the recommended relation for the calculation of the effective second moment of the area about the z-z axis would be:

$$I_{eff} = 0{,}5h_0^2 A_{ch}, \qquad \text{(EN 1993-1-1:2005 [1], relation (6.72))} \qquad (10)$$

By adding these laces, the stiffness of the actual member increased, but according to Eq. (9) and (10) it decreases, this being an inappropriate result, as it would mean that adding connections reduces the stiffness. A correction factor, depending on the stiffness of the battening or lacing system, should be applied on the slenderness factor λ_z like in STAS 10108-0/78 [3], as shown in Eq. (7).

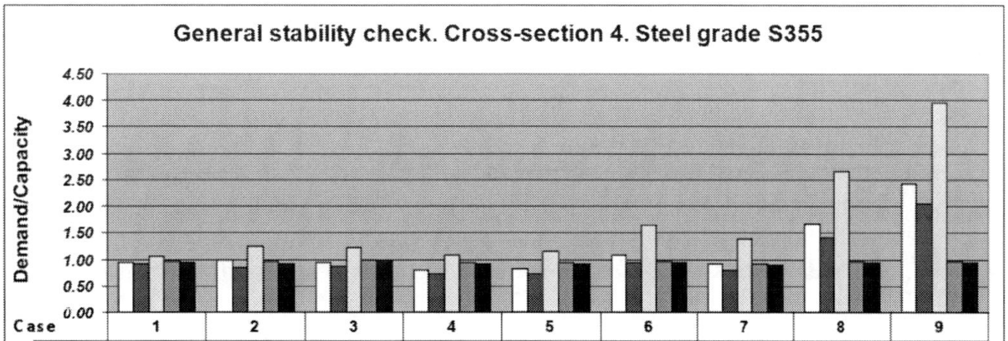

Case	1	2	3	4	5	6	7	8	9
Proc.1	0.947	0.982	0.946	0.811	0.825	1.092	0.923	1.670	2.436
Proc.2	0.917	0.848	0.876	0.741	0.740	0.954	0.804	1.422	2.056
Proc.3	1.070	1.241	1.226	1.088	1.165	1.644	1.398	2.660	3.954
Proc.4	0.976	0.963	0.998	0.940	0.935	0.955	0.922	0.960	0.966
Proc.5	0.941	0.923	0.959	0.908	0.909	0.933	0.905	0.945	0.953

Fig. 6: Cross-section 4; S355; Demand/capacity ratio for different values of the slenderness factor λ_z

For cross-section 4 (buckling curve a), very small buckling resistances values were obtained for $\lambda_z \geq 150$ in case of the procedures 1, 2 and 3 (see the graphics in Fig. 6).

According to EN 1993-1-1:2005[1] the value of the efficiency factor μ, used for the calculation of the moment of inertia of the built-up cross-section formed of two widely distanced chords about the the z-z axis (which does not cross the material) is equal to zero for $\lambda_z \geq 150$.

For cross-section 4 the calculations were repeated for the first 3 procedures considering the efficiency factor $\mu = 1.0$ for all values of λ_z. In this case less scattered values of the buckling resistances were obtained for all analyzed situations (see the values in Table 2 and 3).

Table 2: Cross-section 4, steel grade S355, demand/buckling capacity ratios

λz	N_{Ed} [kN]	$L_{cr,z}$ [mm]	Procedure 1 $e_0=L_{cr}/500$		Procedure 2 $e_0=L_{cr}/600$		Procedure 3 $e_0=L_{cr}/300$		Procedure 4	Procedure 5
			$\mu=1$	$\mu\neq1$	$\mu=1$	$\mu\neq1$	$\mu=1$	$\mu\neq1$		
60	1450	6311	0,947	0,947	0,917	0,917	1,070	1,070	0,976	0,941
75	1200	7889	0,982	0,982	0,848	0,848	1,241	1,241	0,963	0,923
90	1000	9467	**0,911**	0,946	**0,817**	0,876	**1,167**	1,226	0,998	0,959
105	750	11045	**0,738**	0,811	**0,681**	0,741	**0,967**	1,088	0,940	0,908
120	600	12623	**0,671**	0,825	**0,612**	0,740	**0,908**	1,165	0,935	0,909
135	500	14201	**0,673**	1,092	**0,605**	0,954	**0,947**	1,644	0,955	0,933
150	400	15779	**0,546**	0,923	**0,490**	0,804	**0,770**	1,398	0,922	0,905
165	350	17357	**0,616**	1,670	**0,544**	1,422	**0,904**	2,660	0,960	0,945
180	300	18935	**0,612**	2,436	**0,537**	2,056	**0,915**	3,954	0,966	0,953

Table 3: Cross-section 4, steel grade S235, demand/buckling capacity ratios

λ_z	N_{Ed} [kN]	$L_{cr,z}$ [mm]	Procedure 1 $e_0=L_{cr}/500$		Procedure 2 $e_0=L_{cr}/600$		Procedure 3 $e_0=L_{cr}/300$		Procedure 4	Procedure 5
			$\mu=1$	$\mu\neq1$	$\mu=1$	$\mu\neq1$	$\mu=1$	$\mu\neq1$		
60	1450	6311	0,980	0,980	0,955	0,955	1,082	1,082	0,965	0,946
75	1200	7889	0,970	0,970	0,933	0,933	1,117	1,117	0,963	0,940
90	1000	9467	**0,988**	1,018	**0,943**	0,960	**1,217**	1,250	0,992	0,964
105	750	11045	**0,944**	1,009	**0,878**	0,933	**1,205**	1,314	0,971	0,945
120	600	12623	**0,801**	0,909	**0,740**	0,830	**1,046**	1,227	0,920	0,897
135	500	14201	**0,721**	0,907	**0,660**	0,815	**0,966**	1,275	0,905	0,886
150	400	15779	**0,817**	1,382	**0,734**	1,204	**1,152**	2,094	0,960	0,943
165	350	17357	**0,922**	2,500	**0,814**	2,129	**1,353**	3,982	0,992	0,977
180	300	18935	**0,917**	3,870	**0,803**	3,647	**1,371**	5,920	0,994	0,981

4. Conclusions

The main conclusions of our study are:

1. for the considered cross-sections made of S235 and S355 steel grades, procedure 1 (recommended in EN 1993-1-1:2005 [1]), that uses a single value for the initial bow imperfection $e_0 = L/500$ for all situations, is unsafe compared to procedure 5 (continuous smeared connection between the two chords); bigger values than $L/500$ should be used for initial bow imperfection when checking for buckling of built-up members;
2. using different values for the initial bow imperfection, depending on the buckling curve recommended for the chord cross-sections (procedures 2 and 3) conducted to results on the safe side, compared to procedure 1;
3. procedure 3 (taking into account different values of the initial bow imperfection $e_0 = L/350...L/150$, bigger than $L/500$, depending on the buckling curve) and procedure 4 (checking the compressed member as a whole) are more appropriate;
4. the use of the efficiency factor μ for the calculation of the effective second moment of the area should be reconsidered.

References

[1] Eurocode 3, EN 1993-1-1:2005, *Design of steel structures, Part1-1: General rules and rules for buildings*, CEN, 2005.

[2] ENV Eurocode 3, *ENV 1993-1-1:1992, Design of steel structures, Part1-1: General rules and rules for buildings*, CEN, 2002.

[3] STAS 10108-0/78, *Non Industrial, Industrial and Agriculture Buildings, Design of steel structures*, 1978.

WELDING RESIDUAL STRESSES IN HIGH-STRENGTH STEEL. EXPERIMENTAL RESULTS

Benjamin Launert[a], Michael Rhode[b], Hartmut Pasternak[a] and Thomas Kannengießer[b]

[a] *Department of Steel and Timber Construction, Brandenburg University of Technology, Cottbus, Germany*
[b] *BAM Federal Institute for Materials Research and Testing, Berlin, Germany*

Abstract: This article presents the latest results of an ongoing national research project on improved models for the prediction of welding residual stresses of thick-plated welded I-girders. The experimental program is presented and the importance of different influencing factors on the residual stresses is discussed in detail. All results are compared for mild (S355J2+N) and high strength (S690QL) steel. Finally, conclusions for further works are drawn.

1. Introduction

To take into account the effects of welding in capacity design, a geometrically and physically nonlinear analysis of the imperfect structure (also GMNIA) is needed. The quality of such proof significantly depends on whether the imperfections introduced to the numerical model cover the effect of the existent imperfections equivalently. Here are still major gaps in the knowledge [1]. Some recent studies have shown the buckling design to be quite conservative [2,3]. A more precise description of welding imperfections can be realized through welding simulation, which has developed strongly for the last 10 years. However, the modelling and the computational effort are still enormous [2]. The industrial application is therefore limited to some mechanical engineering applications, e.g. in car manufacturing. In steel construction, simplified models are used.

Geometric deviations are controlled by compliance with the tolerance limits prescribed in EN 1090-2. For plated structures, the imperfections can be either local or global. Usually the imperfections are introduced as eigenmodes that are scaled to a proper magnitude. For the section types investigated in this study, the ultimate load is influenced by bow imperfections in each y- and z-direction (typically $L/1500$ to $L/1000$). As the post-weld deformation is also linked to the heat input and the sequencing, a more precise distinction should be made [3].

Systematic measurements are however rarely published [1], so there is a lack of reliable data. The residual stresses are sometimes taken into account by fictitious additional deformations. It is nevertheless possible to include the residual stresses directly in the model with modern FE-codes [1].

A common approach for the residual welding stresses is given by [4]. The comparison of this model with experimental and numerical results in [2] led to adequate agreement for normal strength steel (NSS). However, the parameter field which the model can be applied upon is undefined. Hence, a general review is critical. In addition to the plate thicknesses, the welding method and the parameters are expected to have a strong impact on the residual stresses [3,5]. For high strength steel (HSS), the residual stresses seem to be lower in comparison to the yield strength [6,7]. The influence of different residual stress patterns on centrally loaded columns was studied in [2]. The results indicate that a favourable buckling curve should be assigned for HSS. Therefore, new residual stress models are developed in an ongoing national research project [3].

2. Welding Experiments

The experiments are divided into three parts: (1) a comparative study of different influencing parameters for bead-on-plate specimens, (2) the evaluation of the cutting influence (flame cut) to the chord edges in welded I-girders and (3) the manufacture of component-like welded I-girders under workshop conditions.

2.1 Bead-on-plate

This experiment encompasses a single bead-on-plate run welded on a plate. The aim is the determination and comparison of the residual stresses before and after welding in NSS and HSS. The temperature cycles were measured at different positions during welding. Results are used for the calibration of the later numerical models. In total, eight specimens, four from each S355J2+N and S690QL, were MAG-welded. The varied model parameters are the plate thickness ($t = 10$ and 25 mm) and the weld heat input ($\bar{Q} = 8.3$ and 5.8 kJ/cm) by the variation of the welding speed ($v = 45$ and 65 cm/min). The electrical power for the weld process was kept constant at $I = 260$ A and $U = 30$ V. The filler material was chosen to be in accordance with the characteristics of the base material more or less (Union K56 for S355 and Union NiMoCr for S690). All plates were subjected to a stress-relief heat treatment before welding (500°C for 12 h).

The longitudinal and transverse residual stresses were determined by X-ray diffraction (XRD) after welding. In Fig. 1 the longitudinal residual stresses are plotted exemplary for $t = 10$ mm and $\bar{Q} = 8.3$ kJ/cm. The results are compared for S355J2+N and S690QL. In the left column the absolute residual stress values are given. The right column shows the ratio of measured residual stress values and yield strength of the base material. For this purpose, tensile tests at room temperature were conducted, 5 per plate thickness and per grade. The mean values are on average 100 MPa higher than the standard values. As the magnitude of tensile residual stress is influenced by the strength of the material in particular, it is reasonable not to use the characteristic values in this context. In Fig. 2 the ratios for all samples are added to a diagram originally published by [7].

The results support the assumption that the tensile residual stresses in HSS are lower compared to the yield strength: (0.6-0.7)$\cdot f_y$ (approx. 1.0$\cdot f_y$ for NSS).

In addition to the material grade, it must be noted that the heat input (which is determined by the welding method and the welding parameters) and the plate thickness play an important role as they act on the width of the tensile field. However, the impact of a change in the heat input is reduced with an increase in plate thickness. Therefore, it seems reasonable to use the ratios of the energy per unit length ($\bar{Q}$) and the sum of plate thicknesses in the weld joint ($\sum t$) instead. This would also reduce the number of parameters in a simplified model. Moreover, the widths of the tensile residual stresses for S690 are slightly decreased compared to S355. This can be explained by the increase of the yield strength at elevated temperatures. Assuming that the maximum temperature isotherms are equal (thermo-physical properties are the same), the zone which is subjected to plastic deformations during welding decreases with an increase in the yield strength.

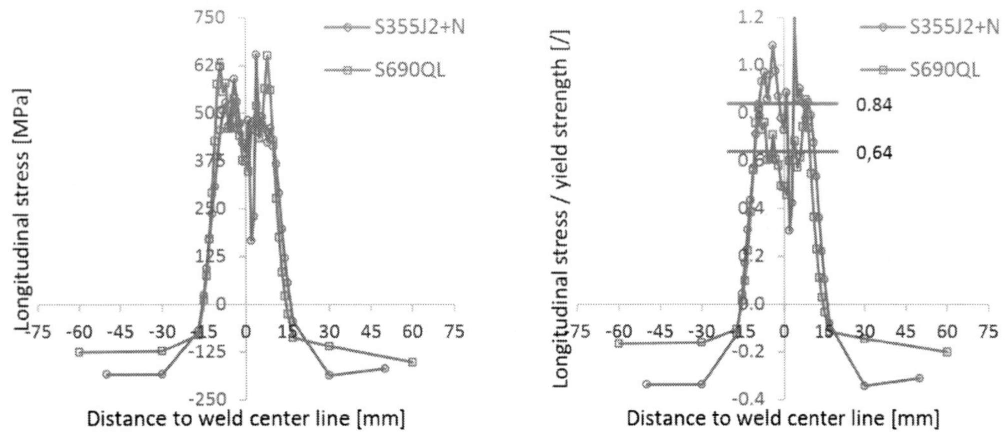

Fig. 1: Influence of the material grade (S355, S690) on the residual stress distribution (by XRD), for t = 10 mm and $\bar{Q}$ = 8.3 kJ/cm

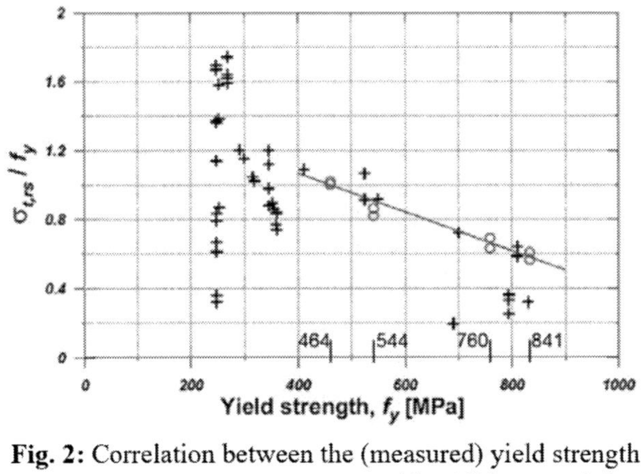

grade	$\bar{Q}$	tensile stress (mean)	
S355J2+N	kJ/cm	ratio	width[mm]
10 mm	8.3	0.84	34.2
	5.8	0.88	25.4
25 mm	8.3	1.00	30.3
	5.8	1.03	26.1
S690QL	kJ/cm	ratio	width[mm]
10 mm	8.3	0.64	29.6
	5.8	0.70	23.2
25 mm	8.3	0.58	23.8
	5.8	0.62	21.2

Fig. 2: Correlation between the (measured) yield strength and tensile residual stress (originally by [7]) with additions made by the authors

It can be concluded that the material, the geometry, especially the plate thickness, and the heat input act on the residual stress field. The influence of an initial stress field on the result is discussed in the following.

2.2 T-joint

It is well known that the cutting method has an influence on the residual stresses [6]. Though, yet no systematic study was published. In steel construction, the cutting is typically realized by flame cuts. The metal is heated by a torch to its kindling temperature and is oxidized by a stream of oxygen from the torch. The heat effect is quite similar to that of a welding torch. As a result, the tensile residual stresses near the cutting edge superpose with the welding residual stresses leading to a reduction or reversal of compressive residual stresses. A numerical study taking into account this effect has shown beneficial effects on the capacity of centrally loaded columns [8].

This study is aimed to systemize the influence of cutting to the chord edges in I-girders. The specimens used in this approach are T-joints with different edge preparations. In a first step the base plate is cut and measured by XRD. In a second step a stiffener is welded in the center line of the chord plate with two fillet welds using MAG process. Afterwards the residual stress measurement is repeated. To reduce the number of specimens, one edge is saw cut, the other is flame cut. The width of the chord plate b_f is 150 mm (in accordance with specimens in 2.1). Fig. 3 shows the results of a preliminary investigation which was conducted to decide on further steps.

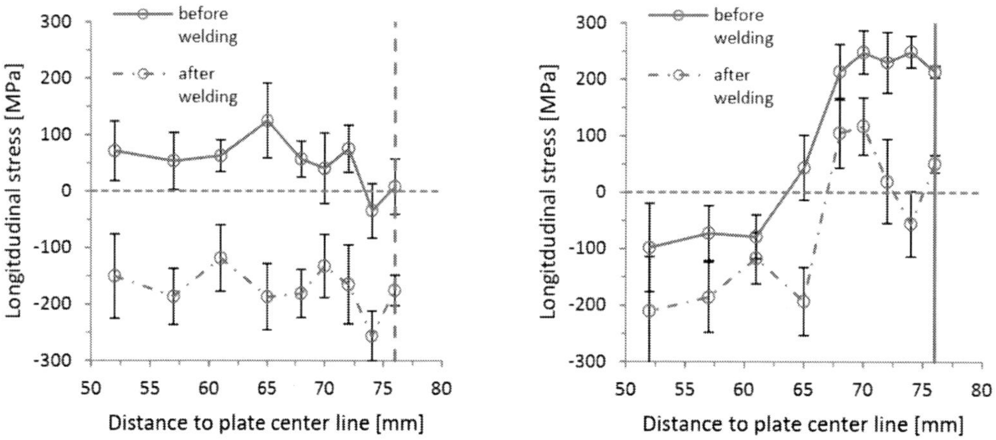

Fig. 3: Influence of the edge preparation on the residual stresses before and after welding S355J2+N, chord $t_f = 20$ mm, stiffener $t_w = 10$ mm, weld heat input $\bar{Q} = 9.2$ kJ/cm per fillet (saw cut (left): indicated by a dashed vertical line, flame cut (right): indicated by a solid vertical line)

It can be seen that a saw cut induces no remarkable residual stresses near the cutting edge. After the welding of the stiffener, the stresses are shifted by a nearly constant value. The compressive residual stress is at approximately -200 MPa. In contrast, the flame cut induces remarkable tensile residual stresses along the cutting edge. As a consequence, the stresses at the edge of the chord remain in the tensile range or close to zero after welding. The affected area is limited to about ~15 mm. In [6] this value is assumed with t_f (in here 20 mm). Further measurements will be conducted with two different weld heat inputs for two combinations of plate thicknesses and for S355 and S690.

Summarized, it is affirmed that an edge preparation with flame cuts induces remarkable longitudinal tensile residual stresses. Those stresses superpose with the stresses resulting from welding.

2.3 Full-scale specimens (I-girders)

The fabrication was realized under workshop conditions by a steel manufacturing company. The base material is again S355J2+N and S690QL. Each series consists of four girders. The flange width is 150 mm and the web height is 220 mm in all specimens. It was shown in 2.2 that the edge preparation by flame cut influences the overall residual stresses in the chord plates. To avoid side effects, all plates were waterjet-cut. The plate thicknesses are 25 or 20 mm in the chords and 15 or 10 mm in the web. The weld type is a one-layer fillet weld. All welds are deposited subsequently after each other by MAG process. Each seam is about 1 m to ensure uniform residual stresses in the middle section. The influence of the weld heat input is studied by changing the welding parameters. The residual stresses were measured for eight girders at BAM after welding. A combination of partial sectioning method and mobile XRD is applied. Representative results for the sectioning of four girders with the plate thicknesses 25 and 15 mm are presented in 3. The corresponding welding parameters are summarized in Table 1.

Table 1: Welding parameters of four MAG-welded girders (chords: 25 mm, web: 15 mm)

Girder-No.	S1	S2	W1	W2
Material	S355J2+N		S690QL	
I [A]	338	327	324	329
U [V]	33,4	33,4	33,7	33,7
v [cm/min]	49	31	49	31

3. Residual stress measurements

3.1 Sectioning method

The measurement of residual stresses by the sectioning method has been used for decades to measure (typically longitudinal) residual stresses in full-size structural members. Therefore, the girders are vertically sectioned by a band saw into smaller slices [6, 9]. To obtain reliable results, it is important to perform a careful preparation of the test piece such as proper location of the test section, strain gauge locations and cutting positions. Every sectioning causes the assimilation of residual stresses in the remaining girder section by relaxation processes. The occurring mechanical elastic relaxation strains are measured at each sectioning by strain gauges applied in longitudinal (and transverse) direction of the girder. From there, the residual stresses are calculated by Eq. (1). The analysis is further simplified by assuming that the transverse stresses are negligible.

$$\sigma_1 = -\frac{E}{1-\upsilon^2}\left(\varepsilon_1 + \nu\varepsilon_2\right) \qquad (1)$$

where, σ is the calculated longitudinal residual stress (in MPa), ε is the measured relaxed elastic strain (/). The index "1" indicates the longitudinal direction in the direction of the weld seam. The index "2" indicates the transverse direction. E is the Young's modulus and υ is the Poisson's ratio.

Results for S355 and S690 referred to the yield strength are shown in Fig. 4 for the inner and outer sides of the chords. Straight graphs correspond to the mean values of measured yield strength (based on min. 5 tensile tests per plate) and dashed graphs to the characteristic yield strength according to Eurocode. It can be noticed that the S690 shows a more favourable behaviour with regard to its yield strength. The values in compression are at approximately half of the corresponding values from S355. Also, the increase in the tensile region is less steep. By extrapolating these values it can be concluded that the tensile residual stresses do not reach the yield strength of the base material (this effect was already stated earlier in 2.1). In addition, it can be seen how the increase in weld heat input (indicated by the index 1 or 2) acts on the magnitudes in compression. An increase of weld heat input with otherwise the same conditions seems to influence less on S690 compared to S355. Furthermore, the stress gradients through the thickness should be taken into account for S690 in a simplified model. Such gradients are also existent for S355. However, for the compressive stresses it seems to be less important in there.

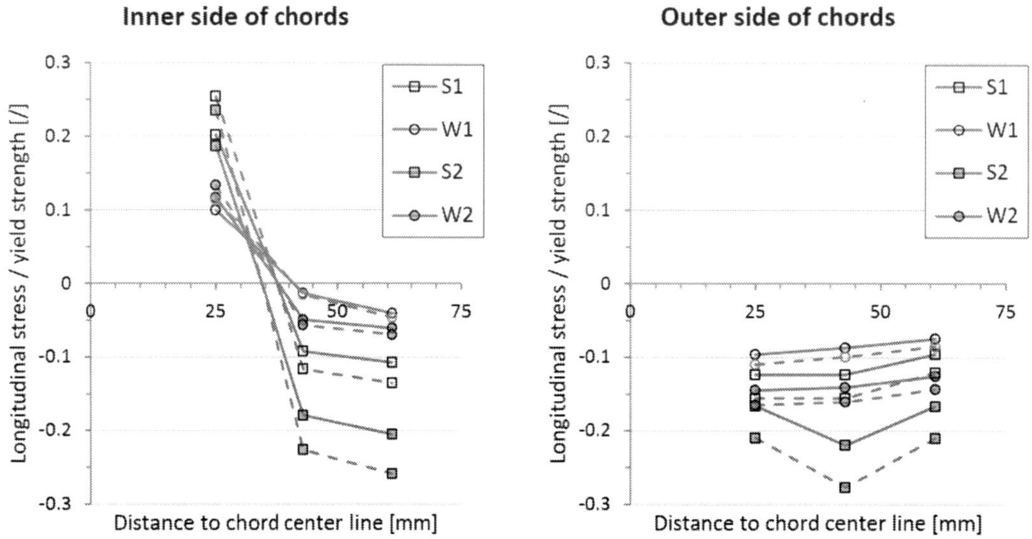

Fig. 4: Longitudinal residual stresses of four girders (Table 1), exemplary for half of the chord

3.2 XRD method

For the future it is planned to use X-ray diffraction (XRD) method for the remaining sections in order to determine any residual stresses present in the material which were not relaxed due to sectioning. For this purpose the chords are additionally separated from the web by saw cuts to obtain free accessibility for the XRD. The residual stresses by XRD are then superimposed to the stresses determined by the sectioning method from where the overall residual stresses are concluded.

Table 2: X-ray measuring and evaluation parameters

Radiation	Cr Kα
hkl	Ferrite: 211
Aperture	3 mm
Exposure time	5 s
ψ step size	-45° to 45° in 10 steps

The XRD is based on the determination of interplanar spacing within the crystal lattice under a certain applied load. The penetration depth is about several µm [9]. Hence, the residual stresses are obtained at the surface near region. The method can be applied upon all reachable surfaces for non-destructive testing and is well established [10]. The accessibility is limited due to the necessary different angles for the measurement of the diffracted X-rays. Table 2 shows the used parameters.

4. Conclusions

This article presents the first steps on improved models for the implementation of welding imperfections in high-strength steel (HSS). The main focus is put on (longitudinal) residual welding stresses. The results obtained to date led to the main conclusions as summarized in the following:

1. The bead-on plate specimens have shown that tensile residual stresses reach more or less the yield strength in case of the S355J2+N (NSS). In contrast, the S690QL shows reduced tensile residual stresses of approximately 60 to 70% of the yield strength.
2. The material grade, the plate thickness and the heat input were identified to be the main influencing parameters on the residual stresses.
3. It can be ascertained too that an edge preparation by flame cut indicates remarkable longitudinal tensile residual stresses. Those stresses superpose with the weld residual stresses. The reduction of the compressive residual stresses on the chord edges acts beneficial in terms of the ultimate load.
4. The residual stress measurements by sectioning method on component-like girders have shown similar tendencies. The S690 shows generally a much more favourable behaviour with regard to its yield strength.
5. Hence, HSS (compared to NSS) have beneficial effects on both: (i) an increased load capacity due to the generally increased strength and (ii) the decreased importance of the longitudinal residual welding stresses. This implies an adjustment of the recent buckling classification.

Acknowledgments

The authors thank the German Federation of Industrial Research Associations (AiF) for the financial support of the research project IGF-No. 18104 BG. This project is carried out under the auspices of AiF and funded by the Federal Ministry of Economics and Technology (BMWi) as part of the program to support Industrial Community Research and Development (IGF).

References

[1] Johansson B, Maquoi R, Sedlacek G, Mueller C, Beg D. *Commentary and Worked Examples to EN 1993-1-5: Plated Structural Elements, Background documents in support to the implementation, harmonization and further development of the Eurocodes (JRC Scientific and Technical Reports),* Luxembourg: Office for Official Publications of the European Communities, Luxembourg, 2007.

[2] Pasternak H, Launert B, Krausche T. "Welding of girders with thick plates - Fabrication, measurement and simulation", *Journal of Constructional Steel Research,* 115, 407-416, 2015.

[3] Pasternak H, Kannengießer T et al. "*IGF 18104 BG - Enhancement of Load Capacity of Welded High-Strength I-Shape Sections Using Improved Design Models for the Consideration of Residual Stress 04/2014-04/2016*" (in German), Project description, 2013.

[4] N.N. "*Swedish Design Rules for Steel Structures*", Boverket - Swedish National Board of Housing, Building and Planning, Karlskrona, Sweden, 2003.

[5] Kannengiesser T, Schroepfer D. "*IGF 17267 N - Influence of heat input on the residual stresses and cold cracking susceptibility in welded components made of high-strength fine grained structural steels*" (in German), Final report, 2015.

[6] Beg D, Hladnik L. "Eigenspannungen bei geschweißten I-Profilen aus hochfesten Stählen", *Stahlbau*, 63(5), 134-139, 1994.

[7] Clarin M. *High Strength Steel: Local Buckling and Residual Stresses* (PhD-Thesis), Lulea: University of Technology, Sweden, 2004.

[8] Pasternak H, Launert B. "Improved design approaches for the load bearing capacity of welded I-profiles from high strength steel considering realistic residual stresses", *Proceedings of the 7th European Conference on Steel and Composite Structures* (Eds.: R. Landolfo, F.M. Mazzaloni), Naples, Italy, 159-160, 2014.

[9] Rossini NS, Dassisti M, Benyounis KY, Olabi AG. "Methods of measuring residual stresses in components", *New Rubber Materials, Test Methods and Processes*, 35, 572-588, 2012.

[10] Rhode M, Kromm A, Kannengiesser T. "Residual stresses in multi-layer component welds", *Trends in Welding Research 2012: Proceedings of the 9th International Conference* (Eds.: t. DebRoy, S.A. David, J.N. DuPont, H.K. Bhadeshia), Chicago (Illinois), USA, 48-56, 2012.

BOW IMPERFECTIONS FOR FLEXURAL BUCKLING ACCORDING TO EUROCODE 3 PART 1-1

Joachim Lindner[a], Adrian Just[b] and Ulrike Kuhlmann[b]

[a] *Formerly: Technical University Berlin, Chair of Steel Structures, Germany*
[b] *University of Stuttgart, Institute of Structural Design, Germany*

Abstract: The resistance of compression members may be calculated by the design buckling resistance based on the reduction factor χ, which depends on the slenderness or by the cross section resistance based on internal forces according to a second order analysis taking into account also equivalent initial bow imperfections. The second way is advantageous especially in the case of axial forces and bending. Values for equivalent initial bow imperfections are given in codes like Eurocode 3 [1] or DIN 18800 [2]. For comparison initial bow imperfections are derived here from the buckling curves for compression members of different profiles, buckling directions and steel grades based on different types of cross sectional interaction. The results are discussed and commented with regard to the design of compression members.

1. Introduction

According to Eurocode 3 Part 1-1, a member subjected to an axial compression force may be verified by two different approaches: a verification based on the European buckling curves given in EN 1993-1-1, sec. 6.3.1.1, or a cross sectional verification according EN 1993-1-1, sec. 6.2 with bending moments according to second order theory assuming initial bow imperfections e_0 according to EN 1993-1-1, sec. 5.3. The consistency of these two approaches were discussed.

In this paper it is presupposed that the buckling curves used for the equivalent member method, which depend on a statistical evaluation of many tests and a mechanical model, are "correct". Assuming that also the cross sectional verification according to second order theory including an initial bow imperfection results in the same load carrying capacity N_u, an imperfection e_0 can be derived depending on the relative slenderness $\bar{\lambda}$. These imperfections are compared to the values given in EN 1993-1-1, Table 5.1 which do not depend on the relative slenderness. This comparison is done for various cross sections, types of interaction, steel grades and buckling directions.

Furthermore FEM calculations are carried out taking into account plastification within the cross section and along the longitudinal axis, residual stresses, linear elastic-ideal plastic constitutive law and $L/1000$ as bow imperfection. As a result these “real” load carrying capacities are also compared to the different approaches for bow imperfections.

2. Derivation of the bow imperfection

The accurate determination of the equivalent imperfection is carried out on the basis of centrally loaded struts, see Fig. 1, where a sine curve for the initial bow with the maximum value of e_0 is assumed.

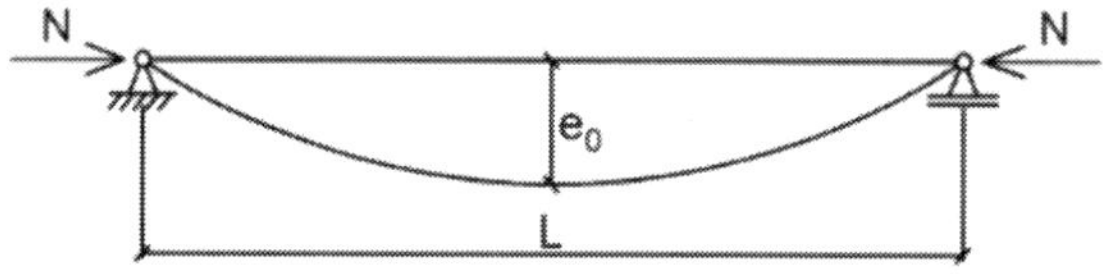

Fig. 1. Centrally loaded member

The equivalent bow imperfection is determined by introducing the moment M, calculated by second order theory, in an appropriate interaction formula. This way is shown exemplarily for the linear imperfection of Eq. (1), [3, 4, 5, 6].

$$\frac{N}{N_{pl}} + \frac{M}{M_{pl}} \leq 1 \tag{1}$$

$$M = N \cdot e_0 \cdot \frac{1}{1 - \frac{1}{\alpha_{cr}}} = N \cdot e_0 \cdot \frac{1}{1 - \frac{N}{N_{cr}}} \tag{2}$$

using $N = \chi \cdot N_{pl}$ (3)

$\chi = f(\bar{\lambda})$ reduction factor for the decisive buckling direction (4)

$\bar{\lambda} = \sqrt{\frac{N_{pl}}{N_{cr}}}$ normalized relative slenderness (5)

leads to

$$e_0 = \frac{(1-\chi) \cdot (1 - \chi \cdot \bar{\lambda}^2)}{\chi} \cdot \frac{M_{pl}}{N_{pl}} \tag{6}$$

Generally this may be written as

$$e_0 = \frac{\psi \cdot (1 - \chi \cdot \bar{\lambda}^2)}{\chi} \cdot \frac{M_{pl}}{N_{pl}} \tag{7}$$

where ψ is the interaction factor for the used interaction formula

$$\psi = \frac{M_{pl,N}}{M_{pl}} \tag{8}$$

For comparison of different influences it is more appropriate not to use the absolute value of e_0 but the non-dimensional value of

$$j = \frac{L}{e_0} \tag{9}$$

In contrast to [3] in this derivation no partial safety factor was taken into account. This means that the amplification factor α_{cr} was taken as a system factor independently from γ_{M0} or γ_{M1}, similar as it was done also for other derivations [6, 7, 8].

3. Interaction factors for different interaction formulae

For the linear plastic interaction it was shown before

$$\psi = 1 - \chi \tag{10a}$$

EN 1993-1-1 [1] gives in chapter 6.2.9.1 various interaction formulae.

For bending of I- or H-profiles about the strong axis y-y (Eq. (6.36))

$$\psi = \frac{1-\chi}{1-0.5\cdot a} \quad \text{but} \quad \psi \le 1.0 \tag{10b}$$

$$\text{where} \quad a = \frac{A - 2\cdot b\cdot t_f}{A}$$

For bending of I- or H-profiles about the weak axis z-z (Eq. (6.37, 6.38))

$$\psi = 1 - \left(\frac{\chi - a}{1-a}\right)^2 \quad \text{but} \quad \psi \le 1.0 \tag{10c}$$

For bending of rectangular hollow sections about the strong axis y-y Eq. (10b) may be used taking t instead of t_f (Eq. (6.39)) and for bending of rectangular hollow sections about the weak axis z-z Eq. (10b) may be used too but with h instead of b and t instead of t_f.

For circular hollow sections CHS:

$$\psi = 1 - \chi^{1,7} \tag{10d}$$

Nonlinear plastic interaction for rectangular solid sections:

$$\psi = 1 - \chi^2 \tag{10e}$$

The German code DIN 18800-1[2] contains the following solutions:

For bending of I- or H-profiles about the strong axis y-y

$$\psi = \frac{1-\chi}{0.9} \quad \text{but} \quad \psi \le 1.0 \tag{10f}$$

For bending of I- or H-profiles about the weak axis z-z

$$\psi = \frac{1-\chi^2}{0.91} \quad \text{but} \quad \psi \le 1.0 \tag{10g}$$

Kindmann/Stroetmann [9] made proposals to modify the rules in EN 1993-1-1, 6.2.9.1:

For bending of I- or H-profiles about the strong axis y-y

$$\psi = \left(\frac{h}{2} - \frac{N_{pl}\cdot(1-\chi)}{4\cdot b\cdot f_y}\right)\cdot(1-\chi)\cdot\frac{N_{pl}}{M_{pl,y}} \quad \text{for} \quad \chi > \frac{t_w\cdot(h - 2\cdot t_f)}{A} \tag{10h}$$

$$\psi = \left(\frac{M_{pl,y}}{\chi\cdot N_{pl}} - \frac{\chi\cdot N_{pl}}{4\cdot t_w\cdot f_y}\right)\cdot\chi\cdot\frac{N_{pl}}{M_{pl,y}} \quad \text{for} \quad \chi \le \frac{t_w\cdot(h - 2\cdot t_f)}{A} \tag{10j}$$

For bending of I- or H-profiles about the weak axis z-z

$$\psi = \left(\frac{b}{2} - \frac{N_{pl}\cdot(1-\chi)}{8\cdot t_f\cdot f_y}\right)\cdot(1-\chi)\cdot\frac{N_{pl}}{M_{pl,z}} \quad \text{for} \quad \chi > \frac{t_w\cdot h + 0.7\cdot r^2}{A} \tag{10k}$$

$$\psi = \left(\frac{M_{pl,z}}{\chi\cdot N_{pl}} - \frac{\chi\cdot N_{pl}}{4\cdot h\cdot f_y}\right)\cdot\chi\cdot\frac{N_{pl}}{M_{pl,z}} \quad \text{for} \quad \chi \le \frac{t_w\cdot h + 0.7\cdot r^2}{A} \tag{10m}$$

For circular hollow sections CHS the exact solution is given by Eq. (10n)

$$\psi = \cos(0.5\cdot\chi\cdot\pi) \tag{10n}$$

4. Influence of the cross section

As it was shown before the interaction factor ψ depends on the reduction factor χ for flexural buckling which is based on the relevant buckling curve. The buckling curve has to be chosen from Table 6.2 [1], for rolled and welded I-sections depending on their dimensions, buckling direction and steel grade.

For the normal steel grade S235 buckling curves a, b and c are relevant for I-profiles up to a flange thickness of t_f = 100 mm. Therefore all "normal" rolled I- and H-profiles profiles were investigated for these three buckling curves, taking into account the following profiles:

- IPE 80 to IPE 600, IPEo 180 to IPEo 600, IPEv 400 to IPEv 600, HEA 100 to HEA 1000, HEB 100 to HEB 1000, HEM 100 to HEM 1000, HEAA 100 to HEAA 1000

The results in this chapter refer only to the interaction formulae given in EN 1993-1-1, sec. 6.2.9.1 [1], this means (Eq. (10b), (10c)) were used.

As typical example the results of the non-dimensional value $j = L/e_0$ (Eq. (10)) are given in dependence on the slenderness for a selection of rolled I-profiles depending on buckling curve "b" in Fig. 2.

Two groups of results may be recognized. One group of profiles associated to buckling direction z-z like e.g. HEB 1000 with a minimum value of $j \approx 196$, the other group of profiles associated to buckling direction y-y like e.g. HEA 160 with a minimum value of $j \approx 270$. The differences between different profiles in one group are small. The value of j proposed in Table 5.1 of EC 3 of $j = 200$ is in the range of the results given here.

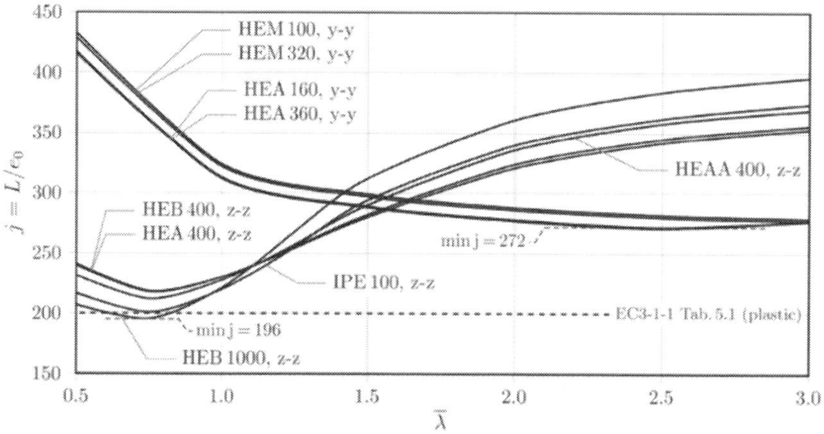

Fig. 2. Non-dimensional representative value $j = L/e_0$ of initial bow imperfections for profiles associated to buckling curve "b", S235, depending on the slenderness

Further calculation show that the results of the non-dimensional value $j = L/e_0$ (Eq. (10) for typical rolled I-profiles circular hollow sections CHS associated to buckling curve "a" are very similar to each other with a minimum value of $j \approx 410$. Additional investigations on CHS [4] show that different diameters D and ratios D/t have only a very little influence on the value j. The value of j proposed in Table 5.1 of EC 3 of $j = 250$ is much smaller than the values derived from the buckling curves depending on the slenderness.

The results for the non-dimensional value $j = L/e_0$ (Eq. (10)) for typical rolled I-profiles associated to curve "c" for the buckling direction z-z lead to a minimum value of $j \approx 165$. Circular hollow sections CHS behave similar with a slightly higher values of $j \approx 180$. The value of j proposed in Table 5.1 of EC 3 for curve "c" of $j = 150$ is in the range of the results given here.

So the results calculated and compared support for a great extend the values given in Table 5.1 of EC3 and would even partly allow for a modest relaxation.

5. Influence of different interaction equations

The type of interaction equation assumed for the cross section verification based on second order analysis has a great influence on the calculated value of the initial bow imperfection e_0 respectively on the relative value $j = L/e_0$. In addition to the interactions mentioned in chapter 3 the mechanically exact interaction is taken into account when calculating the imperfection derived from reduction factor χ_u based on FEM calculations (as described in section 1). In all cases the yield strength of $f_y = 235$ N/mm^2 has been assumed.

As an example of several calculations realised, the results for HEM 120, buckling direction z-z, buckling curve "c" are given in Fig. 3. Results for profiles belonging to buckling curve "a" or "b" are comparable. The interaction of EN 1993-1-1, sec 6.2.9.1 leads to the minimum values of j, comparable values are derived of the interaction of *Kindmann/Stroetmann* in [9], and the values derived of the FEM calculations are more favourable. Linear plastic interaction gives higher values of j on the safe side. If the load carrying capacity is limited to the yield strength then a linear elastic interaction is decisive. In this case the value j of plastic linear interaction may just be multiplied by a factor of M_{pl}/M_{el}. The constant values of EN 1993-1-1, Table 5.1, are suitable in all cases, but may be improved if linear interaction is used to verify the section.

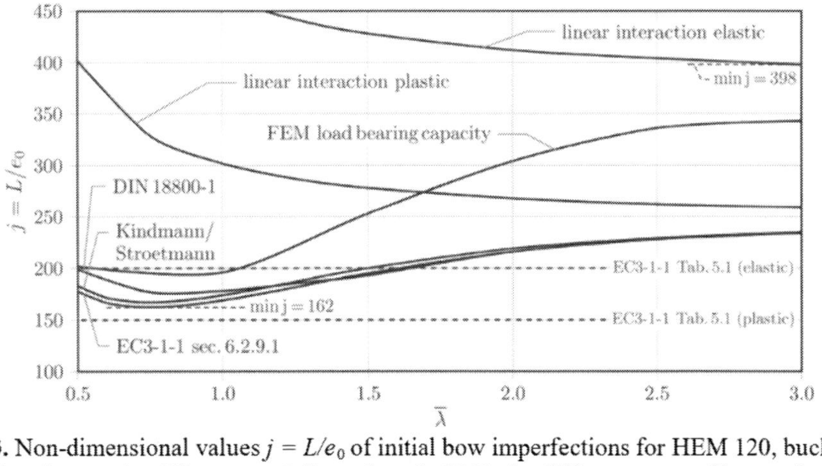

Fig. 3. Non-dimensional values $j = L/e_0$ of initial bow imperfections for HEM 120, buckling direction z-z, buckling curve "c", steel grade S235, for different types of interaction

6. Influence of the steel grade

All previous results are presented in dependence of the non-dimensional slenderness $\bar{\lambda}$. From Eq. (9) the relative bow imperfection j depends on L by $j = L/e_0$, where L can be written as given in Eq. (11).

$$L = \bar{\lambda} \cdot \lambda_1 \cdot i = \bar{\lambda} \cdot i \cdot 93.91 \cdot \sqrt{\frac{235}{f_{y,act}}} \tag{11}$$

where $\bar{\lambda}$ non-dimensional slenderness

$$\lambda_1 = \pi \cdot \sqrt{\frac{E}{f_y}} = 93.91 \cdot \varepsilon \qquad \text{and} \qquad \varepsilon = \sqrt{\frac{235}{f_{y,act}}}$$

i radius of gyration

As a consequence, the values of j for steel grades higher than S235 may be calculated based on the values of j derived from the various interaction formulae of section 3 for S235 by multiplying those by the factor ε, where $\varepsilon = 0{,}924$ for S275, $\varepsilon = 0{,}814$ for S355, $\varepsilon = 0{,}748$ for S420. For S460 special buckling curves are given in EN 1993-1-1 [1]. Therefore for these buckling curves "a_0", "a" and "b" all rolled I- and H-profiles mentioned in chapter 4 are also investigated here. Fig. 4 and Fig. 5 show examples of comparison for selected sections in steel grade S355 and S460.

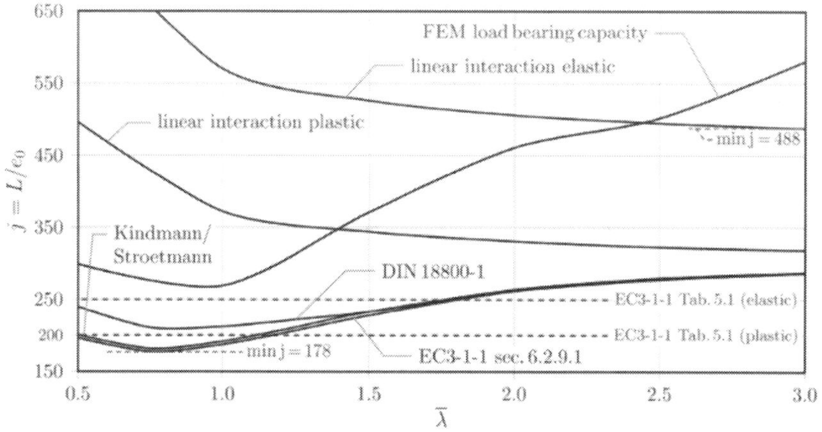

Fig. 4. Non-dimensional values $j = L/e_0$ of initial bow imperfections for HEB 400 buckling direction z-z, buckling curve "b", steel grade S355, for different interaction formulae

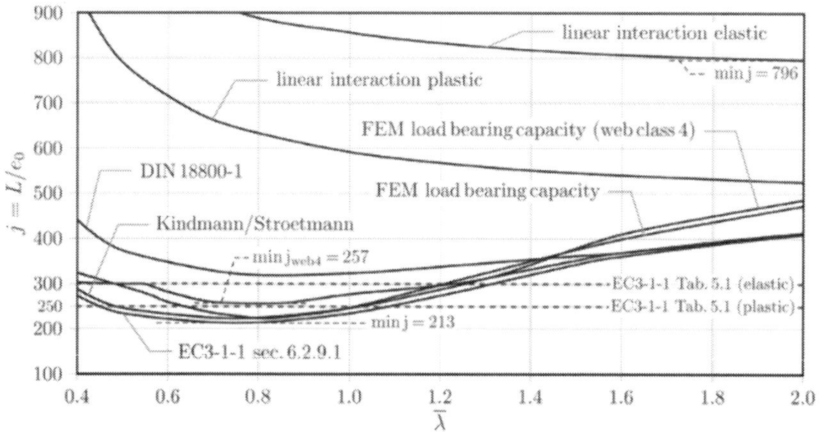

Fig. 5. Non-dimensional values $j = L/e_0$ of initial bow imperfections, for IPE 500, buckling direction z-z, buckling curve "a", steel grade S460, for different interaction formulae

In Fig. 5 results for an IPE 500, buckling direction z-z, in steel grade S460 are given where this profile is classified in buckling curve "a". This considers that the classification in EN 1993-1-1, Table 6.2 is incorrect and must be corrected [6, 10]; it will be changed according to a recent amendment decided already in TC 250/SC 3 [11].

In all cases the results from EN 1993-1-1, sec 6.9.2.1 [1] and *Kindmann/Stroetmann* [9] are close together, whereas the plastic linear interaction leads to much higher j-values. Obviously the constant values of EC3 Table 5.1 for the plastic verification ($j = 200$ for curve "b" and $j = 250$ for curve "a") are not sufficiently covering the results of the nonlinear plastic interaction.

So the comparison with the "real" values derived by FEM calculations are of interest. Regarding the FEM calculations special conditions are present for **rolled** I- and H-sections. It was shown earlier by ECCS, explained in [6], that the residual stresses do not change their absolute values with higher steel grades, but are approximately the same as for S235. Therefore several FEM ultimate load calculations were carried out again for higher steel grades but using the residual stresses of S235. Some of the results are shown in Fig. 4 and Fig. 5.

It may be seen from Fig. 4 that numerically the value j from EC 3 for plastic capacity is underestimated, but the FEM results are more favourable and therefore no reduction of the EC 3 value seems necessary.

The results for IPE 500 for steel grade 460 are calculated for the full cross section, neglecting the fact that the c/t-ratio of the web exceeds the value of EN 1993-1-1, Table 5.2: $c/t = 426/10.2 = 41{,}8 > 27.2$ ($=38\ \varepsilon$). Additionally a FEM calculation was carried out taking into account that the web is categorized as class 4 web. The results are also given in Fig. 5. In fact the differences between both FEM calculations are small and may be neglected. However in this case the j-value from EC 3 Table 5.1 seems to be too optimistic and should be corrected, although the result for the class 4 web confirms in this special case the EC 3 value. Generally it can be determined, that the results of j for all investigated profiles are also valid in case the b/t-ratio of the web exceeds the value for class 2 sections.

7. Axial compression and bending

All results shown in the previous chapters are based on investigations on centrally loaded members. If bending moments are present the results may be more unfavourable, because a larger part of the cross section in longitudinal direction may be plastified, thus leading to a stiffness reduction and therefore higher necessary bow imperfections for simplified calculations. If this is neglected the load carrying capacity may be underestimated. For the German code DIN 18800-2 intensive additional investigations were carried out [5] leading to a maximal deviation of up to approximately 5% for constant moment distribution and very slender members. At the time this was also discussed in a great detail in TC8 of ECCS. At the end it was decided to neglect this effect of axial compression and bending because higher errors occur for very special cases only. This may be assumed in a similar way to be valid now.

8. Summary and conclusions

Investigations were carried out concerning the necessary numerical values for initial bow imperfections e_0 given as non-dimensional representative bow imperfection $j = L/e_0$. The influences of different rolled I-profiles, different interaction equations and different steel grades were looked at. The following conclusions may be drawn:

- For S235 the constant values of initial bow imperfections given in Table 5.1 of EC3 Part 1 are mainly in the range of values calculated here, independent of the assumed interaction formulae. A slight difference for buckling curve "b", buckling direction z-z, may be acceptable.

- The linear elastic interaction leads to much more favourable values so that it is recommended to define improved values for that situation.
- For steel grades higher than S235 the constant values of Table 5.1 of EC3 Part 1 may nevertheless be kept for rolled I- and H-profiles only based on results of refined FEM calculations considering plastification along the members, realistic residual stresses and a geometrical imperfection of $L/1000$. Exceptions were identified for sections in buckling direction z-z and S460, buckling curves "a" and "b".
- In other cases for all profiles with steel grades higher than S235 the j-values given in Table 5.1 of EC3 Part 1-1 have to be multiplied by ε with $\varepsilon = 0{,}924$ (S275), $\varepsilon = 0{,}814$ (S355), $\varepsilon = 0{,}748$ (S420).
- Additional investigations for welded sections and the influence of additional bending moments would allow to confirm and improve the approach.

References

[1] EN 1993-1-1:2009. *Eurocode 3: Design of steel structures – Part 1-1: General rules and rules for buildings*.

[2] DIN 18800-2:2008-11. *Stahlbauten-Teil 2: Stabilitätsfälle - Knicken von Stäben und Stabwerken (Steel structures-Part 2: Stability - Buckling of bars and skeletal structures).*

[3] Taras A, Kuhlmann U, Just A. "Design of Compression Members by 2nd Order Analysis – Imperfection Amplitudes, Material Dependency, Influence of γ_{M1}", *Document ECCS – TC8, TC8-2013-06-005*, June 2013.

[4] Lindner J. "Equivalent initial bow imperfections for hollow sections", *Tubular Structures VI, Proceedings 6th Intern. Symposium on Tubular Structures*, Melbourne/Australia, Balkema, Rotterdam, 159-162, 1994.

[5] Lindner J, Scheer J, Schmidt H. *Stahlbauten, Erläuterungen zu DIN 18800 Teil 1 bis Teil 4 (Steel structures, Commentary to DIN 18800 Part 1 to Part 4),* Berlin, Beuth Verlag/ Ernst & Sohn, 3. Ed., 1998.

[6] Kuhlmann U, Feldmann M, Lindner J, Müller C, Stroetmann R. u. Mitarbeit v. Just A. *Eurocode 3 Bemessung und Konstruktion von Stahlbauten, Band 1: Allgemeine Regeln und Hochbau. DIN EN 1993-1-1 mit Nationalem Anhang. Kommentar und Beispiele* (in German), Bauforumstahl, Berlin: Beuth Verlag / Ernst & Sohn, 2014.

[7] Boissonnade N, Greiner R, Jaspart JP, Lindner J. *ECCS TC 8, No. 119: Rules for member stability in EN 1993-1-1, Background documentation and design guidelines*, ECCS publication, Brussels, 2006.

[8] Sedlacek G. *Consistency of the equivalent geometric imperfection used in design and in the tolerances for geometric imperfection used in execution*, Document ECCS-TC8-2010-06-001, February 2010.

[9] Stroetmann R, Faßl T. *Influence of the interaction formulas for the plastic cross section resistance to the assumption of bow imperfection for flexural buckling.* October 2014, working paper N0019 working group 1-1 of CEN/TC 250/SC 3.

[10] Lindner J. *Classification of rolled I-profiles fabricated in steel grade S460 within table 6.2 of EN 1993-1-1 – rev. 3*, March 2015, working paper N0037 working group 1-1 of CEN/TC 250/SC 3.

[11] CEN/TC 250/SC 3, Document N2164, *Classification of rolled I-Profiles fabricated in steel grade S460 within Table 6.2 of EN 1993-1-1*, Decision 19/2015 on Amendment AM-1-1-2015-001, 2015.

EXPERIMENTS ON THE ROTATIONAL CAPACITY OF BEAMS MADE OF HIGH STRENGTH STEEL

Nicole Schillo and Markus Feldmann

RWTH Aachen University, Institute of Steel Construction, Aachen, Germany

Abstract: In scope of the research project RUOSTE, which is partly funded by the RFCS, the rotational capacity of beams made of high strength steel grades S700 and S960 was investigated. Beams made from high strength steel are supposed to have lower rotational capacity, and are thus excluded from plastic/plastic design according to [1] and [2]. Although the strain hardening of high strength steel material is generally lower in comparison with mild steels, combined with the actual ductility it might be still sufficient to reach the respective cross-section classification. In the research presented here, 8 tests were carried out to assess the limit of cross-section class 1 for high strength steel.

1. Introduction

1.1 Rotational capacity of beams

EC3-1-1 [1] allows the exploitation of plastic resistance of members, defining cross-section class 1 (plastic hinge theory) and cross-section class 2 (punctual exploitation of N_{pl} or M_{pl}). In plastic hinge theory it is assumed that after M_{pl} is reached at one point of a statically indeterminated structure, the system is able to redistribute additional moments to another point until M_{pl} is reached there as well. This is applicable until the system reaches its ultimate static strength. After formation of the last plastic hinge the system becomes kinematic.

However, to reach M_{pl} a certain plastic rotation and thus rotational capacity is necessary: the structural system must be able to endure the displacement controlled loading until the subsequently developing hinges reach their plastic resistance. The required rotation depends not only on the cross-section, but also on the system itself (frame, continuous beam, etc.) and the loading conditions. Exemplarily, rotational capacity requirements for different structures are shown in Fig. 1. One of the most unfavourable systems is considered to be a continuous beam with $R_{req} = 3$, which is taken in [1] in general as a minimum requirement.

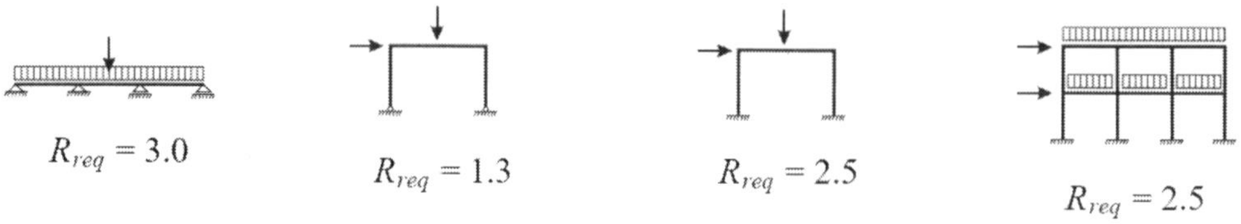

Fig. 1: Required rotational capacity in dependency of structural system [10]

The rotational capacity is usually evaluated in 3-point bending tests, as here the strain concentration is much more localized than in 4-point bending tests or continuously loaded beams. In Fig. 2 a comparison of the consequences of a concentrated and a uniform distributed load on the rotational capacity is shown. The steeper gradient of moment for the concentrated load leads to more localized straining of the structure, and thus to lower or more conservative rotational angles. For uniform loaded structures, the strain affects a larger area, leading to higher rotational angles but lower M_u/M_{pl}-values. The moment-rotation curve can be separated in both cases in a stable (the structure is with increasing angle still able to take more load) and an unstable (with increasing angle the load is decreasing) branch. The maximum point is defined by M_u, after which further loading is applied displacement controlled. Until M_u, the system benefits from the strain hardening of the material. In competition, local buckling effects lead to a geometric softening of the systems stiffness, which dominate in the unstable branch. This effect of accompanying displacement controlled loading allows also to take the falling branch into consideration for the calculation of rotational capacity.

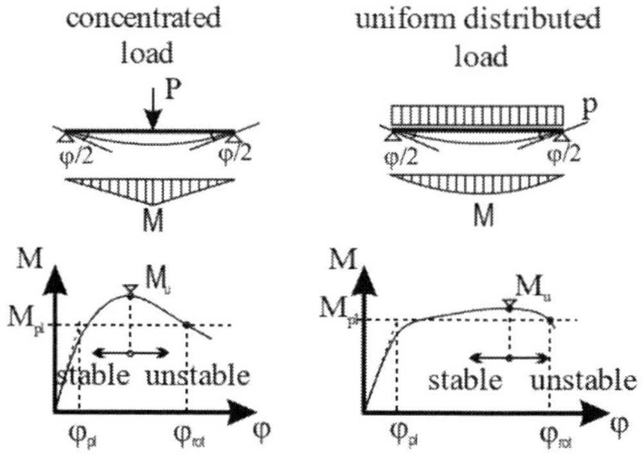

Fig. 2: Comparison R_{act} for concentrated and distributed load

Fig. 3 shows how to assess the rotational capacity from a given Moment-Rotation curve.

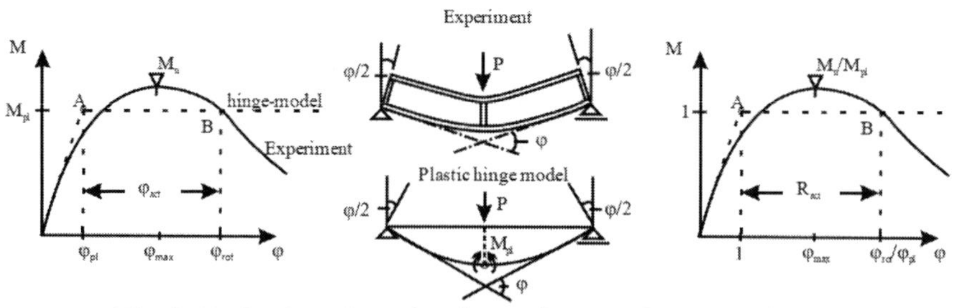

Fig. 3: Evaluation of rotational capacity φ_{act} from experiments

φ_{pl} is calculated theoretically using the plastic-hinge model, assuming elastic rotation in the hinge until reaching the first time the full plastic moment M_{pl} in point (A). When the curve falls later under the M_{pl}-value (B) due to local instability, φ_{rot} is defined and the actual (plastic) rotation can be evaluated:

$$\varphi_{act} = \varphi_{rot} - \varphi_{pl}$$

The normalized requirement R_{act} is then subsequently assessed by drawing M/M_{pl} over φ_{rot} $/\varphi_{rot}$ (Fig. 3, right). To assess the actual rotational capacity, only the plastic rotation of the beam is examined, the elastic part has to be excluded:

$$R_{act} = \frac{\varphi_{rot} - \varphi_{pl}}{\varphi_{pl}}$$

In [3], it is pointed out that the angle measured as the tangent on the member at the plastic hinge under the concentrated load is much lower than the angle measured at the supports. While the plastic hinge model assumes much localized infinite strains with steep rigid body rotation, in reality the strains are finite but distributed over a certain length. This leads to a curvature of the member rather than a rotation. [4] follows this argumentation and uses a fixed strain value of φ_{edge} = 4 % at the edge to evaluate the rotational capacity. This strain based model is used to evaluate the test results of [5], but finds much lower values for M/M_{pl} than given in experiments.

Roik and Kuhlmann investigated in [5] and [6] the influence of web-stiffness, b/t-ratios and moment gradient on the rotational capacity. They varied the span of specimens and found higher rotations and strains when using shorter spans. They suggested to define the b/t limit in dependence of the required rotational capacity.

Petersen [7] performed tests on beams made of St37 and St52 to study the effects of different stiffeners and evaluation processes on the rotational capacity of beams. He found the usage of using lower or upper yield strength in the evaluation process of utter importance, whereas in the present study always the 0.2% proof stress was used. He investigated also the strain development in the compression flange. The experiments were done on specimens with and without stiffeners under the load introduction. The kind of stiffener influenced the rotational capacity, a coupling of upper and lower flange led to increased M/M_{pl}-ratios, especially after first buckling effects occurred. Not only due to this findings, EC3-1-1 [1] requires to stiffen plastic hinges (section 6.3.5.2). As a design check, he suggests to use the fulfilment of the b/t requirement as well as an additional rotational check. [1] allows to neglect the design check for rotational capacity, if the proposed b/t-ratio is fulfilled and thus cross-section class (CSC) 1 is reached.

1.2 Material behaviour of high strength steel

The behaviour of high strength steel material is considerably different compared with mild steels in terms of strain hardening and ductility. While low steel grades develop a significant yield plateau and profit usually from large f_u/f_y-ratios, the stress-strain characteristic of higher steel grades show usually no distinguished yield point. The yield strength is therefore usually assessed using the 0.2% proof stress. Steels grades up to S460 are included in the current EC3-1-1 [1], for the higher grades EN1993-1-12 [2] provides further requirements:

- Ultimate strength to yield strength should be at least 1.05
- Strain at ultimate strength ε_u should be not less than $15 \cdot f_y/E$
- elongation at fracture ε_f on a gauge length of $5.65\sqrt{A_0}$ (where $\sqrt{A_0}$ is the original cross-sectional area) should be at least 10%

Especially the second requirement is hard to achieve, and recent research on wide-plate tension tests [8][9] showed the inappropriateness of that rule: the material properties are derived from coupon tests, where the derived values for strain at uniform elongation and strain at fracture are depending also on the geometry. For structural members, the true stress-strain characteristic is decisive, and the mentioned requirements have to be adjusted to be transferable to other geometries than the one-dimensional coupon test [8]. In the present study, the material properties are assessed therefore to investigate if the requirements are justified.

For beams made of high strength steel (hss) it could be shown that with increasing yield strength the rotational capacity decreases [10]. This is related to the missing yield plateau of HSS-material and the lower f_u/f_y-ratios, implying lower strain hardening effects. However, experimental tests on high strength steels are scarce, and the models used so far for the steels covered by [1] to predict rotational capacity showed to be not applicable when extrapolated to hss [6][10][11].

2. Experimental programme

2.1 Test matrix and denomination

In order to exploit plastic hinge theory and thus achieve CSC 1, [1] requires in clause (5.6) stiffeners in the areas of applied load. However, concern was raised if the welding of such stiffeners could lead to softening in the heat affected zone of high strength steel specimens. As the top flange was very thick and allowed for a large distribution area of the concentrated force, the specimens were designed without stiffeners. It can be assumed that if the welding could be conducted such that the cross section is not weakened, additional stiffener would lead to higher rotational capacity than obtained in the tests described here (see also [7]).

The specimens were designed such that they fulfilled CSC 1 requirements of [1]. The upper flange of the I-section was much thicker to avoid local buckling in the compression zone. Half of the specimens were tested as 3-point bending tests, and the other half as 4-point bending tests. In 3P the strains are very localized: higher strains are achieved, but lower rotational capacity. Due to the constant moment in 4P tests, the strains are spread over a wider range, while the rotations achieved are higher. The test matrix is shown in Table 1, together with the CSC-limits and actual classification. The denomination of the tests include the test sequential number, indication of loading (3P, 4P), the nominal yield strength and the height of the web, the width of the upper flange, as well as the thickness of the web (which equals the thickness of the lower flange). The upper flange was for all specimens 15 mm.

Table 1: Test matrix of test series 1, bending tests to evaluate strain characteristics

Test	$f_{y,web}$ [N/mm²]	$f_{y,up,flange}$ [N/mm²]	Web		Flange	
			c/t, actual	CSC1, limit	*c/t*, actual	CSC1, limit
01_3P_S960_160_140_8	985	1000	19.1	57.3	4.4	4.4
02_3P_S700_160_140_8	738	751	19.7	67.0	4.5	5.1
03_3P_S700_160_110_6	738	751	25.0	67.7	3.4	5.1
04_3P_S960_160_110_6	985	1000	24.7	57.0	3.4	4.8
05_4P_S700_160_110_6	738	751	25.0	67.7	3.4	5.1
06_4P_S960_160_110_6	985	1000	24.7	57.0	3.4	4.8
07_4P_S700_160_140_8	738	751	19.7	67.0	4.5	5.1
08_4P_S960_160_140_8	985	1000	19.1	57.3	4.4	4.4

2.2 Material properties and fabrication

Material was provided by two steel producers, indicated in the Tables 1 and 2 with P1 and P2. In case of P1, the plate material was produced with quenched and tempering process according to [12]. The two materials of P2 fulfilled the requirements of [13], where S500M was produced using thermomecanically rolling processes and additionally, for S960, a direct quenching treatment was used. The beams were assembled in the workshop of the respective producers with matching welds. The main material properties from the material test certificates are summarized in Table 2.

Table 2: Material properties of tested beams

Test	Prod.	t [mm]	f_y [N/mm²]	f_u [N/mm²]	f_y/f_u	ε_u [%]	$\varepsilon_{u,EC}$ [%]	A_5 [%]
01_3P_S960_160_140_8	2	8.07	985	1044	1.06	5.3	7.04	15
02_3P_S700_160_140_8	1	8.12	738	818	1.11	4.7	5.27	15.3
03_3P_S700_160_110_6	1	5.99	746	829	1.11	4.4	5.33	13.7
04_3P_S960_160_110_6	2	6.08	1010	1053	1.04	4.9	7.21	16
05_4P_S700_160_110_6	1	5.99	746	829	1.11	4.4	5.33	13.7
06_4P_S960_160_110_6	2	6.08	1010	1053	1.04	4.9	7.21	16
07_4P_S700_160_140_8	1	8.12	738	818	1.11	4.7	5.27	15.3
08_4P_S960_160_140_8	2	8.07	985	1044	1.06	5.3	7.04	15
Upper Flange, 15mm	1	15	751	812	1.13	-	5.36	17.9
Upper Flange, 15mm	2	14.83	1000	1052	1.05	5.39	7.14	15

The requirements given in EC3-1-12 [2] regarding the A_5 strain at fracture are fulfilled by all steels investigated, even if applied on S960. The f_u/f_y-ratio is fulfilled by the S700 steel grades. From the batches of S960, one would be below the threshold of 1.05 with 1.04. The EC3 requirement regarding the uniform elongation at ultimate load ε_u was in no case fulfilled.

2.3 Test setup and procedure

The basic dimensions and applied measuring equipment used for the 3- and 4-point-bending tests are shown schematically in Fig. 4.

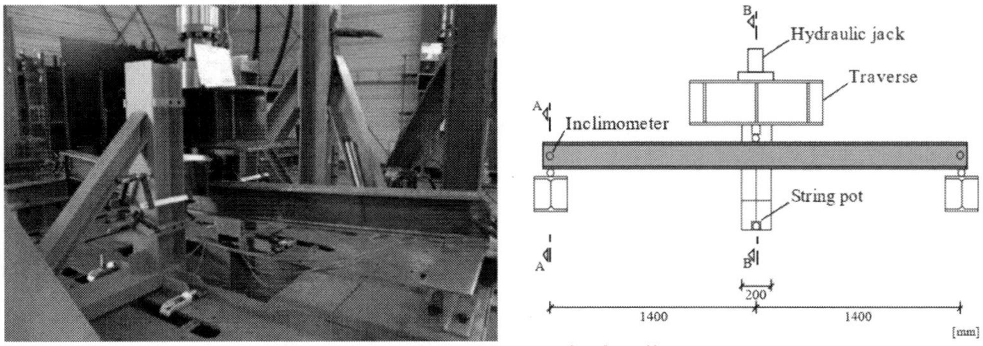

Fig. 4: Test-setup 3-point bending

The prevention of lateral torsional buckling was achieved by steel brackets at mid-span and at the supports. As top and lower flange had different widths only the top flange was restrained laterally at midspan, as this section part was subjected to compression and therefore prone to lateral deflection. At the supports, the lower flanges were constrained additionally by two lateral supports at each side. The testing machine was a 1 MN hydraulic jack.

All specimens were loaded displacement controlled. As the displacement capacity of the hydraulic jack (10 cm) was not sufficient, the specimens were loaded until the limit of the hydraulic jack was reached and then unloaded. The plastic deformation was then used to underput the load introduction with further steel plates. In the next reloading cycle the appliance of further displacement was thus possible.

To determine the end-rotation of the beam, two inclinometers at each end of the beams measured the progressive inclination during testing. The vertical displacement was recorded with a string-pot sensor, applied at mid-span.

Strain gauges were applied on both flanges and the web at a longitudinal distance of 90 mm from midspan, to assess the attained local strains during testing. The strain gauges were not placed directly at midspan to prevent the strain gauges placed on the top flange being damaged by the load applying hydraulic jack. Only for Test 05 and 08 were the strain gauges applied directly at midspan. Two strain gauges are placed on the outer surfaces of the top and lower flange, respectively. Four additional strain gauges were distributed over the height of the web to assess the strain development and the corresponding shifting of the neutral axis.

3. Evaluation of experiments

3.1 Rotational capacity

The evaluation regarding the rotational capacity of the beams is summarized in Table 3, a selection of corresponding normalized moment-rotation curves are depicted in Fig. 5. Specimens made of S700 material showed always higher rotations compared to their S960 counterpart. Three different failure modes could be observed, although they all appeared far in the plastic branch: Test 01, 07 and 08 failed "ductile", meaning that deformations occurred in plane, slight necking effects on the lower flange could be observed. Test 01 was aborted due to testing rig issues, a higher rotational capacity could be expected therefore. Test 02 failed finally due to ductile cracking of the lower flange after considerably rotation. The tests 03 to 06 showed a lateral torsional buckling mode, resulting in a swaying of the lower flange. The plastic hinge was reached by all these specimens but slightly missed by Test 06 (M/M_{pl}= 0.99), in consequence no rotational capacity on M_{pl}-level could be evaluated. An evaluation on a lower load level (e.g. 95% M_{pl}) would have resulted in a considerable rotational capacity. Rotations in 3-point bending tests are in general lower but combined with higher strains, thus the 4-point bending tests show higher rotational capacity although here the M_{max}/M_{pl}-values are lower. The 4-point-bending tests 05, 07 (S700) and 08 (S960) had to be aborted due to testing rig limits.

Requiring a rotational capacity of R_{req} = 3, it can be concluded that the beams made of S700 material are in general capable of fulfilling the requirements for cross-section class 1, while specimens made of S960 do not.

Table 3: Evaluated rotational capacity and ultimate failure mode

Test	M_{max} [kNm]	M_{pl} [kNm]	M_{max}/M_{pl} [-]	$\varphi_{max,left/right}$ [°]	R_{act} [-]	Ult. Failure
01_3P_S960_160_140_8	294.5	278.5	1.06	5.64	1.48	ductile
02_3P_S700_160_140_8	222.1	201.4	1.10	9.85	3.61	Crack
03_3P_S700_160_110_6	172.5	157.7	1.09	9.09	3.40	LTB
04_3P_S960_160_110_6	218.6	212.7	1.03	8.33	1.66	LTB
05_4P_S700_160_110_6	166.5	157.7	1.06	13.55	5.04	LTB
06_4P_S960_160_110_6	210.8	212.7	0.99	10.14	-	LTB
07_4P_S700_160_140_8	217.4	201.4	1.08	13.00	4.97	ductile
08_4P_S960_160_140_8	287.3	278.5	1.03	9.26	2.42	ductile

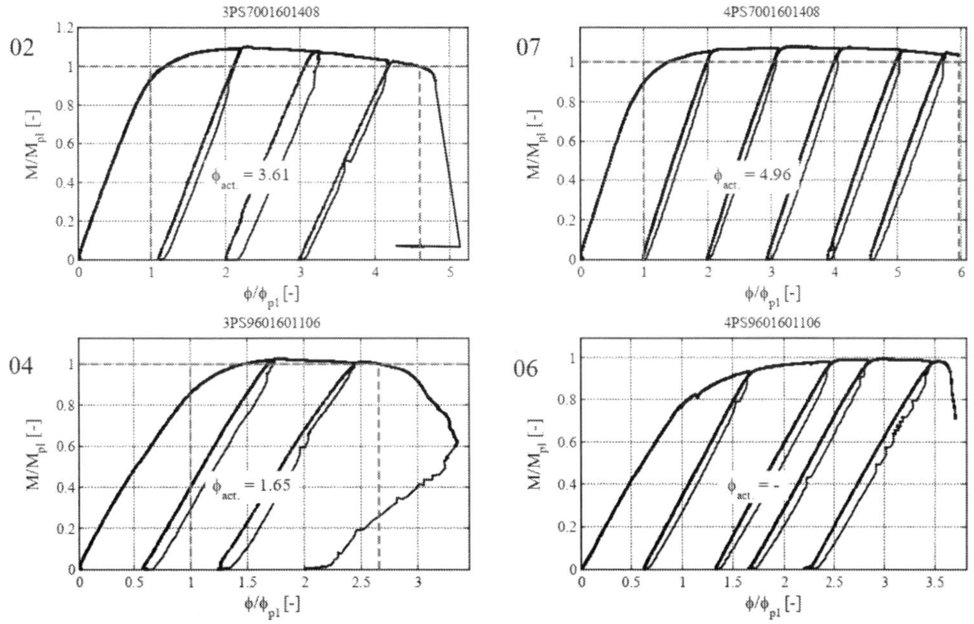

Fig. 5: Normalized Moment-Rotation Curves: 3P (left), 4P (right), S700 (top) and S960 (bottom)

3.2 Strain development during testing

Strains were measured at 90 mm from the load introduction, Test 05 and 08 directly at mid-span. The tensile strains achieved by the S700 specimens varied between 4.72% up to 11.17%. The specimens of 8 mm web thickness failed ductile, while the 6 mm specimens were prone to stability failure and reached thus lower strains. The thinner web results in a torsional restrain on the outstanding flanges, influencing the rotational capacity. This was observed also by [6] and [10]. The strains achieved by the S960 specimens varied between 5.20% and 7.32%.

The strains attained at the edge of the lower flange are at ultimate load (M_u) considerably lower than the strains required for ε_u in [1], but the structural behaviour was not affected. This is due to the different stress-strain state, which is 1D in the coupon test and 3D in the actual structural member. The tests serve as practical example that the aforementioned requirement is not adequate to formulate the actual demands. The background is further explained in [9].

4. Conclusions

1. Presuming a rotational capacity requirement of R = 3, it can be concluded all specimens made from S700 material with single symmetric specimens achieved this value. The S960 specimens did not reach R = 3, but 2 out of 4 had to be aborted due to the large deflections and corresponding testing rig issues. One test (06) did not reach M_{pl}. This test was conducted as 4-point bending test, which implies lower M/M_{pl}-values than to be expected in 3-point bending tests. The specimen was also characterized by a thinner web, implying lower torsional restrain on the outstanding upper flange.
2. For the lower steel grades, S700 and consequently S500, the restrictions given in [2] for plastic design might be considered as too conservative.
3. Current CSC limits for beams in bending seem applicable for S700, provided lateral torsional buckling can be excluded. The S960 specimens did not reach CSC1, although the

usage of stiffeners might improve the rotational capacity. Further experiments are recommended and underway.
4. The material properties regarding f_u/f_y-ratio and elongation at fracture was achieved by all steel grades, one material set of S960 had a slightly lower yield-strength ratio than required with 1.04. The corresponding specimen (Test 04) did reach M_{pl}, but did fail before R = 3 due to lateral torsional buckling.
5. The material characterization in terms of uniform elongation ε_u was in all cases lower than required in EC3, however the overall structural behaviour seemed not to be detrimentally influenced by this fact.

Acknowledgments

The authors want to thank Louis Mertens for his great dedication and assistance during the tests.

References

[1] European Committee for Standardization. *Eurocode 3: Design of steel structures — part 1-1: General rules and rules for buildings*, 27.10.2010.

[2] European Committee for Standardization. *Eurocode 3: Design of steel structures — part 1-12: General - high strength steels*, 28.10.2010.

[3] Petersen C. *Stahlbau: Grundlagen der Berechnung und baulichen Ausbildung von Stahlbauten,* Vieweg, 3 edition, 1992.

[4] Scheer J, Nölke H. "Anmerkungen zur Fließgelenkrotation in stabtragwerken", *Stahlbau*, 72(7):517–526, 2003.

[5] Roik K-H, Kuhlmann U. "Experimentelle Ermittlung der Rotationskapazität biegebeanspruchter I-Profile", *Stahlbau*, 56(12), 353–358, 1987.

[6] Roik K-H, Kuhlmann U. "Rechnerische Ermittlung der Rotationskapazität biegebeanspruchter I-Profile", *Stahlbau*, 56(11):321–327, 1987.

[7] Petersen C. *Rotationskapazität in Fießgelenken als Grenzzustandsgroesse.* In DASt Bericht aus Forschung, Entwicklung und Normung, Volume 17/1990, pages 43–54. Stahlbau-Verlagsgesellschaft, Köln, 1990.

[8] Feldmann M, Schaffrath S, Schillo N. *Derivation of new, optimized strength criteria for steel components*. Düsseldorf.

[9] Schaffrath S, Schillo N, Feldmann M. "Derivation of strain requirements for high strength steel using Johnson Cook Model", *Proceedings of Nordic Steel Conference 2015*, Tampere, 2015.

[10] Spangemacher R. *Zum Rotationsnachweis von Stahlkonstruktionen, die nach dem Traglastverfahren berechnet werden.* PhD thesis, RWTH Aachen, Aachen, 1992.

[11] Feldmann M. *Zur Rotationskapazität von I-Profilen statisch und dynamisch belasteter Träger*. PhD thesis, RWTH Aachen, Aachen, 1994.

[12] European Committee for Standardization. *Hot rolled products of structural steels - part 6: Technical delivery conditions for flat products of high yield strength structural steels in the quenched and tempered conditions*; German version prEN 10025-6:2011.

[13] EN 10149. *Hot rolled flat products made of high yield strength steels for cold forming - part 2: Technical delivery conditions for thermomechanically rolled steels*; German version, 12.2013.

GLOBAL BUCKLING BEHAVIOUR OF HIGH STRENGTH STEEL HOLLOW SECTION COLUMNS

Balázs Somodi and Balázs Kövesdi

Budapest University of Technology and Economics, Department of Structural Engineering, Hungary

Abstract: The application field of high strength steel (HSS) is continuously growing nowadays, but there are only a few available test results for flexural buckling behaviour of HSS hollow section columns. The present study introduces an experimental program where S420, S460, S500, S700 and S960 welded and cold-formed hollow section columns are tested. The out-of-straightness and the residual stress of the members are measured and residual stress models are also proposed. The flexural buckling resistance of the columns are measured, statistically evaluated, and appropriate buckling curves are proposed.

1. Introduction

The application field of high strength steel (HSS) structures is growing nowadays in the civil engineering praxis due to the numerous advantages of the HSS compared to normal strength steel (NSS): economic design, material saving, possibility of creation lighter and more aesthetic structures. However, the application range of the current EN 1993-1-1 [1] for column (flexural) buckling is limited to ordinary steel materials up to the grade of S460. The EN 1993-1-12 [2] gives design background for materials up to grade of S700, however, for the determination of the flexural buckling resistance of hollow section columns the same rules are applicable for material grades between S460 - S700 as for NSS columns.

The determination of the global (flexural) buckling resistance is highly important in the design of HSS structures, because due to the high yield strength smaller cross sections can be used. It results in more slender structural elements, which can be sensitive for stability failure. According to the previous research results [3,4,5,6,7] the flexural buckling behaviour of HSS structures is less severe than in case of NSS structures. The difference comes from the different residual stresses, material properties and geometric imperfections. Several previous investigations and the residual stress measurements in the RUOSTE project [8] also showed that the residual stress amplitudes are smaller for HSS structures compared to the yield strength than in case of NSS members. On the other side the reduction due to the flexural buckling is governed by the ratio of residual stress to the yield strength, rather than the magnitude of residual stress itself. Therefore it is expected that a higher column buckling curve could be used for HSS members than for NSS members, consequently the design using HSS structures could be more economical. The general aim of this paper is to investigate the column (flexural) buckling behaviour of HSS hollow section columns up to S960 steel grade. Based on the experimental investigation our aim is to propose safe and economic design buckling curves for HSS columns having square hollow sections. The current investigation contains cold formed and welded hollow section members as well, which are separately handled, thus different buckling curves can be used for these two column types.

2. Research program

There is only a limited number of column buckling tests available in the international literature dealing with hollow section HSS members. Therefore the objective of the present study is to design and perform a new experimental research program to investigate the global buckling behaviour of welded and cold-formed hollow section columns.

A total of 54 flexural buckling tests are carried out in the Structural Laboratory of the Budapest University of Technology and Economics Department of Structural Engineering. A total of 26 test specimens are investigated having steel grades between S500 – S960, and a complementary investigation is made on 28 columns with S420 and S460 steel grades. The aim of the complementary tests are to give reference and comparison background to the test results made on HSS members (S500 – S960) and to increase the investigated parameter range. The test specimens are produced by three international steel manufacturers (Ruukki, SSAB and VoestAlpine). From the 26 HSS test specimens 17 columns are made from cold-formed, and 9 from welded cross sections. The additional 28 specimens made from S420 and S460 steel grades are all cold formed sections. The investigated specimens cover the relative slenderness range between $0.54 < \lambda < 1.49$. The specimen's geometries are designed to ensure a quasi-uniform distribution by the test results within the studied slenderness region.

The flexural buckling behaviour is highly sensitive on all possible types of imperfection which includes the initial out-of-straightness of the column, the loading end-eccentricities and the residual stress pattern of the tested member. The accurate knowledge of these parameters is especially important in the evaluation process of flexural buckling test results. Therefore the initial out-of-straightness of all the 54 test specimens are measured prior the tests, the end-eccentricities during the loading are also measured using strain gauges applied on bot ends of the columns, and the residual stresses of identically manufactured hollow sections are measured using sectioning technique. 22 different cross-section's residual stress pattern are measured, 9 welded HSS members (S500-S960), 8 cold-formed HSS sections (S500-S960) and 5 cold formed specimens for S420 and S460 steel grades.

Based on the current and the previous experimental results found in the international literature the differences in the flexural buckling behaviour of HSS and NSS members are identified, studied and evaluated. Finally based on the experimental investigations design buckling curves are proposed for cold formed and welded HSS hollow section members separately. The objectives of the present paper are achieved using the following research strategy:

1. literature review on buckling tests of hollow section HSS columns,
2. literature review on HSS residual stress measurements,
3. residual stress measurements on welded and cold formed sections with different *b/t* ratios and different steel grades (S420, S460, S500, S700 and S960),

4. evaluation of the test results and design residual stress model development for HSS hollow section members (separately for welded and cold-formed sections),
5. measurement of the out-of-straightness imperfection on the buckling test specimens,
6. flexural buckling tests on welded and cold-formed HSS members with different global slenderness and different steel grades (S420, S460, S500, S700 and S960),
7. evaluation and documentation of the test results,
8. proposal for design column buckling curve for cold formed and welded HSS hollow section members based on statistical evaluation of present and previous test results.

3. Residual stress measurements

3.1 General research strategy

To investigate the residual stress distribution in HSS members residual stress measurements are carried out on square hollow sections in order to have accurate data for the evaluation of the buckling behaviour. The measurements are carried out on both welded and cold-formed sections. During the measurements the longitudinal residual stress pattern of the analysed cross-sections are determined. The measurements are carried out using the sectioning technique (Fig. 1). The deformations are recorded using electric strain gauges. The residual stress patterns are measured on the outer and the inner surfaces of the sections as well. Due to the symmetry of the cross-sections two sides of the welded and three sides of the cold-formed sections are measured in a detailed way. The typical distributions of the strain gauges are shown in Fig. 1.

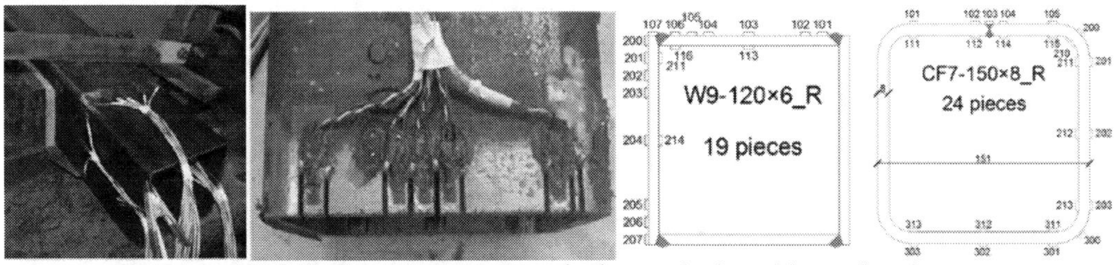

Fig. 1: Residual stress measurement: sectioning method, positions of strain gauges

3.2 Residual stresses on cold-formed specimens

The cold-formed specimens are manufactured using the "continuous forming" process, which includes the following main steps:

- (i) roll-forming of the straight strip first into a circular open tube,
- (ii) joining the edges of the open tube by welding,
- (iii) flattening the tube walls to form the desired rectangular shape.

For all the 13 tested cold-formed specimens the residual stresses are measured on the outer and the inner surface as well. On the outer surface tension, on the inner surface compression residual stresses are measured, as expected based on the previous test results found in the international literature made on cold-formed specimens using the continuous forming method.

At those points where inner and outer residual stresses are also measured the obtained values are averaged. Assuming a linear through thickness stress distribution, the average values give the membrane residual stresses. The measurements proved that these values are close to zero in most of the cases, therefore it can be concluded that the membrane residual stresses are negligible in case of the analysed sections using the studied manufacturing method. In the same time the residual stress measurements showed that the bending-type residual stresses are dominant for the analysed cold-formed specimens. It is also observed that the bending-type residual stress is less severe on the corners than on the flat parts.

The average bending residual stresses on the surface are determined for all the studied specimens considering the flat parts of the cross-section. The average stresses are separately calculated for the outer ($\sigma_{RS,b,out}$) and the inner ($\sigma_{RS,b,in}$) surfaces, as shown in Table 1.

Table 1: Residual stress measurement results on cold-formed specimens

Sign	Producer	Material	f_y [MPa]	b [mm]	t [mm]	b_{lok}/t	$\sigma_{RS,b,out}$ [MPa]	$\sigma_{RS,b,in}$ [MPa]
CF4.2-100×3_R	Ruukki	S420	458	100	3	28.7	320	-292
CF4.2-120×6_R	Ruukki		506	120	6	15.3	331	-326
CF4.2_150×5_R	Ruukki		479	150	5	25.6	351	-331
CF4.6_150×8_R	Ruukki	S460	508	150	8	14.0	467	-489
CF4.6_120×4_R	Ruukki		559	120	4	25.0	297	-349
CF5-120×6_V	VoestA	S500	624	120	6	15.7	518	-488
CF5-130×4_V	VoestA		573	130	4	28.0	557	-452
CF5-200×5_V	VoestA		567	200	5	35.6	408	-387
CF7-150×8_R	Ruukki	S700	742	150	8	13.75	579	-551
CF7-150×4_V	VoestA		799	150	4	32.5	534	-561
CF9-120×6_R	Ruukki	S960	1088	120	6	15.3	452	-386
CF9-150×7_R	Ruukki		1049	150	7.1	16.6	502	-619
CF9-120×4_R	Ruukki		1114	120	4	24.5	391	-353

Based on the current and previous test results a residual stress model (Fig. 2.) is proposed for HSS material grades (S420-S960). According to this new model the bending-type residual stresses in the flat plate parts and in the corner regions can be determined by Eq. (1).

$$\sigma_{RS,b,flat} = \pm\left(1.28 - \frac{f_y}{1270\,MPa}\right)\cdot f_y, \qquad \sigma_{RS,b,corner} = 0.55\cdot\sigma_{RS,b} \tag{1}$$

3.3 Residual stresses on welded specimens

In case of welded specimens it is observed that the residual stresses on the inner and outer surfaces are close to each other, which means that there are only negligible bending residual stresses in the plates, and the dominant residual stress is the membrane type stress. The measured results are in harmony with the expected tendencies based on the literature overview. High tensile residual stresses are measured in the corner zones and a moderate compression residual stresses in the middle part of the plates. The whole residual stress distribution diagram is determined for all test specimens based on the detailed residual stress measurements. These diagrams are studied and evaluated and three values are determined for each investigated sides which are the followings,

- maximum tensile stresses in corner zone ($\sigma_{RS,t}$),
- average compression stresses in the plate middle zone ($\sigma_{RS,c}$),
- length of the compressed plate width (L_c).

The obtained residual stress values are summarized in Table 2. It is observed that the compression residual stress does not increase linearly with the yield strength in case of HSS members. Based on the current and previous test results a residual stress model (Fig. 2.) is proposed for HSS welded hollow sections. According to this model the tension and the compression residual stresses can be calculated by Eq. (2) – (4), where b_0 is the inner width (b_0=b-2t) of the specimen.

Table 2: Residual stress measurement results on welded specimens

Sign	Producer	Material	f_y	b	t	b_{lok}/t	$\sigma_{RS,c}$	$\sigma_{RS,t}$
W5-120×6_R	Ruukki	S500	546	120	6	18.0	-169	517
W5-150×6_R	Ruukki		546	150	6	23.0	-127	420
W5-195×6_R	Ruukki		563	195	6	30.5	-104	473
W5-250×4_S	SSAB		624	250	4	60.5	-109	527
W9-120×6_S	SSAB	S960	1005	120	6	18.0	-71	528
W9-120×6_R	Ruukki		992	120	6	18.0	-145	840
W9-170×6_R	Ruukki		992	170	6	26.3	-104	889
W9-220×6_R	Ruukki		992	220	6	34.7	-124	793
W9-250×6_R	Ruukki		992	250	6	39.7	-76	931

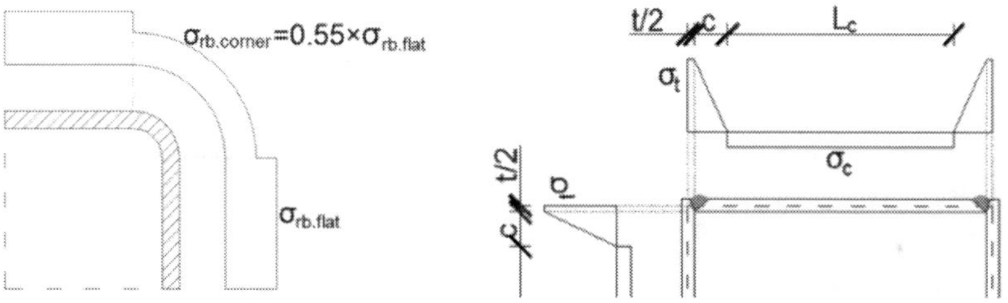

Fig. 2: HSS residual stress model for cold-formed (left) and for welded (right) sections

$$\sigma_{rt} = \mathrm{Min}((1.132 - 0.033 \cdot t) \cdot f_y; f_y). \tag{2}$$

$$\text{If } t \leq 6\text{: } \sigma_{rc} = 20 - 5t - \left(14000 \cdot \frac{1}{t-2} - 800\right) \cdot \left(\frac{b_0}{t}\right)^{-1}, \tag{3}$$

$$\text{If } t \geq 6\text{: } \sigma_{rc} = 20 - 5t - (4620 - 320t) \cdot \left(\frac{b_0}{t}\right)^{-1}, \tag{4}$$

4. Flexural buckling tests

4.1 General research strategy

A total of 54 hollow section columns are tested and the flexural buckling resistances are measured using hinged boundary conditions. The hinged connection is provided by a cylindrical testing rig that allowed free rotation of the end cross-sections. During the tests the axial and lateral displacement of the specimens, the rotation of the hinges, and the stress distribution of the bottom, middle and top cross-section are measured. The test arrangement is shown in Fig. 3. The end-eccentricities of the loading process are calculated based on the measured stress distribution of the bottom and top cross-sections. From the measured load carrying capacities (N_{test}) the buckling reduction coefficients (χ_{test}) are determined. The results are statistically evaluated according to the rules of the EN1990 Annex D [9]. The evaluation is made individually based on the new tests and also together with test specimens taken from the international literature. The evaluation is made by grouping of the test results based on the analysed global slenderness ranges of the specimens. The modified partial safety factor (γ_{M1}^*) is determined following the concept that was used by Sedlacek et al. [10] for the derivation of the column buckling curves for NSS structures of the EN1993-1-1 [1].

4.2 Out-of-straightness measurement

The exact geometry of the test specimens are measured prior to the buckling tests using special equipment based on a moving inductive displacement transducer. The resulted shapes of the 8 cold-formed specimens are shown in Fig. 3. It can be observed that the specimens are bow-shaped. The maximum amplitude of the out-of-straightness is usually much smaller than the manufacturing tolerance given by the Eurocode (L/750). The average magnitude of the out-of-straightness imperfection is L/3200 for the welded and L/8300 for the cold formed specimens. It is observed, that no clear tendencies can be observed depending on the steel grade, which means that for the high strength steel members the same out-of-straightness imperfection shapes are expected than for the normal strength steel members, this means a slight contradiction with the generalized imperfection concept of the European buckling curves.

Despite that the measured out-of-straightness imperfections were significantly small, the total imperfection during the tests was higher due to the applied end-eccentricities. The total imperfections applied in the tests are in average L/1100 for welded and L/1800 for cold-formed specimens. However, it should be mentioned that the exact placement of the specimens into the loading equipment is quite difficult, and therefore it is almost impossible to guarantee the exact L/1000 imperfection magnitude, what would be necessary for column buckling curve development.

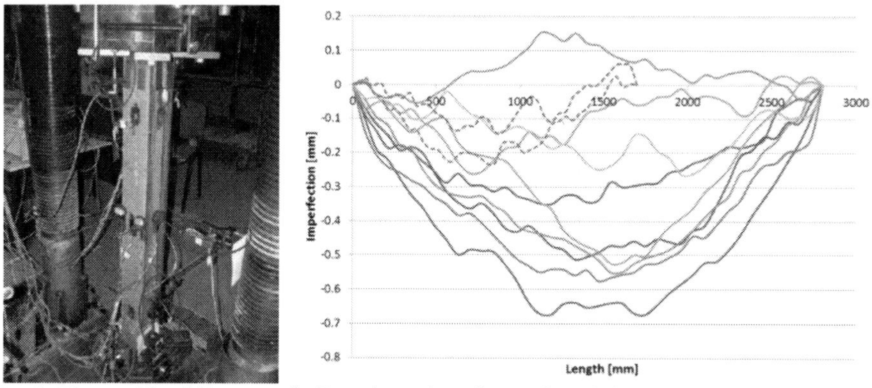

Fig. 3: Flexural buckling test (left) and results of out-of-straightness measurement (right)

4.3 Buckling test results on cold-formed specimens

The calculated buckling reduction factors (χ_{test}) of the new and previous tests [7] are plotted in Fig. 4 in the function of the global slenderness together with the column buckling curves of the EN 1993-1-1 [1]. The statistical evaluation is made separately in three groups; for HSS (i) steel grades (S500-S960), for S460 (ii) steel grades and for S420 (iii) steel grades. The obtained partial safety factors are summarized in Table 3. Based on the calculated values it can be concluded that for S460 or higher steel grades the buckling curve c is too conservative, and the buckling **curve b** is applicable with adequate safety.

Table 3 – Determined values of the modified partial safety factor (γ_{M1}^*)

material	curve a	curve b	curve c
S500-S700-S960	1.09	**0.99**	0.91
S460	1.07	**0.97**	0.88
S420	1.16	1.05	**0.96**

Parallel to the experimental research program additional numerical calculations are also executed to determine applicable buckling curves. In the numerical simulations geometric imperfection ($L/1000$) and the developed residual stress model is applied, which can be used for HSS structures. The buckling resistances are determined by geometrical and material nonlinear analysis using the measured imperfections (GMNIA). The results of the numerical simulations proved the observations of the laboratory tests that the specimens manufactured from S960 steel material provided the largest buckling resistance and for the cold-formed hollow section columns made from S960 steel grade the column buckling **curve a** may be applicable with adequate safety.

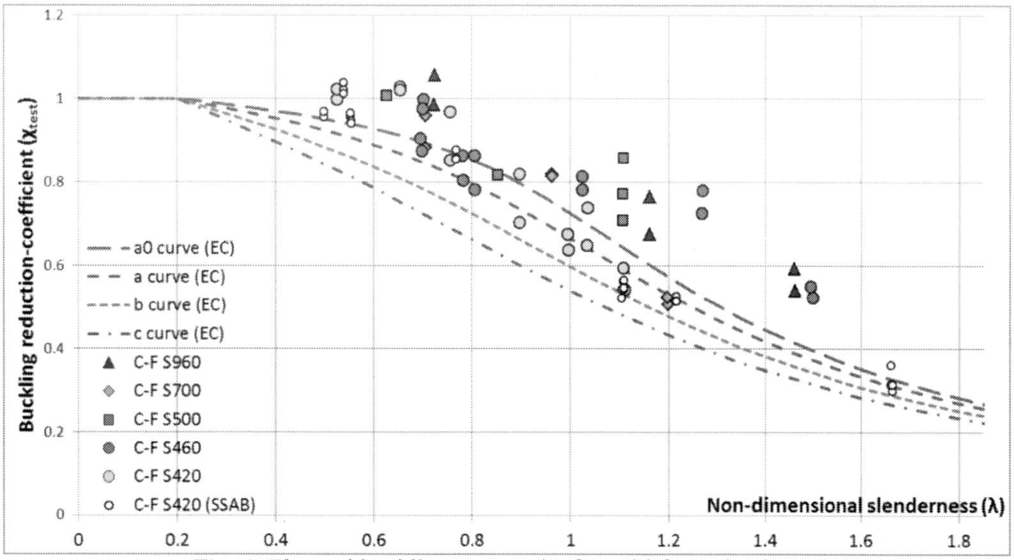

Fig. 4: Flexural buckling test results for cold-formed columns

4.4 Buckling test results on welded specimens

The calculated buckling reduction factors (χ_{test}) of the new and previous tests [4], [5], [6] are plotted in Fig. 5 in the function of the global slenderness together with the column buckling curves of the EN 1993-1-1 [1].

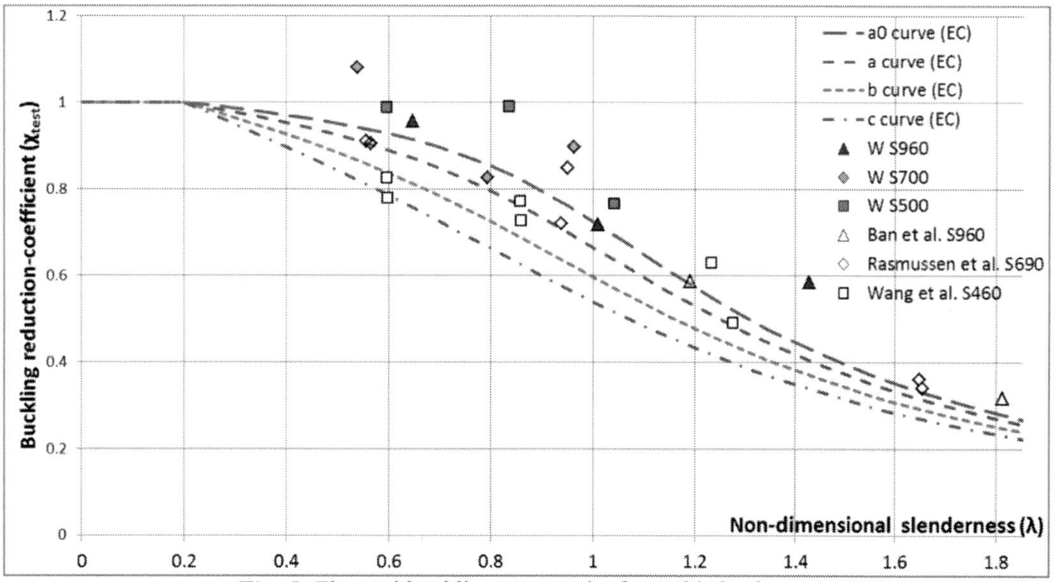

Fig. 5: Flexural buckling test results for welded columns

The statistical evaluation of all the test results showed that the buckling **curve c** is too conservative for HSS structures. The buckling **curve b** is proposed for HSS hollow sections made of S500 or higher steel grades, thus this buckling curve gives the best approximation of the test results with adequate safety. The additional numerical calculations proved, that the applicable buckling curve for welded hollow section columns between S500 – S700 is the **curve b** and for S960 the buckling **curve a**.

5. Summary and conclusions

The paper presents the results of an experimental research program where cold-formed and welded hollow section columns are tested between S420 - S960 steel grades. The buckling resistance of the test specimens are measured, statistically evaluated, and appropriate buckling curves are proposed. The following conclusions are drawn from the experimental and additional numerical results:

- The buckling **curve b** is safely applicable for cold-formed hollow section columns between S460 - S700 and the buckling **curve a** may be applicable for S960 material grades instead if the buckling **curve c**.
- The buckling **curve b** is safely applicable for welded hollow section columns (b/t < 30) between S500 - S700, and the buckling **curve a** may be applicable for S960 material instead if the buckling **curve c**.

Acknowledgement

The presented research results are part of the RUOSTE: Rules On High-Strength Steel; RFCS Project RFSR-CT-2012-00036. The financial support is gratefully acknowledged.

References

[1] EN 1993-1-1. *Eurocode 3: Design of steel structures - Part 1.1: General rules and rules for buildings*. CEN. 2009.

[2] EN 1993-1-12. *Eurocode 3: Design of steel structures - Part 1-12: Additional rules for the extension of EN 1993 up to steel grades S700*. CEN. 2007.

[3] Fukumoto Y, Itoh Y. "Evaluation of Multiple Column Curves using the Experimental Data-Base Approach", *Journal of Constructional Steel Research,* 3(3), 1983.

[4] Rasmussen KJR, Hancock GJ. "Tests of High Strength Steel Columns", *Journal of Constructional Steel Research,* 34, 27-52, 1995.

[5] Ban H, Shi G, Shi Y, Bradford MA. "Experimental investigation of the overall buckling behaviour of 960 MPa high strength steel columns", *Journal of Constructional Steel Research,* 88, 256-266, 2013.

[6] Wang Y-B, Li G-Q, Chen S-W, Sun F-F. "Experimental and numerical study on the behavior of axially compressed high strength steel box-columns", *Engineering Structures,* 58, 79-91, 2014.

[7] Heinisuo M. *Axial resistance of double grade (S355, S420) hollow sections manufactured by SSAB*, 2014.

[8] *RUOSTE: Rules On High-Strength Steel*; RFCS Project RFSR-CT-2012-00036.

[9] EN 1990. *Eurocode: Basis of structural design*. CEN. 2005.

[10] Johansson B, Maquoi R, Sedlacek G. "New design rules for plated structures in Eurocode 3", *Journal of Constructional Steel Research*, 57, 279-311, 2001.

DESIGN IMPERFECTIONS FOR STEEL BEAM LATERAL TORSIONAL BUCKLING

H.H. Snijder, R.P. van der Aa, H. Hofmeyer, B.W.E.M. van Hove

Eindhoven University of Technology (TU/e), Department of the Built Environment, The Netherlands

Abstract: To perform geometrically and materially nonlinear analyses including imperfections for steel beam lateral torsional buckling, the size and shape of the geometric imperfection can be obtained from EN 1993-1-1. The shape is prescribed as an initial bow along the weak axis of the section, excluding torsion of the cross-section. The shape of the imperfection can alternatively be taken equal to the lateral torsional buckling mode, including torsion. Several tables and formulas exist to determine the size of the imperfection. Different imperfection approaches were applied in finite element simulations to evaluate the lateral torsional nonlinear buckling resistances and to compare them to the results obtained with design rules.

1. Introduction

Eurocode 3 provides design rules for the assessment of lateral torsional buckling (LTB) in clause 6.3.2 of EN 1993-1-1 [1]. Alternatively, the code allows LTB to be assessed by performing Geometrically and Materially Nonlinear Analyses with Imperfections of beams. For carrying out such GMNIA calculations, the size and shape of the equivalent geometric imperfection is given in clause 5.3.4(3) of EN 1993-1-1. The shape is prescribed as an equivalent initial bow of the weak axis of the profile considered, excluding torsion of the cross-section. This comes down to the weak axis flexural buckling (FB) mode. The size is prescribed as $ke_{0,d}$ where k is a factor having as recommended value $k = 0.5$ and $e_{0,d}$ is the initial local bow imperfection given in Table 5.1 of EN 1993-1-1. As an alternative, the initial local bow imperfection $e_{0,d}$ can be based on eq. (5.10) of EN 1993-1-1:

$$e_{0,d} = \alpha\left(\bar{\lambda} - 0.2\right)\frac{M_{Rk}}{N_{Rk}} \tag{1}$$

where α is the imperfection factor according to Table 6.1 of EN 1993-1-1, $\bar{\lambda}$ is the non-dimensional slenderness for weak axis buckling, M_{Rk} is the characteristic moment resistance, and N_{Rk} is the characteristic normal force resistance. In case of class 1 and 2 cross-sections $M_{Rk} = M_{pl}$ and $N_{Rk} = N_{pl}$ where M_{pl} is the plastic moment resistance of the cross-section and N_{pl} is its plastic normal force resistance.

In line with clause 5.3.2(11) of EN 1993-1-1, the shape of the imperfection can also be taken according to the LTB mode, including torsion of the cross-section. For the size of the imperfection, Eq. (1) can be used. However, for the next generation of Eurocodes, new LTB design rules [2,3] have been derived as an alternative to those of the current EN 1993-1-1. Using these new LTB design rules and their derivation, a formula has been obtained which, for an LTB mode, describes explicitly the imperfection size:

$$e_{0,d} = \alpha_{LT}\left(\bar{\lambda} - 0.2\right)\frac{M_{Rk}}{N_{Rk}} \tag{2}$$

where α_{LT} is the imperfection factor according to Table 1. In Table 1 h is the section height, b is the section width and $W_{el,y}$ respectively $W_{el,z}$ are the strong and weak axis section modulus.

The three different imperfection approaches for LTB as shown in Table 2 have been considered in this paper.

Table 1: Imperfection factors for Eq. (2)

Cross-section	Limits	α_{LT}
Rolled I-section	$h/b > 1.2$	$0.12\sqrt{W_{el,y}/W_{el,z}} \leq 0.34$
	$h/b \leq 1.2$	$0.16\sqrt{W_{el,y}/W_{el,z}} \leq 0.49$

Table 2: Three imperfection approaches

Approach	Imperfection shape	Imperfection size
1	Weak axis FB mode, excluding torsion	Eq. (1)
2	LTB mode, including torsion	Eq. (1)
3	LTB mode, including torsion	Eq. (2)

A finite element model for LTB was developed and verified using several other models in literature. With that model, the different approaches for imperfection shape and size were applied to evaluate the LTB resistances. Subsequently, these LTB resistances were compared with the LTB resistances according to the appropriate design rules. For the approaches 1 and 2 the design rules according to the clauses 6.3.2.1 and 6.3.2.2 for the so called 'general case' were used as reference while for approach 3 the newly developed design rules according to [2,3] was used. This paper is based on research work as reported in [4].

2. Scope

The imperfection study in this paper is carried out for hot-rolled class 1 and 2 IPE, HEA, and HEB sections. However, the fillet radius between web and flanges is neglected; therefore the sections are denoted by adding an asterisk: e.g. IPE* sections. The section properties have been modified accordingly. Three load cases have been considered as indicated in Fig. 1.

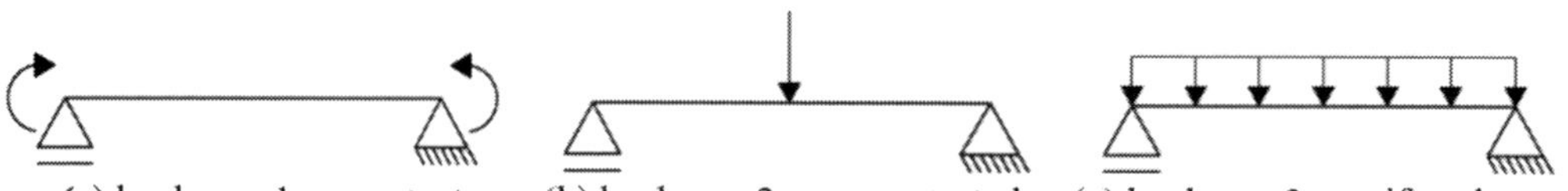

(a) load case 1 – constant bending moment (b) load case 2 – concentrated load (c) load case 3 – uniformly distributed load

Fig. 1: Three load cases (LC) considered

3. Finite element model

For the finite element model, so-called "S8R" Mindlin-Reissner elements are used within Abaqus. These are quadrilateral eight node shell elements with six degrees of freedom per node, four integration points in surface direction, and five integration points in thickness direction. The elements require a regular mesh and are suitable for thick shell applications. Eight elements over the width of the flanges and 16 elements over the height of the web are applied. A bilinear stress-strain diagram, neglecting strain hardening, is used with yield stress f_y = 235 N/mm^2, Young's modulus E = 2.1 x 10^5 N/mm^2, and Poisson's ratio 0.3. The Von Mises yield criterion was applied. With respect to the boundary conditions, Fig. 2 shows the end fork conditions of the beam.

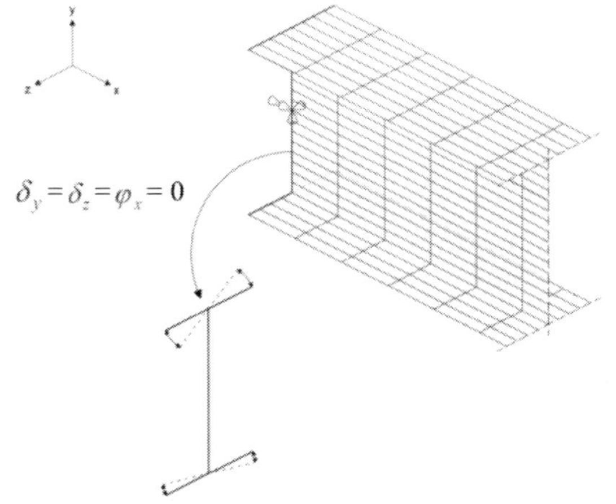

Fig. 2: End fork conditions of the beam

Distortion of the end cross-section is prevented by applying kinematic coupling constraints. This results in an in-plane stiff cross-section (as δ_y and δ_z of all nodes are coupled), which cannot rotate along the beam axis (as $\varphi_x = 0$ for all nodes), but warping is still allowed. Since beam geometry, load cases, imperfection shapes, and failure modes are symmetrical, only half the beam is modelled applying symmetry conditions $\delta_x = 0$, $\varphi_y = 0$, and $\varphi_z = 0$.

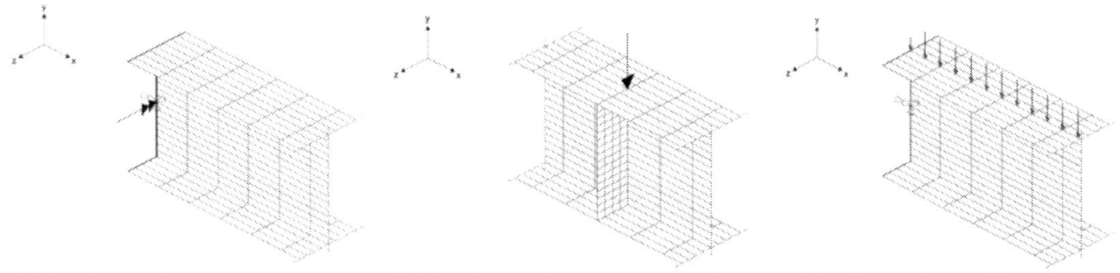

(a) load case 1 – constant bending moment

(b) load case 2 – concentrated load

(c) load case 3 – uniformly distributed load

Fig. 3: Application of loads

Loads are applied as shown in Fig. 3. For the GMNIA, two types of imperfection shapes, i.e. the weak axis FB mode excluding torsion and the LTB mode including torsion are applied after performing the relevant linear buckling analysis (LBA). In Fig. 4 these shapes are shown in the cross-section at mid-span of the beam. For the LTB mode the size of the imperfection is measured at the heart of the top flange.

As a validation of the finite element model, its results obtained from GMNIA calculations were compared with numerical results of [2, 5] and were found to be in good agreement [4].

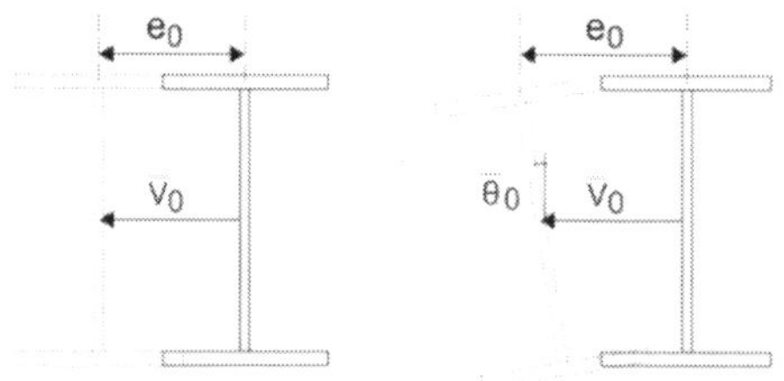

(a) FB mode (b) LTB mode

Fig. 4: Two imperfection shapes at cross-section halfway the beam

4. Simulation and comparison procedure

The comparison of simulations with the finite element method (FEM) and EN 1993-1-1 design rule results is illustrated for an IPE240* section with a length of 3400 mm for load case 1 (constant bending moment) and imperfection approach 2: LTB shape and size of the imperfection according to Eq. (1). First an LBA is performed resulting in an elastic critical bending moment M_{cr} = 84.1 kNm. See Fig. 5 for the corresponding buckling mode which is also used for the shape of the imperfection.

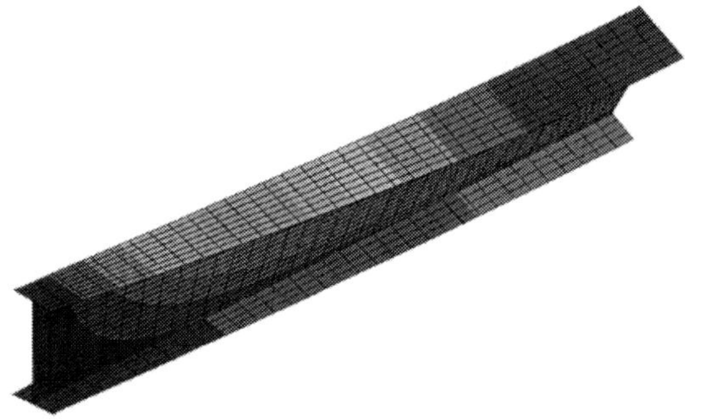

Fig. 5: Half buckling mode shape of 3400 mm long IPE240* beam

The plastic resistance can be calculated as $M_{pl} = W_{pl,y} \cdot f_y = 353.9 \cdot 103 \cdot 235 \cdot 10\text{-}6 = 83.0$ kNm. The non-dimensional slenderness then becomes:

$$\bar{\lambda}_{LT,FEM} = \sqrt{\frac{M_{pl}}{M_{cr}}} = \sqrt{\frac{83.0}{84.1}} = 0.99$$

First the non-dimensional slenderness $\bar{\lambda}$ for weak axis buckling needs to be determined using the elastic critical force for weak axis buckling $N_{cr,z}$:

$$N_{cr,z} = \frac{\pi^2 EI_z}{L_{cr,z}^2} = \frac{\pi^2 \cdot 2.1 \cdot 10^5 \cdot 282.7 \cdot 10^4}{3400^2} = 506.9 \cdot 10^3\,\text{N}$$

$$\bar{\lambda} = \sqrt{\frac{Af_y}{N_{cr,z}}} = \sqrt{\frac{3779 \cdot 235}{506.9 \cdot 10^3}} = 1.32$$

In these calculations I_z is the weak axis second moment of area, $L_{cr,z}$ is the buckling length for weak axis buckling and A is the area of the cross-section. Since h/b = 240/120 = 2 > 1.2 and the flange thickness t_f = 9.8 mm is smaller than 40 mm, buckling curve 'b' applies for weak axis buckling according to Table 6.2 of EN 1993-1-1 and the imperfection factor α =

0.34 according to Table 6.1 of EN 1993-1-1. The imperfection size is then determined using Eq. (1):

$$e_{0,d} = \alpha(\bar{\lambda} - 0.2)\frac{M_{Rk}}{N_{Rk}} = \alpha(\bar{\lambda} - 0.2)\frac{W_{pl,z}}{A} = 0.34(1.32 - 0.2)\frac{72772}{3779} = 7.33\,\text{mm}$$

With $k = 0.5$ the final imperfection size becomes: 3.67 mm. With this imperfection size and the imperfection shape of Fig. 5, a GMNIA is performed, for which load-displacement curves are shown in Fig. 6.

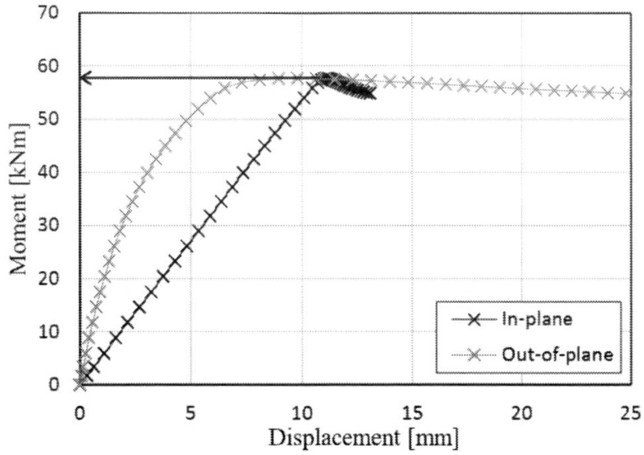

Fig. 6: Load-displacement diagram obtained with GMNIA for a 3400 mm long IPE240* beam

The ultimate load is reached at $M_R = 57.6$ kNm. The reduction factor can then be calculated as:

$$\chi_{LT,FEM} = \frac{M_R}{M_{pl}} = \frac{57.6}{83.0} = 0.694$$

The combination of $\bar{\lambda}_{LT,FEM} = 0.99$ and $\chi_{LT,FEM} = 0.694$ is compared with the current lateral torsional buckling curve of clause 6.3.2.2 of EN 1993-1-1 as shown in Fig. 7 (black arrows). More results for other non-dimensional slendernesses (beam lengths) and other load cases are

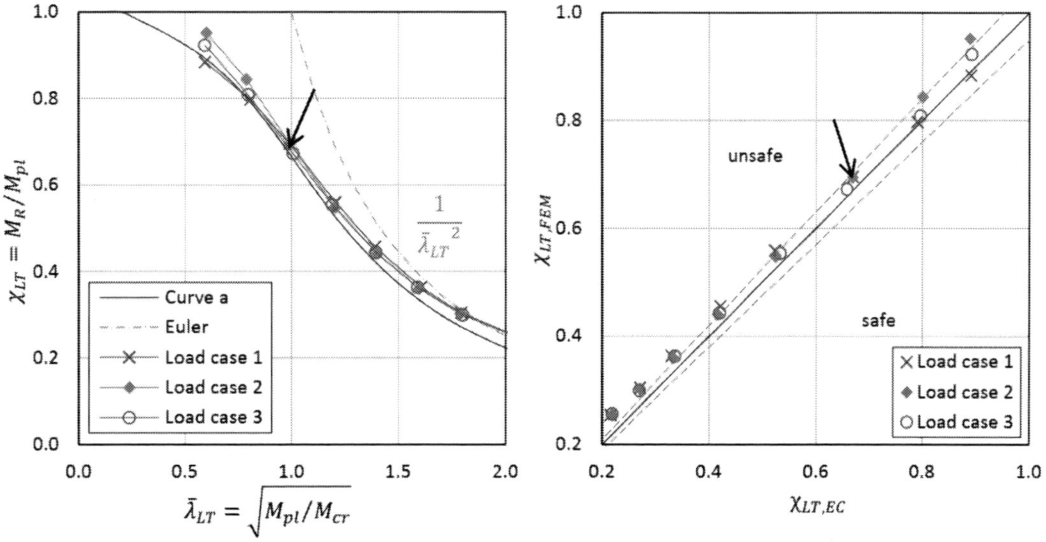

Fig. 7: Comparison of $\chi_{LT,FEM}$ with $\chi_{LT,EC}$ for imperfection approach 2 for an IPE240* beam

also included in Fig. 7. In the left part of Fig. 7 a direct comparison is made between the FEM results and buckling curve of EN 1993-1-1. FEM results above the buckling curve are on the unsafe side, i.e. they are calculated using a too small imperfection size so that $k > 0.5$ should have been chosen. In the right part the reduction factors calculated by FEM are compared with those calculated with EN 1993-1-1, Eurocode 3 (EC3). The diagonal black line represents a perfect match between both reduction factors. If $\chi_{LT,FEM} > \chi_{LT,EC}$ then the FEM results are on the unsafe side. The two dashed lines represent a 5% over- or underestimation.

5. Results

5.1 Imperfection approach 1: weak axis FB mode with Eq. (1) size

The results for imperfection approach 1, with the imperfection shape based on the weak axis FB mode and the imperfection size determined with Eq. (1), are shown in Fig. 8 for an IPE600* beam. Similar results were obtained for IPE240* and HEA300* beams [4].

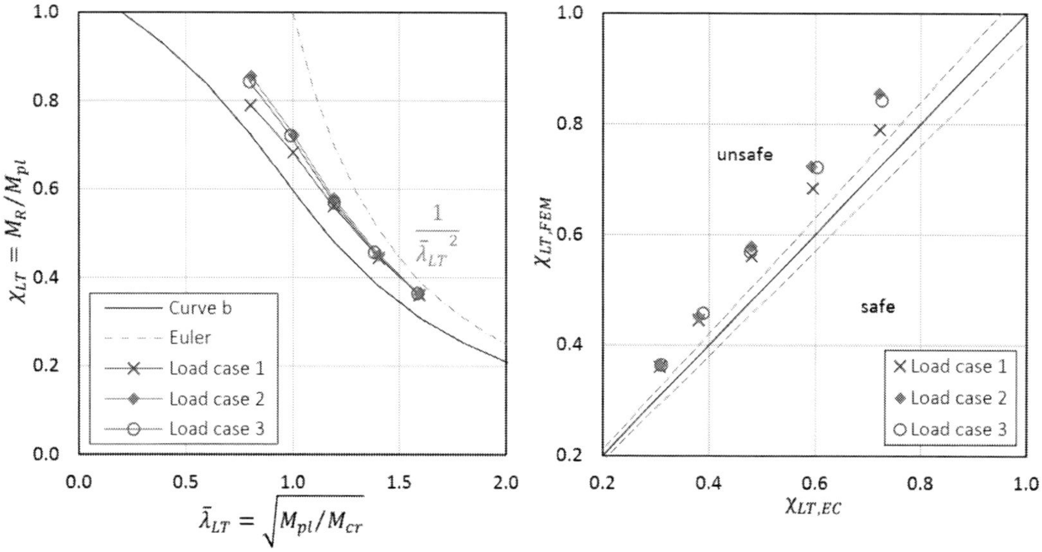

Fig. 8: Comparison of $\chi_{LT,FEM}$ with $\chi_{LT,EC}$ for imperfection approach 1 for an IPE600* beam

As can be seen, load case 2 (concentrated load) gives the largest discrepancies between FEM and EC3 results. All FEM results are largely on the unsafe side of EC3 results meaning that $k = 0.5$ is a far too small value. Therefore, it was decided to determine the k-value more accurately, such that the FEM results are within the 5% limits. The analyses concentrated on load case 2 and on relative slenderness 0.9 where the influence of imperfections is substantial. Fig. 9 shows for IPE beams the required k-values to get $\chi_{LT,FEM}$ to the target value $\chi_{LT,EC}$ as a function of I_y/I_z, where I_y is the strong axis second moment of area. These k-values range from 2.4 to 1.1 and are summarized in Table 3, 3rd column. In a similar way required k-values for HEA beams were obtained [4], also summarized in Table 3.

5.2 Imperfection approach 2: LTB mode with Eq. (1) size

Fig. 7 shows the results for an IPE240* beam for imperfection approach 2, with the imperfection shape based on the LTB mode, and the imperfection size determined with Eq. (1). Similar

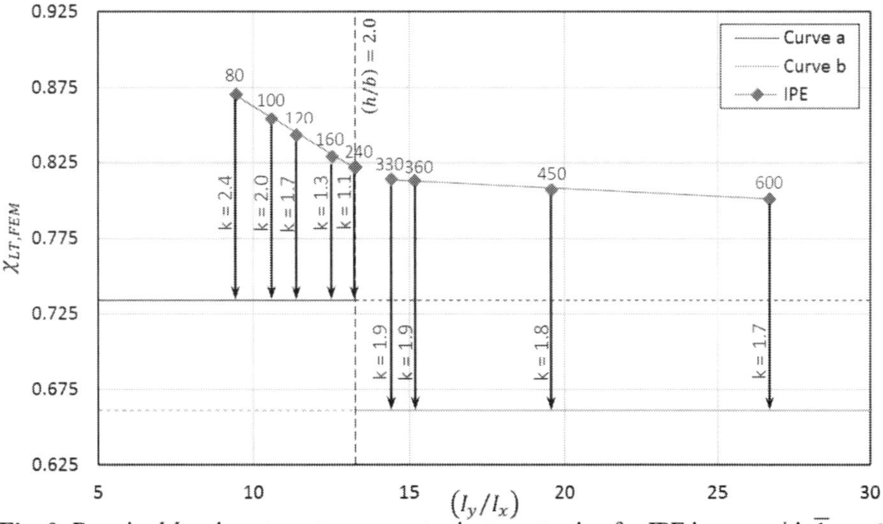

Fig. 9: Required k-values to get $\chi_{LT,FEM}$ to the target value for IPE beams with $\bar{\lambda}_{LT} = 0.9$

Table 3: Required k-values for rolled sections

Section	Limits	Imperfection approach 1	Imperfection approach 2
IPE	$h/b \leq 2.0$	$k = -0.34\frac{I_y}{I_z} + 5.60$	$k = -0.13\frac{I_y}{I_z} + 2.44$
	$h/b > 2.0$	$k = -0.017\frac{I_y}{I_z} + 2.15$	$k = -0.017\frac{I_y}{I_z} + 1.44$
HEA	$h/b \leq 1.0$	$k = -6.90\frac{I_y}{I_z} + 20.40$	$k = -1.69\frac{I_y}{I_z} + 5.39$
	$1.0 < h/b \leq 1.2$	$k = -0.11\frac{I_y}{I_z} + 1.65$	$k = -0.11\frac{I_y}{I_z} + 1.05$
	$1.2 < h/b \leq 2.0$	$k = -0.041\frac{I_y}{I_z} + 1.70$	$k = -0.027\frac{I_y}{I_z} + 0.94$
	$h/b > 2.0$	$k = -0.008\frac{I_y}{I_z} + 2.20$	$k = -0.008\frac{I_y}{I_z} + 1.21$

results were obtained for IPE600* and HEA300* beams [4]. In a similar way as described in section 5.1, required k-values were obtained [4], see Table 3, last column. The required k-values for IPE beams range now from 1.2 to 0.7 [4], still substantially greater than 0.5, but far better than for imperfection approach 1 when neglecting torsion, see Fig. 9.

5.3 Imperfection approach 3: LTB mode with Eq. (2) size

Fig. 10 shows the results for IPE240* and HEA300* beams for imperfection approach 3, with the imperfection shape based on the LTB mode, and the imperfection size determined by Eq. (2). For IPE240* beams, all results are satisfactory below the 5% upper limit. For HEA300* beams with $h/b \leq 1.2$, the results are close to or even slightly above the 5% upper limit. More beams have been investigated to find similar results [4]. It can be concluded that this imperfection approach, as perhaps may have been expected, leads to consistent results, with the additional benefit that no k-value is needed.

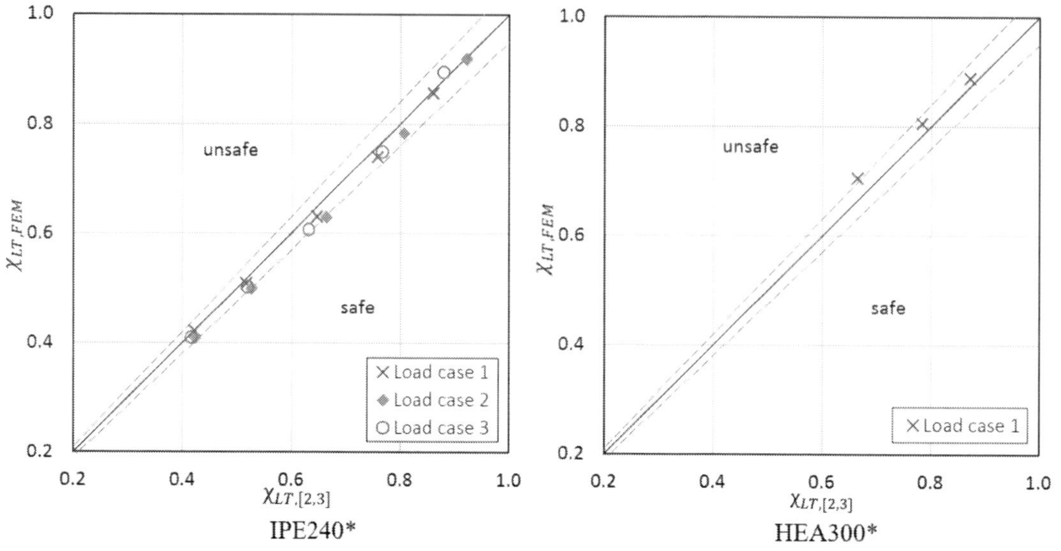

Fig. 10: Comparison of $\chi_{LT,FEM}$ with $\chi_{LT,[2,3]}$ for imperfection approach 3

6. Conclusions

Three different imperfection approaches were studied to be used in GMNIA calculations for LTB of beams in bending. It is shown that imperfection approach 1, using the FB buckling mode as imperfection shape and Eq. (1) for obtaining the imperfection size, requires k-values far greater than the recommended value $k = 0.5$ in cl. 5.3.4(3) of EN 1993-1-1. Also imperfection approach 2, using the LTB buckling mode as imperfection shape and Eq. (1) for the imperfection size, requires k-values substantially greater than $k = 0.5$, the greatest value being 1.2, indicating that this approach is far better than imperfection approach 1. Consistent results were obtained for imperfection approach 3, using the LTB buckling mode as imperfection shape and Eq. (2) for the imperfection size. This imperfection approach does not need a k-value and is advised to be used in the next version of EN 1993-1-1.

References

[1] EN 1993-1-1. *Eurocode 3: Design of steel structures- part 1-1: General rules and rules for buildings*, Brussels, 2011.

[2] Taras A, Greiner R, Unterweger H. *Proposal for amended rules for member buckling and semi-compact cross-section design*, Doc. CEN-TC250-SC3_N1898, 2013.

[3] Taras A. *Contribution to the development of consistent stability design rules for steel members*, PhD thesis, Graz University of Technology, Graz, Austria, 2010.

[4] Aa RP van der. *Numerical assessment of the design imperfections for steel beam lateral torsional buckling*, Master thesis, report 2015.96, Eindhoven University of Technology, Eindhoven, Dept. of the Built Environment, Structural Design, The Netherlands, 2015.

[5] Bruins RHJ. *Lateral torsional buckling of laterally restrained steel beams*, Master thesis, report 2007.7, Eindhoven University of Technology, Dept. of the Built Environment, Structural Design, Eindhoven 2007.

STABILIZING OF I-BEAMS WITH DISCRETE INTERMEDIATE TORSIONAL RESTRAINTS

Richard Stroetmann

Technische Universität Dresden, Institut für Stahl- und Holzbau, Dresden, Germany

Abstract: The stability of beams subject to lateral torsional buckling can be significantly improved by local torsional restraints of the flanges. In the presented paper, the influence of local flange restraints on the stability of beams is described in detail. In addition, calculation models to determine the torsional spring stiffness and the influence of the local buckling effect are treated. The determination of the critical buckling moments and the description of load capacity tests of beams with local torsional restraints follow. From the statistical evaluation of the results from load bearing tests on I-beams with local restraints of the top chords a design approach based on EN 1993-1-1 is presented.

1. Introduction

The construction of the supports and of the connections to restraining elements has often a significant influence on the stability of beams and columns. The realistic modelling of supports and connections is complex even for planar systems. In three-dimensional systems, additional difficulties arise for members prone to lateral torsional buckling because of a higher number of deformational degrees of freedom. In most cases, the boundary conditions in the models representing practical support conditions and restraints can only be approximations with differing degrees of accuracy. The assumptions for these conditions depend on the mechanical model.

The buckling loads of members with thin-walled open cross sections are usually determined using a second-order analysis considering biaxial bending and warping torsion but excluding the distortion of cross sections. For an economic design, it is often necessary to consider the stabilizing effect of restraining structural elements. If one of the flanges of I-beams are restrained by purlins or secondary beams against twist rotations, local plate bending moments occur in the region of the connections, which are not considered directly in the calculation model. The assumption that sections do not distort becomes inaccurate.

2. Behaviour of beams with local flange restraints

2.1 Stability of single beams

The restraining beams often have a bending stiffness that is comparable with the torsional stiffness of the main beam, so a high degree of restraint against torsion and distortion can be achieved (see [1]). However, the restraining effect cannot be used to the full extent for the stabilization of the main beams. The local moments in the connections cause distortions in the cross sections (see above). This effect is of particular importance when calculating the critical load. For systems according to Fig. 1, the effect of the restraining beams on the main beams can be accounted for by single loads and torsional springs. This requires, inter alia, that the restraining beams do not reach the stability limit before the main beams do.

Figure 1 shows in the diagram the ratio of the critical loads $P_{cr} / P_{cr,0}$ as a function of the common logarithm of the spring stiffness C_{xx} for two different beams. The curves were determined with and without considering cross-sectional distortion. The reference values $P_{cr,0}$ are the critical loads that were obtained for the respective beam without restraints ($C_{xx} = 0$). The calculations were performed with the FEM software PROFIL [2]. In Table 1, the critical loads P_{cr} are listed for limit values of the spring stiffness C_{xx}. In addition, the obtained differences are given. It can be seen that significant increases in the critical loads are possible due to the local torsional restraints.

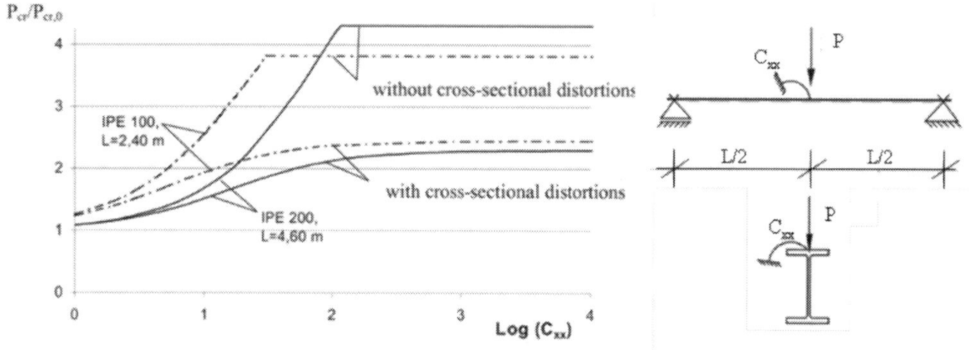

Fig. 1: Critical load as a function of the torsional spring stiffness C_{xx} [kNm]

Table 1: Critical loads P_{cr} [kN], ratios and differences

C_{xx} [kNm]	IPE 100, L=2.40 m			IPE 200, L=4.60 m		
	with c-s distortions	without c-s distortions	difference [%]	with c-s distortions	without c-s distortions	difference [%]
0	14.6	14.7	+1	28.5	28.7	+1
∞	35.9	56.1	+56	65.8	123.6	+88
$P_{cr,\infty} / P_{cr,0}$	2.46	3.82		2.31	4.32	

The influence of cross-sectional distortions (c-s distortions) increases with increasing spring stiffness. The curves of the ratios of the critical loads $P_{cr} / P_{cr,0}$ determined neglecting the cross-sectional distortions change into horizontal lines when certain spring stiffnesses are exceeded. At these points, the eigenmodes of the beams change from a symmetric to an anti-symmetric shape. A further increase of the spring stiffness has no effect on the buckling resistance because of the absence of rotation at its connecting point in the eigenmode. The curves $P_{cr} / P_{cr,0}$ determined considering the cross-sectional distortions are smooth. The eigenmodes of the beams are symmetric for all torsional spring stiffnesses. The critical loads obtained with the assumption of a rigid lateral support are far from being reached.

2.2 Torsional spring stiffness due to connected sections

Assuming in the stability analysis that the cross sections does not distort, the determination of discrete torsional spring stiffness is carried out according to the same principle as for continuous torsional spring restraints. The resultant spring stiffness C_ϑ is calculated from a system of torsional springs in series:

$$\frac{1}{C_\vartheta} = \frac{1}{C_{\vartheta R}} + \frac{1}{C_{\vartheta C}} + \frac{1}{C_{\vartheta D}} \tag{1}$$

C_ϑ characteristic value of the torsional spring stiffness;
$C_{\vartheta R}$ torsional spring stiffness resulting from the bending stiffness of the restraining beam;
$C_{\vartheta C}$ torsional spring stiffness resulting from the deformation of the connection;
$C_{\vartheta D}$ torsional spring stiffness due to the cross section distortion of the restraint beam.

The spring stiffness C_ϑ corresponds to the spring stiffness C_{xx} in Fig. 1. The proportion of the cross-sectional deformation is greatest for beams without stiffeners (Fig. 2). The influence of the deformations of the connection is usually neglected if rolled sections are connected by bolts typical in steel construction, as shown in Fig. 2.

The influence of realistic load applications into I-sections on the lateral torsional buckling has been investigated by Lindner and Schmidt [3,4]. In [3], an equation for determining the torsional spring stiffness resulting from the cross-sectional deformation is given:

$$C_{\vartheta D} = \sqrt{\frac{E \cdot t_w^3}{h - t_f} \cdot GI_{T,ch}} \tag{2}$$

t_w web thickness;
$h\text{-}t_f$ distance of the centroid of the flanges (h = section height, t_f = flange thickness);
$I_{T,ch}$ St. Venant torsion constant of a flange.

High local stresses result when larger torsional moments M_x are applied. This is for example the case for slender beams that are braced mainly by locally concentrated fixations. In many cases, the failure of the beam is induced by plastification at the points of fixation [3].

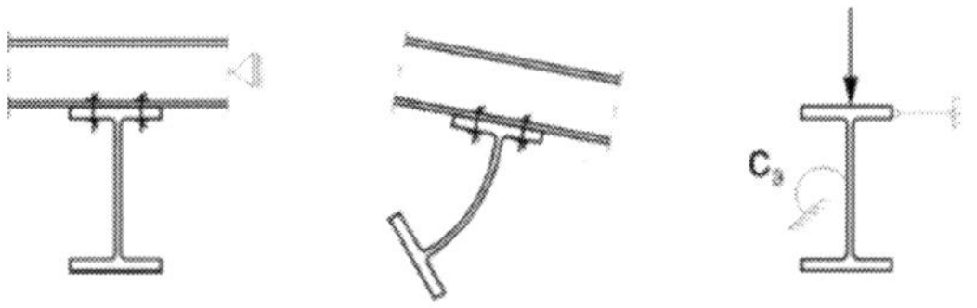

Fig. 2: Restraining of a beam susceptible to lateral torsional buckling by a connected I-section

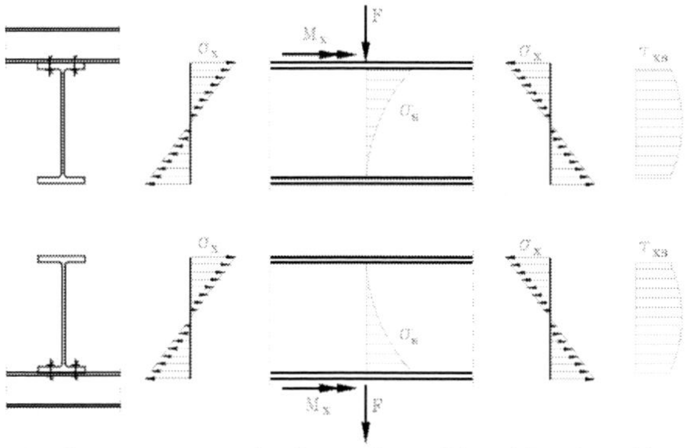

Fig. 3: Membrane stresses in the region of local load applications

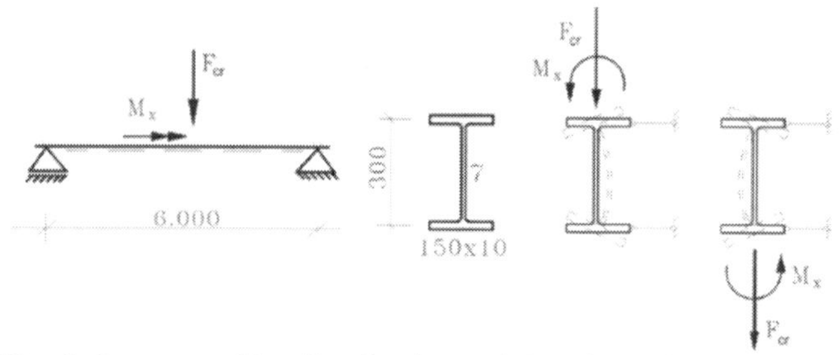

Fig. 4: System and loading for determining the spring value $C_{\vartheta D}$

Table 2: Calculation results for the example shown in Fig. 6

	Eq. (2)	1st-order theory with c-s distortions	2nd-order theory with c-s distortions	2nd-order theory with c-s distortions
point of load application			top flange	bottom flange
LTB F_{cr} [kN]			≅100	≅160
$C_{\vartheta D}$ [kNm]	31.1	38.6	32.4	48.0
difference [%]	-	+24.1	+4.2	+54.3

The equations for the deformations and section forces derived above contain no influences of nonlinear load-bearing behaviour. For slender webs and flanges and large membrane stresses, the influence of local buckling may be significant. The application of locally concentrated forces causes stresses σ_z in the direction of the centre line of the profile, which is superimposed on stresses σ_x and τ_{xz} resulting from the global load transfer (Fig. 3). Therefore, the influence of geometric nonlinearities is significant even for rolled sections.

The influence of the stress state on the spring value $C_{\vartheta D}$ is demonstrated in the following example. A beam with a cross-section similar to an IPE 300 profile and $L = 6$ m is analysed (Fig. 4). The spring value $C_{\vartheta D}$ is determined for three different cases by calculating the entire beam using the program PROFIL [2] considering the cross-sectional distortions:

1. using first-order analysis;
2. using second-order analysis for the case that the load is applied to the top flange;
3. using second-order analysis for the case that the load is applied to the bottom flange.

The spring values according to second-order analysis are determined for systems subjected to the expected lateral torsional buckling load F_{cr}. The calculation results are listed in Table 2. The difference between the spring values when the load is applied to the bottom flange compared to that when the load is applied to the top flange is about 50%. In Table 2, the spring stiffnesses obtained from the more precise calculation are compared to the value given by Eqn. (2). It can be seen that Eqn. (2) gives a conservative result in the present case. It contains certain reserves that cover the influence of second-order theory for the case when the load is applied to the top flange.

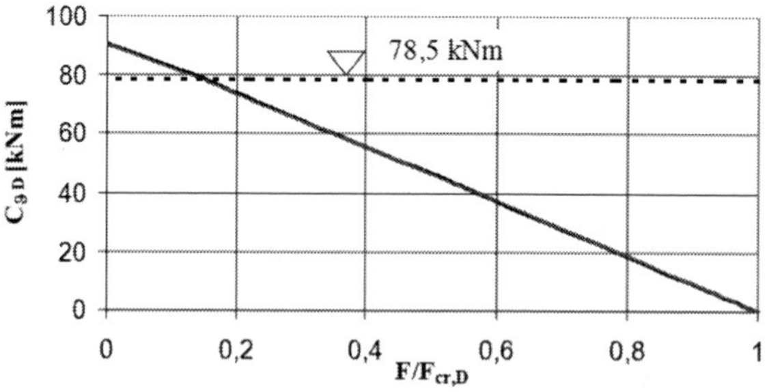

Fig. 5: Dependence of the value $C_{\vartheta D}$ on the degree of utilization with regard to distortional buckling

For a beam IPE 400 subjected to a single point load on the top flange at mid-span, Fig. 5 shows the torsional spring stiffness C_{9D} as a function of the utilization ratio $F/F_{cr,D}$ for the distortional buckling induced by combined longitudinal, transverse and shear stresses. The influences of the different types of stresses on the distortional buckling were varied by investigating various span lengths (L=3.00 m to 8.00 m) and end support conditions (hinged and fixed). From the figure, it can be seen that the stiffness C_{9D} decreases with increasing utilization ratio $F/F_{cr,D}$. This dependence is nearly linear and independent of the ratio of the stress components σ_x, σ_z and τ_{xz}. The spring stiffness calculated using Eqn. (2) gives C_{9D} = 78.5 kNm, which is about 13% smaller than those obtained from PROFIL [2].

3. Buckling loads

For the simplified ULS check on the basis of EN 1993-1-1, Section 6.3.2, the critical load M_{cr} of the lateral torsional buckling problem is required. For beams with locally torsional restrained flanges, the calculation can be performed with different methods varying in complexity. These include:

a) structural-mechanical analysis with shell elements using the FE method,
b) the use of calculation programs specially developed for this type of task (e.g. PROFIL),
c) calculation with beam models, based on second order analysis considering biaxial bending and warping torsion and the local torsional restraint (e.g., DRILL, LTBeam [5,6]),
d) the use of approximation formulas and prepared solutions.

The method a) is comparatively complex. A major advantage of the FEM calculation with shell elements is that the influences of the cross-sectional distortions and local stresses in the region of the load applications and flange restraints are accounted for more accurately. For rolled sections, the influence of the root radius of the section on the intersections of the web and the flanges are to be considered appropriately in the modelling. This is especially true for the torsion constant I_T, but also for the second moment of area I_y.

In [7], a macro element was developed specifically for doubly symmetrical I-beams, which is used in PROFIL [2]. This element considers the flanges as beams and the web as a plate (Fig. 6). By approximating the deformations of the web with fourth-order Hermite polynomials, local buckling effects can be represented approximately.

The methods c) and d) require a suitable method for determining the torsional spring stiffness resulting from the local torsional restraint of the flanges (see 2.2). Using an example, it has been shown that with this equation often a lower estimate of the stiffness is made (Table 2). If the section is susceptible to plate buckling because of its plate thicknesses and stress state, the spring stiffness has to be reduced, e.g., using Eqn. (3).

The methods c) and d) require a suitable method for determining the torsional spring stiffness resulting from the local torsional restraint of the flanges (see 2.2). Using an example, it has been shown that with this equation often a lower estimate of the stiffness is made (Table 2). If the section is susceptible to plate buckling because of its plate thicknesses and stress state, the spring stiffness has to be reduced, e.g., using Eqn. (3).

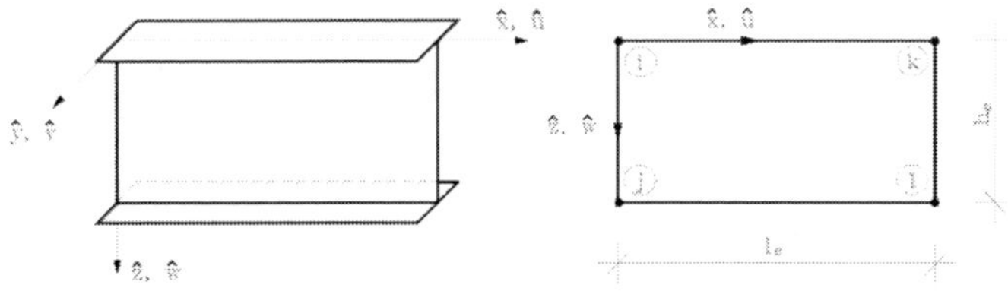

Fig. 6: I-section element, notation according to [7]

$$C_{\vartheta D} = C_{\vartheta D,0}\left(1 - \frac{\alpha_{cr,LT}}{\alpha_{cr,D}}\right) \tag{3}$$

$C_{\vartheta D,0}$ torsional spring stiffness resulting from the cross-sectional deformation (distortion), calculated by a first order analysis;

$\alpha_{cr,LT}$ amplifier to reach the elastic critical load for lateral torsional buckling;

$\alpha_{cr,D}$ amplifier to reach the elastic critical load for distortional buckling of the lateral supported beam (see Fig. 5).

Figure 7 presents the results of a parametric study for the calculation of M_{cr} of a single-span beam with a section IPE 400, a single point load on the top flange, and a local flange restraint at mid-span. The span length, end support conditions and calculation methods were varied in the determination of M_{cr}. FEM calculations with the program ANSYS as well as calculations with programs PROFIL [2] and DRILL [5] was performed. The calculation excluding the cross-sectional distortions was performed with the approach of a torsional spring stiffness $C_{\vartheta D}$ according Eqn. (2). In the calculations with ANSYS and PROFIL, a rigid flange restraint for the rotation φ_x about the longitudinal axis of the top flange was assumed. The cross-sectional distortions were considered by the programs with varying degrees of accuracy.

For small span lengths, the influence of the flange restraints is partially significantly overestimated with programmes PROFIL and DRILL. The deviations with programme PROFIL do not exceed 10% (single span beam with L=3.00 m and both ends fixed) and thus are relatively moderate. The differences related to ANSYS resulting from the simple shape functions in z-direction for the lateral displacements (see Figure 6 and the accompanying explanations). With increasing span lengths, the differences obtained with the three calculation methods become smaller. The influence of the geometric nonlinearity on the value of the spring stiffness becomes smaller, so that programme DRILL generates also useful results with the approach for the spring stiffness according to Eqn. (2).

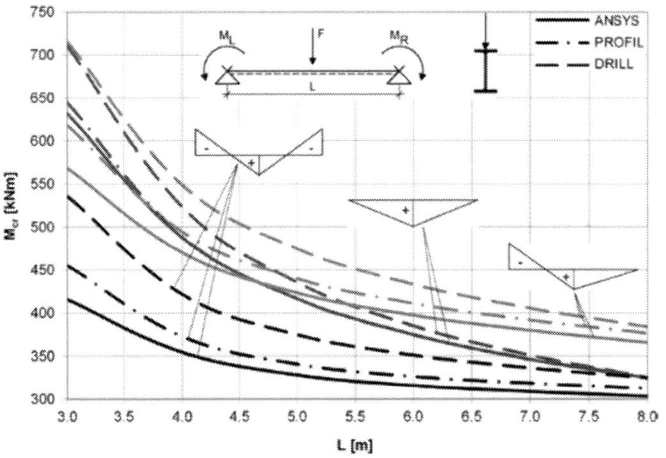

Fig. 7: Dependence of M_{cr} for single span beams IPE400 with different end support conditions and local flange restraints at the top chord at mid span

4. Component tests

As part of a research project, Lindner and Schmidt investigated the lateral buckling problem of I-beams considering realistic load applications [3,4]. The investigations covered systems with I-beams that were positioned perpendicularly to each other. Furthermore, the beam subjected to the load seated freely on the top flange of the beam susceptible to lateral torsional buckling (Fig. 8). A total of 71 tests were conducted, in which the parameters longitudinal and transverse beam section, longitudinal and transverse beam length, type of load application, construction of load application zone, and the support conditions of the transverse beam were varied. To detect the load-deformation behaviour, the global deformations of the longitudinal beams and the local deformations in the load application zone were measured during the tests after applying the individual load steps.

The beam systems shown in Fig. 8 correspond to the so-called load capacity group 1 according to [3,4]. The transverse beams are seated on the longitudinal beams with the entire width of the top flange. No stiffeners are used in the load application zones. For systems according to Fig. 8 b), the transverse beams can move freely in the axial direction at the support (roller bearing). To avoid rotations of the chord line of the beam, the support was lowered during the tests.

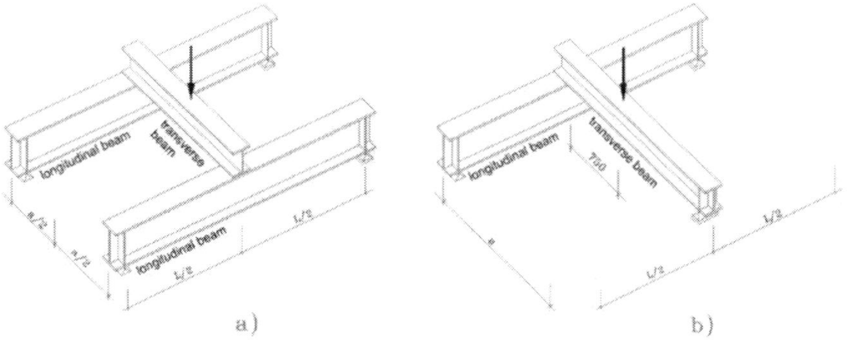

Fig. 8: I-beams positioned perpendicularly

5. Evaluation of test results and design approach

In the development of the design approach in accordance with EN 1993-1-1, different possibilities for the calculation of the buckling loads of beams with local torsional restraint were considered. Using the programmes ANSYS, PROFIL and DRILL, P_{cr} of the structures shown in Fig. 8 were calculated. In the analyses, the dimensions of the cross sections of the tested beams have been considered [3,4]. Furthermore, the different design rules and buckling curves of EN 1993-1-1 were investigated and the results statistically evaluated according to EN 1990 Annex D.

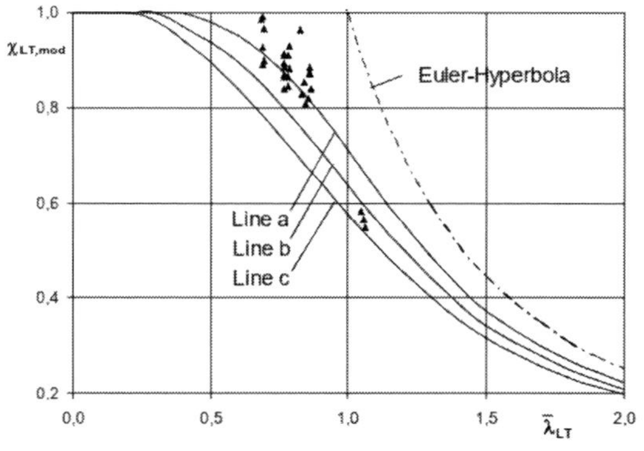

P_{cr} acc.	Buckling line	Number of tests $n = 30$ General case Line b	Line c	Specific case Line c	Line d
PROFIL	R_m	1.108	1.210	1.048	1.163
	V_δ	0.069	0.067	0.074	0.068
	γ_{Rd}	1.115	1.015	1.196	1.059
DRILL	R_m	1.120	1.224	1.059	1.176
	V_δ	0.063	0.064	0.068	0.066
	γ_{Rd}	1,085	0.994	1.165	1.042
DRILL mod	R_m	1.108	1.211	1.049	1.163
	V_δ	0.067	0.066	0.072	0.067
	γ_{Rd}	1.108	1.010	1.190	1.055

Fig. 9: Classification of the test results in the $\chi_{LT,mod}$– $\bar{\lambda}_{LT}$ diagram and statistical evaluation

Fig. 9 shows the classification of the test results in a $\chi_{LT,mod} - \bar{\lambda}_{LT}$ diagram. For the determination of the slenderness $\bar{\lambda}_{LT}$, the buckling loads P_{cr} were calculated using PROFIL, the full plastic resistances P_{pl} were determined from the cross section dimensions and the yield strength of the test specimens. The results of the statistical evaluation of the design rules given in chapter 6.3.2.1 to 6.3.2.3 of EN 1993-1-1 are shown in Fig. 9. They include the general and the specific case of LTB, each with two buckling curves and three methods for calculation of buckling loads. Based on these investigations, the buckling curve c of the general case combined with the modification by using the f-value for the moment diagram according to Eq. (6.58) of EN 1993-1-1 is recommended. Thus a differentiation between medium and larger slendernesses is not necessary. In addition, the simplified calculation of the torsional spring stiffness by Eqns. (1) and (2) is possible for most of the practical cases. In the range of small to medium slendernesses a moderate overestimation of the spring stiffness is compensated by the more conservative assumption of the buckling curve χ_{LT}.

6. Conclusions

The stability of beams can be increased considerably by taking local torsional restraints from purlins or secondary beams into account. For the value of the local torsional spring stiffness, the influence of cross-sectional distortion is significant if no stiffeners between the flanges are inserted. Locally concentrated forces cause stresses σ_z, which are superimposed on stresses σ_x and τ_{xz} resulting from the global load transfer. The effect of local buckling may reduce the torsional spring stiffness significantly. The calculation of M_{cr} is possible by software or with formulas by hand. If programs that exclude the effect of cross-section distortion are used, this effect should be considered in the determination of the value of the local torsional spring stiffness. The development and calibration of a design approach for local restraint beams based on Eurocode 3, Part 1.1 were shown. For this, test results from beam systems of earlier investigations [3,4] were evaluated.

References

[1] Yura JA. "Fundamentals of Beam Bracing", *Engineering Journal*, First Quarter, 11-26, Texas at Austin 2001.

[2] Stroetmann R. "PROFIL – FEM-Program for the structural analysis of space structures with I-Profiles", *Fachgebiet Stahlbau und Werkstoffmechanik, Technische Universität Darmstadt*, Germany, 1999.

[3] Lindner J, Schmidt JS. "Erfassung wirklichkeitsnaher Lasteinleitung in I-Trägern", *Forschungsbericht P77 der Studiengesellschaft für Anwendungstechnik von Eisen und Stahl e. V.*, Düsseldorf, Germany, 1982.

[4] Lindner J, Schmidt JS. "Biegedrillknicken von I-Trägern unter Berücksichtigung wirklichkeitsnaher Lasteinleitung", *Der Stahlbau 51(1982) Volume 9, Verlag Ernst & Sohn*, Berlin, Germany, 257-263, 1982.

[5] Friemann H. "Program DRILL – Lateral torsional buckling of beams and columns" Version 2012.340, *FIDES DV-Partner GmbH*, München, Germany, 2012.

[6] CTICM – Centre Technique Industriel de la Construction Métallique "Program LTBeam – Lateral Torsional Buckling of Beams" Version 1.0.11, Paris, France, 2013.

[7] Stroetmann R. "Zur Stabilitätsberechnung räumlicher Tragsysteme mit I-Profilen nach der Methode der finiten Elemente", *Publication of the Institute for „Stahlbau und Werkstoffmechanik", Technische Universität Darmstadt*, Germany, 1999.

EXPERIMENTAL STUDY ON THE BUCKLING BEHAVIOR OF TAPERED STRUCTURAL MEMBERS

Trayana Tankova[a], João Rodrigues[b], Liliana Marques[a], João Pedro Martins[a] and Luís Simões da Silva[a]

[a] *ISISE – Department of Civil Engineering, University of Coimbra*
[b] *Department of Civil Engineering, University of Coimbra*

Abstract: This article presents the results from experimental study on non-uniform I steel members. These elements allow optimization of the distribution of the material which can result in significant material savings. However, there is a lack of guidance on how to define the critical cross-section, which governs the stability design. There are certain difficulties when non-uniform members shall be designed according to EC3-1-1 such as: choice of buckling curve, critical cross-section, and torsional effect caused by non-symmetric restraints, the effect of imperfections and the actual conditions of loading. The experimental programme is summarized and the results for one of the beams tested are presented in detail.

1. Introduction

The excellent characteristics of the structural steel together with the evolution of the knowledge about the behaviour of steel members and structures in the recent years has led to increased use of members with variable section. Those are particularly used as girders in large span structures, such as stadia and airport roof structures. However, the application of non-uniform members still lacks design methods capable of predicting the actual behaviour and thus reflecting the efficiency of these solutions, resulting in the need to deepen the knowledge of the stability behaviour of steel elements with non-uniform section.

A correct application of tapered elements, whether with variable inertia, with unconventional longitudinal axis or irregular distribution of restraints, allows an optimized distribution of the material using larger sections in the most loaded areas and smaller sections where the material demand is lower.

The tapered elements are usually manufactured in two ways: welding plates or cutting and welding of hot-rolled profiles. The fabrication process induces imperfections (geometric and residual stresses), which highly influence the load bearing capacity of the steel members.

This paper presents one test (Fig. 1) of the experimental programme. Firstly the experimental programme is presented, then the test set-up is detailed. Finally, the obtained results for one of the beams are discussed.

Fig. 1: Experimental test on a web-tapered steel beam

2. Experimental programme

The experimental programme performed within the scope of research project TAPERSTEEL. Its objective was to study the stability behaviour of non-uniform steel elements. The results from the experimental programme are going to serve for calibration of advanced numerical model, which is further used to enlarge the databank of results for calibration of design rules.

Table 1: Experimental programme

N	Member	γ_h	Form	h_{min} [mm]	b [mm]	t_w [mm]	t_f [mm]	L [m]	Failure mode*
1	Column 1	4	V	120	100	12	12	6	*FB In-Plane*
2	Column 2	2	V	185	110	6	12	6	*FB In-Plane*
3	Beam 1	4	L	250	210	12	16	4	*LTB*
4	**Beam 2**	**4**	**L**	**250**	**210**	**6**	**16**	**4**	***LTB***
5	Column 3	3	L	120	100	10	16	6	*FB In-Plane*
6	BC	3	L	120	100	10	16	6	*N+My – LTB*

*

FB In-Plane: in-plane flexural buckling
FB In-Plane + Local: in-plane flexural buckling with local buckling effects
LTB: Lateral-torsional buckling
N+My – LTB: Lateral-torsional buckling of member loaded with major axis bending moment and axial force

The main objectives of the experimental programme are:

- Experiments on non-uniform steel columns, beams and beam-column;
- Analysis of the influence of different tapering ratios;
- Characterization of the material behaviour;
- Measurement of geometrical imperfections;
- Estimation of the ultimate loads;
- Comparison with the estimation of advanced numerical simulations.

The nominal dimensions of the tested elements and the buckling failure mode are given in Table 1.

The paper focuses on the results obtained for *Beam 2*. Comparison is performed with the estimation of advanced numerical simulation GMNIA (Geometrical and Material Non-linear Analysis with Imperfections). The FEA was performed using software product ABAQUS/CAE [1]. The numerical model was built using nominal yield stress f_y=355 MPa. The numerical simulation revealed lateral displacement at the critical location d_{max}=255,8 mm corresponding to maximum load $P_{max,num}$ = 636,38 kN, which itself produces maximum bending moment at the support $M_{max,num}$ = 1215,99 kNm. The maximum displacement at the end of the cantilever is d_{max} = 110mm.

3. Experiment N 4 – Beam 2

3.1 Experimental Layout

The tested beam is simply supported, the test layout is given in Fig. 2. On its small section, it is supported with two-directional hinge. At the end of the non-uniform section, the lateral and vertical displacements as well as the rotation are restrained.

The load was applied as a point load on the top flange of the beam at 8.5cm distance from the end of the cantilever part. The beam was laterally restrained at the intermediate support and at the point of load application (see Fig. 2).

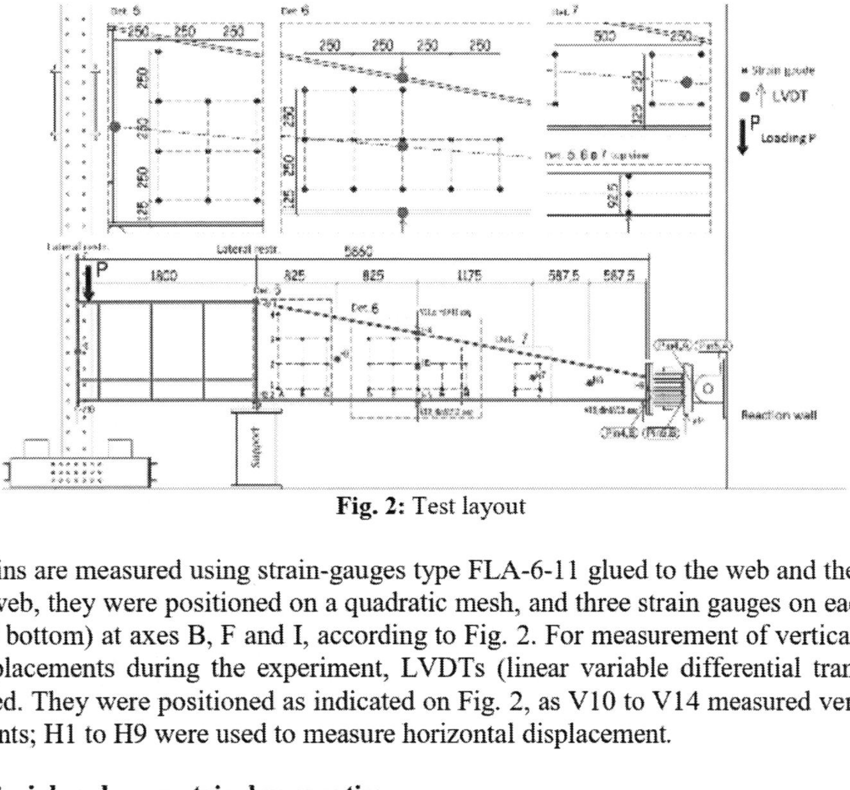

Fig. 2: Test layout

The strains are measured using strain-gauges type FLA-6-11 glued to the web and the flanges. On the web, they were positioned on a quadratic mesh, and three strain gauges on each flange (top and bottom) at axes B, F and I, according to Fig. 2. For measurement of vertical and lateral displacements during the experiment, LVDTs (linear variable differential transformer) were used. They were positioned as indicated on Fig. 2, as V10 to V14 measured vertical displacements; H1 to H9 were used to measure horizontal displacement.

3.2 Material and geometrical properties

All elements were fabricated using welding of hot-rolled plates according to EN 10025-2:2004 (+M) [2], the steel grade of all elements is S355 JR +M. The bolts used were a high strength steel bolts class 8.8. The screws were in class 8.8 with hexagon head M30 Grade A standards EN 15048-1[3] and EN ISO 4017:2008[4], respectively. The nuts and washers were also used in Grade A, in accordance with the standards EN ISO 4032: 2008 [5] and EN ISO 7089: 2008 [6], respectively.

3.2.1 Coupon tests

In order to obtain the real stress-strain curves and, subsequently, the experimentally observed yield and ultimate stresses, tensile coupon tests were performed. Overall, six coupon tests were performed on Beam 2 (three coupon tests from the web and the flanges).
The resulting stress-strain curves are represented in Fig. 3. The obtained stress-strain curves show that the results for the yield stress for 5 coupon tests is higher than the nominal 355 MPa. The test Web_C was excluded, since its result was lower than the guaranteed nominal value. The estimated mean value of the yield stress for all coupon tests performed on Beam 2 is $f_{y,m}$=396 MPa.

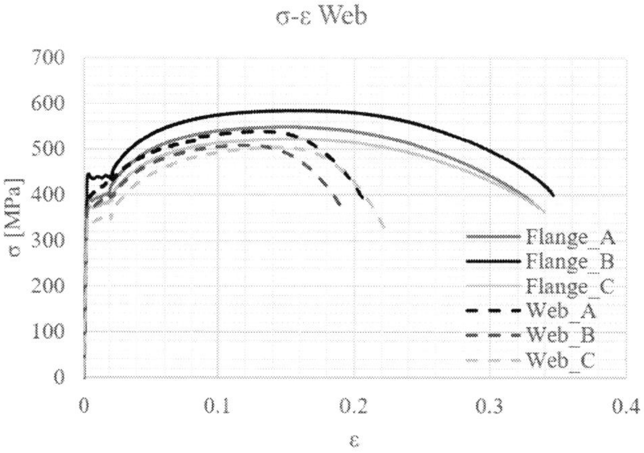

Fig. 3: Stress-strain curves

3.2.2 Geometrical dimensions and deviations

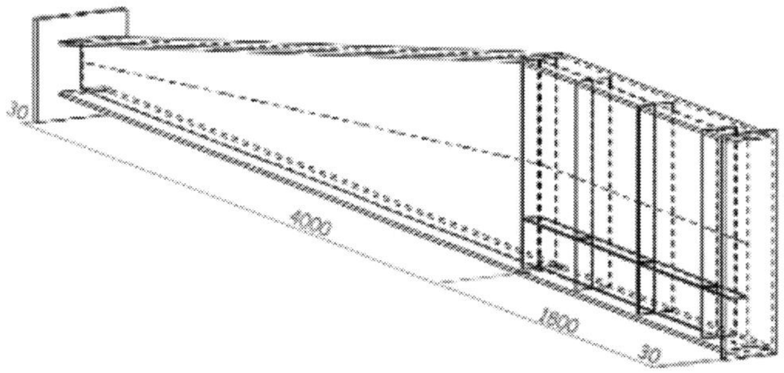

Fig. 4: Tested beam

Table 2: Geometrical properties

h_{man} (mm)	250
h_{max} (mm)	1000
b (mm)	210
T_{flange} (mm)	16
$t_{alma,ABAQUS}$ (mm)	6
t_{alma} (mm)	7,10

The studied beam has variable cross-section in its first 4 meters length and it has 1.8m of uniform section. Due to the slender web, the uniform part is reinforced with web stiffeners in order to avoid local cross-section effects (this part of the beam is aimed to serve only for load transmission).

Geometrical imperfections are one of the most important features determining the global behaviour of slender structures. In this case, imperfections are local and global geometric deviations from the perfect structure, i.e. deviation of the web from the vertical alignment. An advanced laser system (see Fig. 5) was used to gather the information about geometric imperfections. The amplitudes can be observed in Fig. 6.

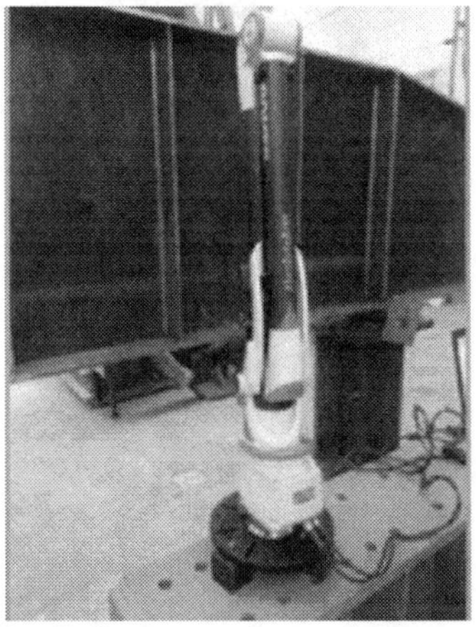

Fig. 5: Laser system used to measure geometric imperfections

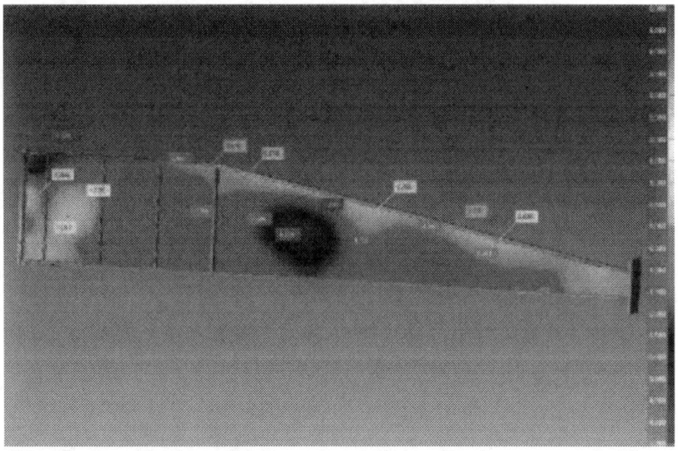

Fig. 6: Amplitudes of the geometric imperfections measured by the laser system

3.1 Supports

Along the reaction wall (see Fig. 2), the beam is supported a two-directional pinned connection, which allows rotation in the two main directions and prevents any translation and torsion about the longitudinal axis. The connection element bearings to be assured by screwing with M30 class 8.8 screws, Grade A, with nuts and washers equivalent. The bearings are in steel S355 with pins class 8.8 and diameter 90mm, the connection is illustrated on Fig. 7.

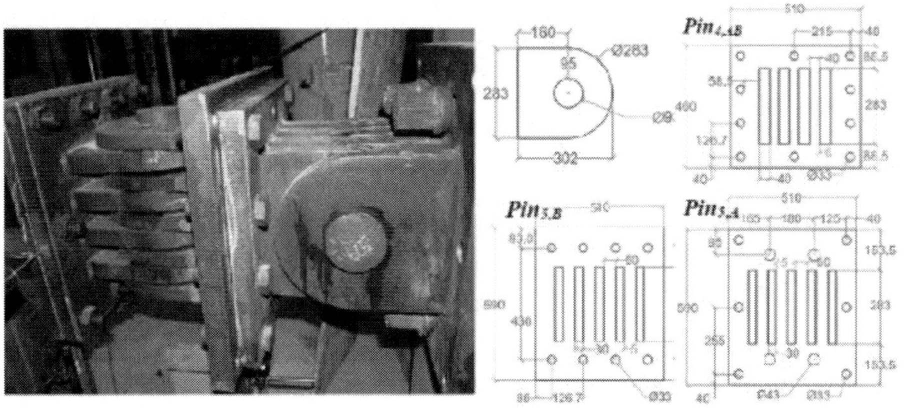

Fig. 7: Pinned connection and its dimensions

The intermediate support, in the transition between the variable and constant height parts, is a simple support, allowing the rotation of the element about the axis of the highest cross-section inertia. A hot-rolled HEB300 profile with welded end-plate was used. The plate has thickness of 40mm in S355 and a Class 8.8 pin with diameter 75 mm (Fig. 8).

The lateral braces used are coupled to roller-bearings. The braces are made of hot-rolled profiles HEB120 in steel S235, inclined at 45 degrees to the free end (one roller-bearing on each side) (Fig. 8b) and parallel to the intermediate support elements (two roller-bearings on each side), preventing the twist beam (Fig. 8c).

a) intermediate support b) Lateral restraint at the free end c) Lateral restraint at the intermediate support

Fig. 8: Support conditions *Beam 2*

3.2 Loading

The test was performed in two stages. Initially, the beam was subjected to point loading with variable intensity. The loading was conducted in successively applied loads of 10%, 20%, 40% and 60% $P_{max,num}$, as between each cycle, total unloading of the member was done. The last cycle reached 40ton ≈ 393 kN.

The application of this load intensity within the elastic range allowed for setting of the lay-out, eliminating deviations, and sliding of the bearings. The measurement of loading and un-loading in elastic range allows to estimate the rigidity of the overall structure. The loading was performed under load control at a rate of 0.20 kN / sec.

In the second stage, after the adjustments resulting from the first, the specimen was loaded at a constant rate of 0.015 mm/s until reaching the ultimate load $P_{max,real}$, which already in-cludes deformations in the plastic range. This experiment resulted in the behaviour curve in terms of load-displacement.

4. Results

4.1 Cycles of loading and unloading

The load displacement curve given on Fig. 9a shows the vertical displacements at the load application point throughout the first stage of the experiment (cyclic loading with load incre-ments). There was particular settlement of the frame in the second loading phase (up to 20% $P_{max,num}$). It is clearly identified on Fig. 9, where three levels that indicate adjustments in ver-tical displacement of 5,62 mm. Table 3 summarizes the adjustments in each cycle.

The location at H6 is where the instability occurred by lateral buckling in the second stage of the experiment. The total horizontal displacement adjustment recorded is 0.90 mm to the right (left positive sense), on the contrary to the instability direction (see Fig. 9b).

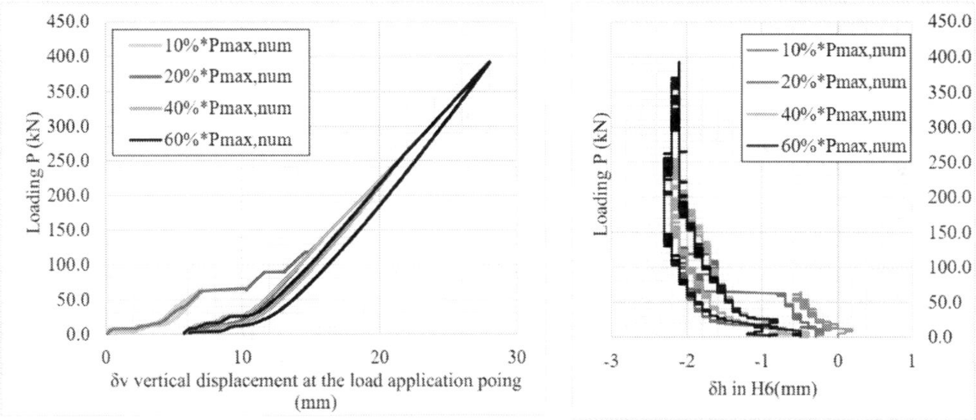

Fig. 9: Evolution of the displacements at V10 (vertical) and H6 (horizontal) during the cycles of loading and unloading

Table 3: Vertical adjustment V10 at the end of each cycle

Cycle	Adjustment V10 (mm)
10%*$P_{max,num}$≈65 kN	0,12
20%*$P_{max,num}$≈130 kN	5,62
40%*$P_{max,num}$≈260 kN	0,00
60%*$P_{max,num}$ ≈400 kN	0,36
Total	6,10

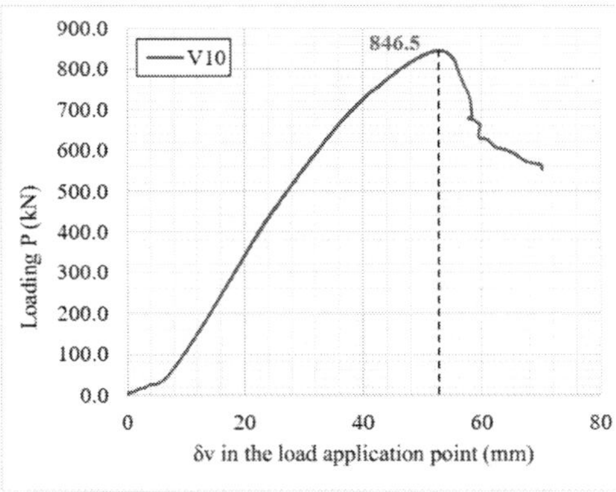

Fig. 10: Vertical displacement during the experiment

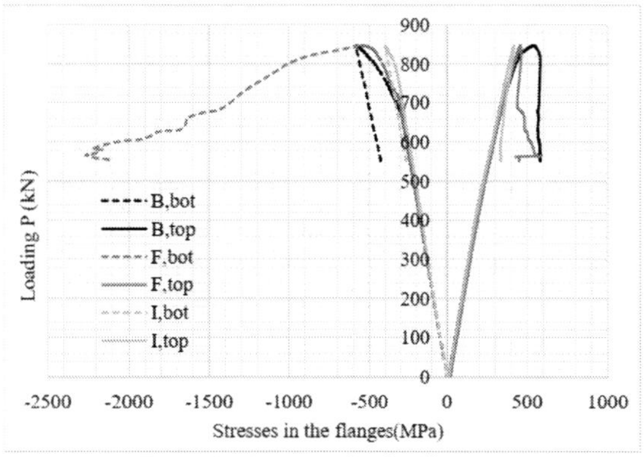

Fig. 11: Stresses in the flanges

4.2 Loading until the ultimate load

The load causing the instability of the beam by lateral buckling occurred at $P_{max,real}$ = 846,5 kN, corresponding to a vertical displacement (downward) in the load application point δ_{V10} = 53,36 mm (Fig. 10). The numerical model predicted twice higher deformation in this section. Along axis F vertical displacements occurred at $\delta_{V11,V12}$ = 25,28 mm and at the smallest cross-section the beam 35,43 mm. Regarding the horizontal displacements, at H6 the maximum horizontal displacement recorded was δ_{H6} = 132 mm.

The maximum load led to a bending moment $M_{max\ Real}$ = 1451.75 kN.m, i.e, 19.40% higher than the estimation of the numerical model. The difference between those two estimations can be explained mainly by the increase in the real strength of the beam whereas the f_{ynom} steel yield stress is higher than 355 MPa considered in the numerical model. In addition, the measured thickness of the web is estimated 7.10 mm, which is different from the prescribed. In the numerical model, the web thickness was set to 6mm, thus imposing differences with the numerical model.

The longitudinal stresses in the flanges were measured in axes B, F and I. A noticeably symmetrical pattern is observed, even though the longitudinal axis of the member is inclined, similar values are obtained when comparing the lower flange to the upper flange at the same cross-section (Fig. 11).

5. Conclusions

This paper presented one of the experiments performed in the scope of experimental programme on tapered members. The study aimed to evaluate the stability resistance of tapered members and the results of the experiments will be further used to calibrate numerical model.

The experimental layout was firstly detailed, including beam's dimensions, supports, and load application procedure. The auxiliary tests were also presented, namely the tensile coupon tests and the laser measurement of geometrical imperfections.

Furthermore, the results of the experiment were detailed according to its two stages. In the first one, adjustments in the experimental layout were observed (total vertical settlement of 6.10mm), thus confirming the necessity of this loading stage. The recorded stresses and displacements are considered acceptable for future calibration of numerical models.

Finally, in the second stage of the test the ultimate load that was obtained was almost 20% higher than the estimation of the numerical model. However, this difference is explained by the higher value of the yield stress (f_{ym}=396 MPa) and the higher web thickness (t_w=7.10 mm).

Acknowledgments

The research leading to these results has received funding from:

- the Portuguese Foundation for Science and Technology (FCT) under grant agreement TAPERSTEEL PTDC/ECM-EST/1970/2012.
- the Portuguese Foundation for Science and Technology (FCT) under grant agreement SFRH/BD/99702/2014.

References

[1] Abaqus (2012). v.6.12, Dassault Systems/Simulia, Providence, RI, USA.
[2] EN 10025: 2004, CEN Brussels. Hot rolled products of structural steels, 2004.
[3] EN 15048-1:2007, CEN Brussels, Non-preloaded structural bolting assembles, 2007.
[4] EN ISO 4017: 2008, Hexagon head screws - Product grades A and B.2008,
[5] EN ISO 4032: 2008, Hexagon regular nuts (style 1) - Product grades A and B, 2008,
[6] EN ISO 7089: 2008, Plain washers - Normal series - Product grade A, 2008.

Plate, shell and space structures

EXPERIMENTAL ANALYSIS OF MONOTONIC AND CYCLIC BEHAVIOUR OF LONGITUDINALLY STIFFENED GIRDERS

Dib Abdelkarim, László Gergely Vigh, Péter Hegyi and Balázs Kövesdi

Department of Structural Engineering, Budapest University of Technology and Economics, Hungary

Abstract: This paper introduces an innovative structural solution for dissipative zones of moment-resisting steel frames and eccentrically braced frames. High local ductility and high performance are intended to be achieved by the application of longitudinally multi-stiffened girder elements. Monotonic and cyclic test series are completed to analyse the influence of different stiffener configurations on the cyclic strength and cyclic ductility of the girder. Based on the experimental results, practical conclusions are drawn on the preferable configuration. The test results further provide solid base for the numerical model development.

1. Introduction

Ductile design of steel structures subjected to seismic effects requires creation of dissipative zones (plastic hinges). In case of moment resisting frames and eccentrically braced frames, in order to achieve high ductility this requirement yields to the application of compact sections (Class 1 as per Eurocode 3 Part 1-5, EC3-1-5 [1]) in the dissipative zone, otherwise premature local plate buckling will govern the design and high ductility cannot be obtained.

Buckling resistance of thin web plates of I girder can be significantly increased by application of stiffeners in the web. Vertical stiffeners are often required in seismic links of eccentrically braced frames, but rarely used in moment resisting frames. Application of longitudinal stiffeners is common in bridge structures only. Such stiffeners are typically designed as relatively stiff and strong stiffener, that is characterized by column-type buckling behaviour, resulting in relatively low ductility.

Vigh [2] and Simon et al [3] stated that using flexible (i.e. with low rigidity relative to the plate rigidity) multiple longitudinal stiffeners enables plate-like overall buckling behaviour, which is beneficial with respects to ductility. Although the behaviour of such stiffened plates subjected to monotonic loading is studied in several researches [2-7], there is no data available on the cyclic ductility.

For this reason, an experimental programme is worked out and completed to study monotonic and cyclic behaviour of flexible multiple longitudinally stiffened plate girders. The possible use of the results are two-fold: a) it provides solid base for numerical model development applicable for virtual experimenting and parametric study; b) by drawing general idea on the actual behaviour, and by the extension of the results with virtual experiment results, it helps developing prototypes and design recommendations.

Note that such configuration may be also beneficial with respect to other extreme loadings such as fire, [8]. Further details on the development of a high-performance frame structure system utilizing such innovative solutions can be found in [9, 10].

2. Experimental programme

In order to simulate the actual circumstances of plastic hinges in moment resisting frames, four-point tests are performed on simply supported girders. The girder shown in Fig. 1 consists of three parts: the inner part (length of 1 m) with varying configurations and being the subject of investigation (i.e. representing the dissipative zone), and the two external parts (each with 3.5 m of length) that shall remain elastic during the test. The internal part is associated with 500×4 mm web and 150×10 mm flanges.

The application of such configuration results in pure bending in the middle panel. Such representation provides feasible approximation for the behaviour of plastic hinges where bending-shear interaction does not play role, i.e. it is assumed that shear force demand-to-capacity ratio is limited.

In the investigated middle panel, 2-3 stiffeners are placed with varying dimensions. The studied configurations are listed in Table 1. The dimensions were selected in order to cover the different type of behaviour in overall plate buckling, including the reference case of no stiffeners. Further details on the preliminary analysis is found in [11, 12].

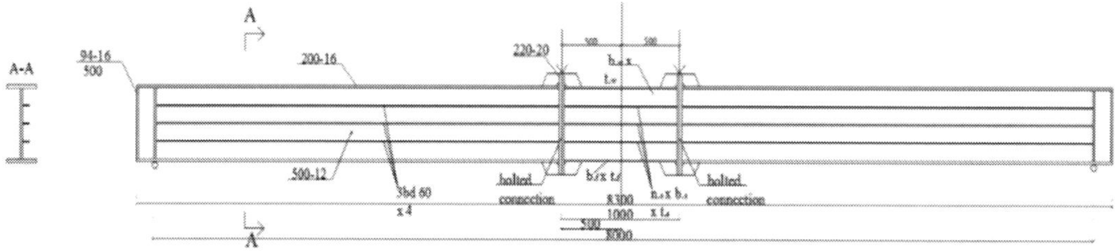

Fig. 1: Configuration of tested specimen

Fig. 2 shows the test configuration. Load and midspan deflection as primary characteristic parameter were measured during tests, as well as strain gages were placed on the external fibers of the flanges and at the upper stiffener for evaluation and verification of expected stress distribution.

Two types of test were carried out for each configuration: a) monotonic test; b) cyclic test. For the cyclic test, test protocol prescribed by EN 15129 [13] is used (Fig. 3). The design (or target) displacement is defined as a deflection of 60 mm. Such a deflection results in a plastic chord rotation that is sufficient for medium to high ductility class.

According to the standard, 5 cycles are repeated at the 25 and 50% of the design displacement, and at least 10 cycles – for confirmation of sufficient behaviour – are needed on the design displacement level. Beyond the 10th cycle, the test was continued with applying the same amplitude until rupture or significant strength degradation was observed.

Fig. 2: Test configuration

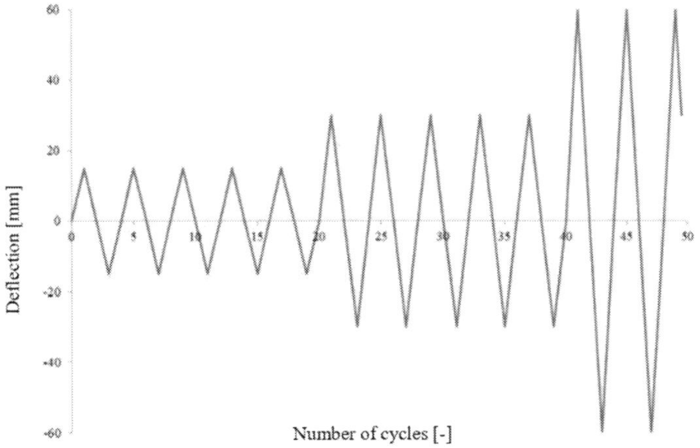

Fig. 3: Cyclic loading protocol

Table 1: Specimen geometries and monotonic test results

Configuration elements	Number of stiff-eners	size of the stiffeners	load		deflection		Ductility [-]
			yield point [kN]	ultimate capacity [kN]	at yield [mm]	ultimate [mm]	
unstiffened	-	-	149	152.65	28	80	2.85
2×40-4	2	40-4	152	179.74	35	130	3.71
3×40-4	3	40-4	n/a	171.65	30*	118	3.93
3×60-4	3	60-4	164	192.46	35	n/a	n/a

* estimated value

3. Monotonic test results

Fig. 4 shows the obtained load vs. midspan deflection curves of the monotonic tests, while Fig. 5 illustrates the observed ultimate failure modes. Buckling failure of the web always occurred in form of overall plate buckling or stiffener local plate buckling. Table 1 summarizes the corresponding results: load at first yield, ultimate load, yield displacement, ultimate displacement and ductility. Yield displacement is defined as the yield displacement of ideal elastic-perfectly plastic behaviour, where the initial stiffness is taken as the secant modulus considering 60% of the ultimate load, and the elastic branch is reaching the ultimate load. The ultimate deflection is taken as the deflection (beyond the ultimate load level, in the descending branch of the curve), at the load level of 90% of the ultimate load. Ductility is then calculated as the ultimate deflection divided by the yield deflection.

Note that the above definition is relatively stringent and does not necessarily represent the actual ductility of the configuration. Nevertheless, it is applicable comparison purposes and for the measurement of tendencies.

Comparing the different test specimens, it is clear that the addition of flexible stiffeners increases the strength as well as the ductility of the system.

4. Cyclic test results

Regarding the cyclic failure (first mechanism and low-cycle fatigue) of the specimens, very clear tendencies can be observed. Ultimate failure mechanism and the corresponding hysteretic curves are shown in Figs. 6-7.

Failure of the reference specimen of unstiffened web plate initiated with web plate buckling accompanied by flange plate buckling, immediately at the level of design displacement (no buckling was observed below this level). Significant strength degradation can be observed with the subsequent cycles at this level, resulting in weld rupture (Fig. 6/a) at cycle #38, due to low-cycle fatigue. The observed significant strength degradation and possible sensitiveness to plate buckling problem is in line with expectations.

In case of the specimen with 2×40-4 stiffeners, similar failure mechanism was observed (Fig. 6/b), and weld rupture initiated in cycle #40. It is also observed that fracture of the lower flange occurred soon after the weld failure. Conceptually different explanations can be given to this observation: a) due to the increased resistance, web failure might be closer to flange failure, or b) the presence of stiffeners and their buckling initiate deformations causing flange buckling.

Specimen with 3×40-4 stiffeners can be considered as the most ductile configuration (Fig 7/c). Plate buckling was not observed neither at the first two (25-50%) amplitudes nor in the first 10 cycles at the design deflection. Beyond this point, combined web-flange buckling was observed, causing gradual strength degradation. The level of strength degradation is much smaller than in case of any other specimens. The ultimate failure is low-cycle fatigue of the web-to-flange weld connection.

The specimen with 3×60-4 stiffeners behaves in a similar way as the previous specimen, however, the fracture of the weld (Fig. 6/c) due to low-cycle fatigue occurred earlier at the 70th cycle. It might be explained by the fact that the applied stiffeners own higher rigidity and strength and thus they dominate the deformation of the web (i.e. tends to column-type buckling), causing larger deformation demands on the welds as well, yielding to premature weld fracture. Accordingly, the large deformation capacity observed during monotonic test cannot be utilized.

Fig. 8/a and 8/b show the absorbed energy per cycle and the total (cumulative) absorbed energy. It is stated the absorbed energy during the first 10 cycles (i.e. at the levels of 25-50% of the design displacement) is negligible in comparison to the values achieved at the design displacement level. Although the specimen with the strongest stiffeners had earlier failure, the absorbed energy is comparable with the one obtained with 3×40-4 stiffeners. The absorbed energy in case of the unstiffened web plate and in case of 2×40-4 stiffeners was approximately half of that of the other two cases.

Note that AISC 341-05 [14] prescribes a requirement for seismic devices to have a minimum value of absorbed energy ratio of 200. It is concluded that specimens with 3×40/60×4 stiffeners satisfy this criteria only.

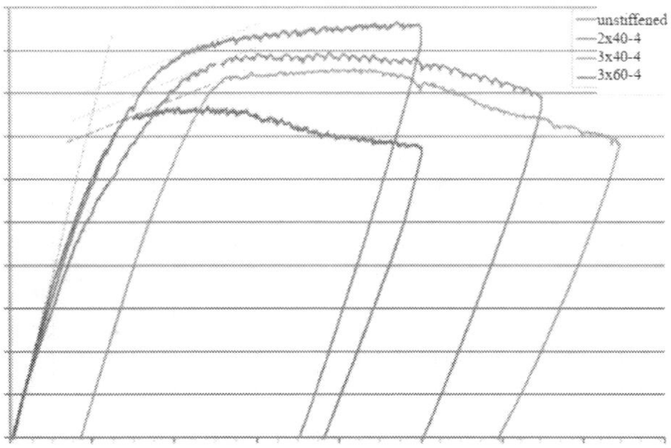

Fig. 4: Monotonic force-displacement curves

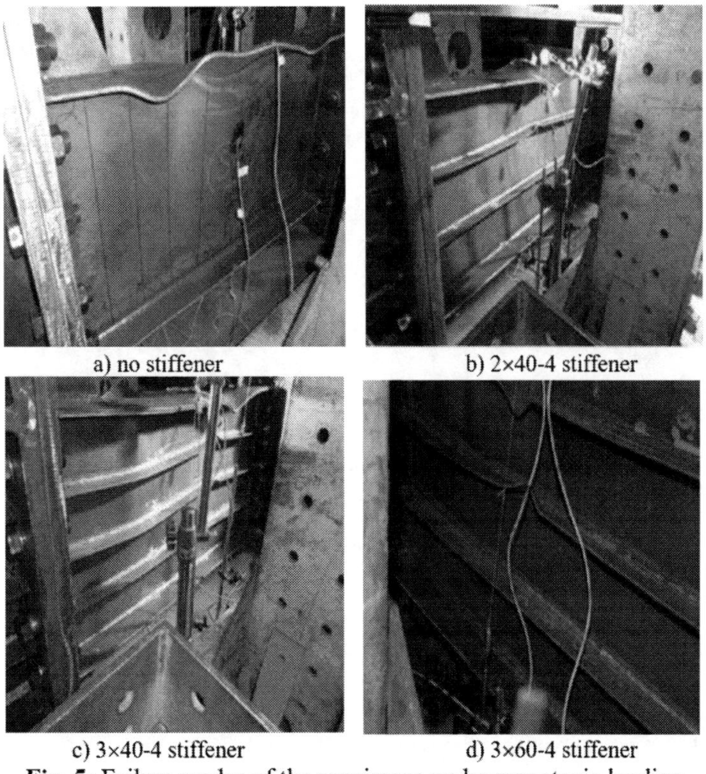

a) no stiffener b) 2×40-4 stiffener

c) 3×40-4 stiffener d) 3×60-4 stiffener

Fig. 5: Failure modes of the specimens under monotonic loading

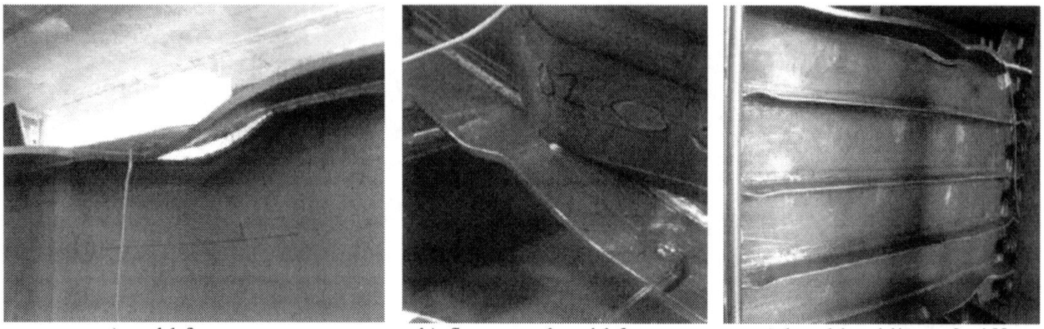

a) weld fracture b) flange and weld fracture c) local buckling of stiffener

Fig. 6: Cyclic failure modes

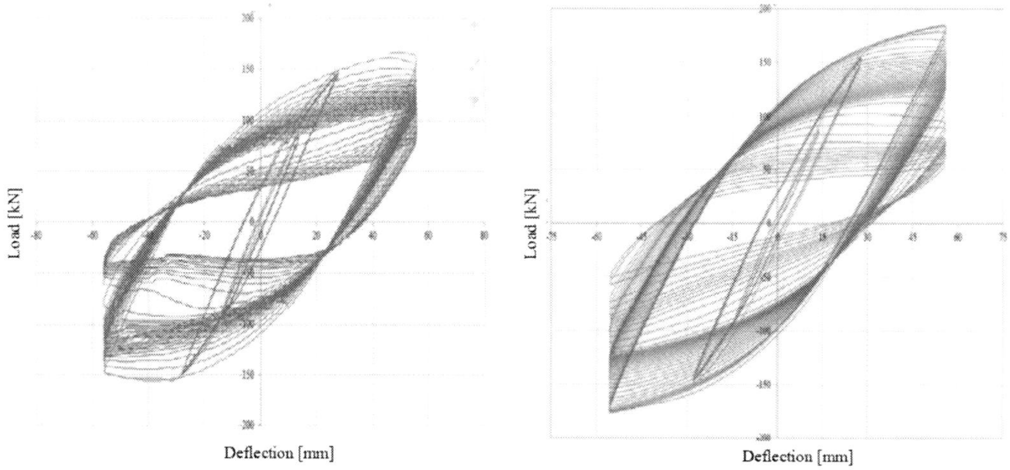

a) unstiffened panel b) 2×40-4 stiffener

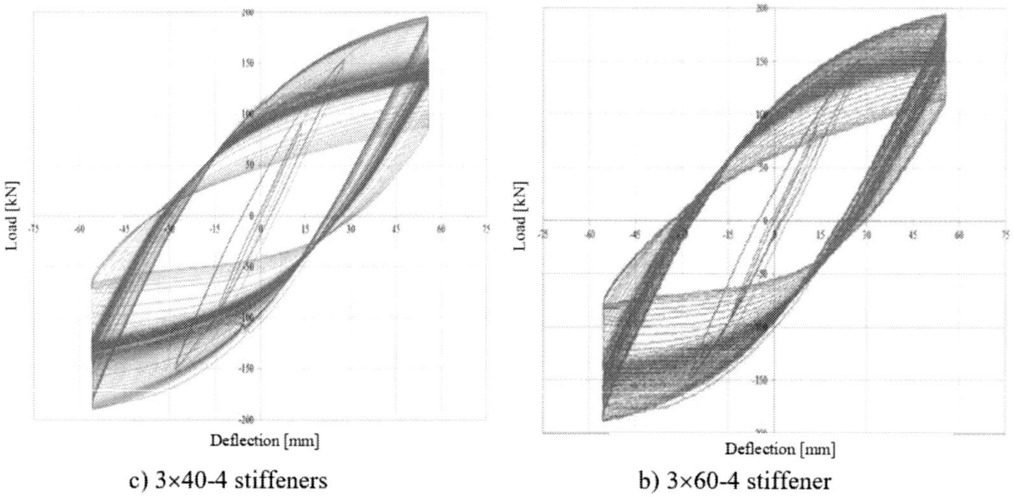

c) 3×40-4 stiffeners b) 3×60-4 stiffener

Fig. 7: Hysteretic curves

EN15129 dealing with seismic devices prescribes a criteria on the strength degradation. Since the behaviour of such plastic hinge is strongly dependent on the plastic deformations, the longitudinally stiffened plates can be considered as non-linear displacement dependent device. To characterize its property, the so-called second-branch stiffness K_2 is defined for nonlinear displacement dependent devices, as follows:

$$K_2 = \frac{V_{Ebd} - V\left(0.5d_{bd}\right)}{0.5d_{bd}} \tag{1}$$

where V_{Ebd} is the design shear force, d_{bd} is the design displacement. As per the code, the cyclic degradation of the K_2 stiffness is limited by the following criteria:

$$\kappa = \frac{\left|K_{2,i} - K_{2,3}\right|}{K_{2,3}} \tag{2}$$

where $K_{2,i}$ represents the second-branch stiffness calculated for the ith cycle on the level of design displacement, i.e. the reference is chosen as the third cycle. Fig. 9 illustrates the calculated k factors for the tested configurations at the different cycle numbers. It can be clearly stated that the unstiffened plate girder as well as the girder with 2×40-4 stiffeners violate the criteria, while the other two specimens show sufficient performance with this respect.

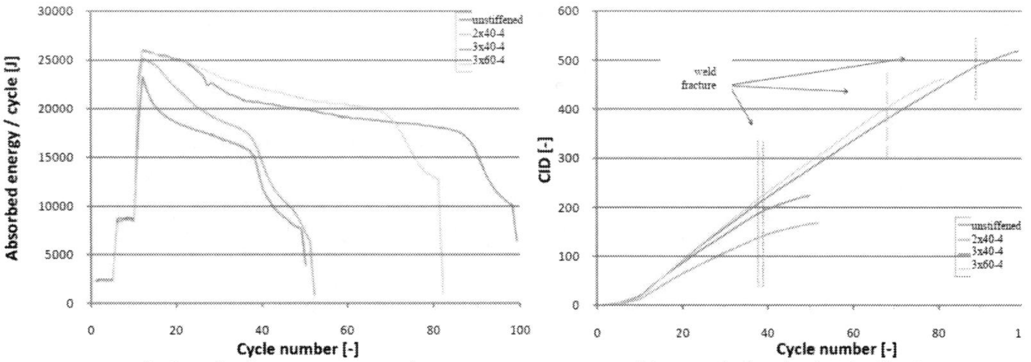

a) absorbed energy per cycle b) cumulative inelastic displacement

Fig. 8: Absorbed energy

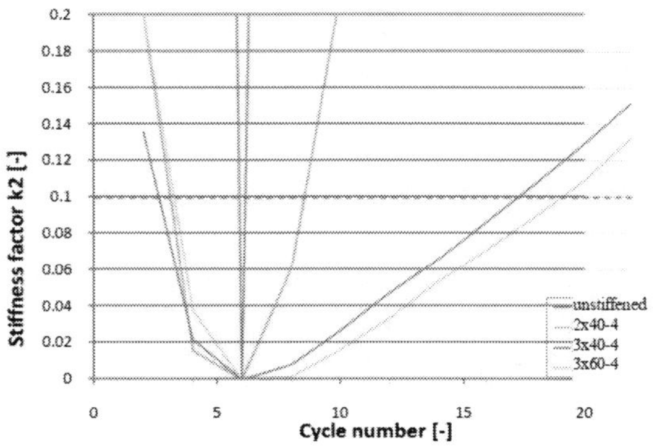

Fig. 9: Stiffness factor

5. Concluding remarks

An innovative structural configuration incorporating multiple flexible stiffeners in webs of plate girders is experimentally studied in the paper. The presented results are valid for plastic hinges subjected to pure bending; however, the results presume sufficient performance for cases of limited bending-shear interaction.

Based on the monotonic and cyclic test results of experimental studies on four different specimens varying the stiffener configurations suggest that the application of flexible stiffener may considerably increase the buckling resistance of the web (leading to better cross-section class) and improve the ductility of the member. An optimal stiffener geometry can be found to achieve sufficient ductility for moderate to high ductility class.

Acknowledgments

This paper was supported by the *János Bolyai Research Scholarship* of the *Hungarian Academy of Sciences*. The presented results are part of a research work partially supported by "*HighPerFrame*" R&D project (in cooperation of BME Department of Structural Engineering and Rutin Ltd.) Nr. GOP-1.1.1-11-2012-0568, supported by the *Új Széchenyi Terv*.

References

[1] EN 1993-1-5:2005, *Eurocode 3 – Design of steel structures – Part 1-5: Plated structural elements*, CEN, Brussels, 2005.

[2] Vigh LG. "Virtual and real test based analysis and design of non-conventional thin-walled metal structures", *PhD dissertation*, BME, Budapest. 128 p. 2006.

[3] Simon J, Kemenczés A, Vigh LG. "Optimal stiffener geometry of stiffened plates: Non-linear analysis of plates stiffened with flat bars under uniaxial compression", *Proc. Eurosteel 2014*. Napoli, Italy, 2014.09.10-12. Napoli: ECCS, 2014. Paper 421. 6 p.

[4] Aalberg A, Langseth M, Larsen PK. "Stiffened aluminium panels subjected to axial compression", *Thin-Walled Structures*, 39, 861-885, 2001.

[5] Jakab G, Dunai L, Macdonald RT. "Behaviour of plate girders with single-sided fillet weld", *Proc. EUROSTEEL 2005*, Vol. B, pp. 2.2.1-2.2.8, Maastricht, 8-10 June 2005.

[6] Dunai L, Jakab G, Néző J, Topping BHV. "Experiments on welded plate girders: fabrication, imperfection and behaviour", *Proc. 1st Intl Conf on Advances in Experimental Structural Engineering (AESE '05)*, Vol. 1, pp. 51-58., Nagoya, 19-21 July 2005.

[7] Jakab G, Szabó G, Dunai L. "Imperfection sensitivity of welded beams: experiment and simulation", *ICMS 2006 Intl Conf on Metal Structures: Steel – A New and Traditional Material For Building*, Brasov, Romania, 20-22 September 2006.

[8] Horváth L, Kövesdi B. "Innovatív módszerek és megoldások acélcsarnokok tűzvédelmi tervezésében", *MAGÉSZ Acélszerkezetek*, 2015/1, Különszám.

[9] Dunai L, Papp Z. "Nagy teljesítményű acélszerkezeti rendszer – A HighPerFrame kutatás-fejlesztési projekt", *MAGÉSZ Acélszerkezetek*, 2015/1, Különszám.

[10] Balogh T, Vigh LG. "Változó keresztmetszetű acél csarnokszerkezetek optimális tervezése szeizmikus hatásokra", *MAGÉSZ Acélszerkezetek*, 2015/1, Különszám

[11] Bartal M, Hegyi P, Apáti J, Kövesdi B, Vigh LG. "Keretszerkezetek innovatív energiaelnyelő szerkezeti kialakításának fejlesztése: Hosszbordával merevített képlékeny csukló numerikus és kísérleti vizsgálata", *Magész Acélszerkezetek*, 12:(1. különszám) pp. 32-40, 2015.

[12] Bartal M.: "Hosszbordákkal merevített acél lemezes szerkezeti elemek statikus és ciklikus vizsgálata", *MSc diplomamunka*, BME Hidak és Szerkezetek Tanszék, 2015.

[13] EN 15129:2010, *Anti-seismic devices*, CEN, Brussels, 2010.

[14] ANSI/AISC 341-05, *Seismic Provisions for Structural Steel Buildings*, American Institute of Steel Construction (AISC); 2005.

PLATE GIRDERS UNDER BENDING

Roland Abspoel

Delft University of Technology, Delft, The Netherlands

Abstract: In a material economy driven plate girder design, the lever arm between the flanges will increase. This leads to higher stiffness and bending moment resistance, but also to an increase of the web slenderness. This means that high strength steels can be used leading to a large reduction of the steel consumption. However, Eurocode 3 [3] restricts the web slenderness based on the formula to avoid flange induced buckling, derived by Basler [2]. Experimental and theoretical research by Abspoel [1] conducted at the Stevin II Laboratory of Delft University of Technology, shows that this formula is too conservative. Ten unstiffened plate girders with high web slenderness's are tested, focussed on flange induced buckling.

1. Introduction

In many steel structures like buildings, industrial halls and bridges, standard hot-rolled sections are used. These sections are divided into specific types, such as IPE, HEA, HEB, HEM, HED and UNP in Europe and similar profiles in the USA. The range of hot-rolled sections is limited and therefore fabricated plate girders are used when the standard hot-rolled sections do not meet the requirements for stiffness, strength, stability and economy.

Such a plate girder is built up from steel plates for the top and bottom flange and for the web, welded together to an I-shape cross-section, single or double symmetric. Using this type of plate girders, a high degree of optimisation of material use is achieved by using different plate thicknesses and widths for the flanges, and thickness and height for the web over the span of the girder adapted to the distribution of bending moments and the shear forces.

Optimisations can be carried out for many aspects, but in the PhD-thesis by Abspoel [1] the ultimate bending moment resistance of a plate girder, given a certain weight per unit length, is the main topic for optimisation. For a long time, this was not or hardly of interest at all in Western countries, especially because the cost of structures was mainly determined by labour cost and hardly by material cost. Due to the increasing automation in the production of plate girders, the material cost becomes more important than the labour cost. However, for nowadays structures, life cycle costs and the environmental impact of structures become of greater influence on the design, beside to the increasing cost of steel by expanding demand by booming economies like China, Brazil, India and other upcoming economies. So, optimisation for minimal use of materials has become highly important.

Using higher steel grades, applying most material in the flanges and increasing the lever arm between both flanges are the main possibilities to maximise the bending moment resistance of a plate girder under pure bending given a certain amount of steel.

In case of a fixed cross-sectional area, by increasing the lever arm, more material is placed in the web, reducing the remaining material for the flanges. However, the lever arm can also be increased without using more material in the web by increasing the web height and decreasing the web thickness. This process is restricted by a limit for a practical thickness of the web, to enable welding of the section and also the handling of the plate girder.

The slenderness of the flanges, expressed in the width to thickness ratio of the flange plate, is restricted such that at least yielding of the outer fibre is possible to ensure that the plate girder exhibits at least a "not brittle-like" post-critical behaviour.

However, EN 1993-1-5 [3] limits the web slenderness by a specific phenomenon called "flange induced buckling". This phenomenon has been studied by Basler [2] and described as "vertical buckling of the compressive flange into the web".

Basler based this maximum web slenderness only on one laboratory test result and so it was considered of interest to perform additional research on this limitation to investigate whether this phenomenon really limits the bending moment resistance of a plate girder and to see if it is possible to increase this maximum web slenderness.

2. Vertical buckling of the compressive flange into the web

Basler [2] developed a model to determine the maximum web slenderness, based on vertical buckling of the compressive flange into the web. The curvature of the test panel of the plate girder under pure bending is constant. It is assumed that the compressive flange yields and because of this curvature a stress perpendicular to the flange σ_n acts on the web, see Fig. 1.

$$\sigma_n = A_{tf} \cdot f_{y.tf} \cdot \frac{2 \cdot \varepsilon_{tf}}{h \cdot t_w} \approx \frac{A_{tf}}{A_w} \cdot 2 f_{y.tf} \cdot \varepsilon_{tf} = \sigma_{cr} = \frac{\pi^2 E}{12 \cdot (1-\upsilon^2) \cdot \left(\frac{h_w}{t_w}\right)^2} \tag{1}$$

The strain of the top flange is assumed to be larger than the yield strain, $\varepsilon_{tf} \geq \varepsilon_{y.tf}$, because it is required that the flange fully yields to obtain some deformation capacity to guarantee that the compressive flange will fully yield. To be sure $\varepsilon_{tf} \geq \varepsilon_{y.tf}$, the strain ε_{tf} of the top flange is taken equal to $\varepsilon_{tf} = \left(f_{y.tf} + \sigma_r\right)/E$, where σ_r is the residual stress in the compressive flange.

$$\beta_{w.\max.I} = \frac{h_w}{t_w} = \sqrt{\frac{\pi^2 E}{24 \cdot (1-\upsilon^2)} \cdot \frac{A_w}{A_{tf}} \cdot \frac{1}{f_{y.tf} \cdot \frac{f_{y.tf} + \sigma_r}{E}}} = 0.67 \cdot \sqrt{\frac{A_w}{A_{tf}}} \sqrt{\frac{E^2}{f_{y.tf} \cdot \left(f_{y.tf} + \sigma_r\right)}} \tag{2}$$

This is the most general equation for the maximum web slenderness $\beta_{w.\max}$. By assuming minimum values for the ratio of area A_w/A_f, but also by assuming a minimum for the residual stress σ_r in the compressive flange, this equation can be simplified.

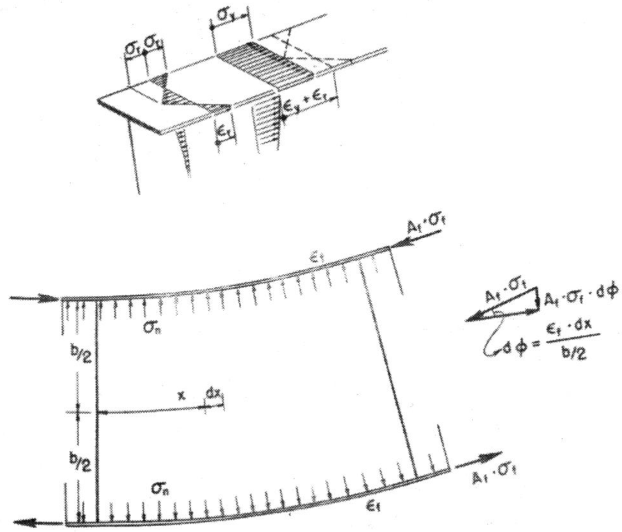

Fig. 1: Determination of the maximum web slenderness according to Basler [2]

Basler [2] mentioned that the ratio of area $\rho = \frac{A_w}{A_f}$ shall not be taken smaller than 0.5. Substitution of the ratio of area $\rho = 0.5$ in Eqn. (2), gives for the maximum web slenderness β_{max}:

$$\beta_{w.max.II} = \frac{0.48E}{\sqrt{f_{y.tf} \cdot (f_{y.tf} + \sigma_r)}} \tag{3}$$

A second simplification of Eqn. (2) is found by assuming a residual stress level of $\sigma_r = \frac{f_{y.tf}}{2}$ for mild steel and so the maximum web slenderness $\beta_{w.max}$ becomes:

$$\beta_{w.max.III} = \sqrt{\frac{\pi^2}{36 \cdot (1-\upsilon^2)}} \cdot \frac{E}{f_{y.tf}} \cdot \sqrt{\frac{A_w}{A_f}} = 0.55 \cdot \frac{E}{f_{y.tf}} \cdot \sqrt{\frac{A_w}{A_f}} \tag{4}$$

In EN1993-1-5 [3] this Eqn. (4) is used. Vertical buckling of the compressive flange into the web is called "flange induced buckling". A third simplification of the maximum web slenderness $\beta_{w.max}$ is given by taking into account a ratio of area $\rho = 0.5$ as well as a residual stress level of $\sigma_r = \frac{f_{y.tf}}{2}$. The maximum web slenderness $\beta_{w.max}$ as given in Eqn. (2), changes into:

$$\beta_{w.max.IV} = \frac{0.40E}{f_{y.tf}} \tag{5}$$

In EUR 8849 [4], a draft Eurocode, the equation for the maximum web slenderness $\beta_{w.max}$ is presented as given in Eqn. (6). For mild steel S235, the maximum web slenderness is 360.

Test specimen G4-T2 failed by vertical buckling of the compressive flange. The actual web slenderness of girder G4 was $\beta_w = h_w / t_w = 1270/3.3 = 387.6$ and the actual ratio of area was $\rho = \frac{A_w}{A_f} = \frac{h_w \cdot t_w}{b_{tf} \cdot t_{tf}} = \frac{1270 \cdot 3.3}{308.9 \cdot 19.7} = 0.69$. The actual maximum web slenderness $\beta_{w.max}$ based on this ratio of area $\rho = A_w / A_f = 0.69$ according to Eqn. (4) is determined with:

$$\beta_{w.max} = 0.55 \cdot \frac{E}{f_{y.tf}} \cdot \sqrt{\frac{A_w}{A_f}} = 0.55 \cdot \frac{206000}{259.2} \cdot \sqrt{0.69} = 368.3$$

The web slenderness β_w of test specimen G4-T2 is larger than the maximum web slenderness $\beta_{w.max}$ and so there will be vertical buckling of the flange into the web. The strain was "slightly" higher than the yield strain ε_y. The actual strengths and areas of test specimen G4-T2 are given in Table 1, including the actual web slenderness β_w and the actual maximum web slenderness $\beta_{w.max.III}$ based on Eqn. (3) and $\beta_{w.max.IV}$ based on Eqn. (4).

Table 1

	A_{tf} [mm²]	$f_{y.tf}$ [MPa]	A_w [mm²]	$\beta_{w.max.III}$ [-]	$\beta_{w.max.IV}$ [-]	β_w [-]
G4-T2	6072.1	259.2	4161.2	368.3	324.3	387.6

3. Delft experiments

In Fig. 2 a schematic presentation of the test setup is given.

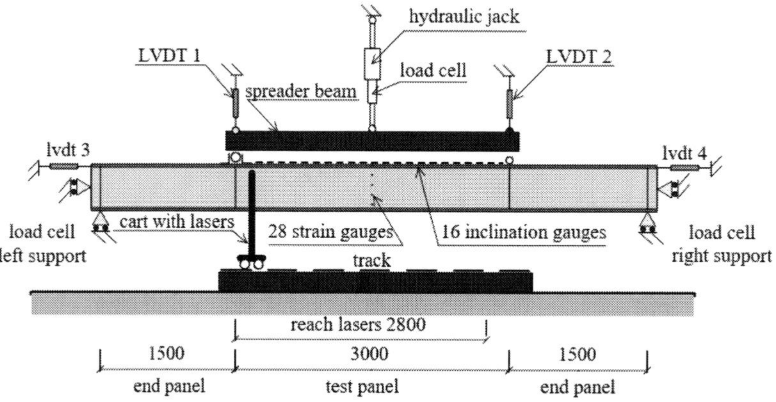

Fig. 2: Schematic presentation of a test specimen and the test rig

The web slenderness β_w of interest lies between 400 and 800, related to commonly used ratios of areas ρ, between ½ and 2 according to Basler [2]. Based on the maximum web slenderness $\beta_{w.max}$, see Eqn. (4), the web slenderness β_w lies between 360 and 720. Because of the shift of the neutral axis due to the use of the effective width method, these web slenderness's are taken higher than described by Basler, namely between 400 and 800.

To make such relatively large objects more suitable for testing in a laboratory, the dimensions are scaled down to a web thickness of 1 mm and a span of 6000 mm for all 10 test specimens. Based on this web thickness $t_w = 1\ mm$, web heights of 400 mm, 600 mm and 800 mm are used, which leads to a web slenderness β_w of 400, 600 and 800 respectively.

Related to the limitation of the ratios of area between ½ and 2 the cross-sectional area of the flanges is taken 200 mm² and 400 mm² and the following flange dimensions are taken: 50×4 mm², 80×5 mm² and 100×4 mm². The last two flanges have the same flange area A_f, but different flange dimensions. Especially, because of the difference in flange thicknesses

t_f, the torsional stiffness's EI_t differ, based on $b \cdot t_f^3$. The actual geometrical dimensions of the test specimens are given in Table 2. The maximum web slenderness is according to Eqn. (4) and based on the actual yield stresses, for test girder 9, 800×80, $f_{y.tf} = 320\ MPa$.

Table 2

	h_w [mm]	t_w [mm]	b [mm]	t_f [mm]	β_w [-]	$\beta_{w.max}$ [-]	ρ [-]
1, 400×50	400	1.0	50	4	400	471.2	2.0
2, 400×80(1)	400	1.0	80	5	400	344.7	1.0
3, 400×80(2)	400	1.0	80	5	400	349.6	1.0
4, 400×100	400	1.0	100	4	400	315.7	1.0
5, 600×50	600	1.0	50	4	600	585.0	3.0
6, 600×80	600	1.0	80	5	600	403.1	1.5
7, 600×100	600	1.0	100	4	600	395.5	2.5
8, 800×50	800	1.0	50	4	800	664.2	4.0
9, 800×80	800	1.0	80	5	800	476.8	2.0
10, 800×100	800	1.0	100	4	800	470.7	2.0

4. Measurements on the test panel of test girder 9, 800×80

The results of test girder 9, 800×80, are shown in this section. For the results of the remaining nine test girders see Abspoel [1]. Fig. 3 shows the P-δ diagram of the test girder, based on the actuator force and the deformations of the piston. It can be seen that the girder finally fails after reaching the maximum force in deformation step H.

The total out-of-plane deflections of the web in the test panel are measured by using lasers on a movable cart, see Figs. 4, 5 and 6 for respectively deformation step 0, H and I. Next to these total out-of-plane deflections of the web, the out-of-plane deformations as function of the load can be seen in Figs. 7 and 8 and finally, Figs. 9, 10 and 11 show the increment of the out-of-plane deflections.

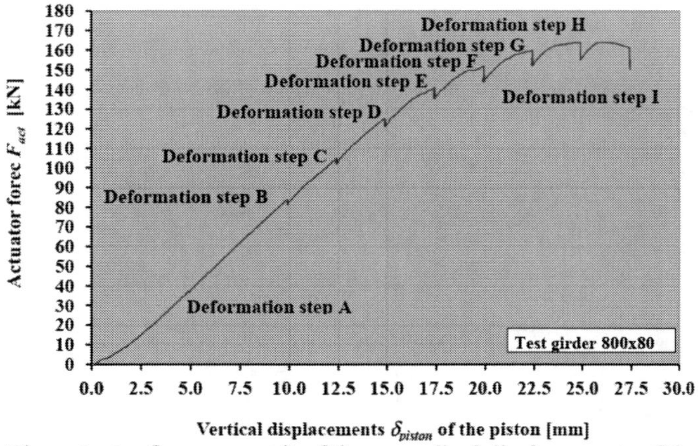

Fig. 3: The actuator force as result of the prescribed displacements of the piston

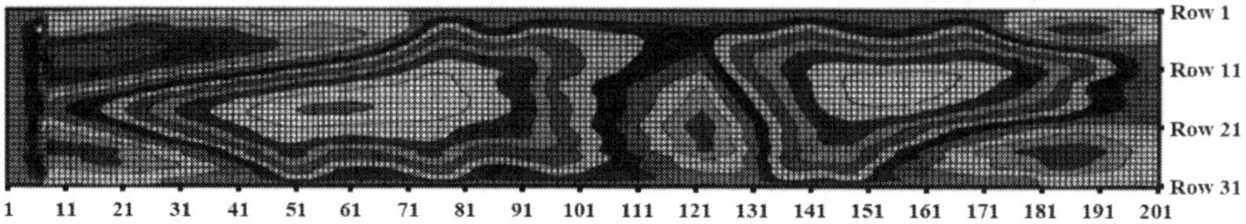

Fig. 4: Initial total out-of-plane displacements, deformation step 0

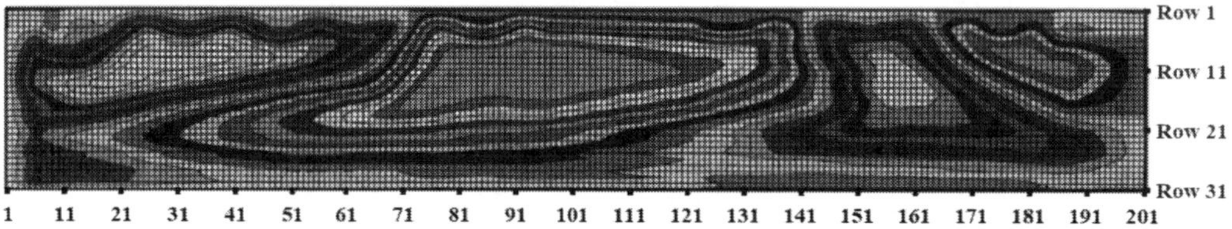

Fig. 5: Total out-of-plane displacements, deformation step H

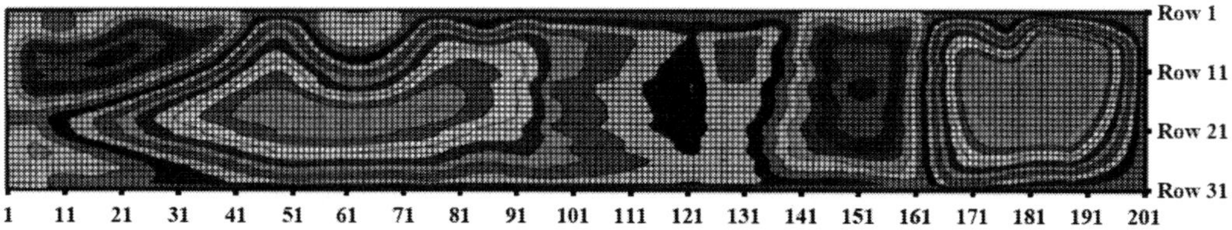

Fig. 6: Total out-of-plane displacements, deformation step I

From Figs. 4 to 6 it can be seen that the total out-of-plane deflections of the web in the test panel is asymmetric over the span of the test panel. The maximum load is reached for deformation step H and in the next deformation step the pattern of the deflections changed suddenly and the girder fails. Figs. 7 and 8 show that at the right side of the test panel one big buckle is subdivided into two buckles, a snap through. Figs. 9 and 10 show that in deformation step I almost the whole web buckles.

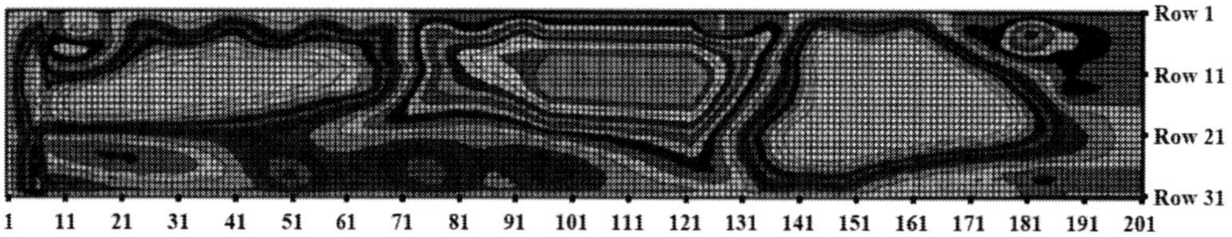

Fig. 7: The out-of-plane deflections as function of the loading in deformation step H

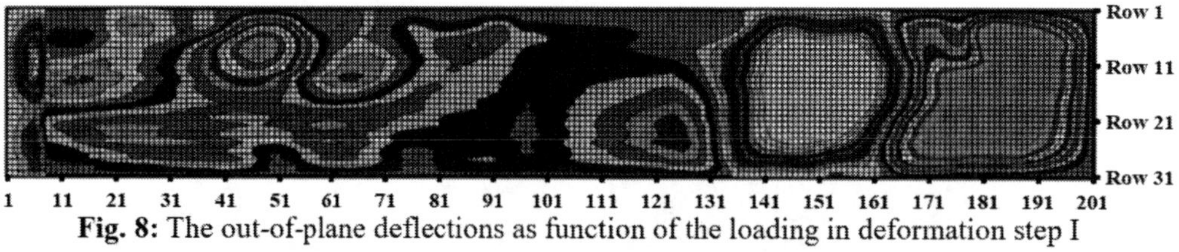

Fig. 8: The out-of-plane deflections as function of the loading in deformation step I

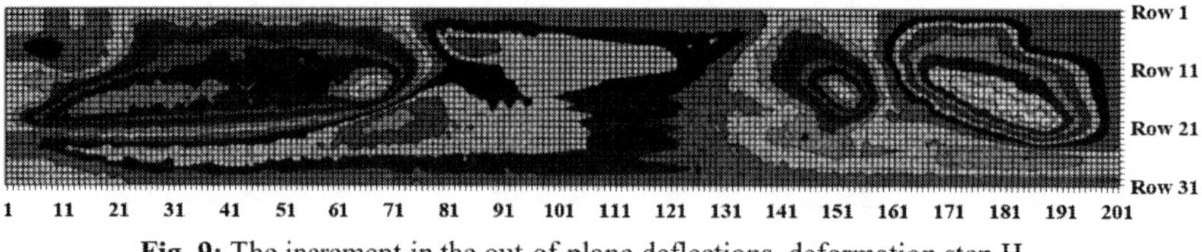

Fig. 9: The increment in the out-of-plane deflections, deformation step H

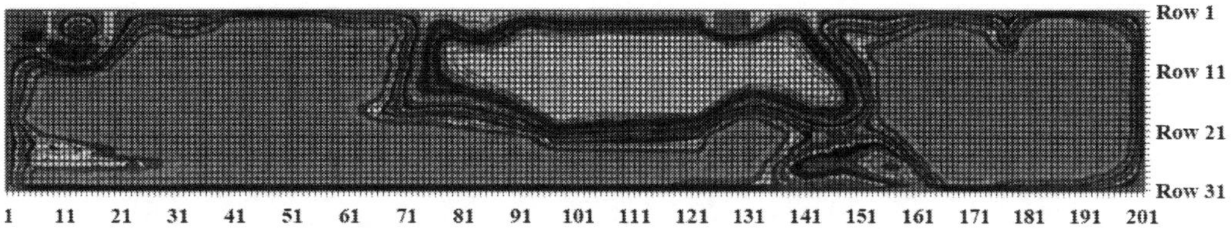

Fig. 10: The increment in the out-of-plane deflections, deformation step I

Table 3 shows the vertical deflections at the load introductions, measured by two Lvdt's at the top of the spreader beam, for all test girders for step the maximum load appears.

Table 3

	$\delta_{z.Lvdt1}$ [mm]	$\delta_{z.Lvdt2}$ [mm]	$\Delta\delta_z$ [mm]
1, 400×50	27.14	26.59	-0.55
2, 400×80(1)	31.86	32.59	+0.73
3, 400×80(2)	40.27	41.61	+1.34
4, 400×100	37.28	36.41	-0.87
5, 600×50	31.30	25.44	-5.86
6, 600×80	22.25	22.29	+0.04
7, 600×100	23.55	23.90	+0.35
8, 800×50	20.87	22.29	+1.40
9, 800×80	18.63	19.76	+2.13
10, 800×100	18.63	20.87	+2.61

From Table 3 it can be seen that there is hardly any difference in vertical deflection at both load introductions and so it can be concluded that flange induced buckling is not determining the maximum load. Even a maximum difference of 5.86 mm is rather small over the length of 3.0 m of the test panel. Fig. 11 shows the vertical deflection over the span of the test panel for every deformation step for test girder 9, 800×80. It can be seen, that the girder deforms asymmetric in deformation step I, so after the load is a maximum in the P-δ diagram in deformation step H.

The bending moment resistance of this test girder 9, 800×80, is rather close to the bending moment resistance based on the cross-section existing of only the flanges, but also close to the effective bending moment resistance taking into account partly yielding of the web. The yield stress of all webs is smaller than the yield stress of the flanges.

5. Conclusions

The following conclusions can be drawn:

1. The model for vertical buckling of the compressive flange into the web as adopted in the EN 1993-1-5 [3] and in the AISC 360-10 [5] is much too conservative according to the test results. This can be caused by the assumption related to the residual stresses, the assumption that the web is simply supported by the flanges, the way the curvature is calculated, the neglect of the influence of the stresses due to bending;
2. It can be concluded that the maximum web slenderness is larger than 800 up to S355, based on actual yield stresses, which is, related to the application of plate girders, very high;

3. For cases where the deflections criterion is not governing, it is concluded that possible material savings obtained by using high strength steel grades and by optimising the cross-section of plate girders are possible up to approximately 50%. This is an important finding for reducing the carbon footprint of steel structures.

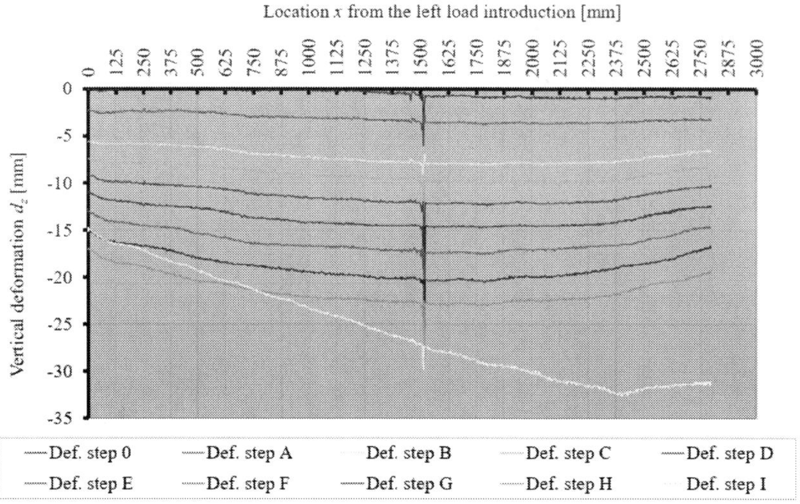

Fig. 11: Vertical deformations between both load introductions

References

[1] Abspoel R. "Optimisation of plate girders", *Dissertation*, Delft, The Netherlands, 2015.

[2] Basler K. "Strength of plate girders", *Dissertation*, Bethlehem, America, 1959.

[3] EN 1993-1-5, *Eurocode 3: Design of steel structures – Part 1-5 Plated structural elements*, 2006.

[4] EUR 8849, *Draft Eurocode 3: Common unified rules for steel structures*, 1984

[5] AISC 360-10, *Specification for structural steel buildings*, American Institute of Steel Constructions, 2010.

DISTORTION AND TORSION OF EXPERIMENTAL BOX-GIRDER CALCULATED USING ANALOGY WITH BENDING

Ivan J. Baláž[a] and Yvona P. Koleková[b]

[a] *Department of Metal and Timber Structures, Faculty of Civil Engineering, STU in Bratislava, Slovakia*
[b] *Department of Structural Mechanics, Faculty of Civil Engineering, STU in Bratislava, Slovakia*

Abstract: Steel experimental trapezoidal box-girder loaded by antisymmetrical action. Calculation of internal forces of torsion and distortion using analogy with bending. Comparison of experimental direct stresses in longitudinal direction in girder and direct stresses in transverse direction in transverse stiffening frames with theoretical ones. Simple, quick and enough exact "hand" calculations used for more complicated phenomena as torsion and distortion are, based on analogies with bending, are very important and useful especially in the age of computer programs. Analogies with solutions of more simple problems make solutions of more difficult problems easily obtainable and well understandable.

1. Introduction

The theory of thin-walled structures was developed in Russia by Vasilij Zacharovič Vlasov, who published basic publications in 1931, 1935, 1936 [1], 1940 and seven monographies in 1933, 1936, 1940 [2], 1949 [3], 1949 and 1958. Vlasov was very close to invention of Generalized Beam Theory (GBT), when he died in 1958. GBT which describes the mechanical behaviour of prismatic structures in the 1st order theory by ordinary differential equations

$$E^kC^kV^{IV}(x)-G^kD^kV^{II}(x)+{}^kB^kV(x)={}^kq(x) \qquad (1)$$

for each mode k, which are not coupled and can be solved independently, was developed in Germany by Richard Schardt in 1966 [4] and by Gerhard Sedlacek in 1968 [5]. Sedlacek gave to this theory its name. Direct application of Vlasov theory [3] in finding of optimal transverse stiffening of 1-, 2-, 3- and 4-cells girders was published in [6]. In [7] it was shown that by using Vlasov`s theory [3] one may obtain the same results as according to GBT. The bible of GBT (VTB – Verallgemeinerte Technische Biegetheorie in German) was published by Schardt in 1989 [8]. GBT was further extended to include second-order effects (local and global buckling) by Schardt in the second and third volume of his fundamental books, which unfortunately were never published. GBT was introduced into English speaking countries in nineties by Davies and Leach from UK [9-11]. In 21st century GBT was heavily developed by D. Camotim from Lisbon and his collaborators [12]. They published a lot of papers in this area. The Vlasov's theory and GBT was taught at Department of Metal and Timber Structures, Slovak University of Technology (STU) in Bratislava already in the end of the seventies [13]. GBT has been taught by the first author from the eighties [14, 15]. Two PhD thesis were successfully defended by Rendek [16] (linear problems of GBT) and Buček [17] (non-linear problems of GBT). More references may be found in [7].

The behaviour of thin-walled members was investigated at STU in Bratislava also experimentally. The results of the tests of small thin-walled cantilever were published in [18]. Big experimental model was tested in [19].

2. Experimental investigation

The following phenomena were investigated:

1. distribution of direct stresses in longitudinal direction and in the transverse stiffener frames under antisymmetrical loading;
2. distribution of direct stresses in longitudinal direction and in the transverse stiffener frames under symmetrical loading;
3. measurement of imperfections of the wide flange in compression and imperfections of the web in combined shear and bending by photogrammetric way;
4. influence of imperfections on buckling resistance of flange and web;
5. resistance of compressed stiffened flange when box-girder was in normal position (the wide flange was in compression, the narrow one in tension);
6. resistance of compressed stiffened flanges when box-girder was in reverse position (the narrow flange was in compression, the wide one in tension);
7. shear resistance of the box-girder webs, which limited the resistance of the box-girder.

In this paper, only results related to 1) are presented.

The total length of experimental box-girder was 8.4 m. Its span was 8 m and weight 2000 kg. Distribution of longitudinal stiffeners is on Fig. 2. The transverse stiffening frames (Fig. 4) were strengthened at the supports (Fig. 5) and in the cross-sections were loading was introduced (Fig. 3). The width of the opening in diaphragma was 470 mm (Figs. 3 and 5). The box-girder was loaded by a) antisymmetrical and b) by symmetrical actions (Fig. 4). The strength properties of the steel were f_y = 262 MPa and f_u = 390 MPa.

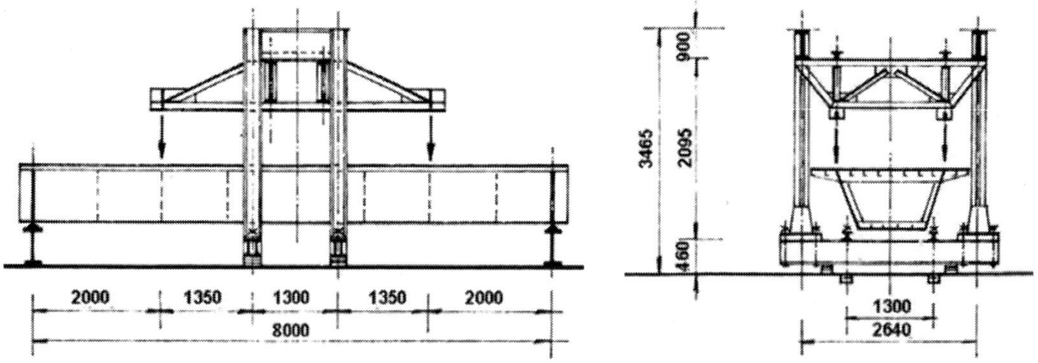

Fig. 1: Arrangement of loading frame. Four hydraulic jacks with capacity 500 kN each

The maximum capacity of every of 4 hydraulic jacks was 500 kN. The stresses were evaluated in 204 points with the help of the logger TSA-63. The deformations were measured with the exactness 0.1 mm in the sections 0, $L/8$, $L/2$, $3L/8$, L.

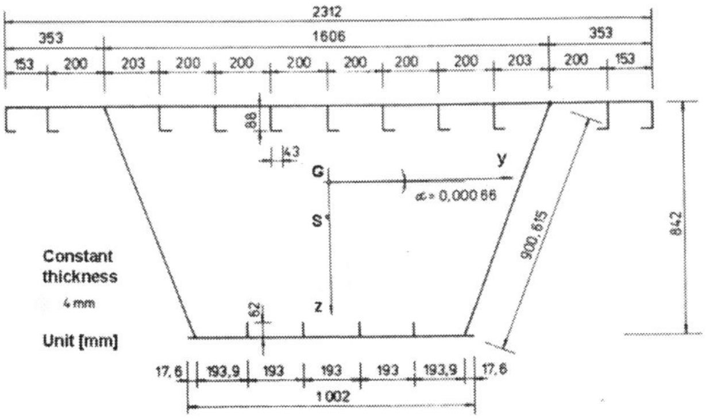

Fig. 2: Cross-section

Fig. 3: Transverse stiffening

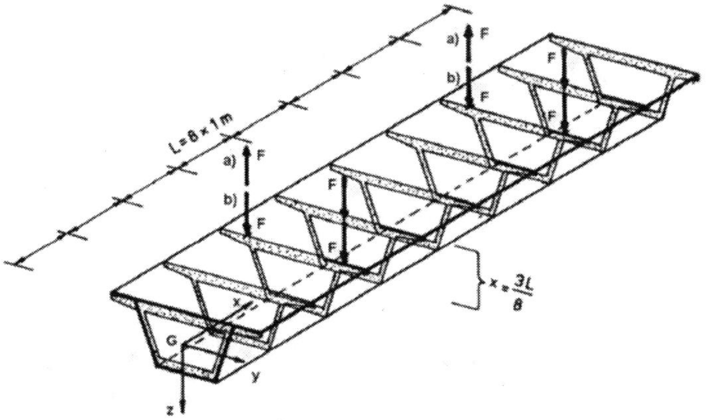

Fig. 4: Theoretical model of experimental box-girder

Fig. 5: Reverse position

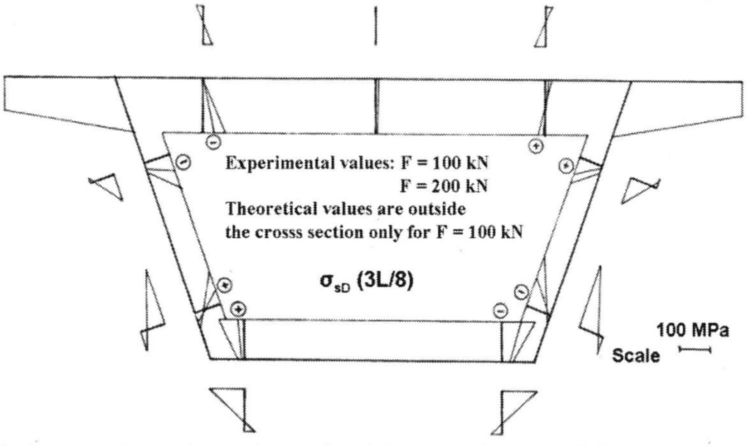

Fig. 6: Graphical comparison of experimental and theoretical values of direct stress σ_{sD} in transverse direction due to distortion on transverse stiffening frame in section $x = 3L/8$

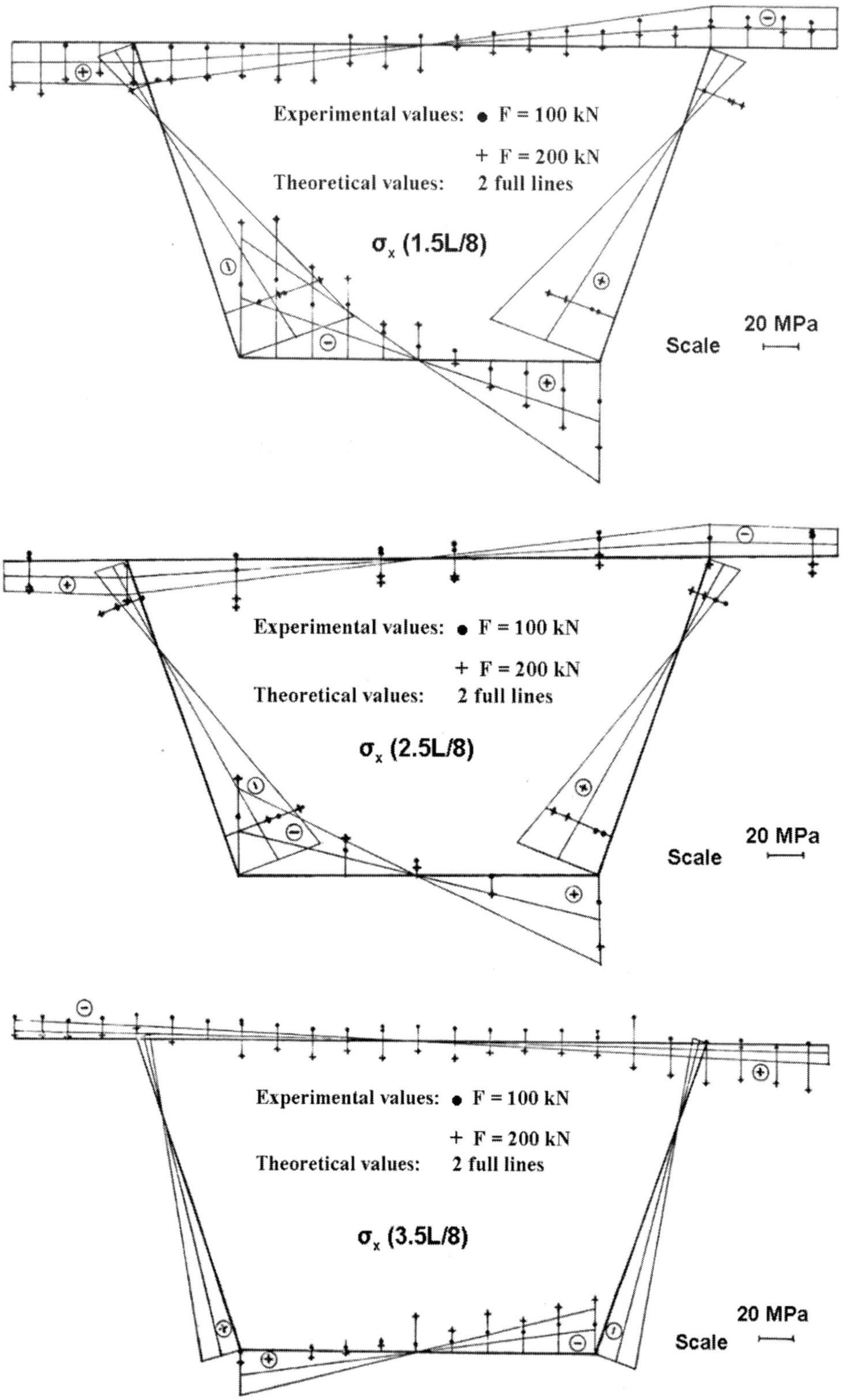

Fig. 7: Graphical comparison of experimental and theoretical values of direct stress σ_x in longitudinal direction due to distortion and torsion in three sections $x = 1.5L/8$, $2.5L/8$, $3.5L/8$

3. Theoretical calculation

The cross-sectional properties of distortion and torsion were calculated by the own computer program SEKTOR. The cross-sectional properties for this type of cross-section may be calculated also from formulae given in [20]. The internal forces were calculated on the bases of analogies. In the case of distortion it is the analogy with the beam in bending loaded at its end by axial tension force. The torsional rigidity GI_B plays the role of axial tension force (Fig. 8). In the case of distortion it is analogy with beam in bending on the elastic foundation and on the discrete elastic springs. The resistance of beam cross-section against its distortion plays the role of elastic foundation (this is for steel beams negligible) and the resistance of transverse stiffening frames against cross-section distortion EI_{RV} plays the role of discrete elastic springs (Fig. 9).

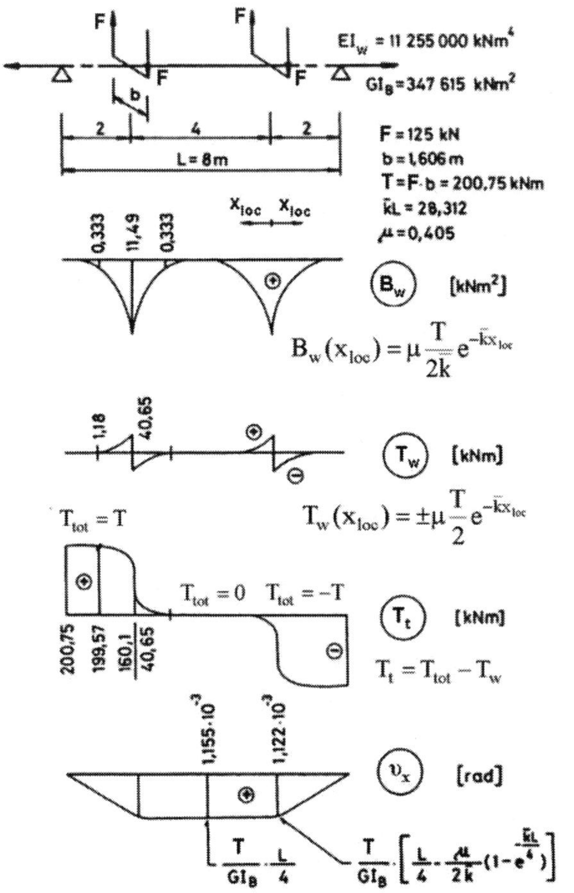

Fig. 8: Internal forces of torsion with taking into account the influence of the shear. The formulae are valid for the beam with the infinitive length on both ends

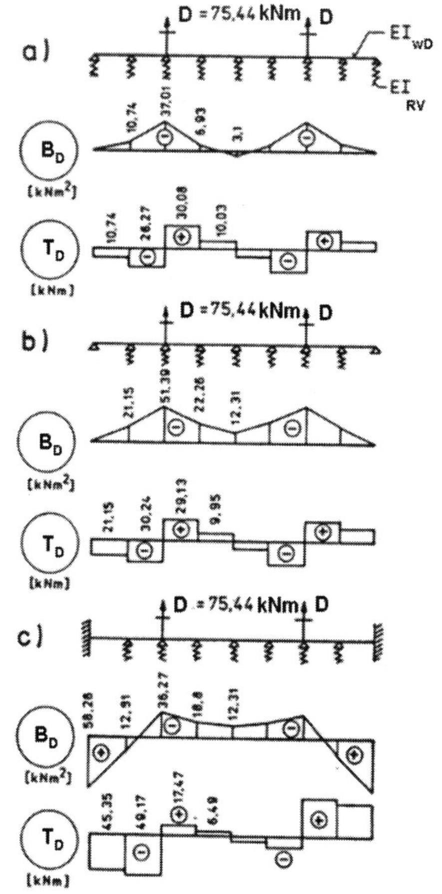

Fig. 9: Internal forces of distortion on the beam with the ends: a) on the elastic spring, b) simply supported, c) fixed

Details of hand calculation of cross-sectional properties of distortion and torsion are given in [15]. Details of hand calculation of internal forces, direct and shear stresses are given in [14]. Theoretical results given here are valid for $F = 125$ kN. They may be used for any value of F by multiplying $F/125$ kN. Fig. 9 shows also the influence of distortion boundary conditions on the internal forces distribution. Case a) was used for comparison with experimental values. Case b) means that distortion of end cross-sections is fully prevented and their warping is free. Case c) means that both are prevented: distortion and warping of end cross-sections.

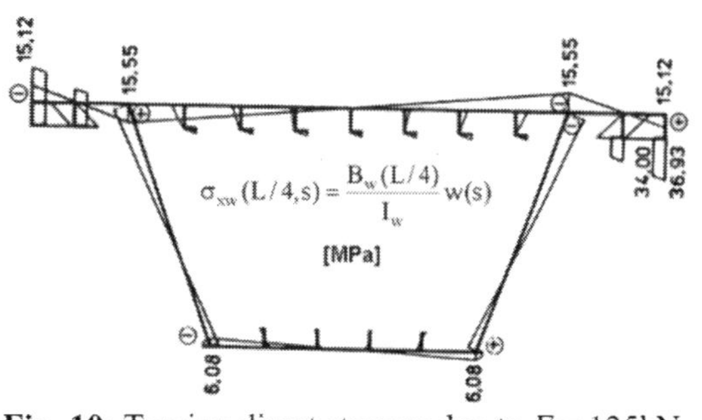

Fig. 10: Torsion direct stresses due to F = 125kN

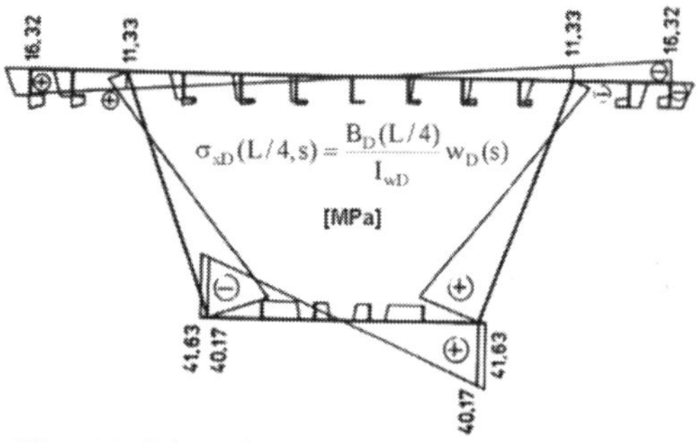

Fig. 11: Distortion stresses due to F = 125kN

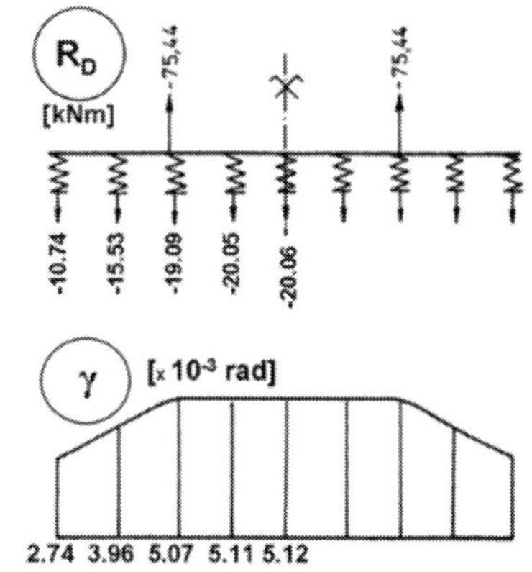

Fig. 12: Distortional reactions R_D and distribution of cross-section distortion γ

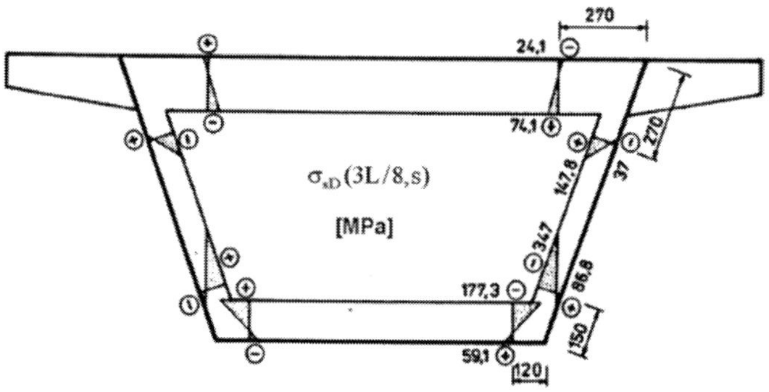

Fig. 13: Distortion stresses in the stiffening frame

4. Conclusions

The experimental and theoretical results show that transverse stiffening was relatively weak and therefore direct stresses due to cross-section distortion (Fig. 11) are greater than those due to torsion (Fig. 10). On the axis of cross-section symmetry the direct stresses are zero (Fig. 10, Fig. 11) due to antisymmetrical loading (Fig. 4, case b)). This was not true for measured experimental values of direct stresses (Fig. 7). The reasons are: a) the experimental beam was at the ends elastic supported in torsion, b) it was necessary to weld additive transverse stiffener outside of cross-section in sections $2L/8$ and $6L/8$ only on the left side to enable to load the box-girder with trapezoidal cross-section shape by hydraulic jack from bottom side. These two reasons are responsible that distribution of theoretical direct stresses were not perfectly antisymmetrical (Fig. 7).

Distortional bimoment B_D changes the sign along the box-girder length (Fig. 9, case a). The same phenomenon may be seen in distribution of experimental direct stresses. Compare direct stresses in sections $1.5L/8$ and $2.5L/8$ with those in section $3.5L/8$ (Fig. 7).

Theoretical direct stresses in transverse stiffening frames are greater than experimental ones (Fig. 6).

Nevertheless, the comparison of theoretical and experimental values show, that behaviour of experimental box-girder loaded by antisymmetrical loading may be described by the values obtained by hand calculation based on analogies with beam in bending. More details about GBT may be found also in the papers [21-28].

Acknowledgments

Project No. 1/0819/15 was supported by the Slovak Grant Agency VEGA.

References

[1] Vlasov VZ. *New method of prismatic beams from thin-walled profiles under normal force, bending and torsion*, Vestnik VIA RKKA, No. 20, 1936.
[2] Vlasov VZ. *Thin-walled elastic beams, strength, stability, vibration*, Goststrojizdat, Moscow, 1940.
[3] Vlasov VZ. *Structural mechanics of thin-walled space systems*, Strojizdat, Moscow, 1949.
[4] Schardt R. „Eine Erweiterung der technischen Biegelehre für die Berechnung biegesteifer prismatischer Faltwerke“, *Der Stahlbau*, 35, H. 6, 161, 1966.
[5] Sedlacek G. „Anwendung der erweiterten Biege- und Verdrehtheorie auf die Berechnung von Kastenträgern mit verformabren Querschnitt“, *Stresse-Brücke-Tunnel*, No. 9, 241-244, No. 12, 329-335, 1971.
[6] Djubek J, Baláž I. Box-Shaped Girders, *Int. Journal of Mechanical Sciences,* 17(10), 617-626, 1975.
[7] Baláž I. „Dünnwandige Stäbe mit öffenem oder geschlossenem deformierbarem Querschnitt“, *Stahlbau*, 68, H.1, 70-77, 1999.
[8] Schardt R. *Verallgemeinerte Technische Biegetheorie. Lineare Probleme.* Springer-Verlag, 1989.
[9] Davies JM, Leach P. „Some applications of generalized beam theory“, *Eleventh International Specialty Conference on Cold-Formed Steel Structures*, St. Louis, Missouri, 479-501, 1992.
[10] Davies JM, Leach P. „First-Order Generalized Beam Theory“, *Journal of Constructional Steel Research*, Vol. 31, 187-220, 1994.
[11] Davies JM, Leach P, Heinz D. „Second-Order Generalized Beam Theory“, *Journal of Constructional Steel Research*, Vol. 31, 221-241, 1994.
[12] Silvestre N, Camotim D. „First-order generalised beam theory for arbitrary orthotropic materials“, *Thin-Walled Structures*, 40:755–789, 2002.
[13] Tesár A. *Metal Structures and Bridges, Modern Steel Bridges,* 2nd part, Textbook, STU in Bratislava, 1977 (in Slovak).
[14] Baláž I. *Thin-walled steel structures. Influence of torsion in thin-walled bridge systems.* Textbook, STU in Bratislava, 1st edition 1984, 5th edition 2004, (in Slovak).
[15] Baláž I, Agócs Z. *Metal structures I, Handbook for design of elements of steel structures, I. part*, Textbook, Alfa Bratislava, 1st edition 1985, 5th edition 1994, (in Slovak).
[16] Rendek S. *Behaviour of thin-walled beams with deformable cross-section*, STU in Bratislava, Supervisor: I. Baláž, 2004.
[17] Buček P. *Stability problems of thin-walled structures used by GBT,* STU in Bratislava, Supervisor: I. Baláž, 2008.

[18] Rendek S, Baláž I. „Computer program employing generalised beam theory", *Proceedings of International Colloquium on Stability and Ductility of Steel Structures*, Budapest, 809-816, 2002.

[19] Tesár A, et al. *Resistance of thin-walled orthotropic bridge systems*, Research report III-8-2/16, STU in Bratislava, 1980.

[20] Steinle A. „Torsion and Profilverformung beim einzelligen Kastenträger". *Beton- und Stahlbetonbau*, H.9, 215-222, 1970.

[21] Baláž I. „Unit deformations of thin-walled cross-sections", *Proceedings of International Sympozium „Desing and reconstructions of steel structures"*, Bratislava, 19-26, 1998, (in Slovak).

[22] Baláž I. „Thin-walled beams with open or closed deformable shape of cross-section", *Inžinierske stavby*, 46, No.8, 279-285,1998, (in Slovak).

[23] Baláž I. „Calculation of internal forces and deformations of thin-walled beams with deformable cross-section with taking into account influence of shear", *International sympozium „Theoretical and technological problems of metal structures"*, Smolenice, 13-22, 1999, (in Slovak).

[24] Baláž I, Koleková Y. „Generalised beam theory. Warping functions and warping constants", *Proceedings of XXXVII. National Workshop of Employees in the Branch of Steel Structures with Participation of Foreign Specialists*, Zochova chata – Modra, 213-216, 2011.

[25] Baláž I, Koleková Y. „Calculation of warping constants", *CD Proceedings of the 9th International Conference on New Trends in Statics and Dynamics of Buildings,* Bratislava, 1-4, 2011.

[26] Baláž I, Koleková Y. „Unit warping functions and unit deformations". *CD Proceedings of the 9th International Conference on New Trends in Statics and Dynamics of Buildings,* Bratislava, 1-6, 2011.

[27] Baláž I, Koleková Y. „Modified generalized beam theory", *CD Proceedings of IX. International Scientific Conference of Faculty of Civil Engineering*, TU Košice, 1-8, 2012.

[28] Baláž I. „Analysis using generalised beam theory and analogy with bending", *Proceedings of 20th Conference Statics of Structures 2015 „Calculations – maintenance – trengthening",* Piešťany, 144-151, 2015 (in Slovak).

INFLUENCE OF STRUCTURAL OPENINGS ON THE BUCKLING STRENGTH OF CYLINDRICAL SHELLS UNDER AXIAL COMPRESSION

Emanuele Brunesi[a], Roberto Nascimbene[a] and J. Michael Rotter[b]

[a] *EUCENTRE, European Centre for Training and Research in Earthquake Engineering*
[b] *Institute for Infrastructure and Environment, The University of Edinburgh*

Abstract: Computational models and numerical assessment strategies were prepared in the framework of European rules for shell buckling (EN 1993-1-6) and were used to study the stability of a large set of cylindrical steel thin-shell tanks with structural openings, predicting behavioural changes in buckling and post-buckling response with variations in the cutout configuration (i.e. shape, size, location and number). Numerical estimates were related to the test data of previous experiments to explore critical design aspects, decoupling the roles played by material nonlinearity and geometric nonlinearity, as well as the contribution of initial geometrical imperfections in the buckling mechanism of axially compressed cylindrical thin-shells.

1. Introduction

For access reasons, cylindrical steel storage silos and tanks often require a structural cutout in the wall surface to provide an access port or door. These structures are thin shells and are very sensitive to failure by buckling, as extensively shown in the literature [1-6]. It is necessary to stiffen the shell around the opening to permit the imposed forces to be carried past it, but even with this stiffening, the local stress condition adjacent to the opening is usually the most critical for the shell. When the shell is loaded with stored product, vertical compressive forces develop in the wall, and the wall thickness is controlled by considerations of buckling under axial compression. The propensity for buckling in the zone in such a shell adjacent to the opening is vital to safe design [7-11], and it is desirable to explore both buckling and post-buckling behaviour [12]. Despite this, a few studies have appeared in the literature so far, investigating the influence of cutouts [7-11] or emphasizing the need for specific design procedures to explicitly account for it [12]. Recent seismic events in Italy (May 20 and 29, 2012, Emilia earthquakes [6,13]) have reaffirmed this critical scenario, considering their moderate size [13]. Fig.

1 shows an example of a buckling mechanism that occurred in correspondence to an elliptical opening. Even though reinforcement was provided in this case, the collapse mode revealed the key role played by the cutout. Large outward bulging began developing at its mid-height and visible stress/strain concentrations suddenly propagated throughout the shell, causing a more prominent deformation to take place between the opening and the base of the tank wall.

Fig. 1: Buckling in correspondence to an opening during the May 2012 Emilia, Italy, earthquakes

The research described in this paper investigated the influence of different numbers of openings, with various shapes, sizes and positions, on the buckling response of axially compressed cylindrical thin-shell structures. High-definition FE models, characterized by different levels of sophistication, were developed to reproduce the experimental behaviour of twenty-eight cutout configurations tested in previous experiments [10], taking this set to provide a comprehensive database against which to validate the numerical evaluations of the study. In order to examine behavioural changes in buckling strength and post-buckling regime as a consequence of parametric variations in the cutout (i.e. shape, size, location and number), the stability of each configuration was studied using each of the analysis processes defined in the European rules for shell buckling [14,15]. Failure modes and axial load-end shortening curves were collected and discussed to identify trends useful for a partial upgrading of the current Eurocode coverage. A paradigm was developed to relate the buckling and post-buckling behaviour of the same geometry using different types of analysis, as well as to relate the result for a cutout to that of the equivalent "perfect" shell structure with no cutout. Estimates were then made of the strength decay caused by the openings, examining the effect of changing the cutout angle and the diameter according to the analysis that was used (i.e. LBA, MNA, GNA, GMNA and GMNIA). Experimental and numerical predictions were correlated and buckling strength ratios were computed to point out peculiar design aspects for these structures, quantifying the influence of plasticity, stability and initial geometrical imperfections.

2. Review of the requirements of EN 1993-1-6

Safety against the most damaging event that may occur in the lifetime of the structure must be ensured, in accordance with many different potential modes of failure [12,16-18]. To this aim, the current version of the European Standard for Shells [14] and European Recommendations on Shell Buckling [15] carefully defined the design process when using different methods of analysis (i.e. hand computations, simple FE simulations, or fully geometrically and materially nonlinear analyses explicitly including initial geometric imperfections). In this framework, the different classes of analysis are planned to lead to progressively less and less conservative estimates of the design resistance, as uncertainties are eliminated, and all static and quasi-static treatments of structural strength are addressed, identifying possible limit states and failure criteria. As a result, the evident gap existing between simple design prescriptions or closed-form analytical expressions and the outcome of sophisticated and time-consuming numerical analysis, which produce a much richer and more complex image of what is desirable or necessary, has been bridged in this Standard, providing clear instructions to designers on how to conservatively interpret the results coming from each type of analysis.

Developed in light of a performance-based approach, this Code permits the user to obtain the most economic structure or to perform the most accurate and rigorous assessment of an existing structure, explicitly addressing questions of plasticity, stability and imperfection sensitivity, in accordance with a well-accepted limit state approach based on partial safety factors on both loading and structural strength. Computer calculations are highly desirable when accurate assessment is needed, and the EN 1993-1-6 [2007] allows different levels of sophistication to be used. In particular, two computational procedures are available. The LBA-to-MNA procedure derives from simple and well-known analyses: linear bifurcation analysis leading to the elastic critical resistance and rigid plastic analysis to predict the plastic reference strength, based on small displacement theory (i.e. no change in geometry). Both simulations introduce uncertainties, related to the pre-buckling stress regime and the phenomenon of snap-through buckling, in the case of LBA, and concerning the strain hardening and the capability of reproducing very large local plastic strain concentrations before the collapse mechanism becomes fully developed, in the case of MNA. In addition, hand calculations are used to give estimates of the imperfection sensitivity and elastic-plastic interaction parameters. By contrast, a second FE process, based on fully nonlinear analysis, is permitted in the EN 1993-1-6 [2007] for investigations of failures and design of major structures. In particular, this method requires four analyses to be carried out; both LBA and MNA must be performed to establish the reference resistances and the global slenderness of the structure. Furthermore, a GMNA must be undertaken so that the effects of the geometric imperfections in the GMNIA can be ascertained for the conditions considered.

Having the capacity to accommodate all forms of static structural analysis within the same design philosophy, the framework of the EN 1993-1-6 [14] constitutes the background of the wide set of simulations presented in the following. More specific and detailed explanations of the rationale behind the European Standard can be found in the literature [2,5,12,15-18].

3. Calibration of FE models

In light of this scenario, computational models and numerical assessment strategies were prepared in the framework of European rules for shell buckling [14,15], as discussed hereafter. A summary of the prevailing modeling assumptions and numerical techniques adopted for each analysis method is given, after a brief description of the experimental database assumed [10].

3.1 Definition of case-study structures

The prototype shells were equipped with cutouts characterized by different shapes (i.e. square, circular and rectangular) and dimensions. The position of the openings along the vertical axis of the shell was also studied. The notation used to parameterize the cutout is defined in Fig. 2, where an example of the initial geometric imperfections measured after the fabrication of the specimens is provided. In addition, Table 1 categorizes the twenty-nine reference shells tested into six cases of study, depending on opening shape, size, position and number.

In circumferential and axial directions, the cutout was described by r^* and h_0, while its position up to the height of the shell is controlled by f or alternatively h_b, as expressed in Eq. (1):

$$h_b = f\frac{l-h_0}{2} \quad \text{and} \quad r^* = \frac{c}{\sqrt{rt}} \tag{1}$$

where r, l and t are radius, length and thickness of the shell, respectively. The values assumed for these quantities are 49.5 mm, 104 mm and 0.175 mm, resulting in a length-to-radius ratio (l/r) and a radius-to-thickness ratio (r/t) of about 2 and 280, respectively.

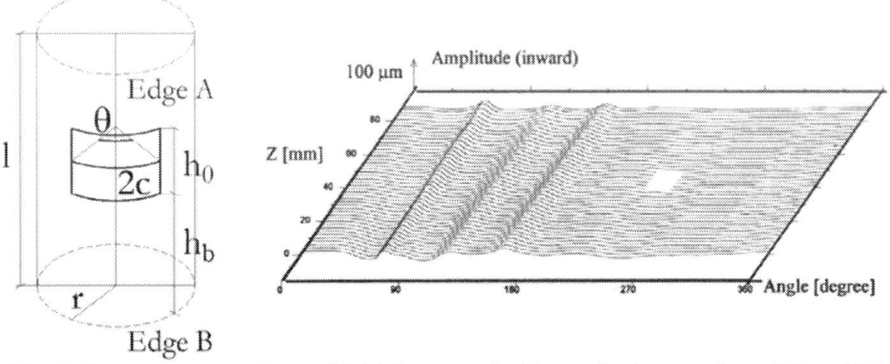

Fig. 2: Parameters, notation and initial geometrical imperfections – adapted from [10]

In Table 1, the numerical values assumed for each prototype are provided.

Table 1: Reference cutout configurations – adapted from [10]

Case	Parameter [-]	a	b	c	d	e	f	g
1				No cutouts				
2	r^*	1.47	2.94	4.40	6.60	8.81	-	-
3	f	1.00	0.75	0.50	0.25	0.00	-	-
4	r^*	1.47	4.40	4.40	8.81	-	-	-
5	h_0/l	0.17	0.25	0.37	0.50	0.75	-	-
6	r^*	2.94	4.40	6.60	8.81	13.23	-	-
7	r^*	0.15	0.28	0.73	1.46	2.88	4.34	5.75

A more comprehensive discussion of the experimental programme, in terms of manufacturing process, inherent initial geometrical imperfections and test setup, can be found in [10].

3.2 Computational model and numerical strategy

High-definition 3D FE models were developed using 2D shell elements based on mixed interpolation of tensorial components [6]. In detail, a mesh of 4-noded single-layered 2-points integration shells in accordance with Reissner-Mindlin plate theory was constructed with a relatively high mesh density in correspondence to the square, rectangular and circular cutouts, in order to capture visible strain-stress gradients. When geometrical nonlinearities are accounted in the set of analyses (i.e. GNA, GMNA and GMNIA), large displacement-large strain kinematics with automatic switching between updated Lagrangian Hencky and updated Lagrangian Jaumann formulations was used to balance the accuracy of the analysis and the computational efforts [19-21]. By contrast, a small displacement-small strain kinematics approach was assumed, whether no changes in geometry were included in the FE simulation carried out (i.e. LBA and MNA). Furthermore, a classical Von Mises yielding criterion was used, if material nonlinearity is considered in the analysis (i.e. MNA, GMNA and GMNIA). By means of average resistance, the true stress-strain relationships [19-21] assumed were calibrated according to the experimental tests [10]. An implicit solution strategy, based on the load-displacement-control method proposed by Bathe and Dvorkin [22] (i.e. arc-length method), was exploited to solve for the nonlinear equilibrium path of the response. Therefore, the twenty-nine prototype shells were studied through five simulation techniques, resulting in a total of 145 simulations.

4. Results and discussion

Numerical results were collected and compared to experimental observations, aiming to quantify the mismatch observed for each case-study and analysis approach, in terms of axial force-displacement curves and post-buckling failure mode, showing changes in the response due to parametric variations in shape, size and position of the cutouts.

The accuracy of GMNIA was quantified in comparison with experimental predictions [10], as shown in Fig. 3 and 4. In particular, Fig. 3 presents the comparison between the experimentally observed and numerically predicted axial load-end shortening capacity curves, while Fig. 4 compares the mechanism at failure in case of a square opening.

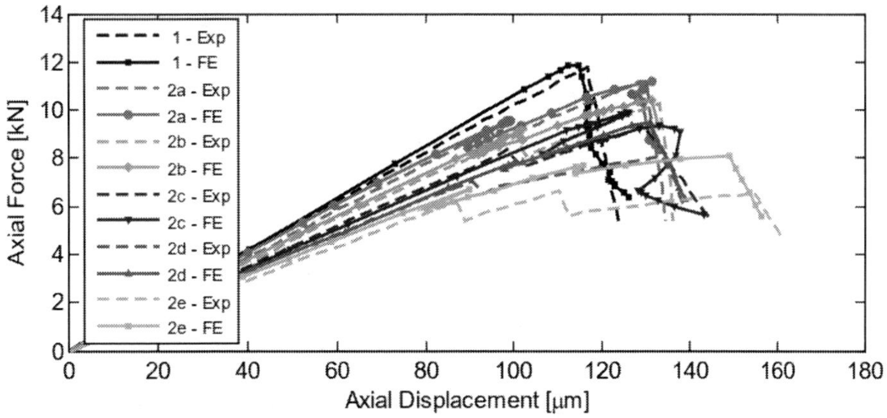

Fig. 3: Experimental vs. FE capacity curves: Case 2 – GMNIA

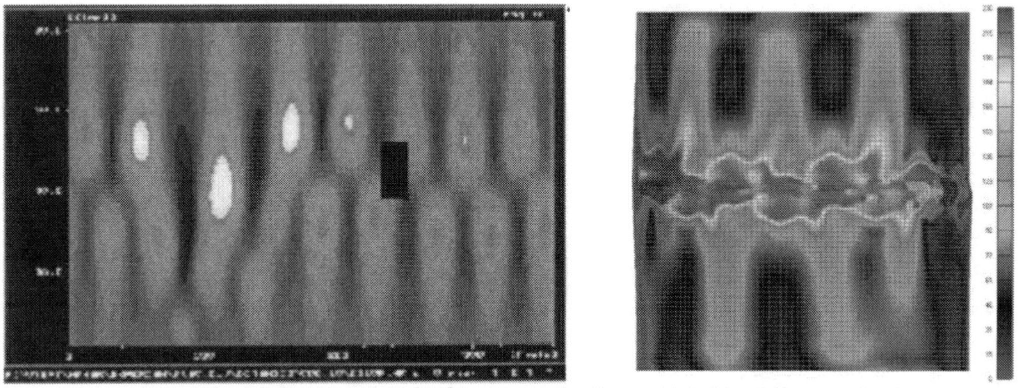

Fig. 4: Experimentally observed [10] vs. numerically predicted buckling mechanism

The force-displacement curve remains linear up to the first local buckling, which occurs close to the contour of the opening. After this point, the capacity curves remain quite linear and stable with a reduced slope until a global collapse buckling, which involves the entire shell, takes place. In the nonlinear pre-buckling state, large radial displacements develop near the edges of the openings, in particular around the rim of the cutout. As the size of the cutout increases, the buckling resistance and the accuracy of the analysis decrease, whilst a negligible sensitivity is observed in terms of failure mode. The analyses reveal that r^* is the key parameter controlling buckling and post-buckling strength, for any opening shape (square, rectangular and circular).

A comparison between different simulation techniques is provided in Fig. 5 and 6, in terms of axial force-axial displacement curves and peak buckling strength estimates.

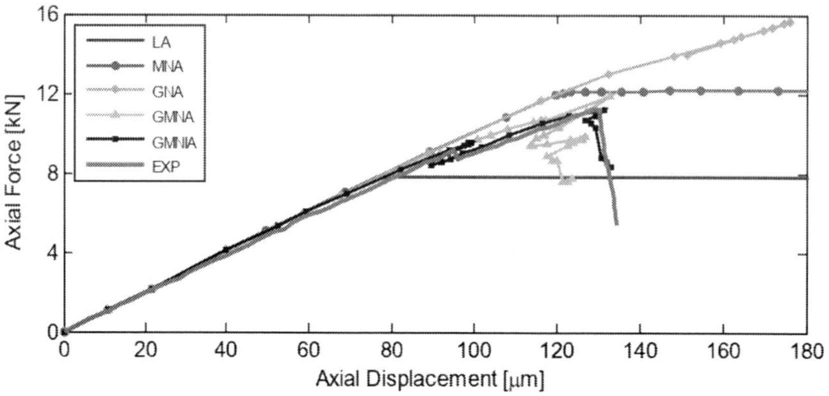

Fig. 5: Comparison between different types of analysis: Case 2a – $r^* = 1.47$

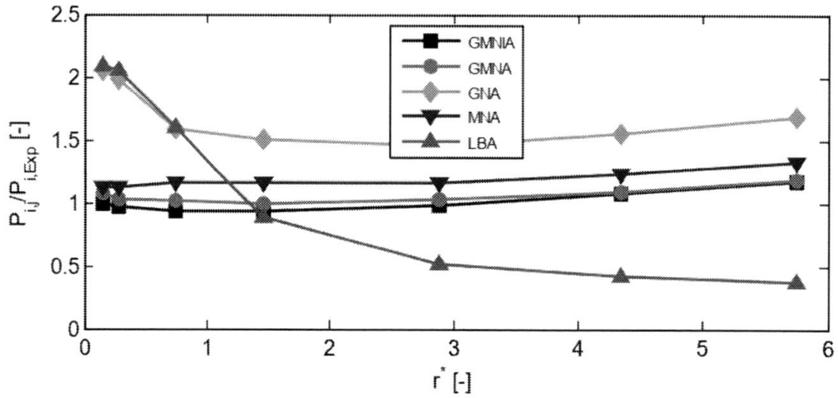

Fig. 6: Accuracy of different analysis methods: FE estimates vs. experimental data – Case 7

GMNIA was reaffirmed to be the most accurate simulation technique, since the average of the numerical-to-experimental peak strength ratios equals 1.06, with a standard deviation of 7.6%. If initial geometric imperfections are omitted in the simulations, their accuracy decreases, as a higher mismatch (13%), with higher standard deviation (9.4%), is predicted in average. GNA is shown to be even less accurate than MNA, since their discrepancy increases up to 60% and 32%, respectively. In addition, LBA underestimates the load carrying capacity in the majority of the cases, but very small circular cutouts, as the linear bifurcation theory predicts buckling in correspondence to the vertical edge of the opening. A summary of FE estimates is provided in Fig. 7, where peak buckling strengths obtained from each numerical method are normalized to the experimental results of Jullien and Limam [10].

Even if GNA is less safe and accurate than MNA, the former technique was observed to be conservative when applied to estimate the peak buckling strength decay caused by a structural opening with increasing cutout angles (see Fig. 8). This trend is even more pronounced in the case of LBA, which is able to account for neither geometric nor material nonlinearity.

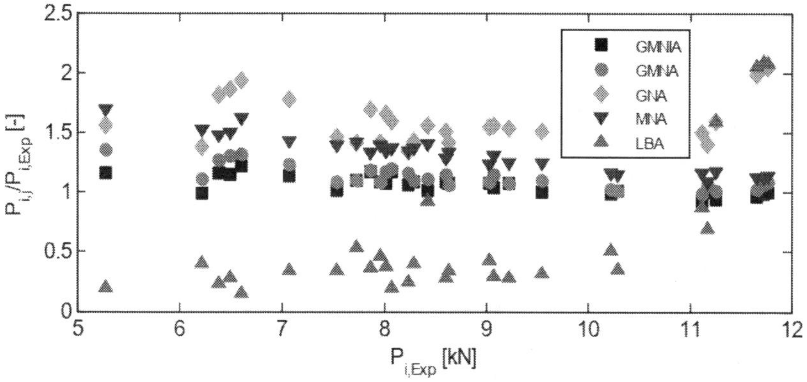

Fig. 7: Accuracy of FE prediction using the entire experimental database [10]

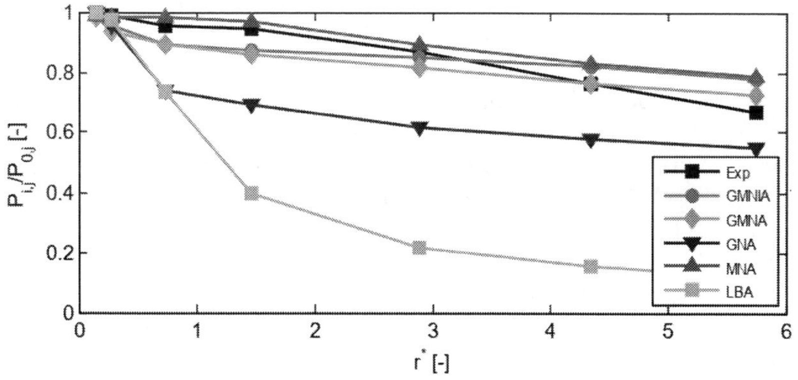

Fig. 8: Buckling strength decay with variations in the opening size: Case 7

Finally, Fig. 8 implicitly reaffirms a moderate sensitivity of buckling mechanism and strength to the initial geometric imperfections, showing that both GMNIA and GMNA are effective in predicting buckling strength reduction with variations in the dimensionless length r^*.

5. Conclusions

In this research, the buckling and post-buckling response of isotropic axially compressed thin cylindrical shells, presenting structural cutouts with various shape, size, position and number, was investigated. A modeling procedure, based on detailed numerical models, was prepared in the computational framework of EN 1993-1-6 and its effectiveness was validated using a wide set of previous experiments. Hence, numerical observations reaffirmed that GMNIA is an attractive tool for buckling and post-buckling response assessment of such structures under axial compression. Even though the other analysis methods revealed a larger mismatch with experimental results, this discrepancy presented specific trends that are predictable at the design stage. The trends emerged may be synthesized in a semi-analytical assessment proposal, after investigation of their sensitivity to radius-to-thickness ratio and boundary conditions, which is currently the focus of the ongoing research.

References

[1] Rotter JM, Teng JG. "Elastic stability of cylindrical shells with weld depressions", *Journal of Structural Engineering ASCE*, 115(5), 1244-1263, 1989.

[2] Rotter JM. "Local collapse of axially compressed pressurized thin steel cylinders", *Journal of Structural Engineering ASCE*, 116(7), 1955-1970, 1990.

[3] Teng JG, Rotter JM. "Buckling of pressurized axisymmetrically imperfect cylinders under axial loads", *Journal of Engineering Mechanics*, 118(2), 229-247, 1992.
[4] Song CY, Teng JG, Rotter JM. "Imperfection sensitivity of thin elastic cylindrical shells subject to partial axial compression", *International Journal of Solids and Structures*, 41(24-25), 7155-7180, 2004.
[5] Rotter JM. "Buckling of cylindrical shells under axial compression", in *Buckling of Thin Metal Shells* (Eds.: J.G. Teng & J.M. Rotter), Spon, London, 42-87, 2004.
[6] Brunesi E, Nascimbene R, Pagani M, Beilic D. "Seismic performance of storage steel tanks during the May 2012 Emilia, Italy, earthquakes", *Journal of Performance of Constructed Facilities ASCE*, 29(5), 04014137, 2015.
[7] Almroth BO, Holmes AMC. "Buckling of shells with cutouts, experiment and analysis", *International Journal of Solids and Structures*, 8(8), 1057-1071, 1972.
[8] Toda S. "Buckling of cylinders with cutouts under axial compression", *Experimental Mechanics*, 23(4), 414-417, 1983.
[9] Nemeth MP. "Buckling behavior of compression-loaded symmetrically laminated angle-ply plates with holes", *AIAA Journal*, 26(3), 330-336, 1988.
[10] Jullien JF, Limam A. "Effects of openings of the buckling of cylindrical shells subjected to axial compression", *Thin-Walled Structures*, 31(1-3), 187-202, 1998.
[11] Hilburger MW, Britt VO, Nemeth MP. "Buckling behavior of compression-loaded quasi-isotropic curved panels with a circular cutout", *International Journal of Solids and Structures*, 38(9), 1495-1522, 2001.
[12] Rotter JM. "Shell structures: The new European standard and current research needs", *Thin-Walled Structures*, 31(1-3), 3-23, 1998.
[13] Belleri A, Brunesi E, Nascimbene R, Pagani M, Riva P. "Seismic performance of precast industrial facilities following major earthquakes in the Italian territory", *Journal of Performance of Constructed Facilities ASCE*, 29(5), 04014135, 2015.
[14] EN 1993-1-6 (2007) "*Eurocode 3: Design of steel structures – Part 1-6: Strength and stability of shell structures*", CEN, Brussels, Belgium, 2007.
[15] ECCS (2008) "*European recommendations for steel construction: Buckling of shells*", 5th ed. European Convention for Constructional Steelwork, Brussels, Belgium, 2008.
[16] Rotter JM. "Design standards and calculations for imperfect pressurised axially compressed cylinders", *Proc. Int. Conf. on Carrying Capacity Steel Shell Structures*, Brno, Czech Republic, 354-360, 1997.
[17] Rotter JM. "The practical design of shell structures exploiting different methods of analysis", in *Shell Structures: Theory and Applications* (Eds.: W. Pietraszkiewicz & C. Szymczak), Taylor and Francis, London, 71-86, 2005.
[18] Rotter JM. "The elastic-plastic imperfection sensitivity of axially compressed cylinders with weld depressions", *Proc. EUROSTEEL 2008*, Graz, Austria, 1497-1502, 2008.
[19] Brunesi E, Nascimbene R, Rassati GA. "Response of partially-restrained bolted beam-to-column connections under cyclic loads", *Journal of Constructional Steel Research*, 97, 24-38, 2014.
[20] Brunesi E, Nascimbene R, Rassati GA. "Seismic response of MRFs with partially-restrained bolted beam-to-column connections through FE analyses", *Journal of Constructional Steel Research*, 107, 37-49, 2015.
[21] Brunesi E, Nascimbene R, Deyanova M, Pagani C, Zambelli S. "Numerical simulation of hollow steel profiles for lightweight concrete sandwich panels", *Computers and Concrete*, 15(6), 951-972, 2015.
[22] Bathe KJ, Dvorkin EN. "On the automatic solution of nonlinear finite element equations", *Computers and Structures*, 17(5-6), 871-879, 1983.

COMPARISON OF FE-CALCULATED BUCKLING STRENGTHS OF LOCALLY SUPPORTED STEEL SILOS TO RELEVANT EUROCODE INTERACTION CURVES

Arne Jansseune[a], Jan Belis[b] and Wouter De Corte[a]

[a] *Department of Industrial Technology and Construction, Faculty of Engineering and Architecture, Ghent University*
[b] *Department of Structural Engineering LMO, Faculty of Engineering and Architecture, Ghent University*

Abstract: In this paper, the buckling strengths derived from a validated shell based FEM model of locally supported axially compressed silos are compared to the interaction curves given in EN 1993-1-6 for axial compression. These buckling strengths were numerically determined for a wide range of geometries and yield strengths, considering two configuration types of locally supported steel silos: with engaged columns and partial-height U-shaped longitudinal stiffeners, respectively. For these, a remarkably good correspondence is found between the numerical results of locally supported steel silos and the interaction curves for uniformly compressed cylindrical steel silos, although unconservative in many cases.

1. Introduction

Generally, the decisive design state of locally supported axially compressed steel silos is elasto-plastic buckling (interaction of instability and yielding). To estimate the elasto-plastic buckling load, the relevant Eurocode (EN 1993-1-6) gives three design methods: (1) the stress design approach, (2) the MNA/LBA approach, and (3) the GMNIA approach [1]; each with varying levels of calculation effort and accuracy. Because such silos exhibit zones of high stress concentrations, e.g. in the silo wall above the local supports, much too conservative results are obtained when the failure load is estimated on basis of a stress design approach. As a result, the first method is less appropriate to apply. On the other hand, two simple numerical shell calculations (MNA and LBA) must be performed for the second method, it is relatively easy to apply. However, because there are currently no meridional buckling parameters available for locally supported stiffened silos, the designer has to guess these parameters in a conservative way, making this method less accessible. Finally, the third and last method can be more accurate, but due to its complexity (material and geometric non-linearity, imperfections, etc.), this method is more demanding and requires much more knowledge than does either of the other type in order to correctly interpret the failure behaviour and to determine the elasto-plastic failure load (different failure criteria) of the imperfect structure.

2. Material and methods

2.1 Scope

Silo configurations

In this contribution, two types of locally supported cylindrical steel silos were considered. For lighter silos, a possible solution is to engage the supporting columns over a short distance along the cylindrical barrel, combined with a light transition ring (Fig. 1 (a)). For medium to heavy silos, a possible solution consists of supporting columns terminating at the transition junction, a partial-height U-shaped longitudinal stiffener above each supporting column, and an upper and a lower ring stiffener (Fig. 1 (b)).

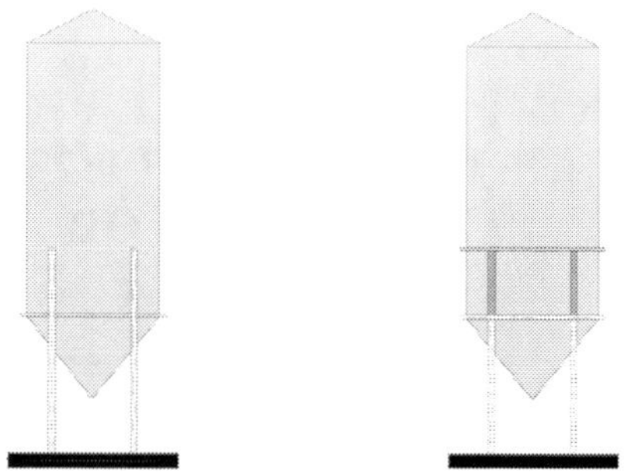

a) Engaged columns. b) Columns to bottom + longitudinal and ring stiffeners.

Fig. 1: Supporting and stiffening arrangements for locally supported silos

Geometry

The radius-to-thickness ratio R/t of the cylindrical barrel was varied between 100 and 1000, while the silo height h was either 2 or 10 times the cylinder radius R, representing either a short or a long barrel (See Fig. 2 (a)). The supporting configuration of a locally supported silo typically has a small number of discrete supports. In this study, the number of supporting columns n_{sup} was chosen equal to four or six, equally spaced over the complete circumference and either eccentrically (engaged columns) or concentrically (U-shaped stiffeners) positioned relative to the silo wall (See Fig. 2 (b) and (c)). The circumferential width d_{stif} was varied between 0.05 and 0.30 times the cylinder radius R, corresponding to a supported proportion of the circumference μ_{sup} between 3.2% and 19.1% (for n_{sup}= 4). The ratio of the radial width to the circumferential width w_{stif}/d_{stif} was equal to 100% or 200% for the engaged columns, and equal to 25% for the U-shaped longitudinal stiffeners. The stiffener's thickness t_{stif} was varied in the interval from the maximum of the silo wall thickness t and $1/40$ of the circumferential width d_{stif} to the minimum of 5 times the silo wall thickness t and $1/10$ of the circumferential width d_{stif}. The ratio of the attached height to the cylinder radius h_{stif}^{sup}/R was varied between 0.5 and 2.0 in steps of 0.5, while a constant value of 4.0 was adopted for the unattached column height h_{stif}^{sup}/R. Both ring stiffeneners have a rectangular cross-section and their thickness was chosen equal as the silo wall. The radial width was equal to 0.10 or 0.20 times the cylinder radius R for the upper and the lower ring, respectively. For a full discussion of the choice of the above scope of geometries, the reader is referred to the doctoral dissertation of the first author [2].

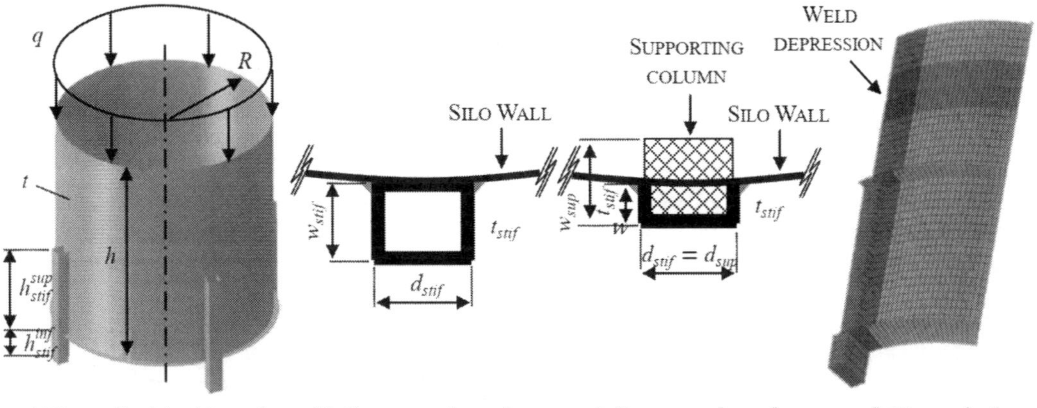

a) The cylindrical barrel. b) Cross-section of an engaged column. c) Cross-section of a U-shaped stiffener. d) Numerical model.

Fig. 2: Geometry and numerical model

2.2 Numerical model

All numerical simulations were performed with the finite element package Abaqus [3] using a shell based numerical model (Fig. 2 (d)), which was validated against experimental results obtained by a small number of well-chosen destructive experiments on scale models [2, 4]. A uniform edge load q was applied on the upper edge of the barrel, generating axial compression in the shell wall (Fig. 2 (a)). Quadrilateral shell elements (S8R5) were used for the entire model. Boundary conditions restricting the out-of-roundness deformations were imposed on the upper edge and the non-supported lower edge of the barrel, representing the influence of the non-modelled roof and hopper, respectively. In addition, the model was reduced to $1/8$ (n_{sup}= 4) or $1/12$ (n_{sup}= 6) of the entire circumference by exploiting symmetry (see Fig. 2 (d)). As a consequence, symmetrical boundary conditions were applied on all free edges created by the symmetry transformation. The local supports were considered as fully clamped supports.

Depending on the shell calculation (LBA, MNA, GNA, GNIA, GMNA, or GMNIA), material non-linearity (M), geometric non-linearity (G), and imperfections (I) were taken into account. The material behaviour either was perfectly elastic or ideal elasto-plastic with a Young's modulus E of 210 GPa and a poisson's ratio v of 0.3. To be able to deduce capacity or buckling curves for silos, the silo slenderness λ was varied by changing the yield stress σ_y from very small values (e.g. 5 MPa) to very large values (e.g. 2000 MPa) [5, 6]. Although these values may seen as unrealistic in practice, they may help with the calculation of the curves.

In this way, the overall slenderness can be varied without influencing the imperfection sensitivity. Based on an imperfection sensitivity study [2], an inward weld depression type A was chosen as imperfection shape, with the half-wavelength equal to the linear elastic bending half-wavelength, equivalent amplitudes according to [1, 7, 8], and located in circumferential direction above the top of the engaged column or the U-shaped longitudinal stiffener (See Fig. 2 (d)).

3. Results and discussion

3.1 Numerical results

In first instance, capacity curves were determined for a wide range of locally supported steel silos (1104 in total) and a variable imperfection quality class (perfect, A, B, and C) by varying the yield stress σ_y of the ideal elasto-plastic material behaviour. In Fig. 3 the capacity curves are depicted for one of the geometries in a dimensionless strength χ versus relative slenderness λ plot. The upper curve represents the silo without imperfections; the other curves represent the imperfect silo with a weld depression in the critical region (See Fig. 2 (d)). Clearly, the dimensionless strength χ decreases when the fabrication quality (i.e. larger equivalent imperfection amplitude) deteriorates.

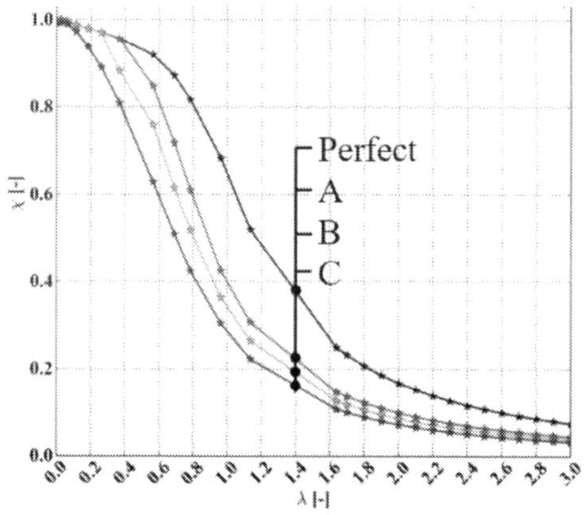

Fig. 3: Illustrative example of a capacity curve

3.2 Comparison with the existing interaction curves

In this part, the large dataset of the above numerical results is reduced to only those combinations of strength χ and slenderness λ which correspond to steel grades used in practice: 235, 275, 355, 460, 690, and 960MPa. This reduced dataset still contains approximately 27000 data points, and are compared with the interaction curves of Eurocode 3 determined on the basis of the meridional buckling parameters mentioned in Section D.1.2 (Annex D) of EN 1993-1-6 [1]. It is important to note that these interaction parameters were originally formulated for unstiffened cylindrical steel silos subjected to uniform meridional compression [9]. According to EN 1993-1-6, these parameters would be rather conservative and may be used for the design of other types of axially compressed silos. In what follows, it will be checked whether or not this assumption is correct for locally supported stiffened cylindrical shells, where the axial stresses exhibit a clear non-uniform distribution in circumferential direction.

For the radius-to-thickness ratios R/t of 200 and 1000, both the data points (Abaqus) and the interaction curves (Eurocode) are displayed in Fig. 4 (a) and (b), respectively. Each subfigure consists of four χ - λ plots: the perfect structure (upper left), quality class A (upper right), quality class B (lower left), and quality class C (lower right). Surprisingly, the lower bounds of the data points are relatively close to the interaction curves. For perfect silos (all values of R/t) and for relatively thick-walled imperfect silos ($R/t < 500$; $\lambda < 1$), a large part of the data points are located below the characteristic interaction curve (20.7% in total), which means that the calculated strength χ_{Abq} is smaller than the lower bound estimated strength χ_{EC}, which is unsafe. In contrast, the design is safe for relatively thin-walled imperfect silos ($R/t \geq 500$; $\lambda \geq 1$) where the calculated strengths χ_{Abq} are equal to or larger than the lower bound estimates of the strength χ_{EC}. However, this unconservative situation of the first category of data points is largely nullified by the presence of a partial coefficient on the resistance γ_{M1} in design. Indeed, the characteristic value of the dimensionless strength χ is divided by 1.10 [1] to obtain the design value of the strength χ, which corresponds with a reduction of approximately 9%.

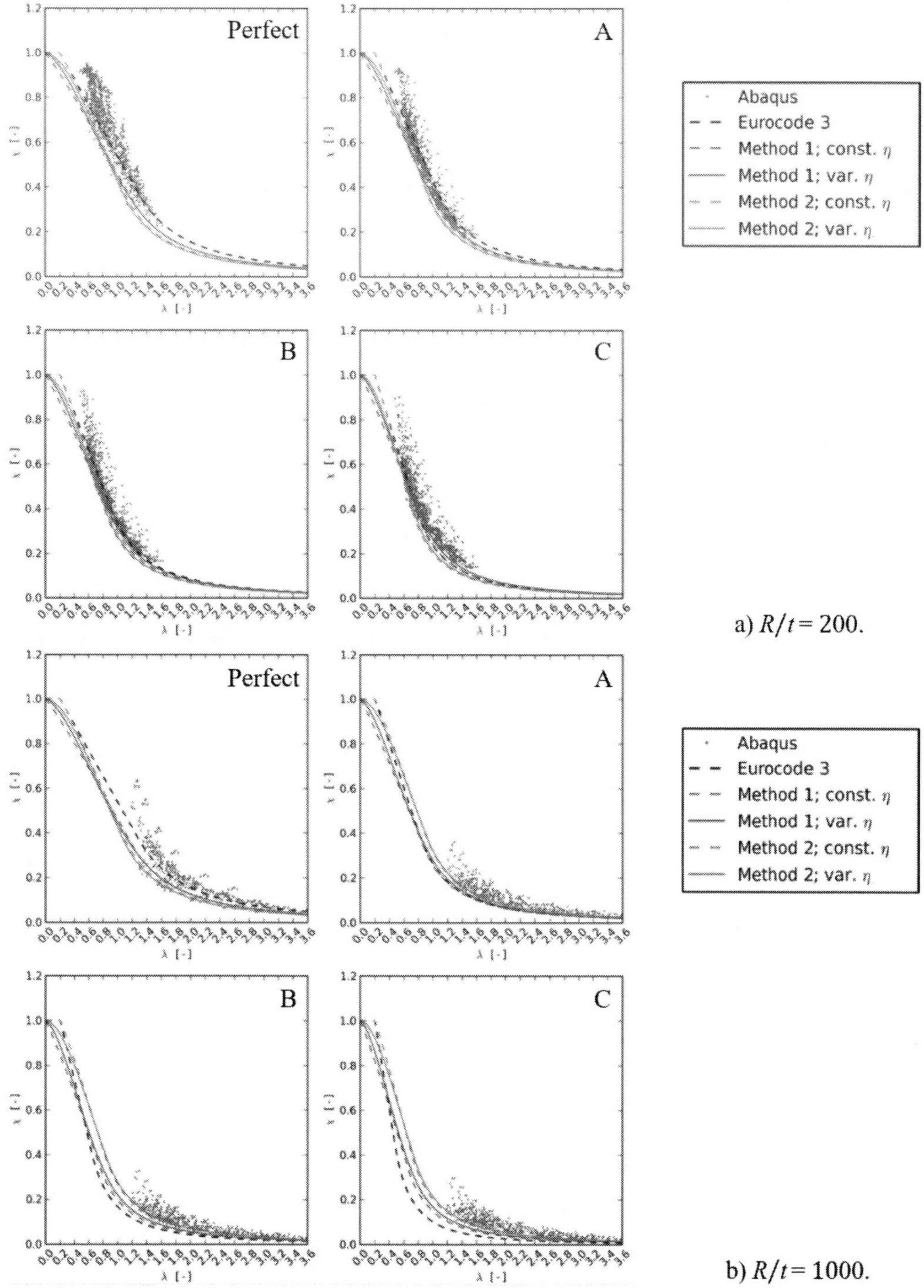

Fig. 4: Comparison of the numerical results, the existing and the proposed interaction curves for locally supported silos with variable radius-to-thickness ratio R/t and quality class.

3.3 Proposal for adapted interaction curves

From the previous section, it was found that 20.7% of the total points in the dataset is located below the Eurocode 3 interaction curves, which corresponds to an unwanted and unsafe situation. Therefore, in this section, it is attempted to find interaction curves which fit better to the lower bound of the dataset (Abaqus) and are safe for almost the complete dataset. The search for these buckling curves is done by varying the value of all buckling parameters (i.e. α_x, β_x, λ_{0x}, and η_x) within a realistic range with a small step size. The plastic range factor β_x and the squash limit load λ_{0x} are constant values, while the interaction exponent η_x can be either constant or vary linearly with the slenderness in the elasto-plastic part of the capacity curve [5, 10]. In addition, two different methods were used for the determination of the (im)perfect elastic reduction factor α_x, (see next paragraphs). The two approaches both for η_x and α_x result in four different methods.

For the determination of the meridional elastic reduction factor α_x, Eq. (1) is used for the first method. Clearly, this formula is very similar to the current expression in [1], except that the constants are replaced by three variable coefficients a, b, and c. The ratio of the imperfection amplitude to the silo thickness $\Delta w_k/t$ is done in the same way as in [1] and depends on the radius-to-thickness ratio R/t and the quality class.

$$\alpha_x = \frac{a}{1+b\cdot(\Delta w_k/t)^c} \tag{1}$$

The second method assumes for each imperfection quality class a constant value of the meridional elastic reduction factor α_x, which is independent of the radius-to-thickness ratio R/t. In Eq. (2) to (5) the coefficients d, e, f, and g are the unknown variables, the second symbol in the subscript of the meridional elastic reduction factor α_x represents the imperfection quality class.

$$\alpha_{x,0} = d \tag{2}$$

$$\alpha_{x,A} = e\cdot\alpha_{x,0} \tag{3}$$

$$\alpha_{x,B} = f\cdot\alpha_{x,0} \tag{4}$$

$$\alpha_{x,C} = g\cdot\alpha_{x,0} \tag{5}$$

After comparing all combinations of buckling parameters and the consequent buckling curves, only one set of buckling parameters was selected for each method of which the corresponding buckling curves fits best to the lower bound of the dataset (Abaqus). An overview of these interaction parameters is given in Table 1, and the best-fitting lower bound interaction curves (and the Eurocode 3 curve) are plotted in Fig. 4 for the radius-to-thickness ratios R/t of 200 (a) and 1000 (b).

From Fig. 4, it is found that all new proposed buckling curves fit better to the lower bound of the dataset compared to the existing Eurocode interaction curve. Furthermore, the allowable percentage of points that are located below the interaction curves is limited to 2.7% to avoid unconservative design. Furthermore, the proposed interaction curves are located below and are more conservative than the existing Eurocode interaction curve for the entire range of slendernesses in the more thick-walled silos: $R/t<500$ (quality class A), $R/t<333$ (quality class B), and $R/t<250$ (quality class C). When the radius-to-thickness ratio R/t increases, the Eurocode interaction curve shifts downwardly and the Eurocode estimate gradually decreases, as a result of which the Eurocode estimate is located below all proposals for larger slendernesses: $R/t>500$ (quality class A), $R/t>333$ (quality class B), and $R/t>250$

(quality class C). The best correspondence (i.e. with the smallest value of "*dev*" in Table 1) is found for the second method with a variable interaction exponent η_x .

Table 1: Overview of the parameters of the best-fitting lower bounds to the dataset.

Parameter		Eurocode (const. η_x)	Method 1 (const. η_x)	Method 1 (var. η_x)	Method 2 (const. η_x)	Method 2 (var. η_x)
α_x [-]	a [-]	0.62	0.5	0.5	-	-
	b [-]	1.91	1.0	1.0	-	-
	c [-]	1.44	1.2	1.0	-	-
	d [-]	-	-	-	0.45	0.43
	e [-]	-	-	-	0.70	0.75
	f [-]	-	-	-	0.60	0.65
	g [-]	-	-	-	0.45	0.50
β_x [-]		0.60	0.64	0.74	0.56	0.62
$\lambda_{x,0}$ [-]		0.20	0.00	0.00	0.20	0.00
$\lambda_{x,p}$ [-]	$=f(\alpha_x; \beta_x)$					
η_x [-]		1.0	1.082	-	1.082	-
$\eta_{x,0}$ [-]		-	-	1.371	-	1.697
$\eta_{x,p}$ [-]		-	-	0.600	-	1.082
under curve [%]		20.7	0.02	0.02	0.02	0.02
dev [-]		-	0.111	0.107	0.102	0.101

Since the above presented results are derived as "lower bound" interaction curves to the scattered dataset, it is important to mention that, for most cases, safe but also rather conservative outcomes are obtained when the corresponding interaction parameters are applied for design. This is the main disadvantage of the approach of fitting a lower bound.

4. Conclusions

The main conclusions are:

1. In general, a relatively good agreement was surprisingly found between the current buckling curves for meridional compression given in Eurocode 3 (based on the buckling parameters for unstiffened uniformly compressed cylindrical steel silos) and the lower bound of the dataset derived from numerical results (locally supported stiffened cylindrical steel silos with a marked non-uniform axial stress distribution in circumferential direction).
2. A number of sets of buckling parameters was proposed of which the corresponding buckling curves fit better to the lower bound of the scattered cloud of numerical results.

References

[1] EN 1993-1-6. *Eurocode 3: Design of Steel Structures, Part 1.6: Strength and Stability of Shell Structures*, CEN, Brussels, 2007.

[2] Jansseune A. *Optimisation of the Stiffening Configuration of Axially Compressed Steel Silos on Local Supports*. PhD thesis, Department of Industrial Technology and Construction, Faculty of Engineering and Architecture. Ghent University, Ghent, 2015.

[3] Abaqus. *ABAQUS version 6.9 documentation*. Dassault Systèmes Simulia Corp., Providence, RI (USA), 2009.

[4] Jansseune A, Belis J, De Corte W. "Experimental validation of a numerical model for locally supported steel silos", *Proceedings of the International Association for Shell and Spatial Structures (IASS)*, Amsterdam, 2015.

[5] Doerich C, Rotter JM. "Generalised capacity curves for stability and plasticity: Application and limitations", *Thin-Walled Structures*, Elsevier 49(9): 1132-1140, 2011.

[6] Rotter JM. "Buckling of shallow conical shell roofs for small diameter tanks and silos", In *Proceedings of the International Conference on Design, Inspection and Maintenance of Cylindrical Steel Tanks and Pipelines*, Prague, Czech Republic: 169-175, 2003.

[7] Rotter JM, Teng, JG. "Elastic stability of cylindrical shells with weld depressions", *Journal of Structural Engineering*, ASCE 115(5): 1244-1263, 1989.

[8] Song CY, Teng JG, Rotter, JM. "Imperfection sensitivity of thin elastic cylindrical shells subject to partial axial compression", *International Journal of Solids and Structures*, Elsevier 41(24-25): 7155-7180, 2004.

[9] ECCS. *Buckling of Steel Shells: European Design Recommendations*, 5th Edition, European Convention for Constructional Steelwork, Brussels, 2008.

[10] Doerich C. *Strength and stability of locally supported cylinders*. PhD thesis, Institute for Infrastructure & Environment, The School of Engineering and Electronics. University of Edinburgh, Edinburgh, 2007.

BUCKLING BEHAVIOR OF STIFFENED PLATE UNDER BIAXIAL COMPRESSION

Joseph Ndogmo[a], Martin Mensinger[a] and Ioan Both[b]

[a] *Technische Universität München, Germany*
[b] *Politehnica University of Timisoara, Romania*

Abstract: The structural elements in steel and composite bridges are sensitive to stability problems due to their large dimensions. Mainly, resulting elements are thin plates falling under the design requirements provided by EN1993-1-5. Due to the results of recent scientific papers regarding the unstiffened plates under biaxial stress, the review of the buckling proof formats for stiffened plates under biaxial loading was needed. The article gives an overview of buckling proof formats according to EN 1993-1-5 for steel and composite bridge and it is limited to the usual method of reduced stresses in Germany. For a given example, this paper discusses whether the proposed approach in the literature is also appropriate for the case of stiffened panels.

1. Introduction

The elements of steel and composite bridges are generally imposed by the buckling verifications. The usual design method of reduced stress was regulated in 2003 by DIN Technical Report 103 (DIN-Fachbericht 103) [1], in Germany. Along with the effective width method, the reduced stress method represents the buckling proof formats according to EN 1993-1-5 [2] for steel and composite bridges.

The results of recent scientific papers regarding the unstiffened plates under biaxial stress [3] suggests that the review of the buckling proof formats for stiffened plates under biaxial loading is needed.

The article gives an overview of buckling proof formats according to EN 1993-1-5 [2] for steel and composite bridges, limited to the usual method of reduced stress, in Germany, introduced in 2003 by DIN Technical Report 103. In this paper it is discussed if the proposed approach in [3] is also appropriate to the case of stiffened panels.

2. Buckling verification according to EN 1993-1-5

The background of the design methods in EN 1993-1-5 [2] is represented by the development of the Eurocodes, which brought together various European design cultures. In most European countries, partly exclusively, the method of effective widths is applied. This describes the load bearing behaviour of buckling plates realistic and takes into account both post-critical load bearing behaviour and stress rearrangements within a cross section. Compared to the method of reduced stresses the concept of effective widths leads to more economical results, but it may also have the consequence that by repeatedly forming small bumps (called web-breathing) fatigue damage occurs.

In Germany the reservations against the method of effective widths come from the higher computational complexity (load case dependent cross-sections) and the lower flexibility in the application (non-rectangular buckling panels are not covered by the method).

The reduced stress method which is applied in Germany considers the post-critical load bearing reserves in a lesser extent and does not especially allow any stress redistribution within the cross-section. The following represents the details of the method of reduced stress.

In EN 1993-1-5 section 10 [2] the method of reduced stress for both single load and combination of loads is regulated. In this method the post-critical load bearing reserves of the cross section are not fully activated and stress redistribution is disregarded. This method is thus not economical in high-rise buildings, but produces robust results and applies to complex geometries. The quintessence of the buckling checks according to method of reduced stresses is the equation (10.5) in [2] reproduced as Eq. (1):

$$\left(\frac{\sigma_{x,Ed}}{\rho_{cx}\, f_{y,k}/\gamma_{M1}}\right)^2+\left(\frac{\sigma_{z,Ed}}{\rho_{cz}\, f_{y,k}/\gamma_{M1}}\right)^2-\left(\frac{\sigma_{x,Ed}}{\rho_{cx}\, f_{y,k}/\gamma_{M1}}\right)\left(\frac{\sigma_{z,Ed}}{\rho_{cz}\, f_{y,k}/\gamma_{M1}}\right)+3\left(\frac{\tau_{Ed}}{\chi_{w}\, f_{y,k}/\gamma_{M1}}\right)^2\leq 1 \qquad (1)$$

The relation corresponds to the verification of the equivalent stress, modified by the corresponding reduction factor ρ_{cx}, ρ_{cz} and χ_w, to avoid buckling. The condition in Eq. (1) is applied for all total panels with their single panels and subpanels in the longitudinal direction of the bridge (see Fig. 1).

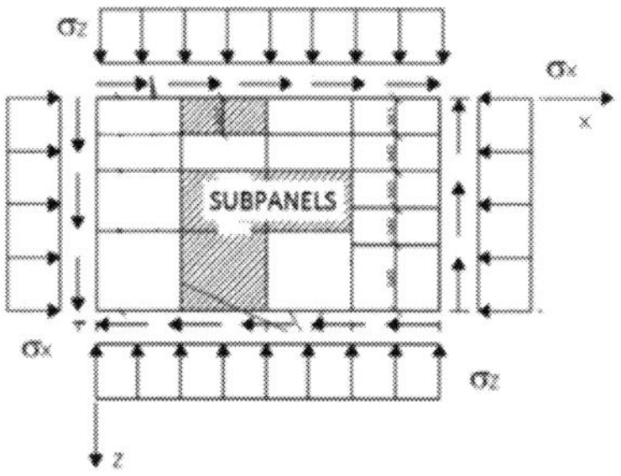

Fig. 1: Buckling panel and stresses [4]

In contrast to the procedure according to DIN 18800-3 [4], the reduction factors are calculated with a "global" slenderness, which takes simultaneously all stress components into account, Eq. (2).

$$\bar{\lambda}_P=\sqrt{\alpha_{ult,k}/\alpha_{cr}} \qquad (2)$$

This procedure is mechanically reasonable because the critical buckling stress depends on the overall state of stress and it is also contemporary due to the fact that, nowadays α_{cr} can be easily and reliable determined either using an approximation relation (equation (10.6) in [1] and [2]) or using special finite element programs.

There is no possibility for elongated buckling panels under lateral compression to develop stabilizing tensile stresses perpendicular to the loading direction. In this case plate buckling becomes column buckling. Since the buckling curves used for the plate buckling verification are considerably favourable with respect to the column buckling curves, this leads to a safety problem. For this reason it is necessary to interpolate in certain aspect ratios between the plate buckling and the unfavourable column buckling.

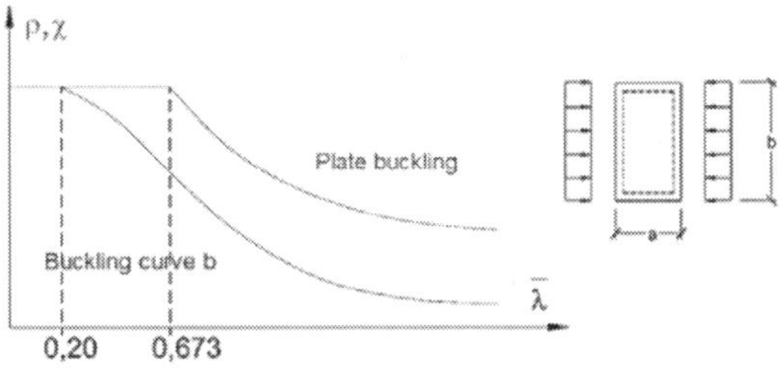

Fig. 2: Column buckling curve vs Plate buckling curve

The interpolation is performed using the formulae in Fig. 3, which is based on the work of Fischer and Harre [5]. It is found in unchanged form in DIN 18800-3 [4], in DIN Fachbericht 103 [1] and in EN 1993-1-5 [2]. From [5] it can be seen that the former studies were concerned only with uniaxial compressed rectangular panels. That means the transferability on plates with multiaxial stress states is not strictly covered. In the standardization today is therefore assumed that for the determination of $\sigma_{cr,p}$ in the expression of ξ, only the stresses "in the direction of the column-like behaviour" are taken into account, while the other stress components are not to be considered. Neither in the DIN Fachbericht 103 [1] nor in EN 1993-1-5 [2] clear evidence can be found, so that misuse is possible and can lead to imprecise results. In later discussed proposal of Braun and Kuhlmann [6], the influence of biaxial loading on the column-like behaviour has been included correctly. Again, the interpolation formulae will continue to be applied considering only the load "in the direction of the column-like behaviour", the unfavourable influence of transverse compression is considered by the modification of the buckling verification formula for unstiffened plates. In the case of longitudinally stiffened panels the effect of the lateral compressive stress is of minor importance.

3. Necessity of the modification of the buckling verification formula for plates under biaxial compression

The root of the buckling verification is the proposal from Scheer and Nölke [7], see also Eq. (1). Using only one comparative slenderness λ_P $(= (f_{y,k}/\ \sigma_{VPi})^{0,5})$ the reduction factors ρ_{cx}, ρ_{cz} and χ_w (in the original source κ_x *, κ_y * and κ_τ*) are determined. In a comparison with DIN 18800-3 [4], it is obvious that the verification can lead, for biaxial compression, to falsely favourable results, because the middle term of Eq. (1) reduces the result, when the stresses have the same sign. This term of the equation is originally from the equivalent stress verification, which is based on the hypothesis that flow is achieved if the deformation energy under multiaxial loads is equivalent to the deformation energy for yielding under uniaxial load. That means, the hydrostatic stress state does not contribute to yielding. Therefore stresses in the x and z direction with "same sign" have a favourable effect. In the case of the plate buckling, however this is not true. In particular, lateral compressive stresses generate additional driving forces and negative influence, depending on the displacement of the edges of the buckling panels, the post-critical load carrying reserves. The middle term in Eq. (1) must therefore be modified so that it becomes small with increasing of the plate buckling risk. With the introduction of the factor $V = \rho_{cx} \cdot \rho_{cz}$ in the verification it is ensured that: with the increasing of the buckling risk ρ_{cx} and/or ρ_{cz} become lower and the mixed term of Eq. (1) vanishes. However, the reduction factors are calculated using the respective equivalent slenderness ratio and not with the global slenderness ratio as described in [1].

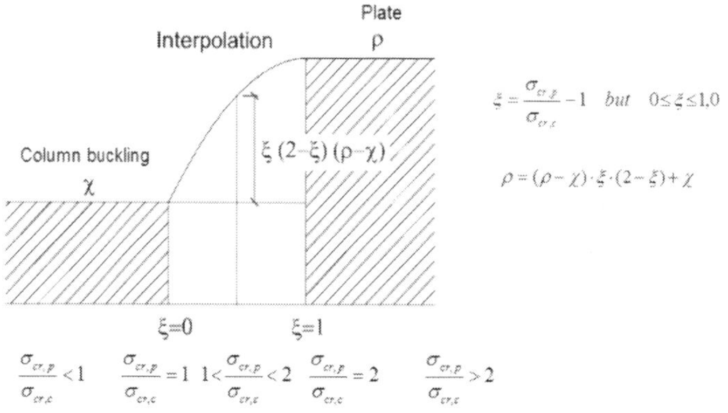

Fig. 3: Interaction between plate and column buckling

The works of Braun [6], [3], Maur [8] and Winterstetter [9] have consistently shown that Eq. (1) can deliver results on the unsafe side in case of biaxial compression. After extensive numerical studies of unstiffened plates which are confirmed in [10], Braun / Kuhlmann made the following proposition, which is different from Eq. (1) by the introduction of the factor V for biaxial compression and $V = 1$ for all other cases.

$$\eta = \sqrt{\left(\frac{\sigma_{x,Ed}}{\rho_{cx} f_{y,k}/\gamma_{M1}}\right)^2 + \left(\frac{\sigma_{z,Ed}}{\rho_{cz} f_{y,k}/\gamma_{M1}}\right)^2 - V\left(\frac{\sigma_{x,Ed}}{\rho_{cx} f_{y,k}/\gamma_{M1}}\right)\left(\frac{\sigma_{z,Ed}}{\rho_{cz} f_{y,k}/\gamma_{M1}}\right) + 3\left(\frac{\tau_{Ed}}{\chi_w f_{y,k}/\gamma_{M1}}\right)^2} \le 1 \quad (3)$$

The proposal of Braun / Kuhlmann is based solely on studies of unstiffened plates and is purposeful for this. The modification is mainly based on a parametric study based on FE calculations (GMNIA) [3] and therefore taken into account the post-critical behaviour. It is currently unclear if this proposal can be transmitted without any conditions to longitudinal stiffened plates (with column-like behaviour), as they are frequently present in bridges, because scientific knowledge is missing. However, it is expected that this proposal leads to more conservative results for the overall behaviour of stiffened plates. The compressive stresses σ_z hardly influence the overall buckling behaviour of the panel in the x-direction, but can lead the subpanel between the webs of stiffeners, which are under higher biaxial compression, to column-like buckling.

From Fig. 4 it can be seen that the transverse compression affects only the sheet, but not the stiffener directly. The buckling of the longitudinal stiffened overall panel (= buckling of the stiffener) is therefore affected by the lateral compression only by the following two effects: a) possibly smaller effective width of the concurred sheet metal part due to lateral compression, b) larger driving forces due to the effects from the second order theory, no stabilizing tensile forces.

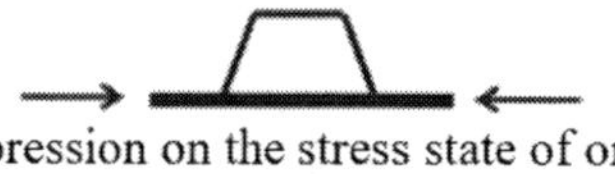

Fig. 4: Influence of transverse compression on the stress state of one stiffener with participatory panel

In articular, for the first of the two points there are currently no normative regulations provided. Both effects are not described by the approach of Braun / Kuhlmann which is based on the studies regarding the buckling of a subpanel.

The buckling in the longitudinal direction of the stiffened buckling plate can be simplified as the buckling of columns, formed from the stiffeners and the cooperating part of the sheet. The corresponding verification is governed by section 9 of EN 1993-1-5 [2]. In this section of the standard, minimum requirements for transverse stiffeners are defined and the stability verification of these stiffeners according to second order theory taking into account their longitudinal stresses and additional drive forces due to the stresses perpendicular to the stiffeners is demanded. In the case of longitudinal stiffened panel with biaxial stress state, this section of the standard takes effect, as always stresses transverse to the stiffener occur. The verification leads automatically to a rigidity of the stiffeners which creates the node lines in the buckling shape at the stiffener place, so that the buckling of the subpanel is decisive for the design of the entire panel. For the modification of Eq. (1) regarding in particular overall plate buckling there is obviously need for more research and it would be advisable to accompany some real buildings scientifically and conduct further investigations. This also applies to subpanels of stiffened plates with strong column-like behaviour. Results of numerical analysis regarding the boundary conditions of plates are given in [11].

4. Buckling verification for the case of bridge launching

In the following section the impact of the introduction of the factor V on the construction is discussed using the example of the bottom plate of a box section of a bridge, see Fig. 5a). During the incremental launching method of a steel bridge with box cross section the webs and the bottom plate are under biaxial compression.

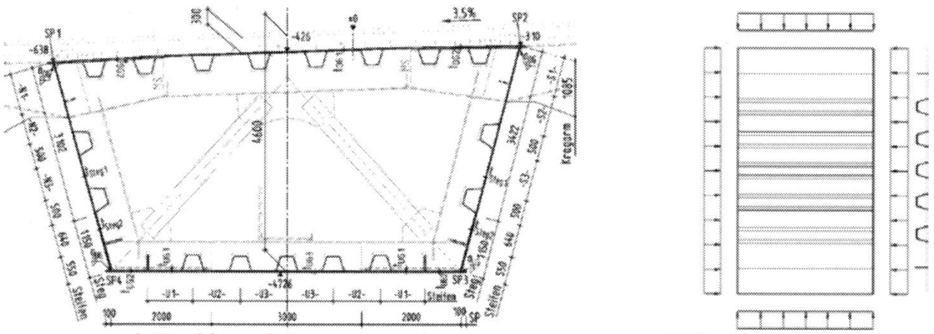

a) Steel box girder cross-section b) Loads on the bottom panel

Fig. 5: Extract of steel box girder

The buckling panel having a thickness of 12 mm has a length of 4000 mm and a width of 7000 mm; the two edge stiffener each having a thickness of 30 mm and a height of 300 mm; the trapezoidal stiffeners each have a height of 300 mm, a width of 250 mm and a thickness of 8 mm; the two edge panels each having a width of 725 mm, the subpanels between the webs of the trapezoidal stiffeners have a width of 450 mm and the subpanels between the trapezoidal stiffeners 475 mm. The steel grade is S355. The longitudinal compression stress is 66 MPa and the transverse one is 50 MPa.

For the determination of the smallest factor α_{cr} for the increasing of the design loads to reach the elastic critical load for the complete acting stress field, there are basically two possible approaches: a) the entire panel is considered, whereby overall or subpanel buckling may occur (verification of the entire panel with decisive subpanel buckling), b) only the buckling of the entire panel is considered and by this way the buckling mode of the subpanels is excluded, afterwards, the buckling of the subpanels is separately verified (verification of the entire panel with decisive global buckling).

In the first case the subpanels must not be verified separately and in the latter case they must be verified separately. The second approach gives a more favourable utilization rate.

4.1 Verification of the entire panel with decisive local buckling of the subpanels

Fig. 6a shows the buckling shape of the largest single subpanels in the edges having six waves and a buckling value $k_{\sigma x}$ = 1236.5 for the sole effect of σ_x. Fig. 6b shows the buckling shape of the single subpanels in the edges having one wave and a buckling value $k_{\sigma z}$ = 143.3 for the sole effect of σ_z. Here it can be seen that the transverse compressive stresses have no appreciable effect on the buckling behaviour of the stiffened bottom plate when the longitudinal stiffeners are rigid enough to form nodal lines. These longitudinal stiffeners must be able to carry longitudinal stresses additionally to the driven forces due to the longitudinal and transversal compressive stresses and moments according to second order theory due to the initial deformation and elastic deflection. Fig. 6c shows the buckling shape under biaxial compression with a magnification factor α_{cr} = 1.530.

The ratio of the critical plate buckling stress $\sigma_{cr,p} = k_{\sigma x}\ \sigma_E$ (σ_E = 190000·(12/7000)^2 in MPa) to the critical column buckling stress $\sigma_{cr,c}$ ($\sigma_{cr,sl}$, see equation (4.10) and Figure A.1 in [2]) gives a value less than 1, indicating a pure column buckling behaviour of the stiffened bottom plate in the x-direction, see Fig. 3. The buckling verification for the total panel without considering the correction factor V would show that the bottom plate is safe according to buckling (η = 0.997 < 1.0), even though local buckling of the subpanels occurs, see Table 1. The buckling verification for the total panel with consideration of the correction factor V, which is justified in this case because local buckling of the subpanels is decisive, will lead to the result that the bottom panel buckles (η = 1.2 > 1.0).

In Table 1, σ_{cr} is calculated with the approximation formula (10.6) in [2]. It is about less than the value 1.53 determined by the FE program EBPlate and is thus on the safe side. The considered values for σ_x and σ_z are 66 N/mm^2 and 50 N/mm^2, respectively.

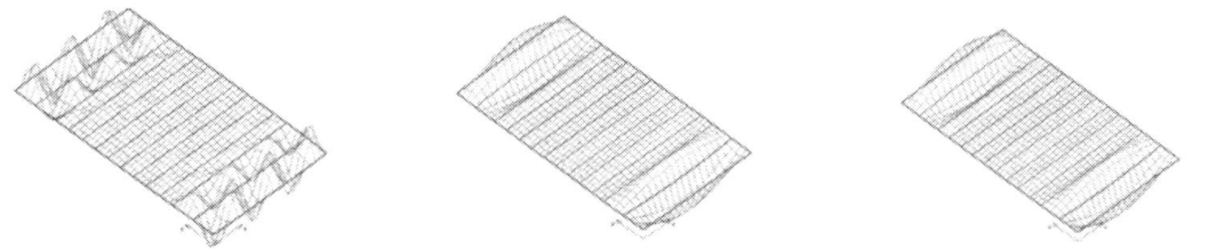

a) with the sole effect of σ_x b) with the sole effect of σ_z c) under biaxial compression

Fig. 6: Buckling shape of the bottom plate with decisive local buckling of the subpanels

Table 1: Buckling verification of the bottom plate with decisive local buckling of the subpanels

Panel	t mm	a m	b mm	α_{ult}	$k_{\sigma,x}$	$\sigma_{cr,x}$ N/mm^2	$\alpha_{cr,x}$	$k_{\sigma,z}$	$\sigma_{cr,z}$ N/mm^2	$\alpha_{cr,z}$	α_{cr}	λ_p	ρ_{cz}	ρ_{cx}	η w/o V	η withV
EF1	12	4	725	5.95	4.0	208	3.16	1.045	54	1.09	0.81	2.71	0.129	0.339	1.043	1.336
EF2	12	4	725		4.0	208	3.16	1.045	54	1.09	0.81	2.71	0.129	0.339	1.043	1.336
EF3	12	4	450		4.0	540	8.19	1.025	138	2.77	2.07	1.70	0.268	0.513	0.513	0.679
EF5	12	4	475		4.0	485	7.35	1.045	127	2.53	1.89	1.78	0.256	0.493	0.536	0.712
GF	12	4	7000		1237	690	10.5	143	80	1.60	1.39	2.07	0.364	0.18	0.997	1.203

4.2 Verification of the entire panel with decisive global buckling

Fig. 7a shows the buckling shape with one wave and $\sigma_{cr,x}$ = 25.24 for the sole effect of σ_x. Fig. 7b shows the buckling shape of the total plate under the sole effect of σ_z with $\sigma_{cr,x}$ = 4.01. The longitudinal stiffeners deform during the buckling process of the bottom plate but should not buckle in order not to lose their stiffening effect. Because the lateral compression exists the stiffeners must be verified in accordance with Section 9 of EN 1993-1-5 [2]. Fig. 7c shows the buckling shape under biaxial compression with a magnification factor of σ_{cr} = 3.58

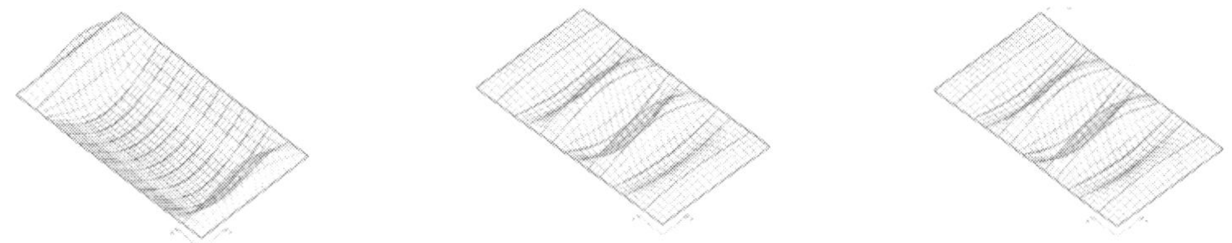

a) with the sole effect of σ_x b) with the sole effect of σ_z c) under biaxial compression

Fig. 7: Buckling shape of the bottom panel with decisive global buckling

The ratio of the critical plate buckling stress $\sigma_{cr,p}$ to the critical column buckling stress $\sigma_{cr,c}$ gives a value greater than 1 but less than 2, indicating a column-like behaviour of the stiffened bottom plate in the x-direction (see Fig. 3). The buckling verification for the total panel without considering the correction factor V would show that the bottom plate is safe according to buckling (η = 0.357 < 1.0, Table 2). The subpanels must be verified separately. Table 2, for EF1 and EF2, shows that the subpanels with a width of 725 mm do not have sufficient safety against buckling (η = 1.043 >1.0). Thus the buckling behaviour of the isolated considered subpanel corresponds exactly to the behaviour of the subpanel within the total panel where local buckling is decisive (see Fig. 6 and Fig. 8).

For the buckling verification of the subpanel in Table 2, σ_{cr} is calculated with the approximation formula (10.6) in [2]. It is for the overall plate about less than the value 3.58 determined by the FE program EBPlate and is thus on the safe side. In the z direction the subpanel has a column-like behavior (ξ = 0,045). If, as shown in Table 3, the verification is performed with σ_{cr} calculated from the FE analysis without taking into account the correction factor V, one comes to the conclusion that the verification is observed (η = 0.81 <1.0). With the consideration of V the verification is just not satisfied (η = 1.06 >1.0). This is only possible because during the FE calculation the nodal lines of the buckled subpanel in x-direction (Fig. 9a) stabilize the subpanel during buckling in z-direction. That leads to a higher σ_{cr} (1.06 in Table 3) or to a smaller global slenderness of 2.366. Carrying out the pure column buckling of the subpanel in the z-direction under the sole effect of σ_z, one finds that the verification does not succeed by far (η = 1.27 >1.0). This means that the buckling shape with many waves due to the stresses in x-direction has positive influence on the buckling behaviour in z-direction. This assumption should not be considered in an assessment, since there is no guarantee that the actual buckling shape or the real imperfection shape also has this supporting effect. In fact, for the given buckling panel geometry the stresses in x-direction only slightly influence the buckling in the z-direction (in negative but not in positive sense) because due to the small curvature of the buckling shape in x-direction they only produce low drive forces. The described effect shows as an example the problem of pure application of FE methods for the buckling verification, which initially led to an underestimation of the buckling risk in this particular case. The first four lines from Table 1 remain unchanged.

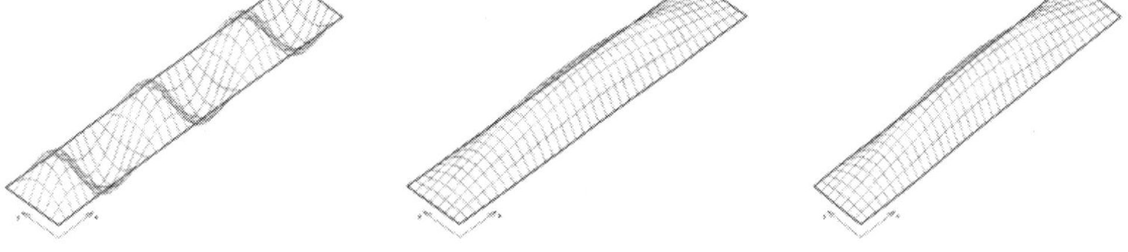

a) with the sole effect of σ_x b) with the sole effect of σ_z c) under biaxial compression

Fig. 8: Buckling shape of the single subpanel with the width 725 mm

Table 2: Buckling verification of the bottom plate with decisive global buckling

Panel	t mm	a m	b mm	α_{ult}	$k_{\sigma,x}$	$\sigma_{cr,x}$ N/mm^2	$\alpha_{cr,x}$	$k_{\sigma,z}$	$\sigma_{cr,z}$ N/mm^2	$\alpha_{cr,z}$	α_{cr}	λ_p	ρ_{cz}	ρ_{cx}	η w/o V	η withV
GF	12	4	7000	5.95	2983	1666	25.24	360	201	4.02	3.47	1.31	0.364	0.504	0.357	0.452

Table 3: Buckling verification of the bottom plate with decisive overall plate buckling

Panel	t mm	a m	b mm	σ_x N/mm^2	σ_z N/mm^2	α_{cr}	λ_p	ρ_{cz}	ρ_{cx}	η withV	η w/o V
EF1	12	4	725	66.0	50	1.0633	2.366	0.166	0.383	0.810	1.059

5. Conclusions

The introduction of the correction factor V for the buckling verification of one plate under bi-axial compression using the reduced stress method is targeted and justified for unstiffened panels. A transfer of this factor on stiffened panels under biaxial compression leads in certain cases to more conservative results of this method as before, in other cases the influence is rather low. The use of the modification factor V for the buckling verification of the entire panel, when local buckling of the subpanel is not decisive, is questionable from the mechanical point of view. The buckling verification of the entire panel is directly related to the verification of the longitudinal stiffeners in accordance with Section 9 of EN 1993-1-5. A scientific review, if the introduction of the correction factor V, the aggravation of the buckling verification for panels without local buckling of the subpanels, is justified or not, is outstanding and desirable.

References

[1] DIN-Fachbericht 103:2009-03: Stahlbrücken. März 2009.

[2] DIN EN 1993-1-5:2010-12: Eurocode 3: Bemessung und Konstruktion von Stahlbauten - Teil 1-5: Plattenförmige Bauteile; Deutsche Fassung EN 1993-1-5:2006 + AC:2009.

[3] Braun B. *Stability of steel plates under combined loading*, Mitteilungen des Instituts für Konstruktion und Entwurf; Nr. 2010-3. Stuttgart 2010.

[4] DIN 18800 Teil 3 Stahlbauten; Stabilitätsfälle, Plattenbeulen. Ausgabe Dezember 2008.

[5] Fischer M, Harre W. "Ermittlung der Traglastkurven von einachsig gedrückten Rechteckplatten aus Baustahl der Seitenverhältnisse a ≤ 1 mit Hilfe von Versuchen (Teil I und II)", *Stahlbau* 47(7&8), 199-204, 239-247, 1978.

[6] Braun B, Kuhlmann U. "Reduced stress design of plates under biaxial compression", *Steel Construction,* 5(1), 33-40, 2012.

[7] Scheer J, Nölke H. "Zum Nachweis der Beulsicherheit von Platten bei gleichzeitiger Wirkung mehrerer Randspannungen", *Stahlbau,* 70(9), 718-729, 2001.

[8] Maur J, Schmidt H, Verwiebe C. "Spannungsbasierter Beulsicherheitsnachweisebener und gekrümmter stählerner Flächentragwerke unter kombinierter Membranbeanspruchung – Eine vergleichende Analyse der Eurocode-Formate", *Stahlbau,* 80(11), 804-813, 2011.

[9] Winterstetter TA. "Zum Nachweis der Beulsicherheit von Platten- und Schalentragwerken aus Stahl bei mehrachsiger Beanspruchung", *Stahlbau*, 71(11), 816-822, 2002.

[10] Unterweger H, Kettler M. "Einzelfeldbeulen – Wirklich große Unterschiede zwischen Eurocode EN 1993-1-5 und DIN 18800-3", *Stahlbau*, 82(8), 597-608, 2013.

[11] Wang X, Huang J. "Elastoplastic buckling analyses of rectangular plates under biaxial loadings by the differential qudrature method", *Thin-Walled Structures*, 47(1), 14-20, 2009.

THE NEW METHOD OF REFERENCE RESISTANCE DESIGN FOR SHELL STRUCTURES

J. Michael Rotter

Institute for Infrastructure and Environment, School of Engineering, University of Edinburgh, UK

Abstract: This paper presents the remedy to a long-standing difficulty in the design process for the buckling of shells. Structural members have long been designed to global measures of resistance (e.g. bending moments), but the local character of most shell buckling modes has led to hand calculations that chiefly use a local stress. Where a structure experiences significant non-uniform plasticity before buckling, the local stress model invariably leads to excessively conservative design. Key situations are global bending, local supports, around penetrations, junctions between shell segments and similar details. These difficulties are now resolved by Reference Resistance Design (RRD) as a shell design conceptual process.

1. Introduction

This paper outlines the new method for thin shell buckling design that has recently been accepted as an amendment to the Eurocode on Strength and Stability of Shells [1]. The paper first outlines the background of some shortcomings in the current Stress Design approach.

The design of thin shell structures against buckling has a long and somewhat chequered history. Most practical thin shells are designed using hand calculations, exploiting algebraic expressions that define the expected buckling resistance. Early rules for the elastic buckling resistance were entirely derived as empirical lower bounds to test results [2,3]. These were necessarily very limited in scope because only a narrow range of geometries and load cases had been tested. The simplest test case is the uniform isotropic elastic cylinder under uniform axial compression (Fig. 1), which is easy to fabricate and easy to test in a classic compression machine. Other widely tested cases are the same cylinder under uniform external or hydrostatic pressure, and under torsion (Fig. 1). The critical aspect of these three load cases is the uniform membrane stress throughout the specimen: the fact that the axial compression buckling mode was found to be relatively local led to the concept that design could be undertaken using the highest local magnitude of the compressive membrane stress. However, this choice prevents the method from accurately dealing with more practical stress states in shells.

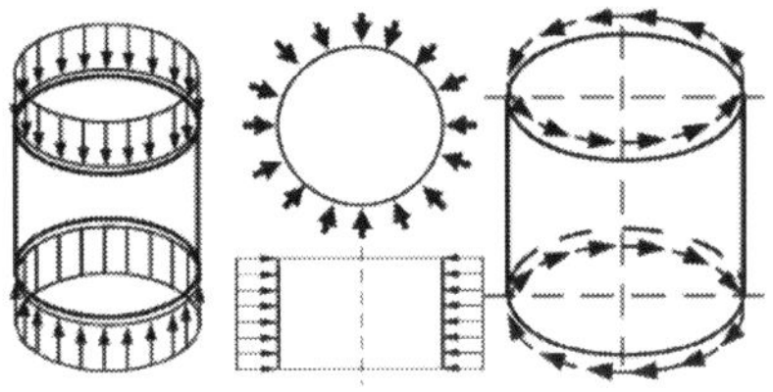

Fig. 1: The three uniform membrane stress states

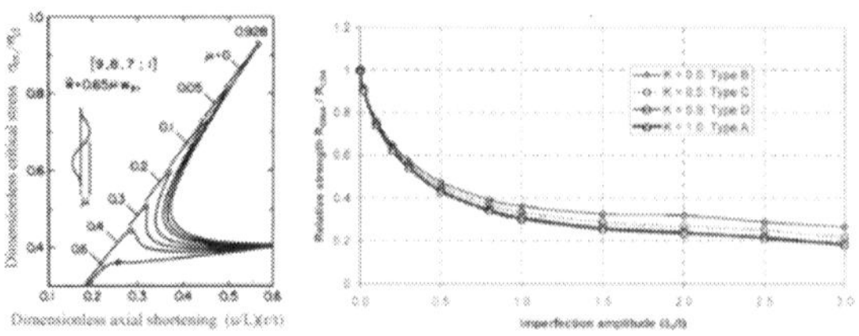

Fig. 2: Typical changing buckling resistance as the imperfection amplitude increases

When more complex conditions are to be assessed, the natural extension of the above concept indicates that it should again be the local peak compressive membrane stress that is deemed to control elastic buckling. For thicker shells, in which buckling occurs after significant plastic deformation, a simple empirical adjustment of the elastic buckling prediction was made to allow for plasticity. This conceptual framework, using the peak calculated membrane stress determined from an elastic calculation to determine when buckling might occur, made it difficult to develop quality design rules of useful accuracy for more complex load patterns, complex stress states and cases involving significant plasticity before buckling.

For all three of the above simple test cases, geometric imperfections played a strong role in reducing the elastic buckling strength (Fig. 2), and it was always difficult to obtain a reliable prediction of the test outcome, whether analytically or computationally [4]. Imperfections can take many different forms and both the form and amplitude of the critical imperfection is vital information for an accurate buckling prediction [5]. A fuller exploration is now possible by using modern computational analyses of shell buckling, but such analyses require that a provisional design has already been performed using hand calculations [6].

Some shells, such as tubes under global bending, buckle after considerable plasticity, so that any quasi-elastic peak stress at the buckling state is far above the yield stress. The only hand calculation method in shell structural design codes until now, known as "Stress Design", cannot address such situations at all. But the new hand calculation method of Reference Resistance Design (RRD) has now been devised to address all shell buckling problems for which appropriate data exist. RRD is similar to the computational method of LBA-MNA, but does not depend on computer calculations. RRD is now defined in EN 1993-1-6 [7] for tubes under global bending and spherical domes under external pressure. It has great potential for application to openings, local supports, wind loading, stiffener terminations, shell junctions, eccentric solids flow in silos, stiffened shells, stepped walls, torispherical heads, spherical and elliptical domes and similar problems of local loads or structural non-uniformity.

2. Stress Design against shell buckling

2.1 Simple designs with uniform stress states

Both experiments and algebraic analyses are easier to perform on shell structures using a loading that produces a uniform membrane stress state in the shell. The easiest shell form is the right circular cylinder, both to fabricate and to analyse, so the literature on this case is particularly extensive. For doubly curved shells, such as partial spheres, a uniform external fluid pressure produces a uniform membrane stress in the body of the shell. For both of these situations, the boundary conditions in an experiment lead to a some additional complexity, but the boundary effect is usually minor relative to the dramatic effect of geometric imperfections.

Because of this background, almost all shell buckling design for general cases is based on the uniform membrane stress condition throughout a shell. The three classic simple cases

(Fig. 1) of uniform axial compression, uniform external pressure and uniform torsion in a cylinder have become the reference cases on which most shell buckling design rules are based. Even where a combination of these three uniform membrane stress states occurs, the rather complicated result is seen as an "interaction" between the three elementary stress states [1].

Unfortunately the buckling modes for these three elementary cases have quite different shapes. Under uniform stress, external pressure usually induces large buckles, shear moderate ones and axial compression very small buckles. If the stresses have a rapid spatial variation, the buckle cannot be large: an axial compression buckle may then not differ much from its uniform stress form, but that is not true where the peak stress is circumferential or shear.

A key feature of a uniform stress state is that changes of geometry before buckling in a thin shell (geometric nonlinearity) play only a small role (less than 15%). However, when a shell is unsymmetrically loaded, changes in geometry affect the result a lot [8], so conclusions drawn from the tests under uniform stress states are significantly misleading.

A further feature of the uniform pre-buckling stress state is that the shape and size of the buckle can vary significantly without much change in the buckling resistance [5,9,10]. This leads to the concept that the interaction between rival modes leads to a high sensitivity to geometric imperfections [9]. However, when the loading produces a stress state that is far from uniform, the potential for different modes to have similar buckling resistances declines considerably, so the uniform membrane stress condition is again a misleading model.

A final troublesome feature of the uniform stress state as a basis for all others occurs when the buckling stress is high, as in thick cylinders. Almost the entire shell reaches the yield stress simultaneously, and the test then unfortunately suggests that a small correction of the elastic buckling stress can account for yield. This useful simplification for uniform stress states turns out to be very unfortunate for many practical shell buckling conditions [11-13].

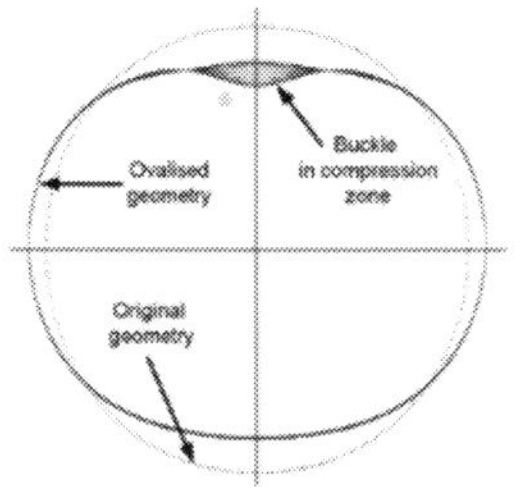

Fig. 3: Ovalisation of a cylinder under bending

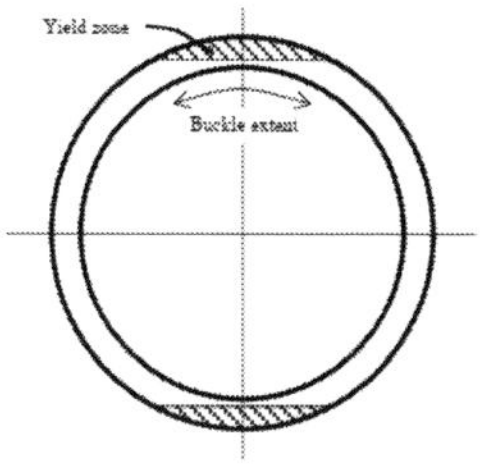

Fig. 4: Partial yield zones under bending

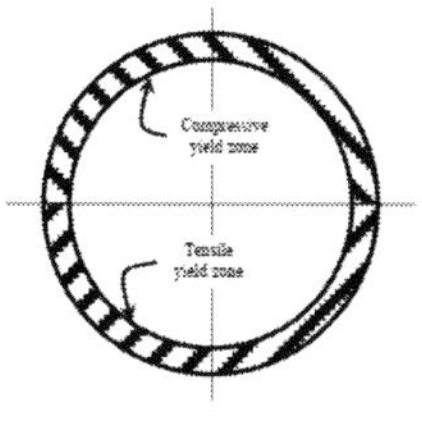

Fig. 5: Full yield under bending

Fig. 6: Large buckles in a stepped wall tank

2.2 Global bending of cylinders

Where a thin elastic cylinder is subjected to global bending, matching compressive and tensile stresses develop on the opposite surfaces, and these lead to ovalisation of the tubular section (Fig. 3). Ovalisation is a severe case of geometric nonlinearity, and leads to an elastic limit load: a condition of snap-through under elastic conditions [14]. Such a snap through is just pre-empted by elastic buckling, but this buckle cannot be predicted without accounting for the reduced lever arm between the tensile and compressive stress peaks, and the increased local radius of curvature in compression: both are geometrically nonlinear phenomena.

Thicker shells are more challenging, since a significantly large zone of yielding can develop before buckling (Fig. 4): at the condition of full plasticity (Fig. 5), the highest stress obtained using an elastic calculation then corresponds to a stress far above the yield stress. The concept that buckling is controlled by the stress at a point is completely inappropriate.

2.3 Stepped walls in tanks and silos

In tanks and silos, the pressure on the inside surface increases with depth below the surface; in silos the axial compression in the wall grows even faster. Consequently the shell wall has an increased thickness with depth below the surface. Metal sheets of constant thickness produce a stepped wall: each uniform zone is termed a strake. The most critical location in each strake is at its base, where the formation of an axial compression buckle is restrained by the thicker strake below. The point of highest membrane stress is thus at a point where the uniform compression resistance is certainly lower than the value required to cause a buckle. Since the buckle lies in a steep stress gradient and has a finite size, the highest stress is a poor measure.

Under external pressure or wind when empty, large buckles develop over a major part of the height (Fig. 6), but the circumferential stress varies with the strake thickness, so it is hard to devise a simple buckling criterion based on the stress concept to cover this case [1].

2.4 Load combinations and complex stress patterns

In almost all structures, it is necessary to consider multiple loads acting simultaneously. Since each load case will produce a different pattern of stress in the structure, a combination will inevitably induce more than one of the three membrane stresses at highly stressed locations. Current standards use the idea that the shell will buckle under the highest combination of stresses, and buckling is assessed on the assumption that this stress state extends throughout the whole shell as a uniform membrane stress state. This is a very conservative treatment, and designers often ignore the low strength predictions that arise from this approach.

Some have argued that it has not been shown that a different high membrane stresses at different locations (axial or circumferential or shear) cannot combine to induce buckling at a lower loading than that of the highest stressed single point. But this would lead to even greater conservatism. The right way forward is to reject the concept of the stress at one point.

2.5 Conical shells under external pressure

A conical shell under external pressure presents a tricky problem for the Stress Design approach especially when combined with axial load. The external pressure induces both circumferential and meridional non-uniform membrane stresses, also controlled by which end is axially supported. Identification of the critical stress location is not simple, the boundary conditions are more complicated, and the buckle may extend over the whole shell or only part of it. The significant challenges posed by this case are outlined in the commentary in [15].

2.6 Openings, cutouts, local supports and eccentric discharge of silos

Where more complex conditions arise, the Stress Design approach fails completely. Local supports (Fig. 7) and brackets usually induce local plasticity before buckling (Fig. 8), and the peak stresses around openings (for human access) or cutouts (for services) are very localised (Fig. 9). A buckle can only form over a finite area, so steep stress gradients lead to much higher peak stresses at buckling. Where a buckle forms adjacent to a boundary, the additional restraint raises the buckling resistance well above that for the uniform membrane stress states above. The eccentric discharge of a silo (Fig. 10) produces very complex patterns of pressure on the wall, causing significant nonlinear pre-buckling changes of geometry and localised stress states that are far from calculable using the Stress Design approach.

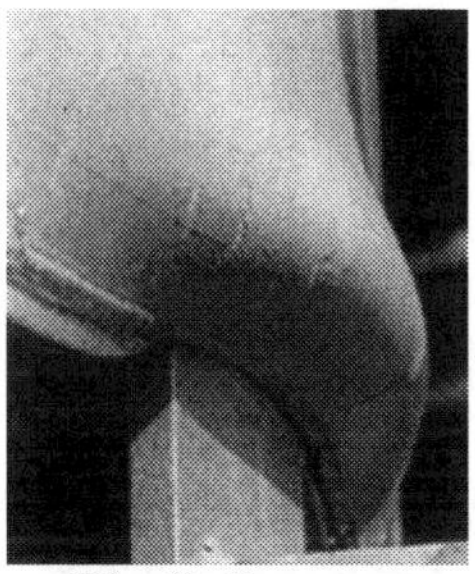

Fig. 7: Support

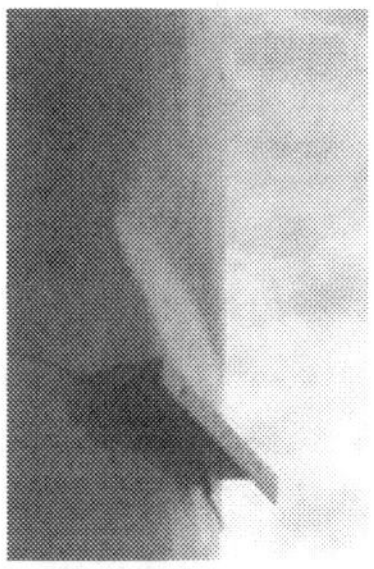

Fig. 8: Bracket

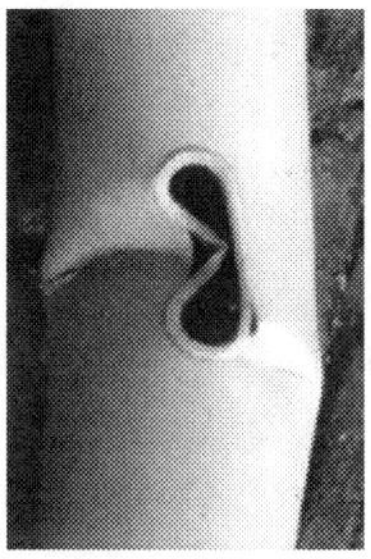

Fig. 9: Cutout

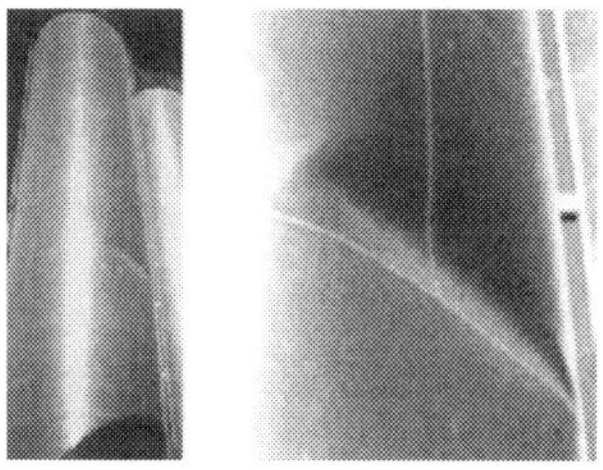

Fig. 10: Eccentric flow

2.7 Spherical, tori-spherical and tori-conical shells

Design rules for partial spherical domes, tori-spherical and tori-conical shells under external and internal pressure were established long ago using empirical approximations to computations [16]. But since such results are not set out in a global context, it is impossible to generalise them. If transformed into the format of RRD, there is great scope to extend the descriptions to deal with variations in geometry, material properties and imperfections.

3. Computational treatments

3.1 Introduction

Computational assessments of shell structures have become very easy to perform with powerful modern machines and programs. But these are not a complete solution to the problem.

The computer can only perform an analysis of a planned and pre-dimensioned structure, but programs to proportion the structure are still relatively rare. For thin shell structures, the range of roles and consequent details of geometry, structural form (e.g. rolled sheets or corrugated or stiffened construction), the specialist load cases and the many load combinations make it difficult to produce a general program for this purpose.

In any global analysis the loading usually cannot be defined by a single value, but consists of a many point loads, line loads and spatially varying surface pressures. This set of loads, at the design values, produce the reference stress state in the structure. A computation of the structure's ultimate resistance can then be calculated as a multiple of these loads, here defined as the characteristic resistance R_k. After applying an appropriate partial factor γ_M, the design resistance R_d is obtained ($R_d = R_k/\gamma_M$). The test of a satisfactory design is simply $R_d \geq 1$.

Since the computer generally only performs an analysis and identifies a failure condition using the criterion of failure for the chosen analysis type, the design process is necessarily iterative and must begin with a hand calculation to produce the trial dimensions [6].

3.2 Fully nonlinear analysis GMNIA

The enthusiastic analyst, working on a one-off special structure with a significant budget, may well choose to use a fully geometrically and materially nonlinear treatment to identify the buckling or collapse condition. But shell structures are peculiarly sensitive to geometric imperfections, which can take many forms and cannot be easily assessed in advance. Other uncertainties, such as the sequence of construction and differential settlements, can make these analyses difficult to use in a design process. EN 1993-1-6 has extensive rules on the use of GMNIA, but it is not really a cost effective design tool, except under special circumstances.

3.3 The short-cut of LBA-MNA

The drafters of EN 1993-1-6 [1] sought to find a simpler method by which computer analyses could be usefully deployed in design. The result is LBA-MNA, in which the proposed design is first subjected to a linear elastic bifurcation analysis (LBA) and then to a materially nonlinear small displacement theory analysis (MNA) to find two reference resistances [6]. A linear elastic analysis can be used to make a conservative estimate of the MNA resistance, but it has been shown that the conservatism of this choice can be extremely variable [12].

To obtain a realistic measure of the true resistance, the two reference values must next be combined, together with estimates of the effect of imperfections, the onset of plasticity, the slenderness at which the full plastic resistance can be attained and the degree to which plasticity affects moderately slender shells [1]. These estimates currently assume that the target shell will have similar characteristics to those of Stress Design, so the shortcomings outlined above of that process affect them. The method is consequently a useful tool for the evaluation of a relatively well understood structure, but is ineffective for many practical situations.

4. Reference resistance design (RRD)

4.1 Introduction

The above descriptions of hand calculations using Stress Design reveal many shortcomings when these are applied to conditions other than completely uniform membrane stresses. The new Reference Resistance Design (RRD) is a hand calculation shell design method that does not have the restrictions of Stress Design, has the advantages of the concepts of LBA-MNA but does not have its shortcoming in which key parameters must be estimated but cannot be evaluated accurately. It offers a significant opportunity to exploit many existing design proposals that have been expressed in empirical form, since they can be transformed into a consistent format. Furthermore, as more problems are included as specific cases of RRD, it will become easier to estimate the parameters required for an LBA/MNA design.

4.2 Outline of the resistance evaluation

Reference Resistance Design (RRD) applies to specific shell buckling cases of geometry, boundary conditions and loading. It uses algebraic expressions, covering the full range of geometries, for the two reference resistances: the linear elastic critical resistance R_{cr} and the accurately determined plastic collapse resistance based on small displacement theory and an ideal elastic-plastic material model R_{pl}. Using the deduced slenderness of the shell $\bar{\lambda} = \sqrt{R_{pl}/R_{cr}}$, RRD then algebraically defines accurately determined values for the effect of geometric nonlinearity α_G, the imperfection sensitivity α_I, the resistance level where plasticity first affects the global resistance β, the slenderness $\bar{\lambda}_o$ at which the reference plastic resistance R_{pl} is achieved, the severity of plasticity effects at each slenderness η, and the potential strength gain due to strain or geometric hardening χ_h. These parameters must all be defined using a comprehensive study of the specific shell buckling case before adoption into RRD.

4.3 Significance of each parameter

Each of the above parameters represents a specific attribute of the shell buckling behaviour. Geometric nonlinearity (pre-buckling deformations) α_G has strong effect on thin shells under unsymmetrical loads, especially when in-plane boundary deformations occur (e.g. settlement). By contrast, the sensitivity to geometric imperfections α_I is more powerful in the uniform membrane stress states that are used as the basis of Stress Design. The load level at which plasticity affects the global bucking behaviour β is often much above the first yield condition, because local bending stresses in shells lead to early yield but do not usually affect the buckling condition. The value of $\bar{\lambda}_o$ may be affected by the geometric imperfection amplitude.

The slenderness at which the simple plastic resistance is exceeded $\bar{\lambda}_o$ is high for structures that experience either geometric or material hardening, but is very low for imperfect shells with either geometric softening or low strain hardening. Some GMNIA calculations with deep imperfections may indicate negative values for $\bar{\lambda}_o$, but such thick shells have small imperfections, so interpret them with care. The plasticity severity factor η can characterise a big range of possible behaviours: anywhere from little interaction between elastic buckling and plastic collapse, through to first plasticity having a big impact. The strain hardening behaviour χ_h is greatly simplified to a linear variation: it is very conservative for thick shells.

The constraints on the values these parameters can take have been identified [17].

4.4 Equations defining the characteristic resistance

These parameters are combined to define the characteristic resistance (Fig. 11) as

$$R_k/R_{pl} = \chi_h - (\bar{\lambda}/\bar{\lambda}_o)(\chi_h - 1) \qquad when \quad \bar{\lambda} \le \bar{\lambda}_o$$

$$R_k/R_{pl} = 1 - \beta\{(\bar{\lambda} - \bar{\lambda}_o)/(\bar{\lambda}_p - \bar{\lambda}_o)\}^{\eta} \qquad when \quad \bar{\lambda}_o < \bar{\lambda} < \bar{\lambda}_p \qquad (1)$$

$$R_k/R_{pl} = \alpha/\bar{\lambda}^2 \qquad when \quad \bar{\lambda}_p \le \bar{\lambda}$$

with $\alpha = \alpha_G \alpha_I$ and $\bar{\lambda}_p = \sqrt{\alpha/(1-\beta)}$.

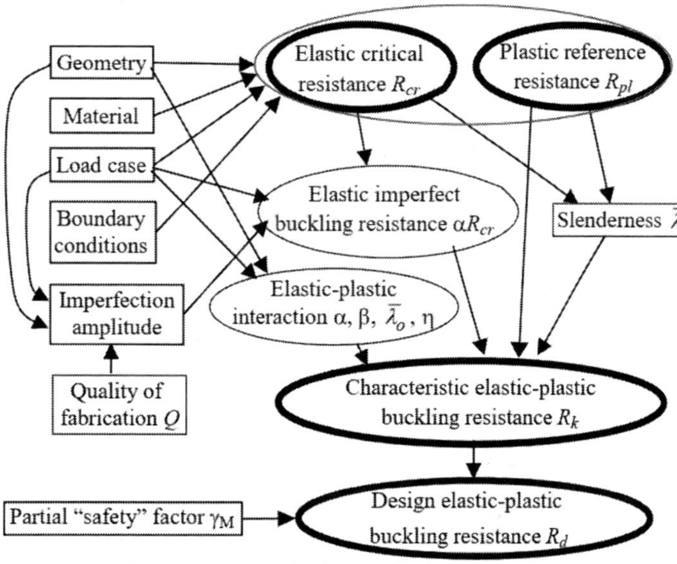

Fig. 11: RRD design calculation process

4.5 Application to other problems

The RRD method has quite general application, so it can be used to accurately describe the outcome of most studies of the static strength of structures. Older buckling studies that produced empirical outcomes only require the discovery of the simple algebraic expressions that define R_{cr} and R_{pl}. More recent shell buckling studies that already produced suitable material for adoption into RRD include local supports, shell junctions, bracket supports, shell penetrations and conical roofs. References to these studies may be obtained from this author.

5. Conclusions

This paper has outlined the new hand calculation Reference Resistance Design (RRD) method of EN 1993-1-6 [1,7]. It has set out some of the many reasons why such a method is needed. Researchers studying the many unsolved shell buckling problems are urged to present their findings in the above parametric form for direct adoption into EN 1993-1-6.

References

[1] EN 1993-1-6. *Eurocode 3: Design of steel structures, Part 1.6: Strength and stability of shell structures*, CEN, Brussels, 2007.

[2] *Buckling of Thin-Walled Circular Cylinders*. NASA SP8007, 1968.

[3] *European Recommendations for Steel Construction: Buckling of Shells*. 1st edition, European Convention for Constructional Steelwork, Brussels, 1980.

[4] Schmidt H, Rotter JM. "Rules for the buckling limit state assessment using global numerical analysis" in *Stability of Steel Shells: European Design Recommendations*: 5th Edn, Europ. Conv. for Constr. Steelwk, Brussels, 89-143, 2013.

[5] Yamaki N. *Elastic Stability of Circular Cylindrical Shells*, North Holland, Elsevier Applied Science Publishers, Amsterdam, 1984.

[6] Rotter JM. "Shell buckling design and assessment and the LBA-MNA methodology", Stahlbau, 80(11), 791-803, 2011.

[7] Rotter JM. "Design of shells using reference resistances", Amendment AM-1-6-2013-05 to EN 1993-1-6, Approved by CEN TC250 SC3, 2013.

[8] Sadowski AJ, Rotter JM. "Study of buckling in steel silos under eccentric discharge flows of stored solids", *J Engrg Mechs ASCE*, 136(6), 769-776, 2010.

[9] Koiter WT. *On the Stability of Elastic Equilibrium*, PhD Thesis (in Dutch), Delft University, Netherlands, 1945.

[10] Calladine CR. *Theory of Shell Structures*, Camb Univ Pr, Cambridge, 1983.

[11] Doerich C, Rotter JM. "Estimating the plastic collapse load of a shell using LA, MNA and GMNA finite element analyses", *Proc. 6th International Conference on Thin-Walled Structures*, Timisoara, Romania, 889-896, 2011.

[12] Doerich C, Rotter JM. "Accurate determination of plastic collapse loads from finite element analyses", *J Pres Ves Tech*, 133(1), 011202-1 to 10, 2011.

[13] Doerich C, Rotter JM. "Behavior of cylindrical steel shells supported on local brackets", *J Struct Engrg ASCE*, 134(8), 1269-1277, 2008.

[14] Houliara S, Karamanos SA. "Buckling and post-buckling of long pressurized elastic thin-walled tubes under in-plane bending", *Int. J. Non-Lin. Mechs.*, 41, 491-511, 2006.

[15] Schmidt H, Rotter JM, Greiner R. "Rules for the buckling limit state assessment using stress design" in *Stability of Steel Shells: European Design Recommendations*: 5th Edn, Europ. Conv. for Constr. Steelwk, Brussels, 145-165, 2013.

[16] Galletly GD, Blachut J. "Torispherical Shells under Internal Pressure", *Proc. I. Mech. E.*, 199(C3) 225-238, 1985.

[17] Doerich C, Rotter JM. "Generalised capacity curves for stability and plasticity: application and limitations" *Thin-Walled Structures*, 49(9) 1132-1140, 2011.

LENGTH EFFECTS IN THE BUCKLING OF IMPERFECT AXIALLY COMPRESSED CYLINDERS

J. Michael Rotter and Hussain Al-Lawati

Institute for Infrastructure and Environment, School of Engineering, University of Edinburgh, UK

Abstract: This paper describes a wide-ranging exploration of the elastic buckling of imperfect cylindrical shells of many lengths under uniform axial compression. It describes the buckling resistance and imperfection sensitivity for the full range of lengths from the shortest to the longest. The circumferential weld depression imperfection is chosen as the most damaging practical form for an axially compressed cylinder. The analyses are based on geometrically nonlinear prebuckling analyses with explicitly modelled imperfections (GMNIA). Several thousand calculations have revealed interesting discoveries about the behaviour at different lengths, and reveals shortcomings in the classical description of imperfection sensitivity.

1. Introduction

The buckling of cylindrical shells under axial compression (Fig. 1) has a very long history, with the fullest literature of any shell buckling geometry or load case. The great difficulty in developing nonlinear algebraic analyses of the buckling of **imperfect** shells and the huge variety of imperfection shapes and amplitudes that can occur have hampered the development of a full understanding. Most experimental, theoretical and computational studies have focussed on the buckling resistance of medium length shells, which is the category that was first studied and which was found to be very sensitive to geometric imperfections. The buckling resistance of **perfect** shells in different length ranges was also explored long ago, but a clear and comprehensive description of the variation of the sensitivity to imperfections in different length ranges does not appear to have ever been carried out.

The study described in this paper used the most damaging practical imperfection form for axially compressed cylinders, the circumferential weld depression. It studied the elastic buckling resistance and imperfection sensitivity over a huge range of lengths. It used geometrically nonlinear analyses with explicitly modelled imperfections (GMNIA), with many thousands of calculations to explore each detail of small variations in buckling resistance.

The Eurocode on the Strength and Stability of Shells [1] identifies three length regimes for axially compressed cylinders and defines their linear bifurcation buckling resistances with simple algebraic expressions. But the imperfection sensitivity is deemed to be the same irrespective of the length and is taken as that for a typical medium length cylinder. This paper shows that some shorter cylinders are more imperfection sensitive than this simple concept supposes, and that other features of the full buckling resistance-length-imperfection relationship show that the current model of EN 1993-1-6 [1] requires some significant improvement.

A more complete version of the introduction in Section 2 may be found in [2]. An extended description of the underlying theoretical work may be found in [3].

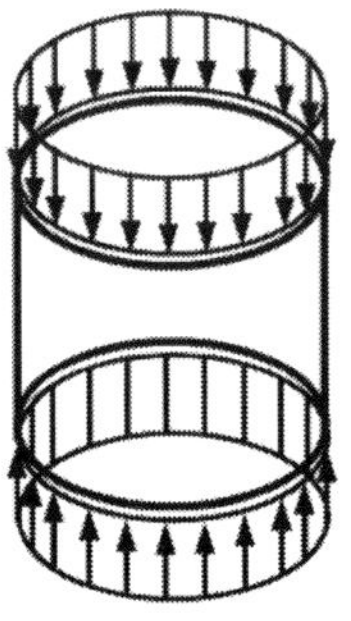

Fig. 1: Cylindrical shell under axial compression

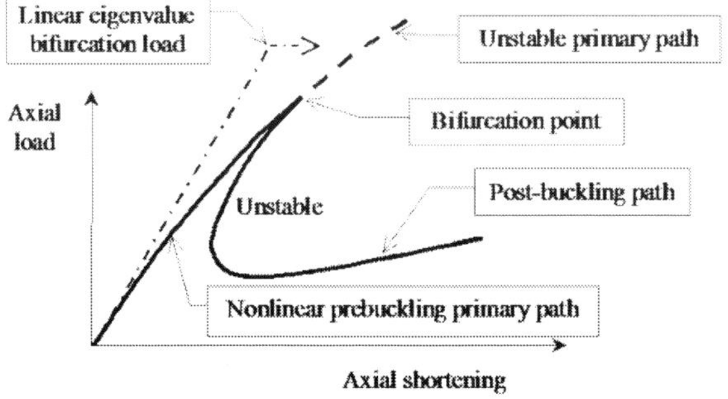

Fig. 2: Typical load-axial displacement relationship for a perfect cylindrical shell under axial compression

2. Buckling of perfect elastic shells under axial compression

2.1 Linear bifurcation in perfect elastic medium length cylindrical shells

The strength of an axially compressed cylindrical shell under any condition is normally compared with the linear bifurcation buckling stress of a perfect shell under the ideal conditions of medium length, with prebuckling stresses unaffected by the boundary conditions and boundaries that restrain circumferential displacements during buckling. This buckling stress was independently first found by Lorenz [4] and is known as the "classical elastic critical stress"

$$\sigma_{cl} = \frac{E}{\sqrt{3\left(1-\nu^2\right)}}\frac{t}{r} \tag{1}$$

in which E is Young's modulus, ν is Poisson's ratio (typically around $\nu = 0.3$), t is the shell wall thickness and r is the cylinder radius.

2.2 Bifurcation accounting for geometrically nonlinear effects

If a geometrically perfect thin elastic cylinder is axially compressed under uniform compression, the load-end shortening relationship is typically as shown in Fig. 2. A very linear prebuckling path (with little for the experimentalist to observe) is suddenly terminated as the shell bifurcates into a non-symmetric mode, with several full buckle waves around the circumference, and often many waves up the height. At buckling, the load falls very rapidly, and the cylinder actually increases in length (Fig. 2) as the displacements normal to the surface grow. As the load falls (Fig. 2), bifurcation after bifurcation occurs as the mode switches from one circumferential wave number n to the next in dynamic jumps [3].

The bifurcation load found using a traditional nonlinear prebuckling analysis is slightly lower than the load found from a linear eigenvalue calculation, because the restraint of radial displacements at the ends leads to local bending whose amplitude grows in a nonlinear manner as the classical critical value is approached [5]. The resulting increased local stresses near the boundaries lead to a slightly lower buckling load. The nonlinear bifurcation load is then usually described as being about 15% below the linear bifurcation load. By contrast, if the boundaries are free to expand radially, the prebuckling condition is a membrane state of stress, and the classical elastic critical stress is reached even in a nonlinear analysis.

At the classical elastic buckling stress, very many different buckling modes or eigenmodes are all simultaneously critical (sometimes over 100 modes with resistances within 1%). Many modes are possible: most involve a periodic variation of deflection in both the axial and circumferential directions, with different wavelengths in the two directions. These modes are termed "chequerboard". The critical stress for axisymmetric buckling, with systematic inward and outward waves of short wavelength only in the axial direction is also given by Eq. 1.

2.3 Cylindrical shells of different lengths

The above description relates to medium length cylindrical shells where buckling is internal to the shell. To provide a consistent representation of all geometries, the length of longer shells must be characterised using the dimensionless length Ω, given by

$$\Omega = \frac{L}{r}\sqrt{\frac{t}{r}} \tag{2}$$

By contrast, the length of shorter shells must be characterised using the dimensionless length ω, given by

$$\omega = \frac{L}{r}\sqrt{\frac{r}{t}} = \frac{L}{\sqrt{rt}} \tag{3}$$

As the medium length cylinder becomes longer in dimensionless terms, or thicker (lower r/t), the critical circumferential mode becomes lower, finally reaching mode 1 (a single sine wave around the circumference) corresponding to Euler buckling. Thus, in longer and thicker cylinders Euler buckling occurs, though this is often omitted in shell buckling standards (e.g. [1]). Euler buckling is a linear bifurcation event, so it is unaffected by whether the prebuckling analysis is linear or nonlinear.

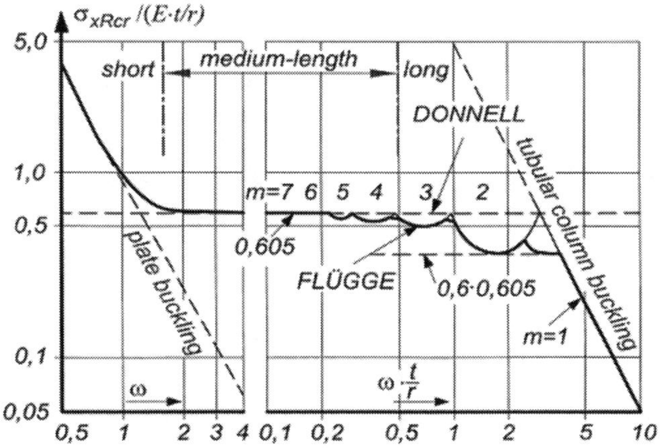

Fig. 3: The effect of cylinder length on linear bifurcation stress of perfect cylindrical shells (after ECCS EDR5, 2013)

A different effect develops when the cylinder is very short. The natural axial wavelength (classical wavelength) for an axisymmetric buckle is given by

$$\lambda_{cl} \cong 1.728\sqrt{rt} \cong 1.728r\sqrt{t/r} \tag{4}$$

In short cylinders, the shell length L is comparable or less than the classical wavelength (Eq. 4), so that the boundary conditions interfere with the formation of a buckle and the buckling resistance increases sharply with reducing length. These descriptions of long and short cylinder behaviour lead to the classic division of shell buckling into the three categories [1,6]. These three categories are illustrated in Fig. 3.

3. Buckling of imperfect elastic shells under axial compression

3.1 Imperfection sensitivity in elastic medium length cylindrical shells

Very minor deviations from the perfect geometry lead to a dramatic loss of buckling resistance in medium length thin elastic cylinders. This phenomenon is well established and has been the subject of much work in the field of shell buckling [2,3]. A typical set of "imperfection sensitivity" curves [3], showing the reduction in buckling strength with the depth δ_0 of a local deviation is shown in Fig. 4 where different imperfection forms have been assumed. The deviation must be seen as relative to the wall thickness t, not the radius r. A silo or tank radius might be 10m, with a thickness loss of 10mm (r/t=1000), but a minor dent in the wall of 10mm can cause up to a 70% loss of strength.

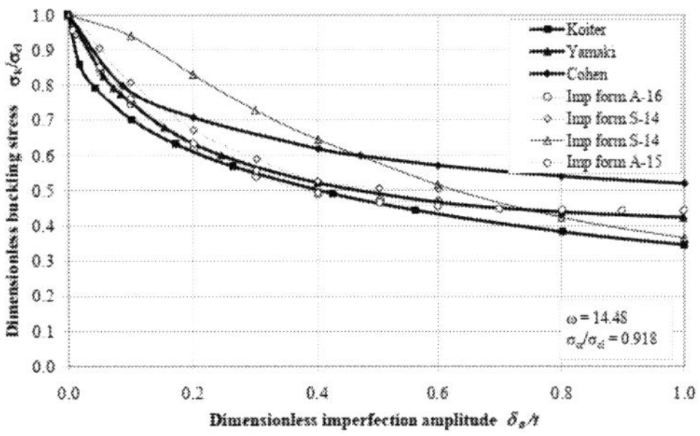

Fig. 5: Classic curves showing the loss of buckling resistance with imperfection amplitude (after Yamaki [3])

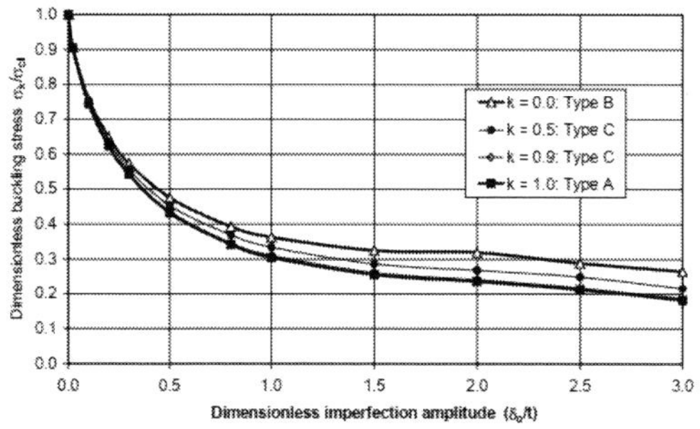

Fig. 6: Imperfection sensitivity for a circumferential weld depressions (after Rotter [7])

Many imperfection forms assumed in previous studies [3] do not relate to practical construction, so the lowest possible curve is not useful. However, it has been shown [7] that a depression at a circumferential weld, which is a natural result of large shell fabrication processes, is effectively the worst that can arise for axial compression buckling. This shape is defined by

$$\delta = \delta_o e^{-\left(\frac{\pi x}{\lambda}\right)}\left(\cos\frac{\pi x}{\lambda} + k\sin\frac{\pi x}{\lambda}\right) \tag{5}$$

in which x is the axial distance from the centre of the weld, k is a factor that depends on the rate of weld cooling and λ is the linear bending half wavelength of the shell given by

$$\lambda = \frac{\pi}{\left[3\left(1-\nu^2\right)\right]^{1/4}}\sqrt{rt} \approx 2.44\sqrt{rt} \tag{6}$$

Rotter and Teng [8] defined two extremes (Type A, k =1: Type B, k =0). The imperfection sensitivity curve is shown in Fig. 6, where Type A is seen to give the lowest strength and has consequently been widely used in other studies.

4. Modelling of the elastic cylindrical shell

The present study examined only thin elastic cylindrical shells with radius to thickness ratios in the range $200 < r/t < 1000$, with a full range of lengths. The weld depression imperfection was modelled as Type A [8] described above. A single imperfection of this form was introduced into the shell geometry at its half height. The imperfection amplitude is everywhere characterised by the ratio δ_o/t indicating a small proportion of the thickness of the shell.

The nonlinear elastic-plastic axisymmetric shell buckling finite element program NEPAS [9] was used throughout this study. It has been widely verified and used for many studies over the last 25 years. The shell was modelled as elastic with modulus $E = 2\times10^5$ MPa and Poisson's ratio $\nu = 0.30$. The program is computationally very efficient, permitting extremely fine meshes to capture the precise form of the imperfection and its effects accurately.

The shells were modelled with prebuckling boundary conditions that give a prebuckling membrane stress state at locations distant from the discrete imperfection. In the prebuckling analysis the ends were free to expand radially, but were restrained against end rotation and circumferential displacement. At buckling, the ends were restrained against radial displacement, but were free to rotate, corresponding to the classical boundary condition S3 [3]. The reasons for changed boundary conditions may be read in [6].

5. Buckling of very slightly imperfect cylindrical shells

The program was first used to undertake a detailed and extensive exploration of the buckling of shells with a tiny geometric imperfection. This was done to explore the proximity of the smallest possible imperfection to the theoretically established conditions. The calculated buckling resistances for $r/t = 200$ and 1000 are displayed in Fig. 7: this large difference in r/t has been chosen to illustrate the independence of r/t. The buckling predictions are plotted against the two dimensionless length parameters ω and Ω on a logarithmic scale. The perfect shell buckling resistances defined by EN 1993-1-6 [1](solid line) are shown for comparison, together with the appropriate Euler buckling curve (dashed line).

A number of observations can be made immediately from Fig. 7. Firstly, Fig. 7(a) shows that the outcome for the two values of r/t is identical for shorter cylinders when the dimensionless length is defined as ω, whilst Fig. 7(b) shows that the same is true for longer cylinders using the dimensionless length Ω. The circumferential mode drops to one wave around the circumference at $\Omega = 3.8$, giving Euler buckling even though the analysis is nonlinear.

Secondly, the comparison of the very slightly imperfect cylinder with the perfect description of EN 1993-1-6 [1] shows that imperfection sensitivity is a phenomenon of medium length cylinders, but is very different in short and long ones. Next, the very tiny imperfection in the medium length cylinders corresponding to only 1% of the shell thickness ($\delta_o/t = 0.01$) produces a strength loss of 7% that is independent of length over an extended range ($8 < \omega$ and $\Omega < 0.1$). The small deviations from this value that occur around $\omega = 3.6$ and $\omega = 7$ are the same for both values of r/t. Similarly the low circumferential mode changes found by Flügge and illustrated in Fig. 4 are identical for both r/t values from $\Omega = 0.15$ upwards.

It is more significant that Fig. 7(b) shows that the reductions in perfect shell resistance are not well represented in [1], since the strength reductions should begin at $\Omega = 0.2$ instead of the current definition of $\Omega = 0.5$. The lower bound in this region used in [1] is clearly necessary as the real behaviour is too complicated to neatly capture in a rule in a standard.

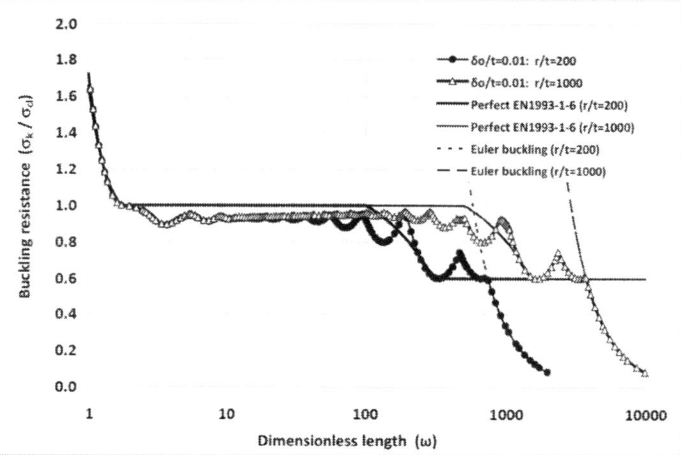

a) plotted against the dimensionless length ω (Eq. 4)

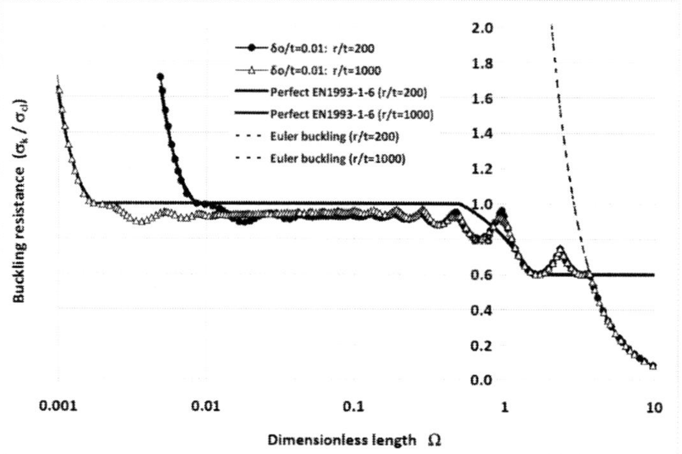

b) plotted against the dimensionless length Ω (Eq. 3)

Fig. 7: Very slightly imperfect shell buckling stress states compared with EN 1993-1-6 [1]

6. Buckling in imperfect cylinders

When a weld depression is introduced in the cylinder, the buckling strength is naturally reduced (Figs 5 & 6). However, the imperfection sensitivity is not uniform across the full range of lengths, as shown in Fig. 8 for the single slenderness $r/t = 1000$.

The following observations are explained in terms of the results for $r/t = 1000$ for reasons of limited space, but the same comments may be made about other shell slendernesses.

The conventional concept of imperfection sensitivity can be seen in Fig. 8 at a length of around $\omega = 150$, where the reduction in buckling strength lies clearly in a zone unaffected by the specific length. At small imperfections, the mode changes shown in Fig. 7 clearly still influence the outcome in longer lengths, but this effect does not persist to affect the buckling resistance of significantly imperfect shells. For all imperfection amplitudes greater than $\delta_o/t = 0.25$, which is effectively most practical construction, the special reduction (indicated as Flügge in Fig. 3) should be omitted for practical design.

All imperfection levels are cut off by the same Euler buckling curve, but their shell buckling resistance is sustained right up to the point where Euler buckling gives a lower resistance. Further, in shorter medium length cylinders, the imperfection sensitivity increases, with significantly reduced buckling strengths for deeper imperfections at lengths around $\omega = 10$. The upper end of the range in which this effect affects the buckling resistance also depends on the amplitude of the imperfection (around $\omega = 17$ for $\delta_o/t = 0.2$; up to $\omega = 35$ for $\delta_o/t = 1.0$ and $\omega = 110$ for $\delta_o/t = 2.5$). The defined "short" range in EN 1993-1-6 [1] does not begin until ω drops below 1.7, so this extra fall in buckling resistance in shorter medium cylinders should certainly be considered. It may be noted that the extensive calculations described in [3] do not extend to deep imperfections and almost all lie in the range $7 < \omega < 23$, though the use of Batdorf's Z as a dimensionless squared length makes the length range appear substantial.

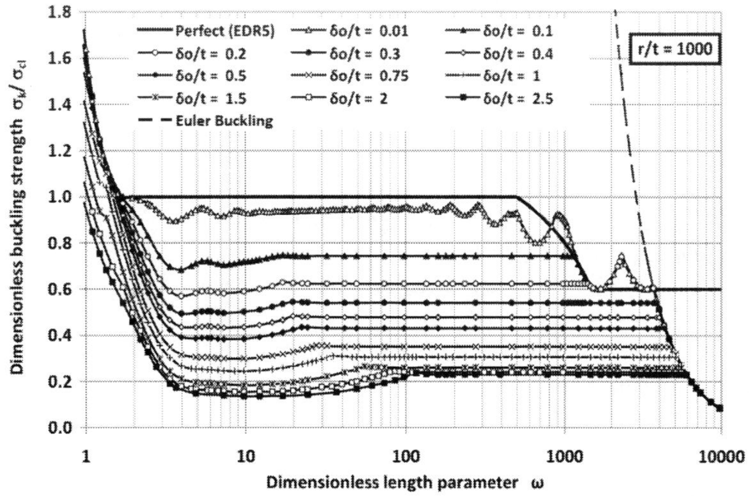

Fig. 8: Variation of elastic buckling strength of imperfect cylinders with dimensionless length ω

In examining short cylinders, it is clear that all practical imperfection magnitudes ($0.1 < \delta_o/t < 2.5$) begin to display short range behaviour at around $\omega = 3.5$, which is far longer than that of the perfect shell (around $\omega = 1.7$). Imperfection sensitivity continues into the short range, but progressively diminishes in severity. The shell with $\delta_o/t < 1.0$ lost 76% of its perfect shell strength at $\omega = 8$, but loses only 17% of the corresponding strength at $\omega = 1$.

The Eurocode on shells [1] relies extensively on the proposition that the elastic critical value for any buckling condition R_{cr} should be reduced by an imperfection factor α that is uniform across a wide range of geometries and conditions. For axial compression, this translates into a uniform imperfection sensitivity that is independent of length throughout all lengths. Both the concept and the standard clearly need more accurate models.

7. Conclusions

The buckling strength of elastic imperfect cylindrical shells has been the subject of a very large number of previous research studies. It is well established that these shells are very sensitive to minor imperfections, but most studies have focussed only on medium length shells. When a fuller range of lengths, with a clearly defined detrimental but practical imperfection form is studied, a thorough examination leads to many unexpected conclusions.

The length ranges defined by the behaviour of the perfect shell do not map onto those of the imperfect shell. Since all shells are imperfect, these ranges should be amended to reflect that situation. There is a special range of lengths at which medium length cylinders are more sensitive to imperfections than the reference condition that pertains for most medium length cylinders. Euler buckling does not intervene over shell buckling until the imperfect buckling resistance is higher than the Euler buckling value.

The general model used in the Eurocode on shells [1] indicates that an elastic critical buckling resistance should be reduced by an imperfection sensitivity factor which depends on the load case but is relatively independent of the geometry. This is evidently a considerable oversimplification. But in oversimplifying this concept, it has also led to more complicated descriptions of perfect shell strength than are appropriate.

Acknowledgments

The authors gratefully acknowledge the financial support of the second author by His Majesty Sultan Qaboos Scholarship Program, through the Ministry of Higher Education of Oman.

References

[1] EN 1993-1-6. *Eurocode 3: Design of steel structures, Part 1.6: Strength and stability of shell structures*, CEN, Brussels, 2007.

[2] Rotter JM. "Buckling of cylindrical shells under axial compression", in *Buckling of Thin Metal Shells*, eds. J.G. Teng & J.M. Rotter, Spon, London, 42-87, 2004.

[3] Yamaki N. *Elastic Stability of Circular Cylindrical Shells*, North Holland, Elsevier Applied Science Publishers, Amsterdam, 1984.

[4] Lorenz R. "Buckling of a Cylindrical Shell under Axial Compression" (in German) Zeitschrift des Vereines Deutscher Ingenieure, 52, 1766, 1908.

[5] Rotter JM. "Stress Amplification in Unstiffened Cylindrical Steel Silos and Tanks", *Civil Engg. Trans., Inst. Engrs., Aust.*, CE31 (3) 142-148, 1989.

[6] Schmidt H, Rotter JM. "Cylindrical shells of constant wall thickness under general loading" in *Stability of Steel Shells: European Design Recommendations*: 5th Edn, Europ. Conv. for Constr. Steelwk., Brussels, 167-216, 2013.

[7] Rotter JM. "Axial Compression: Imperfection Sensitivity, Tolerances and Computational Strength Assessments for EN 1993-1-6", *Proc. Euroshell 71*, Essen, Germany, 1-79, 2013.

[8] Rotter JM, Teng JG. "Elastic Stability of Cylindrical Shells with Weld Depressions", *J. Struct. Engrg., Am. Soc. Civ. Engrs.*, 115 (5) 1244-1263, 1989.

[9] Teng JG, Rotter JM. "Non-Symmetric Bifurcation of Geometrically Non-Linear Elastic-Plastic Axisymmetric Shells subject to Combined Loads including Torsion", *Comps. & Structs.*, 32 (2) 453-477, 1989.

[10] Flügge W. "Die Stabilität der Kreiszylinderschale", *Ing.-Arch.*, 3 (5) 463-506, 1932.

ON THE ELASTO-PLASTIC STABILITY ANALYSIS OF CIRCULAR CYLINDRICAL SHELLS

Rabee Shamass[a], Giulio Alfano[a] and Federico Guarracino[b]

[a] *College of Engineering, Design and Physical Sciences, Brunel University, UK*
[b] *Department of Structural Engineering, University of Naples "FedericoII", ITALY*

Abstract: The stability of thin circular shells under proportional and non-proportional loading, i.e. axial tensile stress and external pressure is extensively investigated. Two plasticity theories are considered; the flow and the deformation theory of plasticity. The results obtained confirm that under over-constrained kinematic assumptions the deformation theory tends to provides lower values of buckling pressure and the discrepancies in the results from the two plasticity theories increase with increasing thickness ratios, tensile stresses, boundary clamping and E/σ_y ratio.

1. Introduction

Plastic buckling of circular cylindrical shells has been the subject of active research for many decades due to its importance to the design of aerospace, submarine, offshore and civil engineering structures. It typically occurs in the case of moderately thick cylinders subjected to axial compression, external pressure, torsion or combinations of such loads.

Generally speaking, the plasticity models that have been proposed for metals in the strain hardening range can be divided into two groups: the 'deformation theory' of plasticity and the 'flow theory' of plasticity. In both of these theories the plastic deformations do not allow volume changes as plastic yielding is governed by the second invariant J_2 of the deviatoric part of the stress tensor, whereby in this respect they are both so-called J_2 theories. However, the deformation theory of plasticity is based on the assumption that for continued loading the state of stress is uniquely determined by the state of strain and, therefore, it is a special class of path-independent non-linear elasticity constitutive laws. According to this assumption, after a strain reversal, rather than recovering the initial elastic stiffness, as is found in physical tests, the initial loading curve is followed. On the other hand, the flow theory of plasticity assumes that an (infinitesimal) increment of stress is uniquely determined by the existing strain and its increment. This leads to a path-dependent relationship in which the current stress depends not only on the value of the current total strain but also on how the actual strain value has been reached, thus making the constitutive relationship path dependent.

There is a general agreement among engineers and researchers that the deformation theory of plasticity lacks physical rigour in comparison to the flow theory. However, use of the deformation theory predicts buckling loads that are less than corresponding loads obtained with the incremental theory, and evidence of comparison between measured and calculated buckling loads points in favour of deformation theory results. This is generally known as the "plate buckling paradox".

Onat and Drucker [1] first pointed out through an approximate analysis that buckling predictions based on the flow theory for long plates supported on three sides tend to those predicted by the deformation theory if small but unavoidable imperfections are taken into account. Restricting attention to the plastic buckling of circular cylindrical shells, Mao and Lu [2] analytically examined simply supported cylinders made of aluminium alloy subjected to axial compression load. They compared the buckling stresses predicted by their analytical formula with the experimental results conducted by Lee [3] and found that the deformation theory provides closer results with the tests while the flow theory significantly over-predicts the critical loads.

Blachut et al. [4] conducted experimental and numerical analyses of 30 mild-steel machined cylinders, of different dimensions, subject to axial tension and increasing external pressure. They showed that agreement between the buckling stresses calculated using the two theories was strongly dependent on the ratio of the length L of the cylindrical shell to its outer diameter D. For short cylinders ($L/D \leq 1$) the plastic buckling pressure predicted by flow or deformation theory coincided only when the tensile axial load vanished. By increasing the axial tensile load, the plastic buckling pressure calculated using the flow theory of plasticity quickly diverged from corresponding values calculated using the deformation theory, which were closer to the experimental values. For specimens with L/D ranging from 1.5 to 2 the results predicted by both theories were very similar for a certain range of combined loading, beyond which the values calculated using the flow theory began to deviate from the corresponding results using the deformation theory and became unrealistic in correspondence of large plastic strains.

Bardi and Kyriakides [5] tested fifteen cylindrical stainless steel tubes, with D/t ranging between 23 and 52, under axial compression and determined the critical stresses and strains at the onset of wrinkling. They reported the buckling modes, including the number and the size of waves. They also calculated the same quantities analytically using the deformation or the flow plasticity theory. The calculations included the effects of assuming both isotropic and anisotropic material behaviour. Bardi and Kyriakides concluded that the flow theory significantly over-predicts the critical stresses and strains while the deformation theory leads to critical stress and strain in better agreement with the experimental results. Moreover, the flow theory grossly over-predicted the wavelength of wrinkles while the deformation theory was in better agreement with the wavelengths measured in the tests.

On the basis of previous investigations led by the present authors [6-9], it is shown that the implicit kinematic constraint in assuming a certain buckling shape as the basis of the analysis seems to be the main reason for the discrepancy between the results from the flow and deformation theory of plasticity – a fact which has suggested the existence of a plastic buckling paradox – and from carefully validated geometrically nonlinear finite element (FE) modelling. In fact, it is shown that with an accurate modelling a very good agreement between numerical and experimental results can be obtained in the case of the physically sound flow theory of plasticity.

2. Constitutive relationships based on the deformation theory of plasticity

The deformation theory of plasticity is based on the assumption that for continued loading the state of stress is uniquely determined by the state of strain and, therefore, it is a special class of path-independent non-linear elasticity constitutive laws.

The deformation theory of plasticity is obtained by extending the Ramberg-Osgood law to the case of a multi-axial stress state using the von Mises formulation (J_2 theory) and results in the following path-independent relationship

$$\boldsymbol{\varepsilon} = (1+\nu)\, dev\,\boldsymbol{\sigma} - (1-2\nu)\, sph\,\boldsymbol{\sigma} + \frac{3}{2}\alpha \left(\frac{\sqrt{\frac{3}{2}}\, dev\,\boldsymbol{\sigma}}{\sigma_y} \right)^{n-1} dev\,\boldsymbol{\sigma} \qquad (1)$$

where $\boldsymbol{\varepsilon}$ and $\boldsymbol{\sigma}$ denote the strain and stress tensors, while dev $\boldsymbol{\sigma}$ and sph $\boldsymbol{\sigma}$ denote the deviatoric and spherical parts of the stress tensor, respectively. n is a constant that depends on the material being considered.

3. Constitutive relationships based on the flow theory of plasticity

The classical J_2 flow theory of plasticity, with nonlinear isotropic hardening and in the small-strain regime [14] is based on the additive decomposition of the spatial rate of the deformation tensor $\dot{\boldsymbol{\varepsilon}}$ into its elastic and plastic parts $\dot{\boldsymbol{\varepsilon}}_e$ and $\dot{\boldsymbol{\varepsilon}}_p$, respectively:

$$\dot{\boldsymbol{\varepsilon}} = \dot{\boldsymbol{\varepsilon}}_e + \dot{\boldsymbol{\varepsilon}}_p \qquad (1)$$

The remaining well-known governing equations are as follows:

$$\dot{\boldsymbol{\sigma}} = 2G\dot{\boldsymbol{\varepsilon}}_e + \lambda \,\mathrm{tr}\, \dot{\boldsymbol{\varepsilon}}_e \mathbf{I} \qquad (2)$$

$$\dot{\boldsymbol{\varepsilon}}_p = \dot{\eta} \left(\frac{\partial f}{\partial s} \right)_{s = dev\,\boldsymbol{\sigma}} \qquad (3)$$

$$f\left(\boldsymbol{\sigma}, \boldsymbol{\varepsilon}_p^{eq}\right) = \mathrm{dev}\boldsymbol{\sigma} - \sqrt{\frac{2}{3}}\,\bar{\boldsymbol{\sigma}}\left(\boldsymbol{\varepsilon}_p^{eq}\right) \qquad (4)$$

$$\dot{\eta} \geq 0, \qquad f\left(\boldsymbol{\sigma}, \boldsymbol{\varepsilon}_p^{eq}\right) \leq 0, \qquad \dot{\eta} f\left(\boldsymbol{\sigma}, \boldsymbol{\varepsilon}_p^{eq}\right) = 0 \qquad (5)$$

where G and λ denote the Lamé's elastic constants, $\mathbf{I}$ is the (rank-2) identity tensor, f is the yield function according to von Mises' yield criterion, η is a hardening parameter. Assuming nonlinear isotropic hardening, the current yield strength, $\bar{\boldsymbol{\sigma}}$ is an increasing function of the equivalent plastic strain ε_p^{eq}:

$$\varepsilon_p^{eq}(t) = \int_{-\infty}^{t} \left| \dot{\varepsilon}_p(\tau) \right| \, d\tau \qquad (6)$$

4. Stability analyses

4.1 Bifurcation analysis

The bifurcation analysis has been conducted by means of an analytical treatment developed ad-hoc on the basis of the approach presented in Chakrabarty [10], by the program BOSOR5 [11] and by the application of the Differential Quadrature Method [12]. The employment of these different approaches has allowed to cover a number of loading conditions, cylinder geometry and boundary conditions. All the treatments have been extended, when necessary, to cover both the use of deformation and flow theory of plasticity.

The bifurcation load in the plastic range and the corresponding buckling mode for axisymmetrically loaded shells is determined in BOSOR5 through a sequence of two consecutive analyses [11]. The first one is a nonlinear pre-buckling analysis which is valid for small strains and moderately large rotations and accounts for material nonlinearity. This nonlinear problem is solved using a strategy in which nested iteration loop are applied at each load level. The inner loop is used to analyse the nonlinear behaviour caused by the moderately large displacements using the Newton-Raphson method. The outer loop is used to evaluate the constitutive matrix and the plastic strain components, and to test loading and unloading condition in the material by means of a sub-incremental strategy. The results from this analysis are used in the subsequent analysis, which is an eigenvalue analysis which yields the bifurcation load and the corresponding axisymmetric or non-symmetric buckling mode, respectively. At the bifurcation load the infinitesimal displacement field, has components in the axial, circumferential and radial direction denoted as δu, δv and δw. They are assumed to vary harmonically around the circumference. It is important to notice that the discretisation in BOSOR5 is only performed in the meridian direction because the resulting displacements are axisymmetric in the pre-buckling phase and the buckling mode is assumed to vary harmonically in the circumferential direction in the bifurcation buckling analysis. BOSOR5, moreover, cannot handle any kind of boundary condition.

Since there are many practical cases of buckling of shells having various combinations of boundary conditions the Differential Quadrature Method (DQM) may offer some advantages over the analytical and BOSOR5 approaches and at the same time a clearer insight into the mechanics of the problem under analysis. In fact, the DQM leaves a certain freedom in dealing with the boundary conditions of the problem. The DQM is routinely employed to provide solutions to partial differential equations arising in various simplified models of fluid flow, diffusion of neutrons through homogeneous media and one-dimensional nonlinear transient heat diffusion and conduction problems. The DQM is an approximation method to calculate the *kth*-order derivative of the solution function $f(x)$ at a grid point i. Consider firstly one dimensional problem. The *kth*-order derivative of the function $f(x)$ is given by a linear weighting of the function values in the domain.

4.2 Nonlinear stability analysis

The nonlinear stability analysis of imperfect cylinders has been numerically simulated by means of the non-linear FE commercial package ABAQUS, version 6.11-1 [13] using both the flow and deformation theories of plasticity. Specific attention has been paid to adopt model parameters which, in the case of proportional monotonic (increasing) loading, result in the same stress-strain curve in both theories, to within a negligible numerical error.

The cylindrical specimens were modelled using a general purpose 4-noded shell element which has six degrees of freedom at each node. This element is named “S4” in the commercial software ABAQUS and is based on a thick shell theory. The shell formulation accounts for finite membrane strains, therefore this element can be used to perform large strain analyses. The element is widely used for industrial applications because it is suitable for both thin and thick shells. The S4 element uses a normal integration rule with four integration points. The enhanced-strains approach is employed to prevent shear and membrane locking. Among the ABAQUS elements, S4 outperforms S4R as the former evaluates more accurately the membrane strains, which plays a key role in the problem at hand.

4.3 Experimental data

As stated in the Introduction, Blachut et al. [4] conducted tests on 30 machined cylinders made of mild steel with outer diameter *34 mm* and length-diameter ratio (*L/D*) of 1.0, 1.5 and 2.0. In the experimental setting, one flange of the specimen was rigidly attached to the end flange of the pressure chamber and the other flange was bolted to a coupling device which in turn was bolted to the load cell, see Fig. 1.

In order to prevent any eccentricity of the axial load exerted on the specimen, the load cell was cantered with respect to the test chamber. The authors pointed out that the maximum initial radial imperfections measured at the mid-length of the specimens were about 1% of the thickness. Giezen et al. [14] tested cylindrical specimens of aluminium alloy 6061-T4. Two sets of specimens were tested, namely Set A and Set B. The average wall-thickness values of the first and second set were 0.76 and 0.71 mm, respectively, and the length-diameter ratio (*L/D*) was equal to one. The maximum initial imperfection was found to be about 0.076mm (10% of the thickness).

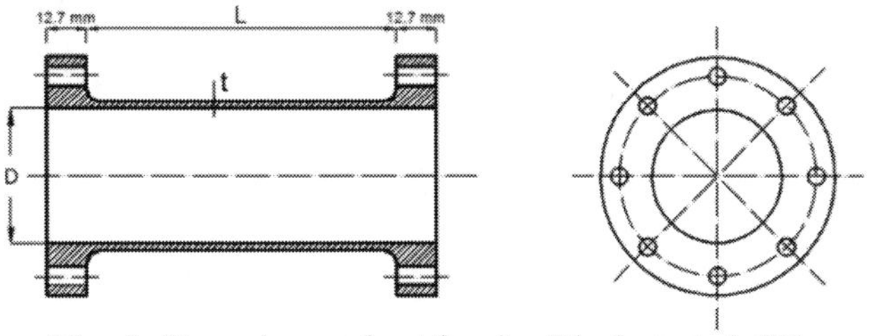

Fig. 1: Experimental setting by Blachut et al. [4]

5. Results and discussion

Table 1 collects the results from experimental tests and BOSOR5 numerical analyses. The results have been chosen to represent cases in which the flow theory of plasticity, according to BOSOR5, does not provide a buckling load or strongly overestimates the ones from tests, and cases in which there is agreement between the flow and deformation theory of plasticity.
Table 2, along with the analytical results, shows the results from non-linear FE analyses obtained by means of the commercial package ABAQUS.

Table 1: Experimental vs. BOSOR5 results (NA=Not Available)

Sp.	Experimental results			BOSOR5 results: Deformation theory		BOSOR5 results: Flow theory	
	Number of waves	Axial tension (N)	External pressure (MPa)	Number of waves	Buckling pressure (MPa)	Number of waves	Buckling pressure (MPa)
S1	NA	17960	4.07	NA	5.65	NA	NA
S2	NA	0	12.76	NA	13.29	NA	13.15
S5	NA	12010	8.28	NA	8.63	NA	NA
M2	NA	10670	8.14	NA	7.75	NA	NA
SP.1-Set B	4	0	5.26	5	5.98	5	6.22
SP.6-Set B	4	11771	3.00	5	3.32	4	6.20
SP.3-Set A	5	2341	6.27	5	6.25	4	6.49

Table 2: Numerical vs analytical results

Sp.	Numerical results (ABAQUS): Deformation Theory		Numerical results (ABAQUS): Flow theory		Analytical results: Deformation Theory		Analytical results: Flow Theory	
	Number of waves	Buckling pressure (MPa)	Number of waves	Buckling pressure (MPa)	Number of waves	Buckling pressure (MPa)	Number of waves	Buckling pressure (MPa)
S1	4	5.53	4	5.64	4	5.29	2	16.24
S2	6	13.14	6	13.15	4	13.24	4	13.28
S5	4	8.73	4	8.83	4	8.56	2	11.02
M2	3	7.84	3	7.87	4	7.75	4	7.91
SP.1-Set B	4	5.09	4	5.15	5	5.32	5	5.44
SP.6-Set B	4	2.91	4	3.22	4	2.75	3	5.36
SP.3-Set A	5	5.25	5	5.28	4	6.00	4	6.27

The analytical treatment, differently from BOSOR5, always provides a value of the buckling pressure, albeit sometimes very different from the experimental results. In fact, BOSOR5 routines do not point to a buckling load in the cases of specimens S1, S2, S5 and M2 tested by Blachut et al. [4]. The numerical analyses conducted by means of the non-linear FE code ABAQUS, instead, lead to the correct determination of the buckling loads, in accordance with the experimental results, both for the deformation and the flow theory of plasticity.

The main finding is that when the buckling modes coincide using either the deformation or the flow theory, i.e. in the case of specimens S2, M2, SP1-Set B and SP3-Set A, the buckling loads result the same and in line with the experimental and FE results. When the buckling modes do not coincide in the case of the deformation or of the flow theory of plasticity, then the buckling loads provided by the flow theory of plasticity result much higher than those provided by the deformation theory, see specimens S1, S5 and SP6-Set B.

By means of the DQM analyses it has also been possible to draw some deductions about the influence of the boundary conditions. It is concluded that, in presence of an overcostrained kinematics, the existence of axial, circumferential and rotational restrains at the edges of the cylinders with increasing the axial tensile stresses can significantly increase the discrepancies between the flow and deformation theories' results in the range 8%-260%.

Finally, in presence of an overcostrained kinematics Fig. 2 shows the Influence of *L/D* ratio on the lateral buckling pressure *q* according both to the flow and the deformation theory of plasticity. σ_t is the tensile axial load.

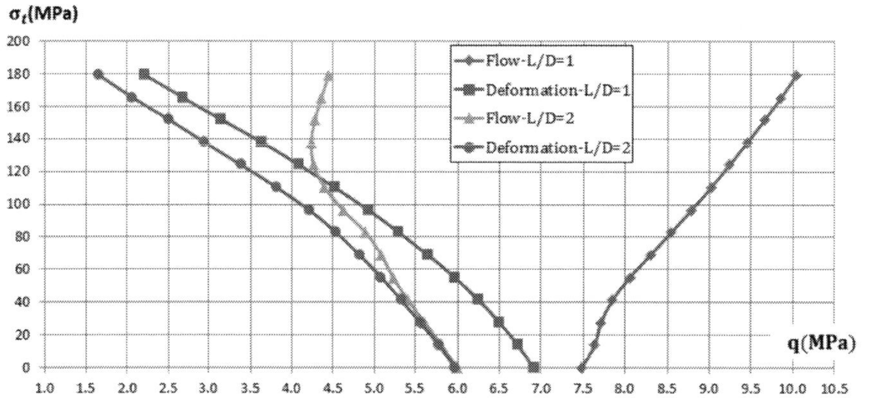

Fig. 2: Influence of *L/D* ratio on the lateral buckling pressure *q* according both to the flow and the deformation theory of plasticity. σ_t is the tensile axial load

6. Conclusions

The main findings of the study are:

1. the root of the apparent plastic buckling paradox in the elasto-plastic analysis of circular cylindrical shells is due to over-constrained kinematic assumptions. This fact leads to overestimate the buckling pressures when the flow theory of plasticity is used, while the deformation theory counterbalances the excessive kinematic stiffness and provides results which are much lower that the flow theory findings [6-9];
2. by conducting geometrically nonlinear finite-element analyses, the flow theory provides physically reliable results, which are in accordance with the deformation theory ones and with the experimental results;
3. implementing the flow theory of plasticity in the elastic-plastic bifurcation analysis may lead to overestimate the buckling pressures. The large discrepancies between flow and deformation theories results observed analytically vanish when using the flow theory in non-linear incremental analysis.

In conclusion, it is recommended to use a geometrically nonlinear finite-element formulation for imperfect shells with carefully determined and validated constitutive laws to avoid the discrepancies between two plasticity theories and to track accurate post-buckling curve in the case of the physically more sound flow theory of plasticity.

References

[1] Onat ET, Drucker DC. " Inelastic instability and incremental theories of plasticity", *Journal of the Aeronautical Sciences*, 20, 181-186, 1953.

[2] Mao R, Lu G. " A study of the plastic buckling of axially compressed cylindrical shells with a thick-shell theory", *International Journal of Mechanical Sciences*, 43(10), 2319-2330, 2010.

[3] Lee L. "Inelastic buckling of initially imperfect cylindrical shells subjected to axial compression", *Journal of Aerospace Science*, 29, 87-95, 1962.

[4] Blachut J, Galletly GD, James S. "On the plastic buckling paradox for cylindrical shells", *Proceedings of the Institution of Mechanical Engineers, Part C*, 210(5), 477-488, 1996.

[5] Bardi FC, Kyriakides S. "Plastic buckling of circular tubes under axial compression - Part I: Experiments", *International Journal of Mechanical Sciences*, 48(8), 830-841, 2006.

[6] Shamass R, Alfano G, Guarracino F. "A numerical investigation into the plastic buckling of circular cylinders", *Shell Structures: Theory and Applications, Vol.3 - Proceedings of the 10th SSTA Conference* (Eds.: W. Pietraszkiewicz, J. Gòrski), Taylor & Francis, 441-444, 2014.

[7] Shamass R, Alfano G, Guarracino F. "A numerical investigation into the plastic buckling paradox for circular cylindrical shells under axial compression", *Engineering Structures*, 75, 429-447, 2014.

[8] Shamass R, Alfano G, Guarracino F. "An analytical insight into the buckling paradox for circular cylindrical shells under axial and lateral loading", *Mathematical Problems in Engineering,* 514267, 1-11, 2015.

[9] Shamass R, Alfano G, Guarracino F. "An investigation into the plastic buckling paradox for circular cylindrical shells under non-proportional loading", *Thin-Walled Structures,* 95, 347-362, 2015.

[10] Chakrabarty J. *Applied Plasticity, 2nd edition*, New York: Springer, 2010.

[11] Bushnell D. "BOSOR5 – Program for Buckling of Complex, Branched Shells of Revolution Including Large Deflections, Plasticity and Creep", *Structural Analysis Systems – Vol. 2* (Ed.: Niku-Lari A.), Pergamon, 25-54, 1986.
[12] Bert CW, Malik M. "Differential quadrature method in computational mechanics: a review", *Applied Mechanics Reviews*, 49(1), 1-28, 1996.
[13] Simulia, "ABAQUS Theory Manual.Version 6.11-1", *Dassault Systems*, Waltham, MA, USA, 2011.
[14] Giezen JJ, Babcock CD and Singer J. "Plastic buckling of cylindrical shells under biaxial loading", *Experimental Mechanics*, 31(4), 337-343, 1991.

TRANSVERSAL LOAD INTRODUCTION INTO UNSTIFFENED GIRDER WEBS – PROPOSAL FOR A SIMPLIFIED EUROCODE–COMPATIBLE DESIGN CHECK

Harald Unterweger and Andreas Taras

Institute for Steel Structures, Graz University of Technology, Austria

Abstract: The introduction of concentrated transversal loads without stiffeners into webs of hot-rolled or welded beams and girders leads to economic solutions. Practical, simple design checks for this detail solution were thus included in national standards (e.g. DIN 18800), but can no longer be found directly in the Eurocode. In this paper the individual rules for transverse load introduction, found in EN 1993-1-8 and EN 1993-1-5, are compared and discussed. Finally, a design recommendation is presented, which is consistent with the different rules given in the Eurocode and which considers all the occurring main effects (buckling of the web, accompanying shear and bending).

1. Introduction

The transfer of concentrated transversal loads into the webs of hot-rolled or welded beams and girders without stiffeners is a common detail, particularly in building structures. Avoiding the stiffeners allows cost-efficient solutions, for example for joints between purlins and roof-girders.

Due to the frequent use of such details, simple design rules were thus included in national standards – e.g. in DIN 18800 [1], as shown in Fig. 1. In this standard, a constant load distribution shape was assumed, leading to an effective length l_y with constant normal stresses σ_z in vertical direction. This leads to the load carrying capacity $F_{z,R}$ given in Fig. 1 and Eq. (1), with web thickness t_w and yield strength f_y of the web. The application of Eq. (1) is limited to stocky webs, which are not susceptible to local buckling. This simple and efficient procedure, however, did not consider all relevant load-carrying mechanism correctly.

$$F_{z,R} = l_y \cdot t_w \cdot f_y \qquad (1)$$

In Eurocode 3, the "general rules" standard EN 1993-1-1 does not contain any design check for this detailing solution. In part EN 1993-1-5 [2], which deals with buckling of plates, rules are given for local buckling under transverse force for the different relevant cases in practice (see Fig. 6). Finally, some "hidden" design rules – i.e. not directly recognizable as treating the same problem for a practitioner – for the studied detail are available in EN 1993-1-8 [3], in the chapter for beam to column joints. The basic joint component "column web in transverse compression", also shown in Fig. 1, essentially covers the same type of load transfer problem.

In this paper, these two sets of design rules in the Eurocode are briefly compared and discussed (chapter 3). At first, the relevant load carrying mechanism at transversal load introduction points are summed up with all relevant interaction phenomena.

The main focus of this paper is the presentation of a design model for details with transversal load introduction into unstiffened girder webs, for all relevant cases in practice (Fig. 6), which is consistent with all different rules in the Eurocode. Although the proposal also covers slender webs with accompanying higher shear and bending are included, rather simple, practical design formulas are suggested.

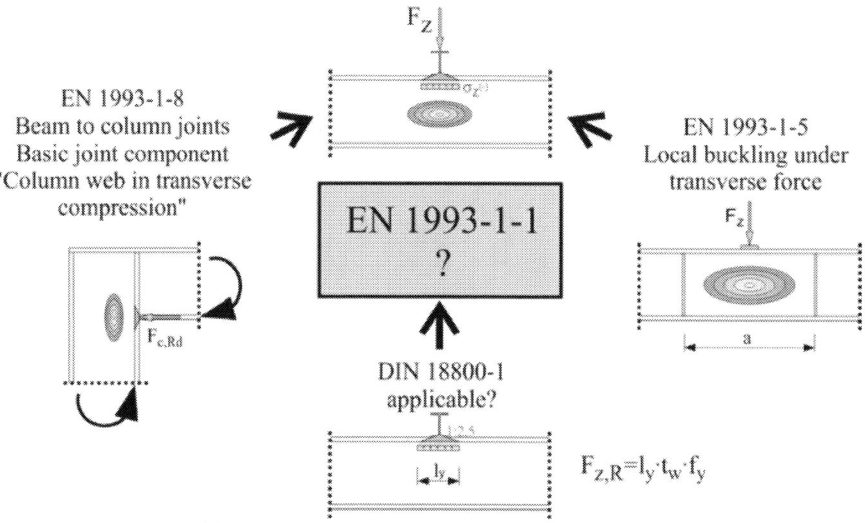

Fig. 1: Transversal load introduction design checks as treated in Eurocode 3

2. Overview of relevant load carrying mechanism for transversal load introduction

In Fig. 2, all the relevant effects (I-IV) are presented that should be considered in order to accurately describe the transverse load carrying capacity F_z:

1. Effect I: local vertical normal stresses σ_z, with maximum value at the web top (constant stress σ_z along effective length l_{eff})
2. Effect II: Local buckling due to those vertical normal stresses for slender webs, considering the individual cases (section 1, 2 in Fig. 2)
3. Effect III: Interaction with normal stresses σ_x in the longitudinal direction, due to girder bending, with the following differentiation:
   a) Both stresses (σ_x, σ_z) are in compression
   b) σ_z in compression and σ_x in tension
4. Effect IV: Interaction with shear stresses τ, due to shear forces V in the web

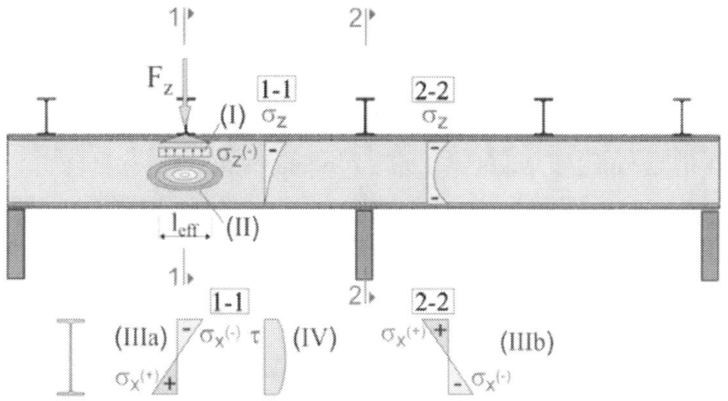

Fig. 2: Relevant effects (I-IV) present at transversal load introduction points

3. Comparison of design rules in the Eurocode for transversal load introduction

In Table 1 the different rules of EN 1993-1-8 [3] and EN 1993-1-5 [2] which consider the transversal load introduction into unstiffened girder webs are summarized. For better understanding, "new" symbols are used, because the given symbols in both parts of the Eurocode are different. Due to space limitations, only a short summary of the differences will be presented.

Table 1: Summary of the design rules for transversal load introduction cases in EN 1993-1-8 and EN 1993-1-5

Effect	"new" symbol	EN 1993-1-8	EN 1993-1-5
Transverse load capacity		$F_{z,R} = w \cdot k_w \cdot r_{p,z} \cdot l_y \cdot t_w \cdot f_y$	
Length of stiff bearing	s_s	smaller than in DIN 18800-1	
Effective loaded length	l_y	$s_s + 5 \cdot \left(t_f + \sqrt{2} \cdot a\right)$	depending on slenderness ($\bar{\lambda}_{p,z}$)
Relative plate slenderness	$\bar{\lambda}_{p,z}$	$\bar{\lambda}_{p,z} = 0{,}932 \sqrt{\frac{l_y \cdot h_w \cdot f_y}{E \cdot t_w^2}}$	$\bar{\lambda}_{p,z} = \frac{1}{\sqrt{0{,}9 \cdot k_F}} \cdot \sqrt{\frac{l_y \cdot h_w \cdot f_y}{E \cdot t_w^2}}$
Reduction factor for local buckling	$\rho_{p,z}$	$\bar{\lambda}_{p,z} \leq 0.72$: $\rho_{p,z} = 1.0$ $\bar{\lambda}_{p,z} > 0.72$: $\rho_{p,z} = \left(\bar{\lambda}_{p,z} - 0.2\right) / \bar{\lambda}_{p,z}^{\;2}$	$\bar{\lambda}_{p,z} \leq 0.50$: $\rho_{p,z} = 1.0$ $\bar{\lambda}_{p,z} > 0.50$: $\rho_{p,z} = \frac{0.5}{\bar{\lambda}_{p,z}} \leq 1.0$
Interaction F_z with shear	ω	$\omega = \sqrt{1 - 0.43 \cdot \left(\frac{V_{Ed}}{V_{Rd}}\right)^2}$	currently not available
Interaction F_z with longitudinal compression	k_w	$\sigma_{x,com} > 0.7 \cdot f_y$ $k_w = 1.7 - \lvert\sigma_{x,com}\rvert / f_y$ (else $k_w = 1.0$)	$\sigma_{x,com} > 0.5 \cdot f_y$ $k_w = 1.4 - \frac{0.8 \cdot \lvert\sigma_{x,com}\rvert}{f_y} \leq 1.0$ (else $k_w = 1.0$)
Interaction F_z with longitudinal tension	k_w	Not intended	von Mises

3.1 Length of stiff bearing s_s

In order to consider all cases in practice, the rules of EN 1993-1-5 will be used, see Fig. 7.

3.2 Effective loaded length l_y

In EN 1993-1-8, the load distribution slope is equal to the rules in [1]. In contrast to this, the rules of EN 1993-1-5 give higher values, also depending of the web slenderness and the ratio b_f / t_w (with flange width b_f). In Fig. 3, for all practical cases (see Fig. 6, case 1, 2, 1R) the differences are presented for a representative girder, with a varying web slenderness h_w / t_w.

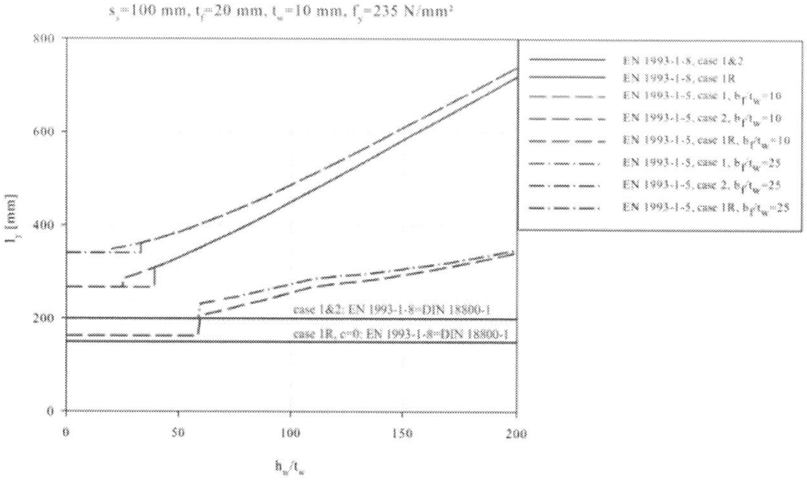

Fig. 3: Comparison of the effective length l_y in EN 1993-1-5 and EN 1993-1-8 for a representative girder and varying web slenderness

3.3 Plate slenderness and reduction factor for local buckling

In EN 1993-1-5, the ideal buckling load - and therefore also the relative plate slenderness - considers the different buckling cases (see Fig. 6) through individual buckling factors k_F. In EN 1993-1-8, a simplified, very conservative value is used (leading to k_F = 1,28 in all cases), and therefore a significant higher plate slenderness must be used. However, the reduction factor for local buckling in EN 1993-1-8 is significant higher, yet finally still always leads to more conservative results for the buckling effect, as will be shown later.

3.4 Interaction of F_z with shear

The reduction factor ω in EN 1993-1-8, which considers the interaction with shear, depends on a factor β, which shows the relationship between shear force V_{Ed} in the web and transverse force $(\beta = V_{Ed} / F_z)$.

It can be shown that the interaction between transverse force $F_{z,Ed}$ and shear force V_{Ed} is based on Eq. (2), with f = 1,3 / 3 = 0,43. The shear resistance V_{Rd} for the girder is based on Eq. (5). If the acting transversal force F_z is expressed as $F_{z,Ed} = \omega F_{z,Rd}$, rearranging of Eq. (2) gives the factor ω, based on the shear force V_{Ed}, see Table 1 and in Eq. (4).

$$\left(\frac{F_{z,Ed}}{F_{z,Rd}}\right)^2 + f\left(\frac{V_{Ed}}{V_{Rd}}\right)^2 = 1.0 \qquad (2)$$

3.4 Interaction of F_z with longitudinal stresses

The interaction with longitudinal compression stresses $\sigma_{x,com}$, expressed with the reduction factor k_w, is different in both parts of the Eurocode and is only relevant for high stresses σ_x.

The interaction with longitudinal tension stresses $\sigma_{x,ten}$ is only mentioned in EN 1993-1-5, by checking a von Mises equivalent stress.

4. Comparison of load carrying capacities F_z in the two Eurocode parts

In Figs. (4) and (5), the load carrying capacities F_z of EN 1993-1-5 [2] and EN 1993-1-8 [3], based on the individual procedures summarized in Table 1, are presented - without any interaction with shear and longitudinal stresses ($\omega = k_w = 1{,}0$).

The results are given for an individual transversal load inside the girder (case 1 in Fig. 4) and at the end of the girder (case 1R in Fig. 5). For a representative girder, the slenderness of the web (h_w / t_w) is varied. The ultimate load carrying capacity $F_{z,Rd}$ is related to the capacity of the pure web (effective length $l_y = s_s$).

It can be seen that the design rules of EN 1993-1-8 are always conservative in comparison with EN 1993-1-5 and coincide with the results based on DIN for stocky webs. The results based on EN 1993-1-5, depending also on the ratio b_f / t_w (with flange width b_f), are showing non-appropriate sudden increases of load carrying capacity if buckling starts (also mentioned in [4]).

Therefore in the future the calculation of the effective loaded length in EN 1993-1-5 should be improved, to eliminate this unfounded behaviour.

For case 2, not shown in the figures, the results of EN 1993-1-8 are also always on the safe side – the results lie in between the two presented cases.

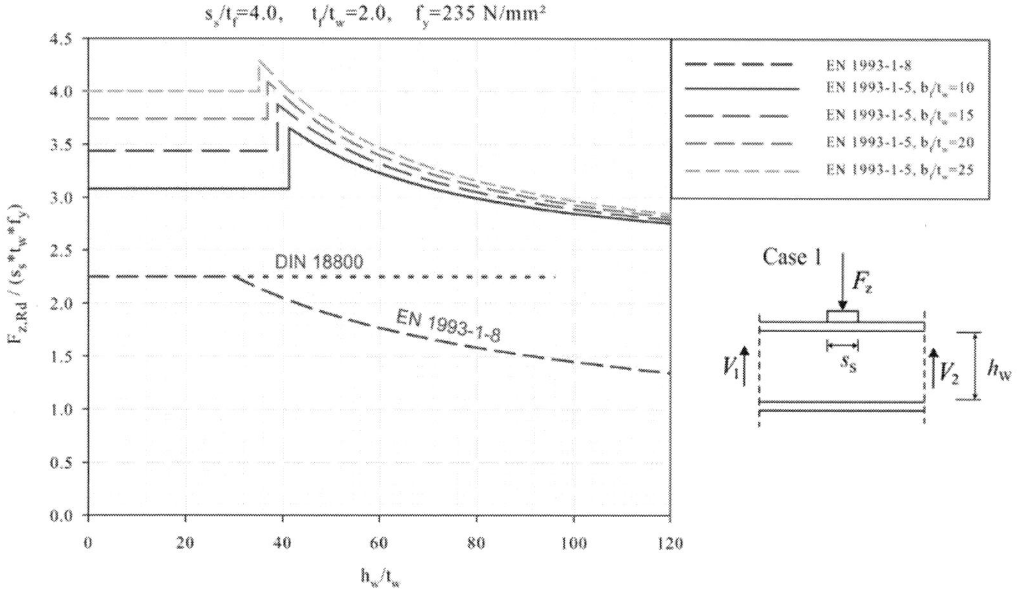

Fig. 4: Comparison of the resistance for transversal load introduction according to EN 1993-1-8 and EN 1993-1-5 for case 1 (with k_F = 6,0 for a very long panel)

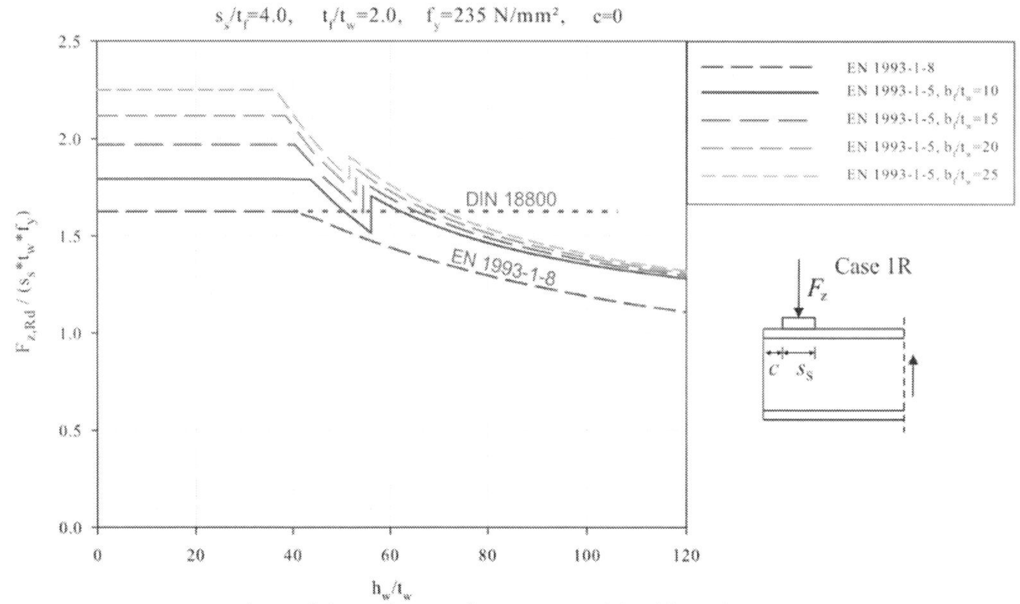

Fig. 5: Comparison of the resistance for transversal load introduction according to EN 1993-1-8 and EN 1993-1-5 for case 1R

5. Design check recommendation for all cases in practice

The recommended design procedure considers all different cases in practice, shown in Fig. 6. The case 2R should be avoided, due to very limited load carrying capacity, and is therefore restricted to very small transversal forces (P_z).

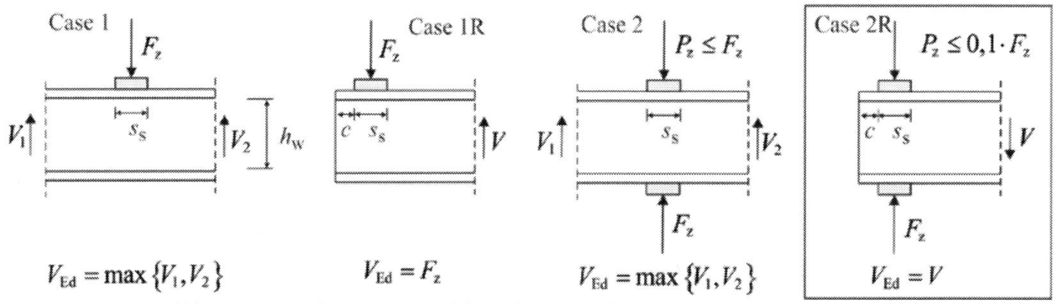

Fig. 6: Different cases for transversal load introduction into unstiffened girder webs and relevant shear forces V_{Ed}

The load carrying capacity $F_{z,Rd}$ is based on Eq. (3), with web thickness t_w and yield strength of the web f_y.

$$F_{z,Rd} = \frac{\omega \cdot k_w \cdot \chi_{Fz} \cdot l_y \cdot t_w \cdot f_y}{\gamma_M} \tag{3}$$

The calculation of the effective loading length l_y is based on Fig. (7).

The reduction factor ω for the interaction with shear V_{Ed} is based on Eq. (4), considering the shear capacity of the girder according to EN 1993-1-1 in Eq. (5).

$$\omega = \sqrt{1 - 0.43 \cdot \left(\frac{V_{Ed}}{V_{Rd}}\right)^2} \tag{4}$$

$$V_{Rd} = A_v \cdot f_y / \left(\sqrt{3} \cdot \gamma_{M0}\right) \tag{5}$$

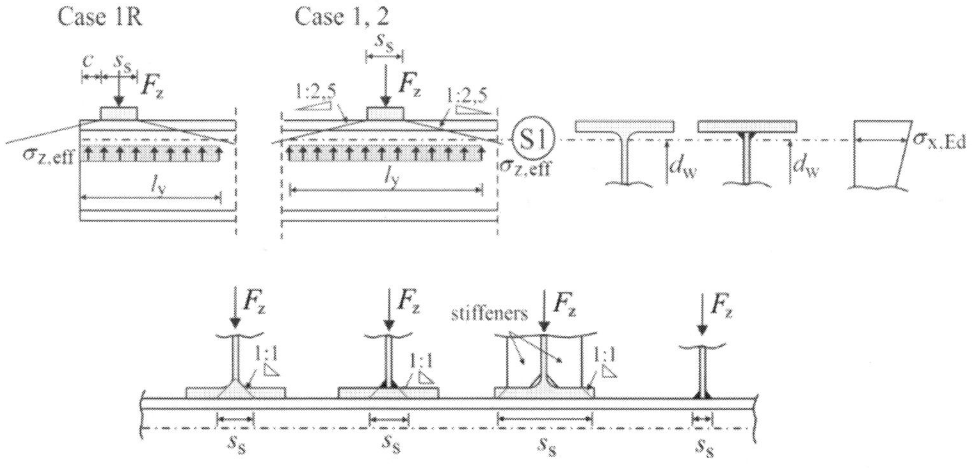

Fig. 7: Calculation of the effective loading length l_y and definition of the web depth d_w

The reduction factor k_w – only relevant if the longitudinal normal stress $\sigma_{x,Ed}$ is in compression (Fig. 7) and $|\sigma_{x,Ed}| / f_y > 0.7$ – is based on EN 1993-1-8, with Eq. (6).

$$k_w = 1.7 - |\sigma_{x,Ed}| / f_y \leq 1.0 \tag{6}$$

Plate buckling is only relevant ($\chi_{Fz} < 1.0$) if the slenderness (d_w/t_w) of the web increases the limit value of Eq. (7), with web depth d_w based on Fig. (7) and factor ε based on Eq. (10).

$$\left(d_w / t_w\right)_{limit} = 23.1 \cdot \varepsilon \cdot \sqrt{d_w / l_y} \tag{7}$$

Then the buckling reduction factor χ_{Fz} is based on Eq. (8).

$$\chi_{Fz} = \frac{\bar{\lambda}_{p,z} - 0.2}{\bar{\lambda}_{p,z}^2} \text{ for } \bar{\lambda}_{p,z} > 0.72 \tag{8}$$

$$\bar{\lambda}_{p,z} = 0.932 \sqrt{\frac{l_y \cdot d_w \cdot f_y}{E \cdot t_w^2}} = \frac{d_w}{t_w} \frac{\sqrt{l_y / d_w}}{32.1 \cdot \varepsilon} \tag{9}$$

$$\varepsilon = \sqrt{235 / f_y} \tag{10}$$

Applicable partial safety factor γ_M:

If $\chi_{Fz} < 1{,}0$; then: $\gamma_M = \gamma_{M1}$

Else: $\gamma_M = \gamma_{M0}$

In case of longitudinal stresses σ_x in tension an additional check based on the von Mises equivalent stress σ_{vw} in the web is suggested:

$$\sigma_{vw} = \sqrt{\sigma_{x,Ed}^2 + \sigma_{z,eff}^2 - \sigma_{x,Ed} \cdot \sigma_{z,eff} + 3\tau_w^2} \leq \frac{f_y}{\gamma_{M0}} \tag{11}$$

$$\sigma_{z,eff} = F_z / \left(l_y \cdot t_w\right) \tag{12}$$

6. Conclusions

In this paper, the transversal load introduction into unstiffened girder webs was analysed and the different rules for this topic of Eurocode 3, found in EN 1993-1-5 and EN 1993-1-8, were summed up and compared. Based on that, a Eurocode-compatible, simple design procedure for all relevant cases in practice was recommended, summed up in Fig. 6. The main conclusions are:

1. The load carrying capacity $F_{z,R}$ in EN 1993-1-5 should be improved, to eliminate the jump up of the load capacity, if the slenderness of the web is increased (see Figs. 4, 5)
2. The simplified load carrying capacity $F_{z,R}$ in EN 1993-1-8 for slender webs, susceptible to local buckling, gives conservative results for all different cases in Fig. 6, when compared with EN 1993-1-5
3. The suggested design procedure is based on the design rules in EN 1993-1-8 with a modified, more generally applicable reduction factor ω for the interaction with shear

References

[1] DIN 18800-1 (2008), *Stahlbauten – Teil 1: Bemessung und Konstruktion*, DIN – Deutsches Institut für Normung, Berlin

[2] EN 1993-1-5, *Eurocode 3: Design of steel structures – Part 1-5: Plated structural elements.*

[3] EN 1993-1-8 + AC (2009), *Eurocode 3: Design of steel structures – Part 1-8: Design of joints.*

[4] Johansson B, Maquoi R, Sedlacek G, Müller C, Bed D. *Commentary and Worked Examples to EN 1993-1-5 Plated Structural Elements*, JRC – ECCS, Brussel, 2007.

ELEPHANT'S FOOT BUCKLING AND IMPERFECTION SENSITIVITY OF INTERNALLY PRESSURIZED THIN-WALLED METAL CYLINDERS SUBJECTED TO BENDING

Maria Vathi and Spyros A. Karamanos

Department of Mechanical Engineering, University of Thessaly, Volos, Greece

Abstract: Upright cylindrical steel storage tanks, subjected to lateral loading, may fail in the form of local buckling near their bottom, due to the simultaneous action of excessive axial compression in the presence of internal pressure, a failure mode known as "elephant's foot buckling". The present paper is aimed at examining this failure mode rigorously, considering material and geometric nonlinearities, accounting for initial imperfections for pressure levels up to 80% of yield pressure. The accuracy of simplified axisymmetric models in modelling this instability mode is also examined. Finally, the numerical results are compared with the predictions from the relevant provisions of EN 1993-1-6 and EN 1998-4.

1. Introduction

Thin-walled bottom-supported steel shells of upright cylindrical shape are subject to internal pressure from stored materials. When subjected to severe lateral loading (e.g. seismic loading), overturning is resisted by axial compressive stresses in the wall. Under this combined loading condition, the shell may exhibit local buckling in the form of a localized outward bulge near its bottom (Fig. 1), which is associated with loss of lateral resistance and shell failure. The biaxial stress state at the critical area of buckling formation (axial compression and circumferential tension) is the key issue for understanding the buckling mechanism. In particular, the presence of internal pressure has a significant effect on the buckling strength. Low values of pressure may have a beneficial effect on shell capacity, whereas high pressure values may result in local yielding near the shell bottom, and precipitate the formation of an elastic-plastic buckling failure [1]. This buckling failure mode is commonly known as "elephant's foot buckling" and has been observed in quite a few post-earthquake inspections of liquid storage tanks. It has been also recognized as a major design issue for storage shells, such as bins, silos and tanks.

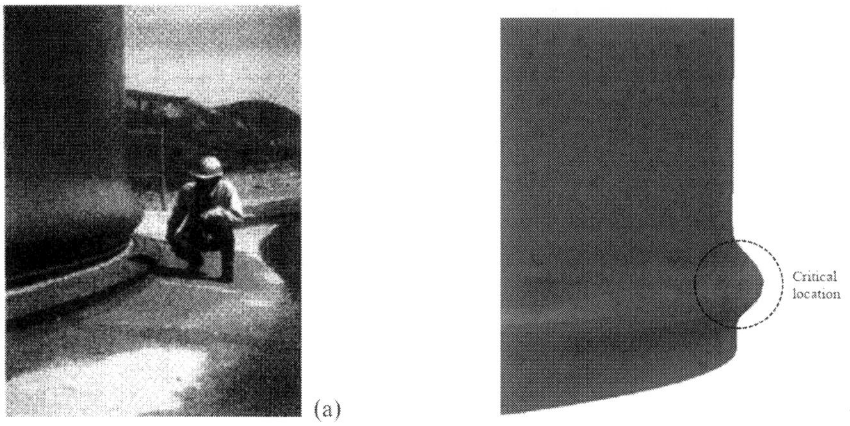

(a) (b)

Fig. 1: Elephant's foot buckling in liquid storage tanks (a) existing tank (Photo: Courtesy of University of California, Berkeley) [1], (b) buckling shape from present finite element analysis

Despite several early works on buckling of internally pressurized cylinders [2][3][4], "elephant's foot buckling" has received relatively little attention. The first systematic study on this subject has been reported in [5], using finite element axisymmetric models. This work resulted in the proposal of a semi-empirical equation, which has been adopted by several design standards and recommendations such as the New Zealand Seismic Tank Design Recommendations [6] and Earthquake Standard NZS 2654 [7]. Furthermore, this equation has been the basis for the recent European standard EN 1993-1-6 for shell design [8], and the Seismic Eurocode for tanks EN 1998-4 [9]. A recent experimental study on buckling of internally-pressurized cylinders has been reported in [10], whereas for a thorough presentation of this subject, in relation with the provisions of EN 1993-1-6, the reader is referred to the paper by Rotter [11].

The present study focuses on the analysis and design of liquid storage tanks under lateral loading, motivated by their damage in severe seismic events. A typical steel shell is considered with height-to-diameter ratio equal to about 1. The shell is considered fixed at its base, and subjected to lateral load, at a certain height, assuming a loading set-up used in the experiments of Mathon & Limam [10]. Different levels of internal pressure are considered; hydrostatic pressure is considered first, whereas constant pressure at levels up to 80% of nominal yield pressure is also considered. In section 2, the numerical finite element model is described and, in section 3, numerical results are obtained for different levels of pressure. In this section, imperfection sensitivity of the buckling strength is examined, whereas the gradual development of local stresses and strains at the buckling (bulging) zone is monitored throughout the shell deformation history. The accuracy of predictions from simplified axisymmetric models is also examined in section 4, whereas the numerical results are compared with the design provisions of EN 1993-1-6 and EN 1998-4 in section 5.

2. Description of the numerical model

The model is developed in ABAQUS/Standard and employs shell elements for the fixed-bottom cylindrical shell. The shell is discretized with four-node reduced-integration shell finite elements, which account for geometric and material nonlinearities. The element size at the critical location in the longitudinal direction of the cylinder is 0.8% of the tank radius. The model is inspired by the experimental set-up described in [10], shown in Fig. 2(a) and Fig. 2(b); the cylinder is bottom-fixed, capped at its top (using a kinematic coupling of the top nodes with a central "reference point") and lateral loading is applied at the top of a fictitious rigid rod, connected with the "reference point" of the top cross-section. Inelastic deformation of the steel material is taken into account through a J_2 flow plasticity constitutive model.

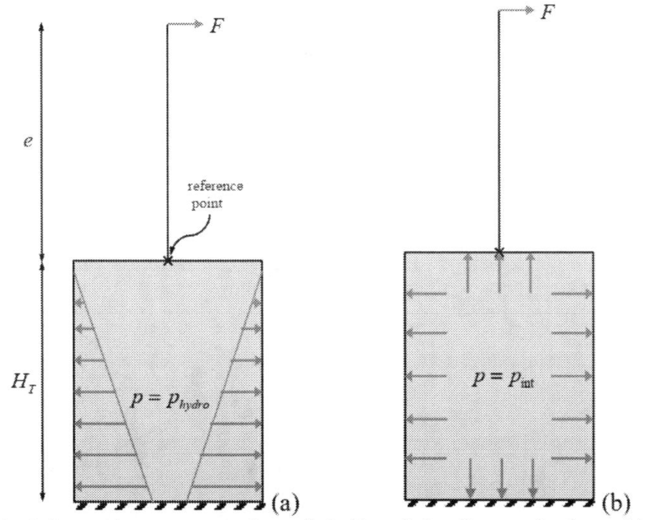

Fig. 2: Schematic representation of shell tank loading under lateral load; (a) hydrostatic pressure (b) internal constant pressure

Initial imperfections are considered in the form of initial bulging for the analysis of imperfect shells. These initial imperfect shapes are obtained from the post-buckling shape of the perfect cylinder, and they are scaled appropriately to obtain the desired imperfection amplitude. In the following the imperfection amplitude w_0 is expressed in terms of shell thickness t.

The analysis is conducted in two steps. In the first step, internal pressure is raised up to a certain prescribed level [Fig. 2(a) and Fig. 2(b)]. Subsequently, keeping the pressure constant, the lateral load at the top of the rigid rod is gradually increased until shell failure in terms of local buckling occurs. The latter step is performed through a Riks continuation algorithm, to overcome snap-back features of the equilibrium path.

3. Numerical results

Results are presented for a fixed bottom cylindrical shell, with diameter, D, equal to 6.3 meters, and constant thickness equal to 6 mm. The shell height, H_T, is 7.4 meters, with liquid fill height, H, equal to 6.9 meters. The shell is completely fixed at its bottom (clamped at its base) and the shell material is mild steel with yield stress 235 MPa. The top section of the shell is considered undeformed and the center is connected to a fictitious stiff rod where lateral loading is applied. Three different rod lengths are chosen; $e = 0$, $e = H_T$ and $e = 5H_T$. Both hydrostatic pressure and constant internal pressure up to 80% of the nominal yield pressure are considered. In the latter case, since the tank is closed (capped), pressure on the top section is taken into account.

The buckling shape is shown in Fig. 1(b) and the consecutive stages of elephant's foot buckling gradual formation in the finite element model are depicted in Fig. 3. The moment-displacement curves for the case of hydrostatic pressure are shown in Fig. 4(a) for different values of initial imperfection amplitude. The moment reported in those graphs corresponds to its value at the tank bottom, whereas the corresponding displacement is the horizontal displacement of the "reference point" at the top of the tank. The sensitivity on the maximum moment sustained by the cylinder in terms of the imperfection amplitude is shown in Fig.

4(b), showing that the height of lateral load may not be a critical factor for the value of maximum moment.

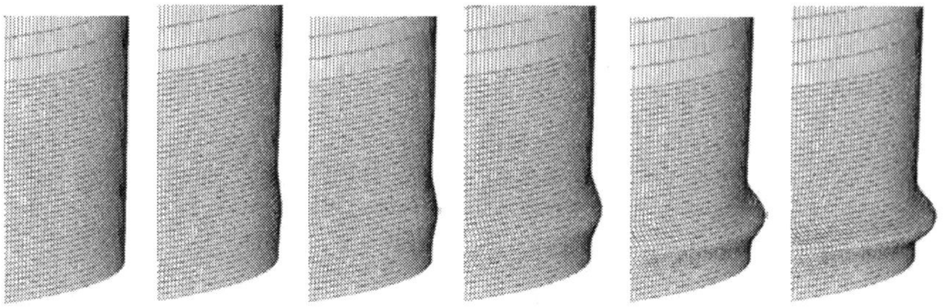

Fig. 3: Consecutive stages of elephant's foot buckling formation in the finite element simulation.

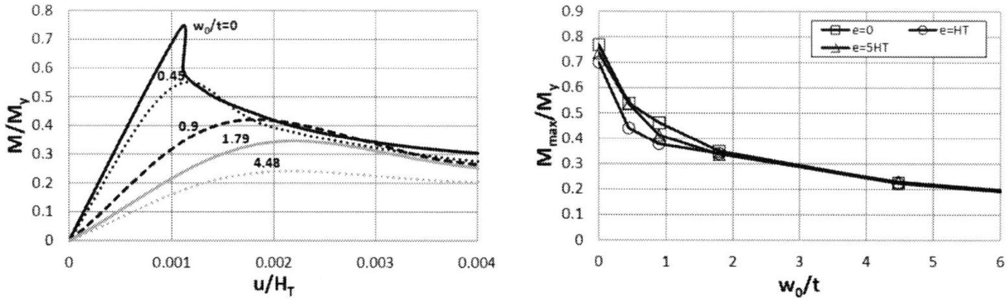

(a) (b)

Fig. 4: (a) Normalized moment-displacement curves for e=5H_T and hydrostatic pressure, (b) normalized maximum moment-imperfection curves for the case of hydrostatic pressure for different lengths of the rigid rod

The evolution of longitudinal (axial) and hoop stresses and strains at the critical location of the buckled area are shown in Fig. 5 and Fig. 6 respectively, in terms of the maximum bulging (outward) displacement of the buckled area, for the case of the initially imperfect cylinder with e=5H_T and w_0/t=0.45. The values of stress in Fig. 5 show that the membrane stresses are quite low with respect to the stresses developed at the inner and outer surface of the shell. The results show that the stresses and strains are larger on the outer surface of the buckled area.

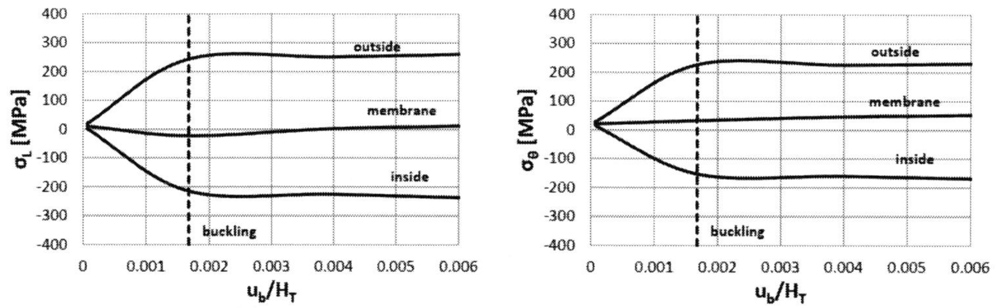

(a) (b)

Fig. 5: (a) Longitudinal and (b) hoop stresses at the critical location of the buckled area in terms of the normalised maximum bulging (outward) displacement of the buckled area

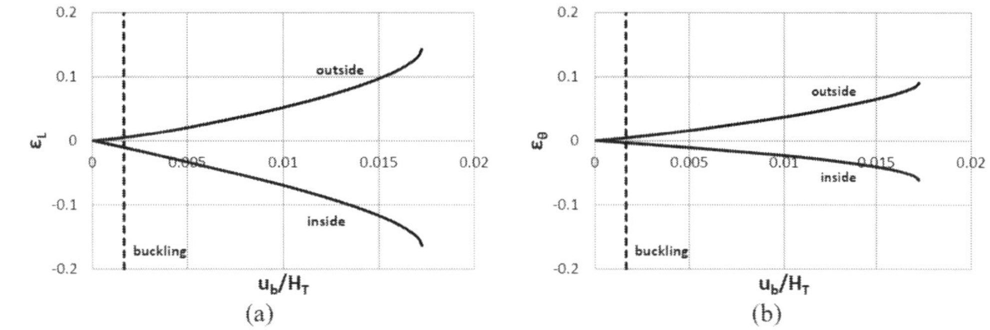

Fig. 6: (a) Longitudinal and (b) hoop strains at the critical location of the buckled area in terms of the normalised maximum bulging (outward) displacement of the buckled area

The cylindrical shell under consideration is also analysed under lateral load, in the presence of internal pressure value of 40% of yield pressure including capped effects on the top. The base moment-lateral displacement diagram is shown in Fig. 7(a), whereas the maximum moment is plotted in terms of the imperfection amplitude in Fig. 7(b). Similar analyses of the tank for different levels of internal pressure show that the presence of a moderate pressure level up to 80% may have a small effect on the lateral resistance of the cylindrical shell. In particular, the maximum overturning moment drops from 85% to 76% of the yielding moment for pressure increase from 15% to 80% of the yielding pressure. The above observations are in good correlation with the experimental results presented in [10], and the discussion offered in the paper by Rotter [11].

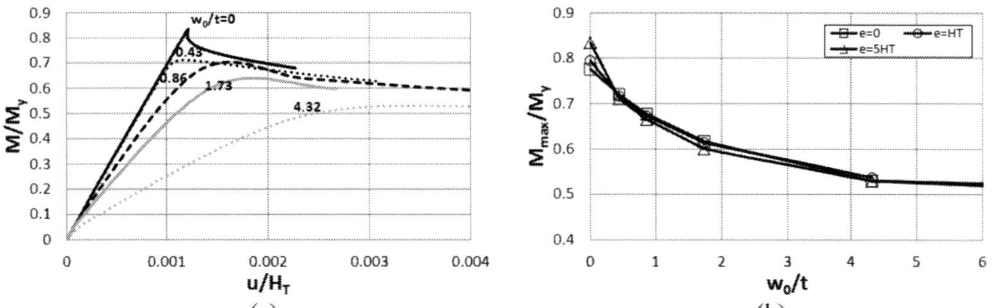

Fig. 7: (a) Normalized moment-displacement curves for $e=5H_T$ and pressure value equal to 40% of yield pressure, (b) normalized maximum moment-imperfection curves for the case of internal pressure value of 40% of yield pressure for different lengths of the rigid rod

4. On the use of axisymmetric shell models

In addition to the three-dimensional models described above, the possibility of using axisymmetric models for buckling analysis of pressurized cylinders under lateral loading is also examined. The shape of the elephant's foot buckle extends over a significant circumferential sector of the cylinder on either side of the most-stressed generator, so that the assumption of axisymmetric conditions seems quite reasonable.

The axisymmetric model employed is shown in Fig. 8. The length of the axisymmetric strip, H_{ax}, is chosen significantly larger than the length of the elephant's foot buckling pattern, obtained in the 3D analyses, and smaller than the total height of the tank for simplification. The bottom end of the model is fixed (clamped), whereas the top end is pinned.

Pressure is applied in the inner surface first and then kept constant, whereas axial loading is distributed in several nodes along the model height, according to the results of the 3D analyses, so that a representative distribution of axial compression along the shell generator exists.

The results from the axisymmetric model are shown in Fig. 9, in terms of the moment-displacement curve. The value of moment is computed by multiplying the membrane force with the bending resistance of the horizontal circular section of the tank (equal to πR^2, where R is the radius of the tank).

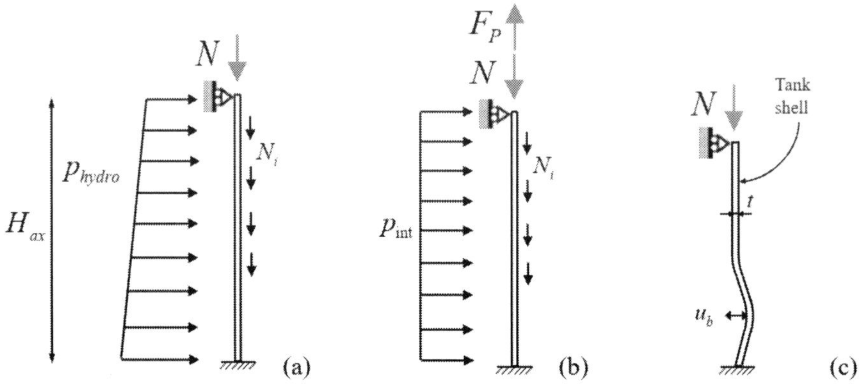

Fig. 8: Schematic representation of the axisymmetric model: (a) buckling analysis under hydrostatic pressure, (b) buckling analysis under internal constant pressure, (c) deformed shape

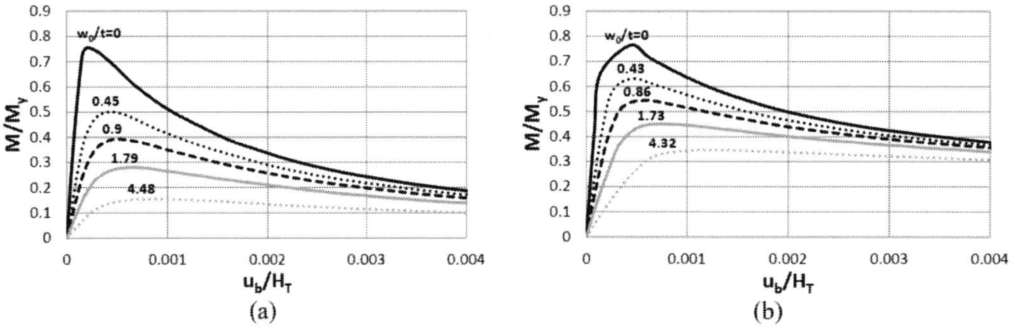

Fig. 9: Normalized moment-displacement curves obtained from the axisymmetric model for different imperfection amplitudes (a) in the case of the hydrostatic pressure (b) in the case of internal pressure equal to 40% of the yield pressure

The results of the axisymmetric model are compared with the ones of the three-dimensional model in Fig. 10. The comparison shows that there exists a good correlation between the two models for small imperfection amplitudes, despite the fact that the axisymmetric analysis results are rather conservative. It is the authors' opinion that axisymmetric models can be used for a simplified model analysis as they provide a reasonable level of accuracy.

5. Comparison with EN design provisions for steel shell design

The results obtained numerically with the three-dimensional model are compared with the provisions of EN 1993-1-6, paragraph 8.5.2 and Annex D (D.1.5) and those of EN 1998-4, paragraph A.10.3 in Fig. 11. To achieve a fair comparison, the imperfection amplitude in EN 1993-1-6 (Eqn. D.15), is taken equal to the value considered in the numerical analyses. The comparison shows good correlation for small imperfection amplitudes. However, this comparison is out of the imperfection range of EN 1993-1-6 (larger than the specified "poor fabrication quality" value shown in Fig. 11 as EN 1993-1-6 limit) and this explains the difference between numerical results and EN 1993-1-6 predictions. Furthermore, the provisions of EN 1998-4 appear to be somewhat conservative for small values of imperfection, but may not provide reliable predictions for imperfection amplitudes of moderate and large values.

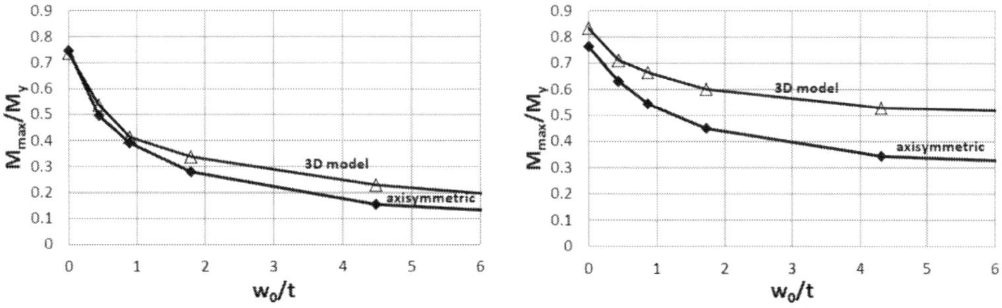

(a) (b)

Fig. 10: Normalized maximum moment in terms of imperfection for the 3D and the axisymmetric model in the case of (a) hydrostatic pressure (b) internal pressure equal to 40% of yield pressure

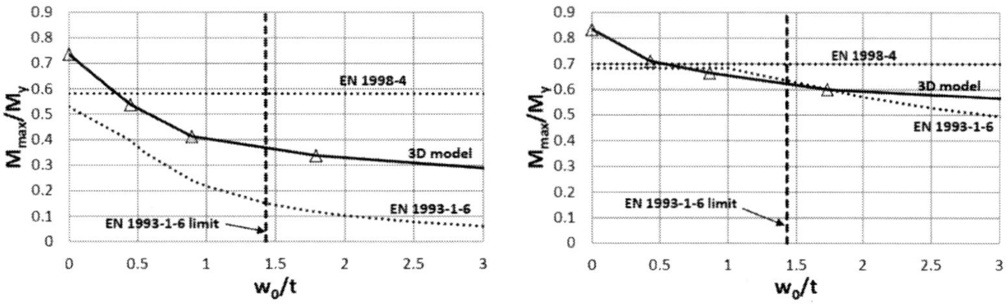

(a) (b)

Fig. 11: Normalized maximum moment in terms of imperfection for three-dimensional analysis and the relevant provisions of EN 1993-1-6 and EN 1998-4 in the case of (a) hydrostatic pressure (b) internal pressure equal to 40% of yield pressure

6. Conclusions

In the present paper, elephant's foot buckling has been examined rigorously using three-dimensional nonlinear finite element tools, in terms of buckling and post-buckling, considering material and geometric nonlinearities, accounting for initial imperfections for hydrostatic pressure and for internal pressure levels up to 80% of yield pressure. The accuracy of simplified axisymmetric models in modelling this instability mode provided fairly good results. Special emphasis has been given on the development of local stresses and strains at the buckled area, in the pre-buckling and the post-buckling configuration. Finally, the numerical results have been compared with the predictions from the relevant provisions of EN 1993-1-6 and EN 1998-4. The comparison showed that the EN 1993-1-6 specification is quite conservative in predicting the bending moment capacity. On the other hand, the empirical equation of EN 1998-4 may not provide reliable predictions of buckling strength in terms of imperfection amplitude.

Acknowledgments

This research has been co - financed by the European Union (European Social Fund – ESF) and Greek national funds through the Operational Program "Education and Lifelong Learning" of the National Strategic Reference Framework (NSRF) - Research Funding Program: THALES. Investing in knowledge society through the European Social Fund. The authors would also like to thank Prof. J. Michael Rotter, University of Edinburgh, for the fruitful discussions on "elephant's foot buckling" during the ECCS TWG 8.4 meetings.

References

[1] Malhotra PK. "Practical Nonlinear Seismic Analysis of Tanks", *Earthquake Spectra*, 16 (2), 473-492, 2000.

[2] Harris LA, Suer HS, Skene WT, Benjamin RJ. "The Stability of Thin-Walled Unstiffened Circular Cylinders under Axial Compression including the Effects of Internal Pressure", *Journal of the Aeronautical Sciences*, 24 (8), 587–596, 1957.

[3] Weingarten VI, Morgan EJ, Seide P. "Elastic Stability of Thin-Walled Cylindrical and Conical Shells under Combined Internal Pressure and Axial Compression", *AIAA Journal*, 3 (6), 1118–1125, 1965.

[4] Hutchinson JW. "Axial Buckling of Pressurised Imperfect Cylindrical Shells", *AIAA Journal*, 3, 1461–1466, 1965.

[5] Rotter JM. "Local Inelastic Collapse of Pressurised Thin Cylindrical Steel Shells under Axial Compression", *Journal of Structural Engineering*, ASCE, 116 (7), 1955–1970, 1990.

[6] Priestley MJN, Davidson BJ, Honey GD, Hopkins DC, Martin RJ, Ramsay G, Vessey JV, Wood JH. *Seismic Design of Storage Tanks*, Recommendations of a Study Group, New Zealand National Society for Earthquake Engineering, 1986.

[7] NZS 2654, *Specification for manufacture of vertical steel welded non-refrigerated storage tanks with butt-welded shells for the petroleum industry*, Standards New Zealand, Wellington, New Zealand, 1989.

[8] EN 1993-1-6, *Eurocode 3: Design of steel structures, Part 1.6: Strength and stability of shell structures*, CEN, Brussels, 2006.

[9] EN 1998-4, *Eurocode 8: Design provisions for earthquake resistance of structures-Part 4: Silos, tanks and pipelines*, CEN, Brussels, 2006.

[10] Mathon C, Limam A. "Experimental collapse of thin cylindrical shells submitted to internal pressure and pure bending", *Thin-Walled Structures*, 44, 39-50, 2006.

[11] Rotter JM. "Elephant's Foot Buckling in Pressurised Cylindrical Shells", *Stahlbau*, 75 (9), 742-747, 2006.

DESIGN REQUIREMENTS FOR INTERMEDIATE RING STIFFENERS ON DISCRETELY SUPPORTED CYLINDRICAL STEEL SHELLS

Özer Zeybek[a], Cem Topkaya[a] and J. Michael Rotter[b]

[a] *Department of Civil Engineering, Middle East Technical University, Ankara, Turkey*
[b] *Institute for Infrastructure and Environment, University of Edinburgh, Scotland, UK*

Abstract: One way of reducing the non-uniformity of axial membrane stresses in discretely supported cylindrical metal silos is to use a combination of a flexible ring beam with an intermediate ring stiffener. Previous work done by the authors has identified the ideal location and the design requirements for the intermediate ring placed at this location. This paper extends the findings to the design of intermediate ring stiffeners placed at locations below the ideal height. The membrane theory of shells and Vlasov's curved beam theory are used to derive closed form expressions that define strength and stiffness requirements for the intermediate ring stiffeners. The details of formulations and their verification are presented herein.

1. Introduction

Discrete supports in cylindrical metal silos introduce local forces into the shell and produce a circumferential non-uniformity in the axial membrane stresses in the silo, which must be taken into account in assessing the stability of the shell. One alternative method of achieving uniform stresses is to use an intermediate ring stiffener as shown in Fig. 1. Greiner [1], Öry et al. [2] and Öry and Reimerdes [3] showed that an intermediate ring stiffener can be very effective in reducing the circumferential non-uniformity of axial stresses in the shell. Studies conducted by these researchers identified the variation of the axial stress distributions up the height of the shell. It was shown that an intermediate ring stiffener can cause a dramatic decrease in the axial stress, producing a more uniform stress state above the intermediate ring. Recently

Topkaya and Rotter [4] showed that there is an ideal location for an intermediate ring stiffener, such that the axial membrane stress above this ring is circumferentially completely uniform. The ideal location (H_I) shown in Fig. 1 was determined analytically and is expressed in terms of basic geometric variables [4].

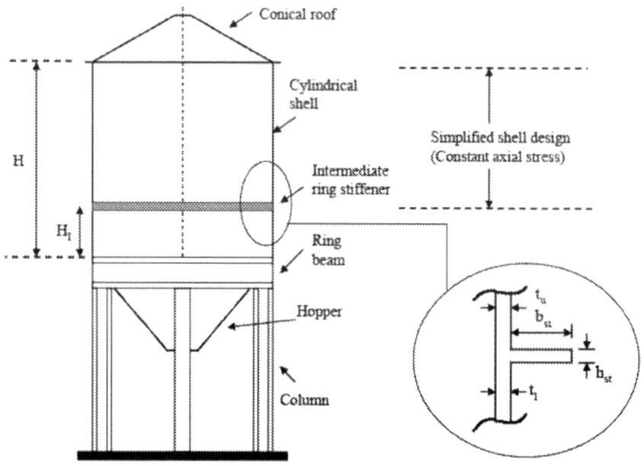

Fig. 1: Typical circular planform silo

The intermediate ring stiffener is expected to have sufficient strength and circumferential stiffness to fulfill its function properly. The key requirement of this intermediate ring stiffener is to prevent or significantly control the circumferential displacements of the cylindrical shell at that level. If the ring stiffener has inadequate stiffness then the circumferential uniformity of the axial stresses above is not achieved. Furthermore, there is an interaction between the cylindrical shell and the ring stiffener which causes stress resultants to develop in the ring. These stress resultants could potentially cause failure of the ring stiffener either by yielding or by instability. To overcome this failure, the authors [5] have developed a design criterion for intermediate ring stiffeners at ideal location for discretely supported cylindrical shells.

In cases where a shell with a large radius rests on a few supports, the ideal location can be quite high and the option of placing the intermediate ring stiffener below the ideal height may provide a viable solution. This paper explores the strength and stiffness requirements for intermediate ring stiffeners placed below the ideal location. In accordance with this goal, the cylindrical shell below the intermediate ring stiffener is analyzed using the membrane theory of shells and the reactions produced by the stiffener on the shell are identified. These reactions are then applied to the intermediate ring stiffener. Vlasov's curved beam theory is used to derive closed form expressions for the circumferential variation of the stress resultants to obtain a strength design criterion for the stiffener. A stiffness criterion is then developed by examining the ratio of the circumferential stiffness of the intermediate ring stiffener to that of the cylindrical shell. These analytical studies are then validated by complementary finite element analyses to identify the requirements for practical design.

2. Stress transfer into intermediate ring stiffeners

Topkaya and Rotter [4] determined the ideal location for an intermediate ring stiffener using the membrane theory of shells [6]. The cylindrical shell is separated into two parts: an upper shell and a lower shell of height, H_L, with the intermediate ring at their junction, as shown in Fig. 2. The fundamental harmonic term of the column support force is deemed sufficient to study the requirements of the ring stiffener, so the support force is represented by

$$P_x = P_{xn} \cos n\theta \tag{1}$$

where P_x = external distributed axial line load applied to the base of the shell; P_{xn} = Fourier coefficient for the nth harmonic of axial line load; n = number of uniformly spaced column supports; and θ= circumferential coordinate.

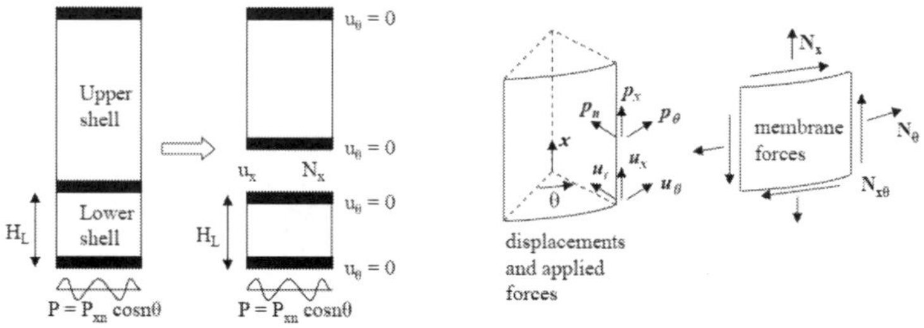

Fig. 2: Boundary conditions used in closed form solution and loading, displacements, and stress resultants in an element of the cylindrical shell

Considering the cylindrical shell element shown in Fig. 2, the equilibrium equations are:

$$\frac{\partial N_x}{\partial x} + \frac{1}{r}\frac{\partial N_{x\theta}}{\partial \theta} + p_x = 0 \tag{2}$$

$$\frac{\partial N_{x\theta}}{\partial x} + \frac{1}{r}\frac{\partial N_{\theta}}{\partial \theta} + p_\theta = 0 \tag{3}$$

$$N_\theta + r\, p_n = 0 \tag{4}$$

where r = middle surface radius; N_x, N_θ, $N_{x\theta}$ = axial, circumferential and shear membrane stress resultants respectively; and p_x, p_θ, p_n = external distributed pressures in the axial, circumferential and radial directions respectively.

The strain displacement relationships can be written as:

$$\varepsilon_x = \frac{\partial u_x}{\partial x} = \frac{1}{Et}\left(N_x - \nu N_\theta\right) \tag{5}$$

$$\varepsilon_\theta = \frac{1}{r}\frac{\partial u_\theta}{\partial \theta} - \frac{u_r}{r} = \frac{1}{Et}\left(N_\theta - \nu N_x\right) \tag{6}$$

$$\gamma_{x\theta} = \frac{1}{r}\frac{\partial u_x}{\partial \theta} + \frac{\partial u_\theta}{\partial x} = \frac{1}{Gt}N_{x\theta} = \frac{2(1+\nu)}{Et}N_{x\theta} \tag{7}$$

where u_x, u_θ, u_r = displacements in the axial, circumferential and radial directions respectively; ε_x, ε_θ = strains in the axial and circumferential directions respectively; $\gamma_{x\theta}$ = shear strain; ν = 0.3 = Poisson's ratio; E = modulus of elasticity; G = shear modulus; and t = thickness of the cylindrical shell.

The solution of governing differential equations for the boundary conditions indicated in Fig. 2 gives the following equation for the axial displacements of the lower shell segment [4]:

$$u_x = \frac{P_{xn}\cos n\theta}{Et}\left[\frac{x^2}{2H_L} - x + \frac{H_L}{3} + \frac{2(1+\nu)r^2}{H_L n^2}\right] \tag{8}$$

Imposing the condition of $u_x = 0$ at $x = H_L = H_I$, leads to the expression for the ideal height of an intermediate ring (H_I):

$$H_I = \sqrt{12(1+\nu)}\frac{r}{n} \tag{9}$$

The reactions in the shell at the boundary between the lower and upper segments can be treated as the loading exerted on the intermediate ring. Solution of differential equations yields the following expression for the shear membrane stress resultant ($N_{x\theta}$)

$$N_{x\theta} = -\frac{r}{H_L n} P_{xn} \sin n\theta \tag{10}$$

The axial membrane stress resultant vanishes at the ideal height but is non-zero for heights below the ideal height. The effect of axial membrane stress resultant on the intermediate ring stiffener will be treated separately in the following parts of the paper using the axial displacements as a measure. Eqn. 10 indicates that the intermediate ring is subjected to a sinusoidally varying circumferential line load whose magnitude is directly related to the discrete support force P_{xn}.

3. Algebraic closed-form solution of stress resultants in the intermediate ring stiffener – Strength criterion

The Vlasov curved beam differential equations [7-8] were used to study the response of the intermediate ring stiffener. The equilibrium equations were first derived for the curved beam element shown in Fig. 3, where three orthogonal internal forces and three internal moments develop at each cross-section. The six basic equilibrium equations can be expressed as follows:

$$\frac{1}{r}\left[\frac{dQ_r}{d\theta} + Q_\theta\right] + q_r = 0 \qquad \frac{1}{r}\frac{dQ_x}{d\theta} + q_x = 0 \tag{11}$$

$$\frac{1}{r}\left[\frac{dQ_\theta}{d\theta} - Q_r\right] + q_\theta = 0 \qquad \frac{1}{r}\left[\frac{dM_r}{d\theta} + T_\theta\right] - Q_x + m_r = 0 \tag{12}$$

$$\frac{1}{r}\frac{dM_x}{d\theta} + m_x + Q_r = 0 \qquad \frac{1}{r}\left[\frac{dT_\theta}{d\theta} - M_r\right] + m_\theta = 0 \tag{13}$$

where r = radius of the ring beam centroid; M_r = bending moment in the ring about a radial axis; M_x = bending moment in the ring about a transverse axis; T_θ = torsional moment in the ring; q_x, q_θ, q_r = distributed line loads per unit length in the transverse, circumferential and radial directions respectively; m_x, m_θ, m_r = distributed applied torques per unit circumference about the transverse, circumferential and radial directions respectively; Q_θ = circumferential tensile force in the ring; Q_x, Q_r = shear forces in the ring in transverse and radial directions respectively.

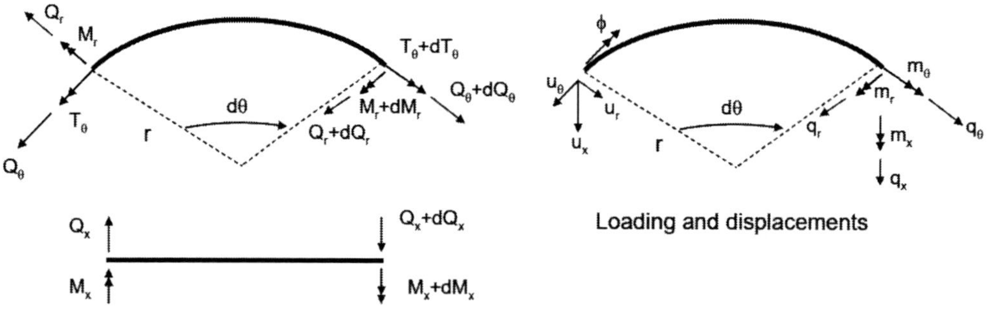

Fig. 3: Differential curved beam element and sign conventions

The following expressions define the force-deformation relationships:

$$Q_\theta = \frac{EA}{r}\left(\frac{du_\theta}{d\theta} - u_r\right) \qquad M_r = \frac{-EI_r}{r}\left(\frac{d^2u_x}{rd\theta^2} - \phi\right) \tag{14}$$

$$M_x = \frac{EI_x}{r^2}\left(\frac{d^2u_r}{d\theta^2} + u_r\right) \qquad T_\theta = -\frac{EI_w}{r^3}\left(\frac{d^3\phi}{d\theta^3} + \frac{1}{r}\frac{d^3u_x}{d\theta^3}\right) + \frac{GK_T}{r}\left(\frac{d\phi}{d\theta} + \frac{1}{r}\frac{du_x}{d\theta}\right) \tag{15}$$

where u_x, u_θ, u_r = displacements in the vertical, circumferential, and radial directions respectively as shown in Fig. 3; ϕ = rotation as shown in Fig. 3; I_r, I_x = bending moment of inertia of the ring stiffener about radial and vertical axes respectively; I_w = warping constant of the ring stiffener; K_T = uniform torsional constant of the ring stiffener; A = cross sectional area of the ring stiffener; and G = shear modulus.

Eqns. 11 through 13 can be solved for the loading condition defined in Eqn. 10 (i.e. $q_\theta = N_{x\theta}$) to arrive at the following relationships:

$$M_r(\theta) = 0 \qquad T_\theta(\theta) = 0 \qquad Q_x(\theta) = 0 \tag{16}$$

$$M_x(\theta) = \frac{P_{xn}}{H_L}\frac{r^3}{n^2(n^2-1)}\cos n\theta \tag{17}$$

$$Q_r(\theta) = \frac{P_{xn}}{H_L}\frac{r^2}{n(n^2-1)}\sin n\theta \tag{18}$$

$$Q_\theta(\theta) = -\frac{P_{xn}}{H_L}\frac{r^2}{(n^2-1)}\cos n\theta \tag{19}$$

The axial displacements and axial stress resultants vanish at the level of intermediate ring stiffener when this stiffener is placed at the ideal height defined in Eqn. 9. When the ring stiffener is placed below the ideal height, however, the axial displacements are non-zero and presence of these deformation demands should be considered in the design of the intermediate ring stiffener. The findings of $M_r(\theta) = 0$, $T_\theta(\theta) = 0$, $Q_x(\theta) = 0$ are valid for an intermediate ring placed at the ideal height. For other locations the bending moment in the ring stiffener about a radial axis can be derived from the axial displacements of the lower shell. The fundamental assumption adopted here is that the shell is very stiff in its own plane and the stiffness of the intermediate ring stiffener is negligible when compared with the stiffness of the shell. Therefore, axial displacements at H=H_L can be imposed as a displacement condition on the intermediate ring stiffener. According to Eqn. 8 the axial displacements at $H = H_L$ can be expressed as follows:

$$u_x = \frac{P_{xn}\cos n\theta}{6EtH_L n^2}\left[12(1+\nu)r^2 - H_L^2 n^2\right] \tag{20}$$

By inserting Eqn. 20 into Eq. 14 and considering that $\phi = 0$ the following relationship can be derived:

$$M_r(\theta) = \frac{EI_r}{r^2}\frac{d^2u_x}{d\theta^2} = \frac{-I_r P_{xn}\cos n\theta}{6H_L r^2 t}\left[12(1+\nu)r^2 - H_L^2 n^2\right] \tag{21}$$

4. Computational verification of the above equations

The accuracy of the equations derived for the stress resultants was verified using numerical analysis. The silo structure analyzed here rests on 4 supports and had a cylinder radius of r = 3000 mm and a height of H = 15000 mm. An intermediate ring stiffener was placed below the ideal height ($H_L = H_I/2$ = 1500 mm where $H_I \approx$ 3000mm based on Eqn. 9) and the modelling technique explained in Zeybek *et al.* [5] was adopted. A lower shell thickness of 8 mm was considered. Two different upper shell thicknesses were considered to study the effect of the stiffness of upper shell. In order to minimize the effects of the upper shell segment its thickness was considered as 1 mm in the first set of analysis. In the second set the upper shell thickness was considered equal to the lower shell thickness of 8 mm. The commercial finite element program, ANSYS [9], was used to perform the numerical analysis. The loading at the bottom of the shell was chosen to give a maximum axial membrane stress of 1.0 GPa above the support based on Eqn. 1. The calculated variations of M_x, M_r, Q_r, and Q_θ around the circumference are shown in Fig. 4, together with the predictions of the closed form solutions (Eqns. 17, 18, 19, and 21). The comparisons show that when the upper shell thickness is considered as 1 mm, the above equations provide acceptably accurate solutions, with the largest differences being 2.30%, 6.87%, 2.59%, and 4.86% for maximum bending moment in the transverse axis, bending moment in the radial axis, shear force and circumferential force respectively. When the upper shell thickness is considered as 8 mm, the above equations provide conservative solutions, with the largest differences being 17.9%, 39.6%, 18.2%, and 22.2% for maximum bending moment in the transverse axis, bending moment in the radial axis, shear force and circumferential force respectively.

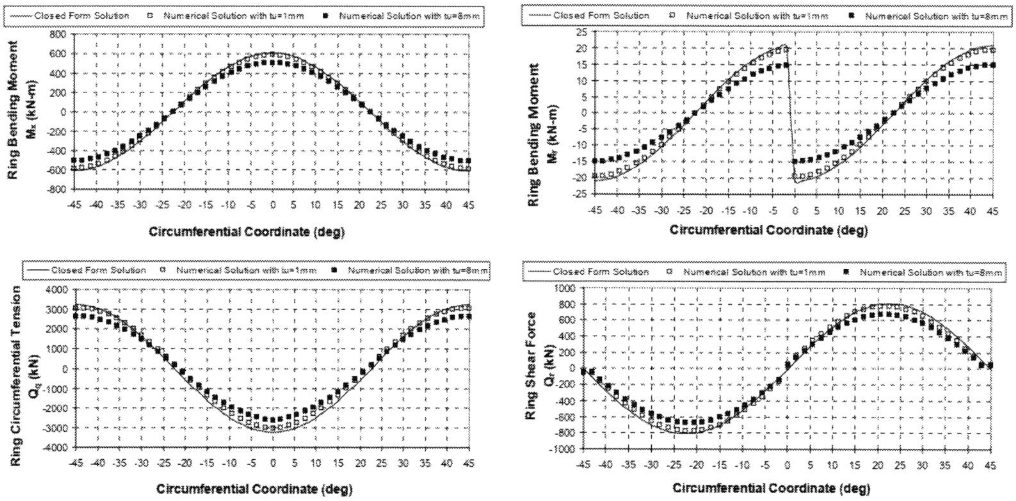

Fig. 4: Comparison of closed form solution with numerical solutions for intermediate ring stiffener

5. Stiffness criterion for the intermediate ring stiffener

The intermediate ring stiffener must have adequate circumferential stiffness to fulfil its function properly. A stiffness criterion was developed by comparing the relative stiffnesses of the intermediate ring and the shell. The intermediate ring stiffness ratio (χ) is expressed here as:

$$\chi = \frac{K_{shell}}{K_{stiffener}} = \frac{N_{x\theta}/u_{\theta,shell}}{q_\theta/u_{\theta,stiffener}} = \frac{u_{\theta,stiffener}}{u_{\theta,shell}} \qquad (22)$$

where $K_{stiffener}$, K_{shell} = circumferential stiffnesses of the intermediate ring stiffener and shell respectively; $u_{\theta,stiffener}$, $u_{\theta,shell}$ = circumferential displacements of the intermediate ring stiffener and shell respectively. The loading condition of Eqn. 10 was adopted to compute the stiffnesses.

The loading and boundary conditions given in Fig. 5 were used to derive circumferential displacements of the shell which can be expressed as follows:

$$u_{\theta(shell)} = -\frac{H_L^2 n}{2Etr} P_{xn} \sin n\theta \tag{23}$$

The circumferential displacements of the intermediate ring stiffener under the same loading condition (i.e. $q_\theta = N_{x\theta}$) can be obtained by solving Vlasov's curved beam differential equations as follows:

$$u_{\theta(stiffener)} = \frac{-P_{xn}}{H_L} \frac{r^3\left(Ar^2 + I_x n^4 - I_x n^2\right)}{AEI_x n^3\left(n^2-1\right)^2} \sin n\theta \tag{24}$$

$$N_{x\theta} = -\frac{r}{H_L n} P_{xn} \sin n\theta \qquad u_x = \frac{P_{xn} \cos n\theta}{6EtH_L n^2}\left[12(1+\nu)r^2 - H_L^2 n^2\right]$$

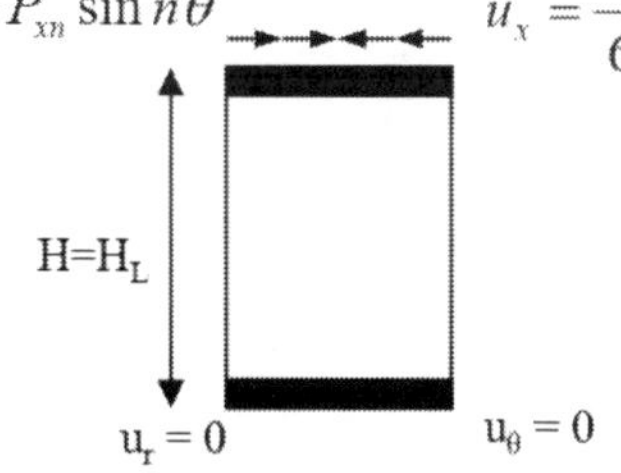

Fig. 5: Boundary conditions used for closed form solution of shell stiffness

Finally the intermediate ring stiffness ratio χ can be expressed as:

$$\chi = \frac{2r^4 t\left(Ar^2 + I_x n^4 - I_x n^2\right)}{AI_x H_L^3 n^4 \left(n^2-1\right)^2} \tag{25}$$

A finite element parametric study was conducted to obtain a practical limit for the intermediate ring stiffness ratio χ. Pursuant to this goal, combinations of shells and intermediate ring stiffeners were analyzed under the fundamental harmonic of the column support. The finite element models described by Zeybek et al. [5] were adopted, with geometries defined as radius $r = 1500$, 3000 and 4500 mm, thickness $t = 4$, 6 and 10 mm, number of supports $n = 4$, with all shells having a height $H = 20000$ mm.

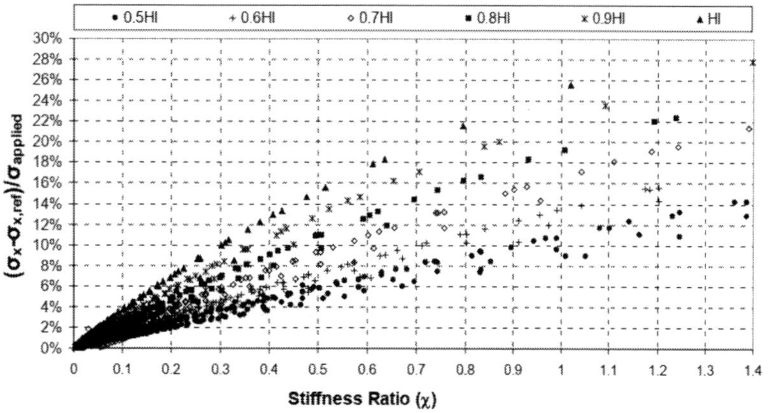

Fig. 6: Assessment of an appropriate value for the intermediate ring stiffness ratio χ

A total of 9 combinations were adopted for the shell geometries with 30 different intermediate ring stiffener dimensions. For simplicity, the intermediate ring stiffener was assumed to have a rectangular cross section with a constant height (h_{st}=20 mm) and a width (b_{st}) varying between 275 and 1000 mm. In each analysis, the axial membrane stress (σ_x) in the shell above the intermediate ring stiffener was recorded. Intermediate ring heights of $0.5H_I$, $0.6H_I$, $0.7H_I$, $0.8H_I$, $0.9H_I$, $1.0H_I$ were considered. A reference axial membrane stress ($\sigma_{x,ref}$) was determined by replacing the intermediate ring with idealized rigid boundary conditions. The differences between the stresses (σ_x-$\sigma_{x,ref}$) normalized by the applied stress at the bottom of the shell ($\sigma_{applied}$ = 1 GPa) are plotted in Fig. 6 as a function of the stiffness ratio χ.

When the stiffness ratio is lower than about 0.2, the differences in axial membrane stresses above the intermediate stiffener is observed to drop to below 6% of the applied value. It should be noted that the effective shell width is not considered in calculation of cross sectional area (A) and bending moment of inertia about the vertical axis (I_x) of the ring stiffener.

6. Conclusions

The forces transferred to the intermediate ring stiffener from the shell were obtained using the membrane theory of shells. The intermediate ring stiffener was analyzed under this loading using Vlasov's curved beam differential equations. Expressions for axial force, shear force and bending moment in the axial and radial direction of the stiffener were devised as a function of the geometrical properties and applied loading.

A stiffness criterion has been developed, based on the relative stiffnesses of the ring stiffener and the cylindrical shell under the harmonic force transfer between the shell and the stiffener. The resulting intermediate ring stiffness ratio (χ) depends on the geometrical properties of both the shell and the stiffener. A complementary finite element parametric study was conducted to determine practical limits for the intermediate ring stiffness ratio. The results indicate that ratios below about $\chi < 0.2$ provide a satisfactory uniform axial membrane stress distribution above the intermediate ring stiffener, so this limit is recommended for practical design.

References

[1] Greiner R., "Zur Ingenieurmässigen Berechnung und Konstruktion zylindrischer Behälter aus Stahl unter allgemeiner Belastung", (Analysis and Construction of Cylindrical Steel Cylinders under non axisymmetric Loading). *Proc. "Wissenschaft und Praxis"*, Vol. 31 FHS Biberach/Riss, StahlbauSeminar 1983, 51 pp., 1983.

[2] Öry H, Reimerdes HG Tritsch W. "Beitrag zur Bemessung der Schalen von Metallsilos" *Stahlbau,* 8, 243-248, 1984.

[3] Öry H. Reimerdes HG. "Stresses in and Stability of Thin Walled Shells under Non-ideal Load Distribution", *Proc. Int. Colloq. Stability Plate & Shell Structs*, Gent, ECCS, 555-561, 1987.

[4] Topkaya C, Rotter JM. "Ideal location of intermediate ring stiffeners on discretely supported cylindrical shells", *ASCE Journal of Engineering Mechanics*, 140 04013001-1, 04013001-10, 2014.

[5] Zeybek Ö, Topkaya C and Rotter JM. "Strength and stiffness requirements for intermediate ring stiffeners on discretely supported cylindrical shells", *Thin Walled Structures*, 96, 64-74, 2015.

[6] Ventsel E, Krauthammer T. *Thin plates and shells: theory, analysis*, and applications. Marcel Dekker, NY, 2001.

[7] Vlasov VZ. *Thin-walled elastic beams*. National Science Foundation, Washington, DC, 1961.

[8] Heins CP. *Bending and torsional design in structural members*. Lexington Books, Lexington, Massachusetts, 1975.

[9] ANSYS, Version 12.1 On-Line User's manual, 2010.

Robustness

T-STUBS RESPONSE TO EXTREME LOADING

Ioan Both[a], Raul Zaharia[a], Ioan Mărginean[a], Calin Neagu[a], Florea Dinu[a,b] and Dan Dubina[a,b]

[a] *Department of Steel Structures and Structural Mechanics, Politehnica University of Timisoara, Romania*
[b] *Laboratory of Steel Structures, Romanian Academy, Timisoara Branch, Romania*

Abstract: The behaviour of steel structures is highly influenced by the connections between elements. In multi-storey steel frame structures using bolted end plate connections, the design of connections is based on the component method, using equivalent T-stubs. The ductility of the connection is related to the ability of the T-stubs in developing large deformations in the tension zone. Besides the normal static loading conditions, robustness of structures considers extreme loadings related to the strain rate aspect. The paper presents the influence of the strain rate and of the elevated temperatures on the resistance and ductility of T-stubs, based on experimental results.

1. Introduction

Structural robustness can be enhanced by continuity, redundancy and ductility [1]. The beam-to-column joints and their capability to withstand significant forces and deformations that may appear in the structure, represent a decisive factor in the development of alternate load path (AP). The AP method, with its emphasis on continuity and ductility, is similar to current seismic design practice [2]. However, there are specific problems which need to be considered when key elements, particularly columns, are lost, i.e. development of the catenary forces in the beams and admissibility criteria to be considered in the design considering the interaction between axial loads and bending moments. It is therefore of interest to study if today design procedures [3] can provide enough robustness for connections under extreme loading conditions.

A large experimental program on connection components, joints and sub-assemblies is in progress within the framework of the CODEC research program at the Politehnica University of Timisoara [4]. Connection components include bolted T-stubs, tested under cvasi-static and high strain rate loading, at normal and elevated temperatures. The geometry of the T-stubs was selected such that the failure is characterised by mode 1 and mode 2 defined by EN1993-

1-8 [3]. The results presented in this paper were obtained for the T-stubs with mode 1 type of failure at normal temperature.

2. Test setup

The T-stubs were tested in the laboratory of Steel Structures and Structural Mechanics of the Politehnica University of Timisoara, using Instron dynamic universal testing machine of 1000 kN. The specimens subjected to elevated temperature were introduced in an environmental chamber therefore the web of the specimens had an increased length. Each test was performed by connecting two identical T-stubs using two bolts M16, grade 10.9 as presented in Fig. 1 a). The generic dimensions of a T-stub are presented in Fig. 1b).

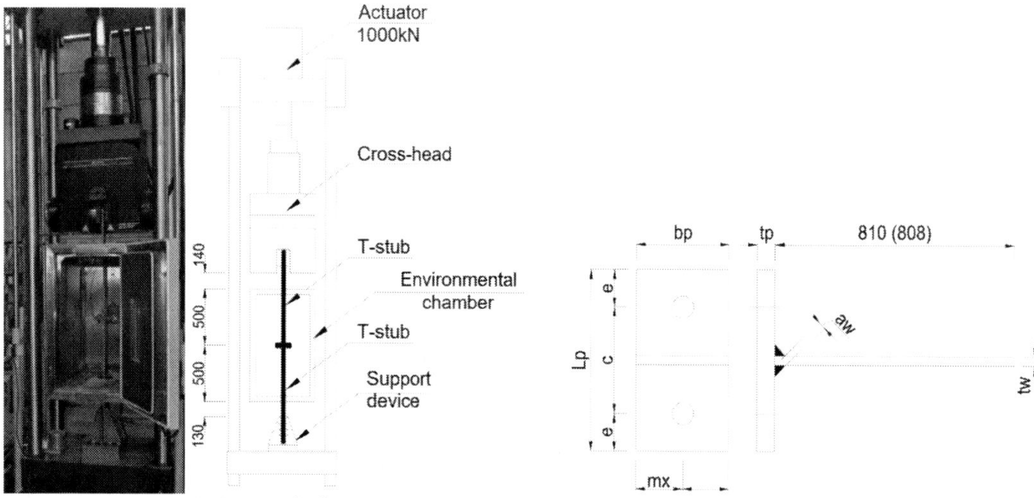

a) General view

b) Specimen

Fig. 1: Experimental setup

The nominal and the measured dimensions are given in Table 1. The name of the specimens indicates, respectively, the type of specimen, flange thickness, bolt diameter, distance between bolts and loading procedure. The suffixes of specimens, i.e. C and CS, stand for the tests performed at normal temperature, quasi-static (0.05 mm/s) and under high strain rate (10 mm/s), respectively. The suffixes of specimens, i.e. T and TS, stand for the tests under elevated temperature, quasi-static and under high strain rate, respectively.

Table 1: Specimens' dimensions

Specimen	Nominal								Measured					
	a_w	t_w	t_p	b_p	L_p	c	e	m_x	t_p	b_p	L_p	c	e	m_x
T-10-16-100C	7	10	10	90	160	100	30	45	9.6	88.9	160	99.9	30.3	43.7
									9.6	89.2	160	100.5	30.5	43.6
T-10-16-100CS	7	10	10	90	160	100	30	45	9.7	89.1	160	99.8	30.5	43.8
									9.7	89.4	160	99.8	30.2	43.9
T-10-16-100T	7	10	10	90	160	100	30	45	9.7	89.4	160	99.9	30.2	44.3
									9.7	89.2	160	100.0	30.9	43.8
T-10-16-100TS	7	10	10	90	160	100	30	45	9.7	89.0	160	99.8	31.1	43.8
									9.7	89.0	160	100.6	30.5	43.9
T-10-16-120C	7	10	10	90	180	120	30	45	9.7	89	180	120.3	30.5	44.4

Specimen	Nominal								Measured					
	a_w	t_w	t_p	b_p	L_p	c	e	m_x	t_p	b_p	L_p	c	e	m_x
									9.7	89.5	180	120.9	30.2	44.4
T-10-16-120CS	7	10	10	90	180	120	30	45	9.7	89.3	180	102.7	30.2	44.4
									9.7	89.3	180	102.9	30.3	44.4
T-10-16-120T	7	10	10	90	180	120	30	45	9.7	89.8	180	120.9	30.1	44.3
									9.7	89.3	180	121.5	29.7	44.0
T-10-16-120TS	7	10	10	90	180	120	30	45	9.7	89.3	180	120.4	30.3	44.5
									9.7	89.3	180	120.7	30.3	44.0
T-10-16-140C	7	10	10	90	200	140	30	45	9.7	89.3	200	141.3	30.2	44.8
									9.7	89.37	200	141.9	30.2	45.9
T-10-16-140CS	7	10	10	90	200	140	30	45	9.7	89.7	200	141.7	29.4	46.1
									9.7	89.5	200	142.5	29.4	46.3
T-10-16-140T	7	10	10	90	200	140	30	45	9.7	89.2	200	141.9	29.3	45.9
									9.7	89.1	200	141.3	30.2	46.1
T-10-16-140TS	7	10	10	90	200	140	30	45	9.7	89.8	200	141.3	29.5	45.9
									9.7	89.1	200	141.3	29.9	44.5
T-12-16-100C	7	10	12	90	160	100	30	45	11.8	88.6	160	101.3	29.6	44.8
									11.8	89.2	160	101.4	28.6	43.5
T-12-16-100CS	7	10	12	90	160	100	30	45	11.8	88.9	160	102.7	27.9	44.7
									11.8	89.2	160	101.5	29	44.8
T-12-16-100T	7	10	12	90	160	100	30	45	11.8	88.9	160	101.5	29.1	44.1
									11.8	89.0	160	101.6	29.2	43.5
T-12-16-100TS	7	10	12	90	160	100	30	45	11.8	89.2	160	101.9	29.4	43.6
									11.8	89.2	160	101.0	29.1	43.9
T-12-16-120C	7	10	12	90	180	120	30	45	11.8	89.7	180	121.5	29.6	46.0
									11.8	89.9	180	121.0	29.5	45.8
T-12-16-120CS	7	10	12	90	180	120	30	45	11.8	88.9	180	120.9	30.2	43.5
									11.8	89.8	180	120.8	29.7	43.8
T-12-16-120T	7	10	12	90	180	120	30	45	11.8	88.2	180	119.9	30.4	43.8
									11.8	87.9	180	120.0	30.3	44.2
T-12-16-120TS	7	10	12	90	180	120	30	45	11.8	89.4	180	121.0	29.4	43.3
									11.8	88.7	180	121.2	29.7	44.3
T-12-16-140C	7	10	12	90	200	140	30	45	11.8	89.4	200	139.9	30.6	45.0
									11.8	89.5	200	140.4	30.5	43.2
T-12-16-140CS	7	10	12	90	200	140	30	45	11.8	90.0	200	140.7	30.8	45.2
									11.8	89.4	200	140.4	28.7	43.4
T-12-16-140T	7	10	12	90	200	140	30	45	11.8	89.7	200	141.4	30.6	45.7
									11.8	89.4	200	140.3	30.0	46.0
T-12-16-140TS	7	10	12	90	200	140	30	45	11.8	88.7	200	140.0	29.4	44.7
									11.8	90.7	200	141.1	28.2	45.8

The flange was chosen as S235 steel grade, while for the web S355 was considered, in order to avoid any plastic deformation in this component. A fillet weld type was used.

The reduction factor for load combination in fire, η_{fi}, takes values function of the proportionality between imposed load and permanent load. As a simplification, EN1993-1-2 [5] allows the use of a value of 0.65. Using this value as the reduction factor for effective yield strength provided in Table 3.1 of EN1993-1-2 [5], it results the temperature of 542°C. Thus, this temperature was used in the experimental program, for which, according to Table D1 of EN1993-1-2 [5], the strength reduction factor for bolts is 0.41 (see Fig. 2).

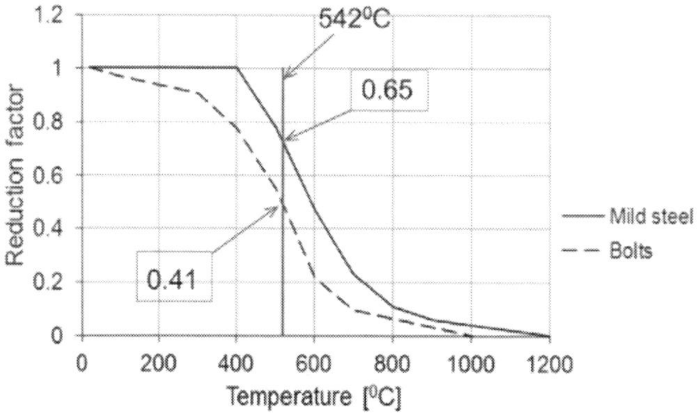

Fig. 2: Strength reduction of mild steel and bolts under elevated temperatures

3. Experimental results

3.1 Tests at normal temperature

Fig. 3 shows the force-displacement curves for quasi-static and high strain rate tests at normal temperature. The increase of the end plate thickness leads to higher resistance, but reduces the ultimate deformation capacity. For an increase of the bolt distance from 100 to 120 mm, there is a reduction of the resistance, but the deformation capacity is higher. For bolt row distance of 140 mm, the deformation capacity increases without any further reduction of the resistance. In all cases the failure is attained due to the fracture of the bolts, see Fig. 4 and Fig. 5. No failure of the welding was recorded. For the most elongated specimens (with 140 mm bolt row distance), some cracks were observed near the welding, in the heat affected zone. There is a small influence of the loading rate, in terms of ultimate resistance, deformation capacity and failure mode.

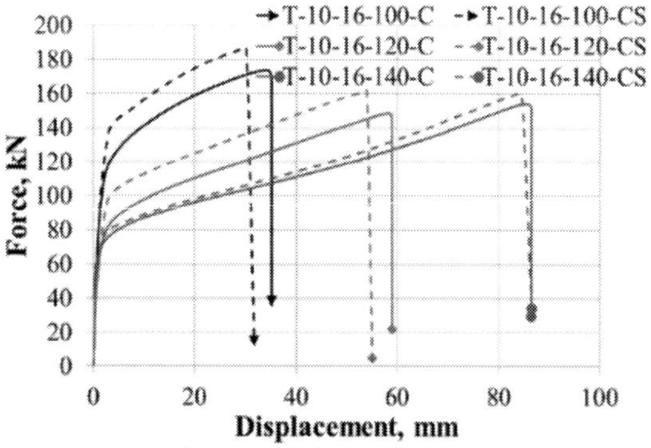

a) Flange thickness 10 mm

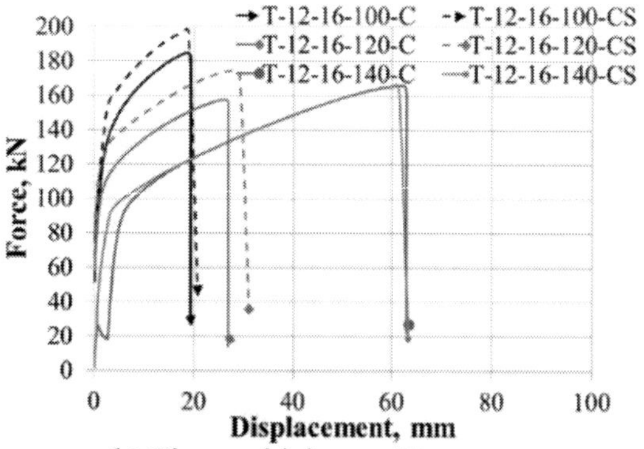

b) Flange thickness 12 mm

Fig. 3: Force displacement curves

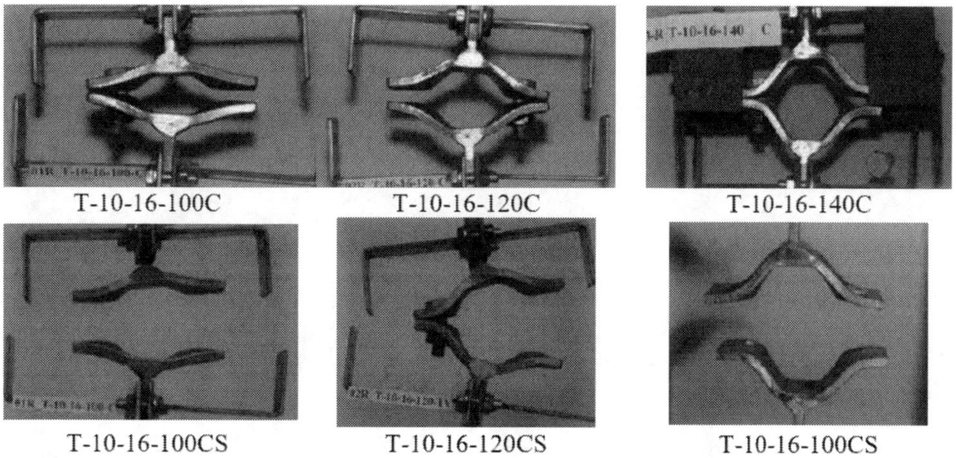

Fig. 4: T-stub specimens with flange thickness 10 mm – normal temperature

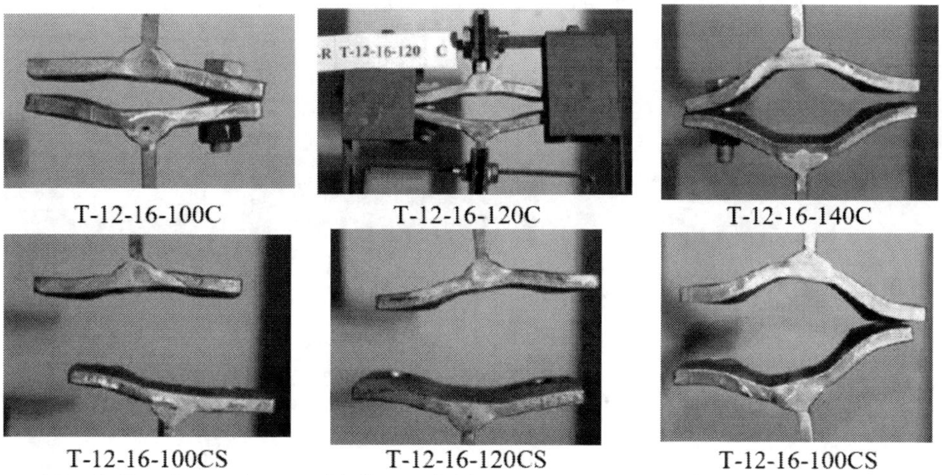

Fig. 5: T-stub specimens with flange thickness 12 mm – normal temperature

3.2 Tests at elevated temperature

Fig. 6 and Fig. 7 present the deformed shape of T-stubs after failure, for flange thicknesses of 10 mm and 12 mm, respectively. No welding damage was observed.

The response of T-stubs under elevated temperature at high strain rate loading presents important differences in both load-carrying capacity and deformability with respect to the same specimens tested quasi-static. The qualitative results that may be observed in Figs. 6-7 are quantified in Fig. 8. For the same geometrical configuration, under elevated temperature, higher resistances are attained at high strain rate loading in comparison with the tests performed quasi-static.

The ratio between the resistance of T-stubs tested at high strain rate and the resistance in case of quasi-static tests presents a reasonable consistent variation, as Fig. 9a suggests (for 542°C). In normal temperature conditions, the difference between the resistance of T-stubs tested at high strain rate and the resistance in case of normal strain rate is neglectable, as shown in Fig. 9a (for 20°C). In Fig. 9b, it is observed that the ratio between the T-stub resistance at elevated temperature and at normal temperature is always bigger for high strain rates. The closer the specimen is to failure mode 3 (bolts fracture), the closer the ratio is to the strength reduction of bolts at 542°C, $k_{y,b,542} = 0.41$ (see Fig. 2).

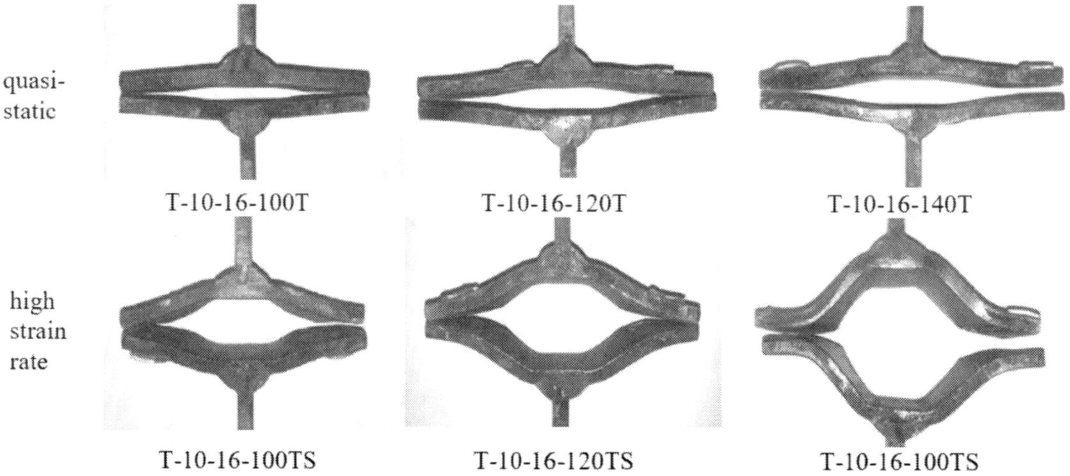

Fig. 6: T-stub specimens with flange thickness 10 mm – elevated temperature

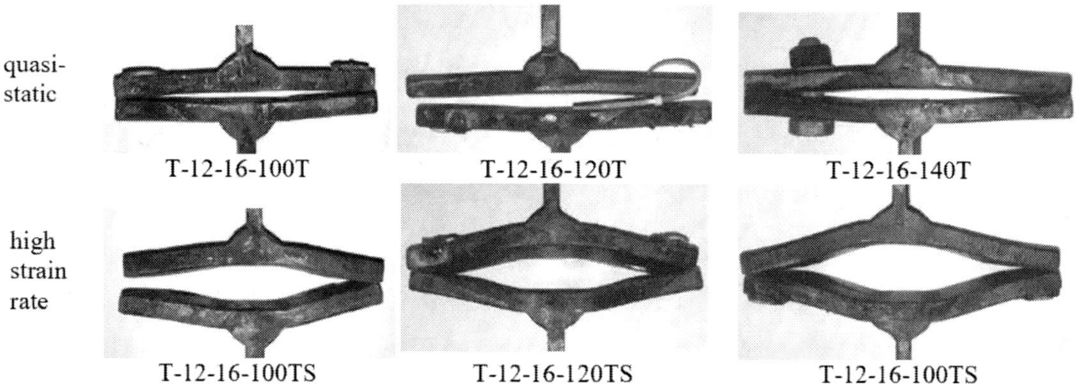

Fig. 7: T-stub specimens with flange thickness 12mm – elevated temperature

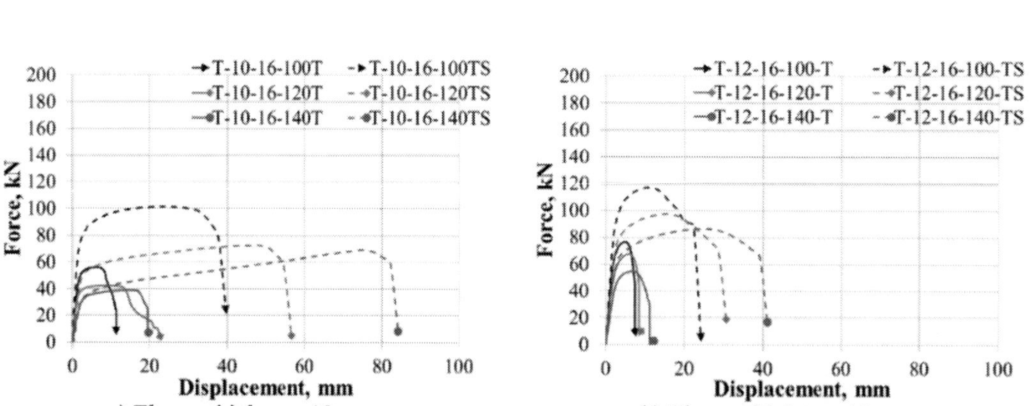

a) Flange thickness 10 mm

b) Flange thickness 12 mm

Fig. 8: Force displacement curves – elevated temperature

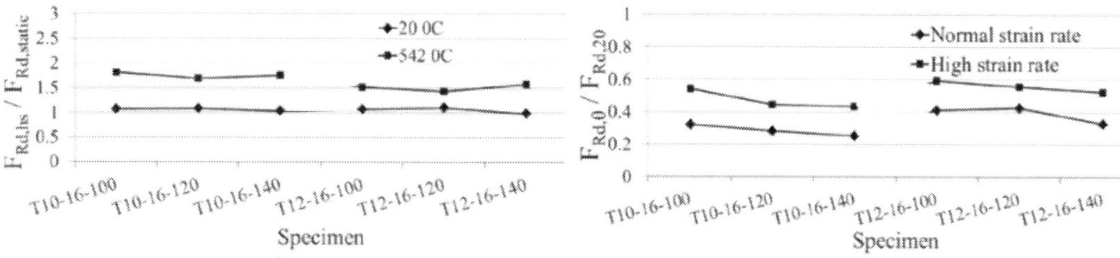

a) b)

Fig. 9: High rate/ quasi-static loading resistance ratio

For the same geometrical configuration, significantly higher deformations are attained at high strain rate loading, in comparison with the tests performed quasi-static. It has to be mentioned that the deformations obtained at 542°C at high strain rate are very similar with the ones obtained at normal temperatures, for both quasi-static and high strain rate loading. In fact, at normal temperature, there were very small differences in terms of deformations between normal and high strain rates, the higher strain rates leading to slightly smaller deformations. In case of tests performed at elevated temperature, the phenomenon is opposite.

The very small deformations obtained quasi-static at 542°C may be explained by the fact that the ratio of the resistances of bolts/ flanges is different at elevated temperatures. At 542°C, the steel resistance in the flanges is reduced to 65%, while the resistance of the bolts suffers a stronger reduction, to 41% (see Fig. 2). Therefore, the relative resistance of the bolt in the T-stub assembly is smaller at elevated temperatures and the failure of the bolt is produced in the quasi-static tests before the flanges may develop important plastic deformations. It should be mentioned that, at normal temperature, only failure Mode 1 was obtained, while at elevated temperature, as it may be seen in Figs. 6-7, the failure modes in the quasi-static tests were closer to Mode 2 or even Mode 3 in case of 12 mm flanges.

On the other hand, the over-strength of the bolts due to the high strain rate loading (T-stubs resistance higher by 40-80%, as shown in Fig. 9a), allow the flanges to develop plastic deformations and to obtain also the Mode 1 type failure, as shown in Fig. 6 and Fig. 7.

Fig. 10 and Fig. 11 compare the force-displacement curves of normal and elevated temperature tests, obtained for quasi-static and high strain-rate loading protocols. For the quasi-static tests, the elevated temperature influences significantly both strength and deformation (Fig. 10), while for the tests performed under high strain rate conditions, the elevated temperature influences significantly the strength, and in a lesser amount the deformation (Fig. 11).

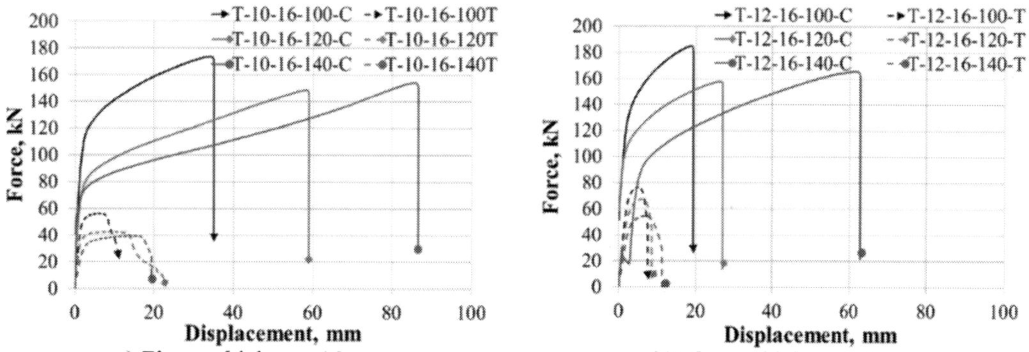

a) Flange thickness 10 mm b) Flange thickness 12 mm

Fig. 10: Force displacement curves in quasi-static tests

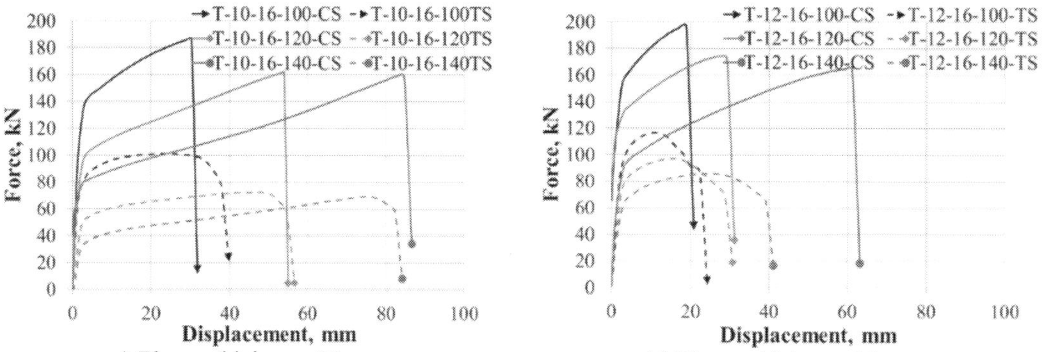

a) Flange thickness 10 mm b) Flange thickness 12 mm

Fig. 11: Force displacement curves in tests performed under high strain rate

4. Conclusions

The behaviour of T-stub components was investigated with respect to their influence on the capacity of beam-to-column joints to resist the loss of a column. Experimental results on bolted T-stub specimens under normal temperature showed that more ductile configurations can be obtained by reducing the end plate thickness or by increasing bolt row distance, while the loading rate has a reduced impact on the ultimate capacity and failure mode on both T-stubs and weld details. On the other hand, under elevated temperatures, a significant increase of both resistance and ductility of the T-subs subjected to high strain rate was emphasised, in comparison with the T-stubs loaded in quasi-static conditions.

Acknowledgments

Partial funding for this research was provided by the Executive Agency for Higher Education, Research, Development and Innovation Funding, Romania, under grant PCCA 55/2012 "Structural conception and COllapse control performance based DEsign of multistory structures under aCcidental actions" (2012-2016) and by the strategic grant POSDRU/159/1.5/S/137070 (2014) of the Ministry of National Education, Romania, co-financed by the European Social Funds – Investing in People, within the Sectorial Operational Programme Human Resources Development 2007-2013.

References

[1] Drysdale D. *An introduction to fire dynamics*, John Wiley & Sons, Ltd, 2011.

[2] NISTIR 7396, "Best Practices for Reducing the Potential for Progressive Collapse in Buildings" National Institute of Standards and Technology, Oakland, CA, 2007

[3] C.E.N. "Eurocode 3: Design of steel structures - Part 1-8: Design of joints", Brussels 2005.

[4] CODEC. "Structural conception and COllapse control performance based DEsign of multistory structures under aCcidental actions", 2012. Contract 55/2012/ UEFISCDI, Romania.

[5] CEN. "Eurocode 3: Design of steel structures - Part 1-2: General rules - Structural fire design", Brussels, 2005.

EXPERIMENTAL EVALUATION OF PROGRESSIVE COLLAPSE RESISTANCE OF STEEL MOMENT FRAME CONNECTIONS

Florea Dinu[a,b], Ioan Mărginean[a], Dan Dubina[a,b] and Calin Neagu [a]

[a] *Department of Steel Structures and Structural Mechanics, Politehnica University Timisoara, Romania*
[b] *Laboratory of Steel Structures, Romanian Academy, Timisoara Branch, Romania*

Abstract: During their lifetime, buildings may be subjected to accidental actions, like blast and impact, which can cause significant local damages. To limit the effects of such unforeseen events, the structure should retain the structural integrity by limiting the extent of damage and prevent its progress. Taking advantage of structure's inherent redundancy and available load paths, seismic resistant steel frames are considered appropriate to resist local damage without collapse. However, there are specific problems, which need to be considered when localized failures, particularly of columns, occur, i.e. large deformations and catenary response of beams. In this study, we investigated the performance of four steel frame beam-to-column connection types affected by the column removal. Acceptance criteria for progressive collapse events were proposed and compared to existing provisions.

1. Introduction

Steel framed structures are widely used in construction of multi-storey buildings, offering many several advantages over other types of constructions. According to EN 1990 [1], relevant design situations must be selected and applied, based on the conditions of use and also on the requirements concerning the performance of the structure. Most of ordinary buildings are designed for persistent and transient design situations, only. If the building may be subjected to seismic events, seismic design situations must be also considered. Even not considered in design, structures must resist accidental actions, e.g. fire, explosions or impact. In such situations, the structure must resist the action without being damaged to an extent disproportionate to the original cause. The disproportionality refers to the situation in which failure of one member causes a major collapse, with a magnitude disproportionate to the initial event; this is also known as progressive collapse [2]. Avoidance or limitation of potential damage may be done by several means, from elimination or reduction of hazard, to selection of a structural form which has low sensitivity to hazards and can survive adequately the accidental removal of an individual member or a limited part of the structure [1]. For the later option to be effective, it is necessary to provide continuity across the damaged area, and thus to allow the development of alternate loads paths. The alternate path method (AP) provides a formal check of the capability of the structural system to resist the removal of specific members, such as columns. The method does not require the characterization of the threat causing loss of the member. The AP method, with its emphasis on continuity and ductility, is similar to current seismic design practice ([2]). Seismic design procedures can be adopted as reference for progressive collapse design, but several modifications are necessary to accommodate the particular issues associated with progressive collapse. One such issue is the response of members and connections and their qualification for progressive collapse consideration. Thus, seismic resistant moment frames should be designed and detailed so that the required plastic deformations occur primarily in the beams or in the beam-to-column connections [3], [4].

If the structure is designed to dissipate energy in the beams, the connections of the beams to the columns should be strengthened and stiffened by using cover plates, haunches. Two examples for such type of connections are Welded Cover Plated Flange connection (CWP) and Haunched End Plate Bolted connection (EPH), respectively. These types of connections are widely used in practice, even they are not the most cost-effective and simple to execute connections. Alternatively, the section of the beam can be reduced at a distance away from the connection to obtain a favorable plastic mechanism. One such connection is the Reduced Beam Section welded connection, RBS.

If the structure is designed to dissipate energy in the connections, their flexural capacity is a fraction of that of the beams, and the connection is classified as partial strength connection. One example is the unstiffened extended end plate bolted connection (EP).

If for evaluation of seismic performance of connections and members cyclic loads with increasing magnitude are applied, without axial loading, in case of progressive collapse, the connection and member experiences one half cycle of loading, often in conjunction with a significant axial load, due to large deformations and catenary response. Unfortunately, only limited experimental research on the performance of steel frame connections under column removal scenarios were conducted so far.

When experimental testing is of concern, the use of connection components and sub-assemblage testing are most common solutions, allowing the complete evaluation of the response and validation of computational models. Yang and Tan [5] tested experimentally the performance of common types of bolted steel beam-column joints under a central-column-removal scenario. The results showed that tensile capacities of beam-column joints after undergoing large rotations usually control the failure mode and the formation of catenary action. Xu and Ellingwood [6] investigated the performance of steel frames with partially restrained connections fabricated from bolted T-stubs following damage to load-bearing columns. They reported that frames with strong T-stub connections can resist collapse in damage scenarios involving the notional removal of one column, whereas the robustness of the frames with weak T-stub connections is questionable. Structural system testing, involving entire building systems or parts of them is expensive and difficult, and therefore very few tests have been performed so far. Astaneh-Asl et al., [7] experimentally studied the ability of a typical steel structure to resist progressive collapse in the event of the loss of a column due to a blast and attempted to establish the failure modes. Another study [8] confirmed that a retrofit scheme in which cables are added to the sides of beams could be used to develop larger catenary action with a higher safety factor. A very recent study ([9], [10]) investigated the capacity of a 3D steel frame structure with extended end-plate bolted beam-to-column connections to support the loss of a column. The results showed that connections possessed sufficient strength to resist the catenary forces developed in the beams, with a rotation capacity over 200 mrad.

In the present study, we investigated the potential of four typical steel frame beam-to-column connections to develop catenary action under column loss and how these connections perform when subjected to large deformation demands. Connections were designed and detailed to meet the seismic design requirements for special moment frame connections. The specimens, having two spans of 3.0 m each, were subjected to a monotonic loading applied to the top of the removed central column until complete failure. The study is part of a research program devoted to the design of structures to sustain extreme loading events without collapse [11].

2. Specimens and test setup

Figure 1.a shows the plan layout of the case-study building. The bays and spans each measure 8.0 m. Structure was calculated for the effect of gravity loads (permanent and variable actions) and lateral loads (wind and seismic actions), using the Eurocodes. The dead and live loads were each 4.0 kN/m^2 and the reference wind pressure was 0.5 kN/m^2. The structures were located in a low seismicity area, characterized by a design ground acceleration a_g of 0.08 g and a control period T_C of 0.7 s. It should be noted that, the seismic intensity and the response spectrum used in design were those given in the Romanian Seismic Code, P100-1/2006 [12]. High dissipative structural behavior was considered using a behavior factor q of 6.5. An inter-story drift limitation of 0.008 of the story height was used for the seismic design at the serviceability limit state. Various loading cases, including seismic design loads, were used for the structural design of members, connections and details, using the relevant Eurocode parts. No particular accidental design situation was considered in design. The highlighted area (Figure 1.a) indicates in the perimeter frame extracted for investigation. Due to space limitation in the laboratory, the frame was scaled down from 8.0 m span to 3.0 m span. Each specimen consists of two steel beams and three columns and have a length of 6.0 m between the center lines of the marginal columns. Figure 1.b shows the experimental frame, with the test set-up, the specimen and the 100 tones actuator. To consider the restraint from the surrounding elements in the reference structure, horizontal restraints were provided by a strong reaction wall, on one side, and by a brace system, on the other side. Out of plane restraints were also provided along the span of the beams and at the central column. Note that the in-plane rotation of the central column was not restrained. In this study, four specimens were constructed and tested, each with a different type of beam-to-column connections but with the same sections for members. In all specimens, beam section was IPE220 while the column section was HEB260 with flanges reduced to 160 mm width. Steel material in plates and profiles was S275J0 and bolts were class 10.9 and 8.8.

First specimen (CWP) used welded cover plated flange connections. The top and bottom beam flange cover plates are 12 mm thick, width is 130 mm and length is 150 mm. Cover plates were welded with complete joint penetration CJP groove welds to the column flange, and with fillet welds to the top and bottom beam flanges. Beam web connection was made using a bolted shear tab, reinforced with filet welds to the beam web and column flange.

Second specimen (EPH) used extended end plate bolted connections and bottom haunches, with 20 mm thick end plate and six rows of bolts M20 class 10.9. The height of the haunch is 110 mm while the length is 150 mm (same length as cover plates in CWP).

Third specimen (RBS) used reduced beam section connections, with circular radius cuts in both top and bottom flanges of the beam. Dimensions of the reduced zones, calculated according to ANSI/AISC 358-10 [4], are a=66 mm, b=150 mm and c=22 mm. Welds of beam flanges to the column are CJP groove welds.

Forth specimen (EP) used unstiffened extended end plate bolted connections, with 16 mm thick end plate and five rows of bolts M16 class 10.9.

Figure 2.a-d shows the construction details of the connections for the specimens and Figure 2.e the moment-rotation characteristics of the connections and the classification according to EN 1993-1-8 [13]. The instrumentation included displacement transducers for vertical deflection and joint rotation. For estimating the axial force in beams, additional transducers were placed on horizontal elements of the in-plane restraining system. As these elements remain elastic, it is easy to estimate the intensity of the axial force in the beams during the test.

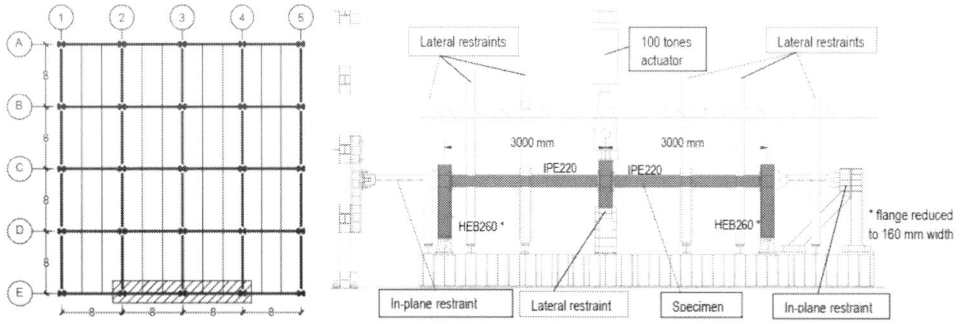

a) b)

Fig. 1: Plan layout of the reference multi-storey building with the position of the specimens extracted for testing (a) and test set-up (b)

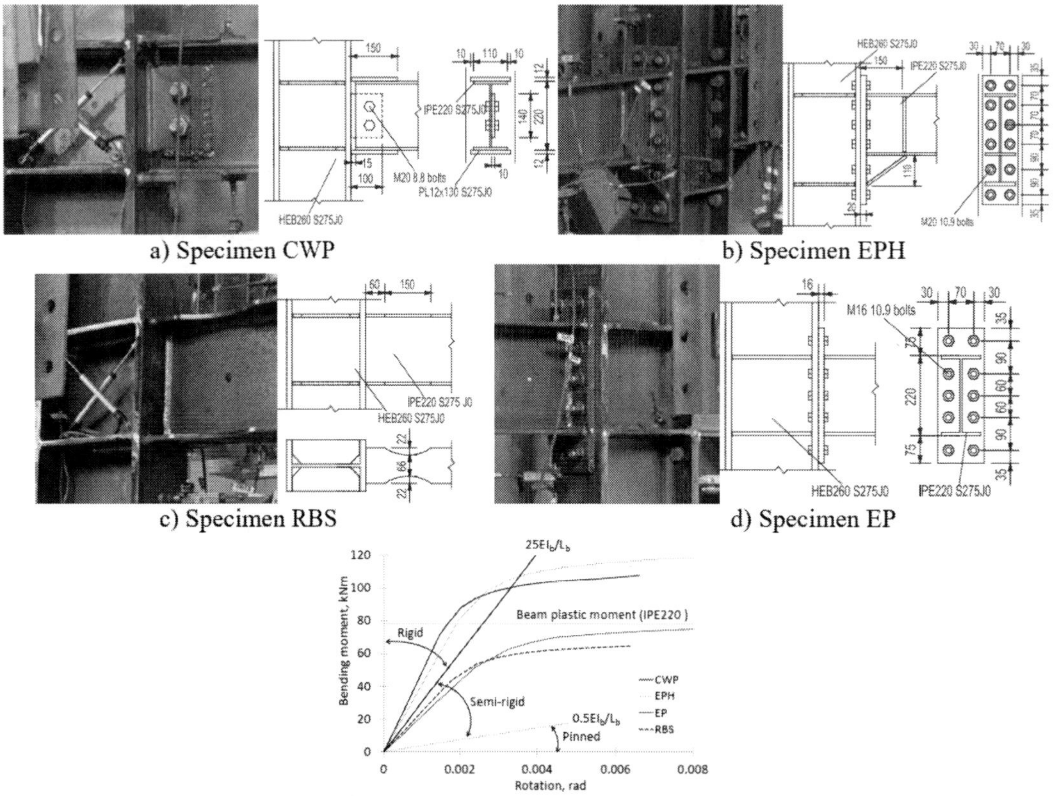

a) Specimen CWP b) Specimen EPH

c) Specimen RBS d) Specimen EP

e) moment-rotation characteristics of the connections

Fig. 2: Details of test specimens (dimensions are in mm)

Before the test, bolts and material coupons from beams, columns, and plates were tested to evaluate the mechanical characteristics of materials. The mechanical characteristics are listed in Table 1. The vertical force was applied on top of the central column and was gradually increased until failure of the specimen was reached. The yield displacement, D_y, and yield force, F_y, were calculated according to the ECCS method. Data from displacement transducers located at beams ends were used to capture the joint rotations R1, R2 and R3, see Figure 3. The results from the four experimental tests are presented in the subsequent sections.

Table 1: Average characteristic values for steel profiles, plates and bolts

Element	f_y (N/mm²)	f_u (N/mm²)	ε_y (%)	A_{gt} (%)
Beam flange IPE220, t = 9.2 mm	351	498	0.17	15.0
Beam web IPE220, t = 5.9 mm	370	497	0.18	15.0
Column web HEB 260, t = 10 mm	402	583	0.19	12.9
Column flange HEB 260, t = 17.5 mm	393	589	0.19	13.3
End plate, t = 16 mm	305	417	0.15	17.1
End plate, t = 20 mm	279	430	0.13	12.7
Cover plate, t = 12 mm	315	455	0.15	16.3
Shear tab, t = 10 mm	314	416	0.15	16.7
Bolt, M20 class 10.9	920*	1085	1.75*	12.2
Bolt, M16 class 10.9	965*	1080	1.76*	12.0
Bolt, M20 class 8.8	672*	825	1.78*	12.3

Note: f_y - yield strength; f_u- ultimate strength; ε_y- yield strain; A_{gt}- Total elongation at maximum stress
* 0.2% offset yield point

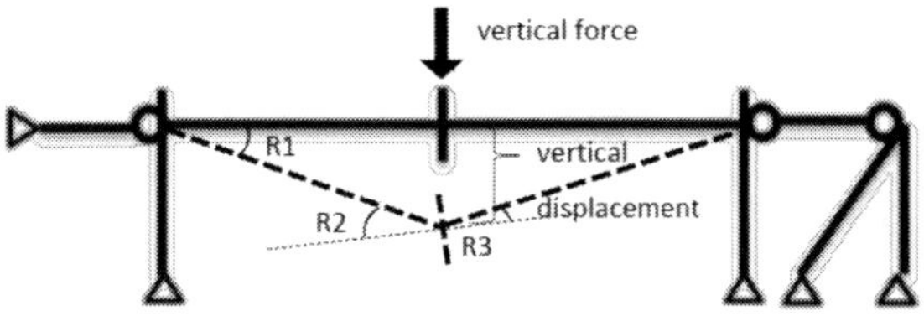

Fig. 3: Schematic representation of the experimental setup and beam rotation notations

3. Experimental results

3.1 Full strength connections

3.1.1 Welded cover plated flange connections CWP

Figure 4.a shows the vertical force vs. middle column displacement for specimen CWP. Three stages can be identified on the curve, i.e. elastic, flexural and catenary, without clear point of demarcation but with some zones of interaction. At the initial loading stage, the specimen was in elastic and the applied load increased almost linearly because the connection was stiff and no slippages were possible in the connection components. The local buckling of the top flange of the right beam, near the connection with the central column, indicated the initiation of yielding and, at this point, the displacement was 35 mm and the applied force was 147 kN. After yielding started, the second stage was initiated and the flexural stiffness started to decrease. The maximum applied force at pure flexural stage (no axial tension in beams) was 201 kN and the corresponding vertical displacement was 115 mm. To note that up to 115 mm vertical displacement, the beams were in compression (axial force was negative), with a maximum compressive force of 17 kN, indicating a very low arching behavior in the structure, see Figure 4.b. Up to a vertical displacement of 210 mm, most of the applied load was still resisted by the flexural capacity but the catenary action started to develop, and the axial force in beams increased to 214 kN. At the end of this stage, which can be called flexural-catenary stage, the bending moment reached the maximum value, i.e. 153 kNm, see Figure 4.b. After this point, the flexural resistance started to decrease while the catenary action became more predominant. The stiffness continued to increase until the vertical displacement and applied force reached 519 mm and 603 kN, respectively. At this point, due to large tensile force in beams, the end connection of the right beam (near the central column) started to fracture, first in the bottom cover plate, which completely separated from the column flange, followed immediately by a large fracture in the shear tab. This fracture was accompanied by a large drop in the applied force. The axial force in beams reached a maximum value of 1230 kN, then started to decrease, see Figure 4.b. Because the fracture was quite violent, the test was halted and the transducers located near the central column were detached for safety reasons. The test was then resumed and the applied load started to increase again, until the top cover plate fractured and the beam was completely separated from the column, see Figure 4.c. When the test was stopped, the ultimate vertical displacement reached 586 mm. Figure 4.d shows the rotation in sections R1, R2 and R3. It may be seen that up to a vertical displacement of 210 mm (end of flexural-catenary stage), the rotations were almost identical. After this point, due to the rotation of the central column, the two beam ends connected to the central column recorded different rotations, i.e. R2 and R3. At the peak applied load, the maximum rotation recorded in the connections was R3=0.193 rad.

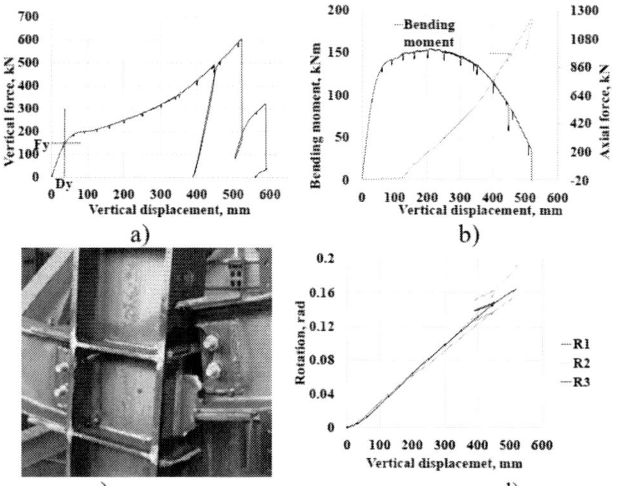

c) d)

Fig. 4: Specimen CWP: a) vertical force vs. vertical displacement; b) bending moment and axial force vs. vertical displacement; c) failure mode; d) beam end rotation vs. vertical displacement

3.1.2 Bottom haunch end plate bolted connections EPH

Figure 5.a shows the vertical force vs. middle column displacement for specimen EPH. As in case of specimen CWP, three stages can be identified on the curve, i.e. elastic, flexural and catenary. At the initial loading stage, the specimen was in elastic and the applied load increased almost linearly because the connection is rigid and there were virtually no slippage in the connection. The local buckling of the top flange of the right beam, near the connection with the central column, indicated the initiation of yielding and, at this point, the displacement was 37 mm while the applied force reached 147 kN. After yielding started, the second stage was initiated and the flexural stiffness started to decrease. The maximum applied force at pure flexural stage was 212 kN and the corresponding vertical displacement was 110 mm. To note that up to 110 mm vertical displacement, the axial force in beams was nearly zero, indicating there is no arching behavior in the structure, see Figure 5.b. Up to a vertical displacement of 171 mm, most of the applied load was still resisted by the flexural capacity but the catenary action started to develop, and the axial force in beams increased from zero to 235 kN. At the end of this flexural-catenary stage, the bending moment reached the maximum value, i.e. 156 kNm, see Figure 5.b. After this point, the flexural resistance started to decrease while the catenary action became more predominant. The stiffness continued to increase until the vertical displacement and applied force reached 440 mm and 477 kN, respectively. At this point, due to large tensile force in beams, the right connection of the right beam failed due to the fracture of first three bolt rows in tension, see Figure 5.d. Next two bolt rows also suffered plastic deformations but did not fracture because the test was stopped due to safety reasons. The axial force in beams reached a maximum value of 1035 kN, see Figure 5.b. Figure 5.c shows the rotation in sections R1, R2 and R3. It may be seen that up to a vertical displacement of 171 mm (end of flexural-catenary stage), the three rotations were very similar. After this point, due to the rotation of the central column, the two beam ends connected to the central column recorded different rotations. At the peak applied load, the maximum rotation recorded in the connections was R1=0.130 rad. To note that when the right connection of the right beam failed due to fracture of the bolts, the beams connections to the central column showed no visible damages, see Figure 5.e. The reason for this behavior is the non-symmetrical arrangement of the connection. Thus, under sagging bending, the connection has more capacity in tension because more bolt rows are engaged. In case of hogging bending, there are less bolt rows active in tension.

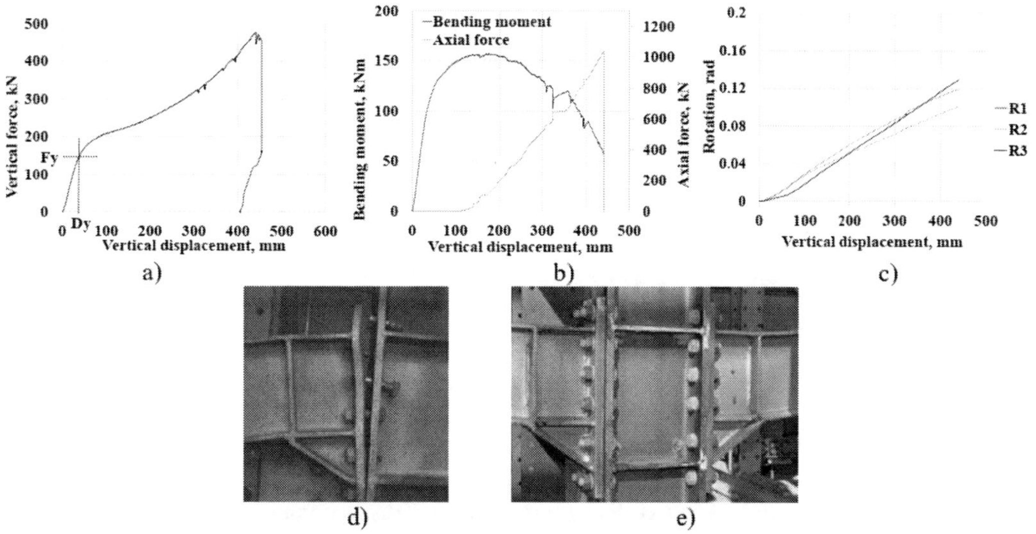

Fig. 5: Specimen EPH: a) vertical force vs. vertical displacement; b) bending moment and axial force vs. vertical displacement; c) beam end rotation vs. vertical displacement; d)- e) failure mode

3.2 Partial strength connections

3.2.1 Reduced beam section connections RBS

Figure 6.a shows the vertical force vs. middle column displacement for specimen RBS. Three stages can be identified on the curve, i.e. elastic, flexural and catenary. At the initial loading stage, the specimen was in elastic and the applied load increased almost linearly because the connection was stiff and no slippages were possible in the connection components. The yielding initiated at a vertical displacement of 33 mm, while the applied force amounted 110 kN. After yielding started, the second stage was initiated and the flexural stiffness started to decrease. The maximum applied force at pure flexural stage (no tensile axial force in beams) was 195 kN and the corresponding vertical displacement was 200 mm. To note that up to this point the axial force in beams was nearly zero, indicating no arching behavior in the structure, see Figure 6.b. Up to a vertical displacement of 250 mm, most of the applied load was still resisted by the flexural capacity, but the catenary action started to develop, and the axial force in beams increased from zero to 100 kN. At the end of this flexural-catenary stage, the bending moment reached the maximum value, i.e. 147 kNm, see Figure 6.b. After this point, the flexural resistance started to decrease while the catenary action became more predominant. The stiffness continued to increase until the vertical displacement and applied force reached 480 mm and 401 kN, respectively, when, at the marginal connection, due to large tensile forces in beams, a crack was initiated in the top flange of the reduced beam zone. The fracture then propagated in the web (see Figure 6.c) and the test was stopped due to safety reasons, because the rupture was quite violent. The maximum axial force recorded in the beams was 753 kN (Figure 6.b). Figure 6.d shows the rotation in sections R1, R2 and R3. It may be seen that the three rotations were almost identical up to the end of the test indicating that the rotation of the central column was negligible. At the peak applied load, the maximum rotation recorded in the connections was R3=0.172 rad.

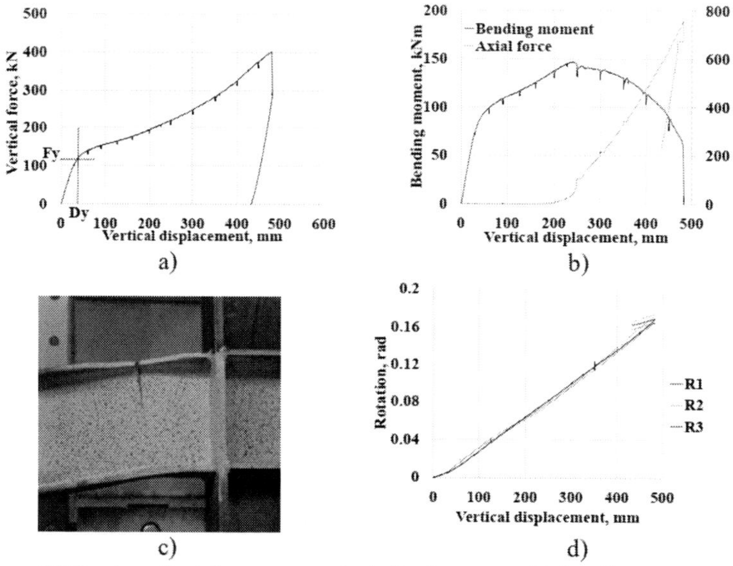

Fig. 6: Specimen RBS: a) vertical force vs. vertical displacement; b) bending moment and axial force vs. vertical displacement; c) failure mode; d) beam end rotation vs. vertical displacement

3.2.2 Unstiffened extended end plate bolted connections EP

Figure 7.a shows the vertical force vs. middle column displacement for specimen EP. Compared to the first three specimens (full strength specimens CWP and EPH and partial strength RBS), this specimen did not show any distinctive catenary behavior. The initiation of yielding was due to bending in the end plate of the right beam, near the connection with the central column, at a vertical displacement of 39 mm and an applied force of 117 kN. After yielding started, the second stage was initiated and the flexural stiffness started to decrease. The maximum applied force at pure flexural stage (no tensile force in beams) was 175 kN and the corresponding vertical displacement was 154 mm. To note that up to 154 mm vertical displacement, the axial force in beams was nearly zero, indicating there is no arching behaviour in the structure, Figure 7.b. At a maximum applied force of 182 kN and a corresponding vertical displacement of 194 mm the specimen suffered first failure due to the fracture of the bottom external bolt row of the right beam connection near the central column. The fracture was caused by the flexural action, with a minor contribution from axial load. The test continued and the failure propagated to second bolt rows within the same connection. In the same time, also the top second bolt row of the right beam connection away from the central column failed due to excessive tensile forces. The specimen finally failed when three bolt rows from left and right connections of the right beam, and two bolt rows from the left connection of the left beam were fractured. At this final stage, the axial force in beams reached 571 kN. The main cause of the premature failure is the insufficient tying resistance of the connection which led to an insufficient rotation capacity required to develop significant catenary action. Figure 7.c shows the rotation in sections R1, R2 and R3. It may be seen that up to a vertical displacement of 175 mm, the rotations R2 and R3 were almost identical, suggesting the central column remained on vertical position. After this point, due to the rotation of the central column, the rotation concentrated in the right connection R3, while R2 started to reduce. At the peak applied load, the rotation recorded in the connections was R3 = 0.079rad. The rotations beyond this point cannot be considered acceptable, because the resistance started to decrease, indicating the progressive collapse is imminent.

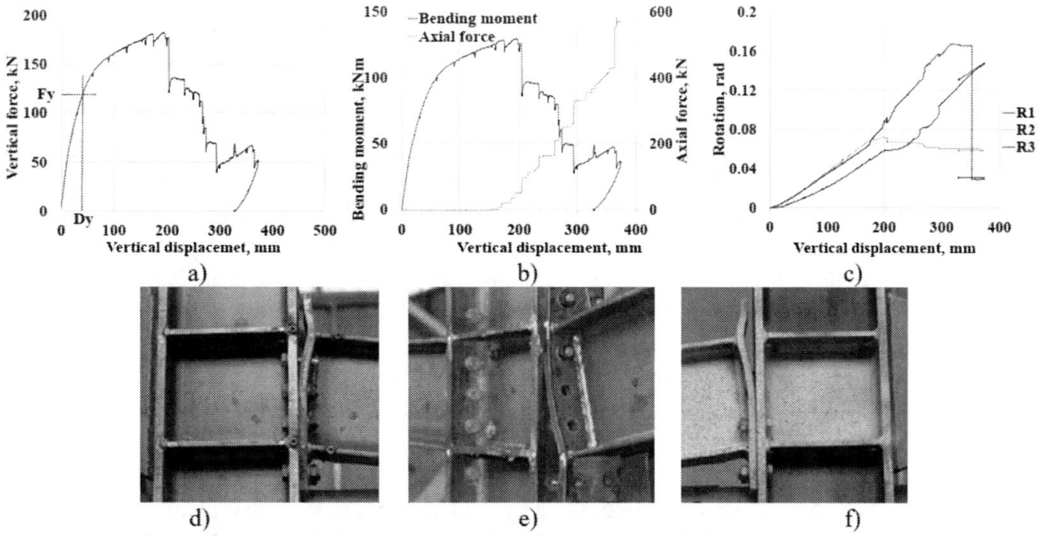

Fig. 7: Specimen EP: a) vertical force vs. vertical displacement; b) bending moment and axial force vs. vertical displacement; c) beam end rotation vs. vertical displacement; d)-f) failure mode

4. Conclusions

An experimental program was conducted at PU Timisoara to investigate the performance of four types of moment resisting connections when a column is removed. The specimens were extracted from a six-story moment resisting frame structure, designed to meet the seismic design requirements for special moment frames. The two-span specimens were tested under a monotonically increased vertical force, applied on top of the central column using a displacement control, until failure of the specimens was reached. The specimens designed with full strength connections (CWP and EPH) showed a good response, experiencing large deformation capacity and catenary response. RBS specimen, where the beam is reduced near the column face, also developed large axial forces in beams before failure, with an ultimate rotation over 170 mrad. The specimen with unstiffened end plate bolted connections showed the poorest behavior and failed before developing significant catenary action in beams. The combined contribution from direct applied axial loads, prying effects and bending due to flexible end plate, finally contributed to bolt fracture, with an ultimate rotation of 79 mrad. Without significant strengthening of the bolts, this connection cannot be used for developing catenary forces and therefore should be used for flexural-based design only.

The results and conclusions presented above are based on a limited number of tests. Therefore, additional studies are necessary to confirm or to adjust the findings. Thus, an extensive numerical parametric study is under development, aiming at investigating the factors that may have a significant impact on the catenary response of connections under extreme loading conditions, e.g. influence of beam depth on the rotation limit, beam web-to-column detail (for welded connections) and strain rate.

Acknowledgments

Funding for this research was provided by the Executive Agency for Higher Education, Research, Development and Innovation Funding, Romania, under grant PCCA 55/2012 "Structural conception and collapse control performance based design of multistory structures under accidental actions" (2012-2016).

References

[1] EN 1990 "Eurocode - Basis of structural design", CEN, 2002.
[2] NISTIR 7396 "Best Practices for Reducing the Potential for Progressive Collapse in Buildings, National Institute of Standards and Technology", Oakland, CA, 2007.
[3] EN 1998-1 "Eurocode 8: Design of structures for earthquake resistance – Part 1: General rules, seismic actions and rules for buildings", CEN, 2004.
[4] ANSI/AISC 358-10, "Prequalified Connections for Special and Intermediate Steel Moment Frames for Seismic Applications", AISC, 2014.
[5] Yang B, Tan KH. "Experimental tests of different types of bolted steel beam-column joints under a central column removal scenario", *Eng. Structures*, 54, 112-130, 2013.
[6] Xu G, Ellingwood B. "Disproportionate collapse performance of partially restrained steel frames with bolted T-stub connections", *Engineering Structures*, 33, 32–43, 2011.
[7] Astaneh-Asl A, Jones B, Zhao Y, Hwa R. "Floor catenary action to prevent progressive collapse of steel structures", (Report No. UCB/CE-Steel-03/2001). Department of Civil and Environmental Engineering, University of California, Berkeley, USA, 2001.
[8] Tan S, Astaneh-Asl A. "Cable-based retrofit of steel building floors to prevent progressive collapse". Berkeley: University of California, 2003.
[9] Dinu F, Marginean I, Dubina D, Petran I. "Experimental testing and numerical analysis of 3D steel frame system under column loss", *Eng. Structures*, 113, 59-70, 2016.
[10] Dinu F, Dubina D, Marginean I. "Improving the structural robustness of multi-story steel-frame buildings", *Structure and Infrastructure Eng.*, 11(8), 1028-1041, 2015.
[11] CODEC. "Structural conception and collapse control performance based design of multistory structures under accidental actions" (2012–2016), Exec. Agency for Higher Ed., Resc., Development and Innov. Funding, Romania, PN II PCCA 55/2012.
[12] P100-1 (2006). Seismic design code. Part 1: "Earthquake resistant design of buildings P100-1/2006" (Buletinul Constructiilor, No. 12-13). INCERC, Bucharest, Romania, Ministry of Transport, Construction and Tourism MTCT (in Romanian), 2006.
[13] EN 1993-1-8 "Eurocode 3 Design of steel structures - Part 1-8: Design of joints", 2005.

EXPERIMENTAL TESTING OF 3D STEEL FRAME WITH COMPOSITE BEAMS UNDER COLUMN LOSS

Florea Dinu[a,b], Ioan Mărginean[a], Dan Dubina[a,b], Ioan Petran[c], Mircea Pastrav[d], Andreea Sigauan[a] and Adrian Ciutina[a]

[a] *Department of Steel Structures and Structural Mechanics, Politehnica University Timisoara, Romania*
[b] *Laboratory of Steel Structures, Romanian Academy, Timisoara Branch, Romania*
[c] *Department of Structures, Technical University of Cluj-Napoca, Romania*
[d] *National Institute for Research and Development in Construction, Urban Planning and Sustainable Spatial Development "URBAN-INCERC", Cluj-Napoca Branch, Romania*

Abstract: The different nature and intensity of accidental loads make difficult the development of design requirements for such situations. Therefore, a better strategy is to limit the extent of damage so that the progressive collapse is not initiated. Features like ductility and continuity provide more deformation capacity and redistribution of loads, so that the structure can bridge over damaged/lost elements. Interaction between the steel beams and the concrete slab is also expected to enhance the resistance after the loss of a column. This paper presents the results of an experimental study that aimed at investigating the contribution of the floor system and beam-floor interaction to the load redistribution capacity in case of a column loss. For this purpose, a 3D steel frame structure, with composite beams and extended end-plate bolted beam-to-column connections, was considered. The specimen was tested under monotonic loading applied to the top of the central column until complete failure. The results showed the system was capable of developing larger capacity to resist the loss of a column but lower deformation capacity when compared to a bare steel frame tested under similar conditions.

1. Introduction

Performance demands for structural and non-structural elements of buildings are linked to the actions that might affect them over time. The limitations in achieving the expected performance can be caused either by underestimation of some of these actions, or by overestimation of the element capacity. When requirements refer to structural characteristics necessary to provide the capacity of the element or the entire structure to remain standing, the consequences are expressed in terms of risk for the public safety (deaths or serious injury), economic or social losses and damage to the environment [1]. In order to reduce this risk, these potential consequences need to be directly addressed by design. For example, EN 1991-1-7 [2] stipulates that a structure shall be designed in such a way that it will not be damaged by events like fire, explosions, impact or consequences of human errors, to an extent disproportionate to the original cause. Similar requirements are stipulated in ASCE Standard 7-05 [3], where buildings and other structures shall be designed to sustain local damage with structural system as whole remaining stable and not being damaged to an extent disproportionate to the original damage. The different nature and intensity of possible accidental load events make difficult the development of design requirements for such situations. Therefore, a better strategy is to limit the extent of damage so that the progressive collapse is not initiated. This approach may include the Alternate Path (AP) method, which requires that the structure should resist the loss of one or more critical load-bearing elements without developing disproportionate collapse of the building, [4]. However, there are concerns that primary load carrying members would not be able to develop the required tie forces because of the significant deformation demands. Some structural features may improve the load redistribution and thus, may reduce the deformation capacity demands. One example is the design that considers the interaction between concrete floor and steel beams. Several studies were conducted to consider the ability of the floor system to provide the necessary load redistribution (Astaneh-Asl et al., 2001 [5]; Stevens, 2008 [6]; Demonceau. and Jaspart, 2010 [7]; Alashker et al., 2010 [8]; Dubina and Dinu, 2012 [9]; Pham and Tan, 2013 [10]). The results showed that the presence of the beam-floor interaction could enhance the resistance to progressive collapse in case of column loss, especially for less redundant structural systems.

This paper presents the results of an experimental study that aimed at investigating the contribution of the floor system to the load redistribution capacity in case of a column loss. For this purpose, a 3D steel frame structure, with composite beams and extended end-plate bolted beam-to-column connections, was considered. The beam-to-column connections were designed as fully restrained connections. The specimen was tested under monotonic loading applied to the top of the central column until complete failure. The test results were compared with another test performed on a similar structure, with same geometric configuration, member sections and connections but with no interaction between the steel beams and concrete floor (bare steel frame structure). The results obtained for bare steel structure were reported in detail elsewhere [11]. The study is part of a research program devoted to the design of structures that can sustain extreme loading events without collapse [12].

2. Specimens and test set-up

This research focuses on the behaviour of two-way steel moment frames, designed with interaction between steel beams and concrete floor, in case of a column loss. For this reason, a 3D specimen was designed and constructed. The specimen has two spans and two bays and was extracted from the first floor of a reference building of four-bay, four-span of 8.0 m, and six-story of 4.0 m steel structure with moment frames in both directions. Because of laboratory restrictions, the 16.0 m by 16.0 m assembly was downscaled, resulting in a 6.0 m by 6.0 m specimen. The tested system included main and secondary beams, acting as composite sections. The number of shear studs was calculated to achieve full shear connection. Fig. 1 shows the framing plan with the dimensions, types, and sizes of element cross sections, types of connections between steel members, headed stud shear connectors used for beams and reinforced concrete floor. Fig. 2.a shows the specimen during construction, before pouring the concrete, while Fig. 2.b shows the specimen ready for testing, with the concrete floor in place. Note that structural steel S275 was used for beams, S355 for columns, concrete C20/25 for concrete floor, reinforcement from S420 steel and steel S235J2 + C450 for headed shear studs. Before the test, bolts, shear studs and materials from members (beams, columns, plates, concrete floor) were tested to evaluate the mechanical characteristics, see Table 1.

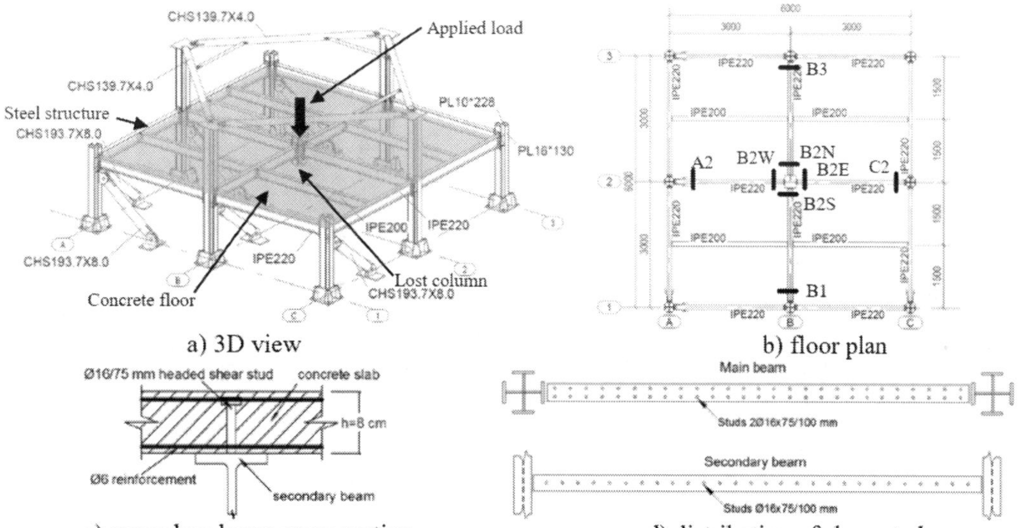

a) 3D view b) floor plan

c) secondary beam cross section d) distribution of shear studs

Fig. 1: Views and details of the specimen

Table 1. Average characteristic values for steel profiles, plates and bolts

Element	f_y (N/mm²)	f_u (N/mm²)	ε_y (%)	A_{gt} (%)	f_{ck}
Beam flange IPE220, t = 9.2 mm	345	464	0.16	28.0	-
Beam web IPE220, t = 5.9 mm	353	463	0.17	30.4	-
Column web HEB 260, t = 10 mm	407	539	0.19	27.0	-
Column flange HEB 260, t = 17.5 mm	420	529	0.20	27.0	-
End plate, t = 20 mm	408	535	0.19	24.4	-
Bolt, M20 class 10.9	905*	1081	-	12.0	-
Concrete floor	-	-	-	-	32.0
Reinforcement ϕ6 mm	346	480	0.16	30.0	-

Note: f_y - yield strength; f_u - ultimate strength; ε_y - yield strain; A_{gt} - total elongation at maximum stress; f_{ck} –characteristic compressive strength of the concrete cube, at 28 days; * 0.2% offset yield point

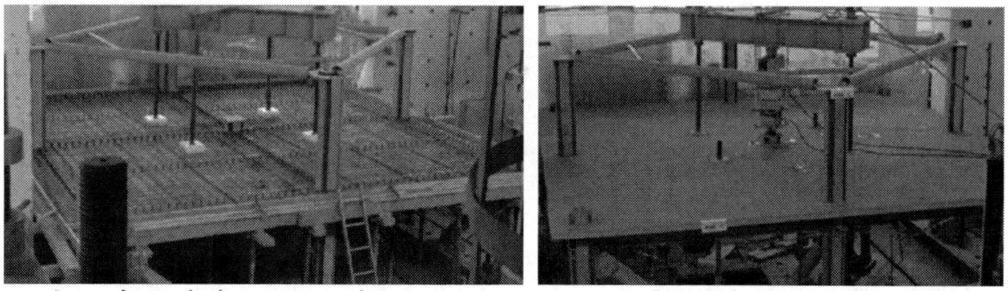

a) specimen during construction b) specimen ready for testing

Fig. 2: Views of specimen and test setup

In the test, measurements of displacements and strains were taken at different locations in the structure. Four transducers were used to measure the vertical displacement below the removed column. These four transducers were aligned in two perpendicular directions to capture the possible spatial movement of the column. Four other transducers were used to measure the horizontal displacements at the mid-height (beam floor level) of the middle perimeter columns (Fig. 3.a). Apart from the vertical displacement below the removed column, beam end rotation is very useful for characterizing the deformation capacity of the structure. Thus, each internal main beam had six transducers, three at each end, to determine the beam end rotations (Fig. 3.a). The strains were measured at one beam end (B1) using the digital image correlation system DIC [13], which employs an optical method to measure deformation on an object's surface [14], see Fig. 3.b. A total station mounted at high elevation was also used to monitor the vertical deflection at different locations on top side of the floor. The system had an accuracy of 1 mm. Monitored points were located on the diagonals of the floor and along the internal beams. Owing to several obstructions, some points were not fully accessible during the testing.

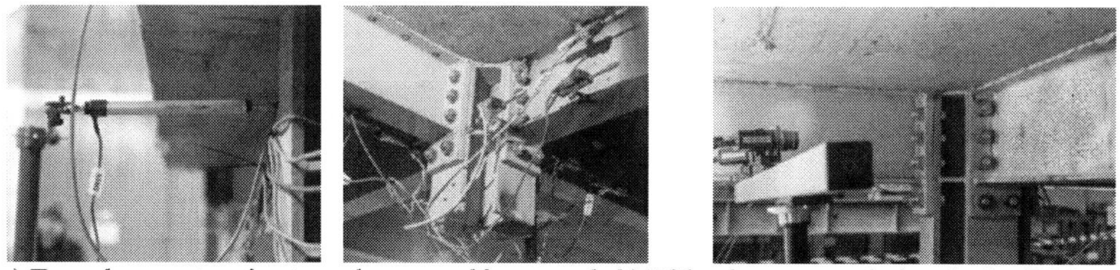

a) Transducers at perimeter columns and beam ends b) Video image correlation device DIC

Fig. 3: Instrumentation of specimen

3. Experimental results

3.1 Test observations

The force was applied statically to the top of the central column and gradually increased until the failure of the specimen using a displacement-controlled loading procedure. Several short pauses were made to allow the inspection of the specimen and make relevant notes. These pauses can be identified by the spikes visible in the force–displacement curve. Because the stroke of a hydraulic jack was not sufficiently long, two stops were made to allow the extension. During these stops, the force was reduced to zero. First cracks in the concrete floor were observed on the top surface, at a vertical displacement of about 30 mm, and followed a circular pattern in the region near the middle perimeter columns (Fig. 4.a). More cracks started to develop afterwards, spreading in and out from the first visible crack line. As seen from the vertical force vs. vertical displacement curve (Fig. 4.b), the maximum applied force was 910 kN, and the corresponding displacement was 279 mm. After the attainment of peak force, the capacity slightly decreased to 899 kN, when the beam B2-B3 fractured at end B2, near the connection with the central column (point A in Fig. 4.b), see Fig. 5.a. The corresponding vertical displacement was 291 mm. The fracture was initiated at the bottom flange, and then extended in the web for approximately 120 mm, causing a significant drop in the force. Full depth cracks developed within the floor in the central region, around the removed column. The test continued, and the force started to increase from 674 kN to 759 kN, until the beam A2-B2 also fractured at end B2, near the connection with the central column (point B in Fig. 4.b). The corresponding vertical displacement was 348 mm. To note that, at this point, the beam B2-B3 suffered complete failure (the beam separated from the end plate), see Fig. 5.b. The concrete was completely crushed and started to detach from the beams top flanges. No permanent deformations were visible in the end plate bolted connections of the main beams to the perimeter columns or to the central column, respectively. The test continued until the concrete floor completely separated from the steel beams in the central region surrounding the removed column (see Fig. 5.c); after that the test was stopped (point C in Fig. 4.b). As can be seen from Fig. 4.a, crack development followed an almost circular pattern. At failure, the concrete slab failed in shear and completely separated from the beams. After the test, the specimen was dismantled and the concrete slab has been removed from the supporting beams. This unveiled that the shear studs did not suffered any visible plastic deformations.

The plastic deformations at beam-ends that withstood hogging moments (on the perimeter) were much smaller than those measured at internal beam ends under sagging moments, see Fig. 6. Thus, Fig. 7 shows the rotation in main beams, at both ends. In point A (first fracture in beams), the ultimate rotation in beam B2-B3 at end B2N amounted 0.121rad. In the beams located on the other direction, the rotation continued to increase, until, at point B, beam A2-B2 also failed in section B2W, at an ultimate rotation of 0.160 rad.

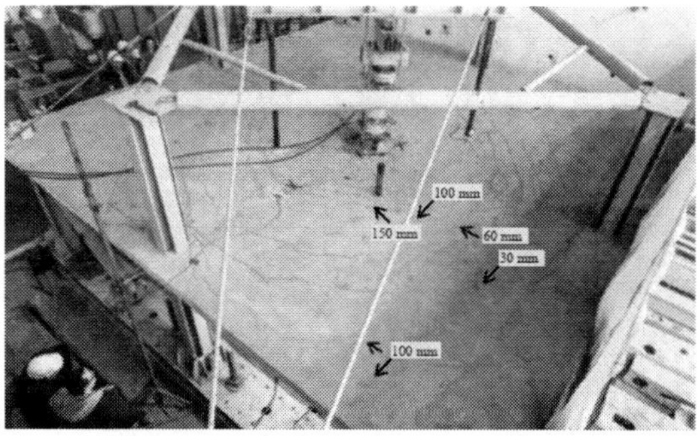

a)

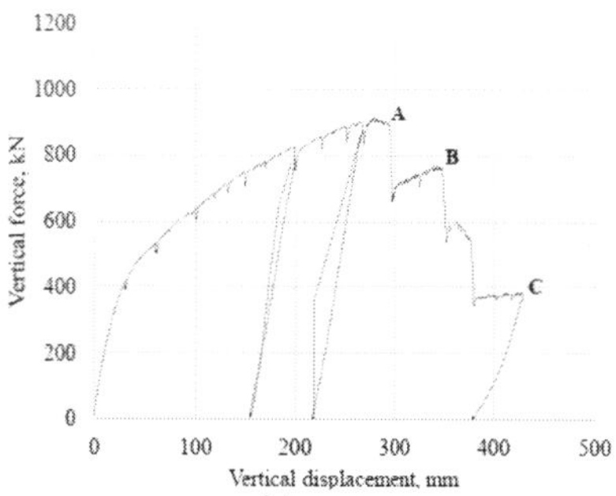

b)

Fig. 4: Specimen after the test (a) and vertical force vs. vertical displacement of central column (b)

a) fracture at beam B2-B3

b) fracture at beam A2-B2

c) central node at the end of the test

Fig. 5: Details with failure mechanism

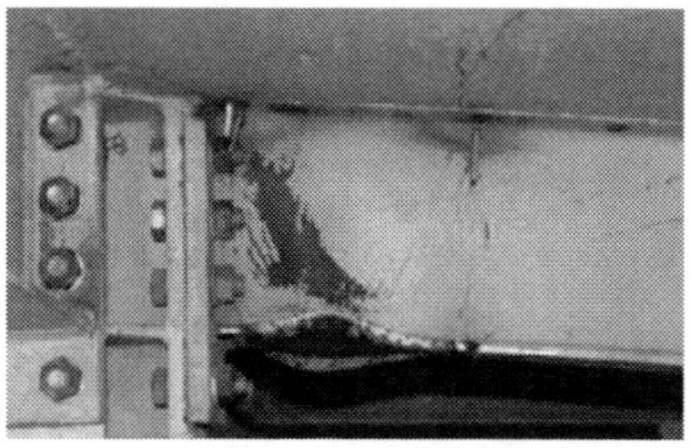

a) beam A2-B2, section A2

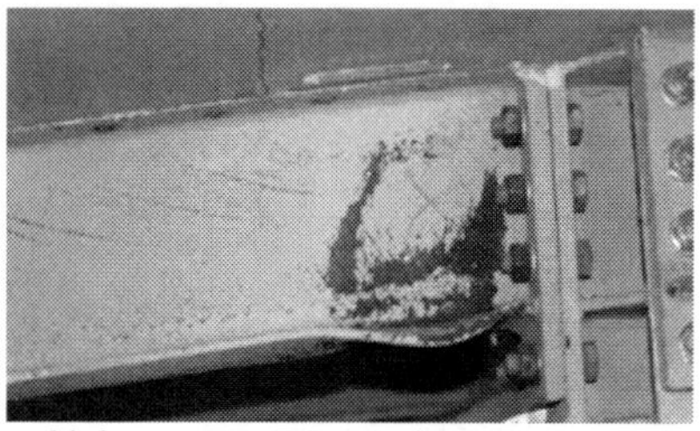

b) beam B1-B2, section B1

Fig. 6: Plastic deformations in beams at point B (see Fig. 4)

Fig. 8.a shows the strain map contour at 200 mm vertical displacement at beam end C2, which withstood hogging moments during the test, while the strain along the height of the beam at 70 mm from same beam end (i.e. C2) is shown in Fig. 8.b. After 250 mm vertical displacements measurements were stopped due to safety reasons (risk of concrete spalling falling on the cameras). The composite action with floor slab causes a shift of 55 mm in the location of neutral axis of the beams at end under hogging moment. To note that the position of the neutral axis does not change until 250 mm vertical displacement, indicating that the axial force in beams has little influence. Fig. 9 shows the displacement profile of the main beams and of the floor slab diagonals, at different load levels. Before 408 kN, the beams are in elastic and exhibit a typical flexural behaviour, with the inflection point at the middle of the span (centre column), see Fig. 9.a. After this load level, the plastic deformations initiate at beam ends and the deflection profile changes, each beam showing distinct inflection points located around the mid-span of each individual beam. As the beam ends under sagging moment exhibit larger plastic deformations compared with the beam ends under hogging moment, beyond 747 kN, the deflection profile of each beam changed and became shaped like a cantilever, almost straight to the tip (central column). A similar behaviour was observed for the concrete floor, with the observation that, at same relative position to the centre of the floor, the displacements measured smaller values when compared with the beams.

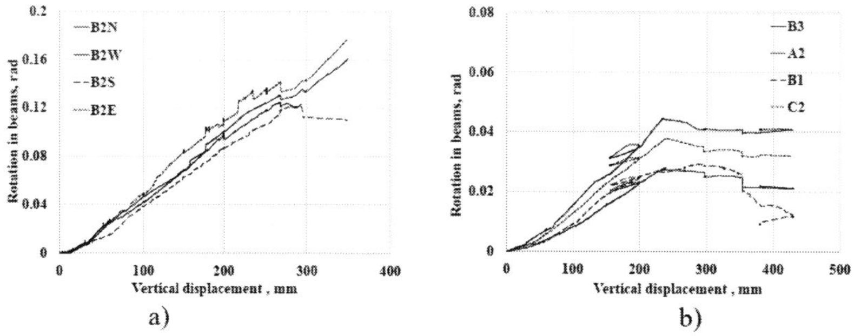

Fig. 7: Rotation in beams: a) beam ends near central column (sagging); b) beam ends near marginal columns (hogging) (for notation of sections, see Fig. 1.b

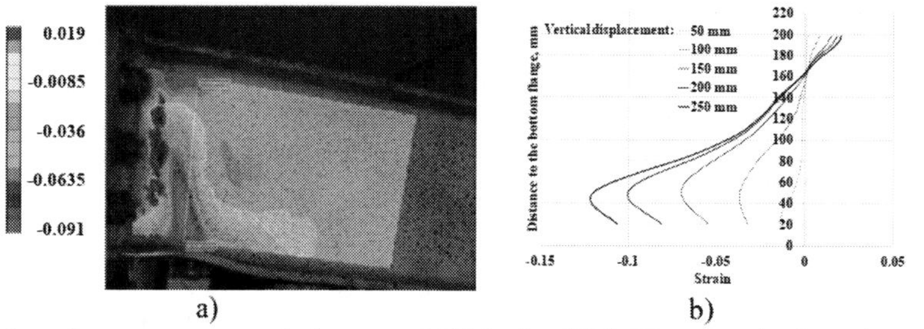

Fig. 8: Optical strain measurements in beam end C2 (using VIC-3D system): a) strain map contour at 200 mm vertical displacement; b) evolution of strain at 70 mm from beam end

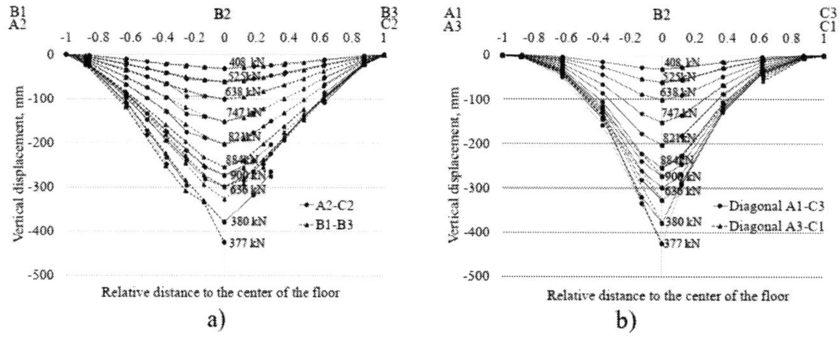

Fig. 9: Vertical displacement along the span of the beam (a) and floor diagonals (b)

3.2 Composite specimen vs. bare steel specimen

In order to observe the effect of the beam-floor interaction on the behaviour of the frame specimen, the results are compared with the bare steel frame results obtained in a previous test [11]. The vertical force versus vertical displacement curves for the central column are shown in Fig. 10. The maximum recorded force was 910 kN, 24% larger than the peak force in case of bare steel specimen, while the corresponding displacement was 279 mm, 50% less when compared with the bare steel specimen. As seen from Fig. 10.a, the composite beam specimen has a larger initial stiffness and yield strength compared with the bare steel specimen. After the elastic stage, both specimens show a significant increase of capacity until the peak load is attained. However, the mechanisms contributing to this increase are different. Thus, for the composite specimen, the main source is the compressive arch action, which can be identified by the outward movement of the four perimeter columns (A2, C2, B1, B3), see Fig. 10.b. The arch action is caused by the asymmetry of composite beam cross section under hogging and sagging moment, respectively. In case of bare steel specimen, a catenary force starts to develop under large deformation conditions, leading to a significant increase in the ultimate deformation capacity. The development of the catenary forces in beams causes the columns to deflect inward, see Fig. 10.c. The ultimate rotation for composite beam specimen, which amounted 0.160 rad, is smaller than that corresponding to bare steel specimen, i.e. $\theta_u = 0.206$ rad [11]. This reduction is mainly caused by the composite action with floor slab and the shift in the location of neutral axis, because for similar beam rotations, larger strains are expected in beams when compared to bare steel beams.

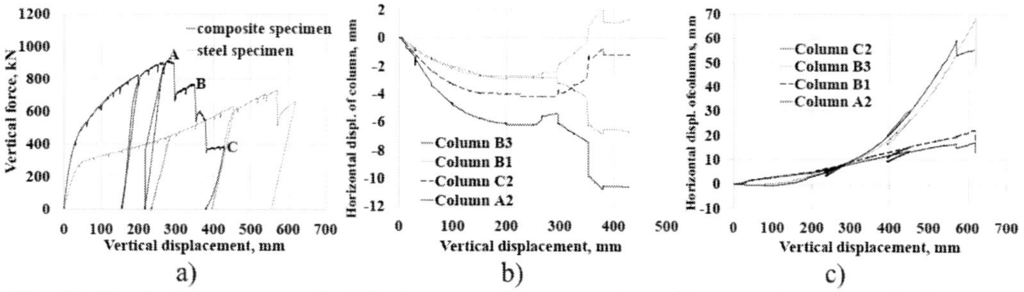

Fig. 10: Vertical force vs. vertical displacement for composite and bare steel specimen [11] (a) and horizontal displacement of middle columns vs. vertical displacement of central column for composite beam specimen (b) and for bare steel specimen [11] (c)

4. Conclusions

In this study, we investigated the capacity of steel frame systems with composite beams to resist collapse under the loss of a column. A specimen with two bays and two spans was extracted from a six-story steel moment-frame building designed to resist seismic forces in a low-seismicity area. The support of the central column was removed, and a hydraulic jack was used to push the column down. As the load increased, plastic deformations occurred at the internal beams ends because of bending effects. First beam fractured at a vertical displacement of 291 mm, and then the capacity started to decrease, until a second beam fractured, at a vertical displacement of 348 mm. Due to the composite action between the steel beam and concrete slab, the peak force was 24% larger than that corresponding to bare steel specimen, while the ultimate rotation of beams reduced correspondingly. The beam-to-column connections showed very good behaviour and had sufficient strength to allow the development of plastic hinges in the beams. A numerical model will be developed and validated against the experimental data. To generalize the findings from this study, a numerical investigation will be carried out. This will include different frame typologies and different initial design conditions, e.g., structures designed for different seismic hazard levels.

Acknowledgments

Partial funding for this research was provided by the Executive Agency for Higher Education, Research, Development and Innovation Funding (UEFISCDI), Romania, under grant PCCA 55/2012 "Structural conception and collapse control performance based design of multistory structures under accidental actions" (2012-2016).

References

[1] NISTIR 7396 "Best Practices for Reducing the Potential for Progressive Collapse in Buildings, National Institute of Standards and Technology", Oakland, CA, 2007.

[2] EN1991-1-7 "Eurocode 1 – Actions on structures – Part 1-7: General actions – Accidental actions", 2006.

[3] American Society of Civil Engineers (ASCE).Minimum Design Loads for Buildings and Other Structures. ASCE/SEI 7-05, Reston, Virginia; 2005.

[4] UFC 4-023-03 – Unified Facilities Criteria: Design of Buildings to Resist Progressive Collapse, 2009.

[5] Astaneh-Asl, A., Jones, B., Zhao, Y., & Hwa, R. "Floor catenary action to prevent progressive collapse of steel structures", (Report No. UCB/CE-Steel-03/2001). Department of Civil and Environmental Engineering, University of California, Berkeley, USA, 2001.

[6] Stevens, D, "Assessment and Proposed Approach for Tie Forces in Framed and Load-bearing Wall Structures". Final rep., Prepared for Security Eng. Working Group, 2008.

[7] Demonceau JF, Jaspart JP. "Experimental test simulating a column loss in a composite frame", *Advanced Steel Construction*, 6, 891-913, 2010.

[8] Alashker Y, El-Tawil S, Sadek F. "Progressive collapse resistance of steel-concrete composite floors", *Journal of Structural Engineering* ASCE 136(10), 1187–1196, 2010.

[9] Dinu F, Dubina D. "Robustness based design of steel building frames under extreme loads", *STESSA 2012 Behavior of Steel Structures in Seismic Areas*, 9-11.01.2012, CRC Press, Taylor & Francis Group, ISBN:978-0-415-62105-2, 905-912, 2012.

[10] Pham XD, Tan KH. "Experimental study of beam–slab substructures subjected to a penultimate-internal column loss" *Engineering Structures* 55, 2–15, 2013.

[11] Dinu F, Marginean I, Dubina D, Petran I. "Experimental testing and numerical analysis of 3D steel frame system under column loss", *Engineering Structures,* 113, 2016.

[12] CODEC. "Structural conception and collapse control performance based design of multistory structures under accidental actions" (2012–2016), Exec. Agency for Higher Ed., Resc., Development and Innov. Funding, Romania, PN II PCCA 55/2012.

[13] VIC-3D Digital Image Correlation Software, Correlated Solutions Inc., 2016.

[14] Sutton MA, Orteu JJ, Schreier H. "Image correlation for shape, motion and deformation measurements: basic concepts, theory and applications", *Springer Science & Business Media*, 2009.

EFFECTS OF CLOSE RANGE BLASTS ON STEEL FRAMES. EXPERIMENTAL TESTING AND NUMERICAL VALIDATION

Florea Dinu[a,b], Ioan Mărginean[a], Andreea Sigauan[a], Attila Kovacs[c], Emilian Ghicioi[c] and Dragos Vasilescu[c]

[a] *Department of Steel Structures and Structural Mechanics, Politehnica University Timisoara, Romania*
[b] *Laboratory of Steel Structures, Romanian Academy, Timisoara Branch, Romania*
[c] *INCD INSEMEX Petroşani, Romania*

Abstract: During last decades, there was an increased interest from research and design professionals to provide effective strategies in protecting buildings and other assets from the direct effects of blasts or other incidents. Experimental tests, conducted over a large range of distances and charge weights, helped at developing analytical approaches and charts which can be used to calculate blast parameters. Due to the lack of test data and inapplicability of common scaling rules, in the last years special attention was devoted to close-in blasts, located in the proximity of the structural elements. Such explosive charges may cause extreme local damage of the elements or even complete loss of load bearing capacity. In the study presented in the paper, two types of beam-column assemblies have been tested under explosive charges detonated close to the specimens. Numerical models, developed using Extreme Loading for Structures software, were validated using the test data collected in the experimental program.

1. Introduction

Over their lifespan, the buildings may be subjected to a variety of actions. Design and execution of buildings should take into account that some of these actions may reach extreme values, much higher than those currently used at initial design. Therefore, the buildings must be able to resist such events without being damaged disproportionately compared to the original cause. One example is the gas explosions that are produced in residential buildings due to failure or improper use of the gas facilities. In such situations, the structure can be severely damaged, leading to partial or complete failure, e.g. the Ronan Point building, in London, in 1968. Explosions caused by gas accumulations have great significance, given the large number of residential buildings (and other types of spaces) where it is currently used. Hazards can be caused both by external distribution networks and the devices that use the gas inside. For example, in the United States, the average probability of producing a gas explosion is about 1.8×10^{-5}/ year [1]. Studies conducted in the UK showed similar values of the order 2.3×10^{-5}/ year to 1.86×10^{-5}/ year. In other countries, e.g. Romania, there are no public data about such events, and this makes it difficult to assess the risk of such events based on the number of dwellings. The second important issue is the maximum pressure in the rooms affected by the explosion. Studies conducted after the accident in the Ronan Point building showed that these pressures rarely exceed the value of 17 kPa. Even if this value is much higher than "ordinary" loads, it is less than the amount recommended in EN 1991-1-7 [2], namely 34 kPa. The pressure resulting from gas explosions is therefore very high compared to permanent or live loads (occupancy, wind, snow), which rarely exceed 4-5 kPa. The presence of voids limits first pulse intensity and pressure fluctuations of the second phase presents a reduced significance for the structural response. Therefore, it can be considered that the effect on the structure is quasi-static (Fig. 1.a).

Unlike internal gas explosions, external explosions are usually produced from intentional causes, the source being both classic explosives (TNT, C-4, Semtex) but also improvised charges, like those based on ammonium nitrates, a chemical fertilizer widely used in agriculture. Because of the intentional nature, these events cannot be modeled using a Poisson distribution model, which is widely used to model the random occurrence of rare events [1]. Explosions caused by explosives, known as detonations, create shock waves that propagate at high speeds from the point of detonation. The initial phase, of positive pressure, is followed by a longer and less intense negative pressure phase, Fig. 1.b. Unlike gas explosions, detonations induce significant dynamic effects, and therefore they must be accounted for in design.

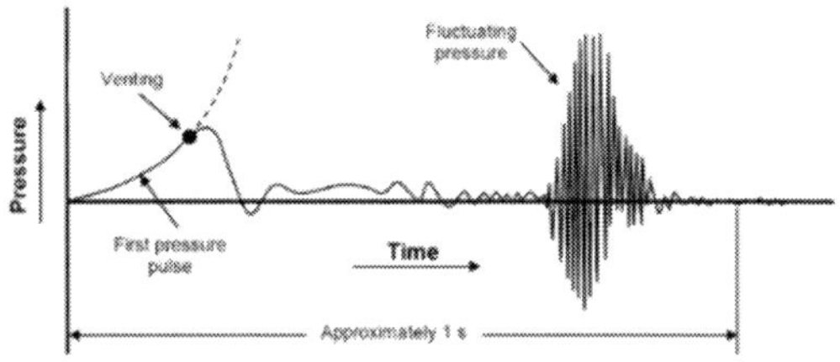

a) gas explosion ([3])

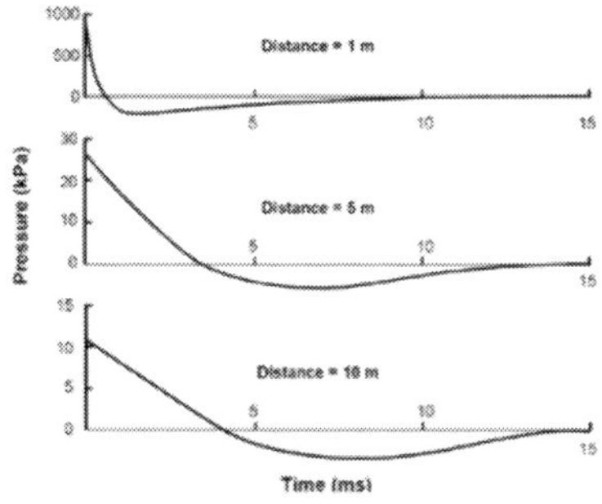

b) blast wave ([4])

Fig. 1: Typical pressure-time records

In terms of the effects on the structure, distance between the source of the explosion and structure plays a very important role. When the explosive charge is placed at greater distance from the structure, failure usually occurs due to excessive bending. The pressure released by the shock wave can be considered as uniformly distributed on the surface of the element. An explosion at a very short distance from the building causes a localized failure, like punching or shear failure. This is caused by the rigidity of the structural element, which produces inertial resistance to blast. Punching or shear-type failure occurs before structural element is able to respond by bending.

The study presented in the paper investigated the response of beam-column steel elements subjected to close-in blasts. The research focused on the evaluation of typical steel frames response when subjected to direct effects of a blast and the direct calculation of blast loads for further numerical simulations. Two 3D frame specimens were tested inside a bunker using different charge weights, located at different distances. A numerical model was also validated using the advanced nonlinear dynamic analysis software Extreme Loading for Structures [5]. The study is part of a research program devoted to the design of structures to sustain extreme loading events without collapse [6].

2. Specimens, test setup and blast loading

2.1 Description of specimens

Two 3D specimens were designed and constructed for testing under blast effects inside a bunker. Specimens were extracted from a typical moment resisting steel frame structure, see highlighted areas in Fig. 2.a. Due to space limitation in the bunker, the specimens were scaled down from 8.0 m span to 3.0 m span. First specimen (see Fig. 2.a, Fig. 3.a) includes a column (blast induces deformations in the major axis direction), two half-span longitudinal beams rigidly connected to the flanges of the column using extended end plate bolted connections and one half-span transversal beam, connected to the column web using a simple clip angle connection.

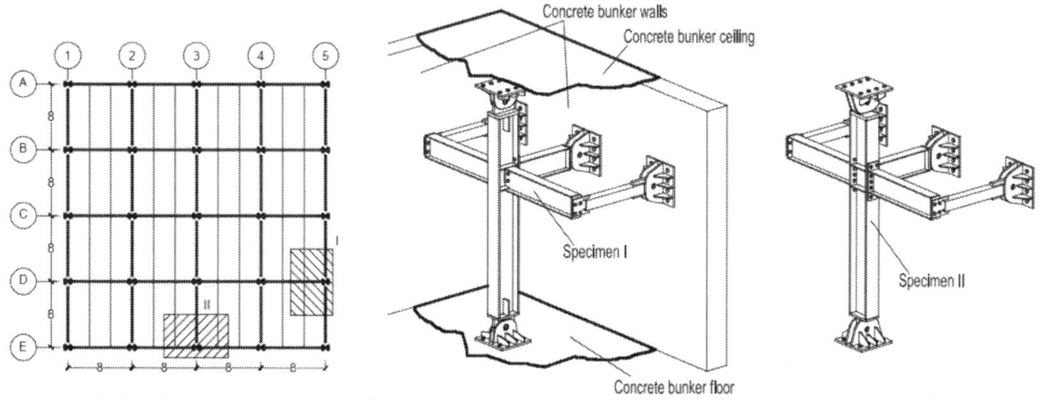

a) plan layout b) specimen I with the bunker details c) specimen II

Fig. 2: Plan layout of the reference multi-story building with the position of the specimens extracted for testing and 3D viws of specimens

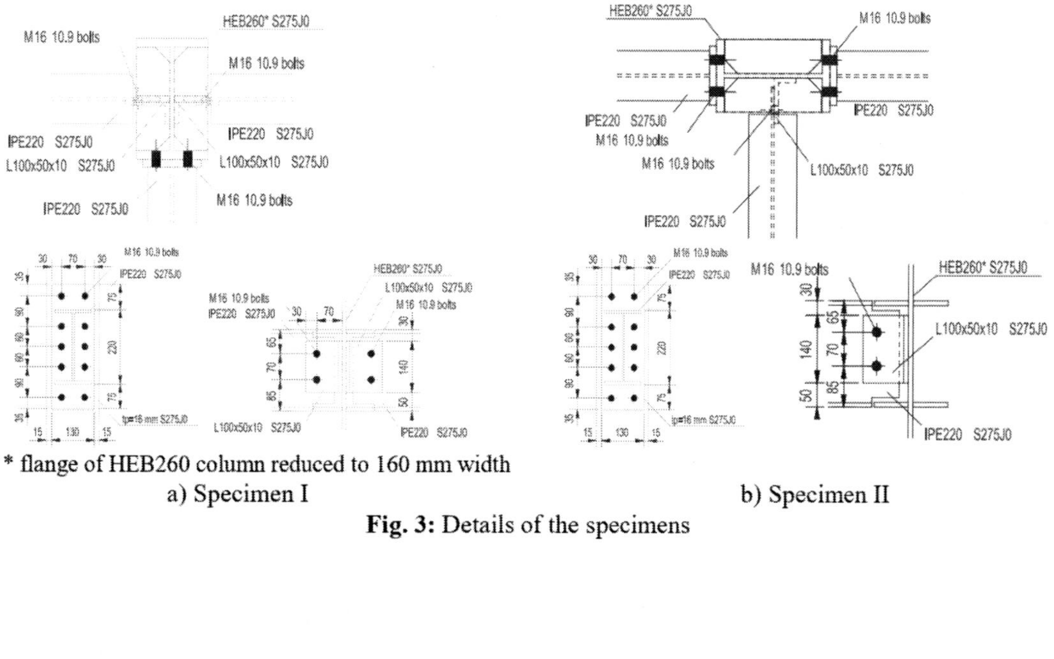

* flange of HEB260 column reduced to 160 mm width

a) Specimen I b) Specimen II

Fig. 3: Details of the specimens

Second specimen (see Fig. 2.b, Fig. 3.b) includes a column (blast induces deformations in the minor axis direction), two half-span longitudinal beams connected to the column web using simple clip angle connections and one half-span transversal beam, connected to the column flange using extended end plate bolted connections. For both specimens, lateral restraints made from tubular profiles were used at the ends of longitudinal beams. IPE 220 section was used for beams, while columns were made from HEB 260, but with flanges reduced to 160 mm width. Steel material in plates and profiles was S275J0 and bolts were grade 10.9. Table 1 summarizes the measured material properties of the specimens.

Table 1: Average characteristic values for materials in steel profiles, plates and bolts

Element	f_y (N/mm²) yield strength	f_u (N/mm²) ultimate strength	ε_y (%) yield strain	A_{gt} (%) Total elongation at maximum stress
Beam flange IPE220, t = 9.2 mm	345	464	0.16	28.0
Beam web IPE220, t = 5.9 mm	353	463	0.17	30.4
Column web HEB 260, t = 10 mm	407	539	0.19	27.0
Column flange HEB 260, t = 17.5 mm	420	529	0.20	27.0
End plate, t = 16 mm	305	417	0.15	17.1
Bolt, M16 class 10.9	965*	1080	1.76*	12.0

Note: * 0.2% offset yield point

2.2 Blast effects

The main hazard components of an explosion are blast (overpressure), fragmentation, and thermal effect. In our study, only first issue has been of interest. As seen in Fig. 1.b, the peak pressure value depends very much on the distance of the detonation point from the structure of interest. Thus, the peak pressure value, and also the velocity of the blast wave, decrease rapidly by increasing the distance between the blast source and the target surface. The effect of distance on the blast characteristics can be taken into account by the introduction of scaling laws [7]. These laws have the ability to scale parameters, which were defined through experiments, in order to be used for varying values of distance and charge energy release [8]. The experimental results are, in this way, generalized to include cases that are different from the initial experimental setup. The most common blast scaling law is the one introduced by Hopkinson [9] and Cranz [10]. According to Hopkinson-Cranz law, a dimensional scaled distance is introduced as described by Eq. (1):

$$Z = R/\left(W^{1/3}\right) \tag{1}$$

where: Z is the scaled distance, in m/kg$^{1/3}$, R is the distance from the detonation source to the point of interest [m] and W is the weight of the explosive [kg TNT or equivalent TNT].

As the scaled distance reduces, the peak overpressure increases. For example, in case of Murrah Federal Building in Oklahoma City, which collapsed in 1995 as a result of a blast produced by a large truck bomb (1814 kg equivalent TNT), located at a distance R of 1.5 m, the scaled distance was Z = 0.12 m/kg$^{1/3}$, which resulted in a very large peak overpressure, estimated at 20 000 de kPa. The scaling law from Eq. (1) should be corrected when blast test is done inside a bunker. A more general expression to evaluate the peak pressure (or peak overpressure) and the variation with the distance (Eq. (2)), is the one proposed by Richards and Moore [11]:

$$P = A \times \left(R/\left(W\right)^{b}\right)^{a} \tag{2}$$

where: P is the peak overpressure (kPa), A is the site constant (evaluated experimentally), a is site exponent (evaluated experimentally, is always negative), b is the site exponent for the charge weight (evaluated experimentally), R is the distance from the detonation source to the point of interest [m] and W is the weight of the explosive [kg TNT or equivalent TNT]. Using experimental calibration tests, Eq. (2) can be used in case of bunker tests. In the study, calibration blast tests were performed first, in order to evaluate the site exponents A, b and a. Then, the specimens were subjected to blast of increasing intensities, obtained by increasing the charge weight and/or reducing the distance from the blast to the specimen.

3. Experimental tests

Fig. 4 shows the first specimen tested in the bunker. In order to evaluate the site exponents A, b and a, and then the pressure inside the bunker, two Kiestler pressure sensors were mounted on a special frame, at 4 meter distance from the specimen, in front of the bunker venting. Explosive material used in the testing has a TNT equivalence of 1. To note that the effects of gravity loads on the columns and beams were not considered in the test.

First four blasts used charges of $m_{1,1}$ = 125 g, $m_{1,2}$ = 250 g, $m_{1,3}$ = 500 g and $m_{1,4}$ = 1000 g, located at 0.27 m from the beams mid-length, freely suspended (see Fig. 4). Last blast used a charge weight $m_{1,5}$ = 1000g, located at the external face of the column, at mid-height, see Fig. 4. The pressure measured after each explosion were $p_{1,max}$ = 20 kPa, $p_{2,max}$ = 40 kPa, $p_{3,max}$ = 80 kPa and $p_{4,max}$ = 175 kPa. The peak pressure during last explosion attained the maximum value, $p_{5,max}$ = 1400 kPa.

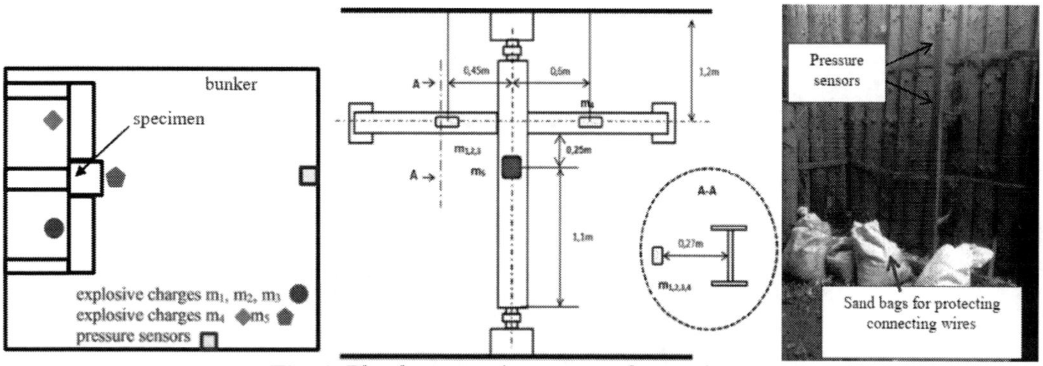

Fig. 4: Plan layout and test set-up for specimen I

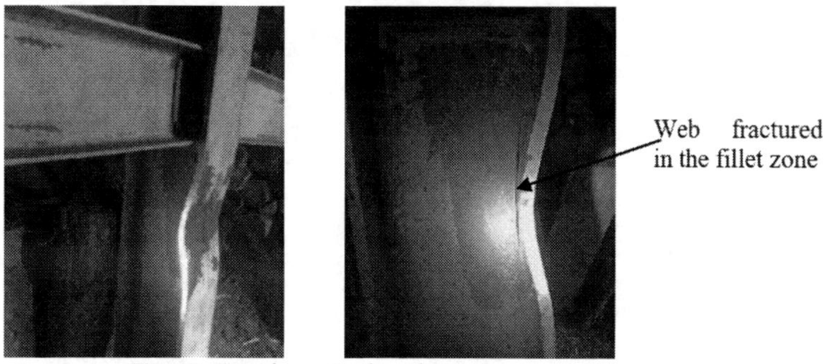

Fig. 5: Specimen I after the test

While for first four explosions there were no visible deformations at the level of the specimen, the last charge, m_5, caused severe local deformations of the column near the source, and the web complete fractured at the contact with the flange, in the fillet zone, for a length of 300 mm.

For calibration of experimental set-up, a charge with weight m = 125 g was detonated at a distance d = 0.27m from the position of the pressure sensors. The value of the peak pressure was $P_{calibration}$ =1400 kPa. With the values of the pressure measured during each detonation, the following specific coefficients of the bunker were determined: A = 3850, a = -0.73, b = 3.87/3 = 1.29. The peak pressure value of the blast wave, decrease rapidly with the distance between the blast source and the target surface, as seen in Fig. 6. Apart from the pressure measurements, also the permanent deflection of the specimen was measured in several points.

Second specimen was tested for three charges. First charge had a weight $m_{2,1}$ = 500 g, located at distance D_1 = 0.5 m from the specimen and H = 1.0 m height from the base of the column. No visible effects were recorded, nor in the beams or in the column. The test continued with detonation of a charge of $m_{2,2}$ = 1815 g, located at the same distance and height (D_2 = 0.5 m, H = 1.0 m). Web of the column has been deformed plastically out of plane for approximately 22 mm, but without any visible cracks (Fig. 7.a). The third detonation was produced by a charge blast of $m_{2,3}$ = 1815 g, located at same height but closer distance, $D3$ = 0.20m. The web of the column has been completely removed within a length of almost 600 mm around the point of detonation, see Fig. 7.b.

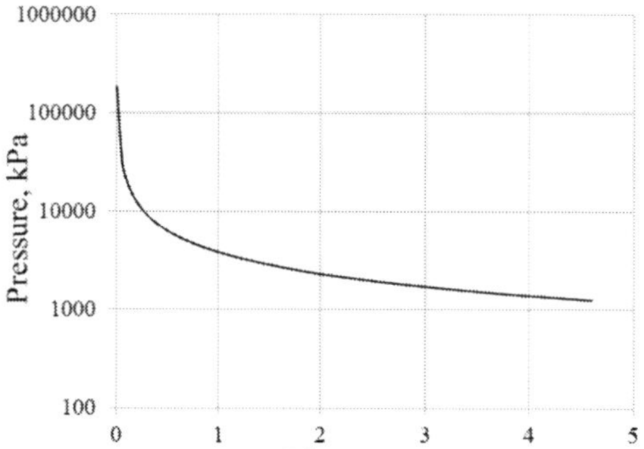

Fig. 6: Peak pressure vs. distance between the blast source and the target surface

a) Plastic deformations in column web after blast $m_{2,2}$

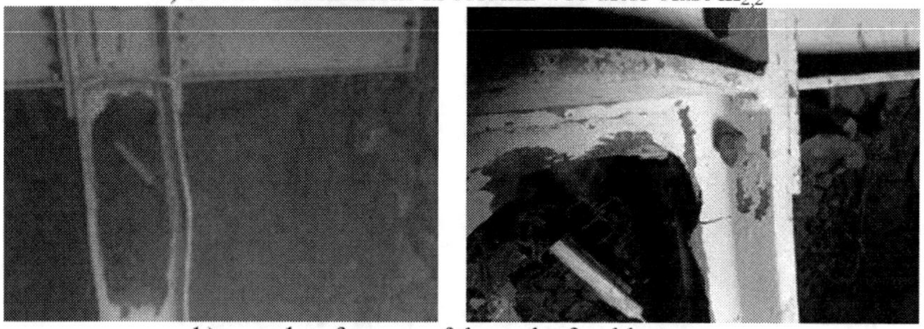

b) complete fracture of the web after blast $m_{2,3}$

Fig. 7: Results of blasts for Specimen II

To note that, because the effects of gravity loads were not considered in the test, the column did not failed. However, in real conditions, the gravity loads are always present in the structure, and therefore such damage would be similar to a complete loss of bearing capacity (removal of the column). The permanent deflections of the specimen were measured in several points after the test and were used for numerical model validation.

4. Calibration of the numerical model

The performance of the steel specimens undergoing close range detonations was also predicted using ELS [5] (see Fig. 8). ELS utilizes a nonlinear solver based on the applied element method [12] which is a derivative of the finite element method and the discrete element method. In ELS, the structure is modeled as an assembly of small elements, which are assumed to be connected by one normal and two shear springs located at contact points distributed around the element edges. The average normal strain is calculated by taking the average of the absolute values of strains on each face. When the average strain value at the element face reaches the separation strain, all springs at this face are removed and elements are not connected any more (until they collide). Columns, beams, and plates were modeled as solid elements and could undergo deformations at the interface between the discretized elements. The constraints, made of tubular sections, were also modeled as solid elements. The bolts were modeled using individual springs: one for normal stresses and two for shear stresses. The column bases and transversal beam end connection to the bunker wall were considered pinned, and all displacements were prevented.

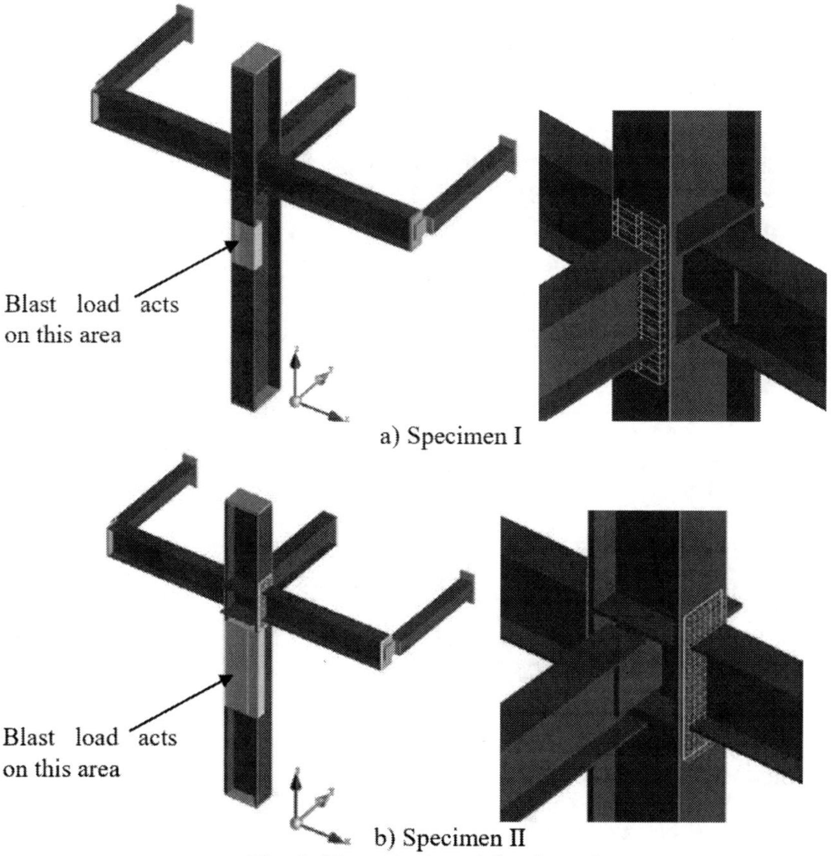

a) Specimen I

b) Specimen II

Fig. 8: Numerical models of specimens

Materials were modelled using characteristic properties presented in Table 1. Blast pressure was modeled using the experimental data presented in previous section. The variation of pressure with distance obtained in the experimental test were introduced in the program, and the point of detonation was the same, i.e. $D = 0.2$ m from the column face. The shaded areas presented in Fig. 8 (left) indicate the surface of the elements loaded with the blast pressure.

Due to the dynamic, impulsive character of the blast load, strain rates effects in the material are very important. The rate dependency has been considered by means of the following relationships[13]:

$$\frac{f_{ysr}}{f_y} = 1 + \frac{21}{f_y}\log\frac{\dot{\varepsilon}}{\varepsilon_0} \quad (3)$$

$$\frac{f_{usr}}{f_u} = 1 + \frac{7.4}{f_y}\log\frac{\dot{\varepsilon}}{\varepsilon_0} \quad (4)$$

where: $\dot{\varepsilon}$ = strain rate, $\varepsilon_0 = 10^{-4}$

f_y, f_u = yield and tensile strength in quasi-static conditions, $\varepsilon_0 = 10^{-4}$

f_{ysr}, f_{usr} = yield and tensile strength at strain rate $\dot{\varepsilon}$.

Because the strain rate is not initially known, the analysis is done first using static material properties, see Table 1. The strain rate is then calculated in the location of interest and material properties are corrected using Eq. (3) and Eq (4). For first specimen, the maximum strain rate recorded in the column web was $\dot{\varepsilon} = 312.23$ (1/sec), while for second specimen $\dot{\varepsilon} = 83.56$ (1/sec).

Fig. 9 and Fig. 10 display the deformed shape of the specimens and the displacement history (along Y axis). Comparisons between numerical and experimental results show a very good agreement. The permanent deflection of the columns is very close to the measurements done after the test. Also, the extension of damage in the columns are very similar, with the same location and extension of fracture lines.

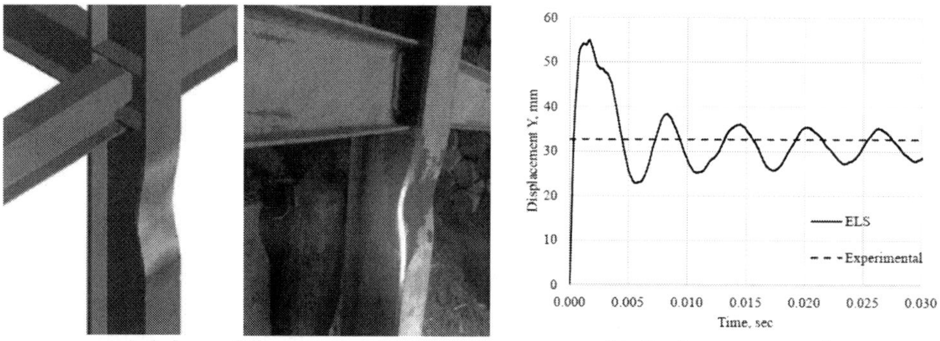

a) deformed shape b) displacement vs. time

Fig. 9: Specimen I, numerical vs. experimental, test $m_{1,5}$

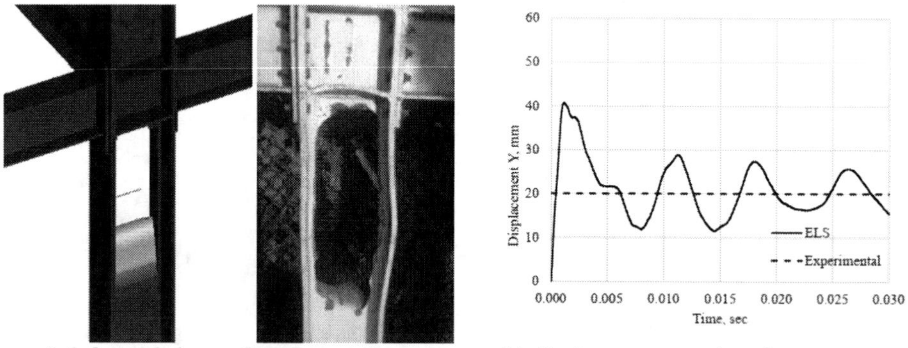

a) deformed shape after test $m_{2,3}$ b) displacement vs. time for test $m_{2,2}$

Fig. 10: Specimen II, numerical vs. experimental

The blast load pressure and therefore the blast effects decrease very rapidly with the distance from the explosive charge. For close-in detonations, the effects can concentrate in the area of the element adjacent to the explosive charge, while the effects on the remaining structure can be neglected. Thus, Fig. 11.a and Fig. 11.b show the shear strains in springs located in the filet zone of the column web, within a length of 50 cm for specimen I and 60 cm for specimen II, respectively, measured from the bottom side of the beams. It can be seen that, even for small charges (maximum weight 1815g equivalent TNT), the local damages are extreme, but localized to the area close to the point of detonation.

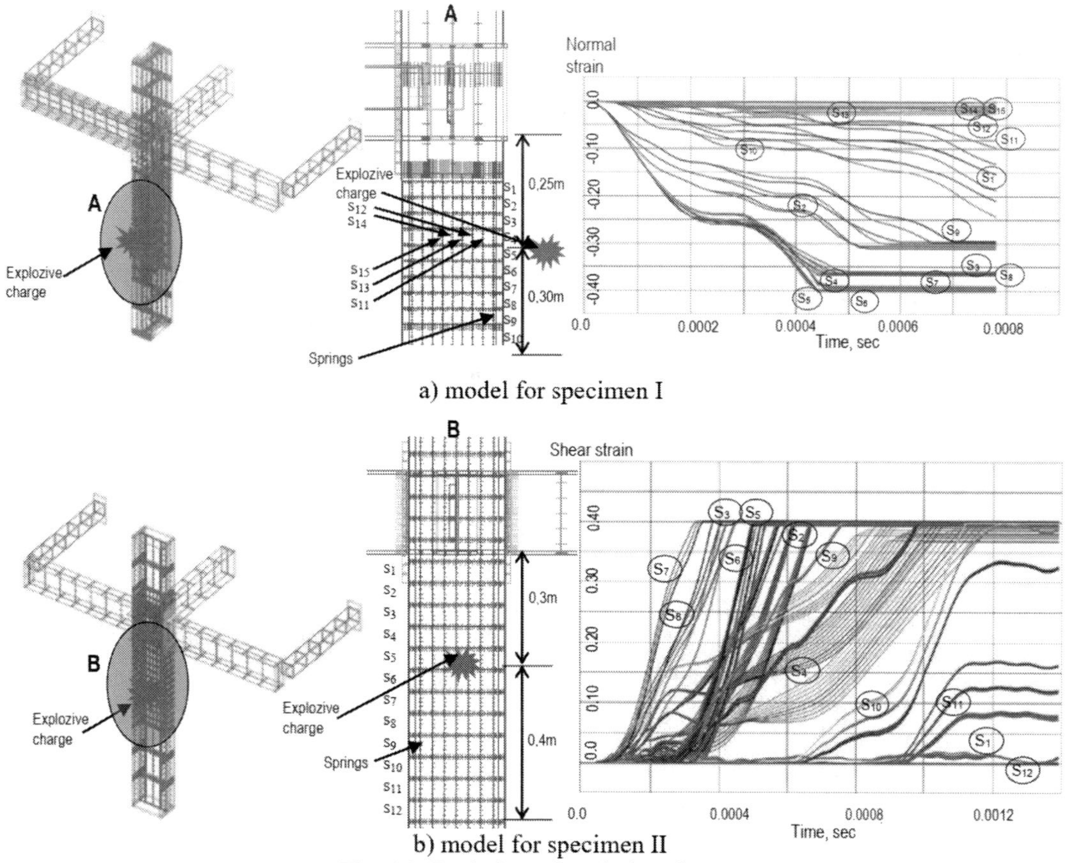

a) model for specimen I

b) model for specimen II

Fig. 11: Evolution of strain in columns

5. Conclusions

The study presented in the paper focused on the evaluation of close-in blast effects against steel frame structures. Two specimens were tested for charges of different weights and located at different distances to the specimens. The specimens were extracted from a six-story moment resisting frame structure, designed to meet the seismic design requirements for special moment frames. The variation of blast pressure with time and distance has been evaluated for bunker test conditions. The results showed that charges located at close distance can produce large damages in the members, with complete fracture of the section walls. Thus, for the specimen loaded against the strong axis (charge normal to the column flange), the blast caused sever local bending of the external flange, and fracture of the web at the contact with the flange, in the fillet zone. In case of specimen loaded against the weak axis (charge normal to the column web), punching or shear-type failure developed, with the web completely removed for a length of 600 mm, before structural element to be able to respond in bending. The later situation can be extremely dangerous, especially for buildings with perimeter steel moment resisting frames and interior gravity frames, if the loads cannot be distributed to the adjacent structural members. In the test, the complete failure of the column was prevented only due to the absence of any gravity loads. Numerical models that were calibrated using test data indicated a very good agreement, giving the possibility to extend the research to full scale structures, using different blast loading conditions.

Acknowledgments

Funding for this research was provided by the Executive Agency for Higher Education, Research, Development and Innovation Funding, Romania, under grant PCCA 55/2012 "Structural conception and collapse control performance based design of multistory structures under accidental actions" (2012-2016).

References

[1] NISTIR 7396. *Best Practices for Reducing the Potential for Progressive Collapse in Buildings*. National Institute of Standards and Technology Administration, U.S. Dep. of Commerce, 2007.

[2] EN1991-1-7: Eurocode 1: Actions on structures – Part 1-7 General actions - Accidental Actions.

[3] Leyendecker EV, Ellingwood BR. "Design Methods for Reducing the Risk of Progressive Collapse in Buildings", *Building Science Series,* No. 98, National Bureau of Standards, Washington, DC, 1976.

[4] Mainstone RJ. The hazards of Explosion. *Impact and Other Random Loadings on Tall Buildings*, Current Paper 64/74, Building Resch. Establishment, Garston, UK, 1974.

[5] ELS (2010). Extreme loading for structures (Version 3.1). Durham, NC: ASI.

[6] CODEC (2012). "Structural conception and collapse control performance based design of multistory structures under accidental actions" (2012–2016), Executive Agency for Higher Ed., Research, Development and Innovation Funding, Romania, PN II PCCA 55/2012.

[7] US Departments of the Army, Navy and Airforce (1990). Technical Manual, Army TM5-1300, Navy NAVFAC P-397, Air Force AFR 88–22, Structures to resist the effects of accidental explosions. Washington, DC: US Dep. of Commerce, Nat. Techn. Inf. Service.

[8] Karlos V, Solomos G. "Calculation of Blast Loads for Application to Structural Components", JRC 32253-2011 with DG-HOME Activity A5 - Blast Simulation Technology Development, EU, Institute for the Protection and Security of the Citizen, 2013.

[9] Hopkinson B. British Ordnance Board Minutes 13565, 1915.

[10] Cranz C. Lehrbuch der Ballistik, Springer-Verlag, Berlin, 1926.

[11] Richards AB, Moore AJ. *Blast vibration course measurement - assessment – control.* Terrock Pty Ltd, 2005.

[12] Tagel-Din H, Meguro K. "Applied element method for simulation of nonlinear materials: Theory and application for RC structures", *Structural Engineering/Earthquake Engineering*, International Journal of the Japan Society of Civil Engineers (JSCE), 17, 137–148, 2000.

[13] Kaneko H. "*Influence of strain-rate on yield ratio, Kobe Earthquake Damage to Steel Moment Connections and Suggested Improvement*", JSSC Technical Report No. 39, 1997.

EXPERIMENTAL ASSESSMENT OF CONNECTION RESPONSE IN COMPOSITE FLOOR CONSTRUCTION FOLLOWING A COLUMN LOSS

Amirhoushang Jamshidi and Robert G. Driver

Department of Civil and Environmental Engineering, University of Alberta, Edmonton, AB, Canada.

Abstract: The susceptibility of structures to extensive collapse when subjected to a localised failure due to an extreme loading event has in recent years gained considerable research attention. The scenario most often used to assess such performance, either experimentally or through computer simulation, is the loss of a column, requiring the floor system to bridge across two building bays with the aid of catenary action. The response of steel shear connections as a component of a composite floor framing system under the column removal scenario is still largely unknown due to the complex interaction of the slab and the steel framework at large deflections. While the slab itself can participate in maintaining the integrity of the overall floor system, its presence also amplifies the demand on the steel connections after experiencing initial flexural action. This paper describes the details of a comprehensive experimental program that has been completed to assess the behaviour of connections in composite floor construction by evaluating the failure mode, load carrying mechanisms, strength, and ductility of the connections and concrete slab. A variety of parameters were varied, including the connection type, connection depth, connection material thickness, and assumed beam span. The research objectives, the experimental protocols, and the most significant conclusions drawn from test observations are discussed.

1. Introduction

When a column is abruptly disengaged by an abnormal load, the resulting double-span beam must bridge over the removed column by developing a new equilibrium path to redistribute the load to the adjacent elements. This severe damage in a steel gravity frame, which is designed to carry primarily gravity loads, creates significant demands on simple shear connections. Adding the large displacement of a removed column to the double-span effect results in the development of catenary action, which changes the nature of the loading transferred to the connections. The connections must then resist combinations of shear and axial forces, and bending moment, concurrently.

Localised failure of a structural component subjected to an extreme load can initiate propagation of the failure to contiguous elements, eventually resulting in the collapse of the entire structure or a major part of it, which is known as disproportionate or progressive collapse. It should be noted that the terms "progressive collapse" and "disproportionate collapse" are used interchangeably throughout this paper because of their general reference to the degree of structural collapse, in spite of different specific meanings. Current North American building codes [1,2] do not specify any prescribed detail-oriented guidance for design of structural components against progressive collapse. However, some non-mandatory standards and guidelines, such as those of the U.S. General Services Administration [3] and Department of Defence [4], outline the alternative path approach to mitigating progressive collapse.

Many studies have focused on the behaviour of steel connections under catenary action in the absence of floor slab effects. Some of these have included physical tests on steel connections considering only the steel members [5,6,7]. However, both the loading and behaviour of a steel connection of a composite beam are different from those of a connection of a bare steel beam. Few research projects have been carried out to investigate the collapse resistance of composite floor systems with shear connections [8]. Using high-fidelity numerical simulations, it was shown by the authors [9,10,11] that the presence of a concrete slab (including composite action) changes the behaviour of connections considerably compared to those in a bare frame.

In order to study the behaviour of commonly-used steel shear connections in composite floor systems under a column-removal scenario, an experimental program consisting of 17 full-scale physical tests has been designed. This paper presents an overview of the experimental program and loading regime, along with the results of the tests. The experiments addressed only the behaviour of connections attached to the lost column – positive bending moment (sagging). However, the behaviour of connections at the other end (hogging moments) were investigated numerically, which is beyond the scope of this paper and will be published in the near future.

2. Experimental Program

2.1 Prototype

It is widely accepted that in order to study progressive collapse behaviour of connections, consideration may be given to the affected spans by imposing appropriate boundary conditions to represent the surrounding structure. Two prototype steel gravity framing systems with a rectangular plan and clear spans of 6.0 m and 9.0 m were designed. Fig. 1 shows the plan layout and elevation of one of the prototypes. Two types of shear connections were considered in this research: shear tab welded to the column and bolted to the beam; and double-angle bolted to both the column and beam.

The flooring system considered includes a 127 mm (5 in.) thick flat concrete slab connected to the beams by shear studs designed for fully-composite action. The central column in the examined area is assumed to be disengaged, so as to represent the removed column under a column-loss scenario. Only connections attached to the removed column were considered in the testing program, with the testing zone depicted in Fig. 1.

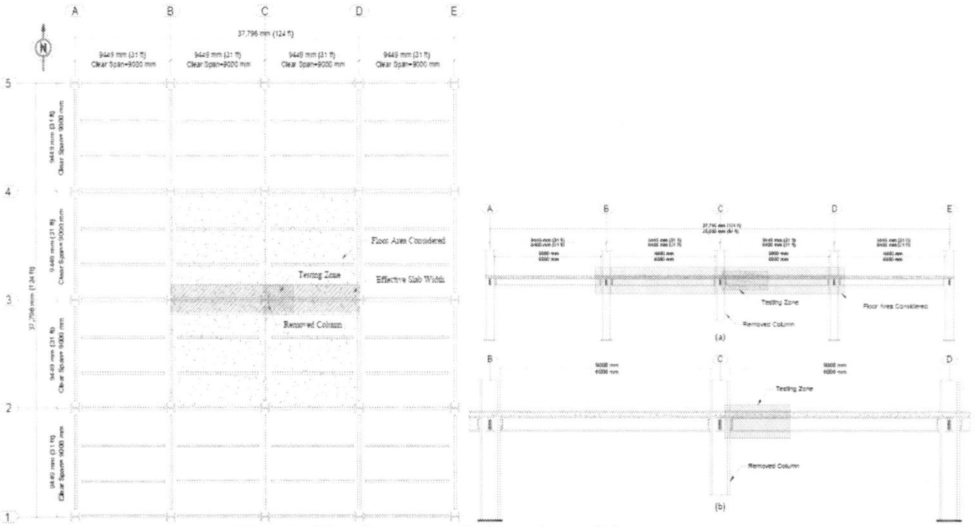

Fig. 1: Plan layout and elevation of the prototype

2.2 Test Specimens

In order to examine the effect of the composite slab on steel connections, a total of 17 full-scale steel shear connections with concrete slab were designed to test under a column removal condition. Table 1 summarizes the matrix of the experimental program. The specimens vary in type and depth of connection (and number of bolts), plate/angle thickness, and assumed span length (with associated concrete slab width). In general, the matrix is classified into two connection types: shear tab and double angle.

Table 1: Test matrix

Connection Type	Thickness	Bolt Dia.	Number of Bolt Rows		Type of Slab*
	mm	mm (in.)	3 Bolts	5 Bolts	
Shear Tab	9.5	22 (7/8)	2 + 1 (Pilot Test)	2	A
	6.4	19 (3/4)	2	2	B
Double Angle	9.5	22 (7/8)	2	2	A
	6.4	19 (3/4)	2	2	B

* Type A: Clear Span Length (L_s) = 6.0 m and Slab Width (b_1) = 1.50 m
* Type B: Clear Span Length (L_s) = 9.0 m and Slab Width (b_1) = 2.04 m

Shear-tab connections with a single plate were welded to the stub column flange and bolted to the beam web. Double-angle connections comprising two similar angles where bolted to both the stub column flange and beam web in a single vertical line at each leg. A vertical pitch of 80 mm and end/edge distance of 35 mm from the centre of the bolt holes to the edges were used. Details and dimensions of the connections are shown in Fig. 2. All connections in this test program were attached to W250×89 column stubs. Angles and plates used in the connections were specified as CAN/CSA G40.21-04 Grade 300W. Beams and columns were specified as Grade 350W and all bolts used in the connections were ASTM A325.

The width of concrete slab was taken as the lesser of 0.25 times the composite beam span and the average distance between beam centrelines. Due to the space limitation in the test set-up, the effective width of 2.04 m was constructed for the 9.0 m span instead of 2.25 m.

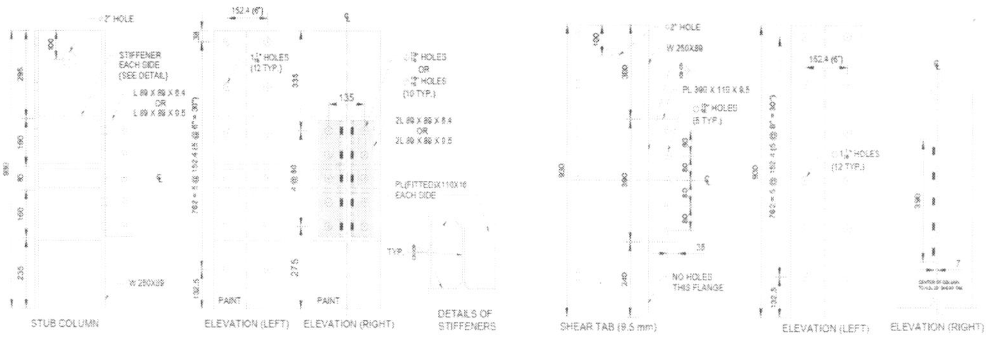

(a) (b)

Fig. 2: Details of shear connections: (a) 5-bolt shear tabs; (b) 5-bolt double angles

3. Loading Regime

In the absence of a predefined loading protocol for testing specimens in a progressive collapse scenario, column removal has been adopted by guidelines and codes as a useful method by which the robustness of structural components is assessed by bridging a two-span frame over the local failure of the middle column. A loading regime based on the column removal approach is believed to provide a gauge of capacity and ductility demands of components such as connections. It is not always feasible to test a full frame, owing to the high associated costs and structural lab limitations. One method to carry out the test, while maintaining full scale, is to isolate the area of interest of the structure and define the load history in advance. Two methods were proposed in this research: equilibrium of forces and geometrical compatibility of displacements. The pilot test and one out of 16 program tests were performed based on the first method [11], while the rest of the experiments were run based on the second approach.

3.1 Equilibrium of Forces

The equilibrium of forces method outlines a basis for proportioning shear and axial forces and bending moment at each loading stage. Fig. 3 shows a symmetrically double-span frame with a central point load and corresponding internal forces in one of the spans. The resulting vertical force carried by each end section, V, in terms of horizontal force, H, bending moments, M^+ and M^-, and the beam chord rotation angle, θ, is given by Eq. (1).

$$V = H\tan\theta + \frac{M^+ + M^-}{L_s} \quad (1)$$

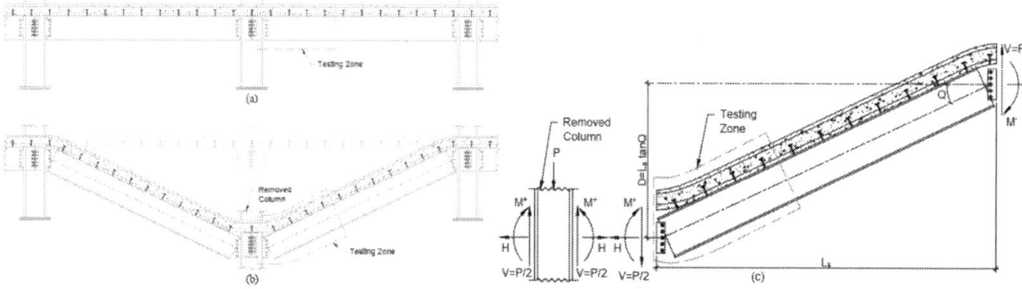

Fig. 3: Double-span frame: (a) before collapse; (b) after column removal; (c) free body diagram

Assuming equal bending moments at the ends simplifies the second term in Eq. 1 to $2M/L_S$, which results in the inflection point being located at the middle of span. However, this assumption is valid only if the two connections are similar and are placed at the centreline of the beam. In the case of asymmetrical connections, such as those in composite construction, owing to the nature of the concrete slab with different rotational stiffnesses and evolutionary behaviours in tension and compression, the assumption of similar moments is no longer valid and the location of the inflection point won't be located at the middle of span. Therefore, an approach based on the compatibility of displacements was introduced to overcome the limitations encountered in this method.

3.2 Geometric Compatibility of Displacements

This technique correlates vertical deflection of the central column, axial deformation of the connections, and beam rotation based on trigonometry, as shown in Fig. 4. Arching action develops as a result of unsymmetrical neutral axes that exist at the two beam ends and axial restraint provided by the surrounding frames. Thus, deformations are computed based on a simple strut and tie model (truss analogy), shown in Fig. 4, to capture the arching/catenary action. The angle developed between the strut and tie members is called the initial angle of arching action (α), which is the key parameter in developing the required axial force in the test. The total axial elongation of the connections is assumed to be entirely attributed to the connection close to the removed column as long as the arching action exists. Once the arching action switches to the catenary phase, the elongation is attributed to both connections.

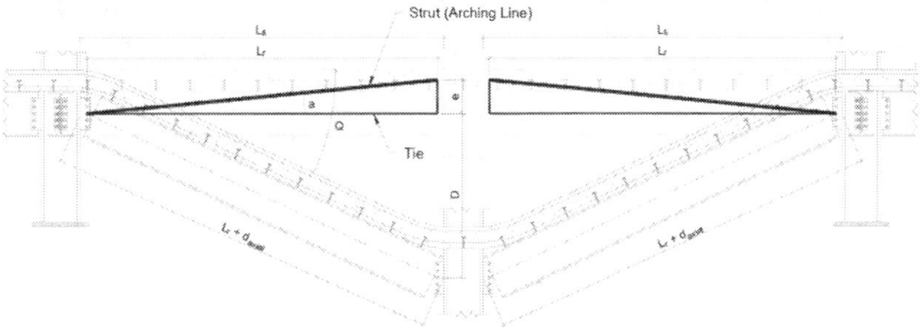

Fig. 4: Geometric compatibilities of displacements

3.3 Test Procedure

Having the target displacements a priori, the test objectives could be achieved by applying loads on a short cantilever beam. Fig. 5 shows the test set-up and instrumentation.

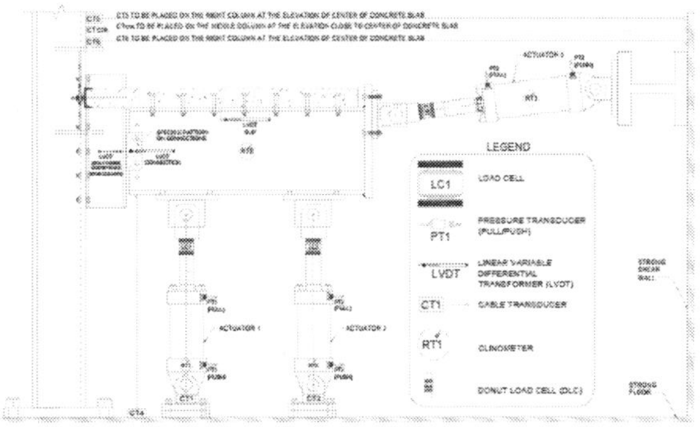

Fig. 5: Test set-up and instrumentation

Rotation and axial deformation were applied quasi-statically to the specimens through the beam using three actuators. Displacements were computed based on the planar geometry compatibilities.

4. Test Results

No failure or significant permanent deformation of bolts was observed in either connection type. Bearing of the top flange of the test beam against the stub column was not observed. A gap of 25 mm and 10 mm, respectively, was placed initially for the shear tabs and double angles. Although at higher rotations the gap increased before contact happened, after the beginning of each test, beam rotation caused the top flange to move toward the stub column flange.

Excessive elongation of the shear tab plate holes under the bearing stress exerted by the bolts was the main factor for ductility of the shear tab connections. Bolt tear-out of the shear-tab plate was the governing failure mode for all specimens. As a result of tensile force developed in the plate, all shear-tab plates failed by bolt tear-out in the direction of the beam axis. Fig. 6 shows the aftermath of two shear tabs with three and five bolt configurations.

The failure mechanism of the double angle connections was different than the shear tabs, mostly because of the ability of angles to unfold, as shown in Fig. 6. Deformation was initially shaped by unfolding of both angles as the angle heels were pulled away from the stub column flange. Since the top of the angles was not as free as the bottom (due to the presence of the concrete slab), the unfolding mechanism was only developed along a partial length of the angles. After unfolding, tearing initiated and propagated from the bottom of the angles where the maximum tensile stress developed.

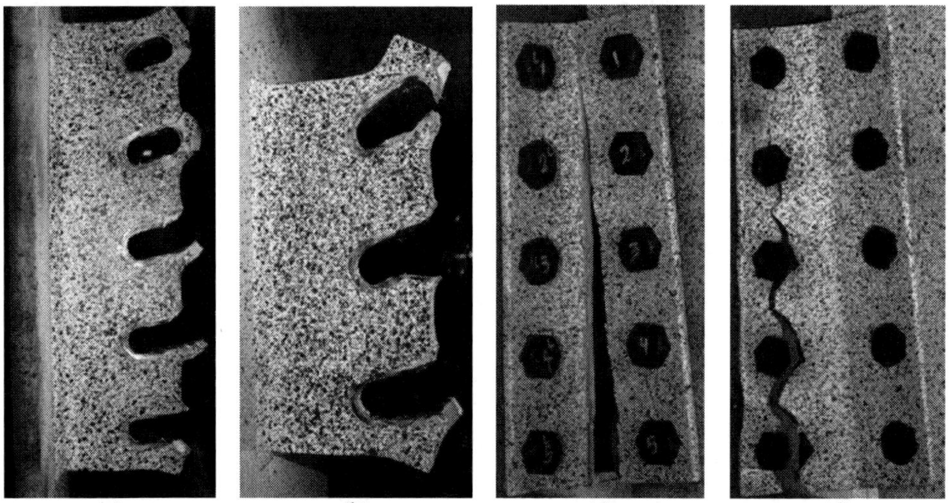

(a) (b) (c) (d)

Fig. 6: Failure modes of (a) five-bolt shear tab; (b) three-bolt shear tab; (c) tearing of gross section near the heel of double angle; (d) fracture of double angle near bolt line attached to the stub column

The failure mode of the concrete slab was similar for all specimens and was characterized by concrete cracking and crushing, typically concentrated around the slab's opening close to the stub column. Cracks in the concrete slab always initiated at a small beam rotation (less than 0.025 rad), around the shear studs closest to the stub column, and propagated in a radial direction, as shown in Fig. 7. Cracks located at shear studs close to the connections were deeper, especially for the specimens with shorter spans (6.0 m). Near the steel connections, the length, depth and number of cracks on the bottom of the concrete slab were more due to the stress concentration. Since the neutral axis was within the concrete slab (based on the composite cross-sectional geometry), cracks always occurred at the soffit of the slab.

General shapes of the load response curves were similar for all connections. Fig. 8 shows the load development in terms of vertical and horizontal loads, and bending moment, of two five-bolt deep connections with a Type A slab (9.5 mm thick shear tab and double angle). As shown, the composite section exhibited a relatively high initial axial and rotational stiffness. The horizontal component of the arching force developed in the cross-section increased rapidly at low rotations and it reached a maximum immediately prior the extreme bolt starting to bulge out or tearing initiation at the angle heels. It then decreased in a smooth manner until switching to the catenary phase. Most of the occurrences of bolt tear-out and angle tearing and fracture occurred prior to the catenary phase.

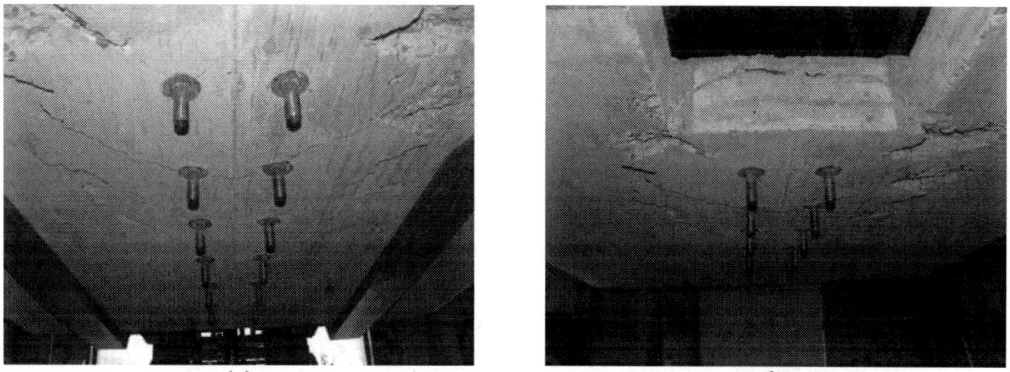

(a) (b)

Fig 7: Failure mode of concrete slab: (a) soffit of slab; (b) opening to the stub column

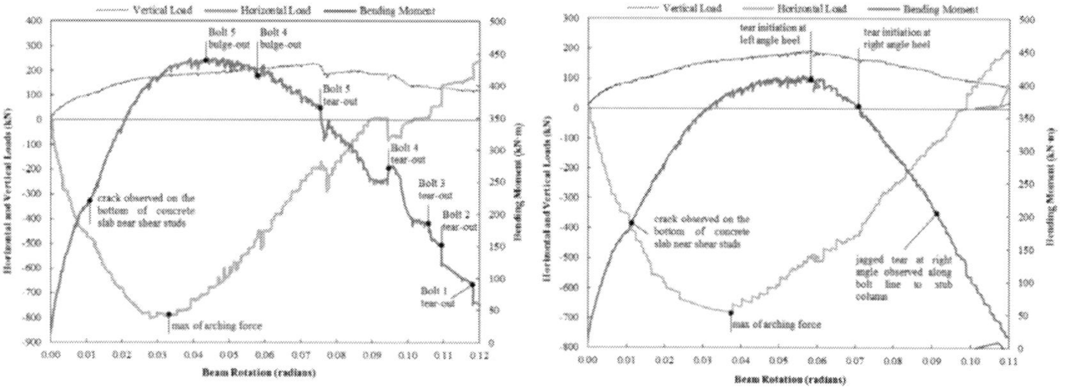

(a) (b)

Fig. 8: Load versus beam rotation of five-bolt deep connections with Type A slab (9.5 mm thick connection elements): (a) shear tab; (b) double angle

5. Conclusions

A full-scale testing program designed to study the behaviour of shear connections of a composite beam in a central-column-removal scenario was described. One-way behaviour has been studied, so any contribution due to two-way action is neglected. Two types of connections with two thicknesses and depths, and with different assumed span lengths, were included in the research program. The test set-up, loading regime, and test procedure were described. Following are the main conclusions reached:

1. The loading regime outlined herein is a practical method for conducting component level tests on most steel shear connections to simulate a column removal scenario;
2. The results indicate that arching action developed initially with a significant horizontal compressive force, while at large deformations the loads were resisted largely by catenary action;

3. The concrete slab affects the behaviour of the connection adjacent to the removed column mainly by shifting of the neutral axis to within the slab, placing the connection into nearly pure tension;
4. Although a higher axial load developed in the composite beam, its rotational capacity was markedly smaller than that obtained by Oosterhof and Driver [5], and Weigand et al [7] for a steel beam due to the prying force induced at an early stage of loading;
5. Bolt tear-out of the shear-tab plate and tearing of the gross section near the heel of the angles were the governing failure modes of the connections. The results demonstrate that the presence of a slab amplifies the demand on the connections considerably and reduces the ductility;
6. Full-scale testing, along with the evaluation of data, is ongoing at the University of Alberta to investigate the influence of a slab on the behaviour of shear connections.

Acknowledgments

The authors would like to acknowledge the financial support provided by the Natural Sciences and Engineering Research Council, concrete donation from Lafarge North America, and Compute Canada for providing advanced research and high performance computing systems.

References

[1] NBCC. *Canadian Commission on Building and Fire Codes*, National building code of Canada, National Research Council of Canada, Ottawa, ON, 2010.

[2] ASCE. *Minimum design loads for buildings and other structures*, ASCE/SEI 7-10, American Society of Civil Engineers, Reston, VA, 2010.

[3] U.S. General Services Administration (GSA). *Progressive collapse analysis and design guidelines for new federal office buildings and major modernization projects,* Washington, DC, 2003.

[4] Department of Defense (DoD). *Design of buildings to resist progressive collapse*, Unified Facility Criteria (UFC) 4-023-03, Washington, DC, 2009.

[5] Oosterhof SA, Driver RG. "Behavior of Steel Shear Connections Under Column-removal Demands", *Journal of Structural Engineering, American Society of Civil Engineers*, 141(4), 2015.

[6] Thompson S. "Axial, shear and moment interaction of single plate "shear tab" connections", *Master's Thesis*, Milwaukee School of Engineering, Milwaukee, WI, 2009.

[7] Weigand J, Francisco T, Meissner J, Fahnestock L, Liu J, Berman J. "Large-scale experimental evaluation of steel gravity framing structural integrity", *Structures Congress*, American Society of Civil Engineers, Pittsburgh, PA, 2013.

[8] Main JA, Sadek F. "Robustness of Steel Gravity Frame System with Single-Plate Shear Connections." *Rep. No. NIST Technical Note 1749* (corrections dated February 2013), National Institute of Standards and Technology (NIST), U.S. Department of Commerce, Gaithersburg, MD, USA, 2012.

[9] Jamshidi A, Driver RG. "Progressive Collapse Resistance of Steel Gravity Frames Considering Floor Slab Effects", *3rd International Structural Specialty Conference*, CSCE 2012, Edmonton, AB, Canada, STR-1003-1-STR-1003-10, 2012.

[10] Jamshidi A, Driver RG. "Structural Integrity of Composite Steel Gravity Frame Systems", *Structures Congress 2013*, American Society of Civil Engineers (ASCE), Pittsburgh, PA, USA, 55-66, 2013.

[11] Jamshidi A, Driver RG. "Full-Scale Tests on Shear Connections of Composite Beams under a Column Removal Scenario", *Structures Congress 2014*, American Society of Civil Engineers (ASCE), Boston, MA, USA, 931-942, 2014.

NUMERICAL MODELING OF DYNAMIC RESPONSE OF STEEL MOMENT FRAMES FOLLOWING SUDDEN COLUMN LOSS

Ioan Mărginean[a], Florea Dinu[a,b], Dan Dubina[a,b], Ioan Petran[c], Mihai Senila[c] and Henriette Szilagyi[d]

[a] *Department of Steel Structures and Structural Mechanics, Politehnica University Timisoara, Romania*
[b] *Laboratory of Steel Structures, Romanian Academy, Timisoara Branch, Romania*
[c] *Department of Structures, Technical University of Cluj-Napoca, Romania*
[d] *National Institute for Research and Development in Construction, Urban Planning and Sustainable Spatial Development "URBAN-INCERC", Cluj-Napoca Branch, Romania*

Abstract: In the last years, significant research effort was devoted to the evaluation of the capacity of steel frame structures to resist progressive collapse after sudden column loss. Due to the complex load-structure interaction and material behavior, it can be very difficult to evaluate the ultimate capacity of structural components by means of currently used analytical methods. Therefore, considerable research effort has been directed to experimental testing and sophisticated numerical simulations. Although the sudden column loss is a dynamic process, most experimental studies on full scale or scaled down specimens were performed under quasi-static loads. In order to incorporate the dynamic effects, increase factors were proposed and incorporated in the analysis and design recommendations. The paper presents the results of a study devoted to the evaluation of steel frame response following the loss of a column. Advanced numerical models are validated using experimental test results and dynamic increase factors are proposed.

1. Introduction

Design of building structures should provide sufficient structural robustness in order to reduce the risk of progressive collapse under extreme loading events. In order to achieve the robustness in a safe and economical manner, it is necessary to determine the structural features that may produce adequate response in the event of an extreme loading that may cause large local damages, i.e. loss of some key members. This may be done by using the alternate path (AP) method to ascertain the capacity of a structure to resist the loss of one or more critical load-bearing elements without causing disproportionate collapse [1]. The complex behaviour prior to failure makes it difficult to predict the ultimate capacity of structural components only by means of analytical methods that are being used in current practice. Therefore, considerable research effort has been directed to experimental testing and sophisticated numerical simulations. In this regard, El-Tawil et al. [2] presented in a recent forum paper the current trends and the research required to address gaps in the understanding of progressive collapse. Among them were highlighted the need for improved modelling and design guidance and also detailed test data.

Testing of full scale building structures subjected to column loss scenarios is expensive and technically demanding. An exception would be the actual buildings that are planned for demolition, which may be instrumented and tested before the demolition. Song et al. [3] investigated the progressive collapse performance of an existing steel frame building in situ by physically removing four first story columns from one of the perimeter frames. Structural system testing, involving planar systems or 3D systems, is a more common approach, as it can produce an accurate prediction of the response with less demands, both technical and economic. In what concerns the structural system testing, although dynamic tests would give the most accurate results, it is difficult to monitor the structural response during the test and make the relevant observations. As the exact load corresponding to the failure point is hard to evaluate, any other distribution and intensity of loads can cause no damages, yielding or rupture of the members or connections. As a result, it is not possible to capture the entire post-yielding behaviour using a single model, only. These limitations and difficulties have led to an increased interest for quasi-static tests to investigate the response of frame building under column loss events. In the quasi-static tests, the response is investigated either by gradually increasing a point load at the missing column location, or by applying a distributed load along the beams while gradually release the support force at the lost column.

Numerical simulations require much less effort and are considerable less expensive than experimental testing. The use of advanced methods increases the accuracy of the analysis but requires higher computational effort and advanced engineering skills. In addition, to reproduce the actual behaviour with sufficient accuracy, models should be validated against experimental data. The transition from original intact configuration to the damaged state is assumed to be instantaneous, thus exposing the structure to a dynamic effect [4].

Whereas non-linear dynamic analysis provides most accurate results, in practice the static nonlinear analysis is the preferred method, as it requires less resources and engineering skills. The benefits come however with shortcomings, for example the consideration of the dynamic effect. When first introduced as a general design method for reducing the progressive collapse potential ([5]; [6]), the alternate path method employed a factor of 2 in the static analysis to increase the loads applied to the bays adjacent to and above the lost member (dynamic effect). Subsequent studies ([7]; [8]; [9]; [10]) revealed some issues with the use of a constant dynamic increase factor DIF. First, a DIF of 2.0 is appropriate for buildings required to remain elastic; however, for progressive collapse events, structures are typically design to respond in the nonlinear range. Second, DIF did not vary with levels of performance and dissipation capacity of the structure. Third, distinction should be made between linear and nonlinear static analysis [11]. Changes incorporated in a revised version of UFC provisions [12]included the evaluation of DIF as a function of the normalized rotation (allowable plastic rotation divided by the rotation at yield of the cross section) and distinct load increase factors for linear and nonlinear static models. More recently, a series of studies were conducted with the aim of investigating the role of catenary action and connection performance on the evaluation of the dynamic behavior of steel frames subjected to a sudden column loss. Liu et al. [13], [14] used experimental tests and numerical simulations to investigate the dynamic behaviour of steel connections subjected to a sudden column removal scenario. By comparing the full dynamic response with its corresponding quasi-static response, both the displacement-based dynamic increase factor (DIF) and force-based DIF were investigated, respectively. In another study, Fu et al. [15] used AP method to study the dynamic performance of two-dimensional steel frames with bolted joint, under a sudden column removal scenario. Dynamic increased factors were obtained by comparing the dynamic responses with the corresponding nonlinear static responses of the bolted-angle steel joints.

The paper presents the results of a study devoted to the evaluation of the response of a steel frame structure following the loss of a column. A scaled-down 3D steel specimen with two bays and two spans was constructed. The beam-to-column end plate bolted connections were designed as fully rigid and fully restrained connections. The specimen was tested under monotonic loading applied to the top of the central column until complete failure. Experimental test results were validated using advanced numerical program ELS [16]. Based on the validated numerical model, dynamic analyses were carried out using different levels of gravity loads until the attainment of failure. Based on the results of the static and dynamic response, both displacement-based (DIF) and force-based DIF were calculated. The study is part of a research program devoted to the design of structures to sustain extreme loading events without collapse [17].

2. Numerical simulations

2.1 Validation of the numerical model

The limitations and costs involved in experimental testing make the numerical simulations very attractive for the complex field of progressive collapse evaluation after the loss of a column. In order to predict the response of the structure with sufficient accuracy, numerical simulations require comprehensive validation of the models. In order to study the response of steel frames following the loss of a column, a numerical model was developed using the advanced nonlinear dynamic analysis software Extreme Loading for Structures [16]. A view of the ELS model, including the geometry, steel elements and connections, is shown in Fig. 1. Numerical model was validated against experimental data obtained in a monotonic quasi-static test [18]. In the test, the response of a two-span and two-bay steel frame specimen after the loss of a central column was investigated. A point load (PL) was applied on top of the central column and increased until failure occurred. The test setup and the specimen are presented in Fig. 2. Structural steel S275 (yield strength of 275 N/mm^2) was used for beams and S355 (yield strength of 355 N/mm^2) for columns. The extended end-plate bolted beam-to-column connections were designed as fully restrained connections.

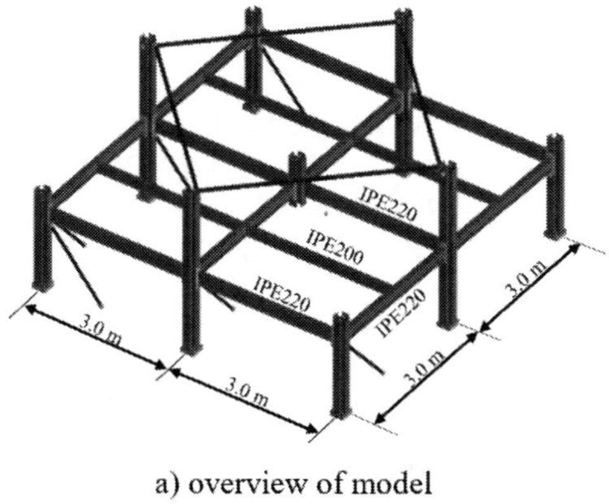

a) overview of model

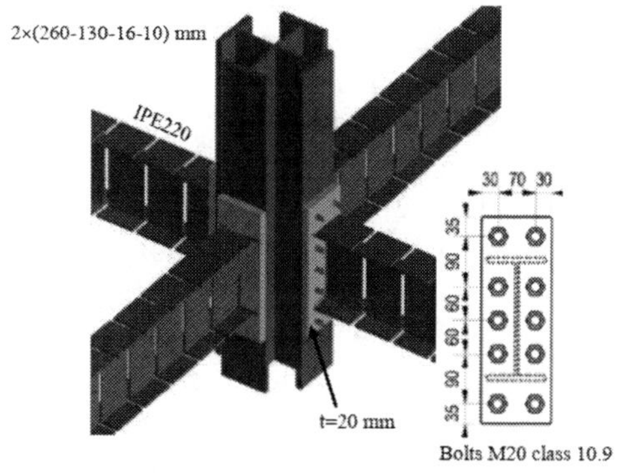

b) detailed view of beam-to-column connection

Fig. 1: Numerical model of structure

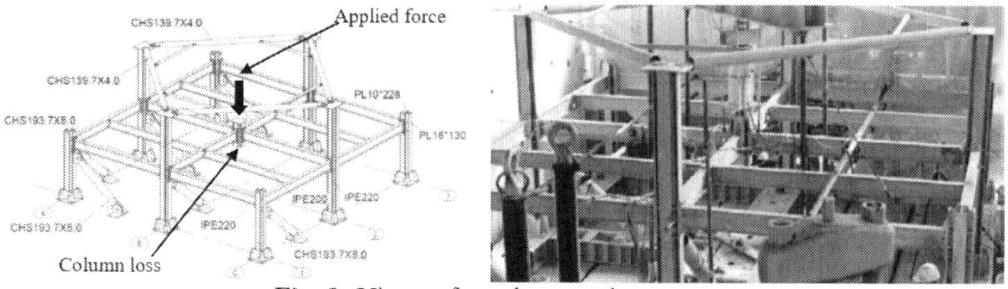

Fig. 2: Views of specimen and test setup

The results of the numerical analysis showed that the model could be used to capture the behaviour of the specimen and the failure mode. The overall relationship between the vertical force and the vertical displacement below the central column is shown Fig. 3.a and Fig. 3.b shows the specimen after the test and the deformed shape of the numerical model at failure.

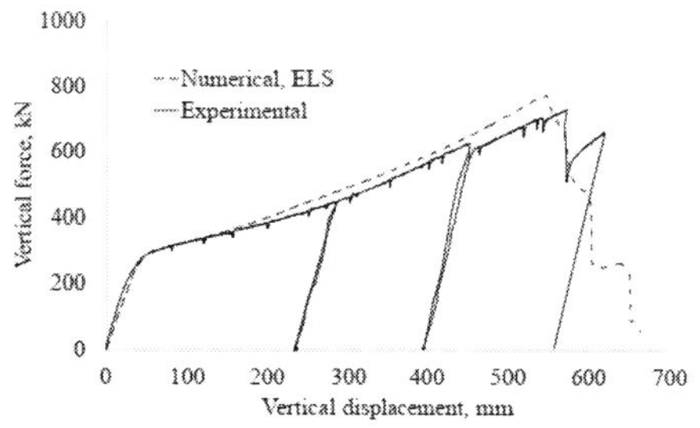

a) vertical force-vertical displacement

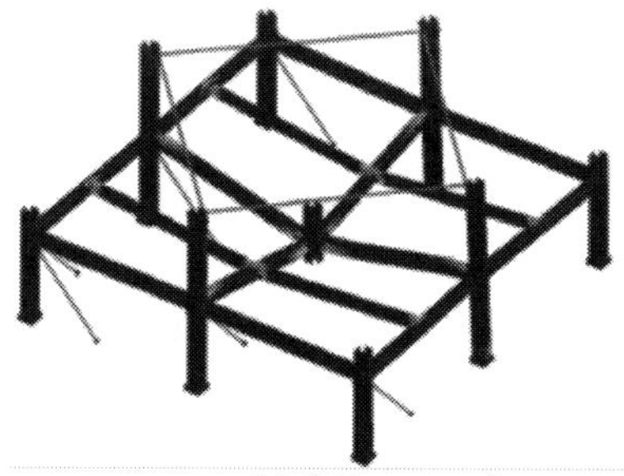

b) failure mode

Fig. 3: Numerical results

2.2 Imposed column displacement vs. column release testing methods

As described in Section 1, the frame response after the column loss can be investigated either by gradually increasing the point load at the missing column location (PL method), or by applying a distributed load along the beams, most common using uniformly distributed load (UDL method) while releasing the support force at the lost column. PL method, which was employed in the test and numerical validation modelling presented described in the previous section, is easier to apply and control and is preferred over the second one. However, the later, also called column release method, is more suitable to represent the response because the interaction forces (between applied UDL and column support force) in the connection region are considered in the analysis [19]. For comparing the results of the two methods, the second method was also applied to the validated model. Uniformly distributed loads with different intensities were applied on the internal beams in a non-linear static analysis. In the first stage, the load was applied on the structure with the central column in place. In the second stage, the column support was released while the vertical reaction of the central column was monitored, see Fig. 4. The decrease in the reaction of the central column represents the load that is undertaken by the beams. The amount of initial gravity load was varied so that on the initial structure the compression axial force in central column would be 33%, 66% and respectively 100% of the maximum vertical force applied on the model (PL method). The models are noted as follows:

- Point load (model validated against test) PL
- UDL of 42kN/m on main beams (equivalent 28 kN/m^2) UDL-1
- UDL of 83kN/m on main beams (equivalent 56 kN/m^2) UDL-2
- UDL of 125kN/m on main beams (equivalent 84 kN/m^2) UDL-3

Fig. 4.c shows the vertical reaction force vs. vertical displacement curves obtained using the two methods. In what concerns the UDL curves, initial reaction forces are always negative as a result of applied distributed loads on beams. If the ultimate force is beyond zero, this indicates the applied UDL is less than the capacity and the structure does not fail at additional point load is required to reach failure. If the ultimate force does not reach zero value, this indicates the applied UDL is larger than the capacity and the structure fails before redistributing all load. Therefore, the determination of the exact value of UDL corresponding to the attainment of failure requires several iterations. In order to compare the results of PL and UDL methods, reaction forces obtained using UDL (Fig. 4.c) were offset to start from zero force, see Fig. 4.d. A uniform distributed load of 33% from the maximum column point load (UDL-1) does not substantially change the response, and its influence may be neglected. In case of UDL-2, the ductility of the system is reduced with 25%. A uniform distributed load that produces a central column reaction equal to its maximum capacity in the reference loading system (UDL-3) will reduce the vertical force capacity and vertical displacement by 40%, and also the yielding force by 17%. The different ultimate capacities obtained by varying the intensity of the UDL are resulting as the effects of different moment distribution before failure, see Fig. 5. To note the M_N represents the maximum bending moment corrected due of the presence of the axial force (M-N interaction).

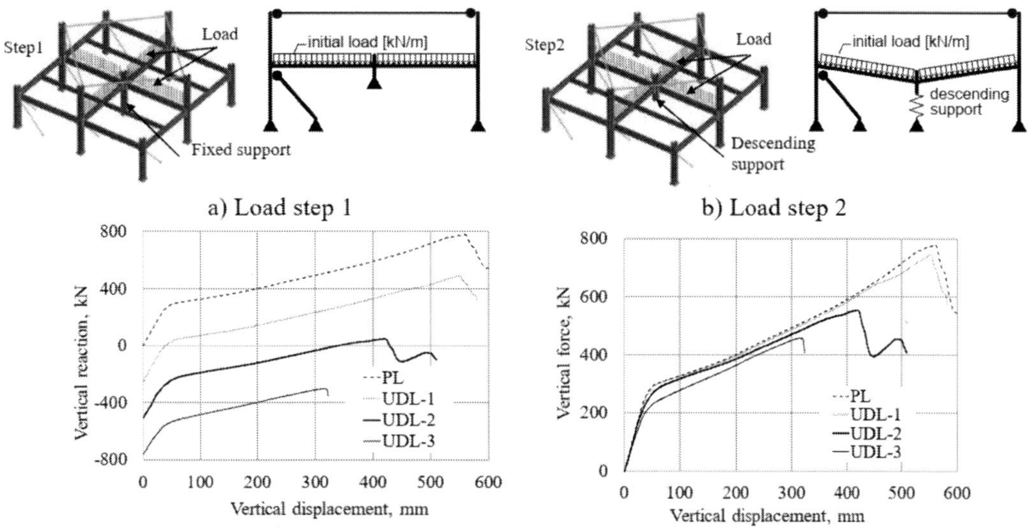

a) Load step 1 b) Load step 2

c) Vertical reaction vs. vertical displacement d) Vertical force vs. vertical displacement

Fig. 4: Uniformly distributed load method

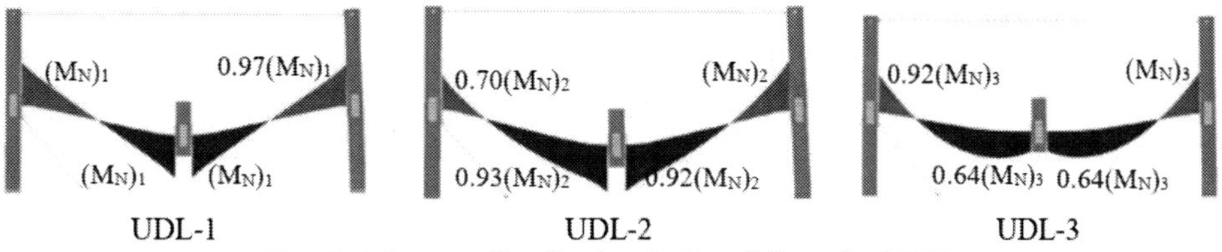

Fig. 5: Moment distribution before failure for UDL

As previously described, the determination of the exact value of UDL corresponding to the attainment of failure requires several iterations. A direct method to obtain the complete path of force-displacement curve under gravity load can employ a pushdown analysis, where the column is removed and the distributed load is incremented up to the attainment of failure. The advantage of the pushdown analysis is the possibility to directly compare the point load results with all UDL at zero reaction force. Fig. 6.a shows the numerical model and distribution of loads in the pushdown analysis. In order to compare the results of the pushdown simulation with the point load test, the total vertical reaction (sum of total applied loads) needs to be allocated to each column based on their tributary zone. For the central column, the corresponding vertical force was 1/4 of the total reaction. Fig. 6.b shows comparatively the vertical force - vertical displacement curves for the reference model (validated against experimental test) and the pushdown analysis. For comparison, the equivalent uniformly distributed load on the floor is also shown on the secondary vertical axis. As seen from the two curves, the initial stiffness and post-yielding behavior are almost identical until the vertical force reaches 380 kN. After this, the model with UDL exhibits an almost constant stiffness, until the failure is attained at a force of 510 kN (or an equivalent UDL of 62 kN/m^2). The ultimate displacement amounted 410 mm. Comparatively, the ultimate resistance and deformation capacity of the model loaded with a point load at the missing column location were much higher, i.e. 773 kN and 564 mm, respectively. As the case with distributed gravity loads on the floors represents a more realistic loading scenario, the prediction of the structural capacity using point load method requires further calibrations to improve the consistency. When compared with the discrete UDL analysis (see Fig. 4.c), it may be seen the ultimate capacity is very close to the UDL-2 (56 kN/m^2).

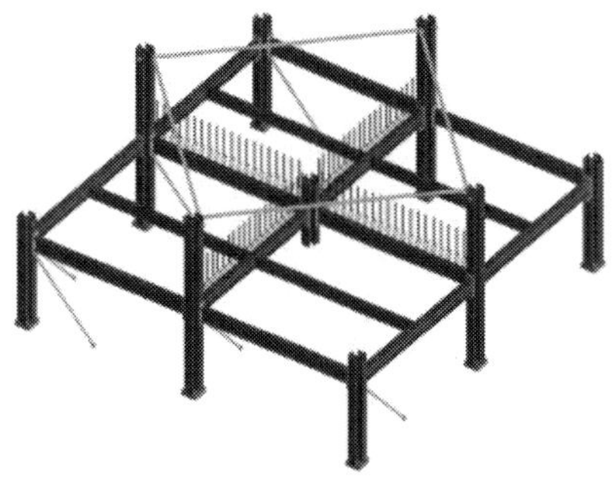

a) Distribution of UDL

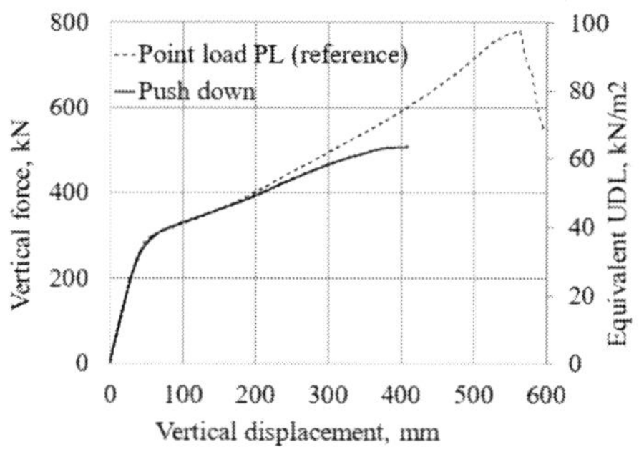

b) Force-displacement curve

Fig. 6: Pushdown analysis

2.3 Non-linear dynamic response

In the event of a sudden column loss, the total force acting on the floor is the summation of directly applied gravity loads and inertial forces developed during dynamic response of the structure. For the study, the model was loaded with a uniformly distributed gravity load (UDL) and then the central column was instantaneously removed, thus exposing the structure to a dynamic effect. The analysis was done for UDL of increasing intensity, up to failure. Fig. 7.a shows the history of vertical displacement for different levels of UDL. The ultimate load before failure amounted 40 kN/m2, at a maximum vertical displacement of 321 mm. The vertical force - vertical displacement envelope obtained in the incremental dynamic analysis is presented in Fig. 7.b and compared with the static pushdown curve. The maximum loading in the dynamic analysis is less than that 2/3 of the static pushdown capacity. In order to quantify the dynamic effect in the structure, either the force or displacement response may be used. The two dynamic factors DIF are defined in Fig. 7.b ([20]). The force-based DIF_F and displacement-based DIF_D, which were calculated based on the numerical analysis results are presented in Fig. 8. The values are compared with DIF calculated using formulas proposed in [12] and [20]. Due to the contribution of post-yield stiffness (increased due to development of catenary action), the force-based amplification factor DIF_F is underestimated for normalized rotations larger than 2 when UFC formula [12] is adopted. A better estimation is given by analytical formula proposed in [20]. The results obtained for DIFD are also in good agreement with the formula proposed in [20].

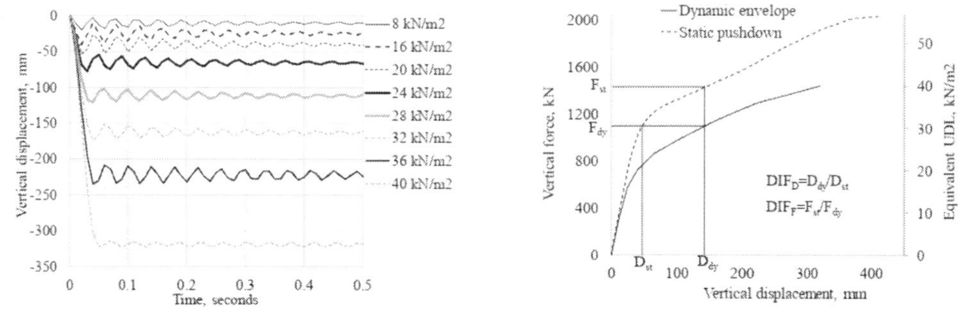

a) Vertical displacement vs time b) Static vs dynamic force displacement curves

Fig. 7: Dynamic analysis results

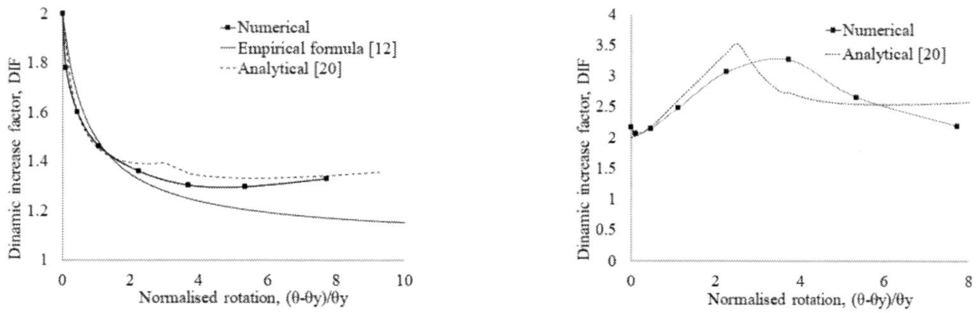

a) Force-based factor DIF_F b) displacement-based factor DIF_D

Fig. 8: Comparison of dynamic increase factors and

3. Conclusions

The study presented in the paper aimed at evaluating the dynamic response of steel frame structures due to a sudden removal of a column using numerical simulations. The model was first validated for static loading regime using test data obtained in a previous study. For evaluating dynamic response factors, nonlinear dynamic analyses were performed and the results were compared with similar results obtained using empirical and analytical formulas. The lower limit of force-based DIF appears to be higher than empirical based results [12]. Similar observations were reported in [20]. This increase may be explained by the increase of post-yield stiffness under large deformations as a result of catenary action development. The displacement-based DIF increases with the increase of the plastic deformation until reaches a maximum value of 3.25 for a normalized rotation of 3.6.

Acknowledgments

Partial funding for this research was provided by the Executive Agency for Higher Education, Research, Development and Innovation Funding, Romania, under grant PCCA 55/2012 "Structural conception and collapse control performance based design of multistory structures under accidental actions" (2012-2016).

References

[1] DoD, U, *Unified facilities criteria: design of buildings to resist progressive collapse.* 2009, UFC 4-023-03. United States Department of Defense.

[2] El-Tawil S, Li H, Kunnath S, "Computational Simulation of Gravity-Induced Progressive Collapse of Steel-Frame Buildings: Current Trends and Future Research Needs", *Journal of Structural Engineering*, 140(8), A2513001, 2014.

[3] Song BI, Sezen H. "Experimental and analytical progressive collapse assessment of a steel frame building", *Engineering Structures*, 56, 664-672, 2013.

[4] Ellingwood BR, Smilowitz R, Dusenberry DO, Duthinh D, Lew H, Carino NJ, *Best practices for reducing the potential for progressive collapse in buildings*. 2007: US Department of Commerce, National Institute of Standards and Technology.

[5] GSA, U, "Progressive collapse analysis and design guidelines for new federal office buildings and major modernization projects". *Washington, DC*, 2003.

[6] DoD, "Design of buildings to resist progressive collapse". *UFC*, 2005.

[7] Ruth P, Marchand KA, Williamson EB, "Static equivalency in progressive collapse alternate path analysis: reducing conservatism while retaining structural integrity", *Journal of Performance of Constructed Facilities*, 20(4), 349-364, 2006.

[8] McKay A, Marchand K, Williamson E, Crowder B, Stevens D. "Dynamic and nonlinear load increase factors for collapse design and analysis". *Proc. Int. Symp. on Interaction of the Effects of Munitions with Structures (ISIEMS)*, 2007.

[9] Foley CM, Barnes K, Schneeman C. "Quantifying and Enhancing Robustness in Steel Structures: Part 2--Floor Framing Systems". *Eng. journal*, 45(4), 267, 2008.

[10] Tsai M-H, You Z-K. "Experimental evaluation of inelastic dynamic amplification factors for progressive collapse analysis under sudden support loss". *Mechanics Research Communications*, 40, 56-62, 2012.

[11] Stevens D, Crowder B, Sunshine D, Marchand K, Smilowitz R, Williamson E, Waggoner M. "DoD research and criteria for the design of buildings to resist progressive collapse", *Journal of Structural Engineering*, 137(9), 870-880, 2011.

[12] DoD, U, *Unified facilities criteria: design of buildings to resist progressive collapse.* 2013, UFC 4-023-03. United States Department of Defense.

[13] Liu, C, KH Tan, and TC Fung, "Investigations of nonlinear dynamic performance of top-and-seat with web angle connections subjected to sudden column removal". *Engineering Structures*, 99, 449-461, 2015.

[14] Liu C, Tan KH, Fung TC. "Component-based steel beam–column connections modelling for dynamic progressive collapse analysis". *Journal of Constructional Steel Research*, 107 24-36, 2015.

[15] Fu Q, Yang B, Hu Y, Xiong G, Nie S, Zhang W, Dai G, "Dynamic analyses of bolted-angle steel joints against progressive collapse based on component-based model". *Journal of Constructional Steel Research*, 117, 161-174, 2016.

[16] ELS, *Extreme loading for structures*. Applied Science International: Durham, 2010.

[17] CODEC 2012, Structural conception and collapse control performance based design of multistory structures under accidental actions CODEC, 2012-2016.

[18] Dinu F, Marginean I, Dubina D, Petran I. "Experimental testing and numerical analysis of 3D steel frame system under column loss", *Eng. Struct*, 113, 59-70, 2016.

[19] Liu C, Tan KH, Fung TC. "Dynamic behaviour of web cleat connections subjected to sudden column removal scenario", *Journal of Const. Steel Research*, 86, 92-106, 2013.

[20] Tsai M-H, "An analytical methodology for the dynamic amplification factor in progressive collapse evaluation of building structures", *Mechanics Research Communications*, 37(1), 61-66, 2010.

PROGRESSIVE COLLAPSE OF MULTI-STOREY STEEL FRAME BUILDINGS: DEVELOPING CATENARY ACTION

Andreea Sigauan[a], Florea Dinu[a,b] and Dan Dubina[a,b]

[a] *Department of Steel Structures and Structural Mechanics, Politehnica University Timisoara, Romania*
[b] *Laboratory of Steel Structures, Romanian Academy, Timisoara Branch, Romania*

Abstract: Progressive collapse resistance depends of several factors, e.g. overstrength and ductility of members and connections, redundancy of structural system enabling for load path redistribution, and ties. There are also other factors, consequent to previous structural properties, which may also influence the structural response, such as the potential strength reserve against gravity loads effect. For instance, structures designed to resist moderate to strong seismic actions may result in large strength reserve when compared with structures designed to resist gravity loads only, thus increasing the progressive collapse resistance. The study presented in the paper investigated the resistance to progressive collapse of steel moment-resisting frame buildings, located in a low seismicity zone, focusing on the influence of beam overstrength, beam span-to-depth ratio and moment capacity of beam-to-column (full strength, partial strength). The applied element method was used to predict the structural response under different failure scenarios by means of nonlinear dynamic analyses.

1. Introduction

Structures should be designed by taking into consideration the conditions under which they are expected to fulfil their basic functions. The relevant design situations shall be selected taking into account the circumstances under which the structure is required to fulfil its function [1]. Considering the different features of possible actions, it could be said that the design of a building structure to withstand a specific load may be ineffective for other loads. However, some design conditions, for example the capacity design employed in seismic design codes, produce more robust structures by virtue of the explicit consideration of the strength ratio between members, failure mechanism, redistribution capacity, and ductility requirements. This assumption is supported by the study reported in [2], which concluded that, if the Murrah Building had been designed to resist seismic action, its progressive collapse would have been precluded. Thus, to achieve a design that is both safe and economical, it is necessary to determine the structural features that may produce adequate response in the event of extreme loading. This may be done by using the alternate path method AP to ascertain the capacity of a structure to resist the loss of one or more critical load-bearing elements without causing disproportionate collapse [3]. The AP method, with its emphasis on continuity and ductility, is similar to current seismic resistant design practice [4].

Although the seismic design philosophy may be considered as a model for controlling the collapse of structures subjected to extreme events other than earthquakes, there are specific issues that should be considered to forestall localized failures, particularly of columns [5]. For example, the development of catenary forces in beams and the admissibility criteria for beam-to-column connections need the consideration of the axial force – bending moment interaction. In a recent study, Dinu et al. [6] have investigated the response of two-way steel moment frame systems under the removal of a central column. For the reference frame, extended end-plate bolted connections were used to connect the beams to the columns. First, an experimental test was carried out, then, a numerical model was validated using nonlinear dynamic analysis program Extreme Loading for Structures [7]. The system has proved its capacity to undergo large deformations associated with catenary response in the beams without failure of the connections, while the beams developed ultimate plastic rotation larger than the limits given in the codes.

The present study continued the investigation by considering different column loss scenarios, varying three parameters, i.e. beam span-to-depth ratio, the level of overstrength in structural elements (demand/capacity ratio for gravity load design), and moment capacity of beam-to-column joints. Bare steel members only have been considered, but the diaphragm effect of the slab was considered in the numerical simulations. Nonlinear dynamic analysis program Extreme Loading for Structures was used in the parametric study.

2. Parametric study

2.1 Description of steel frames

The frame structures considered for the parametric study were four-bay, four-span and six-story two-way steel moment frames, see Fig. 1. For the initial design, the structures were designed for permanent design situations (gravity and wind loads), according to Eurocodes. The dead load was 4.0 kN/m^2, live load was 3.0 kN/m^2 and the reference wind pressure was 0.5 kN/m^2. To simulate the influence of seismic design on sizing the members of the gravity load resisting system, beams were designed for three levels of strength reserve, i.e. 20%, 40% and 60%. These levels represent the reserve in the moment capacity of the beams from the gravity load design. Low levels of strength reserve, e.g. 20% indicate buildings in low seismicity areas, designed primarily against gravity loading. On the contrary, high levels (e.g. 60%) indicate buildings designed to withstand larger earthquakes. The capacity of the columns was accordingly adjusted to achieve strong column-weak beam behaviour. No particular accidental design situations were considered in design.

Table 1 shows the geometrical characteristics of the frames and section of members. The story height is 4.0 m for the first story and 3.5 m for the others. The span and bay is 6.0, 7.5 and 9.0 m respectively. For 6.0 m and 7.5 m spans, the secondary beams are placed at 2.0 m and 2.5 m intervals, respectively, while for 9.0 m span structure, the secondary beams are placed at 2.25 m. For the initial design, equal strength beam-to-column rigid moment joints have been considered ($M_{j,pl} = M_{b,pl}$). The secondary beams were connected to the main beams with shear joints (nominally pinned). As mentioned, diaphragm effect of the slab preventing out-of-plane deformation of beams has been considered even the composite steel-concrete action was not modelled. Steel material in plates and profiles was grade S355, while bolts were class 10.9.

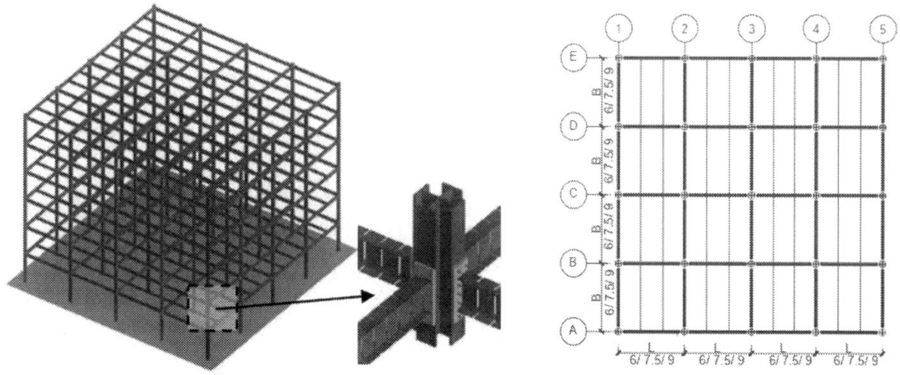

a) b)

Fig. 1: Details of the structures: a) isometric view of the structural model a beam-to-column connection detail; b) plan layout

Table 1: Characteristics of the frames

Frame	*L*, *B* (m)	Strength reserve in beams	Main beam	Secondary beam	Column
1	6.0	20%	IPE330	IPE300	X-HEB340-HEB200
		40%	IPE360	IPE300	X-HEB360-HEB200
		60%	IPE450	IPE300	X-HEB400-HEB250
2	7.5	20%	IPE450	IPE360	X-2HEB450
		40%	IPE500	IPE360	X-2HEB550
		60%	IPE550	IPE360	X-2HEB700
3	9.0	20%	IPE500	IPE400	X-2HEB650
		40%	IPE550	IPE400	X-2HEB700
		60%	IPE750x134	IPE400	X-2HEB800

2.2. Progressive collapse assessment

The assessment of progressive collapse using the AP method is in accordance with the UFC 4-023-03 guidelines [4]. For nonlinear dynamic analysis, the following gravity load combination is applied to the entire structure:

$$G = 1.2D + 0.5L \tag{1}$$

where: G = Gravity loads, D = Dead loads, L = Live loads.

Although the finite element method (FEM) is accurate and reliable for analyzing continuum structures, the onset of element separation is difficult to automate and the modelling of debris collision is time consuming. A more accurate prediction of the structural performance under extreme loading may be obtained by the applied element method (AEM) through ELS software [7]. The AEM can be used to investigate structural collapse behavior through all the stages of the application of the loads, namely, the elastic stage, crack initiation, during propagation in tension-weak materials, during the strain hardening effect in the post-elastic range, the element physical discontinuity stage, the element collision (dynamic contact) stage, and at collision with the ground and adjacent structures. The progressive collapse analysis of the structure can be used to determine how many columns may be lost before the collapse of the structure. It may also be of interest to evaluate the reserve capacity for supporting the gravity loads for a specific column loss scenario, which may be expressed as the ratio of the failure load to the nominal gravity load. This can be calculated using the combinations described earlier.

2.3 Validation of numerical model

The numerical model developed in ELS was validated using experimental test results obtained within the framework of the CODEC research program [6]. The test setup and the specimen are presented in Fig. 2, and also the detailed numerical model utilized by the ELS software.

The numerical results showed that the model can be used to simulate the behavior of the specimen until complete fracture development. Thus, Fig. 3 presents comparatively the deformed shape of the specimen at failure in the experimental test and numerical simulation. The vertical force vs. column displacement is shown in Fig. 4, and may be seen there is a very good agreement between the test data and the numerical simulation. More details about test results and numerical data may be found in the reference [7].

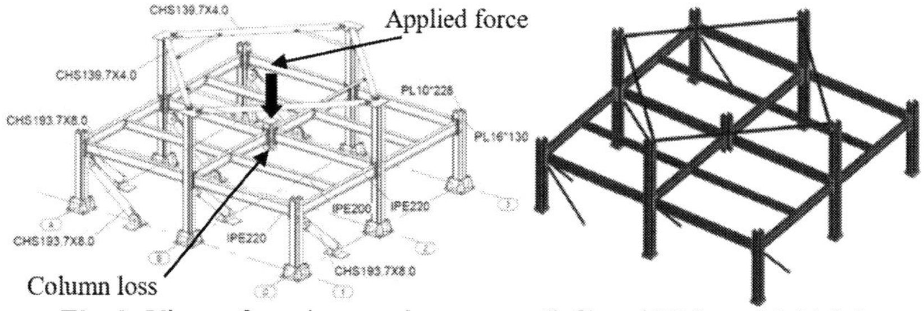

Fig. 2: Views of specimen and test set-up (left) and ELS model (right)

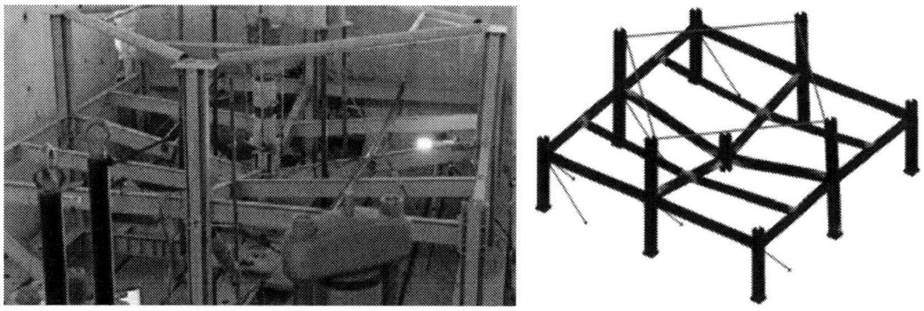

Fig. 3: Failure mode of the specimen, experimental (left) and numerical (right)

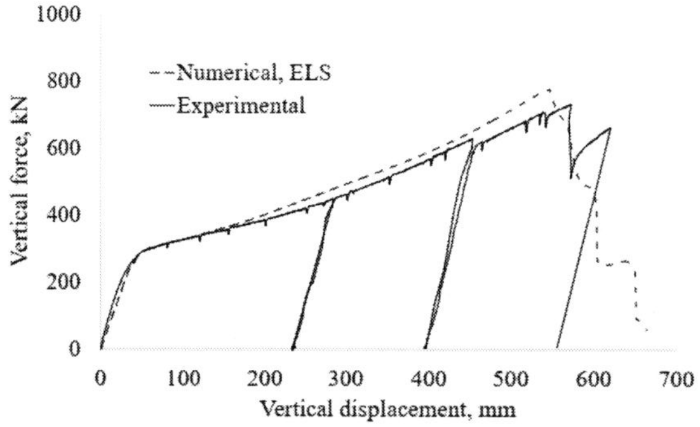

Fig. 4: Experimental vs. numerical results: vertical force-vertical displacement

3. Numerical simulations

The progressive collapse analysis was carried out for four column loss scenarios, namely, a) loss of the corner column (A1), b) loss of one edge column (B1), c) loss of one internal column (B) and d) loss of one internal column (C3), see Fig. 5. The loading procedure consisted of two stages. First, the vertical loads evaluated using Eq. 1 are statically applied to the structure. Once static equilibrium is reached, a column is instantly removed, thus inducing dynamic effects in the structure. The material enhancement (increase of material strength - yielding, ultimate) due to strain rate was neglected. By combining the frame span with the level of strength reserve and column loss scenario, the total number of cases that were analyzed was 36. Each case can be identified by a combination of span, in meters, reserve strength, in % and column loss scenario, e.g. 6-40-C3 represent the structure with 6.0 m span and bay, 40% beam strength reserve and C3 is the position of the removed column. To identify the critical components of the progressive collapse resistance, the gravity loads were incremented until collapse. The robustness criteria were obtained using as reference the ratio of the load at collapse to the initial gravity load.

Based on Eq. 1, the design load for robustness evaluation amounted 6.3 kN/m^2, representing 60% of the load applied in the permanent design situation, i.e. 9.9 kN/m^2.

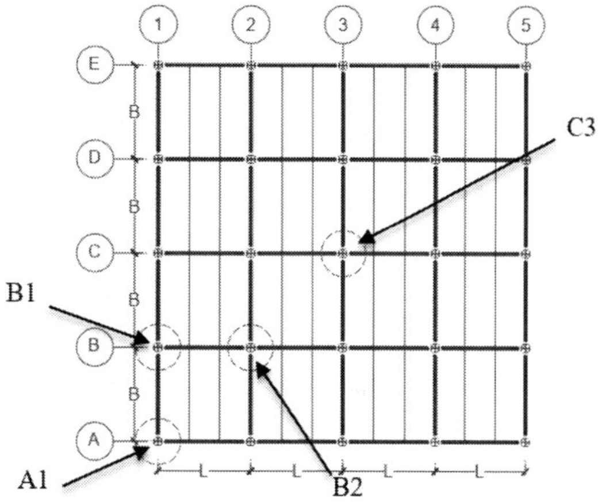

Fig. 5: Column loss scenarios

Fig. 6 presents the three failure modes after the loss of a column obtained in the numerical analyses. First mode is global (Fig. 6.a), and develops in case of internal column loss scenarios (B2, C3). Failure of beams triggers the collapse of adjacent columns, and the damage may extend also outside of the directly affected area (for C3). Second failure mode (see Fig. 6.b) is characterized by the failure of the free edge columns due to the development of large catenary forces in beams. This failure can develop mainly in case of penultimate column loss (B1) but also internal column loss (B2) if the columns are weak. If the beams are connected to stronger columns, the failure mode changes to the global mode. Third failure mode is characterized by the failure of the beam without developing catenary forces in beams (see Fig. 6.c). This type of failure is typical for corner column loss. The ultimate capacity is governed by the flexural capacity of the beams, and can be improved by strengthening the beams (larger strength reserve), tying systems, strong floors or other means to hang the affected area.

Fig. 7 presents the gravity load multiplier vs. beam rotation curves for each column loss scenario obtained in the incremental dynamic analyses. The curves are plotted using the permanent (residual) displacements obtained at different gravity load multipliers, see Fig. 8.

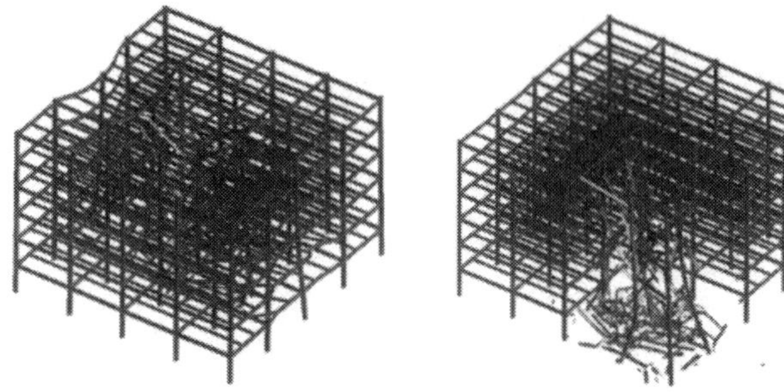

a) global failure mode

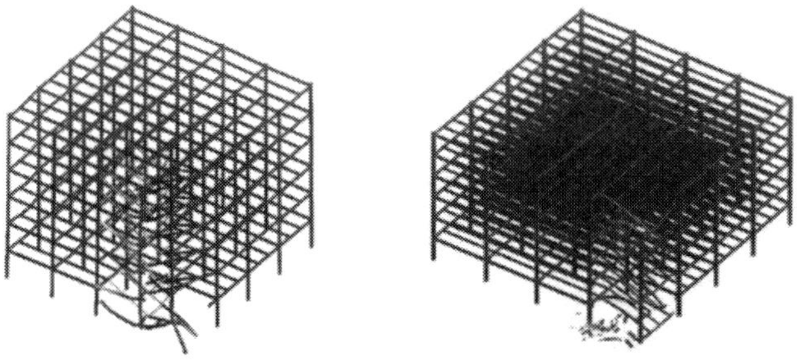

b) column failure c) beam failure

Fig. 6: Failure modes for different column removal scenarios

As may be seen from Fig. 7a-c, larger reserve strength (i.e. 60%) leads to more robust structure, being characterized by largest ultimate gravity load multiplier λ, but less ductile behavior. As the reserve strength diminishes, the ultimate capacity for supporting the gravity loads decreases, but structures exhibit more ductile responses, excepting corner column loss scenario, where the ultimate rotations tend to have similar values, independent of L/d ratio (Fig. 7.d). The complex relation between beam span, strength reserve, ultimate load multiplier and ultimate rotation may be better expressed if the frame span is normalized to the beam depth, i.e. L/d ratio. Thus, the increase of L/d ratio decreases the progressive collapse resistance while increasing the ultimate rotation, independent of column loss scenario. The main reason for this increased deformation capacity is the ability of the frame structure to develop catenary action in beams. This deformation capacity may be limited by the capacity of beams or connections to resist the large axial forces that are produced after undergoing large deformations.

In order to investigate the role of beam-to-column connection resistance in the progressive collapse resistance and development of catenary action, two types of connections were used for scenario C3. The first type is a full strength connection (FS), with the same flexural resistance as the beam and the second type is a partial strength connection, with 80% flexural resistance compared to the beam (PS). Fig. 9 presents the vertical force - beam rotation curves for two beam span to depth ratios, L/d (12 and 18). In case of lower span to depth ratio (L/d=12), the response is dominated by the flexural action, while the contribution from the catenary action to the total load resistance is reduced. For partial strength connection structure, a little increase in catenary action is observed at large deformation stage. For higher span to depth ratio (L/d=18), contribution of catenary action increases to almost 40% for partial strength connection structure. The results show that for internal column loss scenarios, the use of partial strength connections might reduce the progressive collapse resistance, especially for low L/d ratio, where the contribution of flexural resistance is predominant, while for larger L/d ratio, the development of significant catenary action at large deformation stages can compensate the reduction in flexural capacity.

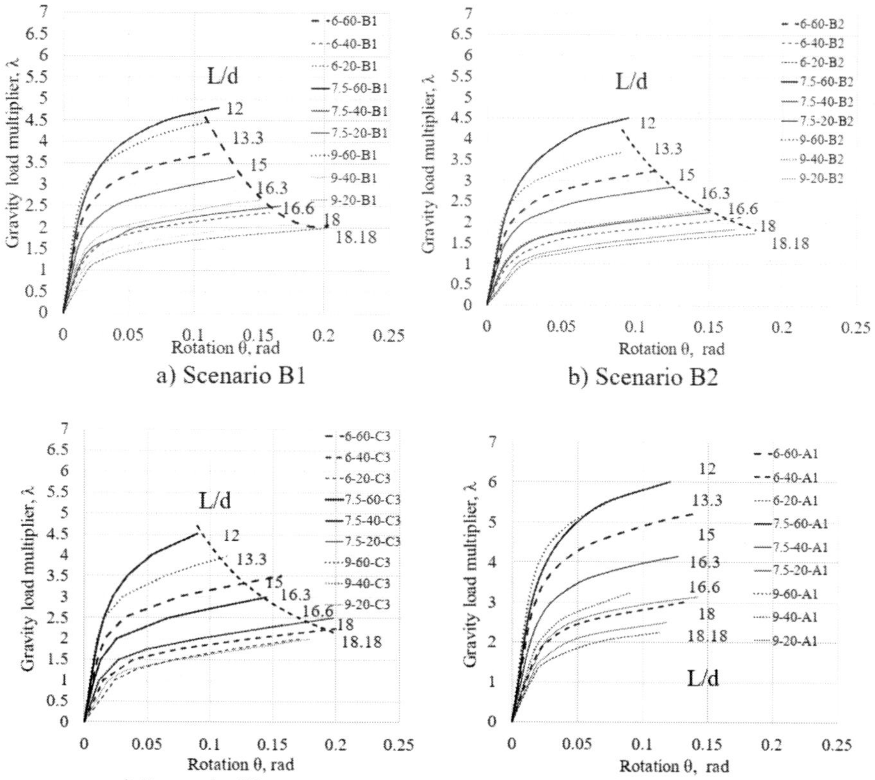

a) Scenario B1 b) Scenario B2

c) Scenario C3 d) Scenario A1

Fig. 7: Gravity load multiplier vs. beam rotation

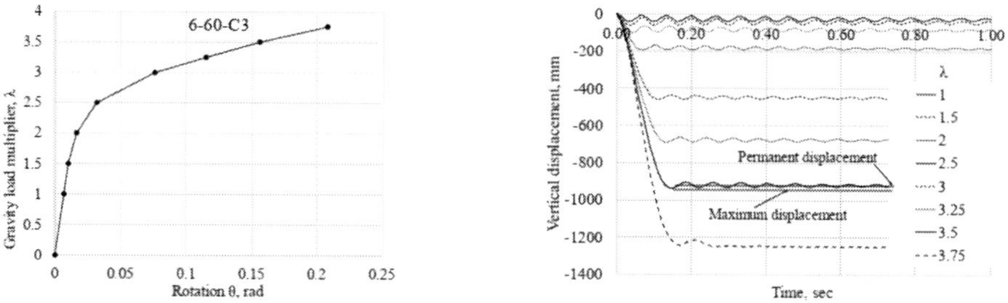

a) Gravity load multiplier vs. beam rotation b) vertical displacement vs. time

Fig. 8: Results for 6-60-C3 structure

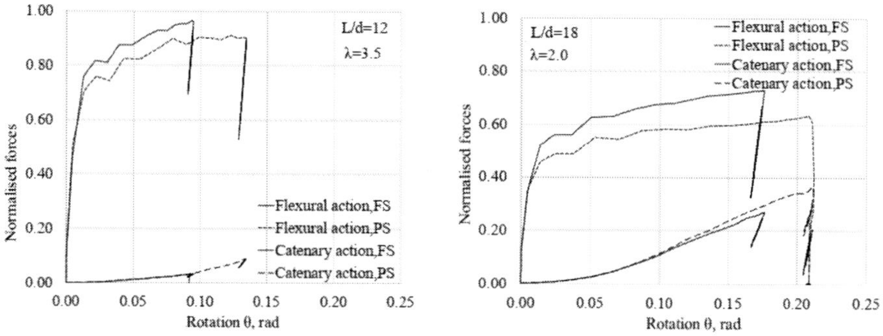

Fig. 9: Development of bending moments and axial forces in beams

4. Conclusions

A numerical study was carried out for investigating the progressive collapse resistance of steel moment frame structures following the loss of a column. By changing the strength reserve of beams and bay/span of frames, different beam depth to span ratio *L/d* were obtained. A total of 36 cases were analyzed using incremental nonlinear dynamic analysis. The numerical models were validated against experimental test data obtained within the framework of the CODEC research program. The results of the numerical analysis showed that the model could be used to capture the behavior of the specimen and the failure mode.

L/d ratio is an important parameter for describing the robustness of the frame structure and the interaction between bending moment and axial force in beams. As *L/d* ratio increases, vertical load resistance decreases while rotation capacity increases, resulting in a larger contribution of catenary action to resist the loss of a column. Exception makes the corner column loss, where only the flexural action contributes to the progressive collapse resistance following the removal of the column. In this latter case, the *L/d* ratio does not affect the rotation capacity, which are almost identical, independent of span to depth ratio.

Acknowledgments

Funding for this research was provided by the Executive Agency for Higher Education, Research, Development and Innovation Funding, Romania, under grant PCCA 55/2012 "Structural conception and collapse control performance based design of multistory structures under accidental actions" (2012-2016).

References

[1] European Committee for Standardization CEN. *EN 1990: Eurocode - Basis of structural design*, 2002.

[2] Federal Emergency Management Agency FEMA *The Oklahoma City Bombing: Improving building performance through multi-hazard mitigation, FEMA 277*, ASCE, 1996.

[3] Department of Defense DoD. *Design of buildings to resist progressive collapse. UFC 4-023-03*, Washington, DC, 2013.

[4] National Institute of Standards and Technology NISTIR. *Best Practices for Reducing the Potential for Progressive Collapse in Buildings NISTIR 7396*, Technology Administration, U.S. Dep. of Commerce, 2007.

[5] Dubina D, Dinu F. "Collapse prevention design of multistorey steel building frames under extreme actions", *Proceedings of Nordic Steel Construction Conference*, 2012.

[6] Dinu F, Marginean I, Dubina D., Petran I. "Experimental testing and numerical analysis of 3D steel frame system under column loss", *Eng. Structures*, 113, 59-70, 2016.

[7] *ELS Extreme loading for structures (Version 3.1)*. Durham, NC: Applied Science International, 2010.

Seismic resistant structures

DEVELOPMENT OF A SEISMIC TESTING FACILITY TO TEST FULL SCALE STEEL STORAGE RACKS

Tanim Ahmed, Mahmoud Shaheen, Kim J.R. Rasmussen and Hao Zhang

School of Civil Engineering, The University of Sydney, Sydney, Australia

Abstract: This paper describes the design and commissioning of an experimental facility for testing heavily loaded steel storage racks under seismic excitation. A one-degree-of-freedom rig was designed to test drive-in racks with 4.4m × 4.4m footprint, up to 5m in height and pallet loads of up to 54 tons. Particular attention was paid to instrumenting the frames with accelerometers and laser or LVDT displacement transducers to capture the 3D cross- and down-aisle deformations and accelerations resulting from the eccentric bracing of the frames. The paper describes the concepts adopted for the design and construction of the test facility as well as the rationale for the instrumentation.

1. Introduction

Steel storage racks are important structures that carry loads much higher than their self-weight and stored goods are by large the main source of load resisted by the frame. For space utilization and providing access to stored goods, their use has increased steadily with a significant amount in areas with public access. Traditionally their design guidelines have been developed primarily focusing on gravity loads [1-3]. Very little information is available on the seismic response of storage racks [4-7].

Experimental verification of the performance of a structure under seismic events can be carried out by shake table tests or pushover tests. A shake table test captures the effects of a varied range of loading frequencies, effects of strain rate and the complete three dimensional response of the structure. To obtain a proper understanding of the seismic response of drive-in racks, full scale seismic tests on a commercial rack sub-assembly have been carried out. This paper describes the development of the seismic testing facility for testing the racks in both down-aisle and cross-aisle directions.

2. Design considerations

A fully loaded, full scale drive-in rack subassembly was to be tested in both down-aisle and cross-aisle directions. It needed to carry a maximum pallet load of 54 tons. In the first stage the rack would be tested in the down-aisle direction. Subsequently, the rack would be dismantled and rotated 90 degrees for testing in the cross-aisle direction.

The test rig had four major components: i) Hydraulic Jack, ii) Roller Support Beams, iii) Skates/Rollers and iv) RHS Load Transfer Frame. The test set-up was specifically designed to be operated on a standard concrete slab without the need for connection to a strong floor. Four I-beams (150 UC-37.2) were anchored to the concrete floor using heavy duty anchor bolts. A rectangular 4×4 grid frame made of RHS members with rollers underneath, was mounted on these I-beams, enabling the RHS-frame to move forwards and backwards along the beams. A hydraulic jack anchored to a 150 mm concrete slab was attached to one end of the rectangular frame, which could move the RHS-frame according to any given earthquake signal. A schematic diagram of the test rig components is shown in Fig. 1.

A finite element (FE) model of the rack subassembly was created using Abaqus. Linear transient dynamic analyses were carried out to obtain the base shears developed due to application of a suite of preselected earthquake records, including El Centro-180, Duzce-180 and Tabas-LN earthquake signals. Table 1 shows the computed base shears and forces for different spine bracing configurations under El Centro-180 loading. The maximum base shear of 350 kN was encountered for the Tabas-LN earthquake in the cross-aisle direction. The base shear was also calculated using the Equivalent Static Method (ESM) of NZS 1170.5.2004 [8] for different site conditions. The maximum base shear from ESM was 220 kN for the cross-aisle direction, considering a 60% effective load of the total pallet weight.

Table 1: Base Shear

	Spine Bracing Configuration	Displacement (DI loading, mm) Down-aisle	Displacement (DI loading, mm) Cross-aisle	Base Shear (kN) (Down-aisle Loading)	Base Shear (kN) (Cross-aisle Loading)
1	No Bracing	330	0	130	200
2	Mid Bay	130	15	135	205
3	Side Bay	132	17	138	205

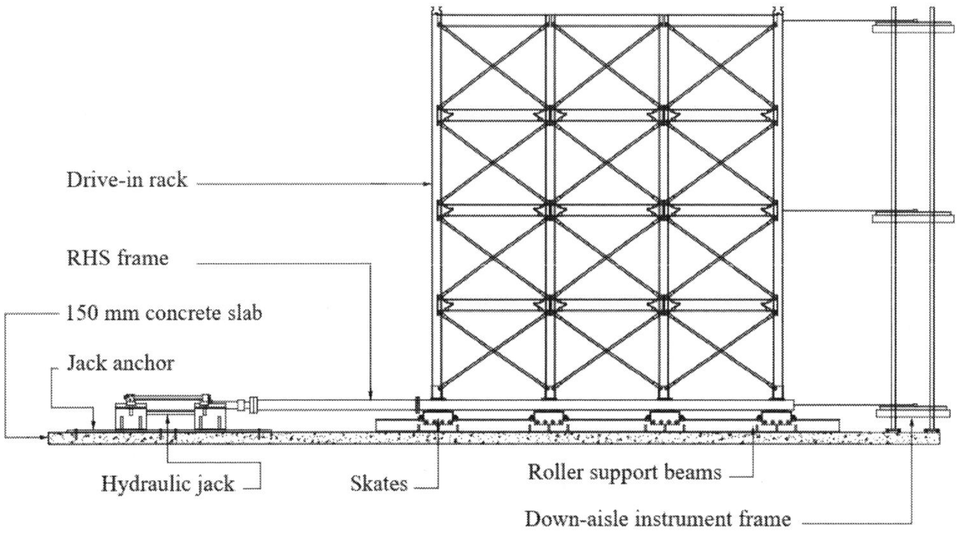

Fig. 1: Schematic diagram of test set-up

Other information including the reaction forces at the upright bases and member forces developed in the test rig were used to design the rig components. Initially, dynamic analyses of an FE model of the rack loaded with 54 ton pallet load were performed. The resulting reaction forces extracted from the model were used for carrying out a static analysis on a model containing only the test rig to determine the member forces. Later, a combined model with the rack attached to the rig was also developed.

3. Components of the test rig

3.1 Hydraulic jack and jack anchor

The hydraulic jack, servo valve and accumulators were designed in cooperation with CMA Electro Hydraulic Engineers. Considering the frequency components of the selected suite of earthquakes and the maximum base shear obtained from the FE analysis, a 600 kN dynamic load rated jack was designed. To run the jack to its full capacity without any pressure fluctuation in the oil supply, accumulators were installed for both supply and return line. Some of the features of the jack are:

- Solenoid controlled main flow isolation valve to the servo actuator. This is a large valve to allow the peak flows up to 1000 l/min anticipated at 500mm/sec piston speed.
- Integrated Digital Position feedback transducer inside the cylinder.
- Integrated servo valve(s), suited for both dynamic and static motion control.
- 900 kN load cell attachment.

Because of the layer of roller beams and skates the test rig level was elevated from the ground by about 315 mm. The jack was placed at the same elevation using vertically standing 70 mm plates as shown in Fig. 2. In the jack-anchor design, both the direct shear force and the moment produced as a result of the elevated jack were taken into consideration. Four vertical plates were welded to a large base plate which was anchored to the concrete slab of the structures laboratory. The vertical plates had threaded holes for bolts to connect the jack through its side-lug holes. On each vertical plate a key-way slot was also cut to accommodate the side lugs of the jack. Although the bolts were sufficient to hold the jack to the anchor frame, the key-way slots were added to transfer the shear load.

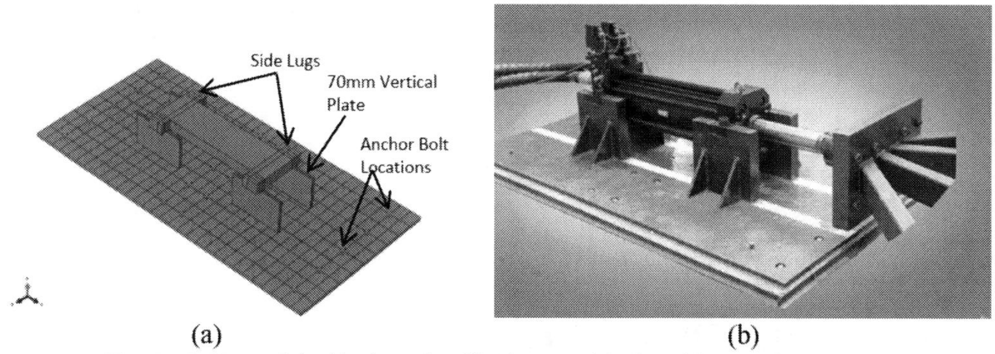

Fig. 2: a) FE model of jack anchor b) The actual jack and its attachments

An FE model (Fig. 2a) was developed to determine the forces on the anchor bolts and also to determine their best possible locations. Based on the FE analysis, a 32mm plate was used for the base plate. For anchoring the jack to the concrete floor, M16 SpaTec Plus safety anchors were used for their high capacity in shear and tension. The anchor bolts were placed in 4 rows, each row containing 5 bolts. Fig. 2b shows the constructed jack and support base.

3.2 Roller support beams

Below the test frame, the roller beams are connected directly to the concrete slab. The whole rack is sitting on a set of rollers and moves along the laid beams. Four 6m long I-beams were laid parallel to each other and fastened to the ground using a total of 96 M16 Spatec safety anchors. The anchor bolts were fitted through 12mm rectangular connector plates welded to the bottom flanges of the beams and screwed into anchors inserted into the concrete floor of the laboratory. The uprights at the back of the rack, where spine bracing is to be located for the down-aisle testing, were expected to experience the maximum uplift. A finite element model was created to determine the stresses developed at different locations of the beam for the applied loads. The loading locations were varied to calculate the maximum reaction forces on the anchoring positions and also to find the maximum stresses in the I-beams. Based on the FE analysis, the I-beam below the uprights featuring spine bracing were designed with extra anchor bolts to withstand an uplift force of 120 kN. The other 3 I-beams were designed for uplift forces of 60 KN.

3.3 Skates

One of the major components of the test rig was the rollers, or skates, running on the I-beams. From the FE simulations it was found that the uprights would face uplift, especially those connected to spine bracing. The maximum uplift occurs when spine bracing is installed from the top to the bottom level of the rack at the mid-bay only. For this reason, the skates under the uprights with spine bracing were designed to transfer the maximum uplift force. The other skates were designed for a lower uplift capacity.

15-OT type Hilman rollers with 15 tons vertical downward loading capacity were chosen for the skates. Bearings were attached to the rollers to prevent the uplift and lateral movement. The rollers expected to experience heavier uplift were fitted with 8 bearings (4 on each side) to restrict the uplift, while the others had 4 bearings (2 on each side). To restrict lateral movement, each roller was fitted with 4 guide bearings, 2 on each side. Fig. 3 shows the bearing attachments on the rollers.

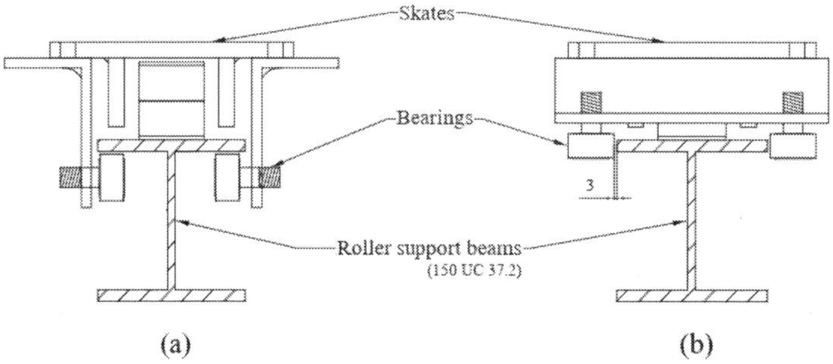

Fig. 3: Views of skates with bearings to restrain (a) uplift and (b) lateral movement

3.4 RHS frame

The dimensions of the RHS frame were governed by both the strength requirement and consideration of the base plate size of the rack. 75 mm wide beams were chosen to facilitate the bolting of the upright base plates of the rack. The frame was designed for a base shear of 600 kN using 125×75×6 RHS members. The frame had a rectangular and a triangular part, as shown in Fig. 4(a). The rectangular part is a grid system having 4 continuous members running in the loading direction. Perpendicular segmented link members connected the continuous members at three equal intervals. All the members in the rectangular part were welded to each other. The end of the rectangular part closer to the jack had extra cross bracings to ensure uniform displacement across the width of the platform. To avoid interference with the base plate connection of the uprights, the cross-bracings could not go all the way to corners. Additional smaller members perpendicular to the diagonal bracings were placed at the corner locations.

A triangular frame made of same RHS sections was connected to the rectangular part using bolted connections, to transfer the load from the hydraulic jack. The connection between the jack and the triangular frame included a spherical alignment coupler to allow for +/-3 mm and +/-3 degrees misalignment. Fig. 4b shows photos of different parts of the rig during construction, including the triangular frame, a skate on a roller beam, jack to frame connection and a partial view of the 4×4 grid frame (clock-wise, starting from top-left).

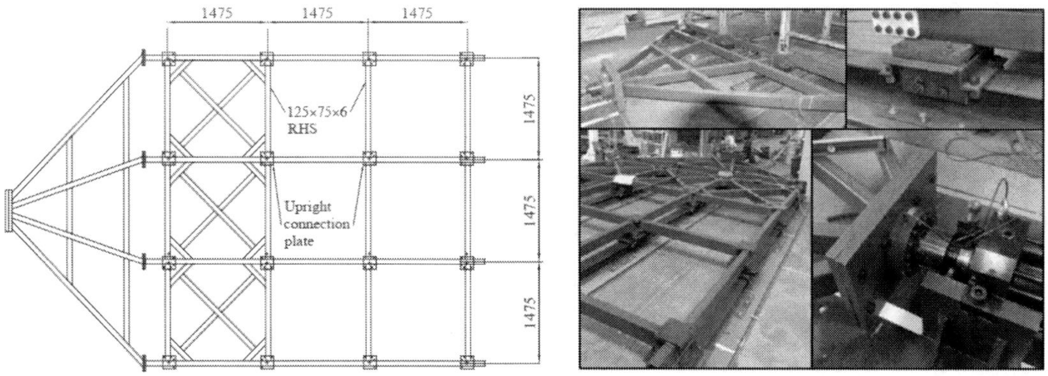

(a) (b)

Fig. 4: a) Test rig drawing for construction, b) different components of the rig

4. Instrumentation

To capture the response in both the loading direction and its perpendicular direction, instruments were placed on both sides of the rack. Two instrument frames, perpendicular to each other, were built on the ground, detached from the moving rig. To record the down-aisle movement, eight transducers (LVDT) with 1400 mm measuring range were placed at the top and mid-level of the rack. Two LVDTs of 600 mm range were used at the base level to measure the down-aisle displacement of the rig. In a similar arrangement, eight laser displacement sensors with a measuring range of 200 mm were used for the cross-aisle response measurement at the top- and mid-levels, and two 20 mm range sensors were placed at the base level. The laser sensors were attached directly to the rack. The movement of the rack in the cross-aisle direction was picked up by the laser beams reflected from the flat surface of a horizontal RHS member on the cross-aisle instrument frame. Fig. 5 shows the two instrument frames.

Strain gauges were placed on uprights, spine and plan braces to monitor internal forces developed during the seismic loading. On the uprights, strain gauges were placed on the neutral axis on both sides of the symmetric axis. They were calibrated for both tension and compression by separate stub column tests and tension tests on separate upright specimens.
Accelerometers were placed on the rack, pallet and concrete blocks on three rail beam levels to observe the possible effect of sliding motion of the pallets and goods (concrete blocks).

(a) (b)

Fig. 5: (a) Transducer frame (down-aisle measurements), (b) laser sensor frame (cross-aisle measurements)

5. Rig performance and seismic tests

The jack performance and rig commissioning process included a varied range of tests starting with the unhooked jack being tested with different simple control signals, including sinusoidal and saw tooth signals. Subsequently, to ensure that the jack can follow complex actual ground motion signals, the system was tested with the scaled suite of earthquakes to be used in the seismic tests. The scaling of the earthquake signals were carried out following the seismic design code of New Zealand [8], considering the zone coefficient for Wellington and structural natural period obtained from the FE frequency analysis. To keep consistency and avoid complications among different tests, the scaling of signals were kept to the one obtained in this stage. Only multiplying factors, (MF=0.1, 0.2 etc.) were used to increase or decrease the signal strength. Different loading scenarios were investigated, where the jack was attached to the rig and gradually concrete blocks were added to the RHS frame, finally reaching a total of 50.5 tons.

To allow for the entry of fork lift trucks, drive-in racks cannot have any bracing at the front. Only the back is braced with spine braces. Tests were carried out to understand the effects of spine bracing configuration on the three-dimensional behaviour of the drive-in racks. Seven different spine bracing configurations were tested. As part of the rack tests, the dynamic properties of the racks were obtained from free vibration tests, using step functions of varied speed and base displacements, explained in more details in [9]. For seismic tests, first, the empty racks were tested under three different earthquake loads including the El Centro-180, Duzce-180 and Tabas-LN. The empty rack was tested with a maximum of 100% (MF=1.0) of the scaled suite of the selected earthquakes. The fully loaded rack was tested under the same suite of earthquakes but with a maximum of 30% (MF=0.3) of the scaled earthquakes. In all these cases the performance of the rig was monitored using different instrumentation. The two 600 mm LVDTs placed at the two outer corners of the rig were used to measure the rack displacement and to check for any differential movement between the two points. The rig performance was found to be consistent in all phases of the tests, both in terms of acceleration and displacement. For the fully loaded rack with 30% of El Centro loading, the control signal had peak ground acceleration (PGA) of 0.10g and the measured PGA was found to be 0.11g.

Fig. 6 shows a comparison of the control signal and the output displacement signal recorded from the ground level LVDTs during a fully loaded test.

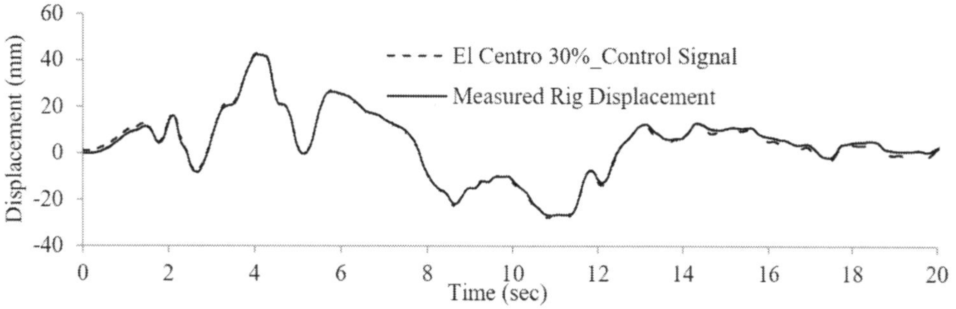

Fig. 6: Control signal and measured displacement comparison

To monitor the torsional behaviour of the rack, the mid- and top-level were instrumented with four transducers each. Fig. 7 shows the displacement response at four locations on the top level of the rack for 30% of the scaled El Centro loading. The displacement gradually reduces from the front to the back because of the spine bracing. Table 2 shows the variation of responses including the base shear, the maximum top displacement and maximum upright forces for the seven spine bracing configuration tested using 20% of the scaled El Centro loading, as reported in [9].

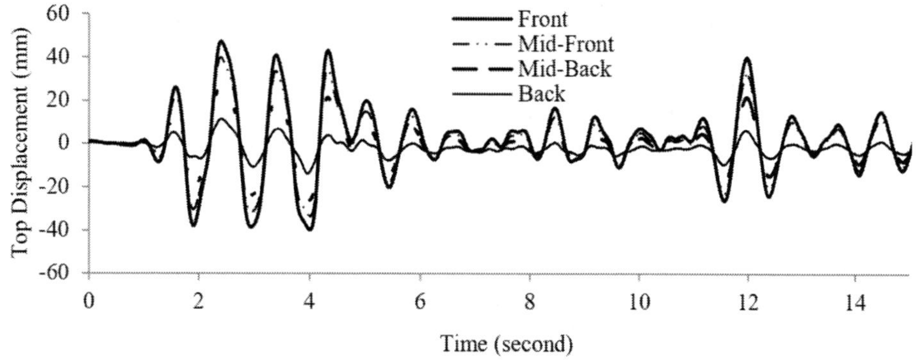

Fig. 7: Down-aisle displacement response at different locations on top level of the rack for El Centro 30% loading

Table 2: Test Results for Fully Loaded Rack (El Centro 20%)

Bracing Configuration	Period of Structure	Base Shear (kN)	Max Top Displacement (DI) (mm)		Max Seismic Axial Force on Upright (kN)	
			Front	*Back*	*Compression*	*Tension*
Config-1	0.68	45	32	7	35	7.2
Config-2	0.81	40	41	20	29	7.2
Config-3	1.04	34	39	27	21	3.8
Config-4	1.28	29	36	30	11	2.1
Config-5	0.81	40	38	24	30	9.0
Config-6	0.73	46	32	14	40	5.9
Config-7	0.91	39	35	25	39	7.0

6. Conclusions

A shake table was designed to test drive-in steel storage racks in both down-aisle and cross-aisle directions. The methodology and principles behind the design of the shake table has been presented along with description of the major components of the rig and the instrument frames.

Tests in the down-aisle direction have been successfully carried out with sample results presented in this paper. The rig performed consistently with different levels of loading conditions including the empty rack and loaded rack with varying levels of earthquake signals.

The instruments performed as intended, specially, the LVDTs and laser sensors were able to successfully capture the displacements at different locations of the rack, and the strain gauges provided reliable recording for determining axial forces in the uprights. Research is underway to test the rack in the cross-aisle direction.

Acknowledgments

This research was supported by Australian Research Council under Linkage Project Grant LP120100458. This support is gratefully acknowledged.

References

[1] Peköz T, Weng CC. "Design of cold-formed steel columns", *9th International Specialty Conference on Cold-Formed Steel Structure*, Missouri University of Science and Technology: St Louis, Missouri, USA, 27-43, 1988.

[2] Sarawit AT, Pekoz T. "Design of industrial storage racks", *Sixteenth International Specialty Conference on Cold-Formed Steel Structures*, Orlando, Florida, USA, 369-84, 2002.

[3] Beattie GJ, Deam BL. "Design guide-seismic design of high level storage racking systems with public access", Porirua, New Zealand: BRANZ Ltd, 2007.

[4] Padilla-Llano DA, Moen CD, Eatherton MR. "Energy dissipation of thin-walled cold-formed steel members", *AISI-Specifications for the Design of Cold-Formed Steel Structural Members, Paper 49,* 2013.

[5] "EUR 23744: Storage racks in seismic areas", *Research fund for coal and steel, final report*, 2009.

[6] Bernuzzi C, Chesi C, Parisi MA. "Seismic behaviour of steel storage pallet racks", *Stessa 2003: Behaviour of Steel Structures in Seismic Areas,* 769-74, 2003.

[7] Filiatrault A, Higgins PS, Wanitkorkul A, Courtwright JA, Michael R. "Experimental seismic response of base isolated pallet-type steel storage racks", *Earthquake Spectra,* 24(3), 617-39, 2008.

[8] NZS 1170.5:2004 Structural design actions - Part 5: Earthquake actions - New Zealand, Standards New Zealand, 2004.

[9] Ahmed T, Rasmussen KJR, Zhang H. "Down-aisle seismic tests of drive-in rack systems", *Proceedings of the International Colloquium on Stability and Ductility of Steel Structures* (SDSS2016), Timisoara, 2016.

DOWN-AISLE SEISMIC TESTS OF DRIVE-IN RACK SYSTEMS

Tanim Ahmed, Kim J.R. Rasmussen and Hao Zhang

School of Civil Engineering, The University of Sydney, Sydney, Australia

Abstract: This paper presents the results obtained from uniaxial shake table tests carried out on fully loaded 3×3 bay drive-in racks subjected to standard recorded earthquakes including the El Centro, Duzce and Tabas time histories. In the tests, the structural response was recorded to understand the 3D behaviour characterizing drive-in rack systems. Various spine bracing configurations are investigated to determine the effect of gradually reducing the spine bracing stiffness of the frame, in terms of increased deflections and increased period of vibration. Details of the test frames, the experimental set-up and the results of full-scale testing are provided and conclusions are drawn regarding the optimal bracing configurations.

1. Introduction

Steel storage racks have different structural components and different structural behaviour to regular steel structures, requiring special design considerations. They are self-standing and more slender structures made of cold-formed sections (CFS), carrying loads many times higher than their self-weight. They are also more likely to carry the full design live load many times in their design life. Uprights of the storage racks are generally made of thin-walled, open sections produced from cold-formed steels. They are usually perforated at fixed intervals to facilitate connection of upright frame bracings, portal beams, rail beams and other rack components. The perforations change the behaviour of the members and reduce the axial and bending capacities [1, 2]. Another limitation of rack structures is that CFS thin-walled sections cannot develop plastic hinges in beams and uprights. This results in relatively low ductility causing reduced seismic resistance [3-4].

In drive-in racks, uprights joined by diagonal bracings create upright frames that provide lateral stability and shear stiffness in the cross-aisle direction. In the down-aisle direction, in order to facilitate loading and unloading, bracing can only be provided at the back, known as spine bracing. Hence the stability in that direction depends significantly on the plan bracing, spine bracing, portal beam-column joint stiffness and base plate connection stiffness [5]. Between the base plates and the plan bracing at the top, the only structural component providing down-aisle lateral stiffness is the spine bracing. The eccentricity of the spine bracing causes an asymmetric, 3D torsional behaviour of the rack, where the down-aisle motion induces coupled movement in the perpendicular cross-aisle direction [6]. The stiffness of the portal beam-to-column connection is highly non-linear, which makes it difficult to understand both the static and dynamic behaviour of the rack.

In addition to the lack of sufficient guidelines in international standards for drive-in racks, available data on experimental tests are also very limited [5-11]. Most studies focus on rack components rather than full scale tests [12, 13]. The experimental study presented herein consists of a series of shake table tests carried out on full scale three-bay wide, three pallets deep and three storage level high drive-in racks. Seven different spine bracing configurations have been experimentally investigated with the objective to examine the effect of various spine bracing configurations on the seismic response of the rack. By controlling the spine bracing configuration, the lateral stiffness in the down-aisle direction can be altered, which changes the natural period of the structure and thus the seismic response can be shifted to a preferred zone in the response spectrum. Fig. 1 shows a typical setup of a fully loaded drive-in rack.

2. Experimental program

Shake table tests provide the most realistic procedure for assessing the seismic behaviour of a structure. The uniaxial shake table facility designed and commissioned at the structures laboratory of The University of Sydney as part of this project was used for carrying out tests on the drive-in rack structures [14].

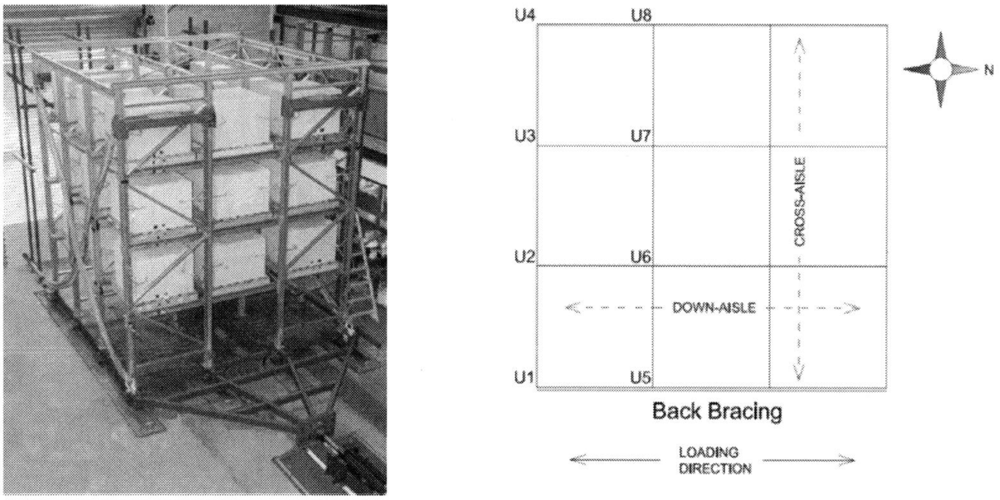

a) b)

Fig. 1: (a) Loaded rack on test rig, (b) upright numbering of tested rack

The dynamic properties of the rack were studied as part of the project. Seismic tests were carried out to investigate the 3-dimensional behaviour of the rack, under down-aisle seismic loading. The rack responses were recorded using instruments including strain gauges, LVDTs, accelerometers and laser displacement sensors. The responses studied include the axial forces developed in the uprights, the maximum base shear, and displacement response history in both down-aisle and cross-aisle directions at three different levels of the rack.

2.1 Rack subassembly

A fully loaded full scale drive-in rack sub-assembly was tested to study its three dimensional responses under horizontal seismic loads. The rack was 4.5m deep, 4.5m wide and 5m high. The plan dimensions were governed by the requirement of the test rig to suit both down-aisle and cross-aisle testing.

The rack was designed to carry a nominal 2 ton load on each pallet. For the tests, concrete blocks with an average measured weight of 1.87 ton were used to replicate the weight from storage goods. A total of 27 concrete blocks were used to load the rack. The horizontal plan bracing at the top covered all three bays and was kept the same for all tests. The spine bracings were changed to vary the lateral stiffness of the rack. The bracings were made of a set of two diagonal members with each set covering a single bay of a single level. Fig. 2 shows the seven different bracing configurations tested. The spine bracing members were made of C250 grade 26.9 × 2 mm pipe sections as shown in Fig. 3.

2.2 Instrumentation

Down-aisle displacements of the rack were recorded at three levels on different locations using LVDTs. Laser sensors were used for recording the cross-aisle movements. Strain gauges were placed on uprights, spine and plan braces to calculate their internal forces. Accelerometers were used on the test rig to record the actual ground acceleration exerted on the rack. To determine the differences in the pallet and rack movement, accelerometers were also placed on the rack, pallet and concrete blocks. For the spine braces in each configuration, the strain gauges were placed at the location where the maximum force was expected. Strain gauges on plan braces were on a fixed location that would develop the maximum plan bracing force.

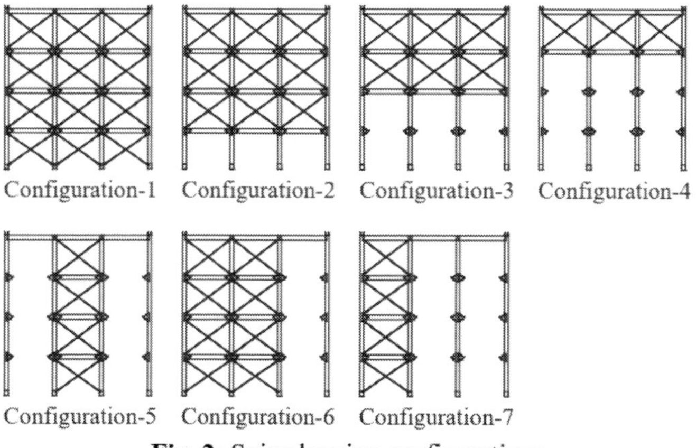

Fig. 2: Spine bracing configurations

Fig. 3: Spine bracing

2.3 Test sequence and ground motion records

For each bracing configuration two series of tests were carried out. First, dynamic properties of the rack were obtained using free vibration tests, where the test rig was given a fixed displacement using a step function of a specific speed. The free vibration response was recorded until the rack came to complete rest. Each test was performed twice to confirm repeatability.

The second series consisted of different ground motion records applied in a gradually increasing scale to the rack in the down-aisle direction. Table-1 shows the earthquakes used in the tests. Among the three ground motions, Tabas has forward directivity. The ground motion records were selected from the suite of records proposed by Oyarzo-Vera et al. [15].

Table 1: Ground motion records and their scaling factors

Record Name	Station	Date	Magnitude	Mecha-nism	Scaling Factor	Scaled PGA (g)
El Centro-180	117 El Centro Array #9	Imperial Valley 19/05/1940	M (7.0)	Strike Slip	1.08	0.34
Duzce-180	Duzce	Duzce, Turkey 11/12/1999	M (7.14)	Strike Slip	0.63	0.22
Tabas-LN	Tabas	Tabas, Iran 09/16/1978	M (7.35)	Reverse	0.50	0.42

Prior to the tests, frequency analyses were carried out on finite element (FE) models of the racks, developed using the FE software Abaqus. The analyses showed that the natural periods range from 0.7s to 1.4s, for different spine bracing configurations. The earthquake signals were scaled following the procedure specified in Section 5.5.2 of NZS 1170.5 [16]. Fig. 4 shows the scaled response spectrum of El Centro-180 earthquake for a structure with a natural period of 0.85 second. Table 1 summarises the earthquakes and the peak ground accelerations (PGA) of the scaled signals.

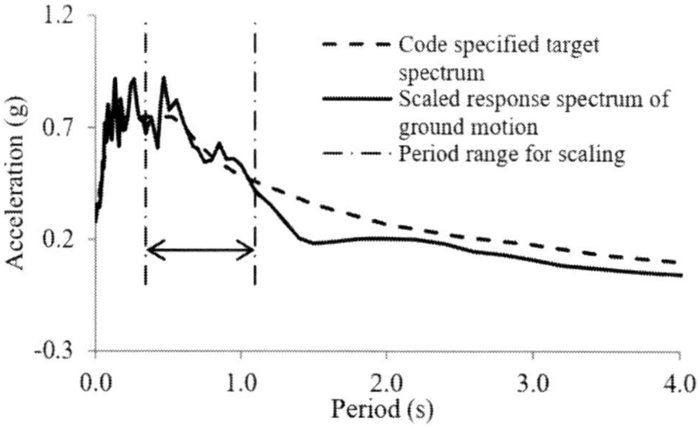

Fig. 4: Scaled response spectrum of El Centro-180 ground motion

3. Tests on fully loaded racks

The fully loaded racks with seven different spine bracing configurations were tested in the elastic range to investigate the effects of spine bracing configurations on the three-dimensional behaviour of the racks. Keeping the loading level within the elastic capacity of the racks enabled the usage of the same rack for multiple tests, modifying the bracing system only. The testing sequence is presented in Table 2.

Table 2: Loaded rack test sequence

Test #	Test ID	Bracing con-figuration	Test type	Speed (mm/s)	% of scaled earthquake	Scaled PGA (g)
1	20 mm/s-B_Config-1	1	Free Vibration	20 (Backward)		
2	30 mm/s-B_Config-1	1	Free Vibration	30 (Backward)		
3	40 mm/s-B_Config-1	1	Free Vibration	40 (Backward)		
4	40 mm/s-F_Config-1	1	Free Vibration	40 (Forward)		
5	El-Centro-10%_Config-1	1	Earthquake		10	0.034
6	El-Centro-20%_Config-1	1	Earthquake		20	0.068
7	El-Centro-30%_Config-1	1	Earthquake		30	0.101
8	Duzce-10%_Config-1	1	Earthquake		10	0.022
9	Duzce-20%_Config-1	1	Earthquake		20	0.044
10	Duzce-30%_Config-1	1	Earthquake		30	0.066
11	Tabas-10%_Config-1	1	Earthquake		10	0.042
12	Tabas-20%_Config-1	1	Earthquake		20	0.084
13	Tabas-30%_Config-1	1	Earthquake		30	0.125

Free-vibration tests on the fully loaded racks were carried out to determine the natural frequencies and critical damping ratios. The seismic tests started with 10% of the scaled earthquakes, i.e. the scaled earthquake signals were multiplied again by a factor (MF) of 0.1, and similarly multiplying factors of 0.2 and 0.3 were applied on the scaled signals to achieve 20% and 30% of the scaled earthquakes, respectively. In order to keep the response within the elastic range of the structures, a maximum of 30% of the scaled earthquakes were used. The same test sequence was followed for all seven bracing configurations. For configuration-4 a maximum of only 20% of the scaled signal was applied due to the high level of sway and strain readings.

4. Test results and discussion

4.1 Natural period and critical damping ratio

The second column of Table 3 shows the natural periods of the structures. The period increases as the spine braces are removed, which has a significant effect on the structural response. In configurations 5-7, the change in the period is not as pronounced and the values are in the same range as for configuration-2.

Table 3: Free vibration and seismic test results

	Free-vibration test results		*Seismic test results*					
Bracing Configuration	Natural Period (s)	Critical Damping Ratio, ξ %	Maximum Base Shear (kN)	Max Top Displacement (DI) (mm)		Max Top Displacement (CI) (mm)	Max SB Force (kN)	Max PB Force (kN)
				Front	***Back***			
Config-1	0.68	2.4	60	47	14	12	1.4	4.7
Config-2	0.81	2.9	52	53	26	9	4.2	5.0
Config-3	1.04	2.3	37	60	40	4	2.9	4.1
Config-4	1.28	1.9						
Config-5	0.81	2.9	50	50	32	5	8.8	3.7
Config-6	0.73	2.6	56	50	22	10	6.7	4.5
Config-7	0.91	3.5	50	52	34	9	5.1	3.0

The decay of vibration amplitude of a structure under free vibration depends on its damping ratio. Damping of a structure depends on multiple mechanisms acting on the structure simultaneously. In mathematical calculation, all these components are idealized as equivalent viscous damping [17]. In the present study, the critical damping ratios were calculated from the displacement response at the top of the rack. The decay of motion has been related to amplitudes 10 cycles apart, starting from the initial maximum free-vibration displacement. Column three of Table 3 shows damping ratios for all the rack configurations as obtained from the tests.

4.2 Seismic response of loaded racks

It follows from Table 3 that for configuration-1 to 3 as spine braces are removed, the down-aisle (DI) displacements increase. However, the difference in down-aisle top displacement at the front and back of the rack reduces. The cross-aisle (CI) displacement also reduces indicating a reduction in the torsional movement generated by the eccentricity of the spine bracing. Fig. 5 shows the top displacement time history at the front and back of the rack with configuration-1, under El Centro 30% loading.

Last two columns of Table 3 show the maximum tension forces developed at the spine (SB) and plan braces (PB). For configuration-5 to configuration-7, the stiffness from spine braces is concentrated in a narrower zone compared to the first four configurations. Hence, the spine bracing force is higher for configuration-4, 5 and 6. The plan bracing force is fairly insensitive to the change of spine bracing configuration, as seen from the results in Table 3.

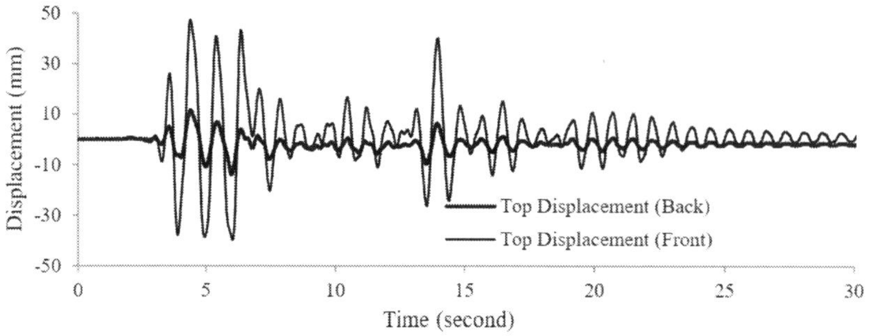

Fig. 5: Top displacement time history for configuration-1 rack

In Fig. 6 and Fig. 7, the response of the spine bracing member on level-3 of the rack (configuration-3) is presented in terms of displacement time history and force-displacement hysteresis behaviour respectively, for El Centro 90% loading. The spine braces are designed as tension only members. Still, they can take some compressive force. The spine braces start to buckle around 10 kN compressive force, as can be seen from the load-displacement curve. The dense region near location-1 in Fig. 7 represents the small displacements, where the member behaves in a linear-elastic manner. Location-2 represents the maximum compressive force developed in the member. The maximum tension force developed in the member is almost twice as the maximum compressive force, occurring with the maximum elongation of the bracing member, marked as location-3 in Fig. 6 and Fig.7.

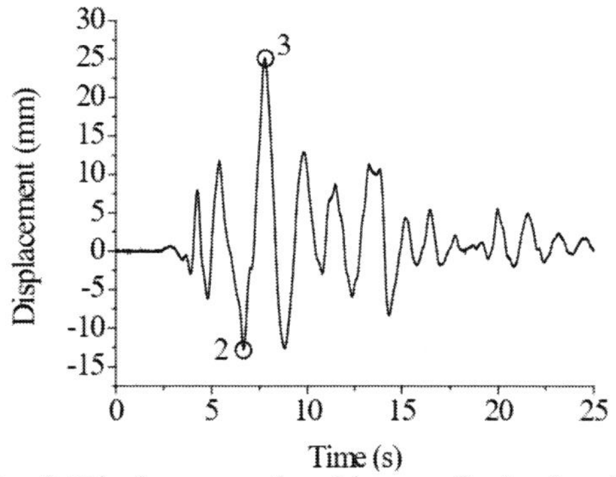

Fig. 6: Displacement time history of spine bracing member from configuration-3 rack

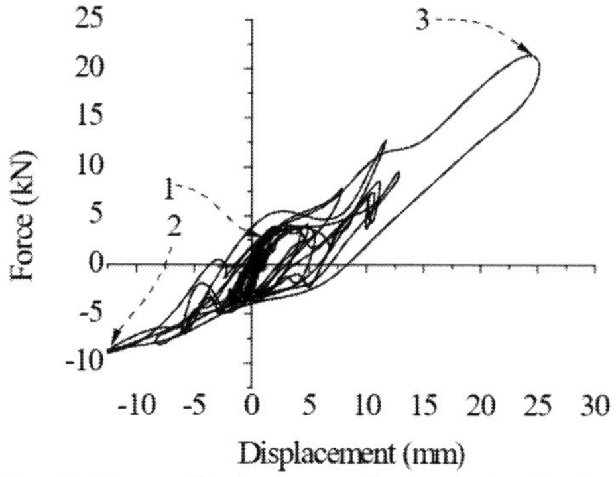

Fig. 7: Force-displacement hysteresis of spine bracing member

Table 4 and 5 show the maximum compressive and tensile forces developed in the uprights during El Centro 30% loading. On a symmetric structure with linear behaviour, deforming in a pure one directional sway mode, the tensile and compressive forces in columns would be same. From the results shown in Table-4 and 5, it is observed that the seismic tensile and compressive forces are not equal and the maximum compressive force occurs in the corner upright U1. The torsional motion of the rack causes the corner uprights (U1 and U13) to experience higher seismic compressive loads. In compression, the baseplates are completely rigid, apart from the initial gap between the upright and baseplate, which closes during the pallet loading process. In tension, the baseplates can experience uplift, causing a reduction of the tensile force in the upright. In configuration-1 to 4 the braces are removed level by level and in configuration-5 to 7 they are changed bay-wise. For configuration-1 to 4, the maximum compressive force develops in upright-1 (U1). In most of the uprights the maximum compressive force gradually reduces as the spine braces are removed. This is associated with increased sway amplitude of the rack. Results for configuration-5 to configuration-7 show that the maximum compressive force develops in the upright connected with the spine braces. In the front uprights, especially uprights U6, U7 and U8, the maximum compressive force does not change significantly with the variation in spine bracing arrangements.

Table 4 Maximum seismic compressive force for El Centro-30% loading

Rack Configuration	U1	U2	U3	U4	U5	U6	U7	U8
	C (kN)	C (kN)	C (kN)	C (kN)	C (kN)	C (kN)	C (kN)	C (kN)
Config-1	-49.9	-18.9	-38.1	-20.4	-6.9	-6.5	-14.8	-12.9
Config-2	-38.3	-9.3	-29.9	-21.5	-14.7	-5.7	-15.2	-11.7
Config-3	-26.5	-3.9	-19.5	-7.4	-13.6	-3.0	-13.5	-7.5
Config-5	-20.3	-7.3	-21.8	-17.5	-41.2	-5.8	-14.9	-11.0
Config-6	-53.2	-11.7	-29.7	-27.3	-7.7	-4.3	-17.5	-11.4
Config-7	-52.5	-7.5	-21.4	-20.4	-28.5	-4.9	-15.7	-8.6

Table 5: Maximum seismic tensile force for El Centro-30% loading

Rack Configuration	U1	U2	U3	U4	U5	U6	U7	U8
	T (kN)	T (kN)	T (kN)	T (kN)	T (kN)	T (kN)	T (kN)	T (kN)
Config-1	13.0	22.4	24.7	17.5	24.5	3.2	8.3	8.0
Config-2	11.9	12.4	24.0	15.4	13.9	3.2	9.8	5.5
Config-3	8.8	9.9	12.8	8.2	9.0	3.2	5.5	5.8
Config-5	11.1	14.8	22.4	13.3	33.3	14.5	7.6	6.3
Config-6	12.1	24.7	24.4	17.7	22.3	4.4	8.8	8.0
Config-7	12.4	22.8	23.9	12.9	28.0	9.3	7.0	7.1

5. Conclusions

The rack responses have been studied for seven bracing configurations under the citations within 30% of the scaled earthquake records. A comparison among configurations-1, 2 and 3 shows that removing one and two rows of spine braces increases the top displacement by 13% and 28% respectively, whereas the corresponding base shear reduces by 13% and 38% respectively.

Torsional motion of the rack is caused by the eccentricity of the spine bracing. Removing the spine braces from the bottom levels reduces the torsional movement but results in increased level of overall rack displacement.

The variation in the tension forces developing in the plan bracing is minimal for different configurations. The spine bracing forces are comparatively higher for configuration-5 due to the narrow area on which it acts.

For configurations-1, 6 and 7, the maximum seismic compressive force develops in upright-1 (U1). However, the most heavily loaded upright for the tested loading arrangement is U7, due to the high initial static pallet loads.

Among configurations-1 to 4, configuration-2 and configuration-3 have less torsional motion and moderate top displacement. Between the two, configuration-3 attracts significantly smaller base shear. Performance wise, configuration-5 to 7 are similar and can be compared to configuration-2. However, they attract higher spine bracing forces. Based on the tests performed in the elastic range, configuration-3 and configuration-5 show the most desirable response, with configuration-3 developing lower spine bracing force and configuration-5 undergoing smaller top displacement.

Acknowledgements

This research was supported by Australian Research Council under Linkage Project Grant LP120100458. This support is gratefully acknowledged. The authors are also grateful to Dematic Pty. Ltd. for supplying all testing materials at no cost.

References

[1] Bernuzzi C, Maxenti F. "European alternatives to design perforated thin-walled cold-formed beam–columns for steel storage systems", *Journal of Constructional Steel Research*, 110(0), 121-136, 2015.

[2] Peköz T, Weng CC. "Design of cold-formed steel columns", *9th International Specialty Conference on Cold-Formed Steel Structure*, Missouri University of Science and Technology: St Louis, Missouri, USA, 27-43, 1988.

[3] EN 1993-1-1. "Eurocode 3: design of steel structures", Part1.1: general rules and rules for buildings, 2005.

[4] ANSI/AISC 341-10. "Seismic Provisions for Structural Steel Buildings", Illinois, American Institute of Steel Construction (AISC), 2010.

[5] Freitas AMS, Souza FT, Freitas MSR. "Analysis and behavior of steel storage drive-in racks", *Thin-Walled Structures*, 48(2), 110-117, 2010.

[6] Gilbert BP, Rasmussen KJR. "Drive-In Steel Storage Racks I: Stiffness Tests and 3D Load-Transfer Mechanisms", *Journal of Structural Engineering-ASCE*, 138(2), 135-147, 2012.

[7] Gilbert BP, Rasmussen KJR. "Recent research on the design and behaviour of drive-in steel storage racking systems", *Steel Construction*, 4(4), 232-241, 2011.

[8] Gilbert BP, Rasmussen KJR. "Bolted moment connections in drive-in and drive-through steel storage racks", *Journal of Constructional Steel Research*, 66(6), 755-766, 2010.

[9] Godley MHR. "The behaviour of drive-in storage structures", *16th international specialty conference on cold-formed steel structures,* Orlando, Florida, USA, 340-352, 2002

[10] Murray NW. "Stability analysis of drive-in racking storage systems", *Structural Stability and Design*, 189-195, 1995.

[11] Salmon MA, Welch RE, Longinow A. "Analysis of drive-in and drive-thru storage racks", *Second specialty conference on cold-formed steel structures*, St. Louis, Missouri, 617-639, 1973.

[12] Anbarasu M, Murugapandian G. "Experimental study on cold-formed steel web stiffened lipped channel columns undergoing distortional–global interaction", *Materials and Structures*, 1-10, 2015.

[13] Trouncer AN, Rasmussen KJR. "Flexural–torsional buckling of ultra light-gauge steel storage rack uprights", *Thin-Walled Structures*, 81(0), 159-174, 2014.

[14] Ahmed T, Shaheen M, Rasmussen KJR, Zhang H. "Development of a seismic testing facility to test full scale steel storage racks", *Proceedings of the International Colloquium on Stability and Ductility of Steel Structures* (SDSS2016), Timisoara, 2016.

[15] Oyarzo-Vera CA, McVerry GH, Ingham JM. "Seismic Zonation and Default Suite of Ground-Motion Records for Time-History Analysis in the North Island of New Zealand", *Earthquake Spectra*, 28(2), 667-688, 2012.

[16] NZS 1170.5. Structural design actions - Part 5: Earthquake actions, New Zealand, Standards New Zealand, 2004.

[17] Chopra AK. *"Dynamics of structures: theory and applications to earthquake engineering"*, 3rd edition, Prentice Hall, Upper Saddle River, NJ, 2007.

EVALUATION OF SEISMIC RESPONSE FACTORS FOR BRBFs USING FEMA P695 METHODOLOGY

Mehmet Bakır Bozkurt and Cem Topkaya

Department of Civil Engineering, Middle East Technical University, Ankara, Turkey

Abstract: Seismic response factors given in ASCE7-10 for buckling restrained braced frames (BRBFs) were evaluated using the Methodology outlined in FEMA P695. Pursuant to this goal three single diagonal BRBF archetypes were designed according to the US Specifications. Time history analyses of these archetypes were conducted in accordance with FEMA P695 and using non-simulated collapse models. Maximum and cumulative brace strains were investigated in detail to evaluate the strain demands recommended by AISC341-10. Furthermore, the seismic response factors were assessed in light of the brace and frame deformation demands.

1. Introduction

Buckling restrained brace frames (BRBFs) are among various lateral load resisting systems for steel structures under seismic loading. A typical steel BRBF is composed of beams, columns, and buckling restrained braces (BRBs). During a seismic event BRBs yield in tension and compression and contribute to energy dissipation. When compared with conventional steel braces, BRBs provide nearly equal tensile and compressive resistances.

In the United States design recommendations for BRBs have been incorporated into AISC 341-10 Seismic Provisions for Structural Steel Buildings [1]. According to AISC341-10[1] buckling restrained braces shall be designed, tested and detailed to accommodate expected deformations. Expected deformations are those corresponding to a story drift of at least 2% of the story height or two times the design story drift, whichever is larger. Qualifying cyclic tests are required for conformance demonstration. In general, uniaxial and subassemblage tests are performed according to the loading protocol recommended in AISC341-10 [1]. The loading protocol is based on the design story drift. Furthermore, the brace test specimen is required to achieve a cumulative axial deformation of at least 200 times the yield deformation under uniaxial testing.

The equivalent lateral force procedure can be used together with a set of seismic response factors to obtain the design story drift. This procedure enables elastic analysis and design which is based on reduced seismic forces. The idea here is that the amount of lateral forces is reduced by taking into account yielding and ductility of the lateral load resisting system. The general structural response shown in Fig. 1 can be considered to develop response factors. Their formulation according to Uang [2] is as follows:

$$\mu_s = \frac{\Delta_{max}}{\Delta_y} \quad R_\mu = \frac{V_e}{V_y} \quad \Omega_o = \frac{V_y}{V_s} \quad R = \frac{V_e}{V_s} = R_\mu \Omega_o \quad C_d = \frac{\Delta_{max}}{\Delta_s} = \mu_s \Omega_o \qquad (1)$$

where, V_e is the ultimate elastic base shear, V_s is the base shear at the first significant yield, V_y is the base shear at the structural collapse level, Δ_s is the drift at the first significant yield, Δ_y is the drift at the structural collapse level, Δ_{max} is the maximum amount of drift, μ_s is the ductility factor, Ω_o is the overstrength factor, R_μ is the ductility reduction factor, R is the response modification factor, and C_d is the deflection amplification factor.

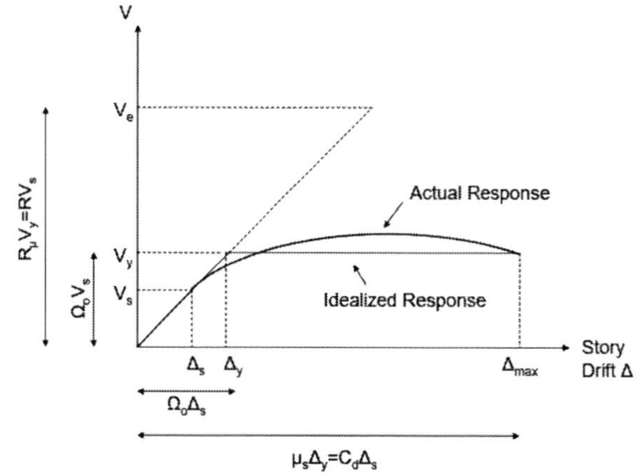

Fig. 1: General structural response

Seismic response factors were developed for various lateral load resisting systems based on observations from past earthquakes and engineering judgment. These factors vary from one specification to the other. In the United States, seismic response factors for BRBFs are given in Minimum Design Loads for Buildings and Other Structures [3] hereafter referred as ASCE7-10. The recommended values of the response modification factor (R), the overstrength factor (Ω_o), and the deflection amplification factor (C_d) are 8, 2.5, and 5, respectively.

Brace deformation demands must be accurately determined at the design stage for satisfactory performance of a BRBF. The design and detailing of a BRB is directly influenced by the design story drift which depends on the seismic response factors. A study has been undertaken to evaluate the seismic response factors for BRBFs using the Methodology outlined in FEMA P695 [4]. Pursuant to this goal three archetype BRBFs were designed and evaluated according to the Methodology. The details of the evaluation are presented herein.

2. Overview of the FEMA P695 Methodology

The Methodology requires nonlinear collapse simulation on the selected archetype models. Collapse simulation is conducted using a far field record set that consists of 22 pairs of ground motions. All 44 ground motion records must be individually applied to an archetype in cases where a two dimensional analysis is performed. The ground motion records are scaled twice. The first scaling is required to anchor the median spectrum of the far field record set to the Maximum Considered Earthquake (MCE) response spectra at the fundamental period of the archetype. The second scaling is applied successively to all far field ground motions until 50 percent of the archetypes exhibit collapse. The amount of scaling that results in the collapse of 50 percent of the archetypes is compared with a variable named the Adjusted Collapse Margin Ratio (*ACMR*). The target *ACMR* values are tabulated in the FEMA P695 document and depend on the total system collapse uncertainty (β_{TOT}), and collapse probability. Two conditions must be satisfied for acceptable performance. The average value of *ACMR* for each performance group should meet the target *ACMR* for 10 percent collapse probability ($ACMR_{10\%}$). Furthermore, individual values of *ACMR* for each index archetype within a performance group should meet the target *ACMR* for 20 percent collapse probability ($ACMR_{20\%}$). While successive scaling approach can be adopted for new structural systems, scaling of all ground motions using a pre-calculated scaling factor is sufficient for evaluation of existing systems. Because individual archetypes are considered in this study, the 20 percent probability of collapse was adopted as a criterion for *ACMR* (i.e. $ACMR_{20\%}$).

The total system collapse uncertainty (β_{TOT}) depends on various factors such as record-to-record collapse uncertainty, design requirements-related collapse uncertainty, test data-related collapse uncertainty, and modelling-related collapse uncertainty. The methodology enables to use non-simulated collapse models for collapse failure modes that cannot be explicitly modelled. Non-simulated collapse modes can be indirectly evaluated using alternative limit state checks on structural response quantities measured in the analysis.

3. Design and selection of archetypes

Different Seismic Design Categories (SDC) can be adopted in the Methodology in order to represent the variation in seismic hazard. In the present study only one seismic design category namely SDC D_{max} was considered which represents the highest seismic hazard level. The MCE, 5 percent damped, spectral response acceleration parameter at short periods adjusted after site class effects (S_{MS}) was taken as 1.50g. The MCE, 5 percent damped, spectral response acceleration parameter at a period of 1 sec adjusted after site class effects (S_{M1}) was taken as 0.90g.

Two geometrical configurations can be adopted for BRBFs where the first one employs single diagonal braces and the second one employs chevron type braces. In the present study performance of only single diagonal type BRBFs were evaluated.

Only one type of floor plan shown in Fig. 2 was considered. The floor plan is rectangular with side dimensions of 36 meters and 22.8 meters. There are a total of four bays with single diagonal BRBs in the long direction of the floor plan which are indicated as BF-1 in Fig. 2. A total of two bays with chevron type BRBs are employed in the short direction of the floor plan which are indicated as BF-2 in Fig. 2. Only BF-1 type frames were designed as a part of this study. All beam-to-column connections of the BRBF were considered simple connections with no moment transfer. A dead load of 5 kN/m^2 and a live load of 2 kN/m^2 which are typical for steel office buildings were considered as loading. Story height was taken as 3.5 meters for all stories except the first story where the height was equal to 3.8 meters. In order to take into account variations in structural periods, 3, 6, and 9 story BRBFs were considered.

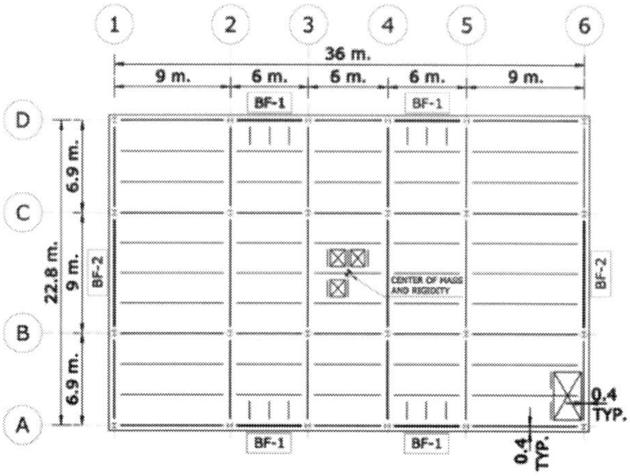

Fig. 2: Floor plan used for the study

A992 grade steel with a yield strength of 345 MPa was considered for all framing members and the core plates of BRBs. It was assumed that the non-yielding portion of a BRB accounts for 33 percent of its total length. Designs were conducted according to ASCE 7-10 [3], AISC 341-05 [1], and AISC 360-05 [5]. Archetypes were designed by minimizing the weight of the framing. Beam, column and brace members of 3 archetypes are given in Table 1. Archetype properties and scaling factors are given in Table 2.

Table 1: Member sizes of archetypes and response quantities

AT	Story	Column	Beam	Brace	*DISD* (%)	*DBAS* (%)	*MISD* (%)	*MBAS* (%)	*CBAS*
1	1	W14x48	W10x30	PL15x135	1.49	1.06	3.14	2.16	198
	2	W14x48	W10x30	PL15x100	2.00	1.31	2.65	1.77	175
	3	W14x38	W10x19	PL15x60	1.98	1.29	2.88	1.96	261
2	1	W14x68	W10x45	PL25x150	0.80	0.57	3.17	2.17	205
	2	W14x68	W10x45	PL25x130	1.57	1.03	2.71	1.72	143
	3	W14x53	W10x45	PL25x120	1.82	1.19	2.02	1.21	125
	4	W14x53	W10x45	PL25x105	2.00	1.30	1.74	0.96	108
	5	W14x38	W10x30	PL25x95	1.96	1.28	1.51	0.81	72
	6	W14x38	W10x30	PL25x65	1.97	1.29	1.41	0.71	96
3	1	W14x132	W10x68	PL35x175	0.52	0.37	2.64	1.78	151
	2	W14x132	W10x68	PL35x155	1.07	0.70	1.90	1.12	89
	3	W14x132	W10x68	PL35x155	1.29	0.84	1.39	0.68	58
	4	W14x132	W10x68	PL35x145	1.47	0.96	1.25	0.55	44
	5	W14x68	W10x68	PL35x135	1.73	1.13	1.48	0.66	44
	6	W14x68	W10x68	PL35x120	1.78	1.16	1.80	0.84	67
	7	W14x53	W10x45	PL35x110	1.88	1.23	1.72	0.77	67
	8	W14x53	W10x45	PL35x105	2.00	1.30	1.44	0.51	31
	9	W14x53	W10x45	PL35x60	1.99	1.30	1.75	0.80	84

DIDS: Design interstory drift, *DBAS*: Design brace axial strain, *MISD*: Median interstory drift, *MBAS*: median brace axial strain, *CBAS*: Median cumulative brace axial strain normalized by the yield strain.

Table 2: Archetype properties and scaling factors

AT	*Ns*	Mass (tons)	*T* (sec)	SF_1	μ_T	β_{RTR}	β_{TOT}	$ACMR_{20\%}$	*SSF*	*CMR* (SF_2)	*SF*
1	3	1255	0.631	2.46	8.06	0.400	0.436	1.443	1.37	1.06	2.60
2	6	2510	1.032	2.57	3.48	0.400	0.436	1.443	1.28	1.12	2.89
3	9	3765	1.386	2.51	2.82	0.382	0.419	1.423	1.29	1.10	2.76

T: fundamental period of vibration, SF_1: First scaling factor for anchoring far-field record set to MCE spectral demand, μ_T: period-based ductility of an index archetype model, β_{RTR}: record-to-record collapse uncertainty, β_{TOT}: total system collapse uncertainty, *SSF*: Spectral shape factor, *CMR:* Collapse margin ratio, *SF:* Ultimate scaling factor.

4. Numerical modelling details and analysis results

Performances of the designed archetypes were evaluated by making use of numerical analysis. The OPENSEES [6] computational framework was used for numerical simulations. Two-dimensional finite element models were used to model the archetypes. The beams and columns of the archetypes were modelled with linear beam elements and the braces were modelled with non-linear truss elements. In general, one of the BRBF bays was modelled and the tributary mass was added to two of the nodes at every story.

The total system collapse uncertainty is dependent on four factors, three of which requires judgment. These factors depend on the knowledge level and modelling capabilities about the system of interest. BRBFs have been studied for over 15 years and have been implemented in the practice. In addition, computational models for BRBFs were also developed and simulation of BRBF behaviour can be conducted with confidence. Therefore, high quality level was assigned to design requirements-related collapse uncertainty (β_{DR}=0.1), test data-related collapse uncertainty (β_{TD}=0.1), and modelling-related collapse uncertainty (β_{MDL}=0.1). The fourth factor that needs to be considered is the record-to-record collapse uncertainty (β_{RTR}) which depends on the period based ductility (μ_T). The μ_T values were determined by conducting nonlinear static (pushover) analysis in accordance with ASCE41-13 [7] and are reported in Table 2. Resulting β_{TOT} and $ACMR_{20\%}$ are reported alongside Spectral Shape Factors (SSF) and ultimate scaling factors (*SF*) for each archetype in Table 2. The archetypes were subjected to 44 ground motion records and the records were scaled by the ultimate scaling factors. A two percent mass and stiffness proportional damping was used in time history analysis.

Evaluation of archetype performance was based on non-simulated collapse models. Buckling restrained braces generally exhibit stable behaviour followed by fracture. Fracture in steel members is difficult to simulate and the Methodology allows for non-simulated collapse models where fracture in members is expected. Furthermore, BRBFs have little redundancy and when one brace fractures the force demand on the fractured brace has to be transferred to all the other braces which eventually results in overloading and fracture in those braces too. In addition, fracture of a brace in any one story triggers soft story mechanism which can potentially trigger collapse of the system.

Interstory drift ratios, brace axial strains and cumulative brace axial strains were investigated in detail. Design demands and median demands obtained from time history analysis are indicated in Table 1. Variation of interstory drift ratios and brace axial strains along the height are given in Figs. 3 through 5. The median of response quantities were used for assessment purposes.

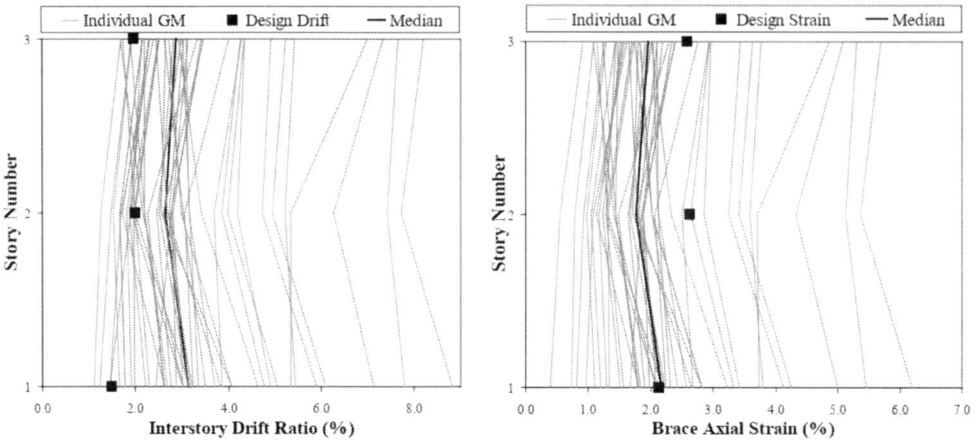

a) b)

Fig. 3: Response of 3 Story Archetype

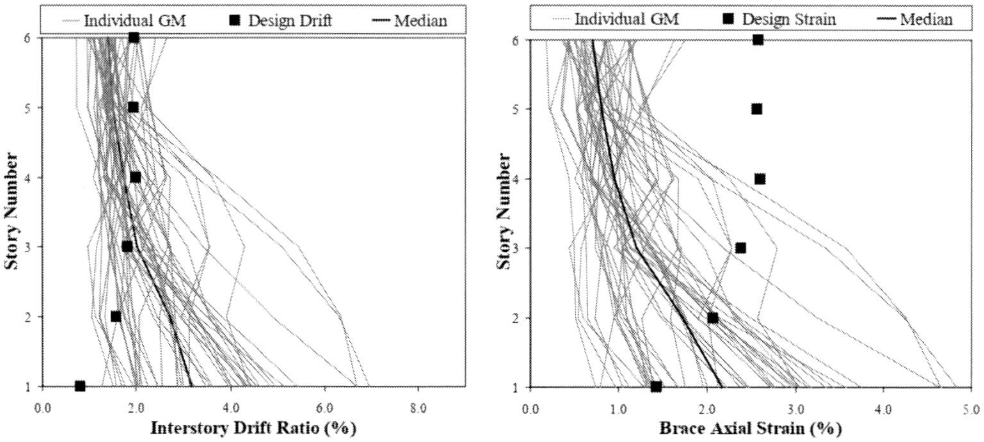

a) b)

Fig. 4: Response of 6 Story Archetype

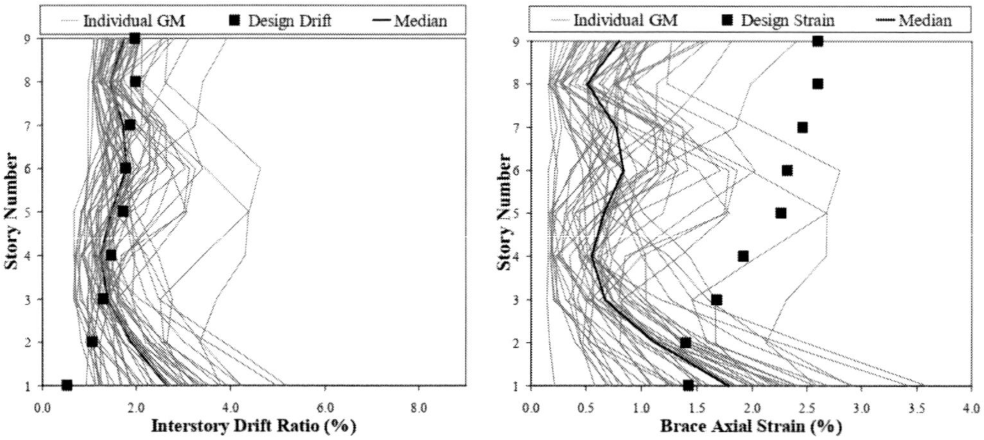

a) b)

Fig. 5: Response of 9 Story Archetype

5. Evaluation of BRBF performance and seismic response factors

Analysis results indicate that calculated interstory drifts and design interstory drifts have different variations along the height. The calculated interstory drifts at MCE level ground motions is expected to be 1.5 times the design interstory drifts which are determined considering design based earthquake. Calculated interstory drifts are significantly higher than the design interstory drifts for lower stories. For the first story of 9-story BRBF archetype the difference is five fold. The differences are more pronounced as the total number of stories increases. On the other hand for higher stories the calculated interstory drifts are observed to be less than the design interstory drifts. The differences observed at the lower stories can be attributable to the differences between the R and C_d factor adopted in design. According to Newmark's equal displacement rule the C_d factor should be taken equal to the R factor to be able to accurately estimate the inelastic demands. For BRBFs, however, the C_d factor is taken lower than the R factor. It should be emphasized that the design of 3 BRBF archetypes were governed by drift limitations. In general, larger sections than required for strength are used at bottom stories to control the drifts at the top of the structure. Considering a C_d factor equal to the R factor would result in significant over-design of BRBFs which may adversely affect the cost of this system. One alternative would be to develop C_d factors that vary over the height as it was developed for eccentrically braced frames by Kuşyılmaz and Topkaya [8].

Excessive interstory drifts which concentrate to bottom stories resulted in brace axial strains that exceeded the expected strains. In fact for 6 and 9 story archetypes the design strains for the bottom story braces are governed by the minimum interstory drift of 2 percent. Despite this provision the median brace axial strains at the first story of 6 and 9 story BRBFs are 1.52 and 1.25 times the design axial strains respectively. The analysis results revealed that the expected deformations of a BRB should be determined by considering a story drift of at least 3 percent if the current response factors are to be used.

The median of cumulative axial strains is generally below the 200 times the yield strain limit except a few cases. For the third story of 3-story BRBF archetype the cumulative strains reached to 261 times the yield strain. It was observed that the cumulative strains do not have a uniform variation over the height of the building. In general the cumulative strains decrease significantly at higher stories.

6. Conclusions

A numerical study on seismic performance factors of BRBFs has been presented. The Methodology outlined in FEMA P695 was applied to BRBFs to evaluate the response factors. Nonlinear time history analyses were conducted for 3 archetypes and the structures were subjected to a set of ground motions in excess of the Maximum Considered Earthquake (MCE) ground motions.

The analysis results indicate that there are marked differences between the calculated interstory drifts and design interstory drifts. These difference stem from the fact that different values are assigned to the deflection amplification factor (C_d) and response modification factor (R). Yielding of BRBs was observed to be non-uniform along the height and significantly higher demands are produced at the bottom stories when compared with top stories. In the past, deflection amplification factors that vary over the height of the structure were developed for other lateral load resisting systems [8] and a similar approach can be taken to arrive at BRBF behaviour that results in more uniform yielding along the height. Future research should focus on developing relationships between response modification factor and deflection amplification factor for BRBFs that vary over the height of the structure.

Brace strain demands are calculated using the interstory drifts at the design stage. Any underestimation of interstory drifts would result in an underestimation of the brace strain demands. In order to safeguard against underestimations the AISC341-10 Specification [1] provides a minimum brace strain demand that corresponds to 2 percent interstory drift. Analysis results showed that the maximum strain in the bottom story of 6- and 9-story archetypes were higher than the design strains in spite of the minimum demand that corresponds to 2 percent interstory drift. The differences can exceed 50 percent which indicate a potential weakness in the design of buckling-restrained braces. Until further research it is recommended to calculate the minimum strain demand based on 3 percent interstory drift.

Examination of cumulative brace axial strain demands also showed a non-uniform behaviour along the height. While the bottom story braces are subjected to higher cumulative strain demands the top stories are subjected to relatively minor demands. The cumulative axial strain limit of 200 times the yield strain set forth by the AISC341-10 Specification [1] was significantly exceeded only in one of the stories of a 3-story archetype.

References

[1] American Institute of Steel Construction (AISC). *Seismic Provisions for Structural Steel Buildings ANSI/AISC 341-10*, American Institute of Steel Construction, Chicago, IL, 2010.

[2] Uang CM. "Establishing R (or R_w) and C_d factors for building seismic provisions", *ASCE Journal of Structural Engineering*, 117(1), 19-28, 1991.

[3] American Society of Civil Engineers and Structural Engineering Institute (ASCE/SEI). *Minimum Design Loads for Buildings and Other Structures ASCE7-10*, American Society of Civil Engineers, Reston, VA, 2010.

[4] Federal Emergency Management Agency (FEMA). "Quantification of Building Seismic Performance Factors", *FEMA P695 ATC-63 Project Report*, Federal Emergency Management Agency, Washington, DC, 2009.

[5] American Institute of Steel Construction (AISC). *Specification for Structural Steel Buildings ANSI/AISC 360-10*, American Institute of Steel Construction, Chicago, IL, 2010.

[6] OPENSEES. Version 2.0 user command-language manual; 2009.

[7] American Society of Civil Engineers. *Seismic Rehabilitation of Existing Buildings*, ASCE/SEI 41-13, American Society of Civil Engineers, Reston, VA, 2014.

[8] Kuşyılmaz A, Topkaya C "Displacement amplification factors for steel eccentrically braced frames", *Earthquake Engineering and Structural Dynamics*, 44, 167-184, 2014.

DESIGN OF CONCENTRICALLY BRACED STEEL FRAMES FOR OPTIMIZED PERFORMANCES IN MODERATE EARTHQUAKE AREAS

Hervé Degée[a], Carlo Castiglioni[b], Alper Kanyilmaz[b], Inigo Calderon[c] and Pierre-Olivier Martin[d]

[a] *Hasselt University, Belgium*
[b] *Politecnico di Milano, Italy*
[c] *Tecnalia, Spain*
[d] *CTICM, France*

Abstract: The present contribution summarizes the part of the outcomes of the currently running European research project ***Meakado*** dealing specifically with Concentrically Braced Frames (CBF). The global objective of the research project ***Meakado*** is to develop specific design procedures for steel and steel-concrete composite structures in regions characterised by a low to moderate seismic activity, with an appropriate reliability level. The final aim is to find an optimal balance between safety and economical concerns.

1. Introduction

Even in the most advanced seismic design methods like performance-based design, the general philosophy is always based on the assumption of global and fully developed plastic mechanisms whatever the seismicity level, together with the use of corresponding capacity design principles. The strict application of these principles for designing steel and steel-concrete composite structures in regions of low to moderate seismicity is however clearly leading to solutions that are on one hand rather difficult to implement and on the other hand often resulting in a significant increase of the building costs. As consequence, for economy reasons, it is often decided to design on the base of q=1,5 only (Ductility Class low, or DCL, design) and to neglect any further provisions aiming at enhancing the seismic performance, a practice which, from a safety point of view, cannot always lead to satisfactory structural solutions.

The aim of the research program ***Meakado*** is therefore to study design options with requirements proportioned to the actual seismic context of constructions in areas characterized by a low or moderate seismic hazard, contrary to most researches aiming at maximizing the seismic performances. More precisely the objective is to propose design rules that are optimised for the actual seismic action, providing the necessary safety level without imposing excessive requirements, and thus limiting the incremental complexity and costs associated with anti-seismic design.

The research ***Meakado*** has chosen to focus essentially on concentrically braced frames (CBF) and moment resisting frames (MRF), as being the most relevant typologies in the European construction market. Frames with dissipative eccentric bracing or other anti-seismic configurations (damping devices, isolators,...) are of scarce relevance for low-to-medium seismicity areas. The present paper presents a broad overview of the developments carried out so far with regard to CBF configurations. Detailed information can be found in the latest updates of the progress reports of the project.

2. General Methodology

As stated above, the objective of the proposed research action is to study an intermediate way of designing steel structures in which reduced but controlled amount of ductility is accounted for, providing thus the necessary safety with respect to uncertainties on the seismic action, but where the local ductility and structural homogeneity requirements are less stringent than required for instance by Eurocode 8 for a Ductility Class Medium (DCM) design in order to focus on intermediate values of behavior factors. These requirements should be tuned according to the actual seismicity level of the area. Two main directions of investigation are to be followed to target the announced objectives.

The first one consists in the exploitation of phenomena that are known to contribute to energy dissipation in steel structures subjected to earthquake action, but whose knowledge is not yet sufficient to quantify them as sources of controlled energy dissipation in the definition of the q-factors. So phenomena like

1. Slip in bolted connections,
2. Plastic ovalization of bolt holes,
3. Post-buckling strength of diagonals in compression.

will be considered in the research and investigated either by means of experimental tests or numerical simulations in order to quantify the energy dissipation that they can provide and to adjust consequently the values of corresponding q factors.

The second direction consists in an investigation on the possibilities of tuning EC8 design rules given for DCM to the actual seismicity zone and to the targeted behaviour factor. The following DCM-EC8 design rules are planned to be re-considered, mainly on the base of numerical simulation tools calibrated and validated by the experimental campaigns.

- Slenderness of diagonals must remain within a limited interval, limiting consequently the type of profiles that can be used for seismic bracings. Boundaries of this interval will be reconsidered. Another possibility could be to impose these restrictions only on a limited number of storeys.
- The overstrength coefficient (ratio between seismic demand and cross-section resistance) of diagonal bracings must not vary of more than 25% all over the height of the entire building. The limit of 25% as well as the number of stories on which the variation has to be limited will be reconsidered.

All the above parameters are planned to be studied first separately then in combination.

The main expected deliverable is a set of recommendations that could be included in the next revision of Eurocode 8 and that would allow better tuned design of steel structures in low-to-moderate seismicity areas, aiming at ensuring the adequate reliability level, improving the economic competitiveness and simplifying the design practice where possible.

3. Assessment of energy dissipation in bolted shear connections

In case of earthquake resistant CBF structures, the horizontal forces induced by inertia when the base motion occurs are absorbed by means of diagonal bracing elements that works in tension and compression, depending on the sense of the load at each time. This cyclic loading could induce local plasticity in connection regions, inducing fatigue but also dissipating energy by hysteresis. The control of this hysteresis behavior is likely to enhance the energy dissipation in a building in a way allowing taking it into account from the very beginning in design stage, letting the structural designer to do a safe relaxation of structural requirements and avoiding unnecessary waste of steel resources.

Considering this, a first Task is defined in the Meakado project to test some of the most typical CBF bracing connections trying to understand the variables conditioning their energy absorption under cyclic loads. The test specimens and the test setup are illustrated at Figures 1 and 2.

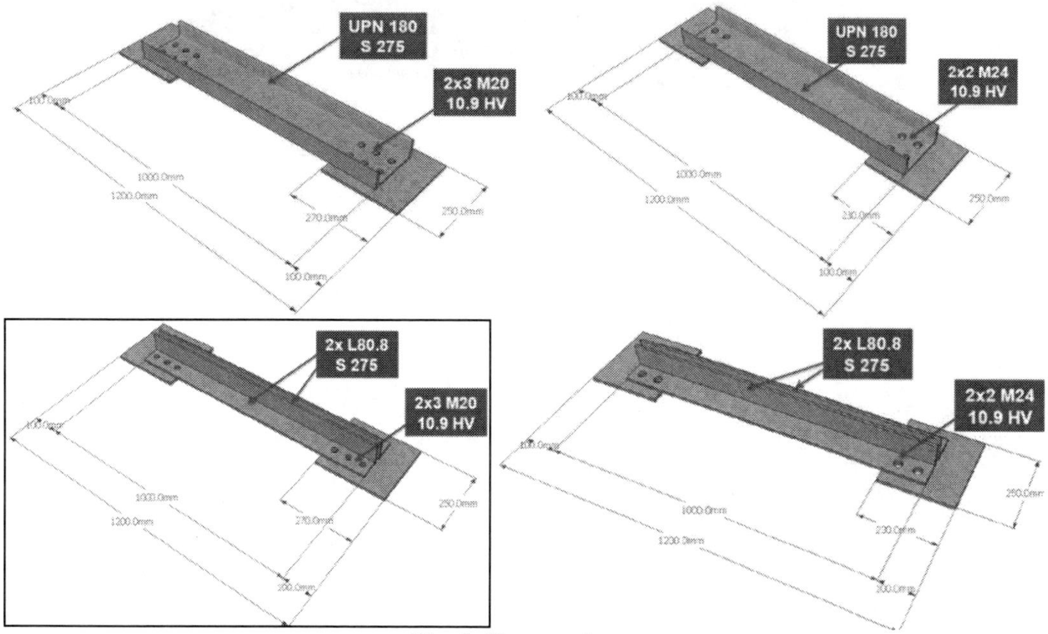

Fig. 1: Test specimens

To this respect, five very relevant parameters have been identified, whose participation in hysteretic behavior could be foreseen.

- The first parameter is the use of prestressed bolts to execute the connection, this enables energy dissipation by friction, thus, in principle, prestressed bolt connections should improve the connection performance regarding energy dissipation, but the quantification of this effect remains unsolved;
- The second relevant parameter is the geometry of the bolt connection, the amount of bolts, the bolt-rows, the metric, the distances to edges and spacing between bolts etc. Initially, this should have influence in the development of local plasticity that leads to energy absorption;

- The third relevant parameter is the gusset thickness, the more thickness, the more rigidity in relation to profile, the less energy absorption developed in the gusset by plastic deformation;
- The fourth parameter is the steel profile. Typically both channels and angles are used in practice for CBF bracing (i.e. UPN and LPN profiles). The profile type will influence in the maximum allowable load, the amount of bolts and the web/flange thicknesses, etc.;
- The last relevant parameter to be assessed is the material properties, whose variations depend on the hot rolling process producing changes in the yield and maximum resistance of profiles changing its dissipative behaviour. The more ductility and plasticity margin, the better energy absorption.

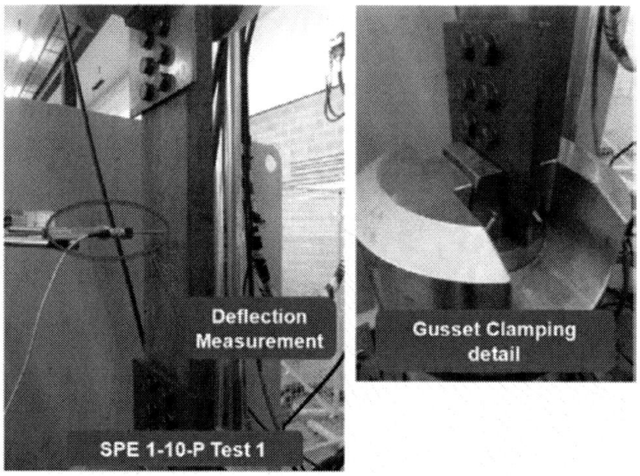

Fig. 2: Test setup

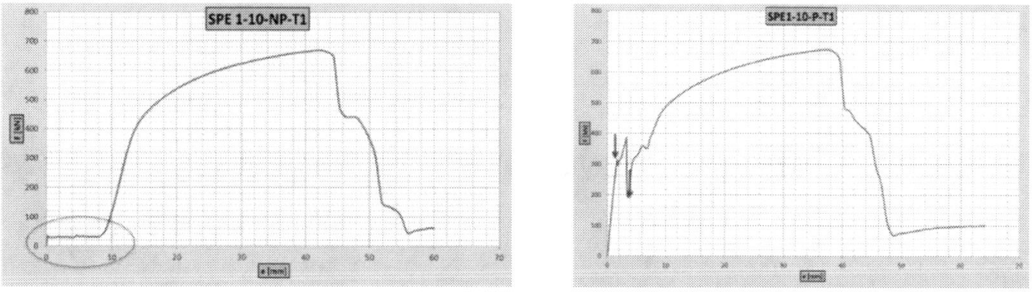

Fig. 3: Monotonic curves – comparison of (a) non- prestressed and (b) prestressed connections

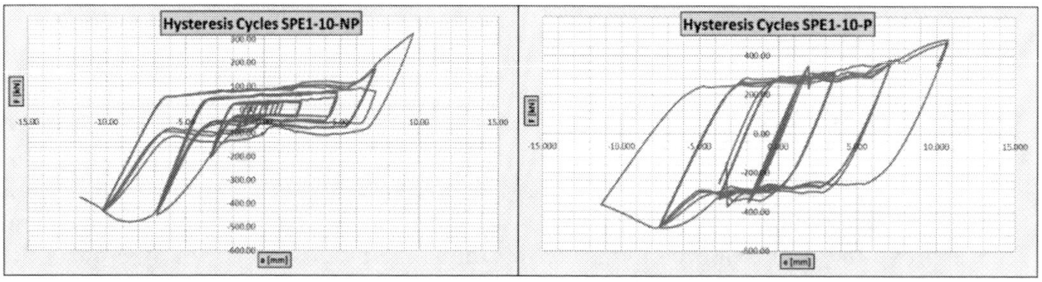

Fig. 4: Cyclic curves – comparison of (a) non-prestressed and (b) prestressed connections

An interesting first observation in non prestressed specimens is illustrated in Figure 3.a and 4.a with monotonic tests of SPE1-10-NPT. The two graphs exhibit an initial region with high displacements at constant low force associated with the adjustment of all bolts in their holes until their complete load-share. As second interesting observation, the curve in tension reaches a maximum force of about 700 kN while in compression it barely reaches 500 kN because of the buckling in compression. It is noteworthy that, during monotonic tests, failure comes from the hole's ovalization and then block shearing failure of profile web, while in compression the failure comes from the permanent buckling of the profile, not of the connection. In comparison, monotonic tests carried out on prestressed connections, such as illustrated in Figure 3.b and 4.b, show out a well fit from the very beginning, with two clear force drops in force-displacement curves corresponding to the points where bolt slip occurred. The ultimate force is approximately the same as for the non-prestressed configuration, but without initial displacement zone without force augmentation. Such a region is, instead, delayed to the point where bolt slip happens and is extended until the point where the specimen recovers its bearing capacity. Once again, the failure of the element in tension occurs by block shearing (Figures 5.a to 5.c) and the failure in compression by unstable buckling. A detailed analysis of the test results can be found in [1].

Fig. 5.a to 5.c: Typical failure mode (Channel connection in tension)

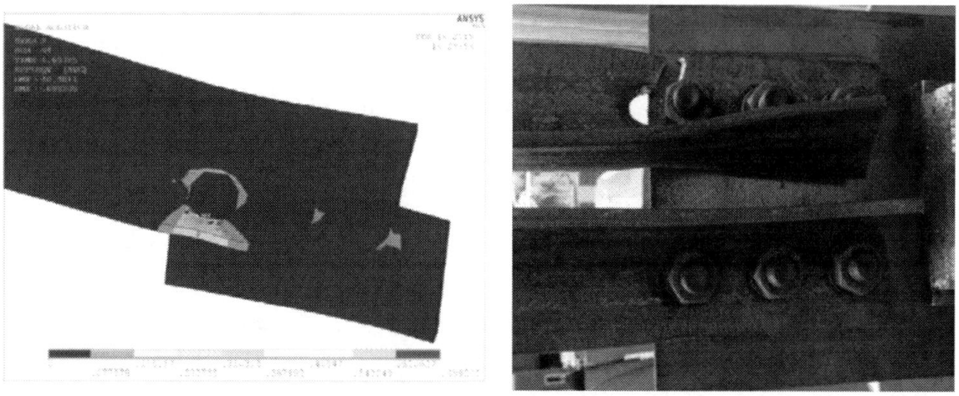

Fig. 6: Failure mode of angle specimens in tension - comparison test/simulation

Beside these experimental activities, specific numerical models have also been implemented to allow the extension of the scope of the test results (also detailed in [1]). Numerical simulations and test results show out to be in reasonably good agreement, as illustrated in Figures 6 and 7.

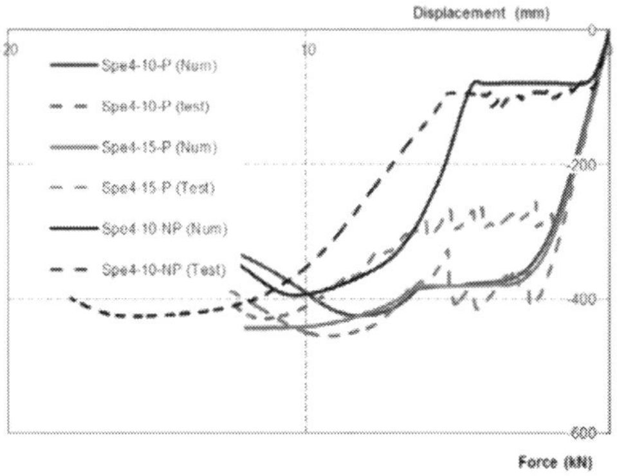

Fig. 7: Comparison of numerical (block lines) and experimental (dash line) results

4. Design and performances of CBF with relaxed design criteria

When considering a CBF structure (concentrically braced frame) designed according to ductility classes DCM or DCH of EN 1998-1, stringent requirements are to be considered regarding the bracing slenderness, the overstrength homogeneity and the overstrength of the joints.

A first step in a global evaluation procedure consists in designing a set of case studies according to these requirements. In order to be consistent with the objectives of the Meakado project, these case studies are evaluated considering a low to medium seismicity. 15 configurations have been designed for 3 structural typologies (Figure 8.a to 8.c) and a number or levels ranging from 4 to 12.

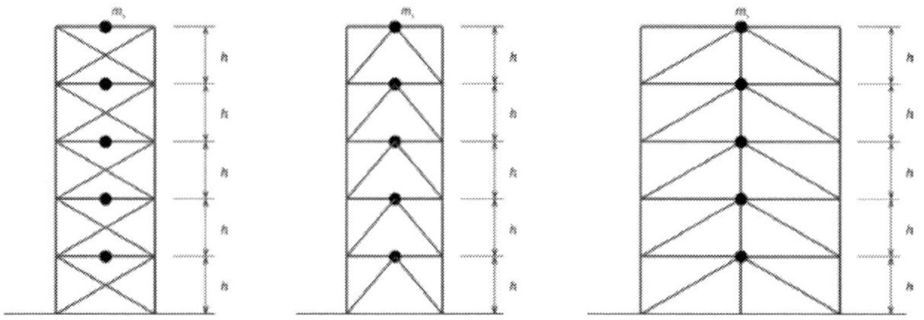

Fig. 8: Structural typologies – (a) X bracings, (b) Inverted V bracings, (c) N bracings

After selection of a reduced set of configurations, for 4 and 8 levels, the steel structures have been re-designed considering successively the variation of following parameters (as summarized in Table 1):

- **Peak ground acceleration** of 0.25g and 0.35g;
- **Relaxed overstrength criterion** for 0.15g, with $\Omega_i \leq 1{,}50\ \Omega_{max}$ (instead of 1.25 Ω_{max});
- **Relaxed slenderness criterion** for 0.15 g, with $\bar{\lambda} \leq 2{,}5$ (instead of 2);
- **Simultaneous reduction on criteria for overstrength and slenderness.**

It can then be showed that relaxation of the EC8 rules permits the adoption of lighter profiles, more in the case of relaxing slenderness than overstrength homogeneity rule.

Incremental dynamic time-history analyses are then used to assess the performances of the structures designed strictly to Eurocode 8 or with relaxed criteria. Figures 9 and 10.a to 10.c illustrate results of such analyses in terms of maximum base shear and of distribution of the yielding within the brace elements.

Table 1: Relaxed design parameters

EN1998-1 rules	Ω_{max} < **1,25** Ω_{min} **1,3 < λ < 2**
Relaxed slenderness (**λ**) **rule**	Ω_{max} < **1,25** Ω_{min} **1,3 < λ < 2,5**
Relaxed overstrength homogeneity (**Ω**) rule	Ω_{max} < **1,50** Ω_{min} **1,3 < λ < 2**

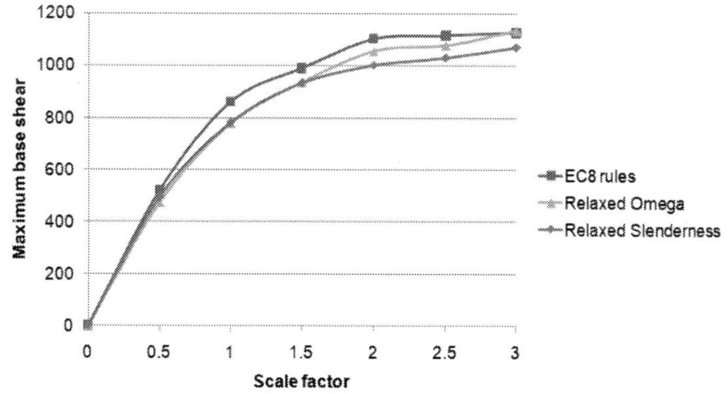

Fig. 9: Example of Incremental Dynamic Analyses curves for a 4-storey N-configuration

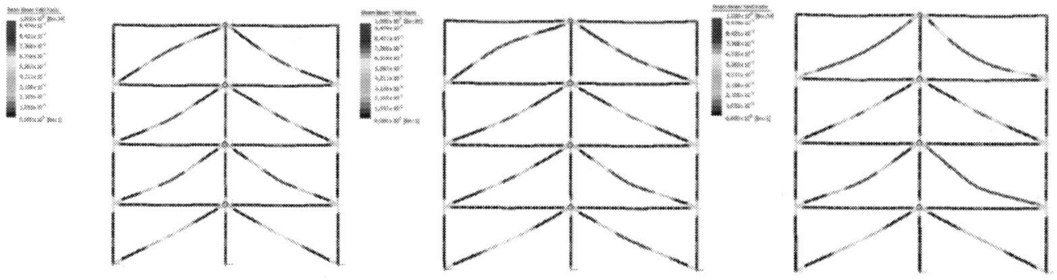

(a) (b) (c)

Fig. 10: Distribution of yielding for a 4-storey N-configuration. (a) EC8 design; (b) relaxed slenderness; (c) relaxed overstrength homogeneity

Based on results considering a limited number of configurations and a limited number of input time-histories, it appears that even ignoring completely the code limits on slenderness and overstrength does not cause the occurrence of soft-storey mechanism and provide a ductility of the structure sufficient to resist the targeted acceleration level with a sufficient safety margin. The conclusions need however to be further elaborated in a broader range of parameters.

5. Conclusions and perspectives

As stated in the description of its general methodology, the final objective of the ***Meakado*** research project is to study an optimized way of designing steel and composite structures in which reduced but controlled amount of ductility is accounted for (intermediate between DCL and DCM), providing thus the necessary safety with respect to uncertainties on the seismic action, but where the local ductility and structural homogeneity requirements are less stringent than for DCM. So the seismic risk can be tackled in a reasonable manner in regions with low or moderate seismic action. The development of this optimized approach is to be based on a deeper understanding of a number of specific mechanisms, with a specific attention dedicated to the energy dissipation in classical bolted shear connections.

Moreover, beside a better control of the local characteristics, the project also focuses on the possibility to release some rules recommended by Eurocode 8 and felt as too stringent by most designers in low seismic country where earthquake is generally considered as a non-critical issue. To this respect, some parameter studies have been initiated and can already show out interesting trends. These investigations will continue all along the second part of the project and be complemented by validation tests, also in view of finally deriving practical design recommendations.

Acknowledgements

The authors acknowledge the support received from the Research Fund for Coal and Steel of the European Commission in the frame of the research grant agreement RFSR-CT-2013-00022 Meakado.

References

[1] Degée H. et al. "Design of steel and composite structures with limited ductility requirements for optimized performances in moderate earthquake areas - Meakado", *RFCS-CT-2013-00022 mid-term technical report*, 2015.

[2] CEN, EN 1998-1, *Eurocode 8 – Design of structures for earthquake resistance – Part 1: General rules, seismic action and rules for building* (2004).

THE BUCKLING-RESTRAINED BRACE HAVING HIGH STRUCTURAL PERFORMANCE

Mamoru Iwata[a], Mitsumasa Midorikawa[b] and Kazuhisa Koyano[a]

[a] *Department of Architecture and Building Engineering, Kanagawa University, Japan*
[b] *Faculty of Engineering, Hokkaido University, Japan*

Abstract: The authors have studied the buckling-restrained brace providing stable hysteretic characteristic even under high-strain conditions. In this paper, our past studies are analyzed and the conditions of high-performance of buckling-restrained braces are extracted. The considered buckling-restrained brace is tested. As a result, the buckling-restrained brace having large cumulative plastic strain energy ratio has been proposed.

1. Introduction

Fig. 1 shows the buckling-restrained brace (hereafter referred to as BRB) that we have been studying. The BRB was prepared by welding together a core plate covered with unbonded material and a pair of steel mortar planks (mortar-filled channel steel). We conducted comparative performance tests on the BRB. Specimens having core plates of varying cross-sectional area, width-to-thickness ratio, and slenderness ratio along with those with restraining parts of varying geometrical moment of inertia, were fabricated for testing purposes.

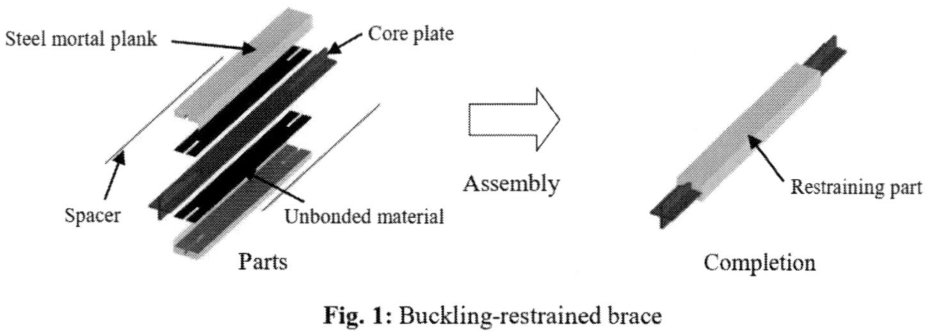

Fig. 1: Buckling-restrained brace

According to Reference [1], structural performance requirements are effectively fulfilled given that the cumulative plastic strain energy ratio ω_u calculated from the test results (experimental value) exceeds the value obtained from formulae that express the minimum performance value ω_r (required value). As illustrated in Fig. 2, the cumulative plastic strain energy ratio ω_u is a parameter found by dividing the buckling-restrained brace's cumulative plastic strain energy E_t, which is calculated from the brace's hysteresis characteristics, by W_y, which is obtained by multiplying the core plate's yield load P_y by its elastic-limit deformation δ_y, and nondimensionalizing the results of the calculation.

Based on the results of our past studies, this study analyzed in detail the factors that determine the experimental value ω_u. Moreover, based on the analytical results obtained, we fabricated specimens using the factors that determined the experimental value ω_u as parameters, and conducted full-scale loading tests to confirm the performance of BRB having a large ω_u. Based on the loading test results, we were able to specify the prerequisites for a feasible BRB having a large ω_u.

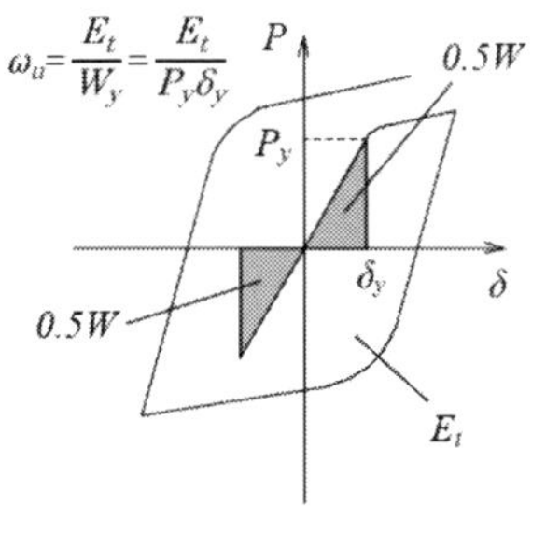

Fig. 2: Conceptual diagram of ω_u calculation

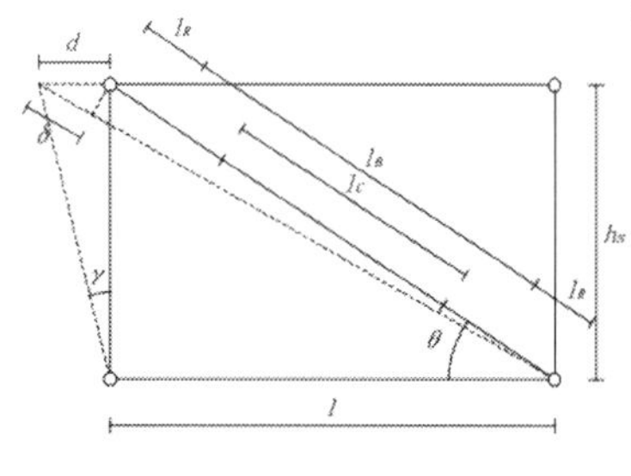

Fig. 3: Test model

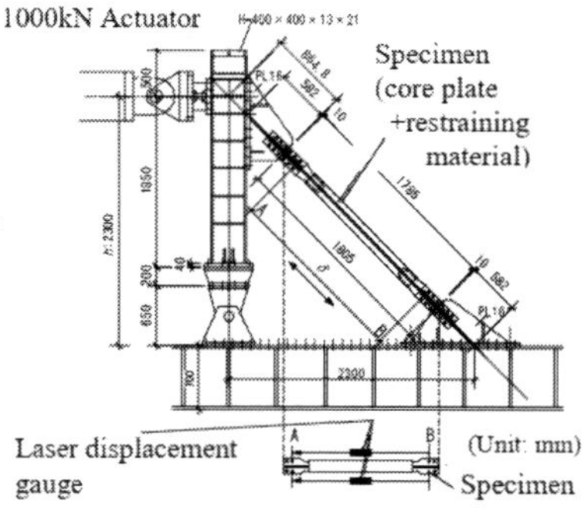

Fig. 4: An example of loading equipment

2. Results obtained from past studies and test model used

2.1 Test model

Fig. 3 shows the test model used in past studies. The model represents a building in which BRBs are incorporated. The testing assumes that shear deformation occurs in the building under the influence of horizontal force. In the figure, the core plate length l_B is obtained by subtracting the length of the rigid zone at both ends l_R from the BRB length.

The ratio of core plate length l_B to core plate plastic zone length l_C is defined as the plastic length ratio (= l_C / l_B).

2.2 Loading plan

An example of the loading equipment is presented in Fig. 4. Using a 1000 kN actuator, tests are conducted by applying increasing loads alternately in the positive and negative directions.

2.3 Test results

Fig. 5 shows the results of tests conducted in our past studies. The vertical axis of the graph represents ω_u, and the horizontal axis, restraining index R. R is the value found by dividing the Euler buckling load of the restraining part P_E by the yield load of the core plate P_y. Core plate length l_B is assigned to buckling length L.

In the test results, the ultimate state of the core plate is categorized as local deformation, strong-axis deformation, or tensile fracture. Deformation around the weak axis is defined as local deformation, deformation around the strong axis as strong-axis deformation, and fracture of the core plate under tension as tensile fracture.

According to Reference [1], ω_u obtained from tests in which the core plate experiences tensile fracture or strong-axis deformation in its ultimate state tends to become larger than that obtained from tests in which the core plate ultimate state is local deformation. For this reason, the following equations are used to evaluate the performance of a BRB regarding the lower limit of ω_u (experimental value) of the specimens in which the core plate ultimate state is local deformation as the required value ω_r of a BRB:

$$\omega_r = 150 \times R \ (R \leq 6.0), \ \omega_r = 900 \ (R > 6.0)$$

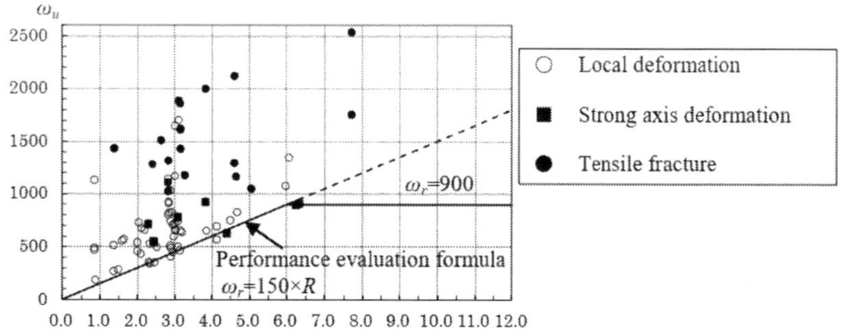

Fig. 5: ω_u - R Relationship

3. Analysis of past studies

3.1 Ultimate state of the core plate

As seen in Fig. 5, some specimens have extremely large ω_u while others have a slightly smaller ω_u value compared to that obtained from the performance evaluation formula.

The ω_u of all specimens where the core plate experienced tensile fracture in the ultimate state is > 1000. This indicates that in order to increase the ω_u of a BRB, it is crucial that the core plate functions properly up to nearly the steel's rupture strength without experiencing local deformation or strong-axis deformation.

3.2 Restraining index *R*

Fig. 5 provides the following information:

1) $R \leq 3$; Local deformation is most likely to occur.

2) $3 < R \leq 6$; Some specimens experience tensile fracture or strong-axis deformation. Local deformation is observed in many specimens. The value of ω_u varies widely in this range.

3) $6 < R$; Some specimens experience local deformation or strong-axis deformation in the vicinity of $R = 6$. Where R is larger, all specimens experience tensile fracture, along with an increase in the ω_u value.

3.3 Compressive-to-tensile strength ratio *α*

Compressive-to-tensile strength ratio α is found by dividing the compressive strength P_c by the tensile strength P_t, where P_c is the maximum compressive load at each strain amplitude and P_t is the maximum tensile load at each strain amplitude. When α increases, the specimen's strength on the compression side increases, leading to potential early-stage local deformation and strong-axis deformation. Reference [2] reports that a high-order buckling mode develops in the core plate plastic zone during compression, and that as the slenderness ratio λ of the core plate plastic zone increases, the number of points of contact with the restraining part increases, causing a rise in the frictional force, and ultimately an increase in α.

3.4 Shape and dimension details

3.4.1 Area of core plate plastic zone is decreased

According to Reference [3], many specimens in their ultimate state experienced local deformation in the core plate end sections when the area of the core plate plastic zone was not decreased.

3.4.2 Reinforcement around strong axis

Reference [4] states that installing a round steel bar on both sides of the core plate is effective in reinforcing the area around the strong axis of the core plate. To further increase ω_u, however, additional reinforcement is necessary. By decreasing the area of the core plate plastic zone and installing a spacer in this narrowed section, the restraining force around the strong axis can be increased.

3.4.3 Plastic length ratio (Core plate plastic zone length / core plate length)

Where the core plate length is the same, λ is increased by increasing the core plate plastic zone length ratio (Reference [5]). When λ is increased, a high-order buckling mode develops in the core plate plastic zone (Reference [6]), and frictional force generated by contact with the restraining part rises to increase α. It is appropriate to set the plastic length ratio at 30–50% in order to suppress an increase in α and to induce tensile fracture of the core plate in its ultimate state.

3.4.4 Width-to-thickness ratio

The ratio of core plate width to core plate thickness is referred to as the width-to-thickness ratio. Reference [4] reports that increasing the width-to-thickness ratio causes early-stage local deformation, resulting in performance degradation. On the other hand, decreasing the width-to-thickness ratio increases strong-axis deformation. In order to reduce local deformation and strong-axis deformation so as to induce tensile fracture, the width-to-thickness ratio should be set at around 6 to 8.

3.4.5 Clearance

When subjected to compression, the cross-sectional area of the core plate increases due to the influence of Poisson's ratio; Poisson's ratio should be considered when determining the clearance. As the amount of deformation found taking into consideration Poisson's ratio is the gain in the core plate width and thickness, the clearance should be larger than the gain. According to Reference [4, 7], a clearance ratio (clearance / core plate thickness) of 10 to 15% is considered appropriate.

3.4.6 Rib penetration length

Reducing the rib penetration length tends to cause drooping rib ends (Reference [8]). It shows the results of testing, conducted in Reference [5], on BRBs with a small rib penetration length ratio (rib penetration length / core plate plastic zone length) of 9%. It is considered appropriate to set the rib penetration length ratio at 10–20%.

4. Test Plan

4.1 Specimens

Table 1 lists the specimens to be used. Fig. 6 shows the view. The data obtained from core plate and mortar material testing is presented in Table 2. The yield load of the core plate P_y is calculated by multiplying the yield stress by the cross-sectional area.

4.2 Specimen parameters

Specimens with varying R, plastic length, and rib penetration length ratio are used. These parameters are presented in Table 2.

4.3 Loading plan

The test equipment and loading pattern are the same as that used in past studies (see Fig. 4). At the maximum axial strain (3%), loading is continued until either the specimen strength drops to 80% of the maximum strength or tensile fracture occurs.

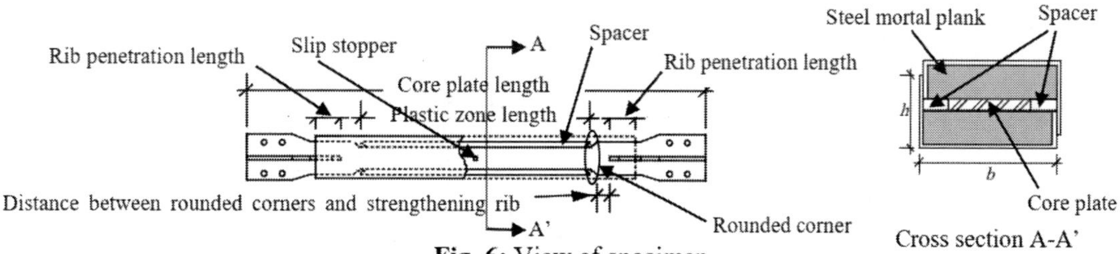

Fig. 6: View of specimen

Table 1: List of specimens

Specimens	Core plate					Restrained part		
	Dimensions (mm)	Cross-sectional area (mm^2)	P_y (kN)	Core plate length (mm)	Plastic zone length (mm)	h (mm)	b (mm)	P_E (kN)
Specimen 1	PL-84×12	1008	308.1	1785	892	82	159.2	1980
Specimen 2					536	106		3427
Specimen 3			298.3		714	82		1980
Specimen 4					536	95		2701
Specimen 5			297.5		892			

Table 2: Material test results and specimen parameters

Specimens	σ_y (N/mm^2)	Tensile strength (N/mm^2)	ε_y (%)	Compressive strength (N/mm^2)	R (=P_E/P_y)	Plastic length ratio (%)	Rib penetration length ratio (%)
Specimen 1	308	420	0.162	72.5	6.4	50	17
Specimen 2					11.0	30	28
Specimen 3	298	415	0.165	64.1	6.6	40	21
Specimen 4					9.0	30	28
Specimen 5		417	0.158	71.0		50	10

5. Test results

5.1 Test process and failure mode characteristics

The test process and the failure mode characteristics of the individual specimens are shown in Table 3. For hysteresis characteristics, the relationship between the yield strength ratio β (P /

P_y; P = axial load) and ε (core plate axial strain (%): the ratio of axial deformation to core plate plastic zone length) of Specimen 1 and Specimen 4 is plotted in Fig. 7.

5.2 Core plate failure mode characteristics

Specimen 1 experienced local deformation in the strengthening rib end. For the other specimens, local deformation or tensile fracture occurred in the center of the core plates.

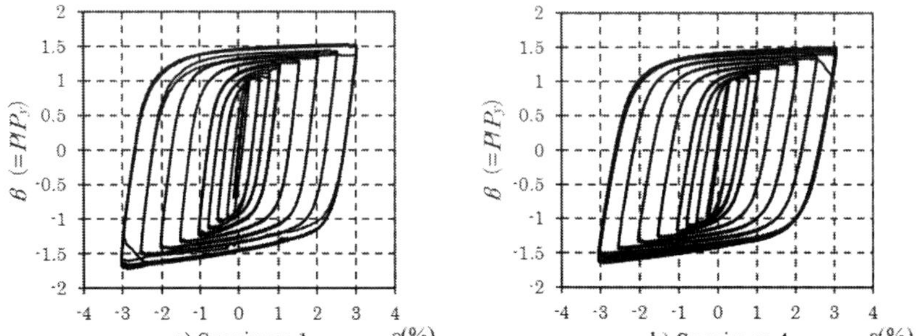

a) Specimen 1 b) Specimen 4

Fig. 7: β - ε relationship

Table 3: Testing process and failure mode characteristics and performance evaluation

Specimens	Testing process	Failure mode characteristics	E_t (kN/m)	ω_u
Specimen 1	5th compression with a 3% strain	Local deformation	400	986
Specimen 2	16th tension with a 3% strain	Tensile fracture	447	1860
Specimen 3	9th compression with a 3% strain	Local deformation	419	1341
Specimen 4	18th tension with a 3% strain	Tensile fracture	507	2162
Specimen 5	8th tension with a 3% strain		491	1264

6. Discussion of test results

6.1 Hysteresis characteristics

All specimens demonstrated stable fusiform-shaped hysteresis characteristics. Yield strength ratio β was found to be approximately 1.5 on the tension side and 1.6–1.7 on the compression side. In all specimens, the yield strength ratio β is larger on the compression side.

6.2 Performance evaluation

Fig. 8 depicts the relationship between ω_u and R obtained from the results of the test performed in this study and that obtained from past studies. Table 3 shows the values of the cumulative plastic strain energy E_t and ω_u of the individual specimens. The ω_u of all specimens, excluding Specimen 1, was >1200

6.3 Restraining index *R*

Specimen 3 experienced local deformation in the central part of the core plate plastic zone in its ultimate state. It is considered that the local deformation occurred due to insufficient R. However, ω_u is large enough; hence, provided that $R > 6$, this specimen would be capable of delivering adequate performance.

6.4 Compressive-to-tensile strength ratio α

The graph in Fig. 9 depicts α at the maximum axial strain for the individual specimens. The α values shown in the graph in Fig. 9 were calculated at each repetition of maximum axial strain (axial strain 3%).

6.5 Influence of core plate end

The ω_u value of Specimen 1 alone that experienced local deformation in the core plate plastic end is 1200 or less. This indicates that avoiding stress concentration in the core plate plastic zone end is crucial for preventing early-stage strength deterioration.

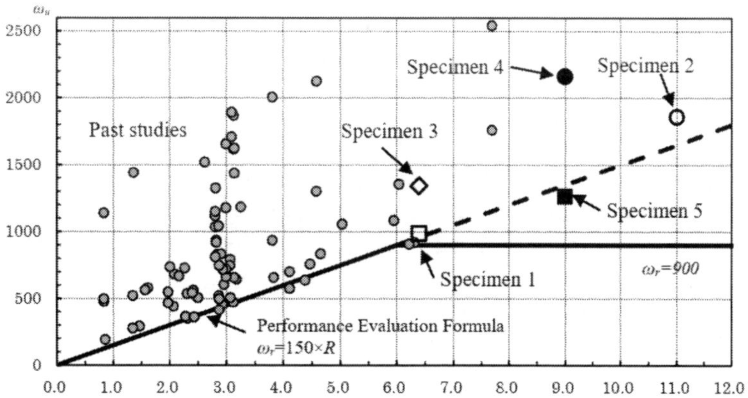

Fig. 8: $\omega_u - R$ relationship

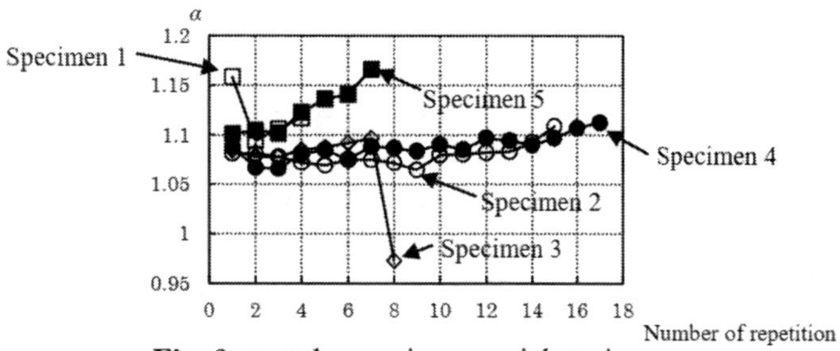

Fig. 9: α at the maximum axial strain

7. Conclusions

The following findings were obtained.

1. The results of analyses performed in past studies revealed that in order to attain a large ω_u, the restraining index should be set at 6 or over, the rise in compressive-to-tensile strength ratio α should be limited, the area of the core plate plastic zone should be decreased, and a spacer should be installed around the strong axis of the core plate.
2. Based on the analytical results, we fabricated a BRB and performed testing. The test results verified that it is possible to develop a BRB having a large cumulative plastic strain energy ratio ω_u, as large as >1200, and a compressive-to-tensile strength ratio α of <1.15.

3. Based on the analytical and testing results, this paper presented the prerequisites for a BRB having a large ω_u. The BRB having a large required value ω_r can be developed by designing a BRB that meets these prerequisites.

Acknowledgments

The authors wish to thank Mr. Masatoshi Murai, Chief Engineer at Kanagawa University (at the time this study was performed) and Mr. Ryota Iizuka (a graduate student at Kanagawa University (Iwata Laboratory) at the time this study was performed) for their cooperation in performing this study.

References

[1] Iwata M, Murai M. "Buckling-restrained brace using steel mortar planks; performance evaluation as a hysteretic damper", *Earthquake Engineering and Structural Dynamics*, 35(14), 1807-1826, 2006.

[2] Midorikawa M, Tanaka Y, Ohtake M, Asari T, Murai M, Iwata M. "Experimental Study on Buckling-restrained Braces Using Steel Mortar Planks: Evaluation of the Buckling Mode Number and the Relationship between Compression-to-tension Strength ratio and Slenderness Ratio", *Journal of Structural and Construction Engineering, Transactions of AIJ*, 76(664), 1153-1160, 2011.

[3] Iizuka R, Wakayama T, Midorikawa M, Iwata M. "Experimental Study on Buckling-restrained Braces Using Steel Mortar Planks: Strong-axis Buckling Mode Due To Difference in Core Plate Slenderness Ratio", *Journal of Structural and Construction Engineering, Transactions of AIJ*, 20(78), 15-21, 2013.

[4] Murase R, Murai M, Iwata M. "Experimental Study on Buckling-restrained Braces Using Steel Mortar Planks -Part 4 Effects of core plate width-thickness ratio, restraining force and cross-section details-", *Journal of Structural and Construction Engineering, Transactions of AIJ*, 620, 117-124, 2007.

[5] Tadokoro A, Midorikawa M, Murai M, Iwata M. "Experimental Study on Bucklin-restrained Braces Using Steel Mortar Planks -Influences of Core Plate Length, Plastic Length Ratio and Rib Plate Length-", *Journal of Structural and Construction Engineering, Transactions of AIJ*, 74(641), 1363-1369, 2009.

[6] Midorikawa M, Wakayama T, Iizuka R, Asari T, Murai M, Iwata M. "Experimental Study on Bucklin-restrained Braces Using Steel Mortar Planks -Evaluation of the Buckling Mode Number, Compression-to-tension Strength Ratio and Friction Force-", *Journal of Structural and Construction Engineering, Transactions of AIJ*, 77(681), 1763-1771, 2012.

[7] Midorikawa M, Sasaki D, Asari T, Murai M, Iwata M. "Experimental Study on Buckling-restrained Braces Using Steel Mortar Planks -Influences of the Clearance Between a Core Plate and a Restraining Member on Compressive Strength and Estimation of the Number of Buckling Modes Related to Compressive Strength-", *Journal of Structural and Construction Engineering, Transactions of AIJ*, 75(653), 1361-1368, 2010.

[8] Takeuchi T, Matsui R, Mihara S, Oya T, Okamoto Y, Ozaki H, Iwata M. "Out-of-plane Mechanical Stability of Buckling-restrained Braces Using Steel Mortar Planks", *AIJ Journal of Technology and Design*, 45, 569-574, 2014.

SPECTRAL-MATCHED GROUND MOTIONS FOR SEISMIC PERFORMANCE ASSESSMENT

Johnn P. Judd

Department of Civil and Architectural Engineering, University of Wyoming, Laramie, Wyoming USA

Abstract: Seismic performance assessment of buildings employs nonlinear response history analysis and often requires spectrally-matched ground motions. The impact of spectral-matched ground motions on the nonlinear response and predicted seismic performance of 4-story and 8-story steel moment-frame buildings is discussed in this paper. The results indicate that measured dispersion in incremental dynamic analysis response was much wider at higher intensities for ground motions spectrally matched to the target spectrum, compared to the original (un-matched) ground motion set. Furthermore, spectral acceleration intensity measures defined by longer periods slightly decreased the conditional probability of collapse.

1. Introduction

An accurate assessment of the seismic performance of buildings usually requires nonlinear response history analysis. Input ground motions can be derived from a variety of sources:

- real strong-motion ground recordings;
- real recordings that have the amplitude linearly-scaled to match a target spectrum, acceleration or other measure;
- real recordings that have been modified in terms of frequency content, duration, amplitude (and so forth);
- synthetic ground motions based on a "seed" real recording, and;
- stochastic ground motions generated using a variety of methods that may or may not be based on physical models.

A set of ground motions (also referred to as an ensemble or suite of ground motions) is then used to capture response variability due to frequency and other attributes of ground motion [1]. Such variability is termed "record-to-record" (RTR) variability. An alternative approach is to use one ground motion and account for record-to-record variability elsewhere, such as in a First Order Second Moment (FOSM) method.

Selection and scaling of a set of ground motions (acceleration records) to be used in analysis may be accomplished in several ways [2,3], with most methods focused on tectonic plate boundary regions, such as the western United States [4].

1.1 Selection of Ground Motions

Ground motions in the central and eastern United States, however, differ from ground motions in the western United States due to geological dissimilarities between the two regions. Namely, the central and eastern ground motions have higher frequency content [5,6], owing to various factors: hazard contributions tend to come from smaller-magnitude earthquakes that have less low-frequency content for a given hazard level; earthquakes have more high-frequency energy; attenuation of motion (especially high-frequencies) is slower due to a high Q factor; and the target uniform hazard spectral shape for a site is shifted towards higher frequencies because the reference ground motion often corresponds to a hard-rock condition, and high-frequency motion is not attenuated as much by near-surface site effects [5].

Thus, various ground motion sets have been proposed for structures located in the central and eastern United States. For example, while developing a ground motion prediction model for the central and eastern United States, Somerville et al. [7] generated a set of ground motions using a physics-based simulation procedure. Fernández [8], Rix and Fernandez [9], developed a synthetic ground motion ensemble for the Mississippi embayment area. Celik [10] used ground motions by Fernández in his analysis of the central and eastern United States in order to capture nonlinear soil behavior (i.e., damping) in site response and resonance in the soil column, and concluded that other ground motions (developed by Wen and Wu [11]) overestimate the ground motion intensity at low periods and do not capture soil column resonance at high periods.

Huang et al. [12] formed a set of ground motions based on real recordings, mostly from the United States, but augmented with records from Canada and elsewhere in the world where there were large intra-plate earthquakes. Some of the recordings were from the western United States where it was determined that the fault mechanism and high-velocity sedimentary layer were appropriate for the eastern United States.

Hines et al. [13] propose an ensemble of ground motions for the eastern United States. Their ensemble does not employ amplitude scaling. Instead, the motions were selected from a United States Nuclear Regulatory Commission (NUREG) database [14]. In the NUREG database, real recordings (mostly from the western United States) were modified (scaled using theoretical transfer functions) as to be appropriate for the tectonic environment of the central and eastern United States. The modified records exhibit higher spectral accelerations and peak in a higher frequency range compared to the original records.

Since high-frequency ground motions are largely stochastic [15,16], there is no perceived problem with modifying ground motions, although synthetic ground motions may lead to differences in the predicted structural response (see for example [17]).

1.2 Scaling of Ground Motions

Acceleration records that are appropriate, in terms of fault mechanism, magnitude, and distance from source-to-site, for a particular location are typically sparse, however, so actual records are often modified by adjusting frequency content, duration, and amplitude, to "match" a target response spectrum. Yet, the impact of such spectral-matched ground motions on the nonlinear response and predicted seismic performance is not well understood.

2. Objectives and Scope

This paper discusses the assessment sensitivity of 4-story and 8-story steel moment-frame buildings to spectral matched ground motions. Ground motions were spectral matched to the target design spectrum corresponding to the highest spectral accelerations for Seismic Design Category (SDC) D (S_{DS} = 1.0 g, S_{D1} = 0.6 g). The probability of building collapse as a function of spectral intensity was predicted using nonlinear response history analysis, following the FEMA P-695 methodology, where the ground motion set was incrementally scaled with respect to the target spectrum until collapse.

3. Methodology

3.1 Building Archetypes

The building plan used 20-ft bays and perimeter moment frames. The first story height (measured to the top of the beam) was 15 ft. Upper story height was 13 ft. 1-story and 2-story moment frame columns were considered pinned at the base, taller columns were considered fixed. The layout, story heights, and gravity loads match the special steel moment framing building archetypes developed in the ATC-76 project [18].

Gravity columns were considered pinned at the base, and were spliced 4 ft above the third and sixth floors. The ATC-76 evaluation [18] did not model gravity framing and, as a result, the gravity framing layout, including the direction and spacing of filler beams every 10 ft, column splices, and gravity column orientation, were developed in this study. The gravity frame columns were oriented with the strong axis in the same direction as the moment frames (the longitudinal direction), and thus the reserve lateral strength observed in this study represents an upper bound on the contribution for this building configuration.

Two types of steel moment frames were considered: (1) a non-ductile moment frame, with fully-restrained flange welded connections, and (2) a ductile moment frame with reduced beam section (RBS) connections. The non-ductile steel moment frames were designed for the lowest spectral accelerations corresponding to SDC B (S_{DS} = 0.167 g, S_{D1} = 0.067 g). The frames were not detailed for seismic resistance. A summary of structural members for the non-ductile moment frame buildings is available [19]. The ductile moment frames were designed for the lowest spectral accelerations corresponding to SDC D (S_{DS} = 0.5 g, S_{D1} = 0.2 g), and were previously developed in the ATC-76 project [18]. The details of the design and member sizes are available in Appendix D of the ATC report.

3.2 Structural Modelling

A structural model of the building (Fig. 1a) was developed using *OpenSees* [20] finite element software. Structural framing members were modeled using an assembly of linear elastic elements and zero-length nonlinear springs to simulate the formation of plastic hinges in beams at the column face, plastic hinges at the top or bottom of columns, and shear yielding of the column panel zone. The gravity framing was explicitly represented, instead of using a "leaning column," and column splices were explicitly included (idealized as pinned connections with no rotational constraint). Second-order (P-Δ) effects were included using the "corotational" approach, where the local element coordinate frame continuously rotated with the element to capture large displacements. The computed fundamental period of vibration was used for both referencing spectral accelerations and for estimating the inherent damping (taken as 2.5%) in the building that was not explicitly modelled through component hysteresis.

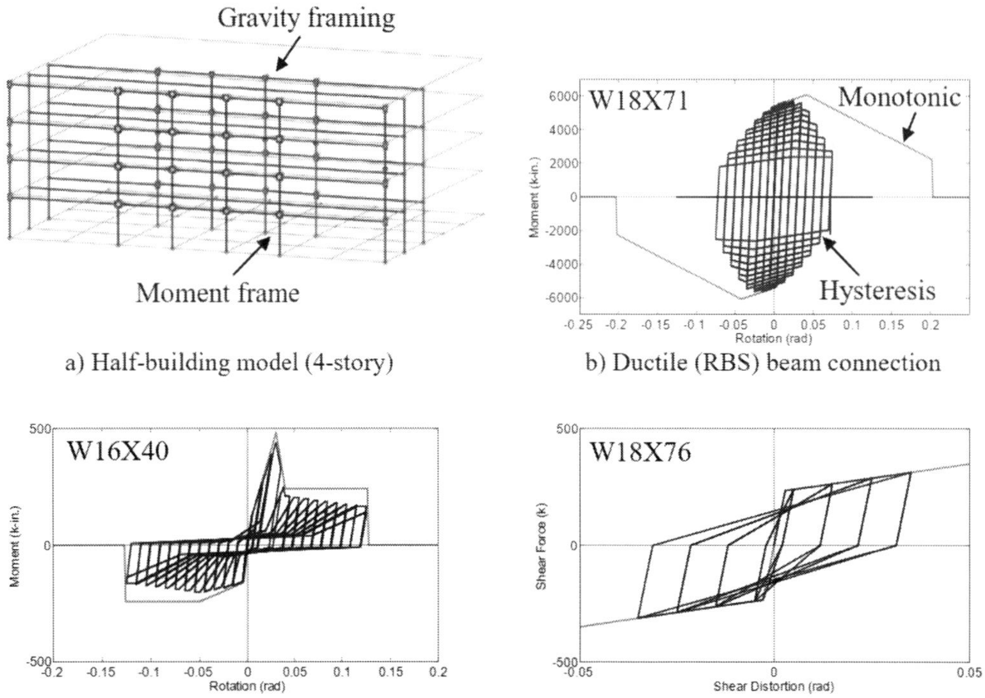

a) Half-building model (4-story)
b) Ductile (RBS) beam connection
c) Shear-tab connection
d) Column panel zone

Fig. 1: Structural model of the building and idealized component behaviour

The moment-rotation behavior of non-ductile beam connections was represented using a tri-linear loading and unloading rule-based hysteresis model [21]. Hysteresis parameters were based on the performance of fully-restrained flange welded connections designed prior to the 1994 Northridge earthquake. The moment rotation behaviour of ductile beam connections (Fig. 1b) was represented using the modified Ibarra-Medina-Krawinkler bilinear model. Hysteresis parameters were based on a statistical analysis of experimental test results [22]. The moment-rotation behavior of the gravity framing shear tab connections (Fig. 1c) was represented using the Lowes et al. [21] model. Hysteresis parameters were based on test data of shear tab connections [23], corresponding analytical models [24,25], and FEMA P-440A [26].

Column behavior was idealized similarly to ductile beam behavior, except column plastic hinge behavior was reduced to account for axial load interaction. The reduction was based on AISC 360-10 Equations H1-1a and H1-1b [27] using a constant axial load resulting from the gravity load plus half the lateral load calculated in a nonlinear static (pushover) analysis. Column panel zones were explicitly represented using the Krawinkler joint model [28,29] which consisted of 8 rigid beam elements connected by hinges and one zero-length spring to represent the combined contributions of the column web and the column flanges. Shear-distortion (Fig. 1d) was modeled using a tri-linear envelope curve and bilinear hysteresis behavior.

3.3 Analysis Approach

Seismic performance was assessed using a sequence of gravity load analysis, frequency (period-determination) analysis, nonlinear pushover analyses (one analysis to estimate axial loads in columns, followed by one analysis that accounted for axial load interaction), and nonlinear dynamic response history analyses that were incrementally scaled in intensity, with respect to the maximum target spectrum for SDC D, until dynamic instability ensued.

The FEMA P-695 Far-Field ground motion set [30] was spectral matched to the target design spectrum by employing an algorithm developed by Adekristi [31], where wavelets and Broyden updating are used to account for a shift in the time corresponding to the peak response. Fig. 2 shows a comparison of the acceleration, velocity, and displacement response histories of the original (un-matched) and spectral matched ground motion for a representative record (FEMA P-695 Far Field ground motion ID No. 1). The response indicates that realistic physical behavior, with zero velocity and minimal displacement at the end of the record, was maintained in the spectral matching process for this ground motion record. This comparison was in line with limits recommended to evaluate the adequacy of the spectral-matching process [32].

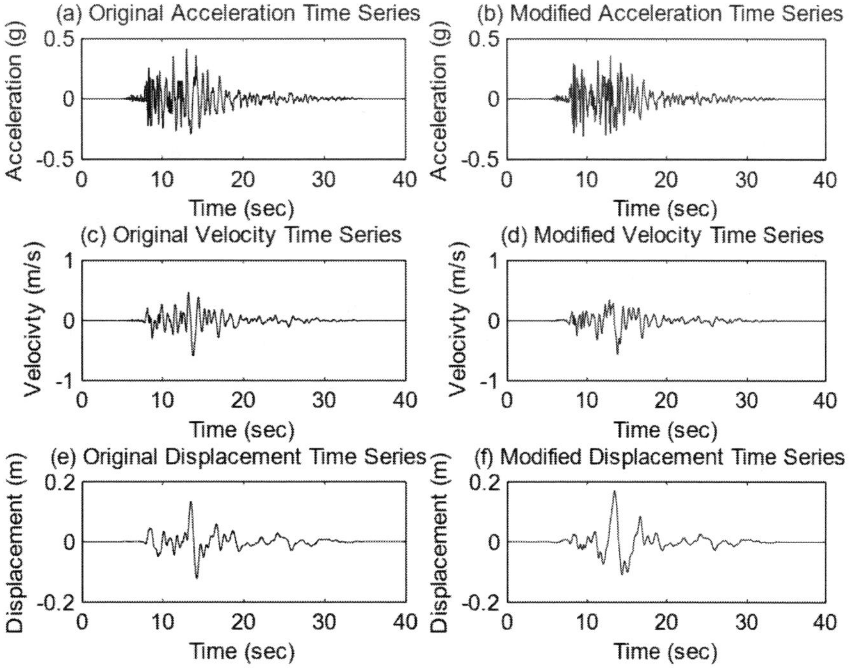

Fig. 2: Response histories representative ground motion
(FEMA P-695 Far Field ground motion ID No. 1)

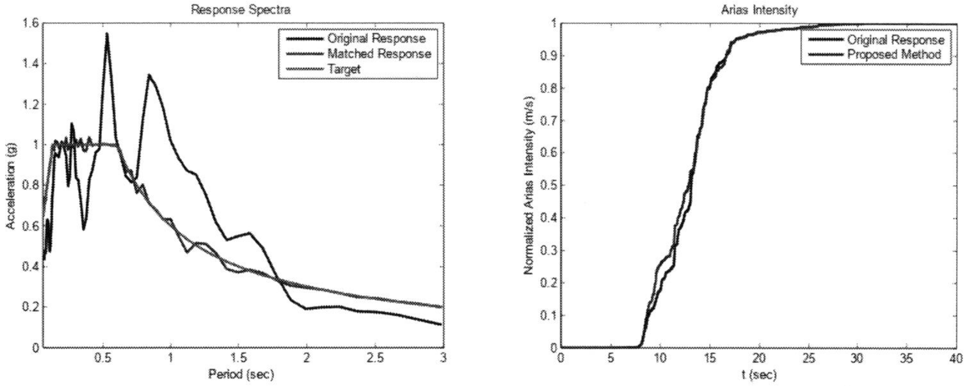

a) Response spectrum b) Normalized arias intensity

Fig. 3: Response of representative ground motion
(FEMA P-695 Far Field ground motion ID No. 1)

For the same ground motion, Fig. 3 shows a comparison of the response spectrum and normalized arias intensity (acceleration of transient waves) of the original (un-matched) and spectral matched ground motion. The response indicates that the matching process also produced a ground motion that was very similar to the original ground motion, in terms of spectral response and ground shaking intensity.

4. Results

The results are summarized in Table 1 (upper rows of the table for non-ductile buildings and lower rows for ductile buildings). The non-ductile 4-story response is shown in Fig. 4 and Fig. 5. The non-ductile 8-story response is shown in Fig. 6 and Fig. 7. The ductile 8-story response is shown in Fig. 8 and Fig. 9. The seismic collapse assessment was sensitive to the ground motion variability (with respect to the target spectrum) and, in general, the matched ground motions led to a higher *CMR* value and consequently a lower probability of collapse.

Interestingly, the dispersion in response was much wider for the matched ground motions. A possible cause for the increased dispersion could be attributed to the fact that the high frequency content of ground motions did not match the target spectrum at low periods. To a lesser extent, the low frequency content also did not always match the target spectrum (see Fig. 4b, for example).

Table 1: Results

Ground Motion Set	T_1 (s)	Static Pushover C_s	Ω	μ_T	Incremental Dynamic Analyses S_{aMCE} (g)	CMR	SSF	ACMR	β_{RTR}	β_{Total}	P_c | MCE
Non-ductile buildings designed for SDC B_{min}, evaluated at SDC D_{max}											
4-Story (MF+GF)											
Original	2.01	0.21	0.41	3.94	0.45	0.93	1.39	1.30	0.37	0.65	34%
Matched						1.08	1.39	1.50	0.34	0.63	26%
8-Story (MF+GF)											
Original	2.85	0.12	0.55	2.35	0.32	0.87	1.27	1.10	0.34	0.63	44%
Matched						1.39	1.36	1.90	0.39	0.66	17%
Ductile buildings designed for SDC D_{min}, evaluated at SDC D_{max}											
8-Story											
Original	2.88	0.05	1.31	3.44	0.31	0.87	1.36	1.18	0.34	0.46	36%
Matched						1.12	1.36	1.52	0.37	0.48	19%

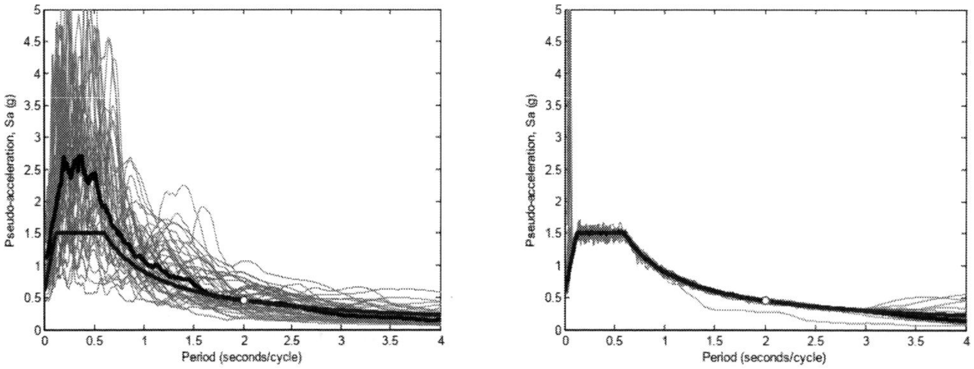

a) Original ground motions b) Spectral-matched ground motions

Fig. 4: Scaled response spectrum for non-ductile 4-story building

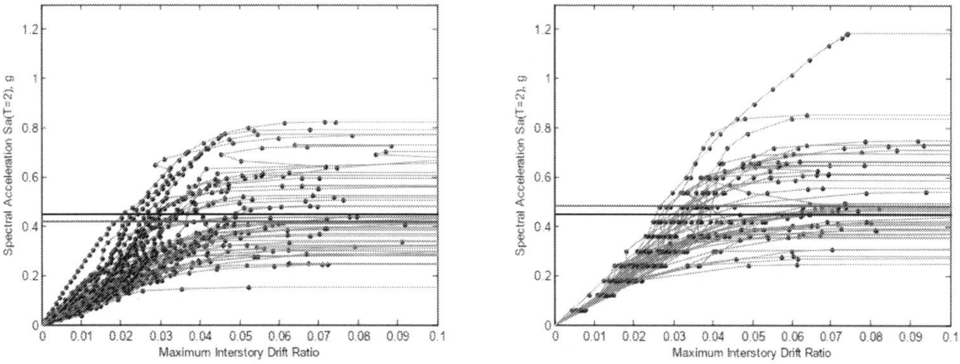

a) Original ground motions b) Spectral-matched ground motions

Fig. 5: Incremental dynamic analyses results for non-ductile 4-story building

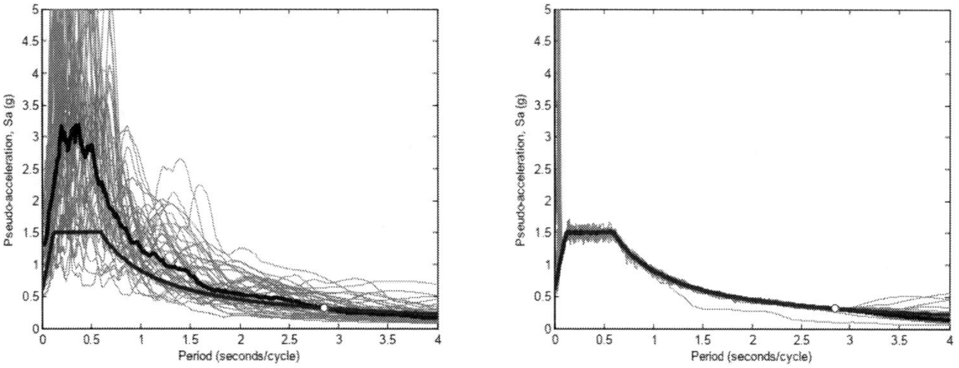

a) Original ground motions b) Spectral-matched ground motions

Fig. 6: Scaled response spectrum for non-ductile 8-story building

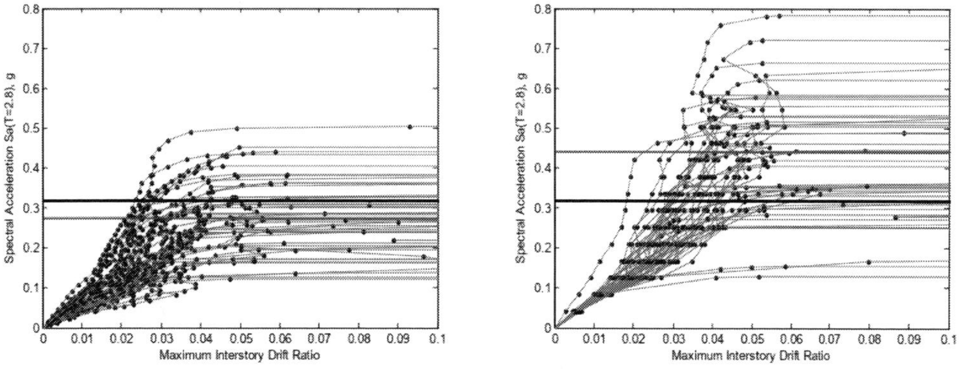

a) Original ground motions b) Spectral-matched ground motions

Fig. 7: Incremental dynamic analyses results for non-ductile 8-story building

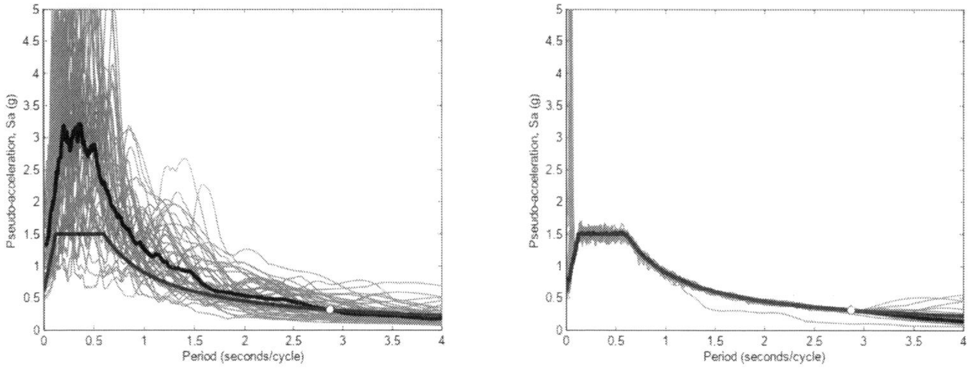

a) Original ground motions b) Spectral-matched ground motions

Fig. 8: Scaled response spectrum for ductile 8-story building

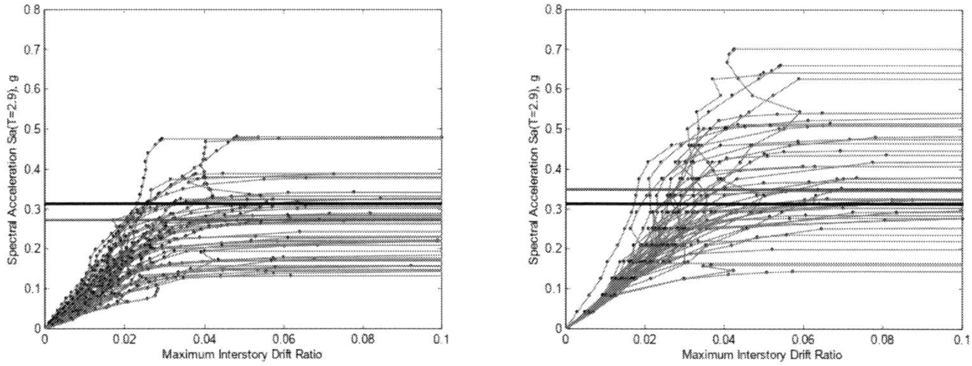

a) Original ground motions b) Spectral-matched ground motions

Fig. 9: Incremental dynamic analyses results for ductile 8-story building

5. Conclusions

The seismic collapse assessment was sensitive to the ground motion variability. Measured dispersion in incremental dynamic analysis curves was much wider for ground motions spectrally matched to the target spectrum, compared to the original (un-matched) FEMA P-695 Far Field set, a probable reflection of high and low frequency content outside the matching bounds. Spectral acceleration intensity measures defined by longer periods slightly decreased the conditional probability of seismic collapse.

References

[1] Liel AB, Haselton CB, Deierlein GG, Baker JW. "Incorporating modeling uncertainties in the assessment of seismic collapse risk of buildings," *Structural Safety*, 31: 197–211, 2009.

[2] Kalkan E, Chopra AK. "Modal-pushover-based ground-motion scaling procedure," *Journal of Structural Engineering*, 137(3): 298–310, 2011.

[3] NIST. *Selecting and scaling earthquake ground motions for performing response-history analysis*. NIST GRC 11-917-15, National Institute of Standards and Technology (NIST), Gaithersburg, Maryland, 2011.

[4] Baker JW, Lin T, Shahi SK, Jayaram N. *New ground motion selection procedures and selected motions for the PEER transportation research program.* PEER Report 2011/03, Pacific Earthquake Engineering Research Center (PEER), University of California, Berkeley, 2011.

[5] NIST. *Selecting and scaling earthquake ground motions for performing response-history analysis.* NIST GRC 11-917-15, National Institute of Standards and Technology (NIST), Gaithersburg, Maryland, 2011.

[6] GEER Team. *Geotechnical reconnaissance of the 2011 Central Virginia earthquake: report of the National Science Foundation sponsored Geotechnical Extreme Events Reconnaissance (GEER) Team.* J. R. Martin, (editor), 2012.

[7] Somerville PG, Collins N, Abrahamson N, Graves R, Saikia C. *Ground motion attenuation relations for the central and eastern United States.* Final Report, United States Geological Survey, Reston, Virginia, 2001.

[8] Fernández JA. *Numerical simulation of earthquake ground motions in the upper Mississippi embayment.* Ph.D. dissertation, Civil Engineering, School of Civil and Environmental Engineering, Georgia Institute of Technology, Atlanta, Georgia, 2007.

[9] Rix GJ, Fernandez JA. "Probabilistic ground motions for selected cities in the upper Mississippi embayment,"http://www.ce.gatech.edu/~geosys/soil_dynamics/research/, 31 July 2006, Georgia Institute of Technology, Atlanta, Georgia, 2006.

[10] Celik OC. *Probabilistic assessment of non-ductile reinforced concrete frames susceptible to mid-America ground motions.* Ph.D. dissertation, Civil Engineering, School of Civil and Environmental Engineering, Georgia Institute of Technology, Atlanta, Georgia, 2007.

[11] Wen YK, Wu CL. "Uniform hazard ground motions for mid-America cities." *Earthquake Spectra*, 17(2): 359–384, 2001.

[12] Huang Y-N, Whittaker AS, Luco N. *Establishing maximum spectral demand for performance-based earthquake engineering: collaborative research with the University at Buffalo and the USGS.* Technical Report: United States Geological Survey, Reston, Virginia. USGS Award Number 08HQGR0017, 2010.

[13] Hines EM, Baise LG, Swift SS. "Ground-motion suite selection for eastern North America," *Journal of Structural Engineering*, 137(3): 358–366, 2011.

[14] McGuire RK, Silva WJ, Constantino CJ. *Technical basis for revision of regulatory guidance on design ground motions.* Report NUREG/CR-6728, U. S. Nuclear Regulatory Commission, 2001.

[15] Hanks T, McGuire R. "The character of high-frequency strong ground motion," *Bulletin of the Seismological Society of America*, 71: 2071–2095, 1981.

[16] Boore D. "Prediction of ground motion using the stochastic method," *Pure and Applied Geophysics*, 160: 635–676, 2003.

[17] Milburn TW. *A comparison of the structural response to synthetic earthquake ground motions compatible with central eastern United States attenuation models.* Ph.D. dissertation, Civil Engineering. University of Louisville, Louisville, Kentucky, 2009.

[18] NIST. *Evaluation of the FEMA P-695 methodology for quantification of building seismic performance factors.* NIST GCR 10-917-8, National Institute of Standards and Technology (NIST), Gaithersburg, Maryland, 2009.

[19] Judd JP, Charney FA. "Seismic collapse prevention system for steel-frame buildings," *Journal of Constructional Steel Research*, published online 2015.

[20] PEER. *Open Systems for Earthquake Engineering Simulation (OpenSees)*, version 2.4.6. Pacific Earthquake Engineering Research Center (PEER), University of California, Berkeley, California, 2015.

[21] Lowes LN, Mitra N, Altoonash A. *A beam-column joint model for simulating the earthquake response of reinforced concrete frames*. PEER Report 2003/10, Pacific Earthquake Engineering Research Center, University of California, Berkeley, California, 2004.

[22] Lignos DG, Krawinkler H. "Deterioration modeling of steel components in support of collapse prediction of steel moment frames under earthquake loading," *Journal of Structural Engineering*, 137(11): 1291–1302, 2011.

[23] Liu J, Astaneh-Asl A. "Cyclic testing of simple connections including effects of slab," *Journal of Structural Engineering*, 126(1): 32–39, 2000.

[24] Liu J, Astaneh-Asl A. "Moment-rotation parameters for composite shear tab connections," *Journal of Structural Engineering*, 130(9): 1371–1380, 2004.

[25] Wen R, Akbas B, Shen J. "Practical moment–rotation relations of steel shear tab connections," *Journal of Constructional Steel Research*, 88: 296–308, 2013.

[26] FEMA. *Effects of strength and stiffness degradation on seismic response*. FEMA P-440A, Federal Emergency Management Agency (FEMA), Washington, D. C., 2009.

[27] AISC. *Specification for structural steel buildings*. ANSI/AISC 360-10, American Institute for Steel Construction (AISC), Chicago, Illinois, 2010.

[28] Krawinkler H. "Shear in beam-column joints in seismic design of frames," *Engineering Journal*, 15(3): 82–91, 1978.

[29] Charney FA, Marshall JA. "Comparison of the Krawinkler and Scissors models for including beam–column joint deformations in the analysis of moment resisting frames," *Engineering Journal*, 43(1): 31–48, 2006.

[30] FEMA. *Quantification of building seismic performance factors*. FEMA P-695, Federal Emergency Management Agency (FEMA), Washington, D. C., 2009.

[31] Adekristi A. *Algorithm for spectral matching of earthquake ground motions using wavelets and Broyden updating*. M.S. thesis, Civil Engineering, Virginia Polytechnic Institute and State University, Blacksburg, Virginia, 2013.

[32] Grant DN, Diaferia R. "Assessing adequacy of spectrum-matched ground motions for response history analysis," *Earthquake Engineering and Structural Dynamics*, 42(9): 1265–1280, 2012.

SEISMIC COLLAPSE RISK OF STEEL-CONCRETE COMPOSITE MOMENT-FRAME BUILDINGS

Johnn P. Judd

Department of Civil and Architectural Engineering, University of Wyoming, Laramie, Wyoming USA

Abstract: This paper examines the seismic collapse risk of 4-story and 8-story buildings with steel-concrete composite moment frame connections and composite shear tab gravity framing connections. The vulnerability of buildings to earthquake-induced collapse (collapse fragility) was predicted using a sequence of vibration, pushover, and response history analyses. Seismic risk was defined using probabilistic ground motions for European sites, topographic data as a proxy for shear wave velocity, and integrating collapse fragility and seismic hazard. The results indicate that composite action between the beam and the concrete slab can significantly enhance the building seismic resistance, compared to the bare steel frame.

1. Introduction

The seismic resistance of buildings with moment-frame steel beams mechanically fastened to a concrete slab on a steel deck can be significant due to composite action. Experimental tests, analytical modeling [1,2], and field reconnaissance (e.g. [3]) have together indicated that composite action significantly improves side-sway collapse capacity of moment frame buildings during an earthquake. Although the effect of composite action between the beam and the concrete slab has been investigated for fully-restrained ductile reduced beam section (RBS) connections, the combined influence of partially-restrained composite shear tab connections on seismic risk has not yet been examined.

This paper discusses the collapse risk of steel-concrete composite moment-frame buildings compared to the bare steel frame buildings. The probability of building collapse as a function of spectral intensity (collapse fragility) was predicted using nonlinear response history analysis, following the FEMA P-695 methodology [4], where the ground motion set was incrementally scaled with respect to the target spectrum until collapse. The risk of seismic collapse was determined by integrating collapse fragility and seismic hazard for selected sites in Europe, accounting for directional effects and soil amplification.

2. Methodology

2.1 Building Archetypes

The building plan used 20-ft bays and perimeter moment frames. The first story height (measured to the top of the beam) was 15 ft. Upper story height was 13 ft. 1-story and 2-story moment frame columns were considered pinned at the base, taller columns were considered fixed. The layout, story heights, and gravity loads match the special steel moment framing building archetypes developed in the ATC-76 project [5]. The ductile moment frames were designed for the highest spectral accelerations corresponding to SDC D (S_{DS} = 1.0 g, S_{D1} = 0.6 g) and soil site class D (V_{s30} = 183 to 366 m/s). The details of the design and the member sizes are available in Appendix D of the ATC project report.

Gravity columns were considered pinned at the base, and were spliced 4 ft above the third and sixth floors. The ATC-76 evaluation [5] did not model gravity framing. The gravity framing shear tab connections, including the direction and spacing of filler beams every 10 ft, column splices, and gravity column orientation, were developed as part of this study. The gravity frame columns were oriented with the strong axis in the same direction as the moment frames (the longitudinal direction), and the reserve lateral strength in the shear tab connections observed in this study represents an upper bound for this building configuration.

2.2 Structural Modelling

A structural model of the building was developed using *OpenSees* [6] finite element software. Structural framing members (Fig. 1a) were modeled using an assembly of linear elastic elements and zero-length nonlinear springs to simulate the formation of plastic hinges in beams at the face of the column, plastic hinges at the top or bottom of columns, and shear yielding of the column panel zone. The gravity framing was explicitly represented, instead of using a "leaning column," and column splices were explicitly included (idealized as pinned connections with no rotational constraint). Second-order (P-Δ) effects were included using the "corotational" approach, where the local element coordinate frame continuously rotated with the element to capture large displacements. The computed fundamental period of vibration was used for both referencing spectral accelerations and for estimating the inherent damping (taken as 2.0%) not explicitly modelled through component hysteresis.

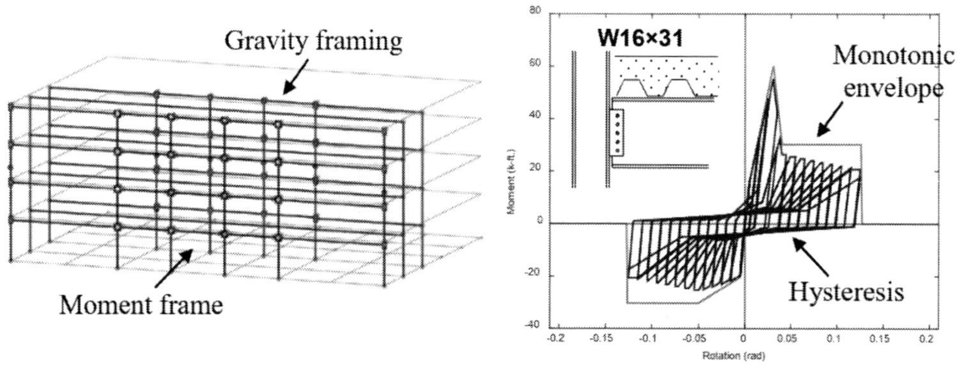

a) Half-building model (4-story) b) Composite shear-tab

Fig. 1: Structural model and idealized gravity beam connection behavior

The moment-rotation behavior of the composite shear tab connections (Fig. 1b) was represented using a hysteretic model with pinching, originally developed for reinforced concrete

[7]. Hysteresis parameters were based on test data of shear tab connections [8], corresponding analytical models [9,10], and FEMA P-440A [11]. The shear tab connection behavior was asymmetric due to the effect of composite action with the concrete slab. The connections exhibit cyclic degradation of both stiffness and strength, until approximately 10% to 12% drift, after which the rotational stiffness and flexural strength of the connection was zero. The negative-moment capacity boundary (monotonic envelope) was based on the behavior of the bare steel beam.

Experimental data and finite element analyses [10] have shown that a linear-elastic stress distribution was developed at the negative yield moment, Eq. (1) and a uniform plastic stress distribution was developed at the negative ultimate moment, Eq. (2) based on the geometry of the shear tab connection, where d_{st} is the height of shear tab (taken as d_{beam} – 6 inches), t_{st} is shear tab thickness (taken as 3/8 inches), R_y = 1.1, and F_y = 50 ksi:

$$M_y^- = (d_{st}^2 t_{st} R_y F_y)/6 \quad (1)$$

$$M_u^- = (d_{st}^2 t_{st} R_y F_y)/4 \quad (2)$$

The positive-moment capacity boundary was based on observations of experimental data [9] that indicate maximum strength for positive moment, Eq. (3) was twice that for negative moment, and positive-moment yield strength, Eq. (4) was a quarter of maximum strength:

$$M_y^+ = 0.25\, M_u^+ \quad (3)$$

$$M_u^+ = 2\, M_u^- \quad (4)$$

The total (ultimate) rotational capacity, Eq. (5) for positive and negative moment was based on the gap, g between the beam flange and the column, and distance from the mid-height of the shear tab to the beam flange, $d_{beam}/2$ [9]:

$$\theta_{Total} = g/(d/2) \quad (5)$$

The displacement ratio was 0.5, reloading force ratio was 0.25, unloading force ratio was 0.05, cyclic strength degradation parameters gK = gD = [0.0 0.1 0.0 0.0 0.2], gF = [0.0 0.4 0.0 0.4 0.9], and maximum cyclic to monotonic energy displacement ratio = 10, as was assumed in FEMA P-440A [11].

The moment rotation behavior of ductile beam connections (Fig. 2) was represented using the modified Ibarra-Medina-Krawinkler bilinear model. Hysteresis parameters for bare steel RBS (Fig. 2a) and composite RBS (Fig. 2b) connections were based on a statistical analysis of experimental test data [1,12].

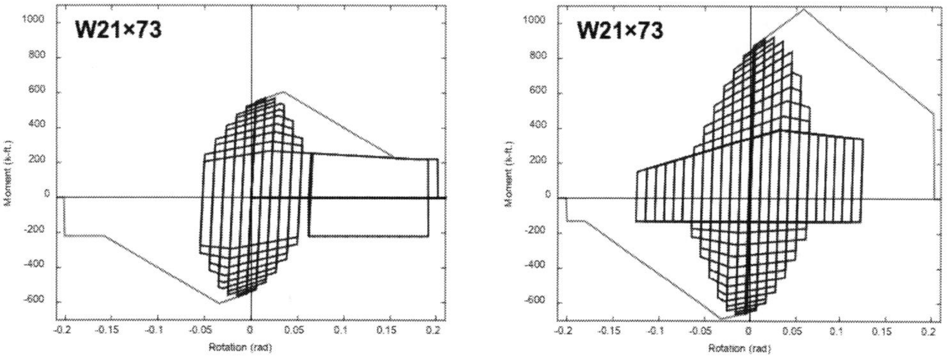

a) Bare steel (non-composite) RBS   b) Composite RBS

Fig. 2: Idealized moment frame beam connection behavior

The plastic moment, M_p was based on the expected yield strength, $F_y R_y$ (where R_y = 1.1) and the plastic section modulus, Z, calculated at the centerline of the RBS cut using AISC

358-10 Equation 5.8-4 [13]. Isotropic hardening was included in the model in an approximate way by increasing the plastic moment to an "effective" yield strength, M_y = 1.1 M_p (see [12]). This strength represented an "average" strain hardening [14] but did not account for cyclic hardening. The post-yield strength ratio (M_c/M_y) was calculated as the ratio of the capping strength to the effective yield strength. This was similar to the strain hardening ratio, as expressed by the C_{pr} factor (see AISC 358-10 page 9.2-6). In this study, the ultimate rotation capacity θ_u was taken as 0.2 [5], but it was recognized that the actual ultimate rotation capacity depends on the loading protocol. The reference cumulative rotation capacity, Λ was based on experimental data, and deterioration parameters used default values (c = 1.0).

Moment-rotation behaviour parameters for RBS connections in Table 1 were modified to account for composite action based on statistical analysis by Elkady and Lignos [1]. The flexure strength of composite beams was predicted to be higher than the strength of the bare steel beams. The rate of cyclic deterioration, D was higher for positive flexure strength. The composite beam pre-capping plastic positive and negative rotations θ_p were higher for positive flexure and lower for negative flexure. The reduction in negative rotation was due to lateral torsion of the bottom flange [15]. The stiffness of composite beams, I_{comp} may be estimated in various ways (see AISC Seismic Design Manual, for example). In this study I_{comp} was based on test data [1] and corresponded, approximately, to the lower-bound elastic stiffness, I_{LB} for the portion of the concrete slab used to balance the shear force transferred by the shear headed stud anchors, $\sum Q_n = 0.25F_yA_s$.

Table 1: Comparison of moment-rotation behaviour parameters

Connection		M_y/M_p	M_c/M_y	M_r/M_y	D	θ_p	θ_{pc}	I_{comp}/I
Bare steel RBS		1.10	1.10	0.4	1.00	(~0.02) [(a)]	(~0.2) [(a)]	--
Composite RBS	M^+	1.35	1.30	0.3	1.15	1.80	1.35	1.4
	M^-	1.25	1.05	0.2	1.00	0.95	0.95	

(a) Regression equation in [12].

Column behavior was idealized similarly to ductile beam behavior, except column plastic hinge behavior was reduced to account for axial load interaction. The reduction was based on AISC 360-10 Equations H1-1a and H1-1b [16] using a constant axial load resulting from the gravity load plus half the lateral load calculated in a nonlinear static (pushover) analysis. Column panel zones were explicitly represented using the Krawinkler joint model [17,18], which consisted of 8 rigid beam elements connected by hinges and one zero-length spring to represent the combined contributions of the column web and the column flanges. Shear-distortion was modeled using a tri-linear envelope curve and bilinear hysteresis behavior.

3. Results

Collapse fragility was assessed using a sequence of analyses. A pre-load gravity analysis was followed by an Eigen-value frequency determination analysis to determine the periods and modes of vibration of the building. The preliminary gravity and vibration analyses were followed by two nonlinear pushover analyses (one pushover analysis to estimate axial loads in columns, followed by one pushover analysis that accounted for axial load interaction). For nonlinear static analysis the finite element system of equations was solved using static load control for gravity analysis, and displacement control [19,20].

Table 2 summarizes the results of the period-determination and pushover analyses. The computed fundamental periods were higher than empirical predictions, but the computed periods were in line with prior analyses of a similar archetype building [5,1]. The effect of composite action was significant. The static over-strength, Ω increased by 8% to 33% compared to the bare frame for the 4-story and 8-story buildings, respectively. Composite action led to a modest increase in period-based ductility, μ_T. Fig. 3 compares the pushover response for the 4-story building. The increase in the initial global stiffness was due to the increase in beam stiffness. Both the bare frame and composite building models developed a collapse mechanism in story 2 under the pushover load profile.

Nonlinear dynamic response history analyses used the FEMA P-695 Far-Field ground motion set. The ground motions were incrementally scaled in intensity, with respect to the target spectrum for the maximum spectral accelerations in SDC D, until dynamic instability and collapse ensued (Fig. 4). For nonlinear response history analysis the system of equations was solved using the Newmark implicit integration method with the special case of average acceleration (integration parameters, $\gamma = 1/2$ and $\beta = 1/4$) and a self-adaptive solution strategy [21].

Collapse fragility was determined by fitting individual collapse points (spectral acceleration at the reference period and corresponding probability of collapse for a one scaled ground motion record) with a curve, assuming a lognormal cumulative distribution function (Fig. 5a). The median collapse spectral acceleration was adjusted (increased) to account for spectral shape and the total dispersion, β_{Total} in collapse capacity was calculated using Eq. (6).

$$\beta_{Total} = \sqrt{\beta_{RTR}^2 + \beta_{DR}^2 + \beta_{TD}^2 + \beta_{MDL}^2} \tag{6}$$

Table 2: Results from period determination and static pushover analyses

	Periods of Vibration		Static Pushover			
Building Model	T_1 (s)	T_2 (s)	C_s	V_{max}/W	Ω	μ_T
		4-story				
Bare Frame	1.46	0.47	0.079	0.15	1.92	6.00
Composite	1.27	0.41		0.20	2.55	6.49
		8-story				
Bare Frame	2.03	0.70	0.046	0.13	2.90	3.54
Composite	1.94	0.67		0.14	3.12	3.57

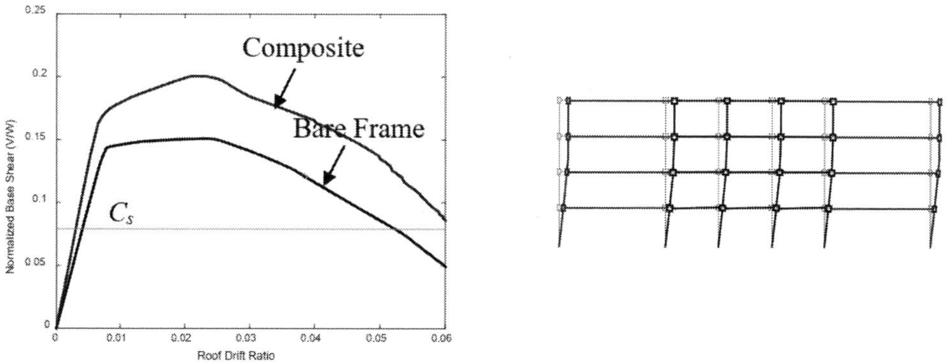

a) Global force-deformation behavior b) Un-deformed and deformed elevation

Fig. 3: Nonlinear static pushover response of the 4-story building

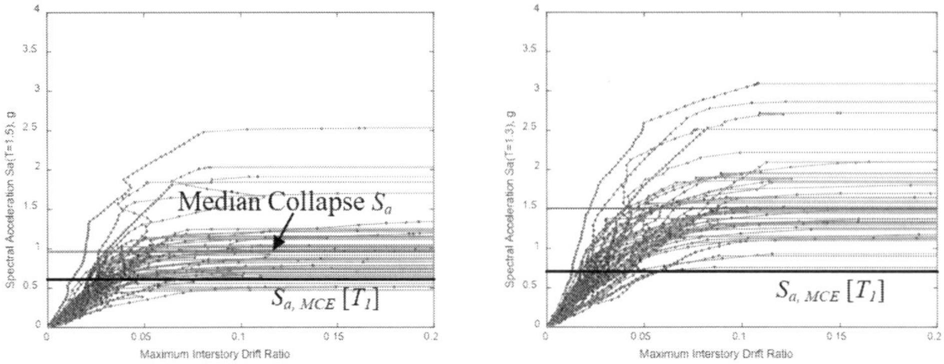

a) Bare frame b) Composite

Fig. 4: Incremental dynamic analyses response of the 4-story building

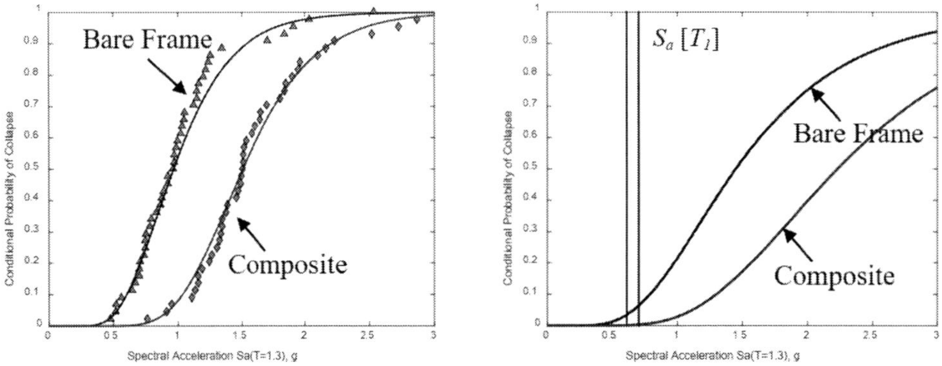

a) Measured points and fitted curve b) Adjusted collapse fragility curve

Fig. 5: Collapse fragility for the 4-story building

The β_{RTR} in Eq. (6) was taken as the measured (record-to-record) dispersion in the nonlinear seismic response history analyses. Epistemic uncertainty was incorporated by assigning dispersion due to design β_{DR}= 0.1, test data β_{TD}= 0.2, and modeling uncertainty β_{MDL}= 0.2 based the FEMA P-695 rubric.

Table 3 summarizes the results from the incremental dynamic analyses. The increase in fundamental period due to composite action led to a small increase in the spectral acceleration, S_a [T_1] corresponding to the Maximum Considered Earthquake (MCE) ground motions for SDC D_{max}. Composite action increased the dynamic over-strength, or Collapse Margin Ratio (*CMR*), of the 4-story building by 36%. The increase in *CMR* for the 8-story building was 4%. Similar increases were measured when the over-strength was adjusted (*ACMR*) for period elongation and strong ground motion for long-periods using the Spectral Shape Factor (*SSF*) based on period-based ductility determined in the static pushover analyses. All buildings satisfied the ASCE 7-10 [22] target conditional collapse probability of 10%.

Seismic collapse risk of the buildings was determined for selected sites in Europe. Seismic hazard in terms of probabilistic bedrock ground motions (i.e. Fig. 6a) provided by the European Facility for Earthquake Hazard and Risk (EFEHR) database (http://www.efehr.org) was adjusted for the maximum-direction spectrum of the FEMA P-695 Far-Field set used to determine the collapse fragility (see [23]). For each site, topographic data was used as a proxy for the average shear wave velocity of the top 30 meters of soil, V_{s30} using the United States Geological Survey (USGS) global V_{s30} map server [24]. Soil amplification of the bedrock motion was based on the ASCE 7-10/NEHRP procedure [22].

Table 3: Results from incremental dynamic analyses

	S_a $[T_1]$ (g)							
Building Model	MCE	Collapse	*CMR*	*SSF*	*ACMR*	β_{RTR}	β_{Total}	P_c | MCE
4-story								
Bare Frame	0.62	0.97	1.57	1.50	2.36	0.37	0.47	3.5%
Composite	0.71	1.51	2.14	1.48	3.16	0.29	0.42	0.3%
8-story								
Bare Frame	0.44	0.61	1.36	1.36	1.86	0.34	0.45	8.6%
Composite	0.46	0.66	1.41	1.37	1.93	0.34	0.45	7.2%

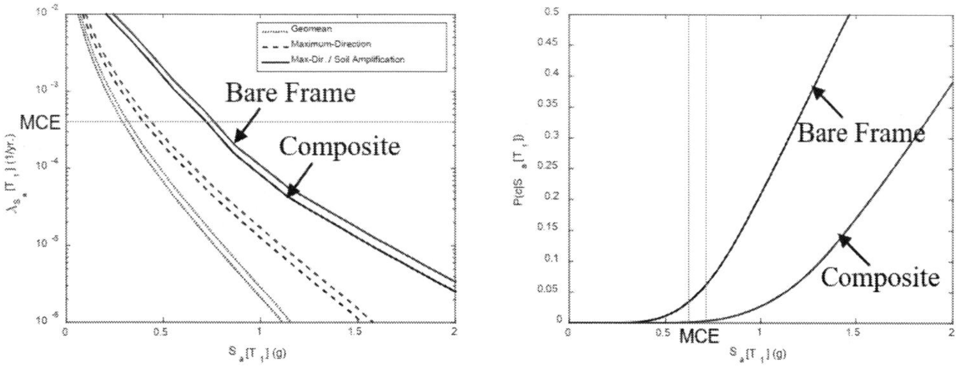

a) Seismic hazard curve (Bucharest site) b) Adjusted collapse fragility

Fig. 6: Probabilistic ground motions and collapse fragility (4-story building)

Seismic collapse risk was calculated by integrating the probabilistic ground motions (i.e. Fig. 6a) and the adjusted collapse fragility (i.e. Fig. 5b). As a result, the risk was applicable only to the archetype buildings described in this study. European-type construction was not investigated. The risk was used solely to illustrate the effects of composite action. The derivative of the fragility curve, the probability density function (PDF), was used to avoid taking the derivative of a polynomial fit of the hazard curve, which could be unstable [23]. Fig. 7 gives results from the risk analysis for the 4-story building at the Bucharest site.

The collapse rate (Fig. 7a) provides a deaggregation of risk to discover spectral accelerations that contribute the most to the annual rate of collapse, λ_c.

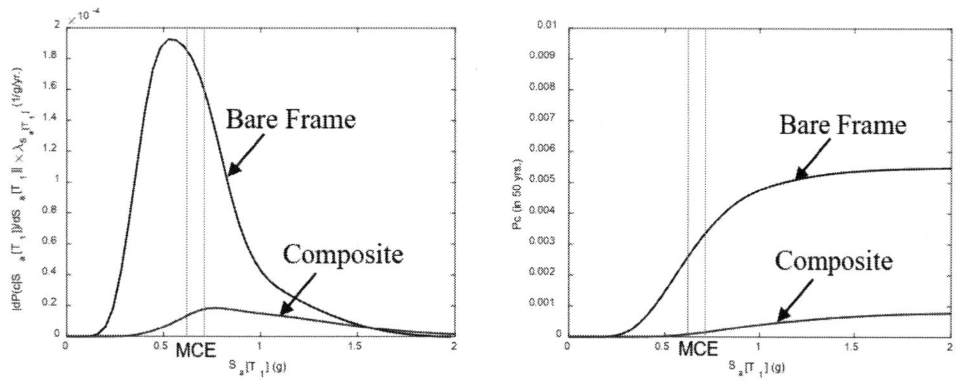

a) Deaggregation of collapse rate b) Cumulative collapse risk (over 50 years)

Fig. 7: Risk analysis for 4-story building at Bucharest site

For example, composite action shifts the dominate hazard for the 4-story building from infrequent events (roughly 500 yr. to 5000 yr. return period) to very rare events (roughly 2,500 yr. to 100,000 yr. return period). The cumulative risk of collapse over the typical lifespan of a building, taken as 50 yrs. (Fig. 7b) was based on characterizing seismic hazard as a Poisson-type process.

Table 4 summarizes the seismic collapse risk. The four locations ranged from high seismic areas (Istanbul) to moderate areas (Brussels). The Bucharest site was closest to the design in terms of spectral accelerations at the computed fundamental period of the buildings. Except for the high-seismic site (Istanbul), all other sites and buildings satisfied the ASCE 7-10 target collapse risk of 1% in 50 yrs. The effect of composite action was significant for the 4-story building, reducing the risk of collapse by half or more, and even satisfying the target risk for Istanbul. Composite action was less significant for the 8-story building risk analysis, where the reduction in structural vulnerability due to composite action was offset by an increase in spectral demands (due to the stiffer building frame) and an increase in the likelihood of the ground motions to actually occur at the site.

Table 4: Results from seismic collapse risk analyses

Building Model	Site				
	City	V_{s30} (m/s)	S_{aMCE} $[T_1]$ (g)	λ_c (collapses/yr.)	P_c in 50 yrs.
			4-story		
Bare Frame	Istanbul	360	1.06	$3.57\text{x}10^{-4}$	1.8%
	Bucharest	240	0.72	$1.10\text{x}10^{-4}$	0.6%
	Lisbon	490	0.47	$4.46\text{x}10^{-5}$	0.2%
	Brussels	360	0.15	$2.02\text{x}10^{-6}$	0.0%
Composite	Istanbul	360	1.16	$1.47\text{x}10^{-4}$	0.7%
	Bucharest	240	0.76	$1.64\text{x}10^{-5}$	0.1%
	Lisbon	490	0.51	$1.57\text{x}10^{-5}$	0.1%
	Brussels	360	0.16	$5.29\text{x}10^{-7}$	0.0%
			8-story		
Bare Frame	Istanbul	360	0.78	$5.92\text{x}10^{-4}$	2.9%
	Bucharest	240	0.47	$1.65\text{x}10^{-4}$	0.8%
	Lisbon	490	0.28	$5.82\text{x}10^{-4}$	0.3%
	Brussels	360	0.08	$2.75\text{x}10^{-6}$	0.0%
Composite	Istanbul	360	0.84	$5.78\text{x}10^{-4}$	2.9%
	Bucharest	240	0.53	$1.83\text{x}10^{-4}$	0.9%
	Lisbon	490	0.32	$6.12\text{x}10^{-5}$	0.3%
	Brussels	360	0.09	$3.05\text{x}10^{-6}$	0.0%

4. Conclusions

Composite action between the bare steel beam and the concrete slab significantly enhanced the seismic resistance of both buildings, especially the 4-story building. The static over-strength of the building increased by 33% for the 4-story building compared to the bare frame. The dynamic over-strength of the 4-story building increased by 36%. The effect of composite action significantly reduced the risk of collapse for the 4-story building. The benefits of composite action were relatively modest for the 8-story building. All buildings satisfied the target conditional collapse probability of 10% and the target collapse risk of 1% in 50 yrs. The computed periods of vibration were closer to empirical estimates due to composite action, and this also led to increased seismic demands. For the 8-story building, the increased demands associated with the stiffer building offset the increase in collapse resistance provided by composite action. Although limited to specific buildings and connection types, the results suggest that an accurate understanding of seismic collapse risk requires inclusion of both the gravity framing and the moment frame and the effects of composite action.

References

[1] Elkady A, Lignos DG. "Modeling of the composite action in fully restrained beam-to-column connections: implications in the seismic design and collapse capacity of steel special moment frames," *Earthquake Engineering and Structural Dynamics*, 43(13): 1935–1954, 2014.

[2] Bozorgmehr A, Leon RT. "Collapse assessment of partial restraint composite connection (PR-CC) moment resisting frames using FEMA P695 (ATC-63) methodology," *Proceedings of the 15th World Conference on Earthquake Engineering*, September 24–28, Lisbon, Portugal, 2012.

[3] Leon RT. "Composite connections," *Progress in Structural Engineering and Materials*, 1(2): 159–169, 1998.

[4] FEMA. *Quantification of building seismic performance factors*. FEMA P-695, Federal Emergency Management Agency (FEMA), Washington, D. C., 2009.

[5] NIST. *Evaluation of the FEMA P-695 methodology for quantification of building seismic performance factors*. NIST GCR 10-917-8, National Institute of Standards and Technology (NIST), Gaithersburg, Maryland, 2010.

[6] PEER. *Open Systems for Earthquake Engineering Simulation (OpenSees)*, version 2.4.6. Pacific Earthquake Engineering Research Center (PEER), University of California, Berkeley, California, 2015.

[7] Lowes LN, Mitra N, Altoonash A. *A beam-column joint model for simulating the earthquake response of reinforced concrete frames*. PEER Report 2003/10, Pacific Earthquake Engineering Research Center, University of California, Berkeley, California, 2004.

[8] Liu J, Astaneh-Asl A. "Cyclic testing of simple connections including effects of slab," *Journal of Structural Engineering*, 126(1): 32–39, 2000.

[9] Liu J, Astaneh-Asl A. "Moment-rotation parameters for composite shear tab connections," *Journal of Structural Engineering*, 130(9): 1371–1380, 2004.

[10] Wen R, Akbas B, Shen J. "Practical moment–rotation relations of steel shear tab connections," *Journal of Constructional Steel Research*, 88: 296–308, 2013.

[11] FEMA. *Effects of strength and stiffness degradation on seismic response*. FEMA P-440A, Federal Emergency Management Agency (FEMA), Washington, D. C., 2009.

[12] Lignos DG, Krawinkler H. "Deterioration modeling of steel components in support of collapse prediction of steel moment frames under earthquake loading," *Journal of Structural Engineering*, 137(11): 1291–1302, 2011.

[13] AISC. *Prequalified connections for special and intermediate steel moment frames for seismic applications*, including supplement No. 1. ANSI/AISC 358-10, American Institute for Steel Construction (AISC), Chicago, Illinois, 2011.

[14] Lignos D. *Sidesway collapse of deteriorating structural systems under seismic excitations*. Ph.D. dissertation, Civil Engineering, Stanford University, Stanford, California, 2008.

[15] Lignos D, Chung Y, Nagae T, Nakashima M. "Numerical and experimental evaluation of seismic capacity of high-rise steel buildings subjected to long duration earthquakes," *Computers and Structures* 89: 959–967, 2011.

[16] AISC. *Specification for structural steel buildings*. ANSI/AISC 360-10, American Institute for Steel Construction (AISC), Chicago, Illinois, 2010.

[17] Krawinkler H. "Shear in beam-column joints in seismic design of frames," *Engineering Journal*, 15(3): 82–91, 1978.

[18] Charney FA, Marshall JA. "Comparison of the Krawinkler and Scissors models for including beam–column joint deformations in the analysis of moment resisting frames," *Engineering Journal*, 43(1): 31–48, 2006.

[19] Batoz J-L, Dhatt G. "Incremental displacement algorithms for nonlinear problems," *International Journal for Numerical Methods in Engineering: Short Communications*, 14: 1262–1267, 1979.

[20] Ramm E. "Strategies for tracing the nonlinear response near limit points," *Nonlinear finite element analysis in structural mechanics, proceedings of the Europe–U. S. Workshop*, New York, New York, 63–89, 1981.

[21] Judd JP. *Multi-hazard performance of steel moment-frame buildings with collapse prevention systems in the central and eastern United States*. Ph.D. dissertation, Civil Engineering, Virginia Polytechnic Institute and State University, Blacksburg, Virginia, 2015.

[22] ASCE. *Minimum design loads for buildings and other structures*. ASCE/SEI 7-10, American Society of Civil Engineers (ASCE), Reston, Virginia, 2010.

[23] Judd JP, Charney FA. "Earthquake risk analysis of structures," *Proceedings of the 9th International Conference on Structural Dynamics (Eurodyn)*, June 30–July 2, Porto, Portugal, 2014.

[24] USGS. *Global V_{s30} map server*. United States Geological Survey (USGS), Reston Virginia, 2015.

THE USE OF REDUCED CROSS-SECTIONS IN ASEISMIC STRUCTURES

Helmuth Köber and Marina Stoian

Technical University of Civil Engineering Bucharest, Romania

Abstract: For eccentrically braced frames with long links, dog-bone configurations at the link ends ensure significant lateral stiffness of the frame and a better control of the plastic deformations along the links. Details with reduced cross-sections near the ends of upper storeys diagonals in concentrically braced frames lead to an adequate slenderness for the braces and to a proper tensile capacity of the diagonals. A favourable global plastic hinge mechanism can be sized by design for concentrically braced frames, by providing configurations with reduced cross-sections near the ends of frame girders. Details resembling the "dog-bone" near the base of bottom-story columns appear to be safer from the point of view of ensuring the general stability of the first-story columns during strong earthquakes.

1. Introduction

The present paper intends to illustrate some advantages of using configurations with reduced cross-sections in several seismic resistant steel structures equipped with different types of braced frames.

2. Configurations with Reduced Cross-sections for Long Links

In a well-designed seismic resistant eccentrically braced frame, the greatest part of the energy induced in the structure by strong earthquakes is dissipated through plastic deformations concentrated in the links. To provide this, the links are sized for code specific lateral loads [1, 2] and all the other structural members of the frames (columns, braces, beam segments outside the dissipative members) are designed for the forces generated by the fully yielded and strain hardened links [1, 2, 3].

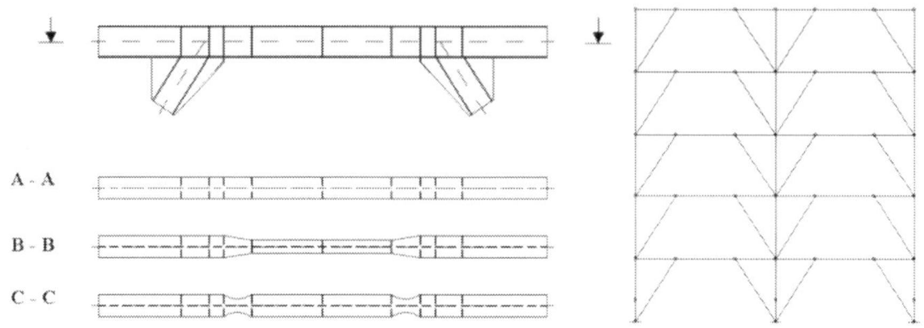

Fig. 1: Constructive solutions for long links in an eccentrically braced frame (solution A-A was used for frame no. 2, solution B-B was used for frame no. 1, solution C-C was used for frame no. 3)

Making the link to be the weakest element of the frame, the designer can force the yielding to occur in the ductile link elements, while preventing inelastic deformations in members with a non-ductile behaviour [1, 2, 3]. Generally this design concept conducts to smaller cross-sections for the links than for the beam segments near them (see configuration B-B in Fig. 1). Especially in eccentrically braced frames with long links the smaller cross-sections of the links leads to smaller values for the lateral stiffness of the structure. On the other hand if the same cross-section is used for the dissipative members and for the beam segments outside the links (see configuration A-A in Fig. 1), the lateral stiffness of the frame increases but the plastic deformations may occur in other members than links too (a favourable general plastic failure mechanism would be difficult to obtain, see Fig. 2). Many plastic hinges appeared in unwanted zones, mainly in the braces and beam segments outside the dissipative members.

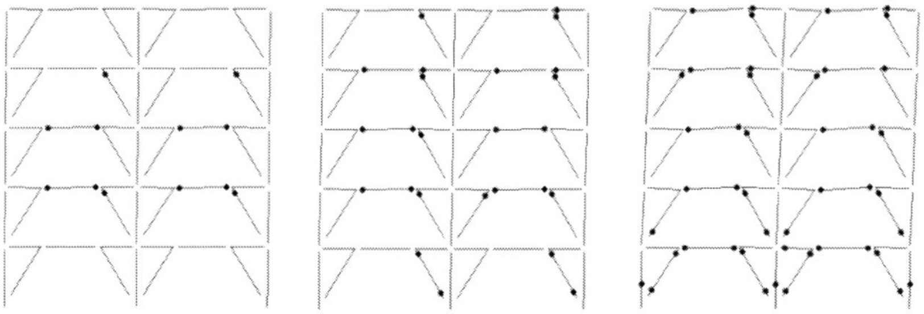

Fig. 2: Some steps of the static nonlinear analysis of frame nr. 2 (configuration B-B for links)

A configuration with reduced beam flanges sections at the member ends is proposed (see configuration C-C in Fig. 1). We tried to outline the advantages of this solution on a designed structure. A five-storey eccentrically braced frame with two spans of 8.0m and a story height of 3.5m was considered. The long links with a length of 3.4m were placed in the central part of the frame girders, between two braces (see Fig. 1). All kind of structural elements had built-up I-shaped cross-sections.

If the same cross-section is used for the whole flexural dissipative member, the positions of the plastic hinges along this element cannot be predicted exactly and so the maximum inelastic link deformations cannot be estimated exactly. The reduced cross-sections near the ends of the links ensure a better control of the positions where plastic hinges can occur along the link. As shown in Fig. 3b), a greater distance between the positions of plastic hinges along the dissipative member, reduces the link axis rotation angle [3].

Numerical tests showed a good behaviour of the structure with reduced beam flanges at the ends of long links. These structures have sufficient stiffness under lateral loads and they develop a favourable global plastic failure mechanism. They are also efficient from the point of view of material consumption

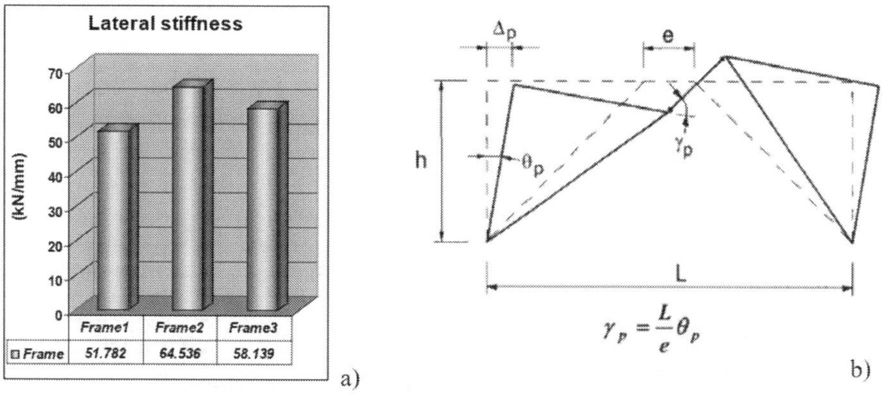

Fig. 3: a) Lateral stiffness values; b) Influence of the distance between plastic hinges e on link axis rotation angle γ_p

3. Configurations Resembling the Dog-Bone for Braces

Configurations resembling the "dog-bone" near the ends of the upper storeys diagonals in concentrically braced frames (see Fig. 4) ensure on one hand, the abidance of the slenderness demand for the braces [1] ($\bar{\lambda} \leq 2.0$) and on the other hand, small values of the overstrength ratio [1, 2] (Ω_i(N)=$N_{plRd,i} / N_{Ed,i}$).

The reduced cross-section of the diagonal in the "dog-bone" detail zones ensures an adequate tensile capacity of the braces and small Ω_i(N) values. At the same time the buckling behaviour and slenderness of the braces is not significantly affected by the reduced cross-sections near the member ends and the slenderness demand of the braces will be satisfied.

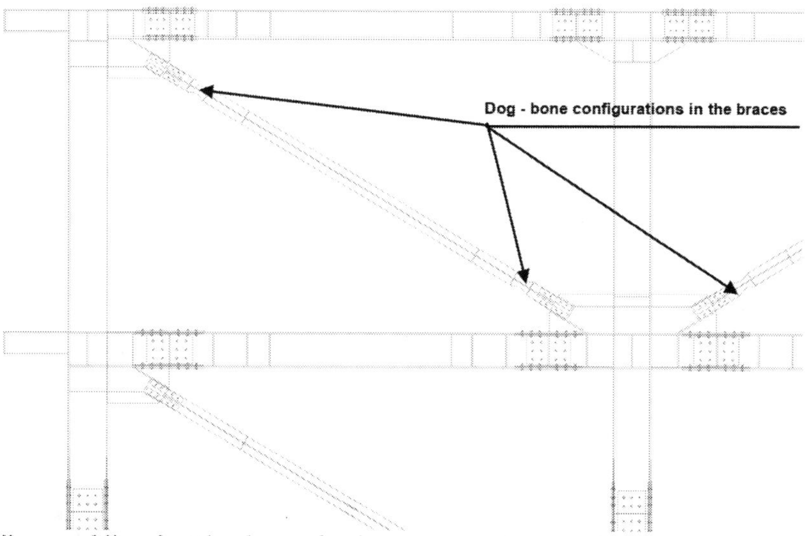

Fig. 4: Details resembling the "dog-bone" for the top storey diagonals in a concentrically braced frame

A ten-storey concentrically braced frame is considered having two spans of 6.0m and a story height of 3.5m, (see Fig. 5.a). The diagonals had built-up I-shaped cross-sections, placed with the web normally to the bracing plane (see Fig. 4). The same current cross-sections are used for the braces located at the storeys 1, 5-6; respectively 2-4; 7-8 and 9-10 (as indicated in Table 1). The non-dimensional slenderness limitation ($\bar{\lambda} \leq 2.0$) is satisfied in all cases.

It can be observed by analyzing Table 1, that the values of $\Omega_8^{(N)}$ and of $\Omega_{10}^{(N)}$ are greater than 1.25-times the minimum $\Omega_i^{(N)}$ value (which is equal to $\Omega_7^{(N)}=1.031$). If, in order to reduce the value of $\Omega_{10}^{(N)}$, a smaller cross-section is used for the braces placed at story 8 and story 10, the slenderness limitation ($\bar{\lambda} \leq 2.0$) will be not fulfilled. By providing dog-bone configurations for the diagonals from story 8 and story 10 (by reducing locally the width of the flanges of the braces cross-section at both ends of the diagonals - or only at one end), the values of $\Omega_8^{(N)}$ and of $\Omega_{10}^{(N)}$ will be reduced and the difference between the maximum and minimum $\Omega_i^{(N)}$-values will be smaller than 25% (see the values in the right part of Table 1). The slenderness demand of the braces will be also satisfied.

Table 1: Reducing the values of the $\Omega_i^{(N)}$ ratios by using dog-bone details at the braces ends

Story	Current brace cross-section				Reduced brace cross-section		
	Web	Flanges	$\Omega_i^{(N)}$ value	$\bar{\lambda}$ value	Web	Flanges	$\Omega_i^{(N)}$ value
1	230 × 8	260 × 15	$\Omega_1^{(N)}=1.249$	1.178	-	-	-
2	250 × 9	280 × 16	$\Omega_2^{(N)}=1.075$	1.102	-	-	-
3	250 × 9	280 × 16	$\Omega_3^{(N)}=1.091$	1.102	-	-	-
4	250 × 9	280 × 16	$\Omega_4^{(N)}=1.144$	1.102	-	-	-
5	230 × 8	260 × 15	$\Omega_5^{(N)}=1.067$	1.178	-	-	-
6	230 × 8	260 × 15	$\Omega_6^{(N)}=1.178$	1.178	-	-	-
7	200 × 8	220 × 12	$\Omega_7^{(N)}=1.031$	1.435	-	-	-
8	200 × 8	220 × 12	$\Omega_8^{(N)}=\mathbf{1.351}$	1.435	200 × 8	160 × 12	$\Omega_8^{(N)}=\mathbf{1.054}$
9	160 × 5	170 × 10	$\Omega_9^{(N)}=1.067$	1.809	-	-	-
10	160 × 5	170 × 10	$\Omega_{10}^{(N)}=\mathbf{1.919}$	1.809	160 × 5	90 × 10	$\Omega_{10}^{(N)}=\mathbf{1.134}$

Remark: The webs and flanges dimensions are given in mm!

On the other hand, if greater cross-sections are provided for the diagonals located at storeys 1-9, the condition that the maximum $\Omega_i^{(N)}$-value does not differ from the minimum $\Omega_i^{(N)}$-value by more than 25% will be also fulfilled. But in this case, a homogeneous dissipative behaviour of all the diagonals will be ensured with much greater material costs. The estimated steel consumption for the diagonals would increase with about 48%, whilst the estimated material consumption for the whole frames will increase with about 14% (see Fig. 5.b). Situation 1 in Fig. 5.b is the one when a homogeneous dissipative behaviour of all the braces is ensured by using dog-bone configurations for the 8- and 10-story diagonals. Situation 2 is the one, when greater braces cross-sections are provided for the storeys 1-9, in order to reduce the differences among the $\Omega_i^{(N)}$-values.

Compared analyses led to the conclusion that the increase of labour cost due to the supplementary cuttings of the flanges at the braces ends is fully compensated by the reduction of overall material cost.

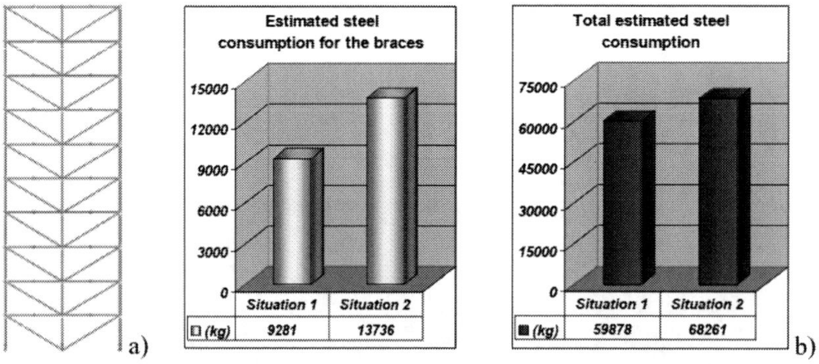

Fig. 5: a) Analyzed concentrically braced frame; b) Estimated steel consumption

4. Dog-Bone Details in Girders of Concentrically Braced Frames

An explication for the poor behaviour of concentrically braced frames during dynamic nonlinear analyses could relay in the fact that according to most seismic design procedures a global plastic failure mechanism is not sized clearly by design for this kind of seismic resistant structures. Generally as long as inelastic deformations are concentrated only in the diagonals, the concentrically braced frames have a predictable behaviour. When the loading level increases and plastic deformations appear in other type of members (columns and girders), the behaviour of a concentrically braced frame is difficult to control. Unfavourable distributions of plastic hinges may lead to the apparition of several local plastic mechanisms (see Fig. 6).

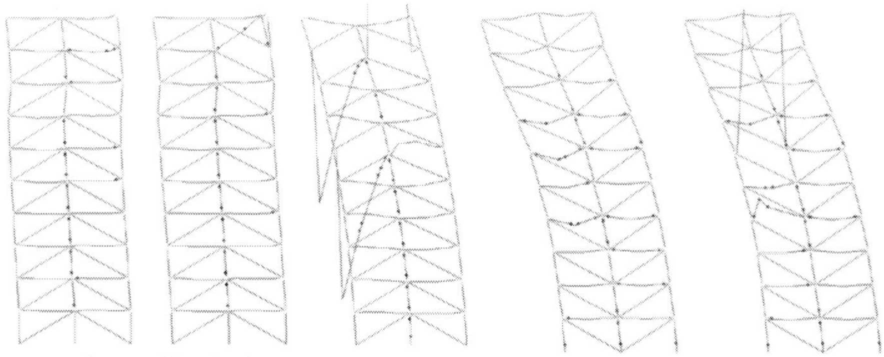

Fig. 6: Unfavourable plastic hinges distributions in CBF during dynamic nonlinear analyses

In order to size a favourable global plastic failure mechanism for a concentrically braced frame potentially plastic zones are provided along the girders [4]. A configuration with reduced beam flanges sections (resembling the dog-bone detail) near the connections with the braces can be adopted for the girders of concentrically braced frames, as indicated in Fig. 7.

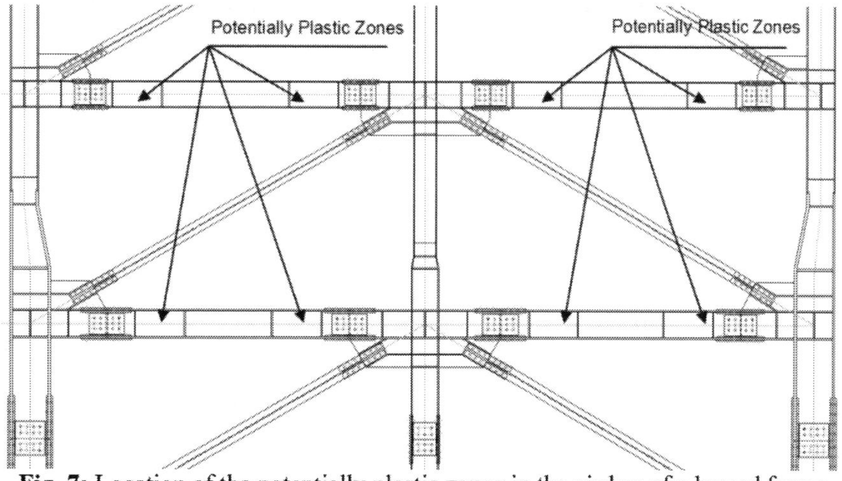

Fig. 7: Location of the potentially plastic zones in the girders of a braced frame

So for concentrically braced frames accepting plastic deformations in the diagonals, at the bottom of first-story columns and in the potentially plastic zones located along the frame girders (near the connections with the braces) can ensure a global plastic failure mechanism. Also a better control of the positions of plastic hinges along the girders of the concentrically braced frames is provided [4]. The same concept can be used for designing a favourable global plastic mechanism for frames equipped with buckling restrained braces.

5. Details with Reduced Cross-sections for First-Story Columns

When a braced steel frame is subjected to strong earthquakes, inelastic deformations can appear also near the bottom of the first-story columns. Several structural details were analyzed for the potentially plastic zones located near the bottom end of the columns considering: reduced flanges cross-sections and/or transverse and longitudinal stiffeners for the columns bottom zone (as shown in Figs. 8 and 9).

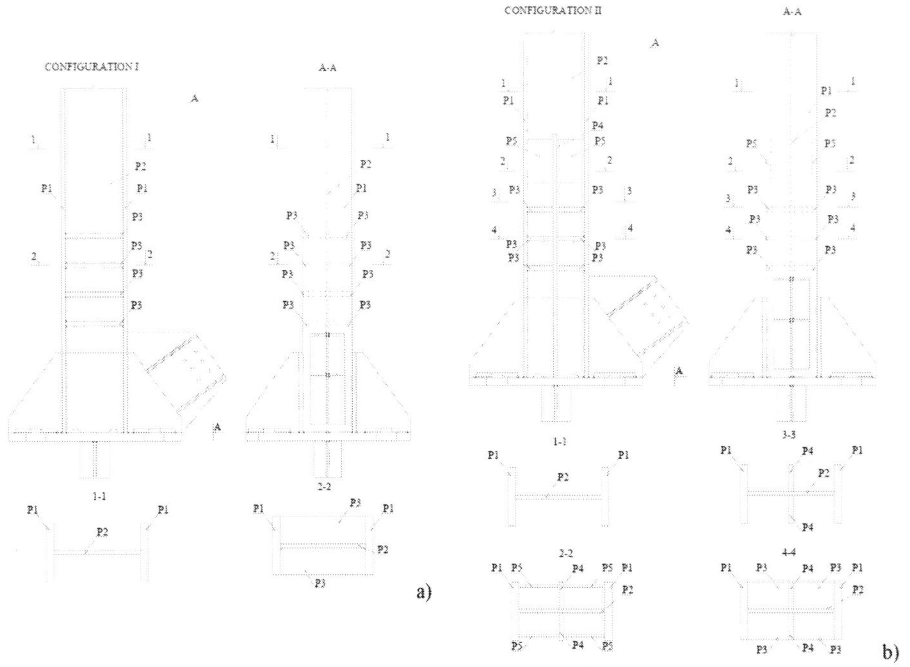

Fig. 8: a) Configuration I; b) Configuration II

Configuration 1 is the reference analysis detail and is shown in Fig. 8.a. The column has the same cross-section on the entire height of the first-story column. Transverse web stiffeners (P3) were used to avoid early local buckling in the potentially plastic zone.

Compared to configuration 1, in the second considered detail an additional pair of longitudinal stiffeners (P4) was placed on the web (see Fig. 8.b). These longitudinal stiffeners (P4) were provided in order to reduce the axial loading level in the flanges, to make room for stresses generated by bending moment. The pair of stiffeners (P4) is placed in the middle of the web in order to not increase significant the flexural capacity of the potentially plastic zone cross-section. Plates P5 were used to facilitate the axial load transfer from the column flanges (P1) to the longitudinal web stiffeners (P4) reducing at the same time load concentration effects. Transverse web stiffeners (P3) are kept in all configurations to reduce the risk of local buckling in the potentially plastic zone (see Figs. 8 and 9).

The third configuration has a reduced flange cross-section in the potentially plastic zone (resembling the dog-bone detail) as shown in Fig. 9.a. The reduced flanges cross-sections ensure a better control of the position of plastic hinges along the first-story column. If the same cross-section is used along the whole member, the positions of the plastic hinges cannot be predicted exactly. The longitudinal web stiffener is kept to assure quite the same axial capacity all along the first-story column height. In configuration 3 the reduced width of the flanges in the potentially plastic zone is about 25% smaller than the flanges width in the rest of the first-story column.

In the fourth considered detail the first configuration column cross-section was kept for the potentially plastic zone, whilst the rest of the column has larger flanges cross-sections in order to increase the buckling capacity of the first-story column (see Fig. 9.b). The reduced flange width in the potentially plastic zone is also about 25% smaller than in the rest of the first-story column.

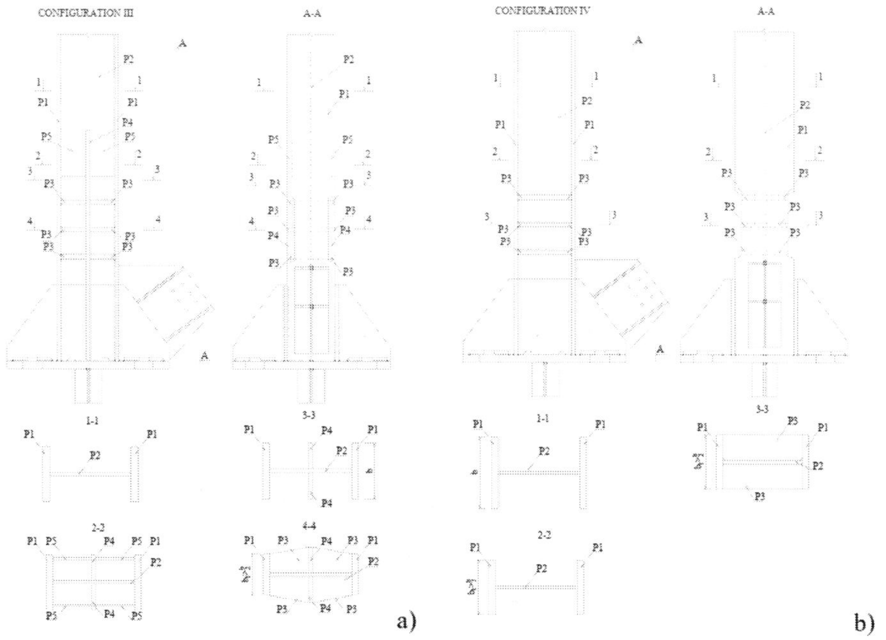

Fig. 9: a) Configuration III; b) Configuration IV

A ten-story eccentrically braced frame with short dissipative members of 1.2m is considered. The frame has two spans of 6.6m and a story height of 3.5m (as shown in Fig. 10.a).

Compared to the other considered construction details, configuration 3 leads to smaller bending moments and greater plastic hinges rotations at the bottom of the columns (see Fig. 10.b). The smaller bending moment values lead to smaller anchor bolts for the columns. The maximum bending moments recorded during the dynamic nonlinear analyses [3] at the bottom of first-story columns in the other considered configurations are nearly the same. Compared to these values, the bending moments registered for configuration 3 are about 20÷26% smaller.

The greatest plastic hinge rotations in the potentially plastic zones of the first-story columns were recorded during dynamic nonlinear analyses [5] for configuration 3. The values of the maximum plastic hinge rotations at the bottom of the columns for the other considered configurations were in the same range. Compared to these values, the plastic rotations for configuration 3 were about 37÷45% greater (see Fig. 10.b).

The buckling resistance of first-story columns was evaluated using relations (6.61), (6.62) and annex B from EN 1993-1-1:2005 [6]. The general stability was verified taking into consideration the most unfavourable loading situations that were recorded in the first-story columns [5] (pairs of bending moment values at the ends of the first-story columns and compressive axial forces that were recorded in the verified element at the same time during the dynamic nonlinear analysis). Generally these unfavourable loading situations occurred when plastic hinges were developed near the bottom of first-story columns [5].

It can be observed from the graphics in Fig. 10.c, that configuration 4 provides the greatest buckling resistance for the situations when inelastic deformations appear in the potentially plastic zones at the bottom of the columns. The general stability criterion is not satisfied in many situations (up to 6%) especially for the marginal first-story columns in case of configuration 1 and 2. Configuration 4 appears to be the safest from the point of view of assuring the general stability of the first-story column in the situation when plastic deformations occur in the potentially plastic zone at the bottom of the column.

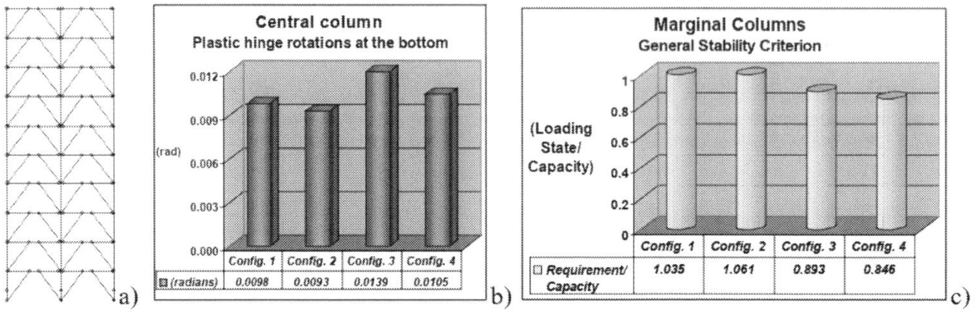

Fig. 10: a) Analyzed eccentrically braced frame; b) Maximum plastic hinge rotations; c) General stability check

6. Conclusions

Constructive details with reduced cross-sections can be used in several situations in order to improve the behaviour of seismic resistant structures during strong earthquakes:

1. in case of eccentrically braced frames with flexural links, a dog-bone configuration at the link ends ensures significant lateral stiffness for the eccentrically braced frame and a better control of the inelastic deformations along the dissipative members;
2. details with reduced cross-sections near the ends of the upper storeys diagonals in concentrically braced frames lead to an adequate slenderness for the braces and to a proper tensile capacity of the diagonals, with small overstrength ratio values;
3. providing configurations with reduced cross-sections for potentially plastic zones along the frame girders, a favourable global plastic hinge mechanism can be sized by design for concentrically braced frames;
4. configurations with "dog-bone" details near the base of bottom-story columns appear to be safer from the point of view of ensuring the general stability of first-story columns in the situations when plastic deformations occur in the bottom story columns.

References

[1] Eurocode 8, EN 1998-1:2004, *Design of structures for earthquake resistance, Part 1: General rules, seismic actions and rules for buildings*, CEN, 2004.

[2] Ministry of Transportations, Public Works and Territory Planning, 2013, *Code for aseismic design - Part I - Design prescriptions for buildings, P100-1/2013*, Bucharest, Romania, 2013.

[3] *Seismic Provisions for Structural Steel Buildings*, American Institute of Steel Construction AISC - 2010.

[4] Köber H, Beţea S. "An Alternative Method for the Design of Centrically Braced Frames", *Proc. of the 5th European Conference on Steel and Composite Structures - Eurosteel 2008*, 03-05 September, Graz, Austria, Vol. B, 1425-1430, 2008.

[5] Tsai KC, Li JW. *Drain2D+ A General Purpose Computer program for Static and Dynamic Analyses of Inelastic 2D Structures*, 1994.

[6] Eurocode 3, *EN 1993-1-1:2005, Design of steel structures - Part 1-1: General rules and rules for buildings*, CEN, 2005.

ECCENTRICALLY BRACED FRAMES EQUIPPED WITH VERTICAL TRUSS ELEMENTS

Marina Stoian and Helmuth Köber

Technical University of Civil Engineering Bucharest, Romania

Abstract: The present paper is intended to relieve the behaviour under severe seismic actions of some eccentrically bracing systems equipped with additional vertical connection elements (vertical truss elements between the ends of all dissipative members that are not connected directly to columns) compared to the seismic behaviour of traditional eccentrically bracings. Each considered structural type (with and without additional vertical truss elements) was subjected to three dynamic nonlinear analyses, using Vrancea earthquakes acceleration records. The values of different parameters recorded during nonlinear analyses were compared.

1. Introduction

The main objective of the present work is to compare the behaviour under seismic actions of two constructive systems: traditional eccentrically braced frames and eccentrically braced frames equipped with additional double-hinged vertical connection elements between the ends of the dissipative members from all storeys. Except for the bottom story, these additional truss elements are connecting the ends of links that are not fixed directly to the columns.

Three eccentrically bracing systems were considered with and without additional truss elements, as indicated in Fig. 1. The analyzed frames had ten storeys of 3.5m and two spans of 6.6m. All the frames (with and without vertical truss elements) had short links with a length of 1.2m. In case of frame K the dissipative members are located in the central part of the frame girders, between to trusses. In case of frame DC and frame DM the links are placed between a brace and a column (for frame DC the links are located on both sides of the central column, whilst for frame DM the dissipative elements are placed near the lateral columns).

The six analyzed frames were sized for the forces produced by the same horizontal seismic load, evaluated according to the Romanian code P100-1/2013 [1] and the provisions of the European standard Eurocode 8, EN 1998-1:2004 [2]. All kind of structural elements (columns, diagonals, vertical trusses, dissipative members and adjacent beam segments) had built-up I-shaped cross-sections, designed according to Eurocode 3, EN1993-1-1:2005 [3].

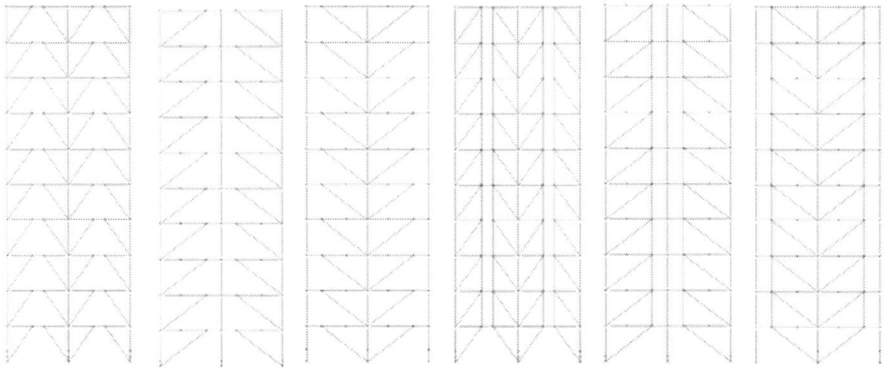

Fig. 1: Analyzed frames

2. Behaviour During Dynamic Nonlinear Analyses

Each considered structural type (with and without additional vertical truss elements) was subjected to dynamic nonlinear analyses using Vrancea earthquakes acceleration records [4]. The N-S components of the acceleration records of Vrancea earthquakes from 1977, 1986 and 1990 (recorded at the INCERC station in Bucharest) were considered. All accelerations records were calibrated for a peak ground value of about 0.3 times the acceleration of gravity. Rayleigh damping was considered [4, 5, 6].

2.1 Extreme Values for Base Shear Forces and Horizontal Floor Displacements

Greater base shear forces could be noticed for the frames with additional trusses compared to the traditional eccentrically braced frames. The maximum differences between the values of the base shear forces were up to 18% (see Fig. 2). These can be explained by the greater lateral stiffness of the eccentrically braced frames equipped with vertical truss elements.

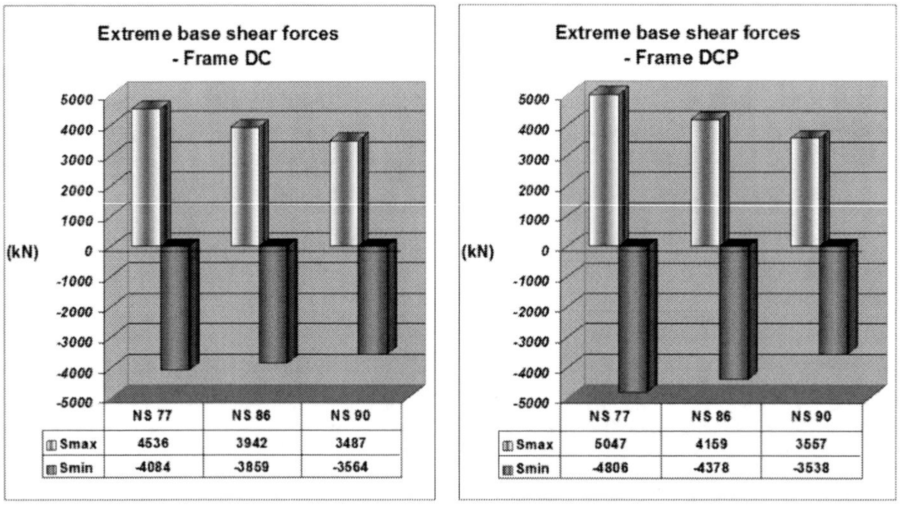

Fig. 2: Extreme values of the base shear forces recorded for frame DC and frame DCP

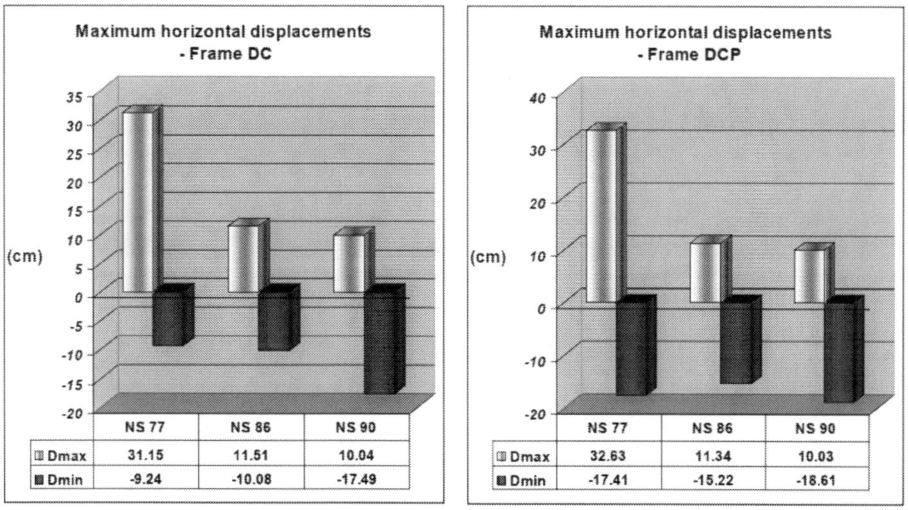

Fig. 3: Extreme values of the lateral floor displacements noticed for frame DC and frame DCP

Although a smaller lateral stiffness was noticed for the traditionally eccentrically braced frames (up to 7%), greater horizontal displacements could be observed during dynamic non-linear analyses (up to 5%) for the eccentrically braced frames equipped with additional vertical trusses (see Fig. 3). The values of the maximum positive and negative horizontal displacements are more balanced for the frames with vertical truss-elements, especially in case of the acceleration record of the Vrancea earthquake from 1977.

2.2 Plastic deformations in the dissipative members

All considered structures had a favourable, predictable behaviour during dynamic nonlinear analyses. Inelastic deformations could be noticed only in the dissipative members and in the potentially plastic zones located near the bottom of first story columns.

For the frames without vertical trusses, the distribution of plastic deformations in the dissipative members along the height of the building is quite different from a dynamic analysis to another (see Fig. 4 and 5). For the frames equipped with additional truss elements, more uniform distributions of inelastic deformations in the links along the height of the frames could be noticed in case of all considered acceleration records (as shown in Fig. 4 and 6).

For example for frame K (see Fig. 4), maximum inelastic deformations could be observed in the dissipative members of story 3 and 4 (in case of the dynamic nonlinear analysis using the acceleration record of the Vrancea 1977 earthquake), in the links of story 6 and 7 (for the Vrancea 1986 earthquake) and in the dissipative members of story 5 (in case of the Vrancea 1990 earthquake). In case of frame KP, as shown in Fig. 4, the maximum plastic deformations in the dissipative members were noticed in the storeys 7, 8 and 9 for the Vrancea 1977 and Vrancea 1990 earthquakes, whilst in case of the Vrancea 1977 earthquake the maximum deformations were observed in the links from the last two storeys. Compared to frame K a much more uniform distribution of the plastic deformations along the height of the structures could be observed for frame KP (compare the values in the graphics of Fig. 4).

For frame DC the greatest inelastic deformations were recorded as shown in Fig. 5: in the links of story 4 and 5 (Vrancea 1977), in the dissipative members of story 8 and 9 (Vrancea 1986) and in the links of story 7 (Vrancea 1990). For frame DM the greatest plastic deformations could be noticed: in the dissipative members of story 3 and 4 (Vrancea 1977), in the links of story 3 (Vrancea 1986) and in the dissipative members of story 6 (Vrancea 1990). It can be observed that the values of the plastic deformations are quite different along the height of the frames without additional truss-elements.

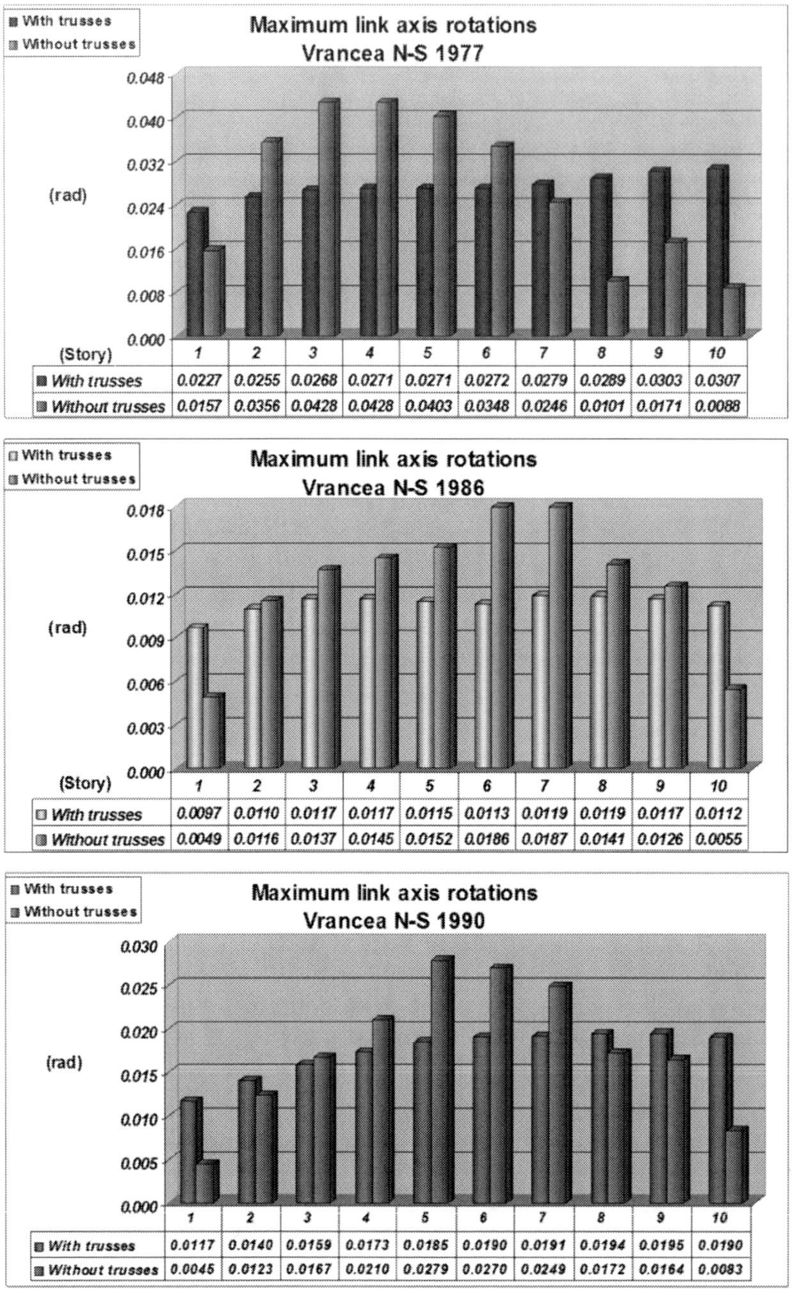

(Story)	1	2	3	4	5	6	7	8	9	10
With trusses	0.0227	0.0255	0.0268	0.0271	0.0271	0.0272	0.0279	0.0289	0.0303	0.0307
Without trusses	0.0157	0.0356	0.0428	0.0428	0.0403	0.0348	0.0246	0.0101	0.0171	0.0088

(Story)	1	2	3	4	5	6	7	8	9	10
With trusses	0.0097	0.0110	0.0117	0.0117	0.0115	0.0113	0.0119	0.0119	0.0117	0.0112
Without trusses	0.0049	0.0116	0.0137	0.0145	0.0152	0.0186	0.0187	0.0141	0.0126	0.0055

	1	2	3	4	5	6	7	8	9	10
With trusses	0.0117	0.0140	0.0159	0.0173	0.0185	0.0190	0.0191	0.0194	0.0195	0.0190
Without trusses	0.0045	0.0123	0.0167	0.0210	0.0279	0.0270	0.0249	0.0172	0.0164	0.0083

Fig. 4: Maximum plastic deformations in the dissipative members of frame K and KP

For the frames equipped with vertical trusses, the maximum observed inelastic deformations are 25÷35% smaller compared to the one noticed for the traditionally eccentrically braced frames. In most cases the maximum plastic link deformations could be noticed in the upper storeys of the frames with additional truss elements. The situation with smaller inelastic deformations in the upper two storeys of the traditionally eccentrically braced frames is avoided (compare the graphics in Fig. 5 and 6).

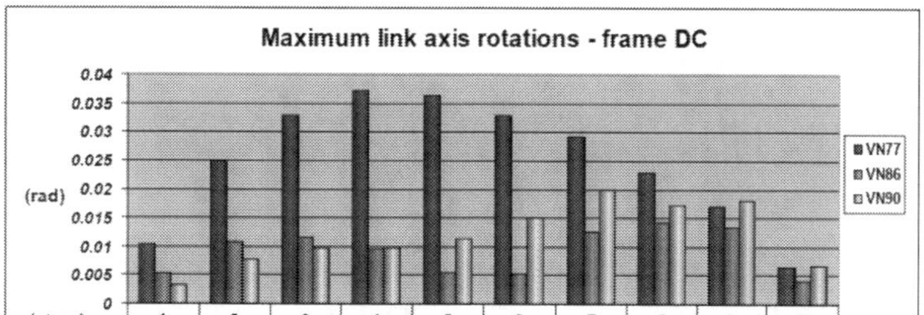

(story)	1	2	3	4	5	6	7	8	9	10
VN77	0.01044	0.02473	0.03293	0.03712	0.03629	0.0328	0.02914	0.02295	0.01699	0.00644
VN86	0.00524	0.01079	0.01157	0.00945	0.00545	0.00518	0.01255	0.01419	0.01331	0.00413
VN90	0.00334	0.00764	0.00971	0.00982	0.01142	0.01507	0.01984	0.01729	0.01814	0.00672

Fig. 5: Maximum plastic deformations in the dissipative members of frame DC

For all the eccentrically braced frames with additional trusses, a more uniform distribution of the plastic deformations along the height of the structures could be observed. The plastic deformation demand of all dissipative members is in the same range and so the links placed at different storeys, participate in a comparable manner at the dissipation of energy through plastic deformations.

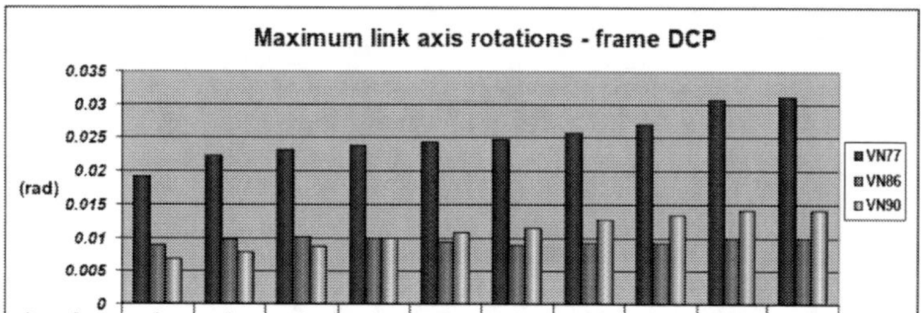

(story)	1	2	3	4	5	6	7	8	9	10
VN77	0.01923	0.02222	0.02321	0.02394	0.02446	0.02474	0.0258	0.02691	0.03075	0.03111
VN86	0.00896	0.00982	0.01012	0.00993	0.00945	0.00884	0.00917	0.00917	0.00992	0.00989
VN90	0.00682	0.00782	0.00878	0.00985	0.01084	0.01156	0.01264	0.01334	0.01411	0.01414

Fig. 6: Maximum plastic deformations in the dissipative members of frame DCP

2.3 Maximum axial forces and bending moments during dynamic nonlinear analyses

For the adjacent beam segments, compared to traditional eccentrically bracings (see Fig. 7), providing additional truss elements led to greater axial forces (about 34% for frame DCP and up to 26% for frame DMP) and to smaller values for the bending moments (up to 27% in case of frame DCP and about 21% for frame DMP). Compared to frame K, in case of frame KP, an average increase of the axial forces with about 28% combined with an average decrease of the bending moment with about 25% could be observed for the beam segments placed outside the dissipative members.

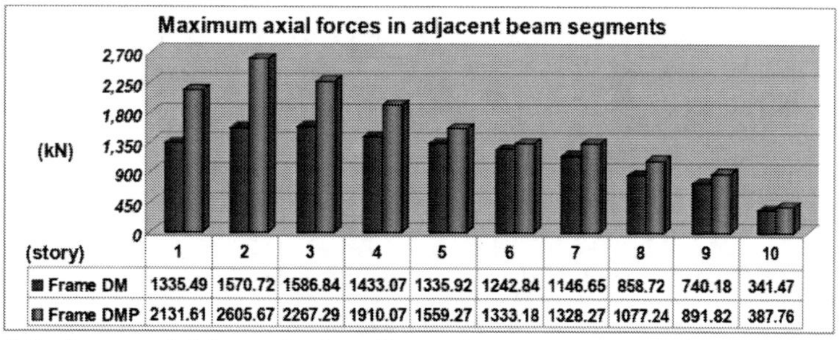

(story)	1	2	3	4	5	6	7	8	9	10
Frame DM	1335.49	1570.72	1586.84	1433.07	1335.92	1242.84	1146.65	858.72	740.18	341.47
Frame DMP	2131.61	2605.67	2267.29	1910.07	1559.27	1333.18	1328.27	1077.24	891.82	387.76

Fig. 7: Maximum axial forces in the adjacent beam segments of frame DM and DMP

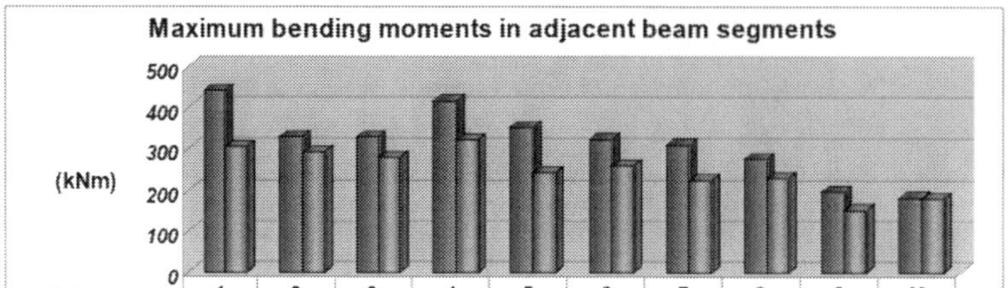

(story)	1	2	3	4	5	6	7	8	9	10
Frame DC	440.92	329.93	328.96	417.07	353.27	324.84	310.73	275.84	197.57	182.25
Frame DCP	307.02	292.37	278.63	323.45	244.07	259.83	224.83	229.96	152.78	179.71

Fig. 8: Maximum bending moments in the adjacent beam segments of frame DC and DCP

Greater axial forces were noticed in the braces of the frames with additional trusses (see Fig. 10). An average increase up to 35% was observed for frame DMP, up to 32% in case of frame DCP and an increase of about 17% for frame KP (compared to the frames without vertical trusses). In addition, greater bending moments were recorded in the diagonals of the frames equipped with vertical trusses (see Fig. 9): about 25% greater in case of frame DMP, respectively about 6% greater in case of frame KP. In case of frame DCP and DC the bending moment values in the diagonals were quite the same, except for the first story. For the bottom story braces, a big increase of the values of bending moments and axial forces (sometimes greater than 100%) could be noticed for all analyzed bracing systems. In the absence of the vertical truss-elements at the first story, the bottom story diagonals had to balance the greatest part of the axial forces from the trusses from the story above. At other storeys (excepting the bottom story), the axial force from a vertical truss element from a story above is balanced mainly by the axial force in the truss from the story below.

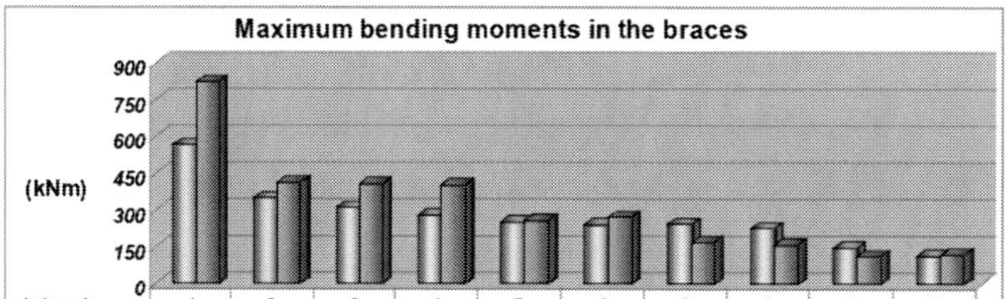

(story)	1	2	3	4	5	6	7	8	9	10
Frame DM	565.84	350.91	310.65	280.65	253.54	242.01	244.63	227.73	145.93	114.22
Frame DMP	821.32	415.28	407.02	403.06	260.18	273.18	169.70	157.34	112.15	121.02

Fig. 9: Maximum bending moments in the braces of frame DM and DMP

The greater axial forces and bending moment values led to greater cross-sections for the diagonals of the frames with additional truss-elements, compared to one without vertical trusses.

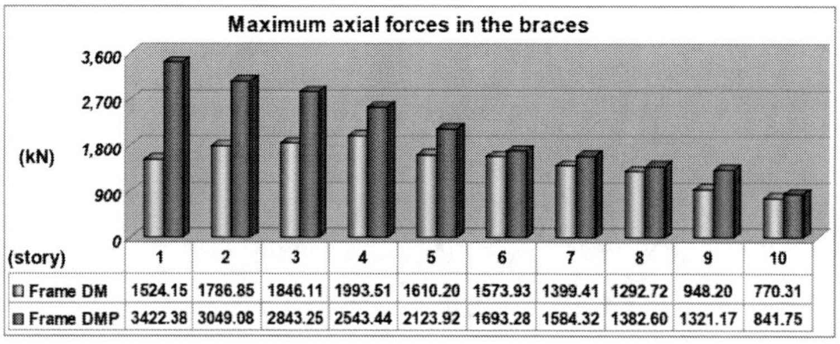

(story)	1	2	3	4	5	6	7	8	9	10
Frame DM	1524.15	1786.85	1846.11	1993.51	1610.20	1573.93	1399.41	1292.72	948.20	770.31
Frame DMP	3422.38	3049.08	2843.25	2543.44	2123.92	1693.28	1584.32	1382.60	1321.17	841.75

Fig. 10: Maximum axial forces in the diagonals of frame DM and DMP

Greater axial forces were noticed in the marginal columns of the frames equipped with vertical trusses, compared to the ones without trusses. The average axial force increase was between 14% in case of frame DCP and 8% in case of frame KP. Except for the first story, smaller bending moments were recorded in the frames with additional trusses. The average bending moment reduction was about 32% for frame DCP, about 26% in case of frame DMP and over 17% for frame KP. At the bottom story of the frames equipped with additional trusses, the absence of vertical truss-elements, led to a big increase of the bending moment values in the columns (sometimes greater than 75%) for all considered bracing systems.

Except for the first story, smaller bending moments were noticed in the central columns of the frames with vertical truss-elements, compared to the frames without trusses. The average axial force reduction was about 24% in case of frame KP, 23% for frame DMP and about 16% in case of frame DCP. In addition, for frame DCP and frame KP smaller axial forces were recorded, compared to traditional eccentrically bracings (about 4% in case of frame DCP and about 3% for frame KP). The axial forces in the columns of frame DCP and frame DC were in the same range.

Providing additional vertical trusses in eccentrically braced frames increases the axial forces in most structural elements. The frames equipped with vertical trusses start to behave more like a vertical truss-girder, which explains the smaller recorded bending moment values observed in most structural elements (except for the diagonals).

3. Estimated Steel Consumption

The estimated steel consumption for the two considered structural types is indicated in Fig. 11. The same design value for the seismic force was used for the six analyzed frames (the dissipative elements of all the analyzed frames were sized for the forces produced by the same design seismic force).

In case of the frames equipped with vertical truss elements, smaller steel consumptions were noticed for the lateral and central columns, whilst greater consumption values were obtained for the diagonals. Almost the same material consumption values could be noticed for the frame girders for both frame types.

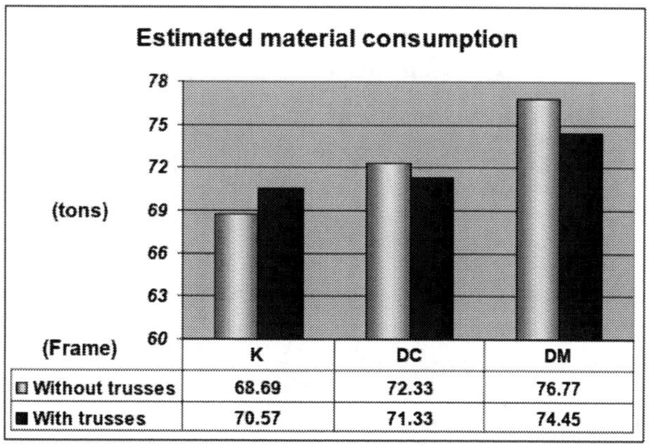

Fig. 11: Estimated material consumption for the six analyzed frames

The constructive solution with additional vertical truss elements led to a smaller material consumption in case of the DC- and DM-bracing systems (with about 1.7% and respectively 3.5%). For these two bracing configurations only one end of the dissipative members is connected to a brace.

For the K- bracing system, with the dissipative members placed between two braces in the central part of the frame girders, traditional eccentrically bracing proved to be more economic (with about 2.7%). This fact can be explained by the greater number of diagonals and additional truss elements used in case of the K-bracing system compared to the DC- or DM-bracing configurations.

4. Conclusions

The main differences in the behaviour under seismic actions of the two analyzed constructive systems (traditional eccentrically braced frames and eccentrically braced frames equipped with additional double-hinged vertical connection elements between the ends of the dissipative members from all storeys) are:

1. the main advantage of the eccentrically braced frames with vertical truss elements consists in the more uniform distribution of inelastic deformations along the height of the structure; the plastic deformation demand of all dissipative members is in the same range and so the dissipative members placed at different storeys, participate in a comparable manner at the dissipation of energy through plastic deformations;
2. the main disadvantage of providing additional vertical trusses in eccentrically braced frames consists in the difficulty of the emplacement of door and window openings in the braced bays and spans;
3. the additional truss elements conduct to a greater lateral stiffness compared to traditional eccentrically bracing;
4. for the frames with vertical trusses, generally greater axial forces were recorded in all kind of structural elements and smaller bending moments could be noticed in the columns and beam segments placed outside the dissipative members;
5. providing additional truss elements on the whole height of the building except for the first story, increases the values of the axial forces and bending moments in the columns and braces of the bottom story;
6. the values of the estimated steel consumption are quite the same for the eccentrically braced frames equipped or not with vertical truss elements.

References

[1] Ministry of Transportations, Public Works and Territory Planning, *Code for aseismic design - Part I - Design prescriptions for buildings - P100-1/2013*, Bucharest, Romania, 2013.

[2] Eurocode 8, *EN 1998-1:2004, Design of structures for earthquake resistance, Part 1: General rules, seismic actions and rules for buildings*, CEN, 2004.

[3] Eurocode 3, *EN 1993-1-1:2005, Design of steel structures - Part 1-1: General rules and rules for buildings*, CEN, 2005.

[4] Tsai KC, Li JW. *Drain2D+ A General Purpose Computer program for Static and Dynamic Analyses of Inelastic 2D Structures*, 1994.

[5] Chopra AK. *Dynamics of Structures. Theory and Applications to Earthquakes Engineering* - 3rd Edition, 2004.

[6] *Seismic Provisions for Structural Steel Buildings*, American Institute of Steel Construction AISC - 2010.

RELATION BETWEEN PLASTIC DEFORMATION CAPACITY AND CUMULATIVE PLASTIC DEFORMATION CAPACITY FOR H-SHAPED BEAMS WITH LOCAL BUCKLING

Atsushi Suzuki[a] and Yoshihiro Kimura[b]

[a] *Graduate School of International Cultural Studies, Tohoku University, Japan*
[b] *New Industry Creation Hatchery Center, Tohoku University, Japan*

Abstract: In general, the seismic capacity of H-shaped beams with local buckling is estimated by two main structural factors. One of them is the plastic deformation capacity in the skeleton curve, and the other is the cumulative hysteresis energy in cumulative hysteresis loops. Their correlation has not been shown even though each one has been individually used as the seismic capacity of the beams. In this paper, the correlation between the cumulative hysteresis energy and the plastic deformation capacity of H-shaped beams are experimentally investigated in 11 test series and empirical equations based on the precious experimental results are suggested.

1. Introduction

Structural capacities of H-shaped beam are evaluated as the plastic deformation capacity in the monotonic loading case such as push over analyses [1], or hysteretic energy absorption and cumulative plastic deformation capacity in the cyclic loading case such as seismic response analyses [2] in design procedures. However, the relationships between the loading hysteresis and the plastic deformation capacity, cumulative plastic deformation capacity of H-shaped beams under the local buckling have not been clarified yet.

In addition, the hysteretic energy absorption is converted from the cumulative plastic deformation capacity through multiplying Bauschinger effect coefficient to explain the extra energy absorption upon the cumulative plastic deformation capacity [1]. However, it is still ambiguous about the connection between the hysteretic energy absorption during the whole loading history and the cumulative plastic deformation capacity obtained from the skeleton curve.

This paper reports the experimental results conducted on the H-shaped beams which possess several flange width-thickness ratios and shear span ratios under monotonic and cyclic loading with different loading histories. Based on the obtained results from the experiments, the plastic deformation capacity, cumulative plastic deformation capacity and Bauschinger effect coefficient at the ultimate strength are collected. Furthermore, this paper clarifies the effect of difference of loading history on the plastic deformation capacity and cumulative plastic deformation capacity at ultimate strength with referring the results of experiment carried out in this study and previous experimental data [3-23]. In addition, an evaluation formula to estimate Bauschinger effect coefficient is proposed with employing the loading amplitude and number of cycles as parameters for the cyclic loading case.

2. Cyclic loading test on H-shaped beam with several flange width-thickness ratios, shear-span ratios, and loading histories

2.1 Summary of specimens

An H-shaped specimen is illustrated in Fig. 1. A loading beam is welded with a test beam by cover plates on upper and lower flange. The loading beam keeps elastic when specimens carry the ultimate strength. The experimental series is tabulated in Table 1, and the specimen designation is illustrated above Table 1. The material of steel is SN400. The web height is 200mm, web thickness is 8mm, and flange thickness is 12mm. Table 2 lists the measured dimensions of each specimen. The experimental parameters are the loading history (monotonic or cyclic), width-thickness ratio (b/t_f= 4.17, 6.25, and 8.33), and shear span ratio (L/H = 4, 5, and 6).

Fig. 2 illustrates the loading protocol. Here, δ_p denotes the displacement at which specimens achieve the full plastic strength, and is calculated from Eq. (1). As given in Eq. (1), δ_p consists of the displacement caused by the bending moment δ_{mp}, and shear force δ_{sp}.

$$\delta_p = \delta_{mp} + \delta_{sp},\ \delta_{mp} = \frac{P_p L^3}{3EI},\ \delta_{sp} = \frac{\kappa P_p L}{AG} \tag{1}$$

where P_p is the shear force at full plastic strength of beam, L is beam length, E is Young's modulus, I is geometrical moment of inertia, κ is shape factor, A is cross-sectional area, and G is shear stiffness modulus.

The results of material test are tabulated in Table 3. Here, σ_y is the yield stress, and σ_u is the ultimate stress.

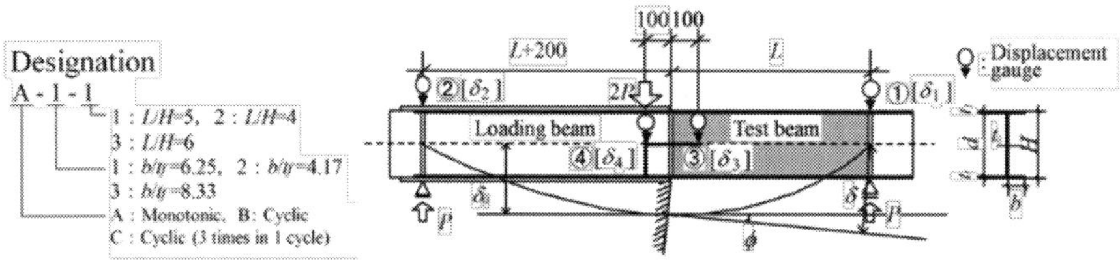

Fig. 1: Specimen

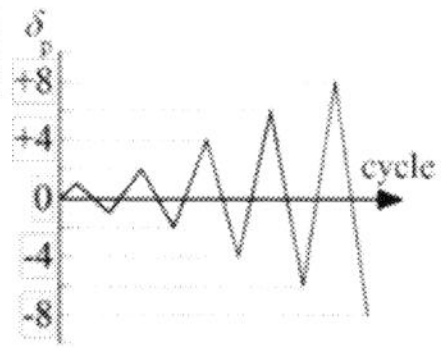

Fig. 2: Loading protocol

Table 1: List of specimens

Specimen	L [mm]	b/t_f [-]	d/t_w [-]	Slenderness ratio	Loading history
A-1-1	1000	6.25	22	24.55	Monotonic
B-1-1					Cyclic
C-1-1					Cyclic (3times)
A-2-1		4.17		37.88	Monotonic
B-2-1					Cyclic
A-3-1		8.33		18.15	Monotonic
B-3-1					Cyclic
A-1-2	800	6.25		19.64	Monotonic
B-1-2					Cyclic
A-1-3	1200	6.25		29.46	Monotonic
B-1-3					Cyclic

Table 2: Measured scales

Specimen	H [mm]	B [mm]	t_f [mm]	t_w [mm]
A-1-1	202	150	11.9	8.2
B-1-1	202	150	12.1	8.1
C-1-1	201	150	12	8.25
A-2-1	201	101	12	8.25
B-2-1	202	100	12.1	8.25
A-3-1	201	200	11.8	8.2
B-3-1	201	201	11.8	8.2
A-1-2	202	150	12	8.25
B-1-2	202	150	11.9	8.25
A-1-3	201	150	11.9	8.35
B-1-3	201	151	12	8.25

2.2 Plastic deformation properties of H-shaped beams under different loading histories

Fig. 3 shows the result of the cyclic loading test conducted on C-1-1. The vertical axis is the normalized shear force P/P_p. Here, P denoted the applied shear force, and P_p denotes the shear force which leads the specimens to the full plastic condition. The horizontal axis is the normalized displacement δ/δ_p. Here, δ is the displacement of beam (*cf.* Fig. 1), and δ_p is the value obtained from Eq. (1). ∇ indicates the ultimate loading.

Fig. 4 gives the cumulative hysteresis curve obtained from Fig. 3. $\alpha_B \eta_{max}$ is the normalized hysteresis energy absorption. $\alpha_B \eta_{max}$ can be calculated by subtracting the elastic energy absorption from the total energy absorption until the specimen achieves the ultimate strength in accumulative hysteresis curve. α_B is the Bauschinger effect coefficient.

Fig. 5 gives the skeleton curve obtained from Fig. 3. The skeleton curve can be drawn by connecting bold lines (skeleton part) indicated in Fig. 3. The plastic deformation capacity μ_{max} is obtained by subtracting the elastic normalized displacement from the normalized displacement at the ultimate strength in the skeleton curve. The cumulative plastic deformation capacity η_{max} is obtained by subtracting the elastic energy absorption from the total energy absorption until the specimen achieves the ultimate strength in the skeleton curve.

Fig. 6 shows the P/P_p - δ/δ_p relationship of specimens of same section and beam length under the different loading histories. Fig. 6(a) gives the result of the cyclic loading test, and

Table 3: Material properties

Thickness [mm]	σ_y [N/mm^2]	σ_u [N/mm^2]	Elongation [%]
8	323	444	21
12	283	428	24

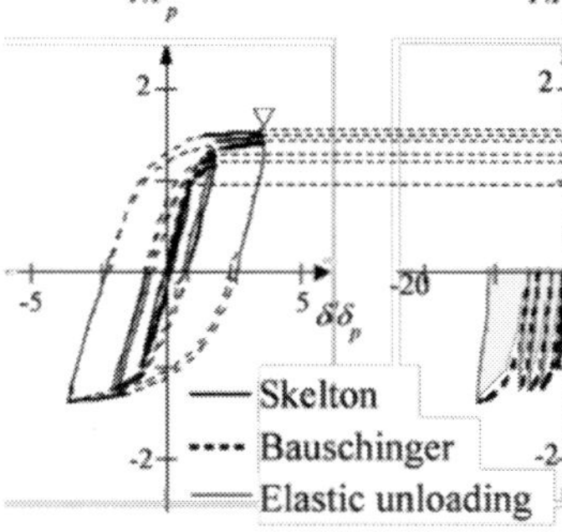

Fig 3: Hysteresis curve

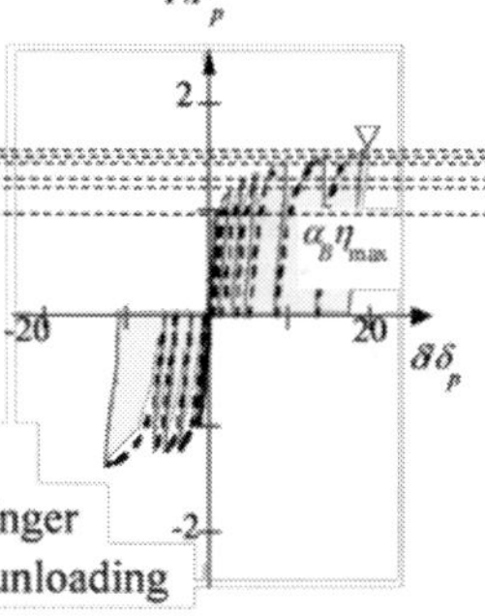

Fig 4: Accumulative hysteresis curve

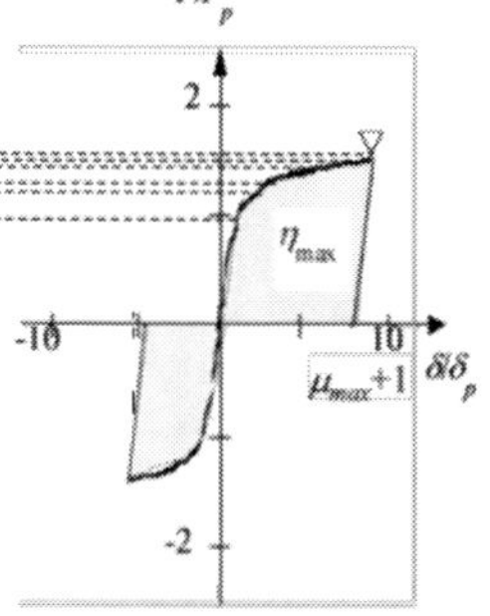

Fig 5: Skeleton curve

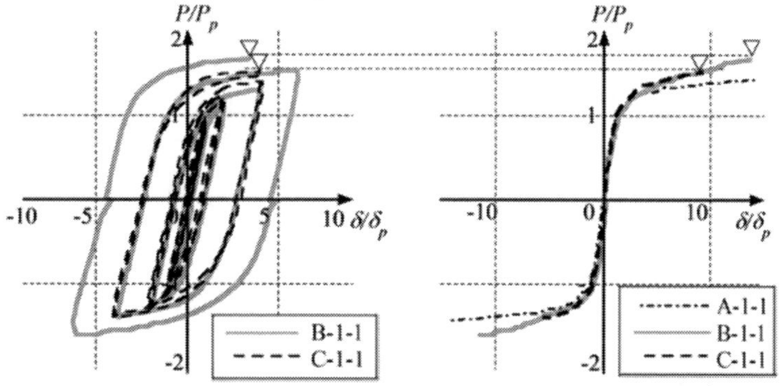

(a) Hysteresis curve (b) Skeleton curve

Fig. 6: Comparison of experimental results between cyclic and monotonic loading

Fig. 6(b) illustrates the comparison between the skeleton curve obtained from Fig. 6(a) and the result of the monotonic loading test. The same hysteresis curve are drawn in both positive and negative fields in the monotonic loading case. Fig. 6(b) confirms that the ultimate strength in cyclic loading is higher than that in the monotonic loading test even for a same beam section and beam length, whereas the plastic deformation capacity in the monotonic loading test is larger than that in the cyclic loading test. This shows that the skeleton curve obtained from the cyclic loading test and monotonic loading test are not equivalent.

2.3 Effect of difference of loading history on structural capacity of H-shaped beam

The results of all specimens are tabulated in Table 4. The contents are the buckling mode, number of cycles when the local buckling occurred, plastic deformation capacity μ_{max}, cumulative plastic deformation capacity η_{max}, and Bauschinger effect coefficient α_B. A-1-2 which is subjected to monotonic loading achieved the ultimate strength due to the local buckling. In contrast, B-1-2 which is subjected to cyclic loading showed a prominent shear deformation, even though it reached the ultimate strength with a local bucking at flange. This confirms that the different loading histories can generate different buckling modes even with the same beam section and beam length.

Fig. 7 shows the comparison of the capacities of H-shaped beams with same section and beam length under the monotonic and cyclic loading cases. Fig. 7(a), (b), (c) explain the normalized ultimate strength, plastic deformation capacity, and cumulative plastic deformation

Table 4: List of experimental results

Specimen	Buckling mode	Cycle at which local buckling is confirmed	μ_{max}	η_{max}	α_B
A-1-1	Local		13.27	15.96	
B-1-1	Local	$+2\delta_p$	12.65	16.16	2.06
C-1-1	Local	$+2\delta_p$(1st loading)	8.17	9.45	2.48
A-2-1	Combined (Local and Lateral)		8.34	7.96	
B-2-1	Combined (Local and Lateral)	$-4\delta_p$	8.64	13.84	1.63
A-3-1	Local		11.86	13.33	
B-3-1	Local, failure at tensile flange	$+2\delta_p$	5.46	10.41	1.61
A-1-2	Local		20.9	26.59	
B-1-2	Local (prominent shear deformation)	$+6\delta_p$	8.88	17.1	2.02
A-1-3	Local		12.84	15.47	
B-1-3	Local	$+2\delta_p$	7.85	14.09	1.82

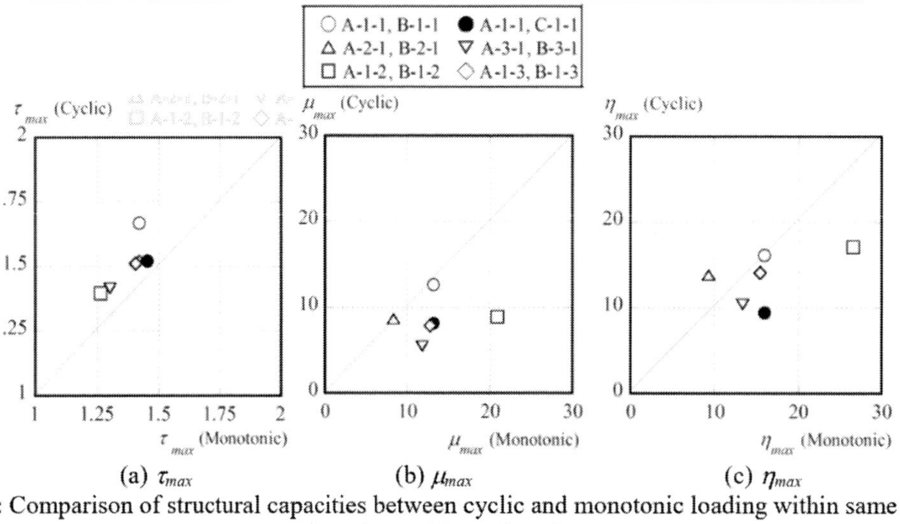

Fig. 7: Comparison of structural capacities between cyclic and monotonic loading within same beam section and beam length

capacity, respectively. The results of cyclic loading tests provide higher normalized ultimate strength than those obtained from monotonic loading tests in all cases, whereas cyclic loading tests give lower plastic deformation capacity and cumulative deformation capacities, except A-2-1 and B-2-1 in which was both local and lateral buckling.

3. Plastic deformation and cumulative deformation properties of H-shaped beams with local buckling under different loading histories

3.1 Ultimate strength, plastic deformation capacity, and cumulative plastic deformation capacity of H-shaped beam with local buckling under different loading histories

The evaluation formulas for assessing the capacities of H-shaped beam are proposed using the database established from the experimental results conducted in this study and previous researches [3-23].

This paper applies the equivalent width-thickness ratio $(b/t)_{eq}$, shear span ratio L/H, number of cycles N, and normalized cyclic amplitude μ_x as parameters with referring previous formulas, which include the flange width-thickness ratio (b/t_f), web width-thickness ratio (d/t_w), slenderness ratio λ_y, and yield strain ε_y as parameters [24].

$$\tau_{max} = \left[1.5 - 0.57(b/t)_{eq} - 0.010 L/H\right]\left[1.1 - 0.1\frac{N}{\sum \mu_{xi}}\right] \tag{2}$$

$$\mu_{max} = \left[-7.2 + \frac{4.5}{(b/t)_{eq}} + \frac{8}{L/H}\right]\left[0.85 + 0.08\frac{\sum \mu_{xi}}{N}\right] - 1 \tag{3}$$

$$\eta_{max} = \frac{(\tau_{max}+1)\,(\mu_{max}+1-\tau_{max})}{2} \tag{4}$$

Here, $(b/t)_{eq}$ can be calculated from Eq. (5) [25].

$$\left(\frac{b}{t_f}\right)_{eq} = \sqrt{\frac{\sigma_{yf}}{E}\left(\frac{b}{t_f}\right)^2 + \frac{\sigma_{yw}}{41E}\left(\frac{d}{t_w}\right)^2} \tag{5}$$

Fig. 8(a) and (b) show the comparison of the plastic deformation capacity and cumulative plastic deformation capacity between the experimental and predicted values, respectively. The hatching field indicates the range where the error rate is less than 30%. The proposed evaluation formulas can estimate the experimental results with 23.1% and 25.2% of error rate in the plastic deformation capacity and cumulative plastic deformation capacity cases, whereas the previous formulas involved larger error rate, 31.3% and 35.5%, respectively.

3.2 Effect of cyclic loading history on Bauschinger effect of H-shaped beam with local buckling

The reference [1] is guiding to set Bauschinger effect coefficient α_B as a constant, 2.0. However, the experimental results obtained in this study and in previous researches imply that α_B varies depending on the number of cycles N and its normalized amplitude μ_x. Therefore, this paper proposes Eq. (6) through trial and error process for assessing α_B with employing both parameters. Here, α_B becomes 1.0 in the monotonic loading case.

$$\alpha_B = 1 + 0.5(N-1)\log\sqrt{\sum_{i=1} \mu_{xi}} \tag{6}$$

where N is the number of cycles, μ_{xi} is the normalized amplitude in i step.

Fig. 9 illustrates the relationship between the second term in Eq. (6) and α_B. The dashed line shows Eq. (6). α_B has a linear relationship with the second term in Eq. (6), and α_B calculated from the proposed formula generally agrees with the experimental results.

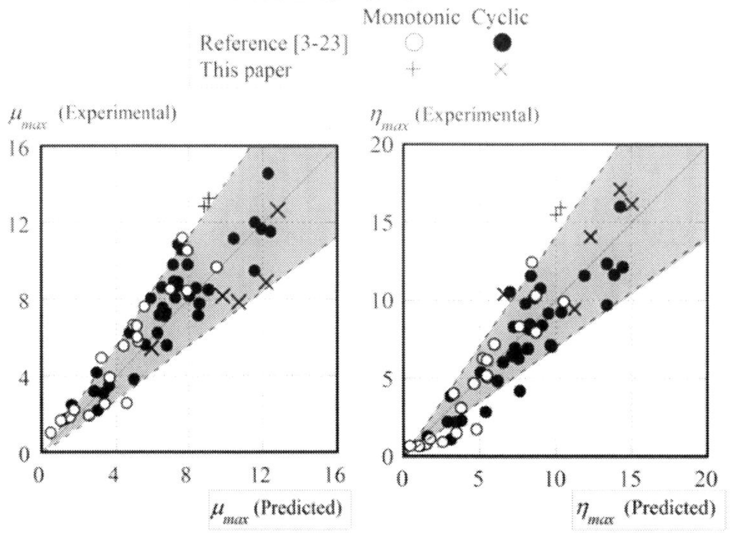

(a) μ_{max} (b) η_{max}

Fig. 8: Comparison between experimental and predicted values

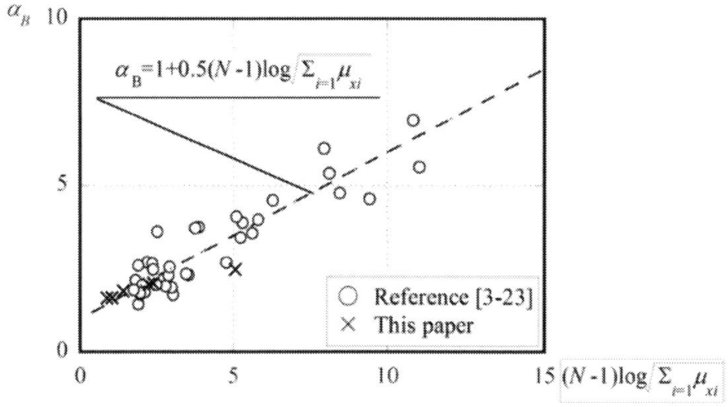

Fig. 9: Relationship between Bauschinger effect coefficient and second term in Eq. (6)

Based on the discussion above, the hysteresis energy absorption can be obtained through multiplying η_{max} obtained from Eq. (4) by α_B calculated from Eq. (6).

4. Conclusion

The obtained findings are summarized as follows:

1) The plastic deformation capacity and cumulative plastic deformation capacity of H-shaped beam are degraded by the larger number of cycles even if they are of the same amplitude in the cyclic loading.
2) Higher plastic deformation capacity and cumulative plastic deformation capacity are achieved under monotonic loading than under cyclic loading for H-shaped beam of the same section and length.

3) The ultimate strength, plastic deformation capacity and cumulative plastic deformation capacity can be estimated by the proposed evaluation formulas, regardless of the loading histories.
4) Bauschinger effect coefficient α_B is enlarged with the increase of number of cycle and amplitude. α_B can be evaluated by Eq. (6), and the hysteresis energy absorption can be assessed as the product of η_{max} and α_B calculated by Eq. (4) and Eq. (6).

References

[1] Architectural Institute of Japan. *Ultimate strength and deformation capacity of building in seismic design (1990)*, 1990.

[2] Architectural Institute of Japan. *Current situations and problems in methods to evaluate ultimate seismic capacity of building*, 2005.

[3] Fukuchi Y, Ogura M. "Experimental studies on local bucklings and hysteretic characteristics of H-shaped beams", *Journal of structural and construction engineering*, 65-71, 1975.2.

[4] Fujikawa T, Fujiwara K. "Research on strength and deformation capacity of H-shaped beam under cyclic loading", *Proceedings of Architectural Institute of Japan Kinki chapter research meeting*, 337-340, 1984.6.

[5] Fujiwara K, Kato S. "A study on deformation capacity of H-shaped steel beams with relative large ratio of web depth to thickness under cyclic loading", *Proceedings of Architectural Institute of Japan Kinki chapter research meeting*, 425-428, 1985.5.

[6] Sakai J, Matsui C, Kuno T. "Effects of the difference of the value of yield ratio of steel and the collapse-types of frame on the ductility of steel frame", *Summaries of technical papers of annual meeting Architectural Institute of Japan*, 1351-1352, 1990.10.

[7] Fujita T, Nakagomi T, Mizusaki Y. "The experimental research concern with structural behaviors of welded beam-to-column connections using cold-formed square steel tube column", *Summaries of technical papers of annual meeting Architectural Institute of Japan*, 1537-1538, 1992.8.

[8] Kai S, Sera K et al. "Structural behavior of H-shaped beam-end connected to RHS-column", *Summaries of technical papers of annual meeting Architectural Institute of Japan*, 1541-1542, 1992.8.

[9] Makishi T, Yamamoto N et al. "Experimental study on the deformation capacity of welded beam end connection without weld access hole", *Summaries of technical papers of annual meeting Architectural Institute of Japan*, 1257-1260, 1993.9.

[10] Ito H, Kimura M et al. "Experimental study on beam-to-column welding connection without beam scallops", *Summaries of technical papers of annual meeting Architectural Institute of Japan*, 1261-1262, 1993.9.

[11] Suzuki T, Motoyuki S et al. "Study on effect of width-thickness ratio on deformation capacity of beams with scallops", *Journal of structural and construction engineering*, 107-114, 1996.8.

[12] Minami K, Fujita T et al. "Experimental study on effect of non-scallops on deformation capacities of beam-to-column welded joints", *Summaries of technical papers of annual meeting Architectural Institute of Japan*, 631-632, 1996.9.

[13] Ishimaru R, Kagami S et al. "Experimental study on the statical characteristics of the WBFW type beam-to-column connections", *Summaries of technical papers of annual meeting Architectural Institute of Japan*, 429-430, 1997.9.

[14] Sawamoto Y, Yoshida H et al. "Experimental study on detail of H-shaped beam-to-box column connection", *Summaries of technical papers of annual meeting Architectural Institute of Japan*, 375-376, 1998.9.

[15] Yasuno T, Tsujioka S, Tada M. "Alternate loading test of haunch-beams made from rolled wide-flange shape", *Summaries of technical papers of annual meeting Architectural Institute of Japan*, 583-586, 2001.9.

[16] Kamaga H, Makino Y et al. "Testing of beam-to-RHS column connections without weld access holes for field welding", *Summaries of technical papers of annual meeting Architectural Institute of Japan*, 919-922, 2001.9.

[17] Nakano T, Masuda H et al. "Experimental study on behavior of WBFW type beam-to-column connection", *Journal of structural and construction engineering*, 139-144, 2002.6.

[18] Nakano T, Masuda H, Tanaka A. "Experimental study on effect of reinforcements at web connected parts of WF beam to SHS column connections", *Journal of structural and construction engineering*, 145-152, 2003.4.

[19] Suzuki T, Ono T, Kanebako Y. "The local buckling and inelastic deformation capacity of steel beams under shear bending", *Journal of structural and construction engineering*, 91-98, 1977.10.

[20] Kato T, Akiyama H, Obi Y. "Experimental study of H-shaped beam with relatively large width-thickness ratio on local buckling characteristics", *Summaries of technical papers of annual meeting Architectural Institute of Japan*, 1075-1076, 1976.10.

[21] Yoda K, Imai K et al. "Bending capacity of thin-walled welding H-section beams", *Journal of structural and construction engineering*, 60-72, 1989.3.

[22] Yoshikawa M, Suzuki T et al. "Investigation on estimation of inelastic behavior of H-shaped beam with circular hole", *Summaries of technical papers of annual meeting Architectural Institute of Japan*, 1473-1474, 1993.9.

[23] Igarashi K, Suzuki T et al. "A study on plastic deformation capacity of H-shaped stiffened beams with web opening", *Summaries of technical papers of annual meeting Architectural Institute of Japan*, 505-506, 2000.9.

[24] Kato B, Akiyama H, Obi Y. "Deformation characteristics of H-shaped steel members influenced by local buckling", *Journal of structural and construction engineering*, 49-57, 1977.7.

[25] Kadono A, Sasaki M et al. "Experimental study on the effect of yield ratio on the bending strength increasing ratio and the ductility of steel structures' members", *Journal of structural engineering*, 673-682, 1994.3.

MODELLING STRATEGY FOR THE SEISMIC RESPONSE OF A MOMENT RESISTING STEEL FRAME BUILDING

Lucia Tirca, Melina Bosco, Nenghui Lin and Lan Lin

Department of Building, Civil and Environmental Engineering, Concordia University, Montreal, Canada

Abstract: Performance-based earthquake engineering (PBEE) requires accurate estimation of the seismic demand and capacity of structures. To tackle this task, incremental dynamic analysis (IDA) is employed herein to investigate the performance of moderately ductile moment resisting frame (MD-MRF) office buildings designed according to the current building code of Canada and steel design standard. For simulating the nonlinear response from yielding to failure, the MD-MRF beams were modelled in the OpenSees software as *beam with finite length hinge* elements. The plastic hinge segment was made of fiber cross-sections. In addition, a proposed low-cycle fatigue model capturing the MD-MRF beams failure is presented.

1. Introduction

Performance-based earthquake engineering (PBEE) requires accurate estimation of the seismic demand and capacity of structures which implies accurate nonlinear finite element models. Hence, the strength and stiffness deterioration corresponding to damage accumulation at the level of plastic hinges developed in beams of moment resisting frames need to be identified. There are several models in the literature which lead to various results. However, the proposed models should be stable, reliable and computationally effective during the time-history analyses which are highly nonlinear.

2. Building Studied

In this study, a 4-storey moderately ductile moment resisting frame (MD-MRF) office building located on Site Class C (firm soil) in Vancouver, B.C. is considered. The plan view of the building, as well as, the magnitude of gravity, snow, and live loads at the roof and typical floor level are given in Fig. 1. The typical storey height *h* is 3.8 m, while the height of ground floor level is 4.5 m. Four MD-MRFs of three bays each are displaced in the N-S direction and two MD-MRFs of five bays each are located in the E-W direction. All MD-MRFs columns are continuous over the building's height. The MD-MRFs columns (e.g. Type 1 and Type 2) are fixed at their base and the gravity columns (Type 3) are pinned. The studied building is designed according to the NBCC 2010 [1] and CSA/S16-09 standard [2] provisions. The static base shear V is calculated with the following equation: $V = S(T_a)M_v I_E W/R_d R_0$, where $S(T_a)$ is the 5% damped design spectral response acceleration corresponding to the fundamental lateral period of vibration T_a, I_E is the importance factor, M_v is the factor accounting for higher mode effect on base shear, W is the seismic weight of the building which includes 25% of snow load, R_d and R_0 are the ductility-related and overstrength-related force modification factors. More specifically, in this case study, M_v and I_E are equal to 1.0, the R_d and R_0 are equal to 3.5 and 1.5, respectively, and the seismic weight is $W = 33452$ kN. The fundamental period corresponding to empirical equation given in [1] is $T_a = 0.085h_n^{3/4}$, where $h_n = 15.9$ m is the total building height. From the above equation it results $T_a = 0.68\ s$. When a dynamic analysis is considered, the code allows to magnify the period by a factor of 1.5 which conducts to $T_a = 1.02\ s$. Thus, from the equivalent static force procedure, the computed base shear is $V = 2102$ kN. In this analysis the P-Δ effect is considered, while the torsional effect is neglected. All the MD-MRF beam and column members are made of W-shapes with $F_y = 350$ MPa and $F_u = 450$ MPa. The storey seismic weight and the cross-section adopted for all the members are given in Table 1. The building frame is symmetric and the orthogonal framing systems are uncoupled. Only the response of the building in the N-S direction is investigated. From dynamic analysis by means of OpenSees [3] the resulted 1st and 2nd mode period of vibration in the N-S direction are 1.387s and 0.403s, respectively. The studied building is not torsional sensitive. Hence, all MD-MRFs members were proportioned to carry the design base shear V_d which was not less than $0.8V$ as recommended in [1].

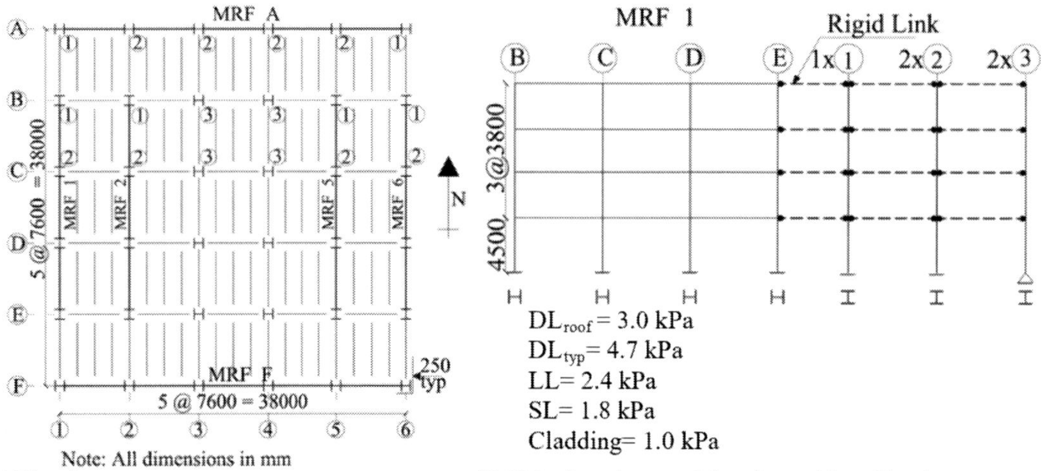

a) Plan layout b) N-S elevation model and considered loads

Fig. 1: Building studied

3. Numerical Analyses

Due to building symmetry, only one quarter of the building is simulated in the OpenSees software as shown in Fig. 1b. The P-Δ effect was considered and the Rayleigh formulation was used to define the mass and stiffness proportional damping coefficients which correspond to the 1st and 3rd mode of vibration. The equivalent viscous damping was equal to 2% and damping was not assigned to elements that are expected to respond in the plastic range [4].

Table 1: Seismic weight and MD-MRF members' cross-sections

Storey	Seismic Weight (kN)	Beams	Column Type 1	Column Type 2	Column Type 3
4	5462	W460x67	W460x113	W530x165	W310x79
3	9330	W530x66	W460x113	W530x165	W310x79
2	9330	W530x66	W460x113	W530x165	W360x122
1	9338	W530x74	W460x113	W530x165	W360x122

3.1 Structural model

Beams of the MD-MRF are modelled by *beam with finite length hinge* elements, whereas the plastic hinge segments were made of fiber cross-sections. Hence, in the plastic hinge zone, each flange is divided into 40 x 4 fibers, while the web is discretised into 20 fibers. The middle part of the *beam with finite length hinges* element behaves elastically, while the plastic hinge segments made of *Steel02* material behave in the nonlinear range. The plastic hinge integration method derives from the Gauss-Radau rule where deformations are integrated over the specified plastic hinge length [5]. In this study, the degradation of stiffness and strength due to local buckling of flanges are simulated by wrapping the low-cycle fatigue material to *Steel02* material assigned to fibers of plastic hinge zone cross-sections. The fatigue material, which is already implemented in OpenSees, uses an accumulative strain model to predict damage in accordance with the Miner's rule. The relation between the plastic strain amplitude experienced at each cycle and the number of cycles to failure is that proposed by Coffin and Manson. The fatigue ductility exponent m is considered constant and is set as $m=-0.5$. To simulate the effect of local buckling and gradual stiffness and strength degradation, the predicted fatigue coefficient ε_0 is assigned to flange fibres according to a linear distribution [6]. A minimum value $\varepsilon_{0,min}$ is assigned to fibers located at the edges of flanges and a maximum value ($\varepsilon_{0,min} + \Delta\varepsilon_0$) is assigned to flange fibers located at the intersection with the web. While the $\varepsilon_{0,min}$ value is related to the initiation of local buckling, the term $\Delta\varepsilon_0$ controls the rate of degradation. The length of the plastic hinges L_{pl} and the parameters $\varepsilon_{0,min}$ and $\Delta\varepsilon_0$ were calibrated against experimental test results. The predicted equations given in [6] are:

$$L_{pl} = 0.22\, L_V \tag{1}$$

$$\varepsilon_{0,\mathrm{min}} = 0.029 \tag{2}$$

$$\Delta\varepsilon_0 = 0.217 - 0.770 \cdot \lambda_f + 0.452 \frac{b_f t_f}{d\, L_v} + 0.902 \frac{L_m}{L_v} \geq 0 \tag{3}$$

where L_v is the shear length, λ_f is the slenderness of the flange, d is the depth of the cross-section, b_f and t_f are the width and the thickness of the flange and L_m is the wave length of buckled beam's flange computed as reported in [7,8]. In Fig. 2 is showed the cyclic response of the proposed model plotted in term of moment-rotation. It is noted that the *beam with hinges* element does not account for shear deformability of the beam. To overcome this drawback, one vertical spring was added at each end of the beam as reported in [9]. Each column belonging to the MD-MRF was modelled with 8 nonlinear beam-column elements with distributed plasticity. An initial camber displacement of 0.1% of the column height is applied at the column mid-length. The *Steel02* material was assigned to these columns. All gravity columns were modelled with elastic elements.

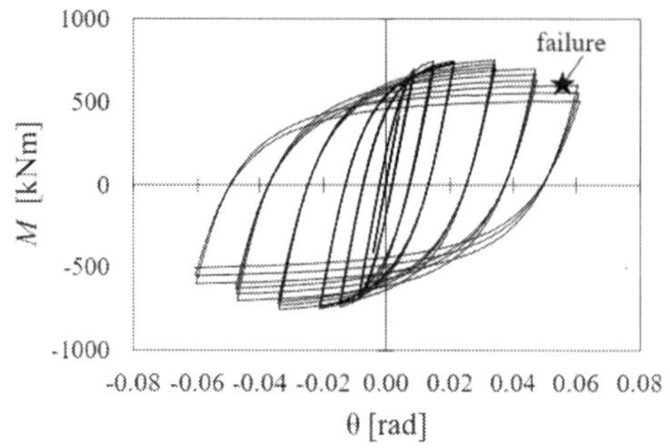

Fig. 2: Moment –rotation cyclic behaviour obtained from the proposed model

3.2 Seismic input

In this study, two sets of ground motions were considered. The first set consists of seven crustal ground motions selected from earthquakes with M_w6.9 and M_w6.7. These records corresponding to Site Class C (firm soil) were selected from the PEER Ground Motion Database website (www.peer.berkeley.edu).The second set of ground motions consists of three records from the M_w9 Tohoku earthquake, Japan. More details are given in [10]. The second set of records was added to investigate the effect of subduction ground motions characterised by longer Trifunac duration. All these records, their duration t and the Trifunac duration t_d are listed in Table 2. The crustal earthquake records (1st set) were scaled according to the ASCE/SEI 7-10 provisions [11], such that the mean spectrum of minimum seven ground motions to match or be above the design spectrum (2% of exceedance in 50 years) over the period of interest $0.2T_1 - 1.5T_1$ (Fig. 3). A scaling factor equal to 1.0 was applied to all subduction ground motions. To evaluate the residual interstorey drifts, 10.0 s of zero acceleration points were added at the end of each accelerogram.

Table 2: Selected ground motions

Set	NGA-Comp	Event	Station	t (s)	t_d (s)
C1	739-250	Loma Prieta	Anderson Dam (Downstream)	40.00	10.40
C2	767-000	Loma Prieta	Gilroy Array #3	39.94	6.37
C3	787-360	Loma Prieta	Palo Alto - SLAC Lab	39.57	11.58
C4	963-090	Northridge	Castaic, Old Ridge Route	45.00	9.08
C5	986-185	Northridge	LA - Brentwood VA Hospital	49.82	11.43
C6	1006-090	Northridge	LA - UCLA Grounds	60.00	11.30
C7	1039-180	Northridge	Moorpark - Fire Station	40.00	14.22
S1	MYG004-EW	Tohoku	MYG004	300.0	85.0
S2	FKS005-EW	Tohoku	FKS005	300.0	92.0
S3	FKS009-EW	Tohoku	FKS009	300.0	74.0

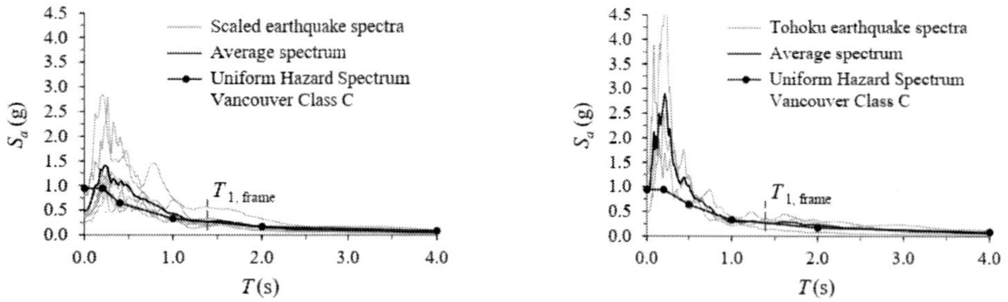

a) Crustal earthquake record set b) Subduction earthquake record set

Fig. 3: Response spectra of scaled records

4. Seismic response of the MD-MRF frame at design level

The seismic response of the MD-MRF building was analysed under the selected earthquake records scaled to match the 2%/50 years design spectrum for Vancouver. The response is expressed in terms of global and local parameters, namely the peak interstorey drift and peak residual interstorey drift at each floor and the rotation demand at the ends of each MD-MRF beam. Fig. 4 shows the height-wise distribution of peak interstorey drift, δ_{max} corresponding to each ground motion and Fig. 5 presents the peak residual interstorey drift, δ_r. Both aforementioned parameters are expressed as a percentage of the storey height *h*. As depicted, the maximum peak interstorey drift along the building height occurred at the 2nd floor level. The peak of the mean interstorey drift corresponding to the crustal record set is *1.55%h* and the peak of the mean plus one standard deviation is *2.15%h*. As resulted, these values are within the code limit which is *2.5%h*. When the studied building was subjected to the crustal record set, the peak of the mean residual interstorey drift and the peak of the mean plus one standard deviation was *0.20%h* and *0.46%h*, respectively. It is noted that according to [12], the *1.0%h* peak residual interstorey drift marks the life safety performance level and the *5.0%h* peak interstorey drift corresponds to the near collapse performance level. The magnitudes of residual interstorey drifts are lower for subduction ground motions. These results are in line with those obtained in [10] for the 4-storey concentrically braced frame building.

5. Assessment of collapse safety using incremental dynamic analysis

In order to assess the seismic performance of the MD-MRF building and to evaluate the collapse capacity, incremental nonlinear dynamic analyses (IDA) are carried out. The IDA

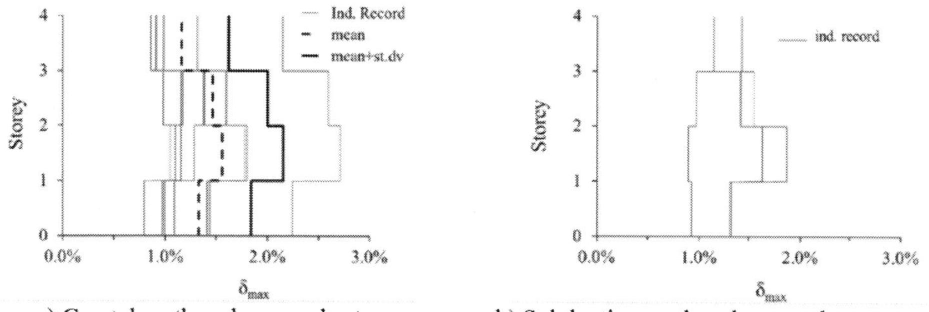

a) Crustal earthquake record set b) Subduction earthquake record set

Fig. 4: Interstorey drift computed at design level

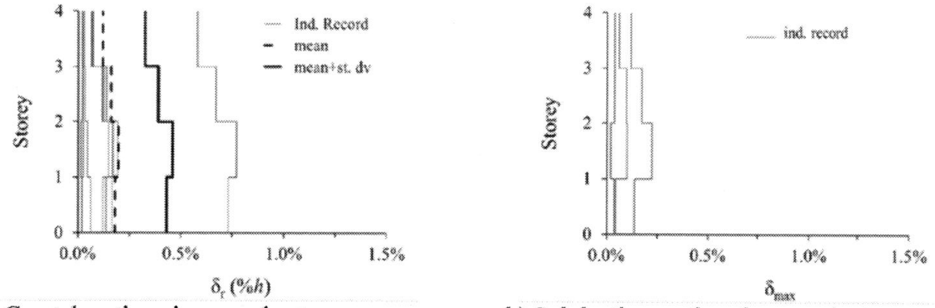

a) Crustal earthquake record set b) Subduction earthquake record set

Fig. 5: Residual interstorey drifts computed at design level curves are computed for each individual ground motion. Each point of the IDA curve obtained for a given ground motion relates the peak interstorey drift resulting from nonlinear time-history analysis to the spectral intensity of the building's first mode period. In order to define the collapse state on each IDA curve, two criteria are considered. The first criterion is intended to evaluate the state of cross-sections within the length of plastic hinge developed in the MD-MRF beams. This criterion is based on the damage index DI calculated for each cross-section as the ratio of the number of flanges fibers in which the Miner's damage index DI_F is equal to one (i.e. the number of fibres reaches fatigue) to the entire number of flanges fibers. When the DI is equal to 1.0, flanges are not able to sustain the bending moment. In [6], a value of DI = 0.375 has been proposed as an index of the local collapse. This value corresponds to approximately 80% of bending moment capacity (e.g. the star mark in Fig.2).

It is noted that the effect of low-cycle fatigue depends on the type of ground motions and is expected to be significant for accelerograms characterized by larger Trifunac duration. In the light of this, a second collapse criterion based on the maximum interstorey drifts δ_{max} is considered. As mentioned above, the maximum interstorey drift assumed acceptable in the near collapse state should not be greater than *5.0%h*.

The obtained IDA curves are shown in Fig. 6. The black circle on each IDA curve marks the spectral acceleration corresponding to a residual interstorey drift of 1.0%*h* and the white circle marks the spectral acceleration corresponding to the first achievement of DI = 0.375 in the cross-section of a plastic hinge zone. The design spectral acceleration is also reported in the figure by a horizontal dashed line. The figure shows that for four crustal ground motions the ultimate behaviour is related to the achievement of a maximum interstorey drift equal to *5.0%h* and in three cases, the local collapse occurs. Furthermore, the maximum interstorey drifts corresponding to the local collapse are only slightly lower than *5.0%h*. Instead, when the subduction earthquake records are considered, the local collapse is achieved for interstorey drifts significantly lower than *5.0%h*. As an example, when the FKS009 record is employed, the interstorey drift corresponding to DI = 0.375 is *3.25%h*. This result confirms the importance of an accurate modelling of stiffness and strength degradation behaviour of plastic hinges when the collapse state of steel MD-MRF buildings is assessed.

To assess the collapse safety of studied building, the following parameters are calculated according to [13]: the median collapse capacity $S_a(\overline{T}_1)$, the collapse margin ratio (CMR) and the adjusted collapse margin ratio (ACMR). The median collapse capacity is the spectral intensity at which the collapse is achieved under 50% of the considered ground motions. In this study, only the crustal records are considered because of the limited number of subduction records used. The computed $S_a(\overline{T}_1)$ value is 1.117 g. The CMR is defined as the ratio of $S_a(\overline{T}_1)$ to the 2%/50 years design spectral acceleration at the period T_1 which is 0.268g.

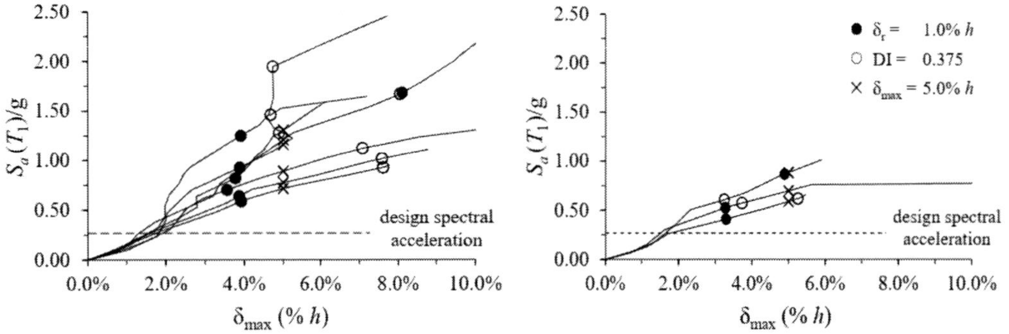

a) Crustal earthquake record set b) Subduction earthquake record set

Fig. 6: The IDA curves

The computed CMR value is 4.165. To account for the effects of spectral shape, the collapse margin ratio is adjusted by multiplying it with the SSF factor calculated according to the procedure described in [13]. The SSF factor is 1.27 and depends on both the period of the structure and the ductility μ_T (5.23). To possess enough collapse safety, the computed ACMR has to be larger than $ACMR_{10\%}$, which is the minimum permissible ACMR corresponding to the 10% probability of collapse under the set of ground motions scaled to the design value. The $ACMR_{10\%}$ depends on the total system collapse uncertainty β_{TOT} given as:

$$\beta_{TOT} = \sqrt{\beta_{RTR}^2 + \beta_{DR}^2 + \beta_{TD}^2 + \beta_{MDL}^2} \tag{4}$$

where β_{RTR} is the record-to-record uncertainty, β_{DR} is the design requirements-related uncertainty, β_{TD} is the test-data related uncertainty and β_{MDL} is the modelling-related uncertainty. The following assumptions are made to calculate the total system collapse uncertainty: $\beta_{DR} = 0.1$ ("A-Superior" quality), $\beta_{TD} = 0.2$ ("B-Good" quality), $\beta_{MDL} = 0.2$ ("B-Good" quality). Further, the record-to-record uncertainty is obtained by a regression analysis and is equal to 0.20. For the computed β_{TOT} value, the $ACMR_{10\%}$ is equal to 1.45 [13]. Thus, the building under investigation satisfies the collapse safety requirement: $ACMR > ACMR_{10\%}$.

6. Prediction of the maximum residual interstorey drift

Since the requirements for direct simulation of residual interstorey drift are computationally expensive, FEMA P-58-1 (2012) [14] provides the following equations to estimate the median peak residual interstorey drift, δ_r, as a function of peak interstorey drift δ_{max}:

$$\delta_r = \begin{cases} 0 & \text{if } \delta_{max} \le \delta_y \\ 0.3 \cdot (\delta_{max} - \delta_y) & \text{if } \delta_y < \delta_{max} < 4\delta_y \\ (\delta_{max} - 3\delta_y) & \text{if } \delta_{max} \ge 4\delta_y \end{cases} \tag{5}$$

where δ_y is the interstorey drift associated with the first yielding in the structure. In this paper, δ_y is evaluated from the IDA curve as the interstorey drift corresponding to a bending moment equal to the factored moment resistance. In Fig. 7, the peak residual interstorey drifts provided by the equations above ($\delta_{r,pred}$) are compared to those obtained from the IDA (δ_r). More specifically, each dot in the figure refers to the results obtained for a single ground motion and a single scaling factor. With two exceptions (963-090 and FKS005 records), Fig. 7 shows that Eq. (5) provides a good prediction of peak residual interstorey drift for values lower than *3.0%h* and a slightly conservative prediction for higher values of δ_r.

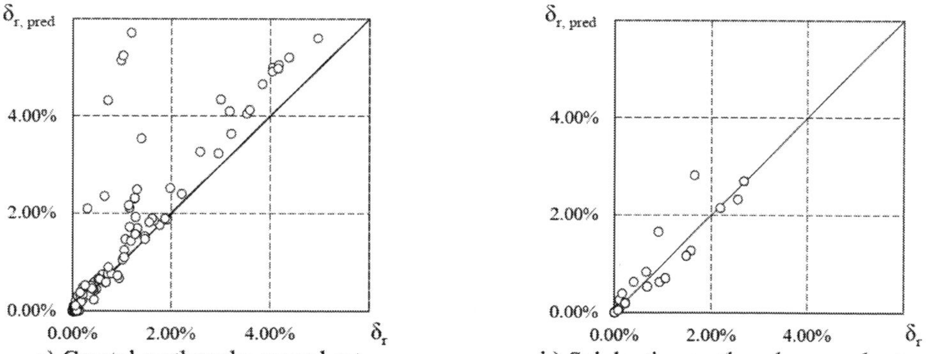

a) Crustal earthquake record set   b) Subduction earthquake record set

Fig. 7: The computed versus predicted peak residual drifts according to FEMA P-58-1

7. Conclusions

The main conclusions of this paper are:

1. The detailed numerical model incorporating the proposed low-cycle fatigue model is able to replicate the nonlinear seismic response of the considered 4-storey MD-MRF building;
2. The importance of modelling accuracy of stiffness and strength degradation due to low-cycle fatigue is higher especially in the case of subduction earthquake records;
3. The adjusted collapse margin ratio of the 4-storey MD-MRF building is larger than the minimum permissible value corresponding to the 10% probability of collapse under a set of crustal ground motions scaled to the design level;
4. The simplified equation given in FEMA P-58-1 provides good estimates of residual interstorey drift when the computed residual interstorey drift is lower than *3%h*.

Acknowledgments

The financial support provided by NSERC is gratefully acknowledged by the authors.

References

[1] National Research Council of Canada (NRCC). "National Building Code of Canada", Associate Committee on the National Building Code, NRCC, Ottawa, Canada, 2010.

[2] Canadian Standards Association (CSA). "Design of steel structures", *S16-09 Standard*, CSA, Toronto, Ontario, Canada, 2009.

[3] McKenna F, Fenves G, Scott M, et al. "Open system for earthquake engineering simulation". *OpenSees software version 2.4.5*, PEER, Univ. of California, Berkeley, 2015.

[4] Charney FA. "Unintended consequences of modeling damping in structures", *Journal of Structural Engineering*, ASCE, 134(4), 581-592, 2008.

[5] Scott M, Fenves G. "Plastic hinge integration methods for force-based beam–column elements", *Journal of Structural Engineering.*, ASCE, 132(2), 244-252, 2006.

[6] Bosco M, Tirca L. "Numerical simulation of steel beam-column members using damage accumulation models", submitted to *J. of Constructional Steel Research,* 2015.

[7] Lay MG. "Flange local buckling in wide-flange shapes", *Journal of the Structural Division*, ASCE, 91(6), 95-115, 1965.

[8] D'Aniello M, Landolfo R, Piluso V, Rizzano G. "Ultimate behaviour of steel beams under non-uniform bending", *Journal of Constructional Steel Research*, 78, 144-158, 2012.

[9] Bosco M, Marino EM, Rossi PP. "Mod elling of steel link beams of short, intermediate or long length", *Engineering Structures Journal*, 84, 406–418, 2015.

[10] Tirca L, Chen L, Tramblay R. "Assessing collapse safety of CBF buildings subjected to crustal and subduction eartquakes", *J. of Constructional Steel Research*, 115, 47-61, 2015.

[11] American Society of Civil Engineers (ASCE), Minimum design loads for buildings and other structures, ASCE/SEI 7-10, Reston, Virginia (2010).

[12] FEMA 356, "Prestandard and commentary for the seismic rehabilitation of buildings", American society of civil engineering, Federal Emergency Management Agency, Washington, D.C, 2000.

[13] FEMA P-695, "Quantification of building seismic performance factors", Federal Emergency Management Agency, Washington, D.C., 2009.

[14] FEMA P-58-1, "Seismic Performance Assessment of Buildings, Vol. 1– Methodology", Federal Emergency Management Agency, Washington, D.C., 2012.

NONLINEAR RESPONSE OF A CBF HOSPITAL BUILDING WITH AND WITHOUT FRICTION DAMPER DEVICES

Lucia Tirca, Masaaki Ohira and Ovidiu M. Serban

Department of Building, Civil and Environmental Engineering, Concordia University, Montreal, Canada

Abstract: In this study, the seismic response of a low-rise moderately ductile CBF hospital building located in Victoria, BC, Canada is analysed by means of OpenSees. The importance category for a hospital building is post-disaster and the investigated parameters were: interstorey drift, residual interstorey drift and floor acceleration. From nonlinear dynamic analysis, the 50th percentiles of interstorey and residual drifts are within the code limit. However, the 50th percentile floor acceleration is larger than 1.1g. When friction damper devices were added in-line with braces (FD-CBF), the results showed that the peak floor acceleration of the FD-CBF system was substantially reduced, while the peak residual interstorey drift increased.

1. Introduction

Concentrically Braced Frames (CBFs) have been frequently employed as seismic force resisting systems (SFRS) due to their high stiffness and moderate ductility.

The CBF system withstands earthquake loading and dissipate the input energy when brace members yield in tension and buckle in compression. However, the CBF system has shown several drawbacks such as the demand concentration within a floor and asymmetric response after braces have buckled in compression. In addition, the CBF system shows high floor accelerations which may be critical for the acceleration-sensitive non-structural components. Although the non-structural components are ignored in the current design methodology, they constitute a major part of losses in the event of an earthquake. Additionally, in critical facilities like hospitals, the cost associated with indirect losses due to damaged equipment, casualties, etc. may be significantly higher than the cost of building replacement.

To overpass the floor acceleration drawback, in this study, it is proposed to install friction damper (FD) devices in-line with diagonal braces (FD-CBF). The proposed FD-CBF system behaves elastically as a traditional CBF before the FD devices are activated and in the nonlinear range after that. Thus, in the case of FD-CBF system, the input energy is dissipated by friction devices and all adjacent members such as: braces, connections, beams and columns of the CBF system are designed to remain in the elastic range. However, the FD-CBF system may undergo large residual deformations under earthquake loading. Past studies [1,2] have shown that peak residual interstorey drift is a key parameter for decision-makers and for a magnitude greater than 0.5%h_s the building is proposed for replacement rather than retrofit.

Thus, the aim of this research is two-fold: (i) to carry out a comparative study for a low-rise moderately ductile CBF hospital building with and without FD devices; and (ii) to establish a numerical model able to simulate the behaviour of brace in-line with friction damper using the OpenSees (Open System for Earthquake Engineering Simulation) framework [3]. The investigated building is a 4-storey hospital located on firm soil (Site Class C) in Victoria, BC, Canada, which is a high-risk seismic zone. According to the current NBCC code [4], the importance category for a hospital building is post-disaster, which implies an amplification of base shear and limits for interstorey drift to 1.0%h_s, where h_s is the storey height.

2. Design of studied buildings

In this study, two SFRS are considered for the hospital building: i) a moderately ductile CBF system (MD-CBF) and ii) a friction braced frame system, FD-CBF. In the first case, brace members were designed to dissipate the earthquake induced energy through yielding and buckling according to the current NBCC code [4] and the CSA/S16-09 standard provisions [5]. In the second case, friction damper devices added in-line with diagonal braces were proportioned to dissipate the input energy through friction, while all adjacent members such as braces, beams, columns and their connections were designed to behave in the elastic range.

2.1 Building description

The studied 4-storey fictitious hospital building located on firm soil (Site Class C) in Victoria, BC, has the same typical floor plan as the Lion Gate hospital located in Vancouver [6].

As illustrated in Fig. 1, the total height of the building is h_n=*18.2 m*, the typical floor height is *4.4 m* and the ground floor height is *5.0 m*. The building is regular and symmetrical in both loading directions, while the floor area is *21.5 m* x *80.4 m*. As depicted in Fig. 1, the selected building is braced in the north-south (N-S) direction by four MD-CBFs with chevron braces (inverted V) and in the east-west (E-W) direction by four MD-CBFs with diagonal tension-compression braces. In this study, only the response of the structure in the E-W direction is presented. As depicted in Fig. 2, all MD-CBF beams and braces are pinned connected at both ends; columns are pinned-connected at their base and are continuous over two storeys. All columns and beams were made of W-shapes with a nominal yield strength F_y= *350MPa* and tensile strength F_u= *450MPa*. All braces were made of hollow structural sections (HSS).

2.2 Seismic design of studied MD-CBF and FD-CBF buildings

According to [4], the empirical fundamental vibration period of the MD-CBF building is given by the expression: $T_a=0.025h_n=0.455$ *s*. A larger first-mode period can be used when a dynamic analysis is employed. However, this value cannot exceed $2T_a$ or *0.91 s* in this case.

To determine the dynamic response of the MD-CBF structure, a numerical model was developed using the ETABS software [7] and the resulted 1st mode dynamic period in the E-W direction was *0.56 s*. Using the equivalent static force procedure [4], the base shear V was computed with the following equation: $V= S(T_a)M_vI_EW/(R_dR_o)$, where $S(T_a)$ is the 5% damped spectral response acceleration corresponding to T_a and is expressed as a ratio to gravitational acceleration, M_v is a factor accounting for higher modes effect (M_v= *1.0*), I_E is the earthquake importance factor of the structure ($I_E = 1.5$ for post-disaster buildings), W is the total seismic weight of the building and R_d, R_0 are the ductility-related and overstrength-related force modification factors equal to 3.0 and 1.3, respectively. However, V shall not be taken less than the base shear calculated according to $V = S(2.0)M_v I_E W/R_d R_o$ and not greater than $V = 2/3[S(0.2)I_E W/R_d R_o]$. For the studied hospital building, the total seismic weight including 25% snow load at the roof level is $W = 32924\ kN$. The site-specific response spectral accelerations for Victoria (Site Class C) at specified periods of 0.2, 0.5, 1.0 and $2.0\ s$ are 1.2, 0.82, 0.38 and 0.18 g, respectively. These values correspond to a probability of exceedance of 2% in 50 years. Considering the 1st mode period of $0.56\ s$, the computed base shear is $V_{dyn} = 9748kN$. Torsion and P-Δ effect were considered in the analysis. After braces of MD-CBF were sized, all beams, columns and their connections were designed according to capacity design method.

To design the FD-CBF building system, there are no straightforward provisions. For example, to estimate the base shear demand, it is proposed to use $R_o = 1.0$ and $R_d = 4.0$, such that the product $R_d R_o$ to be approximately the same for both MD-CBF and FD-CBF systems.

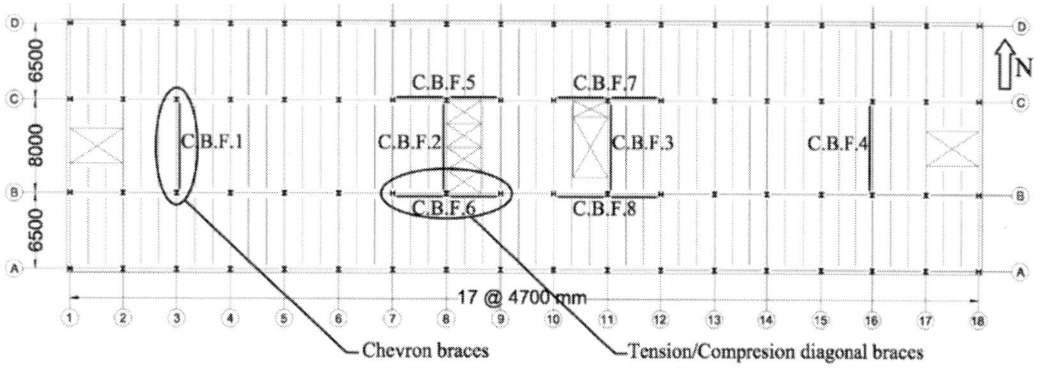

Fig. 1: Plan view of hospital building

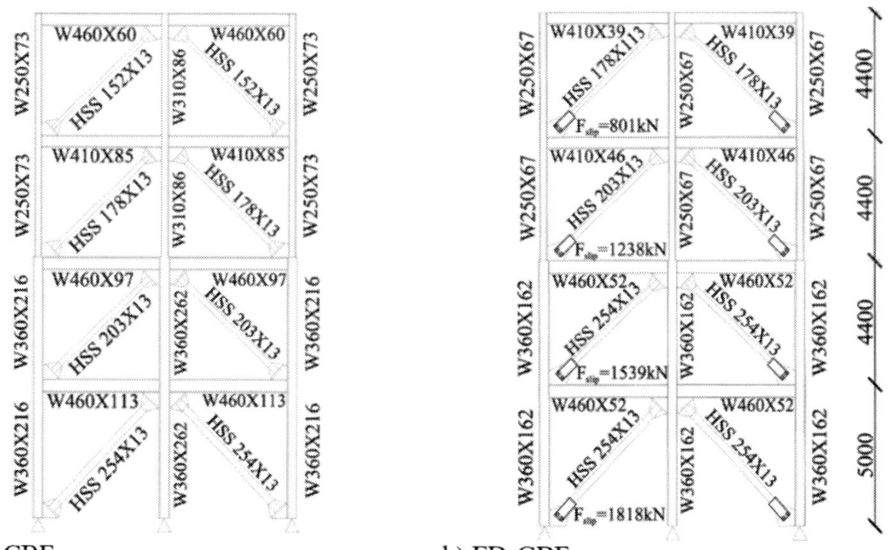

a) MD-CBF b) FD-CBF

Fig. 2: Elevation of MD-CBF and FD-CBF in the E-W direction

As depicted in Fig. 2b, one end of HSS brace was connected to a friction damper and the other end to a traditional gusset plate connection. The FD-CBF system behaves as a traditional

CBF until the first FD device reaches the slip force, F_{slip}. Considering the same lateral force distribution, the slip force in each damper was assessed. Then, according to the NEHRP guidelines [8], each HSS brace was proportioned to carry *130%* F_{slip} of the attached device. Therefore, FD devices preserve braces from buckling and dissipate energy through friction (Fig. 3). Additionally, it was verified that FD devices do not slip under the wind load. In general, the middle plate of FD is welded to the slotted HSS brace, while the external channels are bolted to the gusset plate. The friction occurs when the middle plate slides between the two channels which are connected with pretension bolts [9]. In order to obtain the demanded slip length, one must assess the seismic response of the building through nonlinear time-history dynamic analysis under an ensemble of representative ground motions. The cross-sections of the MB-CBF and FD-CBF system are given in Fig. 2a and Fig. 2b, respectively.

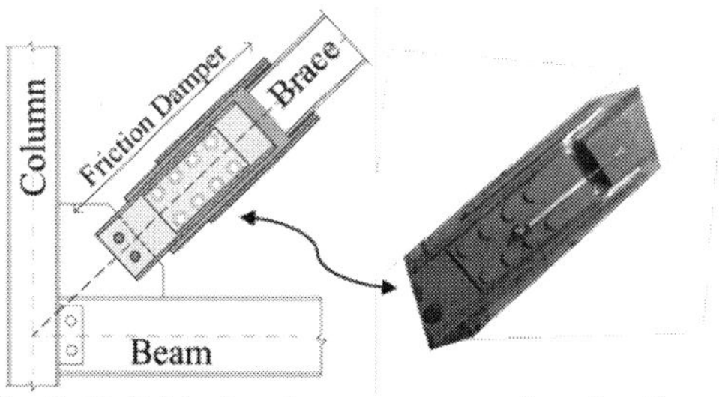

Fig. 3: Pall friction damper connection detail

3. Seismic response

3.1 OpenSees modelling of studied buildings

To investigate the nonlinear response of studied MD-CBF and FD-CBF hospital buildings, detailed OpenSees models were developed. Due to building's symmetry, each OpenSees model loaded in the E-W direction was constructed for one quarter of the building area.

3.1.1 Modelling of MD-CBF hospital building

In the OpenSees model, braces, beams and columns were modelled as non-linear *forced-based beam-column* elements with distributed plasticity, fiber cross-section formulations and three integration points per element.

Each HSS brace was made of 16 elements, 240 fibers and an initial out-of-straightness equal to *1/500* of brace length. The W-shape columns and beams were made of 120 fibers. Each column was divided into 8 elements having an initial out-of-straightness equal to *1/1000* of its length. The *Steel02* material was assigned to all MD-CBF members. To simulate the HSS brace fracture, fatigue material was assigned to parental material of HSS braces (*Steel02*). In this study, the parameters used to define the fatigue material were calculated according to [10]. Brace connections and beam-column connections were modelled by springs assigned in the *Zero-Length* elements. To simulate the out-of-plane behaviour of gusset plate connection attached to the end of HSS brace, a nonlinear rotational spring element made of *Steel02* material was assigned in the *Zero-Length* element. It is noted that the *Zero-Length* element connects the brace member to a rigid link. The stiffness of the rotational spring was based upon the geometry and properties of the gusset plate [11]. For more information regarding the MD-CBF modelling the reader is referred to [12].

3.1.2 Modelling of FD-CBF hospital building

The OpenSees model of the FD-CBF building is very similar with the model created for the MD-CBF building. The difference consists in replacing the rotational spring used to replicate the brace-to-frame connection of bottom end HSS brace with a translational spring located in the same Zero-Length element that is able to simulate the response of friction damper.

Herein, the uniaxial *BoucWen* material was used to simulate the smooth hysteresis behaviour of friction damper replicated by the translational spring displaced in the *Zero-Length* element in the axial direction. The friction damper is activated when the axial force in brace reaches the slip force F_{slip}. The *BoucWen* material is able to simulate the high nonlinear Coulomb friction. The OpenSees model of HSS brace in-line with FD device was calibrated against experimental test results. The slip force and slip length of tested specimen were equal to $F_{slip}=700\ kN$ and $\Delta_{slip}=\pm 50\ mm$, respectively. It is noted that FD can slip when it is loaded either in tension or in compression and the available slip length or stroke should be $1.3\Delta_{slip}$. The testing platform for the HSS brace in-line with FD specimen is depicted in Fig. 4a, while the experimental hysteresis loop is showed in Fig. 4b. As illustrated, the damper was welded to a HSS brace, while the short brace was welded to a *50.8 mm* thick gusset plate at the other end. The brace end was connected to a reaction frame, while the full-size friction damper end was attached to the load cell of a *1500 kN* actuator. The distance between the end of the HSS brace and the upper channel belonging to friction damper is the stroke length. More details regarding the experimental program, loading protocol, and test results obtained on full-scale braces in-line with Pall friction dampers will be presented in future publications.

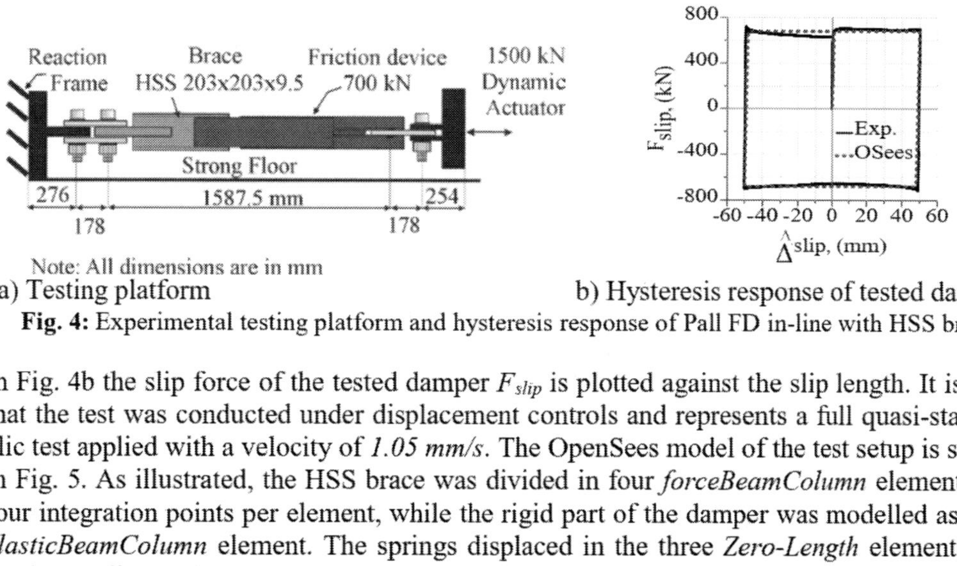

a) Testing platform b) Hysteresis response of tested damper

Fig. 4: Experimental testing platform and hysteresis response of Pall FD in-line with HSS brace

In Fig. 4b the slip force of the tested damper F_{slip} is plotted against the slip length. It is noted that the test was conducted under displacement controls and represents a full quasi-static cyclic test applied with a velocity of *1.05 mm/s*. The OpenSees model of the test setup is showed in Fig. 5. As illustrated, the HSS brace was divided in four *forceBeamColumn* elements with four integration points per element, while the rigid part of the damper was modelled as a stiff *elasticBeamColumn* element. The springs displaced in the three *Zero-Length* elements were used to replicate: the HSS brace-to-frame connection, damper-to-actuator connection and the behaviour of the friction device. Parameters used to define the *BoucWen* material assigned to simulate the FD response are: the ratio of post-yield stiffness to the initial elastic stiffness α, the initial elastic stiffness k_0, the exponent n that influences the sharpness of the model in the transition zones, and γ and β parameters which control the shape of the hysteresis cycle. Other parameters such as: A_0, A, v and η control the degradation process of the model. In this study, it was considered: $A_0 = 1$, $\alpha = A = v = \eta = 0$ while γ and β parameters were considered equal and were calculated based on the formula: $\gamma + \beta = 1/(\Delta_y)^n$. In the latter equation, Δ_y represents the yielding displacement of the system when the first damper starts slipping and can be calculated by using: $\Delta_y = F_{slip}/k_0$ where F_{slip} is the activation slip force and $k_0 = k_{brace} = A_{brace}E/L$. Regarding the sharpness of hysteresis loop parameter, a value of $n=10$ was selected in this study [13]. Using the developed OpenSees model, the replicated FD hysteresis response is given in Fig. 4b against the experimental hysteresis loops.

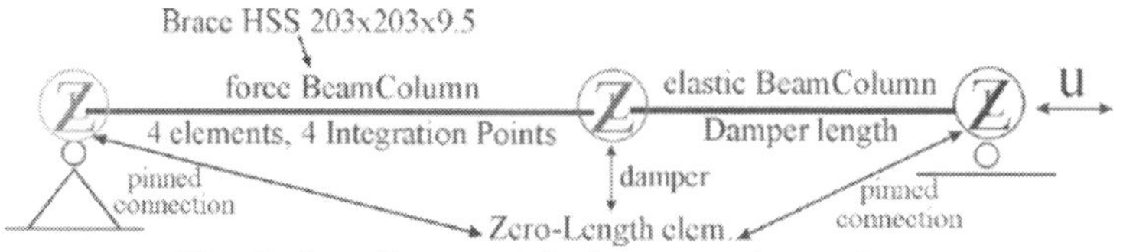

Fig. 5: OpenSees model of the experimental test

3.2 Earthquake ground motions selection and scaling

The 10 selected records for Victoria were from moderate to large crustal earthquakes of magnitude varying from M_w=*6.5* to M_w=*7.5* that occurred on the Pacific coast of North America and were selected from the Peer Ground Motion Database website (www.peer.berkeley.edu). For more information regarding the selected ground motions the reader is referred to [12]. It is noted that scaling procedure complies with ASCE/SEI 7-10 provisions [14] which requires that the mean of 5% damped response spectra of a minimum seven ground motions to match or be above the design spectrum in any point over the period interval $0.2T_1 - 1.5T_1$.

3.3 Seismic response of studied buildings

The hospital building was subjected to nonlinear time-history analysis. An ensemble of 10 crustal ground motions scaled to match the design spectra was considered.

From OpenSees, the resulted first-mode vibration period was *0.54 s* and *0.62 s* for MD-CBF building and FD-CBF building, respectively. The demanded slip length of friction dampers resulted from OpenSees analysis. It was concluded that the maximum demand in term of damper's slip length of *40 mm* was recorded at the 1st floor level under the #796 record (Loma Prieta earthquake), while other ground motions imply the larger demand at the 3rd floor. For this case study, it was proposed to round the slip length demand to Δ_s=*50 mm*. Thus, the employed FD devices should have the stroke equals to $1.3\Delta_{slip}$.

To emphasise the differences in behaviour between the investigated systems, the selected parameters were: i) the interstorey drift, ii) the residual interstorey drift; and iii) floor acceleration. In Fig. 6a to Fig. 6c is depicted the seismic response of the MD-CBF system, while in Fig. 6d to Fig. 6e is showed the results for the FD-CBF building. Moreover, the 50th and 84th fractiles of the recorded engineering demand parameters resulted from nonlinear time-history analysis are also presented in these figures. As illustrated, when the MD-CBF system was employed, the interstorey drift concentration was mainly at the 3rd floor having an average value of $0.85\%h_s$, while the 84th fractiles was $1.02\%h_s$. It is noted that the code interstorey drift limit for post-disaster buildings is $1.00\%h_s$. In the case of the MD-CBF building, the residual interstorey drift values (Fig. 6b) were very small i.e. less than $0.15\%h_s$ for both 50th and 84th fractiles. However, the floor acceleration was larger at the second floor having values ranging from *1.2g* to *1.5g* for the 50th and 84th fractiles, respectively. On the other hand, when a friction device was added in-line with braces, the interstorey drift was more uniformly distributed among the floors, having a value of $1.0\%h_s$ in 84% of the ground motions (at 4th floor), as depicted in Fig. 6d. Another key observation with regard to the residual interstorey drift was that the FD-CBF system undergoes larger residual interstorey drift than the MD-CBF system as shown in Fig. 6e versus Fig. 6b. The summary of residual interstorey drift results of FD-CBF leads to$0.42\%h_s$ and $0.71\%h_s$ for the 50th and 84th fractiles, respectively. Both values were recorded at the 4th floor. Comparing the floor acceleration values of both systems (Fig. 6f versus Fig. 6c), it can be seen that in the case of FD-CBF system the values are considerably reduced.

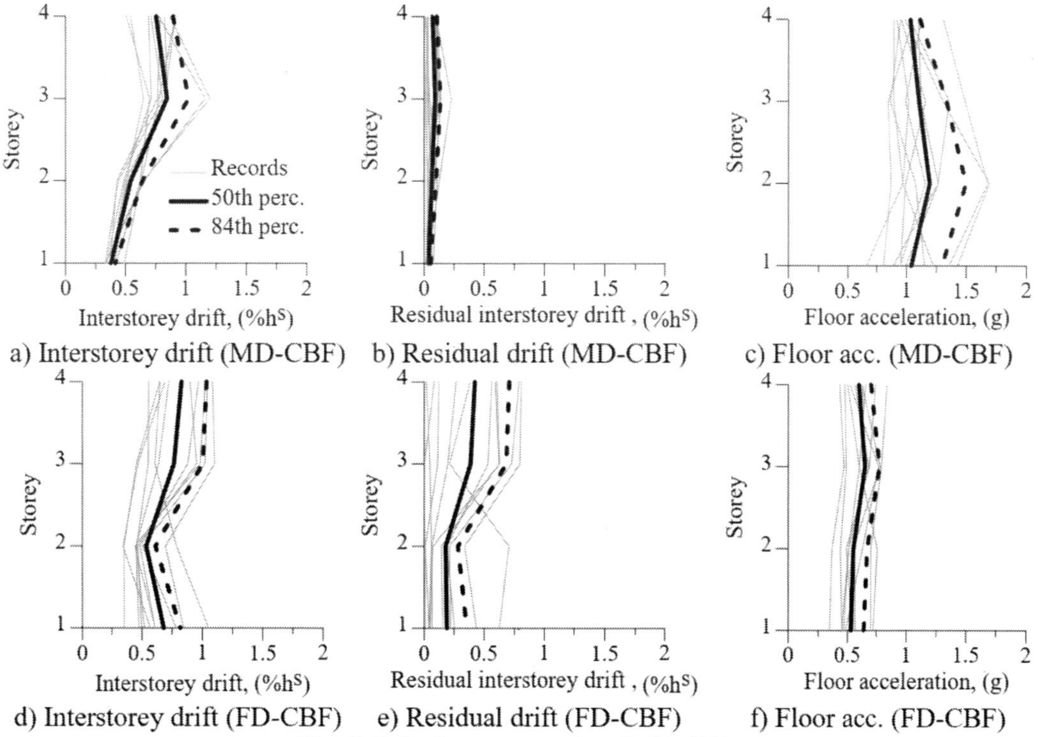

Fig. 6: Seismic response of studied buildings

It can be concluded that FD-CBF systems provide more uniformly distributed demand across the building height by engaging all devices into energy dissipation. Therefore, the floor acceleration was reduced and the failure of acceleration-sensitive non-structural components of this studied post-disaster building was mitigated. However, the residual interstorey drift demand may be greater than $0.5\%h_s$ which imply damage of building contents and displacement-sensitive non-structural components. To overcome this drawback, it is recommended to use a backup moment resisting frame system (MRF) or a self-centering system in conjunction with the FD-CBF system.

4. Conclusions

This study was conducted to evaluate the seismic response of a 4-storey hospital building located in Victoria, BC, Canada. According to NBCC 2010, the importance category for a hospital building is post-disaster. The following findings are reported:

1. Two types of seismic force resisting systems were compared in order to assess their seismic response: i) a traditional MD-CBF with HSS braces designed to yield in tension and buckle in compression and ii) a friction concentrically braced frame (FD-CBF) with friction damper devices installed in-line with HSS braces;
2. A detailed model was developed and a comprehensive nonlinear time-history analysis was carried out by means of OpenSees for both MD-CBF and FD-CBF building systems. Additionally, the smooth hysteresis behaviour of the friction damper was

calibrated against experimental test results. To simulate the response of friction dampers the uniaxial *BoucWen* material was employed;

3. It can be concluded that using friction damper devices installed in CBF systems may lead to large residual interstorey drifts if the seismic demand exceeds the current code. However the floor acceleration was considerably reduced and all dampers were engaged in energy dissipation. To increase the system's performance, it is recommended to use either a backup moment resisting frame or a self-centering system in parallel with the FD-CBF.

Acknowledgments

The financial support provided by the Natural Sciences and Engineering Research Council of Canada is gratefully acknowledged by the authors.

References

[1] McCormick J, Aburano H, Ikenaga M, Nakashima M. "Permissible residual deformation levels for building structures considering both safety and human elements", *Proc. 14th World Conf. Earthquake Eng.*, Beijing, China, Paper ID 05 06-0071, 2008.

[2] Erochko J, Christopoulos C, Tremblay R, Choi H. "Residual drift response of SMRFs and BRB frames in steel buildings designed according to ASCE/SEI 7-05", *J. Structural Engineering*, ASCE, 137(5), 589-599, 2011.

[3] McKenna F, Fenves G, Scott M, et al. "Open system for earthquake engineering simulation", *OpenSees software version 2.4.5*, PEER, Univ. of California, Berkeley, 2015.

[4] National Research Council of Canada (NRCC). "National Building Code of Canada Part 4", Ottawa, Canada, 2010.

[5] Canadian Standards Association (CSA). "Design of steel structures", *S16-09 Standard*, CSA, Toronto, Ontario, Canada, 2009.

[6] Sherstobitoff J, Rezai M, Wong M. "Seismic upgrade of Lions Gate hospital's acute tower south", *Proc. 13th World Conf. Earthquake Eng.*, Vancouver, B.C., 2004.

[7] Computers and Structures, Inc. (CSI). "ETABS, version 9.7.4 (computer program)", CSI, Berkeley, California, 2011.

[8] Federal Emergency Management Agency (FEMA). "Report on the prestandard and commentary for the seismic rehabilitation of buildings", FEMA 356, Washington, 2000.

[9] Pall A, Marsh C. "Response of friction damped braces", Journal of Structures Division, ASCE, Vol. 108, No. ST6, 1313-1323, 1982.

[10] Tirca L, Chen L. "Numerical simulation of inelastic cyclic response of HSS braces upon fracture", *J. of Advanced Steel Construction*, 10(4), 442-462, 2014.

[11] Hsiao PC, Lehman D, Roeder C. "A model to simulate special concentrically braced frames beyond brace fracture", *Journal of Earthquake Engineering and Structural Dynamics*, 42, 182-200, 2013.

[12] Tirca L, Ohira M. "Improved seismic response of braced frames: Use of advanced connections detailing", *Proc. of EUROSTEEL Conf.*, Napoli, Italy, 2014.

[13] Morales Ramirez JD. "Numerical simulations of steel frames equipped with friction-damped diagonal-bracing devices", *MASc Thesis*, Concordia University, Montreal, Quebec, Canada, 2011.

[14] American Society of Civil Engineers (ASCE). "Minimum design loads for buildings and other structures", ASCE/SEI 7-10, Reston, Virginia, USA, 2010.

STRAP-BRACED CFS WALLS: EXPERIMENTAL CHARACTERIZATION OF THE SEISMIC RESPONSE

Maria Teresa Terracciano, Bianca Bucciero, Vincenzo Macillo, Tatiana Pali, Luigi Fiorino and Raffaele Landolfo

University of Naples Federico II, Dept. Structures for Engineering and Architecture, Italy

Abstract: The unconventionality of the CFS structures raised, in recent times, a lot of interest from many national and international companies resulting in the promotion of experimental activity with the main aim of investigating and interpreting the seismic behaviour of these systems. In this framework, an important collaboration between the University of Naples "Federico II" and the Lamieredil S.p.A. company is started in the last years. The main goal of this collaboration is the experimental characterization of global and local seismic response of CFS strap-braced walls. This paper presents the experimental campaign and discusses the main results.

1. Introduction

The adoption of a constructive system in seismic area is related to the clarity and direct applicability of the reference technical code. The use of Cold-Formed Steel (CFS) products for structural applications in seismic area has not been properly contextualized in the current European technical standard. In order to bridge this gap, many national and international companies are promoting experimental activities with the main aim of investigating and interpreting the seismic behaviour of these systems. In this framework, an important collaboration between the University of Naples "Federico II" and the Lamieredil S.p.A. company started in the last years. In particular, the research activity involves the experimental characterization of the seismic response of strap-braced CFS walls by means of monotonic and cyclic tests on full-scale prototypes. Furthermore, in order to evaluate the influence of the main structural components on the global response, the experimental campaign is completed with tests on materials and main connection systems.

2. Experimental activity

The design of the strap-braced CFS wall prototypes represented a preliminary phase to the experimental activity. Two types walls prototypes (elastic light wall, WLE and elastic heavy wall, WHE) were designed according to an elastic design approach [1]. The walls represent the seismic resisting systems of a three-story building in low-medium seismic area (Fig. 1). In particular, the WHE and WLE walls are representative of strap-braced CFS walls placed at the first and last floor of a case study, respectively. The walls were designed according to the procedure described in [2] for resisting to design horizontal loads (H_d) equal to 152 kN (WHE) and to 71 kN (WLE). Furthermore, the theoretical predicted values of the wall strength (H_{th}), evaluated according to the experimental results of wall components and materials, are equal to 186 kN and 95 kN for WHE end WLE walls, respectively. The theoretical predicted values of the wall stiffness are equal to 6.8 kN/mm and 4.2 kN/mm for WHE and WLE walls, respectively.

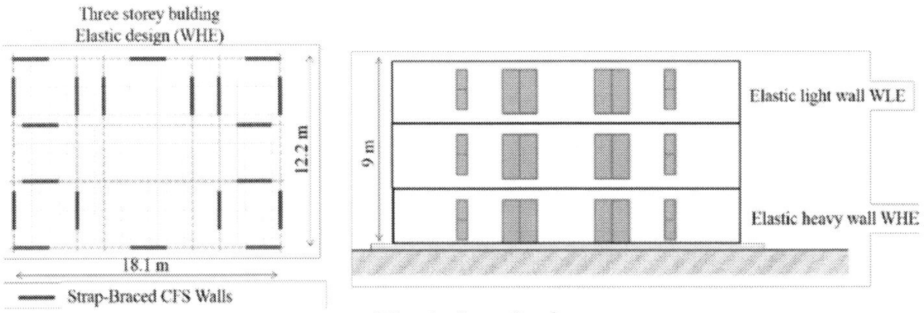

Fig. 1: Case Study

The experimental campaign was carried out in the laboratory of the Department of Structures for Engineering and Architecture of the University of Naples Federico II. In particular, the lateral response of these systems was investigated through the execution of 2 monotonic and 4 cyclic tests on full-scale wall specimens. In order to evaluate the influence of the main structural components on the lateral response of seismic systems, the local response was investigated by means 12 tests on materials, 6 shear tests on connections between gusset plates and strap-braces and 4 tension tests on hold-down.

3. Tests on walls

In order to investigate the lateral response of strap-braced walls 6 tests on full-scale specimens in size of 2400 mm long and 2700 mm high were carried out, of which two monotonic and three cyclic tests on the WHE specimen and one cyclic test on the WLE specimen.

The wall framing was made with stud members, having lipped channel sections (C-sections), spaced at 600 mm on the center and connected at the ends to track members, having unlipped channel sections (U-sections). Chord studs were composed by double C-sections coupled back-to-back. In order to reduce the unbraced length of the chord and interior studs, flat straps were placed at the mid-height of the wall specimens and were screwed to blocking members placed at the ends of walls. The tracks were reinforced by C-section profiles assembled in a box sections. Hold-down devices were used to anchor chord stud ends, while bolts distributed along tracks spaced at 200 mm, were used as shear connections. The wall specimens were completed with strap braced installed in an X configuration on both sides and connected to the wall framing by gusset plates. All the steel members were made with S280 GD steel grade. All connections were made with 6.3 x 40 mm (diameter x length) hexagonal head self-drilling screws. The nominal dimensions and material properties of the wall components are provided in Fig. 2.

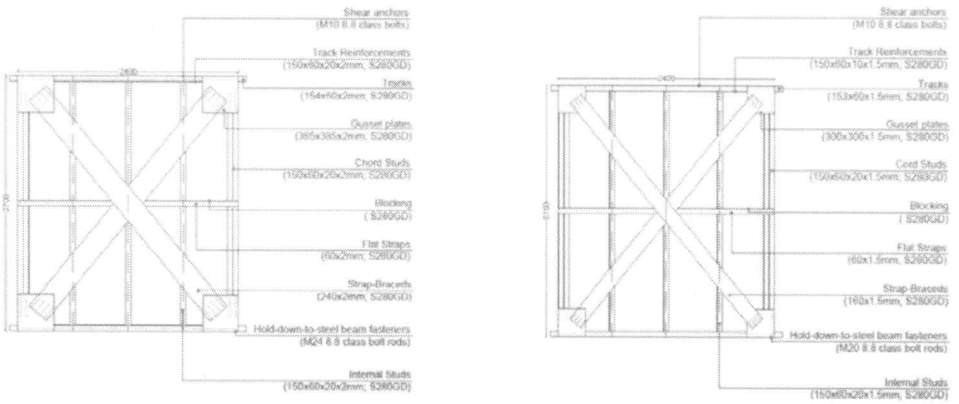

a) WHE specimen b) WLE specimen

Fig. 2: Wall specimens

Tests on full-scale wall specimens were carried out by using a specifically designed testing frame for in-plane horizontal loading. Horizontal loads were transmitted to the upper wall track by means of a steel loading beam with rectangular hollow section. The wall prototype was restrained to the laboratory strong floor by the bottom beam of testing frame. The out-of-plane displacements of the wall were avoided by two lateral supports realized with HEB 140 columns. The tests were performed by using a hydraulic actuator having a stroke displacement of 500 mm and a load capacity of 500 kN. A sliding-hinge was placed between the actuator and the tested wall in order to avoid additional vertical load components. Walls displacements were measured by eight LVDTs, as shown in Fig. 3. In particular, three LVDTs (P_1, P_2 e P_3) were used to measure the horizontal displacements and two LVDTs (P_4, P_5) were used to measure the vertical displacements. The strain of the diagonal straps were recorded by means of three strain-gauges, placed at the centre (SG_2) and at the two ends of the strap (SG_1 and SG_3).

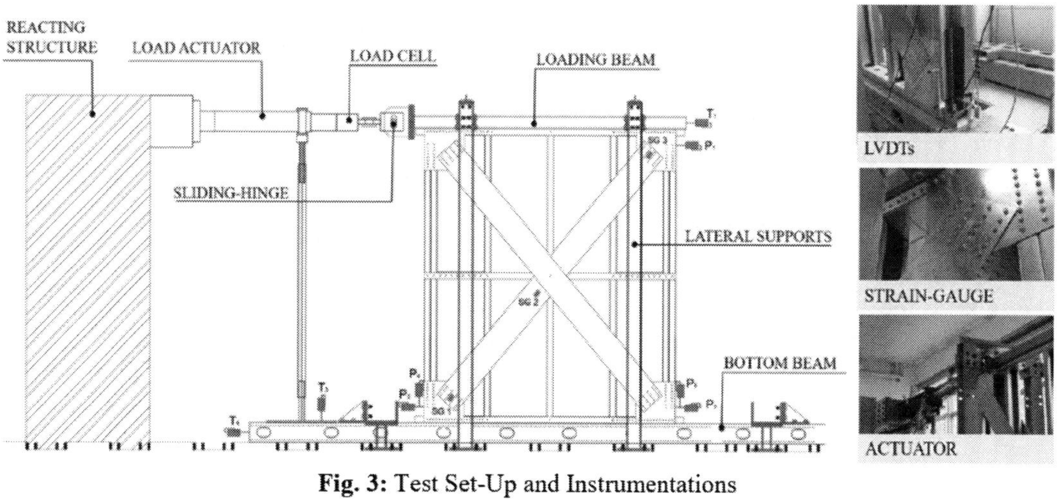

Fig. 3: Test Set-Up and Instrumentations

3.1 Monotonic tests

Two monotonic tests were carried out on the WHE specimen, with displacements imposed at a rate of 0.10 mm/s up to collapse. Table 1 shows test results in terms of: yield strength (H_y); maximum strength (H_p); displacement at the elastic limit (d_y); peak displacement (d_p), corresponding to the maximum strength; elastic stiffness (k_e), corresponding to the slope of the linear portion of the response curve; and observed failure mechanisms.

Table 1: Test results of monotonic wall tests

Type	H_y (kN)	H_p (kN)	d_y (mm)	d_{max} (mm)	k_e *(kN/mm)*	FM
WHE-M1	160.2	187.1	36.6	52.9	3.8	GT
WHE-M2	164.1	185.7	38.9	58.7	4.7	GT

FM: Failure mode; GT: failure of gusset-to-track connection

Fig. 4a shows the response curve in terms of load vs. displacement for WHE-M2 specimen and the above defined parameters. Test results show small differences, within 3% for maximum strength, while for stiffness the difference is 23%. Furthermore, the results highlight that the experimental wall strengths are 0.2% higher with respect to the theoretical predictions, whereas the experimental stiffness is lower of 63% respect to the theoretical value. For both monotonic tests, the observed failure mode was the breaking of the gusset-to-track connection (shear failure of the screws) (Fig. 4b).

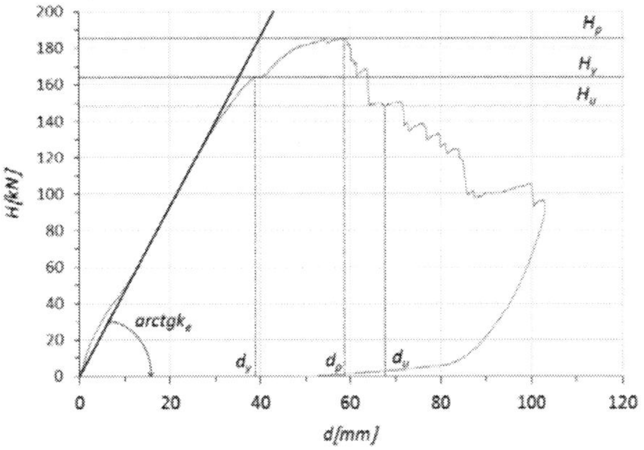

a) response curve of WHE-M2 specimen

b) Failure mode

Fig. 4: Monotonic test

3.2 Cyclic tests

Cyclic tests were carried out by adopting a loading protocol known as "CUREE ordinary ground motions reversed cyclic load protocol" developed for wood walls by Krawinkler et al. [3]. The cyclic loading test protocol consists of a series of stepwise increasing deformation cycles. The displacement amplitudes of each cycle are defined starting from a reference deformation $\Delta = \gamma \, \Delta_m$, where the value of Δ_m is calculated on the basis of monotonic test results, as the displacement corresponding to a load equal to 80% of the maximum load on the post-peak branch of the response curve (conventional ultimate displacement) and γ is assumed equal to 0.60 (Fig. 5a). The results of the cyclic tests are shown in Table 2. Fig. 5b provides the acting load (H) versus measured displacement (d) curve for the WHE-C1 specimen.

Table 2: Test results of cyclic wall tests

Type		H_y (kN)	H_p (mm)	d_p (mm)	k_e (kN/mm)	FM
WLE-C1	Pull/Push	81.8/79.1	102.5/103.7	71.7/59.6	.1.9/2.8	GT+SB
WHE-C1	Pull/Push	191.4/182.0	210.1/199.7	70.8/52.8	3.8/3.8	GT
WHE-C2	Pull/Push	190.1/190.9	199.3/198.7	72.8/52.5	4.5/4.0	GT
WHE-C3	Pull/Push	146.9/187.2	193.1/204.6	135.5/60.0	4.1/4.7	GT+SB

FM: Failure mode; GT: failure of gusset-to-track connection; SB: stud buckling

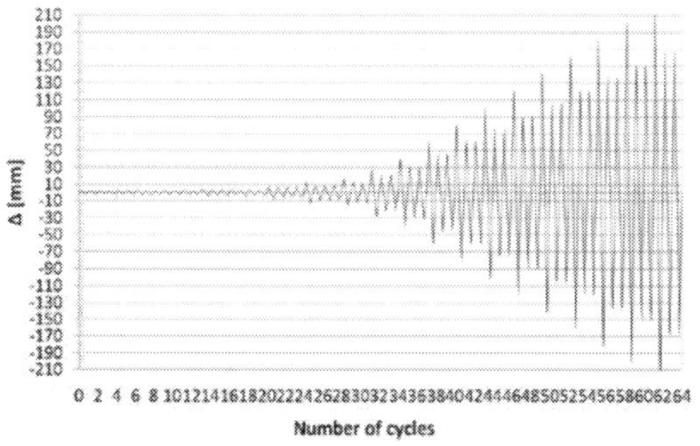

a) CUREE Protocol

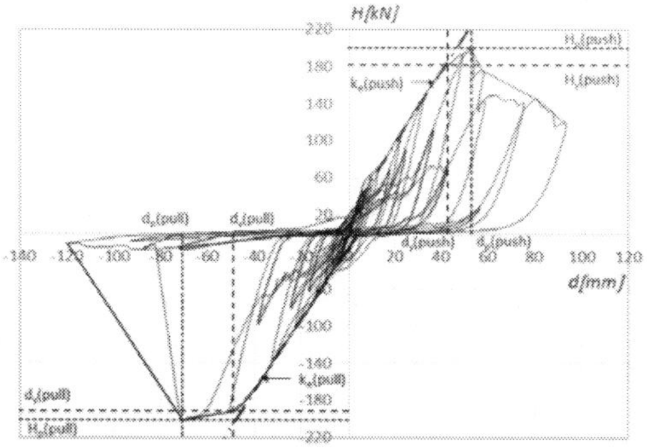

b) Response curve of WHE-C1 specimen

Fig. 5: Cyclic Test

The results show that the strength and stiffness recorded in the pushing phase with respect to the pulling phase have maximum differences of 6% and 15%, respectively, except a variation of 48% for the stiffness of WLE-C1 specimen. The failure mode observed for all tests is due to the interaction of different phenomena. In particular, the first mechanism was the local deformation of the tracks followed by the crushing of the studs and the bearing of the screw holes of the track, while the post-peak response was characterized by the failure of gusset-to-track connections. In the final phase of WHE-C3 and WLE-C1 tests the global instability of the chord studs tests was also observed. The comparison between monotonic and cyclic test results for the WHE walls reveals that the average experimental strength registered under monotonic loads is lower than the one recorded in the cyclic tests with maximum variations of 8%, while the average stiffness for monotonic tests is higher of 2% than the cyclic one.

4. Tests on material and components

4.1 Test on materials

Material coupons of frame members and hold-down devices were subjected to conventional tension tests according to EN ISO 6892-1 [4]. In particular, tests were performed on four specimen types: S280-1.5, S280-2.0, S275-20.0 and S235-8.0. Properties of samples and tension test results of the tension tests are shown in Table 3.

The average experimental values of the yield strength are larger than nominal values (3,5%, 22%, 16%, 14% for S280-1.5, S280-2.0, S275-20.0, S235-8.0, respectively). Also the results in terms of average experimental ultimate strength show an increment respect to nominal values (11%, 16%, 9%, 5% for S280-1.5, S280-2.0, S275-20.0, S235-8.0, respectively).

Table 3: Tests results on material

Type	*Steel grade*	*Thickness* (mm)	*n. tests*	f_y (Mpa)	f_t (Mpa)	$f_{y,av}$ (Mpa)	$f_{t,av}$ (Mpa)
S280-1.5	S280GD	1.5	3	280	360	290	400
S280-2.0	S280GD	2.0	3	280	360	343	420
S275-20.0	S275JR	20.0	3	275	430	320	468
S235-8.0	S235JR	8.0	3	235	360	267	378

f_y: nominal yield strength; f_t: nominal ultimate strength; $f_{y,av}$: average experimental yield strength; $f_{t,av}$: average experimental ultimate strength.

4.2 Tests on frame-to-strap connections

The CFS strap-braced stud walls behaviour is particularly affected by the design of frame-to-strap connections, which in this case takes place through steel gussets. For this reason, a series of tests on connection prototypes reproducing the joints between gusset and strap-bracing was carried out. In particular two types of specimens were tested: CHE connections, adopted in the WHE walls, and CLE connections, adopted in the WLE walls (Fig. 6a). The theoretical strength (F_{th}) values, for the tested connection types, are equal to 83 kN and 177 kN for CLE and CHE prototypes, respectively. The number tests, experimental values of average failure load ($F_{p,av}$), average stiffness ($k_{e,av}$) and failure mechanisms observed for each test typologies are summarized in Table 4. Fig. 6b shows the experimental response curves obtained for all tests. In particular, the experimental strength for CHE connections is lower than theoretical ones with a reduction of 4%, while for CLE connections the experimental value is 3% higher than theoretical ones. For all tests the failure mechanism was the tilting of screws with subsequent net section failure of the plate.

Table 4: Test results on gusset-strap connection

Type	*Steel grade*	*Thick. Plate* (mm)	*no. tests*	*Screw Diameter (mm)*	*no. screws*	$F_{p,av}$ (kN)	$k_{e,av}$ (kN/mm)	FM
CLE	S280GD	1.5	3	6.3	16	85.30	44.04	T+NST
CHE	S280GD	2.0	3	6.3	24	170.34	78.99	T+NST

FM: Failure mode; T: screw tilting; NSF: net section failure of strap-bracing

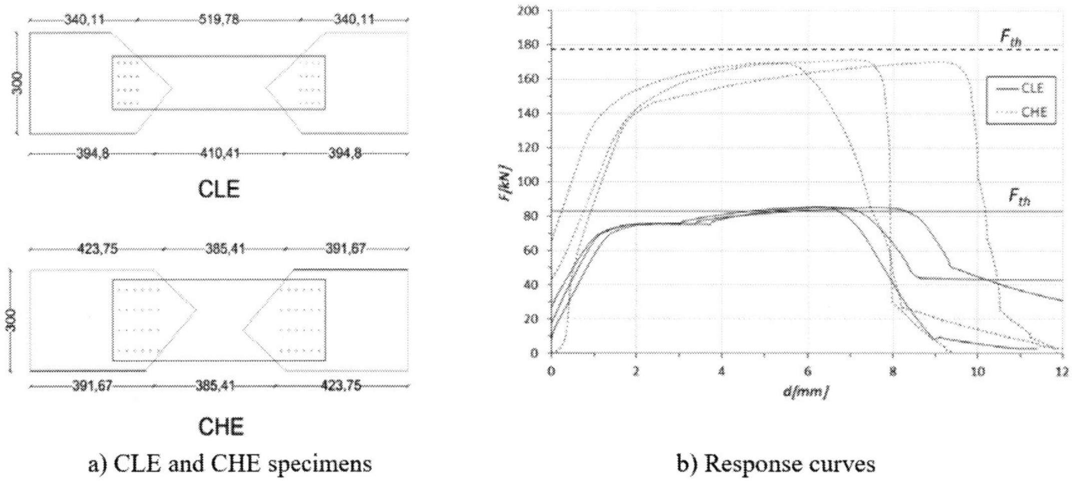

a) CLE and CHE specimens b) Response curves

Fig. 6: Tests on frame-to-strap connections

4.3 Tests on hold-down devices

The local response evaluation of the investigated X-braced CFS systems was completed with four tensile tests on hold-down devices. The two types of hold-down devices adopted for the two different configurations walls were tested: HD-H, adopted in the WHE wall, and HD-L, adopted in the WLE wall (Fig. 7a). Single test was carried out on two specimens connected back-to-back by means of a 30.0 mm thick S355JR steel plate and four M18 or M14 class 8.8 bolts for HD-H or HD-L devices, respectively. The steel plate was clamped to the upper wedge grip of testing machine. Furthermore, each hold-down device was connected at the bottom side to a stiffened T-shaped steel element by means of a M24 or M20 class 8.8 bolt for HD-H or HD-L devices. The T-shaped element was clamped to the bottom wedge grip of testing machine (Fig.7b). The theoretical strength (F_{th}) values for HD-L and HD-H devices are equal to 176.4 kN and 237.7 kN, respectively. Table 5 shows a summary of experimental results, while Fig. 8 shows the experimental force-displacement curves. The limit resistance (F_{lim}) of the device is defined as the force corresponding to a limit displacement (d_{lim}), set equal to 4.7mm, according to the AISI S913 [5].

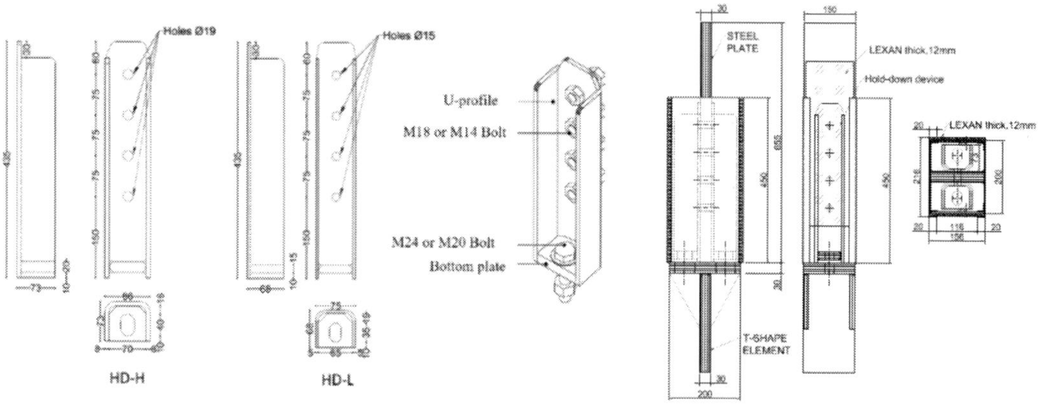

a) HD specimens b) Set-up

Fig. 7: Tests on hold-down device

Table 5: Test results on hold-down device

Type	*Steel grade*	*Thick.* (mm)	*Bolts* (mm)	*no. tests*	$F_{lim,av}$ (kN)	$k_{e,av}$ (kN/mm)	FM
HD-L	Bottom plate S275JR	15	M20 Grade 8.8	2	138.1	353.4	T
	U-profile S235JR	5	M14 Grade 8.8				B
HD-H	Bottom plate S275JR	20	M24 Grade 8.8	2	247.7	89.0	-
	U-profile S235JR	8	M18 Grade 8.8				

FM: Failure mode; T: tension failure of the U-shaped element; F: bending failure of the base plate

Tests of HD-H specimens were carried out to reach the maximum load capacity of the universal testing machine (500 kN). These specimen showed only a strong deformation of the base plate for high values of the applied load, without any rupture (Fig. 9a). HD-L specimens showed two different collapse mechanisms. In particular, during the first test a tensile failure of the U-profile was observed (Fig. 9b), while the second test was characterized by the bending failure of the base plate (Fig. 9c).

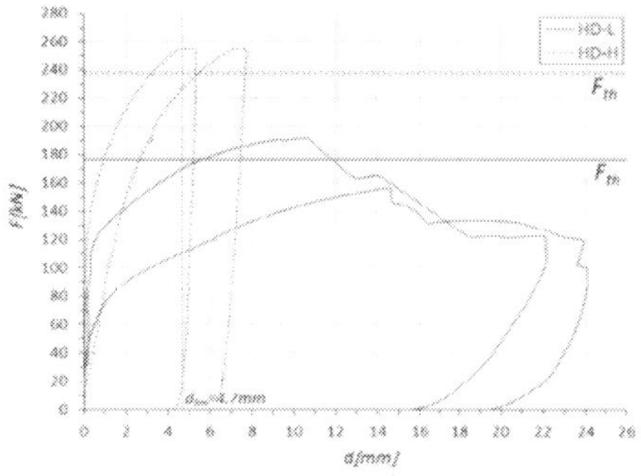

Fig. 8: Response Curves

a) Strong deformation of the base plate b) Tensile failure of the U-profile c) Bending failure of the base plate

Fig. 9: Failure mode of hold-down devices

5. Conclusions

The main conclusions may be summarized as follows:

1. The experimental campaign aimed to study the global response of strap-braced walls designed according to an elastic approach.
2. The local behaviour was also investigated by tests on material, main connection systems and hold-down devices.
3. Results show a good agreement between predicted and experimental behaviour in terms of strength and stiffness.
4. Experimental results may be considered as reference for numerical studies aimed to define seismic design criteria for the investigated systems.

References

[1] Fiorino L, Iuorio O, Macillo V, Terracciano MT, Pali T, Landolfo R. "Seismic Design Method for CFS Diagonal Strap-Braced Stud Walls: Experimental Validation", *Journal of Structural Engineering, ASCE*, 2015.

[2] Macillo V, Iuorio O, Terracciano MT, Fiorino L, Landolfo R. "Seismic response of Cfs strap-braced stud walls: Theoretical study", *Thin-Walled Structures*, 85, 301-312.

[3] Krawinkler H, Parisi F, Ibarra L, Ayoub A, Medina R. "Development of a Testing Protocol for Woodframe Structures", *CUREE*, 2001.

[4] UNI EN ISO 6892-1. *Metallic materials - Tensile testing - Part 1: Method of test at room temperature*. European committee for standardization, 2009.

[5] AISI S913-08, *Test Standard for Hold-downs Attached to Cold-Formed Steel Structural Framing*, United States, 2008.

INFLUENCE OF CYCLIC LOADING FOR STEEL ECCENTRICALLY BRACED FRAMES WITH SHORT LINK UNDER LATERAL LOADS

Adina Vătăman[a], Adrian Ciutina[a] and Daniel Grecea[a,b]

[a] *Department of Steel Structures and Structural Mechanics, Politenica University of Timişoara, Romania*
[b] *Romanian Academy - Timişoara Branch, Romania*

Abstract: Eccentrically braced steel frames are capable of dissipating the seismic energy by plastic hinge formation in the frame's dissipative devices, called 'link elements'. The current study is focused on the calibration of existing experimental monotonic and cyclic results of steel eccentrically braced frames with short links by numerical finite element models in view of improving the overall frame behaviour. A subsequent parametrical study was performed by considering different link lengths and by variation of number of stiffeners disposed on the link. The final results show the influence of cyclic loading in regard to monotonic response. The results are discussed in terms of strength, stiffness and ductility.

1. Introduction

The general design of steel frames subjected to seismic loading considers the lateral stiffness criterion, in order to limit the lateral deformations. As Vayas et al. [1] demonstrated, the deformations of non-structural elements is a concern in the serviceability limit state, but in the ultimate limit state conditions the deformations must be limited in order to avoid the second order instabilities. In addition, the ductility of plastic hinges represents an important factor in case of seismic loading, in order to allow the plastic redistribution of forces [Gioncu et al., 2]. A measure of the ductility is the plastic rotation capacity of the member sections.

Eccentrically Braced Frames (EBF) represent structures designed to withstand horizontal forces by deformation through dissipative mechanisms formed by deformation of link elements. Generally, these are beam segments located outside the triangulated beam-column-brace systems and designed in order to allow the yield in shear or bending. The plastic deformation is mainly dependent on the link element length which can be short (length $e < 1.6M_{pl}/V_{pl}$ according to EN 1998-1 [3]), long (length $e > 3M_{pl}/V_{pl}$) and intermediate links length. Present paper regards the behaviour of short links. They present generally a stiff behaviour and exhibit the plastic energy by shear deformations of the web panel, differently of long link elements for which the plastic hinges form in bending at each side of the links.

The numerical model is based on a large experimental program, previously performed at the Research Centre for Mechanics of Materials and Structural Safety (CEMSIG) of the Politehnica University in Timişoara. The experimental study was concentrated on the behaviour of EBF subjected to lateral loading conditions, similar to seismic loading, which provides the basis for the current numerical approach. The program included eight full-scale EBF tests, which were tested both monotonically and cyclically and included pure steel frames and composite beam solution. The full experimental report can be found elsewhere [4].

The sub-assembly frame chosen for experimental testing was part of a larger four-storey structure designed as a dual frame and represented the ground floor centre bay, as shown in Fig. 1. The dissipative link element has a length of 300 mm and is classified as short link element according to EN1998-1 and consequently the characteristic deformation will be in shear.

2. Monotonic behaviour of EBFs with short links

2.1. Calibration of finite element model and parameters

The numerical FE model was calibrated on the basis of monotonic response of EBF-LF-M frame specimen. This represents a steel structure with the steel sections mentioned in Fig. 1. The beam, including the link element is a HEA 200 European profile, on S235 steel. The system was loaded through a loading actuator at top of the beam.

The numerical finite element model, created using the Abaqus software, considered equivalent geometric and sectional properties as in experimental test. All frame components were modelled as solid finite elements. The assigned material properties corresponds to true stress-strain curve for the beam and dissipative link, built from real the material tests performed on coupons. The Dynamic Explicit type analysis procedure used was with 180 mm applied displacement at tip of the left column, identical to the maximum experimental displacement.

In order to faithfully reproduce the experimental Force-top Displacement behaviour (see Fig. 1a), the slip recorded during testing in brace connections as well as the presence of lateral supporting system of the frame were considered in the FEM analysis. Additional information concerning calibration of monotonic behaviour is given in Vataman et al. [5].The numerical resulting force-displacement curve LS3-H0 closely follows the experimental curve EBF-LF-M, as observed in Fig. 1b. Deformed shapes presented in Figs. 1c and 1d also show a similar failure mode by pure shear of link element.

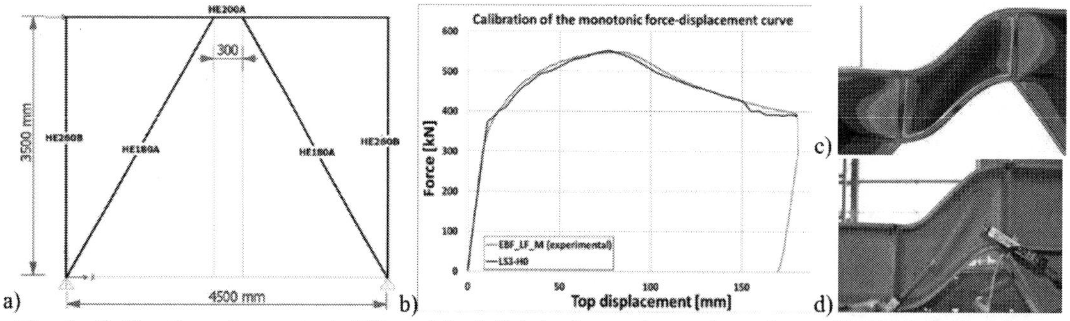

Fig. 1: Calibration of monotonic FE model: a) Original experimental specimen; b) FEM Calibration curve vs. experimental; deformed shape of the dissipative link elements: c) FEM, d) Experimental

2.2. Influence of link length on the monotonic behaviour

The influence of link length in case of monotonic responses was considered through a parametric study, considering different link lengths of 300/400/500/600/750 mm, while preserving the original characteristics of short link elements. The maximum length for the current configuration in order to have a short link was 760 mm.

This resulted in a total of five frame models, as presented in Fig. 2 a). The notation of the EBF models reflect the frame characteristics: LS is the generic name of the series (standing for steel link), first number describes the length of the link (e.g. 3 for 300 mm), H describes the cross-section profile of the beam (HE200A) and the last value indicates the number of stiffeners evenly distributed over the length of the link (0 in these cases).

The results show that link length has no significant influence on elastic and post-elastic ranges, as seen in Fig. 2. There is a slight decrease in elastic and ultimate strengths with the length, but percentage between the resistances is less than 8% for ultimate and 2% in case of elastic limit strength respectively. The maximum strength is obtained in case of 300 mm model, almost coincident with the resistance of the models with the link lengths of 400 and 500 mm. However, in case of larger lengths, such as 750 mm, the web loses the stability earlier, thus having limited ultimate resistance and ductility. The failure modes show that links elements with shortest dimensions (up to 400 mm) are characterized by a global-type shear plasticisation while the longer link elements (500 to 750 mm) are characterised by web crippling on lateral sides of the panel, while the central part remains plane although presenting high levels of shear stresses.

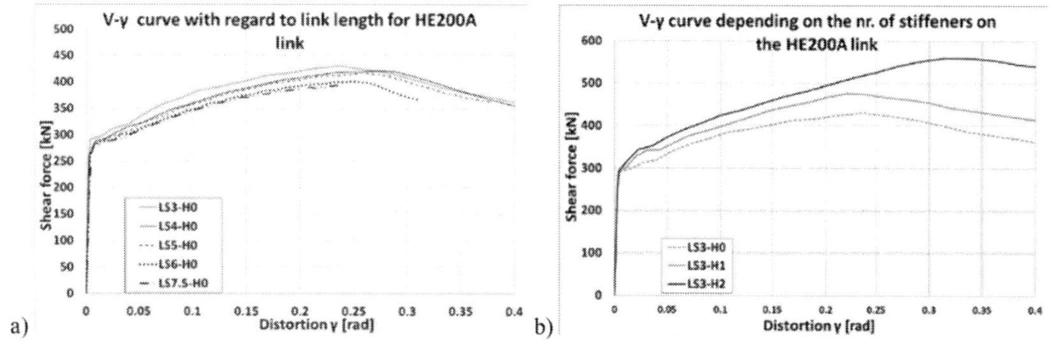

Fig. 2: Monotonic Shear force – link distortion FEM response curves:
a) Influence of link length; b) Influence of number of stiffeners

2.3. Influence of link web stiffening on the monotonic behaviour

In order to study higher levels of shear link stiffening, additional intermediate stiffeners were disposed at equal intervals according to the requirements from the EN1998-1 - maximum distance of 155 mm between stiffeners for HEA 200 profile. The influence of number of stiffeners on the overall shear behaviour was analysed by considering three configurations in case of LS3-H models: (i) models without stiffeners on the dissipative link (ii) models with one stiffener in the middle of the link (representing the normative solution) and (iii) models with 2 stiffeners on the link, at equal intervals from the centre and from the end of the link.

Fig. 2 b) presents the *V-γ* response curves for these cases. The elastic behaviour, initial stiffness and the point of plasticisation are not influenced by the number of stiffeners. However, in the post-elastic range there are some important observations: the stiffening of the link element conducted to linear increases of maximum loads; accordingly, the hardening branch stiffness increased significantly in direct relation to the numbers of stiffeners; the influence on the discharging branch stiffness is not important. For each additional stiffener there is an increase in initial maximum resistance by about 20%. The increase of the ultimate rotation is observed only when using two stiffeners.

Regarding the failure modes, the additional stiffeners have led to the division of the original shear panel into multiple panels, each of these having its own web deformation. In the elastic and initial plastic loading conditions the multiple panels are working together, however in the final loading conditions the plasticisation is concentrated in one of the panels, the other contributing only partially to the global deformation.

3. Cyclic behaviour of EBFs with short links

3.1. Calibration of finite element model and parameters

The calibration considering cyclic loading conditions is based on the same study but considering the response of the pure steel eccentrically braced frame loaded cyclically - specimen EBF-LF-C. The experimental response of EBF-LF-C specimen has shown a ductile behaviour with good dissipation capacity of the link element with high values of distortion exceeding 150 mrad (Fig. 3c). The maximum resistance was of 550 kN, as can be observed in Figure 4. The failure mode was by alternate shear buckling of the link web panel in positive and negative cycles followed by tearing of web panel (see Fig. 4a).

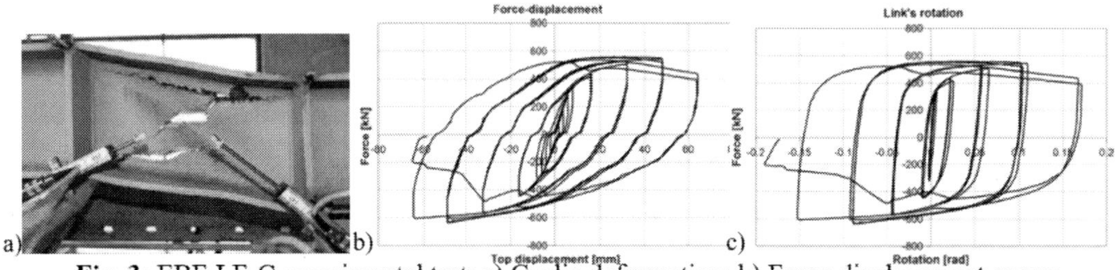

Fig. 3: EBF-LF-C experimental test: a) Cyclic deformation; b) Force-displacement curve; c) Shear force-rotation curve

The numerical model of the tested frame was created and analysed using the finite element (FE) software Abaqus [6]. Whilst taking into consideration a cyclic loading procedure, the geometry and monotonic material properties of the reference specimen were maintained but specific cyclic parameters were accounted:

- plastic material properties were assigned in order to provide a combined isotropic and kinematic hardening for beam steel material, including the dissipative link element [7];
- the cyclic material model adopted considers 5 back-stresses in order to model the material behaviour, as suggested by Chaboche [8].

Solid finite elements were used for the frame members. Hex type elements were used for meshing, considering a sweep technique with mixed median axis and advancing front algorithms. The finite element size ranges from a fine mesh with 6 mm elements on the dissipative area of the frame (link element and adjacent beam area) to larger element sizes of 15 mm for braces, 20 mm for beam and beam stiffeners and 25 mm for columns and column stiffeners.

The loading procedure that followed the ECCS cyclic loading procedure [9] used in experimental testing and based on FE application of cyclic loading [10]. Accordingly, three different amplitudes of $0.25D_y$, $0.5D_y$ and $0.75D_y$ were applied in the elastic range. In the plastic range amplitudes of $1D_y$, $2D_y$, $4D_y$, $6D_y$ and $8D_y$ were applied in order to simulate the cycles of experimental tests. In all above, D_y represents the yield displacement resulted from the interpretation of monotonic tests, in accordance to the ECCS procedure. The load application point is identical to monotonic simulation - at the top of the frame and applied in displacement control.

As in case of monotonic models, the slip in the brace connections should be included in analysis, which means considering a rigid body element and a connector that allows for an elastic spring elongation of the braces. Fig. 4 presents the experimental frame top-displacement curves for experimental response (EBF-LF-C curve) and the FE calibration (LS3-H0-C curve). The change in slip from monotonic loading to cyclic, leads an increased initial stiffness, and also a better accuracy with the restoring stiffness.

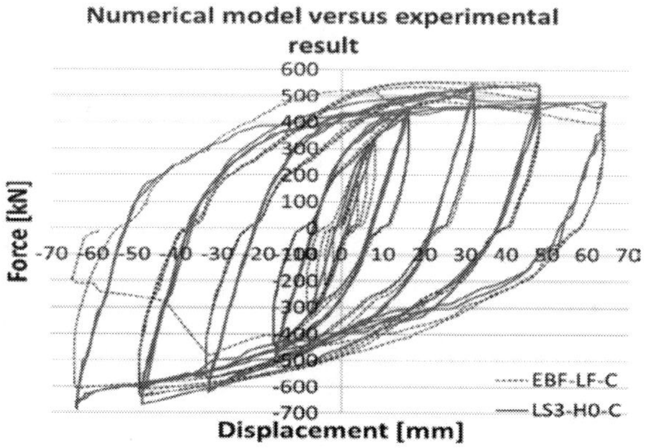

Fig. 4: Calibration curve for cyclic loading

3.2. Influence of link length on cyclic behaviour

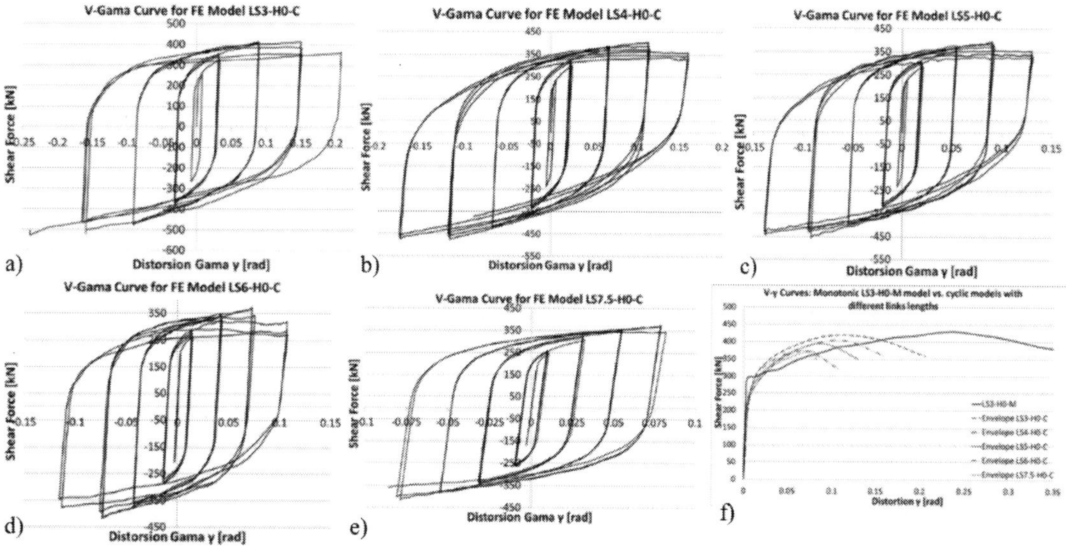

Fig. 5: V-γ response curves for cyclic numerical analysis of EBF models with different link lengths: a) 300 mm; b) 400 mm; c) 500 mm; d) 600 mm; e) 750 mm; f) – envelope curves (positive loading)

Following the parametric study on monotonic FE models, a similar study was performed, taking into consideration different link lengths of 300/400/500/600 and 750 mm. The cyclic results are presented in Fig. 5 in terms of equivalent V-γ (shear force-panel distortion) curves. The values show that the cyclic behaviour introduces a slight limitation in distortion capacity as compared to monotonic responses. However, these values remain higher than the limit value recognised by EN1998-1 of 80 mrad. In case of longer links, the maximum distortion is smaller, as in case of monotonic results. Generally, the cyclic maximum resistances are smaller than the ones recorded in monotonic loading, but the differences are small of up to 5%.

Fig. 5f presents comparatively the V-γ envelope curves from positive quadrants (positive shear force, positive distortion). This shows clearly that in cyclic behaviour, longer links lead to a drop on the maximum cyclic resistance and also on the distortion capacity of the link. Though, the elastic stiffness remains similar for all link lengths.

3.3. Influence of link web stiffening on the monotonic behaviour

A second parameter that has been considered is the stiffening of the link web, with three different configurations being analysed starting from the base model LS3-H0-C without stiffeners on the link: a model with one intermediate vertical stiffener (LS3-H1-C) and respectively a model with two intermediate vertical stiffeners on the link web (LS3-H2-C).

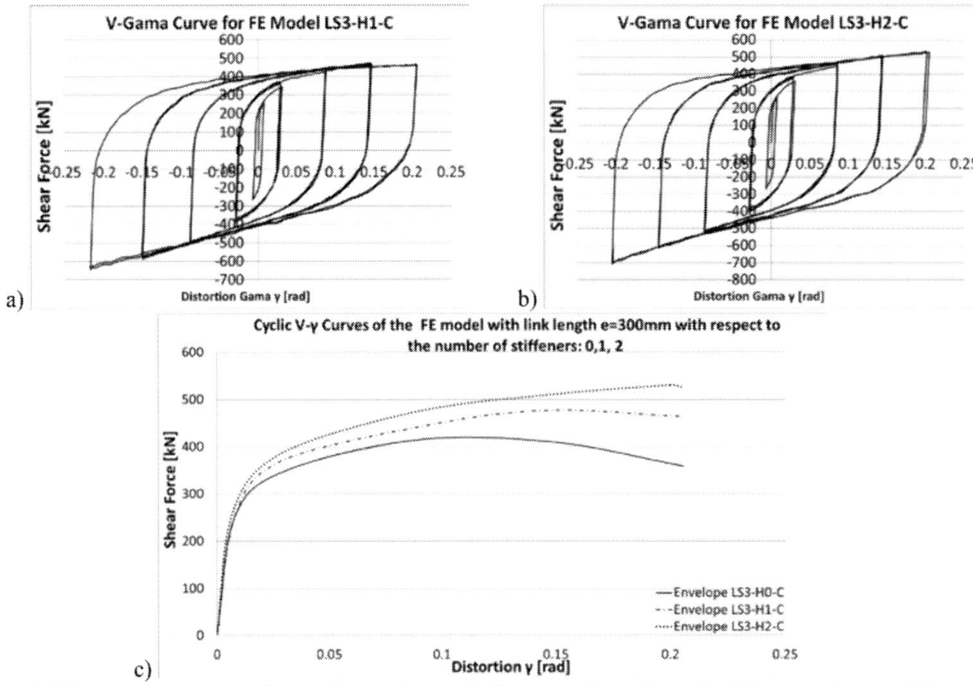

Fig. 6: V-γ response curves for cyclic analyses - influence of number of stiffeners: a) one stiffener; b) two stiffeners c) – envelope curves (positive loading)

According to the shear force-panel distortion V-γ response curves, presented in Fig. 6, the influence of stiffeners is important as they induce important improvements in post-elastic domain:

- the maximum resistance is improved significantly, with more than 10% for each additional stiffener. However, this value is reduced in comparison with monotonic results where the increase is of 20%;
- the distortion capacity remains high, in all the cases over 150mrad. The envelope curves presented in Fig. 6c show that in case of LS3-H2-C model the maximum distortion values should be obtained for cycles superior to $8D_y$.

4. Assessment of cyclic influence

Table 1 presents the analytical values resulted from the V-γ response monotonic and cyclic envelope curves. The values computed from the response graphs include:

- elastic resistance V_y. Represents the shear force value, computed according to ECCS procedure at the intersection of elastic stiffness trend-line and a line of 10% initial stiffness, tangent to V-γ response curve;
- S_{el} – the elastic stiffness of the V-γ response curve;
- V_{max} – maximum shear force;
- γ_{max} – maximum distortion angle computed at 10% drop in V_{max}.

Table 1. Analytical appreciation of LS – H models

Model Name	Link length [mm]	No. of stiffeners	V_y [kN]		S_{el} [kN/rad]		V_{max} [kN]		γ_{max} [rad]	
			M	C	M	C	M	C	M	C
LS3-H0	300	0	283.33	210.67	93689	89764	430.21	416.99	0.23691	0.2093
LS3-H1	300	1	285.56	216.67	95531	93380	476.28	476.6	0.22018	0.20627
LS3-H2	300	2	288.89	222.22	96121	95125	559.77	535	0.31611	0.20530
LS4-H0	400	0	261.11	208.89	86170	85862	421.4	406.9	0.27115	0.17496
LS5-H0	500	0	258.89	213.33	74423	80405	416.73	395.1	0.25768	0.13855
LS6-H0	600	0	261.11	213.33	72456	65379	400.65	372.87	0.24908	0.10762
LS7.5-H0	750	0	261.11	211.11	58385	57707	393.44	378.26	0.23943	0.08189

As in case of monotonic and cyclic experimental tests, all the cyclic models show similar values of elastic stiffness to monotonic loaded models. The values are significantly smaller for longer links (LS7.5 models). No differences exist between models with none, one or two stiffeners on the link.

In exchange, the maximum resistance is directly dependent on the presence of stiffeners and the link length: the stiffeners induce higher maximum loads, while longer links lead to smaller resistances. However, this is not found on the elastic resistance, which is similar in all monotonic and respectively all cyclic cases at different values.

The values of γ_{max}, presented in Table 1 show similar monotonic and cyclic maximum distortions in case of LS3-H0 and LS3-H1 models. However, the differences become higher in case of longer links, recording in case of LS7.5-H0 only 82 mrad in case of cyclic loading starting from a monotonic value of 240 mrad.

5. Conclusions

Present study is focused on the seismic behaviour of EBF with short link elements. Based on a previous experimental study on one-storey EBF, including both monotonic and cyclic loading, two numerical models were developed and calibrated on these experimental results. Further, a parametric study was accomplished based on two parameters: variation of link length and respectively by variation of numbers of stiffeners on the web of the link element. The numerical results show the following:

- both monotonic and cyclic experimental results can be reproduced numerically by careful definition of material laws and refined replication of testing details such as brace elongations; in case of cyclic loading additional material parameters should be considered in order to control the cyclic behaviour of entire frame. The Chaboche material model seems to be adequate in accurately modelling the cyclic material response;
- for both monotonic and cyclic loading, increasing the number of stiffeners will induce higher shear resistances and improves the overall ductility of the frame. However, it does not influence the elastic stiffness of the system;

- the increase of link length while still remaining in range of short links, will conduct to smaller maximum resistances and affects significantly the elastic stiffness of the system: very short links present a stiffness almost double as compared to ones at with length the border to intermediate length links;
- cyclic loading induces both a reduction in maximum shear resistance and distortion capacity of the system. The reduction in shear resistance is up to 5% in all cases, while the reduction in distortions depends on the link length: smaller for shorter link lengths (10%) and reduced to one third in case of longer link;
- in all the monotonic and cyclic models, the maximum link distortions satisfy the minimum normative requirements of 80 mrad, in accordance with EN 1998-1.

References

[1] Vayas I. "Design of braced frames", *Seismic resistant steel structures, Chapter 5, CISM Courses and Lectures*, 420, 241-288, Springer Verlag Wien GmbH, , 2000.

[2] Gioncu V, Mazzolani F. *Seismic design of steel structures*, CRC Press Taylor and Francis Group, 2014.

[3] EN 1998-1 EUROCODE 8, Design of structures for earthquake resistance, Part 1. General rules, seismic actions and rules for buildings, Brussels, CEN, European Committee for Standardisation, 2003.

[4] Danku G. Study of the development of plastic hinges in composite steel-concrete structural members subjected to shear and/or bending, *PhD Thesis*, Politehnica University of Timisoara, Editura Politehnica, Seria 5, Nr. 72, 2011.

[5] Vătăman A, Grecea D, Ciutina A. "Influence of detailing of short links on seismic response of eccentrically braced frames", *The 8th STESSA Conference*, Tongji, Shanghai, 834-841, 2015.

[6] Abaqus 6.11-1, Dassault Systèmes Simulia Corp, 2011.

[7] Arasaratnam P, Sivakumaran KS, Tait MJ. "True stress-strain models for structural steel elements", *International Scholarly Research Notices (ISRN) Civil Engineering*, Volume 2011, Article ID 656401, 2011.

[8] Chaboche JL. "A review of some plasticity and viscoplasticity constitutive theories", *International Journal of Plasticity*, 24(10), 1642-1693, 2008.

[9] European Convention for Constructional Steelwork, ECCS TWS 1.3 N. 45/86, Recommended testing procedure for assessing the behaviour of structural steel elements under cyclic loads, 1986.

[10] Vulcu C, Stratan A, Dubina D. "Numerical simulation of the cyclic loading for welded beam-to-CFT- column joints of dual-steel frames", *Pollack Periodica*, Vol. 7, No. 2, 35-46, 2012.

ENERGY STATE OF MULTI-STORY STRUCTURES EQUIPPED WITH PASSIVE PROTECTION

Pavel Alexa and Aliz Mathe

Technical University of Cluj – Napoca, Romania

Abstract: Proposed contribution deals with energy approach to seismic response of steel multi-story structures. The energy state of seismically acted upon structures is expressed by the amount of seismic input energy and the energy absorption capacity of the structure. To emphasize the versatility of proposed energy approach, the analysed structures are considered in several states of added viscous damping and three states of their lateral stiffness. Also, the structures are acted upon by three seismic actions different from each other by their specific time – variation. Computed results are presented in both, numerical and graphical forms and relevant concluding remarks are inferred regarding the energy approach to seismic response.

1. Introduction

Mitigation of seismic response of structures has reached a status marked by a high technological development and it became a common concern in the day-by-day design activity of structures located in seismic areas. Many times, the attention payed to the reduction of seismic response together with the applied technical solutions are referred as "dissipation of seismic energy". Indeed, dissipation of seismically induced energy into a structure is the key to ensuring the required seismic behaviour and seismic performances. The incentive to this has been put in the form of providing the structures with adequate capability to dissipate seismic energy [1]. Conceiving the structures and sizing their cross sections such that post-elastic behaviour will be induced where and when the designer considered appropriate and according to the provisions of the design codes is, still, a common practice. The ductility state is largely debated and viewed as a possibility to dissipate seismic energy via non-reversible plastic deformations. While the ductility has not lost its design importance and code provided status, the new technologies changed the locations of plastic deformations from the structural members to external equipment. Passive protection of structures seismically acted upon consists, in most of the cases, of "plastic" deformations of installed equipment saving, in this way, the irreversible deformations of structural members and conferring to the structure a high seismic performance capability. The idea of dissipating seismically induced energy via plasticization is, still, valid as in the case of conferring ductility. Energy approach to structural computation refers to the study of seismic energy itself, rather than to the study of (seismic) energy dissipating equipment. Energy based structural computation developed along both, the branch of seismic analysis [2], [3], [4] and the branch of structural design [5], [6]. Expressing seismic response in energy terms requires a large frame of structural mechanical state: elastic (stiffness / flexibility) state, inertia state, inherent damping and added damping state, the seismic action. While in classical structural analysis this complexity is dealt with by vectorial mathematics, energy approach engages scalar (energy) quantities that are both, simpler to deal with and more synthetic in their extent.

In this general frame of energy approach and dissipation of seismic energy, present contribution aims at defining and analysing the energy state of seismically acted upon structures equipped with passive seismic protection consisting of added viscous damping. The general objective – energy state – is dealt with by seismic analyses performed on multi-story steel structures in the presence of supplementary damping. Computed numerical results include seismic input energy and its structural components (kinetic energy, elastic strain energy) and the energy absorption capacity of the structure in several states of lateral stiffness, of damping level and seismic actions. Inclusion in energy state such a large number of structural and seismic action parameters and the scalar nature of energy allows for a synthetic expression of energy based fundamental condition of a seismically performance structure – the energy balance equation in the form:

$$E_{abs} \geq E_i \tag{1}$$

Here, E_{abs} is the required energy absorption capacity of the structure and E_i is amount of seismic energy induced into the structure. Required energy absorption capacity E_{abs} ensures seismic behaviour in the elastic domain of considered structure. Energy balance equation (1) may be viewed as demand – supply relationship: energy absorption capacity E_{abs} (what the structure supplies) has to be greater than E_i - the earthquake demand). How may the energy absorption capacity of the (designed) structure be greater than the induced seismic energy? Adding damping to the inherent damping level of the structure is one way of achieving this goal. Damping and energy are intimately related: damping (as a phenomenon) shows itself only by energy dissipation during motion. Therefore, expressing damping (inherent and added) in energy terms is only natural and appropriate. Relating the energy quantities (seismic input, kinetic, elastic strain) to the level of damping leads to the possibility of a more adequate structural conceiving. Many times, the structural adequacy versus the seismic action is assessed in terms of its kinetic state (mainly concerning the lateral relative displacements – in the case of multi-story structures). The capacity of the conceived / designed structure to absorb seismic energy offers a more comprehensive frame to evaluate the structure and its response to seismic actions than any other components of its mechanical state.

Considering these aspects and viewing the objective of the proposed contribution, it may be easily understood that the present study and its results do not aim at providing conclusions regarding the design of seismically acted upon structures or at proposing more accurate analytical tools for analysis or design of these structures. The proposed contribution focuses on the utility and versatility of energy approach technique to evaluate multi-story structures located in seismic areas. Nevertheless, several sets of seismic analyses have been carried out for the purpose of relating the energy state to mechanical specificities of the structures and to the peculiarities of the seismic actions.

2. Analysed structures and seismic actions

The set of analysed structures comprises several multi-story steel planar frames. The frames are considered in three states of their lateral rigidity (Fig. 1) to allow for assessing the dependence of both, seismic input energy and energy absorption capacity to the elastic state of the structure. Also, several levels of added linear viscous damping to the inherent level of 2% (ratio of critical damping) have been taken into account in order to emphasize the relation of energy absorption capacity with the level of damping. The ratios of critical damping associated to the three levels of added damping are: 5%, 10% and 15%. In their turn, the three recorded earthquakes (Fig. 2) considered in the analyses (El Centro 1940, Vrancea 1977 and Focsani 1986) exhibit three different developments of the seismic actions. Their dynamic vicinity with the analysed structures - in terms of vibration periods (fundamental, for structures and predominant, for earthquakes) – is, also focused.

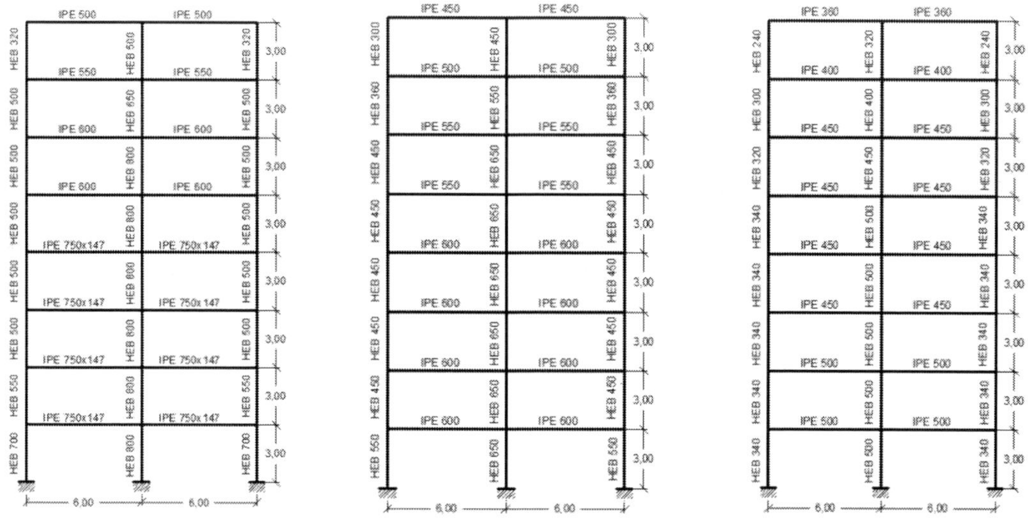

a) Rigid structure b) Moderate rigid structure c) Flexible structure

Fig. 1: Analyzed eight - story structures

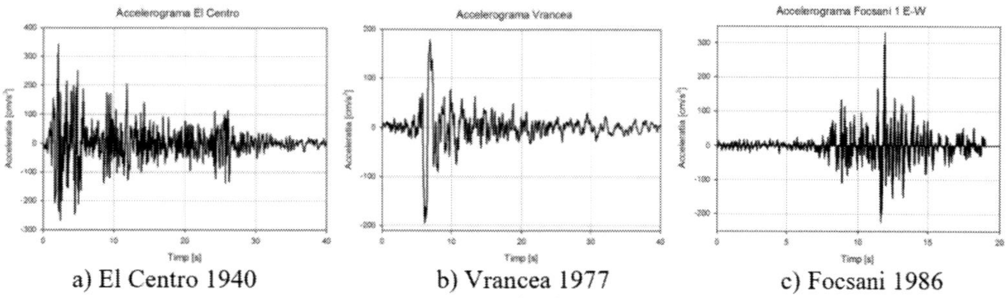

a) El Centro 1940 b) Vrancea 1977 c) Focsani 1986

Fig. 2: Seismic actions

The predominant periods of the three recorded earthquakes are: 0,56 sec. for El Centro, 1,16 sec. for Vrancea 1977 and 0,50 sec. for Focsani 1986. The fundamental periods of natural modes of vibration of the eight – story structure in its lateral stiffness cases are [9]: $T_1 = 0,78$ sec. for the rigid case, $T_1 = 0,897$ sec., for the moderate rigid case and $T_1 = 1,27$ sec. for the flexible case. Presented numerical results exhibit the dependence of seismic input energy E_i and of energy absorption capacity E_d on the seismic action, on the elastic state (lateral stiffness, in this case) and on the damping level. If seismic actions may be viewed as immutable design parameters, the energy absorption capacity constitutes an instrument in the hands of structural engineer in both, the stage of conceiving the structure and in the stage of designing its cross sections. Here lies the importance of E_d – elastic state relationship.

3. Numerical results in energy parameters

Following, are some of the computed results [8] referring to the seismic response of the eight – story frame in terms of variation in time of seismic input energy E_i and energy absorption capacity E_d.

1. The case of Vrancea 1977 seismic action and 2% critical damping ratio.

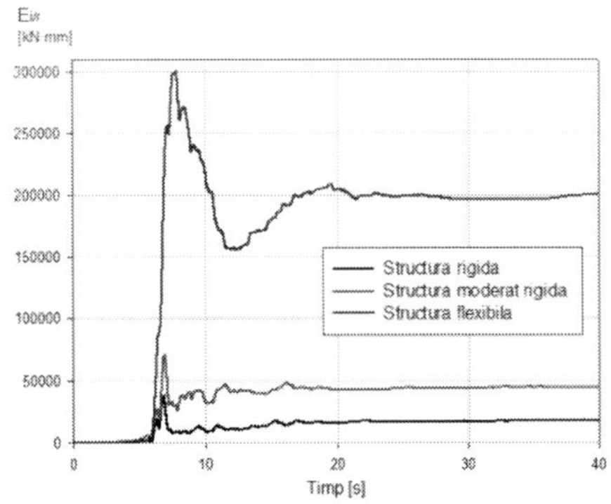

Fig. 3: Relative seismic input energy versus elastic state

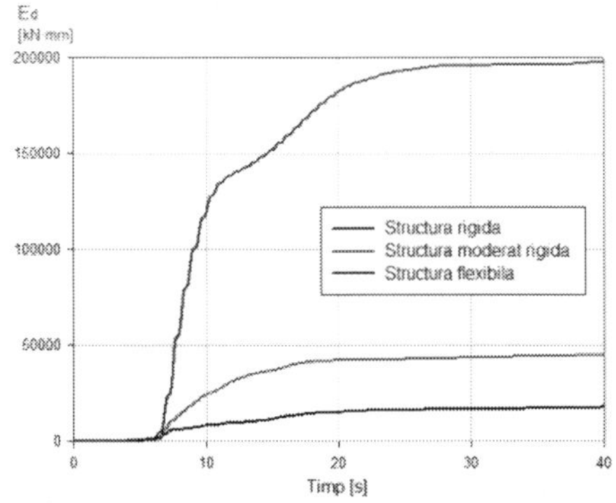

Fig. 4: Energy absorption capacity versus elastic state

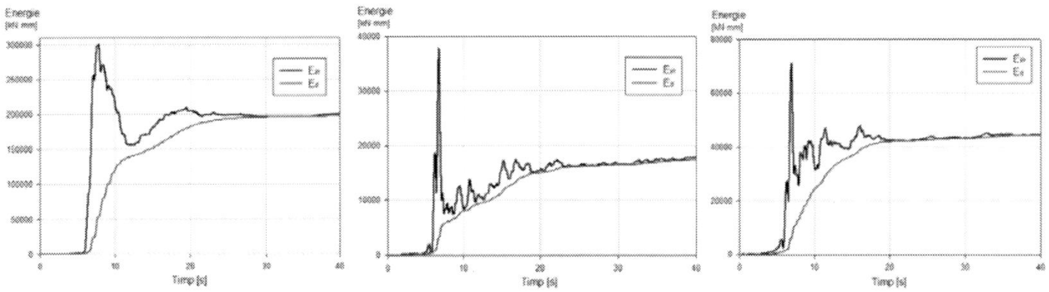

a) Rigid structure b) Moderate rigid structure c) Flexible structure

Fig. 5: Seismic input energy - versus energy absorption capacity

2. The case of Focsani seismic action and 5% critical damping ratio

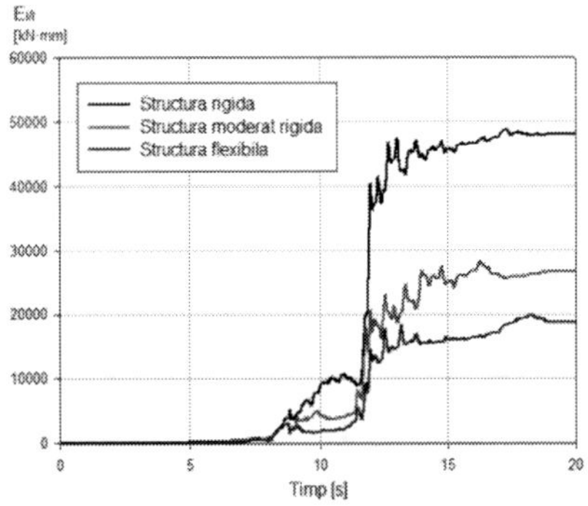

Fig. 6: Relative seismic input energy versus elastic state

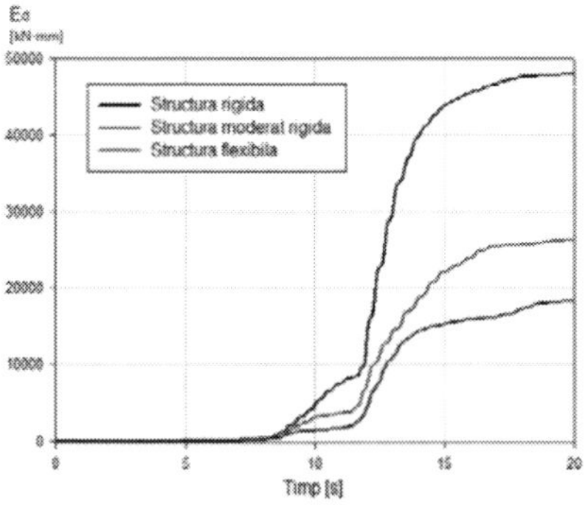

Fig. 7: Energy absorption capacity versus elastic state

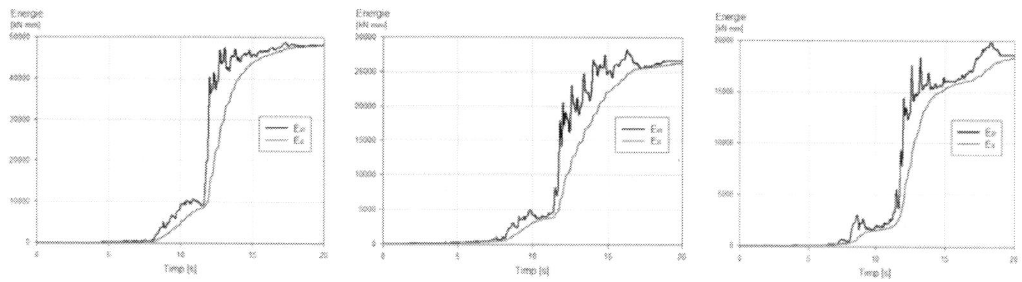

a) Rigid structure b) Moderate rigid structure c) Flexible structure

Fig. 8: Seismic input energy - versus energy absorption capacity

3. The case of El Centro seismic action and 10% fraction of critical damping

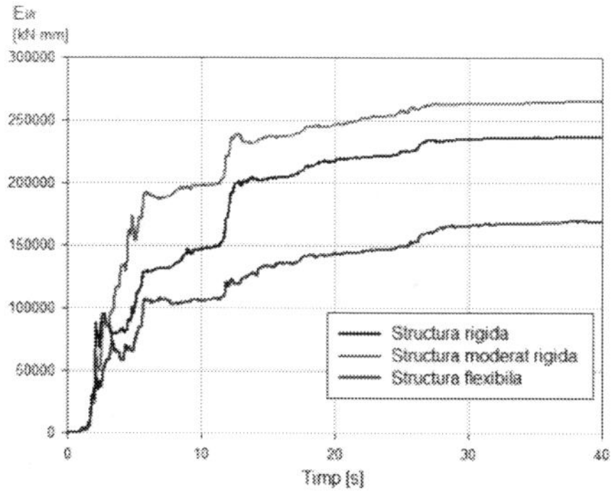

Fig. 9: Relative seismic input energy versus elastic state

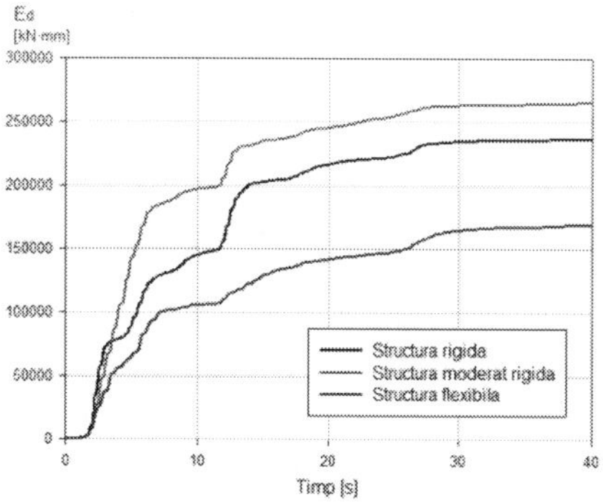

Fig. 10: Energy absorption capacity versus elastic state

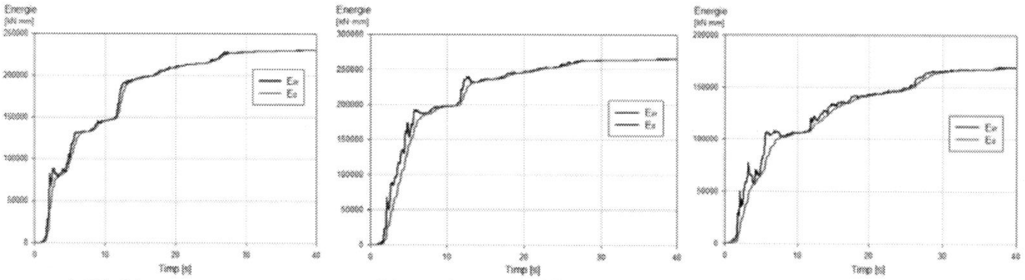

a) Rigid structure b) Moderate rigid structure c) Flexible structure

Fig. 11: Seismic input energy - versus energy absorption capacity

4. The case of El Centro seismic action and 15% fraction of critical damping

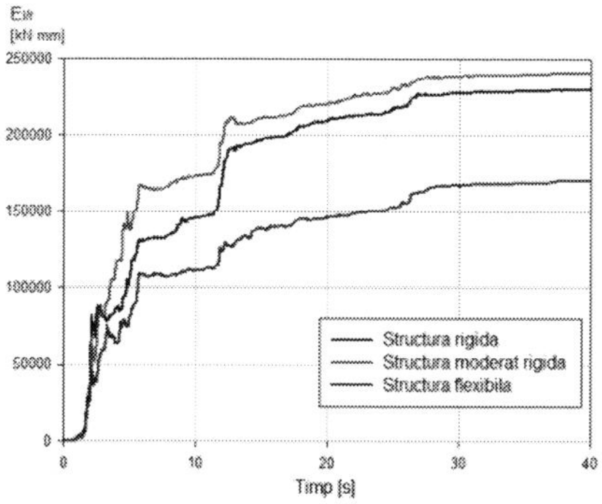

Fig. 12: Relative seismic input energy versus elastic state

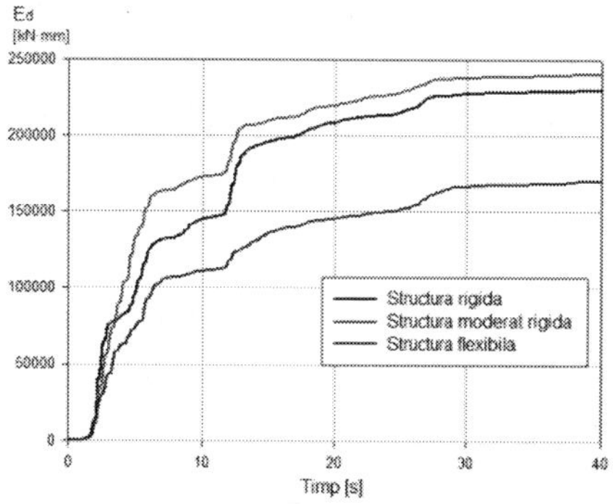

Fig. 13: Energy absorption capacity versus elastic state

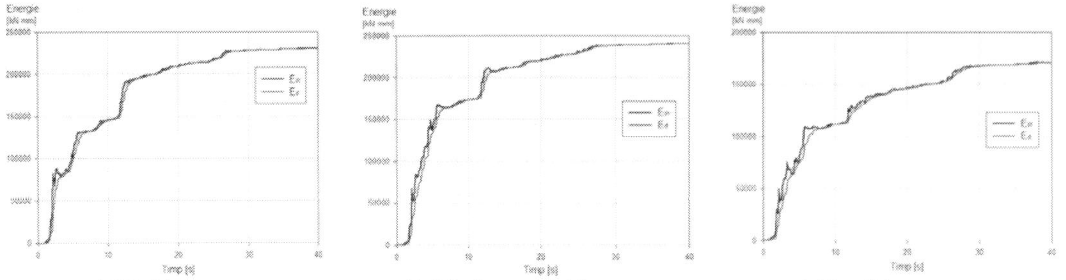

a) Rigid structure b) Moderate rigid structure c) Flexible structure

Fig. 14: Seismic input energy - versus energy absorption capacity

4. Conclusions

Energy state of seismically acted upon structures has been presented in a set of cases emphasizing two energy components: seismic input energy and energy absorption capacity. The computed numerical results are closely associated to the seismic behaviour of multi-story steel structures. Energy amounts (seismic input and absorption capacity) are computed by, already, classical techniques [2], [3] and they refer to the relative motion of dynamic model versus the foundation system of the structure [2], [7]. Dynamic model associated to the energy computation is the usual multi - degree – of – freedom – system with the level lateral displacements as degrees of freedom.

It has to be underlined that the proposed instrument of energy state as part of seismic analysis exhibits an adequate capacity of expressing the seismically induced mechanical state of these structures. Slight variations in the mechanical state of the structure – such as different viscous damping levels – are clearly echoed in the energy term results. For instance, the seismic input energy reaches (Fig. 9) the level of 27,000 (energy units) in the case of 10% ratio of critical damping while, in the case of 15% ratio of critical damping, seismic input energy level (Fig. 12) lowers to 24,000 energy units. Also, this result constitutes an important and reliable information about the influence of damping level on seismic input energy – apart from its influence on the more popular reduction in induced lateral displacements. Concerning the influence of lateral rigidity on seismically induced state of the structure, a prêt – a – porter result is that of reduced lateral displacements in the case of (laterally) stiffer structures. Not so well known result of stiffening is the increase in seismic base shear force. While the reduction in lateral displacements is what is "wanted" by lateral stiffening, the increase in base shear force is, many times, not looked at. Energy state expresses the influence of lateral stiffness in connection with the seismic action (Fig. 3, Fig. 6, Fig. 9, Fig. 12). A flexible structure is more vulnerable (from the point of view of amount of seismic input energy) to Vrancea 1977 seismic action (Fig. 3) while, in the case of El Centro seismic action, the same flexible structure is associated to the smallest amount of seismic input energy (Fig. 9). Regarding energy absorption capacity, energy state exhibits the same power of analysis and of assessment of adequacy of structural conceiving versus location seismicity. While in the case of Vrancea 1977 seismic action, the flexible structure (Fig. 4) expresses the highest absorption capacity, in the case of Focsani 1986 seismic action, the stiff structure (Fig. 7) seems to have the highest absorption capacity. Different situations are encountered in the case of El Centro seismic action. In this case, the flexible structure exhibits the largest absorption capacity of seismic energy, no matter of added damping level (Fig. 10, Fig. 13).

In spite of their usefulness as such, above computed and mentioned results and associated remarks have to be viewed in connection and associated to the power and versatility of proposed tool of assessing the seismic adequacy of multi-story steel structures. Beside the versatility and sensitivity (to slightly different elastic or damping states) of proposed analytical instrument, its simplicity has to be, also, be underlined. The popularity of code imposed design provisions regarding the lateral displacements may, many times, induce amnesia in what concerns other – rather unwanted – results, such as an increase in seismic base shear force (with all its consequences).

In the case of proposed energy instrument, the relation of structural engineer with conceived structure is considered more directly: seismic input energy has to be dissipated without causing structural damages such as permanent plastic deformations. The higher is the amount of seismic input energy, the larger energy absorption capacity has to be provided for the structure. This is a simple and strong conclusion that does not allow for doubts and does not lack unique interpretation. Association of the two energy components (seismic input and energy absorption capacity) with the seismicity of the location may lead to structures capable of high seismic performance. Presented results are only a part of a larger study focused on energy state of seismically acted upon multi-story steel structures. Several techniques of seismic mitigation are considered, such as: added damping, added mass, added stiffness.

References

[1] Housner GW. "Limit design of structures to resist earthquakes", *Proceedings of the 1st World Conference on Earthquake Engineering*, Berkeley, California, 1956.

[2] Uang CM, Bertero VV. "Evaluation of seismic energy in structures", *Earthquake Engineering and Structural Dynamics*, Vol. 19, 1990, pp. 77-90.

[3] Manfredi G. "Evaluation of seismic energy demand", *Earth. Eng. Struct. Dyn.,* 30, 2001, pp. 485-499.

[4] Khashaee P, Mohraz B, Sadek F, Lew HS, Gross JL. "Distribution of Earthquake Input Energy in Structures", *Report NISTIR 6903*, January 2003.

[5] Fajfar P, Vidic T. "Consistent inelastic design spectra: hysteretic and input energy", *Earthquake Engineering and Structural Dynamics,* Vol. 23, 1994, pp. 523-537.

[6] Decanini L, Mollaioli F, Mura A. "Equivalent SDOF systems for the estimation of seismic response of multi-storey frame structures", *Earthquake Resistant Engineering Structures*, III, pp. 101-110, 2001.

[7] Bertero VV, Teran – Gilmore A. "Use of energy concepts in earthquake – resistant analysis and design: Issues and future directions", *Memorias del VIII Seminario Latinoamericano de Ingenieria Sismoressistente y Primeras Jornadas de Ingenieria Estructural*, Merida, Venezuela, 5-8 Julio de 1993.

[8] Ladar I. "Seismic response of steel multi-story structures. An energy approach", *PhD Thesis*, Faculty of Civ. Eng., Technical Univ. of Cluj – Napoca (in Romanian), 2013.

ENERGY APPROACH TO MODAL ANALYSIS

Ioana Ladar

Technical University of Cluj-Napoca, Cluj-Napoca, Romania

Abstract: The contribution proposes an energy approach to the assessment of seismic behaviour of multi-story structures. Two components of energy state are selected to represent the energy state and are being analysed: seismic input energy and energy absorption capacity of the structure. In order to emphasize the capacity and versatility of the energy approach, an eight – story frame is considered in several state of its lateral stiffness and of viscous damping level. Time variations of these two energy components are computed in the case of three recorded seismic actions. The computed results are framed in terms of the proposed energy approach to seismic assessment of multi-story structures.

1. Introduction

Intended contribution reviews the already classical transformations of seismic input energy and dissipated energy by viscous damping into multi-story structures and emphasizes the distribution of these energies into the natural modes of vibrations. The computing process is conducted on steel multi-story frames dynamically modelled as multi degree of freedom systems. The main objective of the intended contribution is to introduce an energy approach to modal based analysis of multi-story steel structures. While the structural engineer enjoys a good command of transformation of seismic input energy into its kinetic, strain, dissipated by viscous inherent and added damping energy components, the distribution of this energy is less exhibited and practiced. A short theoretical part of the present study is dedicated to the distribution of seismic input energy to the natural modes of vibration. It is followed by several numerical results carried on steel multi-story structures seismically acted upon. Computed results - associated to seismic response in energy terms - are presented in comparative numerical and graphical manners. Associated relevant comments and inferred conclusions are presented.

2. Energy state of multi-story structures

Modern technologies of seismic mitigation of structures located in seismic areas aim to dissipating of energy induced into the structures via and during seismic actions. The structures have, therefore, to be provided with an absorption capacity to dissipate seismic energy. Today activity of conceiving seismically acted upon structures and associated sizing of their cross sections cannot avoid the energy state component of design process. Indeed, the energy state of structures under seismic actions became an objective of providing the structures with capabilities to achieve code prescribed seismic performances. Historically, the progress of both, engineering thinking in terms of seismic energy and modern sophisticated technologies to dissipate this energy is huge in spite of the short period since Housner [1] invoked the necessity to equip structures with adequate seismic energy absorption capacity. Though energy absorption capacity is rather a design activity [2], [3], [4], the energy state of structures located in seismically active areas developed as an analysis component as well [5], [6], [7]. Present contribution attempts to cast into modal analysis the energy state of seismically acted upon multi-story structures. Energy approach to seismic analysis and design of structures has to observe the fundamental criterion of energy balance

$$E_{abs} \geq E_i \tag{1}$$

where, E_{abs} is the capacity of structure to absorb seismically induced energy while, E_i is the energy input into the structure by seismic action. Above condition has to be fulfilled jointly by all mechanical state of the structure: the static (equilibrium of applied loads and stress - resultants) state, kinematic (compatibility of displacements and deformation) state, elastic (flexibility / rigidity) state and damping state [8]. In what concerns the elastic state, present development includes only elastic behaviour under seismic action. To fulfil condition (1), the structure has to be provided with capabilities regarding its mechanical, material and geometrical states. Decomposition of seismically induced energy E_i into the structure may offer a better view on the possibilities of providing for E_{abs}:

$$E_i = E_s + E_k + E_d + E_{ad} \tag{2}$$

Here E_s is the elastic strain (deformation) energy, E_k is the kinetic energy, E_d is the capacity of the structure to dissipate energy by its inherent damping state and E_{ad} is the capacity of the structure to dissipate seismic energy via added damping. Detailing the structural components (E_s, E_k, E_d and E_{ad}) of seismic energy requires a dynamic model of seismic analysis. In the case of a multi-story structure, such model may be the traditional MDOF (multi – degree – of – freedom) system (see Fig. 1) that emphasizes only the relative displacement of structure versus its base.

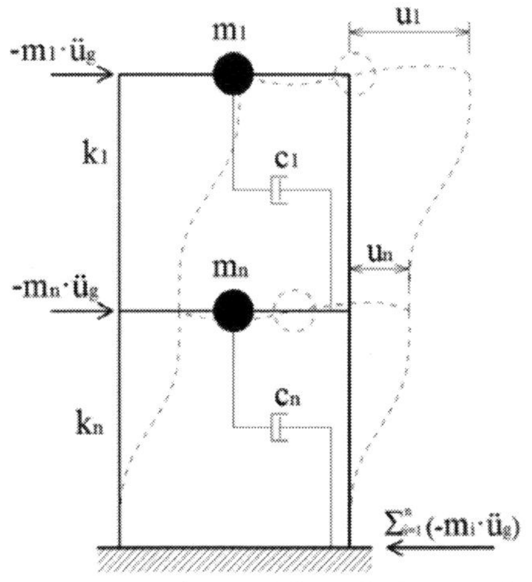

Fig. 1: Dynamic model – relative motion

Associated references to the dynamic model and seismic analysis are the classical ones:

m_i lumped level masses
k_i relative level stiffness
u_i lateral displacements along the degrees of freedom
c_i critical damping rations (assumed linear viscous)
$\ddot{u}_g$ ground acceleration seismically induced
$\boldsymbol{u}$ n vector of degrees of freedom (lateral displacements u_i)
$\boldsymbol{M}$ inertial / mass matrix ($n \times n$) associated to vector $\boldsymbol{u}$
$\boldsymbol{K}$ stiffness matrix ($n \times n$) associated to vector $\boldsymbol{u}$
$\boldsymbol{C}$ damping matrix ($n \times n$) associated to vector $\boldsymbol{u}$.

The structural energy components from Eqn. (2) may, now, be expressed in the following forms [2], [3], [10]:

- Input seismic energy:

$$E_i = -\int_0^t \ddot{u}_g^T(t) \cdot M \cdot \dot{u} \cdot dt = -\int_0^t \ddot{u}_g^T \cdot M \cdot d \cdot g(t) \tag{3}$$

- Relative kinetic energy:

$$E_k = \int_0^t \ddot{u}^T(t) \cdot M \cdot \dot{u} \cdot dt = \int_0^t \ddot{u}^T(t) \cdot M \cdot du \tag{4}$$

- Energy absorption capacity conferred by inherent viscous damping:

$$E_d = \int_0^t \dot{u}^T(t) \cdot C \cdot \dot{u} \cdot dt = \int_0^t \dot{u}^T \cdot C \cdot du \tag{5}$$

- Elastic strain energy:

$$E_s = \int_0^t u^T(t) \cdot K \cdot \dot{u} \cdot dt = \int_0^t u^T(t) \cdot K \cdot du \tag{6}$$

3. Orthogonality of eigen vectors in the presence associated to inertia, stiffness and damping states

The following is a theoretical development that refers to the orthogonality properties of classical eigen vectors $\mathbf{\Phi}_j$ (j = 1 - n) with respect the damping matric **C** as it is presented in [9].

The well-known seismic modal analysis leads to the specific parameters: T_j - the eigen vibration periods, ω_j - the eigen circular frequencies, $\mathbf{\Phi}_j$ - eigen n - vectors, $\mathbf{\Phi} = [\mathbf{\Phi}_1 \cdots \mathbf{\Phi}_j \cdots \mathbf{\Phi}_n]$ – modal ($n \times n$) matrix. The orthogonality of eigen vectors $\mathbf{\Phi}_j$ with respect mass **M** matrix is, also, a common place and leads to the results:

$$(\mathbf{\Phi}_j)^T \cdot \mathbf{M} \cdot \mathbf{\Phi}_k = 0$$
$$(\mathbf{\Phi}_j)^T \cdot \mathbf{M} \cdot \mathbf{\Phi}_j = \mathcal{M}_j$$

where $\mathcal{M}_j$ is the modal mass associated to natural mode j of vibration.

The orthogonality with respect stiffness matrix **K** ends in the following results:

$$(\mathbf{\Phi}_j)^T \cdot \mathbf{K} \cdot \mathbf{\Phi}_j = 0$$
$$(\mathbf{\Phi}_j)^T \cdot \mathbf{K} \cdot \mathbf{\Phi}_k = \mathcal{R}_j$$

where $\mathcal{R}_j$ is the modal stiffness associated to natural mode j of vibration.

Two matrices may, now, be built up with parameters $\mathcal{M}_j$ and $\mathcal{R}_j$ respectively:

$$\boldsymbol{\mathcal{M}} = \mathbf{\Phi}^T \cdot \mathbf{M} \cdot \mathbf{\Phi} \tag{7}$$

and

$$\boldsymbol{\mathcal{R}} = \mathbf{\Phi}^T \cdot \mathbf{R} \cdot \mathbf{\Phi} \tag{8}$$

The orthogonality of eigen vectors $\mathbf{\Phi}_j$ confers to the two matrices $\boldsymbol{\mathcal{M}}$ and $\boldsymbol{\mathcal{R}}$ the diagonal property and undoubted analytical advantages [9].

Extension of orthogonality of eigen vectors over the damping ($n \times n$) matrix **C** requires a specific development of linear viscous damping associated to MDOF systems. Considering damping matrix **C** of Rayleigh type and based on several sets of experimental and site measured results of damping [9], the damping matrix may be cast into such a form as to allow the eigen vectors to emphasize their orthogonality property with respect this damping matrix. The diagonal form of associated damping matrix is crucial to the mathematical model involved in modal analysis in what concerns the solutions of differential equations and the computed results [10]. As it has been proved [9], the orthogonality of eigen vectors $\mathbf{\Phi}_j$ with respect damping matrix **C** depends on the distribution of elastic state, of inertia state and of damping state into the analytical dynamic model. The dynamic model of Figure 1 underlines the interconnectivity of lumped level masses m_i via elastic and damping properties. Such connectivity together a proper distribution of linear viscous damping to the structural elements / levels may, finally, lead to the much wanted orthogonality property of eigen vectors with respect damping matrix **C**. A homogeneous structural built–up along the height of multi-story structures, a homogeneous geometry of the structures (i.e., the same structural type – shear walls, frames, etc.), preservation of structural material (steel, concrete, etc.) and the preservation of connecting type (welding, bolts, semi-rigid beam – column connections, etc.) of structural elements assure a realistic Rayleigh type model of inherent damping. Therefore, a linear viscous damping proportional to both, mass and stiffness may be considered while the computed numerical results may be regarded as adequate [3], [4], [9]:

$$\mathbf{C} = \alpha \cdot \mathbf{M} + \beta \cdot \mathbf{K} \tag{9}$$

The orthogonality of eigen vectors $\mathbf{\Phi}_j$ with respect damping matrix **C** given by Eqn. (9) takes the form:

$$(\mathbf{\Phi} j)^T \cdot \mathbf{C} \cdot \mathbf{\Phi}_k = 0; \; j \neq k \tag{10}$$

Now, the extension of classical modal computation over matrix **C**:

$$\mathbf{\Phi}^T \cdot \mathbf{C} \cdot \mathbf{\Phi} = \mathscr{C} \tag{11}$$

yields into the diagonal form of damping matrix $\mathscr{C}$ given by Eqn. (11) with its entries $\mathscr{C}_j$ given by:

$$\mathscr{C}_j = (\mathbf{\Phi}_j)^T \cdot \mathbf{C} \cdot \mathbf{\Phi}_j \tag{12}$$

With this, the problem of free vibrations of MDOF systems with viscous damping reduces to the traditional form of eigen values.

4. Modal energy state

The differential matrix equation of vibrations of MDOF systems induced by seismic action takes the form:

$$M \cdot \ddot{u} + C \cdot \dot{u} + K \cdot u = -M \cdot 1 \cdot \ddot{u}_g \tag{13}$$

where **1** is the n unit vector.

The usual linear coordinate transformation is applied:

$$\mathbf{u} = \mathbf{\Phi} \cdot \mathbf{x} \tag{14}$$

that allows the transition from generalized (global) coordinates **u** to modal coordinates **x**. Here, n vector x has the (transposed) form:

$$\mathbf{x}^T = [x_1, ..., x_j, ..., x_n]^T$$

with x_j – the modal coordinate associated to natural mode j of vibration. The coordinate transformation in Eqn. (13) has to be extended over the derivatives $\dot{u}$ and $\ddot{u}$ of $\boldsymbol{u}$ as well:

$$\mathbf{u}^T = \mathbf{x}^T \cdot \mathbf{\Phi}^T; \; \dot{u} = \mathbf{\Phi} \cdot \dot{x} \; ; \; \ddot{u} = \mathbf{\Phi} \cdot \ddot{x} \; ; \mathbf{du} = \mathbf{\Phi} \cdot \mathbf{dx} \; ; \mathbf{dx} = \dot{x} \cdot \mathrm{dt} \tag{15}$$

By virtue of Eqns. (14) and (15), and taking into account the orthogonality of eigen vectors $\mathbf{\Phi}_j$ with respect matrices **M**, **K** and **C**, the differential Eqn. (13) in **u** becomes a system of n uncoupled differential equations in x_j:

$$\mathcal{M}_j \cdot \ddot{x}_j(t) + \mathcal{C}_j \cdot \dot{x}_j(t) + \mathcal{K}_j \cdot x_j(t) = -\phi_j^T \cdot M \cdot 1 \cdot \ddot{u}_g(t) \quad (16)$$

The scalar differential Eqn. (16) in x_j may be dealt with in the usual way [9], [10] and its solution is:

$$x_j(t) = -\frac{\Gamma_i}{\omega_i} \cdot \int_0^t \ddot{u}_g(\tau) \cdot e^{-\zeta_j \cdot \omega_j (t-\tau)} \cdot \sin \omega_j (t-\tau) \cdot d\tau \quad (17)$$

where, Γ_j and ζ_j are given by Eqn. (18), while τ is the time interval:

$$\Gamma_i = \frac{\phi_j^T \cdot M \cdot 1}{\mathcal{M}_i} = \frac{\phi_j^T \cdot M \cdot 1}{\phi_j^T \cdot M \cdot \phi_i} \quad (18)$$

$$\xi_i = \frac{\beta_j}{\omega} = \frac{\mathcal{C}_j}{2 \cdot \mathcal{M}_i \cdot \omega_i}$$

The right side member of Eqn. (17) emphasizes the modal distribution $\ddot{u}_g^{(j)}(t)$ of seismic action $\ddot{u}_g(t)$ (recorded accelerogram). That is, the distribution of the "effect" (seismic response in x_j) lead to the modal distribution of the "cause" [10]. These partial results allow the re-expressing the energy components in terms of x_j.

- Input seismic energy E_i

$$E_i = -\int_0^t \sum_{j=1}^n m^T \cdot \Phi_j \cdot \dot{x}_j \cdot \ddot{u}_j \cdot dt \quad (19)$$

By applying the property of double summing:

$$\int_0^t \sum_{j=1}^n = \sum_{j=1}^n \int_0^t \quad (20)$$

it yields into:

$$E_i = -\sum_{j=1}^n \int_0^t m^T \cdot \Phi_j \cdot \dot{x}_j \cdot \ddot{u}_g \cdot dt \quad (21)$$

The concept of modal input energy [10] is, now, introduced in the form:

$$E_i^{(j)} = -\int_0^t m^T \cdot \Phi_j \cdot \dot{x}_j \cdot \ddot{u}_j \cdot dt \quad (22)$$

and the total seismic input energy is given by:

$$E_i = \sum_{j=1}^n E_i^{(j)} \quad (23)$$

- Kinetic energy E_k

$$E_k = \int_0^t \dot{u}^T \cdot M \cdot \dot{u} \cdot dt \quad (24)$$

and its expression in terms of modal coordinates x_j:

$$E_k = \frac{1}{2} \sum_{j=1}^n \mathcal{M}_j \cdot \dot{x}_j^2 \quad (25)$$

The new concept of modal kinetic energy:

$$E_k^{(j)} = \frac{1}{2} \sum_{j=1}^n \mathcal{M}_j \cdot \dot{x}_j^2 \quad (26)$$

and the total kinetic energy being given by the sum over all its modal components:

$$E_k = \sum_{j=1}^n E_k^{(j)} \quad (27)$$

Following the same procedure, the elastic strain energy is:

$$E_s = \sum_{j=1}^n E_s^{(j)} \quad (28)$$

where, its modal components are:

$$E_s^{(j)} = \mathcal{K}_j \cdot x_j \cdot \dot{x}_j \cdot dt \quad (29)$$

and, finally, the energy absorption capacity via viscous inherent damping:

$$E_d = \sum\nolimits_{j=1}^{n} E_d^{(j)} \tag{30}$$

With its modal component given by:

$$E_d^{(j)} = \int_0^t \mathscr{C}_j \cdot \dot{x}_j^2 \cdot dt \tag{31}$$

Concluding, it may be underlined that the seismic input energy and its structural components have been decomposed into their modal components.

5. Numerical results

In the following, a set of numerical results obtained on an eight level multi-story steel planar frame are presented in what regards the modal distribution of seismic energy.

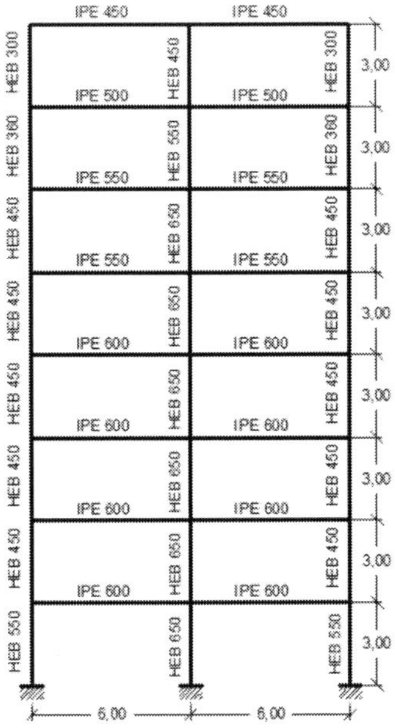

Fig. 2: Analysed structure

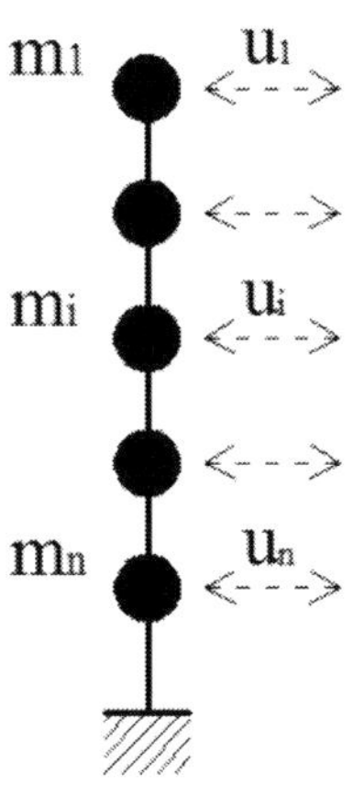

Fig. 3: Dynamic model

The structure (Fig. 2) has been designed according to Romanian code design provisions associated to a seismic location of ground motion $a_g = 0.20g$ and a control period $T = 0.7$ sec. The dynamic / seismic model is a lumped mass system (Fig. 3) of $n = 8$ degrees of freedom. Three seismic actions have been applied in the form of recorded accelerograms: El Centro 1940, Vrancea 1977 and Focsani 1986 (Fig. 4). A ratio of critical inherent (linear viscous) damping of 5% has been considered.

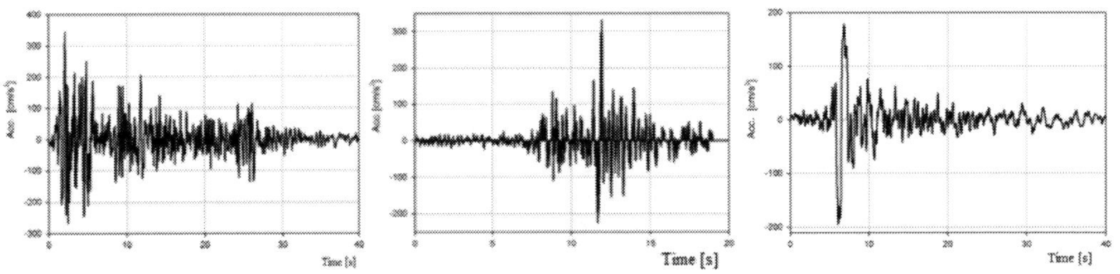

Fig. 4: El Centro 1940, Vrancea 1977 and Focsani 1986 recorded accelerograms

The eight eigen periods of the dynamic model are:
$T_1 = 0.897$ sec; $T_2 = 0.315$ sec; $T_3 = 0{,}176$ sec; $T_4 = 0.116$ sec; $T_5 = 0.104$ sec ;
$T_6 = 0.085$ sec; $T_7 = 0.084$ sec; $T_8 = 0.083$ sec.

Presented results refer to the total seismic (relative) input energy E_i computed by (19) and – aiming at comparison – by (23) as the sum of its modal components $E_i^{(j)}$ (Fig. 5 ÷ Fig. 9).

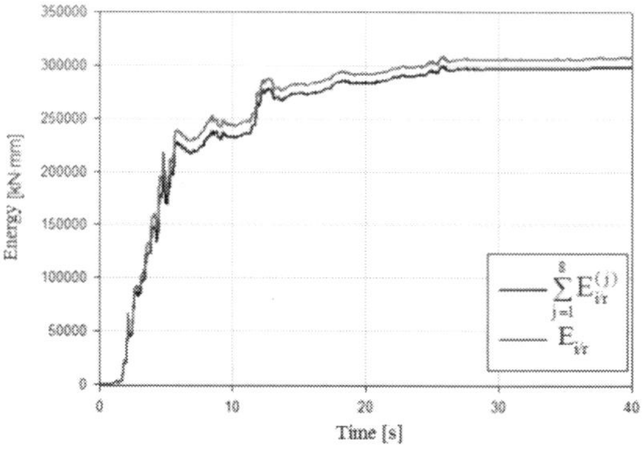

Fig. 5: Seismic (relative) input energy

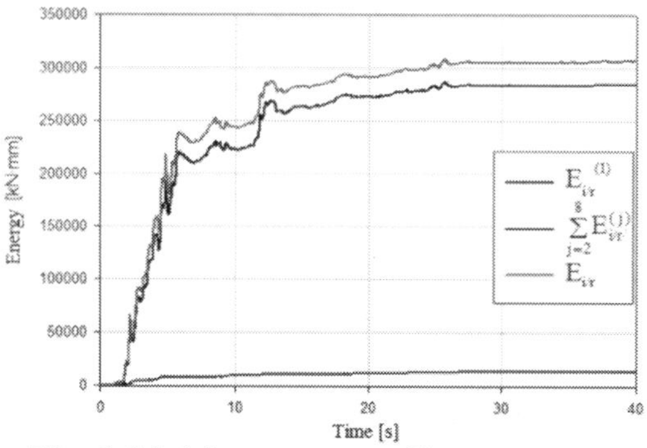

Fig. 6: Modal components of input energy

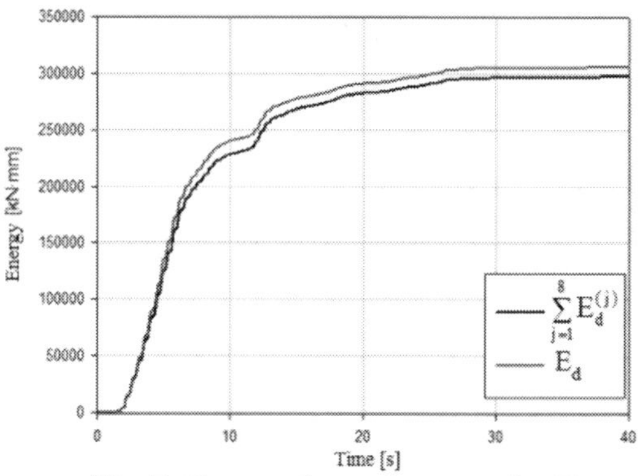

Fig. 7: Energy absorption capacity E_d

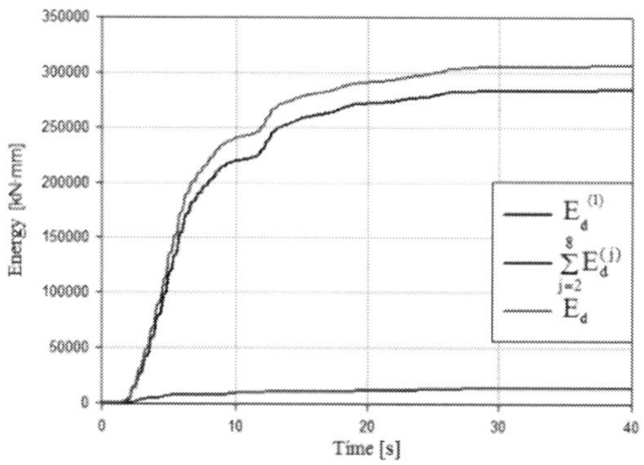

Fig. 8: Modal components of E_d

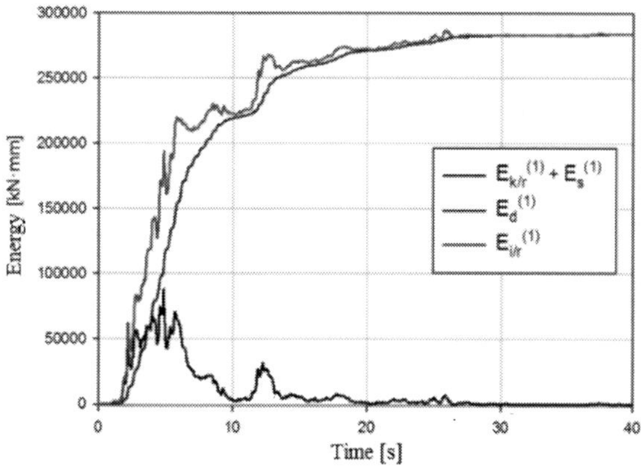

Fig. 9: Modal components of kinetic energy E_k

Comparing energy quantities is a relevant and simple process. Its relevance confer by the scalar nature of energy and the fact that energy includes a large spectrum of mechanical states.

6. Conclusions

A modal decomposition of seismic response in energy terms has been presented. The concept of modal response has been extended from its classical parameters (displacements, stress – resultants, seismic shear force) to the energy response. The theoretical presentation exhibits the simplicity of the process while the numerical results underline the validity of modal decomposition in terms of seismic energy.

References

[1] Housner GW. "Limit design of structures to resist earthquakes", *Proceedings of the 1st World Conference on Earthquake Engineering*, Berkeley, California, 1956.

[2] Uang CM, Bertero VV. "Evaluation of seismic energy in structures", *Earthquake Engineering and Structural Dynamics*, 19, 77–90, 1990.

[3] Fajfar P, Vidic T. "Consistent inelastic design spectra: hysteretic and input energy", *Earthquake Engineering and Structural Dynamics*, 23, 523 – 537, 1994.

[4] Decanini L, Mollaioli F, Mura A. "Equivalent SDOF systems for the estimation of seismic response of multi-storey frame structures", *Earthquake Resistant Engineering Structures*, III, 101 – 110, 2001.

[5] Manfredi G. "Evaluation of seismic energy demand", *Earth. Eng. Struct. Dyn.*, 30, 485 – 499, 2001.

[6] Khashaee P, Mohraz B, Sadek F, Lew HS, Gross JL. "Distribution of Earthquake Input Energy in Structures", *Report NISTIR 6903*, January 2003.

[7] Kalkan E, Kunnath SK. "Relevance of Absolute and Relative Energy Content in Seismic Evaluation of Structures", *Advances in Structural Engineering*, 11(1), 1-18, 2008.

[8] Bertero VV, Teran – Gilmore A. "Use of energy concepts in earthquake – resistant analysis and design: Issues and future directions", *Memorias del VIII Seminario Latinoamericano de Ingenieria Sismoressistente y Primeras Jornadas de Ingenieria Estructural*, Merida, Venezuela, 5-8 Julio de 1993.

[9] Chopra AK. *Dynamics of Structures. Theory and Applications to Earthquake Engineering*, Prentice Hall International, 1995.

[10] Ladar I. *Seismic response of steel multi-story structures. An energy approach*, PhD Thesis, Faculty of Civ. Eng., Technical University of Cluj – Napoca, 2013 (in Romanian).

Steel-concrete composite members and structures

STRENGTH OF SQUARE LACED STEEL-CONCRETE COMPOSITE COLUMNS USING GFRP AND SILICA FUME

Mohamed A. El Aghoury[a], Essam A. Amoush[b], Ihab M. El Aghoury[a] and Mohamed A. Ghanem[c]

[a] *Department of Structural Engineering, Ain Shams University, Abassia, Egypt*
[b] *Department of Structural Engineering, Higher Technological Institute, Tenth of Ramadan City, Egypt*
[c] *Industrial Education, Suez University, Egypt*

Abstract: Laced Steel-Concrete Composite (LSCC) column is a relatively new form of steel composite systems. In this study, the (LSCC) system is composed of double cold-formed channels connected by bolted lacing bars. The volume between the two channels comprising the square box section of the columns is filled with concrete. A total of eleven composite columns using filling concrete with different additives (silica fume and GFRP rebars) are tested under axial compressive load. Composite actions are maintained in the tested specimens by using interior shear connectors. The additive silica fume & GFRP rebars improved the strength of the medium & short composite columns by about (8-20%) & (4-8%); respectively.

1. Introduction

There are many advantages associated with the use of composite columns such as small cross-sections that are capable of withstanding high loads. Similarly, sections with different resistances, but identical external dimensions, can be produced by varying the steel thickness, the concrete strength or by adding additional reinforcement. The outer dimensions of a column can be held constant over a number of floors in a building, thus satisfying the architectural demands easily. Economic efficiency of this system also results from the use of concrete (as a low cost material) and from the time saved by using the highly developed connection techniques of the steel work construction. With concrete filled profiles, the steel section serves as formwork during casting, therefore, provides the opportunity to erect the steel frame of a building and afterwards fill the cross-sections by the concrete. By doing so, the time of erection can be reduced. The protective steel casing also allows the concrete to achieve greater strength; especially in the case of concrete filled circular hollow profiles. That is because the effect of confinement by the steel leads to an increase in the overall compressive resistances. The influence of creep and shrinkage of the concrete can usually be neglected for these sections. However, these effects must be considered for the concrete encased profiles. The complete encased steel section, generally fulfills the technical requirements for the higher classes of the fire protection without any additional measures. For partially encased sections, as well as for concrete filled sections, these requirements can be achieved using additional reinforcement. The partially encased sections have the advantage that it can be produced quite simply by casting the concrete whilst the steel section lies horizontally. Twenty Four hours later, the column can be turned around and further concrete could be added. Another important advantage of these partially encased sections is that there is still a considerable area of steel available for connections, even after pouring the concrete.

There is a need for developing an alternate and simpler structural system which has the properties required for resisting loads, namely, ductility and structural integrity. Arjang Tavassoli [1] tested nine large-scale circular concrete columns reinforced with longitudinal and transverse glass fiber-reinforced polymer (GFRP) rebars. The tested specimens were subjected to lateral cyclic quasi-static loading and simultaneously with a constant axial load. Based on the measured hysteretic loops of the moment versus curvature and shear versus tip deflection relationships, a series of parameters related to ductility and flexural strength are used to evaluate the seismic behavior of each column. The results showed that the concrete columns reinforced with GFRP rebars have stable post-peak branches and can achieve very high levels of deformability. Longitudinal GFRP rebars, maintained their stiffness at high strains and transverse GFRP spirals, provided increasing confinement for the entire duration of the test without any spiral damage. The tests showed that, as an innovative material with excellent corrosion resistance GFRP rebars can be successfully used as internal reinforcement in ductile concrete columns. Mahbuba Begum et al. [2] presented a wide range of numerical investigations on partially incased composite columns (PEC) based on finite element analysis. The PEC columns are fabricated from thin-walled built-up H-shape steel sections with links provided between the opposing flanges to improve the resistance to local buckling. The regions between the flanges and web are filled with concrete. They were applied successfully model to simulate the behavior of 34 PEC columns from five experimental programs. Thirumalaiselvi et al. [3] studied the behavior of a laced beam which consisted of two cover plates and connected by lacings and cross rods. There was also a concrete infilled between the cover plates. In their numerical investigations, both steel and concrete of the laced structural components were required to be modeled as separate entities. However, finite element modeling of the laced structural component was tedious and time consuming due to the presence of lacings besides longitudinal and transverse reinforcement steel bars. Therefore, they developed a simplified approach for the analysis of the laced structural components. They adopted an effective moment of inertia approach in determining the load-deflection response of LSCC beams under monotonic loading. However, with limited available numerical results, an attempt had been made to find out the effective moment of inertia, which can be used to obtain deflection. In addition, they recommended that further research is needed to verify that model.

In this study, a total of eleven specimens are tested for partially encased lacing steel-concrete composite columns under axial load. An investigation of the behaviour of short and medium composite columns is conducted. The strength and the failure modes of the laced steel-concrete composite columns are also discussed.

2. Experimental program

2.1 Section size

Laced Steel-Concrete Composite (LSCC) system, which is composed of Lacing Steel (LS) cage and filled with concrete acting in a composite actions, is tested. The LSCC system consists of two thin steel cold-formed channels, which are connected using lacing bars and transverse bolts and infilled with concrete. Fig. 1 shows a schematic of the elements comprising the LSCC columns along with their dimensions.

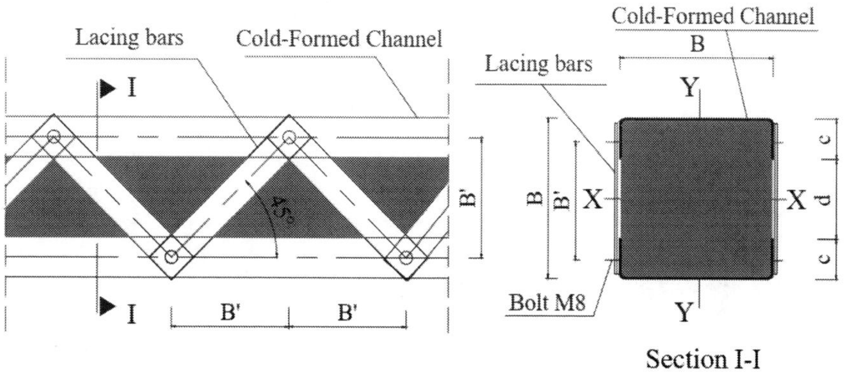

Fig. 1: Elements comprising the proposed LSCC column

The dimensions of the LSCC panel are chosen based on similar LS panels. Cold-formed steel channels are manufactured from sheets with a length of 1200 mm and 2400 mm to simulate short and medium columns; respectively. The specimens consist of two steel plates of sizes 1200 mm × 150 mm and 2400 mm × 150 mm with a thickness of 1.5 mm. The cover plate and lacings are fabricated in a factory environment. The cold-formed channels are placed 100 mm apart. In additions, the lacing bars and end batten plates are used to connect the double cold-formed channels and a single unit is assembled by keeping the channel vertically connected by bolts. Fig. 2 shows the lacing steel column specimen before filling the concrete.

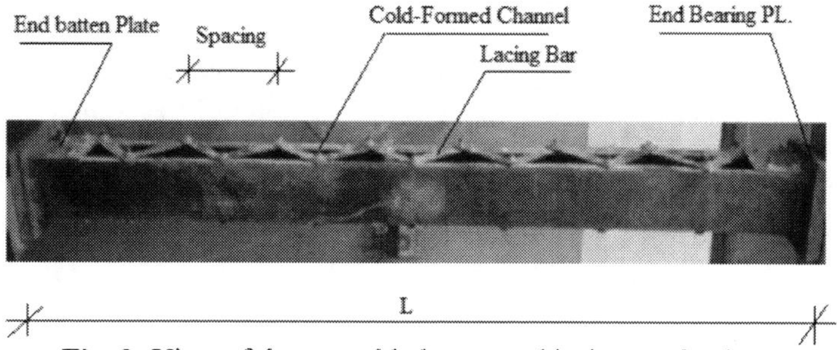

Fig. 2: View of the assembled proposed lacing steel column

Longitudinal steel-to-concrete cross-section area ratios are determined according to Hanswille and Eurocode [4, 5]. Concrete is poured from the top and the channels act as shutters. Composite action is maintained in the tested specimens by using interior shear connectors. Table 1 lists the specimens used in the experimental program.

Table 1: Test specimens dimensions and description

Specimen	Length (mm)	Breadth (B) (mm)	Description
CC-35	1200	100	Short Control Concrete Column
L-35	1200	100	Short Laced Steel Column
L-70	2400	100	Medium Laced Steel Column
LC-35	1200	100	Short Composite Laced Steel Column
LC-70	2400	100	Meduim Composite Laced Steel Column
LSC-35	1200	100	Short Composite Laced Steel column with additives (10% Silica fume)
LSC-70	2400	100	Meduim Composite Laced Steel column with additives (10% Silica Fume)
LSC-GFRP-35	1200	100	Short Composite Laced Steel column with additives (10% Silica Fume & GFRP rebars)
LSC-GFRP-70	2400	100	Meduim Composite Laced Steel column with additives (10% Silica Fume & GFRP rebars)
LSC-rft-35	1200	100	Short Composite Laced Steel column with additives (10% Silica Fume & rft bars)
LSC-rft-70	2400	100	Meduim Composite Laced Steel column with additives (10% Silica Fume & rft bars)

2.2 Test setup and procedure

A total of eleven square LSCC short and medium columns specimens subjected to axial compression load are tested. The investigated columns were chosen to have two overall slenderness ratios; (λ_c=35&70). Two specimens have concrete without additives, and two specimens having 10% silica fume additives: four specimens reinforced with glass fiber reinforced polymers (GFRP) and steel reinforced high strength rebars (rft) are tested. The tested specimens were loaded vertically by a testing machine as shown in Fig. 3. Two cold-formed channels ($B \times c \times t$), 100mm × 25mm × 1.5mm, are used to form box (square) specimens (100mm × 100mm) which are assembled by bolting lacing bars (125mm × 20mm × 1.5mm) bolted at equal intervals (S=2B'=150 mm) throughout the specimen length. At both ends of the specimen, the two channels were assembled by end batten plates of dimensions 100mm × 75mm × 3mm. End bearing plate 150mm × 150mm × 20mm was connected with end batten plates by welding. A theodolite is used to measure the verticality of the specimen. In spite of that, there was a little small eccentricity at the ends of all tested specimens, which was very difficult to avoid. This setup lets the specimen's ends to act as globally pinned; however, the longitudinal warping strains due to torsion are not allowed due to the presence of the thick end plate, 20 mm. The load, shortening and lateral displacement are recorded by using a data acquisition system. Two (LVDT) transducers were placed at mid height of specimen to measure the lateral displacement. One (LVDT), was attached to the end of the specimen to measure the axial shortening. The initial axial load was approximately equal to (1/10–1/15) of the estimated ultimate load and was applied to the specimen to make full contact with the end bearing plate.

3. Material properties

3.1 The cover steel plates

Tensile coupons for the cover plates were prepared and tested according to the ASTM-E8 to determine the yield stress (F_y), tensile strength (F_u), Young's Modulus (E) and percentage elongation at fracture (e_u). Average yield stress and ultimate stress are obtained as 420 MPa and 480 MPa, respectively. In addition, Young's Modulus (E) and percentage elongation on a 10 mm gauge length at fracture (e_u) are 210 GPa and 20%, respectively.

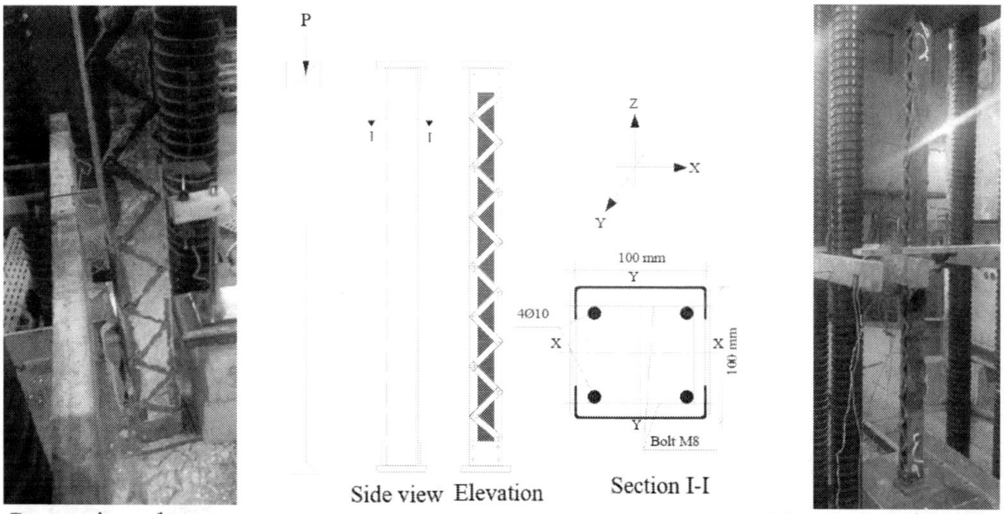

Composite columns   Non-composite column

Fig. 3: Test setup

3.2 Concrete in-filling

The concrete mix design was prepared with (cement: sand: aggregate) ratio = 1:1.25:2.5 with water-cement ratio of 0.45 to give a compressive strength f_{cu}=25 MPa. Quantity of cement, sand, 12 mm aggregates and water are 29 kg, 36.35 kg, 72.5 kg and 13 liters respectively. Concrete is transported in trolleys and filled in the Lacing Steel cage and thirty five standard control cubes are cast. Needle vibrators are used for compaction of the concrete. At the top, needle vibrators are placed horizontally so that the concrete will get filled up at the corners and in between the steel channels. Their compressive strengths are obtained as listed in Table 2. It is clear from the characteristic strength of the 35 tested cubes that the best results were obtained using a percentage of silica fume = 10%.

Table 2: Control cube characteristic compressive strength (MPa)

Cube	Plain concrete (PC)	Additives					
		(rft)	GFRP	10% SF	15% SF	20% SF	25% SF
1	9.5	17.1	12.6	28.5	28.5	24	21.15
2	14.25	20.7	18.75	24.3	20.7	25	25.5
3	23.7	21.3	25.5	28.65	24.15	23	27
4	22.2	20.5	21.3	26.6	23.3	22.2	24.5
5	24.3	19.4	16.9	28.3	25.9	26	25.1
Average	18.84	19.8	19.01	27.27	24.51	24.06	24.65

rft = steel reinforcement rebars, GFRP= Glass Fibre Reinforced Polymer rebars & SF= ratio of (Silica fume /cement) weight

3.3 GFRP rebars preparation

Despite the advances in the quality of the FRP rebars over the last few years, many designers are still reluctant to replace steel with FRP as the main reinforcement in concrete members. However, this hesitation has been addressed to a great degree in the case of reinforced concrete beams and slabs with large volumes of data available from tests performed on large specimens in different research institutions around the world. Contrary to the current understanding of the GFRP bars in tension, the compressive response of these rebars is not well-understood. Therefore, many design codes in North America such as the ACI440 prevent designers from using GFRP bars in members under compression while the CSAS6-06 does not have any provisions regarding this application. Also, few studies have regarded the behavior of the concrete columns reinforced with GFRP rebars under pure axial load, while no experimental data exist in the response of concrete columns reinforced longitudinally and transversely with GFRP under combined axial load and flexure.

In this paper, a manual fabrication of the Glass Fiber Reinforced Polymer (GFRP) rebars, is carried out in the lab. The GFRP matrix composition consists of fiberglass, cobalt, polyester and peroxide. Each 100 gram of polyester is mixed with 1 ml of peroxide and 0.60 ml of cobalt. Fixed ends are installed with nails to a wooden panel. In addition, the polyethylene sheets are attached between the fixed ends after the fiber yarns warping process is completed, the fibers are immersed in the polyester resin (mixture of polyester and peroxide) by using brushes. A torque load is applied to the bar end until it is formed in the aligned position. During the fabrication of the GFRP rebars, the final diameter of the rebars depends on the number of rolls as shown in Fig. 4. Eventually, the GFRP is used in LSCC system subjected to axial compression to study and understand the behaviour of the GFRP rebars in compression.

Bar diameter (mm)	No. of rolls
6	24
8	32
10	40
12	48
16	54
18	64

Fig. 4: Preparation of GFRP rebars and the number of rolls required for each rebar diameter

4. Test Results

For axially loaded short LSCC specimens, the local buckling of the cold-formed channel is observed in short composite columns, as a result of lateral deformation of the infill concrete due axial compression load. The observed deformed shapes are consisted of a series of local buckling waves with half-wave lengths approximately equal to the spacing between the intersections of the lacing bars. In addition, crushing occurred in the infill concrete section (especially when the confined cold-formed channel section with end batten plate is changed to only cold-formed channel with lacing bars) at the ends of the specimen. Also, medium columns failed due to overall bending in addition to the local buckling waves as shown in Fig. 5. After these buckling waves had appeared, the specimens continued carrying additional loads until failure. At failure, local yielding occurred in the cold-formed channels where the infill concrete is behaving as a compression member at the point of intersection with the lacing bars. The results of the tested specimens and their failure modes are listed in Table 3. The composite columns are capable of withstanding high loads, where the composite column strength for LC-35 equals 241 kN is higher than the sum of concrete column, C-35, strength and the strength of the lacing steel column, L-35, which equalled 209.6 kN. Forces in the two cold-formed and lacing bars together act as a couple and resist the applied axial load. In addition, the lacing steel also provides integrity to the infill concrete even at large deformations.

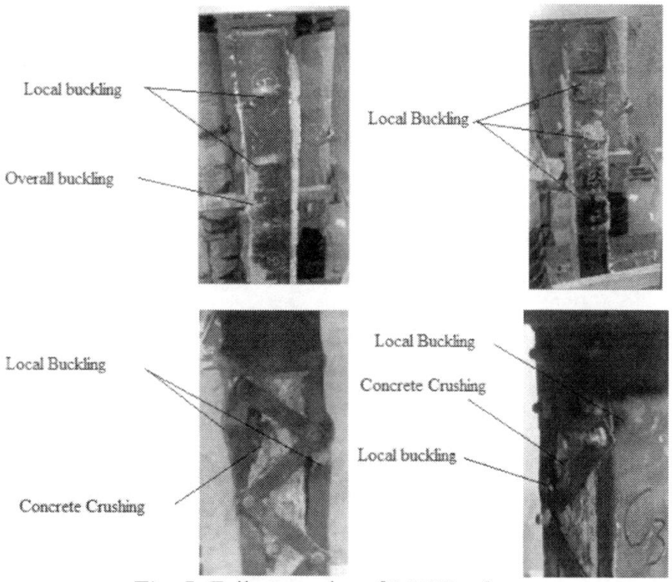

Fig. 5: Failure modes of LSCC column

Fig. 6 shows the axial load vs. shortening relationships for the tested specimens. The control concrete specimen (CC-35) has a lower strength and it fails due to the concrete crushing. However, the control steel specimen (L-35) failed due local buckling and it has an ultimate strength more than the control concrete specimen.

Table 3: Results of tested specimens

Specimen	Ultimate load (kN)	Failure Mode	Specimen	Ultimate load (kN)	Failure Mode
CC-35	81.6	C	LSC-70	272	CF-C
L-35	128	LF-C	LSC-GFRP-35	324	LF-C
L-70	113	CF-C	LSC-GFRP-70	251	CF-C
LC-35	241	LF-C	LSC-rft-35	460	LF-C
LC-70	251	CF-C	LSC-rft-70	288	CF-C
LSC-35	289	LF-C			

C= concrete crushing; LF-C= Local failure and concrete crushing; CF-C= Coupled local-overall failure and concrete crushing

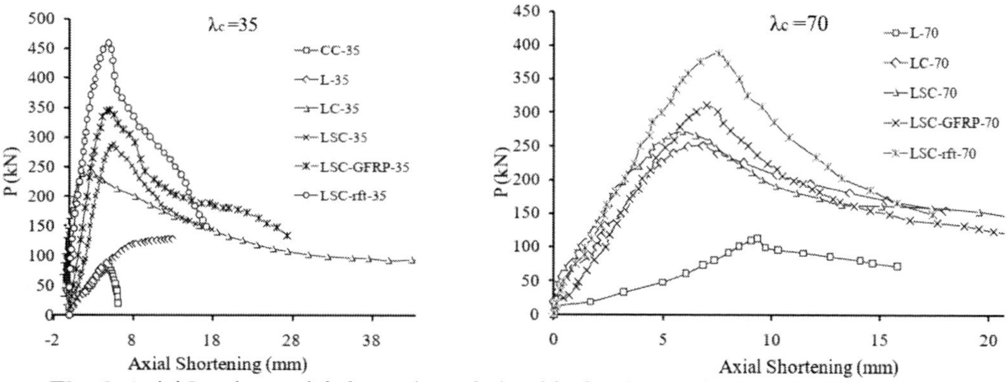

Fig. 6: Axial Load vs. axial shortening relationship for short and medium LSCC columns

The ultimate strength of the composite columns (LC-35 & 70) is obviously affected by the composite action as well as its ductility is higher than the control specimens (CC-35 & LC-35). The additives improved the behaviour and strength of the LSCC columns. Therefore, the strength of (LSC-35 & 70) specimens (which contains 10% silica fume) increases the ultimate capacity of the concrete compressive strength by (8-20%) for medium and short columns; respectively. The reinforced steel bars contribute in increasing the strength of LSCC columns with about 8% for the medium composite columns and by 15% for the short composite columns. The GFRP increases the strength of the LSCC columns. Test results indicate the effect of reinforced steel rebar (rft) improves the compressive strength of the LSCC column by about (42- 54%) more than the GFRP; for the short and medium composite columns; respectively.

5. Conclusions

Behavior and strength of laced steel-concrete composite (LSCC) columns are investigated in this study. A total of eleven specimens are tested for (LSCC) columns under axial load, having overall slenderness ratios (λ_c = 35 & 70). The additive silica fume (10% of silica fume /cement weight content) increases the strength of the concrete compressive strength of the medium and short composite columns by about (8-20%); respectively. The composite action between the steel channel and concrete improves the strength of the laced steel-concrete composite columns. The reinforced steel bars contribute in increasing the strength of LSCC columns with about 8% for medium and 15% for short composite columns. However, the GFRP increases the strength of the medium and short composite columns by about (4-8%); respectively. The effect of reinforced steel rebar (rft) improves the compressive strength of the LSCC column by about (42- 54%) more than the GFRP; for short and medium composite columns; respectively. The failure modes of LSCC columns are local buckling in cold-formed channel and concrete crushing in short composite columns. Also, combined local-overall buckling in cold-formed and crushing of the infill concrete is observed in medium columns. In addition, the lacing steel provides integrity to the infill concrete even at large deformations.

References

[1] Tavassoli A. "Behaviour of GFRP-reinforced concrete columns under combined axial load and flexure" *Master's of Applied Science*, Department of Civil Engineering, University of Toronto, 2013.

[2] Begum M, Driver RG, Elwi AE. "Numerical simulations of the behavior of partially encased composite columns", University of Alberta, Department of Civil & Environmental Engineering, *Structural Engineering Report,* No. 269, July 2007.

[3] Thirumalaiselvi A, Anandavalli N, Rajasankar J, Iyer NR." Prediction of static response of laced steel-concrete composite beam using effective moment of inertia approach", *IOSR Journal of Mechanical and Civil Engineering (IOSR-JMCE),* e-ISSN: 2278-1684, p-ISSN: 2320-334X, PP 01-06.

[4] Hanswille G. "Eurocodes Background and Applications", *Dissemination of information for training*, Institute for Steel and Composite Structures, University of Wuppertal-Germany, 18-20 February 2008, Brussels.

[5] EN 1994, *Eurocode 4: Design of composite steel and concrete structures*, CEN.

NONLINEAR INELASTIC ANALYSIS OF 3D COMPOSITE STEEL-CONCRETE FRAMEWORKS WITH PARTIAL SHEAR CONNECTION

Cosmin G Chiorean and Stefan M. Buru

Technical University of Cluj-Napoca, Cluj-Napoca, Romania

Abstract: This paper presents an efficient computer method for nonlinear inelastic analysis of 3D composite steel-concrete frameworks. The proposed formulation is intended to model the geometrically nonlinear inelastic behaviour of composite beam-column elements using only one element per member. Nonlinear effects of partial shear connection of composite beams are taken into account innovatively by introducing the internal axial force of the concrete slab as a function of the degree of shear connection. The proposed nonlinear analysis formulation has been implemented in a general nonlinear static purpose computer program, NEFCAD. The studies show that the proposed analysis compares very well to advanced finite fibre element solution (ABAQUS) with much less computational effort.

1. Introduction

In recent years, have witnessed significant advances in nonlinear inelastic analysis methods and computer programs for steel and composite steel-concrete framed structures and integrate them into the new and more rational advanced analysis and design procedures [1-4]. Currently the available tools for such analysis are general purpose FE programs that require very fine-grained modelling that is often impractical to the structural engineer. In spite of the availability of such nonlinear inelastic algorithms and powerful computer programs, the nonlinear inelastic analysis of *real large-scale frame structures* still posses huge demands on the most powerful of available computers and still represents unpractical tasks to most designers.

2. Mathematical formulation of the proposed method

The present work attempts to develop accurate yet computational efficient tools for the nonlinear inelastic analysis of real large-scale 3D semi-rigid composite steel-concrete frameworks with partial shear connections fulfilling the practical and advanced analysis requirements. The following assumptions are adopted in the formulation of analytical model: (1) Plane section remain plane after flexural deformation when full shear connection is assumed; when partial shear connection between the concrete slab and steel beam is assumed the plane sections remain plane independently with a linear strain distribution along steel and concrete sections heights with a discontinuity at the steel-concrete interface (Fig.2b); (2) Partial shear connection is taken into account considering the concept of the degree of shear connection; (3) Small strain but arbitrarily large displacements and rotations are considered; (4) Shear deformations are taken into account on the element stiffness but are neglected in the plastic constitutive relationships. The cross-section may assume any shape with multiple polygonal or circular openings. Two constitutive relations to model the concrete under compression and tension are adopted in the proposed formulation [4]. Multi-linear elastic-plastic stress-strain relationships, both in tension and in compression, are assumed for the structural steel and the conventional reinforcing bars. Gradual yielding of cross-sections and elasto-plastic tangent flexural and axial rigidities, when full shear connection is assumed, is described through basic equilibrium, compatibility and material nonlinear constitutive equations and the procedure is described in detail in [3].

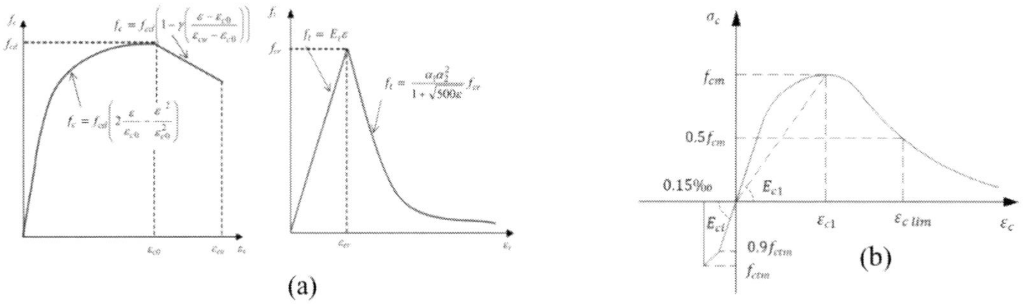

Fig. 1: Nonlinear constitutive equations for concrete (a) Model 1 (b) Model 2 CEB-FIP 1990

2.1 Elasto-plastic flexural rigidity of beam cross-sections with partial shear connection

The composite beam cross-section considered here consists of a concrete slab connected to a steel beam as presented in Fig. 2. Usually, for a given composite beam, the *full shear connection* is defined as the least number of shear connectors (n_f) such that the bending resistance of the beam would not be increased if more shear connectors are provided, whereas *partial shear connection* occurs when the number of connectors (n) used in a beam is lower than n_f. In order to analyse and design the composite beams, simplified methods are very useful and such methods are proposed in international literature and in some codes [5]. For instance in the Eurocode 4 [5] the concept of the degree of shear connection $\eta = n / n_f$ has been introduced. A simplified method to evaluate inelastic response of the composite steel-concrete beam cross-sections having partial shear connection is developed herein. Nonlinear effects of partial shear connection of composite beams are taken into account innovatively by introducing the internal axial force of the concrete slab as function of the degree of shear connection η already defined. Let us consider the cross-section subjected to the action of the external bending moment (M), and axial force (N) as shown in Fig. 2. The origin of the reference axis is considered in the geometric centroid of the structural steel. Under the above assumptions, the resultant strain distribution, corresponding to the curvature ϕ and the axial strains u_c and u_s in concrete and steel respectively, can be expressed in a linear form as:

$$\varepsilon_c = u_c + \phi \cdot (y - r); \ \ \varepsilon_s = u_s + \phi \cdot y + \varepsilon_r \tag{1}$$

where ε_c and ε_s represents the strains in concrete and steel respectively and ε_r represents the initial deformation produced by residual stresses, and is taken into account in above equation for the structural steel only.

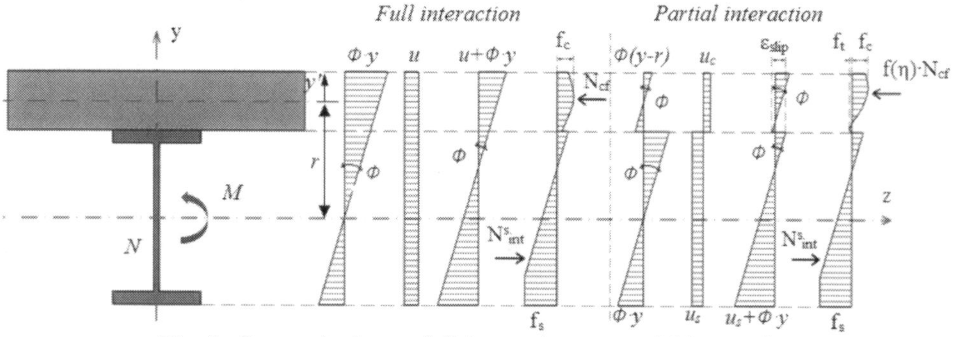

Fig. 2: Composite beam: full interaction vs. partial interaction

The equilibrium is satisfied when the external forces (*N*, *M*) are equal to the internal ones. These conditions can be represented mathematically in terms of the following nonlinear system of equations as:

$$\left\{\int_{A_s}\sigma(\varepsilon_s(u_s,\phi))dA_s+\int_{A_c}\sigma(\varepsilon_c(u_c,\phi))dA_c-N=0;\ \int_{A_s}\sigma(\varepsilon_s(u_s,\phi))ydA_s+\int_{A_c}\sigma(\varepsilon_c(u_c,\phi))ydA_c-M=0\right. \quad (2)$$

In order to solve the above nonlinear system of equations we assume that the axial force in the concrete slab (N^c_{int}) can be computed as function of the axial force in the concrete slab under the assumptions of full composite action (N_{cf}) as (Fig. 2): $N^c_{\text{int}}=f(\eta)\cdot N_{cf}$ where $f(\eta)$ represents a function of the degree of shear connection:

$$f(\eta)=c\eta\left[1+(c\eta)^n\right]^{-\frac{1}{n}} \quad (3)$$

and where the coefficients *c* and *n* are determined by numerical calibrations. In the proposed approach the following values are used: *c*=5; *n*=1.5. In this way with $f(\eta)\cdot N_{cf}=N_c$ and $-f(\eta)\cdot N_{cf}=N_s$ the nonlinear system (2) becomes:

$$\left\{\int_{A_c}\sigma(\varepsilon_c(u_c,\phi))dA_c-N_c=0;\ \int_{A_s}\sigma(\varepsilon_s(u_s,\phi))dA_s-N_s=0;\ \int_{A_s}\sigma(\varepsilon_s(u_s,\phi))ydA_s+\int_{A_c}\sigma(\varepsilon_c(u_c,\phi))ydA_c-M=0\right. \quad (4)$$

The above system can be solved numerically using the iterative Newton method and taking into account the fact that the stresses are implicit functions of the axial strain and curvature through the resultant strain distribution given by the Eq. (1) and then the axial and flexural rigidity of the cross-section can be computed [4].

2.2 Second-order inelastic member analysis

Flexibility-based method is used to formulate the distributed plasticity model of a 3D frame element (12 DOF) (Fig.3a). The spread of inelastic zones within an element is captured considering the variable section flexural EI_y and EI_z and axial *EA* rigidity along the member length, depending on the bending moments and axial force level, cross-sectional shape and nonlinear constitutive relationships as already described. Fig. 3b shows the deformed shape of a 3D beam-column element in local system attached to the initially straight center line, with the rigid body modes removed. Non-linear analysis by the stiffness method requires incremental loading, i.e. the inelastic behaviour is approximated by a series of elastic analysis. The element incremental flexibility matrix $\mathbf{f}_r$ which relates the end displacements to the actions $\Delta\mathbf{s}_r$ can be derived by applying the Maxwell-Mohr rule.

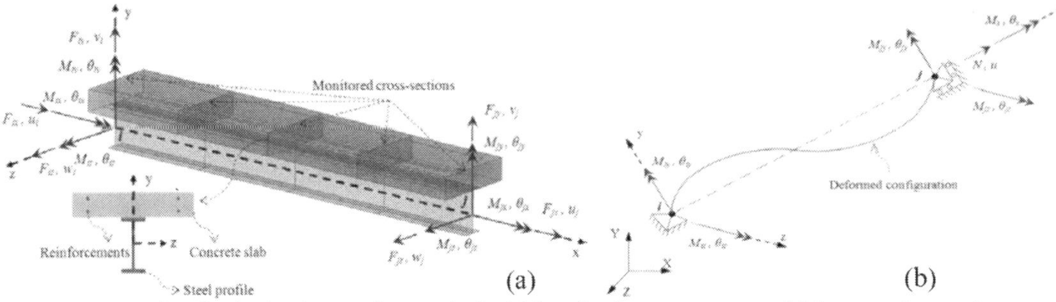

Fig. 3: Second-order inelastic member analysis (a) local reference system; (b) beam-column element with rigid body modes removed.

Assuming elastic behaviour within a load increment, and no coupling of axial and flexural responses at the section level, the generalized displacement in point i of the member produced by the force ΔP ($\Delta M_{iy(z)}, \Delta M_{jy(z)}, \Delta T_{iy(z)}, \Delta T_{jy(z)}$) could be expressed as in Eq. (5) where with superior indices (I) and (II) are marked the first order and second order efforts respectively.

$$\Delta\delta_{i,\Delta P} = \underbrace{\int_0^L \frac{\partial \Delta M^{I}_{y(z)}(x)}{\partial \Delta P}\cdot\frac{\Delta M^{II}_{y(z)}(x)}{EI_{ty(z)}(x)}dx}_{\text{bending moment component}} + \underbrace{\int_0^L \frac{\partial \Delta T^{I}_{z(y)}(x)}{\partial \Delta P}\cdot\frac{\Delta T^{II}_{z(y)}(x)}{GA_{y(z)}(x)}dx}_{\text{shear force component}} \tag{5}$$

The second term in Eq. (5) introduces the additional effect of shear deformations. Explicitly for nodal displacements (rotations at ends i and j) ($\Delta\mathbf{u}_r$) the above relationships could be expressed in a detailed manner as:

$$\Delta\mathbf{u}_r = \begin{bmatrix}\Delta\theta_{iy}\\ \Delta\theta_{jy}\\ \Delta\theta_{iz}\\ \Delta\theta_{jz}\end{bmatrix} = \int_0^L \begin{bmatrix} \frac{\partial\Delta M^I_y(x)}{\partial\Delta M_{iy}} & \frac{\partial\Delta T^I_z(x)}{\partial\Delta M_{iy}} & 0 & 0\\ \frac{\partial\Delta M^I_y(x)}{\partial\Delta M_{jy}} & \frac{\partial\Delta T^I_z(x)}{\partial\Delta M_{jy}} & 0 & 0\\ 0 & 0 & \frac{\partial\Delta M^I_z(x)}{\partial\Delta M_{iz}} & \frac{\partial\Delta T^I_y(x)}{\partial\Delta M_{iz}}\\ 0 & 0 & \frac{\partial\Delta M^I_z(x)}{\partial\Delta M_{jz}} & \frac{\partial\Delta T^I_y(x)}{\partial\Delta M_{jz}} \end{bmatrix} \cdot \begin{bmatrix}\frac{\Delta M^{II}_y(x)}{EI_{ty}(x)}\\ \frac{\Delta T^{II}_z(x)}{GA_{ty}(x)}\\ \frac{\Delta M^{II}_z(x)}{EI_{tz}(x)}\\ \frac{\Delta T^{II}_y(x)}{GA_{tz}(x)}\end{bmatrix} dx = \int_0^L \mathbf{b}^T(x)\Delta\boldsymbol{\varepsilon}(x)dx \tag{6}$$

Solving the following second-order differential equation (with the unknown the bending moment) which expresses the equilibrium of the element by taking into account the second-order geometrical effects, the second order bending moments and shear forces can be evaluated in function of the nodal bending moments and uniform distributed loads q as:

$$\frac{d^2M(x)}{dx^2} + \frac{v^2}{l^2}M(x) = -q\ ;\ v = L\sqrt{\frac{N}{EI_{tm}}} \tag{7}$$

and then the relationship (6) between nodal displacements ($\Delta\mathbf{u}_r$) and the nodal efforts ($\Delta\mathbf{s}_r$) could be further expressed by defining the element flexibility matrix ($\mathbf{f}_r$) and nodal displacements given by the uniform distributed loads ($\boldsymbol{\delta}_r$) as:

$$\Delta\mathbf{u}_r = \mathbf{f}_r\Delta\mathbf{s}_r + \boldsymbol{\delta}_r \tag{8}$$

where:

$$\mathbf{f}_r = \int_0^L \mathbf{b}^T\mathbf{f}_{st}\mathbf{B}dx = \begin{bmatrix}\mathbf{f}_{ry} & \mathbf{0}\\ \mathbf{0} & \mathbf{f}_{rz}\end{bmatrix};\ \boldsymbol{\delta}_r = \int_0^L \mathbf{b}^T\mathbf{f}_{st}\Delta\mathbf{q}dx = \begin{bmatrix}\boldsymbol{\delta}_{ry}\\ \boldsymbol{\delta}_{rz}\end{bmatrix}\ \mathbf{f}_{st} = diag\begin{bmatrix}\frac{1}{EI_{ty}} & \frac{1}{GA_{ty}} & \frac{1}{EI_{tz}} & \frac{1}{GA_{tz}}\end{bmatrix} \tag{9}$$

$$\mathbf{b} = \begin{bmatrix} \frac{x}{L}-1 & \frac{x}{L} & 0 & 0 \\ \frac{1}{L} & \frac{1}{L} & 0 & 0 \\ 0 & 0 & \frac{x}{L}-1 & \frac{x}{L} \\ 0 & 0 & \frac{1}{L} & \frac{1}{L} \end{bmatrix} = \begin{bmatrix} b_1(x) & b_2(x) & 0 & 0 \\ b_3(x) & b_4(x) & 0 & 0 \\ 0 & 0 & b_1(x) & b_2(x) \\ 0 & 0 & b_3(x) & b_4(x) \end{bmatrix}$$

$$\mathbf{B} = \begin{bmatrix} \frac{\sin \nu_y\left(\frac{x}{L}-1\right)}{\sin \nu_y} & \frac{\sin \nu_y\left(\frac{x}{L}\right)}{\sin \nu_y} & 0 & 0 \\ \frac{\frac{\nu_y}{L}\cos \nu_y\left(\frac{x}{L}-1\right)}{\sin \nu_y} & \frac{\frac{\nu_y}{L}\cos \nu_y\left(\frac{x}{L}\right)}{\sin \nu_y} & 0 & 0 \\ 0 & 0 & \frac{\sin \nu_z\left(\frac{x}{L}-1\right)}{\sin \nu_z} & \frac{\sin \nu_z\left(\frac{x}{L}\right)}{\sin \nu_z} \\ 0 & 0 & \frac{\frac{\nu_z}{L}\cos \nu_z\left(\frac{x}{L}-1\right)}{\sin \nu_z} & \frac{\frac{\nu_z}{L}\cos \nu_z\left(\frac{x}{L}\right)}{\sin \nu_z} \end{bmatrix} = \begin{bmatrix} B_{1y}(x) & B_{2y}(x) & 0 & 0 \\ B_{3y}(x) & B_{4y}(x) & 0 & 0 \\ 0 & 0 & B_{1z}(x) & B_{2z}(x) \\ 0 & 0 & B_{3z}(x) & B_{4z}(x) \end{bmatrix}$$

and, for instance, the flexibility sub matrix $\mathbf{f}_{ry}$ could be detailed as:

$$\mathbf{f}_{ry} = \int_0^L \begin{bmatrix} \frac{b_1(x)B_{1y}(x)}{EI_y(x)} + \frac{b_3(x)B_{3y}(x)}{GA_z(x)} & \frac{b_1(x)B_{2y}(x)}{EI_y(x)} + \frac{b_3(x)B_{4y}(x)}{GA_z(x)} \\ \frac{b_2(x)B_{1y}(x)}{EI_y(x)} + \frac{b_4(x)B_{3y}(x)}{GA_z(x)} & \frac{b_2(x)B_{2y}(x)}{EI_y(x)} + \frac{b_4(x)B_{4y}(x)}{GA_z(x)} \end{bmatrix} dx = \begin{bmatrix} \frac{L}{3EI_{0y}}c_1 + \frac{1}{LGA_{0z}}g & \frac{L}{6EI_{0y}}c_3 + \frac{1}{LGA_{0y}}g \\ \frac{L}{6EI_{0y}}c_3 + \frac{1}{LGA_{0y}}g & \frac{L}{3EI_{0y}}c_2 + \frac{1}{LGA_{0y}}g \end{bmatrix} \tag{10}$$

To produce the deformational-stiffness relation, the Eq. (8) is inverted, resulting the following deformational-stiffness equation:

$$\Delta \mathbf{s}_r = \mathbf{k}_r \Delta \mathbf{u}_r - \Delta \mathbf{q}_r; \qquad \Delta \mathbf{q}_r = \mathbf{k}_r \boldsymbol{\delta}_r \tag{11}$$

where the vector $\Delta \mathbf{q}_r$ is the incremental equivalent load vector, whereas $\mathbf{k}_r$ represents the instantaneous element stiffness matrix of the beam-column element without rigid body modes, determined by matrix inversion of the flexural matrix $\mathbf{f}_r$:

$$\mathbf{k}_r = \begin{bmatrix} \mathbf{k}_{ry} = \mathbf{f}_{ry}^{-1} & \mathbf{0} \\ \mathbf{0} & \mathbf{k}_{rz} = \mathbf{f}_{rz}^{-1} \end{bmatrix} \tag{12}$$

The resulting stiffness matrix is a 4×4 matrix, and does not include torsional and axial degrees of freedom. Torsional and axial stiffness components are then added to result in the required 6×6 stiffness matrix. To include rigid body modes, the stiffness matrix is pre- and post multiplied by a transformation matrix to result in the required 12×12 matrix [3]. Although the present study concerns mainly on frames with rigid joints and the effect of the floor slab action is ignored, in the proposed approach can be easily implemented the effects of the nonlinear behaviour of semi-rigid connections, with proper nonlinear moment-relative rotation models for composite connections, and the floor slab action. The effects of finite-size joints and semi-rigid composite connections on the stiffness matrix of the beam-column element and equivalent nodal loads could be included in the proposed formulation. A detailed discussions concerning this issue could be find in [4].

3. Advanced finite element modeling using Abaqus software

The advanced numerical simulation is conducted here by using the specialized software for nonlinear analysis of structures, ABAQUS v. 6.11 [6]. A numerical model considering a combination of 3-D solid elements (for concrete volumes) and shell elements for steel elements has been developed. In this paper, the general purpose S4R shell element is selected for modelling the steel members. Eight-node brick elements with reduced integration, C3D8R, are used to model the concrete slabs. The beam – column joint area is modelled utilizing discrete rigid finite elements R3D4. The reinforcing bars are modelled as two-node three-dimensional linear truss elements, T3D2, which are connected to the concrete matrix with the *Embedded option. The connector elements CONN3D2 are used in the proposed approach for modelling the connection between the steel beam and the concrete slab (Fig. 4). The constitutive law, that describes the in-plane translational behaviour of the connector elements, it can be described by Ollgaard's [7] constitutive model.

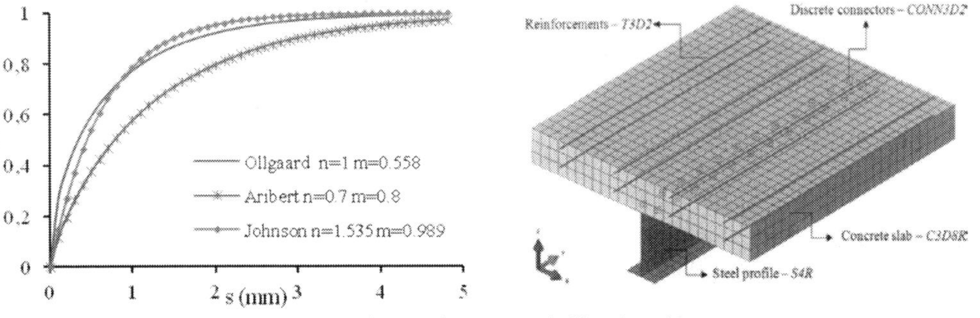

Fig. 4: Advanced FEM modelling in Abaqus

4. Computational examples

In order to trace the equilibrium path, for proportionally and non-proportionally applied loads, the proposed model has been implemented in a simple incremental (work-load) matrix structural-analysis program (NEFCAD) [3]. The accuracy of the analytic procedure and the computer program developed here has been evaluated using several benchmark problems. In the present approach, one element has been used to model each column and beam in all computational examples.

4.1 Simply supported composite beam with partial shear connection

The proposed numerical model is validated by comparisons against Chapman and Balakrishnan [8] experimental tests on E1 simply supported composite beam, as well as against the results predicted by ABAQUS software and those published by other authors [9,10]. The geometry and section properties of the analyzed beam are depicted in Fig. 5.

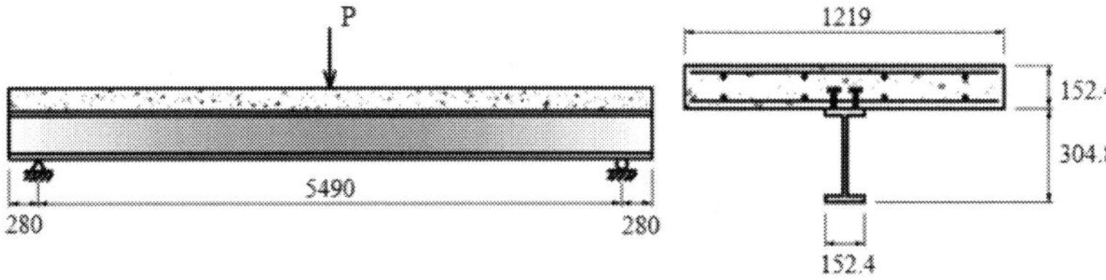

Fig. 5: Simply supported composite beam with partial shear connection

The CEB-FIB Model Code 90 has been used to model the concrete in compression and tension. The mid-span deflection vs. applied load comparative curves is plotted in Fig. 6a. As it can be seen the behaviour of the composite beam predicted by the proposed analysis procedure (with η=1.36 and shape parameters c=5 and n=1.5) and the finite element model described in Section 3 is in close agreement with that of experimental test. Fig. 6b presents the comparative load-deflection curves obtained by the proposed approach when different degrees of shear connections are taken into account. By comparing the curves depicted in Fig. 6b, it can be observed, that the proposed method (NEFCAD) predicts fairly well the nonlinear behaviour and ultimate load capacity of the system when lower levels (PSC=47%, 71%) of shear connection are considered.

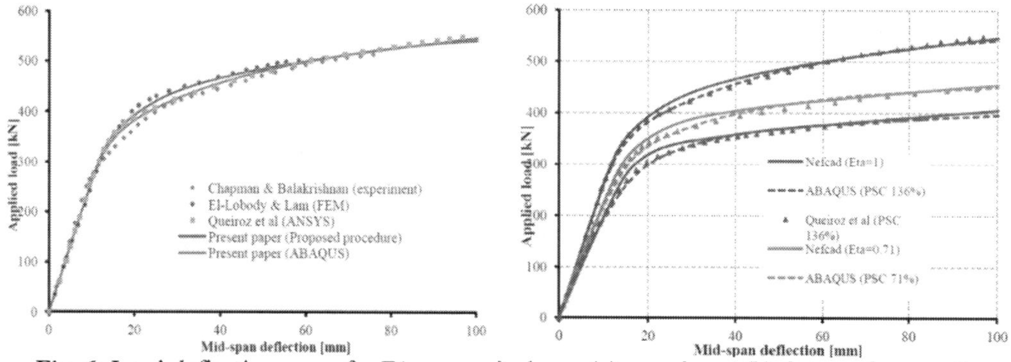

Fig. 6: Load-deflection curve for E1 composite beam (a) experiment (b) degree of connections

4.2 Six-story composite frame

The Vogel's six-story two-bay frame with proportionally applied distributed gravity loads and concentrated lateral loads, is presented in Fig. 7a. The Vogel's frame, studied previously by other researchers, has been modified and included in this verification study.

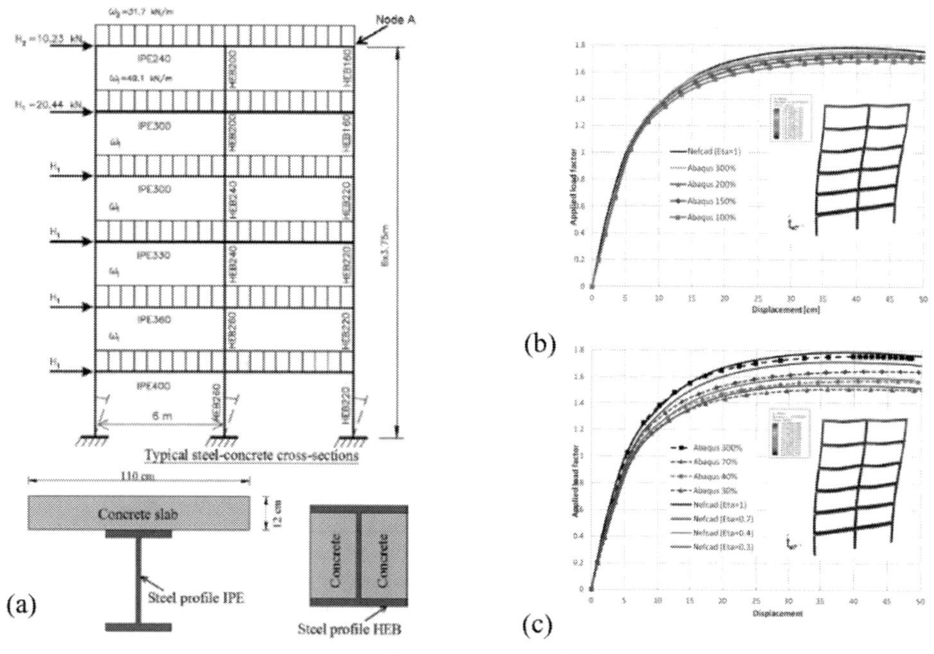

Fig. 7: Six-story composite frame

The geometry, member and material properties, and the applied loading are shown in Fig. 7. The yield strength of all steel members is 235 MPa and Young Modulus is E=20500 MPa and the shear modulus G=7885 MPa. The compressive yield strength of concrete for columns is f_c=26 MPa and for concrete slab f_c=16 MPa. In order to evaluate the effects of the finite size of the joints and to make possible comparisons with more advanced nonlinear FEM solutions (ABAQUS), in the proposed approach (NEFCAD) the frame has been modelled considering member end offsets and assuming effective rigid joint size of one-half the true size of the joint. The diameter of the shear studs has been taken as d=16 mm for which the corresponding ultimate shear capacity, calculated according to Eurocode 4 provisions, is Q_u=64.34 kN. The parameters that describe the shape of the shear connectors constitutive law have been selected as proposed by Ollgaard: n=1 and m=0.558. Fig. 7b presents the comparative load-deflection curves obtained by the proposed approach when the function of degree of shear connection $f(\eta)\rightarrow 1$ and those obtained with the advanced finite element modelling considering different levels of shear connections. In this study levels ranging from 100% to 300% are analyzed using FEM model, by varying the number of shear connectors. It can be observed that, as expected, increasing the level of shear connection the system became more rigid with increased strength and stiffness and the inelastic behaviour becomes similar with the response predicted by the proposed approach for $f(\eta)\rightarrow 1$ (full shear connection), Fig. 7b. The effectiveness of the proposed procedure is further assessed by varying the degree of shear connection (partial shear connection-PSC) and comparing the predicted curves with those obtained with more complex finite element analysis, as shown in Fig.7c. It can be seen the present method predicts fairly well the nonlinear behaviour and ultimate load capacity.

5. Conclusions

A reliable and robust nonlinear inelastic analysis method for composite steel-concrete space frames has been developed. Gradual yielding is described through basic equilibrium, compatibility and material nonlinear constitutive equations. The proposed formulation has been found to be effective in predicting the behaviour of composite beams and frames with partial shear connection and, the numerical results agrees fairly well to those obtained by advanced nonlinear FEM approaches but with much less computational effort.

References

[1] Chen WF, Toma S. *Advanced Analysis of Steel Frames,* CRC Press London, 1994.

[2] Fang L, Chan SL, Wong YL. "Numerical analysis of composite frames with partial shear-stud interaction by one element per member", *Engineering Structures*, 22, 1285-300, 2000.

[3] Chiorean CG. "A computer method for nonlinear inelastic analysis of 3D composite steel-concrete frameworks", *Engineering Structures*, 57, 125-152, 2013.

[4] Chiorean CG, Buru SM. "Advanced nonlinear inelastic analysis of 3D composite steel-concrete frameworks with partial shear connection", *submitted for publication*, 2015.

[5] EN 1994-1-1:2004. *Eurocode 4: Design of composite steel and concrete structures - Part 1-1: General rules and rules for buildings,* ECS, 2004.

[6] Hibbitt, Karlsson & Sorensen Inc. ABAQUS/standard user's manual, version 6.11, 2011.

[7] Olgaard JG, Slutter RG, Fisher JW. "Shear strength of stud connectors in lightweight and normal weight concrete", *AISC Eng. J.*, 8, 495-506, 1971.

[8] Chapman JC, Balakrishnan S. "Experiments on composite beams." *The Structural Engineer*, 42, 369-383, 1964.

[9] Queiroz FD, Vellasco PCGS, Nethercot DA. "Finite element modelling of composite beams with full and partial shear connection", *Journal of Constructional Steel Research*, 63, 505-521, 2007.

[10] El-Lobody E, Lam D. "Finite element analysis of steel–concrete composite girders", *Advances in Structural Engineering,* 6(4), 267–81, 2003.

THEORETICAL AND EXPERIMENTAL INVESTIGATION ON PERFORMANCES OF COMPOSITE STEEL CONCRETE SHEAR WALLS WITH STEEL FIBER REINFORCED CONCRETE

Daniel Dan, Valeriu A. Stoian, Sorin-Codrut Florut, Alexandru Fabian and Emanuela Boita

Department of Civil Engineering, Politehnica University of Timisoara, Timisoara, Romania

Abstract: The paper presents the results of an experimental program started at the Politehnica University of Timisoara, studying the behaviour of composite steel reinforced concrete shear walls (CSRCW) with partially encased profiles. Use of such shear walls represents a widespread solution for providing a lateral load resisting structural system, especially in the case of medium and high-rise buildings placed in seismic areas. Since previous theoretical and experimental studies have proved the performance of such a system, the foremost objective of the experimental program consisted in identifying innovative solutions for composite steel-concrete shear walls with enhanced performance, in the shape of steel fibre reinforced concrete which was used in order to replace traditional reinforced concrete. The effective stiffness, strength and ductility is evaluated using the experimental results.

1. Introduction

In the last decades the need for high-rise buildings has increased due to the industrialization process and the expansion of cities. In order to obtain a lateral load resisting structural system for high-rise buildings, especially in seismic areas, the most applied solutions are composite steel-concrete shear walls with boundary encased steel profiles. Different types of composite shear wall systems have been studied for the purpose of improving the ductility of the conventional reinforced concrete shear walls. Recent research on composite shear walls were conducted by Liao et al. [1], Ji et al. [2], Hossain and Wright [3], Su and Wong [4].

This solution proved to be very efficient and it is successfully used in designing and execution of tall buildings. A new direction towards this domain of research is the use of different types of fibres made from different materials: steel (SFRC), carbon (CFRC), glass (GFRC), plastic or others, added in the concrete mixture in order to improve the mechanical characteristics of materials, the bearing capacity of the structural elements and to reduce the amount of work in execution.

Although most studies were developed by analysing small scale elements, for example steel fibre reinforced concrete beams, [5, 6], there are also studies regarding the performances of wall elements, [7], subjected to seismic actions. In the case of high-performance concrete elements, [8], the addition of different ratios of steel fibre to the concrete mix can lead to an increase of the initial cracking force, while the stiffness and energy dissipation capacity may be increased by up to 40%. Many studies have been developed on self-compacting concrete, [9, 10], (SCSFRC) reinforced with steel fibres and the results are promising. Thus, adding a large amount of fibre in the mixture can change the sudden opening of a crack, as compared to plain concrete, in a more stable process. In addition to crack open some elements SCSFRC exhibited a greater tension stiffening than those without fibres.

Plain concrete has a rather low tensile strength, this drawback being annulled by the introduction of reinforcements. They are continuous and specific localized in structure to optimize performances. fibres are discontinuous and are generally distributed dispersed in the concrete mixture. Thus, fibres are still used in structural applications, along with traditional reinforcement. In general, improvement of capacity is observed from the inclusion of the steel fibre in concrete mixture, [6].

The main objective of this research project is to investigate the behaviour of composite steel-concrete walls after replacing traditional field reinforcement with steel fibres. In order to reveal the behaviour of composite steel - concrete walls reinforced by steel fibres, subjected to combined axial and quasi-static lateral cyclic loads, a theoretical and experimental program was developed at the *Politehnica* University of Timisoara, Civil Engineering Department, Romania, [11, 12].

2. EXPERIMENTAL SPECIMENS AND TEST SET-UP

2.1 Wall specimens characteristics

The experimental program consists of six 1:3 scale elements (CSFRCW-1 to 6), designed using the principles from the existing codes, Eurocode 2, Eurocode 4 and Eurocode 8, [13, 14, 15], applied to composite steel-concrete elements. The steel fibre contribution ratio is determined according to the Technical Regulation from 21/04/2003. The design details of all six types of composite steel fibre reinforced concrete walls are presented in Fig. 1. In this paper are presented the results obtained for the first specimen (CSFRCW-1).

The experimental specimens have 3000 mm height, 1000 mm width and 100 mm in thickness, and represent a three storeys and one bay element from the base of a lateral resisting system made by shear walls. The wall panels were embedded in a heavily reinforced concrete foundation with 1500 mm length, 400 mm height and 350 mm width. The structural steel profiles were connected with the concrete web by headed shear stud connectors with d=13 mm diameter and h=75 mm length placed every 150 mm. The reinforcements from the edges of the web panel consists of 8Ø10 mm vertical bars and Ø8/75 mm horizontal stirrups. Both the steel profiles and the reinforcements were embedded into the reinforced concrete foundation block to ensure anchorage The materials used for elements construction were tested in laboratory in order to determine their mechanical properties. The parameters of the steel sections used in CSFRCW specimens are presented in Table 1.

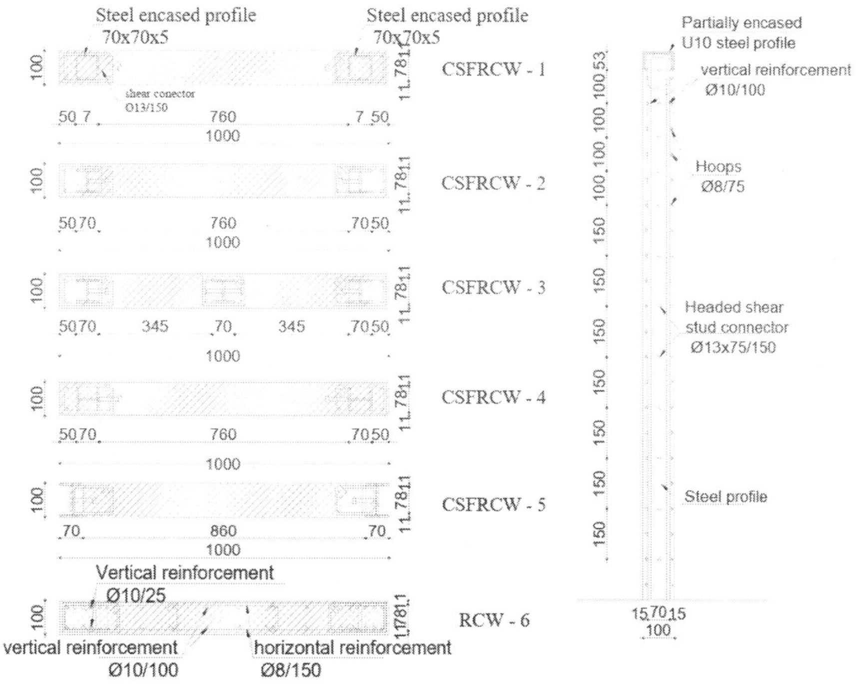

Fig. 1: Details of experimental specimens

Table 1: Parameters of encased steel sections

Specimen label	Steel shape	Encasement level	b_f [mm]	t_f [mm]	h_w [mm]	t_w [mm]
CSFRCW-1	2□	fully	70	5	70	5
CSFRCW-2	2I	fully	70	7	56	7
CSFRCW-3	3I	fully	70	7	56	7
CSFRCW-4	2H	fully	70	7	56	7
CSFRCW-5	2I	partially	70	7	86	7

2.2 Material properties

The design concrete quality was C30/37 class, the reinforcements Bst500S and the structural steel S355 JR (OL52-2k). The steel profiles were manufactured by welding the steel plates. The mixing formula of the concrete used to fabricate the specimens is the following: Cement 320 kg; Water: 170 kg/m^3; Sand: 725 kg; Aggregate: 1087 kg; Filler: 70 kg; Additive: 3.5 l. To determine the compressive and tensile strength of the concrete, 6 samples concrete cubes and 3 samples of prismatic elements were provided. The results of the tests obtained at the age of 28 days are presented in Table 2. For disperse reinforcement were used hooked steel fibres with geometry and mechanical characteristics presented in Table 3.

2.3 Test set-up and loading procedure

The experimental test was performed at the laboratory of Civil Engineering Department, at *Politehnica* University of Timisoara. The specimen was tested under constant vertical load and quasi-static reversed cyclic lateral loads. The lateral loads were applied alternatively from left and right. The test specimen was placed in the same plane as the loading frame, Fig.2, and was anchored with steel bolts into the laboratory reaction floor.

Table 2: Material properties of concrete

Specimen label	Number of samples (compression)	f_{cm} [N/mm^2]	Number of samples (tension)	f_{ctm} [N/mm^2]
CSFRCW-1	6	38.83	3	5.1

Table 3: Characteristics of steel fibres

Fibre Type	Diameter [mm]	Length [mm]	Tensile strength [N/mm^2]	Modulus of elasticity [N/mm^2]	Density [kg/m^3]
Hooked	0.8	60	Min. 1100	210x10^3	7850

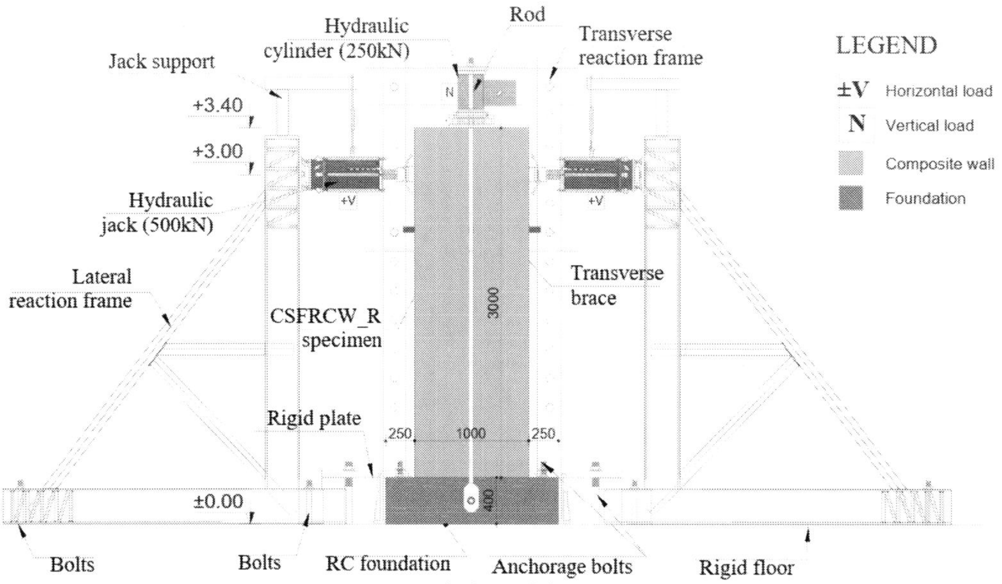

Fig. 2: General view of the test set-up

The loading frame consist of two steel braced frames, placed symmetrically. The lateral force was applied using two 400 kN hydraulic jacks, whereas the vertical force was induced by a 250 kN hydraulic jack. Initially, a constant vertical load of 100 kN was applied to the specimen and was maintained constant during the test. The horizontal forces were applied at 400 mm below the top of the elements, providing thus sufficient anchorage length above the load application level for the reinforcing bars and steel profiles. A transverse brace system was used in order to avoid any out of plane displacements of the specimens.

The recommended ECCS short testing procedure, [16], was used for the cyclic tests. The tests were performed in displacement control model. Minimum four cycles were performed before the elastic limit of the element was reached. After the elastic limit $\Delta_y \cong 20$mm, three cycles were performed at each displacement level. The horizontal forces were applied under controlled cyclic displacements, until the strength of the specimens decreased to 85% of the peak horizontal load.

In order to monitor the behavior of the experimental specimen, pressure transducers, displacements transducers (D) and strain gauges (G) glued on the reinforcements bars and on the steel profiles, were used. The position and the main characteristics of the measuring equipment are presented in Fig. 3.

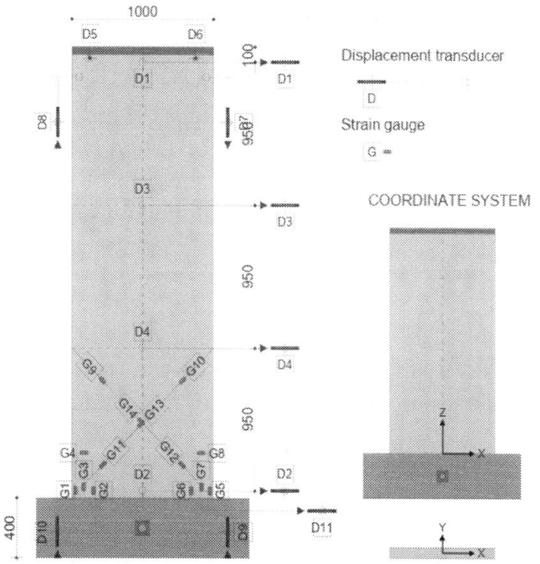

Fig. 3: Position of the displacement transducers and strain gauges

3. Experimental results

3.1 General behavior and failure modes of fiber reinforced elements

Generally, the behavior mode of a reinforced concrete wall subjected to in-plane lateral loading can be referred to as either flexural or shear. The flexural mode is characterized by horizontal cracks extending from the lateral edge in tension toward the opposite compressed side. The shear mode is marked by inclined cracks running across the web and/or straight sliding cracks at the linear joints. Nevertheless, it is quite frequent to observe cracks starting horizontally at the edge, and then sloping toward the compressed toe in the web; this situation can be regarded as flexural-shear behavior. As we predicted, experimental specimen CSFRCW-1 presented a shear mode failure, a behaviour that is not accepted in seismic areas. The behavior aspects of the elements during the test are presented in Table 4 at each displacement level.

Table 4: Response characteristics of the specimen

Displacement stage Δ (mm)	Behaviour aspects
Δ < 5mm	No visible cracks appeared
5 < Δ < 10mm	Diffuse inclined cracks appeared
10 < Δ < 15mm	Diffuse horizontal cracks forming; old inclined cracks developed
15 < Δ < 20mm	Inclined cracks developed along element
Δ > 20mm	Cracks developed towards failure

3.2 Cracking analysis

During the experimental test, the crack development on the surface of the concrete was monitored. The cracks appear when the tensile strength of the concrete is exceeded and the developing direction is generally perpendicular to the direction of the tension stresses. The failure starts with diffused inclined cracks which appeared in the tensioned zone. This cracks are caused by the stresses transfer from the steel profile to the concrete. The first crack appeared approximately at 30 cm from the bottom line. After that, new horizontal cracks and the extension of the existing ones were observed upon further loading. In the cycle Δ_y, diagonal cracks developed towards failure, in cycle 2Δy. The collapse of the element occurred with the development of the existing inclined cracks until the width of the crack was between 0.3-0.6 mm, and yielding of vertical reinforcement occurred. The tested steel fiber reinforcement concrete shear walls with encased steel profiles showed a shear failure, with the damage of the concrete web followed by steel profile and stirrups, from the tensioned zone, yielding.

3.3 Strain analysis

During the experimental tests, the strains were measured using strain gauges placed along the longitudinal direction of the reinforcing bars and on the steel encased profile, see Fig. 3. In Fig. 4 are presented the diagrams of lateral load P versus unit strain (ε) in the steel profile, vertical bars and stirrups.

From the experimental results it was observed that yielding occurred first in vertical reinforcements at a drift of 17.81 mm. The maximum strain recorded in vertical bars was 29.53‰. During the last cycle (2Δy), when concrete failure occurred, yield strain reached in steel profiles and stirrups in the compression zone. The yield values of 2.58‰, for steel profiles, and 2.59 ‰, for stirrups, are attained for total drifts of 26.46 mm and 31.6 mm.

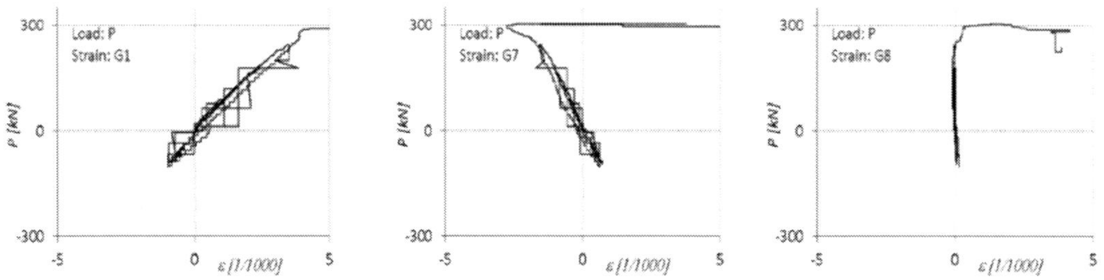

a) $P\text{-}\varepsilon_{steel}$ b) $P\text{-}\varepsilon_{sv}$ c) $P\text{-}\varepsilon_{sh}$

Fig. 4: Lateral load versus steel strain (ε): a) vertical reinforcement (G1); b) steel profile (G7); c) stirrups (G8)

4. Discussion

The main objective of this research project is to investigate the behaviour of composite steel-concrete walls after replacing traditional reinforcement with steel fibre. The results obtained from this program will be compared with the results from another research program developed also at the Politehnica University of Timisoara, that studied six structural Composite Steel Reinforcement Concrete Walls (CSRCW), [11, 12]. In order to provide an accurate comparison, the geometrical dimensions and the cross-section are similar in both studies.

In this paper are presented some comparative results for the first specimens of each research program (CSFRCW-1 and CSRCW-1). The strength of the specimens are defined in terms of load bearing capacity and represents the maximum applied horizontal force, P_{max}. The absolute values of the load bearing capacity for both specimens are presented in Fig. 5.

Although the values of the maximum force are closer, the failure mode is totally different. The concrete wall reinforced with steel fibre had a shear failure at a small displacement capacity compared with the element from the past research, and as it can be observed in Fig. 6 the displacement ductility is very low in the case of composite walls with steel fibres. Also, to describe the major hysteretic characteristics under cyclic behaviour, an envelope curve was generated for each tested specimen, Fig. 7. The behaviour of the tested elements is characterized by the four main points denoted as: initial cracking, element yielding, limit stage and failure stage. In Table 5 are presented the correspondence between the forces and the displacements at each characteristic point from the tests. It can be observed that for the CSFRCW-1 wall initial cracking occurs at a higher force and displacement than for the element CSRCW-1.

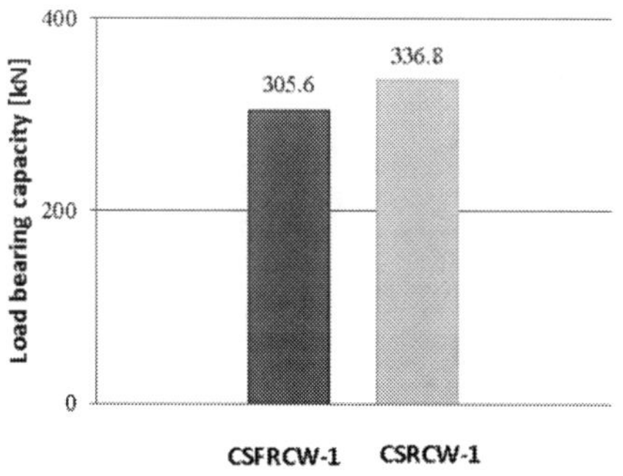

Fig. 5: The load bearing capacity of the specimens

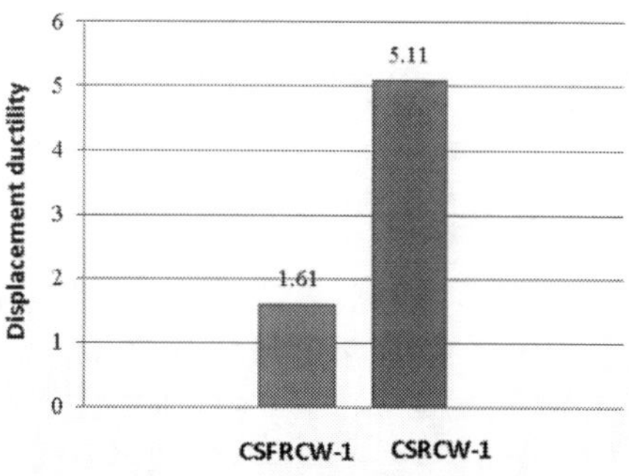

Fig. 6: Displacement ductility

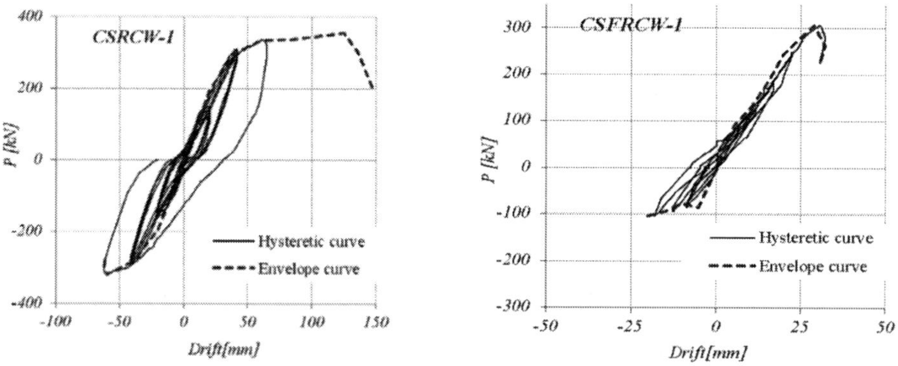

Fig. 7: Cyclic load drift envelope curves of specimens

Table 5: Force and displacement at different characteristic stages

Specimen	Initial cracking		Element yielding		Limit stage		Failure stage	
	P_{cr} [kN]	Δ_{cr} [mm]	P_y [kN]	Δ_y [mm]	P_{max} [kN]	Δ_{max} [mm]	$P_{85\%}$ [kN]	$\Delta_{85\%}$ [mm]
CSFRCW-1	125.6	11.02	245.2	22.84	305.6	28.97	262.5	32.18
CSRCW-1	80.5	7.54	228.2	26.5	354.4	125.1	301.5	135.4

5. Conclusions

The paper presents the experimental results obtained on the first composite steel reinforced concrete wall with steel fibres tested in a specific research program. Adding steel fibres in concrete mixture can lead to improvement of the mechanical characteristics of materials, and to reduce the amount of work in execution.

Using the information presented above and the information from specific literature the following conclusions can be formulated:

1. The results shown the failure mode of the element, with a diagonal crack developing towards failure and yielding of vertical reinforcements and structural steel;

2. Comparing with a traditional composite steel reinforced concrete wall it can be observed that although the maximum load bearing capacities are quite similar, failure was brittle and occurred at a smaller displacement, the displacement ductility significantly smaller;
3. The connection between steel encased profile and reinforced concrete using welded studs behaved as full connection as was assumed in the designing process.

Acknowledgments

The presented work was partially supported by SC Somaco Grup Prefabricate SRL Timisoara. We thank to all the persons for assistance during the design phase and casting the specimens.

References

[1] Liao FY, Han LH, Tao Z. "Experimental behavior of RC shear walls framed with steel reinforced concrete (SRC) columns under cyclic loading", *Proceedings of the 4th International Conference on Steel & Composite Structures*, 233-238, 2010.

[2] Ji X, Sun Y, Qian J, Lu X. "Seismic behaviour and modelling of steel reinforced concrete (SRC) walls", *The Journal of the Int. Assoc. for Earthquake Engineering*, 2014.

[3] Hossain AKM, Wright HD. "Experimental and theoretical behavior of composite walling under in-plane shear", *Journal of Constructional Steel Research*, 61(6):59-83, 2004.

[4] Su R, Wong S. "Seismic behavior of slender reinforced concrete shear walls under high axial load ratio", *Engineering Structures*, 29: 1957-1965, 2007.

[5] Bencardino F, Rizzuti L. "Implications of test methodology on post-cracking and fracture behaviour of Steel Fibre Reinforced Concrete", *Composites Part B: Engineering*, 46, 31-38, 2013.

[6] Ghaffar A, Chavhan AS. "Steel Fibre reinforced concrete", *International Journal of Engineering trends and Technology*, 9(15), 2014.

[7] Carrillo J, Alcocer SM, Pincheira J. "Shaking Table Tests of steel Fiber Reinforced concrete Walls for Housing", *15th World Conference on Earthquake Engineering*, Lisbon, 2012.

[8] Ganesan N, Indira PV, Seena P. "High performance fibre reinforced cement concrete slender structural walls", *Advances in Concrete Construction*, 2(4), 309-324, 2014.

[9] Jansson A, Flasbjer M. "Experimental investigation of surface crack initiation, propagation and tension stiffening in self-compacting steel-fibre-reinforced concrete", *Materials and Structures*, 45, 1127-1143, 2012.

[10] Frazao C, Barros J. "Durability of steel fiber reinforced self-compacting concrete", *Construction and Building Materials*, 80, 155-166, 2015.

[11] Dan D, Fabian A, Stoian V. "Theoretical and experimental study on composite steel–concrete shear walls with vertical steel encased profiles", *Journal of Constructional Steel Research*, 67, 800–813, 2011.

[12] Dan D, Fabian A, Stoian V. "Nonlinear behavior of composite shear walls with vertical steel encased profiles", *Engineering Structures*, 33, 2794-2804, 2011.

[13] EN 1992-1-1. Eurocode 2: Design of concrete structures, part 1-1, general rules and rules for buildings.

[14] EN 1994-1-1. *Eurocode 4: Design of composite steel and concrete structures, part 1-1, general rules and rules for buildings*, CEN.

[15] EN 1998-1. *Eurocode 8: Design of structures for earthquake resistance*, CEN.

[16] ECCS 1999. *Recommended testing procedure for assessing the behaviour of structural steel elements under cyclic loads*, CEN.

BUCKLING RESISTANCE OF HYBRID STEEL-CONCRETE COLUMN

Pisey Keo, Hugues Somja, Quang Huy Nguyen and Mohammed Hjiaj

Structural Engineering Group - INSA de Rennes, France

Abstract: In high-rise reinforced concrete buildings, columns or concrete walls containing multiple embedded profiles are often used, because the loads are such that standard reinforcement with rebars is not sufficient. Those composite elements belong to structures defined as "hybrid", which means that they are neither reinforced concrete structures in the sense of EC2, nor composite steel concrete structures in the sense of EC4. It is unclear whether these hybrid elements behave as reinforced concrete elements or as composite ones. To address this aspect, this paper evaluate the pertinence of the simplified methods proposed in EC2 and EC4 for the design of slender hybrid columns. To do so, a specific FE model which has been developed in [6] in which the geometrical/material nonlinearities are taken into account, will be used to treat different cases of 1140 hybrid columns in order to cover the various possible situations. The parametric study shows that these simplified methods are not satisfactory. Based on the extensive parametric study with 2960 data sets, a new version of bending moment magnification method for slender hybrid columns is proposed.

1. Introduction

In high-rise buildings, slender RC columns containing multiple encased profiles as reinforcement are often used to resist horizontal loads by bending about their strong axes when standard reinforcement with rebars is not sufficient to sustain such large loads. Those composite steel–concrete columns are called "hybrid column" because they are neither RC columns in the sense of EC2 [1] or ACI 318 [2], nor composite columns in the sense of EC4 [3] or AISC 2010 [4]. In the latter, the design rules are provided only for a single encased steel profile. Although such elements are used regularly, practitioners have no regulation applicable to their design. The RFCS research project SMART COmposite COponents [5] aims to fill this gap by providing a design guide for hybrid components. This paper presents the project results regarding hybrid column designs against instability. The Eurocode format will be kept and new expressions of the flexural stiffness along with other parameters involved in the magnification moment method are provided.

2. Slender column designs of Eurocodes 2 and 4

To design a column against instability, Eurocode 2 as well as Eurocode 4 permits to use a nonlinear geometrical and material analysis. The assumptions and safety factors of the two standards are consistent and will be used in the finite element model. EC2 and EC4 propose both a simplified method, called "moment magnification method", in which the first-order bending moment M_{Ed} is multiplied by a magnification factor k to obtain the second-order bending moment. The factor k largely depends on the flexural stiffness and the equivalent moment distribution. Hence, the procedure involves two steps. The first stage is to estimate the first-order moment magnification factor based on the shape of the bending moment diagram and the second one is to compute the effective stiffness *EI*.

3. Parametric study and assessment of simplified methods of EC2 and EC4

In this section, a specific FE model [6] is used to conduct an extensive parametric study in order to assess the applicability of moment magnification methods of EC2 and EC4 to hybrid column designs. To do so, the ultimate load of slender hybrid columns with different types of cross-sections are computed using our FE model and the Eurocode simplified methods. The results obtained with Eurocode methods and FE model are compared against each other to assess the applicability of these simplified methods to hybrid column design.

3.1 Description of the parameters

Five different hybrid cross-section configurations (HSRCC1–5) are considered. The cross-sections HSRCC1 and HSRCC2 are built with 3 steel profiles HEB120. In the first configuration (HSRCC1) the weak-axis of the profiles is parallel to the bending axis whereas in the second configuration (HSRCC2) they are orthogonal (see Fig. 1a and Fig. 1b). Hybrid cross-sections HSRCC3 and HSRCC5 correspond to the so-called mega-column which contains 4 steel profiles located at each corner of the cross-section (see Fig. 1c and Fig. 1d). The cross-section (HSRCC4) has also 3 steel profiles but with larger steel cross-section.

Table 1: Summary of case-studies

Section	*HSRCC1-5*
Concrete	C35/45 ; C60/75 ; C90/105
f_{yk}	500 MPa
f_a	355 MPa
$\bar{\lambda}$	0.5; 1.0; 2.0
e/h	0.0 to 3.0
δ	0.2 to 0.62
φ_{ef}	0; 1.5
r_m	-1; 0; 1

For all cases considered in this study, the limit of elasticity for steel profile is 355MPa and 500 MPa for reinforcement bar. Three concrete classes are considered: C35, C60 and C90. Note that hybrid columns HSRCC4 and HSRCC5 with a significantly high value of steel contribution ratio δ are modeled with concrete class C35. Although the hybrid column HSRCC4 is not totally realistic, it is considered here as a limit case with a large value of δ. For columns subjected to axial compression and bending moment, three different relative slenderness $\bar{\lambda}$ are considered for each crosssection configuration. To access the effect of creep on the ultimate load, both situations (with and without creep) are considered for each case. From the value of relative slenderness and geometry of the cross-section, the column length can be deduced. For columns subjected to compressive load only, the whole range of possible relative slenderness is covered. The parametrical study is summarized in Table 1.

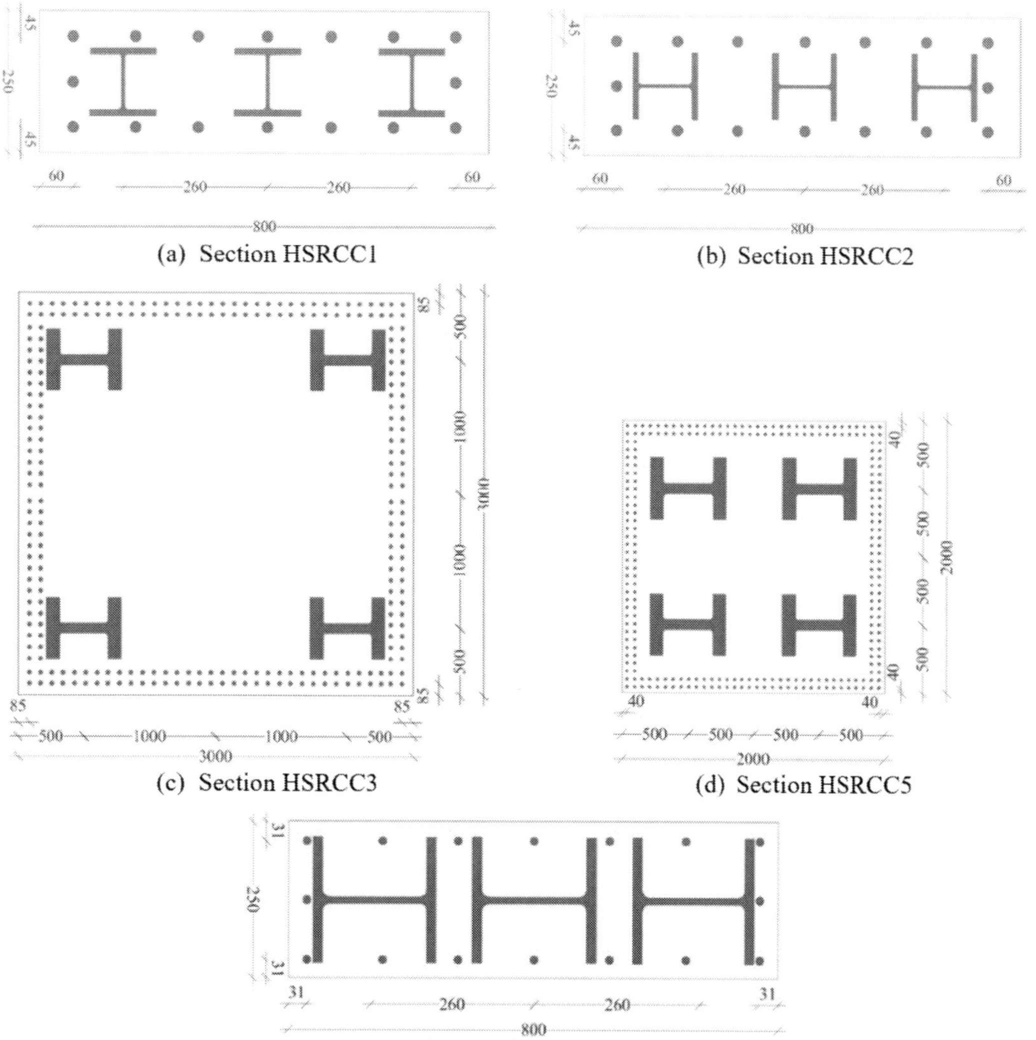

(a) Section HSRCC1

(b) Section HSRCC2

(c) Section HSRCC3

(d) Section HSRCC5

(e) Section HSRCC4

Fig. 1: Cross-sections considered in parametric study

3.2 Material laws

In order to evaluate the accuracy of the safety level when applying the simplified design methods proposed in EC2 and EC4 for hybrid column design, the general design methods (using nonlinear FE analysis) suggested by the Eurocodes should be adopted. Nonlinear material models as well as the safety format have to be properly described. The comparison of the results provided by simplified method of Eurocodes against FE analysis is readily achieved by using the stress–strain relationship presented in Fig. 2.

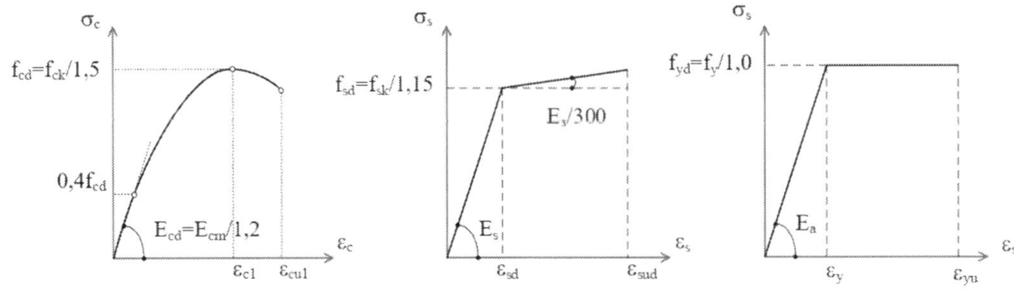

Fig. 2: Material constitutive laws used for FE analysis

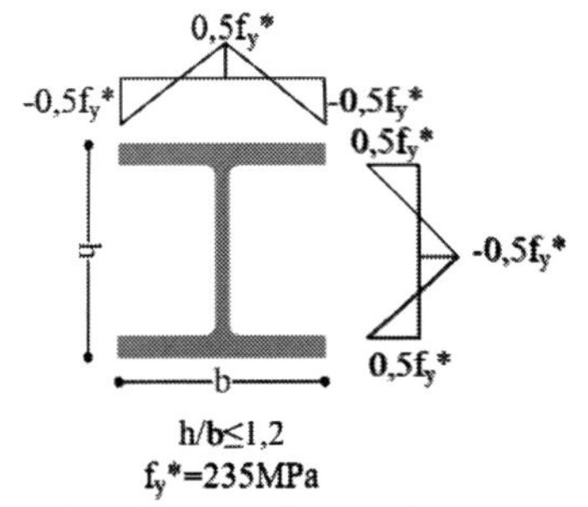

Fig. 3: Residual stress distribution of steel profile

Second-order analysis requires the definition of an imperfection. Those imperfections found their sources in both the geometric imperfection as well as the residual stresses. The definition of this initial deformed shape strongly affects the behaviour of slender columns. For concrete columns, EC2 recommends to consider a geometric imperfection equal to $l/400$ whereas for steel columns EC3 not only suggests to adopt a geometric imperfection equal to $l/1000$ but also to take into account the effects of the residual stress distribution (see Fig. 3). The hybrid column being built as a concrete column; we adopt an initial imperfection w_0 equal to $l/400$ combined with an explicit representation of the residual stress distribution in order to ensure the best accuracy of the results.

3.3 Assessment of the Eurocode moment magnification method

In the present section, the applicability of the EC2 version of the moment magnification method to hybrid columns is assessed by comparing its predictions against FE analysis results for hybrid column with cross-section HSRCC1 (see Fig. 1a). The concrete class is C60 and the effect of creep is taken into account ($\varphi_{ef} = 1.5$). It can be seen from Fig. 4a that in case of pure compression, the moment magnification method of EC2 gives unsafe results for low-to-moderate relative slenderness whereas the method provides conservative results for high relative slenderness. For column subjected to single curvature bending and regardless of the load eccentricity (see Fig. 4b and c), EC2 method overestimates the ultimate load for low-to-moderate relative slenderness ($\bar{\lambda}$=0.5 to 1.0). The same conclusion applies for columns bent in double curvature under antisymmetric bending moment (see Fig. 4d) except for very high load eccentricities (close to pure bending). For high relative slenderness, EC2 method gives safe results except for columns bent in single curvature under large bending moment. The same comparison has been made to evaluate the EC4 version of the moment magnification.

For a hybrid column subjected to pure compression (see Fig. 5a) where the ultimate load of the column is characterized by the resistance in axial compression, the simplified method of EC4 gives safe results regardless of column relative slenderness. Apart from the latter case, this method gives unsafe results for a large number of cases. For low load eccentricity, the ultimate load given by EC4 formulation is safe regardless of column relative slenderness (see Fig. 5b to d). For moderate load eccentricity, EC4 method always overestimates the ultimate load. Under large bending moment, EC4 method gives safe results, particularly for column under symmetric single curvature bending in the zone of nearly pure bending. The conservative nature of the results can be attributed to the equivalent moment factor β, which, in the present case, is equal to 1.1. To evaluate the EC2 and EC4 variant, the ratio R defined below has been selected:

$$R = \frac{R_{FE}}{R_{SM}} \text{ with } R_{FE} = \sqrt{\left(\frac{N_{FE}}{N_{pl,Rd}}\right)^2 + \left(\frac{M_{FE}}{M_{pl,Rd}}\right)^2} \text{ and } R_{SM} = \sqrt{\left(\frac{N_{SM}}{N_{pl,Rd}}\right)^2 + \left(\frac{M_{SM}}{M_{pl,Rd}}\right)^2}.$$

(N_{FE}, M_{FE}) is the ultimate load obtained from FE model; (N_{SM}, M_{SM}) is the one obtained with the simplified methods, and $N_{pl,Rd}$, $M_{pl,Rd}$ are resistance of the cross-section. When R is greater than 1, the simplified approach is safe. Histograms of R obtained for the two simplified methods are presented in Fig. 6. It was found that the mean value and the standard deviation are respectively equal to 0.996 and 0.104 for EC2 simplified method and 1.010 and 0.112 for EC4 simplified method. The percentage of R below 0.97 is 41.84% and 34.86% for EC2 and EC4 simplified method, respectively. As a general conclusion, it can be pointed out that mean estimations of both design codes seem to be correct but that their shortcomings lead to a large scatter of the results.

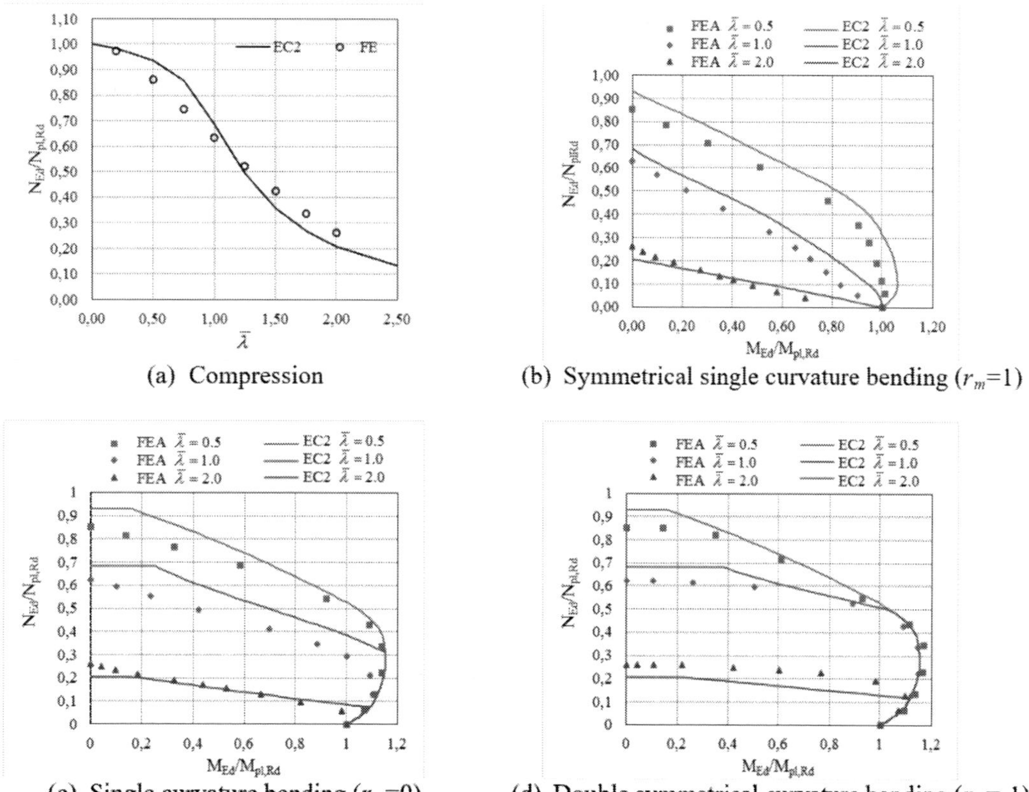

(a) Compression (b) Symmetrical single curvature bending (r_m=1)

(c) Single curvature bending (r_m=0) (d) Double symmetrical curvature bending (r_m=-1)

Fig. 4: Comparison of simplified method of EC2 against FE analysis results – HSRCC1

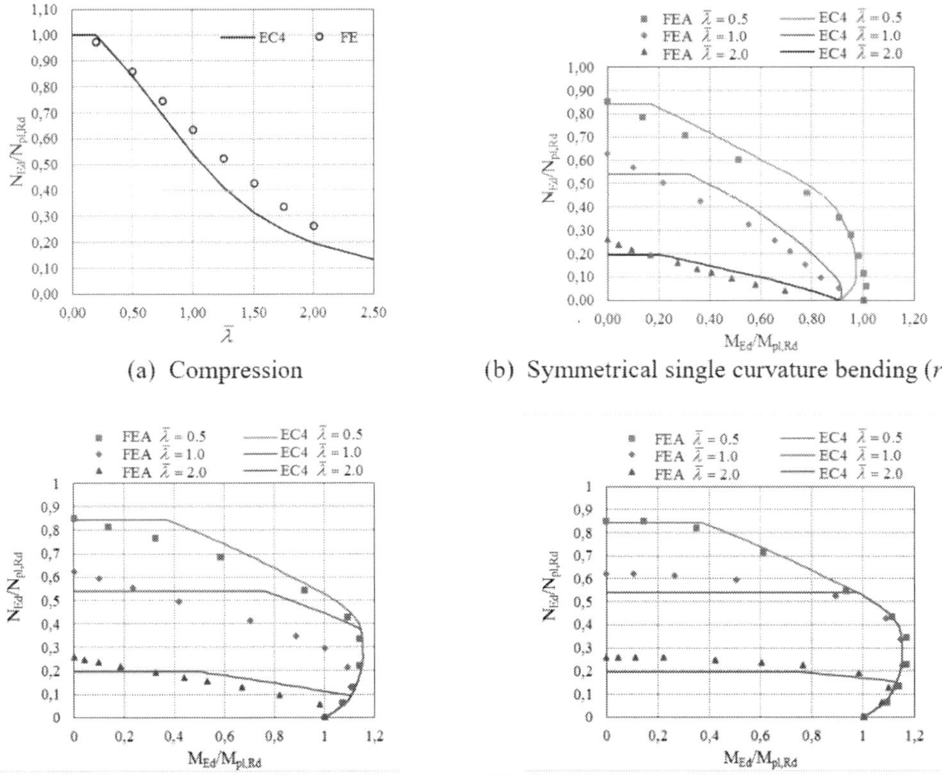

(a) Compression (b) Symmetrical single curvature bending (r_m=1)

(c) Single curvature bending (r_m=0) (d) Double symmetrical curvature bending (r_m=-1)

Fig. 5: Comparison of simplified method of EC4 against FE analysis results – HSRCC1

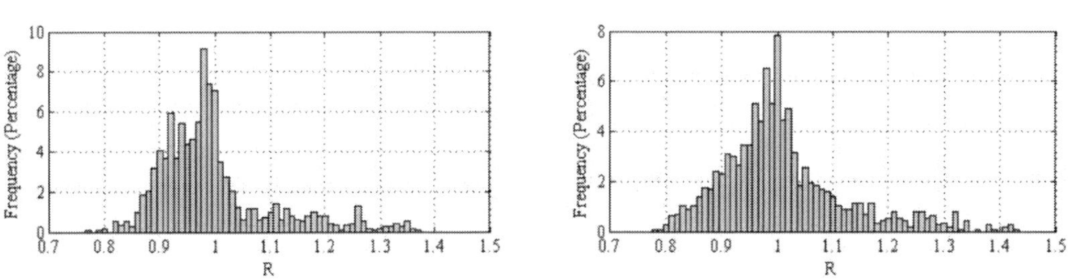

(a) Eurocode 2 (b) Eurocode 4

Fig. 6: Histograms of R obtained for the two simplified Eurocode 2 and Eurocode 4

4. Proposal of a moment magnification design method for hybrid columns

4.1 Further insight into the physical behavior of hybrid column

To determine how to define the most appropriate simplified method, the impact of various parameters has been studied in detail. The most prominent result concerns the effect of residual stresses. It appears that they have no effect on the ultimate load. The profiles being eccentric relative to the centre of gravity of the concrete section, they are subjected to a significant normal force which reduces the effect of residual stress, and it can be concluded that they behave more like rebars. Therefore, it is more appropriate to develop the new simplified method based on the EC2 method, rather than EC4, since EC4 explicitly integrates, through the definition of equivalent imperfections, the effect of residual stresses. Then, the general parametric study was increased to 2964 cases, by varying the yield strength of steel profiles. Additional cases were added with S235 and S355 steel profiles. It is shown that at the ultimate load, there exist a strong dependence of the state of steel profiles to the slenderness, creep coefficient and the ratio of resistance of concrete and steel. Indeed, increasing the deformation of the concrete due to creep results in a partial plasticization of the profiles, and therefore a reduction in their stiffness. This effect is less pronounced for slender columns, for which the deformation when the instability is produced, are limited, and therefore in this case the profile does not yield.

4.2 A proposal for the expression of the flexural stiffness *EI*

Based on the above findings, and on the work of Bonet et al. [7], a new moment magnification method is developed. The second order moment, which will be compared to the moment resistance of the section at the applied normal force N_{Ed}, $M_{pl,N,Rd}$, is obtained by multiplying the first order moment by:

$$k = \frac{\beta}{1 - \dfrac{N_{Ed}}{N_{cr}}} \tag{1}$$

where $\beta = 0,6r_m + 0,4 \geq 0,4$ and N_{cr} is the elastic critical load calculated using the following expression of the effective bending stiffness:

$$EI = K_c E_{cd} I_c + K_s E_s I_s + K_a E_a I_a \tag{2}$$

with

$$K_c = k_1 k_2 / (1 + \varphi_{ef}); k_1 = \sqrt{\frac{f_{ck}}{20}}; k_2 = n\frac{\lambda}{170} 0.2; n = \frac{N_{Ed}}{N_{pl,Rd}}$$

$$K_a = \frac{0.76\left(\dfrac{f_y}{f_{ck}}\right)^{0.0124}}{1 + 105\varphi_{ef}\exp(-0.078\lambda)} \leq 1; \quad K_s = 1$$

All variables subscripted with c belong to the concrete section and those with a belong to the steel beam. Quantities with subscript s are associated with the reinforcing bars.

4.2 Validation of the proposed simplified method

The histogram of the R ratio calculated for the new simplified method is shown in Fig. 7. The mean of the distribution is 1.0022, its standard deviation 0.0459. The 5% fractile is obtained for $R = 0.93$. Its overall variability gives a good estimation of the mean value of the ultimate load with relatively small deviation.

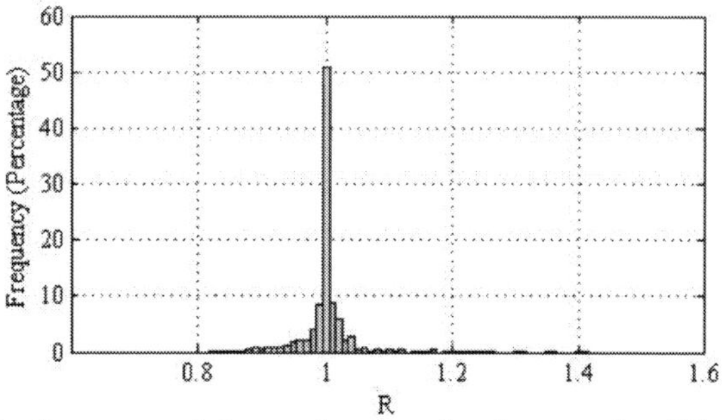

Fig. 7 Performance of the results given by the new simplified method

5. Conclusion

The main objectives of this study was to evaluate the bending moment magnification method proposed in EC2 and EC4 when applied to hybrid columns. To thoroughly analyze the applicability of EC2 and EC4 variants of the moment magnification method to hybrid columns, an extensive parametric study with 1140 data sets has been carried out. The comparison between the results obtained with Eurocode simplified methods and with FE analyzes shows that simplified method of EC2 and EC4 leads to a wide scatter where half of case-studies are unsafe. It means that the proposed effective flexural stiffness *EI* of EC2 and EC4 are not appropriate for the hybrid column design. It was observed that the secant modulus of compressed part of the steel section varies as a function of the creep coefficient φ_{ef} and the geometric slenderness λ. Therefore, in addition to the previous cases already analyzed, a further investigation of the effect of the steel yield stress on the ultimate load of hybrid columns has been carried out. This later investigation was based on an extensive numerical parametric study with 2960 data sets. A simplified method has been proposed for hybrid column design. This method is developed within the context of Eurocodes, i.e. moment magnification approach. In the proposed method, new expressions for the correction factors are proposed in order to take into account the creep effect and the effect of plastification of the steel profiles. The comparisons between proposed simplified method and FE analyzes show that the developed method provides the ultimate load for typical slender hybrid column with adequate accuracy.

Acknowledgments

The authors gratefully acknowledge financial support by the European Commission (Research Fund for Coal and Steel) through the project SMARTCOCO (SMART COmposite COmponents: concrete structures reinforced by steel profiles) under grant agreement RFSRCT-2012-00039.

References

[1] *EN 1992-1-1: Eurocode 2: Design of Concrete Structures: Part 1-1: General Rules and Rules for Buildings.* CEN, 2005.

[2] *ACI 318, Building Code Requirements for Structural Concrete (ACI 318-05) and Commentary (ACI 318R-05),* ACI Committee 318, American Concrete Institute, Farmington Hills, MI, 2005.

[3] *EN 1994-1-1: Eurocode 4: Design of composite steel and concrete structures: Part 1-1: General Rules and Rules for Buildings.* CEN, 2005.

[4] *AISC 2010: Specifications for structural steel buildings*, American Institute for Steel Construction, 2010.

[5] Plumier et al. *Tentative design guide for Hybrid elements*, SMARTCOCO Report, 2015.

[6] Keo P, Nguyen Q-H, Somja H, Hjiaj M. "Geometrically nonlinear analysis of hybrid beam–column with several encased steel profiles in partial interaction", *Engineering Structures*, 100, 66-78, 2015.

[7] Bonet J, Romero M, Miguel P. "Effective flexural stiffness of slender reinforced concrete columns under axial forces and biaxial bending", *Engineering Structures*, 33(3), 881-893, 2011.

EXPERIMENTAL AND NUMERICAL INVESTIGATIONS ON THE FLEXURAL BEHAVIOUR OF BOXED STEEL BEAMS ENCASED IN CONCRETE

S. Peltonen[a], M. V. Leskela[b], P. Kiriakopoulos[c], A. Iliopoulos[a], I. Vayas[d] and K. Spyrakos[c]

[a] *Peikko Group Corporation, R&D department, Finland*
[b] *Structural Consulting Engineer, Oulu, Finland*
[c] *National Technical University of Athens, Laboratory of Earthquake Engineering, Greece*
[d] *National Technical University of Athens, Institute of Steel Structures, Greece*

Abstract: The article summarizes the main results from a longstanding research campaign on the flexural behaviour of a certain type of slim floor beam, generally known as Deltabeam™. Load-displacement curves, slip values and damage patterns are reported for different levels of imposed deflections from three point bending tests of 7.2 m long simply supported beams with various geometrical configurations and depths. The experimental results are compared with those obtained from non-linear FE analyses. The tests showed that the ductility of beams depends not only on the classification of the top flange but also on the ability of the concrete section to contribute to the bending capacity still with high rotations of plastic hinge.

1. Introduction

Deltabeams are boxed shaped built-in steel beams with regularly spaced web openings, to be completely filled with concrete on-site. The geometry of the steel cross-section is shown in Fig. 1a. Its torsional stiffness is high enough to allow the assembly of heavy precast elements on the ledges of the bottom flange without propping (Fig. 1b). After the assembly of precast elements and all necessary reinforcements, the concreting phase follows. The final product is a rigid composite beam with high shear and bending capacity (Figs. 1c and 1d). Deltabeams are delivered on site with the necessary precambering so that deflections are easier to control. The shear connection between steel and concrete is ensured through the dowel action of the connector bars that pass through the web openings. Research on the resistance and slip characteristics of such shear connection is reported in [6 and 7].

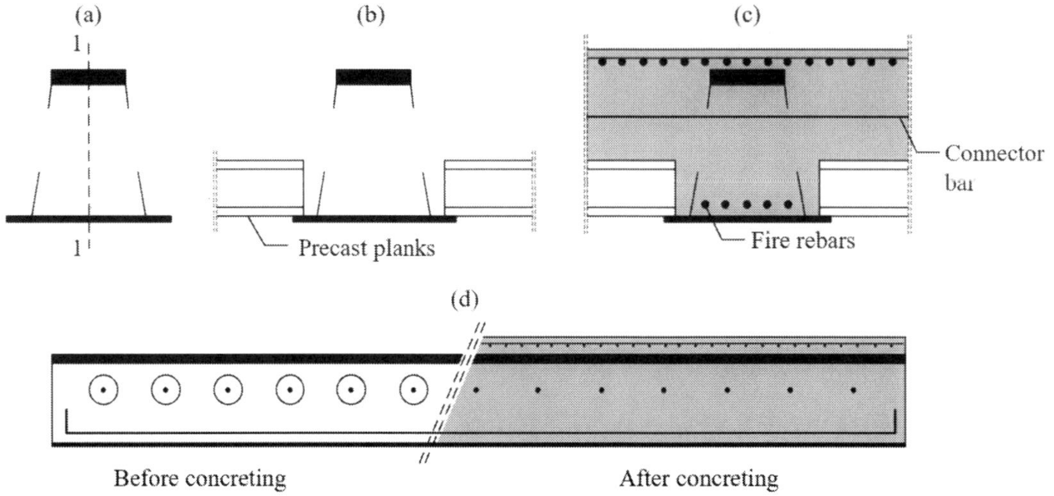

Fig. 1: (a) Steel cross-section, (b) Beam at erection stage, (c) Beam at final stage, (d) Section 1-1

The vertical shear resistance of the composite cross-section is considerably higher than that of the steel part due to the contribution of the core concrete. The main findings from shear tests are reported by *Leskela et al.* in [5]. Deltabeams can also reach a fire resistance rate higher than R180, depending on the thickness of bottom plate and amount of fire rebars inside the boxed section. Frequently the contribution of fire rebars is higher than 50 ~ 70 % of the bending capacity in typical beam sections. Experimental investigations on the fire resistance of Deltabeams equipped with hollow core slabs are found in [3]. Due to their high shear and bending capacity, Deltabeams are effective solutions for buildings that need to be designed against column loss scenarios. Numerical investigations on progressive collapse of buildings with Deltabeams are published in [4].

2. Sagging bending tests

2.1 Scope of the tests

In 2015 a new research campaign on the sagging bending behaviour of Deltabeams started in the institute of steel structures of National Technical University of Athens and is still on going. The aim of this research is to draw conclusions regarding the flexural resistance, the rotational capacity, the shear connection and the damage progression of Deltabeams with different geometries and reinforcing details.

2.2 Test set-up and materials

The experimental set up shown in Fig. 2a refers to 3-point bending tests of simply supported beams with a total length of 7.7 m. The distance between the roller supports is 7.2 m. The displacements in the middle of the beam were imposed by a hydraulic jack of 2.5 MN capacity, servo-driven as displacement controlled. The loading protocol is demonstrated in Table 1. The first 3 cycles refer to a maximum deflection of 25 mm, approximately equal to L/300, which corresponds to a common value for limiting deflections at the stage of serviceability. After three cycles, the displacement is increased to a maximum value of 430 mm.

The two composite sections that were tested are shown in Fig. 3. Specimen DB1 is heavily reinforced with dense stirrups Φ8/100 and 4Φ16 top rebars for achieving the maximum possible ductility. Specimen DB2 is an edge girder with a similar reinforcing as that of DB1.
The concrete used in the specimens is C30/37 according to EN 1992-1-1 [1]. The material for the steel part is S355J2+N according to EN 10025-1:2004 [2]. The mean concrete strength measured after the beam tests was 36.2 N/mm^2. For structural steel, the mean yield limit R_e measured was 410 N/mm^2 and the tensile strength R_m 565 N/mm^2. For reinforcing bars, the mean values for the yield limit and tensile strength were 493 and 650 N/mm^2.

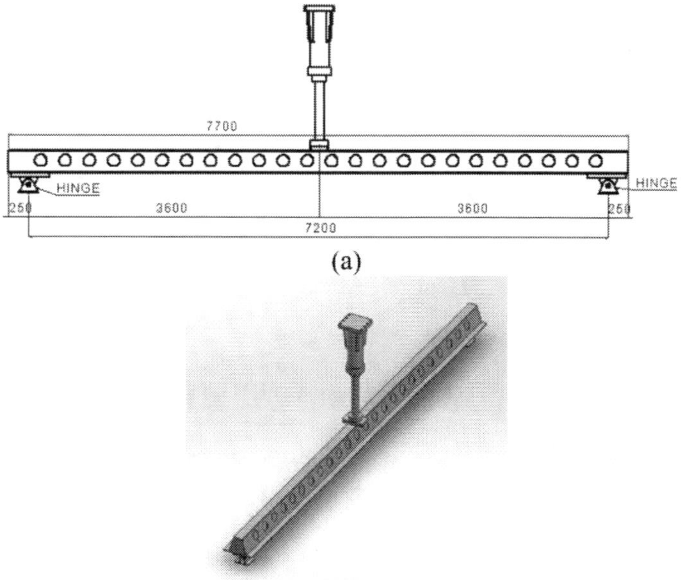

(a)

(b)

Fig. 2: (a) Side view and (b) 3D-view of the test frame

Table 1: Imposed deflections vs. time

Time (mins)	Imposed deflections (mm)
0	0
7.5	25
10	0
17.5	25
20	0
27.5	25
30	0
90	100
150	430

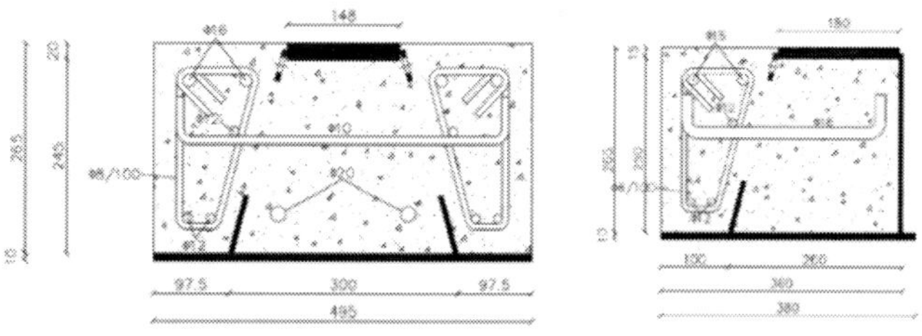

(a) (b)

Fig. 3: Cross-sections of the tested specimens; (a) DB1 and (b) DB2

2.3 Capacity curves and experimental observations

The force-displacement curves are shown in Fig. 4. Both specimens DB1 and DB2 demonstrated a hardening behaviour with extreme ductility values. Indeed, considering 75% of the maximum force as the yield load, the displacement ductility values for DB1 and DB2 become equal to 6.4 and 6.0. The corresponding rotational capacities measured are 124 mrad for DB1 and 117 mrad for DB2. It is pointed out that both tests were terminated when the target displacement, approximately equal to 430 mm, was reached while the diagrams are still ascending. This indicates that both specimens are able to achieve even higher ductility values.

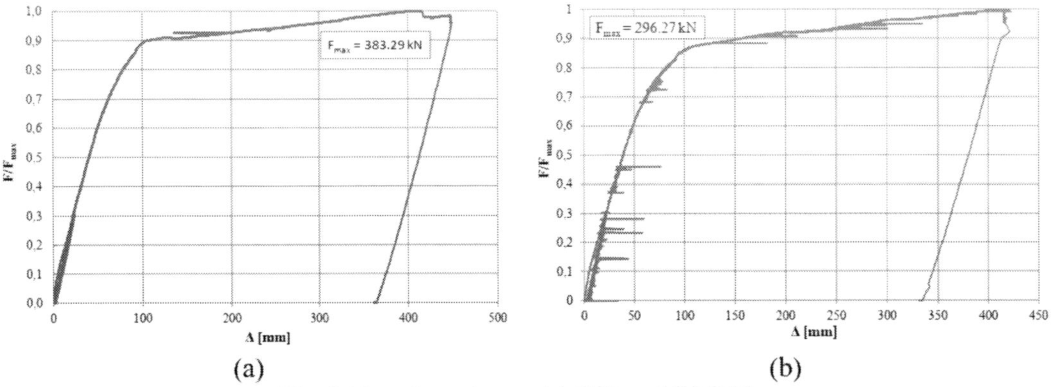

Fig. 4: Tested specimens (a) DB1 and (b) DB2

The damage patterns of the specimens are shown indicatively for three different phases in Fig. 5. At third cycle of the imposed deflection equal to 25 mm, a small number of hairline cracks emerged at the span of DB1 (Fig. 5a). Hairline cracks at the top surface of DB2 were observed, presumably due to torsion (Fig. 5b). The cyclic repetition of the loads at this phase has practically no influence on the stiffness of the specimens that can be considered as non-cracked. For deflections even higher than L/40, which corresponds to ULS-deflection limit [12], typical flexural cracks are found at a length approximately equal to 2.0 m (Figs. 5c and d). Cracks at the top surface of concrete near the loading area indicate initiation of concrete crushing. At the end of both tests, a small part of the concrete cover is lost (Figs. 5e and f). Both specimens DB1 and DB2 kept their integrity and remained at an excellent condition without unwanted phenomena such as excessive spalling and rebar or plate buckling. The main reason for this positive behaviour is attributed to the confinement effect in the core concrete inside the steel section and the dense stirrups that keep the external concrete intact.

2.4 Shear connection

The initial flexural design of the specimens was based on the partial shear connection theory for shallow floor beams [8], omitting the bond between the external concrete sections and the basic beam with concrete infill. However, as the bond was not eliminated, full connection behaviour was evident throughout the tests, as verified by very small slips (Figs. 6a and b) observed in the shear connection interfaces. The Eurocode 2 stress-strain model of the concrete for the design of concrete cross-section cannot cover the whole range of deformation behaviour shown in Fig. 4. However, applying the Sargin model [6] for the concrete and strain hardening properties of reinforcement and structural steels, the experimental ductility characteristics shown comply well with the computed numerical bending moment - curvature response.

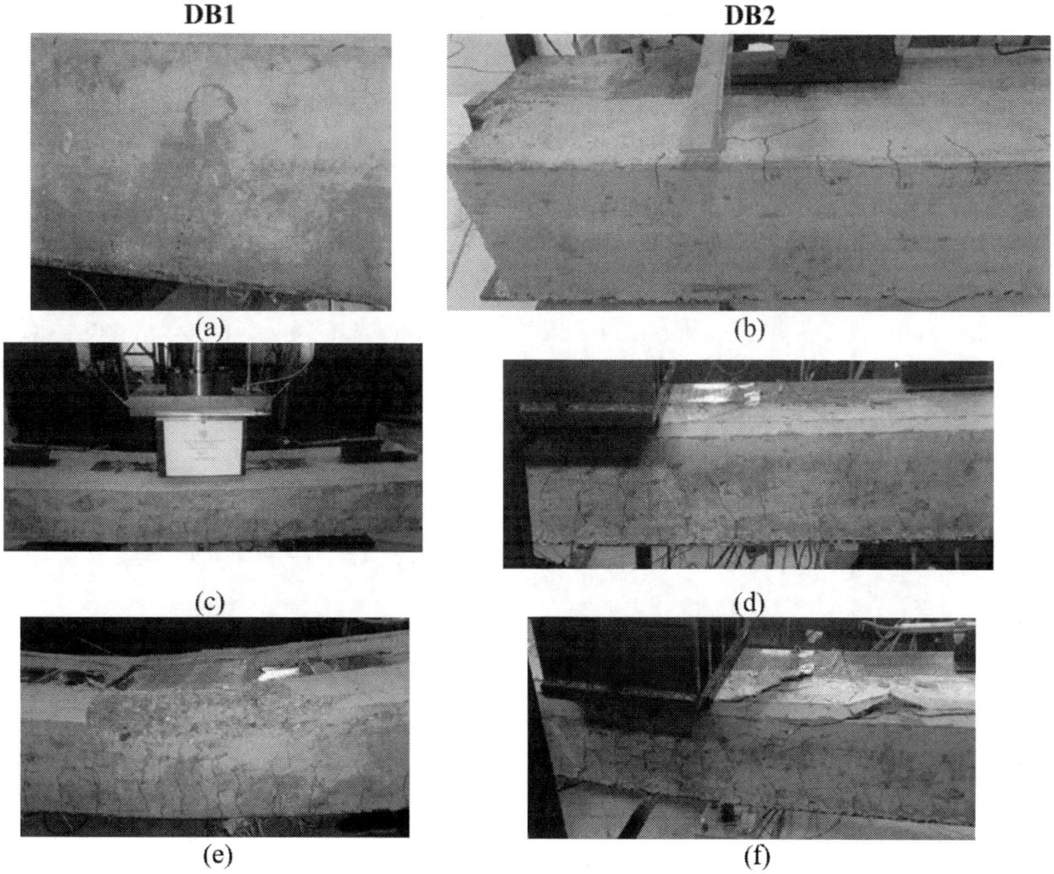

Fig. 5: Damage patterns (a), (b) at third cycle of 25 mm deflection, (c), (d) at 200 mm deflection and (e), (f) at the end of the test

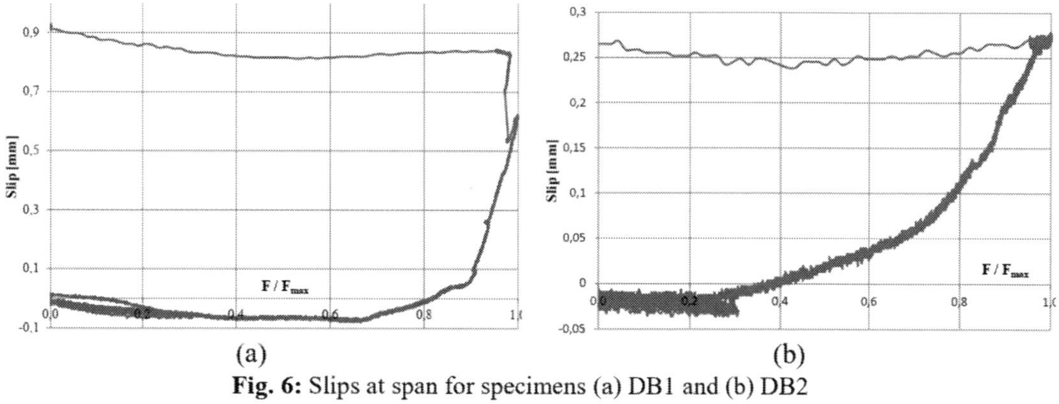

Fig. 6: Slips at span for specimens (a) DB1 and (b) DB2

3. FE-Modelling

In order to simulate the behaviour of the specimen DB1, finite element analysis software Abaqus [10] was used. The model consists of three parts (steel beam, concrete part, steel reinforcement) combined in one assembly. The mesh of the assembly is created with 15540 linear hexahedral elements of type C3D8R for the steel and concrete parts and 3706 linear line elements of type T3D2 for the reinforcement.

Concrete damaged plasticity is used for concrete and for conducting non-linear analysis, the mechanical parameters were taken from [11]. For the steel member, an elastic- plastic material model is used. The reinforcement bars are modelled as a bi-linear material with strain hardening. A general contact with friction factor equal to 0.4 is applied between concrete and steel parts. The two parts are not able to penetrate each other but they can separate after contact. The reinforcement bars are fully embedded in concrete.

The load-deflection responses measured in the middle of the beam span are compared with the numerical results in Fig. 7a; the two curves appear to be quite similar. Compressive and tensile strains at the top and bottom flange of the steel part monitored at span during the tests are also compared with the values calculated. The strain versus load graphs are presented in Fig. 7b. There is a good compatibility between the finite element analysis and the experimental data both for the compressive and tensile strains. One can also observe that despite the asymmetry of the steel cross-section, the strains at bottom flanges are quite symmetrical.

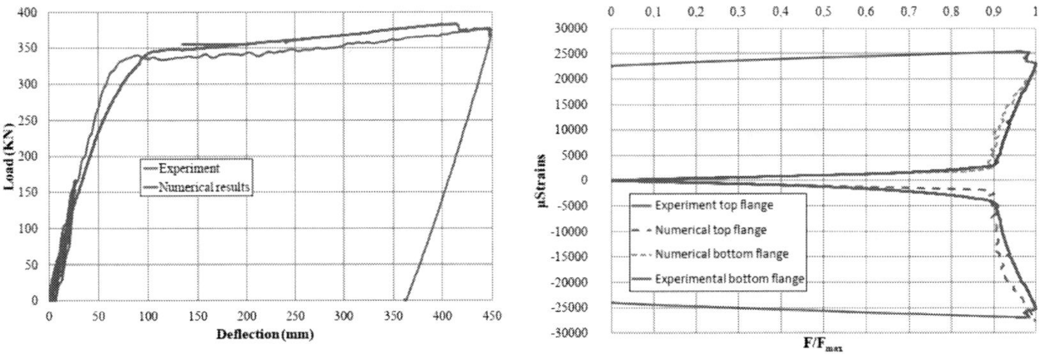

(a) (b)

Fig. 7: Experimental vs. numerical results for (a) Capacity curve and (b) strains at top and bottom plates of the specimens

Stresses in structural steel, concrete and reinforcing bars for deflections $L/40$ = 180 mm and $L/16$ = 450 mm are demonstrated in Fig. 8. Von Mises stresses in structural steel (Figs. 5a and 5b) are above the nominal value of the yield limit (> 355 N/mm^2). It can be seen that even for the extreme deflection of 450 mm, there are parts of the webs at span that remain elastic; this proves that the steel part of the composite cross-section still possesses non-negligible capacity reserves. In Fig. 8c it can be observed that the minimum principal stresses in concrete for 180 mm deflection remain below 30 N/mm^2; indeed for such deflections no concrete crushed (see Figs. 8c and d). At 450 mm deflection, the minimum principal stresses reached 38 N/mm^2 around the loaded area. A better view of the concrete stresses at span is shown in Figs. 9a and b. Concrete inside the webs and stirrups develops high compressive stresses due to confinement. It can be observed that the external concrete in the compression zone outside the stirrups shows zero stresses due to failure. This was observed during the spalling of the concrete cover in Fig. 5f.

The contribution of the longitudinal and transverse reinforcement in the resistance of the composite beam can be assessed from Figs. 8e and 8d. For both deflection values, 180 and 450 mm, the normal stresses in the longitudinal rebars exceeded 500 N/mm^2. The rebars in the compression zone were able to reach their maximum capacity without buckling, exactly the same way as in the experiment. A close look in the stirrups at span offers the information that their upper part is also overstressed; this is due to their support to the longitudinal rebars.

For deflection 180 mm　　For deflection 450 mm

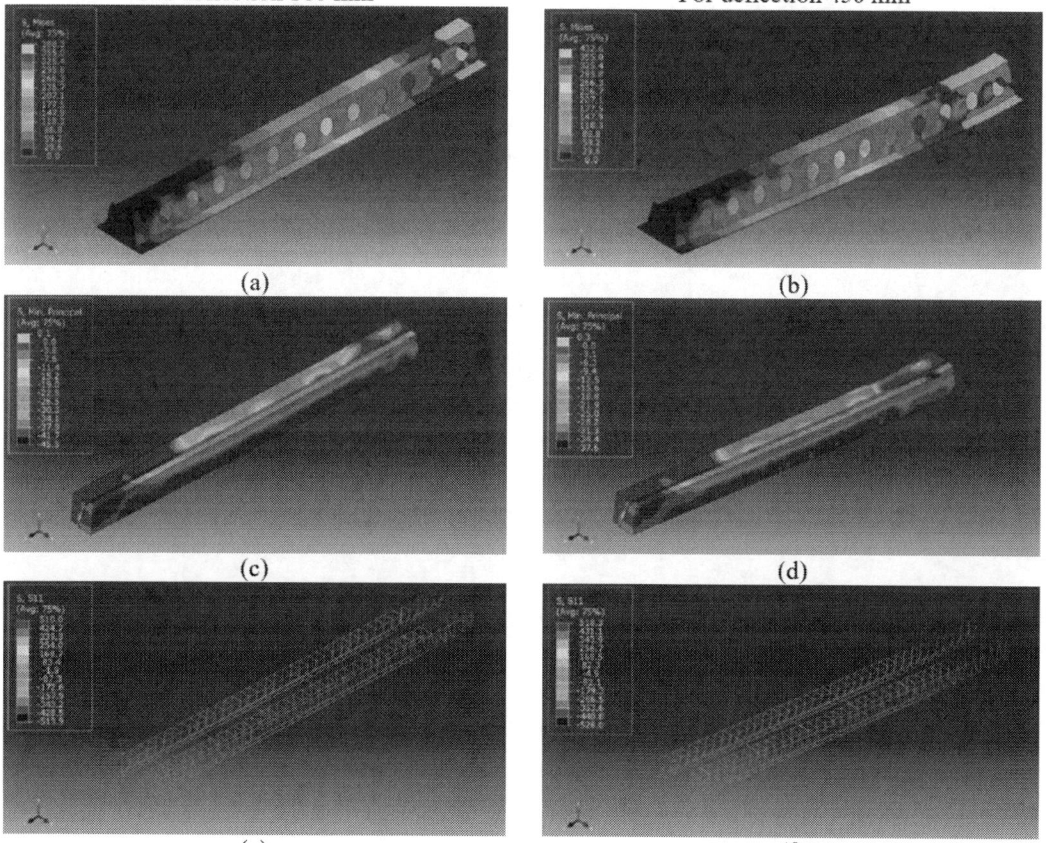

(a), (b), (c), (d), (e), (f)

Fig. 8: Stresses in Structural steel: (a), (b) Concrete: (c), (d) and Reinforcing bars: (e), (f)

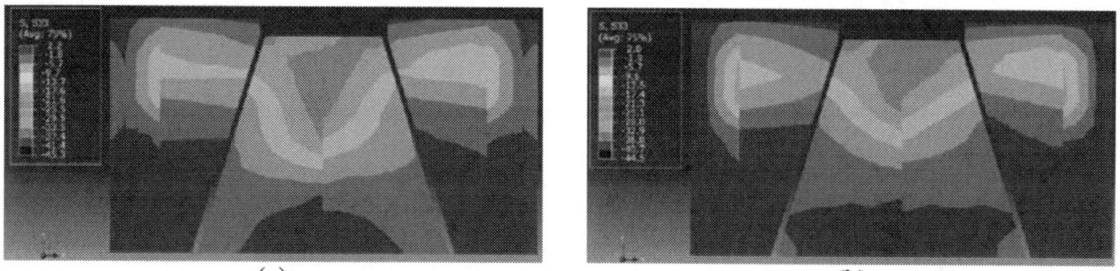

(a)　　(b)

Fig. 9: Normal stresses in the concrete at span for 180 mm (a) and 450 mm (b) deflections

(a)　　(b)

Fig. 10: Deformed member after the end of the test: (a) specimen and (b) its FE-model

As mentioned previously, typical ultimate limit states for static loading refer to a maximum deflection limit close to $L/40$. Higher deflections are associated with extreme situations, such as column loss scenarios, accidental loadings and progressive collapse. Figs. 10a and b show the deformed shape of the specimen DB1 after the test. The deformation capacity of the specimen and its hardening response indicate that such beams can be implemented also for such unexpected situations.

4. Conclusions

In this article, two bending tests of boxed shaped steel cross-sections encased in concrete are presented. The geometry of the steel part in combination with proper reinforcing offers an extremely ductile behaviour without strength degradation. The slips between concrete and structural steel remained very small and the integrity of the specimens was maintained up to the end of the tests. Spalling of concrete was limited and no reinforcing bar or plate buckling occurred. The beneficial structural response indicates that such properly reinforced slim floor beams can be implemented by the designers not only for typical ultimate states but also for the extreme cases, such as progressive collapse and accidental loadings.

References

[1] EN 1992-1-1, *Eurocode 2: Design of concrete structures. Part 1-1: General rules and rules for buildings*. CEN 2004.

[2] EN 10025-1, *Hot rolled products of structural steels - Part 1: General technical delivery conditions*. CEN 2004.

[3] Peltonen S, en Plum CM. "Brandveiligheid - Druckboog en schurfweerstand helpen onbeschermde ligger bij brand". *Bouwen met staal* 215, 50-52, Juni 2010.

[4] Peltonen S, Iliopoulos A, Kiriakopoulos P. "Progressive Collapse Analysis of Composite Framed Buildings with Encased in Concrete Steel Beams". *IABSE Workshop Tuusula, 144-151, Finland 2013.*

[5] Leskela MV, Peltonen S, Iliopoulos A, Kiriakopoulos P. "Numerical and experimental investigations on the vertical shear resistance of boxed steel cross-sections with concrete infill (Deltabeams)". *Eurosteel 2014 Conference Papers* /08/27-039, 6 p.

[6] Peltonen S, Leskela M. "Connection Behaviour of a Concrete Dowel in a Circular Web Hole of a Steel Beam". *Composite Construction in Steel and Concrete V*, 544-552, ASCE 2006.

[7] Leskela MV. "Shear connection by transverse rebars - Shallow floor composite beams". *Eurosteel 2008 Proceedings*, Volume A, 273-278, 2008.

[8] Leskela MV, Peltonen S, Obiala R. "Composite action in shallow floor beams with different shear connections". *Steel Construction*, V. 8 No 2, 90-95, 2015

[9] Sargin M. "Stress-Strain Relationships for Concrete and the Analysis of Structural Concrete Sections", *SM Study No. 4, University of Waterloo, Canada, 1970.*

[10] ABAQUS Theory Manual, "Version 6.14," ABAQUS Inc., USA, 2014.

[11] Kmiecik P, Kaminski M. "Modelling of reinforced concrete structures and composite structures with concrete strength degradation taken into consideration". *Archives of civil and mechanical engineering, Vol. XI, No. 3, 2011.*

[12] Huber G. "Non-linear calculations of composite sections and semi continuous joints". *Ernst und Sohn, p. 137, 2000.*

LAYER ARRANGEMENTS INFLUENCE ON BUCKLING, POST-BUCKLING AND COLLAPSE OF CHANNEL SECTION COMPOSITE PROFILE

Adrian Gliszczynski and Tomas Kubiak

Department of Strength of Materials, Lodz University of Technology, Lodz, Poland

Abstract: The paper deals with buckling and postbuckling behaviour of thin-walled composite beams subjected to pure bending and columns subjected to uniaxial compression. The object of the analysis were structures with C-shaped cross-section made of eight-layer GFRP laminate. The analysis was performed on six layer arrangement: $[0/-45/45/90]_S$, $[90/-45/45/0]_S$, $[90/0/90/0]_S$, $[0/90/0/90]_S$, $[45/-45/45/-45]_S$, $[45/-45/90/0]_S$. Numerical analysis was conducted using ANSYS software. The estimates of load capacity were based on the following failure criteria: Inverse of Tsai-Wu criterion, criterion of maximum stress and Hashin's criterion. The obtained results were compared with the results of the experiment.

1. Introduction

Laminates are currently the most common type of composites and for many years have been applied in girder structures of aircraft wings [1] or helicopter blades [2]. The papers dealing with buckling, postbuckling and failure of thin walled columns with different cross-section can be very easy found in worldwide literature. Papers focusing on steel structures are public e.g. by Schafer [3], Mulligan and Pekoz [4] or Ungureanu et al. [5] and many others. The investigations dealing with buckling and postbuckling of composite columns are presented for example by Barbero and Tomblin [6] or Debski et al. [7]. In world literature there are still not enough paper presenting experimental investigations on buckling, postbuckling and failure of thin-walled FRP structures, especially those which are manufactured in autoclaving process with different layer arrangements [8-10]. The thin-walled channel section profile made of an epoxy resin laminate reinforced with glass fibers (GFRP) has been taken into consideration. The results presented in this paper concern the six arrangements of layers. The comparison of buckling loads, postbuckling equilibrium paths and values of failure load was conducted, depending on the layer arrangement and the type of load.

2. Subjects of the analysis

The specimens were C-shaped profiles with following cross-section dimensions (width x height x thickness of the wall): 82 × 40 × 2.08 mm (see Fig. 1). The considered profiles were produced by autoclaving technique, which allows to manufacture high-efficient composite structures [11-13]. The analyzed C-shaped profiles were subjected to pure bending (beams) or to uniform compression (columns).

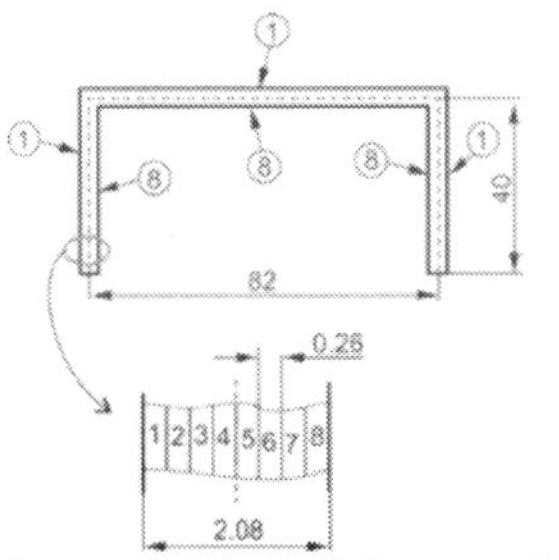

Fig. 1: Dimensions and ply numbering of the tested channel section profiles

Columns and beams had a length of 250 mm and 275 mm, respectively. All specimens were made of glass–epoxy unidirectional prepreg band (SE70/EGL/300g/400mm/35%/PoPa).

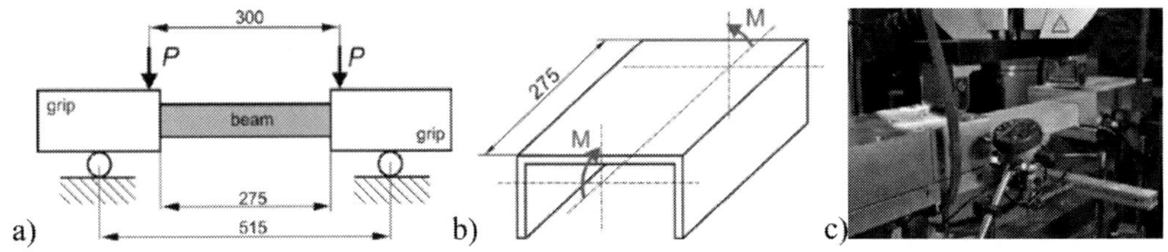

Fig. 2: The diagram of the support and the load with characteristic dimensions (a, b) and the test stand (c) of beams subjected to pure bending

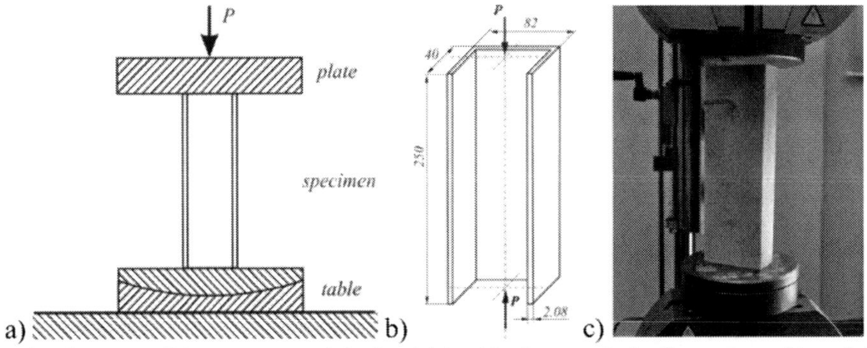

Fig. 3: The diagram of the support and the load (a) with characteristic dimensions (b) and the test stand (c) of columns subjected to compression

A diagram of the performed bending tests with dimensions, describing the span of support and the span of load, is presented in Figs. 2a-2b. Fig. 2c presents the beam fixed on the test stand. The beams were bent in the plane with lowest second moment of area. The load and the support were applied indirectly by additional grips. Similarly, in the case of columns subjected to compression Fig. 3a shows a diagram of the support and the load, Fig. 3b presents the geometrical dimensions of the analyzed channel sections and in Fig. 3c the composite C-shaped column on the test stand can be observed. In the case of the stand for column subjected to compression a plate and a table were used. The plate was mounted to the upper jaw and has only one degree of freedom (vertical movement). The table with spherical bearing which is mounted in the bottom jaw has three degree of freedom (rotations about three perpendicular axes). The loading force applied to the system and the displacement in points, where the load was applied, were obtained directly from the machine sensors. In this testing stand no additional plastic-pads were used. It was assumed that the examined profiles had walls consisting of eight layers with six different symmetric ply arrangements denoted as follows: C_1: $[0/-45/45/90]_S$, C_2: $[90/-45/45/0]_S$, C_3: $[90/0/90/0]_S$, C_4: $[0/90/0/90]_S$, C_5: $[45/-45/45/-45]_S$, C_6: $[45/-45/90/0]_S$. It was assumed that all layups under consideration had their general axis of orthotropy parallel to the wall edges (0^0 – fibres along the longitudinal direction or 90^0 – fibres in the transverse direction) or inclined to the wall edges by 45^0. Material properties for a single lamina were determined in strength tests. The Young's modulus in two orthogonal directions E_1 and E_2, (index "1" denotes the fiber direction and index "2" – transverse to the fiber direction), the Poisson's ratio ν_{12}, the ultimate strength in tension for both orthogonal directions (T_1 and T_2) were determined in accordance with the ASTM D 3039 standard [14]. The ultimate strength in compression (C_1 and C_2) was determined in accordance with the CRAG 400 standard [15]. The maximum shear stress S_{12} and the Kirchhoff modulus G_{12} were determined in accordance with the ASTM D 3518 standard [16]. The determined material properties are presented in Table 1.

Table 1: Mechanical characteristics of the SE70/EGL/300g/400mm/35%/PoPa glass–epoxy composite

E_1 [GPa]	E_2 [GPa]	G_{12} [GPa]	ν_{12} [-]	T_1 [MPa]	T_2 [MPa]	S_{12} [MPa]	C_1 [MPa]	C_2 [MPa]
38.5	8.1	2.0	0.27	792	39	108	679	71

On the basis of the obtained numerical and experimental results, the critical values of buckling load (bending moment M_{CR} and compression force F_{CR}) were determined, in accordance with the following methods [17]: the load vs. square of strain difference P-$(\varepsilon_1 - \varepsilon_2)^2$ method or the load vs. square of deflection method – denoted as M-1 and the load – deflection P-$(\varepsilon_1 - \varepsilon_2)$ curve inflection point method – denoted as M-2.

3. FEM models with assumed boundary conditions

Numerical calculations were conducted in the environment of Ansys® software based on the finite element method. In order to check the impact of layer arrangements on buckling load, postbuckling behaviour and to estimate the load-carrying capacity of composite structures the geometrically non-linear analysis was performed and failure criteria were implemented. The nonlinear problem was solved using the Newton-Raphson algorithm. In presented cases the small initial geometrical imperfection corresponding to buckling mode with amplitude equal to 1/10 of thickness of the profile walls were assumed. In the case of the C-shaped beam, the geometry of the model contains the tested channel section beam with grips. To create an FEA model of the grip, some simplifications were used – a geometric model accounted for a C-section with a length corresponding to the sample together with both grips (Fig. 4a). For the part of the C-shape model corresponding to the grip, a material with higher stiffness was assumed and a diaphragm in the place of support and loading was added. The vertical force was applied to one point of the lines (at each grip - Fig. 4b) corresponding to the line loaded in the real test (Fig. 2a, Fig. 2c). The system of the channel section beam with grips was simply supported at two ends along the line on the bottom surface of the handle. The geometry of the model contained the tested channel section column with assumed boundary condition and load (Fig. 5) as close as possible to the test stands.

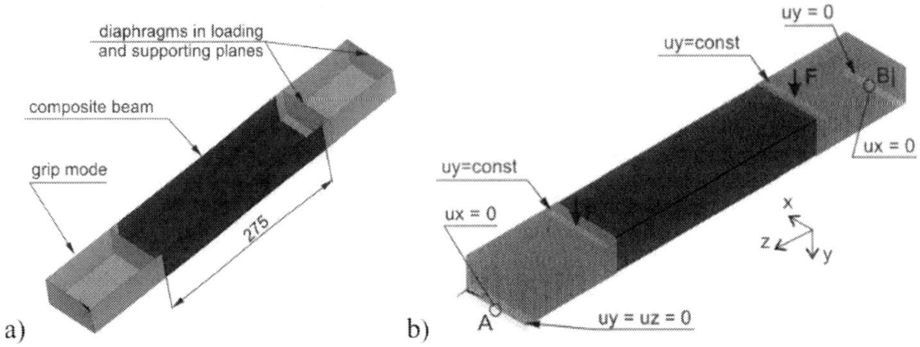

Fig. 4: Geometrical (a) and discrete (b) model of the beam with the grips and the assumed boundary conditions

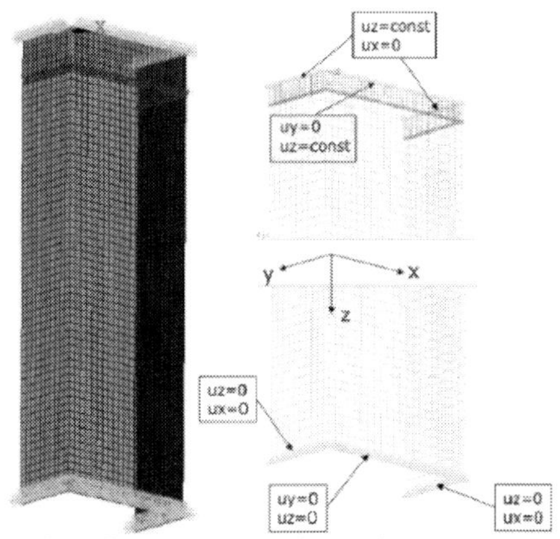

Fig. 5: Discrete model of compressed channel section column with assumed boundary conditions corresponding to the test stand.

A four-node multi-layered shell element with six degrees of freedom at each node was taken for discretisation. According to the ANSYS® documentation [18] this finite element may be used for layered applications for modeling of laminated composite shells or sandwich construction. The accuracy in modelling of composite shells is governed by the first order shear deformation theory (Mindlin-Reissner shell theory).

4. Results of the numerical and experimental investigations

The comparison of buckling loads, postbuckling equilibrium paths and values of failure load was conducted, depending on the layer arrangement and the type of load. The buckling modes for the analyzed profiles with different lay-ups observed on both test stands are shown in Table 2. It can be noted that in both test stands buckling modes are similar (number of half-waves m in the most cases are the same). However, for the column with lay-ups denoted as C_5 two different buckling modes have appeared (see Table 2). In the case of beams with lay-ups denoted as C_2 we can observe the same number of half-waves (m=5) but the experimental studies have shown non-symmetrical distribution of half-waves along the beam (see Table 2). It should also be noted that a different number of sub-steps in nonlinear analysis can lead to different buckling mode and the increasing number of sub-steps does not necessarily guarantee the closer correlation of experimental and numerical results (compare modes for beams with C_6 lay-up).

Table 2: Buckling mode – *m* number of half-waves along the profiles under consideration

	C_1	C_2	C_3	C_4	C_5	C_6
	Beams subjected to pure bending					
NLA – FEM	1	5	5	3	1	$3/5^{*2}$
Experimental	1	5^{*1}	5	5	1	4
	Columns subjected to uniform compression					
NLA – FEM	2	3	2	3	2	3
Experimental	2	3	2	3	3	3

NLA – nonlinear analysis
^{*1}) non-symmetrical distribution along the beam
^{*2}) depending on the number of sub-steps

The differences in the buckling load determined from the numerical simulations and the results of the experimental tests obtained with the conventional strain gauge measurement methods are in the range of: 0.4% (C_5) up to 11.9% (C_6) for the beams subjected to bending and 0.9% (C_6) up to 22.5% (C_5) for the columns subjected to compression (see Table 3). In the case of C-shaped beams it can be said that the highest buckling bending moment M_{CR} was obtained for the beam with the layer arrangement C_1 and C_6 and the lowest - for the cases C_3 and C_5. For similar profiles subjected to compression, the highest buckling load is obtained for columns with the layer arrangement denoted as C_3 and C_6 and the lowest - for the cases C_2 and C_5. These results show the importance of having knowledge of the type of load. The same profile subjected to different schemes of loads can be characterized by various behaviour.

Table 3: Buckling load determined on the basis of the numerical and experimental investigations

	C_1	C_2	C_3	C_4	C_5	C_6
	Beams subjected to pure bending - M_{CR} [N·m]					
FEM M-1	390	296	237	277	257	339
FEM M-2	438	306	262	290	-	417
FEM MV.	414	291	255	283	257	378
EXP (S-G) M-1	354	249	218	264	227	317
EXP (S-G) M-2	378	267	246	270	287	349
EXP (S-G) MV.	366	258	232	267	257	333
	Columns subjected to uniform compression - F_{CR} [kN]					
FEM MV.	6,7	6,8	8,5	8,5	12,0	11,4
EXP (S-G) M-1	8,0	7,9	11,0	9,3	8,9	11,3
EXP (S-G) M-2	9,0	8,9	10,6	9,7	9,7	11,7
EXP (S-G) MV.	8,5	8,4	10,8	9,5	9,3	11,5

EXP (S-G) – experimental results obtained from strain gauges
MV - mean value

The influence of the layer arrangement on the postbuckling stiffness was investigated on the basis of the experimental tests (Figs. 6a-6b). The lowest stiffness in the postbuckling range has the beam with the layer arrangement C_5: $[45/-45/45/-45]_S$ and the highest stiffness are observed for the beam with the layer arrangement C_2: $[90/-45/45/0]_S$. It should also be noted that for the lay-up C_1 and C_5, the stiffness decreases with an increasing load and for the cases C_3 and C_4, the stiffness in the beginning of the postbuckling range decreases and then increases with an increasing load. Analysing the postbuckling equilibrium path of the structures under consideration, it can be easily observed that the columns, which have the lowest stiffness in prebuckling state are characterized by the highest postbuckling stiffness.

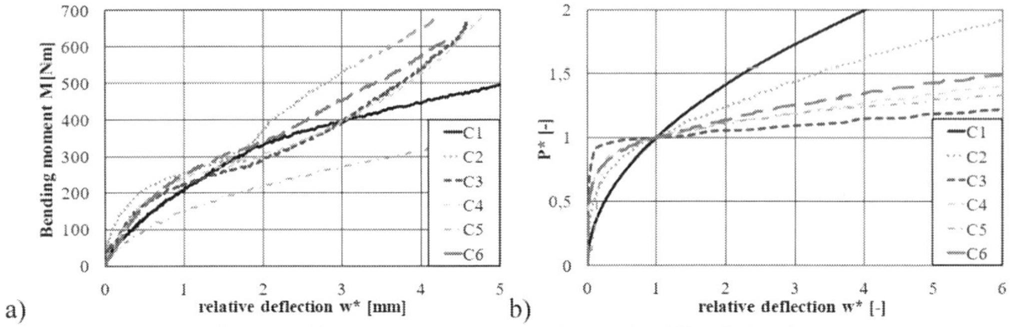

Fig. 6: Influence of layer arrangements on the postbuckling behaviour determined experimentally for beams (a) and columns (b).

The purpose of the numerical studies was also to determine the load carrying capacity of composite profiles subjected to bending and compression. The failure loads (M_f, F_f) leading to the profiles failure for all analyzed cases have been determined with the assumption that the beam fails if the failure criterion is fulfilled in all layers. To find failure load it was assumed that strength factors should be equal or greater than unity in all layers. Analysing the experimental values of failure loads of the columns it can be easily noticed that regardless of the ply arrangement the values of failure loads are nearly identical (Table 4). The exception are the columns with ply system denoted as C_5 for which the failure loads were the lowest.

Table 4: Experimental results and estimated failure loads for the considered criteria

Ply arrangement:	C_1	C_2	C_3	C_4	C_5	C_6
	Beams subjected to pure bending					
Criterion	Failure load M_f [N·m] (layer no.)					
Inverse of Tsai-Wu	$397_{(1)}$	$539_{(5)}$	$761_{(5)}$	$749_{(3)}$	$163_{(1)}$	$540_{(5)}$
Maximum stress	$444_{(1)}$	$576_{(5)}$	$777_{(5)}$	$773_{(6)}$	$168_{(1)}$	$577_{(5)}$
Hashin	$564_{(1)}$	$575_{(5)}$	$761_{(2)}$	$759_{(1)}$	$343_{(8)}$	$569_{(4)}$
Experiment (mean value)	618	655	663	747	499	676
	Columns subjected to uniform compression					
Criterion	Failure load F_f [kN] (layer no.)					
Inverse of Tsai-Wu	$25.7_{(3)}$	$29.2_{(5)}$	$42.8_{(4)}$	$24.6_{(4)}$	$23.5_{(5)}$	$29.8_{(5)}$
Maximum stress	$25.9_{(3)}$	$32.5_{(5)}$	$38.6_{(5)}$	$27.3_{(3)}$	$23.4_{(4)}$	$33.5_{(5)}$
Hashin	$22.7_{(7)}$	$21.5_{(8)}$	$23.4_{(8)}$	$26.5_{(7)}$	$18.4_{(8)}$	$22.7_{(8)}$
Experiment (mean value)	24.6	24.8	24.8	23.8	17.6	24.6

In the case of beams subjected to pure bending it should be noted that the load capacity of the structures is determined mainly by the strength of the composite fibres. After exceeding the compressive strength (C_1) or tension strength (T_1) in the direction of the fibres the composite beams were characterized by a high stiffness degradation, leading to rapid destruction of the structures. Beams with ply arrangements denoted as C_3 and C_4 were characterized by the highest values of failure loads, both experimental and numerical. This was explained by the 50% content of laminates in the layers with longitudinal orientations of the fibres (0^0).

5. Conclusions

The purpose of the studies presented by the authors was to determine and compare the buckling loads, the load carrying capacity and postbuckling behaviour of composite columns subjected to uniaxial compression and beams subjected to pure bending. Numerical analyses have been conducted in the Ansys® program, based on the finite element method. The loads of carrying capacity of the analyzed structures were determined using the following failure criteria: Inverse of the Tsai-Wu criterion, criterion of Maximum stress and Hashin criterion. Based on the performed experimental and numerical studies it has been concluded that:

1. The main parameter determining an increase in the correlation of the experimental and numerical results is the identical buckling mode (the number of half-waves);
2. Factors that can improve the convergence of the numerical results with the experimental results are initial geometrical imperfection and a number of load steps that could identify phenomena such as asymmetry in the form of buckling (cf. the beam C_6);
3. The experimental difference in the values of the load capacity for the beams with ply arrangement denoted as C_3 and C_4 in relation to the lay-up marked as C_5 was approximately 33%. However, in the case of channel sections profiles subjected to uniform compression five out of six analyzed arrangement of layers (except the ply system denoted as C_5) were characterized by the almost identical failure loads. Thus, the orientation of the main directions of orthotropy of the individual layers in laminates can influence in different way on the values of failure loads;
4. The largest convergence of numerical and experimental results was obtained with the implementation of the Hashin criterion, mainly for ply systems for which the other applied failure criteria do not lead to the accurate estimation of failure loads (for beam with ply system denoted as C_5 and columns ply arrangement marked as C_5 and C_6).
5. The load capacity of the structures is determined mainly by the strength of the composite fibres. After exceeding the compressive strength (C_1) or tension strength (T_1) in the direction of the fibres the composite beams were characterized by a high stiffness degradation, leading to rapid destruction of the structures.

Acknowledgments

The paper has been written as a part of the research project financed by the National Centre for Science - decision number DEC-2011/03/B/ST8/06447.

References

[1] York CB. "Buckling analysis and minimum-mass design procedures for composite wing-box structures", *Journal of Aircraft*, 43(2), 528-536, 2006.

[2] Debski H, Ostapiuk M. "Numerical FEM analysis for the part of composite helicopter rotor blade", *Journal of Kones,* 19(1), 71-77, 2012.

[3] Schafer BW. "Local, distortional, and Euler buckling of thin-walled columns", *J. Struct. Eng.*,128(3), 289–299, 2002.

[4] Mulligan GP, Pekoz T. "Locally buckled thin-walled columns", *J. Struct. Eng.*, 110(11), 2635–2654, 1984.

[5] Ungureanu V, Kotelko M, Mania RJ, Dubina D. "Plastic mechanisms database for thin-walled cold-formed steel members in compression and bending", *Thin-Walled Structures*, 48(10), 818–826, 2010.

[6] Barbero E, Tomblin J. "A phenomenological design equation for FRP columns with interaction between local and global buckling", *Thin-Walled Structures*, 18(2), 117–131, 1994.

[7] Debski H, Kubiak T, Teter A. "Experimental investigation of channel-section composite profiles behavior with various sequences of plies subjected to static compression", *Thin-Walled Structures*, 71, 147–154, 2013.

[8] Urbaniak M, Teter A, Kubiak T. "Influence of boundary conditions on the critical and failure load in the GFPR channel cross-section columns subjected to compression", *Composite Structures*, 134, 199–208, 2015.

[9] Gliszczynski A, Kubiak T. "Load-carrying capacity of thin-walled composite beams subjected to pure bending", *International Journal of Structural Stability and Dynamics* (article in print), World Scientific Publishing Company.

[10] Kubiak T, Kolakowski Z, Swiniarski J, Urbaniak M, Gliszczynski A. "Local buckling and postbuckling of composite channel-section beams – numerical and experimental investigations", *Composites Part B* (article in print).

[11] Gliszczynski A, Kubiak T, Urbaniak M. "Influence of boundary conditions and material properties on the stability of bent composite beams" (in Polish), *Modelling in Engineering*, 50(19), 25-32, 2017.

[12] Kubiak T, Kaczmarek L. "Estimation of load-carrying capacity for thin-walled composite beams", *Composite Structures*, 119, 749-756, 2015.

[13] Bienias J, Gliszczynski A, Jakubczak P, Kubiak T, Majerski K. "Influence of autoclaving process parameters on the buckling and postbuckling behaviour of thin-walled channel section beams", *Thin-Walled Structures*, 85, 262-270, 2014.

[14] ASTM D3518: *Standard test method for in-plane shear response of polymer matrix composite materials by tensile test of a ±45° laminate*, Annual Book of ASTM Standards, 100 Barr Harbor Drive, West Conshohocken, PA 19428, USA, Vol. 15.03, 151–157, 1997.

[15] Hodgkinson JM. *An Experimental Comparison of ASTM, BSI and CRAG Standard Test Methods for the Determination of Mechanical Properties of Composite Materials*, The Centre for Composite Materials, Technical Report 90/02, Imperial College, London, 1990.

[16] ASTM D3039M, *Standard test method for tensile properties of polymer matrix composite materials*, American Society for Testing and Materials, 100 Barr Harbor Drive, West Conshohocken, PA 19428, USA, Vol. 15.03, 1997.

[17] Paszkiewicz M, Kubiak T. "Selected problems concerning determination of the buckling load of channel section beams and columns", *Thin-Walled Structures*, 93, 112–121, 2015.

[18] User's Guide ANSYS 16.2, Ansys®, Inc., Houston, USA.

Tubular constructions

GBT-BASED ASSESSMENT OF THE VIBRATION AND BUCKLING BEHAVIOUR OF COLD-FORMED STEEL CHS MEMBERS

Cilmar Basaglia[a], Dinar Camotim[b] and Nuno Silvestre[c]

[a] *Department of Structures, School of Civil Engineering, Architecture and Urban Design, University of Campinas, Brazil*
[b] *CEris, ICIST, DECivil, Instituto Superior Técnico, Universidade de Lisboa, Portugal*
[c] *IDMEC, LAETA, DEM, Instituto Superior Técnico, Universidade de Lisboa, Portugal*

Abstract: This paper reports an ongoing investigation on the use of Generalised Beam Theory (GBT) to perform vibration and buckling analyses of cold-formed steel circular hollow section (CHS) steel members. After presenting a brief overview of the GBT formulation, illustrative numerical results concerning fixed-ended members are presented and discussed, namely their (i) load-free vibration behaviour and (ii) buckling and vibration behaviour under axial compression – in the last case, the member compressive load is expressed as a fraction of its critical buckling value. For validation purposes, some GBT results are compared with values obtained from shell finite element analyses carried out in the code ANSYS.

1. Introduction

Due to the high global and local (wall) slenderness, the structural behaviour and strength of cold-formed steel circular hollow section (CHS) members are invariably governed by local (wall flexure) and/or global (flexural) buckling phenomena (*e.g.*, [1]). Furthermore, given the well-known mathematical similarity between the stability and vibration analyses (eigenvalue problems), and the high relevance of the local (cross-section) deformations on the buckling behaviour of thin-walled members, it is logical to expect a comparable influence on the vibration behaviour of such members. Moreover, since most thin-walled steel members are subjected to simultaneous static and dynamic actions, designers are faced with the need of assessing their dynamic response. Therefore, they must be equipped with numerical tools capable of determining the member local and global vibration behaviour, possibly for arbitrary support conditions and in the presence of more or less significant internal forces and/or moments. However, this task can only be rigorously performed at the moment by resorting to shell finite element analysis, a highly computer-intensive approach that requires a substantial modelling effort, including data input and (mostly) result interpretation.

A very promising alternative to the above approach is the use of a one-dimensional model based on Generalised Beam Theory (GBT), a beam theory incorporating genuine plate theory concepts, *i.e.*, accounting for the member cross-section in- and out-of-plane deformations. In this context, the authors have been employing GBT-based beam finite element formulations to analyse the (i) buckling behaviour of CHS members (*e.g.*, [2]) and (ii) buckling and vibration behaviour of members exhibiting open and/or closed flat-walled cross-sections (*e.g.*, [3]).

The objective of this work is to present the development and numerical implementation of a novel GBT formulation to analyse the vibration and buckling behaviour of thin-walled CHS members, as well as illustrate its application and capabilities. The paper starts with a brief overview of the GBT formulation employed, which is followed by the presentation and discussion of illustrative numerical results involving fixed-ended members, namely results concerning their (i) load-free vibration behaviour and (ii) buckling and vibration behaviour when subjected to axial compression. In the latter case, the member compressive load is expressed as a fraction of its critical buckling value and the investigation focuses mainly on the influence of the compressive load magnitude on the fundamental frequency and vibration mode nature. For validation purposes, some GBT results are compared with numerical values obtained from shell finite element analyses carried out in the code ANSYS [4].

2. GBT formulation

Because the cross-section displacement field is expressed as a linear combination of structurally meaningful deformation modes, GBT analyses involve solving equilibrium equation systems written in a very convenient modal form and lead to solutions that provide in-depth insight on the member structural response under consideration. The performance of a vibration or buckling analysis involves (i) a *cross-section analysis*, to obtain the deformation modes and corresponding mechanical properties, and (ii) the *member analysis*, *i.e.*, the solution of the appropriate eigenvalue problem and the (modal) interpretation of the results determined (natural frequencies or buckling loads and vibration or buckling mode shapes).

The cross-section analysis comprises a set of rather complex sequential operations, already described in detail in [2,5,6]. Figs. 1, 2 and 3 show the deformation mode shapes of the circular hollow cross-section dealt with in this work, which comprise (i) a "*shell-type*" mode family, originally considered by Schardt [5], (ii) an *axis-symmetric* mode and a *torsion* mode, both first introduced by Silvestre [2], and (warping) *shear modes*, due to the authors [6].

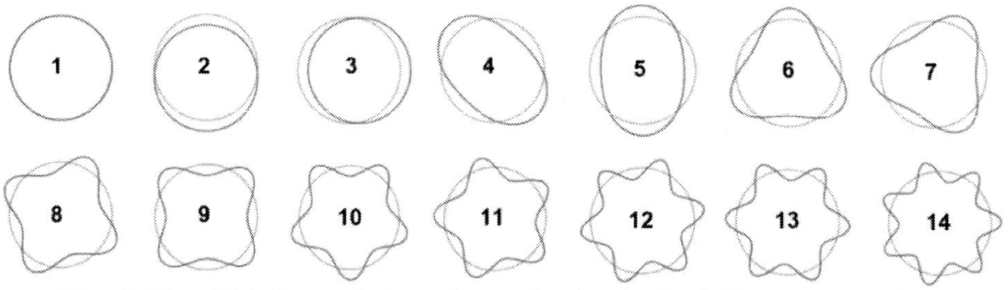

Fig. 1: First *14* shell-type deformation modes for circular hollow cross-sections

After determining the deformation modes, the equilibrium equation system defining the dynamic problem can be readily established as

$$C_{ik}\zeta_{k,xxxx} - D_{ik}\zeta_{k,xx} + B_{ik}\zeta_k - Q_{ik}\zeta_{k,xxtt} + R_{ik}\zeta_{k,tt} - X_{jik}\left(W_j^0\zeta_{k,x}\right)_{,x} = 0 \qquad (1)$$

where (i) $\zeta_k(x,t)$ are functions describing the deformation mode amplitude variation both along the longitudinal direction and with time t, (ii) C_{ik} are stiffness components concerning *generalised* warping effects, (iii) D_{ik} are stiffness components dealing with *generalised* twisting, (iv) matrix B_{ik} is related to the cross-section in-plane deformation (wall transverse bending), (v) Q_{ik} (out-of-plane) and R_{ik} (in-plane) are the inertia matrices, and (vi) X_{jik} are geometric stiffness components associated with the applied stress resultants W_j^0 – all these components are defined in [6].

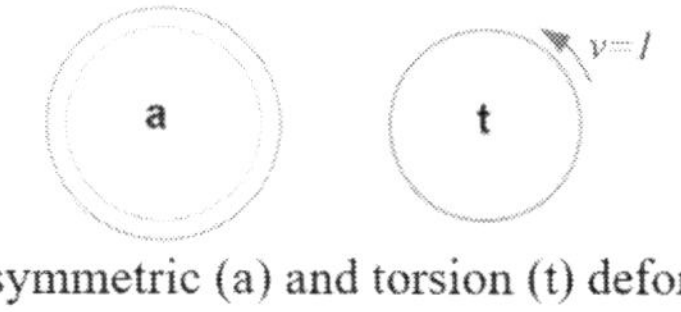

Fig. 2: Axis-symmetric (a) and torsion (t) deformation modes

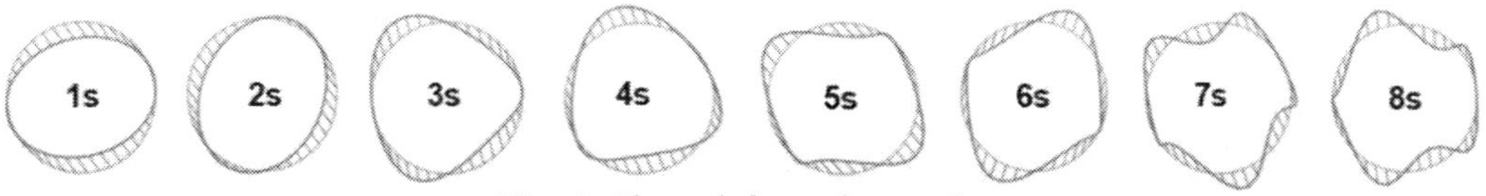

Fig. 3: Shear deformation modes

For members moving in synchronous configurations, the amplitude functions ζ_k can be written as

$$\zeta_k(x,t) = \phi_k(x)Y(t) \tag{2}$$

where $\phi_k(x)$ is a longitudinal shape function and $Y(t)$ is a time-dependent function satisfying the free vibration harmonic equation of motion $Y_{,tt}+\omega^2 Y=0$ (ω is the angular frequency of vibration) – thus, they may be expressed as

$$\zeta_{k,tt} = -\omega^2 \zeta_k \tag{3}$$

It is worth noting that, in the context of geometrically linear static analyses, one has

$$Y(t) = 1 \quad \Leftrightarrow \quad \zeta_k = \phi_k \tag{4}$$

The incorporation of (3) into (1) leads to the final system of equilibrium equations,

$$C_{ik}\phi_{k,xxxx} - D_{ik}\phi_{k,xx} + B_{ik}\phi_k - \omega^2\left(R_{ik}\phi_k - Q_{ik}\phi_{k,xx}\right) - X_{jik}\left(W_j^0\phi_{k,x}\right)_{,x} = 0 \tag{5}$$

which depends exclusively on derivatives with respect to the longitudinal coordinate x.

The methods that have been employed to solve the GBT eigenvalue problems stemming from the above equation are fairly standard. They include (i) Galerkin's method, used by Silvestre [2] to perform CHS member buckling analyses, and (ii) the finite element method, using a beam element formulation specifically developed for GBT analyses and employed so far to analyse members with flat-walled (open and/or closed) cross-sections (*e.g.*, [7]). At this point, it is worth mentioning that the GBT results presented in this work were obtained by means of a beam finite element specifically developed for CHS members – the main steps and procedures involved in the formulation of this finite element are briefly described next [6]:

(i) Approximate the modal amplitude functions $\phi_k(x)$ by means of linear combinations of (i_1) Lagrange cubic polynomial primitives, for the axial extension and warping shear deformation modes (those involving solely warping displacements u), and (i_2) Hermite cubic polynomials, for the remaining shell-type, axis-symmetric and torsion deformation modes. Then, one has

$$\phi_k(x) = \Psi_\alpha(\xi)d_{k\alpha}^{e.i} = \Psi_1 d_{k1}^{e.i} + \Psi_2 d_{k2}^{e.i} + \Psi_3 d_{k3}^{e.i} + \Psi_4 d_{k4}^{e.i} \tag{6}$$

where (i_1) Ψ_α are either Lagrange cubic polynomial primitives or Hermite polynomials and (i_2) $d_{k\alpha}^{e.i}$ are the generalised displacements of finite element i.

(ii) The various sub-matrix components of the finite element linear stiffness $[K_e]$, geometric stiffness $[G_e]$ and mass $[M_e]$ matrices are determined by means of the expressions

$$K_{ik\alpha\beta}^{(e)} = C_{ik} \int_{L_e} \Psi_{\alpha,xx} \Psi_{\beta,xx} dx + B_{ik} \int_{L_e} \Psi_\alpha \Psi_\beta dx + D_{ik}^1 \int_{L_e} \Psi_{\alpha,x} \Psi_{\beta,x} dx + $$
$$+ D_{ik}^2 \int_{L_e} \Psi_{\alpha,xx} \Psi_\beta dx + D_{ki}^2 \int_{L_e} \Psi_\alpha \Psi_{\beta,xx} dx \quad (7)$$

$$G_{ik\alpha\beta}^{(e)} = W_j^0 X_{jik} \int_{L_e} \Psi_{\alpha,x} \Psi_{\beta,x} dx \qquad M_{ik\alpha\beta}^{(e)} = Q_{ik} \int_{L_e} \Psi_{\alpha,x} \Psi_{\beta,x} dx + R_{ik} \int_{L_e} \Psi_\alpha \Psi_\beta dx \quad (8)$$

where the roman (i, j, k) and greek (α, β) subscripts identify the *deformation mode* and *degree of freedom* (modal generalised displacement), respectively.

(iii) Taking into account the member *support conditions*, expressed in terms of modal degrees of freedom, perform the assembly procedure leading to the discrete vibration eigenvalue problem

$$([K]-[G]-\omega^2[M])\{d\} = \{0\} \quad (9)$$

where (iii_1) $[K]$ and $[G]$ are the member linear and geometrical stiffness matrices, (iii_2) $[M]$ is the member mass matrix and (iii_3) $\{d\}$ is the generalised modal amplitude vector – its components are the (unknown) values and/or derivatives of the GBT deformation mode amplitudes at the member nodes (finite element end cross-sections).

3. Numerical results

In this section, numerical results concerning the vibration and buckling behaviour of fixed-ended CHS steel (E = 210 GPa, ν = 0.3 and ρ = 7.85 g/cm^3) members with radius r = 60 mm and wall thickness t = 1.2 mm are presented and discussed. Initially, the vibration behaviour of load-free members is investigated: the variation of the lower natural frequencies and vibration mode shapes with the member length are determined. Next, the buckling behaviour of uniformly compressed columns is addressed, focusing on the determination of the critical buckling loads and mode shapes of columns with various lengths. Finally, the vibration behaviour of axially compressed members is studied, with the compressive load P expressed as a percentage of the corresponding critical buckling value P_{cr}. Most GBT results are compared with values obtained by ANSYS shell finite element analyses.

3.1 Load-free member vibration

The curves depicted in Fig. 4 provide the variation of the first three natural frequencies (ω_f, ω_2, ω_3 – f stands for "fundamental") with the length L, as well as the corresponding vibration mode longitudinal half-wave numbers n_s – note that there is always a set of two vibration modes corresponding to the same frequency (multiplicity 2 eigenvalues), which implies the existence of a vibration mode sub-space. In order to enable a clear visualisation of the curves in Fig. 4, different vertical scales are adopted for 10 cm<L<100 cm and 100 cm<L<1000 cm. The ANSYS values, providing validation for the GBT results, are identified by white circles (ω_f), squares (ω_2) and triangles (ω_3). Fig. 5 concerns members with L=10, 60, 300 and 500 cm, and shows (i) ANSYS 3D vibration mode shapes and (ii) the GBT deformation modes participating in the vibration modes. The comparison between the GBT and ANSYS vibration results prompts the following remarks:

(i) The GBT and ANSYS frequency values agree perfectly, as the differences never exceed *2.5%*. Moreover, there is also very close agreement between the vibration mode shapes shown in Fig. 5 and those provided by the GBT analyses (not shown here).

(ii) The three first natural frequencies decrease monotonically with L and tend to zero as L goes to infinity. While ω_f and ω_2 are very close for $L \approx 25, 55$ and $180\ cm$, the ω_2 and ω_3 values are virtually identical for $L \approx 20, 40, 100$ and $300\ cm$.

(iii) The fundamental vibration modes always exhibit a single half-wave, regardless of the member length. On the other hand, the second and third vibration modes may have more than one half-wave, depending on the member length.

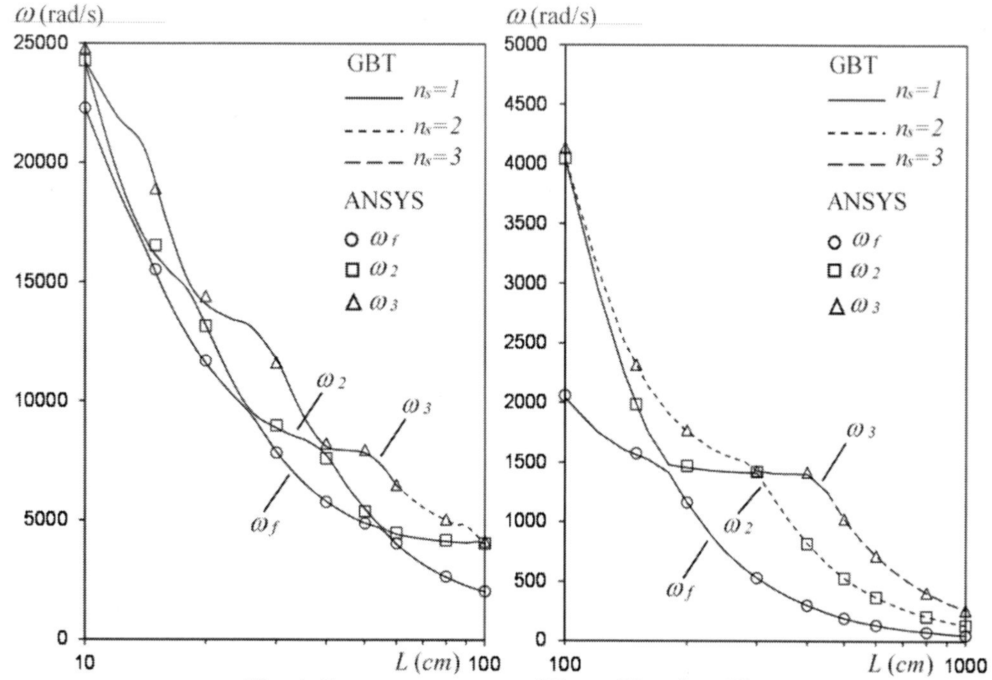

Fig. 4: Frequency curves $\omega_f(L)$, $\omega_2(L)$ and $\omega_3(L)$

ANSYS vibration mode shapes											
L = 10 cm			L = 60 cm			L = 300 cm			L = 500 cm		
ω_f	ω_2	ω_3	ω_f	ω_2	ω_3	ω_f	ω_2	ω_3	ω_f	ω_2	ω_3

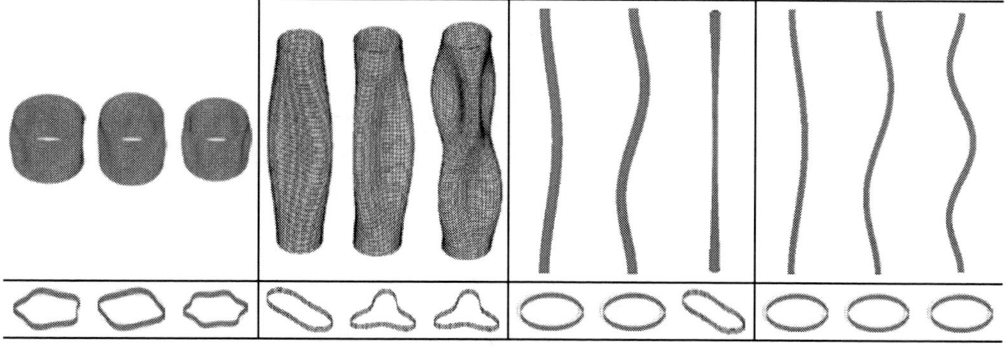

GBT deformation modes (see Figs. 1 and 3)											
10+9s	8+7s	12+11s	4+3s	6+5s	6+5s	2+1s	2+1s	4+3s	2+1s	2+1s	2+1s

Fig. 5: ANSYS vibration mode shapes and GBT deformation modes participating in the vibration mode

(iv) Unlike the column buckling modes, the load-free member vibration modes always have a contribution from *shear modes* (*i.e.*, modes involving non-linear warping displacements), as was already noted by Bebiano [8], in the context of the vibration of open cross-section members.

(v) For L<50cm, the fundamental vibration occurs in a *shell-type* GBT deformation mode of an order higher than that corresponding to the second vibration mode – *e.g.*, for L=10cm the fundamental and second vibration modes have contributions from modes 10 and 8, respectively. However, note that the contribution of the *shear modes* increases with the vibration mode order.

3.2 Fixed-ended column buckling

Fig. 6 shows GBT column buckling results, namely the variation of the critical buckling load P_{cr} (kN) and mode shape with the length L (cm). For several L values, ANSYS shell finite element values are also displayed. The following conclusions can be drawn from the observation of these column buckling results:

(i) The GBT and ANSYS critical buckling loads are extremely close – all differences below *3%*. In addition, there is an excellent agreement between the ANSYS and GBT buckling mode shapes (in the latter case, only buckled cross-sections are shown).

(ii) The critical buckling load P_{cr} always decreases with L – however, for L>*300 cm*, corresponding to global buckling, this P_{cr} decrease takes place at a much higher rate.

(iii) For L≤*60 cm*, the column buckles in critical shell-type modes with *5* and *6* circumferential half-waves (modes 12 and 10) and *2* to *7* longitudinal half-waves. For *60*<L≤*120 cm*, buckling occurs in a critical shell-type mode with *3* and *4* circumferential half-waves (modes 6 and 8) and *3* to *6* longitudinal half-waves. For *120*<L≤*300 cm*, the column buckles in shell-type mode 2 with *2* to *4* longitudinal half-waves. For L>*300 cm*, the columns buckle in single half-wave flexural modes.

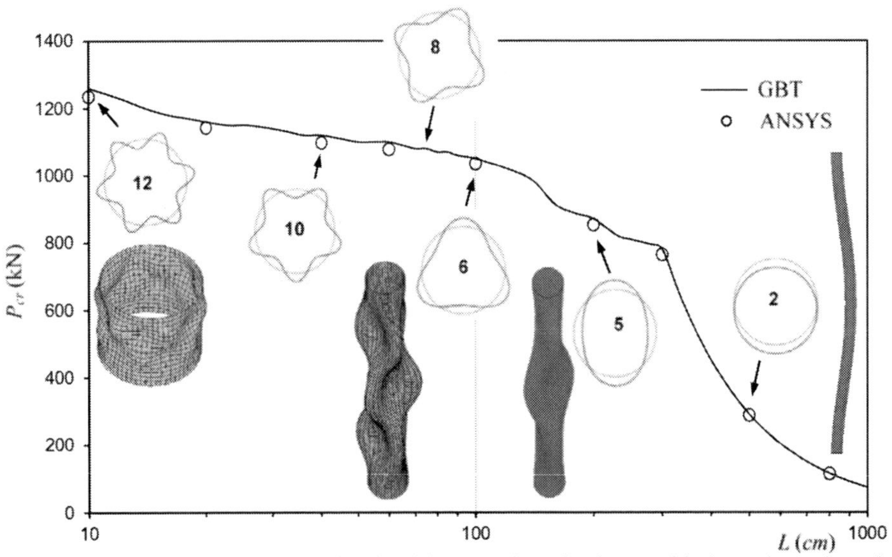

Fig. 6: Variation of the critical buckling load (P_{cr}) and mode shape with the column length (L)

3.3 Axially compressed member vibration

Finally, the vibration of axially compressed members (columns) is investigated – the axial compression level is defined by the parameter $P^0 \equiv W_1^{\,0}$, taken as fraction (α) of the critical buckling value (*i.e.*, $P^0 \equiv \alpha P_{cr}$). Fig. 7 makes it possible to assess the influence of the axial compression on the member fundamental frequency ω_f. Moreover, the 3D views of the buckling and vibration modes yielded by the ANSYS analyses, presented in Fig. 8 for $L=40cm$ and $\alpha=0$, 0.6, 0.95, 0.99, provide information on the evolution of the fundamental frequency and vibration mode nature as the axial compression increases. The results presented in these two figures prompt the following remarks:

(i) The presence of axial compression only causes a visible fundamental frequency drop (*i.e.*, more than *5%*) for $\alpha>0.1$ – a substantial fundamental frequency drop (*i.e.*, more than *30%*) occurs for $\alpha>0.6$. In order to confirm these findings, the corresponding ANSYS values were determined and are also shown in Fig. 7 – an excellent agreement is observed, both in terms of the frequency values and vibration mode shapes (those yielded by the GBT analyses are not shown here).

(ii) The fundamental vibration mode shapes for $\alpha=0$ and $\alpha\neq0$ are very similar up to $\alpha=0.95$. On the other hand, for α very close to 1.0 the vibration mode shape changes drastically and practically replicates the column critical buckling mode shape – indeed, for $\alpha=0.99$ the fundamental vibration and buckling modes virtually coincide, as clearly shown in Fig. 8.

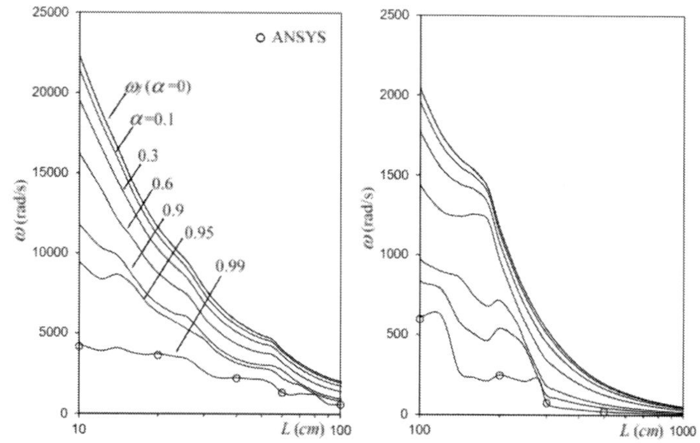

Fig. 7: Variation of the fundamental frequency ω_α with the member length and axial compression level

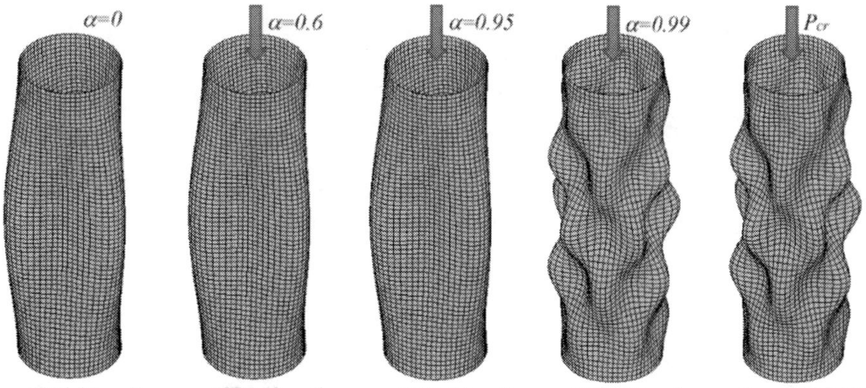

Fig. 8: Members with $L=40cm$: fundamental vibration modes for $\alpha=0$, 0.6, 0.95, 0.99 and critical buckling mode

4. Conclusion

The paper reported the available results of an ongoing investigation on the use of a GBT-based beam finite element approach to analyse the vibration and buckling behaviour of CHS members.

After a brief overview of the most relevant concepts involved in performing a CHS member GBT analysis, the paper described the procedures required to obtain the member overall linear, mass and geometric stiffness matrices. Then, the application and capabilities of the developed methodology were illustrated by means of the presentation and discussion of numerical results involving fixed-ended cold-formed steel members, namely results concerning their (i) load-free vibration behaviour and (ii) buckling and vibration behaviour when subjected to axial compression. In particular, the last illustrative example focused on the influence of the axial compression level on the natural frequency value and vibration mode shape nature – for instance, it was numerically confirmed that the fundamental vibration mode shape tends to the column critical buckling mode shape as the axial compression level approaches P_{cr}. For validation purposes, some GBT results were compared with values obtained by means of ANSYS shell finite element analyses – an excellent agreement was observed in all cases, both for the natural frequencies and vibration mode shapes.

Acknowledgments

The first and third authors gratefully acknowledge the financial support provided respectively by (i) *Conselho Nacional de Desenvolvimento Científico e Tecnológico* (CNPq – Ministry of Science, Technology and Innovation of Brazil), through project nº 446679/2014-3, (ii) FCT (*Fundação para a Ciência e a Tecnologia* – Portugal), through IDMEC, under LAETA, project UID/EMS/50022/2013.

References

[1] Teng JG. "Buckling of thin shells: recent advances and trends", *Applied Mechanics Reviews (ASME)*, 49(4), 263-274, 1996.

[2] Silvestre N. "Generalised beam theory to analyse the buckling behaviour of circular cylindrical shells and tubes", *Thin-Walled Structures,* 45(2), 185-198, 2007.

[3] Camotim D, Basaglia C, Bebiano R, Gonçalves R., Silvestre N. "Latest developments in the GBT analysis of thin-walled steel structures", *Proceedings of International Colloquium on Stability and Ductility of Steel Structures* (SDSS'Rio 2010 – Rio de Janeiro, /9) E. Batista *et al.* (eds.), vol. 1, 33-58, 2010.

[4] Swanson Analysis Systems (SAS) Inc. *ANSYS Reference Manual*, version 12, 2009.

[5] Schardt R. *Verallgemeinerte Technishe Biegetheorie*, Springer-Verlag, Berlin, 1989. (in German).

[6] Basaglia C, Camotim D, Silvestre N. "Buckling and vibration analysis of cold-formed steel CHS members and frames using Generalized Beam Theory", *International Journal of Structural Stability and Dynamics*, 15(8), 1540021-1-25, 2015.

[7] Bebiano R, Gonçalves R, Camotim D. "A cross-section analysis procedure to rationalise and automate the performance of GBT-based structural analyses", *Thin-Walled Structures*, 92 (July), 29-47, 2015.

[8] Bebiano R. *Stability and Dynamics of Thin-Walled Members: Application of Generalised Beam Theory*, PhD Thesis in Civil Engineering, IST, Technical University of Lisbon, 2010.

DESIGN OF EQUAL-WIDTH RHS TRUSS X-JOINTS FOR SIDE WALL BUCKLING IN THE PRESENCE OF CHORD PRE-LOAD

Shanshan Cheng and Jurgen Becque

Department of Civil and Structural Engineering, the University of Sheffield, United Kingdom

Abstract: The paper presents the results of theoretical and numerical studies on the side wall buckling of equal-width welded RHS X-joints, aiming to investigate the effect of chord compression. Previous research has established that the CIDECT rules provide overly conservative predictions for full-width RHS X-joints with high chord slenderness values (h_0/t_0). A new design method is therefore proposed, based on the elastic local buckling stress of the side wall plate, which accounts for the effect of a pre-load in the chord member. The proposed model yields good results when compared with numerical and experimental data.

1. Introduction

Research on welded RHS X-joints started in Europe in the 1960s [1] and the current CIDECT design rules for welded joints [2] reflect many decades of research. For X-joints with equal width of the chord and the brace members, side wall buckling of the chord members is the critical failure mode when the joint is loaded in compression. In the current CIDECT design rules, the design model for side wall buckling isolates a vertical strip in the chord side wall and considers it as a column [1]. While defendable because of its simplicity, this approach obviously ignores the two-dimensional character of the side wall buckling as a plate. Furthermore, it has been known for some time that the current CIDECT design rules [1] for chord side wall failure are quite conservative, and more so as the chord wall slenderness h_0/t_0 increases [2].

Becque and Wilkinson [3] carried out a study to develop an alternative design equation for chord side wall buckling, founded on a rational plate buckling model. The model showed excellent agreement with the experimental results. The authors also presented a clear rationale for limiting the joint capacity to the buckling load, rather than the peak load or 3% deformation limit obtained from an isolated joint test.

In this paper the model developed in [3] is extended to include the effect of a chord pre-load on the side wall buckling of full width RHS X-joints. Numerical models, validated against the experiment, are used to carry out parametric studies and a design model is developed based on a slenderness parameter which is founded on the elastic plate buckling stress.

2. Analytical model

The chord side wall is idealized as an infinitely long plate, simply supported along both longitudinal edges and subject to a uniform localized load σ_1 (originating from the brace members) and a longitudinal compressive stress σ_2 (Fig. 1).

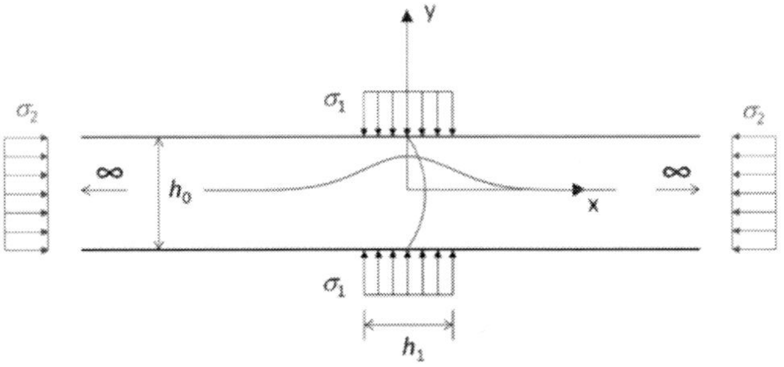

Fig. 1: Idealized model of the chord side wall

The critical buckling stress of the plate is approximated using a Rayleigh-Ritz approach with the following proposed deformed shape:

$$w = \Delta \cos\left(\frac{\pi y}{h_0}\right) e^{-2Bx^2} \tag{1}$$

In Eq. (1), w is the out-of-plane displacement of the plate, while Δ and B are (presently undetermined) parameters. Δ determines the amplitude of the displacements, while B is related to the width of the buckle.
The elastic strain energy U contained in the deformed shape of the plate is given by (e.g. Timoshenko and Gere [4]):

$$U = \frac{D}{2} \int_{x=-\infty}^{x=\infty} \int_{y=-h_0/2}^{y=h_0/2} \left\{ \left(\frac{\partial^2 w}{\partial x^2}\right)^2 + \left(\frac{\partial^2 w}{\partial y^2}\right)^2 + 2v\left(\frac{\partial^2 w}{\partial x^2}\right)\left(\frac{\partial^2 w}{\partial y^2}\right) + 2(1-v)\left(\frac{\partial^2 w}{\partial x \partial y}\right)^2 \right\} dx\, dy \tag{2}$$

In the above equation, D is the flexural rigidity of the plate, given by:

$$D = \frac{Et^3}{12(1-v^2)} \tag{3}$$

In Eq. (3), E is the modulus of elasticity, v is the Poisson's ratio and t is the chord wall thickness. Substitution of Eq. (1) into Eq. (2) eventually leads to:

$$U = \frac{\Delta^2 D}{2} \sqrt{\frac{\pi}{B}} \left(3B^2 h_0 + B\frac{\pi^2}{h_0} + \frac{\pi^4}{4h_0^3} \right) \tag{4}$$

On the other hand, the potential energy of the applied stresses is given by:

$$V = -\frac{\sigma_1 t}{2} \int_{x=-h_1/2}^{x=h_1/2} \int_{y=-h_0/2}^{y=h_0/2} \left(\frac{\partial w}{\partial y}\right)^2 dx\, dy - \frac{\sigma_2 t}{2} \int_{x=-\infty}^{x=\infty} \int_{y=-h_0/2}^{y=h_0/2} \left(\frac{\partial w}{\partial x}\right)^2 dx\, dy \tag{5}$$

Substituting Eq. (1) into Eq. (5) yields:

$$V = -\frac{\Delta^2 \sigma_1 t \pi^2}{4h_0} h_1 - \frac{\Delta^2 \sigma_2 t h_0}{4} \sqrt{\pi B} \tag{6}$$

provided that:

$$\frac{h_1^2 B}{3} << 1 \tag{7}$$

Eq. (6) is obtained from Eq. (5) while neglecting the higher order terms in h_1. It will be shown, once an expression for B has been determined, that Eq. (7) is indeed a reasonable assumption. We now set the derivatives of the total energy $U+V$ with respect to B and Δ equal to zero:

$$\frac{\partial(U+V)}{\partial B} = 0, \quad \frac{\partial(U+V)}{\partial \Delta} = 0 \tag{8}$$

Eqs. (8) eventually lead to:

$$B = \frac{1}{18}\left\{\frac{\sigma_2 t}{2D} - \left(\frac{\pi}{h_0}\right)^2 + \sqrt{\left[\left(\frac{\pi}{h_0}\right)^2 - \frac{\sigma_2 t}{2D}\right]^2 + 9\left(\frac{\pi}{h_0}\right)^4}\right\} \tag{9}$$

and:

$$\sigma_{cr} = \frac{1}{2th_1}\sqrt{\frac{\pi}{B}}\left\{12D\left(\frac{h_0}{\pi}\right)^2 B^2 + 4\left[D - \frac{\sigma_2 t}{2}\left(\frac{h_0}{\pi}\right)^2\right]B + D\left(\frac{\pi}{h_0}\right)^2\right\} \tag{10}$$

If $\sigma_2 = 0$, indicating that there is no chord pre-load, Eq. (9) reduces to:

$$B = \left(\frac{\sqrt{10}-1}{18}\right)\left(\frac{\pi}{h_0}\right)^2 = \frac{1.186}{h_0^2} \tag{11}$$

which is consistent with the value obtained in [3].

After introducing $\frac{\sigma_2 t}{2D} = a\left(\frac{\pi}{h_0}\right)^2$, Eqs. (9) and (10) can be converted into:

$$B = \frac{\sqrt{9+(1-a)^2} - (1-a)}{18}\left(\frac{\pi}{h_0}\right)^2 \tag{12}$$

$$\sigma_{cr} = \frac{D}{2th_1}\sqrt{\frac{\pi}{B}}\left[12B^2\left(\frac{h_0}{\pi}\right)^2 + 4(1-a)B + \left(\frac{\pi}{h_0}\right)^2\right] \tag{13}$$

in which $a = \frac{\sigma_2 t}{2D}\left(\frac{h_0}{\pi}\right)^2 = \frac{6(1-\nu)^2}{\pi^2 E}\sigma_2\left(\frac{h_0}{t}\right)^2$.

Since the brace load is assumed to be uniformly distributed over the brace width h_1, the critical elastic buckling load of the X-joint is calculated as:

$$P_{cr} = 2th_1\,\sigma_{cr} = D\sqrt{\frac{\pi}{B}}\left[12B^2\left(\frac{h_0}{\pi}\right)^2 + 4(1-a)B + \left(\frac{\pi}{h_0}\right)^2\right] \tag{14}$$

The condition in Eq. (7) can now be evaluated. Substituting Eq. (12) into Eq. (7) yields:

$$\left(\frac{h_1}{h_0}\right)^2 << \frac{54}{\pi^2\left[\sqrt{9+(1-a)^2} - (1-a)\right]} \tag{15}$$

The parameter a typically ranges from 0 to 2 in practical applications, with a=2 yielding the most critical value for Eq. (15):

$$\frac{h_1}{h_0} < 1.13 \tag{16}$$

Given that the chord member is typically the larger member compared to the braces (or at most of equal size), h_1/h_0 usually satisfies Eq. (16) and, consequently, Eq. (7).

3. Finite element analysis

There is a profound lack of experimental data related to RHS X-joints with a compressive load applied to the chord. Therefore, the Finite Element (FE) method was adopted in this study to investigate the effect of a chord pre-load on the side wall buckling of equal-width X-joints. The finite element model was developed in Abaqus [5] and was first benchmarked against nine experimental results involving RHS full width X-joints with no chord preload, obtained from [2, 3] and labelled X1-X9 (Table 1). The measured geometric dimensions, imperfections, weld sizes and material properties were incorporated into the models. Hexahedral elements were used throughout the model, except for the welds where tetrahedral elements were employed due to the complexity of the geometry. A 'tie contact' was used to fuse the surfaces between the welds and the brace and chord members together. Three elements were used in the through-thickness direction of the RHS members. A general mesh size of twice the chord wall thickness was used, while a finer size was chosen for the region of the chord side wall under the brace members, where side wall buckling was expected to occur (Fig. 2). The influence of the mesh size, the element type and the analysis type were investigated in a sensitivity study using test X7. A total of 13 models were run, covering mesh sizes ranging from 2mm to 15 mm (in the most refined region), 8-noded as well as 20-noded hexahedral elements, and 'General Static' versus 'Riks' analyses. It was concluded that, as long as the mesh size is smaller than the chord wall thickness in the most refined region, the results are quite insensitive to either the analysis type or the number of nodes in the elements. However, a 20-noded quadratic hexahedral element significantly increased the running time and was therefore not used in the analysis.

The FE results and the experimental data are compared in Table 1 with respect to the ultimate load P_u. Fig. 3 compares the FE predicted load-shortening behaviour of specimen X7 with the test data. A good agreement was generally achieved between the FE models and the test data.

4. Effect of chord preload

To investigate the effect of a chord pre-load, three levels of pre-compression were applied to X1-9, amounting to 25%, 50% and 75% of the chord squash load. A 'General Static' step was first used to apply the chord compression, followed by a 'Riks' step under increasing brace compression. A total of 36 models were analysed, as listed in Table 2. As an example, Fig. 4 shows the load-displacement curves of joint X1, subject to various levels of chord pre-load. It is seen that the chord pre-load has only a very minor effect on the ultimate capacity and behaviour of the joint. This is consistent with the work previously conducted by Platt [6].

Table 1. FE model validation

Label	Ultimate load		
	test	FEA	test/FEA
X1	176	182	0.97
X2	302	285	1.06
X3	373	421	0.89
X4	560	492	1.14
X5	783	751	1.04
X6	409	448	0.91
X7	828	862	0.96
X8	1804	2071	0.87
X9	1289	1339	0.96
		Avg.	0.98
		St. dev.	0.09

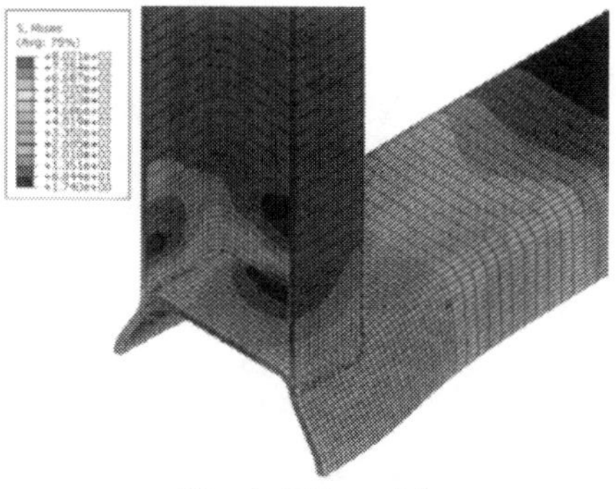

Fig. 2: FE model

As previously explained, the proposed design methodology is based on the initial buckling load of the side wall being the limiting factor in the determination of the joint capacity (in order to avoid non-linear interactive effects between member instability and joint buckling [4]) . Therefore, the buckling loads P_b were determined from all 36 FE models and are listed in Table 2. The exact point of buckling is not always easy to pinpoint, especially when buckling takes place in the inelastic range. However, the buckling load was determined by the divergence point between a geometric non-linear analysis and a linear analysis (both including material non-linearity) in the load vs. axial shortening diagram (Fig. 5).

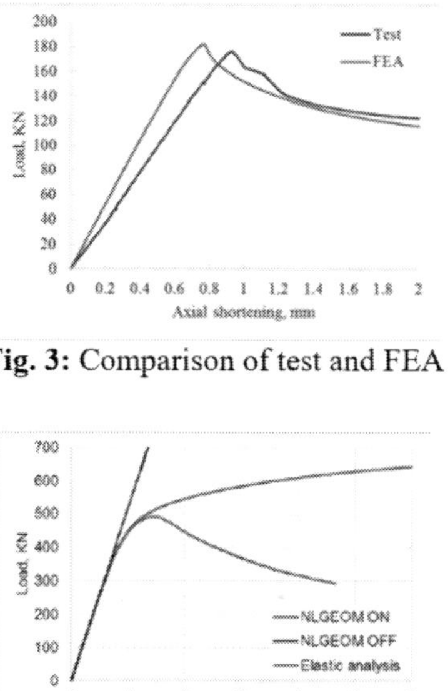

Fig. 3: Comparison of test and FEA

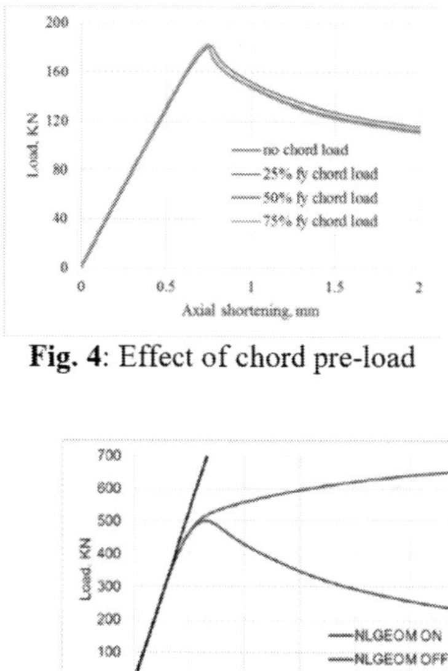

Fig. 4: Effect of chord pre-load

a) No chord pre-load b) 50% chord pre-load level

Fig. 5: Determination of the buckling load of specimen X4

It is seen from Table 2 that, in general, the chord pre-load does not have a major effect on the buckling load. Only the very slender side walls ($h_0/t = 50$) show a distinct reduction in buckling load. No results are provided for X6, X8 and X9 at elevated levels of chord pre-load, since these slender chord sections simply cannot sustain these pre-loads and would collapse in local buckling. Interestingly, a slight increase in buckling load is typically observed in the most stocky sections, which buckle inelastically. This can be attributed to the state of biaxial compression in the side wall actually delaying yielding.

Table 2. FE results and design model predictions

Label	h_0/t	Pre-load level	f_y MPa	P_y KN	$P_{b,FEA}$ kN	P_{cr} kN	P_{CIDECT} kN	$P_{pred,1}$ kN	$P_{pred,2}$ kN	$P_{b,FEA}/P_y$	$P_{CIDECT}/0.9P_{b,FEA}$	$P_{pred,1}/P_{b,FEA}$	$P_{pred,2}/P_{b,FEA}$
X1	36.7	0%	330	232	170	125	61	114	114	0.78	0.40	0.67	0.67
		25%	330	232	168	116	59	106	114	0.77	0.39	0.63	0.68
		50%	330	232	168	106	57	98	114	0.77	0.37	0.58	0.68
		75%	330	232	168	95	53	89	114	0.77	0.35	0.53	0.68
X2	26.7	0%	330	305	277	285	122	228	228	0.93	0.49	0.82	0.82
		25%	330	305	286	272	119	222	228	0.96	0.46	0.78	0.80
		50%	330	305	286	259	114	215	228	0.96	0.44	0.75	0.80
		75%	330	305	286	246	107	208	228	0.96	0.41	0.73	0.80
X3	21.2	0%	400	469	369	594	236	401	401	0.82	0.71	1.09	1.09
		25%	400	469	400	575	229	397	401	0.88	0.64	0.99	1.00
		50%	400	469	399	555	220	392	401	0.88	0.61	0.98	1.00
		75%	400	469	399	535	205	387	401	0.88	0.57	0.97	1.00
X4	18.1	0%	370	513	477	995	319	475	475	0.98	0.74	1.00	1.00
		25%	370	513	482	974	310	474	475	0.99	0.71	0.98	0.99
		50%	370	513	483	953	298	473	475	0.99	0.68	0.98	0.98
		75%	370	513	484	932	278	472	475	0.99	0.64	0.97	0.98
X5	12.9	0%	345	654	672	2465	520	632	632	1.05	0.86	0.94	0.94
		25%	345	654	651	2438	505	632	632	1.02	0.86	0.97	0.97
		50%	345	654	658	2411	485	632	632	1.03	0.82	0.96	0.96
		75%	345	654	666	2384	452	632	632	1.04	0.75	0.95	0.95
X6	50.0	0%	463	834	270	243	75	231	231	0.32	0.31	0.86	0.86
		25%	463	834	213	183	73	175	231	0.26	0.38	0.82	1.08
		50%	—	—	—	—	—	—	—	—	—	—	—
		75%	—	—	—	—	—	—	—	—	—	—	—
X7	25.6	0%	451	953	791	652	285	573	573	0.83	0.40	0.72	0.72
		25%	451	953	801	613	277	544	573	0.84	0.38	0.68	0.72
		50%	451	953	781	572	266	514	573	0.82	0.38	0.66	0.73
		75%	451	953	751	530	248	481	573	0.79	0.37	0.64	0.76
X8	35.3	0%	468	2778	1550	1364	482	1254	1254	0.56	0.35	0.81	0.81
		25%	468	2778	1490	1199	468	1114	1254	0.54	0.35	0.75	0.84
		50%	468	2778	1410	1024	450	961	1254	0.51	0.35	0.68	0.89
		75%	—	—	—	—	—	—	—	—	—	—	—
X9	50.5	0%	481	2745	682	604	227	579	579	0.25	0.37	0.85	0.85
		25%	481	2745	590	445	221	430	579	0.21	0.42	0.73	0.98
		50%	481	—	—	—	—	—	—	—	—	—	—
		75%	481	—	—	—	—	—	—	—	—	—	—
										Avg.	0.52	0.82	0.87
										St. dev.	0.18	0.15	0.13

5. Design

Table 2 shows, for each FE model: the elastic critical buckling load P_{cr} obtained using Eq. (13), the buckling load $P_{b,FEA}$ and the squash load P_y, which is taken as:

$$P_y = 1.2 \times 2 f_y h_1 t = 2.4 f_y h_1 t \tag{18}$$

The factor 1.2 thereby takes into account that some of the load follows an alternative load path through the chord top and bottom faces and then spreads out into the chord side walls. This factor agrees well with the ultimate load observed in the stockiest joint (X5). The non-dimensional slenderness is then defined as:

$$\lambda = \sqrt{\frac{P_y}{P_{cr}}} \quad (17)$$

We also define:

$$\chi = \frac{P_b}{P_y} \quad (17)$$

The non-dimensional buckling loads $P_{b,FEA}/P_y$ obtained from all FE models are plotted against the calculated slenderness values λ in Fig. 6a, which also shows the elastic critical buckling load in dashed line:

$$\chi = \frac{P_{cr}}{P_y} = \frac{1}{\lambda^2} \quad (19)$$

It is seen that, generally, good agreement is obtained between the data and the above equation in the elastic range (λ >> 1). In order to account for the gradual transition of the elastic curve into the inelastic range at lower slenderness values, an equation is proposed which is aligned with the current CIDECT practice of referring, in the design for side wall buckling, to the code specifications for column buckling (e.g. EN1993-1-1 [7]):

$$P_b = \chi P_y \quad (20)$$

with:

$$\chi = \frac{1}{\phi + \sqrt{\phi^2 - \lambda^2}} \leq 1.0 \quad (21)$$

$$\phi = \frac{1}{2}\left[1 + \alpha(\lambda - 0.2) + \lambda^2\right] \quad (22)$$

The value of the imperfection factor α was determined to be 0.08 by conservatively fitting a lower bound curve to the data. This result is also consistent with [3]. Eq. (20) is plotted as a solid line in Fig. 6a. It is seen that Eq. (20) agrees well with the data, both for slender side walls buckling elastically and stocky side walls buckling inelastically, although the proposed equation is conservative in the intermediate slenderness range for high chord pre-loads. A reliability analysis to EN1990 [8] will be carried out as future research in order to improve the reliability of the proposed design equations and make them acceptable for inclusion in design standards. Since the buckling loads were found not to vary all that significantly with different levels of chord pre-load (Table 2), it is worth investigating whether the effect of the chord pre-load can simply be ignored in design. With a=0 in Eq. (13), the methodology reduces to the one previously proposed in [3]. The corresponding design curve is presented in Fig. 6b. Good agreement with the numerical data is observed. It is even noted that much of the conservatism in the intermediate slenderness disappears. However, the predictions are now slightly unsafe in the high slenderness range, with the buckling loads being overestimated by 8% for X6 with h_0/t=50 and a chord pre-load of 25% of the squash load.

Table 2 lists the ratios of the predicted buckling capacities over the numerical results, using the current CIDECT rules, as well as Prediction 1 (Fig. 6a) and Prediction 2 (Fig. 6b). An average ratio of 0.82 was obtained with a standard deviation of 0.15 for Prediction 1, while an average ratio of 0.87 with a standard deviation of 0.13 was achieved by Prediction 2. Both systems strongly outperform the current CIDECT design rule for side wall buckling, which over the same data set features an average ratio of predicted to measured capacity of 0.52 with a standard deviation of 0.18. It is noted that, in the comparison, the CIDECT prediction was divided by a factor of 0.9 since the CIDECT design rules provide *factored* joint resistances which already include an implicit safety factor [1]. Importantly, it is also seen that the CIDECT rule does not offer a consistent margin of safety, but is more conservative for side walls with high h_0/t_0 values. The applicability of the current CIDECT rule is also limited to an h_0/t ratio of 40, while the proposed rule is based on experimental and numerical data including sections with h_0/t ratios of up to 50.

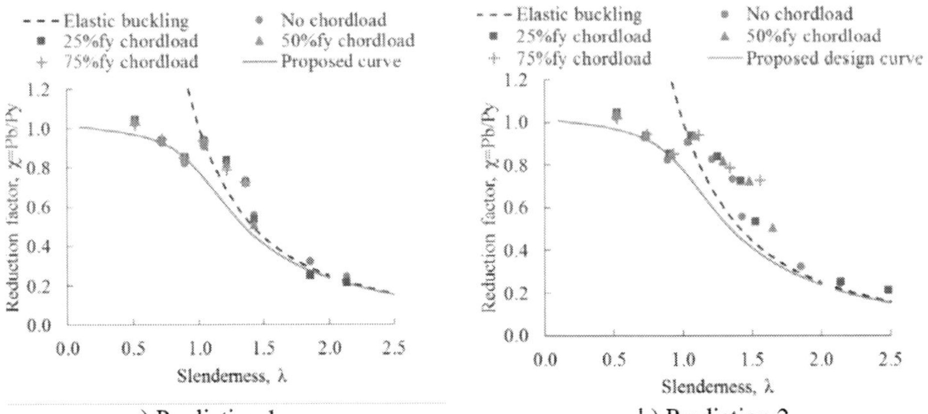

a) Prediction 1 b) Prediction 2

Fig. 6: Proposed design equation

6. Conclusions

This paper presents a study of the side wall buckling of equal width RHS X-joints subject to simultaneous brace and chord compression. As the current CIDECT design rule is known to be overly conservative, especially for large h_0/t_0 ratios, a new design equation is proposed. The approach is founded on a rational analysis of an infinitely long plate subject to a localized distributed load and a longitudinal compressive stress. A Rayleigh-Ritz approximation is used to obtain the critical elastic buckling stress, which is subsequently used in the definition of a slenderness parameter. The new design equation is compared to numerical results including X-joints made of SHS and RHS of widely varying sizes and wall slenderness values and subjected to various levels of chord pre-load. The proposed design equation was found to be in good agreement with the data.

References

[1] Packer JA, Wardenier J, Zhao XL, v. d. Vegte GJ, Kurobane Y. *Design guide for rectangular hollow section (RHS) joints under predominantly loading*, 2009.

[2] Becque J, Wilkinson T. "Experimental investigation of the static capacity of grade C450 RHS T and X truss joints," *Tubular structures XIV*, L. Gardner, Ed., ed: CRC Press, 2011.

[3] Becque J, Wilkinson T. "A new design equation for side wall buckling of RHS truss X-joints," in *Tubular Structures XV*, ed: CRC Press, 2015, pp. 419-426.

[4] Timoshenko SP, Gere JM. *Theory of Elastic Stability (2nd edition).* New York: McGraw-Hill, 1961.

[5] ABAQUS, "ABAQUS Analysis Users' Manual, Version 6.14-1," ABAQUS INC., USA, 2014.

[6] Platt JC. "Side-wall behaviour in full-width rectangular hollow section joints," PhD PhD, University of Nottingham, University of Nottingham, 1984.

[7] BSI, "EN1993-1-1: Design of steel structures," in *Part 1-1: General rules and rules for buildings*, ed. Brussels: BSI, 2005.

[8] BSI, "EN1990: Basis of structural design", BSI, 2002.

EXPERIMENTAL AND NUMERICAL INVESTIGATION OF THE BENDING CAPACITY OF SPIRAL-WELDED STEEL TUBES

Sjors H.J. van Es[a], Arnold M. Gresnigt[a], Daniel Vasilikis[b] and Spyros A. Karamanos[b]

[a] *Delft University of Technology, Delft, The Netherlands*
[b] *University of Thessaly, Volos, Greece*

Abstract: To investigate the structural behaviour of large-diameter spiral-welded steel tubes under bending, a full scale experimental program has been performed, consisting of thirteen 42-inch diameter, spiral-welded steel tubes with D/t ratios ranging between 65 and 120. Additionally, numerical studies have been conducted, using nonlinear finite element simulations. A model has been developed to simulate the performed tests, using the actual material properties and initial geometric imperfections obtained from the tested tubes. Finally, a parametric investigation has been performed that investigates the influence of initial geometric imperfections and presence of girth and coil connection welds on the bending behaviour of the considered tubes.

1. Introduction

The spiral-welding manufacturing process (HSAW) offers an economical method to manufacture relatively thin-walled, large-diameter steel tubes. An overview of this manufacturing process is presented in Fig. 1a. Common tube diameters (D) range from 500 mm to 3000 mm with wall thicknesses (t) between 9 mm and 25 mm. The process is continuous; a steel coil that runs out is connected to a new coil by means of a butt weld without interruption of the spiral-welding process. This weld, running perpendicular to the spiral welds, is denoted as a coil-connection weld (CCW). Aside from this coil connection weld, girth welds (GW) occur in structures when two separate tubes are joined (see Fig. 1b).

Large-diameter spiral-welded tubes are employed in for example hydrocarbon and water pipelines, but also in structural applications. The present research mainly focuses on the behaviour of spiral-welded tubes for application in combined walls. In these soil retaining structures, large diameter, thin-walled tubes are used as primary structural elements to resist horizontal soil loads. Steel tubes used in these applications are generally manufactured with the spiral-welding process. Due to the slenderness of the tubes that are typically applied in these structures, local buckling is the dominating failure mode.

The pure bending response of metal tubes has been investigated numerous times in the last decades, yet investigations focusing on spiral-welded tubes are scarce. In [1], four bending tests are performed on spiral-welded tubes with D/t=48 and 82, intended for linepipe use. Therefore, two of these tests were conducted with internal pressure in the tube. In [2], a similar testing programme is conducted on four spiral welded tubes with D/t=46 and 64. Both studies deem the performance of the tubes to be satisfactory but neither includes the presence of girth welds or coil connection welds. Eight bending tests on more thin-walled spiral-welded tubes have been performed by Reinke et al. [3], as part of the same project as the present research. Also in the tests by Reinke et al., the spiral welded tubes performed well. The spiral weld was deemed to have no influence on the maximum bending moment capacity of the specimens; deformation capacity was not investigated.

The current research is part of a research project with acronym COMBITUBE, which examines the bending behaviour of large-diameter steel tubes, specifically for application in combined walls [4].

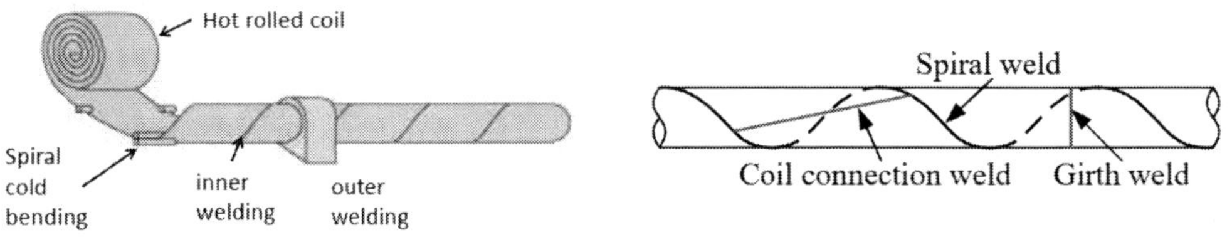

a) Spiral-welding manufacturing process b) CCW and GW in a spiral-welded tube

Fig. 1: Overview of spiral welding manufacturing process and resulting welds in spiral-welded tubes

2. Experimental work

2.1 Overview

The testing program featured investigation of the initial geometry and geometrical imperfections, extensive material testing and four-point bending tests on thirteen 42-inch diameter (1067 mm) spiral-welded carbon steel tubes.

The length of all specimens was 16500 mm. The D/t ratios varied from 65 to 120 and the steel grades varied from X52 to X70 (see Table 1). The experimental program includes plain tubes that feature only a spiral weld, but also tubes with additional welds such as girth welds and coil connection welds. If present, these welds divided a specimen in two or more sections which are referred to as 'specimen parts' and are identified by adding the indicators 'left', 'middle' and 'right' to the specimen identification.

2.1 Measurement of initial geometry and geometrical imperfections

The tube diameter and wall thickness have been measured at five cross sections along the tube in 45 degree increments of axial tube rotation. Additionally, the wall thickness variation across the width of the original coiled steel plate has been investigated by taking measurements along a line perpendicular to the spiral weld. The wall thickness measurements were performed using an ultrasonic thickness measurement device. The geometric profile of the tube wall has been scanned using a laser trolley driving on rails below the specimen. An overview of this test setup is depicted in Fig. 2.

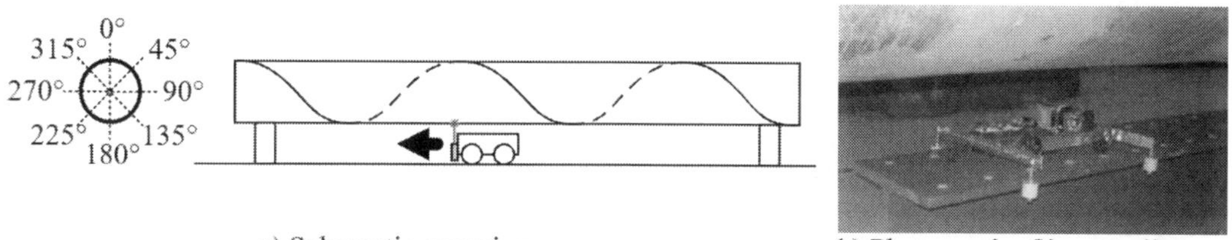

a) Schematic overview b) Photograph of laser trolley

Fig. 2: Overview of test setup for measurement of geometrical imperfections

2.2 Four-point bending tests

To determine the maximum bending capacity of the tubes, the specimens were loaded in four-point bending by applying an upward displacement at the two outer supports (see Fig. 3). Due to limited availability of hydraulic actuators in the laboratory, different actuators were used on each side of the tube. Despite this difference, the loading on the specimen itself was symmetric. After local buckling failure, deformation was further increased until the full stroke of the actuators was used.

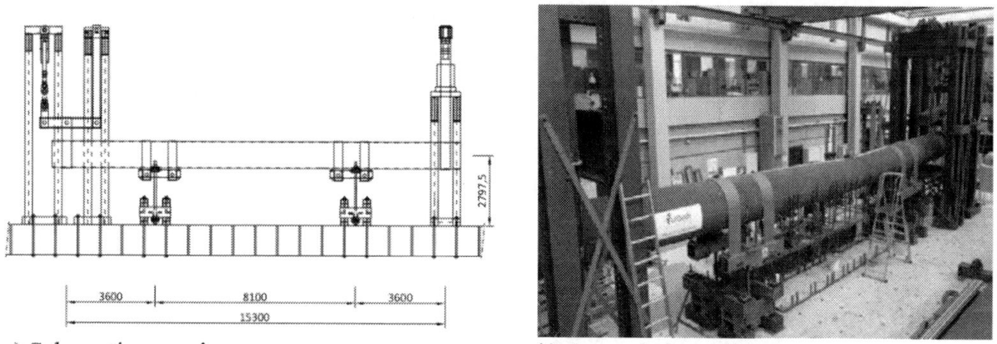

a) Schematic overview b) General view of test setup

Fig. 3: Overview of four-point bending test setup

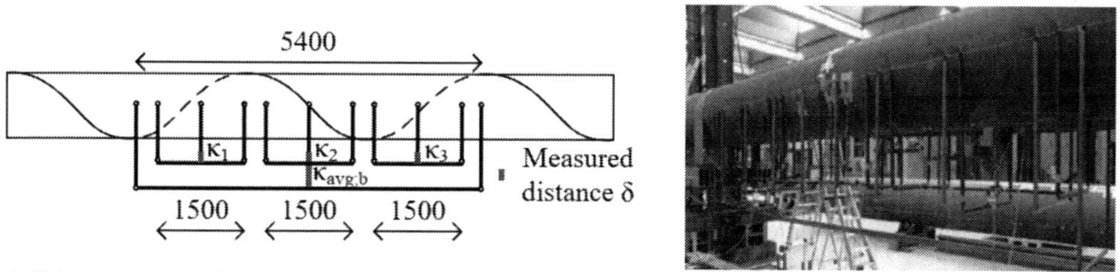

a) Schematic overview b) Photograph of measurement brackets

Fig. 4: Overview of curvature measurements using brackets

During the tests, the applied displacements and the resulting forces were measured continuously. Furthermore, the curvature (κ) of the tube was measured over three tube segments with a length L of 1500 mm, and over the full tube segment that was loaded in constant bending moment. Each curvature measurement used three brackets, of which the outer two were coupled (see Fig. 4). From the measured distance δ between the centre bracket and the coupler bar, the curvature of the examined tube section was calculated using Eq. (1). A more elaborate description of the performed measurements can be found in [5].

$$\kappa = \frac{8 \cdot \delta}{L^2} \tag{1}$$

3. Results of experiments

3.1 Results of measurements of tube geometry and initial geometrical imperfections

The scans of the initial geometry of the tubes show a very regular imperfection pattern. Between the spiral welds a regular pattern of wave-like shapes occurs in all investigated tubes (see Fig. 5). In some of the tubes, the outer surface showed markings, caused by the rolling equipment used to bend a tube from the coiled steel plate. Since these markings align with the amplitudes in the imperfection profiles, the imperfections profile is assumed to originate from the spiral-welding process. Near girth welds and coil connection welds, more local and generally larger geometrical imperfections occur. A more thorough discussion of all geometrical imperfections and their origins can be found in [5] and [6].

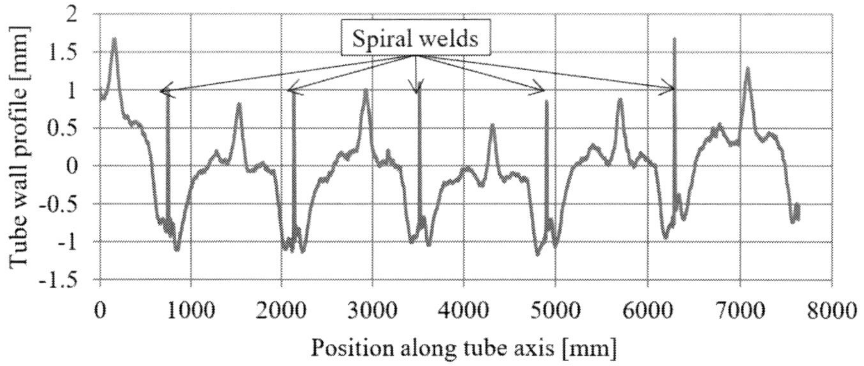

Fig. 5: Example of measured initial geometrical imperfection profile (Tube T2)

The average wall thickness of each specimen is listed in Table 1. The variation of the wall thickness within the specimen parts was found to be very small and is therefore neglected in the following analyses. The measurements over a line perpendicular to a spiral weld have shown a decrease in wall thickness towards the spiral welds of about 1%. This is also considered to be negligible.

The diameter of the tubes was very close to the specified value of 1067 mm. Aside from the average diameter, the initial ovalisation of the tube is recognized as a relevant parameter. The initial ovalisation f is defined as $(D_{max}-D_{min})/D_{ave}$. This parameter was estimated using the diameters measured in 45° increments (see section 2.2). The results show that only one of the tubes exceeds an initial ovalisation of 0.5%, which is considered to be excellent.

3.2 Results of four point bending tests

As a result of the exerted bending loads, all tubular specimens failed in the form of local buckling. The specimens with D/t=120 failed suddenly and violently. In contrast, the transition between a stable and unstable situation was much more gradual for the thicker walled tubes. The violent failure of the thin walled tubes in some cases resulted in failure of the measurement equipment. If this occurred, graphical representations of measurements are truncated at the moment of failure. Examples of failed specimens are presented in Fig. 6.

The main output of the bending tests is the moment-curvature (M-κ) diagram. Since four different curvature measurements were used, each test results in four diagrams. Results of two typical tests are presented in Fig. 7. The results show that in the inelastic part of the moment-curvature response, the four curvature measurements show different results. For example, the curvature measured by measurement κ_1 is significantly higher than the others in Fig. 7a. Apparently, the curvature is not constant within the tube segment that is loaded by a constant bending moment, caused by variations in bending resistance over the length of the tube. The moment-curvature diagrams further show that these tubes did not fail in elastic buckling. Instead, an elasto-plastic buckling mode develops.

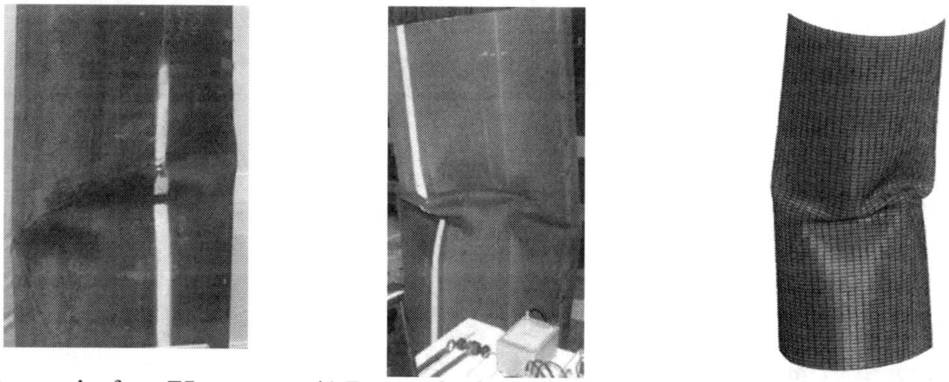

a) Test result of test T7 b) Test result of test T2 c) FEA result of test T2

Fig. 6: Examples of local buckles in spiral-welded tubular specimens

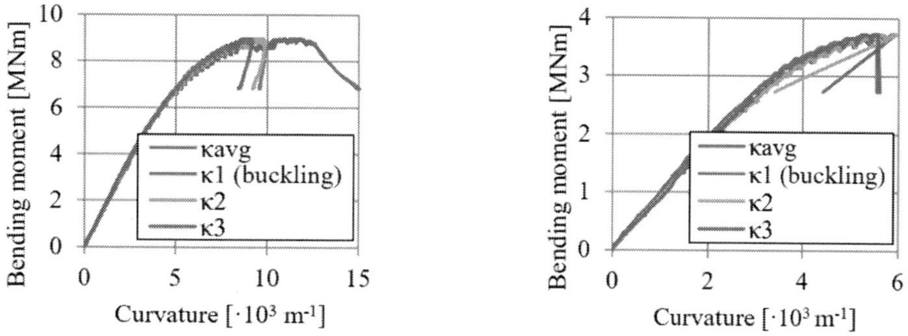

a) Specimen T9 (D/t=65.4) b) Specimen T4 (D/t=116)

Fig. 7: Moment-curvature diagrams of typical specimens. Local buckling takes place within the local measurement marked with '(buckling)'

The critical curvature κ_{cr} is defined as the measured average curvature at the occurrence of the maximum bending moment. An overview of the reached critical compressive strains (ε_{cr}), determined from the critical curvatures, is presented in Fig. 8a. The figure shows that specimens containing a girth weld or coil connection weld exhibit a slightly lower critical strain. It is assumed that is the result of e.g. misalignments at the weld and discontinuities in bending moment capacity, which result in concentration of deformation. The maximum resisted bending moment, normalized by the full plastic bending moment (M_{pl}), is presented in Fig. 8b for each specimen. The slenderness is normalized by $D/(t\varepsilon^2)$, with $\varepsilon^2=235/\sigma_y$, as suggested by [7]. Specimens containing a girth weld or coil connection weld appear to have a lower or equal bending moment capacity than plain specimens. Exception to this trend is specimen T10. Local buckling occurred at one side of the girth weld, on the opposite side of the weld a 50% higher yield strength was measured. Apparently the negative effect of the girth weld was counteracted by a positive effect of this yield strength difference.

In all plain specimens, local buckling occurred at the manufacturing-related geometrical imperfections. None of the buckles formed directly at a spiral weld. The presence of a girth weld or coil connection weld resulted in local buckling at this weld in four out of six cases.

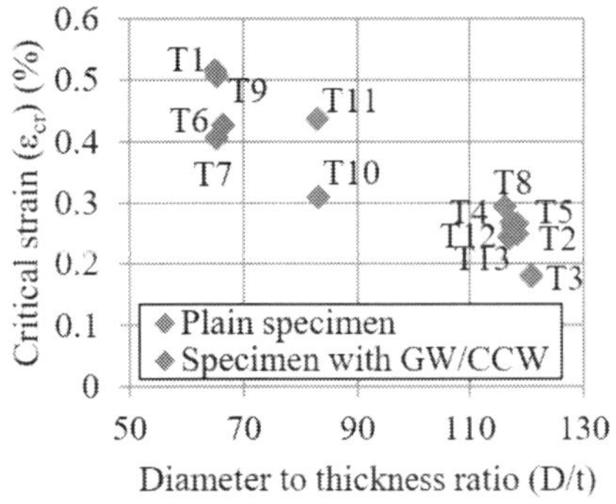

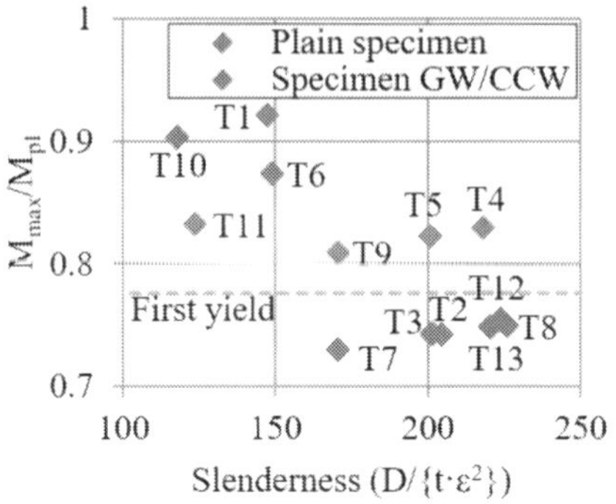

a) Critical strains b) Normalized maximum bending moments

Fig. 8: Overview of test results of four point bending tests

4. Numerical simulation of four point bending tests

4.1 Results of four point bending tests

A numerical model has been developed in ABAQUS/Standard [8]. The tubes are modelled with four-node reduced integration shell elements (S4R). Inelastic material behaviour is considered using a Von Mises plasticity model with isotropic hardening. Residual stresses are inserted in the model to simulate the effects of the manufacturing process of spiral welded tubes [9].

For the simulation of plain tubes, a dense rectangular mesh of elements is employed (Fig. 9a), with an element size equal to 50 mm. The tube is initially imperfect, with a wavy-type (wrinkling) geometric imperfection, in the form of the first buckling mode (Fig. 9b) with amplitude δ_0. Seven elements are present per half-wavelength. In case of a tube with a girth weld, different material properties are taken into account on each side of the weld.

Four reference nodes have been introduced to represent the bending test set-up; two reference points refer to the supports and two to the locations where forces are applied (see Fig. 10). An arc-length continuation algorithm (Riks) is employed to trace the load-displacement path.

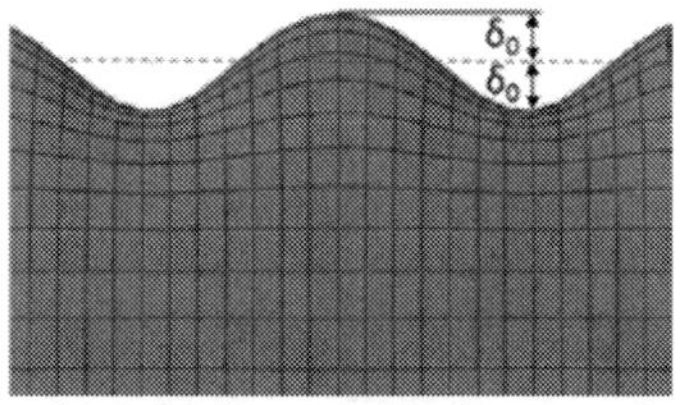

a) Rectangular finite element mesh of initially wrinkled tube (magnified for visualization purposes)

b) Initially wrinkled pattern in the compression side

Fig. 9: Overview of FEA model for plain tubes and definition of geometrical imperfections

Fig. 10: Numerical model of four-point bending in ABAQUS

In addition to the rectangular mesh of Fig. 9, a "spiral mesh" has also been employed (see Fig. 11a) with similar density. In that mesh, the spiral weld is modelled as a separate part, with different material properties, so that over-matching of the spiral weld is considered. In case of tubes containing girth and/or coil connection welds, the tube is divided in parts with separate geometric and material properties (Fig. 11b).

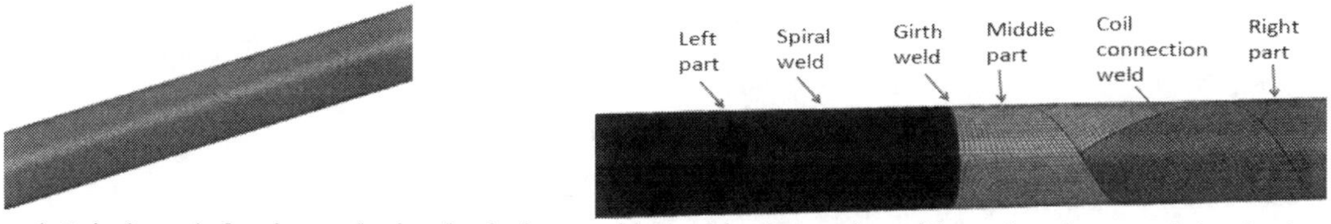

a) Spiral mesh for the analysis of spiral-welded tubes

b) Partitioning and meshing of a tube containing both girth and coil connection weld

Fig. 11: Overview of FEA model with spiral mesh, suitable for modelling of coil connection welds

4.2 Numerical results

The comparison between numerical and experimental results for all tests is presented in the last four columns of Table 1, whereas moment-curvature diagrams for two representative tests are depicted in Fig. 12. The bending curvature reported in the numerical results is calculated as the sum of the two rotations at the middle supports, averaged over the initial distance between the two supports. Moment values are normalized by the plastic bending moment, and bending curvature is normalized by the curvature-like quantity $\kappa_I = t/D_m^2$. The results from Fig. 12 and Table 1 show a very good comparison between the numerical results and the experimental data, and indicate that the developed models are able to accurately predict the bending response of spiral welded tubes in terms of both ultimate bending moment capacity and deformation capacity. In all cases, the buckled shapes obtained numerically are in excellent agreement with the shapes observed in the experiments (e.g. see Fig. 6b and c).

Table 1: Summary of bending tests; comparison between numerical and experimental results

Test	Type	Geometric data			Yield stress	Initial wrinkle	Moment (M_{max}/M_{pl})		Curvature κ_{cr}/κ_I	
		D [mm]	t [mm]	D/t [-]	σ_y [MPa]	δ_0/t [%]	Test [-]	FEM [-]	Test [-]	FEM [-]
T1	Plain	1066	16.4	65.1	541	1.8	0.907	0.940	0.646	0.694
T2	Plain	1067	9.02	118	392	4.5	0.769	0.842	0.559	0.669
T4	Plain	1065	9.16	116	420	3.0	0.871	0.845	0.761	0.713
T5	Plain	1070	9.04	118	403	3.6	0.814	0.852	0.681	0.762
T8	Plain	1068	9.10	117	435	5.8	0.781	0.807	0.697	0.691
T9	Plain	1069	16.30	65.4	571	4.6	0.869	0.935	0.696	0.819
T11	Plain	1068	12.90	83.0	341	3.1	0.865	0.945	0.726	0.728
T3	GW	1069	9.03	118	392	10.0	0.735	0.771	0.432	0.510
T13	GW	1070	9.18	116	443	12.8	0.744	0.772	0.633	0.612
T6	CCW	1066	16.30	65.3	528	9.3	0.876	0.941	0.586	0.631
T7	CCW+GW	1068	16.30	65.4	614	5.5	0.730	0.903	0.533	0.705
T10	CCW+GW	1070	13.1	81.6	333	13.7	0.897	0.916	0.480	0.498
T12	CCW+GW	1069	9.13	117	431	6.7	0.783	0.815	0.690	0.706

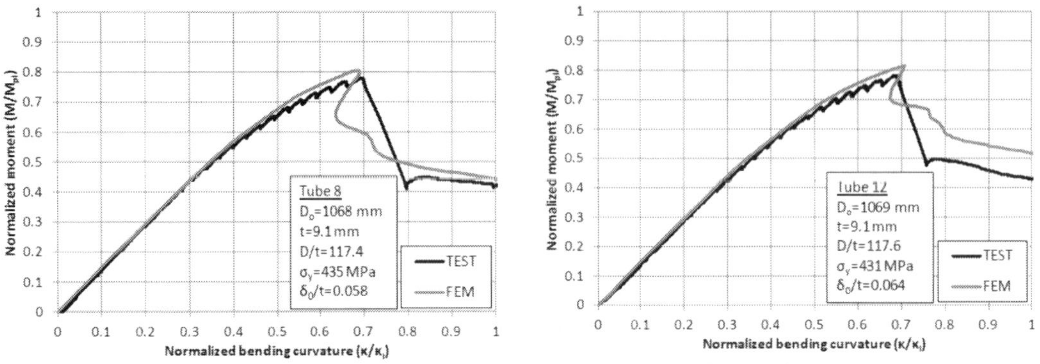

a) Specimen T8 b) Specimen T12

Fig. 12: Comparison between numerical and test results in terms of moment-curvature curves

5. Numerical parameter study

5.1 Wrinkling imperfections sensitivity

Numerical results that investigate the imperfection sensitivity of the bending response are presented in Fig. 13 for 42-inch-diameter tubes with D/t=67 and 119 respectively, in the absence of residual stresses. The results show that both the maximum bending moment and critical curvature are sensitive to the presence of initial wrinkles. The first tube (D/t=67) exhibits substantial inelastic deformation. For rather small values of initial imperfection amplitude (δ_0/t<0.5%), the maximum moment exceeds 95% of M_{pl}. For larger amplitudes of the initial wrinkles, the moment and deformation capacity decrease significantly. In any case, the response is characterized by a clear limit point on the moment-curvature diagram, well into the inelastic range. The thin-walled tube (D/t=119) exhibits a maximum moment less than 90% of M_{pl}. For values of δ_0/t greater than 30%, the "snap-back" in the M-κ diagram is alleviated and the response becomes smoother, characterized by a limit point.

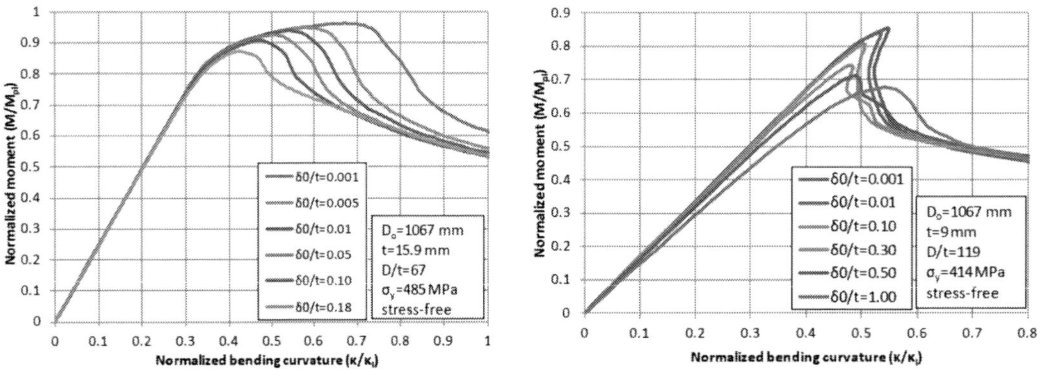

a) Tubes with D/t=67 b) Tubes with D/t=119

Fig. 13: Imperfection sensitivity of 42-inch diameter stress-free tubes subjected to bending

5.2 Effect of girth and coil connection welds

At girth welds and coil connection welds, the possible discontinuity in material properties of the two connected tubes may result in stress and strain concentrations, facilitating the formation of a local buckle. In the model, the yield stress of one part of the tube is chosen equal to 485 MPa, while the yield stress of the other part is considered as a variable. The strength difference between the two parts is expressed as a percentage. The results in Fig. 14 indicate that significant differences in yield stress between the two connecting tube parts may decrease the critical bending curvature. This is attributed to the fact that the discontinuity in material properties acts as an imperfection, causing localization of stresses and deformation and, eventually, premature buckle formation, always in the weakest part of the tube.

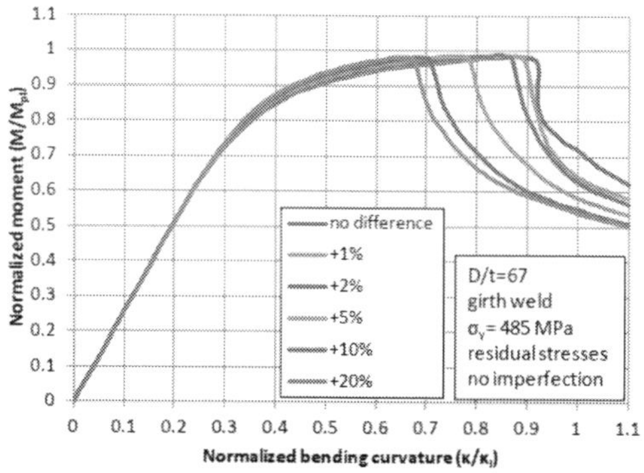

a) Tubes with girth welds

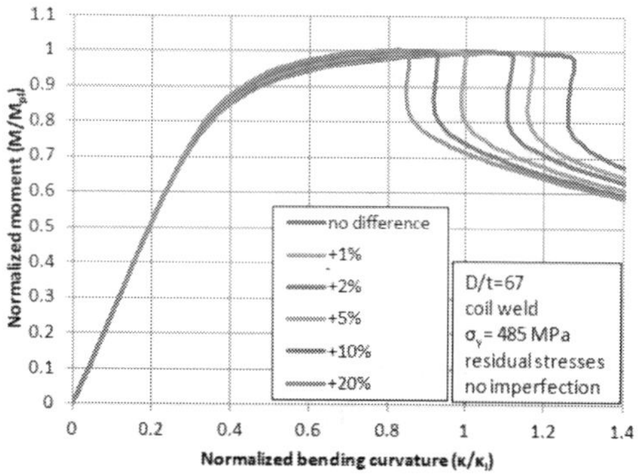

b) Tubes with coil connection welds

Fig. 14: Normalized moment-curvature diagrams of a 42-inch diameter tube (D/t=67, grade X70) for different yield stress values of connected tubular parts (girth welds and coil connection welds)

5.3 Effect of girth weld misalignment

At a girth-weld, there is a possibility for eccentricity between the connected tubes. This is also referred to as “misalignment” or “high-low”. The calculated normalized critical curvature for different levels of girth weld misalignment is presented in Fig. 15. The misalignment has a considerable effect on the bending behaviour, decreasing the maximum moment and critical curvature. During bending, deformation concentrates in the girth weld region due to this misalignment, and buckling always occurs in this area in the FE-analyses. In practice, the buckling location may depend on the misalignment size and overstrength of the girth weld.

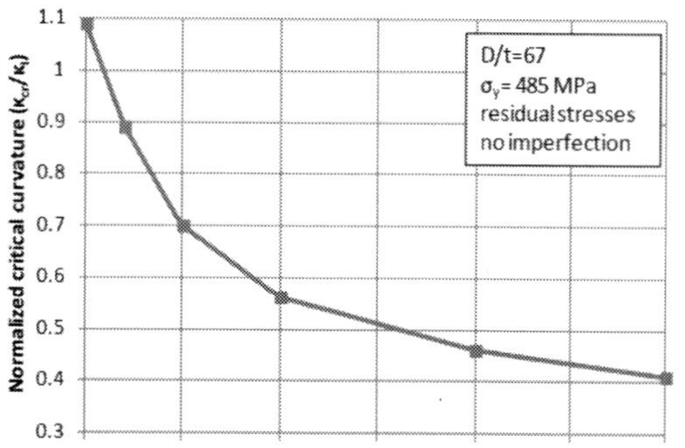

Fig. 15: Variation of normalized critical curvature in terms of the girth weld misalignment

6. Conclusions

This paper presents a summary of the results of thirteen full-scale four-point bending tests on spiral-welded steel tubes. Furthermore, a numerical study of the mechanical behaviour of these tubulars is offered using finite element models. The main findings in this paper are:

1. The tested spiral-welded tubes show regular geometrical imperfections. These imperfections are associated with the spiral-welding manufacturing method. In absence of girth welds and coil connection welds, local buckles form at such imperfections;
2. Curvature was found to be not constant over a length of tube with constant bending moment. Especially beyond the elastic range of the tube, deformation localized;
3. The presence of girth welds and coil connection welds had a penalizing effect on the critical strain and maximum resisted bending moment of the tested tubes;
4. The developed numerical model shows good comparison with experimental data in terms of moment-curvature behaviour (κ_{cr} and M_{max});
5. The tubes under consideration were found to be sensitive to initial geometrical imperfections. Even the relatively thick-walled tubes (D/t=67) show to be sensitive to realistic imperfections that were observed in the laboratory.
6. The effect of girth welds and coil connection welds is visible through the influence of a yield strength difference on either side of the weld and through misalignments at the weld itself. Both these factors can have significant influence on the critical curvature.

In [10], an analytical model is presented that is able to determine structural behaviour of these tubes, taking into account relevant parameters that determine the deformation capacity and bending moment capacity.

Acknowledgments

Funding for this work has been provided by the Research Fund for Coal and Steel (RFCS) of the European Commission, project COMBITUBE: "Bending Resistance of Steel Tubes in CombiWalls", Grant Agreement No. RFSR-CT-2011-00034.

References

[1] Zimmerman T, Xie J, Timms C, Asante J. "Buckling resistance of large diameter spiral welded linepipe", *Proceedings of the Int. Pipeline Conf. (IPC)*, Calgary, 2004

[2] Zimmermann S, Karbasian H, Knoop FM. "Helical Submerged Arc Welded Line Pipe Engineered For Strain Based Design", *Proceedings of ISOPE*, Anchorage, 2013.

[3] Reinke T, Sadowski AJ, Ummenhofer T, Rotter JM. "Large Scale Bending Tests of Spiral Welded Steel Tubes", *Proceedings of Eurosteel*, Naples, Italy, 2014.

[4] Combitube Research Group, "Bending resistance of steel tubes in CombiWalls – COMBITUBE – Final Report", European Comission – RFCS, Brussels, 2015.

[5] Van Es SHJ, Gresnigt AM, Vasilikis D, Karamanos SA. "Ultimate Bending Capacity of Spiral-Welded Steel Tubes – Part I: Experiments", *Thin Walled Structures, IN PRESS*

[6] Sadowski AJ, Van Es SHJ, Reinke T, Rotter JM, Gresnigt AM, Ummenhofer, T. "Harmonic analysis of measured initial geometric imperfections in large spiral welded carbon steel tubes", *Engineering Structures*, 85, 234-248, 2015.

[7] European Committee for Standardization, "Eurocode 3: Design of steel structures – Part 1-1: General rules for buildings", CEN, Brussels, 2006.

[8] ABAQUS, 2013. *User' Manual*, version 6.12, Hibbitt, Karlsson and Sorensen, Inc.

[9] Vasilikis D, Karamanos SA, Van Es SHJ, Gresnigt AM. "Ultimate Bending Capacity of Spiral-Welded Steel Tubes – Part II: Predictions", *Thin Walled Structures, IN PRESS*

[10] Gresnigt AM, Van Es SHJ, Vasilikis D, Karamanos SA. "Strain-based design procedures for spiral-welded steel tubes in combined walls", *The International Colloquium on Stability and Ductility of Steel Structure,* Timisoara, Romania, 2016.

STRAIN-BASED DESIGN PROCEDURES FOR SPIRAL-WELDED STEEL TUBES IN COMBINED WALLS

Arnold M. Gresnigt[a], Sjors H.J. van Es[a], Daniel Vasilikis[b] and Spyros A. Karamanos[b]

[a] *Delft University of Technology, Delft, The Netherlands*
[b] *University of Thessaly, Volos, Greece*

Abstract: Spiral-welded steel tubes with diameter to wall thickness ratios between 60 and 140 are often employed in combined wall systems with local buckling as governing failure mode. The current design rules in Eurocode 3 (EN 1993-5 and EN 1993-1-6) are not capable to obtain a good estimate of the real strength and deformation capacity. In a European RFCS project called Combitube, the structural behaviour of spiral-welded steel tubes has been investigated. An analytical model has been developed that gives more economic designs with a better balanced safety level.

1. Introduction

Spiral-welded steel tubes are often employed in combined wall systems, e.g. for quay wall systems as indicated in Fig. 1. In these applications, the diameter to wall thickness ratio (D/t) ranges between 60 and 140. The main loading is bending in combination with normal force, earth loads and tensile or compressive loads from the infill sheeting. The governing failure mode is local buckling in the inelastic range of the steel.

The current design rules in Eurocode 3 (EN 1993-5 and EN 1993-1-6) follow a design approach based on stress resultants, rather than strains and deformations. It has been shown that these current design rules are not well suited for tubes in combined wall systems. They are not capable to obtain a good estimate of the real strength and deformation capacity and lead to uneconomic designs. Within the framework of a European RFCS project called Combitube, the structural behaviour of spiral-welded tubes for application in combined walls has been investigated. Full-scale four-point bending tests and extensive numerical parametric studies have been performed [1][7][8][9]. Based on the analytical formulation initially developed for buried pipelines [3][6] and the results of the Combitube research, an analytical model has been developed for tubes in combined walls including the effect of the following parameters:

- Diameter to wall thickness ratio (for tubes in CombiWalls usually between 60 and 140).
- Stress-strain properties of the steel, in particular the strain hardening properties.
- Presence of normal (axial) force and shear force in the tube.
- Transverse forces from infill sheeting.
- Cross-sectional ovalisation.
- Initial pipe wall short-wave wrinkling patterns in longitudinal direction.
- Other geometrical Imperfections: offset at girth welds ("high-low"), dimples.
- Residual stresses due to the manufacturing cold-bending process.

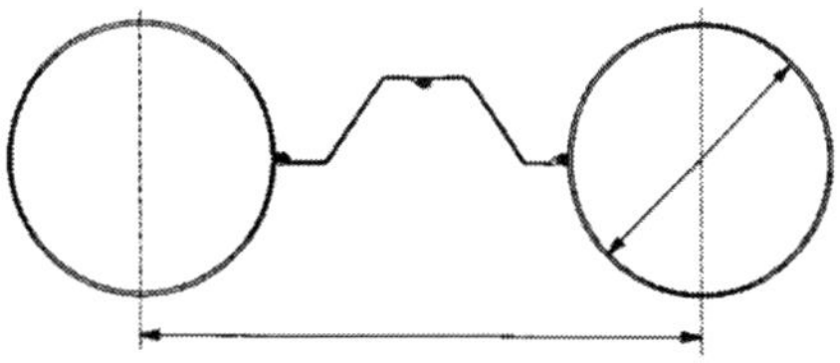

Fig. 1: Schematic and photo of a combined wall of tubes and infill sheeting, also called CombiWall; the infill sheeting is connected to the tubes with welded slots

2. Bending moment curvature diagram

First the bending moment curvature diagram is constructed in pure bending without the effect of other influences such as ovalisation, imperfections, residual stresses, etc. Thereafter the effect of other influences is considered. The equations are set up for thin walled tubes. This allows simplified equations for the cross-section quantities with sufficient accuracy for the present D/t range.

2.1 Moment curvature without other influences

A bilinear stress strain diagram is assumed (no strain hardening). For the elastic part it gives:

$$M = \kappa \cdot EI \quad \text{with} \quad EI = E\pi r^3 t \quad \text{and} \quad \kappa = \frac{\varepsilon_{\max}}{r} = \frac{\sigma_{\max}}{Er}$$
$$M_y = \sigma_y \cdot \pi r^2 t \qquad \kappa_y = \frac{\varepsilon_y}{r} = \frac{\sigma_y}{Er} \quad \text{with} \quad r = (D-t)/2 \tag{1}$$

And for the elastic-plastic part [6]:

$$M = M_p \cdot 0,5\left(\frac{\theta}{\sin\theta} + \cos\theta\right) = 4r^2 t\sigma_y \cdot 0,5\left(\frac{\theta}{\sin\theta} + \cos\theta\right) \;;\; \kappa = \frac{\varepsilon}{r} = \frac{\varepsilon_y}{r \cdot \sin\theta} \tag{2}$$

where:

$$\theta = \arcsin(\varepsilon_y / \varepsilon) \quad \text{with} \quad \varepsilon \geq \varepsilon_y \tag{3}$$

Fig. 2 gives the stress distribution in the elastic-plastic part and the bending moment curvature diagram without other influences. The curvature is normalized by dividing it by the following quantity:

$$\kappa_I = \frac{t}{D_m^2} = \frac{t}{(D-t)^2} \tag{4}$$

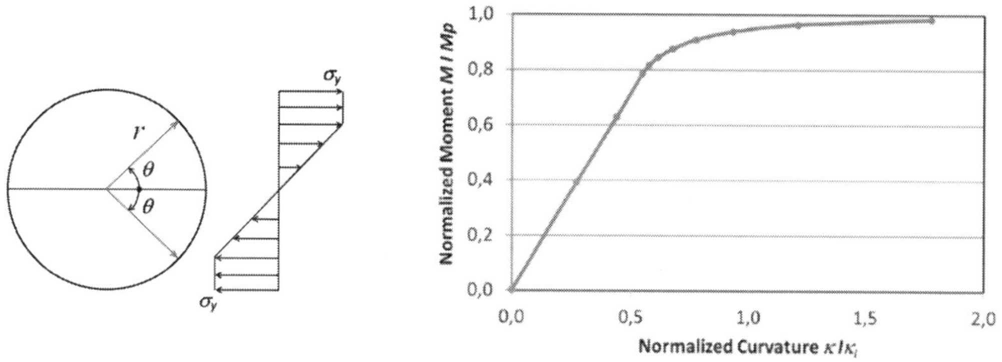

Fig. 2: Stress distribution in bending and bending moment curvature diagram

2.2 Ovalisation

2.2.1 Elastic behaviour

The ovalisation in the elastic part of the bending moment curvature diagram is [6]:

$$a = \frac{r^5}{\rho^2 t^2} = \kappa^2 \frac{r^5}{t^2} \tag{5}$$

This equation can be derived with the model in Fig. 3.

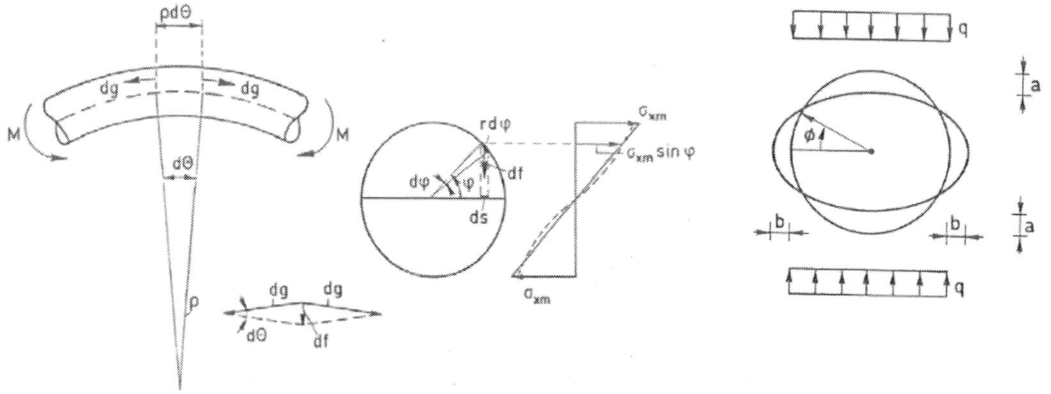

Fig. 3: Ovalisation forces due to linear elastic bending - comparable to a uniform load q [6]

The uniform ovalisation load q and the ovalisation a follow from:

$$q = \frac{df}{\rho \, d\theta \, ds} = \frac{r t E}{\rho^2} = \kappa^2 r t E \quad ; \quad a = \frac{q r^4}{12 E I_{wall}} = \frac{q r^4}{12 E \cdot \frac{1}{12} t^3} = \kappa^2 \frac{r^5}{t^2} \tag{6}$$

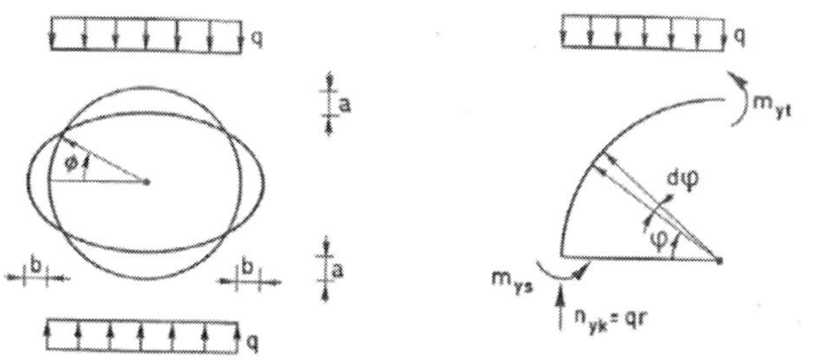

Fig. 4: Ovalisation forces give plate bending moments and plate normal forces

The ovalisation forces cause plate bending moments and plate normal forces in the tube wall (Fig. 4). The model for determining the ovalisation and the plate bending moments enables to take into account and combine these plate bending moments and normal forces with the effect of other loads perpendicular to the tube wall, e.g. soil loads, internal or external pressure and loads from infill sheeting.

2.2.2 Elastic-plastic behaviour

The calculation of the ovalisation in the elastic-plastic part of the bending moment curvature diagram is more complicated. For the full plastic cross section, the normality principle can be applied for the relation between curvature, ovalisation and plate bending moments [6].

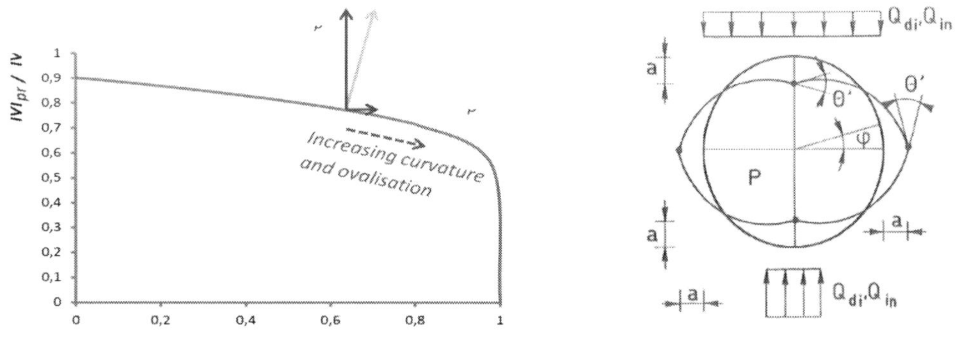

Fig. 5: Yield surface with normality principle and fully plastic cross section

In Fig. 5 the yield surface is given. The plastic moment capacity M_{pr} decreases with increasing values of the plate bending moments. In the right figure, the full plastic cross section with plastic hinges loaded by ovalisation forces and soil loads as in buried pipelines [6] is given. Because the slope of the yield surface is not a constant, a step wise procedure is followed.

$$\delta a = -\frac{r^3}{t}\left(2\psi \cdot \delta\mathrm{K}\right) \quad \text{with} \quad \psi = \frac{dM / M_p}{dm_y / m_y} \tag{7}$$

In this equation, ψ is the slope of the yield surface.

For practical applications an approximate equation is proposed. It is based on the "elastic" model for ovalisation as in Eq. 6. In the elastic part of the bending moment curvature diagram, the ovalisation forces increase due to increasing curvature and increasing axial bending stresses. In the elastic-plastic part, the increase of the ovalisation forces due to bending stresses in axial direction becomes smaller and therefore the ovalisation forces increase slower. For the elastic-plastic part the following equation can be applied for the ovalisation a_p:

$$a_p = \kappa^{1,5} \cdot \kappa_e^{0,5} \cdot \frac{r^5}{t^2} \quad \text{with} \quad \kappa_e = \frac{\varepsilon_y}{r} \tag{8}$$

The area of high plate moments at the sides of the cross section, mainly remains elastic. The plate bending stiffness at the top is governed by the normality principle.

2.3 The effect of ovalisation on the bending moment capacity

Ovalisation gives a reduction in the section modulus and it causes plate bending moments that reduce the tube bending moment capacity. The effect is expressed in next equations:

$$M_{\max,\text{oval}} = g \cdot h \cdot M_p \quad \text{with} \quad h = 1 - \frac{2}{3}\frac{a}{r} \tag{9}$$

where g expresses the effect of the plate moments and plate normal forces.

The stresses due to the plate bending moments and plate normal forces are given in Fig. 6. They are optimized to obtain the highest tube bending moment (optimal distribution of stresses according to the theory of plasticity). Plate bending moments in circumferential direction m_y give also plate bending moments in longitudinal direction called m_x (Poisson's ratio ν is assumed as 0,3). Plate normal forces in circumferential direction are n_y and in longitudinal direction n_x. The plate normal forces n_x give the tube bending moment M.

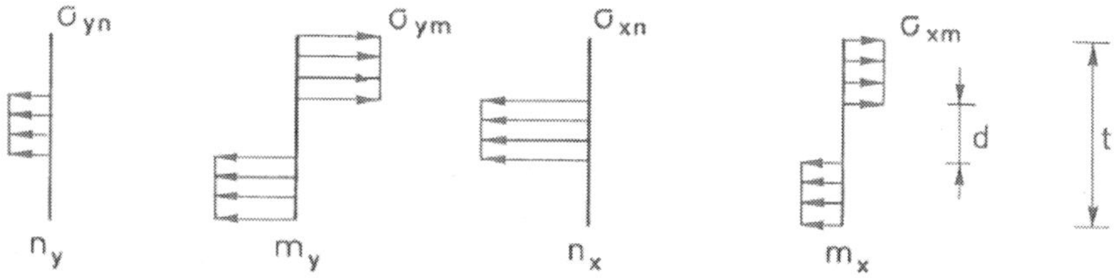

Fig. 6: Plate bending moments and plate normal forces with optimal stress distribution in the tube wall

Once the plate bending moments and plate normal forces in circumferential direction are known, the stresses σ_{xn} and plate normal force n_x can be calculated using the Von Mises yield criterion (Fig. 7). The factor g for tube bending moment capacity can be calculated with:

$$g = \frac{c_1}{6} + \frac{c_2}{3} \tag{10}$$

where

$$c_1 = \sqrt{4 - 3\left(\frac{n_y}{n_p}\right)^2 - 2\sqrt{3}\frac{m_y}{m_p}} \quad \text{and} \quad c_2 = \sqrt{4 - 3\left(\frac{n_y}{n_p}\right)^2} \tag{11}$$

with

$$n_y \le n_p = t\, f_y \quad \text{and} \quad m_y \le m_p = 0,25\, t^2 f_y \tag{12}$$

The equations for n_y and m_y depend on the type of loading and can be found in [6][3][7]. For pure bending, n_y and m_y are:

$$n_y = 0,20 \frac{M_{\max}\,\kappa}{r} \quad \text{and} \quad m_y = 0,071 \cdot M_{\max} \cdot \kappa \tag{13}$$

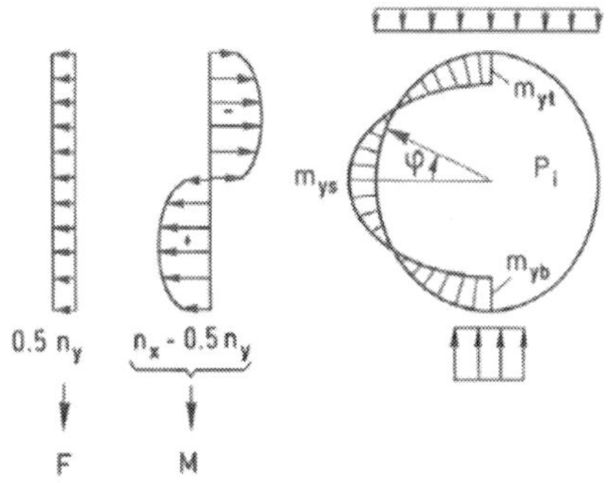

Fig. 7: Plate normal forces in axial direction that give the bending moment in the tube - in the figure also plate normal forces as a result of a normal force in the tube are given

Safe estimates for n_y and m_y can be found by assuming $M_{max} = M_p$ and $\kappa = t/D^2$. The effect of n_y is very small for tubes without internal or external pressure. The effect of ovalisation (reduction of the section modulus) and plate bending moments on the bending moment curvature diagram is depicted in Fig. 8.

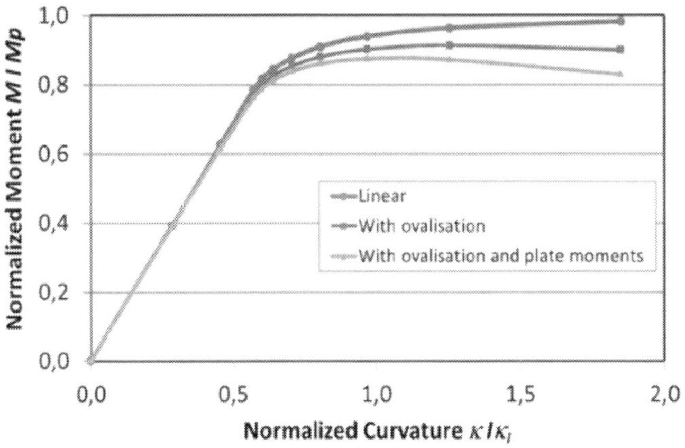

Fig. 8: Effect of ovalisation and plate bending moments on the bending moment curvature diagram

2.4 The effect of residual stresses on the bending moment curvature diagram

Depending on the forming process of the tubes, residual stresses will be present in the tube walls. These residual stresses result in yielding of the tube wall almost from the start of bending. From a bending moment of about $0{,}5M_e$ the effect can be clearly visible. The effect of residual stresses can be taken into account using a modified expression for the curvature.

$$\kappa_{res} = \alpha_{res} \frac{\varepsilon}{r} = \alpha_{res} \frac{\varepsilon_y}{r \cdot \sin\theta} \tag{14}$$

with

$$\alpha_{res} = 1{,}0 \qquad \text{for} \quad M \le 0{,}5M_e$$

$$\alpha_{res} = 1{,}0 + \left(\frac{M}{M_p} - \frac{0{,}5M_e}{M_p} \right)^2 \qquad \text{for} \quad M > 0{,}5M_e \tag{15}$$

The resulting bending moment curvature diagram is given in Fig. 9. It is noted that where the ovalisation is limited, e.g. near endplates, the bending moment capacity will be larger. However, this advantage is only relevant for (very) short tubes and not relevant for tubes in structural applications. Another issue to be taken into account near endplates and connections are the distortions due to welding and uneven introduction of stresses due to variations in stiffness of the supporting structure. In tests it often happens that local buckling occurs near endplates.

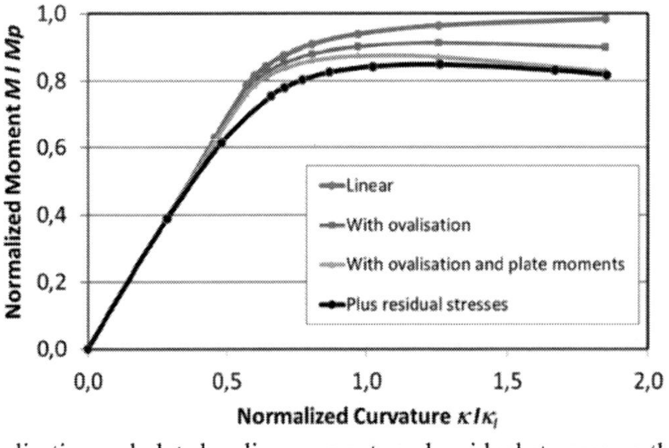

Fig. 9: Effect of ovalisation and plate bending moments and residual stresses on the bending moment curvature diagram

3. Critical curvature for local buckling

3.1 General equations

Starting from the critical strain according to [6][4][3][10], equations have been developed to take into account the various influences that have an effect on the strain at which local buckling occurs. The starting equations are:

$$\varepsilon_{cr} = 0,25\frac{t}{r_o} - 0,0025 \qquad \text{for } \frac{r_o}{t} \leq 60$$
$$\varepsilon_{cr} = 0,10\frac{t}{r_o} \qquad \text{for } \frac{r_o}{t} > 60 \tag{16}$$

where ε_{cr} is the critical compressive strain and r_o is the local radius in the compressed part of the cross section as is indicted in Fig. 10.

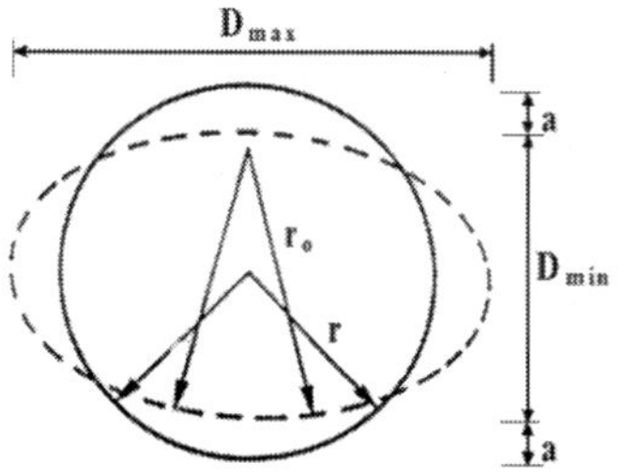

$$r_o = \frac{r}{1-\frac{3a}{r}} \tag{17}$$

$$\kappa_{cr} = \frac{\varepsilon_{cr}}{r_o} \tag{18}$$

Fig. 10: Definition of ovalisation, radius r_o and critical curvature in an ovalised cross section

For pure bending the effect of ovalisation can be neglected because the validation of these equations is done on pure bending tests where ovalisation due to pure bending was included. Therefore, in this case: $r_o = r$. The next equations take into account the other influences.

$$\varepsilon^*_{cr} = \varepsilon_{cr} \cdot \alpha_{geo} \cdot \alpha_{sh} \cdot \alpha_{sand} \quad ; \quad \kappa^*_{cr} = \frac{\varepsilon^*_{cr}}{r_o} \tag{19}$$

where

ε_{cr} is the critical compressive strain in pure bending

α_{geo} is the smallest of the following geometrical imperfections effects:

- α_{un} is the effect of tube surface undulations
- $\alpha_{high-low}$ is the effect of misalignment (high-low) at welds
- α_{dple} is the effect of dimples or dents
- $\alpha_{local-load}$ is the effect of local deformations due to local loads, e.g. waling beams

α_{sh} is the effect of strain hardening

α_{sand} is the effect of sand fill

Below a summary is given of the set of equations to determine these factors.

3.1.1 Surface undulations

For surface undulations the reduction factor for the critical curvature is:

$$\alpha_{un} = a_{un119} - \frac{D/t-119}{52}(a_{un67} - a_{un119}) \tag{20}$$

with $$\alpha_{un67} = 1,37 - 1,2\left(\frac{\delta_{un}}{2t}\right)^{0,3} \quad \text{and} \quad \alpha_{un119} = 1,15 - 0,5\left(\frac{\delta_{un}}{2t}\right)^{0,3} \tag{21}$$

In these equations is δ_{un} the depth of the undulation (difference between valley and adjacent tops). For undulations δ_{un} equal to 0,04t, the factor α_{un} =1,0. This is considered normal quality fabrication. For larger surface undulations, the critical curvature becomes smaller.

δ_{un}/t	0,02	0,04	0,08	0,12	0,16	0,20	0,30
α_{un} (D/t = 67)	1,07	1,00	0,91	0,86	0,81	0,77	0,69
α_{un} (D/t = 93)	1,05	1,00	0,94	0,89	0,86	0,83	0,78
α_{un} (D/t = 119)	1,02	1,00	0,96	0,94	0,92	0,90	0,87

The above equations are based on parameter studies in the Combitube project. Canadian research reported by the Center for Reliable Energy Systems (CRES) [12] has resulted in:

$$\alpha_{un-CRES} = 1,84 - 1,6\left(\frac{\delta_{un}}{t}\right)^{0,2} \tag{22}$$

This factor gives about the same reduction for D/t ratios till about 40. It does not depend on the D/t ratio. For high D/t ratios the CRES reduction factor is much lower than the Combitube result. The field of application in CRES is for D/t between 22 and 104.

3.1.2 Misalignment at girth welds (high-low)

For misalignment at girth welds the following reduction factor can be applied:

$$\text{the smallest of } \alpha_{high-low} = 2,0 - 1,6\left(\frac{\delta_{hl}}{3t}\right)^{0,2} \quad \text{and} \quad \alpha_{hi-lo} = 1,0 \tag{23}$$

Where δ_{hl} is the misalignment at the girth weld as indicated in next table.

δ_{hl}/t	0,096	0,15	0,20	0,30	0,40	0,50
$\alpha_{high-low}$	1,0	0,905	0,840	0,742	0,668	0,607

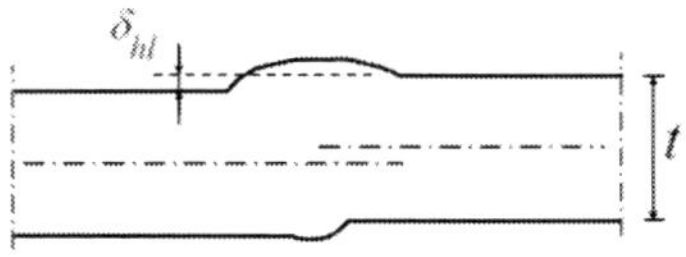

The negative effect of misalignments till δ_{hl} =0,096t is assumed to be compensated by the higher yield strength at the weld (overmatched weld metal) and the extra thickness of the tube wall at the weld (the weld cap). The proposed design rule is based on research in [11] and [12]. It was found that for weld high-low misalignments up to 0,50t, the effect of the misalignment on the compressive strain capacity is equivalent to a 0,15t geometry imperfection (surface undulation) in a plain pipe with zero pressure. Parameter studies in the Combitube project have shown that without the extra thickness of the weld cap and without the overmatched weld strength the effect of misalignments is much larger, as could be expected.

3.1.3 Dimples

For dimples the same reduction factor may be applied as for surface undulations.

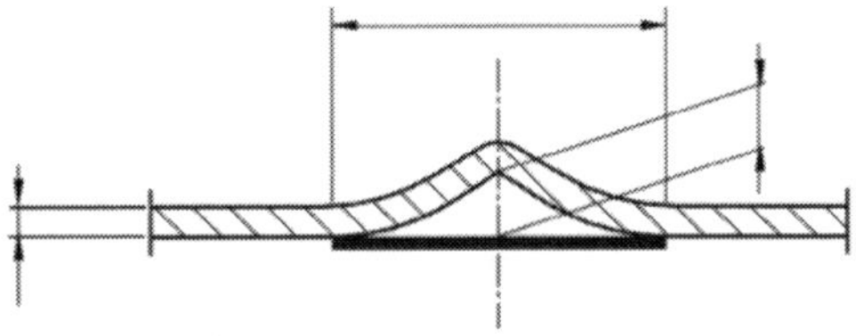

$$\alpha_{dple} = 1,84 - 1,6\left(\frac{\delta_{dple}}{t}\right)^{0,2} \tag{24}$$

The measurement length is according to EN1993-1-6:

$$l_{dpl} = 4\sqrt{rt} \tag{25}$$

3.1.4 Effect of strain hardening

Strain hardening has a positive influence on the local buckling behaviour. The σ_y/σ_t ratio is a measure for the strain hardening.

$$\alpha_{sh} = 1,33 - \frac{0,04}{1 - \sigma_y / \sigma_t} \tag{26}$$

σ_y/σ_t ratio	0,82	0,84	0,86	0,88	0,90	0,92	0,94
α_{sh}	1,11	1,08	1,04	1,00	0,93	0,83	0,66
alternative: $\alpha_{sh\text{-}DNV}$	1,11	1,07	1,03	1,00	0,97	0,93	0,91

The factor α_{sh} is based on parameter studies in the Combitube project. The value of α_{sh} is normalised to 1,00 for σ_y/σ_t =0,85. In the DNV rules [13] the following equation is given:

$$\alpha_{strain-hard} = \left(\frac{\sigma_y}{\sigma_u}\right)^{-1,5} \text{, normalised to 0,85 it gives } \alpha_{sh-DNV} = 0,825\left(\frac{\sigma_y}{\sigma_u}\right)^{-1,5} \tag{27}$$

Another formulation is on basis of the strain hardening modulus E_{st}. According to the parameter studies in the Combitube project:

$$\alpha_{sh-combitube} = 1,8 \cdot h^{-0,14} \quad \text{with} \quad h = E / E_{st} \tag{28}$$

h	40	60	80	100	120	140
$\alpha_{sh\text{-}combitube}$	1,07	1,02	0,98	0,95	0,84	0,90

3.1.5 Effect of sand fill

Sand fill has a positive effect on the critical compressive strain, the bending moment capacity and the post buckling behaviour [14][15].

$$a_{\text{sand}} = a_{\text{empty}} \frac{k_{\text{steel}}}{k_{\text{steel}} + k_{\text{sand}}}$$

$$k_{\text{steel}} = \frac{12\,EI}{r^4}\left[\frac{\text{N}}{\text{m}^3}\right] \quad with \;\; EI = \frac{1}{12} E_{\text{steel}}\, t^3 \qquad k_{\text{sand}} = \frac{E_{\text{sand}}}{r}\left[\frac{\text{N}}{\text{m}^3}\right] \tag{29}$$

4. Comparison with test results

In Fig. 11 the calculated and measured bending moment curvature diagrams are compared for two of the 14 tested tubes. The column κ_{cr}/κ_I *excl.* in the table on next page gives the critical curvature without the effect of surface undulations and the effect of strain hardening. The column with κ_{cr}/κ_I *incl.* gives the critical curvature including these effects. The lines with K1, K2 and K3 give the measured curvatures along the test tubes [1]. Variations in the bending resistance along the tubes cause variations in the measured curvatures.

Local buckling occurs in the section with the largest curvature. The critical curvature also depends on the measuring length. Test T6D10 had a "coil-connection weld" where the plates from two coils were welded during the spiral-tube production. At that spot large undulations were measured, resulting in a low value of the calculated critical curvature [1] and [7].

Test	D (mm)	t (mm)	D/t	ε_{cr} [19]	κ_{cr}/κ_I ***excl.***	f_y (MPa)	f_u (MPa)	f_y/f_u	α_{sh} [30]	δ_{un} (mm)	α_{un} [24]	κ_{cr}/κ_I ***incl.***
T4D4	1065	9,16	116	0,00184	**0,424**	441	539	0,818	1,11	0,6	0,968	**0,455**
T6D10	1066	16,3	65	0,00526	**0,678**	527	609	0,865	1,03	3,0	0,777	**0,544**

5. Concluding remark

An analytical model is presented that is capable of taking into account many more parameters that have an influence on the bending moment capacity and the deformation capacity than the present equations in the European standards. Addressing these influences enables to much better predict the structural behaviour and lead to less scatter in statistical evaluations of test data and so enhance economic and safe designs with better balanced safety level.

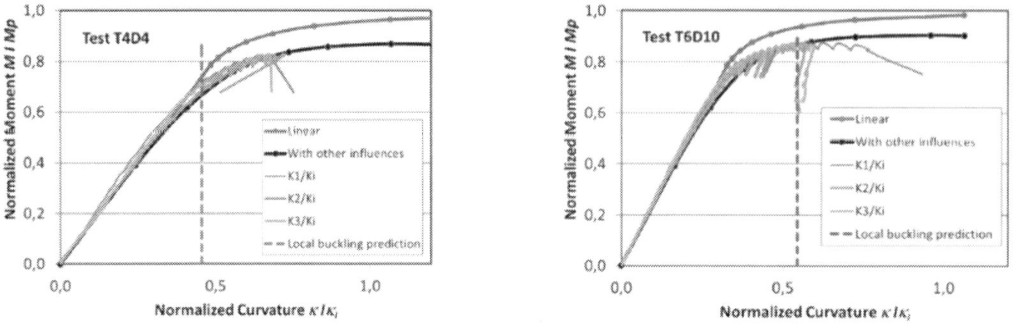

Fig. 11: Test T4D4 and test T6D10 compared with the calculated behaviour, showing good agreement between the tests and the calculations - also showing safe estimates for the local buckling curvature

Acknowledgment

Funding for this work has been provided by the Research Fund for Coal and Steel (RFCS) of the European Commission, project COMBITUBE: "Bending Resistance of Steel Tubes in CombiWalls", Grant Agreement No. RFSR-CT-2011-00034.

References

[1] Van Es SHJ, Gresnigt AM, Vasilikis D, Karamanos SA. "Experimental and Numerical Investigation of the Bending Capacity of Spiral-Welded Steel Tubes", *The International Colloquium on Stability and Ductility of Steel Structures,* Timisoara, Romania, 2016.

[2] EN 1993-1-6 "Eurocode 3: Design of steel structures, Part 1-6: General, Strength and Stability of Shell Structures." *CEN,* 2007.

[3] EN 1993-4-3 "Eurocode 3: Design of Steel Structures, Part 3-4: Pipelines." *CEN,* 2007.

[4] ENV 1993-5 "Eurocode 3: Design of Steel Structures, Part 5: Piling." *CEN,* 1998.

[5] EN 1993-5 "Eurocode 3: Design of Steel Structures, Part 5: Piling." *CEN,* 2007.

[6] Gresnigt AM. "Plastic design of buried steel pipes in settlement areas", *HERON 1986-4.*

[7] European Commission. "Bending resistance of steel tubes in CombiWalls – Final Report", *Research Programme of the Research Fund for Coal and Steel,* Brussels, 2016.

[8] Van Es SHJ, Gresnigt AM, Vasilikis D, Karamanos SA. "Ultimate Bending Capacity of Spiral-Welded Steel Tubes – Part I: Experiments", *Thin Walled Structures,* Vol 102, 286-304, May 2016.

[9] Vasilikis D, Karamanos SA, Van Es SHJ, Gresnigt AM. "Ultimate Bending Capacity of Spiral-Welded Steel Tubes – Part II: Predictions", *Thin Walled Structures,* Vol 102, 305-319, May 2016.

[10] CSA Z662-11. "Oil and Gas Pipeline Systems". *Canadian Standards Association,* 2014.

[11] Nader Yoosef-Ghodsi, Istemi Ozkan, Quishi Chen. "Comparison of compressive strain limit equations*". Paper IPC2014-33182, International Pipeline Conf.,* Calgary, 2014.

[12] Ming Liu, Yong-Yi Wang, Fan Zhang, Kunal Kotian. "Realistic Strain Capacity Models for Pipeline Construction and Maintenance". *Transportation Pipeline and Hazardous Materials Safety Administration Office.* CRES Dublin OH 43017, USA, 2013.

[13] DNV-OS-F101. "Submarine Pipeline Systems". *Det Norske Veritas,* Norway, 2012.

[14] SBRCURnet (2013). "Handbook Quay Walls 2nd Edition." *Centre for Civil Engineering Research and Codes,* Delft, 2013.

[15] Peters DJ, Broos EJ, Gresnigt AM, Van Es SHJ. "Local Buckling Resistance of Sand-filled Spirally Welded Tubes". *Proceedings ISOPE Conference,* Hawaii, 2015.

EXPERIMENTAL INVESTIGATION OF COLD-FORMED FERRITIC STAINLESS STEEL TUBULAR X-JOINTS

Yuner Huang[a] and Ben Young[b]

[a] *School of Engineering, The University of Edinburgh, Edinburgh, United Kingdom*
[b] *Department of Civil Engineering, The University of Hong Kong, Pokfulam Road, Hong Kong, China*

Abstract: An experimental investigation on cold-formed ferritic stainless steel X-joints of rectangular hollow sections is performed. A total of 24 tests was conducted with angle between chord and brace members at either 45° or 90°. Axial compression force was applied to the brace members of the X-joint specimens between fixed ends. Chord face indentation and lateral deformation of chord side wall were measured during testing. Test results were compared with the design strengths predicted by existing design rules. Reliability analysis was performed. It is shown that the existing design rules are generally not suitable to predict the strength of ferritic stainless steel X-joints in this test program.

1. Introduction

Cold-formed ferritic stainless steel has various advantages as construction material, including shining appearance, high strength, ease of construction and maintenance as well as a relatively low price compare with other types of stainless steel materials. It has a relatively lower price and higher strength compare with austenitic stainless steel material, which offer a great potential for ferritic stainless steel to be used in construction. Tubular X-joint connection is widely used in frameworks, truss girders, façades, offshore platforms, claddings, roof in long span structures and so on. However, research on cold-formed ferritic stainless steel tubular connections is not found in literature. Hence, it is necessary to investigate the structural behaviour of cold-formed ferritic stainless steel X-joints, and to facilitate application of ferritic stainless steel X-joints in construction.

The existing design rules for steel connections are developed mainly based on experimental and numerical results of carbon steel sections. Stainless steel material demonstrates different stress-strain behaviour than carbon steel material. Therefore, the existing design rules may not be suitable for cold-formed ferritic stainless steel X-joints. Experimental and numerical investigations on cold-formed austenitic and duplex stainless steel X-joints have been performed by Feng and Young [1 - 4]. It is shown that the CIDECT [5] is quite unconservative for specimens subjected to chord face failure, and quite conservative for specimens subjected to chord side wall failure. X- and K-joint tests have been conducted for square hollow sections [6] circular hollow sections [7]. Zhao [8] assessed deformation limit and ultimate strength of cold-formed carbon steel rectangular hollow section (RHS) T-joints. A failure criteria based on deformation limit was proposed for cold-formed carbon steel RHS T-joints, and the proposed deformation limit was modified from Lu et al. [9]. Despite of previous research on carbon steel and stainless steel X-joints, it should be noted that there is no research on ferritic stainless steel X-joints. Moreover, investigation on X-joints with angle between chord and brace members of 45° is also limited. This paper presents an experimental investigation of cold-formed ferritic stainless steel X-joints, with angle between chord and brace members at either 90° or 45°. The test strengths were compared with design strengths predicted by European Code (EC3) [10]. It should be noted that the design equations of joints in CIDECT [5] is identical with those in EC3 Part 1.8 [10] for carbon steel structures, and the Australian/New Zealand Standard [11] for stainless steel structures adopted the design equations from CIDECT. Reliability analysis was conducted to assess the reliability of the existing design rules for cold-formed ferritic stainless steel X-joints.

2. Experimental Investigation

A total of 24 X-joint tests was conducted in this study, among which 12 specimens had the angle between chords and brace members of 90°, and 12 test specimens had the angle between chords and brace members of 45°. Axial compression force was applied to the brace members of the X-joints. The tests specimens were carefully designed to cover a wide range of geometrical influence factors, including ratio (β) of brace width to chord width (b_1/b_0), ratio (τ) of brace thickness to chord thickness (t_1/t_0), and ratio (2γ) of chord width to chord thickness (b_0/t_0). The brace and chord of the joints were rectangular hollow sections with nominal dimensions ($D\times B\times t$) of 60×40×4, 80×60×4, 100×40×2 and 120×80×3, where D, B, t are the depth, breadth and thickness of the sections, respectively. The end of chord member should be 2.5 times the overall depth of chord member, in order to avoid any failure at the end of the chord. Thus, the nominal chord length (L_0) was taken as $5h_0+h_1$ and $6h_0+\sqrt{2}h_1$ for joints with angle between chord and brace members of 90° and 45° respectively, as shown in Fig. 1, where h_0 = overall depth of chord member and h_1 = overall depth of brace member. The nominal brace length (L_1) was taken as the larger value of $2h_1$ or $2b_1$ to avoid overall buckling. The cross-section dimensions of the specimens as well as width of weld (w_1) and height of weld (w_2) were measured and reported in Table 1. The specimens are labelled so that their joint configuration and nominal cross-section dimensions of chord and brace members can be identified. The letter "C" indicates chord member, which is followed by nominal cross-sectional dimension of chord member ($h_0\times b_0\times t_0$). The letter "B" represents brace member, followed by the nominal cross-sectional dimension of brace members ($h_1\times b_1\times t_1$). The number after letter "A" indicates the angle between chord and brace member in degree. If a test is repeated, then a superscript "#" indicates the repeated test. The chord and brace members were cold-rolled from ferritic stainless steel type EN 1.4003.

A hydraulic testing machine was used to apply axial compressive load to the X-joint specimens. A special ball bearing was used at the top end of the brace member to eliminate any possible gap between the specimen and support. For specimens with angle between chord and brace members of 90°, four vertical linear variable displacement transducers (LVDTs) with extension arms at the tip were used to measure chord face indentation (u). The extension arms were used to ensure that the chord face indentation was measured at 5 mm away from brace members. Another four LVDTs were positioned at the two ends of chord member near the corners, in order to measure the overall vertical displacement of chord member, and to ensure the chord member remains horizontal throughout testing. Two horizontal LVDTs were located at the middle of chord side walls of the member to measure the chord side wall deformation. As the maximum deformation of the chord side wall may varies during testing, especially when the ratio (β) of brace width to chord width (b_1/b_0) is less than 1.0, plastic plates were connected to the tips of the two horizontal LVDTs to measure the maximum chord side wall deformation (v) during testing (Fig. 2). For specimens with angle between chord and brace members of 45°, two L-shaped platforms were attached at mid-line of the two brace members, and moves in the same magnitude of brace members along their mid-line. Two vertical LVDTs were positioned at the platforms to measure chord face indentation (u). Similarly, two horizontal LVDTs with plastic plates connected to the tips were used to measure chord side wall deformation (v) during testing, as shown in Fig. 2.

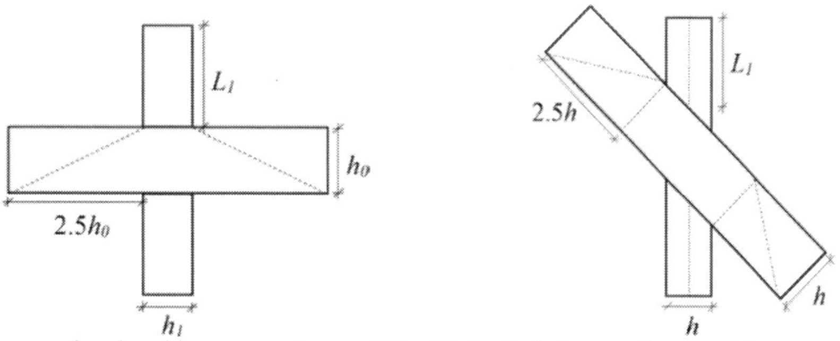

a) Angle between chord and brace members = 90° b) Angle between chord and brace members = 45°
Fig. 1: Dimension of X-joint specimens

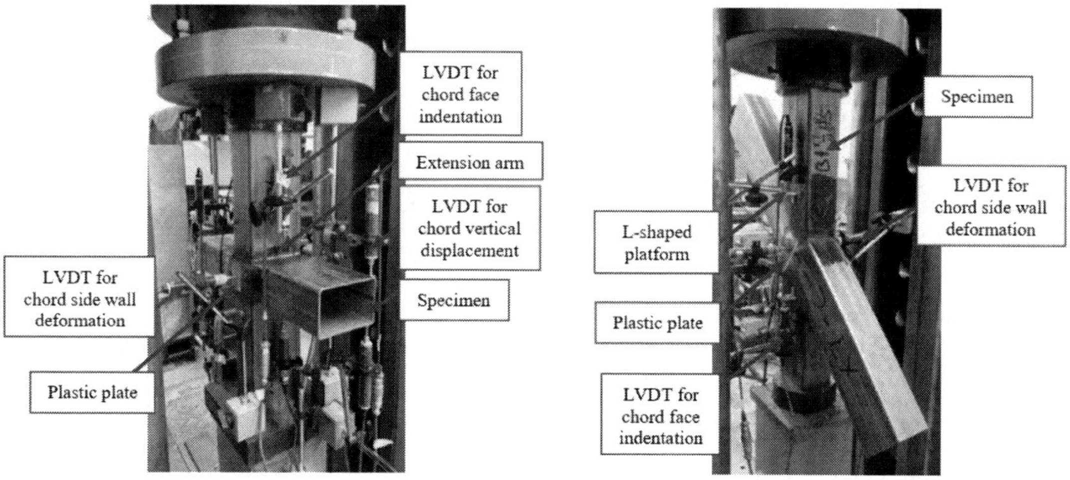

a) Angle between chord and brace members = 90° b) Angle between chord and brace members = 45°
Fig. 2: Setup of X-joint tests

The X-joint test results are summarized in Table 2. According to European Code [10] and CIDECT [5], chord side wall failure (S) and chord face failure (F) occur when the brace to chord width ratio equals to 1.00 and less than or equal to 0.85, respectively. However, it is observed from the test results that chord face indentation (u) and chord side wall deformation

(v) at ultimate load can be in a comparable magnitude in these two categories ($\beta = 1.00$ and $\beta \leq 0.85$). Therefore, failure mode of combination of chord side wall and chord face failure (S+F) should also be defined. In this study, the chord side wall failure mode (S) is defined when the chord side wall deformation (v) is greater than 2.5 times of the chord face indentation (u) by each brace member at ultimate load, while the chord face failure mode (F) is defined when the chord face indentation by each brace member is greater than 2.5 times of the chord side wall deformation at ultimate load. The other situations are classified as combination of chord side wall and chord face failure (S+F). Four failure modes were observed from the test specimens, namely the chord side wall failure (S), chord face failure (F), combination of chord side wall and chord face failure (S+F), and combination of chord side wall and chord face failure as well as local buckling of brace member (S+F+L).The test strengths are defined using four different criteria: (1) ultimate load recorded during tests (N_u), (2) deformation limit proposed by Zhao [8] (N_f), (3) chord face indentation reaches 1% of the chord width ($N_{s,1\%}$), and (4) chord face indentation reaches 3% of the chord width ($N_{s,3\%}$). The test strengths of each specimen defined by these four criteria are summarized in Table 2. Specimens C40×100×2B80×60×4A45, C80×120×3B80×60×4A45 and C80×120×3B60×40×4A45 with angle between chord and brace members of 45° cannot reach peak load. Therefore, the ultimate load (N_u) cannot be obtained. The maximum loads recorded during testing for these three specimens are equal to 60.3, 95.4 and 72.4 kN, respectively. It should be noted that the deformation limit proposed by Zhao [8] was meant for T-joints, instead of X-joints. Suitability of the deformation limit for X-joints is assessed in this study.

Table 1: Measured specimen dimensions

Specimen	Chord				Brace				Weld	
	h_0	b_0	t_0	L_0	h_1	b_1	t_1	L_1	w_1	w_2
C80×60×4B60×40×4A90	80.3	60.0	3.74	463.9	59.7	40.0	3.81	119.1	2.81	3.45
C80×60×4B80×60×4A90	80.3	59.9	3.73	480.3	80.0	60.0	3.72	158.9	3.25	2.98
C60×80×4B60×80×4A90	60.0	80.3	3.75	363.9	59.9	79.9	3.72	159.7	3.34	3.35
C100×40×2B100×40×2A90	99.9	40.2	1.95	605.4	99.6	40.4	1.94	199.7	3.17	3.12
C40×100×2B40×100×2A90	40.2	99.7	1.94	243.5	40.3	99.6	1.94	199.7	2.93	3.02
C40×100×2B80×60×4A90	40.2	99.6	1.94	284.8	80.0	60.0	3.73	158.2	3.12	2.98
C40×100×2B80×60×4A90#	40.2	99.7	1.94	284.4	79.9	59.9	3.73	158.7	2.83	2.79
C120×80×3B120×80×3A90	120.0	79.9	2.82	724.9	120.0	79.9	2.78	239.4	3.50	3.41
C80×120×3B80×60×4A90	79.9	120.2	2.82	478.8	80.0	60.0	3.74	159.7	3.46	3.32
C80×120×3B60×40×4A90	79.9	120.0	2.84	464.8	59.7	39.9	3.81	119.7	2.95	2.89
C80×120×3B60×40×4A90#	79.8	120.1	2.82	464.7	59.7	39.9	3.82	119.8	3.02	3.11
C80×120×3B80×120×3A90	79.8	120.1	2.82	484.7	79.6	119.5	2.82	239.4	2.87	2.69
C80×60×4B60×40×4A45	80.2	60.0	3.74	575.2	60.0	40.1	3.82	119.4	2.01	3.54
C80×60×4B80×60×4A45	80.1	59.9	3.72	599.7	80.1	60.0	3.73	159.6	1.98	2.98
C80×60×4B80×60×4A45#	80.2	60.0	3.72	599.6	80.2	60.0	3.74	159.2	2.21	3.13
C60×80×4B60×80×4A45	59.9	80.1	3.74	453.9	60.0	80.1	3.74	159.5	2.16	3.24
C100×40×2B100×40×2A45	99.9	40.2	1.95	754.2	99.7	40.3	1.95	199.4	2.23	3.05
C100×40×2B100×40×2A45#	99.7	40.3	1.95	754.5	99.8	40.3	1.95	199.4	2.45	3.25
C40×100×2B40×100×2A45	40.2	99.7	1.94	305.0	40.2	99.9	1.95	200.2	2.31	3.13
C40×100×2B80×60×4A45	40.2	99.8	1.95	354.1	80.1	59.9	3.74	159.5	2.10	3.09
C120×80×3B120×80×3A45	119.9	79.9	2.83	899.2	120.0	79.9	2.82	239.4	2.08	3.26
C80×120×3B80×60×4A45	79.9	120.1	2.82	599.9	80.2	59.9	3.75	159.7	2.19	3.13
C80×120×3B60×40×4A45	79.9	120.0	2.83	574.8	59.9	40.0	3.81	119.7	2.10	3.22
C80×120×3B80×120×3A45	79.8	120.0	2.82	604.8	79.9	119.9	2.81	238.3	2.20	2.89

(Unit: mm)

The 1st criterion considers the ultimate load recorded during the tests (N_u), which is controlled by the strength of the joint specimens. The 2nd criterion of deformation limit proposed by Zhao [8], which is modified from Lu et al. [9], combines strength and deformation as controlling factors in determining the test strength. The 3rd and 4th criteria to determine the test strengths are governed by the serviceability limit state. The CIDECT [5] recommends that carbon steel joint deformation (u) under service loads should be limited to 1% and 3% of the chord width (b_0) as $N_{s,1\%}$ and $N_{s,3\%}$, respectively. It should be noted that the 2nd, 3rd and 4th criteria are developed for carbon steel joints only instead of stainless steel joints. However, there is no design rule for stainless steel X-joints. Therefore, suitability of the existing failure criteria for ferritic stainless steel X-joints is assessed.

3. Comparison with Design Strengths

The nominal strengths (N_{EC3}) of the X-joint specimens are compared with the test strengths, as shown in Table 2. Reliability analysis was performed for the current design rules in EC3 Part 1.8 [10]. In this study, the reliability analysis as detailed in the Commentary of the ASCE Specifications [14] was used. The design rules are considered to be reliable if the reliability index is greater than or equal to target reliability index (β) of 2.5. Resistance factor (ϕ) of 1.00 and load combination of 1.35DL+1.5LL in EC3 were used in the reliability analysis, where DL is the dead load and LL is the live load. The Eq. 6.2-2 in the ASCE Specification [14] was adopted in calculating the reliability index. The statistical parameters M_m = 1.10, F_m = 1.00, V_m = 0.10 and V_F = 0.05, which are the mean values and coefficients of variation for material properties and fabrication factors were adopted in this study. The mean value (P_m) and coefficient of variation (V_P) of tested-to-predicted load ratios are shown in Table 2. In calculating the reliability index, Eq. F1.1-3 in the North American cold-formed steel Specification AISI S100 [15] was used to calculate the correction factor, in order to consider the influence of small number of data.

When the ultimate load (N_u) is used as the test strength, the mean value of the test-to-design strength ratio for all specimens equals to 2.88 with the corresponding coefficient of variation (COV) of 0.527, and the reliability index is 2.38. It is shown that the design rules provide the most conservative predictions when the 1st criterion was used to determine the test strengths. The predictions for specimens with $\beta = 1$, which is classified as chord side wall failure according to EC3, is much more conservative and scatter than the other specimens with $\beta \leq 0.85$. The mean values of test-to-design strength for the specimens with $\beta \leq 0.85$ and $\beta = 1$ equal to 1.95 and 3.35 with COV of 0.247 and 0.495, respectively, in Table 3. The 2nd criterion (N_f) is based on both strength and deformation. For all specimens, the mean value of test-to-design strength equals to 2.28 with a very large COV of 0.787. The reliability index is 1.41, due to the large value of COV. The mean values of test-to-design strength for the specimens with $\beta \leq 0.85$ and $\beta = 1$ equal to 0.80 and 3.35 with COV of 0.383 and 0.495, respectively. Therefore, EC3 cannot provide good prediction for the failure criterion proposed by Zhao [8]. When the test strengths are defined by serviceability limit state using the 3rd and 4th criteria, it is shown that EC3 generally provides conservative and very scatter predictions. The mean values of test-to-design strength for all specimens are 1.23 and 1.41 with COV of 0.763 and 0.770 for chord face indentation reaching 1% and 3% of b_0, respectively, as shown in Table 2. When the test strengths are determined by the 3rd criterion, the mean values of test-to-design strength for the specimens with $\beta \leq 0.85$ and $\beta = 1$ equal to 0.35 and 1.85 with COV of 0.384 and 0.395, respectively. When the test strengths are determined by the 4th criterion, the mean values of test-to-design strength for the specimens with $\beta \leq 0.85$ and $\beta = 1$ equal to 0.67 and 1.94 with COV of 0.247 and 0.600, respectively. It is shown that the design rules in EC3 [10] are considered to be not reliable for ferritic stainless steel X-joints, when the test strengths are defined by the four criteria considered in this study.

It is clearly shown that the $1\%b_0$ and $3\%b_0$ deformation limits are not appropriate for ferritic stainless steel X-joints. The deformation limit of $3\%b_o$ was suggested by Lu et al. [9], base on the observation that the chord face indentation corresponding to peak load of carbon steel joints varies between 2.5 – $4\%b_o$. The chord face indentation with respect to chord width (u/b_o) at peak load (N_u) of ferritic stainless steel X-joints in this study was plotted in Fig. 3(b). It is shown that u/b_o at peak load ranges from 0.51-3.15% for specimens with $\beta = 1$, and 5.1-29.5% for specimens with $\beta \leq 0.85$. Therefore, the $3\%b_0$ deformation limit is too small for ferritic stainless steel X-joints with $\beta \leq 0.85$. It is also observed from Fig. 3(b) that the u/b_o ratio decreases as β increases for specimens with $\beta \leq 0.85$. Therefore, the deformation limit for ferritic stainless steel X-joints should varies with β value. It is suggested that further research is needed to propose a suitable deformation limit and design rules.

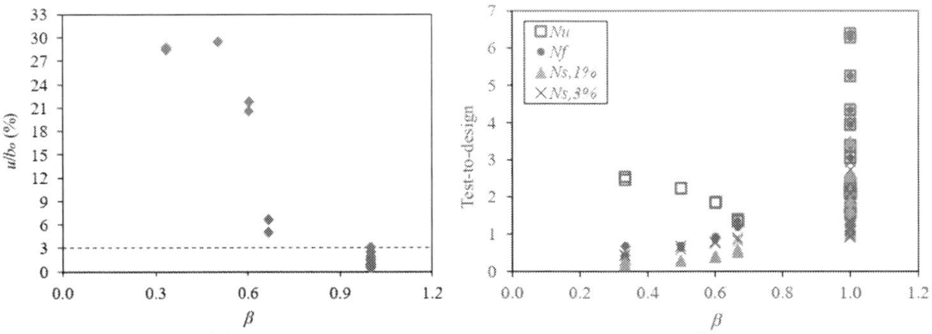

Fig. 3: (a) Relationship between test-to-design ratio with β; (b) Relationship between u/b_o with β

4. Conclusions

Experimental investigation on cold-formed ferritic stainless steel tubular X-joints has been performed. The test specimens cover a wide range of geometrical influence factors. The test specimens had angle between chord members and brace members of 45° and 90°. The failure modes and test strengths defined by four different failure criteria are reported. The test strengths corresponding to different failure criteria were compared with the design strengths calculated by EC3 Part 1.8 [10]. It is shown that the design rules are generally not suitable for cold-formed ferritic stainless steel X-joints. The test strengths were compared with design strengths for specimens with brace width to chord width ratios smaller than or equal to 0.85 ($\beta \leq 0.85$) and equal to unity ($\beta = 1$) separately. This is due to the fact that the current EC3 considers specimens of $\beta \leq 0.85$ (chord face failure) and $\beta = 1$ (chord side wall failure) separately. It is shown that the design rules in EC3 provides quite conservative and scatter predictions for specimens with $\beta = 1$, especially when failure load is defined by peak load or deformation limit proposed by Zhao [8]. For specimens with $\beta \leq 0.85$, the EC3 provides quite conservative predictions when failure load is defined by peak load, but unconservative predictions when the deformation limit proposed by Zhao [8] as well as deformation limit reaching 1% and 3% of the chord width are used. It is clearly shown that the current deformation limit proposed for carbon steel joints are not suitable for ferritic stainless steel X-joints. Further research is needed to propose a suitable deformation limit and design rules for cold-formed ferritic stainless steel tubular X-joints.

Table 2: Comparison of test strengths with design strengths

Specimen	β	Failure mode	N_u (kN)	N_f (kN)	$N_{s,1\%}$ (kN)	$N_{s,3\%}$ (kN)	$\frac{N_u}{N_{EC3}}$	$\frac{N_f}{N_{EC3}}$	$\frac{N_{s,1\%}}{(N_{EC3}/1.5)}$	$\frac{N_{s,3\%}}{(N_{EC3}/1.5)}$
C80×60×4B60×40×4A90	0.67	S+F	119.9	116.4	78.0	116.4	1.38	1.34	0.60	0.90
C80×60×4B80×60×4A90	1.00	S+F	333.9	333.9	333.0	260.9	2.06	2.06	1.37	1.07
C60×80×4B60×80×4A90	1.00	S+F	300.9	300.9	299.5	244.3	1.70	1.70	1.13	0.92
C100×40×2B100×40×2A90	1.00	S	90.5	90.5	79.5	85.4	4.34	4.34	2.54	2.73
C40×100×2B40×100×2A90	1.00	S+F+L	105.1	105.1	99.6	61.0	2.39	2.39	1.51	0.93
C40×100×2B80×60×4A90	0.60	S+F	33.1	16.7	11.1	20.7	1.85	0.93	0.42	0.77
C40×100×2B80×60×4A90#	0.60	S+F	32.8	15.7	10.4	20.3	1.84	0.88	0.39	0.76
C120×80×3B120×80×3A90	1.00	S	200.9	200.9	200.3	161.0	3.94	3.94	2.62	2.11
C80×120×3B80×60×4A90	0.50	S+F	60.3	18.7	12.4	23.8	2.22	0.69	0.31	0.59
C80×120×3B60×40×4A90	0.33	F	51.8	8.3	5.5	12.9	2.46	0.39	0.17	0.41
C80×120×3B60×40×4A90#	0.33	F	52.6	8.4	5.6	14.1	2.54	0.41	0.18	0.45
C80×120×3B80×120×3A90	1.00	S+F	222.2	222.2	178.9	132.8	3.05	3.05	1.64	1.22
C80×60×4B60×40×4A45	0.67	S+F	195.3	171.6	114.4	181.8	1.34	1.18	0.52	0.83
C80×60×4B80×60×4A45	1.00	S+F	362.0	362.0	282.4	343.2	2.12	2.12	1.10	1.34
C80×60×4B80×60×4A45#	1.00	S+F	362.7	362.7	347.6	329.0	2.12	2.12	1.36	1.28
C60×80×4B60×80×4A45	1.00	S+F	312.7	312.7	308.6	289.8	1.55	1.55	1.02	0.96
C100×40×2B100×40×2A45	1.00	S	132.1	132.1	82.3	130.0	6.28	6.28	2.61	4.12
C100×40×2B100×40×2A45#	1.00	S	135.1	135.1	59.4	135.1	6.39	6.39	1.87	4.26
C40×100×2B40×100×2A45	1.00	S+F	106.3	106.3	105.8	81.3	2.25	2.25	1.49	1.15
C40×100×2B80×60×4A45	0.60	F	---	25.2	16.8	35.3	---	0.85	0.38	0.79
C120×80×3B120×80×3A45	1.00	S	273.1	273.1	272.4	229.7	5.25	5.25	3.49	2.94
C80×120×3B80×60×4A45	0.50	S+F	---	27.2	18.1	43.8	---	0.62	0.28	0.67
C80×120×3B60×40×4A45	0.33	F	---	22.5	15.0	26.8	---	0.69	0.31	0.55
C80×120×3B80×120×3A45	1.00	S+F	255.0	255.0	244.4	235.0	3.38	3.38	2.16	2.08
Mean (P_m)							2.88	2.28	1.23	1.41
COV (V_p)							0.527	0.787	0.763	0.770
Resistance factor (ϕ)							1.00	1.00	1.00	1.00
Reliability index							2.38	1.41	0.77	0.91

Note: S = chord side wall failure, F = chord face failure, L = local buckling of brace member

Table 3: Comparison of test strengths with design strengths

	$\beta \le 0.85$				$\beta = 1$			
	$\frac{N_u}{N_{EC3}}$	$\frac{N_f}{N_{EC3}}$	$\frac{N_{\varepsilon,1\%}}{(N_{EC3}/1.5)}$	$\frac{N_{\varepsilon,3\%}}{(N_{EC3}/1.5)}$	$\frac{N_u}{N_{EC3}}$	$\frac{N_f}{N_{EC3}}$	$\frac{N_{\varepsilon,1\%}}{(N_{EC3}/1.5)}$	$\frac{N_{\varepsilon,3\%}}{(N_{EC3}/1.5)}$
# of data	7	10	10	10	14	14	14	14
Mean (P_m)	1.95	0.80	0.35	0.67	3.35	3.35	1.85	1.94
COV (V_p)	0.247	0.383	0.384	0.247	0.495	0.495	0.395	0.600

References

[1] Feng R, Young B. "Tests and behaviour of cold-formed stainless steel tubular X-joints", *Thin-walled structures*, 48, 921-934, 2010.

[2] Feng R, Young B. "Design of cold-formed stainless steel tubular T- and X-joints", *Journal of Constructional Steel Research*, 67, 421-436, 2011.

[3] Feng R, Young B. "Design of cold-formed stainless steel tubular joints at elevated temperatures", *Engineering Structures*, 35, 188-202, 2012.

[4] Feng R, Young B. "Stress concentration factors of cold-formed stainless steel tubular X-joints", *Journal of constructional steel research*, 91, 26-41, 2013.

[5] Packer JA, Wardenier J, Kurobane Y, Dutta D, Yeomans N. "Design guide for rectangular hollow section (RHS) joints under predominantly static loading." *In: Comité International pour le Développement et l'étude de la Construction Tubulaire*, Cologne (Germany): Verlag TüV Rheinland, 1992.

[6] Rasmussen KJR, Young B. "Tests of X- and K-joints in SHS stainless steel tubes", *Journal of Structural Engineering, ASCE*, 127(10), 1173-1182, 2001.

[7] Rasmussen KJR, Hasham AS. "Tests of X- and K-joints in CHS stainless steel tubes", *Journal of Structural Engineering, ASCE*, 127(10), 1183-1189, 2001.

[8] Zhao XL. "Deformation limit and ultimate strength of welded T-joints in cold-formed RHS sections". *Journal of constructional steel research*, 53, 149-165, 2000.

[9] Lu LH, de Winkel GD, Yu Y, Wardenier J. "Deformation limit for the ultimate strength of hollow section joints", *Tubular structures VI*, Rotterdam, Balkema, 341–347, 1994.

[10] Eurocode 3 (EC3). "*Design of steel structures—Part 1-8: Design of joints*". European Committee for Standardization, EN 1993-1-8, CEN. Brussels, Belgium, 2005.

[11] Australian/New Zealand Standard. "*Cold-formed stainless steel structures*". AS/NZS 4673. Standards Australia, Sydney, Australia, 2001.

[12] EC3. "*Design of steel structures – Part 1.4: General rules – Supplementary rules for stainless steels*". European Committee for Standardization, EN 1993-1-4, Brussels, 2006.

[13] Huang Y, Young B. "Experimental investigation of cold-formed ferritic stainless steel tubular T-joints." *International Conference on Advances in Steel Structures*, Lisbon, Portugal (CD-ROM Proceeding).

[14] ASCE. "Specification for the design of cold-formed stainless steel structural members." SEI/ASCE 8-02; Reston, VA: *American Society of Civil Engineers*, 2002.

[15] American Iron and Steel Institute (AISI). "North American Specification for the design of cold-formed steel structural members", AISI S100, *North American Cold-Formed Steel Specification*, Washington, DC, 2012.

ANALYSIS OF PRE-COMPRESSED STEEL TUBULAR MEMBERS UNDER LATERAL IMPACT: A PARAMETRIC STUDY

Tiago Manco[a], João Pedro Martins[a], Constança Rigueiro[b] and Luís Simões da Silva[a]

[a] *ISISE, Department of Civil Engineering, University of Coimbra, Portugal*
[b] *ISISE, Department of Civil Engineering, Polytechnic Institute of Castelo Branco, Portugal*

Abstract: This paper deals with steel tubular members of offshore structures under lateral impact. In this sense, in the present study the resistance of these members is calculated through a wide parametric variation using numerical simulations assuming the same and a higher (more realistic) level of impact energy than the one defined in standards. The tubular members are considered with different slenderness of cross sections, several pre-compression levels and the influence of the adjacent structure is simulated by springs at the column ends. The size of the indenter is also assessed. The results are discussed in terms of collision forces, indentations, global displacements, kinetic and strain energies.

1. Introduction

1.1 Background

The collision of ships against offshore platforms has been one of the major hazards faced by the sector of Oil&Gas. For the past three decades, the standard collision events were considered with the impact from a supply vessel of 5000 ton and a speed of 2 m/s. According to Storheim and Amdahl [1] a considerable increase in requirements regarding ship collisions is expected to be present in the next version of NORSOK N-003, mainly due to grow in number and size of the vessels servicing the Norwegian Shelf (up to 7500-10000 ton) and a higher velocity than 2 m/s (documented by Kvitrud [2]). In this sense, this paper aims to take conclusions about the safety of actual jacket structures (built with tubular circular elements) when impacted by ships with higher kinetic energies than those for which they were designed. For this, a broad parametric study will be carried out in order to assess the influence that the main geometric and loading variables have in the response of the elements. Circular tubes with different slenderness of cross-section subjected to axial pre-loading will be impacted by rigid indenters with different sizes and energies. The effect of the adjacent structure will be simulated through axial and rotational springs placed at the tubes ends. Some authors (e.g. [1]) have concluded that the consideration of rigid indenters with simplified geometries may in some cases do not represent very accurately the real situation of impact. However, the importance of the obtained results is not jeopardised once the main objectives of this study are related with the influence that the variables in study and the change in its values have in the response of these elements.

1.2 State-of-the-art

The dissipation of energy in tubular circular elements has deserved historically a considerable amount of studies. Some of the most relevant are highlighted next. Søreide and Amdahl [1] dealt with impact loads on tubular members. The authors carried out experimental tests to assess the influence of membrane forces and dynamic loading. Ellinas and Walker [4] developed a pioneer work to predict the resistance to indentation suggesting methods for estimating the damage and its impact in the reduction of resistance. Wierzbicki and Suh [5] performed an important theoretical analysis of the indentation of tubes under combined loading. Zeinoddini et al. [6], [7] studied the capacity of axially preloaded tubular steel members under lateral dynamic loads. The authors concluded that the effect of the axial preload has a considerable effect on the lateral collapse load and on the level of energy dissipated. Qvale [8] studied numerically the impact of supply vessels against columns in offshore structures. The contact location and the deformability of the ship were aspects taken into account. Travanca and Hao [9] developed a numerical study of steel tubular members subjected to impact of non-rigid bows assessing the influence of the geometry, boundary and loading conditions. Finally, Storheim and Amdahl [3] carried out an extensive work on ship collisions against offshore structures. This study assesses scenarios of bow and stern collisions against floating platforms and jacket legs and braces.

Despite the existence of a considerable amount of studies dealing with the impact on circular tubes it is verified a scarcity of studies dealing with the relation between the variables involved in the problem and its impact on the obtained results through a broad parametric study. Consequently, the purpose of this paper is, on a certain level, to fill this gap.

When a tubular element is struck by a vessel, usually, it begins to deform by local denting. If the plastic bending resistance is exceeded by the denting force, the collapse occurs (normally through a three hinge mechanism). In NORSOK N-004 [11] the resistance of tubular elements to denting (R) is function of the plastic resistance (R_c), geometric factors (c_1 and c_2) incorporating the width of contact zone, a parameter (k) accounting with axial loads, the diameter (D) and the dent depth (W_d) and is given by:

$$\frac{R}{R_c} = k\, c_1 \left(\frac{W_d}{D}\right)^{c_2} \tag{1}$$

When the axial restraint of the adjacent connection is small the plastic resistance is practically constant due to the inexistence of membrane forces. On the other hand if the connections are sufficiently stiff in the axial direction membrane forces have a significant beneficial effect in the load bearing capacity. For the case of a clamped beam (with full axial restraint) subjected to a concentrate load (P) the load-deflection behaviour can be deducted through rigid-plastic methods of analysis in function of the plastic bending resistance (P_c), the deformation (w) and the diameter (d) as follows:

$$\frac{P}{P_c} = \sqrt{1-\left(\frac{w}{d}\right)^2} + \frac{w}{d}\sin^{-1}\left(\frac{w}{d}\right) \quad , \quad \frac{w}{d} \le 1 \tag{2}$$

$$\frac{P}{P_c} = \frac{\pi}{2}\frac{w}{d} \quad , \quad \frac{w}{d} > 1$$

2. Numerical modelling

2.1 Type of analysis, geometry, boundary conditions and application of the loading

Taking into account the nature of the analysis an explicit dynamic analysis was used through ABAQUS/Explicit. This type of analysis allows to represent efficiently problems with relatively short response times and discontinuous events. Moreover, it is possible to use the large-deformation theory once the models in analysis undergo large deformation and rotations.

The contact properties between the indenter and the tubes were defined using the general contact algorithm available in ABAQUS assuming "hard" contact in the normal direction.

The length of the tubes was fixed to 20 m and the external diameter to 1.5 m. The thicknesses (represented posteriorly by "T") were defined according the cross-section classification from EN 1993-1-1 [10] as 0.06, 0.04, 0.03 and 0.01 m, respectively for cross-sections with class 1, 2, 3 and 4. Although the thickness of 0.01 m have little practical applicability it is useful to draw conclusions about the influence that the thickness has in the results. Rigid indenters ("Ind") with rectangular cross sections were considered with a length of 2 and 4 m (corresponding to 10 and 20% of the length of the tube), a width of 3 m and a height of 0.15 m.

The influence of the adjacent structure is important and consequently the boundary conditions should incorporate its effect. In order to assess its influence infinite and finite stiffness were considered for the axial displacements and rotations on the plane of the tubes. Besides cases with axial fully restrained conditions ("STinf") another cases were defined by axial linear elastic springs with a stiffness of 100 MN/m ("ST100"). For the rotational case, besides infinite rotational stiffness (the considered case unless otherwise stated), 100 MNm/rad were considered for rotational springs ("R100"). These are reasonable values representative of the effect of the real adjacent structures. The remaining degrees of freedom were considered restrained.

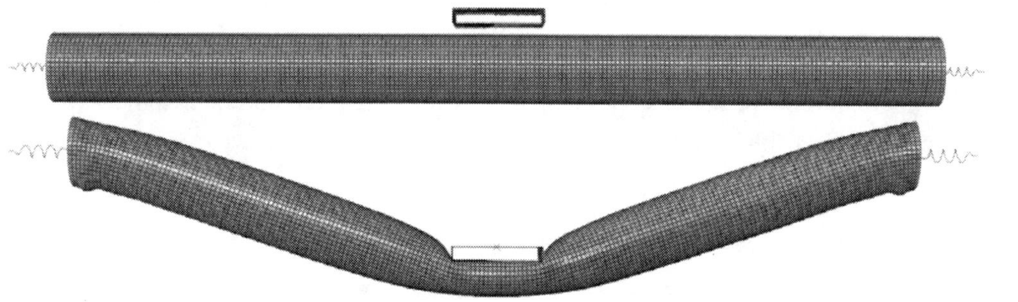

Fig. 1: Example of the deformation of a column with axial flexibility at ends due to an impact

Besides cases where the tubes were considered without axial preload (represented by "AP0"), cases with an axial preload corresponding to 50% of the plastic resistance of the tubes ("AP50") were considered. When the axial displacements were considered restrained ("STinf") the tubes are loaded by the shortening of the tube applying a corresponding displacement at the ends. The new position of the end remain fixed until the end of the analysis. In the case of models with axial flexibility ("ST100") a concentrated load was applied at the column ends. The applied loads were corrected due to the force developed at springs (shortening of the tube).

Two different levels of kinetic energy ("KE") were applied to the indenters in order to compare the effects of the energy specified in standards with a more realistic situation: *Case 1)* Kinetic energy of 10 MJ ("KE10") specified in NORSOK N-004 [11] corresponding to a mass of 5000 ton and a speed 2 m/s; *Case 2)* Higher energy of impact of 60 MJ ("KE60") corresponding to a mass of 7500 ton and a speed of 4 m/s.

2.2 Behaviour of the material

The material was defined with a density of 7800 kg/m^3. The modulus of elasticity and the Poisson's ratio were taken to be 200 GPa and 0.3, respectively. The yield stress was considered equal to 345 MPa (S355 – $t > 16$ mm) with an ultimate stress equal to 470 MPa at an extension of 0.14 using isotropic strain hardening. As in the present case, when strain rates ($\dot{\varepsilon} = d\varepsilon / dt$) are high the rapid increase in strain leads to a considerable increase in the yield stress. The strain rate effect was simulated in ABAQUS using the Cowper-Symonds law which is defined as:

$$\dot{\varepsilon}^{pl} = D\left(\frac{\overline{\sigma}}{\sigma_0} - 1\right)^n \Leftrightarrow \frac{\overline{\sigma}}{\sigma_0} = \left[1 + \left(\frac{\dot{\varepsilon}^{pl}}{D}\right)^{\frac{1}{n}}\right] \tag{3}$$

where $\dot{\varepsilon}^{pl}$ is the equivalent plastic strain rate; σ_0 is the static yield stress; $\overline{\sigma}$ is the yield stress at nonzero strain rate; D and n are empiric material parameters. The standard DNV RP-C208 [12] proposes for common offshore steel materials, in the lack of more accurate data, values of $D = 4000\ s^{-1}$ and $n = 5$ which were the ones used in this study. The progressive damage and failure are necessary to fully represent the behaviour of the material. The tensile failure initiation was used in this study. This means that the stiffness of the material is degraded after the damage initiation according to a determined damage evolution law. The initiation of the ductile damage begins at a corresponding equivalent plastic strain ($\overline{\varepsilon}_D^{pl}$) which is function of the equivalent plastic strain rate ($\dot{\overline{\varepsilon}}^{pl} = 10$ [s^{-1}]) and stress triaxiality ($\eta = -p / q = 0.7$) where p is the pressure stress and q is the Mises equivalent stress. The fracture strain ($\overline{\varepsilon}^{pl}$) was considered equal to 0.15. The considered value of the strain rate corresponds to the maximum value measured in a preceding analysis carried out in ABAQUS (in which the strain rates were neglected).

After the damage initiation the damage evolution can be assumed through a linear relationship function of the plastic displacement u^{pl}. The complete failure is assumed to occur when the total plastic displacement at failure u_f^{pl} is reached:

$$u_f^{pl} = L\varepsilon_f^{pl} \tag{4}$$

Where ε_f^{pl} is the plastic strain at failure and L is the length of the finite element. The assumed plastic strain at failure $\varepsilon_f^{pl} = 0.2$ (a reasonable value for this type of steel) corresponding to $u_f^{pl} = 0.02$ m taking into account that $L = 0.1$ m.

2.4 Type of element, mesh convergence study and FE model validation

Taking into account the high diameter to thickness ratio shell elements were used for the tubular elements - linear four node shell with reduced integration (S4R). S4R is a robust general-purpose element suitable for dynamic situations. The indenter was modelled with rigid elements. The mesh size of the tubes was defined from a mesh convergence study as 0.1×0.1 m^2, corresponding to 201 nodes in the longitudinal direction and 48 nodes in the cross-section.

The numeric models were validated through the study of Travanca and Hao [9]. For this the same conditions were replicated. The achieved values are about 10% smaller than the ones from the referred study. This error can be considered reasonable once different software were used (different algorithms, different elements, etc.).

3. Results

The results will be presented in terms of membrane forces ("MF") and transversal forces ("TF") which are plotted in function of indentation (given by the difference between the top and bottom displacement) and top displacements at the mid-span. The remaining notation presented in the following charts was defined previously in Section 2. The results plotted and the ones not presented (due to space reasons) are also discussed next.

From Fig. 2 and Fig. 3 it is possible to conclude that the indentation is highly affected by the thickness of the tube. As expected, the thinner tubes present a much larger indentation than the thicker ones. Tubes with lower slenderness (small D/t ratio) can reach high bending moments (plastic bending moment) and maintaining it for large rotation. On the other hand, the resistance of tubes with high slenderness are limited by occurrence of local buckling. As an example, the transversal forces are on average 40% larger for thicknesses of 0.06 m than for 0.04 m.

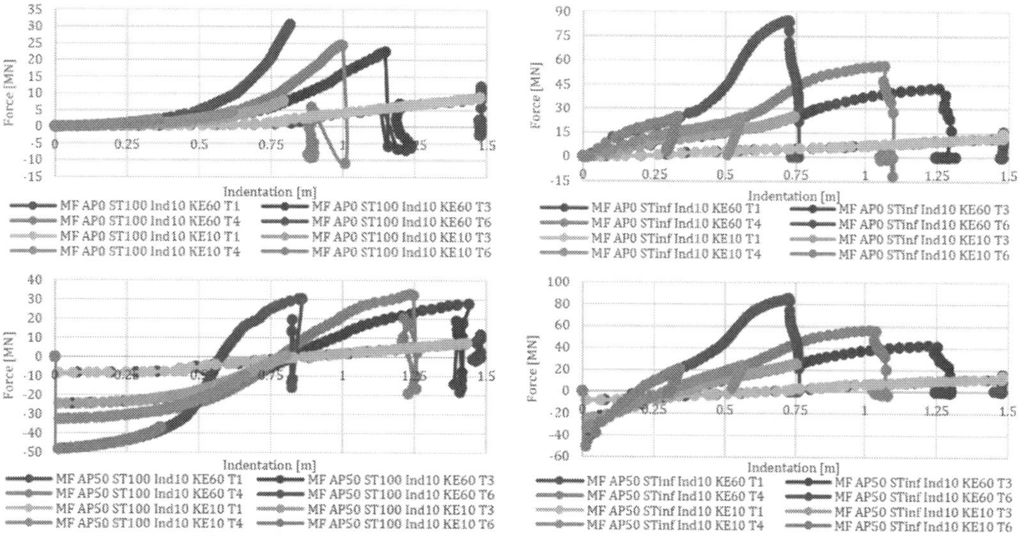

Fig. 2: Membrane force vs indentation curve - tubes with infinite rotational stiffness

The size of the indenter has influence not only in the maximum developed force (wider indenter lead to transversal forces about 10% larger and membrane forces about 10% smaller) but mainly in the indentation for which those forces occur (for the wider indenters the indentation corresponding to the maximum forces are considerable smaller, especially for the thicker tubes).

The axial stiffness of the column ends has an enormous impact in the development of membrane forces, once this is the reason for the development of this force. For example, when fully axial restrained conditions are assumed the maximum membrane forces are about 2.7 times larger than considering an axial stiffness of 100 MN/m for the thickness of 0.06 m. Moreover, the results obtained for the transversal forces are also considerably distinct (1.3 times larger for the fully axial restrained tubes).

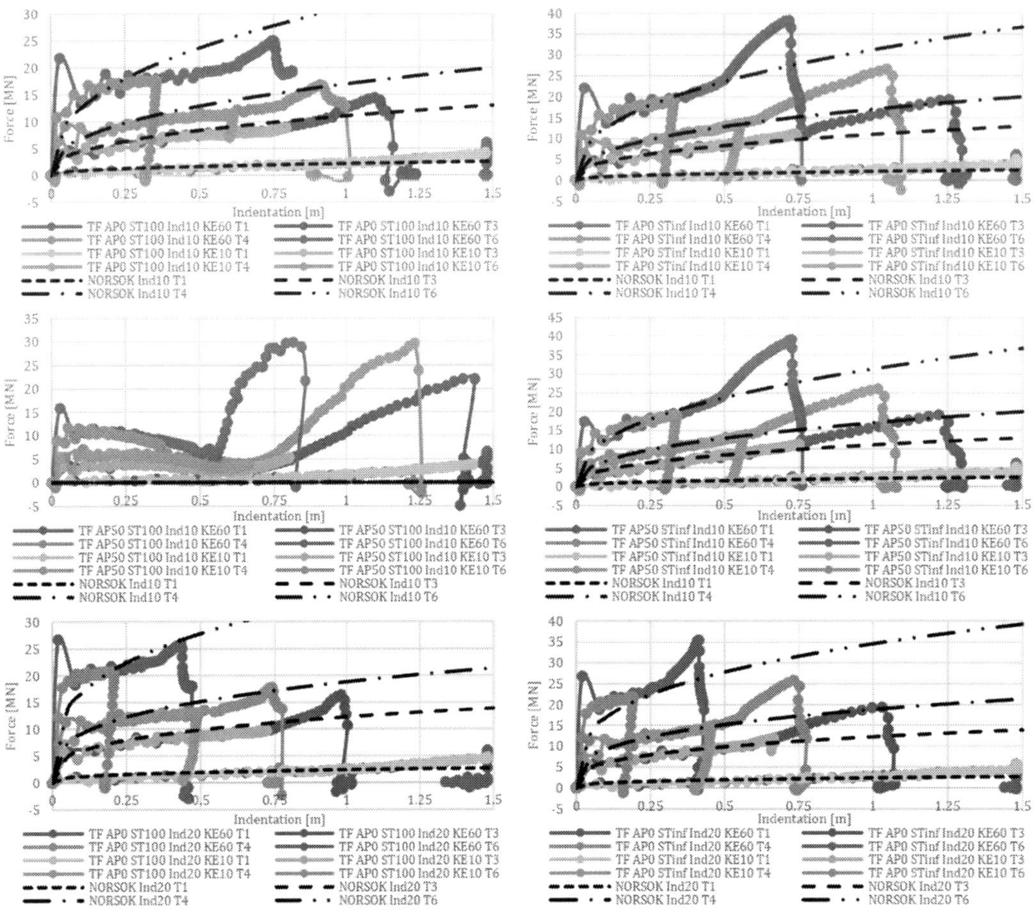

Fig. 3: Transversal force vs indentation curve - comparison with NORSOK N-004 - infinite rotational stiffness

The axial preloading of the tubes causes an initial negative axial force (due to the compression) which vanishes with the increase of the transversal displacements and the development membrane forces. In fact, the effect of the preload is very detrimental as seen in Fig. 3 for AP50. The behaviour of these curves presents an initial positive stiffness and then they begin progressively to lose stiffness due to buckling of the compressed side (the impacted side). With the development of the displacements the compression vanishes and at the point when the membrane forces is positive (see Fig. 2 for AP50 and ST100) the system gains rapidly stiffness and the transversal forces increase significantly. This behaviour is not presented in the cases of infinite stiffness of the column ends because the ends are fixed against inward displacements and consequently the effects of the buckling are limited (see Fig. 3 for AP50 STinf).

The plotted curves from NORSOK in Fig. 3 are given by equation (1) representing the denting of tubular members. It should be noted this curves have the correction factor k applied which accounts for the axial loads. In this case, once axial preloading is present the N_{Sd}/N_{Rd} ratio is larger than 0.6 (N_{Rd} has into account phenomena like local and global buckling) and consequently the factor k is equal to 0 leading in this case to conservative values. From the presented results the curves from NORSOK are generally well adjusted but when axial flexibility is present the analytical curves seems to be slightly non-conservative especially for larger indentations. From Fig. 3 (for AP0 STinf, for example) it is possible to verify that the element is initially governed by bending (affected by local denting), then with the development of membrane tension forces the load carrying capacity increase considerably. Finally, the energy dissipation capacity is limited by tension failure.

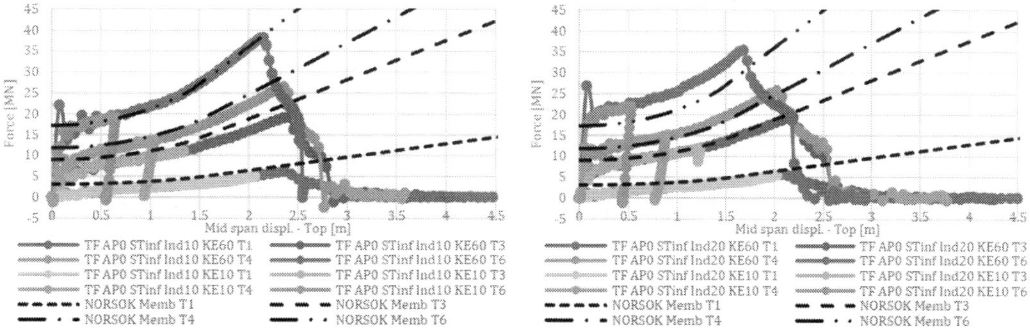

Fig. 4: Transversal force vs mid span (top) displacement curve - comparison with NORSOK N-004 - infinite rotational stiffness

Equations (2) are plotted in Fig. 4 representing the effect of the membrane forces for fully axially restrained ends. The progress of the numerical and analytical curves is very similar. It should be noted that the analytical curves do not have present the effect of the reduced plastic moment capacity of a tube due to indentation. When thicker tubes are impacted by a 10 MJ indenter they show a considerable reserve for the dissipation of energy. The length of the indenter has not a major impact in the maximum transversal load obtained. However, the maximum force is reached earlier for the larger indenter.

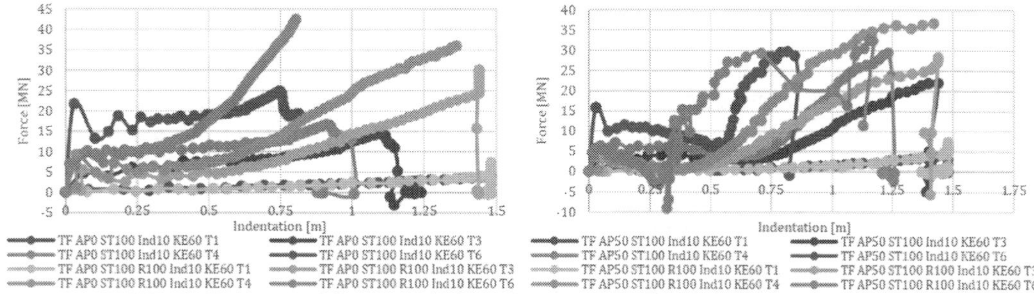

Fig. 5: Transversal forces vs indentation curve - effect of rotational stiffness

The effect of the rotational stiffness is substantial and is more evident when the thickness is larger (see Fig. 5). Once the beam plastic resistance in bending is smaller when rotations are not fully restrained, for small indentation the transversal forces are also smaller, but taking into account that these tubes are able to develop larger transversal displacements, the effect of membrane forces is more evident and takes a greater role for larger indentations.

The force-displacement curves obtained by the two levels of kinetic energy have the same progress, however the indenters with 60 MJ lead to higher indentations and consequently to higher transversal and membrane forces. In the case of 10 MJ (not presented here) the thinner tube is unable to dissipate all the energy in all conditions. In addition when axial preloading is present with inward displacements the first three thicknesses fail to dissipate all the energy. In all other cases the kinetic energy is dissipated totally. The tubes with the greatest thickness (0.06 m) have the capacity to dissipate the most of the 60 MJ (Fig. 6) impact energy: on average 97% when not considering the tubes with ST100 and AP50, in this case the tubes only dissipate about 75% of the 60 MJ.

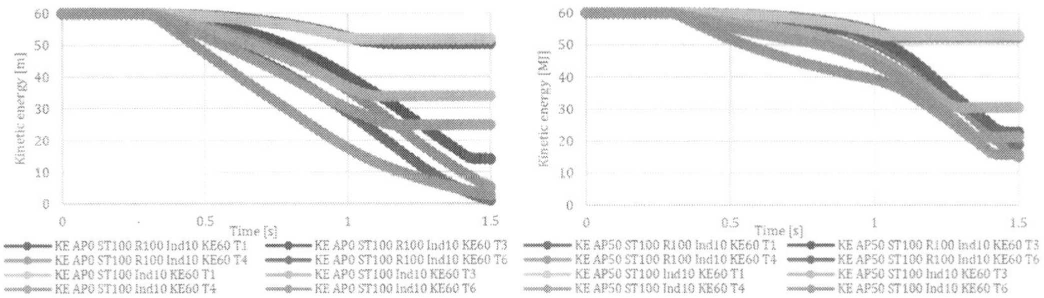

Fig. 6: Dissipation of kinetic energy (60 MJ) – effect of axial preloading and rotational stiffness

On the other hand for the thickness of 0.04 m the dissipation of energy is much smaller. In this case the tubes are able to dissipate on average 66% when not considering the tubes with ST100 and AP50 and 58% in that case. From this it is possible to conclude that *i)* a small increase in thickness leads to a significant improvement of the dissipated energy and *ii)* in the presence of axial preload and axial displacements not fully restrained (as happens in real cases) the dissipated energy is considerably smaller than assuming simplistically fully axially restrained displacements or neglecting the presence of axial preload. The rotational stiffness affects, not only, the rate of energy dissipation (tubes with higher rotational stiffness leads to faster dissipations) but also the value of total dissipated energy. In this case, it is shown that higher rotational stiffness dissipate in some cases much less energy due to rupture of the element (compare the intermediate thicknesses). Finally, the results seem to indicate that the size of the indenter does not have an important influence on the amount of dissipated energy.

4. Conclusions

From the previous discussion of results it is possible to conclude that the consideration of simplified assumptions may lead in some cases to rather distinct results from the ones obtained in real cases. Consequently, one should be aware if the assumed simplifications are, in fact, conservative. It was possible to conclude that the axial preloading of the tubes leads to drastic reductions of the bearing capacity to transversal loads and should not be neglected. The influence of adjacent structure plays an important role in the development of membrane and transversal forces and consequently on the resistance to transversal impacts. It was shown that neglecting the flexibility of the tubes ends may not be on the safe side (in terms of energy dissipation). Moreover, the requirements of elements capable to withstand collisions of 60 MJ are much higher than the requirements of actual standards and consequently actual jacket structures may not be prepared to withstand the additional forces and displacements. The provisions of NORSOK N-004 are generally safe but when the assumed conditions deviate from the conditions of the real problem non-conservative errors can be committed.

References

[1] Storheim M, Amdahl J. "Design of offshore structures against accidental ship collisions". *Marine Structures*, 37, 135-172, 2014.

[2] Kvitrud A. "Collisions between platforms and ships in Norway in the period 2001–2010". *30th International Conference on Ocean, Offshore and Arctic Engineering OMAE2011*, 2011.

[3] Søreide T, Amdahl J. "Deformation characteristics of tubular members with reference to

impact loads from collision and dropped objects". *Norwegian Matimime Research*, No. 2, 1982.

[4] Ellinas C, Walker A. "Damage of offshore tubular bracing members". *Procedings IAB-SE Colloquium on Ship Collision with Bridges and Offshore Structures*, Copenhagen, 1983.

[5] Wierzbicki T, Suh M. "Indentation of tubes under combined loading". *International Journal of Mechanical*, 30(3–4), 229–248, 1988.

[6] Zeinoddini M, Harding G, Parke G. "Effect of impact damage on the capacity of tubular steel members of offshore structures". *Marine Structures*, 11, 141-157, 1998.

[7] Zeinoddini M, Harding G, Parke G. "Dynamic behaviour of axially pre-loaded tubular steel members of offshore structures subjected to impact damage". *Ocean Engineering*, 26, 963-978, 1999.

[8] Qvale K. "Analysis and Design of Columns in Offshore Structures subjected to Supply Vessel Beam Collisions". *Master thesis, Norwegian University of Science and Technology Trondheim*, 2012.

[9] Travanca J, Hao H. "Numerical analysis of steel tubular member response to ship bow impacts". *International Journal of Impact Engineering*, 64, 101-121, 2014.

[10] EN 1993-1-1. "EN 1993-1-1:2005 - Eurocode 3: Design of steel structures - Part 1-1: General rules and rules for buildings". *European Committee for Standardization*, Brussels, 2005.

[11] NORSOK N-004. "NORSOK N-004 - Design of steel structures". Ed. 3, *Standards Norway*, 2013.

[12] DNV-RP-C208. "DNV-RP-C208 – Determination of structural capacity by non-linear FE analysis methods". *Recommended Practice, Det Norske Veritas*, 2013.

STUDY ON DEFORMATION CAPACITY OF SQUARE STEEL TUBULAR COLUMNS

Kazuya Mitsui and Atsushi Sato

Nagoya Institute of Technology, Japan

Abstract: Square steel tubular columns have been widely used in Japan; seismic loads are significant in designing a column. The design formulas for columns that are specified in current recommendation are based on the test results of H-shaped steel columns; therefore, it is questionable that these formulas can be directly applied in square tubular columns. From this background, it is necessary to clarify the deformation capacity of square steel tubular columns under constant axial force with bending moment. For this purpose, full scale tests of square steel tubular columns are conducted. The effects of axial force ratio, material property and loading pattern on the elasto-plastic behaviour of the columns are clarified in this paper.

1. Introduction

In Japan, square steel tubular columns have been widely used as main structures because square steel tubular columns have equal rigidity and strength in two directions at orthogonal angles. Seismic load is significant for column design; combined loadings (i.e., axial force with bending moment) will be subjected. To evaluate seismic performance of columns properly, it is necessary to confirm elasto-plastic behaviours of the square steel tubular columns under constant axial force with bending moment.

A Recommendation for Limit State Design of Steel Structures so called LSD [1] has the design limitations, and square steel tubular columns that are expected to form plastic hinge at the end of the columns must satisfy followings;

$$n_y \leq 0.75 \tag{1}$$

$$n_y \cdot \lambda_{c,E}^{\ 2} \leq 0.10(1 + M_2 / M_1) \tag{2}$$

where: n_y is axial compression force ratio; $\lambda_{c,E}$ is bending slenderness ratio $(= 1/\pi \cdot \sqrt{\sigma_y / E} \cdot L / i_x)$; σ_y is yield stress; E is Young's modulus; L is length of column; i_x is radius of gyration of column in bending axis ; M_2/M_1 is end moment ratio ($|M_1| \geq |M_2|$). These limitations are set for the purpose of ensuring sufficient plastic deformation capacity. These limitations are based on test results of H-shaped steel columns [2]; therefore, it is questionable that these limitations can be directly applied in square steel tubular columns. Although a lot of studies about square steel tubular column (e.g., local buckling strength, elasto-plastic behaviour and the influence of the initial deflection) are investigated [3-5], a test data that will verify the current limitations for square steel tubular columns does not exist. From these backgrounds, it is necessary to confirm the plastic deformation capacity, the maximum strength and deformation state of the Box column.

In this paper, to clarify these relationship, full-scale tests are conducted. Axial compression force ratio, properties of steel, and loading pattern are selected as test parameters.

2. Test program

In order to confirm elasto-plastic behaviours of the square steel tubular columns under constant axial force with bending moment, full-scale tests were conducted. Cold forming square steel tubular columns are used in this study, and the cross section is Box-150×150×9. Width-to-thickness ratio B/t of this section is 16.7, and this value satisfies the LSD limitation which is set for the purpose of ensuring sufficient plastic deformation capacity. The length of specimen L is 2900 mm. The test specimen is shown in Fig. 1. Material properties of the specimens are tabulated in Table. 1.

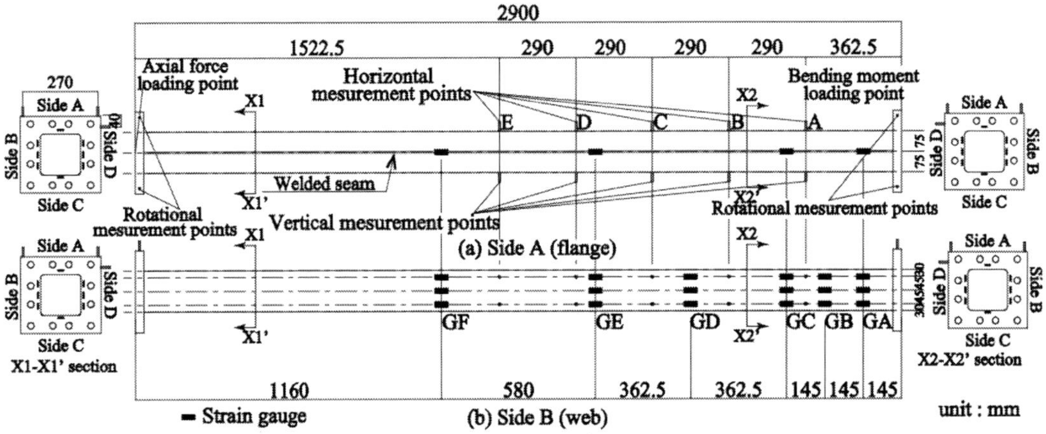

Fig. 1: Test specimen

Table 1: Material properties from coupon tests and compressive test results of stub column

Steel grade	E [kN/mm^2]	σ_y [N/mm^2]	σ_u [N/mm^2]	$Y.R.$ [%]	E_{long}. [%]	${}_sN_{max}$ [kN]	${}_sN_y$ [kN]	${}_{cal}N_y$ [kN]	${}_sN_y/{}_{cal}N_y$
STKR400	211	387	459	84	35	2088	1781	1832	0.972
BCR295	214	420	492	86	32	2384	2008	2102	0.956

σ_u is ultimate stress; $Y.R.$ is yield ratio; E_{long} is elongation; ${}_sN_{max}$ is the maximum strength of stub column; ${}_sN_y$ is yield strength of stub column; ${}_{cal}N_y$ is yield strength based on calculation ($=A\times\sigma_y$)

The compressive tests of stub column were conducted, and these results are also tabulated in Table. 1. The length of stub column is three times as long as the column width and the length is 450 mm. Boundary condition of both sides is fixed support. In this paper, yield strength of stub column is defined as the point where the tangent stiffness decreased to one third of the initial elastic stiffness. Measured geometrical dimensions of the cross section are shown in Table. 2. Geometrical dimensions of specimens are shown in Table. 3. Test parameters are axial compression force ratio, properties of steel and loading pattern. The target axial compression force ratio n_y are 0.20 and 0.30. To confirm the effect of properties of steel, STKR400 and BCR295 grade steels are used for specimens. These grade are one of the specification materials for square steel tubular columns. STKR400 is equivalent to S275JR Grade in European standards. BCR295 Grade has more ductility than STKR400 Grade. Loading patterns are monotonic loading and cyclic loading.

Table 2: Geometrical dimension of cross section

Steel grade	t [mm]	B [mm]	R_{out}/t	A [mm^2]	I [cm^4]	Z_P [cm^3]	M_P [kN ·m]
STKR400	8.67	150.2	1.81	4713	1540	246.5	95.3
BCR296	9.16	150.3	1.69	4953	1606	258.1	108.5

t is measured thickness of column; B is width of column; R_{out} is outer radius at the corner; A is area of cross section; I is moment of inertia; Z_p is plastic secsion modulus; M_p is full plastic bending moment

Table 3: Geometrical dimensions of specimens

Name	Steel grade	λ	${}_F\lambda_{c,E}$	$\lambda_{c,E}$	n_y	N_t [kN]	M_2/M_1	M_{pc} [kN ·m]	θ_{pc} [rad]
ny020_M_S	STKR400	50.7	0.56	0.69	0.20	366	0.0	90.0	0.0286
ny030_M_S	STKR400	50.7	0.56	0.69	0.30	550	0.0	83.4	0.0275
ny030_C_S	STKR400	50.7	0.56	0.69	0.30	550	0.0	83.4	0.0275
ny030_M_B	BCR295	50.9	0.62	0.72	0.30	531	0.0	94.2	0.0310

λ is slenderness ratio; ${}_F\lambda_{c,E}$ is flexural slenderness ratio based on specified yield stress; $\lambda_{c,E}$ is flexural slenderness ratio based on measured yield stress (See table1) ; N_t is target axial compression force; M_{pc} is full plastic bending moment that is considering the effect of axial force; θ_{pc} is elastic rotation corresponding to full plastic bending moment that is considering the effect of axial force

The comparison between limitations in LSD and test specimens is shown in Fig. 2(a). Boundary condition and plan view of test setup are shown in Figs. 2(b-c). As seen from Fig. 2(a) and Table 3, when specimens are evaluated by $\lambda_{c,E}$, ny020_M_S only satisfies limitation Eq.(2), and others do not satisfy limitation Eq. (2). When specimens evaluated by ${}_F\lambda_{c,E}$, ny020_M_S, ny030_M_S and ny030_C_S satisfy limitation Eq. (2), and ny030_M_B does not satisfy limitation Eq.(2). Following limitation can be derived from elastic analysis to guaranty the maximum moment is equal to the applied end moment (i.e., P-δ effect is not dominant)

$$n_y \cdot \lambda_{c,E}^2 \le 0.25(1 + M_2/M_1) \tag{3}$$

As seen from Figs. 2(b-c), in order to observe the displacement history and deformation of cross section, measurement devices are set. Also, in order to observe the rotation of both sides of the column, measurement devices are set. Strain gauges are attached to observe the strain distribution of cross sections and the strain distribution in the longitudinal direction. Specimens are set on the side A (welded seam side) of the specimen upward. In this paper, Side A and Side C are defined as web, and, side B and side D are defined as flange (see Fig.1 and Fig.2).

As seen from Figs. 2(b-c), the left end of the column is supported as roller supporting, and the right side of the column is supported as pin supporting. Rotation axes about y-axis and z-axis at the both ends are restrained. Bending moment is applied at the right end to the orthogonal angles by using center-hole-type jacks after applying constant target axial compression force N_t at the centroid of the column. The hydraulic jack is used to apply target axial compression force N_t, and the target axial compression force is controlled automatically to have constant force during testing. When monotonic loading is applid, testing is conducted until the strength is lost completely or the rotation reaches its stroke (i.e., 0.30 rad). When cyclic loading is applied, alternately increasing cyclic loading pattern is applied, and cyclic loading pattern is expressed as follows;

$+2\theta_{pc}$, $-2\theta_{pc}$, $+2\theta_{pc}$, $-2\theta_{pc}$, $+4\theta_{pc}$, $-4\theta_{pc}$, $+4\theta_{pc}$, $-4\theta_{pc}$, $+6\theta_{pc}$, $-6\theta_{pc}$, $\cdots$

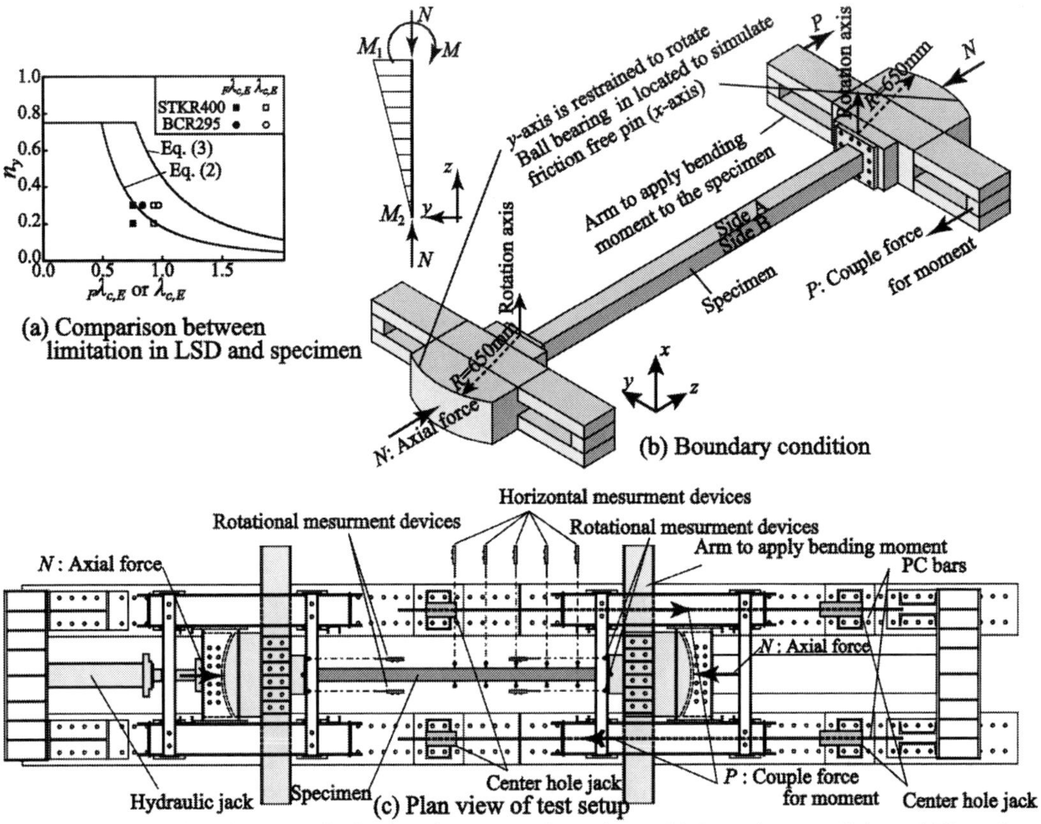

Fig. 2: a) Comparison between limitations in LSD and specimen: b) Boundary condition; c) Plan view

3. Test results

3.1 Relationship between bending moment and rotation

Test results are shown in table. 4. Relationship between bending moment and rotation are shown in Fig. 3. In this paper, the rotation that the maximum bending moment decreased to 90% or bending moment decreased to the full plastic bending moment M_{pc} is defined as the ultimate rotation θ_u. The circle legend in Fig. 3(a) stands for the point where M–θ curve reaches at the full plastic bending moment M_{pc}. The triangle legend in Fig. 3(a) stands for the point where M–θ curve reaches at the maximum bending moment. The square legend in Fig.

3(a) stands for the point where 90% of the maximum moment or the full plastic bending moment M_{pc}. Deformed shapes after the testing are shown in Fig. 4. Plastic deformation capacity of ny020_M_S is 0.179 rad, and it shows significant plastic deformation capacity. Compared ny030_M_S which does not satisfy limitation Eq. (2) with ny020_M_S which satisfies limitation Eq. (2), initial rigidity of ny030_M_S decrease early. Also, the maximum bending moment and deterioration behaviour after the maximum bending moment decrease extremely. As seen from figs. 4(a, b and d), the maximum deformation point differ depending on axial compression force ratio. The maximum deformation point of ny030 series are formed middle point of columns. On the other hand, that of ny020 are formed around the bending moment loading point. These effects were caused by P-δ effect with increasing axial compression force.

Compared BCR295 with STKR400, yield stress of ny030_M_B is larger than that of ny030_M_S. Therefore, the maximum strength of ny030_M_B is larger than that of ny020_M_S. However, plastic deformation capacity of ny030_M_B is smaller than that of ny030_M_S because ny030_M_B does not satisfy limitation Eq. (2).

Although specimens used in this paper were not enough, it is confirm that limitation Eq. (2) can ensure enough plastic deformation capacity.

Table. 4: Test results

Name	M_{max} [kN·m]	M_{min} [kN·m]	θ_{max} [rad]	θ_u [rad]	θ_u/θ_{pc}	${}_{pl}\theta_u$	$\Sigma\theta_{pl}^+$ [rad]	$\Sigma\theta_{pl}^-$ [rad]
ny020_M_S	102.3		0.121	0.225	7.87	0.179		
ny030_M_S	87.4		0.091	0.122	4.45	0.080		
ny030_C_S	94.8	-102.0					0.335	0.268
ny030_M_B	95.77		0.087	0.104	3.46	0.061		

M_{max} is the maximum bending moment; M_{min} is minimum bending moment; θ_{max} is rotation at the maximum bending moment; θ_u is ultimate rotation; ${}_{pl}\theta_u$ is rotation which is removed the elastic rotation from ultimate rotation; $\Sigma\theta_{pl}^+$ is cumulative plastic deformation capacity on the positive side; $\Sigma\theta_{pl}^+$ is cumulative plastic deformation capacity on the negative side

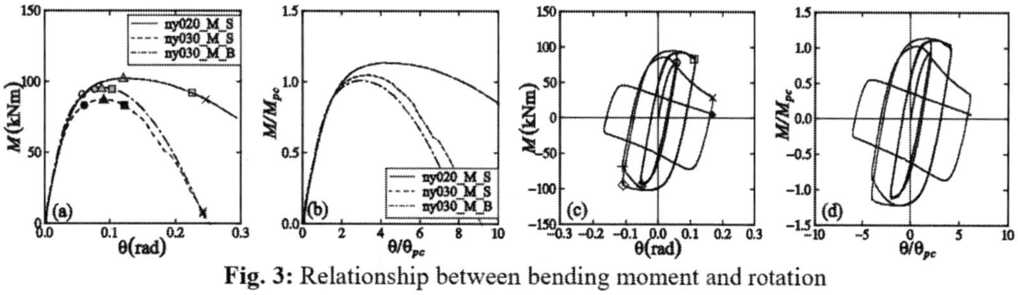

Fig. 3: Relationship between bending moment and rotation

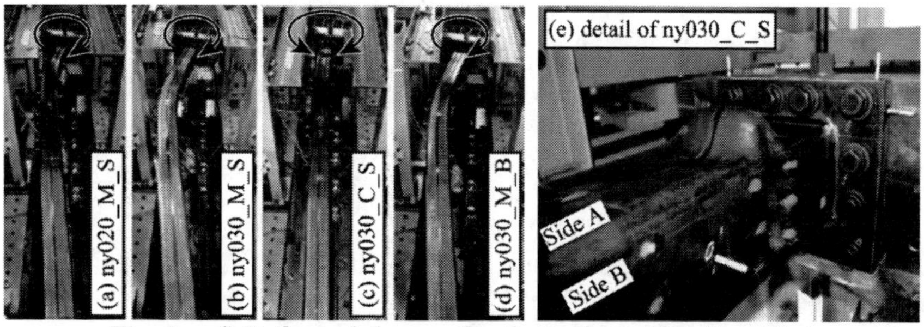

Fig. 4: a-d) Deformed shapes after the testing; e) local buckling

Figs. 3(c,d) show the relationship between bending moment and rotation of cyclic loading pattern. Because specimen which was subjected to cyclic load was influenced by strain hardening, the maximum bending moment of ny030_C_S are larger than that of ny030_M_S. Compared the positive loading side with the negative loading side, the maximum bending moment on the negative loading was larger than the maximum bending moment on the positive loading. Fig. 5(a) shows the relationship between bending moment and cumulative rotation curve, and Fig. 5(c) shows the relationship between bending moment and backbone curve which was illustrated by following Fig. 5(b). As seen from Fig. 5(a) and Table 4, the cumulative plastic deformation capacity on the positive side was 4.18 times as the rotation $_{pl}\theta_u$ which was removed the elastic rotation from ultimate rotation θ_u. And, the cumulative plastic deformation capacity on the negative side was 3.35 times as the rotation $_{pl}\theta_u$. It was confirmed that the plastic deformation capacity of the column which was subjected to cyclic load was larger than that of the column which is subjected to monotonic load. The same effect was confirmed in Fig. 5(c).

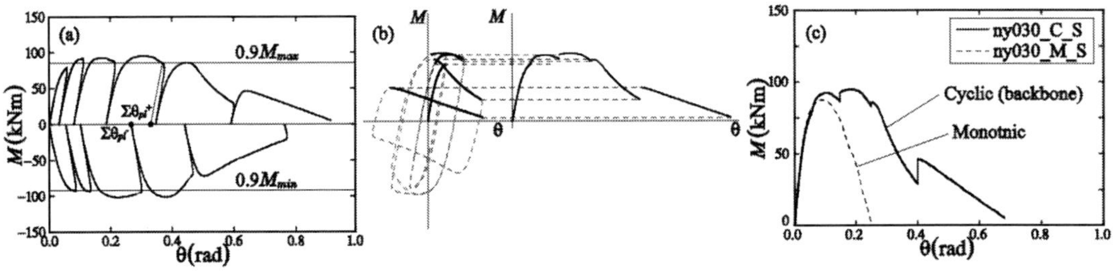

Fig. 5: a) Relationship between bending moment and cumulative plastic rotation: b) backbone curve extraction method: c) relationship between bending moment and backbone curve

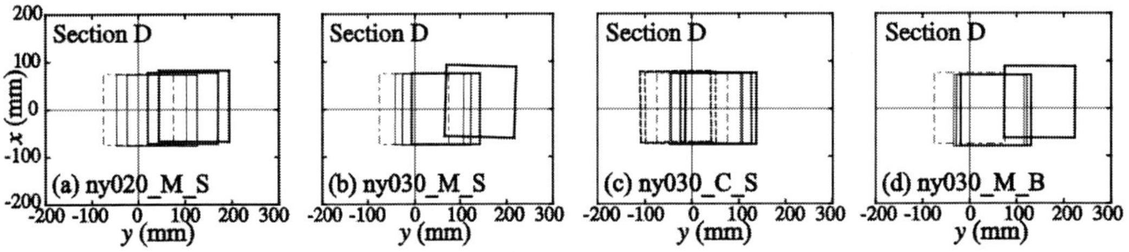

Fig. 6: Deformation history in the cross section D

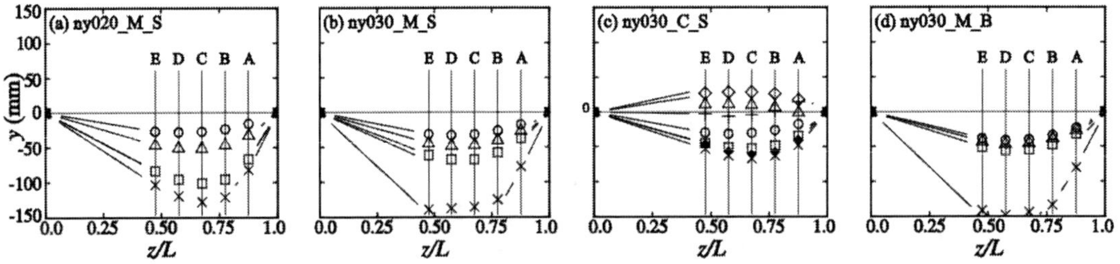

Fig. 7: Deformation history in the longitudinal direction

3.2 Deformation history

Fig. 6 shows the deformation history at the location of cross section D (See fig. 1). Fig. 7 shows the deformation history in the longitudinal direction. These history were computed from measurement results. Dashed-dotted lines stand for an original shape of the cross sections. The line width in Fig. 6 corresponds to the rotation angel, larger angle have thickness line (see Fig. 3). In Fig 6(c), solid lines stand for positive cyclic loading, and dashed line stand for the negative cyclic loading. Each symbol which were shown in Fig. 7 corresponds to each symbol which was shown in Fig. 3.

As seen from Fig. 6, the deformation of the cross section was deformed in-plane direction. Compared ny020_M_S with ny030_M_S, deformation of ny020_M_S was almost at the same level as deformation of ny030_M_S at the full plastic bending moment M_{pc} and at the maximum bending moment M_{max}. However, as seen from Fig. 7, the maximum deformation position of ny020_M_S was different from that of ny030_M_S. With increasing *P-δ* effect, the maximum deformation position moves to the middle point of columns. The same effect was confirmed in Fig. 7(d). The deformation history of the specimen which was subjected to cyclic loading was totally different from the deformation history of the specimen which was subjected to monotonic loading and the same level of axial compression force ratio. The deformation on the negative loading side was smaller than the deformation on the positive loading side. As mentioned before, the maximum deformation point of ny030 series was formed around the middle point of columns. On the other hand, the maximum deformation point of ny020_C_S formed around section C.

3.3 Relationship between *P-δ* moment and rotation

Fig. 8 shows the relationship between *P-δ* moment and rotation. *P-δ* moment curves were illustrated by thin solid lines or dashed lines. *P-δ* moment was determined by multiplying axial compression force on the horizontal deformation which was measured by horizontal measurement devices.

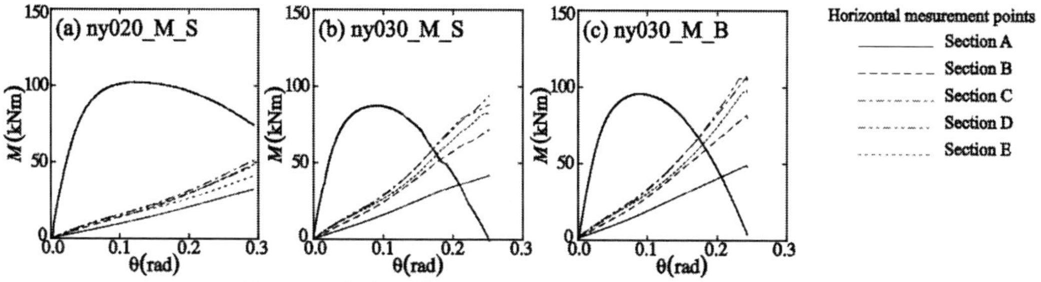

Fig. 8: Relationship between *P-δ* moment and rotation

Compared ny020_M_S with ny030 series, the increase rate of *P-δ* moment of ny030 series with increasing rotation were larger than ny020_M_S. In the range that *M-θ* curve decreases, this tendency was remarkable. Increasing of *P-δ* moment decreases plastic deformation capacity of column. Compared with each *P-δ* moment curve of ny030 series, *P-δ* moment at the section A was the smallest in all section. *P-δ* moment increases with closing to middle point.

4. Conclusions

In this paper, the full scale tests were conducted to clarify the plastic deformation capacity, the maximum strength and the deformation state.
The main conclusions are:

1. Compared ny030_M_S which does not satisfy limitation Eq. (2) with ny020_M_S which satisfies limitation Eq. (2), initial stiffness of ny030_M_S decreases early. Al-

so, the maximum bending moment and deterioration behaviour after the maximum bending moment decrease significantly.

2. The maximum deformation points differed depending on axial compression force ratio. With increasing P-δ effect, the maximum deformation position move to the middle of the column.
3. The maximum strength of ny030_M_B was larger than that of ny020_M_S because the yield stress of BCR295 is larger than that of STKR400. On the other hand, plastic deformation capacity of ny030_M_B was smaller than that of ny030_M_S, because ny030_M_B does not satisfy limitation Eq. (2).
4. It is confirm that limitation Eq. (2) can ensure enough plastic deformation capacity.
5. Because specimen which was subjected to cyclic load was influenced by strain hardening, the maximum bending moment of ny030_C_S was larger than that of ny030_M_S.
6. The plastic deformation capacity of column which was subjected to cyclic load was larger than that of column which was subjected to monotonic load.
7. The deformation of cross section was deformed in-plane direction.
8. The increase rate of P-δ moment of ny030 series with increasing rotation angle were larger than ny020_M_S. And in the range that M-θ curve decreases, this tendency was remarkable. Increasing of P-δ moment decreases plastic deformation capacity of column.

Acknowledgements

Funding for this research was supported by "*The Hibi Science Foundation*".

References

[1] Architectural Institute of Japan. *Recommendation for Limit State Design of Steel Structures*, Japan, 2010.

[2] Suzuki T, Ono T. "A Study of the Plastic Deformation Capacity of H-Shaped Steel Beam-Column (Part I)", *Transactions of the Architectural Institute of Japan*, Japan, 23-29, 1980.

[3] Sato K, Ikarashi K. "Evaluation of coupled local buckling strength of square hollow section member under biaxial bending shear force and axial force", *Journal of Structural and Construction Engineering*, Japan, 1909-1918, 2014.

[4] Tsuda K, Matsui C. "Strength of Square Steel Tubular Beam-Columns under Constant Vertical and Horizontal Loads", *Journal of Structural and Construction Engineering*, Japan, 149-156, 1998.

[5] Ikarashi K, Sato K. "Effect of Initial Imperfection by Manufacturing on Large Deformation Behavior of Cold-formed Square Steel Tube Member", *summaries of technical papers of annual meeting*, Japan, 1237-1238, 2013.

CRUSHING BEHAVIOR OF THIN-WALLED FRUSTA – EXPERIMENTAL STUDY

Artur Mołdawa and Maria Kotełko

Department of Strength of Materials, Łódź University of Technology, Poland

Abstract: In the paper the crushing behaviour of thin-walled, spot-welded prismatic frusta subjected to axial impact load is investigated. Results of experimental quasi-static and impact tests are presented. Validation of FE model is carried out. Comparison of experimental and FE results are presented in the form of load – shortening diagrams. Some conclusions concerning influence of spot welds and load velocity upon the crushing behaviour of frusta are derived.

1. Introduction

Dynamic response of structures in the plastic range is a significant problem in the case of energy absorbers. Such a structural member converts totally or partially the kinetic energy into another form of energy. One of the possible design solutions is the conversion of the kinetic energy of impact into the energy of plastic deformation of a thin-walled metallic structural member. There are numerous types of energy absorbers of that kind cited in the literature [1].

Thin-walled metal tubes are widely used as energy absorption systems in automotive industry due to their high energy absorption capability, easy to fabricate, relatively low price and sustainability at collapse.

A designer of any impact attenuation device must meet two main requirements. The initial collapse load (peak load) must not be too high in order to avoid unacceptably high impact velocities of the vehicle. On the other extreme, the main requirement is a possibly highest energy dissipation capacity, which may not be achieved if the collapse load of the impact device is too low. Thus, maximizing energy absorption and minimizing peak to mean force ratio by seeking for the optimal design of these components are of great significance. Many efforts have been made so far by several researchers to reduce the peak load at initial stage of the crushing process and simultaneously to increase energy absorption capacity in metal tubes, acting as energy absorbers.

One of the solutions are flaws or dents acting as triggers. A trigger may induce the most desirable crushing (collapse) mode, leading to higher energy absorption and mean to peak crushing force ratio. Ferdynus [2] performed the dynamic analysis of axially compressed square section tube with dents in the corners. A parametric study into an optimal dents situation and geometry with respect to the energy absorption capacity of tubes with dents subjected to axial impact was performed by the same author in [3].

In last few years research was carried out into another solution, namely an application of multi-cornered thin-walled tubes as energy absorbing members. Abbasi et al. [4] investigated crushing behavior and energy absorption of hexagonal, octagonal and 12-edge thin-walled columns. Reddy et al. [5] continued to analyze the similar problem for 12-edge columns of different geometry. Ali et al. [6] analyzed axial crushing of pentagonal and cross-shape thin-walled tubes. Another solution, bitubal energy absorber was investigated experimentally under quasi-static load by Sharifi et al. [7].

An alternative solution are compressed thin-walled frusta (truncated circular cones or prisms) [1], currently used as impact attenuation members in car structures, mainly due to the reduction of the peak crushing load in relation to parallelepiped. There are few published results, concerning this type of energy absorbing members. Very recently, Sarkabiri et al. [8] published the results of multi-objective crashworthiness optimization of thin-walled conical groove tubes filled with polyurerthane foam. They stated, that the conical angle reduces the peak crushing force, however it reduces also the absorbed energy.

In [9] the authors of the present paper presented results of the parametric study into the energy absorption capacity, ultimate load and mean crushing load of thin-walled prismatic, spot welded frusta on square foundation subjected to axial impact compressive force. Particularly, the influence of the angle of inclination of the frustum wall was examined.

There are several crashworthiness indicators [10] used to evaluate the crashworthiness of the energy absorbing structure (energy absorber). In the analysis presented in [9] the following indicators have been applied:

- energy absorption (EA);
- mean crushing force (MCF);
- crash load efficiency (CLE).

The results of the parametric study indicated significant decrease of peak crushing force with the increase of frustum angle. It was stated that the influence of frustum angle on the crashworthiness indicators mentioned above is different for different wall thickness values and that relation of the energy absorption in terms of frustum angle is non-linear. The parametric study was carried out using commercial code Abaqus(*) and explicit dynamic analysis [11].

The present paper deals with the experimental study into crushing behavior of steel thin-walled prismatic, spot welded frusta. The study has been performed in order to validate results of the theoretical analysis presented in [9], namely to analyze an actual crushing behavior of frusta under investigation, among others to study an influence of spot welds on the crushing behavior and mode of failure.

2. Specimens under investigation

The subject of investigation was a thin-walled prismatic frustum on square foundation of constant edge length a = 90 mm and height h =160 mm, shown in Fig. 1, subjected to axial, dynamic (impact) compressive load. Parallelepipeds and frusta of angle of inclination (frustum angle) $\alpha = 5°$ were investigated. Specimens were manufactured from steel sheet of thickness t = 1 mm. The material used was the high quality steel Dual Phase DP 800, with main strength parameters: yield stress σ_0 = 590 MPa and Young modulus E = 210 GPa.

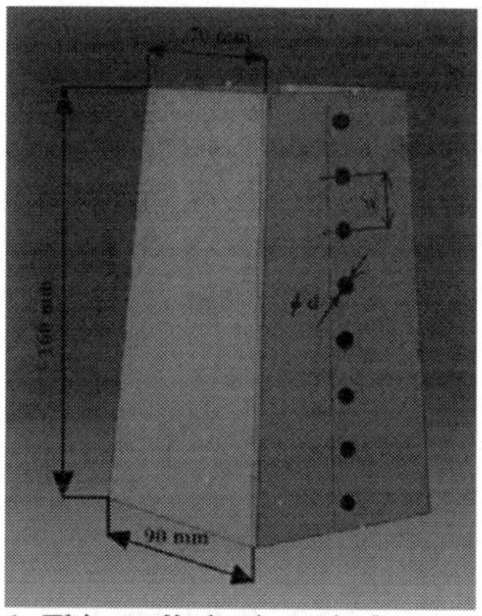

Fig. 1: Thin-walled prismatic frustum

The columns were fabricated from two face – to face channel sections (produced on the folding machine) and joint by means of two seams of one – point spot welds. The distance between spot welds was w = 20 mm.

3. Test results

3.1 Quasi-static tests

Preliminary quasi-static tests were performed on the testing machine Instron, with the load velocity v_0 = 600 mm/min. parallelepiped and frustum specimens of two different spot welds diameter d = 4, 8, [mm] were tested. Typical failure modes, initialized by folding mechanism were observed. At some specimens of spot welds of diameter d = 4 mm, failure by seam tearing was observed, although it did not change significantly the final failure mode.

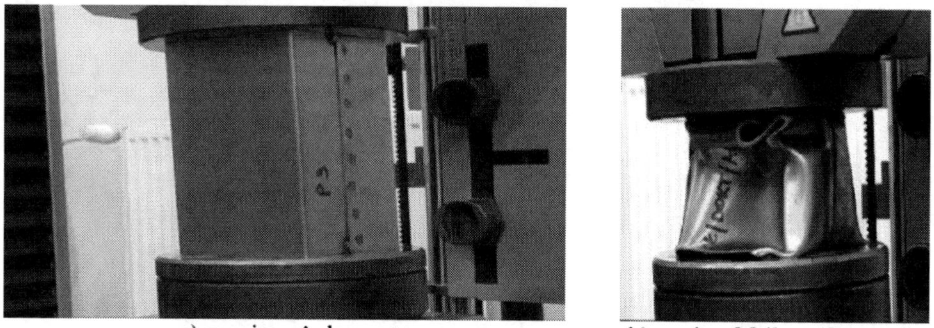

a) specimen's lay-out b) mode of failure (frustum)

Fig. 2: Quasi-static tests

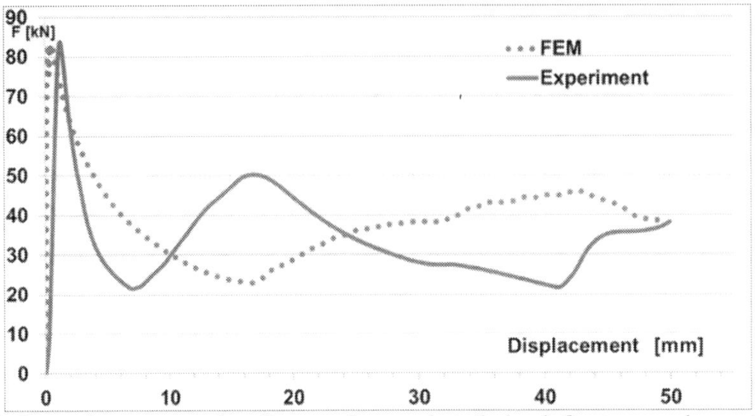

Fig. 3: Load-shortening comparative diagram for quasi-static load: frustum angle $\alpha = 5°$, $d = 8$ mm

Fig.3 shows the comparison of experimental and theoretical load – shortening diagram for frustum. The FE model is described in details in par. 4. FE curve in Fig. 3 was obtained for elasto-plastic material model, neglecting strain rate effect.

3.2 Impact tests

Impact tests were performed on the drop-hammer rig, described in details in [12]. Drop hammer rig consists of the rigid steel frame (12), to which two linear guides are fixed. Impactor (7) is kept between guides by the wheels on ball bearings. For impactor lifting the rig is equipped with two lifts: lower (3) and upper (6). Lower lift is propelled by the cable winch. Special pneumatic lock/unlock mechanism is located inside the lower lift. This mechanism is used for impactor dropping. When additional spring drive of impactor is used, an upper lift has to be used.

Separate computer with specific data acquisition software records data from transducers via signal conditioning system.

1 – cable winch,
2 – screw jack drive,
3 – upper lift,
4 – springs,
5 – linear guide,
6 - lower lift,
7 – impactor (hammer),
8 – piezoelectric accelerometer,
9 – specimen,
10 – bumper,
11 – piezoelectric force transducer,
12 – frame

Fig. 4: Experimental stand for impact tests – drop hammer rig

Dynamic (impact) tests were performed on frusta with spot welds of diameter $d = 8$ mm only. The specimens were situated between two rigid steel plates, as shown in Fig. 5a. Failure modes observed did not differ from those spotted in quasi-static tests.

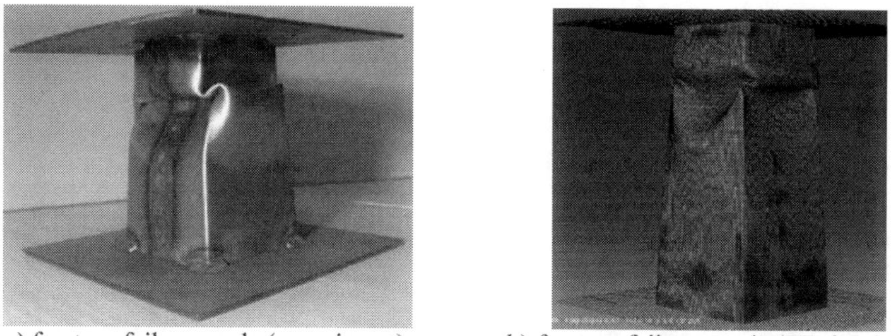

a) frustum failure mode (experiment) b) frustum failure mode (FE simulation)

Fig. 5: Frustum failure mode under impact

4. Validation of FE simulations

Finite element calculations were carried out using commercial code Abaqus and explicit dynamic analysis [11]. An FE model of the column was created using 4-node shell elements S4R. A column model was situated between two rigid elements R3D4. Between upper and lower rigid elements and the column Tie links were applied. In the analysis the residual stress coming from forming process of the metal sheet were taken into account. FE simulation of sheet folding process was performed. Then, the equivalent residua stresses obtained in this simulation were used as initial conditions in the explicit analysis of impact behavior. The spot welds were modelled using the rigid beam elements, situated between two overlapping shell elements. Using the FASTENER elements nodes in the area of the spot weld were connected [11]. Three material models taking into account the strain rate and strain hardening have been applied in the FE analysis, namely: the rate-dependent elasto-plastic material model, Cowper-Symonds and Johnson- Cook model. Equivalent stress and equivalent plastic strain were calculated using Huber-Mises yield criterion.

Fig. 5 shows the comparison of experimental and theoretical failure patterns. The diagrams, showing the comparison of FE simulations based on three material models, together with experimental results are presented in Fig. 6. Curve "strain rate" concerns the rate dependent elasto-plastic material model. Fig. 7 shows experimental and theoretical load-shortening diagrams for frustum. The theoretical diagram corresponds to the Cowper-Symonds material model.

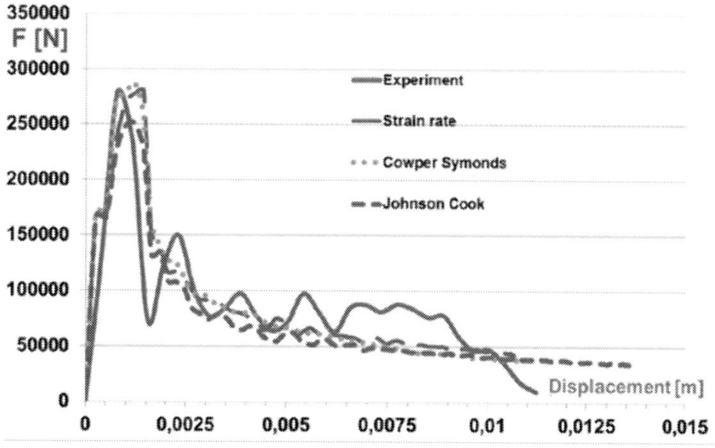

Fig. 6: Comparative load – shortening diagrams (parallelepiped, d = 4 mm); load velocity v_0= 6.19 m/s

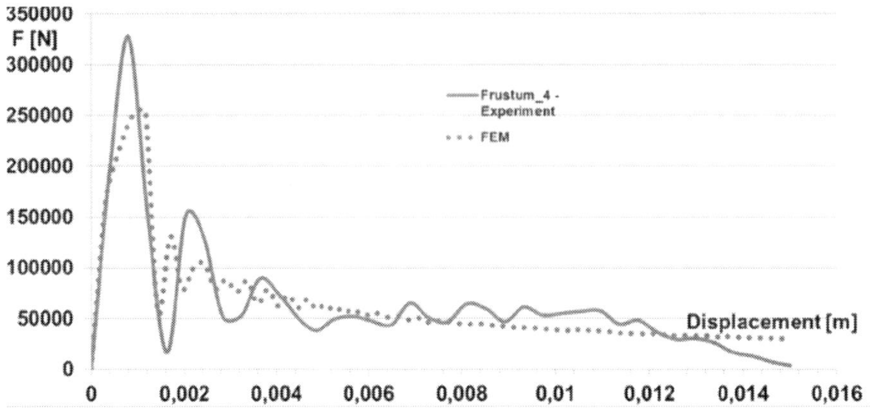

Fig. 7: Comparative load – shortening diagrams (frustum, α = 5°, d = 8 mm); load velocity v_0= 6.19 m/s

5. Conclusions

The main conclusions concerning the crushing behaviour of examined frusta are as follows:

1. the crushing behaviour did not depend significantly from spot welds diameter (even in the case of seam tearing);
2. observed failure modes were similar for quasi-static loads and impact loads.

The conclusions concerning the validation of FE simulations are:

1. an agreement of experimental and FE results was satisfactory and confirmed the correctness of the applied FE model;
2. applied materials models delivered very close results (Fig.6), the best agreement with the experimental results was obtained for Cowper-Symonds model;
3. for specimens with spot welds of diameter d=4 mm (Fig.6) ultimate loads obtained experimentally and theoretically were nearly the same, while for specimens with spot welds of diameter d=8 mm (Fig.7) experimental results were about 15% higher than theoretical; the reason of that discrepancy was the actual stiffness of spot weld seams higher, than assumed in the theoretical FE model.

Results presented in the paper are a part of wider research project. Further research is planned into the investigation of impact load velocity on the frusta crushing behaviour and into increasing the energy absorption capability by means of foam filling.

Acknowledgments

Finite Element calculations carried out in Abaqus by CYFRONET AGH. MNiSW/IBM_BC_HS21/PŁódzka/013/201

References

[1] Alghamdi, "Collapsible impact energy absorbers: an overview", *Thin-Walled Structures*, 39, 189-213, 2001.

[2] Ferdynus M. "An energy absorber in the form of a thin-walled column with square cross-section and dimples", *Eksploatacja i Niezawodność (Maintenance and Reliability)*, 15(3), 253-258, 2013.

[3] Ferdynus M, Kotełko M. "Dynamic axial crushing of flawed thin-walled square section tubes", *Stability of Structures XIV Symposium, Department of Strength of Materials*, Lodz Univesity of Technology, 39-40, 2015.

[4] Abbasi M, Reddy S, Ghafari – Nazari A, Fard M. "Multiobjective crashworthiness optimization of multi-cornered thin-walled sheet metal members", *Thin-Walled Structures*, 89, 31–41,2015.

[5] Reddy S, Abbasi M, Fard M. "Multi-cornered thin-walled sheet metal members for enhanced crashworthiness and occupant protection", *Thin-Walled Structures*, 94, 56-66, 2015.

[6] Ali M, Ohioma E, Kraft F, Alam K. "Theoretical, numerical, and experimental study of dynamic axial crushing of thin walled pentagon and cross-shape tubes", *Thin-Walled Structures*, 94, 253-272, 2015.

[7] Sharifi S, Shakeri M, Ebrahimi Fakhari H, Bodaghi M. "Experimental investigation of bitubal circular energy absorbers under quasi-static axial load", *Thin-Walled Structures*, 89, 42-53, 2015.

[8] Sarkabiri B, Jahan A, Rezvani M. "Multi-objective crashworthiness optimization of thin-walled conical groove tubes filled with polyurethane foam", *Proceedings 3rd Polish Congress of Mechanics, 21st International Conference on Computer Methods in Mechanics*, 947-948, 2015.

[9] Kotełko M, Mołdawa A. "Impact behaviour of spot-welded thin- walled frusta", to be published in Acta Mechanica et Automatica.

[10] Jones N, "Structural Impact". *Cambridge University Press*, 2003.

[11] Dassault Systems, "Abaqus 6.13 Documentation", *Abaqus Online Documentation*, 2013.

[12] Kotełko M, Mołdawa A, Jankowski M. "Axial impact of open-section TWCF columns experimental study", *Romanian Journal of technical sciences - applied mechanics* (former Revue Roumaine des sciences techniques, série de mécanique appliquée), (vol. 59, no. 1-2, 2014); Special Issue "Stability and nonlinear analysis of steel structures – research advances", 59, 72-86, 2014.

INITIAL IMPERFECTIONS FOR FEA OF TUBULAR BEAMS BASED ON VALIDATED EXPERIMENT RESULTS

Marko Pavlović and Milan Veljković

Delft University of Technology, Faculty of Civil Engineering and Geosciences (CITG), Delft, The Netherlands

Abstract: The experiment data available on 4-point bending experiments performed in Australia and Europe on tubular section beams are validated by FEA. Large scattering of available experiment results is investigated by detailed analysis of different geometrical imperfections. Loading set-up is analysed through modelling of welding induced imperfections. It is found that the welding induced imperfections lead to more accurate prediction of the local buckling failure modes and the rotation capacity than the traditionally modelled geometrical Eigen-mode imperfections. Also the influence of measured residual stresses is examined.

1. Introduction

Resistance and rotation capacity of tubular beams depend on strength of the material but also the local buckling which is influenced by slenderness of the cross section and real imperfections: geometric, residual stresses and inhomogeneity of the material. Large scattering of experiment data on the resistance and rotation capacity is present in published experiment results, such as [1] for I beams, which is used as basis for current cross section classification rules in the Eurocode both for open and closed cross sections. Results of 4–point bending experiments on tubular beams [4], [5] and [8] also show large scattering, especially in terms of rotation capacity, see Fig. 1. Reasons for such large scattering are differences in set-ups: cross-section and beam scale and loading point detail, but also on initial imperfections which depend on production process and type of the material: cold-formed, hot-finished, mild steel, HS steel, etc. The most unfavourable equivalent wave-like shape geometric imperfections obtained in linear buckling Eigen-mode analysis are used in state-of-the-art buckling FEA to validate the experiment results and for parametric studies. However, measurements of imperfections on mild steel and high strength steel hollow section beams [4], [6], [8] and [9] report constant longitudinal distribution of mostly convex, “bow-out” geometric imperfection in a wide amplitude range, approximately 1/100 – 1/1000.

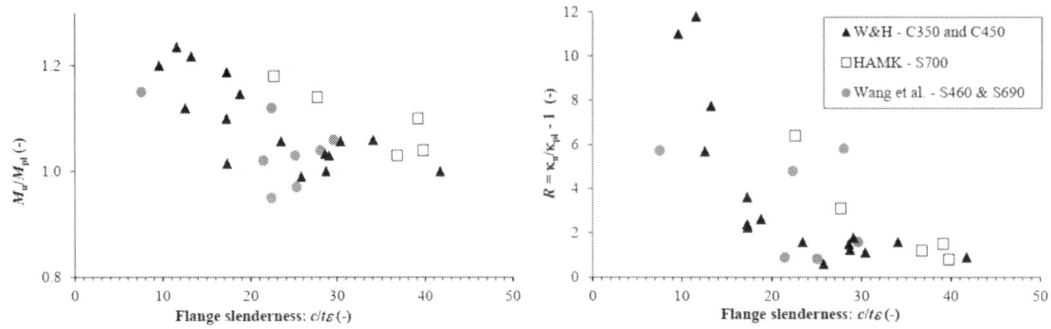

a) Non-dimensional bending resistance b) Rotation capacity

Fig. 1: Scattering experiment results from [4], [5] and [8]

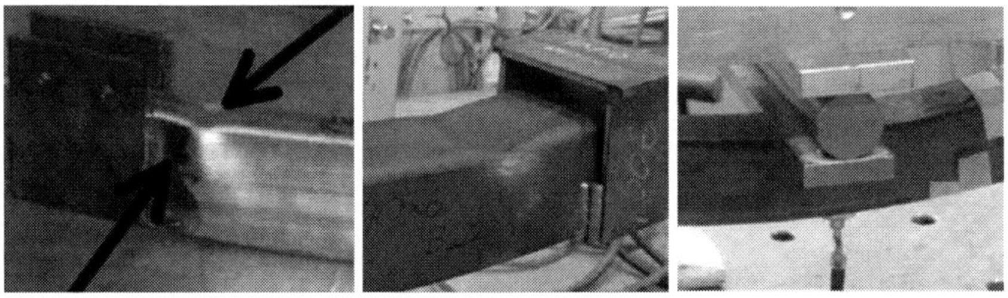

a) Wilkinson and Hancock [4] b) HAMK experiments [5] c) Wang et al. [8]

Fig. 2: Local web and flange buckling near the loading point in bending experiments

Scattering of Eigen-mode imperfections amplitudes used to validate experiment results of different research is also large and do not coincide in shape, nor in amplitude with the measured imperfections [7], [8]. EN 1993-1-5 [2] recommends equivalent bow imperfection of the local subpanel in a cross section as 1/200, either as one bow or a wave-like shape imperfection.

Local buckling in 3-point and 4-point bending experiments occurs in the vicinity of the loading point, see Fig. 2. Wilkinson and Hancock [4] used loading detail by welding side plates over the full height of the web. In HAMK experiments side plates are welded only in the bottom half of the web in order to prevent welding imperfections in the compression zone of the hollow section. Wang et al. [8] applied force by direct contact over the thick plate.

Residual stresses dependent on the forming process (continuous forming, direct forming) might also affect the local buckling. Almost zero membrane residual stresses are obtained in several independent experimental research, see [13]. However, significant bending residual stresses, between 50% and 90% of the yield strength, are observed in the longitudinal and transverse direction of the tubular beam.

Welding imperfections are suspected to have major influence on local buckling in analysed experiments. Therefore they are validated by FEA where welding induced imperfections and residual stresses due to forming are applied. Afterwards, the influence of welding induced imperfections is compared to currently used equivalent geometric imperfections.

2. FE model of reference experiments

The finite element modelling (FEM) in ABAQUS finite element software package [3] is validated by 4-point bending experiments of hollow section beams: on shorter beams having spans between 1.3 and 1.7 m made of S460 performed by Wilkinson and Hancock [4] and on longer span beams made of S700 having spans from 3 m to 6.8 m performed by HAMK [5]. Cold-formed hollow sections are produced in continuous forming (CF) process in both cases.

Totally 9 cases are analysed having wide range of spans, cross-section dimensions, thicknesses and aspect ratios, see [14].

Experiments are modelled using shell elements S3D4r at the outer surface for the beam and solid elements C3D8R for loading plates and welds with proper coupling by tie constraint surface pairs, see Fig. 3. Two vertical plane symmetry boundary conditions is exploited in FEA. Loading point details differences between [4] and [5] experiments, in terms of height of the weld on the web, are carefully taken into account. Exact material properties obtained in coupon tensile tests of each specimen in bending experiments are considered in FEA. Different material properties are used for the flange and the web, according to coupon tests results. General/Static solver with load application by displacement is used in Abaqus [3].

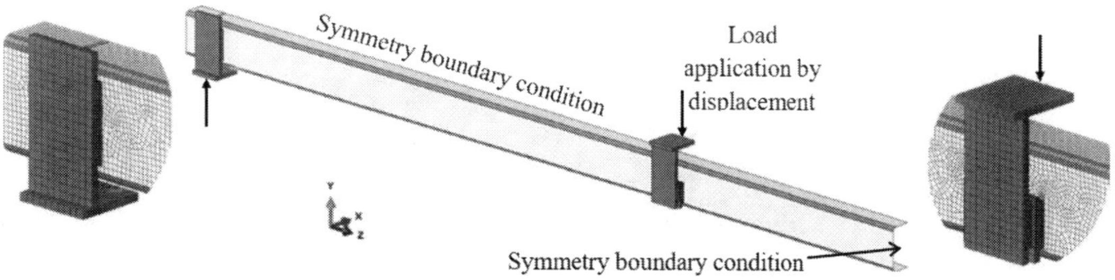

Fig. 3: FE model of 4-point bending experiments

3. Initial imperfections

3.1. Welding induced imperfections

Welding of steel plates causes local deformations and residual stresses in heat affected zone (HAZ). Angular distortion, also called the transverse bending (TB) [10], shown in Fig. 4a, is the most important in the case of side plates welded to the web of the hollow section beam analysed in this study because it produces the critical local geometrical imperfection. The amount of angular distortion due to welding depends on many parameters, such as: type of the weld and welding technique, input heat, thickness of the plates and the weld, etc. However, research results published in [10], [11], combining experiments and thermo-elastic-plastic (TEP) analysis of bead on plate and T-joint fillet welds showed that the angular distortion has an limiting value of $TB_{\text{lim}} = 0.022$ rad, depending mostly on the input heat and thickness of the plate see Fig. 4.

Prediction of welding induced deformations is employed by inherent strain method (ISM). Conservatively, the extreme angular distortion due to welding available in literature $TB = 0.022$ rad is used in FEA of 4-point bending as geometrical imperfection. TEP analysis is performed on plain strain 2D model of the longitudinal section in the middle of the height of the weld in Abaqus Welding Interface (AWI) [12], see Fig. 5. The aim was to obtain the model of distribution of the plastic strains in the HAZ and weld, producing the angular distortion. Effective width of longitudinal plastic strains in the web below the weld is estimated as given in Eq. (1), where a_{w} is the weld throat thickness. For the prescribed value of the angular distortion TB, see Fig. 5c, the induced strain ε_{w} at the outer face of the web is given in Eq. (2).

$$b_{\text{w,eff}} = 1.5a_{\text{w}}\sqrt{2} \quad (1)$$

$$\varepsilon_{\text{w}} = TB \cdot t/b_{\text{w,eff}} \quad (2)$$

Based on the model of distribution the effective width $b_{\text{w,eff}}$ and the induced strain ε_{w} are calculated for each case considered in FE validation of 4-point experiments and applied to the

4-point bending FE model, see Fig. 6. Strains are applied only in longitudinal direction of the beam using orthotropic expansion properties of the material and distribution shown in Fig. 5b.

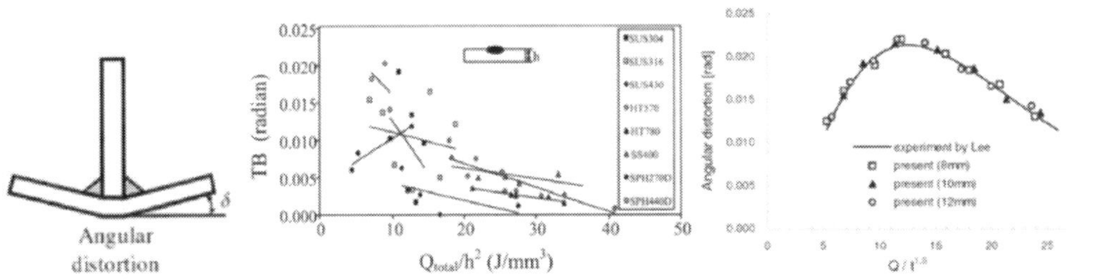

a Shape of distortion b) bead-on-plate welds, 3-5 mm [11] c) fillet welds, plates 8-10 mm [10]

Fig. 4: Transverse bending (angular distortion) due to welding

a) TEP analysis of fillet weld between the web and the loading plate

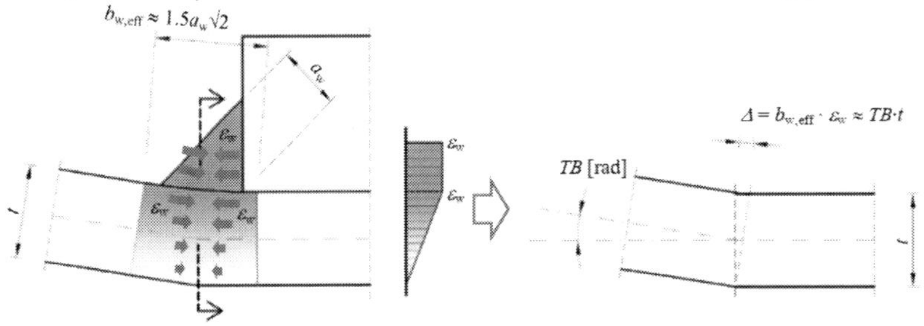

b) induced strain distribution c) induced strain vs. angular distortion *TB*

Fig. 5: Model of welding induced plastic strain distribution.

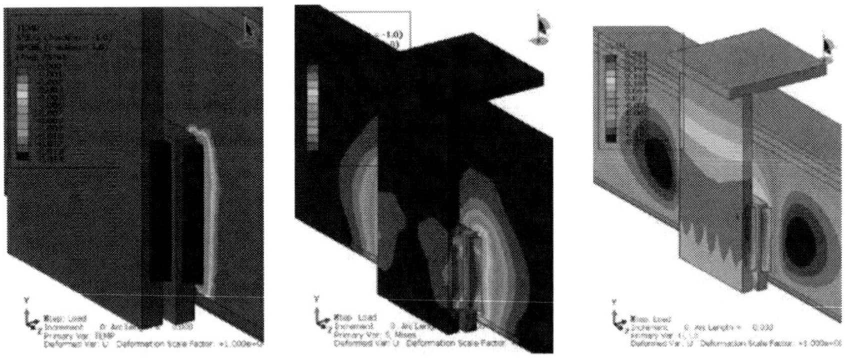

a) strains applied by temperature b) von Mises stresses c) out of plane deformations

Fig. 6: Welding induced strains applied in FE model of RHS 140×140×6

3.2. Residual stresses

Bending residual stresses are indirectly applied in FEA by equivalent temperature strains using orthotropic thermal expansion material properties to define different signs of the longitudinal and transverse direction, see [14]. Application of approximately 50 % of the yield strength residual stresses is achieved.

3.3. Equivalent geometric imperfections

Influence of geometric imperfections in form of Eigen-shape and bow-out imperfections, shown in Fig. 7, are analysed with the aim to show their influence on the scattering of rotation capacity obtained in experiments, in comparison to welding induced imperfections. Square hollow sections from the larger scale experiments [5] and nominal material properties of S500, S700 and S960 materials are used, see [14]. Two sets of imperfection amplitudes are analysed: a) 1/700 – comparable to the imperfections obtained by applying welding induced angular distortions and b) 1/200 – according to EN 1993-1-5 [2] recommendation.

a) Eigen-mode due to bending load b) Eigen-mode due to axial load c) "bow-out" imperfections

Fig. 7: Typical equivalent geometric imperfections used in nonlinear analysis

4. Results and discussion

4.1. Welding induced imperfections vs. residual stresses

Influence of different imperfections are presented in Fig. 8 and Fig. 9: a) without any imperfections, b) with welding induced imperfections only, c) with residual stresses only and d) with welding induced imperfections and residual stresses. Ultimate resistance is well estimated by FEA even if no imperfections are applied. However, the rotation capacity is overestimated and in most of the cases the obtained failure mode in FEA does not reflect experiments because the local buckling occurs in the mid-span of the beam.

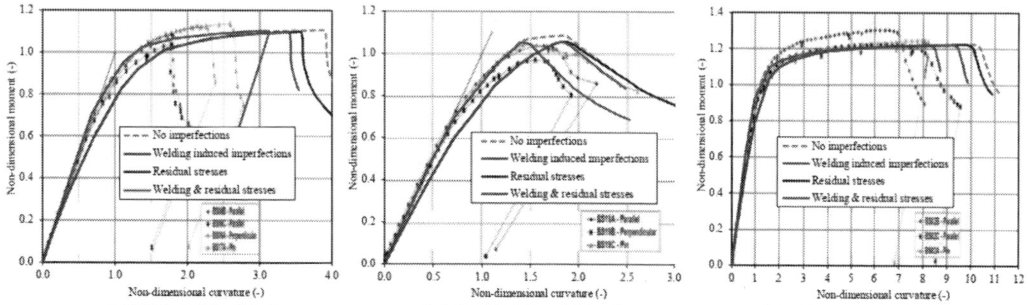

a) RHS 100x50x2 b) SHS 100x100x3 c) RHS 150x50x4

Fig. 8: Experiment [4] vs. FE results

Applying welding imperfections gives very good results in terms of prediction of rotation capacity, see Fig. 8 and Fig. 9. Also the characteristic failure mode obtained in experiments, local buckling near the loading plates, is obtained in each case in FEA, see Fig. 10. If only the residual stresses are applied as defined in section 0 the obtained rotation capacities are overestimated. The square hollow sections showed the same behaviour as if no imperfections are applied and in case of the rectangular hollow sections only up to 10% reduction of the rotation capacity is obtained. If both the welding imperfections and residual stresses are applied the rotation capacity is increased compared to the case where only welding imperfections are applied. Finally, it is concluded that best agreement between the experiment and FEA results is obtained if only welding induced imperfections are applied in FEA of 4- point bending of hollow section beams.

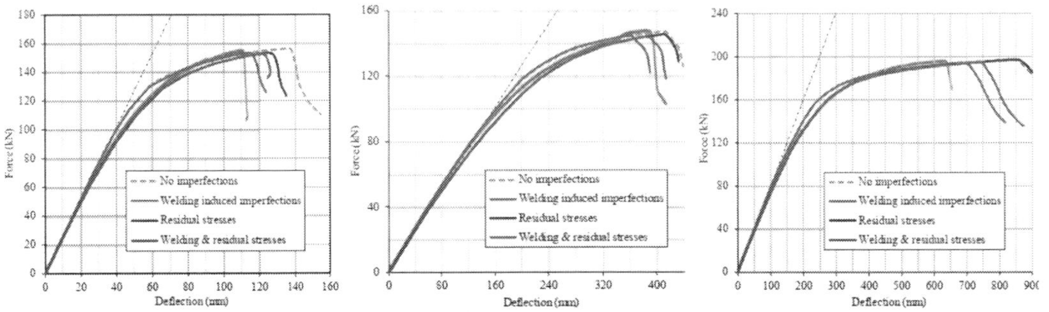

a) RHS 150x100x4 c) SHS 150x150x6 d) SHS 150x150x8

Fig. 9: Experiment [5] vs. FE results

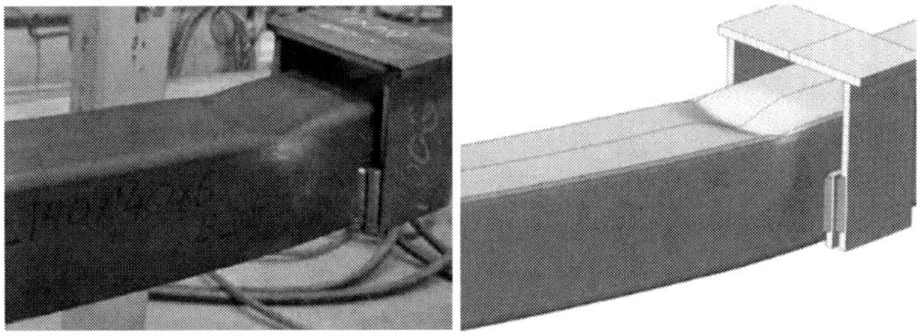

Fig. 10: Failure mode in experiments vs. FEA with welding induced imperfections

4.2. Welding induced imperfections vs. equivalent geometric imperfections

Large scattering of non-dimensional bending resistance and rotation capacity is obtained with the analysed imperfection parameters, see Fig. 11. The average variation ranges of the obtained results are 0.15 and 3 for the bending resistance and the rotation capacity, respectively. Similar scattering can be observed in experiments, see Fig. 1.

Higher resistance and capacity is obtained with bow-out imperfections having amplitude 1/200, compared to the welding induced imperfections having equivalent amplitude of 1/700. This leads to the conclusion that in absence of loading plates welded to the hollow section, higher ductility of the hollow section beams can be obtained.

Much lower bending resistance and rotation capacity is obtained with equivalent Eigen-mode imperfections, compared to the welding induced and bow-out imperfections. The influence of the Eigen-mode imperfection shape (caused either by bending or axial loads), is insignificant compared to influence of the imperfection amplitude. It is concluded that applying the Eigen-mode equivalent geometric imperfection with amplitude 1/200, according to EN 1995-1-5 [2], is unrealistic compared to behaviour obtained with more realistic welding induced imperfections.

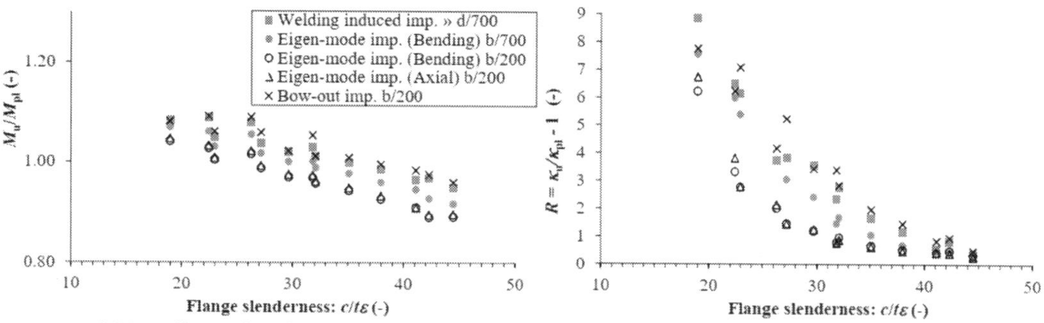

a) Non-dimensional bending resistance b) Rotation capacity

Fig. 11: FEA results for high-strength steel with different imperfections

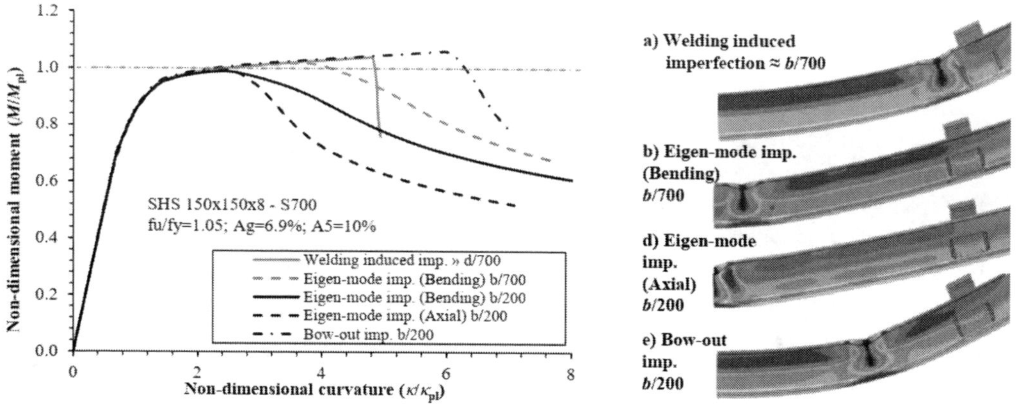

Fig. 12: Moment-rotation curves and location of local buckling with different imperfections

Influence of the imperfection type on post buckling behaviour is shown in Fig. 12 by non-dimensional moment-rotation curves and by location of the local buckling. The local buckling near the loading point in case of the welding induced and bow-out imperfections results in sudden reduction of the bending moment and therefore lower rotation capacity. This is related to the fact that the curvature is calculated based on differences of the deflections at the loading points and at the mid-span of the beam. Once the plastic hinge is formed at the loading point in case of the welding induced imperfections, the three points have constant increase of the deflections and the calculated curvature is almost constant. In case of the bow-out-imperfections the local buckling occurs a bit towards the mid-span and therefore the described phenomenon is less pronounced. In case of the Eigen-mode imperfections the local buckling in the mid span influence the increase of difference in deflections of the loading points and the mid-span and therefore the curvature rapidly increases.

It is concluded that the rotation capacity in the 4-point bending experiments is influenced by position of the local buckling between the loading points (the region of the constant bending moment) which is the additional reason for large scattering of experiment results.

5. Conclusions

Following conclusions are drawn from the study:

1. FEA validation of series of experiment results clearly showed the dominating influence of welding induced imperfections compared to the influence of residual stresses.

2. Least scattering of the experiment results is obtained in experiments where the loading plates are welded only to the part of the cross section which is in tension. This loading detail is recommended for the 4-point bending set-up.
3. The Eigen-mode equivalent geometric imperfection with amplitude 1/200, according to EN 1995-1-5 [3], leads to a bit pessimistic results compared to calculations using the welding induced imperfections.

Acknowledgments

This work was partially supported by RFCS through the Research Project RUOSTE. The authors gratefully acknowledge Prof. Markku Heinisuo for providing valuable information for the project.

References

[1] Sedlacek G, Feldmann M. "The b/t ratios controlling the applicability of analysis models in Eurocode 3, Part 1.1", *Background Document 5.09 for chapter 5 of Eurocode, Part 1.1*, Aachen, Germany 1995.
[2] EN-1993-1-5: *Design of steel structures, Part 1-5: Plated structural elements*, Brussels, Belgium, European Committee for Standardization, 2006.
[3] *ABAQUS Version 6.13 Analysis User's Guide*, 3DS Simulia, 2014.
[4] Wilkinson T, Hancock G. *Tests for the compact web slenderness of cold-formed rectangular hollow sections*, Research Report R744, The University of Sydney 1997.
[5] Zhongcheng M, Havula J. *Bending tests of S700 tubes*, Report 2015-56, University of Applied Sciences HAMK, Hämeenlinna, Finland 2015.
[6] *RUOSTE Project - Background Document on Deliverable 2.2 – Local buckling*. RFCS 2016.
[7] Wilkinson T, Hancock G. *Finite element analysis of plastic bending of cold-formed rectangular hollow section beams*, Research Report R792, The University of Sydney 1999.
[8] Wang J, Afshan S, Gkantou M, Theofanous M, Baniotopoulos C, Gardner L. "Flexural behaviour of hot-finnished high strength seet square and rectangular hollow sections", *Journal of Construction al Steel Research,* 121, 97-109, 2016.
[9] Ma JL, Chan TM, Young B. "Experimental Investigation on Stub-Column Behavior of Cold-Formed High-Strength Steel Tubular Sections", *Journal of Structural Engineering ASCE*, 2015.
[10] Jang CD, Lee CH, Ko DE. "Prediction of welding deformations of stiffened panels", *Proc. Instn. Mech. Engrs., Part M: J Engineering for the Maritime Enviroment,* Vol. 216, 133-143, 2002.
[11] Wang R, Zhang J, Serizawa H, Murakawa H. "Study of welding inherent deformations in thin plates based on finite element analysis using interactive substructure method", *Materials and Design,* 30, 3274-3481, 2009.
[12] *ABAQUS Extensions: Abaqus Welding Interface*, 3DS Simulia, 2015.
[13] *RUOSTE Project - Background Document on Deliverable 2.1 – Residual Stress Measurements*. RFCS 2016.
[14] *RUOSTE Project - Background Document on Deliverable 1.5 – Rotation capacity of beams*. RFCS 2016.

AXIAL BEHAVIOUR OF PRESTRESSED HIGH STRENGTH STEEL TUBULAR MEMBERS

Jie Wang[a], Sheida Afshan[b] and Leroy Gardner[a]

[a] *Imperial College London, London, UK*
[b] *Brunel University London, London, UK*

Abstract: The axial behaviour of high strength steel tubular elements with internal prestressing cables, representing the chord members in prestressed trusses, is investigated herein. Experiments on tensile and compressive members were carried out, with the key variables examined being the steel grade (S460 and S690), the initial prestress level and the presence of grout. FE models were developed to replicate the experiments and generate parametric results. The presence of cables was shown to enhance the tensile load-carrying capacity of the members while the application of prestress extended the elastic range. In compression, prestressing was detrimental, and a modified Perry-Robertson design approach was examined.

1. Introduction

Prestressed steel trusses can offer high material efficiency for long span applications such as aircraft hangars, stadia and warehouses [1, 2]. Cable-in-tube systems, whereby the steel cables are housed within hollow structural sections, are one way to achieve these structural forms. The performance of cable-in-tube systems in normal strength steelwork has been studied in [3, 4]. This paper focuses on the behaviour of such systems in high strength steels (HSS).

2. Experiments

2.1 General

To assess the response of cable-in-tube systems with different prestress levels and with addition of grout, a total of 22 specimens were tested under axial loading - 12 in tension (6 non-grouted and 6 grouted) and 10 in compression (4 non-grouted and 6 grouted). Fig. 1 depicts a typical test specimen. The steel tubes were hot-finished S460 and S690 SHS 50×50×5, which encased Y1860S7 cables. Their material and geometric properties are given in Table 1, where E, f_y and A are the Young's modulus, yield stress and cross-sectional area, respectively. The strength of the grout was derived from cube tests performed on the same day as the member test. The nominal grout strength was 50 N/mm^2.

The key variables of the cable-in-tube specimens were the steel grade (S460 and S690), initial prestress level (No cable, P_{nom}, $0.5P_{opt}$ and P_{opt}) and presence of grout. Fig. 2 gives the labelling system employed throughout the paper. The prestress P_{nom} is a nominal prestress level (around 5 kN) applied to ensure that the cables were taut during grouting and testing. The prestress P_{opt}, as defined in [3], is the optimum initial prestress force that causes the cable and the tube to yield simultaneously when the system is under tension, which maximises the extent of the elastic range. The value of P_{opt} depends on the material and geometric properties of the tube and cable, and can be calculated from Eq. (1) [3], where A, E and f_y are as given in Table 1 and the subscripts c and t denote cable and tube, respectively. Connecting collars were employed at quarter points to ensure that the members did not buckle under the initial prestress [3]. From the measured properties, it was determined that P_{opt} = 189 kN for the S460 members and 167 kN for the S690 members. Note that the actual levels of prestress achieved in the experiments differed slightly from these target values and are reported in the following sections.

$$P_{opt} = \left(\frac{A_c A_t}{A_t E_t + A_c E_c}\right)\left(f_{cy} E_t - f_{ty} E_c\right), \text{ but } P_{opt} \le f_{ty} A_t \text{ and } f_{cy} A_c \quad (1)$$

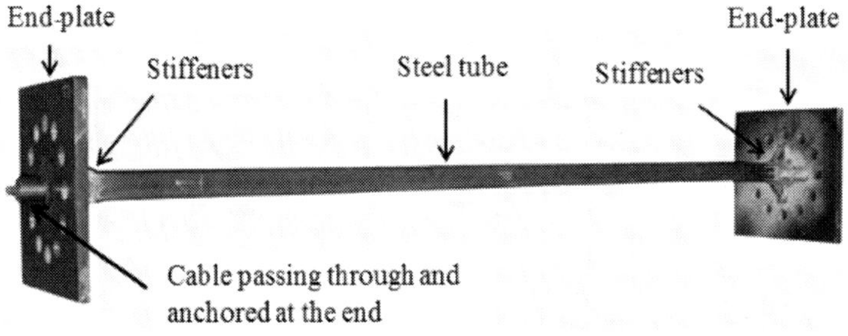

Fig. 1: Configuration of the cable-in-tube specimens

Table 1: Measured material and geometric properties of steel tubes and cable

Component	E (N/mm^2)	f_y (N/mm^2)	A (mm^2)
S460 SHS 50×50×5	210000	505	858
S690 SHS 50×50×5	208000	759	841
Cable	130000	1703	151

2.2 Tensile member tests and results

A total of 12 members were tested under tensile loading. The measured dimensions and prestress levels of the tensile specimens are presented in Table 2, where L_0, h, b, t and r_{ex} are the initial member length before prestressing, depth, width, wall thickness and average outer corner radius of the tube members, respectively, and P_i is the actual initial prestress level achieved. An Intron 2000 kN machine was employed for the tensile member tests. The end-plates of the specimens were bolted onto the base and moving head of the machine, and displacement control was used to load the specimens at a rate of 0.5 mm/min. The applied load, axial displacement and four longitudinal strains at the mid-height of the tubes were recorded during the tests at 1 s intervals. The obtained axial load-displacement curves are plotted in Figs. 3(a) and 3(b) for the S460 and S690 specimens, respectively. The results are discussed in Section 4.

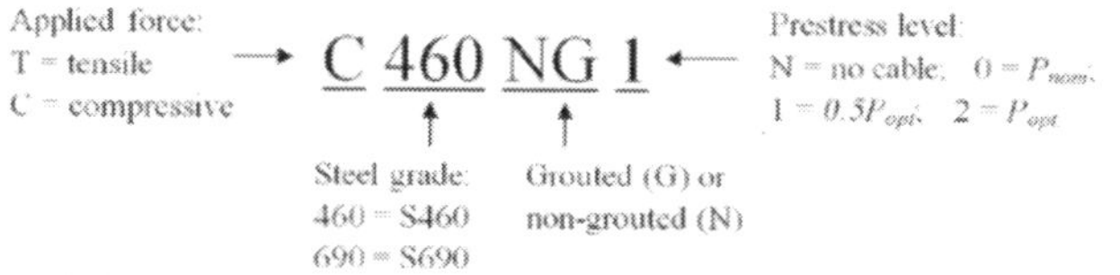

Fig. 2: Labelling convention for cable-in-tube test specimens

Table 2: Measured geometric dimensions and prestress levels of tensile members

Specimen	L_0 (mm)	h (mm)	b (mm)	t (mm)	r_{ex} (mm)	P_i (kN)
T460NGN	1954	50.07	50.36	5.01	5.63	N/A
T460NG0	1997	50.27	50.35	5.01	5.63	4.7 (P_{nom})
T460NG1	2001	50.28	49.96	5.01	5.63	77.3 ($0.41P_{opt}$)
T460NG2	2000	50.32	50.15	5.01	5.63	153.7 ($0.81P_{opt}$)
T460G1	2002	50.11	50.25	5.01	5.63	84 ($0.44P_{opt}$)
T460G2	2001	50.14	49.88	5.01	5.63	150.7 ($0.80P_{opt}$)
T690NGN	2000	50.18	50.43	4.91	5.88	N/A
T690NG0	2000	50.27	50.37	4.91	5.88	5.7 (P_{nom})
T690NG1	1999	50.11	50.36	4.91	5.88	65 ($0.38P_{opt}$)
T690NG2	2003	50.50	50.12	4.91	5.88	131.0 ($0.78P_{opt}$)
T690G1	2001	50.26	50.40	4.91	5.88	86.2 ($0.52P_{opt}$)
T690G2	1954	50.38	50.28	4.91	5.88	124.3 ($0.74P_{opt}$)

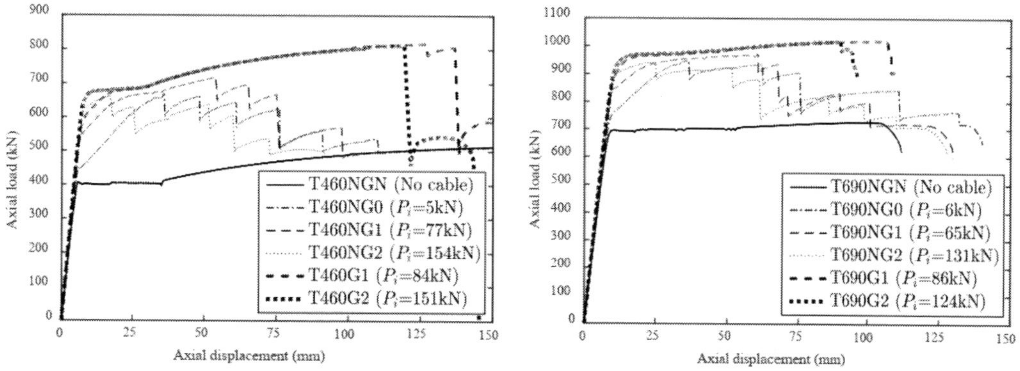

a) S460 members b) S690 members

Fig. 3: Axial load vs. displacement relationship of tensile members

2.2 Compressive member tests and results

The dimensions and prestress levels of the compressive specimens are reported in Table 3, where ω_i is the measured initial bow imperfection amplitude derived from strain gauge readings [5]. Fig. 4 illustrates the test setup. Pin-ended boundary conditions were applied through hardened steel knife edge supports, which allow only in-plane rotation of the members about one axis. Cylinders were used to connect the knife edges to the end-plates, encasing the anchoring system at both ends. The distance between the top and bottom knife edges was taken as the buckling length L_{cr} of the specimens. The monitored variables during testing included all those in the tensile member tests plus the lateral deflection at mid-height, which was recorded using a string pot. The load-lateral deflection curves for the S460 and S690 specimens are shown in Figs 5(a) and 5(b), respectively.

Table 3: Measured dimensions and prestress levels of compressive members

Specimen	L_0 (mm)	h (mm)	b (mm)	t (mm)	r_{ex} (mm)	ω_i (mm)	P_i (kN)
C460NG0	1003	50.32	50.19	5.01	5.63	1.28	6.8 (P_{nom})
C460NG2	1002	50.11	50.31	5.01	5.63	1.58	125.6 ($0.67P_{opt}$)
C460G0	1001	50.13	50.42	5.01	5.63	0.92	7.1 (P_{nom})
C460G1	1002	50.36	50.10	5.01	5.63	1.61	70.8 ($0.37P_{opt}$)
C460G2	1002	50.11	50.40	5.01	5.63	1.03	152.7 ($0.81P_{opt}$)
C690NG0	1000	50.42	50.14	4.91	5.88	0.82	3.7 (P_{nom})
C690NG2	1000	50.38	50.20	4.91	5.88	0.81	120.8 ($0.72P_{opt}$)
C690G0	1000	50.11	50.35	4.91	5.88	0.44	5.0 (P_{nom})
C690G1	1002	50.14	50.39	4.91	5.88	1.88	65.2 ($0.39P_{opt}$)
C690G2	1001	50.26	50.26	4.91	5.88	0.06	123.5 ($0.74P_{opt}$)

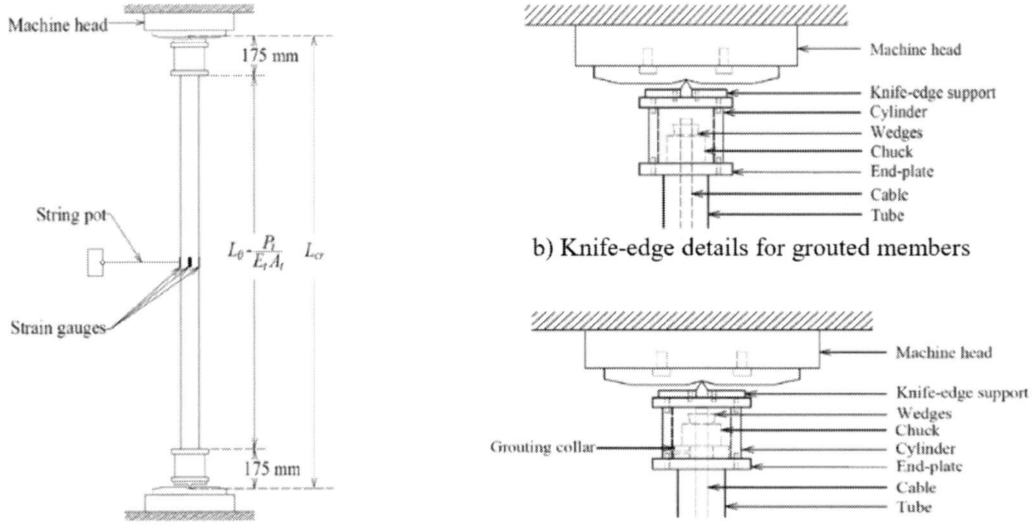

a) Schematic setup

b) Knife-edge details for grouted members

c) Knife-edge details for non-grouted members

Fig. 4: Compressive test setup

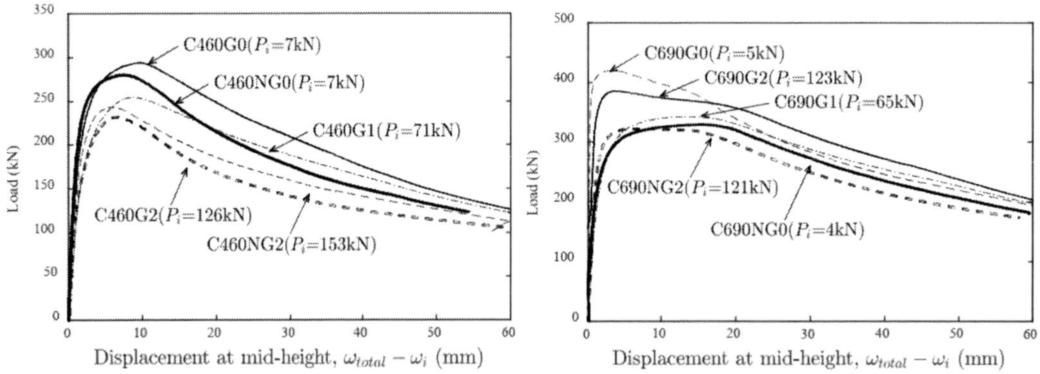

a) S460 members

b) S690 members

Fig. 5: Load-lateral deflection curves compressive members

3. FE validation and parametric study

Numerical modelling was conducted initially to replicate the experimental results and subsequently to generate parametric results for prestressed elements of different geometries. Only compressive members were examined in the numerical study since the tensile behaviour of the members could be accurately captured through simple analytical models.

3.1 Modelling assumptions

The components of the numerical models were the SHS tube, prestressing cable, connecting collars and confined grout for the grouted specimens. A fine mesh of three-dimensional eight node (C3D8) solid elements was used to model all the components. The mesh size reported in [6] was adopted herein. The measured material properties of the cables and steel tubes were employed in the models. The material model of the grout was derived based on the grout cube strength, following the approach described in [6]. A friction coefficient of 0.25 was used to define the tangential behaviour of the contact faces of the tube–grout and grout–cable interface elements. An initial imperfection in the form of the first eigenmode with the amplitude reported in Table 3 was assigned to the models. In the nonlinear analysis, the prestressing was achieved by assigning initial stresses to the elements of each component. The loaded models were in compression via displacement control.

3.2 FE validation and parametric study

The FE models were able to capture accurately the observed the load–deformation history of the prestressed members under compression. This is shown in Fig. 6, in which the results of the C460NG0 and C460G0 models are given as examples. Based upon the validated numerical models, an extensive parametric study was conducted for the development and assessment of design proposals. A total of 192 FE models, comprising 8 member slenderness values (0.5-2.25), 2 cable sizes (A_c=100 or 150 mm^2), 3 prestress levels (P_{nom}, 0.5P_{opt} and P_{opt}), 2 grout conditions (grouted and non-grouted) and 2 steel grades (S460 and S690), were simulated. An initial imperfection of $L/1000$ was applied to all models.

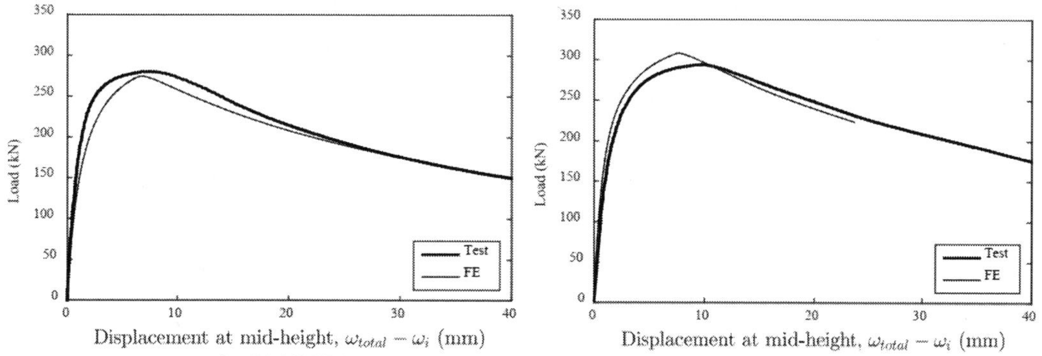

a) C460NG0 b) C460G0

Fig. 6: Validation of FE models for compressive specimens

4. Analysis of results and design recommendations

4.1 Tensile members

The tensile response of the prestressed members could be predicted analytically. By combining the measured geometries (i.e. section sizes and member length) with assumed bi-linear stress-strain material models of the cables and steel tubes, analytical axial load-displacement relationships can be derived [3]. Figs. 7(a) and 7(b) demonstrate that the analytical results match closely the test results of the non-grouted members. The T460NGN and T690NGN results have also been included in Figs 7(a) and 7(b) for comparison purposes.

From Figs. 7(a) and 7(b), it is clear that the introduction of prestressing cables can significantly improve the strength of the cable-in-tube systems. Furthermore, the addition of prestress can be seen to extend the elastic response of the cable-in-tube system. The presence of the grout, as shown in Fig. 3, had a relatively small effect on the strength of the system, but was found to improve the ductility (i.e. axial displacement at ultimate load).

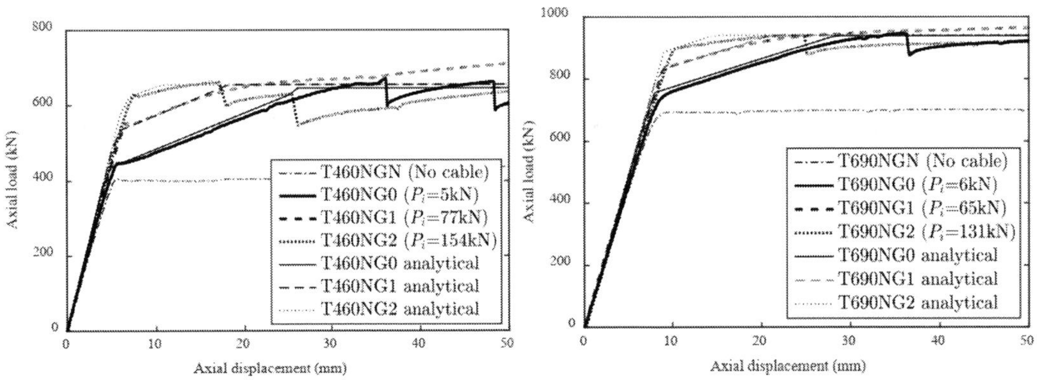

a) S460 non-grouted tensile members b) S690 non-grouted tensile members

Fig. 7: Analytical and test results of non-grouted tensile members

4.2 Compressive members

Prestressing is detrimental for self-anchored cable-in-tube systems in compression. However, the observed reduction in resistance was not as large as the applied initial prestress. This is illustrated in Figs. 5(a) and 5(b), where the C460NG2 (P_i = 126 kN) had only a 70 kN resistance reduction in comparison with the C460NG0 member, and the capacity of the C690NG2 specimen (P_i = 121 kN) was almost the same as the C690NG0 specimen. This is due to the absence of second order bending moments induced by prestressing: i.e. the prestress force remains aligned with the centroidal axis of the member, even in the deformed configuration. The presence of grout was shown, as expected, to increase the compressive capacity - see Figs. 5(a) and 5(b).

The design of cable-in-tube systems under compression has been studied in Gosaye et al. [4], where a method based on the Perry-Robertson approach [7] for conventional column design was developed, and termed the modified Perry-Roberston approach. The method employs the same framework as Eurocode 3 [8] for column design, but accounts for the prestressing cables and grout, as expressed in Eqs. (2)-(7), where the symbols are consistent with those used in Eurocode 3 [8]. The compressive capacity of the system is expressed as the plastic resistance of the cross-section multiplied by a reduction factor to account for member instability:

$$N_{u,cal} = \chi_p N_{pl} \tag{2}$$

where N_{pl} is defined in Eq. (3) for non-grouted and grouted members (denoted with g),

$$N_{pl} = A_t f_{ty} \,;\; N_{pl,g} = A_t f_{ty} + A_g f_{gk} \tag{3}$$

and χ_p is related to the member slenderness through Eqs. (4) and (5).

$$\chi_p = \frac{(1 - P_i / N_{pl})}{\alpha_k \left[\phi_p - \sqrt{\phi_p^2 - \frac{(1 - P_i / N_{pl})}{\alpha_k} \bar{\lambda}^2} \right]} \tag{4}$$

$$\phi_p = \frac{(1 - P_i / N_{pl})\bar{\lambda}^2 + \left[\alpha_k + \alpha(\bar{\lambda} - 0.2)\right]}{2\alpha_k} \tag{5}$$

$\bar{\lambda}$ and α_k are defined in Eqs. (6) and (7), respectively, for members without and with grout, where α is the imperfection factor, with a value of 0.13, corresponding to buckling curve a_0 as recommended by EC3 [8] for S460 and S690 hot-finished tubular members.

$$\bar{\lambda} = \sqrt{\frac{N_{pl}}{N_{cr}}} \,;\quad \bar{\lambda}_g = \frac{L_{cr}}{\pi} \sqrt{\frac{N_{pl}}{E_t I_t + 0.6 E_{gm} I_g}} \tag{6}$$

$$\alpha_k = \frac{K_t}{K_c + K_t} \,;\; \alpha_{k.g} = \frac{K_t + K_g}{K_c + K_t + K_g} \tag{7}$$

In Eq. (7), K_c, K_t and K_g are the axial stiffness of the cable, tube and grout (A_cE_c/L, A_tE_t/L and A_gE_{gm}/L), respectively. Note that for the case of very small initial prestress (i.e. $P_i < N_{b,t}K_c/(K_c+K_t)$ for non-grouted members and $P_i < N_{b,t}K_c/(K_c+K_t+K_g)$ for grouted members, where $N_{b,t}$ is the buckling resistance of the steel tube), the cable will slacken before the member fails, hence the design is equivalent to the case without prestressing, which leads to $P_i = 0$ and $\alpha_k = 1$in Eqs. (4) and (5).

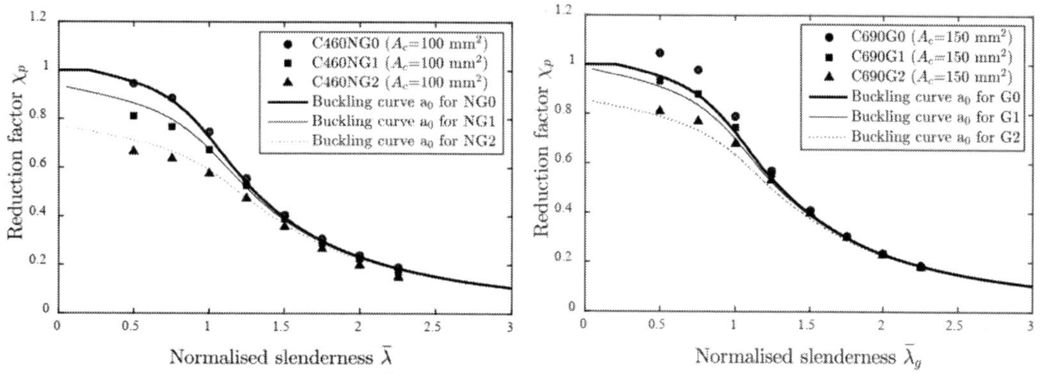

a) C460NG (A_c=100 mm^2) b) C690G (A_c=150 mm^2)

Fig. 8: Assessment of the modified Perry-Robertson approach

The C460NG (A_c=100 mm^2) and C690G (A_c=150 mm^2) FE results are plotted in Figs 8(a) and 8(b), respectively, to assess the design curves determined from the modified Perry-Robertson approach [4]. In general, this approach can provide a good estimation of the compressive resistance of cable-in-tube systems. However, buckling curve a_0 slightly overestimates the strength of the S460 $0.5P_{opt}$ and P_{opt} models, as shown in Fig. 8(a). It also underestimates the S690 results for the complete range of prestress levels considered, especially for the stocky members, as demonstrated in Fig. 8(b). Alternative buckling curves may therefore need to be considered for prestressed HSS tubes.

5. Conclusions

The tensile and compressive behaviour of cable-in-tube systems has been examined through a series of experimental and numerical investigations. Tensile resistance was shown to be enhanced by the addition of the cable while the application of prestress was needed to extend the elastic range and hence the overall performance of the cable-in-tube system. Prestressing was found to reduce the compressive resistance of the system, but the reduction in strength was shown to be much less than the applied prestress due to the absence of second order bending moments. FE models, validated against the experiments, were employed in order to generate parametric results for the assessment of a modified Perry-Robertson design method [4]. Buckling curve a_0 adopted in EC3 [8] was shown to yield some unsafe predictions for the S460 $0.5P_{opt}$ and P_{opt} members, but was safely applicable to the S690 members.

Acknowledgments

The research has received funding from the Research Fund for Coal and Steel (RFCS) under grant agreement No. RFSR CT 2012-00028. V&M DEUTSCHLAND GMBH is acknowledged for the supply of the test specimens. The authors are indebted to Mr Gordon Herbert, Mr Paras Shah and Mr Stephen Okeghie for their assistance during the tests.

References

[1] Li H, Schmidt L. "Post-tensioned and shaped hypar space trusses", *Journal of Structural Engineering, ASCE 1997*, 123(2), 130-137, 1997.
[2] Schmidt LC, Li H. "Studies on post-tensioned and shaped space-truss domes", *Structural Engineering and Mechanics*, 6, 693-710, 1998.
[3] Gosaye J, Gardner L, Wadee A, Ellen M. "Tensile performance of prestressed steel elements", *Engineering Structures*, 79, 234-243, 2014.
[4] Gosaye J, Gardner L, Wadee A, Ellen M. "Compressive behaviour and design of prestressed steel elements", *Structures*, 5, 76-87, 2016.
[5] Zhao O, Rossi B, Gardner L, Young B. "Behaviour of structural stainless cross-sections under combined loading – Part I: Experimental study", *Engineering Structures*, 89, 236-246, 2014.
[6] Tao Z, Wang Z, Yu Q. "Finite element modelling of concrete-filled steel stub columns under axial compression", *Journal of Constructional Steel Research*, 89, 121-131, 2013.
[7] Ayrton WE, Perry J. "On struts", *The Engineer (London)*, 62, 464-465, 1886.
[8] EN 1993-1-1:2005. "Eurocode 3: Design of steel structures – Part 1-1: General rules and rules for buildings", *Brussels, European Committee for Standardization (CEN)*, 2005.

AUTHOR INDEX